ELSEVIER'S DICTIONARY OF BIOLOGY

ELSEVIER'S DICTIONARY OF BIOLOGY

compiled by

RAUNO TIRRI, JUHANI LEHTONEN, RISTO LEMMETYINEN, SEPPO PIHAKASKI and PETTER PORTIN

Department of Biology
University of Turku, Finland

ELSEVIER
Amsterdam – Lausanne – New York – Oxford – Shannon – Singapore – Tokyo

ELSEVIER SCIENCE B.V.
Sara Burgerhartstraat 25
P.O. Box 211, 1000 AE Amsterdam, The Netherlands

First edition 1998
Second impression 2000

Library of Congress Cataloging-in-Publication Data

Elsevier's dictionary of biology / Rauno Tirri ... [et al.]. -- 1st ed.
p. cm.
Includes bibliographical references.
ISBN 0-444-82525-8 (alk. paper)
1. Biology--Dictionaries. I. Tirri, Rauno. II. Elsevier Science Publishers.
QH302.5.E47 1998
570'.3--dc21 98-31199
CIP

ISBN: 0-444-82525-8

⊗The paper used in this publication meets the requirements of ANSI/NISO Z39.48-1992 (Permanence of Paper).
Printed in The Netherlands.

Preface

This dictionary provides concise explanations of words and expressions central to the language of biology. It is intended for use by all who are interested in biology, in education, in the application of biology, or in the pursuit of biology for pleasure.

Biology is an extensive subject that has undergone a vast expansion and so comprises not only such conventional aspects as taxonomy, morphology, biochemistry, functional physiology, and ecology, but also the rapidly expanding new fields of cell biology and molecular biology. This dictionary attempts to provide a comprehensive coverage of biological terms, recognising also that the interface between biology and other sciences, such as chemistry and physics as well as medicine, is becoming less distinct and includes terms from these other areas.

In the preparation of this dictionary the authors have relied heavily on their collective experience of the needs of the students they have taught. However, it will also serve well all who have an interest in biology and its related sciences, particularly medicine and environmental science. It will be of use not only to students and their teachers, but also to journalists and translators, to biologists in industry and the administrators of biological programmes.

We are grateful to many persons who have encouraged and helped us in many forms such as in choice of entries, formulation and review of explanations, checking of language, and drawing of pictures. We especially acknowledge Prof. Ken Bowler, Dr. and Mrs Leslie Barnes, Dr. Esa Lehikoinen, Dr. Timo Vuorisalo, M.Sc. Leena Laitinen, M.Sc. Maija Mustonen, Mr Tuomas Lemmetyinen and Lab. technician Kurt Ståhle.

Turku, Finland, July 1998

Rauno Tirri
Juhani Lehtonen
Risto Lemmetyinen
Seppo Pihakaski
Petter Portin

Information for the Users

Etymology
The origin of many words in this book is in Latin (L.) or Greek (Gr.) and we have introduced etymons, although short, because they are usually of interest to readers and perhaps also a method of remembering words. In cases where subsequent entries have the same origin, the origin is not repeated.

Explanations
The explanations of entries are of various lengths. This is partly because we have wished to explain the most crucial words in detail.

Within many of the explanations the reader will find an arrow to indicate a reference to other words of essential importance in order to relate one term with another.

Choice of entries
One of the most difficult tasks has been the choice of entries. Our choice was influenced by many aspects, such as the fields of the five authors and their colleagues and the existence of special dictionaries available for instance in physics, chemistry, microbiology, and medicine. Many terms have one synonym or more, and although most of them are presented as entries, the explanation is found only for the term most commonly used.

In science, the systematics of the living organisms is under constant change, because improved methods particularly in nucleic acid analysis give new information to this issue. For plants and fungi two different classifications are given. The first one (older) is classical, still often generally accepted. The newer classification for plants is presented mainly according to that in "Biology of Plants" (Raven et al., Worth, 1992), and that for fungi according to "Dictionary of Fungi" (Hawksworth et al., CAB International, 1996). The classification of animals is based mainly, but not exclusively on the book "Synopsis and Classification of Living Organisms" by Sybil P. Parker (McGraw-Hill Company, New York 1982). It has been somewhat intriguing to choose the taxa and therefore we perhaps could not avoid a certain amount of subjectivity. Anyhow, the principle has been to include at least all phyla (protozoans excluded) and the essential classes, and all orders of the tetrapod vertebrates also. Further, we have explained many other taxa mainly because of general interest. Also the data in geological time scales are highly varying in different sources. In this dictionary we have used the data presented by B. F. Windley in table 1.1 of his book "The Evolving Continents" (3rd Edition, 1995, John Wiley & Sons).

This dictionary has about 10,000 entries. Moreover there are a great number of terms, which are not chosen as entries but are included in the explanations of other words. They can be found by logical thinking of the context or by reading this book systematically.

Several annexes follow providing a table of international measure, a listing of chemical elements, the systematics of organisms, and geological periods.

Contents

A

A, symbol for **1.** alanine; **2.** adenine, adenosine; **3.** adenylic acid; **4.** ampere.

***A*,** symbol for **1.** absorbance; **2.** activity; **3.** mass number.

***A*.,** symbol for **1.** arteria (also a.); **2.** anode (also a.).

a, symbol for **1.** thermodynamic activity; **2.** total acidity; **3.** atto, 10^{-18}; **4.** year.

***a*,** **1.** symbol for specific absorption coefficient; **2.** acceleration.

a., 1. anode (also A.); **2.** artery (also A.); **3. aa.** (L. *arteriae*), arteries.

a-, (Gr. L., before h or vowels, **an-**), denoting not, without.

AA, abbreviation for amino acid.

aardvarks (Afr. *aard* = earth, *vark* = pig), → Tubulidentata.

AAS, atomic absorption spectroscopy, → atomic absorption spectrophotometer.

ab-, abs- (L. *ab* = from), denoting from, away, off.

ABA, → abscisic acid.

A-band, anisotropic band, *see* I-band.

abaxial, abaxile (L. *ab* = from, away, *axis* = stem), situated outside of, or facing away from the axis of a structure, as of a stem, body, or part; e.g. in plants, the lower surface of a leaf, facing away from the stem during early development. *Cf.* adaxial.

abdomen, pl. **abdomina,** Am. also **abdomens** (L.) **1.** the region of the vertebrate body that comprises the posterior part of the → coelom and the viscera, i.e. the internal organs other than the heart, lungs, and thymus; in tetrapod vertebrates the abdomen is situated between the thorax and pelvis (in mammals separated from the thorax by the diaphragm), and in lower vertebrates between the cardiac and caudal regions of the body; **2.** in invertebrates, especially in arthropods, a distinct posterior part of the body including a great part of the digestive organs and gonads with the genital orifices.

abdominal (L. *abdominalis*), pertaining to the → abdomen.

abdominal cavity, 1. in mammals, the ventral part of the body cavity (coelom) caudally from the → thoracic cavity, bounded anteriorly by the diaphragm and lined by the → peritoneum; contains the **intraperitoneal organs** (inside the → peritoneal cavity), i.e. the stomach and the main part of the → intestine, the liver, and the spleen, and the **retroperitoneal organs** (between the body wall and peritoneum), i.e. the duodenum, pancreas, kidneys, adrenal glands, and the ascending and descending colon; **2.** sometimes used to refer to the → coelom of lower vertebrates (especially of birds) which is not divided into two sections like in mammals.

abducent nerve (L. *nervus abducens*), *see* cranial nerves.

abduct (L. *abducere* = to lead away), to move or draw away from the midline of the body, as e.g. a limb away from the body; **abduction**, the process of such a movement; **abductor**, any muscle that abducts. *Cf.* adduct.

aberrant chromosome behaviour (L. *aberrare* = to go astray), the deviation of chromosome behaviour from that occurring in normal → mitosis or → meiosis, sometimes having a recognized genetic basis; includes e.g. chromosome loss and non-random segregation of chromosomes.

aberration, a deviation from the normal, e.g. **1.** an abnormal position of an organ; **2.** chromosome aberration (*see* chromosome mutation); **3.** an individual with exceptional features due to environmental, not inherited, factors; **4.** in entomology, a term used instead of form (*forma*); **5.** a mild disorder of the mind; **6.** unequal refraction or focalization of a lens or a mirror. *Adj.* **aberrant**.

abiogenesis, pl. **abiogeneses** (Gr. *abios* = lifeless, *gennan* = to produce), origination of living organisms from non-living matter, e.g. **1.** a suggested origin of life on Earth; **2.** spontaneous generation of life; origination of organisms such as lice and fleas from dirty lifeless material, an earlier belief abandoned in the 19th century. *Adj.* **abiogenetic**.

abiosis, 1. the absence of living phenomena or life; **2.** nonviability. *Adj.* **abiotic**, non-living, not involving living organisms. *Cf.* biotic.

abiotic factors, physical and chemical factors, i.e. **extrinsic factors**, operating outside living organisms, e.g. humidity, climate, inorganic nutrients, and edaphic factors, affecting organisms in the environment (**abiotic environment**). *See* biotic factors.

ablation (L. *ablatus* = carried away), the separation or detachment, especially the removal of an organ by a surgical procedure, or destruction of its structure and function by a noxious substance.

abneural (L. *ab* = away, Gr. *neuron* = nerve), **1.** away from the neural axis; **2.** abnerval, i.e. away from a nerve, e.g. in a muscular fibre away from the point of the nerve entrance.

abnormal (L. *norma* = rule), differing from the usual state or condition; not normal, subnormal, exceptional, aberrant.

ABO blood groups, the major system of the blood group types based on presence or absence of two antigenic glycoprotein structures on the surface of human red blood cells. These structures (**surface antigens, agglutinogens**) are called A and B antigens, i.e. A and B blood group factors (*see* antigen). More than one third of all people have A antigen (**A group**), about 10% have B antigen (**B group**), less than 10% have both these antigens (**AB group**), and more than one third have neither of these (**O group**, sometimes called zero blood); the frequencies vary greatly between different human populations.

ABO blood groups are called natural because a person having no A and B antigens, or only one of these, develops → antibodies against that antigen which is absent in his blood (in the O blood against both antigens, i.e. **anti-A** and **anti-B**). Thus, an individual with A blood has in his plasma antibodies against B blood, those with B blood have correspondingly antibodies against A blood, and people with O blood have antibodies against both A and B blood, i.e. against A and B antigens. People of AB type have no antibodies, otherwise the agglutination (aggregation) of their own red blood cells would occur. In practice, the blood grouping is based on this agglutination reaction when different types of blood are mixed in the test, i.e. a blood sample (erythrocytes) to be tested, is exposed to a known specific antiserum. Typing of ABO blood groups is especially important in blood transfusions; O individuals are common donors, because their cells cannot be agglutinated in any recipient blood. AB individuals are correspondingly universal recipients. *See* rhesus factor.

abomasum, *see* rumination.

aboral (L. *ab* = away, *os* = mouth), opposite to, or away from the mouth; **aborad**, away from the mouth.

abortion (L. *aboriri* = to miscarry, disappear), **1.** the arrested development of an embryo or organ at an early stage; anything that fails to develop; **2.** in mammals, the premature expulsion of the foetus; **abortus**, any product of an abortion. *Verb* to **abort**. *Adj.* **abortive**, incompletely developed, causing abortion.

abrasion (L. *abradere* = to scrape off), rubbing or wearing away by friction; the scraping or rubbing of the surface layer of the skin or the mucous membrane; an act of abrasing (erosion). *Adj.* **abrasive**, producing abrasion.

abscise (L. *abscidere*), to cut away, off.

abscisic acid, ABA, a natural plant hormone, sesquiterpene, $C_{15}H_{20}O_4$; occurs in different plant organs, buds, leaves, seeds, roots, etc. The naturally occurring → isomer, dextrorotatory form (+)-ABA, and the synthetic laevorotatory form (—)-ABA), are physiologically equally active. The biologically active +ABA has the *cis*-configuration but the *trans* form is inactive. ABA promotes the abscission of leaves and flowers and inhibits e.g. the germination of seeds and the opening of stomata. ABA maintains the dormancy of seeds and buds being an antagonist to → gibberellins. Different stress conditions increase the ABA concentration in plant tissues. The terms abscisin and dormin were previously used for ABA.

abscission, 1. cutting off, removal; **2.** shedding of leaves, flowers, or fruits from a plant.

abscission layer/zone, a cell layer which develops at the base of a leaf, flower or fruit, mostly in deciduous trees and bushes; the cells of the layer break down and cause the abscission of the structure. A **protection layer** of cork cells develops over the wound to cover the tissue in the base. Both layers together form the **abscission zone**. *Syn.* separation layer.

absolute density, in ecology, describes the population density as actual numbers of individuals; e.g. the census of a human population in industrial countries, or a population of some rare large-sized animals, such as the white-tailed sea eagle in Europe. In nature, a reliable count is often difficult to perform, and therefore it is more common to use relative density indices. *See* relative density.

absolute zero, 0K (zero kelvin), the zero of thermodynamic temperature, *T*, i.e. the lowest possible temperature, (—273.16°C), at which the thermal movement of atoms in any matter ceases. *T* is expressed as kelvins (K) starting from the absolute zero.

absorbance, also **absorbancy** or **absorbency** (L. *absorbere* = to suck up, absorb), symbol ***A*** or ***OD*** (optical density); the ability of a medium to **absorb** radiation. It expresses the quantity of absorbed radiation into a medium (solid, liquid, or gas), in a laboratory measurement, most usually the absorption of light (ultraviolet, visible, or infrared light) into a solution. Absorbance of solutions is measured by optical instruments such as a → **spectrophotometer**, and is expressed as the negative logarithm of **transmittance**. Absorbance = $\log(I_0/I)$, where I_0 is the intensity of light coming into a sample solution (or light passing through a pure solvent), and I the intensity of light transmitting the sample. Absorbance is dependent on the concentration (e.g. dye concentration) and thickness of a solution layer. **Absorption spectrum** is formed when different wavelengths of light (or other electromagnetic radiations) are absorbed differentially by a solution. Each compound, which can be identified and determined on the basis of the spectral properties, has its specific absorption spectrum. *See* Lambert—Beer's law.

absorbent, 1. having capacity or tendency to absorb; **2.** any material that absorbs into itself radiation or substances, e.g. a liquid absorbing heat, light, radioactive waves, or chemical agents.

absorption (L. *absorptio*), **1.** incorporating or taking in gases, liquids, or radiation; **2.** the entering of substances into blood, lymph, or cells through walls of the intestine or kidney tubules, the skin epithelium, etc.; **3.** in psychology, the devotion of mental activity to one object only; **4.** uptake of nutrients and water by plants. *Verb* to **absorb**. *Cf.* adsorption.

abstinence (L. *abstinere* = to restrain, refrain, hold back), refraining from the use of stimulants, dietary enjoyments, or sexual intercourse; **abstinence symptoms** (withdrawal symptoms), psychic, neuronal and other physiological symptoms occurring especially in humans after ceasing the intake of a stimulant (addictive agent), like morphine, caffeine, or alcohol.

abundance, in ecology, either **absolute abundance** describing the total number of individuals of a taxon (usually a species) per unit area or volume, or **relative abundance** which is the total number of individuals of a taxon in relation to all other studied taxa, usually expressed as percentages. Abundance emphasizes especially the proportions of → constant and dominant species. *See* RNA abundance.

abyssal zone (Gr. *abyssos* = bottomless), the bottom of oceans in the depth below 2,000 m beyond the continental shelf. The organisms in the abyssal zone are adapted to continual darkness, cold, and high pressure.

abyssopelagic zone (Gr. *pelagos* = sea), an oceanic zone above the bottom at the depth between 4,000 and 6,000 m.

Acanthocephala (Gr. *akantha* = spine, thorn, *kephale* = head), **acanthocephalans, spiny-headed worms**; a phylum of endoparasitic worms living in the intestine of vertebrates. They have a cylindrical invaginable proboscis with several curved hooks for attaching to the intestine of the host. The worms are pseudocoelomate and have no digestive tract, and food is absorbed through the body wall. The life cycle requires a crustacean or insect as an intermediate host. The phylum includes more than 500 species throughout the world; many of them are harmful often causing tissue damage in their hosts, e.g. in many fish species, birds, and mammals; normally, no species are human parasites.

Acanthopterygii (Gr. *pterygion* = fin, small wing), a large superorder of ray-finned fishes with spiny fins and scales, e.g. **perch**, **mackerel** and **tuna** (order Perciformes), and **flatfishes** (Pleuronectiformes).

acanthosoma, pl. **acanthosomas** or **acanthosomata** (Gr. *soma* = body), a late larval stage in the life cycle of some decapod crustaceans.

Acari(na) (Gr. *akari* = mite), **acarines**, including **mites** and **ticks**; a large and varied order of arachnids (some authors consider as a subclass) including usually small spherical animals found throughout the world, many of which are medically and economically significant; about 30,000 species are described. The cephalothorax and abdomen are completely fused. Several species are free-living in terrestrial or aquatic environments, but many are endoparasitic or ectoparasitic, e.g. living on birds and mammals. These may cause various diseases, such as dermatitis, scabies, and itching. Some species are vectors of many viral and bacterial diseases in humans and other mammals (e.g. borreliosis, *see* Borrelia). The chelicerae of acarines are adapted for stinging, sucking, or biting. Development from the egg to an adult includes several larval and nymphal stages.

acariasis, also **acaridiasis,** or **acarinosis,** any disease caused by mites; e.g. scabies, dermatitis.

acaricide (L. *caedere* = to kill), any chemical substance used to kill ticks and mites.

acarology (Gr. *logos* = word, discourse), a branch of biology studying mites and ticks (Acari).

acceptor (L. *accipere* = to accept, receive), an atom or a molecule that captures an element (e.g. electron, atom, or an atom group such as amine or methyl groups) from another substance (donor); e.g. hydrogen acceptor, oxygen acceptor.

acceptor splicing site, in genetics, the boundary between the right end of an → intron and the left end of the following → exon in the gene; the → heterogeneous nuclear RNA is cleaved here in → splicing.

accessory bud, a bud situated on either side of the actual → axillary bud.

acclimation (L. *ac* < *ad* = additional, Gr. *klima* = climate), **1.** physiological, phenotypic adaptation of an organism to a certain environmental factor in laboratory circumstances; e.g. acclimation to temperature changes when other environmental factors remain steady. Organisms originating from cold and temperate climates can be acclimated to resist low temperatures when exposed gradually to these temperatures, plants also to a shortened day length; *see* cold hardening; **2.** especially in America, the term is used as a synonym for → acclimatization.

acclimatization, physiological, phenotypic adaptation of organisms to changes in their natural environment; involves processes of physiological, reversible adaptation to many environmental factors simultaneously (*cf.* acclimation), like to temperature, moisture, pressure, illumination, acidity, nutrition, various harmful chemicals, and to the presence of other organisms. These conditions vary seasonally and thus also the acclimatization of organisms. For complete acclimatization several weeks are usually required, and the process is mostly adaptive helping organisms to survive in changed circumstances. The genetic properties determine the biochemical, physiological, and behavioural strategies used by an organism in changed situations. Hormonal systems, in animals also the nervous system, control the activity of biochemical and physiological processes in cells and tissues. Generally, those organisms which during evolution have lived in steady environmental conditions have a poor capability, but organisms living naturally in varying conditions, like in arctic regions, have a good capability of acclimatization.

accommodation (L. *accommodare* = to fit to, adapt), adjustment; in biology, especially the adjustment of the eyes for various distances. In the eyes of mammals as well as of most birds and reptiles, the accommodation for seeing near is adjusted by contraction of the ciliary muscle. This causes the relaxation of the suspensory ligaments around the lens, allowing the curvature of the lens to increase on its anterior surface. At rest the ligaments pull the lens to become more flattened. The eyes of fish and amphibians, and of cephalopod molluscs, accommodate by moving the lens forward or backward in the visual axis, i.e. like in the camera. *Verb* to **accommodate**. *Adj.* **accommodative**.

accretion (L. *ad* = to, *crescere* = to grow), the growth together, or an abnormal adhesion of anatomical structures; e.g. the growth of the placenta into the mucosa of the uterus.

acellular (L. *a* = not, *cellula*, dim. of *cella* = cell), not made up of cells, non-cellular.

acelomate, acelomatous, → acoelomate.

acentric (Gr. *akentrikos* = not centric), **1.** not located in the centre, without a centre, not central; **2.** pertaining to a chromosome or a fragment of chromosome lacking the → centromere.

acet(o)- (L. *acetum* = vinegar), pertaining to acetic acid or a two-carbon fragment of acetic acid.

acetabulum, pl. **acetabula** (L. originally = "vinegar cup"), a cup-shaped structure, e.g. **1.** a cup-shaped depression on the surface of the hip bone in which the femur articulates in tetrapod vertebrates; **2.** a sucker in trematodes (flukes), hirudines (leeches), and cephalopods.

acetaldehyde, acetic aldehyde, ethanal, CH_3CHO; the intermediate formed from pyruvate by decarboxylation e.g. in → fermentation; it is produced also in animals by intestinal microbes and in oxidation of ethanol in liver cells by the action of → alcoholdehydrogenase. Acetaldehyde is oxidized to acetic acid (ethanoic acid) and reduced to ethanol.

acetic (L. *acetum* = vinegar), **1.** pertaining to vinegar, or to the presence of the two-carbon fragment (acetyl group) of → acetic acid; **2.** sour.

acetic acid, ethanoic acid; a saturated aliphatic fatty acid, CH_3COOH; a weak acid which dissociates incompletely to $CH_3COO^- + H^+$. Its salts and esters are **acetates** which are formed in cellular metabolism of organisms. Acetic acid is the effective component in vinegar.

acetic acid fermentation, the reaction pathway in which acetic acid is produced from ethanol by *Acetobacter* bacteria in the presence of oxygen. *See* fermentation.

acetic aldehyde, → acetaldehyde.

acetic ether, → ethyl acetate.

acetone (L. *acetonum*), dimethyl ketone, propanone, CH_3COCH_3, a colourless, inflammable organic liquid, a good lipid solvent that also dissolves in water. It is formed in animals in incomplete oxidation of fatty acids; e.g. in mammals, increased concentrations (acetonaemia and acetonuria) are found during a long fasting period, in diabetes, and especially in cattle during the postparturient period. Some bacteria, such as *Clostridium acetobutylicum*, produce acetone as a result of → fermentation.

acetonitrile, ethanenitrile, methyl cyanide, CH_3CN; occurs in coal tar and is found as a metabolite in plants (*see* indole-3-acetonitrile); a poisonous liquid with an ether-like odour. Acetonitrile is used as a solvent and for the synthesis of many organic compounds, such as α-naphthalene acetic acid, thiamine, and acetamidine.

acetylcholine, ACh, the acetic ester of → choline, $CH_3COOCH_2CH_2N^+(CH_3)_3$; a → neurotransmitter of the nervous system of animals; its synthesis is catalysed by → **choline acetyltransferase** (**c. acetylase**) and its decomposition by → **choline esterase**. Acetylcholine acts in **cholinergic synapses** between different neurones in the central and peripheral nervous system, or between neurones and effector cells (muscle cells or glandular cells), acting through two receptor types, → **muscarinic receptors**, and → **nicotinic receptors**. The pharmacological substances which act like acetylcholine are called **cholinomimetics**, and their antagonists, **cholinolytics**.

acetylcholinesterase, *see* choline esterase.

acetyl coenzyme A, acetyl CoA, the active acetate symbolized by CoAS-$COCH_3$, formed from → coenzyme A and acetic acid. It acts in cell metabolism in the transfer of an acetyl group, CH_3CO- (cleaved e.g. from fatty acids) into the → citric acid cycle, and is thus a link between → glycolysis or → beta-oxidation and the citric acid cycle. Acetyl CoA is also an intermediate in the → glyoxylate cycle and is involved in some syntheses, e.g. of fatty acids and terpenoids.

acetyltransferases, transacetylases; an enzyme sub-subclass, EC 2.3.1., transferring acetyl groups, CH_3CO-, from a compound to another; e.g. → phosphotransacetylase.

achene (Gr. *a* = without, *chainein* = to yawn, gape), an indehiscent, one-seeded → fruit, a subtype of the nut; typical of the families Asteraceae and Cichoriaceae.

achiasmate, pertaining to → meiosis or → bivalents where no chiasmata are formed.

Achilles tendon (Gr. *Achilleus*, a mythical hero), the calcaneal tendon (*tendo calcaneus*) of tetrapod vertebrates; attaches the triceps surae muscle to the back of the heel.

achlamydeous (Gr. *a-* = without, *chlamys* = mantle), pertaining to a flower without sepals and petals.

A chromosome, a → chromosome belonging to the normal chromosome set; opposite to → B chromosomes which are accessory chromosomes.

acid (L. *acidus* = sour), any compound that liberates protons (hydrogen ions, H^+) in its water solution; in an **acidic** solution pH is less than 7; forms a salt when reacting with a base (proton acceptor). Inorganic acids are e.g. hydrochloric acid (HCl), nitric acid (HNO_3), and sulphuric acid (H_2SO_4); organic acids are e.g. acetic acid (CH_3COOH) and many fruit acids. Cells and tissues contain several acids which are the end products or intermediates of metabolism. Some acids are amphoteric, i.e. substances which can act as acids or bases depending on their environment. In cells and tissues, acids are buffered so that **acidity**, **pH**, does not considerably change in organisms. *See* acid-base balance, pH.

acid-base balance, the maintenance of a certain → pH value (acidity) in cells and intercellular fluids of organisms. Important factors for maintaining the balance are pH → buffers, the excretion of acidic metabolites, the removal of carbon dioxide produced in cell respiration, in animals also the reabsorption of bicarbonate by the kidney. Acidity varies only slightly in the cell cytoplasm being usually between pH 7.0 and 7.4. The vacuolar solution of plant cells is normally rather acid depending on organic acids and acid salts accumulated in the vacuole. In animals, pH of intercellular tissue fluids and blood may change much more than cytoplasmic pH.

acid-base titration, *see* titration.

acid deposition, the settling of an acidic material from the atmosphere onto the ground and surface water; occurs especially as a consequence of the use of fossil fuels when atmospheric sulphur oxides form sulphuric acid and sulphates, and nitrogen oxides form nitric acid. These deposit on the ground either as a **wet deposition** which is also called **acid rain**, or as a **dry deposition**, having usually pH values less than 6. At present, acid deposition is one of the worst global environmental threats.

acid dyes, a group of dyestuffs containing a negatively charged chromophore ion, e.g. organic sulphonate. Acid dyes are mainly salts of organic acids and are used for staining natural fibres such as silk and wool; some dyes may contain a metal chelate at the negative site. Of acid dyes e.g. eosin and aniline blue are in use for staining of biological materials. *Cf.* basic dyes.

acid growth hypothesis, a theory explaining that an increased H^+ ion concentration in the vicinity of cell walls caused by → auxin activates enzymes (e.g. pectinases and cellulases) which hydrolyse cell wall material; the softened cell walls allow cells to enlarge. Additional evidence for the hypothesis gives the action of → fusicoccin that induces rapid cell elongation triggering pumping of protons out of sensitive cells causing wall loosening. The actions of fusicoccin and auxin can be blocked by permeating the cell wall with buffers that prevent the extracellular pH from becoming lowered.

acidity, the quantity of acid in a solution, i.e. the concentration of hydrogen ions. *See* pH.

acidophile, also **acidophil** (Gr. *philein* = to love), **1.** a cell or other histological element staining with acid dyes; **2.** acidophilic → granulocyte, a type of white blood cells also called eosinocyte or eosinophil(e). *Adj.* **acidophilic**, pertaining to acidophiles, or to microbes growing in an acid medium.

acidosis, pl. **acidoses,** a state of increased acidity (decreased alkalinity) in animals due to weakened removal of carbon dioxide (**respiratory acidosis**), or disturbances in metabolism (**metabolic acidosis**) or in excretion (**renal acidosis**), resulting in accumulation of acid substances, or depletion of alkaline reserves, especially bicarbonates.

acidotrophic (Gr. *trophe* = nourishment), pertaining to a water body where water is highly acidic owing to sulphates or humic acids (pH below 5.5).

acid phosphatase, a phosphatase enzyme (EC 3.1.3.2) having a low pH optimum; catalyses the cleavage of inorganic phosphate from different organic compounds, occurring in cells e.g. in → lysosomes, cisternae of the → Golgi complex, and dictyosomes.

acid rain, *see* acid deposition.

acinar, acinic (L. *acinus* = grape), pertaining to the → acinus.

acinous, acinose, aciniform, resembling a grape-shaped structure.

acinus, pl. **acini** (L. = berry, grape), **1.** a minute grape-shaped secretory lobe of a compound exocrine gland (**acinous gland**), such as salivary glands or the exocrine part of the pancreas; **2. liver acinus**, the smallest functional unit of the liver tissue; **3.** a sac-like cavity of the termination of a passage, e.g. **pulmonary acinus**, the respiratory lobule comprising a respiratory bronchiole and all its branches.

Acoelomata (Gr. *a* = not, *koilos* = hollow), **acoelomates**; a group of multicellular animals comprising all the phyla in which the animals have no → coelom but a well-defined embryonic mesoderm; includes Platyhelminthes, Nemertea, and Gnathostomulida; sometimes also the more primitive phyla, Porifera, Cnidaria, and Ctenophora are included. *Cf.* Coelomata, Pseudocoelomata.

acoelomate, Am. **acelomate, 1.** not having a true → coelom, i.e. body cavity; also **ac(o)elematous**; **2.** an animal lacking the body cavity. *See* Acoelomata.

acoelous, Am. **acelous,** pertaining to a vertebra

type with a flat-ended vertebral body (*corpus vertebrae*). *Cf.* amphicoelous, opisthocoelous, procoelous.

aconitic acid (*Aconitum napellus*, monkshood or wolfsbane), *cis*-aconitic acid, a tricarboxylic acid, $HOOCCH_2CCH(COOH)_2$; an intermediate in the → citric acid cycle of cell respiration, formed by dehydration of → citric acid. The reaction is catalysed by **aconitase** (EC 4.2.1.3) which also catalyses the incorporation of water into aconitic acid and converts it to isocitric acid. The salts, esters, and the ionic form are called **aconitates**.

aconitine, a very poisonous plant alkaloid, $C_{34}H_{47}NO_{11}$, occurring in *Aconitum* species, formely used as a cardiac sedative.

acontium, pl. **acontia,** also **acontia filaments, acontia threads** (Gr. *akontion* = small javelin, dart), a thread-like organ or prolonged part of septal filaments in some anthozoans (e.g. sea anemones), provided with nematocysts and gland cells. The animals use acontia for catching prey or for defence by protruding them from the mouth or through pores in the body wall.

acorn worms, → Enteropneusta.

acoustico-lateralis system, *see* lateral line.

acoustic organs, *see* hearing, ear.

acquired character, a character modified solely by environmental factors during the development of an organism; acquired characters are not heritable. *See* Lamarckism.

acquired immune deficiency syndrome, → AIDS.

acquired immune response, adaptive immune response; the specific secondary, tertiary, etc. response of the vertebrate immune system to an antigen; it is more rapid and effective than the primary response achieving a higher antibody concentration, especially with immunoglobin G (IgG). The response is based on the presence of special memory cells (→ lymphocytes) activated during primary invasion of the same antigen. *See* antibody, immune system.

acquired tolerance, the → tolerance to a drug or other exogenous chemicals induced in an animal by repeated intake of the drug; occurs especially in the nervous system, e.g. morphine tolerance.

Acrania, (Gr. *a* = not, *kranion* = skull), a group of chordates with no skull (*cranium*); includes two subphyla, **Tunicata** (tunicates) and **Cephalochordata** (lancelets). In some classifications Acrania comprises only the subphylum Cephalochordata. *Syn.* Protochordata.

Acrasiogymnomycotina, a subdivision in the fungal division Gymnomycota; contains the class → Acrasiomycetes.

Acrasiomycetes, cellular slime moulds, the only class of the subdivision Acrasiogymnomycotina, division → Gymnomycota, kingdom Fungi; are simple, motile, unicellular soil microorganisms reproducing through spores which change on germination into motile myxamoebae, creeping together and forming a **pseudoplasmodium** without fusion. The pseudoplasmodium develops into a fructification with a cylindrical stalk and an ovate head, producing new haploid spores. The best known species is the commonly used laboratory research object *Dictyostelium discoideum*. In some systems the group Acrasiomycetes is classified into the division **Acrasiomycota**; the dictyostelids are thus separated into the division Dictyosteliomycota.

acridine, $C_{13}H_9N$, a synthetic mutagenic compound that causes changes in DNA by adding or deleting nucleotide base pairs. Acridine mutagens used in research are known e.g. by the trade names ICR 170 and ICR 191.

acridine orange, 3,6-bis(dimethylamino)-acridine hydrochloride, tetramethyl acridine; a dye used for staining nucleic acids, both DNA and RNA. It can attach to the double-stranded DNA giving fluorescence at 530 nm; if it is bound to phosphate moieties it fluoresces at 640 nm.

acro- (Gr. *akron* = extremity, tip), pertaining to an extremity, height, summit, top, end, or tip.

acrocarpic, acrocarpous (Gr. *karpos* = fruit), pertaining to a plant having the male and female → gametangia at the tip; e.g. the gametophyte of a typical moss.

acrocentric (Gr. *kentron* = centre), pertaining to a type of chromosome which has its → centromere close to the extremity, and hence one arm of the chromosome is very short and the other much longer.

acrolein, acrylaldehyde, propenal, $H_2C{=}CH$-CHO; a colourless volatile liquid used as a well-penetrating fixative of biological material for electron microscopy, suitable even for hard plant tissues; polymerizes readily into resins. Acrolein is an atmospheric pollutant liberated from motor traffic combustion, especially irritating respiratory tracts. *Cf.* acrylic acids.

acromegaly (Gr. *megas* = large, great), an abnormal condition of an adult vertebrate animal, especially known in man, caused by hypersecretion of → growth hormone in adult age, resulting in enlargement of distal parts of the skeleton, i.e. the ears, jaws, nose, toes; the converse state is called **acromicria**.

acromion (L. *omion*, dim. of *omos* = shoulder), the process of the scapula articulating with the clavicle in mammals and some reptiles.

acron, the unsegmented anterior part (head) of the body of a metameric animal, found e.g. in some insects and annelids.

acropetal (L. *petere* = to seek), directing towards a top, apex, or extremity; e.g. pertaining to the youngest part of a shoot situated closest to the apex, or to a substance transported towards the apex of a plant shoot or root; e.g → auxins move acropetally in roots, i.e. towards the root apex. *Cf.* basipetal.

acrosome, acrosomal cap (Gr. *soma* = body), a cap-like, membranous structure at the tip of a spermatozoon covering the anterior part of the nucleus; contains hydrolytic enzymes like **acrosin** and **hyaluronidase**, i.e. enzymes which in fertilization makes the entrance of the spermatozoon possible by decomposing surface structures of the egg cell. Acrosome is lacking in species which have → micropyle in the ovum, as in insects.

acrylamide, propenamide; as a monomer (C_3H_5NO) it is a very poisonous crystalline substance; when polymerized with methylene bisacrylamide (ammonium persulphate as a catalyst) **polyacrylamide** gel is formed. The properties of the polymer can be controlled by changing the concentrations and ratios of the monomers. Acrylamide polymers are used e.g. in → electrophoresis.

ACTH, adrenocorticotrop(h)ic hormone, → corticotropin.

actin-, actini-, actino- (Gr. *aktis* = ray), denoting relationship to a ray, a ray-shaped structure, or to some form of radiation.

actin (L. *actus* = motion), a protein in cells of eukaryotic organisms participating in the motile activity of cells; especially abundant in muscle cells (an essential protein in thin filaments). Actin molecules are primarily globular (**G-actin**), but are then polymerized into fibrous form, filamentous actin (**F-actin**), two of which coil around each other to form an actin filament. In certain conditions F-actin can depolymerize back into G-actin. Actin filaments are cellular **microfilaments** forming an essential part of the → cytoskeleton of the cell. In the presence of calcium ions and ATP, the filaments of actin and → myosin combine into **actomyosin** resulting in the contraction of a cell or its parts. *See* muscle contraction.

actinin, alpha-actinin, a protein component crosslinking actin filaments in the cell → cytoskeleton; in muscle cells it is a major component of the Z-disc, anchoring actin filaments (thin filaments) to the disc.

Actinistia (Gr. *aktis* = ray), → Coelacanthini.

actinomorphic, actinomorphous (Gr. *morphe* = form), radially symmetrical; pertaining to a structure with more than two planes of symmetry, e.g. an actinomorphic flower is divisible vertically into similar halves by each plane.

Actinomycetales (Gr. *mykes* = fungus), **actinomycetes**, an order in the division → Bacteria; they are Gram-positive bacteria, usually growing as branched filaments, mostly without septa between cells. Actinomycetes live in soil, water, and e.g. in the alimentary canal of animals. Some species are important producers of antibiotics (*Streptomyces*); some are pathogenic in mammals, such as *Actinomyces bovis* in cattle (actinomycosis) and → *Mycobacterium* species in humans (leprosy and tuberculosis).

actinomycin, a group of antibiotics used against Gram-positive bacteria and fungi; produced by some *Streptomyces* (*Actinomyces*) species; particularly **actinomycin D** (dactinomycin) is widely used in biological and biochemical studies to inhibit the protein synthesis in cells. Actinomycin D binds to DNA, blocks the proceeding of RNA polymerase, and thus prevents RNA synthesis in both prokaryotes and eukaryotes. **Actinomycin C** (cactinomycin) is a mixture of actinomycins C1, C2 and C3 produced by *Streptomyces chrysomallus*, used as an antineoplastic and immunosuppressive agent.

actinopharynx (Gr. *pharynx* = gullet), the gullet of some anthozoans.

actinopilin, a protein compound in chitin effectively refracting light; therefore an important feature in identification and classification of arachnids.

Actinopoda (Gr. *pous* = foot), a superclass in the subphylum Sarcodina comprising protozoans which have a spherical cell body and radial axopods; includes → Heliozoa and → Radiolaria. In some present classifications Actinopoda is considered as a phylum.

Actinopterygii (Gr. *pteryx* = fin, wing), **actinopterygians, ray-finned fishes**; a subclass in the class Osteichthyes; fishes with paired fins supported by dermal rays, the fleshy lobe of the fins is absent, and the nostrils are not connected to the mouth cavity. They have been dominant fishes from the Devonian period to the present and comprise about 20,000 species which are divided into three infraclasses: **Chondrostei** (sturgeons and bichirs), **Holostei** (e.g. *Amia* and *Lepisosteus*), and **Teleostei** (most living bony fishes).

actinostele, *see* stele.

actinotrocha (Gr. *trochos* = wheel), a free-swimming, ciliated larva of phoronids.

Actinozoa, → Anthozoa.

actinula, pl. **actinulae,** also **actinulas,** an advanced larval stage of some hydrozoans; a real polyp with a short stem in some species, e.g. in the genus *Tubularia*.

action potential (L. *potentia* = power), a transient, quickly propagating change (all-or-none response) in the electric activity of an activated cell, serving in transfer of information especially in the nervous and muscular systems of animals. The action potential (**impulse**) measurable on the cell membrane is elicited by different types of stimulation, such as electrical, chemical, mechanical etc., and at the moment of induction the negative membrane potential of the cell, i.e. the → resting potential (–70 to –90 mV) discharges locally depolarizing the cell membrane, often even becoming positive. The action potential mostly results from **influx of sodium ions** (Na^+) into the cell (Na^+ channels open), although calcium (Ca^{2+}) and chlorine ions (Cl^-) are important cofactors. **Efflux of potassium ions** (K^+) out of the cell returns the potential quickly to the resting level.

Action potential is elicited locally, but propagates immediately along the cell membrane when adjacent Na^+ channels open. In nerve and skeletal muscle cells, action potentials (**nerve** and **muscle impulses**) are short-lasting (1—2 ms, about 100 mV), but action potentials in smooth muscles may be lower and often last much longer. In cardiac muscles, duration is exceptionally long corresponding to the duration of the heart contraction.

In plants, action potentials have also been found, although they are not common and are different from those in animals; often they are connected to injuries of different kinds. Action potentials are generated by transport of calcium ions across the cell membrane, but also spontaneous action potentials are known. *See* ion channels.

activation (L. *activatio*), the process of becoming active, e.g. **1.** the activation of a proenzyme to form the active enzyme usually

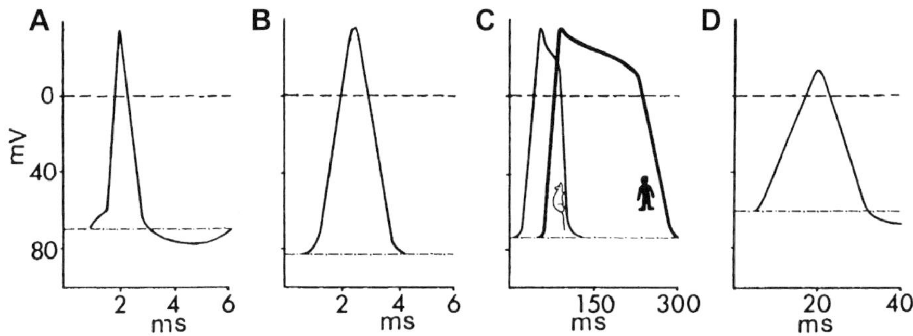

Examples of action potentials in mammalian cells: nerve cell (A), skeletal muscle cell (B), cardiac muscle cell (C), and smooth muscle cell (D), recorded with one electrode inside and the other outside the cell. --- resting power level, -·-·- firing level. In cardiac muscle cells, the duration of action potential varies depending on the size of an animal. In smooth muscle cells, duration, shape, and amplitude vary much in different organs.

by the action of another enzyme; **2.** the stimulation of an egg cell for division by fertilization; **3.** the stimulation of the cerebral cortex by the → reticular formation of the vertebrate brain stem, or corresponding activation of neurones in any brain area; **4.** an increase of the energy content of a molecule or an atom; **5.** the process of inducing radioactivity. *Verb* to **activate.**

activation energy, E_a, the minimum energy required to form a transition complex, i.e. an activated complex of molecules that precedes the formation of end products. This can be expressed by the following equation: A + B = C + D, in which the molecules A and B form first the transition complex AB, having a greater potential energy than that of A + B. The rate of reaction is dependent on the concentration of AB. The complex will yield the end products C + D, the potential energy of which is lower than that of A + B. The activation energy is lower for enzymatic than for uncatalysed chemical reactions; E_a is usually given as kJ/mol. *See* Arrhenius plot.

activation level, in ethology, the internal condition of an animal with an increased need to perform a certain function; the activation level is difficult to distinguish from the term → urge.

activator, a substance that increases the activity of a process; e.g. activates an enzyme reaction, or stimulates the development of a certain structure in an embryo. *See* gene activation.

active site, 1. the region of an enzyme molecule to which a substrate binds forming a transient enzyme-substrate complex during the catalytic conversion of a substrate into a reaction product; more than one active site may exist in an enzyme. Some enzymes require a prosthetic group, such as a metal ion, bound to the active site. In competitive inhibition of the enzyme activity, an inhibitory compound, structurally close to a true substrate, binds to the active site thus reducing or totally inhibiting the enzymatic reaction; **2.** the antigen-binding site of an → antibody.

active transport, the movement of material across the cell membrane or epithelial cell layer by an active process using metabolic energy from ATP; e.g. ion transport by → ion pumps, ATPases. *See* cell membrane transport.

activin, *see* inhibin, transforming growth factors.

actomyosin, a protein complex formed when → actin and → myosin are combined in contraction; the formation and decomposition of actomyosin is responsible for contraction-relaxation processes in muscle cells, but also in many other cell types. *See* muscle contraction.

acu- (L. *acus* = needle) denoting relationship to a needle, sharpness, or sharp needle-like structures in anatomy.

acute toxicity (L. *acutus* = sharp, Gr. *toxikon* = arrow poison), a rapidly developed toxicity in organisms, especially in animals; results from exposure to a poisonous substance or a microbiological element in such high quantities that quick toxication and often death occurs within hours or a few days. Acute and chronic **toxicity tests** are ordered by law to be performed on animals for new compounds and preparations (e.g. drugs) meant for the use of humans. *Cf.* chronic toxicity.

acyl, an organic radical derived from an organic acid by the removal of the hydroxyl group; e.g. acylcoenzyme A (**acyl-CoA**), RCH_2CO-SCoA, an important intermediate of carboxylic acid and → coenzyme A in the oxidation of fatty acids in cells; **acyltransferases** (transacylases), EC subclass 2.3., are enzymes that catalyse the transfer of acyl groups from acyl-CoA to various acceptors.

ad- (L. *ad* = to), **1.** towards, near, adjacent to; **2.** denoting relationship to an increase, or adherence.

adamantine (Gr. *adamas* = utterly hard substance, diamond), pertaining to a hard substance, especially to the enamel (*substantia adamantina*) of the teeth.

adaptability (L. *adaptare* = to adjust, adapt), the ability of an individual to adapt to changes of the environment. *See* adaptation.

adaptation, any change in a structure, function, or behavioural trait of an organism that is evolved by → natural selection. As a result of adaptation, structural and funtional changes occur in the organism contributing to enhanced fitness in new environmental conditions.

Adaptation can be achieved in two ways, either by **phenotypic adaptation** (→ acclimatization), or by **genotypic adaptation** (evolutionary adaptation). In the former process, physiological functions of the organism adapt usually within a few days or weeks to changed environmental conditions. This includes e.g. activation or inactivation of certain enzyme systems at cellular level, usually in the control of the hormonal system, in animals also of the nervous system. The processes will reverse when conditions change back. In phenotypic adaptation the genotype of the organism determines to what extent, and in what ways, the organism is capable of compensating for environmental changes, but the direction and extent of adaptive responses are due to the physiological condition of the organism.

In genotypic adaptation, a change in genes leads to a new genetic constitution which in an organism results in a new way to respond. To the organism and its offspring this change can give better capability (adaptation) to live in changed environmental conditions. Adaptations arise as a co-operation of combinations of genes working in a harmonious balance. The maintenance of new adaptive characteristics is achieved through natural selection.

A special case of phenotypic adaptation is → **sensory adaptation**, i.e. a rapid adjustment of the function of the sense organs to the strength of the stimulus.

adapter, also **adaptor,** one that adapts, e.g. a connective part of an apparatus for adjusting two pieces to fit together, such as adapters for centrifuge tubes to fit the rotor. *See* adaptor.

adaptive, having a tendency or capability of → adaptation.

adaptive divergence (L. *divergere* = to bend away), an evolutionary situation where some external barriers divide a population into two or more distinct subpopulations, and by preventing the gene flow between them makes genetic isolation and adaptation to local environmental factors possible. *See* adaptive radiation.

adaptive enzyme, an old term for → inducible enzyme.

adaptive field, adaptive landscape; a three-dimensional, symbolic, topographic presentation of the significance of changes in gene frequencies for the adaptability of populations.

In the graph the genotypes fit for adaptation inhabit adaptive peaks and less suitable genotypes the adaptive valleys. *See* group selection.

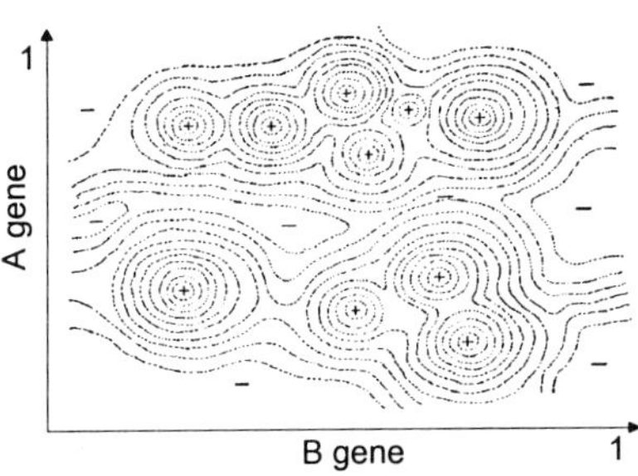

Adaptive field. (+ = adaptive peak, - = adaptive valley). Adaptive peaks represent favourable combinations of gene frequencies and adaptive valleys unfavourable ones. (After Wright - Proc. Sixth Int. Cong. Genetics 1, 1932).

adaptive immune response, → acquired immune response.

adaptive norm, the constitution of → gene frequencies of a population adapted to its environmental circumstances.

adaptive radiation, the evolutionary divergence of a group of organisms having the same phylogenetic origin; through selection, this often leads in a relatively short time interval to the formation of many new evolutionary lines. The beginning of adaptive radiation is often preceded by extensive changes in environmental conditions and consequently by large scale extinction of species, or by the movement of the group to completely new areas or new ecological zones (*see* adaptive zone).

adaptive value, → fitness.

adaptive zone, an evolutionary concept connected with extensive changes in → mega-evolution of organisms, like the shift from aquatic life to terrestrial life or from land to air. According to the concept, the environment is divided into adaptive zones which differ dramatically from each other, e.g. an aquatic environment as opposed to dry land. Certain basic preadaptations in organisms, as e.g. the lungs for air respiration in animals, are conditions for conquering a new adaptive zone. Each adaptive zone is divided into **ecological zones**, as e.g. oceans versus fresh water in aquatic environments, or tundra versus coniferous forests in terrestrial environments, and these further into → **niches**.

adaptor, 1. in genetics, a short two-stranded

DNA molecule that includes the binding sites for → restriction enzymes by means of which DNA molecules can be combined in a controlled way; **2.** adaptor protein; **3.** → adapter.

adaptor protein, also **adapter protein,** a protein that can attach other proteins together, e.g. → actinin or → vinculin.

Ada-response (Ada < adaptive), a response of a bacterial cell during exposure to an → alkylating agent; in the response several genes are activated to protect the cell against the mutagenic effect of the agent.

adaxial (L. *ad* = to, Gr. *axon* = axle), directed towards the axis, facing the axis or stem, situated on the same side as the axis; e.g. pertaining to the upper surface of a leaf. *Cf.* abaxial.

addiction (L. *addicere* = to favour), a reversible psychophysiological condition generated in an animal with highly developed brain, resulting in a habit beyond voluntary control, particularly induced by a habit-forming stimulant such as a drug or alcohol; addiction is especially marked in humans, comprising both physical (physiological) and psychologic (emotional) **dependence** on the drug.

Addison's disease (*Thomas Addison,* 1793—1860), chronic adrenocortical insufficiency; a human disease due to the hypofunction of the adrenal gland with decreased secretion of adrenocortical hormones, especially → glucocorticoids, resulting in e.g. weakness, fatigue, decreased blood pressure, and decreased tolerance to stress situations. *Cf.* Cushing's disease.

additive (L. *additivus*), characterized or produced by addition; in biology, e.g. **1.** pertaining to a process characterized by addition, as the **additive effect** of two substances in an organism; **2.** a substance deliberately added into foodstuffs e.g. as a preservative. *See* additive genes.

additive genes, genes interacting but showing no → dominance if they are alleles, or showing no → epistasis if they are non-alleles; i.e. they are genes whose effects can be summed.

additive genetic variance, the variance between individuals in a given population caused by the additive effect of allelic and/or non-allelic genes. *Cf.* variation.

adduct (L. *adducere* = to draw toward), **1.** to move or draw toward the median plane or axial line of the body; **adduction**, the process of such a moving; **adductor**, any muscle that adducts; *cf.* abduct; **2.** DNA adduct, a covalent bond between a chemical mutagen and DNA.

aden(o)- (Gr. *aden* = gland), denoting relationship to a gland.

adenine, 6-aminopurine, $C_5H_5N_5$, one of the two major purine bases (the other is guanine) formed from one six-atomic and one five-atomic heterocyclic ring; occurs in → nucleosides, → nucleotides, → nucleic acids (DNA and RNA), and → cytokinins.

adenoblast (Gr. *blastos* = germ), an embryonic gland cell in animals developing into a mature secretory cell, i.e. **adenocyte**.

adenocarcinoma, glandular cancer, carcinoma; a malignant neoplasm of epithelial cells growing in a glandular or gland-like pattern.

adenohypophysis, the anterior and intermediary parts of the → pituitary gland (hypophysis) of vertebrates, i.e. the glandular tissue which embryonally derives from → Rathke's pouch; includes the pars distalis, pars intermedia, and pars tuberalis. *Cf.* neurohypophysis.

adenoid, resembling a gland.

adenoma, a benign tumour derived from a glandular epithelium.

adenosine, A, a nucleoside composed of → adenine and ribose.

adenosine diphosphate, ADP, *see* adenosine phosphates.

adenosine monophosphate, AMP, *syn.* adenosine phosphate, adenylic acid, adenylate. *See* adenosine phosphates.

adenosine nucleotide, a → nucleotide composed of adenine, pentose sugar (ribose or deoxyribose), and a phosphate group.

adenosine phosphates, → nucleotides composed of adenine, D-ribose (or deoxyribose), and one, two, or three phosphate groups, forming adenosine monophosphate (**AMP**, or **dAMP** with deoxyribose), adenosine diphosphate (**ADP**, or **dADP**), or adenosine triphosphate (**ATP**, or **dATP**). AMP (adenylate, adenylic acid) bears a phosphate group at position 5' of the ribose moiety, but 2' and 3' monophosphates also exist. The cyclic form of it, i.e. adenosine 3',5'-cyclic monophosphate (*see* cyclic AMP), acts as a → second messenger in cells. Addition of one further phosphate group to AMP, leads to the synthesis of ADP. This is an important agent in energy metabolism of all living cells, for it is converted via photosynthetic phosphorylation (*see* photosynthesis) and → oxidative phosphorylation into → adenosine triphosphate, ATP, which is the most important high-energy compound in cells.

adenosine triphosphatases, ATPases, adenosine 5'-triphosphatases, an enzyme subsubclass EC 3.6.1.; enzymes catalysing the hydrolysis of adenosine triphosphate (ATP) to adenosine diphosphate (ADP) and orthophosphate (P_i). ATPases are large multimeric proteins embedded in the cell membranes or in membranes of cell organelles. The energy liberated from ATP by ATPase is used for the active ion pumping across the cell membranes or contractile activity of intracellular filaments. **P-type ATPases** are the class of ion-pumping enzymes including Na^+,K^+-ATPase, H^+-ATPase and Ca^{2+}-ATPase. They are also called **E_1E_2-ATPases**, referring to the fact that they pass through two different conformational states during each catalytic cycle. **V-type ATPases** are found in tonoplast, lysosomes and storage vesicles. The ATPases in the inner membrane of → mitochondria and thylakoid membranes in → chloroplasts are of **F-type (F_1F_0 type) ATPases**. These enzymes are driven in reverse by the high proton gradient and proton motive force, and thus synthesize, rather than use ATP, and are therefore called **ATP synthases**. *See* adenosine phosphates, ion pumps, proton pump, oxidative phosphorylation.

adenosine triphosphate, ATP, an energy-rich compound produced from ADP in energy-yielding processes, mainly by → cell respiration and → photosynthesis; used by cells as their direct energy source. The hydrolysis of ATP yields adenosine diphosphate (ADP) and an inorganic phosphate residue (P_i). The enzymic transfer of this terminal residue to many different substances means transferring of free chemical energy (30.6 kJ/mole) to catabolic reactions. When ATP is hydrolysed into AMP, pyrophosphate (PP_i) is liberated in the reaction. *See* adenosine phosphates, creatine, adenosine triphosphatases.

adenovirus, any virus of the family Adenoviridae; adenoviruses are double-stranded DNA viruses (class I viruses) developing in cell nuclei of infected birds and mammals, causing infections especially in upper respiratory tracts. Also called adenoidal-pharyngeal-conjunctival virus (A-P-C virus).

adenylate, adenosine monophosphate, AMP. *See* adenosine phosphates.

adenylate cyclase (Gr. *kyklos* = circle), an enzyme, EC 4.6.1.1, of lyase type located in the cell membrane of eukaryotic cells. It catalyses the synthesis of cyclic AMP (cAMP), i.e. the reaction: **ATP —> adenosine 3',5'-cyclic monophosphate** (cyclic AMP, cAMP) + **pyrophosphate**. The enzyme is activated by the linkage of a hormone or neurotransmitter to its receptors on the cell membrane. cAMP acts then as an important → second messenger inside the cell. The enzyme activity is modulated by a → G protein that dissociates from the receptor complex when the hormone (or neurotransmitter molecule) binds and either activates (Gs protein) or inhibits (Gi protein) adenylate cyclase. Earlier the enzyme has been called **adenyl cyclase**. An analogous enzyme, **guanylate cyclase**, catalyses the formation of another second messenger, cyclic GMP, from GTP (*see* guanosine phosphates).

adenylic acid, adenosine monophosphate, AMP. *See* adenosine phosphates.

adequate stimulus, specific stimulus, i.e. that energy form to which the sensory receptor is most sensitive, e.g. light stimulus to photoreceptors of the eye, or mechanical stimulus (vibration) to hair cells of the ear. Also **inadequate stimuli** (if strong enough) may elicit the specific response, as e.g. a strike on the eye produces sensation of light.

ADH, 1. antidiuretic hormone, → vasopressin; **2.** → alcohol dehydrogenase.

adhesion (L. *adhaerere* = to stick to), adhesiveness; e.g. **1.** the ability of particles or molecules to stick or **adhere** to each other (physical attraction); **2.** the adhesion of water on the walls of vessels and tracheids in plants; **3.** → cell adhesion; **adherence**, act or state of adhering or growing together; **adherent**, adhering, sticking.

adhesive proteins, adhesive molecules for cell-cell adhesion; e.g. → cadherins, selectins, and immunoglobulins.

adip(o)- (L. *adeps*, gen. *adipis* = fat), denoting relationship to fat or adipose tissue in animals.

adipocyte (Gr. *kytos* = cavity, cell), fat cell (in animals). *See* adipose tissue.

adipose tissue, 1. white adipose tissue, white fat; a type of areolar connective tissue of animals, the cells (**adipocytes**) of which are enlarged by gathering neutral fats, → triacylglycerols, so much that the whole cytoplasm may be filled with fat pushing the nucleus flattened against the plasma membrane. The adipose tissue usually locates between the inner organs serving as an energy store, but in many homoiothermic animals also under the skin (subcutaneous fat) serving as a heat insulator (especially in aquatic mammals). Many hormones, such as → adrenaline, glucocorticoids, glucagon, and growth hormone, are involved in the regulation of fat deposits and release in tissues; → **lipases** hydrolyse triacylglycerols to fatty acids and glycerol; **2.** → **brown adipose tissue**.

adiposis, excessive accumulation of the body fat; obesity.

aditus, pl. **aditus** (L.), an entrance or opening to an organ, cavity, or channel.

adjuvant (L. *adiuvare* = to aid), **1.** aiding, assisting; **2.** a substance or mixture of different substances which aid another substance to a better function; e.g. Freund's adjuvant enhancing the production of antibodies by an antigen.

ad libitum (L.), abbr. **ad lib.**, according to desire, as food and water given freely to laboratory animals during an experimental period.

administration, in physiology and pharmacology, the application of a substance, e.g. a drug, to an animal.

A-DNA, a quite rare form of deoxyribonucleic acid (DNA) conformation being a right-handed double helix coil; less hydrated than → B-DNA.

adolescence (L. *adolescere* = to grow up), the transitional period between puberty and adulthood when the somatic growth ceases and sexual maturity has developed. *Adj.* **adolescent**.

adoral (L. *ad* = near, to, *os*, gen. *oris* = mouth), near or towards the mouth.

ADP, adenosine diphosphate. *See* adenosine phosphates.

adrenal (L. *ren* = kidney), located near the kidney, or pertaining to the → adrenal gland.

adrenal cortex, *see* adrenal gland.

adrenal gland (L. *glandula suprarenalis*, Gr. *epinephros*), **suprarenal gland, epinephric gland**; a pair of endocrine glands of vertebrates located close to the kidneys. In tetrapods, the gland is developed by fusion of two different embryonal tissues: the **cortex** derived from the coelomic mesoderm, and the ectodermal **medulla** differentiated as a sympathetic ganglion from the neural crest. In cyclostomes and most fish species, fusion of the parts has not occurred but the tissues form two separate organs. The adrenal cortex comprises three tissue zones secreting steroid hormones, **corticosteroids**; the outermost zone (*zona glomerulosa*) produces → mineralocorticoids, the middle zone with the columns of cells (*zona fasciculata*) secretes → glucocorticoids, and the innermost zone (*zona reticularis*) produces small amounts of → sex hormones. Secretion is regulated mainly by → corticotropin (ACTH) released from the anterior → pituitary gland. The adrenal medulla contains **chromaffin cells** which originally are postganglionic sympathetic neurones but have lost their axons. The cells are activated by the → sympathetic nervous system, one cell type secreting → **adrenaline**, and the other → **noradrenaline**, into the blood; the proportion of the cell types varies greatly between animal species. These hormones strongly increase the effectiveness of the sympathetic nervous system and prolong its actions. *See* stress.

adrenaline, Am. **epinephrine**; 3,4-dihydroxy α-(methylaminoethyl)benzyl alcohol; a monoamine hormone secreted with → noradrenaline by chromaffin cells in the adrenal gland and other → **chromaffin tissues** in vertebrates. In lower vertebrates, adrenaline acts also as a neurotransmitter functioning in certain brain → synapses and in nerve endings of the sympathetic nervous system. Together with → noradrenaline and dopamine, adrenaline belongs to **catecholamines**; phenylalanine is the precursor of their biosynthesis.

Adrenaline is secreted quickly in → **stress** conditions when the sympathetic nervous system is activated (fight or flight), and improves the capacity of an animal to cope with changed conditions; e.g. adrenaline increases the heart rate and contraction force, causes vasodilatation in brain, heart, and skeletal muscles, enhances breathing capacity and energy metabolism; it also raises the arousal level and decreases the sense of pain. Adrenaline is evolutionally an old messenger substance functioning also in the nervous system of most invertebrates, and is e.g. involved in the regulation of the embryonic development already prior to the development of the nervous system. *See* adrenergic receptors.

adrenal medulla, *see* adrenal gland.

adrenergic (Gr. *ergon* = work), pertaining to nerves, nerve fibres, and receptors involved in actions of → adrenaline and noradrenaline in animals, i.e. neuronal and hormonal functions of these substances or drugs which mimic actions of the → sympathetic nervous system. *Cf.* cholinergic.

adrenergic receptors, adrenoreceptors, adrenoceptors, protein molecules embedded in the cell membranes of certain neurones and effector cells (muscles, glands) in animals, mediating actions of → **adrenaline** and **noradrenaline**. The receptors are divided into two main groups, **alpha receptors** (α-receptors) and **beta receptors** (β-receptors), mainly on the basis of their pharmacological properties in vertebrates. The former types mediate e.g. vasoconstriction in the skin, the latter e.g. acceleration of the heart, bronchodilatation, and increase of metabolic activity. Both types have 2 or 3 subtypes (α_1, α_2, β_1, β_2, β_3), which have further several subtypes. Although adrenaline and noradrenaline are able to act through all these receptors, their affinity to different receptors varies. In many tissues, the effects of adrenaline are mediated mainly by beta receptors, and those of noradrenaline by alpha receptors.

In synapses, adrenergic receptors usually locate posterior to the synaptic cleft (**postsynaptic receptors**) transmitting nerve impulses from neurone to neurone, or to an effector cell. Some receptors may locate prior to the cleft on the presynaptic membrane (**presynaptic receptors**, autoreceptors), and when noradrenaline binds to them, its own release reduces, i.e. through these receptors, noradrenaline inhibits the function of the adrenergic synapse. For medical use many selective **adrenergic agonists** and **antagonists** (alpha and beta blockers) have been developed synthetically.

adrenocortical, pertaining to the adrenal cortex, or to hormones secreted by its cells. *See* adrenal gland.

adrenocorticotrop(h)ic hormone, ACTH, adrenocorticotrop(h)in, → corticotropin.

adrenocorticotropic hormone-releasing hormone/factor, ACTH-RH, ACTH-RF, → corticoliberin.

adrenolytic (Gr. *lysis* = dissolution), inhibiting or antagonizing adrenergic influence in tissues, i.e. inhibiting the action of → adrenaline or noradrenaline, or of similar sympathomimetic drugs.

adrenoreceptors, → adrenergic receptors.

adrenosterone, 4-androstene-3,11,17-trione; an androgenic steroid hormone produced by the cortex of the → adrenal gland.

adsorbent (L. *ad* = to, *(ab)sorbere* = to suck, absorb) a substance on which some other substance, **adsorbate** (a gas, liquid, or dissolved substance), is attracted and held, i.e. adsorbed; e.g. kieselguhr (diatomaceous earth) in column → chromatography, or an antigen or antibody in immunochemistry. *Verb* to **adsorb**. *See* adsorption.

adsorption, gathering of substances onto the surface of any structure or substance; e.g. the adsorption of nutrient ions on the negatively charged surface of soil → colloids, or the adsorption of impurities on the particles of charcoal. The adsorption technique is used e.g. in → chromatography. *See* adsorbent. *Cf.* absorption.

adventitia (L. *adventicius* = foreign, coming from outside, outermost), the outermost coat of loose connective tissue covering an organ and connecting it to its surroundings; specifically **tunica adventitia**, the outermost coat of various tubular structures, covering e.g. the alimentary canal, respiratory tracts, and blood vessels of vertebrates. *Adj.* **adventitial**, pertaining to the tunica adventitia. *Cf.* adventitious.

adventitious, 1. arising from an external source; **2.** occurring spontaneously or accidentally in contrast to natural or hereditary; **3.** adventitial; *see* adventitia.

adventitious bud, a plant → bud formed in an unusual place often from epidermal tissue due to some inductive factor, e.g. removal of the apical bud. The propagation of plants by cuttings is based on the formation of adventitious buds.

adventitious embryony, *see* agamospermy.

adventitious root, a plant root not arising from the primary root, being formed in an abnormal site e.g. from a node of the stem or leaf cutting. *Cf.* aerial root. *See* root.

adventive organs, organs which grow on unusual positions on a plant and develop later than the normal organs; e.g. adventitious roots may develop in stem nodes.

aecidiospore (Gr. *aikia* = injury, torture, *sporos* = spore), one spore type of rust fungi, formed in a cup-shaped **aecidium**. *See* Uredinales.

aer(o)- (Gr. *aer* = air), pertaining to air, atmosphere, aviation, or a gas containing oxygen.

aerenchyma (Gr. *enchyma* = cast, poured in), a type of plant → parenchyma tissue, characterized by large intercellular air spaces; most typical of water plants in which it transports air and gases to and from the submerged parts.

aerial plankton, air plankton, aeroplankton (Gr. *planktos* = wandering), *see* aerobiology.

aerial root, an adventitious root formed in plant shoots above the ground, common especially in tropical plants; e.g. in *Philodendron* and many other aracean plants the aerial roots grow to the ground and turn into normal roots. In many epiphytic orchids the aerial roots are hanging and absorb water from the air, never growing down to the ground.

aerial shoot, a plant shoot (stem with leaves) growing totally above the ground.

aerial stem, a plant stem growing above ground. *Cf.* rhizome.

aerobe (Gr. *bios* = life), an organism, usually a microbe, which requires molecular oxygen for living (**obligate aerobe**), or can live in the presence of molecular oxygen (**facultative aerobe**). *Syn.* aerobiont. *Cf.* anaerobe.

aerobic, 1. pertaining to the presence of molecular oxygen; **2.** living in the presence of air or molecular oxygen; **3.** pertaining to the oxidative respiration (aerobic respiration).

aerobic respiration, the part of → cell respiration in which reactions takes place only in the presence of oxygen; comprises the → citric acid cycle, → electron transport chain, and → photorespiration in plants.

aerobiology (Gr. *bios* = life, *logos* = word, discourse), the study dealing with organic aerial particles which are floating in airstreams; the particles, living or dead, include e.g. pollen grains, spores, seeds, bacteria, and viruses. The research topics in aerobiology are the origin, dispersal, and biological and environmental effects of these particles (aerial plankton, aeroplankton, air plankton).

aerobiont, → aerobe.

aeroplankton, aerial plankton, air plankton, *see* aerobiology.

aerosol (sol = a fluid colloidal system), fine liquid or solid particles in air or gas, such as smoke, fog, mist, or a solution or suspension dispersed as a fine mist out of a container by means of a propellant gas.

aestivation, Am. **estivation** (L. *aestas* = summer), **1.** in zoology, **summer dormancy**; the dormant state of certain animals, such as some amphibians and lungfish during summer or dry seasons; characterized by decreased activity of biochemical, physiological, and behavioural functions. A similar state is also the → **diapause** of many arthropods; *cf.* hibernation; **2.** in botany, a pattern in which sepals and petals are arranged in a flower bud before the opening of the flower. Aestivation types are numerous and they have been used in systematics; nowadays, however, they are considered to be unreliable as taxonomic characters.

aethalium, pl. **aethalia** (Gr. *aithalos* = thick smoke, soot), a structure formed by slime moulds (*see* Myxomycetes); a single, large → sporangium formed of the whole plasmodium, producing spores.

aetiology, etiology (Gr. *aitia* = cause, *logos* = word, discourse), the study of the causes of diseases; **2.** the cause or origin of a disease. *Adj.* **(a)etiologic**.

afferent (L. *afferre* = to bring to), bringing to or leading towards the centre of an organ or organ system, as an afferent nerve passing to the central nervous system, or an afferent blood or lymphatic vessel leading to an organ. *Cf.* efferent.

affinity (L. *affinis* = neighbouring), in biology, an attraction for binding, as **1.** attraction of certain atoms and molecules to unite, as in binding of an enzyme to its substrate, a hormone to its receptor, or an antigen to its antibody; **2.** the selective staining of a tissue with a certain dye; **3.** affinity → chromatography, with an immobile phase which has a special biochemical affinity for a substance to be separated.

affinity index, in ecology, an index presenting the measure of the similarity in species composition in two communities. The index is measured from the equation $A = C/(a + b)^{1/2}$, where A is the affinity index, a is the number of the species in community 1, b is the number of the species in the community 2, and C is the number of species found in both communities.

aflatoxins (afla < *Aspergillus flavus*, Gr. *toxikon* = arrow poison), very toxic coumarin-like substances produced by *Aspergillus* fungi. Foodstuffs like cereals and peanuts may be contaminated by aflatoxins under bad storage conditions, and they may cause severe poisoning, **aflatoxicosis**, in domestic animals and even in man. For example, poultry, particularly turkeys, die sometimes in large numbers because contaminated crushed peanuts accidentally have been used as their food. Aflatoxins have a rapid toxic effect on liver cells, and in addition, they are shown to be teratogenic and carcinogenic. Aflatoxins are characterized as aflatoxin groups B, G, and M. Type B_1 is one of the most potent environmental mutagens and carcinogens known; its lethal dose in day-old ducklings is 30—40 µg/100 g body weight. Aflatoxins are metabolized by liver microsomal systems, yielding products that bind covalently to DNA and induce DNA adducts.

after-ripening, a process in which the plant seed stays dormant although the seed and embryo are fully developed; the dormancy breaks after a period during which the after-ripening takes place. The seed is able to germinate after this process which is related to chemical changes in the structures, possibly to hormones or growth inhibitors.

agam(o)- (Gr. *agamos* = unmarried), asexual. *Adj.* **agamic, agamous**.

agamete, any non-copulating germ cell, usually a spore; agametes are produced as products either of meiosis or mitosis.

agamogony (Gr. *gone* = birth), a form of asexual reproduction in which a new individual arises from a single cell. Agamogony occurs among unicellular organisms in three forms: a) in the form of a simple division of an individual into two similar new individuals, b) in the form of → budding, c) in the form of cell division in which the nucleus first divides many times and then the cells divide into equal numbers of daughter cells, i.e. new individuals.

agamospecies, a species including only asexually reproducing individuals. *See* apomixis.

agamospermy (Gr. *sperma* = seed), a type of → apomixis that may occur in higher plants; reproduction occurs without fertilization, and it includes → **vegetative reproduction** and **agamospermy**. Agamospermy comprises asexual reproduction through seed formation and it is divided into **adventitious embryony** and **gametophytic apomixis**. In adventitious embryony the plant embryo develops directly from a diploid cell of the → nucellus or of an integument of the → ovule, in gametophytic apomixis from the → embryo sac. Adventitious embryony is common in the genera *Citrus* and *Allium*. Gametophytic apomixis may occur on haploid or diploid level. A haploid embryo may be formed without fertilization from an → egg cell or some other haploid cell of the embryo sac; the phenomenon is called **apogamy**. The vitality of haploid individuals is mostly weak and this form of apomixis is not a permament type of reproduction.

Diploid gametophytic apomixis produces successful descendants, the type being common in many plant groups. The diploid embryo sac formed may be derived from a vegetative cell of nucellus or integuments; the phenomenon is called **apospory**, occurring e.g. in *Hieracium* species (hawkweed) and in *Ranunculus auricomus* (goldilocks). The diploid embryo sac may also be formed from its mother cell in an abnormal way, i.e. by **diplospory** in which → meiosis may be totally lacking (**gonial apospory**) or the chromosome number may not be reduced in meiosis as usually (**aneuspory**); e.g. *Taraxacum* (dandelion).

Some plants need → pollination, although the embryo is formed without fertilization. For example in *Ranunculus auricomus*, the formation of the → endosperm presupposes the fertilization of the polar nucleus in the embryo sac.

The ability to form seeds without fertilization is considered to be a useful property. In some plant species agamospermy is regular (e.g. *Taraxacum* and *Alchemilla*, lady's mantle), but some species have also preserved their ability for sexual reproduction, e.g. *Potentilla* (cinquefoil) and *Hieracium* species. The latter characteristic is called **facultative apomixis**. *Cf.* parthenogenesis.

agamospory (Gr. *sporos* = seed), formation of a plant embryo and seed without fertilization, asexually.

agar, also **agar-agar** (Singhalese = seaweed), gelose; a gelatin-like polymer of → galacturonic acid extracted from some seaweeds, used e.g. for solidifying the culture media for bacteria, fungi, or in tissue and cell cultures.

Agaricales, *see* agarics.

agarics, a group of basidiomycete fungi, the order Agaricales in the class Basidiomycetes, subclass Holobasidiomycetidae, consisting of well-known fungal groups: mushrooms, toadstools and boletes. These fungi have stalked fruiting bodies, in which the spores are formed on the surface of gills or in pores.

agaritin, a substance occurring in minute quantities in the champignon mushroom (*Agaricus bisporus*); possibly has weak mutagenic properties.

agarose, a neutral fraction isolated from → agar; a large polysaccharide that forms a transparent gel used in → electrophoresis and column chromatography, e.g. for analysing nucleic acids.

age distribution, age structure; the distribution of the individuals of a population by its age-groups. In population ecology, the groups are divided in relation to reproduction, i.e the **prereproductive**, **reproductive**, and **postreproductive age-groups**. A **stable age distribution** prevails in the population which grows exponentially (*see* exponential growth) so that the proportions of different age-groups remain unchanged. A **stationary age distribution** prevails in a population which has reached a constant size and where birth rate and death rate are equal. In general, the stationary age distribution is hypothetical and is determined in accordance with death rates of different age-groups.

ageing, Am. **aging,** senescence; the gradual progressive decline in fitness of individuals of multicellular organisms and of their somatic organs and cells, ultimately leading to → death. Most unicellular organisms are potentially immortal, and no ageing phenomena can be observed in them. Ageing does not involve the cells of the → germ line of multicellular organisms. The basis for ageing phenomena are changes in individual cells; the chemical reactions of the cells slow down, and the frequency of cell divisions decreases.

Ageing of the cells and tissues varies greatly according to species, and is thus linked to the → life cycle of the organism. Actually, ageing begins only after the reproductive age has passed. Because different organisms age in different ways and at different rates, ageing must have a genetic basis. However, it is not equally clear whether the differences in ageing between different individuals of the same species are genetically programmed or due to stochastic events. The causes may involve the accumulation of changes in individual cells. The changes may include slowing down of chemical reactions, accumulation of errors in biosynthesis, and reduction in the frequency of cell division, leading to a failure in cellular homoeostasis.

The **theories of ageing**, more than 300 presented, can be divided into theories of programmed ageing, stochastic theories of ageing, and evolutionary theories of ageing. According to the programmed theories, ageing is controlled by genes in the same manner as development from the fertilized egg to adulthood. Stochastic theories emphasizes that ageing is due to more or less random wear and tear phenomena which cause errors in cellular macromolecules. Errors begin to accumulate when the reproductive age has passed, and there is no more need to maintain structures by wasting energy for the repair of errors in somatic cells. Evolutionary theories of ageing explain the differences in the ageing between species, linking ageing to the rest of the life cycle and ecology of the species. *See* programmed cell death.

agenesis, pl. **ageneses** (Gr. *a* = not, *gennan* = to produce), absence of development; incomplete or imperfect development of an organ or its part.

agent (L. *agere* = to act, do, drive), any substance or factor that is able to produce a physical, chemical, or biological effect, which may be either stimulatory (**activating agent**) or inhibitory (**blocking agent**).

ageotropism, → apogeotropism.

age pyramid (Gr. *pyramis* = pyramid), a graphical figure showing the number or percentage of individuals in a population by different age-groups. The graph is divided horizontally into three parts so that the young, prereproductive individuals are at the bottom, the reproductive ones in the middle, and the postreproductive ones on the top of the pyramid. The pyramid of an **increasing population** has a broad base owing to the large portion of young individuals; in a **stable population** birth rate and death rate are equal and the two youngest age-groups are nearly of the same size. The pyramid of a **decreasing population** has a narrow base owing to the small proportion of young individuals.

age structure, → age distribution.

agglomeration (L. *ad* = together, *glomus* = mass, ball), a mass formed of particles gathered together, especially an agglomeration of cells. *Syn.* aggregation.

agglutination (L. *agglutinare* = to glue to), the process of being glued together; e.g. **1.** in immune reactions, the clumping of antigen-bearing cells or particles by the action of specific antibodies (**agglutinins**). A protein antigen on the cell surface eliciting this kind of antibody reaction is called **agglutinogen**; the agglutination of red blood cells by certain plasma antibodies is used for determination of → blood groups; *see* lectins; **2.** the clumping of tissue in the healing of wounds.

aggregated dispersion, clumped distribution, *see* dispersion.

aggregate fruit (L. *aggregare* = to add to), a fruit formed from an apocarpous gynaecium of a single flower, the fruitlets arising from separate ovaries, as in the raspberry. One type of aggregate fruit is the → infructescence or multiple fruit, in which the fruitlets are formed from separate flowers of an inflorescence, as in the pineapple.

aggregation, a mass or group of distinct things, e.g. of cells or organisms; e.g. the aggregation of conspecific individuals of the same population without any social organization, or the process of such grouping. *Syn.* agglomeration.

aggressin (L. *aggressus* = attack), a substance produced by pathogenic microbes postulated to inhibit the resistance of a host without actually being toxic.

aggression (L. *aggredi* = to approach, attack), an offensive behavioural action or threat of an animal, usually directed against a competitive individual (competitor) to force it to abandon something which is useful to the attacker. Aggression is not connected with predation. According to some theories, aggression consists mainly of drive actions (aggression drive), but it is also influenced by learning through experience. *Adj.* **aggressive**.

aggressive mimicry, *see* mimicry.

aggressiveness, an ability or inclination of an animal to behave offensively against other individuals.

Agnatha (Gr. *a* = not, *gnathos* = jaw), **agnathans, jawless vertebrates**; a superclass (class by some authors) of primitive aquatic vertebrates which are jawless and have poorly developed paired fins which may be absent in some species. The superclass includes two classes: **Cephalaspidomorphi** (e.g. lampreys) and **Pteraspidomorphi** (Myxini, e.g. hagfishes).

agnathous, jawless; pertaining to jawless vertebrates. *See* Agnatha.

agonist (Gr. *agonistes* = combatant), one that is engaged in a struggle; in biology, e.g. **1.** a contracting muscle with reference to its opposing muscle, **antagonist**; **2.** a substance, such as a drug, which can occupy physiological receptors producing a similar reaction as does the natural messenger, and is opposed by some antagonist. *Verb* to **agonize.** *Adj.* **agonistic**.

agonistic behaviour, a type of behaviour, usually found in aggressive interactions of animals, which besides attacking elements have also features of a retreating or → appeasement behaviour (submissive behaviour). The agonistic behaviour is common especially in dense populations where environmental resources are scant and a part of the individuals (usually young) are forced to escape from the population.

agony (Gr. *agonia* = contest, struggle, anguish), extreme suffering; the death struggle.

agranulocytes (L. *a-* = not, *granulum*, dim. of *granum* = grain, Gr. *kytos* = cavity, cell), non-granular leukocytes; the term sometimes used for those white blood cells of vertebrates which have no secretory granules in the cytoplasm, thus including the → lymphocytes and → monocytes but excluding the → granulocytes.

agranulocytosis, an abnormal condition in vertebrates characterized by a decreased number of → granulocytes in the blood.

Agrobacterium, a genus of bacteria, the best known species being *A. tumefaciens* which causes → crown gall tumours in plants by transferring the tumour-inducing DNA plasmid (Ti plasmid) into plant cells. The Ti plasmid becomes integrated into the DNA of the host cell where it is capable of independent replication and therefore the plasmid is widely used as a DNA → vector in gene manipulation of plants, e.g. in the production of transgenic plants.

ahemerobe (Gr. *hemeros* = tame), an original

environment which has not been changed by man.

ahnfeltan, a colloidal matter in cell walls of some red algae, → Rhodophyta.

A-horizon, the uppermost, often bleached layer of soil, e.g. of podzol. *Cf.* B-horizon, C-horizon. *See* soil formation.

AIDS, acquired immunodeficiency syndrome; a disease caused by the human immunodeficiency virus (HIV) transmitted by exchange of body fluids, such as blood, semen, or transfused blood products. The disease is characterized by opportunistic infections particularly due to the depletion of helper → lymphocytes.

air bladder, 1. → swim bladder; **2.** in plants, a roundish air bag in the thallus of some brown algae, e.g. in the bladder wrack (*Fucus vesiculosus*) and in the knotted wrack (*Ascophylum nodosum*), or an air bag occurring in pairs in pollen grains of many coniferous trees, e.g. in pines. *See* pneumatophore, buoyancy.

air plankton, aeroplankton, aerial plankton, *see* aerobiology.

air pressure, → atmospheric pressure.

air sacs, 1. thin-walled sac-like extensions of respiratory organs in birds, i.e. originally nine membranous bags (the paired anterior thoracic, cervical, posterior thoracic, abdominal air sacs, and the unpaired interclavicular air sac), that branch from the lungs to different parts of the body, also into the cavities of larger bones; in some species one or more pairs may be fused; **2.** thin-walled expansions of tracheae as parts of the respiratory system in some insects, e.g. in grasshoppers; from the sacs the insect pumps air into the tracheal system by alternate contractions and relaxations of the abdominal body wall.

airway, air passage; in anatomy, any route for passing air into respiratory organs of air-breathing animals; e.g. the → trachea.

akinesia, (Gr. *a-* = not, *kinein* = to move), **1.** the loss of activity, or impairment of power of the voluntary muscles; **2.** the postsystolic phase (rest interval) of the heart. *Syn.* akinesis.

akinesis, pl. **akineses,** absence of motion, immovability; e.g. **1.** → amitosis; **2.** → akinesia.

akinete, a thick-walled resistant resting cell in → cyanobacteria and some → green algae for retaining the organism over unfavourable periods.

akinetic (Gr. *akinetos* = motionless), **1.** immovable, motionless; **2.** pertaining to → akinesia or akinesis; **3.** → acentric, defin. 2.

akinetic skull, *see* cranium.

alanine, abbr. Ala, ala; symbol A; an amino acid occurring in two forms: 2- or α-aminopropionic acid (α-alanine, L-α-alanine, *see* isomers), $CH_3CH(NH_2)COOH$, occurring commonly in proteins of organisms, and 3- or β-aminopropionic acid (β-alanine), NH_2CH_2-CH_2COOH, that does not occur in proteins. α-Alanine is synthesized in cells by the reaction of pyruvate and glutamate, β-alanine is a decarboxylation product of aspartic acid. Alanine is a non-essential amino acid for animals.

albinism (L. *album* = white), a recessive genetic character of birds and mammals in which the pigmentation of the plumage, fur, and skin is totally absent because the production of tyrosinase enzyme is lacking. Thus, these animals are completely white with red eyes; the individual is called **albino**. Albinism is rare in natural conditions because the survival of albinos up to reproductive age is difficult. Many experimental animals are selectively albinos. The number of albinos varies much in human populations, being in average only a few individuals per 100,000. E.g. in humans also some other disturbances may change the production of pigment, sometimes causing white areas on the skin, like in vitiligo.

Also plants with pigmentless, white flowers are called albinos. *Cf.* leucism.

albuginea (L. *albugo* = white spot), a white layer of fibrous tissue, **tunica albuginea**; e.g. **1.** *albuginea oculi*, the sclera of the vertebrate eye; **2.** *albuginea testis*, albuginea of the testis, the immediate covering of the vertebrate testis; **3.** *albuginea ovari*, the tunica covering the vertebrate ovary.

albumen, egg-white; the white outer part in the eggs of birds, many reptiles, and some invertebrates such as gastropods (*see* albumin gland); a viscous protein solution containing chiefly → **albumin** (egg albumin, ovalbumin) and a glycoprotein, called **ovomucoid,** with certain minerals and vitamins. *See* egg.

albumin, a group of water-soluble proteins occurring e.g. in → blood plasma (serum albumin), milk (lactalbumin), and eggs (ovalbumin) of vertebrates, and as reserve protein in seeds of many plants; **albumoses**

are breakdown products of albumin produced by proteolytic enzymes. Bovine serum albumin (BSA) is commonly used in biological and biochemical laboratory work.

albumin gland, albumen gland; an accessory gland in the female reproductive system of gastropods, connected to the ovotestis by the hermaphroditic duct; secretes albumen around fertilized eggs.

albuminoid, albumoid, 1. resembling → albumin; **2.** any of insoluble animal proteins (scleroproteins) occurring especially in horny and cartilaginous tissues; e.g. keratin, collagen, elastin, and spongin.

albuminous cells, cells containing starch in the phloem area of conifer needles. *Syn.* Strasburger's cells.

alcohol (Arab. *al-kuhl* = powder of antimony, later the term was applied to anything impalpable such as spirit), any of several simple organic compounds chiefly formed in → fermentation, yielding carbon dioxide as a side product. An alcohol molecule contains one or more -OH groups, i.e. alcohols are monohydric, dihydric, trihydric, etc. (polyhydric alcohols); e.g. → **methanol** (methyl alcohol), **ethanol** (ethyl alcohol), **propanol** (propyl alcohol), and **butanol** (butyl alcohol) are monohydric, **ethylene glycol** is dihydric, and **glycerol** is trihydric. Alcohols are named according to their hydrocarbon moiety, e.g. methanol, CH_3OH, ethanol, CH_3CH_2OH, propanol, $CH_3CH_2CH_2OH$, butanol, pentanol, hexanol, etc.

The water solubility of alcohols decreases and lipid solubility increases when their hydrocarbon chain becomes longer. Alcohols have isomerism in their molecular structures (*see* isomers) depending on how many carbon atoms are bound to the carbon which has the hydroxyl group, i.e. one, two, or three carbon atoms, the alcohols being named respectively **primary**, **secondary**, or **tertiary alcohols**.

All alcohols are somewhat **toxic** to animals rendering the cell membranes more fluid that results in many additional cellular effects; they probably also have specific effects e.g. on receptor proteins. Most sensitive to alcohols are nerve cells in which alcohol especially changes the activity of interneuronal junctions, → synapses, e.g. activates inhibitory GABA receptors and inhibits excitatory glutamine receptors (NMDA receptors). Symptoms of effects of ethanol, the intoxicating component in alcohol drinks, appear especially in the brain cortex and so in the behaviour of higher animals. The toxicity is not always fatal because many animals (e.g. humans) have a natural detoxification system for ethanol, decreasing the blood alcohol level quite efficiently. For other alcohols this detoxication system is weak or lacking. Also plant tissues may contain low amounts of ethanol, concentrations are highest in plants growing in anaerobic conditions, e.g. in ice incasement. *See* alcohol dehydrogenase, anaerobic respiration.

alcohol dehydrogenase, ADH, alcohol NAD^+ oxidoreductase, EC 1.1.1.1; an enzyme which in the presence of NAD^+ catalyses the oxidation of ethyl alcohol or ethylene glycol into acetaldehyde or ketone, or the reverse reaction. ADH acts in microbial fermentation, and many animals have this enzyme mainly in the liver. It also occurs in plants, being important especially during → anoxia, when ethyl alcohol production occurs in → anaerobic respiration. There are also other alcohol dehydrogenases which use other hydrogen acceptors than NAD^+, e.g. **alcohol $NADP^+$ oxidoreductase**, EC 1.1.1.2.

Alcyonaria (Gr. *alkyonion* = a kind of sponge resembling the nest of the kingfisher, Gr. *alkyon*, L. *-aria* = connected with), **alcyonarian corals**; a cnidarian subclass in the class → Anthozoa; primitive animals with eight tentacles and unpaired complete septa. All species are colonial and the polyps are connected by a spiny skeleton and gastrodermal tubes (**solenia**); many species are known for their colourfulness; e.g. **soft corals**, **sea fans**, **sea pens**. Because of their octomerous symmetry the subclass is also called Octocorallia.

aldehyde (L. *alcohol dehydrogenatum* = alcohol from which hydrogen is removed), an organic compound having a carbon atom bound to one oxygen and to one hydrogen atom, i.e. the molecule has a -HC=O group. Common aldehydes are e.g. acetaldehyde (ethanal) oxidized from ethyl alcohol (ethanol), and → formaldehyde (methanal) oxidized from methyl alcohol (methanol).

aldehyde-lyases, enzymes of the sub-subclass EC 4.1.2. in the lyase class; catalyse the cleavage of a C-C bond in a molecule that contains a hydroxyl group and a carbonyl group, thus forming two smaller molecules (aldehydes or ketones). Called also aldolases.

aldolase, 1. generic name for → aldehyde-lyase; **2.** fructose bisphosphate aldolase, EC 4.1.2.13, an enzyme of the lyase class that in carbohydrate metabolism of cells catalyses the reaction D-fructose 1,6-bisphosphate —> dihydroxyacetone phosphate + D-glyceraldehyde 3-phosphate, in plant cells the formation of sedopheptulose 1,7-bisphosphate from erythrose 4-phosphate and dihydroxyacetone phosphate in the Calvin cycle of → photosynthesis; an important enzyme also e.g. in alcohol fermentation.

aldose, a monosaccharide having potentially an aldehyde group, -CHO; e.g. glucose (aldohexose), or glyceralhehyde.

aldosterone, the main mineralocorticoid hormone produced by the outermost layer, zona glomerulosa, in the cortex of the → adrenal gland in vertebrates; this steroid hormone regulates the mineral and water balance of the body by facilitating potassium (K^+) exchange for sodium (Na^+) in the distal renal tubules. This results in sodium reabsorption and potassium and hydrogen loss.

aleurone cells (Gr. *aleuron* = flour), a cell layer (**aleurone layer**) under the seed coat around the → endosperm in the → caryopsis of grasses (Poaceae), containing **aleurone grains**. The aleurone cells are metabolically active with high activity of hydrolytic enzymes; the aleurone grains contain storage proteins. During the germination of the seeds the cells secrete alpha-amylase to the endosperm, which causes the mobilization of the energy store for use of the growing embryo.

aleuroplast (Gr. *plastos* = moulded), a modified, colourless plastid, storing proteins in many seeds.

algae, eukaryotic green plants, which may be unicellular, filamentous, or form thalli. In the Plant kingdom algae form numerous divisions: → Chlorophyta (**green algae**), Charophyta (**stoneworts**), Phaeophyta (**brown algae**), Rhodophyta (**red algae**), Chrysophyta (**golden algae**), Xanthophyta (**yellow-green algae**), Euglenophyta (**euglenoids**), Pyrrophyta (**dinoflagellates**), and Cryptophyta (**cryptophytes**). Most algae are aquatic (marine or freshwater), but they may also live in damp soil, on rocks, and on tree trunks. Some species are utilized for producing useful materials, as e.g. → agar; some species, especially red algae, (e.g. *Porphyra*) are edible. The cyanobacteria have been called "blue-green algae", but they are more related to Bacteria than algae.

algal bloom, *see* bloom.

algal fungi, *see* Mastigomycota.

algin, sodium alginate; the sodium salt of → alginic acid.

alginic acid, a linear polysaccharide synthesized from β-(1→ 4)-D-mannosyl-uronic acid and α-(1→ 4)-L-gulosyluronic acid residues; occurs in red and brown algae, such as in giant kelp. Salts of alginic acid are **alginates** (acetylated polymers, e.g. algin), and some of them are also produced by some microorganisms, e.g. *Azotobacter vinelandii*. In the presence of calcium, alginates form viscous gels which are widely used for the immobilization of microorganisms, as fire retardants, in food processing, and for textile printing.

algology, a branch of science studying → algae. *Syn.* phycology.

alimentary (L. *alimentum* = food, nourishment), pertaining to food or nutrition.

alimentary canal, digestive tract, gut; the tubulous organ system where nutrients are digested into simpler chemical compounds and from which they are then absorbed into cells, blood, lymph (vertebrates), or haemolymph and interstitial fluids.

In primitive animals, as in cnidarians, ctenophores and turbellarians, the canal is a branched sac (→ gastrovascular system) with only one opening; in more developed animals, the canal includes both the **mouth** and **anus**. In some endoparasites, as in cestodes, the digestive tract is totally absent. In more developed invertebrates, the tract comprises both the mouth and anus and a long canal between them divided into several sections: e.g. in insects the **foregut**, **midgut** and **hindgut**, or as in molluscs the **oesophagus, stomach** and **intestine**. The anterior part of the canal may include also a modified **crop** and gizzard, as in annelids.

In vertebrates, the alimentary canal comprises the **mouth** (oral cavity), **pharynx**, **oesophagus**, **stomach**, and **intestine**. In some birds, a part of the oesophagus is expanded to form the crop, and the stomach may be formed from different sections, as the **proventriculus** and **gizzard**. In some mammals, as in ruminants, the oesophagus is divided into several sections (*see* rumination). The intestine of higher vertebrates is differentiated into small and large intestine.

The histological structure of the vertebrate alimentary canal is principally similar in all parts. The **mucous coat** (**mucous tunic**, *tunica mucosa*), lining the inner surface of the whole canal, has a stratified epithelium in the mouth, pharynx, oesophagus, and anal regions, but a single-layered (simple) columnar epithelium in the stomach and intestine. The connective tissue layer below the epithelium contains smooth muscle cells, and numerous digestive glands (→ gastric glands, intestinal glands) derived from epithelial cells. The next layer under the mucosa is **tunica submucosa**, the thick loose connective tissue layer with some glands. Under the submucosa lies **tunica muscularis** (muscular tunic) with two sublayers of smooth muscles (circular and longitudinal), in the stomach also a third layer, the oblique smooth muscle layer. The mouth, anterior oesophagus, and anal areas have, instead of the smooth muscle layers, the corresponding skeletal muscle layers. The outer surface is covered by connective tissue forming **tunica adventitia** which on free surface of the canal is covered by a thin epithelium, forming **tunica serosa** (serous tunic). In all layers beneath the epithelium are dense blood and lymph vessel nets, lymph nodes, and autonomic nerve fibres from two nerve plexuses.

The ducts from the → salivary glands open into the mouth, and the hepatic and pancreatic ducts into the anterior end of the intestine. The structure and the length of the canal correlates well with the quality of food, herbivores having a much longer and more complex canal than carnivores.

aliphatic compounds (Gr. *aleiphar* = fat, oil), organic compounds made up of a straight (acyclic) carbon chain. *Cf.* cyclic compounds.

alkali, pl. **alkalis, alkalies** (Arab. *al-qaliy* = soda ash), a strongly basic substance yielding hydroxyl ions (OH^-) in water solution, e.g. the hydroxides or carbonates of alkali metals, such as sodium hydroxide, NaOH, potassium hydroxide, KOH, and calcium hydroxide, $Ca(OH)_2$, which react basically and neutralize acids to form salts. *Adj.* **alkaline**, having the properties of an alkali; **alkalic**, containing large proportions of alkalis.

alkali metals, a group of metallic elements comprising the **alkali metals**: lithium (Li), sodium (Na), potassium (K), rubidium (Rb), caesium (Cs), and francium (Fr), which form univalent electropositive ions, M^+, and **alkaline earth metals** (earth alkali metals): magnesium (Mg), calcium (Ca), strontium (Sr), barium (Ba), and radium (Ra), which form bivalent electropositive ions, M^{2+} (M = metal). Most alkali metals are essential elements in organisms.

alkaline phosphatase, APHOS, *see* phosphatases.

alkalinity, an alkaline condition, describing particularly the property of water determined by the acid consumption when titrated to a certain pH value. In natural waters the alkalinity is mainly caused by bicarbonate ions (HCO_3^-). *See* pH.

alkaloids, basic nitrogen-containing organic compounds which occur in 30% of the vascular plants, particularly in angiosperms, usually the nitrogen atoms being members of a heterocyclic ring. Alkaloids are synthesized from common → amino acids, such as aspartic acid, lysine, phenylalanine, tyrosine, tryptophan, and ornithine. Many alkaloids are **harmful** or **poisonous** to man and many animals, and in plants they probably have **defensive properties** against herbivorous animals.

Some alkaloids are used in pharmaceutical preparations, some are drugs which affect specifically the nervous or circulatory system. Alkaloids may disturb transport mechanisms of cellular membranes, transmission of stimuli between cells, protein synthesis, enzyme activity, or processes of cell division. Hundreds of alkaloids are known, among which are e.g. → caffeine, theobromine, ephedrine, morphine and other opiates (e.g. codeine), nicotine, atropine, cocaine, quinine, strychnine, mescaline, and colchicine. Also colourful betalaine pigments of many flowers and fruits are alkaloids occurring in those plants which do not contain anthocyans.

alkalosis, a pathologic condition in vertebrates characterized by loss of hydrogen ions (H^+) or accumulation of basic substances in body fluids (**metabolic alkalosis**) due to changed cellular metabolism, or loss of carbon dioxide (CO_2) as a result of hyperventilation (**respiratory alkalosis**). *Cf.* acidosis.

alkanes, paraffins; saturated acyclic (aliphatic) hydrocarbons with the general formula

C_nH_{2n+2}; their names end with **-ane**, e.g. methane, ethane, propane, butane, etc.; with 1—4 carbons they are gases, with 5—15 carbons, liquids (paraffin oil), and with more than 16 carbons, solid (paraffin wax).

alkenes, olefins; a homologous series of unsaturated acyclic (aliphatic) hydrocarbons with the general formula C_nH_{2n}; have one double bond (unsaturated bond) and are thus more reactive than → alkanes (both have several isomeric forms). Their names end with **-ene**, e.g. methene (methylene, $=CH_2$), ethene (ethylene, $CH_2{:}CH_2$), or propene (propylene, $CH_3CH{:}CH_2$).

alkyl, 1. a univalent saturated hydrocarbon radical of the general formula C_nH_{2n+1}, e.g. methyl (CH_3-) or ethyl (CH_3CH_2-); formed from an alkane by removal of one hydrogen atom; **2.** alkide; a compound in which a metal is combined with one or more alkyl radicals; e.g. tetraethyl lead.

alkylating agents, substances which cause **alkylation**, i.e. introduce an alkyl radical into hydrocarbon chains or aromatic rings (*see* alkyl); many are strong chemical mutagens which transfer an alkyl group to DNA bases changing the pairing properties of the bases. Ethylmethylsulphonate (EMS) is the most common alkylating substance in mutation studies; mutagenic → formaldehyde has become an environmental pollutant.

allantochorion, → chorioallantoic membrane.

allantoin, 5-ureidohydantoin; glyoxyldiureide; an oxidation product of → uric acid forming an end-product of purine metabolism in mammals, other than primates.

allantois, pl. **allantoides** (L. *allas* = sausage, *eidos* = form), **allantoic membrane**; a vascular extraembryonic membrane in reptiles, birds, and mammals, i.e. in the Amniota group of vertebrates; derived as a pouch from the hindgut expanding outside the embryo forming the **allantoic bladder**. In reptiles and birds, the allantois expands chiefly between the → amnion and → chorion close beneath the porous egg shell to serve as a **respiratory organ**, its cavity forms a store for embryonic **excretions**. In most mammals, the allantois contributes to the formation of the → **placenta** and **umbilical cord**. In many reptiles, birds, and mammals, the allantois fuses with the chorion to form the **allantochorion**.

allele (Gr. *allelo-* = one another), allelomorph; one of the alternative forms of a gene, i.e. one of the series of a gene (or genes) occupying the same position or locus on a chromosome; this means that one gene may exist in many alleles which are located on homologous chromosomes. Alleles are mutually exclusive, and arise through gene mutation, and they all affect the same biochemical process or developmental event. In a → haploid organism or stage of life cycle, only one allele of each gene is present in its genome; in a → diploid there are two, and in a → polyploid several alleles. Within a population or a species, tens or even hundreds of different alleles (**multiple alleles**) may exist.

Regarding a given → locus, the organism may be homozygous or heterozygous, in the former the homologous chromosomes are carrying identical alleles, but in the latter, different alleles. *Adj.* **allelic**.

allele-specific oligonucleotide, ASO, an oligonucleotide → probe used for detecting the presence of a given mutant allele of a particular gene the sequence of which has been defined.

allelic complementation, *see* genetic complementation.

allelic exclusion, a phenomenon observed in the lymphocytes in which only one of the two gene alleles, encoding → immunoglobulins, is functional and expressed in the phenotype while the function of the other allele is suppressed.

allelism, the relationship between → alleles, or the existence of alleles.

allelomorph (Gr. *morphe* = form), → allele; one of the alternative forms of a gene. *Adj.* **allelomorphic**.

allelopathy (Gr. *pathos* = passion, suffering), in botany, the phenomenon in which some plants control the seed germination and growth of other plants by excreting substances such as aldehydes and terpenes into the air or soil. *Adj.* **allelopathic**.

Allen's rule (*Grant Allen*, 1848—1899), a zoogeographical rule stating that the mammalian species and subspecies which are evolutionary adapted to cold climate, tend to have relatively shorter extremities (e.g. ears, legs, and tails) than those relatives adapted to warmer climate. Thus, heat loss is smaller in animals living in cold regions. *Cf.* Bergmann's rule, Gloger's rule.

allergy (Gr. *ergon* = work), an immunological hypersensitivity found in higher vertebrates, especially in humans; induced after a certain latent period by repeated exposure to a given antigen (**allergen**), which may be e.g. a bacterial, metal, or drug allergen, or a substance (often a protein) released from other organisms. In allergy, the antigen-antibody reaction usually occurs in the skin and mucous membrane; e.g. in topic allergy immunoglobins involved are of E class (IgE). *Adj.* **allergic**. *See* anaphylaxis.

allo- (Gr. *allos* = other), **1.** denoting other; not normal, unusual; **2.** in chemistry, a prefix used to describe organic compounds, e.g. amino acids, with an asymmetric carbon in a side chain but having the same empirical formula; in sterols and related substances, usually indicates that two rings are in the *trans* position to each other.

alloantibody, *see* alloantigen.

alloantigen, an → antigen which occurs in alternative forms in individuals of a species, as blood-group antigens; also called **isoantigen**. An → antibody which is specific for an alloantigen is called **alloantibody** (isoantibody).

allocation hypothesis, (L. *al-* < *ad* = to, *locus* = place), a postulate presented by *G.C. Williams* in 1966, stating that the resources which are available to an organism are limited and have to be shared or allocated to different competing functions. The hypothesis comprises the conception of → **trade off**, which presents that the allocation of resources to some functions decreases resources available in some other functions, e.g. the resources allocated to reproduction, decrease resources for growth and maintenance.

allochronic (Gr. *chronos* = time), describing a change in timing of different developing organs during embryogenesis, often leading to the evolution of a new species.

allochronic speciation, a sympatric speciation which occurs when a population living in a certain area is divided into two or more reproductive isolates according to their different mating time.

allochthonous (Gr. *chthon* = ground, earth), outside a system, not original, exotic; not originated in the place where found, e.g. **1.** allochthonous material, organic matter which is brought to a water system from surrounding land areas; **2.** allochthonous species, a species dispersed to a certain area from somewhere else; e.g. the waterweed in Europe (native in America); **3.** in geology, pertaining to a mineral formed elsewhere than in the area where found, also called allogenic. *Cf.* autochthonous.

allocyclic, pertaining to differences in the → chromosome coiling between different chromosomes of the chromosome set, or even between whole chromosome sets; may be environmental, genotypic, or cellular in origin. Cell regions which most frequently are subjected to allocyclic behaviour (heteropycnosis) are → centromeres, → telomeres, and → nucleolus organizing regions, while the chromosomes most often allocyclic, are the → sex chromosomes and → B chromosomes. In certain insects, whole chromosome sets derived from different parental sexes show allocyclic behaviour. *Noun* **allocycly**.

alloenzymes, → allozymes.

allogamy (Gr. *gamos* = marriage), **1.** cross-fertilization or exogamy; **2.** cross-pollination or xenogamy. *Adj.* **allogamous**, reproducing by cross-fertilization.

allogenic (Gr. *gennan* = to produce), **1.** pertaining to a change or event which is caused by an external factor; **2.** pertaining to different gene constitutions within the same species (also allogeneic); **3.** in geology, pertaining to a mineral or other constituent that is formed elsewhere than in the rock where it is found, also called allochthonous.

allogenic succession, a temporal → succession of populations, communities, or ecosystems in an area where conditions are changed by factors originating from outside the area. *Cf.* autogenic succession.

allograft, a homologous → graft, i.e. a tissue which is transplanted between genetically different individuals of the same species. *Syn.* homograft.

allogrooming, a behavioural pattern found among some animals; the cleaning of the body (skin, feathers, or hairs) by another individual of the same or another species, i.e. conspecific or interspecific grooming. A special form of allogrooming is the **cleaning symbiosis** in which cleaners help customers e.g. by feeding on ectoparasites living on them. *See* grooming.

allometric growth, a type of growth in which growth rates in different parts of an individual organism differ from each other.

allometry (Gr. *metron* = measure), a study concerning → allometric growth.

allomone, a substance secreted by an organism as signals to individuals of other species. *Cf.* pheromone.

allomorphism (Gr. *morphe* = form, shape), **1.** a change in the cell shape, e.g. due to → metaplasia; **2.** a change of the crystalline form without changing chemically.

allomorphosis, pl. **allomorphoses** (Gr. *morphosis* = shaping), evolution of a phylogenetic line characterized by a rapid increase of specialization; it is common in situations where a taxon is dispersed to a new adaptive zone. Also called **allomorphy**. *Adj.* **allomorphic**.

allopatric (L. *patria* = home country), pertaining to populations or species which live in different geographical areas. The exchange of genes between allopatric populations is restricted or totally obstructed. *Cf.* sympatric.

allopatric speciation, *see* speciation.

alloploid (Gr. *-ploos* = -fold), pertaining to a hybrid individual which is born as the result of natural or artificial crossing of two species. Alloploids carry the chromosome sets of the parental species that differ genetically and structurally, and both sets occur either in a single dose (**allodiploid**) or in multiple doses (**allopolyploid**). Alloploid individuals are usually sterile. The phenomenon leading to an alloploid individual is called **alloploidy**. *See* allopolyploidy.

allopolyploidy (Gr. *polys* = many, *-ploos* = -fold), the existence of duplicate (allotetraploidy, amphidiploidy) or manifold (allohexaploidy, allooctoploidy, etc.) chromosome number in natural or artificial hybrids of different species or genera. Allopolyploidy is a fairly common form of speciation in plants. It always requires crossing between species.

If the haploid chromosome sets of the parental species are designated by A and B, the chromosome constitution of the original allodiploid hybrid is AB. This hybrid is almost always sterile due to disturbances in → meiosis, but because of this disturbance the hybrid may produce unreduced gametes (AB) whose union leads to allopolyploid (allotetraploid) species AABB. This hybrid is fully fertile because meiosis now proceeds normally, each chromosome having its homologue.

all-or-none response, the way of certain excitable cells to respond to a stimulus always at full strength or not at all, i.e. the intensity of the response does not depend on the strength of the stimulus; originally observed in the contraction of the cardiac muscle; seen also e.g. in the generation of an → action potential in a neurone or muscle cell.

allosematic (Gr. *sema* = sign), pertaining to a harmless animal whose colour or structure resembles (mimics) the warning colour or structure of another, usually dangerous species; e.g. hoverflies (Syrphidae) mimic wasps or bees. *Cf.* aposematic, mimicry.

allosterism, allostery (Gr. *stereos* = firm, strong), a property of proteins, particularly enzymes which have two (or more) binding sites, i.e. an → **active site** and **allosteric site** (regulatory site) distinct from each other. The occupation of the allosteric site by **effectors** (activators or inhibitors) may alter the affinity of the enzyme for its substrate (non-competitive inhibition). Effector binding induces a conformational change in the tertiary or quaternary structure of the protein modulating enzymatic activity, e.g. the maximum rate of reaction. Allosterism occurs also on the gene level. A regulatory protein that controls the function of an operon coding for the synthesis of an → inducible enzyme, may be **allosteric**. *See* competitive inhibition.

allotetraploid, amphidiploid. *See* allopolyploidy.

allotopic (Gr. *topos* = place), pertaining to sympatric populations of two or more species which inhabit different habitats in the same distribution area.

allotrophic (Gr. *trophe* = nourishment), **1.** having an altered, especially lowered nutritive value; **2.** pertaining to the influx of nutrients from outside into a water body of an ecosystem; **3.** heterotrophic; *see* heterotroph.

allotropy, allotropism (Gr. *trope* = turning), the occurrence of certain chemical elements in different physical forms; e.g. the forms of carbon are graphite, diamond, and → fullerene (allotropes of carbon). *Adj.* **allotropic**.

allotype, any of the antigenic specifities present in the → immunoglobulins and other proteins of different groups of individuals of the same species and determined by different → alleles of the same gene.

allozyme (Gr. *zyme* = state of fermentation), alloenzyme; one of the forms of the same enzyme encoded by a different allele in the same locus; allozymes often differ from each other e.g. functionally (temperature dependence, pH optimum) or by their electrophoretic mobilities. *Cf.* isozymes.

alpha, A, α, the first letter of the Greek alphabet. Greek letters are used in biological terms as a letter mark or written, e.g. α-cells or alpha cells.

alpha-actinin, α-actinin, *see* actinin.

alpha amylase, *see* amylase.

alpha-blocker, α-adrenergic blocking agent; an agent that competitively blocks alpha-adrenergic receptors in the vertebrate nervous system and its effector organs, antagonizing certain effects of → noradrenaline. *See* adrenergic receptors.

alpha diversity, a biological diversity as a result of competition between species in which the species must have specialized and adapted to a narrowing ecological niche. *Cf.* beta diversity.

alpha foetoprotein, AFP, Am. **alpha fetoprotein,** *see* foetoproteins.

alpha helix, pl. **alpha helices** (L., Gr. *helix* = coil, spiral), α-helix; the right-handed helical configuration of the → polypeptide backbone (primary structure) in protein molecules. The primary structure is formed by → peptide bonds between the carboxyl group of one amino acid and the α-amino group of the next. The helical bending of the backbone (**secondary structure**) is due to the → hydrogen bonds formed between the side groups of amino acid residues extending outward from the helical backbone. Hydrogen bonds are formed between carbonyl oxygen and nitrogen atoms, stabilizing the helix. In the alpha helix, 3.6 amino acid residues exist per each turn, which corresponds to 0.54 nm on the helical axis. This configuration contributes to the secondary structure of fibrous proteins. *See* beta-pleated sheet, double helix.

alpha ketoglutaric acid, α-ketoglutaric acid, an intermediate, $HOOCCH_2CH_2COCOOH$, in the → citric acid cycle of the respiratory metabolism in cells; formed from isocitric acid via decarboxylation and further converted to succinyl CoA. Its salts, esters and ionic form are called **alpha ketoglutarates**.

alpha particles, α-particles; positively charged subatomic particles composed of two protons and two neutrons, being identical to helium nuclei (^4He). The particles are liberated with high velocity (about $2 \cdot 10^7$ m/s) from many radioactive compounds, i.e. from unstable isotopes with a high atomic number, producing **alpha radiation** (alpha rays), e.g. from radon (^{222}Rn), or radium (^{226}Ra), or when thorium (^{234}Th) is formed from uranium (^{238}U). Alpha radiation is very energetic and penetrating, and therefore highly dangerous for organisms. *See* ionizing radiation, radioactivity. *Cf.* beta emission.

alpha receptor, *see* adrenergic receptors.

alpine zone, *see* height zones.

alternating cultivation, in agriculture, the cultivation of different plant species in an area in successive years for growing a better crop. Especially valuable for the soil are leguminous plants, which are able to fix free nitrogen and thus naturally increase the nitrogen content and its availability in the soil.

alternating leaf arrangement, *see* phyllotaxy.

alternation of generations, the alternation of two or more → generations of organisms, reproducing themselves in different ways, e.g. one generation reproducing sexually and the other parthenogenetically, or one being diploid and the other haploid. The alternation between haploid and diploid generations is typical of plants, in which haploid → gametophytes and diploid → sporophytes follow each other.

altricial (L. *altrix* = nourisher), pertaining to a young animal, usually born naked, blind, and immobile, requiring care and nursing after hatching or birth: e.g. a mouse pup, or nidicolous young of birds. *See* nidifugous, nidicolous. *Cf.* precocial.

altruism (L. *alter* = other), unselfishness; a self-jeopardizing, self-exposing, or self-sacrificing behaviour of animals, i.e. any unselfish behavioural pattern which increases the fitness of the object at the expense of the **altruistic** individual. The cost of the behaviour may be e.g. an increasing risk of predation during the altruistic acts. Altruism is rather common e.g. among birds whose young individuals from previous broods may take part in feeding their siblings in the succeeding brood. *See* Hamilton's rule, kin selection.

Alu family (Alu < *Arthrobacter luteus*, a bacterial species), a group of similar DNA sequences in the human genome having a cleavage site for an Alu → restriction enzyme at both ends of each sequence. The term is commonly used to refer to the most frequent type of human → short interspersed element (SINE).

aluminium, Am. **aluminum** (L. *alumen*), a metallic element, symbol Al, atomic weight 26.982, density 2.699; occurs in various compounds in nature, e.g. aluminium silicates and hydroxides ($Al_2O_3 \cdot nH_2O$) in clay. On acidification in nature, aluminium ions are released to the soil and passed on through plants to animals. Into humans it comes chiefly from aluminium dishes and various chemicals (e.g. drugs). Cells do not require aluminium but neither is it clearly toxic to any organism. There are some doubts, however, that long exposure of animals to high aluminium concentrations might harm especially the nervous system.

alveolus, pl. **alveoli** (L. dim. of *alveus* = hollow, cavity) **1.** pulmonary alveolus; *see* lung; **2.** a small sac-like part formed by secretory cells in the compound exocrine glands (**alveolar glands**, e.g. salivary or pancreatic glands); hundreds or thousands of alveoli in one gland release their secretions via ducts which unite into larger ducts (finally usually one) passing out of the gland; **3.** a tooth cavity or socket in the jaw; **4.** a minute chamber on the cell wall of diatoms Bacillariophyceae; **5.** in Indian culture, a piece of tilled land.

Alzheimer's disease (*Alois Alzheimer*, 1864—1915), Alzheimer's dementia; a type of presenile or senile dementia with mental deteriation seen as memory loss, disorientation, and confusion of mind. The disease is caused by neuronal degeneration in the brain tissue, especially in the cerebral cortex and hippocampus; first appearing as the degeneration of cholinergic synapses and death of cholinergic neurones (nerve cells), later also of noradrenergic and serotoninergic neurones. Histologically, the atrophic brain tissue is characterized by senile plaques composed of filamentous or granular argentophilic masses with a myeloid core, and by the distortion and thickening of neurofibrils inside the nerve cells. A slight retardation of the beginning of the deteriation has been achieved with cholinesterase inhibitors which increase the acetylcholine level in the brain.

amacrine cell (Gr. *a-* = non, *makros* = large, long, *is* = fibre), a special neurone type in the retina of the vertebrate eye; unlike the neurones in general, amacrine cells have no axon. The cells regulate the function of optic neuronal pathways by means of dendrite-like processes.

Amastigomycota, a division in → Fungi. The group is considered to be more developed than the other divisions, Gymnomycota, Mastigomycota. Among Amastigomycetes no freely swimming stages exist. The group is divided into three subdivisions: → Zygomycotina, Ascomycotina, and Basidiomycotina. In some systems the group Amastigomycota is divided into the divisions Zygomycota, Ascomycota, Basidiomycota, and Deuteromycota.

ama(ni)toxins (*Amanita* = a fungus genus, Gr. *toxikon* = arrow poison), cytostatic substances found in e.g. *Amanita phalloides*, *A. virosa*, *Galerina marginata*, and some *Macrolepiota* and *Conocybe* species. Amatoxins (e.g. amanin and α-, β-, γ-amanitins) are the most effective fungal poisons causing in animals irreversible damages to tissues, especially in the kidneys, liver, lungs, cardiac muscle, and brain cells. The toxins block RNA synthesis by inhibiting RNA polymerase II and III activity. For humans, a lethal dose is 0.05—0.1 mg/kg body weight; symptoms of poisoning include nausea, diarrhea and severe pain. Liver and kidney damages may develop within 48 hours. One fruiting body of these mushrooms contains the lethal dose manifold. *See* phalloidin(e).

amber codon, amber termination codon, UAG; one of the → termination codons found in all organisms. *See* ochre codon, opal codon.

amber mutation (L. *mutare* = change), a gene mutation in the messenger RNA resulting in the → termination codon UAG (uracil-adenine-guanine). Amber mutation is one of the three different → nonsense mutations all resulting in a terminator codon; other nonsense mutations are "ochre" (UAA) and "opal" (UGA). Amber was named after its discoverer *Bernstein* (means amber in German), and thereafter other mutations were named after different yellowish colours. As the ribosomes translate the messenger RNA into a polypeptide structure in → protein synthesis, translation stops at the terminator codon and protein synthesis ceases because of amber mutation.

ambi-, ambo- (L. *ambo* = both), on both sides, around.

ambivalence (L. *valentia* = strength), a

situation in which opposite forces prevail; e.g. the simultaneous existence of conflicting emotions, wishes, or attitudes toward the same object. *Adj.* **ambivalent**.

ambivalent behaviour, a type of behaviour occurring in a conflicting situation where an animal is simultaneously subjected to two or more motivations; e.g. a gull defending its territory against a strong intruder may show behaviour characteristics of both escape and attack.

ambulacrum, pl. **ambulacra** (L. *ambulare* = to walk), **1.** the ultimate segment of a mite leg, the → apotele being the central part; **2.** ambulacrum system, i.e. five radiating grooves (**ambulacral grooves**) on the body surface of echinoderms. The grooves are formed from calcareous radial plates, each groove with a pair of **ambulacral plate** rows, a pair of **interambulacral plate** rows located between the grooves. In sea stars the grooves reach from the mouth to each arm and in sea urchins from the mouth to anus. The tube feet of the → water-vascular system are projecting out through the pores on the ambulacral plates.

ameboid, → amoeboid.

amebocyte, → amoebocyte.

ameloblast (Old E. *amel* = enamel, Gr. *blastos* = germ), a columnar epithelial cell in the inner layer of the enamel organ in a developing vertebrate tooth; ameloblasts make up enamel to cover the crown of the tooth. *Syn.* adamantoblast, enamel cell, enameloblast.

amensalism (L. *a-* = non, *mensa* = table), an interaction or a symbiosis of two species (populations), in which one is adversely affected and the other (amensal) is unaffected; e.g. an interaction between a bacterium and fungi producing antibiotics. *Cf.* commensalism, mutualism.

Ames test/assay (*Bruce Ames*, b. 1928), an assay for testing the mutagenicity of a chemical compound; utilizes a histidine-requiring mutant strain of *Salmonella typhimurium*. This strain is inoculated into a culture medium deficient in histidine but containing the test compound. Only the back mutants, produced if the compound is mutagenic, are able to grow in this medium and can be counted (this is possible because the back mutants do not require histidine). The assay is applied also for testing the carcinogenicity of chemicals since about 90% of carcinogens are also mutagens. *See* back mutation.

Ametabola (L. *a-* = non, *metabole* = change) → Apterygota.

ametabolous, pertaining to animals which have no → metamorphosis. *Cf.* hemimetabolous, heterometabolous, holo-metabolous.

amide, a nitrogenous compound derived from ammonia (NH_3) by substitution of one or more hydrogen atoms by an acyl radical; general formula R-$CONH_2$ where -$CONH_2$ is the amide group, e.g. acetamide, CH_3CONH_2. Some metals may link to ammonia forming inorganic amides such as sodium amide, $NaNH_2$. Biological amides can be derived also from carboxylic and amino acids by replacing the -OH of the carboxyl group with an amino group. The amide derivatives of aspartic and glutamic acids, i.e. asparagine and glutamine, may occur in proteins. In plants, especially glutamine is an easily transported nitrogenous compound and an important storage compound in many seeds; is used by cells via metabolic interconversions in biosynthetic processes or energy metabolism.

amine, a nitrogenous organic compound formed from ammonia by replacing one ore more hydrogen atoms by a hydrocarbon radical (R), resulting in a primary (RNH_2), secondary (R_2NH), or tertiary amine (R_3N).

amine oxidases, enzymes of the oxidoreductase class catalysing the reaction: $RCH_2NH_2 + H_2O + O_2 \longrightarrow RCHO + NH_3 + H_2O_2$; e.g. **amine oxidase (copper-containing)**, EC 1.4.3.6, also called diamine oxidase or histaminase, acts on primary monoamines and diamines such as histamine in animal tissues, or **amine oxidase (flavin-containing)**, EC 1.4.3.4, also called monoamine oxidase (MAO), acts on primary, secondary, and tertiary amines, e.g. on adrenaline, noradrenaline, dopamine, and serotonin.

aminerotrophic bog (L. *a* = not, *minera* = mine, ore, Gr. *trophe* = nutrient), a combined mire type the central part of which is higher than the edges which are collecting runoff water. Often the centre of the bog forms a water-logged fen, whereas the edges are dryer dwarf-shrub areas. In Europe, the aminerotrophic bogs are common in northern Fennoscandia and Russia, in the lowlands of northern Germany, in the foothills of the

Alps, in the western parts of Scotland, and throughout Ireland. *Syn.* ombrotrophic raised bog.

Amino acids. Abbreviations, one-letter symbols and mRNA codes of amino acids.

Amino acid	Abbreviations	Symbol	mRNA code designation
Alanine	Ala, ala	A	GCU, GCC, GCA, GCG
Arginine	Arg, arg	R	CGU, CGC, CGA, CGG, AGA, AGG
Asparagine	Asn, asn	N	AAU, AAC
Aspartic acid	Asp, asp	D	GAU, GAC
Cysteine	Cys, cys	C	UGU, UGC
Glutamic acid	Glu, glu	E	GAA, GAG
Glutamine	Gln, gln	Q	CAA, CAG
Glycine	Gly, gly	G	GGU, GGC, GGA, GGG
Histidine	His, his	H	CAU, CAC
Isoleucine	Ile, ile	I	AUU, AUC, AUA
Leucine	Leu, leu	L	UUA, UUG, CUU, CUC, CUA, CUG
Lysine	Lys, lys	K	AAA, AAG
Methionine	Met, met	M	AUG
Phenylalanine	Phe, phe	F	UUU, UUC
Proline	Pro, pro	P	CCU, CCC, CCA, CCG
Serine	Ser, ser	S	UCU, UCC, UCA, UCG, AGU, AGC
Threonine	Thr, thr	T	ACU, ACC, ACA, ACG
Tryptophan	Trp, trp	W	UGC
Tyrosine	Tyr, tyr	Y	UAU, UAC
Valine	Val, val	V	GUU, GUC, GUA, GUG

amino acid, an organic acid in which a non-acid hydrogen is replaced by an **amino group** ($-NH_2$), thus forming a compound which contains one or more amino groups and one or more **carboxyl groups** (-COOH), i.e. they are → ampholytes. In solutions such as cell and tissue fluids, amino acids show both basic and acidic properties, hence both groups are dissociating. The structure and length of the carbon chain vary as well as the position of amino and carboxyl groups. In α-amino acids, R is linked to α-carbon next to the carboxyl group (R is any group of atoms, e.g. a carbon chain, aromatic ring, or sulphur).

Amino acids are **constituents of proteins**, 20 different amino acids occurring in natural proteins in L-configurations (*see* isomers). In addition, organisms have also many amino acids which are not incorporated in proteins.

The linear sequence of amino acids in a polypeptide or protein molecule is called **amino acid sequence**, forming the primary structure of these compounds. It is genetically determined in the ribosomal translation of messenger RNA in → protein synthesis. *See* essential amino acids.

amino acid sequence, the sequence of amino acids forming the primary structure of a polypeptide or protein molecule; genetically controlled in the translation of → messenger RNA. *See* protein synthesis.

aminoacyl-tRNA, the complex formed by an amino acid and its specific transfer RNA (tRNA) in → genetic translation. The binding reaction is catalysed by **aminoacyl-tRNA synthetase**: amino acid + tRNA + ATP —> aminoacyl-tRNA + AMP + PP_i.

δ-aminolaevulinic acid, 5-amino-4-oxo-pentanoic acid, $C_5H_9NO_3$, an intermediate compound in the synthesis of porphyrins, such as → haem of haemoglobin, or chlorophyll; formed from the condensation of glycine and succinyl-CoA.

amino site, → A-site.

amitosis, pl. **amitoses,** a nuclear division by a process other than → mitosis; in typical cases involves a dumbbell-shaped cleavage of the nucleus in which no chromosomes nor any mitotic spindle are recognizable; occurs e.g. in ciliates and some other protists.

amitrole, aminotriazole; 3-amino-1,2,4-triazole; a non-selective herbicide which causes chlorophyll deficiency so hampering photosynthesis in plant cells. It also inhibits root growth and germination. *See* triazole.

amixis (L. *a-* = non, Gr. *meignynai* = to mix), a reproduction mechanism in haploid organisms, such as bacteria, in which the characteristics of sexual breeding, i.e. meiosis, and the fusion of nuclei (karyogamy), are lacking, but occasionally some pre- and postmeiotic phenomena may occur. In higher plants it is called → apomixis.

ammocoete, also **ammocete** (Gr. *ammos* = sand, *koite* = bed), the freshwater larval stage of lampreys; in many respects, such as by anatomical and morphological structures, or life habits, it resembles → amphioxus. The name was given before the ontogeny of lampreys was known and the larvae were mistakenly regarded as another species.

ammonia (salt of *Ammon*, an Egyptian god, formerly obtained near the temple of Ammon in Libya), a colourless gas with a pungent smell, NH_3, soluble in water. Ammonia is one of the nitrogen end products of cellular metabolism and biological decomposition in organisms, playing an important role in the → nitrogen cycle in nature and e.g. in the → urea cycle in animals. Ammonia is alkaline and easily attracts hydrogen ions (H^+) from water or acids, forming ammonium ions (NH_4^+). Ammonia is a base and reacts therefore with acids, e.g. with hydrochloric acid forming the salt, ammonium chloride, NH_4Cl. This dissociates to NH_4^+ and Cl^- ions, and further to NH_3 and H^+. Thus, aqueous solution of ammonium chloride produces hydrogen ions (protons) and is acidic; therefore ammonia may act as an acidifying agent in nature. *See* ammonotelic.

ammonification (L. *facere* = to do), the oxidation of organic nitrogen compounds into → ammonia and ammonium ions by some soil microbes. *Cf.* denitrification, nitrification.

ammoniotelic, → ammonotelic.

ammonites, *see* Ammonoidea.

ammonium ion, NH^+, a reduced form of nitrogen; behaves as a univalent metal in forming ammonium compounds. Plants can take ammonium ions from soil and utilize it in their metabolism. Together with hydroxyl ions (OH^-), ammonium ions form ammoniumhydroxide (NH_4OH) which is a weak base. *See* ammonia.

Ammonoidea (*Ammon* = a god in Egyptian mythology represented with ram's horns), **ammonites, ammonoids**; an extinct molluscan subclass of spirally wound cephalopods (class Cephalopoda); the animals were widely prevalent in the Mesozoic era but became extinct by the end of the Cretaceous period. Many species are important → index fossils.

ammonotelic (ammonia + Gr. *telos* = end), pertaining to animals which excrete ammonia as the main end product of nitrogen metabolism, e.g. many aquatic invertebrates and amphibians. *Cf.* ureotelic, uricotelic.

amni(o)- (Gr. *amnion* = membrane around the foetus, caul), denoting relationship to the → amnion.

amniocentesis (Gr. *kentesis* = punction), a method by which a sample is taken from the amniotic fluid by punctuating an injection needle through the uterine wall and foetal membranes.

amnion, pl. **amnia** (Gr.), amniotic sac; the innermost of the → extraembryonic membranes in reptiles, birds, and mammals (Amniota group), forming a protecting sac, filled with **amniotic fluid**, around the embryo. The amnion is usually formed from different embryonal elements, i.e. from folds of the yolk sac, the ectoderm, and the mesoderm. These grow around the embryo, histologically comprising an epithelium tissue layer (ectoderm) on the internal surface, and a mesenchymal tissue layer (mesoderm) externally.

Amniota, amniotes; a group of vertebrates (some authors consider as a superclass) including reptiles, birds and mammals, i.e. animals whose developing embryo is surrounded by the → amnion. *Cf.* Anamniota.

amniotic egg, an egg type characteristic of reptiles, birds, and egg-laying mammals (prototherians); contains a large amount of yolk and is surrounded by a leathery or calcified shell, and within the egg the → extraembryonic membranes are formed during the embryonic development.

Amoeba (Gr. *amoibe* = change), a protozoan genus in the superclass Rhizopoda (phylum Sarcomastigophora); single-celled animals, **amoebae** (amoebas, amebas) which live in large numbers especially in moist soil and small fresh-water bodies. Some species are parasitic causing infections (**amoebiasis, amebiasis**) in many animals, also in mammals. Amoebae change their shape and move and feed by stretching pseudopods from the cell body (**amoeboid motion**, *see* pseudopodium).

amoebocyte (Gr. *kytos* = cavity, cell), a cell type in many invertebrates capable of moving and changing shape like an amoeba; e.g. sponges (Porifera) have five amoebocyte types: 1) → archaeocytes develop into other cell types such as gametes, 2) → scleroblasts and 3) → spongoblasts act as supporting cells, 4) → collencytes correspond fibroblasts of vertebrate connective tissues, and 5) primitive myocytes (muscle cells) are specialized for motility.

amoeboid, Am. **ameboid,** resembling an amoeba; moving like an amoeba. *See* pseudopodium.

amorphous (Gr. *a-* = non, *morphe* = form), formless, undetermined in form, structureless; e.g. **1.** pertaining to → mutations or → alleles (**amorphs**, silent alleles) in which gene function is totally lacking or is completely faulty; in structural genes amorphous alleles or mutations are usually recessive; **2.** not crystallized, e.g. amorphous ice.

AMP, adenosine monophosphate. *See* adenosine phosphates.

ampere, A (*André Ampère*, 1775—1836), the unit of electric current. *See* Appendix 1.

amph(i)-, ampho- (Gr. *amphi* = around, *ampho* = both), denoting both sides, double, surrounding.

amphetamine, α-methylphenethylamine; a synthetic amine which has strong → sympathomimetic action in the vertebrate nervous system; e.g. stimulates the brain decreasing the sense of fatigue. Amphetamine is used in medicine e.g. to reduce appetite in obesity, and is also used illegally as a narcotic psychoactive drug.

amphiatlantic (Gr. *Atlas* = a mythological Titan), occurring on both sides of the Atlantic Ocean.

Amphibia (Gr. *bios* = life), **amphibians**, a class of vertebrates comprising semiaquatic poikilothermic animals whose larvae breathe by gills, but adults have usually lungs and a specified cutaneous respiration (skin respiration); the heart is three-chambered with two atria and one ventricle. Each forelimb usually has four digits; the skin is moist and thin with many glands secreting slime and/or poisons. About 4,200 species are known and they are divided into three orders: **caecilians** (Gymnophiona, or Apoda), **salamanders** and **newts** (Caudata), and **frogs** and **toads** (Anura).

amphibian, 1. pertaining to an organism which is adapted both to terrestrial and aquatic environments; also **amphibious**; **2.** an animal of the class → Amphibia.

amphiblastula (Gr. *blastos* = bud), a larval stage of some sponges, the anterior part comprising flagellated → micromeres, and the posterior part granular → macromeres. *Cf.* parenchymula.

amphibolic (Gr. *amphiballein* = to throw round), **1.** capable of being directed either forward or backward, as e.g. the amphibolic outer toe of an owl; **2.** in biochemistry, pertaining to a catabolic pathway that, however, provides precursors for anabolic pathways.

amphicoelous, Am. **amphicelous** (Gr. *koilos* = hollow), **biconcave**, concave on both surfaces; e.g. amphicoelous vertebrae of fish. *Cf.* acoelous, opisthocoelous, procoelous.

amphicribral (L. *cribrum* = sieve), pertaining to the siphonostele type in plants in which the phloem surrounds the xylem. *See* stele. *Cf.* amphiphloic.

amphid, a small, paired sense receptor (**sense pit**) anteriorly in the skin of nematodes, probably reacting to chemical stimuli; it is usually reduced in parasitic nematodes. *Cf.* phasmid.

amphidiploid, allotetraploid. *See* allopolyploidy.

amphidromous (Gr. *dromein* = to run), pertaining to fish species migrating from salt water to fresh water and vice versa. *Cf.* anadromous, catadromous.

Amphineura (Gr. *neuron* = nerve), *see* Polyplacophora.

amphioxus, pl. **amphioxi,** Am. **amphioxuses** (Gr. *oxys* = sharp), an original generic name for **lancelets**; after the genus was renamed *Branchiostoma*, the term amphioxus has been used as a convenient common name for all the species in the subphylum → Cephalochordata.

amphipathic (Gr. *path-* < *paschein* = to experience, suffer), amphiphilic, amphiphobic; pertaining to a molecule that contains chemical groups with different properties, such as hydrophilic and hydrophobic qualities; e.g. membrane lipids like phosphoglycerides and sphingolipids, or a detergent.

amphiphloic (Gr. *phloos* = bark), pertaining to a siphonostele type in plants in which there is phloem both outside and inside the xylem.

Syn. solenostele. *See* stele. *Cf.* amphicribral.

Amphipoda (Gr. *pous*, gen. *podos* = foot), **amphipods**; a crustacean order in the class Malacostraca including mainly laterally compressed animals which have sessile compound eyes and one pair of maxillipeds; the thoracic and abdominal limbs differ from each other in form and function. Amphipods are mainly free-living aquatic animals from fresh to marine waters; some species are parasitic. The order includes about 6,000 species, e.g. beach fleas and sandhoppers.

Amphisbaenia (Gr. *bainein* = to walk), **amphisbaenians, worm lizards**; a reptilian suborder in the order → Squamata including about 130 species; the long, cylindrical body is short-tailed and usually limbless (except one Mexican genus whose forelimbs are well developed), the eyes and ears are hidden under the soft skin. Worm lizards live in the soil in South America and tropical Africa.

amphistylic skull (Gr. *stylos* = pillar), *see* cranium.

amphitopic (Gr. *topos* = place), pertaining to a population or species which have large tolerance to various environmental factors and thus can live in many different areas.

amphitropic distribution (Gr. *trope* = turning), a distribution of organisms restricted to both margins (north and south) of the tropical zone.

amphitropous ovule, a type of → ovule inverted so that the funicle is in the middle of one side.

ampholyte, amphoteric electrolyte; a compound that can produce or accept a → proton (hydrogen ion, H^+) and thus act as either an acid or a base; e.g. amino acids and proteins. *Adj.* **ampholytic**.

amphophilic, amphophilous (Gr. *philein* = to love), having an affinity to both basic and acid dyes; e.g. a tissue structure.

amphoteric (Gr. *amphoteros* = each of two), having two opposite characteristics, such as a compound which can react as an acid and a base; e.g. proteins, their acidic and basic side groups balancing at the isoelectric point.

ampicillin, a penicillin-like antibiotic with a broad-spectrum activity, widely used as a marker for plasmid transfer.

amplexicaul (L. *amplecti* = to entwine, *caulis* = stem, cabbage), surrounding the stem, such as the base of a sessile leaf.

amplicon, a chromosomal region which takes part in → gene amplification, becoming amplified in it.

amplification (L. *amplificare* = to increase, amplify), enhancement, strengthening; e.g. amplification of a stimulus, electrical signal, or genetic material in → gene amplification.

ampulla, pl. **ampullae,** (L. = two-handled bottle), **1.** an enlargement of a canal or duct; e.g. three **osseous ampullae** (anterior, posterior, and lateral) in one end of each semicircular canal in the vertebrate inner ear; a muscular sac at the inner end of the podium in sea stars; **2. ampule, ampul**, a sealed, sterile container for a medicinal preparation. *Adj.* **ampullar(y)**. *Dim.* **ampullula**. *See* Lorenzini's ampullae.

ampullary organ, *see* electric sense.

amygdala, pl. **amygdalae,** also **amygdaloid nuclei/bodies** (Gr. *amygdale* = almond), **1.** a brain nucleus (*corpus amygdaloideum*) of grey matter consisting of a large group of nerve cells in the temporal area underneath the olfactory cortex in the telencephalon; forms a part of the → limbic system; **2.** → tonsilla.

amyl alcohol, pentanol, $C_5H_{11}OH$, a colourless liquid with a characteristic smell, obtained by fractionation of fuel oil; toxic to organisms. It occurs in several isomers of which → **isoamyl alcohol** is widely used in biochemistry.

amylases (Gr. *amylon* = starch), diastases; enzymes of the sub-subclass EC 3.2.1., catalysing the cleavage of starch, glycogen, and other 1,4-glucans, i.e. 1,4-α-glycosidic bonds in polysaccharides. Three types are found in organisms. The hydrolysis of starch (both amylose and amylopectin) and glycogen (animal starch) by the action of **α-amylase** (EC 3.2.1.1) which hydrolyses inner 1,4-α-glycosidic bonds from the reducing end of the polysaccharide chain. The hydrolytic products of starch are → maltose (a disaccharide) and dextrins; glycogen breaks down into α-glucose. In animals, α-amylase (former name ptyalin) is chiefly produced by salivary glands and the exocrine tissue of the pancreas, the enzyme being important in the digestion of starch and glycogen in the alimentary canal. In plants, α-amylase is particularly produced by cells of germinating seeds and storage tissues.

The hydrolysis of starch is catalysed also by **β-amylase** (EC 3.2.1.2) that cleaves maltose from the non-reducing end of the substrate.

This enzyme is produced especially by plants and microorganisms, occurring abundantly e.g. in cereal seeds. The third enzyme in this group is γ-**amylase** which is glucan 1,4-α-glucosidase produced by microorganisms. Amylases are economically important enzymes used e.g. in brewery.

amyloid, 1. any starch-like substance, usually becoming stained blue with iodine; **2.** translucent extracellular granules or sheets composed of aggregated fibrillary proteins or proteoglycans in subcutaneous tissues of mammals, due to a disturbance in protein synthesis; it stains dark brown with iodine; **amyloidosis** is a disease characterized by extracellular accumulation of amyloid; **3.** a gelatinous, hydrated cellulose produced in paper industry.

amylopectin, one of the two main components of → starch; polymerized from glucose by α-1,4- and α-1,6-bonds which give the molecule a branched structure. Amylopectin dissolves very poorly in water. *See* amylose.

amyloplast(id) (Gr. *plastos* = formed), a colourless starch-containing → plastid in plant cells.

amylose, an unbranched, water-soluble polysaccharide synthesized from glucose by α-1,4-bonds; forms starch together with amylopectin. Amylose stains blue with iodine reagents.

ana- (Gr. *ana* = up), denoting **1.** up, toward; **2.** back, backward; **3.** again, anew; **4.** in chemistry, a compound with two fused 6-membered rings having substituents in positions 1 and 5.

anabiosis, → cryptobiosis.

anabolic (Gr. *anabole* = improvement), pertaining to synthetic metabolism, i.e. → anabolism.

anabolic steroid, usually a synthetic → steroid which has high anabolic efficiency, especially increasing the growth of muscles.

anabolism, the synthesizing part of → metabolism in organisms, resulting in the building of cell and tissue structures. *Cf.* catabolism.

anadromous (Gr. *ana* = up, again, *dromos* = running, course), pertaining to an animal which migrates from salt water to fresh water; e.g. some fish species, such as some salmon species and cyclostomes (lampreys) which migrate from the oceans to spawn in lakes and rivers. *Cf.* amphidromous, catadromous.

anaemia, Am. **anemia** (Gr. *an-* = not, *haima* = blood), a condition in which the blood haemoglobin content is less than normal; may be caused by a decreased number of red blood cells or by a reduced haemoglobin synthesis. Basic reasons may be iron deficiency (**hypoferric anaemia**), increased rate of erythrocyte destruction (**haemolytic anaemia**), disturbance in absorption of vitamin B_{12} (**pernicious anaemia**), or infection (**infectious anaemia**). *Adj.* **anaemic** (anemic).

anaerobe (Gr. *aer* = air, *biosis* = way of living), an organism that lives under conditions where oxygen is absent; may be an **obligate** anaerobe which is unable to live in the presence of even low oxygen concentrations, or a **facultative** anaerobe, i.e. able to live also in aerobic conditions. *Adj.* **anaerobic.**

anaerobic respiration, the part of → cell respiration in which oxygen is not consumed; includes the reactions of the **anaerobic → glycolysis** which gradually decomposes glucose into pyruvic acid which then passes to the aerobic part of respiration. In the absence of oxygen this cannot occur and pyruvate is metabolized further to **lactic acid**, and in some organisms to **ethanol**. The anaerobic part produces only less than 10% of ATP energy in comparison to aerobic respiration, but it may be sufficient for critical periods in the life of organims.

Usually certain cell types, especially muscle cells (anaerobic muscles), can temporarily use anaerobic respiration for their energy need. Because of the accumulation of lactic acid, anaerobic muscles become fatigued and the oxygen debt must be paid by aerobic respiration. Many invertebrates and some fish species (e.g. crucian carp, goldfish), can live for rather long periods using anaerobic respiration. Under anaerobic conditions their metabolism produces lactic acid or ethanol which are later metabolized back to glucose or used in aerobic conditions for ATP production. However, during long anaerobic periods the animals have to excrete these substances from the body with the loss of energy.

Anaerobic reactions also take place in plant tissues, e.g. in seedlings underneath a tight ice cover (ice incasement). In these conditions the cells liberate abundantly carbon dioxide, and accumulate ethanol and lactic acid in their cells.

anaesthesia, Am. **anesthesia** (Gr. *an-* = not,

aisthesis = sensation), a condition of an animal characterized by loss of sensation as a result of pharmacologic depression of the nervous system, or of a disease; e.g. local, spinal, epidural, or general anaesthesia.

anagenesis, pl. **anageneses** (Gr. *ana* = up, *genesis* = birth, origin), **1.** the regeneration of a tissue or a lost part of the body; **2.** a form of progressive → evolution which is connected with the origin of new organs, and types and building plans of their structures. Anagenesis leads to formation of new species without splitting of stem species into two or more species (*cf.* cladogenesis). According to *Julian Huxley*, anagenesis is a general term for all types of evolution that lead to the improvement of biological adaptation. This begins with the adaptation of the species and ends with the general reforms of biological organization which are followed by the origin of new, higher taxa.

anal, pertaining to the → anus or the end of the rectum.

analogous (Gr. *analogia* = proportion), pertaining to structures of different organisms which have the same function but are not evolutionarly related; e.g. the wings of butterflies and birds. *Cf.* homologous.

Anamniota (Gr. *an-* = not, *amnion* = a foetal membrane), **anamniotes**; a group of vertebrates (some authors consider as a superclass) including **cyclostomes**, **fishes**, and **amphibians** whose embryonic development takes place in water and the embryos have no → amnion. *Cf.* Amniota.

anamorphosis, pl. **anamorphoses** (Gr. *anamorphoun* = to transform), **1.** a gradually ascending progression in → phylogeny of an organism; **2.** the larval development in the course of → metamorphosis in some arthropods, such as proturans, chilopods and diplopods, when the number of body segments increases in moults; **3.** in physics, an image generated by a distorting optical system. *Cf.* epimorphosis.

anandrous (Gr. *an-* = not, *aner* = man), pertaining to a flower without stamens.

anaphase (Gr. *ana* = up, again, *phasis* = appearance), a phase of cell division; i.e. the phase of → mitosis or → meiosis in which the chromosomes move towards the poles of the dividing nucleus. In the anaphase of mitosis and the second meiotic division, the daughter chromosomes separate from each other; in the anaphase of the first meiotic division the chromosomes, forming the → bivalent or → multivalent in the preceding metaphase, separate from each other.

anaphylaxis (Gr. *phylaxis* = protection), a kind of immunological hypersensitivity to an antigen (toxin), usually characterized by a transient, strong allergic reaction; may lead to shock (anaphylactic shock).

anaplerosis (Gr. *anapleroein* = to fill up), **1.** the replacement or repair of defective or lost parts, as e.g. the transplantation of tissue to fill a wound; **2.** in biochemistry, denoting enzymatic reactions which replenish the concentration of an important intermediate, allowing the reaction or reaction cycle to continue. *Adj.* **anaplerotic**.

anapsid (Gr. *apsis* = arch), **1.** pertaining to a skull type with no temporal openings; *cf.* diapsid, parapsid, synapsid; **2.** an animal in the subclass → Anapsida.

Anapsida, anapsids; a reptilian subclass characterized by a skull type without temporal apertures; includes the extant order → Chelonia (turtles, tortoises) and some extinct groups of primitive reptiles (e.g. cotylosaurs). *Cf.* Diapsida, Synapsida.

anastomosis, pl. **anastomoses** (Gr. = opening), an intercommunicating or inosculating structure, e.g. **1.** a communicative vessel between two blood vessels or other tubular organ systems; **2.** a connective opening created by surgical or traumatic means between two normally separated organs, as a fistel or stoma; **3.** a connecting branch between leaf veins.

anatomy (Gr. *ana-* = up, back, again, *temnein* = to cut), **1.** the branch of morphology dealing with structures of cells, tissues, and organs of various organisms, e.g. anatomy of animals or anatomy of plants; originally denoted only human anatomy; **2.** the morphological inner structure itself. *Adj.* **anatomical**. *Cf.* morphology.

anatropous ovule (Gr. *tropos* = turn), *see* ovule.

ancestrula, pl. **ancestrulae** (L. *antecedere* = to go before), the first zooid which by budding gives rise to a series of other zooids in an initial bryozoan colony.

Ancylus Lake (L. *Ancylus* = a generic name of some gastropods), a freshwater stage in the development of the Baltic Sea, from about 9,000 to 8,000 years ago, following the

Yoldia Sea.

Andreaeaidae, granite mosses, rock mosses; a subclass in the class Bryopsida (Musci), division Bryophyta, Plant kingdom; in some classifications a class in the division Bryophyta. The species are small and calcifuge; the columella does not extend to the top but is capped by the spore sac. The group is small, containing only the genus *Andreaea*.

andro- (Gr. *aner*, gen. *andros* = man), denoting relationship to a male organ or organism, as to male reproductive organs.

androconium, pl. **androconia** (Gr. *konis* = dust), one of the modified wing-scales of some male lepidopterans; olfactory glands are located at their base secreting pheromones to attract females for copulation.

androdioecious (Gr. *dis* = twice, *oikos* = house), pertaining to a plant species having male and bisexual flowers in separate individuals. *Cf.* andromonoecious.

androgens (Gr. *gennan* = to produce), substances, usually steroid hormones, stimulating the development of male sexual characteristics in animals. In vertebrates, androgens are secreted by → Leydig cells of the testis and some cells of the adrenal cortex; e.g. → testosterone, androsterone, and their derivatives.

androgenic, 1. pertaining to an → androgen; **2.** having masculinizing effect on animals.

androgenesis, 1. development from a male gamete, i.e. male → parthenogenesis; **2.** the development of an egg which contains only paternal chromosomes, due to the failure of the nucleus of the female gamete to participate in fertilization. *Adj.* **androgenetic**.

androgenote, an individual that can be experimentally produced by removing the maternal → pronucleus from the fertilized egg and substituting it with a paternal pronucleus taken from another embryo; hence the androgenote carries two paternal genomes.

androgyny, also **androgynism** (Gr. *gyne* = woman, female), **1.** female pseudohermaphroditism, false → hermaphroditism, in which an individual (**androgyne**) has the gonads of one sex but contradictory other criteria of sex; **2.** having both masculine and feminine characteristics, e.g. seen in the behaviour; **3.** in botany, having both staminate and pistillate flowers in the same inflorescence. *Adj.* **androgynous**.

andromonoecious (Gr. *monos* = single, *oikos* = house), pertaining to a plant species having male and bisexual flowers in the same individual. *Cf.* androdioecious.

androsterone, 3α-hydroxy-5α-androstan-17-one, $C_{19}H_{30}O_2$; a steroid metabolite of → testosterone and some other androgens produced chiefly by the testes of vertebrates; has a weak androgenic effect.

anemochore (Gr. *anemos* = wind, *chorein* = to spread), an organism which is spread by the wind. *Cf.* anemohydrochore, hydrochore.

anemohydrochore (Gr. *hydor* = water), an organism which is spread by the wind and water. *Cf.* anemochore, hydrochore.

anemophily (Gr. *philein* = to love), wind pollination; the pollination of flowers by the wind. *Adj.* **anemophilous**.

anemotaxis (Gr. *taxis* = arrangement), a tactic movement of an organism the direction of which is determined by the wind. *See* taxis.

anencephaly, anencephalia (Gr. *enkephalos* = brain), a congenital malformation of the developmental brain occurring rarely in vertebrates; the cerebral and cerebellar hemispheres are totally lacking and other parts largely reduced. *Adj.* **anencephalous**.

anesthesia, → anaesthesia.

aneugenic, pertaining to a chemical or physical agent that causes aneuploidy.

aneuploid (Gr. *an-* = not, *eu* = genuine, good, *haploos* = onefold), a cell or an individual whose chromosome number differs by one, two, or several from the normal basic chromosome set. Thus, **aneuploidy** includes a genome mutation in which certain chromosomes appear in too few or too many copies. In diploid species, aneuploidy is regarded as **nullisomy** if both parts of a given chromosome pair are lacking, and as **monosomy** if only one partner is missing.

In chromosomal supernumerary, aneuploidy is called **trisomy** if a given chromosome appears in three sets, and **tetrasomy** if it appears in four sets. If the supernumerary involves more than one chromosome pair, the possibilities are a twofold, threefold, etc. trisomy, or tetrasomy. Respectively, the subnumerary may involve more than one chromosome pair, and is then called twofold, threefold, etc. monosomy. In diploid species, the most common cause of aneuploidy is a → non-disjunction of chromosomes during meiosis. *Syn.* heteroploid.

aneurin(e) (Gr. *a-* = not, *neuron* = nerve), vitamin B_1, thiamine; the name was given because the deficiency of the vitamin causes bad neuronal disturbances. *See* vitamin B complex.

aneuspory, *see* agamospermy.

angi(o)- (Gr. *angeion* = vessel), denoting relationship to blood vessels, or lymph vessels.

angiogenesis (Gr. *gennan* = to produce), development of blood vessels; occurs during the embryonic development of animals and later in the regeneration of tissues.

angiokinesis (Gr. *kinein* = to move), vasomotion; constriction and dilatation of blood vessels.

Angiospermae (Gr. *sperma* = seed), **angiosperms**, a subdivision of the division Spermatophyta (seed plants) in the Plant kingdom. Angiosperms have their ovules in a closed → ovary of the gynaecium. The ovules develop into seeds after fertilization, and the ovary forms a closed fruit around the seeds. Angiosperms have actual flowers and the → sporophyte is the dominating stage in their life cycle; the → gametophyte is extremely reduced, being included in the structure of the ovule (female gametophytes) or pollen grains (male gametophytes). Angiosperms are divided into two classes: → **Monocotyledonae** and → **Dicotyledonae**. In some systems Angiospermae is replaced by the group Anthophyta, which is considered to be a division within the Plant kingdom.

angiotensin (L. *tensio* = stretch), any of several peptides in the blood of vertebrates regulating renal function; **angiotensin I** (decapeptide) is formed when **renin** (an enzyme liberated from the kidneys) catalyses the cleavage of **angiotensinogen** (a tetradecapeptide formed by the liver). Angiotensin I is physiologically inactive but is activated into **angiotensin II** (octapeptide) by enzymatic action of a peptidase. Angiotensin II regulates the renal function by constricting blood vessels in the kidneys and by stimulating the secretion of → aldosterone from the adrenal cortex. Also **angiotensin III** is found in the blood; it is a heptapeptide derivative of angiotensin II and has similar but weaker effects on aldosterone secretion.

angiotonin (Gr. *tonos* = tension), a vasoconstrictor polypeptide found in vertebrate tissues; formed in the partial hydrolysis of **hypertensinogen** (angiotensinogen).

angstrom, Å (*A.J. Ångström*, 1814—1874), an old unit of length; 0.1 nm = 10^{-10} m; originally used as a unit of wavelength.

Anguilliformes (L. *anguilla* = eel, *forma* = shape), **eels**; an order of bony fishes with a long and slender body; unpaired fins are continuous, pelvic fins and girdle absent, and cycloid scales minute or absent. Their life cycle includes reproduction and larval stages (leptocephalus) in oceans but they reach maturity in brackish or fresh waters. Males stay in brackish water of estuaries, and females continue up the rivers, i.e. are catadromous.

angular collenchyma (Gr. *kolla* = glue, *enchyma* = cast, poured in), a type of collenchymatous tissue in plants (→ collenchyma), in which the corners of the cell walls are thickened.

anhydrases (Gr. *an-* = not, *hydor* = water), called formerly hydrases; enzymes of the EC sub-subclass 4.2.1. catalysing the removal of water from a compound; usually called **dehydratases** (hydro-lyases) when the reaction is not reversible, and **hydratases** when the reverse of the reaction is dominant; e.g. fumarate hydratase catalysing the fumarate-malate interconversion. → Carbonic anhydrase (carbonate dehydratase) catalyses the decomposition of carbonic acid into carbon dioxide and water. In animals it is an important factor in the transport of carbon dioxide in tissues.

anhydride, a compound derived from a substance by abstraction of water, e.g. the anhydrides of bases are oxides, those of alcohols are ethers.

aniline (Arab. *annil* = indigo blue), phenylamine, aminobenzene; an aromatic, colourless, slightly water soluble liquid, $C_6H_5NH_2$; forms coloured substances together with chlorates. Aniline dyes are widely used in industry. In biological studies aniline is used for staining cells and tissues. A careless use can cause a disease called anilism (e.g. with muscular weakness and methaemoglobinemia).

animal, a unicellular or multicellular organism which needs oxygen and organic food, i.e. is heterotrophic (animal cells are non-photosynthetic and have no cell walls). Animals are usually capable of voluntary movements although some are → sessile, taking food e.g. with the aid of tentacles. In animals, meiosis leads directly to the formation of gametes (terminal meiosis), different in size, i.e. either → macrogametes or microgametes. Typical of most multicellular animals are sensory receptors and the nervous system. *See* Animal kingdom.

animal ecology (Gr. *oikos* = house, *logos* = word, discourse), a branch of biology studying interaction between animals and their environment. *See* ecology.

Animalia, → Animal kingdom.

Animal kingdom, Animalia, the highest taxon of animals, comprising about 30 phyla with about 1.4 million described species; divided in some classifications into three subkingdoms: → **Phagocytellozoa**, **Parazoa**, and **Eumetazoa**. Also heterotrophic unicellular → protists are conventionally included in Animalia as the fourth subkingdom, **Protozoa**. Animals form a large and old branch of eukaryotic organisms first appearing over 600 million years ago in Precambrian seas.

animal physiology (Gr. *physis* = nature, *logos* = word, discourse), **zoophysiology**; a branch of biology studying the function of animals, i.e. development, interaction, regulation, and physiological adaptation of animal cells, tissues, organs, and organ systems; it also deals with neurophysiological functions controlling the behaviour. *See* physiology.

animal pole, that pole of the egg (ovum) which contains the nucleus, opposite to the vegetal pole; in freely floating eggs the animal pole turns upward.

anion (Gr. *anienai* = to go up), a negatively charged → ion, e.g. chloride (Cl^-), sulphate (SO_4^{2-}), or phosphate (PO_4^{3-}). *Cf.* cation.

anisocytic stoma (Gr. *anisos* = unequal, *kytos* = cavity, cell), *see* stoma.

anisogamy (Gr. *gamos* = marriage), a type of sexual conjugation in which the fusing gametes differ in size and/or motility; found in most animals. *Syn.* heterogamy.

anisophylly (Gr. *phyllon* = leaf), the situation in which a plant permanently has leaves of two or more shapes and sizes in the same zone of the shoot, often at the same node; e.g. in *Selaginella* and *Thuja* species. *Adj.* **anisophyllous**. *Cf.* heterophylly.

anisosmotic, pertaining to an osmotic concentration that differs from another, comparable solution. *Syn.* anisotonic. *See* osmosis.

anisospore (Gr. *sporos* = seed), a spore of such plants in which → sexual dimorphism occurs, i.e. they have larger macrospores and smaller microspores. *Cf.* isospore.

anisotomy (Gr. *tome* = cutting), a branching type of plant roots, a subtype of → dichotomy. Branching is **anisotomic** when one of two branches is stronger and better developed than the other; in isotomic branching both branches are equal.

anisotonic (Gr. *tonos* = tension), **1.** not having equal tension; **2.** → anisosmotic.

anisotropic, also **anisotropous** (Gr. *trope* = turning), **1.** having unequal properties in different directions; **2.** pertaining to a structure which is birefractive in electromagnetic radiation; e.g. anisotropic band (→ A-band) in a myofibril. *Cf.* isotropic.

ankle, ankle joint; the tarsal joint in the hindlimb between the foot and leg in tetrapod vertebrates.

anlage (Ger. *Anlage* = embryo, germ), *see* primordium.

annealing (ME *anelen* = to set on fire), the heating and controlled cooling of a material, usually metal or glass, to release stress; in biochemistry and genetics, the process in which two single-stranded polynucleotides form a double-stranded molecule. Heating results in the separation of the individual strands of any double-stranded, nucleic acid helix. Cooling leads to the pairing of the complementary base pairs by hydrogen bonds, and results in the linkage of the polynucleotide strands. Annealing takes place between two complementary polynucelotides of DNA or RNA, producing double-stranded DNA or RNA, or RNA-DNA hybrids. The process is also called nucleic acid → hybridization. *See* polymerase chain reaction.

Annelida, (Annulata) (L. *annulus* = ring, Gr. *eidos* = form), **annelids**, **segmented worms**; a phylum of metamerically segmented worms having a long body, true coelom, and paired setae covered with a thin cuticle (setae are absent in leeches). The nervous system consists of dorsal cerebral ganglia and a long ventral nerve cord with segmental ganglia. The circulatory system is closed and paired nephridia are segmental; respiration occurs through the skin. The phylum includes more than 12,000 species, most of which are marine but many are terrestrial or live in fresh waters. Annelids are divided into three classes: **oligochaetes** (Oligochaeta, e.g. earthworms), **polychaetes** (Polychaeta, e.g. ragworms, lugworms), and **leeches** (Hirudinea).

annual (L. *annus* = year), pertaining to an organism which lives only one year; the term is used especially for plants. An annual plant may germinate in spring and produce seeds in autumn, or the seeds may germinate in autumn and new seeds ripen next spring. *Cf.* biennial, perennial.

annual growth, a part or zone of an organism grown during one year. In some plants and animals the annual growth can be clearly detected in the structure. *See* annual ring.

annual ring, annulus; **1.** in zoology, a marking in an organ or tissue indicating one year's growth (**growth ring**), showing the age of an animal; for example, one of the ridges on the scales of some fish species or similar growth structures seen e.g. in the otolith, or in the shell of a mussel; the rings are markings due to uneven rate of seasonal growth; **2.** in botany, the cell layers formed by the → cambium during one year in the stem or root of a plant (tree). The annual rings can be discerned in the structure because the cells borne in the spring time are larger than those produced later in summer; in autumn and in winter no new cells are produced. The annual rings also show the age of a tree. In tropical areas without clear climatic seasons there are separate growing seasons depending e.g. on rains, resulting in **seasonal rings** in plant structures.

annular thickening, any of ring-like thickenings in the inner walls of xylem vessels or tracheids; provide mechanical support and help the water columns to adhere to the walls of the water-conducting structures.

Annulata, → Annelida.

annulus, pl. **annuli,** Am. also **annuluses** (L.), a ring or ring-shaped anatomical structure; e.g. **1.** → annual ring; **2.** abdominal ring, *annulus abdominalis*, the opening through which the ductus deferens and the gonadal blood vessels enter the inguinal canal in mammals; **3.** a ring-like body segment (not truly segmented) of some annelids, e.g. leeches; **4.** a row of specialized thick-walled cells involved in the opening of moss capsules or sporangia of ferns; **5.** a ring in the stalk of some fungi in the group → Basidiomycota.

anode (Gr. *ana* = up, *hodos* = way), **1.** the positive electrode in an electrolytic bath liberating positive ions (cations) and attracting negative ones (anions); **2.** the positive electrode in a battery; **3.** the positive plate in an electron tube. *Cf.* cathode.

anoestrus, *see* oestrus.

anomaly (Gr. *anomalia*), a marked deviation in the average anatomical structure, especially as a result of congenital developmental disturbance.

anomocytic stoma (Gr. *anomos* = illegal, *kytos* = cavity, cell), *see* stoma.

anonymous group, a group of animals not identifying each other individually. In an **open anonymous group**, the members admit individuals of other species to the group, as in mixed bird swarms. In a **closed anonymous group**, the animals identify each other as a member of the group, but not individually, e.g. on the basis of smell, and drive strangers away.

Anoplura (Gr. *an* = without, *hoplon* = weapon, *oura* = tail), **sucking lice**; an order of small wingless insects (ectoparasites) whose body is depressed, head narrow, and mouthparts specialized for sucking blood from homoiothermic animals. Many species are vectors of various diseases; e.g. the human louse (*Pediculus humanus*) which may transmit typhoid fever. About 2,400 species are described. *Syn.* Siphunculata.

anoxemia (Gr. *an-* = not, ox < oxygen, *haima* = blood), lack of oxygen in arterial blood.

anoxia, total lack of oxygen in inspired gas, blood, and tissues of an organism. *Cf.* hypoxia.

Anseriformes (L. *anser* = goose, *forma* = shape), **swans**, **geese**, and **ducks**; an order of aquatic birds which have a broad beak covered by cornified epidermis with sense pits; the margin of the beak is provided with filtering ridges, the legs are short and the front toes webbed. The order includes 150 globally distributed species.

antagonism (Gr. *antagonizesthai* = to struggle against), a mutual opposition between structures (as muscles), substances, physiological processes, etc.; something that opposes or resists is called an **antagonist**. *Cf.* synergism. *See* agonist.

Antarctic kingdom (Gr. *anti* = opposite, *arktos* = bear, north), an area in plant geography, comprising the Antarctic, islands of Antarctic Ocean, and the southernmost tip of South America (Patagonia).

ANT-C, → antennapedia complex.

antenna, pl. **antennae, antennas** (L. = sail yard), a paired sensory appendage on the head of arthropods, excluding arachnids; long, jointed organs specialized for receiving tactile and olfactory stimuli. Crustaceans have two pairs of antennae; the first pair, **antennules,** are uniramous, the **second antennae** are biramous and, in some species like water fleas and cyclops, enlarged for swimming.

antenna complex, the acceptor of → photons in plant cells; the absorbed energy is transferred from antenna pigments to photosynthetic reaction centres of → chloroplasts. *See* photosynthesis.

antennal gland, an excretory organ at the base of the second antennae of crustaceans. *See* green gland.

antennapedia complex, ANT-C, a gene complex in fruit flies consisting of five homoeotic → selector genes which control the development of head and thoracic segments. *See* bithorax complex.

antennule (L. *antennula*, dim. of *antenna* = sail yard), the first pair of antennae on the head of crustaceans serving as sensory organs; may be absent in some branchiopods, such as water fleas and cyclops. *See* antenna.

anth(o)- (Gr. *anthos* = flower), denoting relationship to flowers.

anther (Gr. *antheros* = flowery), the terminal part of the → stamen, producing → pollen.

anther culture, a branch of cell and tissue culture in order to produce haploid plants, starting from anther tissue or pollen mother cells.

antheridiophore (Gr. *antherix* = spike, *phorein* = to carry along), the stalk that bears an → antheridium; occurs e.g. in some liverworts.

antheridium, pl. **antheridia** (Gr. *-idion* = dim.), the male → gametangium in plants, producing male gametes.

antherozoid (Gr. *zoon* = animal), a freely swimming male gamete of many algae and mosses, developing in male gametangia, i.e. in the antheridia. *Syn.* spermatozoid. *See* gametangium.

anthocarp, → infructescence.

Anthocerophyta, *see* Anthocerotae.

Anthocerotae (Gr. *keras* = horn), **hornworts**; a class in the division Bryophyta, Plant kingdom; in some classifications separated from Bryophyta into a separate division Anthocerophyta; a few small calcifuge species, occurring especially in tropical areas. The gametophyte is disc-shaped and structurally simple, being anchored by rhizoids to the soil. Containing only one chloroplast in each cell, they differ from other bryophyte cells.

anthocyan (Gr. *kyanos* = blue), a red, purple, or blue flavonoid pigment located in the vacuole of a plant cell. The name anthocyan includes both the pigments **anthocyanidins** and their glycosides, **anthocyanins**. Red and blue colours in flowers are due to anthocyans; they occur also in green plant parts but are not visible until chlorophylls are decomposed in autumn. Stress conditions, such as strong wind, low temperature, or high illumination, stimulate the synthesis of anthocyans.

Anthophyta, Magnoliophyta, a division in Plant kingdom in some systems, corresponding to the subdivision Angiospermae of the division Spermatophyta in other classifications.

Anthozoa (Actinozoa) (Gr. *anthos* = flower, *zoon* = animal), **anthozoans**; a cnidarian class; primitive invertebrate animals which have a life cycle without medusa stage, and so a flower-like → polyp is dominating. The gullet (*stomodaeum*) connects the mouth and internal body cavity (*enteron*), and six or eight mesenteries (*septae*) divide the body into radial compartments. Based on the number of the septae, the class is divided into the subclasses **Alcyonaria** (or Octocorallia) and **Zoantharia** (Hexacorallia). In some classifications also a third subclass **Ceriantipatharia** (tube anemones, thorny corals) is described. Known orders are e.g. **sea pens** (Pennatulacea), **stony corals** (Scleractinia, formerly Madreporaria), **sea fans, sea whips** (Gorgonacea), and **sea anemones** (Actiniaria).

anthracene (Gr. *anthrax* = charcoal), anthracin; a tricyclic hydrocarbon obtained from coal tar, originated from → anthraquinones of ancient plants; an important material in chemical industry, e.g. in producing alizarin dyes.

anthracite, a hard coal containing more than 93% of carbon.

anthraquinones, a large group of cyclic compounds, yellow, orange, or red in colour, occurring in fungi, lichens, and higher plants, as well as in some insects (probably originated from plants). One of the simplest is

9,10-anthracenedione, $C_{14}H_8O_2$ (anthraquinone by common name), manufactured from benzene for industry. The reduction product of anthraquinone is → anthrone. Anthraquinones also exist as dimers, and they can be brominated, chlorinated, or nitrated; anthraquinone derivatives are e.g. **anthraquinone dyes**.

anthrone, 9,10-dihydro-9-oxoanthracene, $C_{14}H_{10}O$, prepared from → anthraquinone by reduction; used as a colour reagent for the quantitative determination of sugars and starch.

anthropo- (Gr. *anthropos* = human), denoting relationship to man.

anthropochore (Gr. *chorein* = to spread), a species which is not original in a certain area but transported there by man. *Syn.* hemerochore. *Cf.* anemochore, hydrochore.

anthropogenic (Gr. *genes* = produced), caused or produced by man; e.g. anthropogenic climate.

anthropogeny, anthropogenesis (Gr. *gennan* = to produce), the origin and evolution of man.

Anthropoidea (Gr. *eidos* = form), **anthropoid apes**; a suborder of the order Primates including **New World** and **Old World monkeys** (superfamilies Ceboidea and Cercopithecoidea), **gibbons**, **great apes**, and **man** (superfamily Hominoidea). Extinct anthropoid genera are e.g. *Aegyptopithecus*, *Dryopithecus*, and *Parapithecus*. *Adj.* **anthropoid**.

anthropology (Gr. *logos* = word, discourse), a branch of science dealing with origin, evolution, and social and cultural relationships of man.

anthropomorphism (Gr. *morphe* = shape, form), a pseudoscientific view which explains biological phenomena by terms referring to man, human action, or imaginary human-like features. *Adj.* **anthropomorphic**.

anthropomorphous, human-like, resembling man, shaped like man.

anti- (Gr. *anti* = before, against), denoting against or opposite.

antiadrenergic, antagonistic to → adrenaline or noradrenaline, or to an action of sympathetic nerves and the neurones in which noradrenaline or adrenaline acts as a neurotransmitter.

antiauxin (L. *auxanein* = to give rise to growth), the common name for any compound antagonizing the effects of → auxins in plants.

antibiotic, a natural, synthetic, or semisynthetic substance with a relatively low molecular weight, inhibiting the growth of microorganisms or destroying them. Many antibiotics are derived from bacteria of the genus *Streptomyces* (an exeption is e.g. → penicillin). Antibiotics block the synthesis of the bacterial cell wall (e.g. penicillins) or protein synthesis (e.g. neomycin, streptomycin, and tetracyclines) especially in microbes. Antibiotics are used to control microbial diseases in man and animals. They are also used in culture media in order to prevent contamination by foreign bacteria or to select certain cell lines in cultures, and as markers of gene transfer (e.g. → kanamycin).

antibiotic-resistance element, a genetic element, composed of DNA and often carried by a → plasmid (antibiotic-resistance plasmid, drug-resistance plasmid), rendering its carrier bacterium resistant to an antibiotic.

antibody, Ab, immunoprotein; a globular protein produced by lymphocytes usually by induction of an → antigen in the vertebrate immune system. Each antigen type can then combine with this specifically induced antibody type by **lock-and-key** binding, producing secondary reactions such as precipitation and → agglutination of antibody-antigen complexes. Supposedly, antibodies may also exist naturally without antigen induction (*see* immunity, lymphocytes).

Antibodies are located either on the lymphocyte cell membrane as **receptor antibodies** (cell-bound antibodies), or they are secreted by activated B lymphocytes, **plasma cells**, into the blood and lymph as **soluble antibodies**, → immunoglobulins. Antibodies specific to one single epitope (determinant) of an antigen molecule, are called **monoclonal antibodies** which are produced by a lymphocyte cell line derived from one single lymphoblast. Enormous numbers of different lymphocyte lines exist in each animal and therefore plasma always contains a mixture of antibodies, i.e. **polyclonal antibodies**; monoclonal antibodies can be obtained in cell cultures started from one single cell.

Antibodies are essential elements in the induction of → **immunity** to foreign substances and microorganisms. The immune system can be rendered more efficient by →

vaccinations with antigens that induce new antibodies and thus facilitate the immune defence in the body; also specific antibodies can be injected. Specific monoclonal antibodies are useful tools in biological studies e.g. for recognizing and localizing molecules and cellular and tissue structures.

More narrowly, the term antibody is used only as a synonym of → **immunoglobulin**, i.e. an antibody present in the blood plasma or other body fluids capable of eliciting the immune response. *See* blood groups, opsonins.

antibody diversity/variation, production of different antibodies by different → lymphocytes (*see* antibody, immunoglobulin). Light and heavy chains of the antibody protein are encoded by different genes, the so called **assembled genes**, consisting of codes for constant and variable parts of the antibody. These are inherited in the germ line as different **genelets** but brought together to form a functional gene by somatic recombination during the maturation of the lymphocyte. In the germ cell line there are a few different genelets for the constant part and many for the variable part. Thus millions of different antibodies can be formed in their different combinations.

anticholinergic, antagonistic to → acetylcholine, or to the action of para-sympathetic nerves and the neurones in which acetylcholine acts as a neurotransmitter.

anticipation (L. *anticipare* = to anticipate), in genetics, the earlier appearance of a certain, usually dominant genetic disease (or its appearance in a more severe form) in successive generations. Anticipation was earlier regarded as an apparent phenomenon because only the most severe cases of genetic diseases reach researcher's knowledge; consequently, in families which are involved in examinations anticipation is more likely to be found. From the beginning of 1990's it has become clear that anticipation is a real phenomenon, and also its molecular basis is known in certain cases.

anticlinal (Gr. *klinein* = to bend, lean), inclining in opposite direction, being oriented at right angles to a surface; e.g. describes a possible plane of cell division. *Cf.* periclinal.

anticoagulant (L. *coagulare* = to clot), an agent preventing → blood coagulation; may affect directly on biochemical coagulation processes like → heparin, or indirectly like inhibitors of → vitamin K. Anticoagulants are used in medicine and as pesticides especially against rodents; the most used pesticides so acting are warfarin, bromadiolon, difenacum, and brodifacum.

anticodon (L. *codex* = book), a nucleotide triplet in → transfer RNA, complementary to the → codons residing in → messenger RNA. The anticodon recognizes the codon according to the → base-pair rule, and hence each amino acid carried by transfer RNA will be precisely placed into its right location in the polypeptide chain built up during → protein synthesis.

antidiuresis (Gr. *diourein* = to urinate), reduction of urine volume, suppression of urinary secretion.

antidiuretic, 1. suppressing the rate of urine formation; **2.** an agent which reduces the formation of urine.

antidiuretic hormone, ADH, → vasopressin.

antidote (Gr. *dotos* = what is given), a substance which counteracts the effect of a poison, e.g. by neutralizing it.

antidromic (Gr. *dromos* = running), running in the opposite direction; e.g. pertaining to propagation of a nerve impulse in a direction opposite to normal; may occur in ephaptic transmission (*see* ephapse).

antifreeze substances, → cryoprotectants.

antigen, Ag (Gr. *genos* = origin), a substance which specifically binds to an → antibody, and usually induces the antibody production in lymphatic tissues of vertebrates, thus acting as an **immunogen**; foreign cells, microbes, viruses, and macromolecules (as proteins) may act as antigens, sometimes also the animal's own tissue structures (*see* autoimmunity). All parts of an antigen are not equivalent but certain atom groups in it act as **epitopes** (**determinants**) serving as recognition and binding sites for antibodies. Also small molecules may act as antigens by binding first to certain larger substances in tissues (*see* hapten). *See* immunity, lymphocytes.

antigen-antibody reaction, the binding of an antigen to an antibody, specific to it; in the primary reaction the non-covalent binding unites antigen and antibody molecules together which usually leads to secondary reactions such as → precipitation and agglutination, and activation of the → complement system and phagocytosis.

antigenic variation, the ability of some pathogenic microbes to change their coat antigens during infection.

antigen-presenting cells, APCs, a heterogeneous population of vertebrate cells, such as → monocytes and other macrophages, B lymphocytes, and certain epithelial and endothelial cells called **interdigitating dendritic cells**, IDCs, becoming active especially when stimulated by → cytokines. APCs are found in the blood, skin (Langerhans cells), lymph nodes, spleen, and thymus. The cells take up protein → antigens, move to lymph nodes, and break the antigens into peptide fragments which form complexes with MHC (class II) molecules found abundantly in APCs. Within the paracortex of the lymph node, APCs are recognized by T lymphocytes (T cells) specific for the complex that is displayed on the T cell surface (**antigen presentation**).

Another specialized APC population are the follicular dendritic cells, found in follicles of the B cell areas of the lymph nodes and spleen. They present antigen to B cells and instead of MHC molecules have other molecules for interaction of immune complexes.

antigibberellins, the term usually pertaining to synthetic → growth retardants which inhibit the growth of the plant stem, having effects opposite to → gibberellins. Also a natural plant hormone, → abscisic acid, could be regarded as an antigibberellin.

antiherbivory effect (L. *herba* = plant, *vorare* = to eat), a result of defensive functions caused by the → chemical defence of plants against herbivores.

antihistamine, a drug which antagonizes the effect of → histamine; used in the treatment of allergy symptoms.

antimere (Gr. *meros* = part), an opposite part; e.g. **1.** one of the several identical portions in radially symmetrical organisms that can be divided by any transverse plane of symmetry; **2.** one of the corresponding opposite parts of a bilaterally symmetrical organism, such as the right or left part of the body.

antimetabolite, a substance which antagonizes or replaces a certain → metabolite.

antimorphic (Gr. *morphe* = form), in genetics, pertaining to mutations or → alleles the effect of which is opposite to the corresponding → wild allele. It is believed that **antimorphism** is due to the fact that the gene product of such an allele competes for the enzyme substrate with the enzyme produced by the wild allele, but cannot accomplish the enzyme reaction itself. Antimorphism may also be connected with the phenomenon that the gene product of such an allele competes with the wild type allele in other different interactions of molecules but is unable to carry out the normal gene function. *Noun* **antimorph**.

antimycin A, an antibiotic produced by most species of *Streptomyces* bacteria; inhibits the electron transfer to cytochromes in cell respiration.

anti-oncogene, a tumour-suppressing gene, controlling cell growth; the inactivation of an anti-oncogene leads to the loss of growth control and consequently the development of a tumour. *See* oncogene.

antioxidants, natural or synthetic substances which prevent oxidative processes. Biological antioxidants occur in all organisms for protecting cells and tissues against the harmful action of → **oxygen radicals** produced in normal cell metabolism and e.g. by ultraviolet radiation. The most important antioxidants in organisms are vitamins C and E, beta-carotene, and flavonoids. Also selenium is considered to be an antioxidant, because it is an essential cofactor of an antioxidative enzyme, glutathione peroxidase. Antioxidants are added to foodstuffs for improving their nutritional value and preservation properties.

antipodal cells, antipodals (Gr. *pous*, gen. *podos* = foot), three cells in the typical → embryo sac of a plant, situated on the opposite side to the egg cell and the synergid cells. The antipodal cells e.g. store nucleic acids and transmit nutrition from the chalaza to the other structures of the embryo sac before fertilization.

antiport (L. *portare* = to carry), a membrane protein transporting two different ions or molecules simultaneously in opposite directions across the cell membrane; one ion or molecule moves down the → electrochemical gradient and it can co-transport the other one against a gradient. *Cf.* symport, uniport. *See* cell membrane transport.

antipyretic (Gr. *pyretos* = fever), **1.** reducing fever, antifebrile, febrifugal; **2.** an agent which reduces fever, a febrifuge.

antisense RNA, a small RNA molecule which is complementary to a → messenger RNA (mRNA), and hence by hybridizing with the latter can prevent its function; occurs naturally but can also be produced by → gene manipulation in order to prevent the function of a given gene.

antisense strand, that DNA strand which functions as the template for RNA in → genetic transcription. *Syn.* template strand of DNA. *Cf.* sense strand.

antisepsis (Gr. *sepsis* = putrefaction), a process of inhibiting the multiplication and growth of microorganisms, so preventing infections.

antiseptic, 1. pertaining to → antisepsis; **2.** an agent that inhibits the growth of a microorganism.

antiserum, a blood serum which contains an → antibody (or antibodies) to a certain antigen (or antigens), e.g. for testing blood groups.

antlers, deciduous projections growing on the skull of the animal in the deer family (Cervidae); in elks (mooses) and most deer species, the antlers occur only in males but in reindeer and caribou also in females. Antlers are usually branched and composed of dense connective tissue which develops into a calcified bone for the breeding season, and are shed annually. During growth the antlers are covered by skin in the form of "velvet" which the deer has to remove rubbing them e.g. against tree trunks (**antler velvet shedding**). *See* horn.

anulus, → annulus.

Anura, (Salientia) (Gr. *an* = not, *oura* = tail), **anurans**, **toads** and **frogs;** an amphibian order comprising semiaquatic or terrestrial animals; aquatic larvae (**tadpoles**) are gill-breathing, but adults breathe mainly with lungs and through the skin. Adult anurans have no tail and the caudal vertebrae are fused into the → urostyle, the hindlegs have specialized for jumping. The order includes about 3,400 species in 21 families and 301 genera.

anus, pl. **anus** or **ani,** Am. also **anuses** (L. = ring, anus), the posterior opening of the alimentary canal; in → Deuterostomia it is embryonally derived from the → blastopore, but in → Protostomia the blastopore develops to form the mouth, and the anus arises at the other end of the embryonic intestine, archenteron.

anvil, in anatomy, the → incus.

aorta, pl. **aortae,** Am. also **aortas** (L. < Gr. *aorte*), in vertebrates, the elastic main trunk of the systemic arterial circulation, rising toward the head from the single ventricle of the fish and amphibian hearts and from the left ventricle of the four-chambered heart of other vertebrates. In fish, the **ventral aorta** (*aorta ventralis*) branches into 4—6 paired branchial aortae (gill aortae); in amphibians two **aortic arches** turn caudally and unite behind the heart. In birds only the right aortic arch, and in mammals the left aortic arch, remain during embryogenesis. The more caudal parts are called **thoracic aorta** and **abdominal aorta**. In fish, the aorta carries deoxygenated blood, in amphibians and reptiles partly mixed blood, and in birds and mammals oxygenated blood.

Annelids have anteriorly five pairs of aortic arches between the dorsal and ventral vessels; in insects, the dorsal aorta is the only blood vessel, moving → haemolymph forward from the tubular heart.

aortic arches, *see* aorta.

aortic bulb, → bulbus arteriosus.

aortic body (L. *glomus/corpus aorticum*), a small globular sense organ of vertebrates in the small branch of the aortic arch; it is similar to the → **carotid body**. Both these organs contain chemoreceptors (glosso-pharyngeal nerve endings) which are sensitive to lack of oxygen and to increase in carbon dioxide and hydrogen ion concentrations (pH) in the blood. Aortic and carotid bodies are involved in the control of respiration.

apatite, any calcium phosphate containing some other elements, $Ca_5(PO_4)_3X$, where X may be fluorine (F), chlorine (Cl), or a hydroxyl group (OH). Apatite is an important mineral occurring as granular masses or hexagonal crystals in phosphate-containing rocks, but is formed also in some organisms; in vertebrates it is the chief mineral in bones and teeth. The -OH group is easily changed with F^- and the more acid-resistant **fluoroapatite** is formed in the enamel of the teeth.

aperture (L. *apertus* = open), **1.** in anatomy, an opening or an open hole or gap, e.g. the pelvic outlet (*apertura pelvis inferior)* and the pelvic inlet (*a. p. superior*), the lower and upper openings of the → pelvis; **2.** the opening angle of a lens; **3.** in electron microscopy, the opening angle of the electron beam, or the gap regulating it.

apes, *see* Primates.

apetalous, pertaining to a flower having no →

petals, e.g. birch, willow.

apex, pl. **apices, apexes** (L. = tip), the top or extremity of an anatomical structure, especially of a conical or pyramidal structure; e.g. the apex of the lung or heart, the apex of the snail shell, and the shoot or root apex in plants.

aphids, → Aphididae.

Aphididae (*Aphis* = the type genus), **aphids, plant lice**; a family of homopteran insects that are usually wingless, some species have a pair of wings; they often reproduce also parthenogenetically. Most aphids live by sucking sap from plants, a part of which is secreted as honeydew to be used by ants in a symbiotic interaction. Many aphids are vectors of plant diseases.

aphidiphage (Gr. *phagein* = to eat), an animal feeding on aphids.

aphodus, pl. **aphodi** (Gr. *aphodos* = departure), a short duct in the water-canal system of sponges, leading from the flagellated chamber to the excurrent canal.

aphotic (Gr. *a* = not, *phos* = light), lightless; **aphotic zone**, a deep layer of a water body into which light does not penetrate. *Cf.* photic, euphotic, dysphotic.

aphyllous (Gr. *phyllon* = leaf), leafless, pertaining to a plant without foliage leaves.

apical, situated at the tip, → apex; pertaining to the side of an epithelial cell opposite to the basal side.

apical bud, a plant → bud situated at the tip of the shoot or branch.

apical dominance, a type of plant growth and branching, regulated hormonally by an → apical bud. The bud prevents development of lateral (axillary) buds into branches; if the apical bud is removed, branching takes place. Apical dominance is typical of many trees and shrubs.

apical flower, a flower located at the tip of the shoot, a shoot branch, or of an inflorescence.

apical growth zone, *see* apical meristem.

apical meristem, the → meristem tissue at the tip of plant shoots, roots, and their branches. The cells remain embryonal, retaining their ability to divide, and the **apical growth zones** formed by them support the longitudinal growth of the plant, forming also the primordia of leaves and axillary → buds in the stems. In seed plants the apical meristem is divided into **promeristem** with initial cells, and → **derivative meristems**. On the other hand, the apical meristem can be divided into **tunica** and **corpus**, the former consisting of 1—3 cell layers in the surface and the latter of several cell layers below. These structures have different tasks in plant growth and development, the tunica, for example, being responsible for the formation of leaf primordia. In many pterophytes there is only one continuously dividing apical cell, which produces the apical growth zone.

Apicomplexa (L. *apex* = tip, *complexus* = twisted around, surrounding), a protozoan phylum; endoparasitic protists which live in animals of many phyla; the **apical complex** with several cell organelles (e.g. → micronemes, micropore) in the anterior part of the cell body is characteristic of certain developmental stages; usually cilia and flagella are absent. The phylum is divided into two classes, Perkinsea and Sporozoea; pathogenic in man are e.g. the genera *Monocystis*, *Gregarina*, *Plasmodium*, and *Toxoplasma*.

aplasia (Gr. *a-* = not, *plassein* = to form), congenital absence or defective development of an organ or tissue. *Adj.* **aplastic**.

apo- (Gr. *apo* = away from, off), denoting separated from, derived from.

apocarpous (Gr. *karpos* = fruit), pertaining to a plant → gynaecium having separate carpels; opposite to syncarpous. *Noun* **apocarpy**.

apocrine (Gr. *krinein* = to separate), pertaining to a type of exocrine secretion of animals in which the apical parts of secreting cells are cast off and incorporated into the secretion; e.g. a type of → sweat glands. *Cf.* eccrine, holocrine, merocrine.

Apoda (Gr. *a-* = without, *pous,* gen. *podos* = foot), → Gymnophiona.

apodema, pl. **apodemata,** or **apodemas** (Gr. *apo* = away, *demas* = body), **apodeme**; an inner projection or a nodule in the cuticle of arthropods on which the muscles are inserted.

Apodiformes (Gr. *apous* = footless, L. *forma* = form), an order of short-legged and long-winged birds which are known as excellent flyers. The order includes 429 small-sized species in three families: **swifts** (Apodidae), **tree swifts** (Hemiprocnidae), and **hummingbirds** (Trochilidae). Hummingbirds are very small birds, found in America and the West Indies, characterized by the brilliant plumage of the male. A Chinese species, **edible-nest swiftlet** (*Aerodramus fuciphagus*), builds the nest of saliva which is commonly used by man as food in a soup.

apoenzyme (Gr. *apo* = away, from, *en zyme* = in state of fermentation), the protein part of an → enzyme to which a smaller component, coenzyme, is bound to form an active enzyme (holoenzyme).

apoferritin (L. *ferrum* = iron), a protein present in the intestinal tissue of vertebrates combining with ferric hydroxide-phosphate to form **ferritin**, the absorptive iron carrier found especially in the liver, spleen, bone marrow and gastrointestinal mucosa; a single apoferritin molecule can assimilate thousands of ferric ions.

apogamy, *see* agamospermy.

apogeotropism, ageotropism (Gr. *ge* = earth, *tropos* = turn), negative geotropism, e.g. the growth of roots up from the earth. *See* gravitropism.

apomict, a plant produced or reproducing by → apomixis.

apomixis, pl. **apomixes** (Gr. *mixis* = act of mixing), a type of reproduction of higher plants without fertilization (*cf.* parthenogenesis in animals). Apomixis is divided into two main types: → **agamospermy** and **vegetative reproduction**. In agamospermy seeds and embryos are formed, but without → fertilization. In vegetative reproduction new plants are formed from existing plants (roots, rhizomes, runners, etc.) without seed formation. Apomixis is often thought to include only agamospermy. The descendants formed through apomixis are genetically identical with their mother plants unless special → autosegregation occurs.

apomorphic, apomorphous (Gr. *morphe* = form), in evolution, pertaining to a characteristic which is more developed in relation to its ancestral form; e.g. the feather is apomorphic in relation to the reptilian scale. *Noun* **apomorphy**. *Cf.* **plesiomorphic**.

apophysis, pl. **apophyses** (Gr. *phyesthai* = to grow), **1.** a process or an outgrowth of a bone usually for attachment of muscles; **2.** in plants, a small outgrowth, e.g. in the capsule of mosses.

apophyte (Gr. *phyton* = plant), an original plant species which has benefitted from human activity, such as soil turning.

apoplast (Gr. *plastos* = mould), the continuum formed by all cell walls and intercellular spaces in an individual plant; water and various solutes are passively transported in the apoplast. *Cf.* symplast.

apoprotein, a protein molecule having no → prosthetic group; e.g. an → apoenzyme.

apoptosis (Gr. *ptosis* = falling), → programmed cell death.

apopyle (Gr. *pyle* = gate), an internal opening on the surface of the spongocoel (in syconoid sponges) through which water flows from a flagellated chamber to the spongocoel; or a pore between the flagellated chamber and the excurrent canal in leuconoid sponges through which water leaves the chamber passing through the excurrent canal to the osculum. *Cf.* prosopyle.

aporogamy (Gr. *a* = without, *poros* = a small opening, *gamos* = marriage), growth of the pollen tube from the stigma to the → ovule laterally through the integuments after pollination. *Cf.* porogamy, chalazogamy.

aposematic (Gr. *sema* = sign), describing an organism which by its warning colours, shapes, sounds, or behaviour informs a predator of being noxious or dangerous; e.g. many insects, such as bees and wasps, are aposematic. *Noun* **aposematism.** *Cf.* allosematic.

apospory, *see* agamospermy.

apostatic (Gr. *statos* = standing), differing from normal.

apostatic selection (Gr. *statos* = standing), natural selection that favours exceptional morphs or species; a selective predation which directs against the most common → morph (form) of the prey population: e.g. thrushes usually catch the most common morph of a snail population, and thus the selection favours the other morphs and consequently their proportion increases. Apostatic selection usually leads to → balanced polymorphism in the prey population.

apotele (Gr. *telos* = end), the most distal segment, usually provided with a claw, in the legs of some arthropods. *Syn.* pretarsus.

apothecium, pl. **apothecia** (Gr. *apotheke* = storehouse), a type of fruiting body in → Ascomycetes; an open, bowl-like structure, its surface being covered by asci (*see* ascus). Apothecia are common also in lichens in which they are formed by the fungal partner which mostly is an ascomycete.

apotracheal parenchyma (Gr. *tracheia* = rough (artery), *para* = beside, *enchyma* = poured in), a → parenchyma tissue within the → secondary xylem of trees, located separately from water-transporting vessels. *Cf.* paratracheal parenchyma.

apparatus, pl. **apparatus** (L.), **1.** an instrument usually made up of several parts; **2.** a system of organs performing a special function; may be composed e.g. of glands, blood vessels, ducts, and muscles, as the alimentary apparatus (digestive system), or it may be formed of cell parts, as e.g. Golgi apparatus (→ Golgi complex).

apparent competition, an interaction between two prey species in which a predator can feed on both of them. First, the predator gains advantage by preying only on one species and the population of the predator increases. Hence, the bigger population begins to feed also on the other prey species which becomes threatened. Therefore, through the predator the prey species indirectly have negative effects on each other which is called apparent competition. In ethological studies this is difficult to distinguish from the true interspecific → competition.

appeasement behaviour/display, an intraspecific behavioural pattern occurring in agonistic behaviour of animals in situations when escape is disadvantageous for a subordinate individual. With an appeasement gesture the animal tends to arouse a conflict situation in a dominant attacker and thus inhibit its aggressiveness. Appeasement usually operates either with a **sexual presentation posture** of the subordinate individual or with signals and gestures associated with **infantile behaviour**. In both situations the appeasing individual may be either male or female. The second type of appeasement relies on gestures and signals which are as different as possible from threating postures. *Syn.* submissive posture. *Cf.* threat signals.

appendage, a structure attached to the body; **1.** in vertebrate anatomy, → appendix; **2.** in invertebrates, usually paired projections from the trunk of an animal, such as different types of legs and mouthparts. In arthropods, each body segment has primarily one pair of appendages, which usually are jointed organs specialized for different functions; e.g. the crayfish has eight pairs of head appendages (two antennae, three mouth part appendages, and three maxillipeds), five pairs of thoracic appendages (one cheliped pair and four pairs of walking legs), and six pairs of abdominal appendages (pleopods and one pair of uropods). The appendages of crustaceans may be homologous with the parapods of annelids. Fins and limbs of lower vertebrates are sometimes called appendages. *See* biramous appendage.

Appendicularia (L. *appendicula* = small appendage), → Larvacea.

appendicular skeleton, a part of the vertebrate skeleton comprising the pectoral and pelvic girdles and the limbs.

appendix, pl. **appendices** (L. *appendere* = to hang on), in anatomy, a supplementary structure attached to the main structure; e.g. the **vermiform appendix** (*appendix vermiformis/caeci*), the worm-like diverticulum of the caecum in humans.

appetitive behaviour, an instinctive behavioural pattern which forms the first phase in the goal-orientated behaviour of an animal. The behaviour supposes the preceding increase of motivation level; e.g. a hungry animal begins to seek food (**exploratory behaviour**). Appetitive behaviour may either consist of random movements without any distinct goal, as occurs in many primitive animals, or of well-directed and goal-oriented movements, as those of hungry gulls which fly to a dumping field. The appetitive behaviour is followed by → consummatory acts.

apposition growth, a growth type of the plant cell wall in which cellulose microfibrils deposit one on top of the other, typical of the formation of the → secondary wall. *Cf.* intussusception growth.

aprim(a)eval area (L. *a-* = not, *primus* = first, *aevum* = age), a pristine area or a forest usually under some conservation.

apterium, pl. **apteria** (Gr. *a-* = not, *pteron* = wing), a naked area on the bird's skin, bare of contour feathers and lying between pterylae, e.g. the incubation patch of the female. *Syn.* apteryla. *Cf.* pteryla.

apterous, wingless, like some insects.

Apterygiformes (Gr. *pterygion* = little wing, L. *forma* = form), **kiwis**; a threatened bird order which includes one family, Apterygidae, with three species in the genus *Apteryx*, all living in forests in New Zealand. They are flightless night birds with stunted wings and tail, and a long sensitive beak. *See* ratites.

Apterygota, Ametabola (Gr. *pterygotos* = winged), a subclass of primitively wingless insects which probably is a polyphyletic taxon; have long antennae, and a jointed abdomen with caudal appendages, styli and cerci; → metamorphosis is reduced or absent. The subclass includes four primitive orders: Protura (**proturans**), Diplura (**japygids**), Collembola (**springtails**), and Thysanura (**silverfish** and **bristletails**). *Cf.* Pterygota.

apteryla, pl. **apterylae,** → apterium.

aquatic (L. *aqua* = water), pertaining to water, living in water. *Cf.* semiaquatic.

aquatic ape theory, a theory stating that the evolutionary development of man has included a long phase in a semiaquatic environment. The theory is based on exceptional features of man as compared to other primates, e.g. lack of fur, abundant subcutaneous fat, large numbers of sebaceous glands located evenly in the skin, descendent larynx, efficient control of respiratory movements and → diving reflex, weak sense of smell, the location of the uterus deep in the abdominal cavity, and → neoteny. Corresponding traits exist only in aquatic mammals, such as whales. Furthermore, of all primates only man has eccrine sweat glands which act in temperature regulation but also secrete large quantities of salts. According to the theory these glands have originally been salt glands, secreting extra salt in sea-shore conditions, and probably the exceptionally efficient lacrimal glands had the same function. The theory has not been generally accepted, and many researchers prefer the savanna theory, stating that man has developed in grassland conditions.

aqueductus, pl. **aqueductus** (L.), in anatomy, a canal filled with tissue fluid; e.g. *aqueductus cerebri* (a. Sylvii), the aqueduct of cerebrum, a canal inside the mesencephalon of vertebrates, connecting the third and fourth brain ventricles.

aqueous humour, Am. **aqueous humor** (L. *humor* = liquid), intraocular fluid occupying the anterior and posterior chambers of the camera eye; in vertebrates, it is secreted by the ciliary processes, passes to the anterior chamber and is then reabsorbed at the iridocorneal corner into the venous system.

arabinose, a pentose sugar. *See* monosaccharide.

arachidonic acid, 5,8,11,14-eicosatetraenoic acid; an unsaturated fatty acid, $CH_3(CH_2)_4$-$(CH{=}CHCH_2)_4(CH_2)_2COOH$, produced by plants; in animals, it is an → essential fatty acid occurring as a constituent of phospholipid molecules in cellular membranes, also found in depot fats. Arachidonic acid (like some other unsaturated acids) exists more abundantly in poikilothermic animals living in cold regions, rendering cell membranes more fluid. → Prostaglandins, thromboxanes, and leukotrienes are synthesized from arachidonic acid in many animal tissues. Many plants such as algae, mosses, and ferns are rich in arachidonic acid.

Arachnida (Gr. *arachne* = spider), **arachnids;** a class in the subphylum Chelicerata, phylum Arthropoda, including mainly terrestrial animals that have four pairs of walking legs and strong mouth appendages (→ chelicerae and pedipalps). They respire through book lungs or tracheae, and excrete by Malpighian tubules or coxal glands. Most species are predators, some ticks are parasites, harvestmen feed on plants. The number of described species is about 73,000, and familiar orders are e.g. **spiders** (Araneae), **scorpions** (Scorpiones), **false scorpions** (Pseudoscorpiones), **solifugids** (Solifugae/ Solpugida), **harvestmen** (Opiliones/ Phalangida), and **ticks** and **mites** (Acari).

arachnoid (membrane), arachnoidea, *see* meninx.

arachnology (Gr. *logos* = discourse), a branch of zoology studying arachnids.

Araneae (Araneida), spiders; an order of arachnids; most species have abdominal spinning glands producing silk, and poison glands with ducts to the tip of the chelicerae; the prosoma and the opisthosoma in the body are distinct. About 35,000 species in 30 families are described.

arbor, pl. **arbores** (L. = tree), in anatomy a tree-shaped structure, as *arbor vitae* (*a.v. cerebelli*), a tree-like appearance of white matter in the median section of cerebellum.

arboreal, pertaining to a tree, living in trees; e.g. arboreal apes.

arboretum, a park-like area with various cultivated tree species.

arboviruses (ar < arthropod, bo < borne), a group of RNA viruses found in arthropods but can infect also other animal species, also mammals, such as bats, rodents, and primates. About 50 arbovirus species can produce a disease in man, usually transmitted by arthropods. Arboviruses transmitted by ticks are often classified in a separate category, tick-borne viruses.

arch-, arche-, archi-, archae(o)-, archeo- (Gr. *archein* = to begin), denoting beginning, original, ancestral, primitive, primary, first, or chief.

Archaean, Archean, Arch(a)eozoic (Gr. *archaios* = ancient), an earlier geological eon (period) in the Precambrian era about 4,500 to 2,400 million years ago, preceding Proterozoic; first living systems with anaerobic metabolism emerged. Earlier used as a synonym of Precambrian.

Archaebacteria (Gr. *bakterion* = little rod), organisms resembling ordinary → bacteria but differing biochemically in many ways from them; e.g. they have a novel 16S-like ribosomal RNA component. DNA analyses have shown that the Archaebacteria clearly differ from both typical prokaryotes and eukaryotes and therefore some biologists classify them as a separate kingdom. Many species live in extreme environmental conditions, e.g. in very hot and acid circumstances. Some species can produce methane and they have probably played an important role in formation of the coal and oil stores on the Earth. Archaebacteria are very common especially in plankton of cold seas and in the forest soil of northern areas and thus they seem to be ecologically very important.

archaeocyte, also **archeocyte** (Gr. *kytos* = cavity, cell), an amoeboid cell of sponges which are phagocytic and play a role in digestion; if needed, archaeocytes are capable of forming other types of cells. *Cf.* amoebocyte.

archaeophyte, also **archeophyte** (Gr. *phyton* = plant), a plant species which has been transported to an area by man.

Archaeopteridae (Gr. *arche* = beginning, *pteris* = fern), an exctinct subclass in the class → Pteropsida, Plant kingdom; in some classifications included in the division → Progymnophyta. The group has been small and occurred during Devonian.

Archaeopteryx (Gr. *pteron* = wing), a genus of extinct birds in the subclass Archaeornithes ("lizard birds") which lived in the late Jurassic and early Cretaceous about 150 million years ago. Many of their characteristics were primitive and reptile-like, as teeth, solid bones, a saurian pelvis and long tail with several vertebrae, and three fingers with claws; the pygostyle was absent. Bird-like characteristics were e.g. feathers and the → furcula. The remnants of the known species *Archaeopteryx lithographica* (*lithographica* = inscribed in stone) are found in limestone quarries in Central Europe.

Archaeornithes (Gr. *ornis* = bird), an extinct subclass of birds which lived in the late Jurassic and early Cretaceous. *See* Archaeopteryx.

Archaeozoic (period) (Gr. *zoe* = life), → Archaean.

archegoniophore, (Gr. *gone* = birth, *phorein* = to carry along), a stalk bearing an → archegonium in some plants, e.g. some liverworts.

archegonium, pl. **archegonia,** the female → gametangium in mosses, pteridophytes, and gymnosperms; in other groups the corresponding structure is called → oogonium.

archencephalon (Gr. *enkephalos* = brain), the anterior part of the embryonal brain of vertebrates; develops into the telencephalon and diencephalon.

archenteron, pl. **archentera** (Gr. *enteron* = intestine), the embryonal cavity formed by the invagination of → blastula in the animal groups which have three embryonal germ layers in their gastrula stage. Also called gastrocoel(e) or gastrocele.

archeocyte, → archaeocyte.

archeophyte, → archaeophyte.

Archeozoic (period), → Archaean.

archesporium, pl. **archesporia** (Gr. *sporos* = seed), **archespore**; the original mother cell from which the → macrospores or → microspores of angiosperms develop. The female archesporium in the gynaecium (in nucellus) generally divides into two cells, i.e. the parietal and sporogenic cells. The latter develops into the macrospore mother cell which divides meiotically and produces finally one macrospore. The male archespores in the anther (in stamen) divide as described above, and the sporogenic cells develop into microspore mother cells, which through → meiosis further develop into microspores.

archetype, *see* prototype.

archistome (Gr. *stoma* = mouth), → blastopore.

Archosauria (Gr. *archon* = ruler, *sauros* = lizard), **archosaurs**; a reptilian subclass (in some classifications a superorder in the subclass Diapsida) including many specialized species, mostly extinct; most archosaurs were terrestrial but some specialized for flight. At present, the only extant order is → Crocodilia (25 species), but e.g. the orders Thecodontia, Pterosauria, Saurischia, and Ornithischia are all extinct.

arctic zone (Gr. *arktos* = bear, northern), a zone surrounding the North Pole, the climatic southern line of which coincides with the isotherm of 10°C for the warmest month; in general, between 66.5 ° and 72 ° N. The arctic zone comprises the northern coastal regions in America and Eurasia and the arctic arhipelago. The vegetation consists of low shrubs, perennial herbs, lichens, and mosses (tundra vegetation).

Arctogaea, also **Arctogea** (Gr. *ge* = earth), a zoogeographical area including the Palaearctic, Nearctic, Ethiopaean, and Oriental regions. *Syn.* Megagaea. *Cf.* Neogaea, Notogaea, Palaeogaea.

area-species curve, → species-area curve.

arista, → awn.

arginase, arginine amidase; an enzyme (EC 3.5.3.1) acting in the → urea cycle of animals catalysing the hydrolysis of arginine to urea and ornithine; found in the liver of vertebrates, and in plants.

arginine, abbr. Arg, arg, symbol R, $H_2NC{=}NHNH(CH_2)_3CHNH_2COOH$; a basic amino acid that occurs in proteins of all organisms; one of the → essential amino acids in animals, occurring also as an intermediate in the → urea cycle. Its high energy form, **arginine phosphate**, acts as an energy store in cells of many invertebrates. *Cf.* creatine. *See* arginase.

aridity (L. *aridus* = dry), extreme dryness; very dry climatic conditions in an area or a habitat with less than 250 mm annual rainfall and minor vegetation. *Adj.* **arid.**

aril (L. *arillus* = raisin), fleshy, often coloured covering around some seeds; in some plants it is edible with a certain flavour and used as a spice, such as litsh, rambutane, moscat.

Aristotle's lantern (*Aristoteles*, 384—322 BC), a complex, lantern-shaped chewing apparatus in some echinoids; comprises five strong, sharp teeth and five pairs of retractor and protractor muscles to move the teeth which are supported by a five-sided framework inside the test of the animal.

aromatic (Gr. *aromatikos* = odorous), **1.** pertaining to benzene or related hydrocarbons having a ring structure; **2.** pertaining to a substance which gives off a pleasant, spicy smell, e.g. aromatic amino acids.

aromorphosis, pl. **aromorphoses** (Gr. *ara* = straightway, *morphosis* = forming), an evolution of a phylogenetic line in which the levels of organization and integration increase but specialization seldom occurs. Aromorphosis is usually found in situations where a species is dispersing to a new adaptive zone. *Cf.* allomorphosis.

arousal, a state of responsiveness of an animal to sensory stimulations; wakefulness, vigilance, neural activation.

Arrhenius plot (*Svante Arrhenius*, 1859—1927), a graphic presentation expressing the temperature dependence of a biochemical reaction on the coordinate system, in which the logarithm of a reaction rate (log*V*) on the ordinate is presented as the function of the reciprocal of the absolute temperature (1/K) on the abscissa. The plot gives a straight line for the activity of most enzymes, and on the basis of the plot the → activation energy of the reaction can be calculated.

arrheno- (Gr. *arrhen* = male), denoting relationship to male organisms.

arrhenotoky, male haploidy; a form of → parthenogenesis where unfertilized eggs develop into males and fertilized eggs into females.

arrhythmia (Gr. *a-* = not, *rhythmos* = rhythm), loss of rhythm, having no regular rhythm, e.g. the irregularity of heartbeat.

arrow worms, → Chaetognatha.

arsenic (L. *arsenicum* = poison < Gr. *arsen* = strong), a chemical element, symbol As, atomic weight 74.92; naturally occurs free or as a substitute in different compounds, e.g. arsenic trioxide (As_2O_3). Arsenic is very toxic to organisms as it inhibits aerobic cellular respiration. Arsenic compounds are used in insecticides and herbicides, in conservation of animal samples, and in medicine. The salts of arsenic acid are **arsenates**.

artefact, → artifact.

arteria, pl. **arteriae** (L. < Gr. = originally the windpipe), → artery.

arteriole (L. *arteriola*, dim. of *arteria* = artery), a minute artery branching to capillaries.

artery (L. *arteria*), a blood vessel conveying blood from the heart. *See* blood vessels.

arthr(o)- (Gr. *arthron* = joint), denoting relationship to a joint, articulation.

arthrobranch, arthrobranchia, pl. **arthrobranchs, arthrobranchiae** (Gr. *branchia* = gills), a gill attached to the coxopodite of thoracic appendages in some arthropods, such as crayfish. *Cf.* pleurobranch, podobranch.

Arthropoda (Gr. *pous*, gen. *podos* = foot), **arthropods**; a very large phylum which comprises over 1 million known species and more than 80% of all the animal species living in the world. The body is more or less metamerically segmented, covered by chitin cuticle, with jointed appendages. Primarily, one pair of appendages, specialized for particular functions, are located on each body segment. The nervous system includes the ventral nerve cord with segmental ganglia. Arthropods in the subphylum Trilobita have been extinct from Permian about 200 million years ago. Extant arthropods are divided into four subphyla: **chelicerates** (Chelicerata), **crustaceans** (Crustacea), **insects** and **myriapods** (Uniramia), and **pentastomids** (Pentastomida); previously all other than chelicerates were classified to the common subphylum Mandibulata. At present, some authors classify the phylum Arthropoda as a superphylum with several phyla.

articulation, (L. *articulare* = to divide into joints), **1.** (L. *articulatio*), a joint or a juncture between bones or cartilages in the skeleton of vertebrates, or a joint between rigid, usually chitinous segments of the body or legs in many invertebrates; **2.** the act of joining, connecting, or the state of being jointed; **3.** the enunciation of words and speech.

Articulata (L. *articulus* = small joint), **1.** a class of brachiopod species having the shell valves connected by a hinge; **2.** a subclass of the class Crinoidea (sea lilies); **3.** an old name used by *Georges Cuvier* (1769—1832) for annelids and arthropods.

artifact, also **artefact** (L. *ars* = art, *facere* = to make), anything made by man; e.g. a structure or outcome not naturally present in an object but caused by the technique used, e.g. in a histological preparation.

artificial fertilization, *in vitro* fertilization; the fertilization of the ovum artificially outside the organism and separated from it.

artificial insemination, the transfer of sperm into female genital ducts artificially without copulation. After learning to store the sperm of many animals in deep freeze, the use of artificial insemination specifically for breeding of some domestic animals has almost completely replaced copulation. On medical grounds, artificial insemination is in use also for humans.

artificial seed, an → embryoid originated from a somatic cell covered with endosperm substance and a capsule. Artificial seeds can be used when the natural seed production of a plant species is very low.

artificial selection, a selection process in breeding of domestic species in which the best animal and plant individuals are selected for reproduction; aims to improve economically important characteristics in them.

Artiodactyla (Gr. *artios* = even, *daktylos* = finger, toe), **even-toed ungulates, artiodactyls**; a mammalian order of large or middle-sized herbivores with only two (rarely four) functional toes on each foot. In most artiodactyls the third and fourth toes are covered by a cornified hoof (cloven hoof), the first, second, and fifth toes are reduced. Many species have antlers or horns. The order includes 185 species in three suborders, Suiformes (**pigs, hippopotamus**, etc.), Tylopoda (**camels, llamas, alpacas**), and Ruminantia (**deer, antelopes**, etc.).

aryt(a)enoid cartilage (Gr. *arytaina* = ladle, *eidos* = form), a paired, small pyramidal cartilage of the → larynx; the posterior part of the vocal ligament with some laryngeal muscles attached.

asafoetida, Am. **asafetida** (L. *asa* = gum < Per. *aza* = mastic, L. *foetidus* = fetid), a brown, soft gum resin obtained from the roots of some umbelliferous plants (Apiaceae); has a bitter taste and an obnoxious, alliaceous odour; used as a spice and formerly in medicine as an antispasdomic.

ASAT, AST, → aspartate aminotransferase.

ascariasis, pl. **ascariases** (Gr. *askaris* = an intestinal worm), an infection caused by a roundworm species (*Ascaris*) in many vertebrates, in man by *Ascaris lumbricoides*. Infection is possible usually when worm eggs are ingested with insufficiently washed or uncooked vegetables. In the digestive tract of a host the larvae hatch and move through the circulation to the lungs. If the infection is heavy, pneumonia is possible in the host. Finally the larvae pass to the intestine where they use intestinal contents causing abdominal symptoms and allergic reactions.

Aschelminthes (Gr. *askos* = bladder, *helmis* =

parasitic worm), **aschelminths**; a group of heterogeneous worm-like invertebrates living in oceans and fresh waters; common characteristics are e.g. a complete digestive tract and pseudocoelom. The group were earlier considered to be a separate phylum, Aschelminthes. At present, their systematic position is more problematic: some authors consider them as a superphylum, some like to reject the group as a taxonomic category. The group includes the phyla **gastrotrichs** (Gastrotricha), **rotifers** (Rotifera), **kinorhynchs** (Kinorhyncha), **nematomorphs** (Nematomorpha), and **nematodes** (Nematoda).

Ascidiacea, (Gr. *askidion* = little bag), **ascidians, sea squirts**; a class of tunicates (phylum Chordata); sessile animals living widely in oceans either solitarily or in colonies; some species are compound in which form many individuals share the same test. Solitary individuals have **incurrent** and **excurrent siphons** protruding outside the body. The common characteristics of chordates, i.e. the notochord, nerve cord, branchial sac, and tail, are found only in the larval stage; in adults only the enlarged branchial sac is left. The class comprises about 1,250 species in three orders: Aplousobranchia, Phlebobranchia, and Stolidobranchia.

ascites, pl. **ascites** (L., Gr. *askites* < *askos* = bladder, belly), hydroperitonia, hydroperitoneum; accumulation of tissue fluid in the peritoneal cavity.

ascocarp (Gr. *karpos* = fruit), the fruiting body of ascomycete fungi containing asci with ascospores. *See* Ascomycetes.

ascogenous hyphae, *see* ascogonium, Ascomycetes.

ascogonium, pl. **ascogonia** (Gr. *gone* = birth), the female → gametangium in the fungal group → Ascomycetes. The reproduction takes place through **gametangiogamy** in which whole gametangia are fused: the male gametangium (antheridium) and ascogonium are connected with a special projection, **trichogyne**, along which the nuclei of the antheridium are transported to the female gametangium. After this the ascogonium grows **ascogenous hyphae**, which finally form spore-bearing **asci** after a special → crozier (hook) formation.

Ascolichenes (Gr. *askos* = bag, bladder), → lichens whose fungal partner belongs to the group → Ascomycetes. *Cf.* Basidiolichenes.

ascoma, pl. **ascomata** (L.), *see* Ascomycetes.

Ascomycetes (Gr. *mykes* = fungus), **sac fungi**; a class in subdivision Ascomycotina, division Amastigomycota, kingdom Fungi. The species have septate hyphae and the sexual reproduction involves → **gametangiogamy**, in which there is no formation of → gametes but the whole gametangia fuse. After this the female gametangium, **ascogonium**, grows **ascogenous hyphae** which produce sac-like **asci**, containing **ascospores**. The formation of asci takes mostly place in fruiting bodies (ascocarps); the fruiting body may be a roundish **cleistothecium**, bottle-like **perithecium**, bowl-like **apothecium**, legume-like **hysterothecium** or disc-shaped **ascoma**. The asexual reproduction takes place through non-motile **conidia**. About 33,000 species are described; they may be terrestrial or aquatic, many are microscopically small. Well-known macroscopic genera are *Gyromitra* and *Morchella*. Also many moulds and yeasts belong to this group. In some systems the group Ascomycetes is classified into a separate division, Ascomycota, which is divided into 46 orders.

Ascomycotina, a subdivision in the division Amastigomycota, kingdom Fungi; the only class is → Ascomycetes.

ascon grade, asconoid type, the simplest type of the canal system of sponges in the class Calcarea; water enters through small dermal pores into the spongocoel lined with choanocytes; from there the water emerges through the osculum. Asconoids are found only in the class Calcarea. *Cf.* leucon grade, sycon grade.

ascorbic acid (Gr. *a-* = not, L. *scorbutus* = scurvy), *see* vitamin C.

ascospore (Gr. *askos* = bag, bladder, *sporos* = seed), a spore-type in → Ascomycetes, formed in the → ascus.

ascus, pl. **asci** (L.), a sac-like structure containing eight **ascospores** in ascomycete fungi (→ Ascomycetes); formed from **ascogenous hyphae** grown by the → ascogonium after → crozier (hook) formation.

-ase, a suffix meaning an enzyme, as in → saccharase, lactase, phosphatase, ATPase, etc.

asepsis, pl. **asepses** (Gr. *a-* = not, *sepsis* = putrefaction), a condition free from living pathogenic microbes; keeping sterility. *Adj.* **aseptic**.

aseptate (L. *septum* = cross-wall), without a

septum, not septate; pertaining to filamentous structures of plants or fungal hyphae lacking cell walls between cells.

asexual (L. *asexualis* = sexless), **1.** without sex, sexless; *see* asexual reproduction; **2.** without sexual desire.

asexual reproduction, → reproduction without a union of gametes, i.e. the development of a new individual from either a single cell (**agamogony, agamospermy**) or from a group of cells (**vegetative reproduction**) in the absence of any sexual processes, i.e. without → meiosis and/or → fertilization. *Syn.* agamic reproduction. *See* parthenogenesis, budding.

-asis, also **-osis,** a suffix denoting state or condition; often together with a scientific name of a species or a genus referring to a particular disease caused by a parasitic species, e.g. **ascariasis**, an infection caused by a roundworm, *Ascaris*.

A-site (A < acyl), the binding site on the → ribosome for a charged tRNA molecule in → protein synthesis. *Syn.* amino site.

ASO, → allele-specific oligonucleotide.

asparagine (Gr. *asparagos* = asparagus), abbr. Asn, asn, or Asx, symbol N or B, mol. weight 132.12, chemical formula H_2NCOCH_2-$CHNH_2COOH$; amide of → aspartic acid, occurring in proteins. Both plants and animals can synthesize asparagine, and e.g. the asparagus and pea seedlings are rich in it. In plants asparagine is an important transportable nitrogen compound.

asparaginic acid, → aspartic acid.

aspartame, N-L-α-aspartyl-L-phenylalanine 1-methyl ester; asp-phe methyl ester produced synthetically to be used as an artificial low-caloric sweetener.

aspartate, a salt or ion of → aspartic acid; in animals acts e.g. as a neurotransmitter binding to N-methyl-D-aspartate receptors (NMDA receptors) in glutaminergic → synapses of the central nervous system. *See* glutamate receptors.

aspartic acid, aminosuccinic acid; abbr. Asp, asp, or Asx, symbol D or B, mol. weight 133.10, $HOOCCH_2CHNH_2COOH$, an amino acid occurring in proteins of all organisms, synthesized also by animals; its salts and the ionic form are called **aspartates**. *Syn.* asparaginic acid.

aspect (L. *aspectus* = view), **1.** a look or appearance; **2.** nature, quality, category, character; e.g. in anatomy a surface facing in any designated direction, such as dorsal or ventral aspect; **3.** all the plant species in a plant community at a certain moment; e.g. spring and summer aspects in a deciduous forest.

asphyxia (Gr. *a-* = not, *sphyzein* = to throb), the impaired exchange of respiratory gases due to lack of oxygen in respired air or interruption of respiration; includes → hypoxia or even anoxia, and hypercapnia. *Adj.* **asphyxial**.

aspiration, sucking of gas, fluid, etc., e.g. from a body cavity.

assimilation (L. *assimilare* = to make similar), **1.** the conversion of nutrients into organic molecules by an organism. The most important assimilation is → photosynthesis carried out by plants, in which process carbon dioxide is assimilated into carbohydrates by means of solar energy; **2.** in psychology, a cognitive process by which former experience is applied to a new situation.

association (L. *associare* = to join), union; connection of events; e.g. **1.** genetic association, the non-random occurrence of two genetically separate traits in a population; has its basis in mechanisms other than genetic → linkage; **2.** in ethology, a learned connection between a certain event and a stimulus which has earlier been neutral to the event; *see* conditioned reflex; **3.** in botany, a plant community with a certain species composition, occurring in certain physical and chemical conditions; association is the basic unit in → plant sociology; e.g. in floristic systematics of Zurich—Montpellier (Braun—Blanquet); important in the description of associations are character and separation species; **4.** in chemistry, association (combination) of two or more molecules together; opposite to dissociation; **5.** in psychology, combining two or more ideas in the mind; e.g. free associations, dream associations.

associative learning, a type of → learning of animals in which an initially neutral stimulus evokes a particular response when it occurs with another stimulus that normally evokes this response; e.g. a predatory hawk can learn to associate the presence of a broken eggshell with the presence of a gull chick nearby. In man, the greatest part of learning is of this type but much more complicated, i.e. → insight learning. *See* latent learning.

assortative mating/breeding, a type of non-random mating; the influence of relativeness or similarity of characters of a female and male in mating. In **genotypic assortative mating** (*see* genotype), if positive, the mating individuals are more related than in random mating, and if negative, the mating individuals are less than on average related to each other. Genotypic assortative mating leads to deviation of frequencies of genotypes, but not of frequencies of genes from that expected on the basis of the → Hardy—Weinberg law equilibrium. Correspondingly, in **phenotypic assortative mating** (*see* phenotype), if positive, the mating individuals are phenotypically more similar, and if negative, less similar than on average.

aster (Gr. *aster* = star), a tuft of → microtubules around the → centrosome in animal and some lower plant cells; a star-shaped cytoplasmic structure radiating from the centriole. Aster is associated with the → mitotic spindle. Also called sperm-aster.

Asteroidea (Gr. *eidos* = form), **asteroids, sea stars (starfish)**; a class (subclass) of echinoderms; invertebrates with a five-radiated, star-shaped flattened body, often brightly coloured and composed of a central disc which merges gradually with the five flexible arms. Sea stars are familiar in littoral zones of oceans.

asterosclereid (Gr. *skleros* = hard, *eidos* = form), a type of thick-walled → sclereid in plants.

astrocyte (Gr. *kytos* = cavity, cell), a large neuroglial cell characterized by numerous protoplasmic branches. *Syn.* macroglial cell; collectively called astroglia or macroglia. *See* neuroglia.

astrotaxis (Gr. *tassein* = to arrange), a tactic movement of an animal the direction of which is determined by the position of the stars; e.g. migration routes of some migratory birds. *See* taxis.

asymmetry, lack of symmetry; the difference between the left and right side of an organism or part of it; **directional** and **fluctuating** asymmetry are distinguished, in the former the asymmetry is non-random, in the latter random. *Adj.* **asymmetric(al)**.

asymmetric competition, the competition between two organisms, populations, or species, effects of which are directed more on one of the two competitors. *See* competition.

asymmetric flower, a flower without any planes of symmetry; e.g. *Canna*. *Cf.* bilateral flower, symmetric flower, zygomorphic flower.

atactostele (Gr. *ataktos* = irregular, *stele* = column), *see* stele.

atavism (L. *atavus* = ancestor), a phenomenon of inheritance in which a characteristic that has been hidden for a long time appears again; e.g. the so-called Darwin's tip on the ear lobe and the ability to move ears are suggested as atavistic characteristics in man.

atlas (in Greek mythology, a Titan supporting the Earth on his shoulders), **1.** the first cervical vertebra of tetrapod vertebrates articulating anteriorly with the occipital bone and posteriorly rotating around the tooth (dens) of the axis; **2.** a collection of illustrations of a subject, e.g. a brain atlas.

atmosphere (Gr. *atmos* = vapour, *sphaira* = ball, sphere), **1.** a mixture of gases surrounding the Earth, i.e. the air. The atmosphere includes nitrogen N_2 ca. 78%, oxygen O_2 21%, argon Ar 0.9%, carbon dioxide CO_2 0.03%, and varying small amounts of hydrogen H_2, ozone O_3, methane CH_4, carbon monoxide CO, helium He, neon Ne, krypton Kr, and xenon Xe. The composition of the atmosphere is substantially the same as at the ground level up to at least 88 km, although the air pressure is lower at higher levels. Modern problems associated with the atmosphere are the decrease of → ozone and the → greenhouse effect; **2.** in astronomy, the gaseous envelope surrounding any celestial body; **3.** any gaseous medium; **4.** a conventional unit of pressure, i.e. the air pressure at sea level. *See* pascal.

atmospheric pressure, air pressure, the pressure against the Earth's surface; equals the weight of an air column divided by the area under the air column. The pressure units used in biological literature are **millibar** (1 mbar = 0.001 bar) and **hectopascal** (1 hPa = 100 Pa = 1 mbar; *see* Appendix 1). The normal atmospheric pressure (1 ATM) at sea level is 101.3 kPa (1.013 bar), decreasing vertically so that at 5,500 m above the sea level it is about 50 kPa, at 16,000 m about 10 kPa, and at 32,000 about 1 kPa. On climate maps, places with the same air pressure are joined to form curves, **isobars**, with cyclons to indicate pressure areas where the pressure is lower than that of the surrounding areas. Many organisms are adpated to live in a low pressure with correspondingly lowered oxygen pressure, e.g. in high mountain areas. *See* mountain sickness.

atoll (native name in the Maldive Islands), a coral reef that encircles a lagoon but not an island; usually shaped like a horseshoe.

atom (Gr. *atomos* = indivisible), the smallest particle of any chemical compound (molecule) once believed to be indivisible; made up of subatomic elementary particles: centrally is located the **atomic nucleus** in which the atomic mass is concentrated, and surrounding **electrons** revolve in well defined orbitals around the nucleus. The nucleus comprises varying numbers of positively charged **protons** and electrically neutral **neutrons**. The atoms of different elements have different nuclear charges, $+Z \cdot e$, where Z = whole number of protons (atomic number), and e = an electrical elementary charge. The periodical system of the elements is based on the *Z*-numbers (*see* Appendix 2). Atoms are linked together by different chemical bonds forming molecules. *See* covalent bond, coordinate bond.

atomic absorption spectrophotometer, an apparatus used for quantitative spectroscopic determination of chemical elements, i.e. in **atomic absorption spectroscopy** (**AAS**). A sample dissolved in acid is burned in AAS that identifies different elements based on the typical colour of the flame, and the intensity of the colour is proportional to their concentrations. In biology, AAS is used to determine elements (as e.g. sodium, potassium, calcium, iron, lead) in plant and animal cells and tissues as well as in water and soil samples. The method is suitable for determination of all metals, alkali and alkaline earth metals and of some so called hemimetals (arsenic and selenium), nearly 70 different elements altogether.

atomic weight, relative atomic mass. *See* Appendix 1.

ATP, → adenosine triphosphate.

ATPase, → adenosine triphosphatase.

atrazine, 2-chloro-4-ethylamino-6-isopropylamintriazine, a contact and soil herbicide that inhibits e.g. light reactions of photosynthesis as well as development of chloroplasts. Corn is atrazine resistant, metabolizing the herbicide to its peptide conjugates.

atrial cavity, → peribranchial cavity.

atrial natriuretic peptide/factor, ANP, ANF, a peptide hormone secreted by cardiac muscle cells in vertebrates, mainly in the atrial muscle of the heart; dilates blood vessels decreasing the blood pressure, and stimulates the excretion of sodium ions in urine. ANP belongs to the group of → natriuretic peptides.

atrichia (Gr. *a-* = not, *thrix* = hair), hairlessness; *adj.* **atrichous** or **atrichic**, having no flagella or cilia.

atrichoblast (Gr. *blastos* = bud), a cell type in the plant root epidermis (rhizodermis) without a root hair. *Cf.* trichoblast.

atriopore (L. *atrioporus* < L. *atrium* = entrance hall, Gr. *poros* = pore), an opening between → atrial cavity and the surface of the body, found e.g. in lancelets.

atrioventricular valve, the valve between the atrium and ventricle in the vertebrate heart; one valve in the two-chambered fish heart and two valves in the three-chambered heart of amphibians and four-chambered heart of higher tetrapods, i.e. the **left atrioventricular valve** (**mitral valve, bicuspid valve**) with two cusps on the left side of the heart, and the **right atrioventricular valve** (**tricuspid valve**) with three cusps on the right side of the heart. Thin tendons from → papillary muscles of the inner ventricular surface are attached onto the valve edges controlling the closure of the valves.

atrium, pl. **atria** (L. = entrance hall), a cavity or chamber forming a passage to another cavity or several cavities; e.g. **1.** the atrium of the heart (*atrium cordis*), the left atrium and right atrium (*a. c. sinistrum, a.c. dextrum*) in tetrapod vertebrates; **2.** the atrium of the nose; **3.** a large cavity containing the pharynx in tunicates and cephalochordates. *Adj.* **atrial**.

atrophy, atrophia (Gr. *a-* = not, *trophein* = to nourish), the decrease in size of a tissue, organ, or entire organism; a wasting of tissues or organs. *Adj.* **atrophic**. *Verb* to **atrophy**.

atropine, an alkaloid, $C_{17}H_{23}NO_3$, obtained from the plants *Atropa belladonna*, *Datura stramonium*, and other Solanaceae plants; antagonizes muscarinic cholinergic receptors in animal tissues, i.e. has an anticholinergic effect e.g. in smooth muscles of vertebrates. Atropine is used in medicine e.g. as a mydriatic and antispasmodic drug. *Syn.* hyoscyamine, belladonna. *See* muscarinic receptors.

atropous ovule (Gr. *a-* = not, *trope* = turning),

see ovule.

attenuation, 1. loss of virulence; **2.** in genetics, the termination of transcription in the → operon caused by an → **attenuator**, which is a DNA sequence in the operon upstream of the structural genes. It can lead to the termination of → genetic transcription because a → palindrome in the DNA can form a loop that breaks transcription. Transcription is terminated in the cell if there is enough end product of a reaction which is catalysed by the enzyme encoded by the operon with the attenuator. Such an end product is usually an amino acid, and its → transfer RNA recognizes certain → codons in the attenuator and stops transcription.

audition (L. *audire* = to hear), → hearing.

auditory capsule, the cartilaginous capsule of the vertebrate embryo that surrounds the developing **auditory vesicle** in the inner ear. The capsule develops into the → bony labyrinth, and the vesicle forms the membranous labyrinth of the ear. *Syn.* otic capsule.

auditory membrane, → tympanic membrane.

auditory organ, *see* hearing.

auditory ossicle, one or three small ossicles in the tympanic cavity (middle ear) of the → ear of tetrapod vertebrates forming the bony bridge from the tympanic membrane to the oval window; three ossicles in mammals, i.e. → **malleus**, **incus**, and **stapes**; one ossicle (columella) in other vertebrates corresponding to the stapes of mammals.

auditory sense, sense of → hearing.

auditory tube (L. *tuba auditiva*), a tube in tetrapod vertebrates leading from the tympanic cavity to the nasopharynx, equalizing air pressure between the cavity and the outside air. Also called Eustachian tube, pharyngotympanic tube, otopharyngeal tube, salpinx.

Auerbach's plexus (*Leopold Auerbach*, 1828—1897), myenteric plexus; a nerve plexus in the muscular coat of the intestine in vertebrates; comprises postganglionic autonomic nerve cell bodies and unmyelinated nerve fibres, chiefly controlling muscular activity of the alimentary canal. *Cf.* Meissner's plexus.

auricle (L. *auricula*, dim. of *auris* = ear), **1.** external ear; the pinna (flap) of the ear; **2.** atrial auricle, atrial auricula (*auricula atrii*), a muscular pouch projecting from the anterior portion of either atrium of the heart in tetrapod vertebrates. *Adj.* **auricular**.

auricularia, pl. **auriculariae,** also **auricularias,** a planktonic larval stage of sea cucumbers (Holothuroidea).

Australian region, a realm of biogeography; **1.** in zoology, the Australian region comprises Australia, the Moluccas, Sulawesi, Papua—New Guinea, Tasmania, New Zealand, and the oceanic islands of the South Pacific. The region comprises the **Australian**, **Austro—Malayan**, **New Zealand**, and **Polynesian subregions**. The fauna is characterized by many endemic species, especially monotremes and marsupials; **2.** in plant geography, the Australian region includes Australia and Tasmania, and is characterized by a wide range of endemic species (ca. 80%); *see* zoogeographical region, floristic kingdoms.

australopithecines, (L. *australis* = southern, Gr. *pithekos* = ape), a group of extinct hominids whose ancestor probably was *Australopithecus afarensis*, living in Africa from about 4 to 3 million years ago. This bipedal hominid was short with upright posture, modified dentition, and his face and brain resembled those of the chimpanzee. The fossils of australopithecines are found in Ethiopia. *A. afarensis* was an ancestral species of all human and human-like forms. A known fossil is the skeleton of "Lucy", a 20 year old female who lived in Ethiopia about 3,2 million years ago. Later the species diverged into different lineages, one of which may have evolved to the genus *Homo*. Other species among australopithecines have been e.g. *A. aethiopicus*, *A. africanus*, *A. robustus* and *A. boisei*, who lived from about 2.5 to 1 million years ago.

aut(o)- (Gr. *autos* = self), denoting self or same.

autecology (Gr. *autos* = self, *oikos* = household), a branch of → ecology studying reactions and adaptations of individuals and populations of a species to different environmental factors; important supporting sciences are e.g. physiology, biochemistry, and biophysics. *Cf.* synecology.

author, in biology, a person who first describes a new species or other taxon and gives a scientific name to it.

autoantibody, an → antibody produced in an animal responding against its own normal tissue constituent.

autoantigen, any normal tissue constituent of

an animal which evokes an immune response in the animal's own body, i.e. acts as a self-antigen. *See* antigen.

autocatalysis, pl. **autocatalyses** (Gr. *katalysis* = decomposing), self-catalysis, e.g. **1.** the autoactivation of a chemical reaction that speeds up gradually when some reaction products begin to act as enhancers or catalysts; **2.** the ability of biological macromolecules to control their own metabolism, e.g. the ability of DNA to control its duplication; the molecule catalysing its own production is called **autocatalyst**.

autochthonous (Gr. *chthon* = ground, earth), aboriginal, indigenous, native; pertaining to materials or organisms produced or borne by the original system, or found in a locality where they originate; e.g. **autochthonous species**, a native species being original in a region. *Cf.* allochthonous.

autoclave (L. *clavis* = bar, latch), a steam pressure container used to sterilize water, solutions, and laboratory instruments, as well as to preserve e.g. pharmacological preparations and food; the temperature of the steam reaches 121°C, killing microbes and their spores.

autocrine (Gr. *krinein* = to separate), pertaining to a secretion of a cell which has specific receptors for its own secretory product, resulting in self-stimulation.

autoecious (Gr. *oikos* = house), pertaining to fungi being parasitic on one host plant only during their life cycle, however producing different spore types; typical especially of rust fungi (→ Uredinales). *Cf.* heteroecious.

autogamy (Gr. *gamos* = marriage), **1.** the → self-pollination of a plant; **2.** the obligatory self-fertilization of one undivided cell in which the haploid nuclei unite in pairs. Autogamy is a common way of reorganization of nuclei in *Paramecium* when the → micronucleus and → macronucleus of this unicellular animal unite. *Adj.* **autogamous**. *Cf.* automixis.

autogenesis, pl. **autogeneses** (Gr. *genesis* = production), self-generation; the origin of living matter within the organism itself. *Adj.* **autogenetic** or **autogenic, autogenous**.

autogenic succession, a temporal → succession of populations, communities, or ecosystems in an area where conditions are changed by factors originating inside the area. *Cf.* allogenic succession.

autogenous control, in genetics, the control of a gene caused by a gene product in the cell by either inhibiting (negative autogenous control) or activating (positive autogenous control) the expression of the gene coding for this product.

autograft, autotransplant, autoplast; a → graft of a tissue or organ transferred to a new site in the body of the same individual.

autogrooming, *see* grooming.

autoimmunization, the induction of an immune response in an individual against its own tissue constituents, leading to autoimmune diseases.

autolysis (Gr. *lysis* = decomposition), self-dissolution; the decomposition of cells and tissues by their own enzymes; occurs normally during → metamorphosis, apoptosis, and after the death of an organism. *See* lysosomes.

automimicry (Gr. *mimikos* = imitating), intraspecific → mimicry; occurs in situations where individuals of a species are polymorphic in relation to some characteristics which have different selection value; e.g. the palatability for predators may be different, the palatable individuals mimicking the unpalatable members which thus also become avoided by predators. In ethology, the automimicry means intraspecific behaviour when an animal imitates e.g. the behaviour of the opposite sex or of an individual in another developmental stage. The appeasement behaviour includes often automimicry when a submissive male, facing a dominant individual, mimics postures and displays of a female or a young.

automixis, obligatory self-fertilization by → autogamy, paedogamy, or parthenogamy; opposed to amphimixis (cross-fertilization). *Adj.* **automictic**. *Cf.* autogamy.

autonomic nervous system, that part of the nervous system which innervates involuntary functions of organs, such as endocrine and exocrine glands and smooth and cardiac muscles; in vertebrates comprises → sympathetic nervous system and → parasympathetic nervous system. *Syn.* vegetative, visceral, or involuntary nervous system.

autophagia (Gr. *phagein* = to eat), **1.** biting or eating one's own flesh; **2.** nutrition for the whole body by metabolic consumption of some of one's own tissues; **3.** → autophagy.

autophagosome (Gr. *soma* = body), an intracellular vesicle formed from the endoplasmic reticulum and involved in digestion of worn-out and damaged cell structures. The autophagosomes fuse with → lysosomes which contain hydrolytic enzymes for the digestion. *See* phagocytosis.

autophagy, segregation and digestion of a cell's own structures; the process is necessary for disposal of damaged cells and cell organelles, where → autophagosomes play an essential role. Sometimes also called autophagia.

autoploid, → autopolyploid.

autopolyploid (Gr. *polys* = many, *-ploos* = -fold), a cell, individual, or species that carry more than two sets of → homologous chromosomes. The autopolyploid derivatives of a → diploid species differ usually only quantitatively from the parental species. Autopolyploids are often impaired in fertility, mainly due to disturbances in → meiosis because of occurrence of → multivalents. Vegetatively autopolyploid species are more productive than diploid species. The terms autodiploid, autotriploid, autotetraploid, etc. are used for cells which carry two, three, four, etc. sets of homologous chromosomes, respectively. *Syn.* autoploid. The phenomenon is called **autopolyploidy** (autoploidy).

autoradiography (L. *radius* = ray, Gr. *graphein* = to write), an analytic technique in which a (biological) specimen containing radioactive atoms is overlaid with a photographic film, incubated, and then developed. The location of the radioactive substance is seen as silver grains on the film. The technique is used widely e.g. for localization of cellular macromolecules (e.g. during DNA synthesis) with radioactive-labelled precursors as the radioactive marker. Usually low-energy isotopes (β-emitters) are used, e.g. tritium (H-3), sulphur (S-35), or phosphorus (P-32) isotopes.

autoregulation, → self-regulation.

autosome (Gr. *soma* = body), any chromosome other than the → sex chromosome. Genes residing in the autosomes follow normal → Mendelian inheritance, while the genes located on the sex chromosomes follow → sex-linked inheritance. There are 22 pairs of autosomes in the human chromosomal complement. *Adj.* **autosomal**.

autostylic skull, *see* skull.

autotomy (Gr. *tomein* = to cut), a defensive mechanism of self-amputation occurring in a trapped or damaged part of the body of an animal, when a predator tries to catch it; such parts are e.g. the tail of a lizard, a leg of a spider or crab; also some snail species can self-amputate a part of their foot when e.g. predatory coleopterans are pursuing them.

autotroph (Gr. *trophe* = nutrient), an organism which is able to assimilate carbon dioxide using energy obtainable in its environment, i.e. light energy, like **photoautotrophs** in → photosynthesis, or chemical energy as **chemoautotrophs** in → chemosynthesis, e.g. by oxidation of hydrogen sulphide, H_2S, or ammonia, NH_3. Thus the autotrophs need no organic nutriments. Typical autotrophs are green plants, some bacteria, and many protozoans. *Adj.* **autotrophic**. *Cf.* heterotroph.

autotrophic succession, a temporal → succession of populations, communities, or ecosystems in an area which is mainly affected by production of its own vegetation. *Cf.* allogenic succession, autogenic succession.

autumn tints, the bright colour formation in plant leaves in autumn due to the decomposition of chlorophyll while other colours, especially → anthocyans, become visible. The phenomenon is connected to → cold hardening and is genetically regulated.

aux-, auxano, auxo- (Gr. *auxanein* = to increase), denoting relationship to increase or addition, e.g. in size, intensity, or speed.

auxiliary cells, cells adjoining guard cells of plant stomata; their number and situation vary and different stomatal systems are named according to these cells. *Syn.* subsidiary cells.

auxins, a group of natural or synthetic plant hormones stimulating cell divisions and growth of cells. Auxins, like other hormones act in very low concentrations and have a certain optimum concentration for the maximum effect. The best known natural auxin is **indole-3-acetic acid (IAA)** that is synthesized from an amino acid, tryptophan, in Brassicaceae plants also from → thioglycosides. The primary effect of auxin occurs at gene level inducing synthesis of some enzymes, such as cellulases and other hydrolases, which act as cell wall-softening enzymes; the effect of auxin is mediated by a → signal transduction mechanism. IAA plays an important role in → tropism and leaf → abscission, and together with cytokinins in →

apical dominance; it induces the formation of adventitious roots, causes → parthenocarpy and is an important hormone for the development of mature pollen.

CH_2-COOH
N
H
O-CH_2COOH
Cl
Cl

Examples of natural (IAA, left) and synthetic (2,4-D, right) auxins.

The other well known auxins are e.g. some synthetic hormones, such as → 2,4-dichlorophenoxyacetic acid (2,4-D), → naphthalene acetic acid (NAA, widely used as a herbicide), and indole-3-butyric acid (IBA). Their effects are rather similar to those of IAA.

auxospore (Gr. *sporos* = seed), a resting fertilized egg cell in → Bacillariophyceae; formed in insufficient conditions and germinates into a new individual.

auxotroph (Gr. *trophe* = nutrition), a cell or tissue dependent on additional nutrients. The term is used to describe a cell strain in a culture when the strain requires growth factors or nutritional additives not needed by a normal strain. Those nutrients are added into a minimum medium during cultivation. *Adj.* **auxotrophic**, opposite to → prototrophic.

Aves, (L. *avis* = bird), **birds**; a vertebrate class comprising homoiothermic, feathered animals whose body structure has in many ways developed for flying, i.e. the skeleton is light but sturdy, the forelimbs are evolved into wings, and the high keel in the sternum anchors the flight muscles of the wings. The most distinctive feature of the vertebral column of birds is its rigidity due to the fusion of several vertebrae into the → **synsacrum**. Together with the pelvic girdle the vertebral column forms a stiff framework to support the legs and provide rigidity for flying. The number of bones in the forelimbs are reduced, several bones being fused together. The respiratory system is adapted to high metabolic demands of flying (*see* lung). In addition to the lungs, the nine interconnecting **air sacs** extend between the inner organs in the body cavity and into cavities of the largest bones. Fertilization is internal and reproduction oviparous, eggs are large and contain plenty of yolk. The class Aves includes 27 orders of living birds in two superorders, Palaeognathae and Neognathae; 9,881 species have been described. *See* Archaeopteryx. *Adj.* **avian**.

avidin, antibiotin; an antibacterial glycoprotein in non-denaturated egg white; binds strongly biotin and can prevent its absorption from the intestine. Therefore a daily diet containing raw eggs may lead to biotin deficiency e.g. in man.

avitaminosis, hypovitaminosis, vitamin deficiency; a condition in which the supply of vitamin is inadequate.

Avogadro number (*Amadeo Avogadro*, 1776—1856), the number of molecules (= $6.022 \cdot 10^{23}$) in one mole of any substance. *Syn.* Loschmidt number.

awn, a thin, hooked projection in the lemma of a → floret in grasses. *Syn.* arista.

axenic (Gr. *a* = not, *xenos* = foreign, strange), pertaining to a sterile condition, such as to a pure culture or to germ-free laboratory animals born and raised in sterile circumstances behind a barrier. *See* gnotobiota.

axi(o)- (L. *axis*), denoting relationship to an axis, as the central line of the body or its parts.

axial, 1. pertaining to an axis; axile; **2.** situated in the central part of the body trunk or of the long axis of an organ.

axial gland, an organ in the perihaemal system of echinoderms which is composed of spongy tissue and may produce amoebocytes.

axial skeleton, a part of the skeleton of vertebrates including the skull, vertebral column, ribs, and sternum. *Cf.* appendicular skeleton.

axial system, the cells in the → secondary xylem and → secondary phloem in plant stems and roots, produced by → fusiform initials, i.e. cells in the → vascular cambium giving rise to cells of actual secondary connective tissues. The axial system consists of layers located parallel with the main axis of the growing structure.

axifugal (L. *fugere* = to flee), directed away from an axle, axis, or axon.

axil (L. *axilla* = armpit), the angle between the plant stem (axis) and the leaf petiole.

axillary bud, a bud type in seed plants, formed in each axil in connection of every leaf. Axillary buds form shoot branches or flower shoots, and support the branching of seed plants.

axipetal, also **axiopetal** (L. *petere* = to seek), directed toward an axle, axis, or axon.

axis, pl. **axes** (L. < Gr. *axon* = axle), **1.** a straight line through a spherical body (organ, organism) between the two poles, e.g. the central line around which the body parts are arranged, sometimes denoting the vertebral column or the central nervous system; **2.** the second cervical vertebra in the vertebral column of tetrapods; in most reptiles and birds it is modified with the → atlas to permit very movable articulation of the skull; in mammals with two condyles (odontoid processes) the axis is attached and fused with the atlas, and has usually strong spines for attachment of ligaments supporting the skull; *syn.* **epistropheus**, in mammals called also odontoid vertebra; **3.** the main stem of a plant; **4.** one of the reference lines of the coordinate system, i.e. x-axis, y-axis. *See* axial.

axocoel (Gr. *koiloma* = hollow), a pair of coelomic vesicles in the echinoderm embryo, giving rise to the → stone canal and perihaemal channels. *Cf.* hydrocoel, somatocoel.

axolotl, a larval stage of tiger salamanders (*Ambystoma*, Amphibia) living in Mexico and southwestern regions of the Unites States; it develops into an adult only if the water pool it lives in, happens to dry. Also thyroxine treatment induces its further development into the adult stage. Usually, axolotls gain their sexual maturity in their larval stage, and therefore these non-metamorphic species are said to be **perennibranchiate** ("permanently gilled"). Erroneously, the larva was first named as a separate species. *See* neoteny.

axon (Gr. = axis), the single branch of the nerve cell which usually conducts impulses away from the cell body; a chief element in the → **nerve fibre**. In contrast to other neuronal branches, dendrites, the axon is usually long and mostly forms branches only in its terminal end. Axons may be less than one millimetre long e.g. in the brain, but in many nerves they can be as long as the nerve itself, i.e. even several metres long in large animals, such as in whales. The axon is covered by different types of sheath. *Adj.* **axonal**. *See* giant axon.

axoneme (Gr. *nema* = thread), the central complex array of microtubules in the core of a → cilium and flagellum in eukaryotic organisms.

axopodium, pl. **axopodia** (L. < Gr. *pous*, gen. *podos* = foot), **axopod**; a type of → pseudopodium in some protozoans, e.g. heliozoans; it is supported by long axial rods of microtubules surrounded by cytoplasm. Axopods are arranged radially around the cell body and can be extended or retracted, apparently by addition or removal of microtubular material.

azeotropic mixture, azeotrope (Gr. *a-* = not, *zein* = to boil, *tropos* = turn), constant-boiling mixture; a liquid mixture, when distilled reaches the state in which the composition of liquid phase and vapour phase are equal, and therefore the separation of the substances by distillation is difficult or impossible.

azide, a compound that contains the group $-N_3$; e.g. the poisonous and highly explosive salts of hydrazoic acid (HN_3, the ionic form N_3^-), such as sodium azide (NaN_3). Sodium azide inhibits the growth of microbes, acts as an electron transport inhibitor blocking the electron flow from cytochrome oxidase to oxygen. It is used in biological studies e.g. to inactivate hydrolysing enzymes. Azides are also used in the explosion industry.

azo- (Fr. *azote* = name for nitrogen proposed by Lavoiser), in chemistry, indicating the presence of the group -N=N-, **azo group**.

azo dyes, a large group of compounds having the -N=N- chromophore (azo group); important dye stuffs prepared from diazo compounds, and used e.g. as acid dyes for staining biological material, e.g. soft tissues, wool, cotton, and many artificial fibres; also used as additives in food stuffs.

azoic (Gr. *a-* = not, *zoe* = life), containing no living organisms; without organic life.

azonal (Gr. *a-* = without, *zone* = belt), not arranged in zones; e.g. describes an exceptional soil structure without any layers or horizons. This kind of soil is mostly recently deposited and exposed, e.g. arisen from the sea, and the local climatic factors have not had time to cause → soil formation. The vegetation of an azonal area is also different from that of the surroundings.

B

B, symbol for **1.** boron; **2.** asparagine; **3.** aspartic acid.

***B*,** magnetic flux density.

b, symbol for base in designating the length of a nucleic acid sequence; e.g. 500 b = 500 nucleotides long; 5 kb (kilobase) = 5,000 nucleotides long. *See* base pair.

Bacillariophyceae (L. *bacillus* = little rod, *phykos* = seaweed), **diatoms**; subdivision in the division → Chrysophyta, Plant kingdom. Diatoms are microscopic unicellular algae characterized by a double shell rich in silica, and chlorophyll and carotenoid pigments. The cells reproduce by divisions, while the halves of the double shells separate and form themselves into new individuals. The original size is restored by formation of special → auxospores. Deposition of diatoms during geological periods has produced diatomaceous earth which is inert material sedimented in water, and used e.g. in the production of dynamite. Bacillariophyceae is divided into the orders Centrales and Bacillariales.

bacillus, pl. **bacilli**, a rod-shaped bacterium from the family Bacillaceae; bacilli are spore-forming and mostly aerobic, motile, Gram-positive cells, the most important genus being *Bacillus*. The term was formerly used to refer to any rod-shaped bacterium. *See* bacterium.

backbone, → vertebral column.

backcross, in experimental genetics, a cross-breeding (mating) in which a heterozygous hybrid individual is crossed with either of its parents or a genotypically identical individual. *See* heterozygote.

back mutation, a heritable change in a mutant gene resulting in a reversion of the original wild type gene function.

Bacteria, *see* bacterium.

bactericide, also **bacteriocide** (L. *caedere* = to kill), an agent that kills bacteria; e.g. antibacterial proteins produced by animals, or chemicals produced by man against bacteria; an antibacterial substance (e.g. antibody) of the immune system is called **bacteriolysin**. *Adj.* **bactericidal**, **bacteriocidal**, causing death of bacteria.

bacteriochlorophyll (Gr. *chloros* = green, *phyllon* = leaf), the group of photosynthetic pigments (bacteriochlorophylls a, b, c, d, e, and g) occurring in photosynthetic bacteria; the chemical structure is close to that of plant chlorophylls but differs from them in the substituents around the tetrapyrrole nucleus and in the absorption spectra. The absorption maxima of bacteriochlorophylls are near the infrared region between 900 and 1000 nm.

bacteriocin, an exotoxin protein produced by bacteria killing species or strains of other bacteria by inhibiting nucleic acid and protein synthesis; e.g. staphylococcin produced by *Staphylococcus* bacteria, or colicin produced by *Eschericia coli*. Bacteriocins may be enzyme proteins such as colicin E2 which is a DNase, or colicin E3 which is an RNase.

bacterioid, *see* bacteroid.

bacteriophage (Gr. *phagein* = to eat), **phage**; a virus that parasitizes a bacterial cell; consists of genetic material (either DNA or RNA) and a protein coat surrounding it. Phages exhibit differences in their life cycles and may be grouped into **virulent** or **temperate** phages. The introduction of the genetic material of virulent phages into a susceptible host bacterium, results in the death and dissolution of the bacterial cell and the release of new phage particles (**lytic cycle**). Temperate phages can integrate themselves onto specific points of the bacterial chromosome and replicate with it as one unit, the viral functions then continue indefinitely unexpressed (**lysogenic cycle**). In the integrated stage the viral material is called → **prophage**, and correspondingly, the bacterium carrying a prophage is a **lysogenic bacterium**. The lysogenic cycle can convert into a lytic type spontaneously or by → induction. In that case the prophage is released from the bacterial chromosome, multiplying independently, and the cycle ends with a lysis of the host cell when infectious phages are liberated.

bacteriorhodopsin (Gr. *rhodon* = rose, *opsis* = sight, vision), a purple, light-driven proton-pumping protein in the plasma membrane of the bacterium *Halobacterium halobium*, quite similar to → rhodopsin of animals. Its light-absorbing → chromophore is retinal (retinaldehyde), and the light absorption maxima are 558 nm (dark-adapted form) and 568 nm (light-adapted form). Two protons are transported by the energy of one photon from the cytoplasm to the extracellular side, and the proton gradient formed drives the chemiosmotic synthesis of ATP.

bacteriostatic, inhibiting the reproduction of bacteria but not killing them.

bacterioviridin (L. *viridis* = green), a chlorophyll-like pigment that occurs in green photosynthetic bacteria; its absorption spectrum is similar to that of chlorophyll a. *See* chlorophyll.

bacterium, pl. **bacteria** (L. < Gr. *bakteria* = rod, staff), any of **prokaryotic**, unicellular microorganisms in the group (division) **Bacteria**, classified by some taxonomists into the kingdom Monera. Bacteria are often less than 1 µm in size, but have a true cellular structure. The genetic material is in a ring-shaped nucleic acid chain (DNA) which forms a nuclear body without a nuclear envelope. Special cell organelles typical of → eukaryotes are lacking, the most important physiological processes take place in the cell membrane and cytoplasm. Bacteria usually possess cell walls, the properties of which are used in the classification of bacteria (*see* Gram staining).

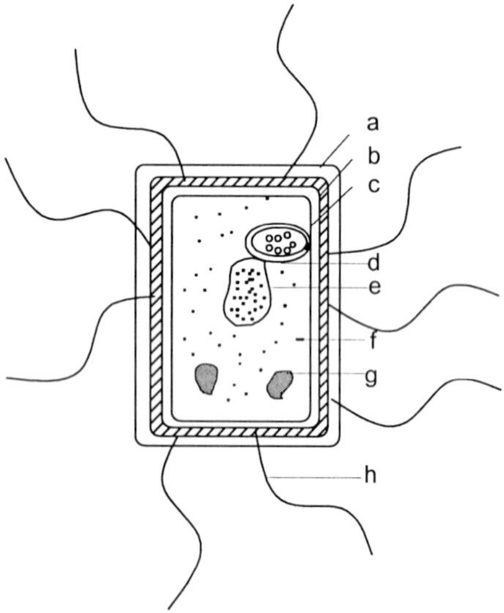

Typical bacterial cell. capsule (a), cell wall (b), cell membrane (c), mesosome (d), nuclear body (e), ribosome (f), storage granules (g), flagellum (h)

Bacteria may be anaerobic or aerobic, actively moving or non-motile, and they reproduce asexually through cell divisions (fissions); they can develop asexual spores. Bacteria can also transfer genetic material to other bacteria by processes corresponding to sexual processes of other organisms (*see* conjugation, F factor, plasmid), and undergo → genetic recombination. Most bacteria are saprophytes or parasites, some are free-living, some autotrophic bacteria producing energy by photosynthesis. The assimilation system, however, is different from that of higher plants. Many parasitic bacteria cause infectious diseases in higher organisms.

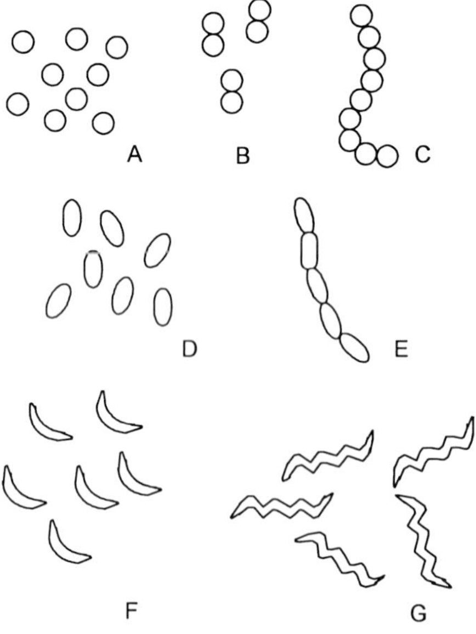

Bacterial forms. cocci (A), diplococci (B), streptococci (C), bacilli (D), chain of bacilli (E), vibrios (F), spirilla (G).

According to their form, bacteria are divided into several groups: **bacilli** (rod-shaped), **cocci** (spherical), **vibrios** (short curved cell), and **spirilla** (long curved cells). The division Bacteria is divided into subclasses → Archaebacteria and Eubacteria. The latter includes, in addition to the typical "true" bacteria, also → rickettsiae, → mycoplasmata, and → Cyanobacteria. Also another, more thorough going classification is presented in Appendixes. *Adj.* **bacterial**, pertaining to, or caused by bacteria; **bacteriogenic** (bacteriogenous), caused by bacteria; **bacteroid** (bacterioid), resembling bacteria.

bacteroid, bacterioid, 1. resembling bacteria; **2.** a small, often rod-shaped bacterium; in plant roots, a group of 46 nitrogen-fixing bacteria, surrounded by a peribacteroid membrane; *see* nitrogen fixation.

baculum, pl. **bacula,** also **baculums** (L. = rod), the penis bone (*os priapi*), supporting the penis of many mammals, such as carnivores, insectivores, bats, and nearly all primates, excluding man.

bag leaf, a bag-like, modified plant leaf, often functioning as an insect trap of → carnivorous plants.

bait trap, an apparatus used for catching insects, especially nocturnal moths; the bait is dipped into a sweet, fermented liquid alluring moths to drink, and then the intoxicated animals fall down into a bottle with poison. *Cf.* light trap.

bakanae disease, a disease of rice caused by *Gibberella* (*Fusarium*) fungi and seen as a giant growth of rice plants; known in Japan since the end of the 19th century. Plants infected by fungi are pale green and produce no seeds. *See* gibberellins.

balanced lethal system, a genetic mechanism which maintains lethal genes in high frequencies in a population. The most common case is when double → heterozygotes are viable carriers of the lethal genes, but the genes are lethal in homozygotic individuals; yet the two lethal genes persist in the population through survival of heterozygotes. The system is used in maintaining lethal genes in various laboratory strains.

balanced polymorphism, *see* genetic polymorphism.

balancers, → halteres.

balancing selection, *see* selection.

Balbiani rings (*Edouard Balbiani*, 1825—1899), large chromosomal → puffs, i.e. swellings, found in → giant chromosomes of chironomid larvae; are characterized by a big size and a ring-shaped structure. Like in other dipteran puffs, RNA synthesis is observed in Balbiani rings indicating that they are transcriptionally active genes. Balbiani rings, like other puffs, are borne when the DNA of the → polytene chromosome, organized into a → chromomere, opens from its packaged form for transcription. Balbiani rings are so large that they can be isolated by a micromanipulation technique, and are therefore specifically suited for the study of → genetic transcription.

baleen (L. *balaena* = whale), whalebone; any of the cornified sheets of the mouth mucous membrane hanging from the palate of the baleen whales (Mysticeti) who filter plankton food through the comb-like structure of the baleens.

ballast plant, a plant species transported to a new area within ballast used in ships.

Baltic Ice Sea, a stage in the developmental success of the Baltic Sea soon after the Ice Age, about 12,000—10,200 years ago; was a cold freshwater lake separated by a glacier from the sea. *See* Ancylus Lake, Yoldia Sea, Littorina Sea.

bar, a unit of pressure. *See* Appendix 1.

barb (L. *barba* = beard), *see* feather.

barbels, small tentacles surrounding the mouth in some fish species, such as hagfish.

barbicels, *see* feather.

barbule (L. *barbula*, dim. of *barba* = beard), *see* feather.

bark, plant → periderm.

barnacles, → Cirripedia.

baroreceptor, also **baroceptor** (Gr. *baros* = weight), pressoreceptor; a stretch receptor in blood vessels for sensing changes in blood pressure; in vertebrates, the receptors are sensory nerve endings reacting to stretch of the wall of the heart auricle, vena cava, aortic arch, and sinuses of the carotid arteries. The nerve fibres from the receptors ascend to the vasomotor and cardioinhibitory centres in the medulla oblongata, controlling blood pressure through autonomic nerves to the heart and blood vessels.

barren, incapable of producing offspring; unproductive, unfruitful, sterile.

Barr body (*Murray Barr*, b. 1908), → sex chromatin; a body in the nuclei of somatic cells in mammalian females formed by the → inactive X chromosome; consists of facultative → heterochromatin and is attached to the nuclear envelope. Because only females have the Barr body, its occurrence in a mammalian cell has been used to indicate the sex of an individual. The number of Barr bodies in the cell is the number of X chromosomes subtracted by one.

barrier, an obstruction; in biology e.g. **1.** an artificial protective barrier made for plants and animals against certain environmental factors, such as radiation, chemical pollution, or microbes; e.g. aseptic conditions for organisms require special systems, such as air-filtration, aseptic substratum, sterilized nutrition, etc.; **2.** physiological barriers in animal tissues, such as → blood-brain barrier, blood-testis barrier, and placental barrier.

Bartholin's gland (*Caspar Bartholin*, 1655—1738), L. *glandula vestibularis major*; greater vestibular gland; a small gland of mammals located on each side of the vaginal orifice, producing mucus; homologous with the → bulbourethral gland in males.

basal (Gr. *basis* = base), pertaining to, or located near, or forming the base or basis of a structure; basic, fundamental. *Cf.* basial.

basal body/granule, a cylindrical cytoplasmic cell organelle with nine sets of → microtubules, located at the base of each → cilium and flagellum acting as an organizing centre for them; resembles the → centriole, from which it originates. Also called kinetosome or basal corpuscle.

basal body temperature, BBT, a term used especially in medicine, denoting the body temperature of a resting person in the morning.

basal cell, the early keratinocyte in the basal layer of the → epidermis in vertebrates; by dividing continuously the cells produce keratinocytes for renewal of the skin. Also called foot cell.

basal disc, a "foot" on the lower end of the cylindrical body of hydrozoans; its cells secrete mucus for attaching the animal to the substratum, and for locomotion.

basal ganglia/nuclei, large brain nuclei at the base of the telencephalon in vertebrates; the nuclei consist of a mass of grey substance (nerve cells) lying deep in the cerebral hemispheres. Basal ganglia comprise e.g. → **corpus striatum**, **corpus amygdaloideum**, **claustrum**, **capsula externa**, and **capsula interna**. They function in the control of motor activity forming the essential part of the → extrapyramidal system.

basal lamina, 1. the membrane structure (50—80 nm in thickness) seen in electronmicrographs around certain cells, such as muscle cells, fat cells, Schwann cells; also a constituent of the thicker structure, the → **basement membrane** in the base of the epithelial cells; sometimes the terms basal lamina and basement membrane are used as synonyms; **2.** the ventral plate-like lamina of the neural tube in vertebrate embryos, also called basal plate; its cells, neuroblasts, give rise to somatic and visceral motor neurones.

basal metabolic rate, BMR, the metabolic rate of a resting animal, especially of a homoiothermic animal; expressed e.g. as oxygen consumption or production of carbon dioxide. The BMR level is controlled by certain hormones, especially by → thyroid hormones, and it usually changes during temperature acclimatization of the animal.

basal salt solution, BSS, a physiological salt solution for vertebrate cells and tissues; contains the most important salts (sodium, potassium, calcium, and magnesium used as chloride, bicarbonate and phosphate salts) in proportions corresponding to the tissue fluid of most vertebrates. BSS is used especially in cell and tissue cultures. *See* physiological solution.

base (L., Gr. *basis*), **1.** basement, bottom, basis; **2.** in chemistry, a substance that liberates OH^- ions on dissociation in an aqueous solution, or a substance that acts as a proton acceptor; reacts with an acid to form a salt and water. Common bases are e.g. sodium hydroxide (NaOH), potassium hydroxide (KOH), and many organic bases in cells, e.g. purines, pyrimidines present in nucleotides and nucleic acids. The pH of a **basic** (alkaline) solution is over 7. *Cf.* acid.

Basedow's disease (*Carl von Basedow*, 1799—1854), a disorder of the thyroid gland described in man; characterized by excessive secretion of → thyroid hormones resulting in several symptoms, such as nervousness, irritability, heat intolerance, etc. Also called **Graves' disease**. Similar disorders are also possible in other vertebrates.

basement membrane, the thin extracellular layer underlying the epithelium against the connective tissue; produced by both epithelial and connective tissue cells and composed of proteoglycans and proteins of laminin and collagen types. It is thick enough to be seen in the light microscope. The basement membrane in tubular or alveolar organs (glands, blood vessels, alimentary canals, respiratory pathways, etc.) mainly determines what substances and cells can pass through the wall. *Syn.* **basilemma**. *See* basal lamina.

base pair, bp, two nitrogenous bases that form a pair in the double-stranded structure of DNA or RNA, with hydrogen bonds holding the structure of the two-stranded molecule together. In DNA, adenine forms a base pair with thymine, and cytosine with guanine; in RNA, thymine is replaced by uracil.

base pairing, hydrogen binding between

appropriate purine and pyrimidine bases in double-stranded DNA or RNA molecules, in → genetic transcription between DNA and RNA strands; pairing occurs according to the → base pair rule.

base pair rule, base pairing rule; a rule of purine and pyrimidine pairing in → deoxyribonucleic acid (DNA); in DNA, composed of two complementary strands, thymine always occurs opposite to the adenine base, and guanine opposite to the cytosine base. DNA → replication and → genetic transcription are based on this fact.

base ratio, the ratio of (A + T) : (G + C) in the double-stranded DNA molecule; base ratio is a species-specific characteristic. (A = adenine, T = thymine, G = guanine, C = cytosine).

base rosette, a cluster of leaves at the base of a stem in → rosette plants.

basial, pertaining to the basion, i.e. the area at the midpoint of the ventral cranium near the foramen magnum of vertebrates. *Cf.* basal.

basibranchial (Gr. *branchia* = gills), *see* gill arch.

basic dyes, cationic dyes which contain a basic organic group that acts as an actively staining part, being usually linked to an inorganic acid; basic dyes are used in microscopy, e.g. for staining nucleic acids which are thus **basophilic**.

basic fibril of chromosome, a fibrillar (string of beads) basic structure of the chromosome formed by → nucleosomes; a long DNA molecule that reaches from end to end of the chromosome, and is twisted around the octameric protein core particle formed from → histones H2a, H2b, H3 and H4, histone H1 holding the structure together.

basic stem, a plant shoot used in plant → grafting.

Basidiolichenes, → lichens whose fungal partner belongs to the group → Basidiomycetes. Basidiolichenes are rare and mostly tropical; only about twenty species are known. Most lichens belong to the group → **Ascolichenes**, their fungal partners belonging to → Ascomycetes. *Syn.* Hymenolichenes.

Basidiomycetes (Gr. *mykes* = fungus), a fungal class in the division Amastigomycota, subdivision Basidiomycotina; in some systems a class in the division Basidiomycota. Basidiomycetes have septate hyphae and they reproduce through formation of **basidiospores** on the surface of a special cell, **basidium**, mostly in a → **fruiting body** as e.g. in bracket fungi, mushrooms and toadstools. The smut fungi (Ustilaginales) and rust fungi (Uredinales) do not form fruit bodies. The total number of species is ca. 30,000. *See* clamp mycelium.

Basidiomycotina, a fungal subdivision in the division Amastigomycota, the only class being → Basidiomycetes. In some systems the group Basidiomycotina is classified into a separate division, **Basidiomycota**, comprising the classes Basidiomycetes, Teliomycetes, and Ustomycetes.

basidiospore (Gr. *sporos* = seed), a spore type of → Basidiomycetes, formed by meiotic divisions on the surface of a special cell, **basidium**. Except in smut fungi (Ustilaginales) and rust fungi (Uredinales), the basidia and basidiospores are formed in fungal fruiting bodies.

basidium, pl. **basidia,** *see* basidiospore.

basilar membrane (of cochlear duct) (L. *lamina basilaris ductus cochlearis*), a membranous structure between the cochlear duct and tympanic canal of the cochlea in the inner ear of vertebrates; the sensory cells (hair cells) and supporting cells of the organ of Corti lie on it.

basilemma, → basement membrane.

basipetal (L. *petere* = to reach), directing to, or developing in the base of an anatomical structure, as the transport of a substance towards a base, e.g. auxin transport in plants; opposite to → acropetal.

basipodite (Gr. *pous*, gen. *podos* = foot), one of the two segments in the basal protopodite of crustacean biramous appendages. *Syn.* basis. *See* coxa, coxopodite.

basis, pl. **bases** (L., Gr.), **1.** → base; **2.** → basipodite.

basket cell, 1. a neurone type in the cerebellar cortex of vertebrates; axons of these nerve cells form a basket-like nest around the soma of → Purkinje cells; **2.** a → myoepithelial cell with many branches occurring basally beside the secretory cells e.g. in lacrimal glands.

basophil(e) (Gr. *basis* = base, *philein* = to love), a cell or other histological element staining with alkaline dyes; specifically referring to **1.** basophilic → granulocyte (basophilic leukocyte); **2.** a cell type (beta cell) in the anterior pituitary gland (adenohypophysis). *Cf.* acidophile.

basophilic, 1. pertaining to → basophiles; **2.** describing microbes which grow in a highly alkaline medium.

bast, a thick fibrous layer formed from phloem fibres in some trees, as in linden.

Batesian mimicry (*H. W. Bates*, 1825—1892), *see* mimicry.

bathyal zone (Gr. *bathys* = deep), in oceanic seas, the bottom area of the continental slope between the continental shelf and abyssal zone at a depth from 200 to 2,000 m, in some coasts to 4,000 m.

bathypelagic zone (Gr. *pelagos* = sea), in oceanic seas, a dark marine depth zone where light does not penetrate, situated above the abyssopelagic zone at a depth of 1,000 to 4,000 m.

bathyplankton (Gr. *planktos* = wandering), plankton organisms having daily vertical migration, i.e. wandering to surface water by night and to deeper water by day. *Cf.* epiplankton.

bats, → Chiroptera.

B cell, 1. B lymphocyte; *see* lymphocytes; **2.** a rarely used term for beta cells of the → pancreatic islands; **3.** β-cell in the anterior pituitary gland.

B chromosomes (Gr. *chroma* = colour, *soma* = body), heterogeneous accessory chromosomes present in the cell nucleus additional to the normal chromosome set; are usually heterochromatic, seldom euchromatic (→ euchromatin). B chromosomes are found especially in plants but also e.g. in insects. Their number per cell may vary between individuals but also inside an individual. B chromosomes are often unstable, and can form chiasmatic bivalents between each other but they do not pair with → A chromosomes. The function of B chromosomes is unknown, but they may have adaptive significance because their number and occurrence vary within the species often according to their living conditions. The existence of B chromosomes in the chromosome set is indicated by adding the symbol +B or (B) after the chromosome number. Their number is not indicated because it varies between individuals and cells, and therefore B chromosomes are not counted into the chromosome number of the species.

B-DNA, the most common form of DNA conformation in living cells; the classical Watson—Crick double helix having a right-handed coil, and being more hydrated than the → A-DNA. *Cf.* Z-DNA.

beak, any pointed horny mouthpart; specifically **1.** the **bill** (neb) of the bird, i.e. the structure formed from the elongated lower mandible and upper maxilla; the beak of modern birds is completely toothless and covered by horny (keratinous) skin, whereas the ancient *Archaeopteryx* had both jaws with teeth set in sockets characteristic of archosaurs; **2.** the jaws of some reptiles (as turtles) and some mammals (monotremes, dolphins) resembling the beak with a more or less horned end; **3.** an appendix of some fruits; e.g. the long beak of the fruit of *Geranium* species formed from the persistent style.

beard worms, → Pogonophora.

becquerel, Bq (*Antoine Henri Becquerel*, 1852—1908), the SI unit of activity (radioactivity). *See* Appendix 1.

behaviour models, *see* instinctive behaviour.

behavioural ecology, a branch of biology emphasizing evolutionary explanations for the reciprocal dependence between animal behaviour and the environment. The main object is an individual and the genes regulating its behaviour. Most studies concern communication between animals, foraging behaviour (*see* optimal foraging theory), sexual selection, and sociobiology. *Cf.* classical ethology.

behaviourism, a psychological theory evolved by *John B. Watson* in the 1910's and 1920's claiming that the behaviour of animals is explainable only as a response to stimuli and as physiological reactions activated by them. Behaviourism has been favoured especially in the United States. *Cf.* ethology, behavioural ecology, classical ethology.

belly, 1. abdomen; the ventral or under part of the vertebrate body containing the abdominal viscera; **2.** the fleshy part of a muscle (*venter musculi*).

belt desmosome, *see* desmosome.

Benedict's test (*S. R. Benedict*, 1884—1936), a qualitative test for determination of reducing sugars; cupric ion Cu^{2+} is reduced to red-yellow copper(I) oxide, Cu_2O, in an alkaline citrate solution.

benign, benignant, (L. *benignus* < *bene* = well), pertaining to the mild character of a disease; non-malignant as referring to a tumour.

benthos (Gr. *benthos* = depth), the organisms living on the bottom of any water body; *adj.* **benthic, benthal, benthonic**.

benzene, benzol, a hydrocarbon, C_6H_6, with

one aromatic six-carbon ring (hexagonal benzene ring) in its molecule; found in coal tar, now made from petroleum. Benzene is a highly volatile poisonous liquid with carcinogenic properties. It is an important solvent in chemical industry as well as in physiological and biochemical laboratory work.

benzoic acid, benzenecarboxylic acid, phenylformic acid; the simplest aromatic carboxylic acid having one COOH group bound to the benzene ring; common in many plants, in high quantities e.g. in lingonberry (*Vaccinium vitis-idaea*) and in the resin of the benzoin tree (*Styrax*). Because of its antimicrobial properties, benzoic acid and its salts, **benzoates**, are used as a preservation substance.

benzyladenine, BA, a synthetic plant hormone having cytokinin-like properties. *Syn.* benzylaminopurine, BAP.

Bergmann's rule (*Karl G. Bergmann*, 1814—1865), an ecogeographical rule stating that the body size of homoiothermic animals of a race living in cold regions tends to be larger than that of races of the same species living in warmer regions. The physiological explanation for the rule is based on the fact that the volume of the body increases in the cube but the surface in the square of a linear dimension, i.e. the larger a body, the relatively smaller is its heat-exchanging surface. *Cf.* Allen's rule, Gloger's rule.

beriberi (Singhalese "I cannot", meaning weakness), a disease caused by a deficiency of vitamin B_1 (thiamine) in humans, characterized by dysfunction of the nervous system (e.g. pain, paralysis), heart failure, and oedema; the endemic form is found especially in southeast areas of Asia. *Adj.* **beriberic**.

berry, a fleshy indehiscent → fruit with several seeds.

beta-amylase, *see* amylases.

beta blockers, drugs inhibiting specifically adrenergic beta receptors, e.g. propranolol. *See* adrenergic receptors.

beta-carotene, β-carotene (L. *carota* = carrot), a yellow carotenoid pigment occurring abundantly in carrots and many vegetables and fruits; it is converted to → vitamin A in animal tissues. *See* carotenoids.

beta cell, β-cell, 1. a cell type of the → pancreatic islets secreting insulin; earlier also called B cells; **2.** B cell, a cell type found in the anterior pituitary gland.

betacyanins (L. *Beta* = the genus of beet, Gr. *kyanos* = blue), reddish flavonoid pigments occurring in ten plant families of the order Centrospermae, e.g. in beetroot; the pigments belong to alkaloids, and are not related to → anthocyans.

beta emission, β-emission, beta radiation; a type of the radioactive particle emission; either electron (β^-) or positron (β^+) radiation. When an atom nucleus is overloaded by neutrons, one neutron is converted to a proton, and one electron is liberated as beta emission. When a nucleus sends a β-particle its charge changes by 1, but the mass remains practically the same. If there are too many protons in the nucleus, one proton can convert to a neutron by liberating a positron (β^+) or catching an electron. The β-particles are small, they have a very high velocity (the velocity of light) and a rather great penetration into biological material causing ionization and cell and tissue damage; e.g. glass is quite a good obstruction to β-emission. Common β-emitters are e.g. ^3H, ^{14}C, ^{32}P and ^{35}S; β-particles can be detected by autoradiography, or by using a Geiger-Muller tube or a scintillation counter.

beta-galactosidase, → β-galactosidase.

beta-globulins, β-globulins, a class of blood plasma proteins in vertebrates, fractioned as its own group in an electrophoretic assay; includes e.g. many lipoproteins and carrier proteins, such as → transferrin. *See* globulins.

beta oxidation, β-oxidation, a cellular reaction pathway where β-carbon (carbon 3) of a fatty acid is oxidized forming a β-keto (β-oxo) acid analogue. Two end carbons in a fatty acid molecule split successively into two-carbon β-keto moieties binding to coenzyme A to form acetyl coenzyme A (acetyl CoA). The acetyl residues of the fatty acids pass then into the → citric acid cycle. Beta oxidation takes place in → mitochondria of all organisms and in → glyoxysomes of plant cells.

beta-pleated sheet, the spatial arrangement of polypeptide molecules in which the laterally associated β-strands, fairly short extended stretches of the polypeptide chain, form a sheet-like pleated conformation. This structure is held together by → hydrogen bonds between carbonyl oxygen atoms and amide hydrogen atoms in two adjacent β-strands of the same or different polypeptide chains, forming a secondary structure or quaternary structure, respectively. Beta conformation is a major type of protein secondary structure especially in globular proteins, often mixed with stretches of → alpha helices. *See* protein.

beta receptors, β-receptors, *see* adrenergic receptors.

bet-hedging theory, a theory concerning evolution of life cycles interpreting why the most usual clutch size in many iteroparous species is smaller than the most productive clutch size. The most usual clutch size is determined by the compromise between the most productive clutch size and the reproductive costs to the parents. In very unpredictable environments, the parent does not invest all of its resources in one breeding season only, because production of too large a brood may decrease the survival of both offspring and parents. In that way the parents have to "hedge their bets" and not release all of their offspring into the same environment. The bet-hedging theory is composed to complement the K- and r-selective theories and to compensate for their deficiencies. *See* K-selection, r-selection, Lack's principle.

B-form helix, B-DNA; the classical right-handed conformation of the DNA → double helix; the most frequent conformation of DNA in living cells. *Cf.* A-DNA, Z-DNA.

B-horizon, the soil layer below the topmost layer, A-horizon. *See* soil formation.

bi- (L. *bis* = twice), a prefix meaning twice, two, or double; before vowels **bin-**.

bicarbonate, an acid salt or ion, HCO_3^-, of a weak dibasic acid, **carbonic acid**, H_2CO_3; e.g. sodium bicarbonate $NaHCO_3$, forming Na^+ and HCO_3^- and $Na_2CO_3 + CO_2$ (alkaline solution). Bicarbonate ions are spontaneously formed in organisms from carbon dioxide and water by the catalytic activity of → carbonic anhydrase, and the ions are important in the transport of carbon dioxide and as a pH buffer in tissues. *See* buffer.

bicentric (Gr. *kentron* = centre), having two centres; e.g. a taxon that has two centres of evolution, or a plant species that has two centres of distribution.

biceps (L. *caput* = head), a muscle having two heads at one end; e.g. *biceps brachii* on the front side of the arm.

bichirs, *see* Polypteriformes.

bicollateral bundle, a type of plant → vascular bundle in which there are two → phloems on opposite sides of the → xylem; is typical of many climbing plants and lianas.

bicorn, bicornate, bicornuate, bicorn(u)ous, (L. *cornus* = horn), **1.** two-horned; **2.** an anatomical structure having two projections or processes; mostly the terms bicornate or bicornuate are used.

bicuspid valve (L. *valva bicuspidalis*), the left atrioventricular valve (*valva atrioventricularis sinistra*) of the four-chambered vertebrate heart. Called also mitral valve.

bicyclic plant (Gr. *kyklos* = ring), a plant having two stages in its life cycle, such as → biennial plants.

biennial (L. *biennium* = period of two years), pertaining to a plant which lives for two years, forming a leaf rosette in the first year and flowering and producing seeds during the second year. A biennial plant (also called biennial) is bicyclic; e.g. carrot, many burrs. *Cf.* hapaxanthous plant, annual, perennial.

bifacial (L. *facies* = form, face), pertaining to a flattened structure with upper and lower surfaces; e.g. typical of leaves.

big-bang reproduction, *see* semelparity.

bilateral (L. *lateralis* < *latus* = side), having longitudinally two sides, pertaining to right and left sides of a structure, plane, etc., or located on both sides, e.g. **bilateral symmetry** of organisms, i.e. their body can be divided by the vertically longitudinal plane into right and left halves which are almost mirror images of one another; the horizontal plane divides the bilateral body into dorsal and ventral halves.

bilateral flower, a flower with two planes of symmetry (e.g. the family Brassicaceae). *Syn.* disymmetric flower. *Cf.* asymmetric flower, symmetric flower, zygomorphic flower.

bile (L. *bilis* or *galla*), **gall**; an alkaline, yellow-greenish fluid secreted by liver cells in vertebrates. Bile is transported from the liver to the small intestine (duodenum in mammals) through → bile ducts. In most vertebrates, the **hepatic bile** is first concentrated in the **gall bladder** to form **cystic bile** (gall-bladder bile). → **Bile salts** and phospholipids emulsify lipids, fatty acids, and mono- and diglycerides in a micellar solution, and thus make the digestion and absorption of these substances possible. **Bicarbonate ions** in the bile render the fluid alkaline allowing the optimal activity of pancreatic and intestinal enzymes. Also cholesterol, lecithin, and many waste products, synthesized or processed by the liver cells, are excreted in bile, and e.g. cholesterol, lecithin, and bile salts are absorbed with fatty acids and glycerides to circulation in the intestine. *See* enterohepatic cycle, micelle.

bile acids, *see* bile salts.

bile ducts, biliary ducts, gall ducts; a ductal system in vertebrates conveying bile from the liver to the intestine (duodenum in mammals); in most animals include bile capillaries (bile canaliculi), interlobular bile ducts, hepatic duct, cystic duct, and the **common bile duct** (*ductus choledochus*).

bile pigments, breakdown products of → haem, originated from → haemoglobin, myoglobin, and cytochromes; are produced in the liver and excreted in bile. One of them is **bilirubin**, red in colour and taken up from plasma by liver cells; it is conjugated with glucuronic acid into **bilirubin diglucuronide**. Other pigments are the green **biliverdin** (dehydrobilirubin, i.e. oxidation derivative of bilirubin), the colourless **urobilinogen**, and its oxidized form **urobilin** (the brownish pigment found also in urine). Bile pigments colour the faeces.

bile salts, salts of bile acids in the bile; formed when bile acid (cholic acid) is conjugated to glycine or taurine to form glycocholic acid (cholylglycine) and taurocholic acid (cholyltaurine), their sodium and potassium salts being respectively **clycocholate** and **taurocholate**. Bile salts are important water-soluble substances for emulsification of lipids in the intestine. *See* bile.

bilharziosis, also **bilharziasis** (*T.M. Bilharz*, 1825—1862), a tropical disease caused by some digenean trematodes of the genus *Bilharzia* (*Schistosoma*); outside Africa the disease is also called → **schistosomiasis**. About 300 million humans suffer from the disease and several millions die from it every year.

bili- (L. *bilis* = bile), denoting relationship to bile or the gall bladder.

biliproteins, chromoproteins containing red (phycoerythrins) and blue (phycocyanins) pigments; occur in algae and cyanobacteria. *Syn.* phycobiliproteins.

bilirubin, *see* bile pigments.

biliverdin, *see* bile pigments.

bilomentum, an indehiscent → fruit which resembles siliqua but breaks into separate parts with one seed in each.

binary fission, a form of vegetative reproduction in unicellular organisms, occurring when a single cell divides into two equal parts.

binary vector, one type of tumour-inducing plasmid (Ti plasmid) used as a vector when the *Agrobacterium* T-DNA sequences and the Ti plasmid virulence genes are cloned into separate plasmids. Foreign DNA to be delivered can be cloned into T-DNA located on a small plasmid. *See* plasmid.

binocular (L. *bini* = pair, *oculus* = eye), **1.** pertaining to both eyes, especially to vision with both eyes (**binocular vision**); **2.** having two optical devices for the use of both eyes, as e.g. in a microscope.

binomial (L. *nomen* = name), in nomenclature, pertaining to a scientific name of a species consisting of two terms, the former designating the genus and the latter the species, e.g. *Homo sapiens*, man. Also used as a noun. *Cf.* trinomial, principle of priority.

bio- (Gr. *bios* = life), denoting relationship to life or living organisms.

bioaccumulation, the accumulation of toxic substances, especially heavy metals and organic pollutants into organisms, as a result of direct uptake from surrounding air, water, or soil, e.g. through the gills. Especially mussels and many fish species are sensitive to the bioaccumulation in water systems. *Syn.* bioconcentration. *Cf.* biomagnification.

bioacoustics (Gr. *akouein* = to hear), a branch of biology dealing with vocalization and sound production of animals. The sounds are recorded on tape and transformed into a visible **sonogram** by using a **sound spectrogram**. The sonogram expresses e.g. sound strength, amplitude, and frequency. Bioacoustics also examines organs which are involved in sound production and reception.

bioactivity, the term describing the efficiency of an agent, such as an enzyme, hormone, vitamin, drug, etc. in a cell, tissue, organ, or organism. The efficiency of a certain substance is measured quantitatively in controlled conditions by comparing its effect with the efficiency of a standard preparation. As a unit of bioactivity is often used **IU** (International Unit) in which case the standard preparation to be compared is described and generally accepted. *Adj.* **bioactive**, eliciting a response in a cell or tissue. *See* biotest.

bioassay, → biotest.

biocatalyst (Gr. *katalyein* = to dissolve), → enzyme.

biochemical oxygen demand, BOD, → biological oxygen demand.

biochemistry (Gr. *chemeia* = alchemy), the branch of science dealing with chemical reactions occurring in organisms and the structure and properties of compounds involved in these reactions.

biocides (L. *caedere* = to kill), toxic, often synthetic substances, which kill living organisms; e.g. → herbicides, fungicides, insecticides.

bioclimatology (Gr. *klima* = climate, *logos* = word, discourse), a science of climate in relation to life and life conditions of organisms.

biocoenosis, biocenosis (Gr. *koinos* = in common), → community.

bioconcentration, → bioaccumulation.

biocybernetics (Gr. *kybernesis* = steerage), a field of science dealing with regulation and communication mechanisms of organisms.

biodegradation, the decomposition of substances by living organisms, especially by bacteria and fungi. *Adj.* **biodegradable,** pertaining to substances which are easily broken down by living organisms.

biodiversity, polymorphism in the living nature; the term refers to the genetic variation of organisms, the polymorphism of the species in a community (→ species diversity), and to the spectrum of different communities. Also the conceptions of dominance and deviation are included in the term. The greater deviation a species has from other species, the more significant is its effect on the biodiversity.

biodynamic (Gr. *dynamis* = force, might), **1.** pertaining to → biodynamics; **2.** relating to a special kind of plant cultivation, **biodynamic cultivation**, which means the application of anthroposophy in agriculture. The system is holistic, **non-scientific**, and it takes into consideration all terrestric and cosmic factors. Artificial fertilizers or pesticides are not used, but e.g. the moon phases are noticed. *Cf.* natural cultivation.

biodynamics, the research branch of science dealing with the force and energy of living matter.

bioenergetics, a branch of ecology studying energy flow through ecosystems.

bioethics, a branch of ethics especially involved in the ethical implications of biology and biological inventions, e.g. giving priority in medical care, protecting the integrity of individuals in human genetic engineering, and reservation of nature. Also the care and handling of experimental and farm animals belongs to the realm of bioethics.

biogas, a mixture of gases that is composed mostly of methane and carbon dioxide; produced by bacterial decomposition from organic material, e.g. from farm, household, and industrial wastes. Biogas is gathered and used as a renewable energy source e.g. in many farms.

biogenesis (Gr. *gennan* = to produce), a term coined by *Thomas Huxley* in the 19th century comprising the idea, generally accepted now, that life originates only from pre-existing life; the theory was based particularly on the work of *Louis Pasteur*.

biogenetic law (Gr. *genetikos* = concerning origin), a theory stating that the main stages of evolution of a species are repeated in the development of the individuals of that species, i.e. → ontogeny repeats → phylogeny. The theory has been developed from Haeckel's postulate (→ recapitulation theory), and is already outdated as a causal theory. The theory was further developed by *Karl von Baer* who suggested that during the embryonal development of an individual, the general features of the group appear first, and subsequently the special features of the species, i.e. ontogeny is specialization.

biogenic (Gr. *-genes* = born), resulting from the activity of an organism, e.g. from an enzymatic function; having its origin in biological processes.

biogeochemical cycle, a cyclic system of chemical elements in various forms between organic substances of living organisms (biosphere) and inorganic compounds, as minerals, air, and water in the soil and ground (geosphere); e.g. cycles of → carbon, oxygen, nitrogen, and phosphorus, or the hydrological cycle (water cycle).

biogeocoenosis, biogeocenosis (Gr. *koinos* = in common), → ecosystem.

biogeography (Gr. *ge* = earth, *graphein* = to write), a branch of biology and geography dealing with the geographical distribution of plants (**plant geography, phytogeography**) and animals (**zoogeography**).

bioindicator, a living organism or a species

sensitive to certain environmental changes, such as to a chemical element, thereby indicating the condition of its environment. At present, bioindicators are much used for monitoring the state of environmental pollution. Demands for species used as a bioindicator are, that their biological features have been scientifically examined, they have a large distribution range, their ability to migrate is rather weak, their tolerance range to the environmental factor is scientifically defined, and they react to a given factor rapidly and repeatedly in the same manner. Examples of bioindicators are many plants and lichens, reacting susceptibly to air pollutants, and the top carnivores, such as eagles, peregrines, or seals.

biolistic method, a direct gene transfer method in which DNA is bombarded into a cell in a special instrument by means of gold or tungsten particles.

biological clock, internal factors which control the → biorhythms of organisms; some are directly determined by specific **clock genes**, some are involved in the function of certain anatomical structures, such as certain hypothalamic brain nuclei in vertebrates. *See* biorhythms, melatonin.

biological control, → biological pest control.

biological cultivation, *see* natural cultivation.

biological half-life, *see* half-life.

biological information flow, the complex of phenomena in which the → genetic information residing in DNA is transcribed in the nucleus into → messenger RNA which guides → protein synthesis in the cytoplasm. Proteins then regulate the characteristics of the cell and the whole individual, i.e. the → phenotype of the individual. *Cf.* central dogma.

biologically efficient dose, a dose of an environmental agent sufficient to cause a → genotoxic effect.

biological oxygen demand, biochemical oxygen demand, BOD, the quantity of oxygen (O_2) used in the aerobic decomposition of organic material by microorganisms in a water system; usually presented as mg O_2/dm^3 water. BOD is used for estimation of environmental loading of waters by organic waste products. Measurements of the oxygen consumption are usually made in standard conditions, i.e. in darkness at 20°C during 5 days (BOD_5), or 7 days (BOD_7). Also called chemical oxygen demand, COD.

biological pest control, the reduction or total elimination of undesirable pest species by using biological but not chemical methods, which may be: **1.** the introduction of other living organisms, such as predators, parasites, bacteria, or viruses, to fight against the pest; **2.** the use of → pheromones attracting the pest animals, especially insects, or of metabolic hormones to reduce reproduction and metamorphosis of the pest species; **3.** the sexual sterilization of pest males by radiation and then introducing them back to nature; **4.** the use of allelopathic chemicals (*see* allelopathy) or photoperiodism in the control of weeds.

biological rhythms, → biorhythms.

biological species, *see* species.

biological wastewater treatment, a method to remove or degrade an organic load from wastewaters using bacteria and protozoans to destroy or mineralize dissolved or suspended organic materials. A part of the material is oxidized into carbon dioxide, a part precipitated to a solid organic mass. The most effective biological treatment is the **activated sludge method** by which it is possible to remove about 90% of the organic load. Beside biological treatment, also mechanical and chemical treatment methods are used in cleaning of wastewaters.

biologism (Gr. *logos* = word, discourse), a doctrine which explains psychological and social events in terms of biology. *Cf.* sociobiology.

biology, a branch of science studying living systems, such as organisms, their parts, and communities. Biology is divided into various sectors on the basis of organism types, organization levels, or of groups of functional phenomena occurring in organisms. The term biology was coined by *Lamarck* in 1802.

Based on organism types, biology is divided into **zoology**, **botany**, **mycology**, and **microbiology**, and based on the organization level: **molecular biology**, **cell biology**, **histology**, **anatomy**, **morphology**, and **ecology**. On the basis of functional phenomena, biology is divided e.g. into **genetics**, **animal physiology** (zoophysiology), **plant physiology**, **embryology**, **ecology**, **ethology**, **systematics**, and **taxonomy**. The most important theories in biology are those concerning evolution, genetics, and cytology. Essential applied fields of biology are e.g. medicine, agriculture, forestry, fishery, and environmental protection.

bioluminescence (L. *lumen* = light) chemiluminescence occurring in living organisms; the phenomenon in which a chemical reaction releases its energy as light without heat production (cold light). Bioluminescence occurs e.g. in many bacteria, algae, protozoans, molluscs, crustaceans, and fishes, especially in species living in deep waters of oceans where the light is used as a lure for prey capture, defence, and communication. Also many insects especially in tropics produce light for intraspecific communication. A well-known phenomenon in tropical areas is the **bioluminescence of the sea**, a blue-white glow caused by some very abundant unicellular algae of the division Cryptophyta, e.g. *Noctiluca miliaris*. Animals produce cold light usually by specific light organs, → **photophores**, where a reactive **luciferin** and ATP complex is oxidized in the reaction catalysed by a **luciferase** enzyme. The structure of the photophores often includes lenses and filters for directing and colouring the light. *See* luminescence.

bioluminescence assay, a quantitative method using light-emitting biomolecules for determination of certain chemical substances, e.g. ATP concentration in a biological sample. The technique is based on the bioluminescence reaction and uses luciferase/luciferin complex prepared from fireflies. *See* luminescence.

biolysis, pl. **biolyses** (Gr. *lyein* = to dissolve), **1.** chemical decomposition of organic material by the action of living organisms; **2.** dissolution of an organism; death. *Adj.* **biolytic**.

biomagnification, an increase of the concentration of an environmental pollutant, as mercury, chlorinated hydrocarbons etc., in a → food web; the higher the position of an animal in a food chain is, the higher is the concentration of pollutants the animal gets into its tissues with food; e.g. concentrations in birds at the end of the food chain are from tens to many hundreds of times higher than they are in the animals farther down in the food chain. *Syn.* biological magnification *Cf.* bioaccumulation.

biomanipulation, the use of various methods for controlling natural biological processes, e.g. the accumulation of plant nutrients or pollutant contents in a water system, usually in a lake. The process includes the use of biological, chemical, or physical approaches; e.g. the removal of phosphates from a water pool by reducing fish stocks through massive fishing, or the introduction of a new animal species to consume the overgrowth of algae. *See* gene manipulation.

biomarker, 1. → bioindicator; **2.** a change in a biological macromolecule or cell organelle caused by environmental factors and used as a marker in analyses.

biomass, the total amount of biological material in a specified area; usually expressed in grams of dry or fresh weight per square metre, and the production of a biomass during a certain time can be estimated by successive determinations. *See* biomass removing.

biomass pyramid, *see* ecological pyramid.

biomass removing, especially in aquatic ecology, the removing or reducing of organic matter (especially vegetation and economically unimportant fishes) from an eutrophic water system; the aim of the process is to remove nutrients, e.g. phosphorus, and thus to improve conditions in the water for certain other organisms, such as an economically important fish species. *Cf.* biomanipulation.

biome, an expansive terrestrial complex of several → ecosystems, which is characterized by similar soil and climatic conditions and vegetation types. The biomes may extend over several continents; e.g. the tundra, taiga, coniferous forests, savanna.

biometrics (Gr. *metrein* = to measure), **1.** statistics applied to biological problems; **2.** → biometry.

biometry, a branch of biology dealing with metrical traits of organisms and populations.

biomonitoring, the continuous observing of environmental changes, such as pollution, using → biomarkers.

biophage (Gr. *phagein* = to eat), an organism that uses other living organisms for nourishment, e.g. an animal, *Cf.* heterotroph, saprophage.

biophysics (Gr. *physis* = nature), the branch of science dealing with physical phenomena in organisms.

biopoiesis, pl. **biopoieses** (Gr. *poiein* = to make, create), the origin of the first living thing, including the chemical history that preceded it. *See* origin of life.

bioreactor (L. *reagere* = to react), a container used for performing biological reactions, e.g. for cultivation of bacteria and other cells in liquid media to produce some chemicals, such as enzymes or pharmaceutical products. Also columns packed with immobilized cells or enzymes are used; immobilized biocatalysts are used to carry different reactions, as e.g. isomerization, hydrolytic reactions, or decolorization.

bioremiditation, a process in which, by using living organisms, pollutants and other harmful substances are destroyed from water and soil.

biorhythms (Gr. *rhythmos* = rhythm), a cyclic occurrence of biochemical, physiological and behavioural functions of organisms. Rapid frequency rhythms are called → **ultradian** rhythms, one-day rhythms are → **circadian** rhythms, and slow frequency rhythms are → **infradian** rhythms. The research branch that deals with biorhythmic phenomena is called **chronobiology**. Human biorhythms based on the exact time of birth, and determined as rhythms of physical power, mental power, and emotional sensitivity are unscientific. The term **biological rhythms** should denote the scientifically verified rhythmic phenomena only.

biosciences, the group name for all fields of sciences dealing with biological phenomena; in addition to different branches of biology, includes also biochemistry, biophysics, and many applied research fields.

biosensor (L. *sensus* = sense), a device installed into a living organism to record a quantitative measure of a reaction to be studied; the response of an organism, such as mechanical, electrical, or chemical change in its tissues, to a physical or chemical stimulus (light, heat, pressure, pH, etc.) can be measured *in vivo* e.g. using a telemetric monitoring of biosensors.

biosphere (Gr. *sphaira* = globe), the zone around the surface of the Earth inhabited by living organisms; includes **hydrosphere**, **atmosphere**, and **lithosphere**. *Syn*. ecosphere.

biosynthesis, pl. **biosyntheses** (Gr. *syntithenai* = to put together), the formation of a substance in living organisms, e.g. → protein synthesis. *Adj*. **biosynthetic**.

biosystem, a functional system formed by a cell organelle, cell, tissue, organ, organism, or a group of organisms.

biota (Gr. *biote* = way of life), the entity of all the living organisms of a certain region at a certain time.

biotechnics, → biotechnology.

biotechnology (Gr. *techne* = skill, craft), technological applications of biology and the study dealing with these aspects; uses cell organelles, cells, tissues, organisms (especially microorganisms), or knowledge obtained from those applied to technology, serving e.g. medicine and drug industry, brewery and food industry, animal and plant breeding, or management of wastes. → Gene manipulation is an important area in biotechnology. Often biotechnics is considered as a synonym of biotechnology.

biotest, bioassay; the determination of a compound by its effect on biological test material, such as living cells, tissues, or organisms. In plant physiology, the best known biotest is the quantitative determination of auxin with oat coleoptiles which are cultivated in auxin solutions. The growth curve is drawn as the function of the known auxin concentrations and the unknown auxin content is read on the curve. The test also gives information if unknown samples have auxin activity or not. Correspondingly, bacteria cultures may be used to determine different kinds of substances. In animal physiology, pharmacology and biochemistry, compounds like hormones or vitamins are often determined using biotests by comparing the effect of a test sample on an animal or its certain organ with an internationally known standard product.

biotic, pertaining to life or living organisms; produced by a living organism.

biotic environment, the part of the environment of an organism composed by biotic factors including e.g. competitors, predators, parasites, and prey organisms. The biotic environment can be divided into various sections on the basis of interaction forms, such as **social**, **competitive**, **sexual**, and **parent-offspring environments**. *See* biotic factors.

biotic factors, environmental factors formed by the activity of living organisms. *Cf.* abiotic factors.

biotic potential, in population ecology, intrinsic rate of natural increase; the maximal growth capacity of a population obtained through maximal birth rate and minimal death rate. *See* innate capacity for increase.

biotin, hexahydro-2-oxo-1H-thieno[3,4-d]imidazole-4-pentanoic acid, $C_{10}H_{16}N_2O_3S$; acts in cells as a coenzyme of carboxylases (**coenzyme R**), i.e. as a carrier of activated carbon dioxide, CO_2. Biotin was first found in the liver extract of mammals but later in cells of all organisms. Many animals (man included) need it in food, and therefore it was regarded as a member of the vitamin B complex, named **vitamin H**. Biotin tends to bind to → avidin, and the complex formed is used in immunological assays for marking antibodies. *See* biotinylation.

biotinylation, the binding of → biotin to a macromolecule; e.g. DNA can be biotinylated by → nick translation, and the biotinylated sample is then linked to an avidin enzyme complex. The enzyme-catalysed (e.g. peroxidase) reaction results in a coloured product indicating the location of a biotinylated probe. *See* avidin.

biotomy (Gr. *tomein* = to cut), scientific examination of structures of organisms by using dissection.

biotope (Gr. *topos* = place), the environment of organisms where the dominant environmental factors are similar; e.g. grassy meadows are nesting biotopes for grassland birds. In botany, the biotope is the growing place, the sum of those environmental conditions in which certain species can succeed. *Cf.* habitat, microhabitat.

biotoxin (Gr. *toxikon* = arrow poison), a poison produced by living organisms.

biotransformation, biodegradation; biochemical conversion of the molecular structure of a compound in a living organism; often associated with change of pharmacological activity of a substance (e.g. drug), or with decreased toxicity and increased excretion of a pollutant in animals.

biotrophic (Gr. *trophe* = nourishment), pertaining to a parasitic organism (**biotroph**) which uses a living host as its nutrient source.

biotype (Gr. *typos* = pattern), **1.** a group of organisms which have similar genotypes; **2.** the former term for → biovar.

biovar (< variation), an intraspecific category of microorganisms distinguished from other strains of the same species on the basis of physiological or biochemical characteristics.

bipinnaria (L. *bis* = twice, *pinna* = feather), the first larval stage of sea stars; a free-living planktonic animal with two ciliated bands. *See* brachiolaria.

bipolar (L. *polus* = pole), having two poles; e.g. **1.** pertaining to an anatomical structure with two poles, as to a nerve cell with two branches; **2.** relating to the polar regions of the Earth, e.g. a bipolar distribution of a species.

bipyridine, dipyridine, dipyridyl with basic formula $C_{10}H_8N_2$; a fungal poison produced e.g. by some *Cortinarius* species; some bipyridines, such as Diquat or Paraquat (methyl viologen), are used as contact herbicides which act as cations inhibiting electron transport in → photosynthesis. Also called bipyridylium.

biramous appendage (L. *ramus* = branch), the two-branched type of crustacean appendage; proximately it consists of the **protopodite**, typically with two joints (**coxopodite**, **basipodite**), branching into the medial **endopodite** and the lateral **exopodite**. The protopodite may have one or more additional processes, **epipodites**, the exopodite one **exite**, and the endopodite one **endite**. In general, the endopodite is larger than the exopodite and forms with the protopodite an axis (**corm**) on which the exopodite stands laterally. The endopodite is primarily formed from five segments, named from the base distally: **ischiopodite**, **meropodite**, **carpopodite**, **propodite**, and **dactylopodite**. Structural modifications of the biramous appendages are very different but two basic types are named, i.e. the bibranched **stenopodium** and leaf-like **phyllopodium**.

birds, → Aves.

birds of prey, *see* Falconiformes, Strigiformes.

birth rate, → natality.

birth site tenacity/fidelity, a tendency of many animal species to return to the birth site for

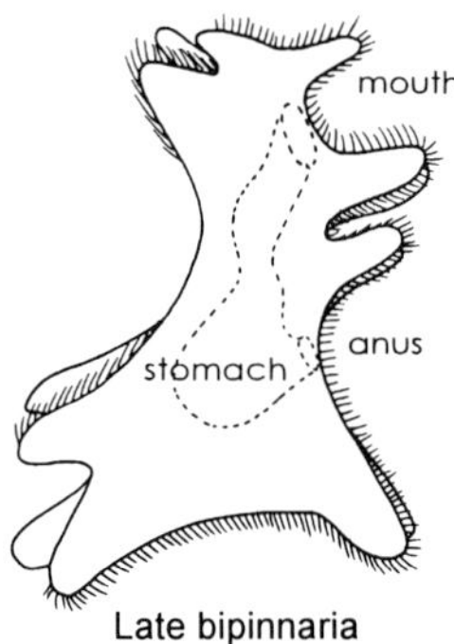

Bipinnaria.

reproduction, e.g. many migration birds. *Syn.* philopatry.

bisaccate (L. *bis* = twice, *saccus* = bag), pertaining to pollen grains with two air sacs, helping their dispersion in the wind; typical of many coniferous trees, e.g. pines.

bisexual, pertaining to **1.** → hermaphrodite, i.e. an individual which has the sexual organs of both sexes; **2.** an individual whose sexual drive is directed to both sexes; **3.** a population comprising individuals of both sexes. *Noun* **bisexuality**. *See* heterosexuality, homosexuality.

bisporic embryo sac, a type of the plant → embryo sac developing in an exceptional way from one of the two macrospores after an incomplete meiosis in the macrospore mother cell; eg. in *Allium* and *Endymion* species. *See* macrospore.

bisymmetric(al), symmetric in two directions, usually radially and bilaterally, as e.g. sea jellies.

bithorax complex, BX-C, a gene complex in the fruit fly consisting of three homoeotic → selector genes specifying to control the development of the thoracic and abdominal segments. *Cf.* antennapedia complex.

biuret, imidodicarbonic diamide; carbamylurea, $H_2NCONHCONH_2$, derived from urea by heating; in an alkaline solution it binds copper ions, staining the solution violet. *See* biuret test.

biuret test, a quantitative staining test for the determination of peptides and proteins; the **biuret reaction** is based on the binding of copper ions to peptide bonds. The test was first developed for the determination of → biuret, later modified for determination of urea and some amino acids.

bivalent (L. *bi-* = double, *valens* = of some value), **1.** divalent; describing an atom, ion, atom group, or a molecule having combining power (valence) of two; e.g. an antibody binding to antigens with two binding sites; **2.** bivalent chromosome; the temporary structure formed by alignment and attachment of two homologous chromosomes in the first meiotic division (*see* meiosis). Bivalents begin to form during zygotene, and are fully formed during pachytene. In chiasmatic meiosis, the chiasmata hold the members of the bivalent together during diplotene, diakinesis, and the first metaphase until the beginning of the first anaphase. In achiasmatic meiosis, the corresponding structures perform the same function. The formation of the bivalents is a necessary condition for the reduction of the chromosome number in meiosis.

Bivalvia (Pelecypoda, Lamellibranchia) (L. *valva* = folding door, valve), **bivalves**, e.g. **clams** and **mussels**; a class of molluscs including animals which have a bivalved, laterally compressed, calcareous shell. The **valves** are held together dorsally by a hinge ligament that causes the valves to gape ventrally. The slender muscular foot extends out between the valves; the head is reduced, the narrow mouth with labial palps takes in food mostly by filter feeding. The sexes are usually separate. Bivalves may be either free-living or sessile, mainly living in oceanic waters, several species also in fresh water; only a few species are adapted to brackish water. About 20,000 species including **mussels**, **clams**, **scallops**, **oysters**, and **shipworms**, are divided into 2—6 extant subclasses in various classifications; e.g. Protobranchia, Pteriomorphia, Palaeoheterodonta, Heterodonta, Anomalodesmata.

black earth/soil, *see* soil formation.

black nucleus, *see* extrapyramidal system.

black wart disease, a potato disease caused by the fungus *Synchytrium endobioticum*, which belongs to the group → Mastigomycota.

bladder, a sac-like structure (L. *vesica*, *cystis*); e.g. **1.** a membranous sac, usually serving in storing of secretions or excretions; e.g. urinary bladder, allantoic bladder, or several types of pathogenic bladders or cysts in animals; **2.** → swim bladder; **3.** in plants, → air bladder.

bladderworm, → cysticercus, a larva of some cestodes (tapeworms).

blast 1. (Gr. *blastos* = germ, shoot), an immature precursor stage of a cell, usually used as a suffix, **-blast**, e.g. osteoblast, fibroblast; *cf.* -plast; **2.** (Old English *blaest*), the wave of air pressure produced by an explosion.

blast(o)- (Gr. *blastos* = germ, shoot), denoting relationship to an immature cell or to the → blastula.

blastema, pl. **blastemas** or **blastemata** (Gr. = offspring, offshoot), the primordial cell mass giving rise to an organ or its part in a developing animal (embryo, larva), or in a regenerating tissue of an adult animal; e.g. the nephroblastema from which the kidney develops.

blastocoel(e), Am. also **blastocele** (Gr. *koilos* = hollow), the fluid-filled cavity of an early embryonic stage, → blastula, occurring in most multicellular animals. *Adj.* **blastoc(o)elic**.

blastocyst (Gr. *kystis* = bladder), blastodermic vesicle; the early stage of a mammalian embryo; a special type of → blastula which has an eccentric embryoblast (inner cell mass) for the development of the embryo, and a surface layer of cells forming the extraembryonic → trophoblast.

blastoderm (Gr. *derma* = skin), germinal membrane; the thin cell mass of an early animal embryo, forming the hollow sphere of the → blastula; usually ball-shaped or cap-shaped, in birds and mammals a disk-shaped structure.

blastogenesis (Gr. *gennan* = to produce), **1.** reproduction of primitive organisms by budding; **2.** the morphological transformation of small lymphocytes into larger blast-like cells in a tissue culture containing certain antigens; **3.** the transmission of inherited characteristics through the germ blast; *cf.* pangenesis theory. *Adj.* **blastogenetic** (blastogenic).

blastoglobulus (L. *globus* = ball), an electron-dense, roundish structure composed of lipids, occurring in plant chloroplasts.

blastomere (Gr. *meros* = part), cleavage cell, i.e. one of the animal cells produced during the first cleavages (divisions) of a fertilized ovum.

blastopore (Gr. *poros* = opening, pore), the opening of the cavity (→ archenteron) of the gastrula stage of an animal embryo; formed by invagination of the → blastula. *Syn.* archistome, protostoma. *See* gastrula.

blastostyle (Gr. *stylos* = pillar), a special structure in the middle capsule (**gonotheca**) of a → **gonozooid** of some colonial hydroids (e.g. *Obelia*); several medusae are produced by budding from one blastostyle.

blastozooid (Gr. *zoon* = animal, *eidos* = form), an individual animal developed by budding, e.g. a zooid in a hydroid colony. *Cf.* oozooid.

blastula, pl. **blastulae,** Am. also **blastulas** (L.), a hollow ball of cells forming an early stage of an animal embryo; arises from the morula stage when the fluid-filled cyst, **blastocoele**, is formed inside the cell group. Simultaneously the cells (**blastomeres**) are arranged in two germ layers to form the **endoderm** and the **ectoderm**. In amniotic vertebrates which often have abundant nutritive reserve in eggs (e.g. birds) the blastula stage forms a disc-like cellular plate, **blastoderm**. The formation of blastula is called **blastulation**. The next stage after blastula is → gastrula.

blepharoplast (Gr. *blepharon* = eyelid), **1.** the term used earlier for → centriole; **2.** the → basal body of flagella.

blight, a potato disease caused e.g. by the fungus *Phytophthora infestans* of the group → Mastigomycota.

blind spot (L. *discus/papilla nervi optici*), the spot where neurones of the optic nerve leave the retina in the vertebrate eye; the area has no sensory cells and so it forms a blind spot. It is located about 15 degrees medially from the visual axis and forms only a very fractional part of the whole visual field. In addition, due to rapid eye movements the blind spot does not interfere with normal vision.

blood, the body fluid circulating in a closed system in all vertebrates and many invertebrates (*see* blood circulation). Its embryonic development is close to that of the connective tissue, and it is composed of yellow liquid, → **blood plasma**, and different types of → **blood cells**. In a respiratory system blood takes oxygen and releases carbon dioxide (**oxygenated blood**), and in tissues around the body it releases oxygen for cells, taking simultaneously carbon dioxide (**deoxygenated blood**) from the tissues (*see* blood pigments). The blood also transports nutritional substances, waste products, and different messengers, like hormones. It maintains osmotic and ion balance, defends physiologically (*see* immune system, blood coagulation), and distributes heat about the body in → thermoregulation.

Other words pertaining to blood refer to *haem-*, *haima* (Gr.), or *sanguis*, *cruor* (L.), all meaning blood. *Cf.* haemolymph.

blood-brain barrier, BBB, an obstruction between the blood and brain tissue for many substances circulating in the blood. In vertebrates it is due to numerous → tight junctions between the capillary endothelial cells and by flattened endings of neuroglial cells. For example, most hormones, neurotransmitters, and toxins cannot penetrate this barrier, and apparently the barrier is involved also in the control of ion balance in the brain tissue. In the vomiting centre in the medulla oblongata the barrier is not as effective as elsewhere and therefore toxic substances can elicit the vomiting reflex. *Cf.* blood-placenta barrier, blood-testis barrier.

blood capillaries, *see* capillaries.

blood cells, blood corpuscles; cells circulating normally in the blood, of which the → **red blood cells** (erythrocytes) of vertebrates function in the transportation of respiratory gases (O_2, CO_2), the → **white blood cells** (leucocytes) in the immunological defence, and the → **blood platelets** (thrombocytes) in blood coagulation. Apparently all blood cells have a common stem cell type from which different types develop as their own lines, first in the embryonic yolk sac, later especially in the bone marrow.

Blood cells are circulating also in the blood or haemolymph of invertebrates serving defence and/or respiration, although → blood pigments (respiratory pigments) often occur extracellularly in the plasma.

blood circulation, the transportation system in the body of most multicellular animals, acting in the transport of respiratory gases and a vast number of nutritional, excretory, and messenger substances (*see* blood, haemolymph). Sponges, cnidarians, ctenophores, flatworms, and roundworms have no circulatory system, and transportation is mainly performed by the → gastrovascular system. Other multicellular animals usually have blood circulation, although in some cases it is partly or totally reduced (e.g. in many parasitic worms, echinoderms). In roundworms (nematodes) and rotifers, the tissue fluid circulating in the → pseudocoelom transports nutritional substances.

Many invertebrates, as molluscs and arthropods, have an **open circulation**, i.e. blood vessels open into the body cavity, bathes the inner organs, and the circulating fluid mixes with tissue fluid forming **haemolymph**. The → heart is either a long, segmentary tube (**tubular heart** e.g. in many arthropods), or a roundish muscle forming a **chambered heart** (in molluscs). In echinoderms the → water-vascular system functions as a circulatory system.

Annelids have a **closed circulation** with beating vessels, but no heart. Also vertebrates have a closed circulation with different types of → blood vessels and a chambered heart. The circulation is either **single-circuit** (two-chambered heart in fishes), or **double-circuit** (three-chambered heart in amphibians, or four-chambered in reptiles, birds and mammals). In the latter type, the blood first circulates into the lungs to be oxygenated (→ **pulmonary circulation**) and then to other tissues to deliver oxygen (→ **systemic circulation**). The circulation may be **incomplete**, when the oxygenated and deoxygenated blood mix in the heart (amphibians and many reptiles), or it may be **complete**, when no mixing occurs (birds and mammals). *See* blood pressure.

blood coagulation/clotting, the clotting process of certain proteins in the blood plasma. It is best known in vertebrates, in which it is a long chain reaction activated by certain substances released from an injured tissue. This cascade of reactions in which inactive enzymes are activated, produces different **clotting factors** some of which activate the final reactions in the plasma causing coagulation, i.e. an inactive protein, **prothrombin** (produced by the liver), is cleft into an active enzyme, **thrombin**, which then converts a soluble protein, **fibrinogen**, into insoluble **fibrin**. This quickly forms a fibrillar net-like structure in the injured area. Some leucocytes and especially → **platelets** (thrombocytes) aggregate in the wound to form the final clot (*see* thromboxane); in small wounds thrombocytes almost alone form the clot. More than ten different clotting factors, numbered I, II, III etc., are found; one is the calcium ion, Ca^{2+}, and by complexing it e.g. with citrates blood coagulation can be inhibited. Heparin, released from → mast cells (heparinocytes) and basophilic granulocytes in blood and other tissues, inhibits normally harmful clotting in the tissues. Hydrolytic decomposition of fibrin occurs enzymatically in fibrinolysis. *See* anticoagulant.

blood corpuscles, → blood cells.

blood formation, → haemopoiesis.

blood group/type, the type of an individual blood defined by an immunological property of one or more → **antigens** present on the surface of red blood cells. The group is genetically determined by gene alleles located on autosomal chromosomes. Blood grouping is intensively studied in humans and some other mammals, but a similar system is found also in other vertebrates. The blood-group factor is a **glycoprotein** protecting the surface of red blood cells. In the immune system this glycoprotein acts as an antigen (**agglutinogen**), and can induce the formation of a specific → **antibody** (**agglutinin**) in the plasma of such a person who does not possess this particular glycoprotein. If the blood group factor contacts with its antibody they attach to each other and thus fix red blood cells together (**agglutination**). This reaction is used for determination of blood groups. Of many human blood groups the most important in clinical practice are the → ABO and Rh groups.

blood picture, the determination (**blood count**) of the quantity and quality of blood cells; indicates the health condition of an animal because the blood picture changes e.g. in illness and stress conditions.

blood pigments, respiratory pigments; pigments occurring in the → blood or → haemolymph. In vertebrates and some invertebrates, blood pigments are found in blood cells but in most invertebrates they are extracellular in the plasma; in many invertebrate groups blood pigments are totally lacking. The pigments function in the transportation of respiratory gases, especially of oxygen. Different types of pigment molecules are all composed of a **protein** molecule and a **metal**, usually in the form of coloured **metalloporphyrin** (*see* porphyrin). In most pigments the metal atom is iron, as in → haem. Of the blood pigments, → **haemoglobin** occurs in all vertebrates, and → **chlorocruorin** and **haemerythrin** in many invertebrates; → **haemocyanin** (with copper) is found in molluscs and many arthropods, → **haemovanadium** (with vanadium) in tunicates.

blood-placenta barrier, an obstruction in mammalian tissues between the maternal blood and the foetal blood, separating the two circulatory systems from each other. During the embryonic development some structures between the uterine and placental walls in many mammalian species disappear. However, there are at least the capillary endothelium of the foetus and a uterine epithelium with some connective tissue separating the circulations from each other. *Cf.* blood-brain barrier, blood-testis barrier.

blood plasma (L. *plasma*), the liquid portion of the blood, the osmotic concentration of which in most vertebrate groups is equal to 0.9% NaCl solution, i.e. about 300 mOsm/l. Also in many invertebrates its concentration is regulated to about the same value as in vertebrates, but in other groups the concentration changes according to the environment (*see* osmotic regulation). **Sodium chloride** (NaCl) is generally a main component in the plasma and in lesser quantities also potassium, calcium, and magnesium **salts**, and about 70 g/l of **plasma proteins** (e.g. albumin, globulins, fibrinogen). There are also various substances necessary for cells, such as **nutrients**, **trace elements**, **vitamins**, **hormones**, and other messenger substances and their metabolites, and many **waste products** such as urea and uric acid. *Cf.* blood serum.

blood platelet, → platelet.

blood pressure, hydrostatic pressure against the walls of blood vessels; highest in large arteries decreasing gradually towards thinner blood vessels. The capillary pressure is usually less than one tenth of the initial arterial pressure. In the veins the pressure changes greatly depending e.g. on the activity and position of an animal. The blood pressure is generally dependent on the action of the heart, the elasticity of the walls of the vessels, and the viscosity and the volume of blood. In tetrapod vertebrates, arterial pressure is manifold in the systemic (greater) circulation as compared to the pulmonary (lesser) circulation, and changes according to contraction phases and the resting phase of the heart. The **systolic pressure**, the maximum arterial pressure of the stroke output of the heart (left ventricle in the four-chambered heart) is about one third higher than the **diastolic pressure**, i.e. the minimum pressure in the resting phase of the heart; in an adult man the pressures are normally about 120/80 mm Hg. Correspondingly, in the pulmonary circulation driven by the right ventricle, arterial pressure is much lower, only about 25/10 mm Hg. The average of these pressure levels is called **mean blood pressure**. The blood pressure is controlled by the autonomic nervous system and humoral factors in the blood (e.g. hormones, O_2), which change the activity of the heart and resistance of the vessels (dilatation, constriction). → Baroreceptors found in the walls of large arteries act as pressure sensors.

blood serum, a clear, yellowish liquid separating from the blood on clotting, i.e. the →

blood plasma without fibrinogen (fibrin).

blood-testis barrier, an obstruction for various substances in the vertebrate testis between the blood circulating in capillaries around seminiferous tubules and the tissue fluid around the germ cells in the tubules. The barrier is formed by the endothelial cells of the capillary wall and by tight junctions between adjacent → Sertoli cells. *Cf.* blood-brain barrier, blood-placenta barrier.

blood typing, blood grouping. *See* blood group.

blood vascular system, the system of → blood vessels.

blood vessels, closed or open channels in organs of most animals for carrying blood to tissues; comprise **arteries** (*arteriae*) with many small branches (**arterioles**) transporting blood from the heart to the peripheral tissues, **veins** (*venae*) beginning with small branches (**venules**) carrying blood back to the heart, and → **capillaries** between arterioles and venules. Also→ anastomoses of various size may exist between arteries (or arterioles) and veins (or venules). In vertebrates, the histological structure of arteries and veins is similar, i.e. against the lumen is a layer of the **tunica intima**, composed of endothelium and connective tissue; in the middle is located the **tunica media** composed of smooth muscle and connective tissue, and outermost is the **tunica externa** composed of connective tissue. The walls of arteries are always much thicker and stronger than those of corresponding veins, and often contain thick elastic elements. The capillaries form a netlike structure, a capillary bed, between the arterioles and venules. Capillary walls are thin, lined by endothelium, i.e. a single layer of flat cells, between which blood plasma can be filtered.

bloom, 1. the flower of a plant, or flowers collectively; **2.** the mass occurrence of planktonic algae (especially → Cyanobacteria) during late summer in eutrophic water bodies, causing harmful changes in taste and smell of the water; some species produce toxins in the water; **3.** the whitish, powdery coating of deposit such as on the surface of certain fruits.

blubber, the fat of whales and other cetaceans occurring mainly under the skin; whale oil is made of blubber.

blue-green algae, → Cyanobacteria.

blue-staining fungus, a fungus causing dead wood (lumber) to turn blue. The fungi do not affect the strength of the wood because they do not dissolve cellulose or lignin, but they use only organic matter of dead cells. Blue-staining fungi occur in many fungal groups.

B lymphocyte, B cell (B < *bursa Fabricii* = a lymphatic gland of birds), a type of → lymphocyte responsible for humoral → immunity. *See* bursa.

BMR, → basal metabolic rate.

B-nine, daminozide; N-(dimethylamino)succinamic acid, used as a plant growth retardant.

BOD, → biological oxygen demand.

body, 1. the physical structure or the trunk of an animal or plant; **2.** a corpse or cadaver; **3.** the main mass or the most important part of an organ, or of a thing; **4.** any mass of material.

body cavity, 1. → blastocoel, i.e. the primary body cavity in an early animal embryo; **2.** → coelom, i.e. secondary body cavity; **3.** → pseudocoelom in more primitive animals.

body cell, a → somatic cell as distinguished from reproductive cells in animals.

body fluids, *see* fluid.

body temperature, temperature prevailing in a living animal; in poikilothermic animals (**poikilotherms**) it mainly follows the ambient temperature, but in homoiothermic animals (**homoiotherms**) the thermoregulatory system maintains the temperature of internal organs (**core temperature**) at a certain, fairly stable level (about 35—37°C in mammals, 39—41°C in birds). The temperature of the peripheral parts of the body may be much lower; e.g. the skin temperature changes drastically, especially in the limbs. *See* thermoregulation.

bog, a wet plant community where bog mosses (→ Sphagnidae) produce a thick, acid peat layer, mostly growing dwarf shrubs and stunted pines; bogs are common especially in Scandinavia, northern Russia and in North America. *See* mire.

bog mosses, → Sphagnidae.

bog (iron) ore, limonite; an impure form of hydrated iron(III) oxide, $Fe_2O_3 \cdot H_2O$, occurring in bogs and marshes.

Bohr effect (*Christian Bohr,* 1855—1911), the dependence of oxygen dissociation from oxyhaemoglobin on a change of acidity or the partial pressure of carbon dioxide in blood. The affinity of oxygen to haemoglobin decreases when acidity or the partial pressure of carbon dioxide increases. Consequently, haemoglobin in the blood releases oxygen more easily in peripheral tissues with higher concentration of carbon dioxide, and binds a higher load of oxygen in the lungs when carbon dioxide escapes from the blood.

bolus (Gr. *bolos* = lump), **1.** a rounded mass of food material passing through the alimentary canal; **2.** a pharmaceutical preparation, such as a dose of drug, to be swallowed.

bombesin, a peptide first isolated from the skin of a frog (*Bombina*), later found in parasympathetic nerve endings of the gastrointestinal mucous membrane in vertebrates; probably acts as a neurotransmitter e.g. stimulating gastric acid secretion, pancreatic secretion, and contraction of the gall bladder.

bond, *see* covalent bond, disulphide bond.

bone, 1. bone tissue (osseous tissue); a hard connective tissue which forms the skeleton of most vertebrates; consists of living cells (→ **osteoblasts**, **osteocytes**, and **osteoclasts**) embedded in the **matrix** containing the framework of collagen fibres and the ground substance impregnated with a mineral component, mostly **hydroxyapatite**, composed chiefly of calcium phosphate with some calcium carbonate and fluoride. Altogether on average, the bone tissue is composed of 15—25% organic material and 75% inorganic material (*see* apatite);

2. a part of the bony skeleton, formed from bone tissue; the surface is covered by a hard membrane (**periosteum**) composed of fibrous connective tissue. The bones may develop directly from the soft fibrous connective tissue (**dermal bones**, membranous bones) or from cartilages (**endochondral bones**). The structure of the bone is usually harder near the surface or in the shafts of long bones (**compact bone**) than deeper in the bone tissue (**trabecular bone**, cancellous bone, spongy bone). The compact bone is organized into cylinders around the central blood vessels called osteons or Haversian systems. The trabecular bone makes up trabeculae lining the marrow cavity. *See* bone marrow.

bone marrow, the soft connective tissue in the medullary cavities of most bones in tetrapod vertebrates; produces different types of blood cells (*see* haemopoiesis). Especially the bone marrow of the long bones of the limbs tends to become filled with fat with age (yellow bone marrow).

bonitation, 1. evaluation of numerical distribution of a species locally or seasonally; **2.** evaluation of the ability of woodland (e.g. of a certain forest type) to produce wood.

bonsai, pl. **bonsai** (Japanese), a tree, grown artificially in dwarf size.

bony fishes, → Osteichthyes.

book lung, lung book; one of the two types of respiratory organs in arachnids, the other being the tracheal system. The book lungs are situated in the front part of the opisthosoma and are formed by paired organs with several flat tissue plates resembling book leaves. The plates have capillaries for gas exchange.

bordered pit, a type of → pit occurring in plant cell walls; bordered pits are common in the → xylem of higher plants, especially in water conducting tracheids of gymnosperms, where they connect separate tracheid cells.

Bordetella (*Jules Bordet*, 1870—1961), a genus of aerobic, Gram-negative bacteria many of which are pathogens living in respiratory organs of many mammals, e.g. rodents, domestic animals, and man; *Bordetella pertussis* causes whooping cough in man.

boreal (L. *borealis* = northern), pertaining to the north, northern biogeographical regions, or north wind.

Boreal Age; the last period of the → Ice Age about 8,800—7,500 years ago, when the climate was warm and continental.

boreal zone, the zone south of the → arctic zone, characterized by continuous coniferous forests.

boreoalpine (L. *alpinus* = living in the Alps), pertaining to a species which lives in the → boreal zone and in the → alpine zone of mountain areas.

boron, symbol **B**, a non-metal element, atomic weight 10.81; occurs naturally only in compounds, such as boric acid and its salts, **borates**. Boron is one of the micronutrients required by cells, e.g. in plants for the metabolism and transport of carbohydrates, and for the development of cell walls and the growth of a pollen tube. Boron deficiency results in rotten fruit and root crops, and inhibition of the growth of the root and shoot apex.

Borrelia (*Amédée Borrel*, 1867—1936), a genus of Gram-negative, anaerobic, helical bacteria (order Spirochaetales) containing many pathogens of many animals, also humans; some are transmitted to birds and mammals by arthropod bites, especially of ticks, causing **borreliosis**, known as Lyme disease in man.

bostryx, pl. **bostryces,** also **bostryxes** (Gr. *bostrychos* = curl), an → inflorescence type, subtype of → cyme.

botany (Gr. *botane* = pasture, herb), a branch of biology studying plants; branches of botany are e.g. plant anatomy, morphology, systematics, ecology, and physiology, newer areas e.g. plant biotechnology and molecular biology.

bothridium, pl. **bothridia** (Gr. *bothros* = pit), a cup-like structure which is either chitinous, as some ciliated pits of acarines, or consists of muscle tissue, as a cup-like sucker in the scolex of some platyhelminths (tapeworms).

bottleneck effect, *see* genetic drift.

botulism (L. *botulus* = sausage), a severe intoxication caused by ingestion of the neurotoxin (**botulin**) produced by *Clostridium botulinum* bacteria in improperly preserved food; botulin causes paralysis in vertebrates, although many species, such as canines and swines, are somewhat tolerant to it. Botulism may cause problems in breeding of domestic animals, and is very toxic in man.

bouquet stage, a meiotic prophase stage in which the chromosomes form a bouquet-shaped appearance lasting from leptotene to pachytene; occurs in the cells of most eukaryotic organisms.

Bovidae (Gr. *bos* = ox), a large mammalian family in the suborder Ruminantia, order Artiodactyla; includes cattle, bison, musk-ox, sheep, goats, and antelopes.

bovine serum albumin, BSA, a protein used as a reference substance in protein determinations and as a protective agent in the isolation of cells and organelles.

bowel, → intestine.

Bowman's capsule (Sir *William Bowman*, 1816—1892), capsule of glomerulus, *capsula glomeruli*; the double-walled, expanded beginning of the nephron tubule in the vertebrate kidney, surrounding the → glomerulus. Also called Müller's capsule or Malpighian capsule. *See* nephron.

bp, → base pair.

Boyle--Mariotte's law (*Robert Boyle*, 1627—1691; *Edme Mariotte*, 1620—1684), a rule expressing that at a constant temperature the volume of a given quantity of an ideal gas varies inversely with its absolute pressure, and the pressure varies inversely with its volume.

Bq, → becquerel.

brachial (L. *brachium* = arm), pertaining to the arm or the forelimb of an animal, or to an arm-like anatomical structure.

brachial plexus (L. *plexus brachialis*), a nerve network in tetrapod vertebrates comprising the ventral rami of the fifth cervical to the first thoracic spinal nerves; innervates the brachial skin and muscles, i.e. those moving the forelimb.

brachiolaria, pl. **brachiolariae** (L. *brachiolum* = small arm), the second larval stage of some sea stars, living free-swimming in the oceanic plankton; developed from the → bipinnaria.

Brachiopoda (Gr. *pous*, gen. *podos* = foot), **brachiopods**, **lamp shells**; an invertebrate phylum, most species of which are extinct; thousands of fossil species are known, and about 300 species still live mainly in oceanic shallow waters. Brachiopods have a true → coelom, many tentacles (arms) of the → **lophophore**, a calcareous **shell** with ventral and dorsal **valves**, and a fleshy stalk (**pedicel**, **peduncle**) which in many species is protruded forward for attaching the animal to a substratum. The phylum is divided into two classes, **Articulata** and **Inarticulata**, which differ from each other by the existence of the hinge connecting the valves (lacking in Inarticulata). A familiar genus is *Lingula*, the "living fossil".

Brachiopterygii (Gr. *pterygion* = fin), a group (in some classifications a subclass) of bony fishes comprising the order → Polypteriformes (bichirs and reedfishes).

brachium, pl. **brachia** (L. = arm), **1.** the arm or forelimb of tetrapod vertebrates, especially the upper part between the shoulder and the elbow; **2.** a process in an anatomical structure resembling the arm. *Adj.* **brachial**.

brachy- (Gr. *brachys* = short), denoting short, brief.

brachypterous (Gr. *pteron* = wing), having short wings, especially referring to certain insects.

brachysclereid (Gr. *skleros* = hard, *eidos* = form), a roundish thick-walled → sclereid (stone cell), common e.g. in the flesh of pear.

brackish water, water with salinity between fresh water and seawater, i.e. the concentration of dissolved salt ranging from 0.5 to 30 g/l. On the basis of the salt content, brackish water can be classified to **oligo-**, **meso-** or **polyhaline**. The largest brackish water bodies are the Baltic Sea, Black Sea, Caspian Sea, and Hudson Bay; in general, also ocean estuaries contain brackish water.

bract (L. *bractea* = thin gold plate, gold leaf), **1.** a modified, often small plant leaf from the axis of which a branch of the stem, a flower, or an inflorescence grow; **2.** a kind of upper leaf within an inflorescence; **3.** a modified, mostly small plant leaf, being often connected with normal leaves by transitional forms. Bracts often follow the leaves of the vegetative stem, being associated with the flowers and branches within the inflorescence. Small bracts are called **bracteoles**. *Cf.* scale leaves.

bracteate inflorescence, an → inflorescence which in addition to flowers has scale-like → bracts. If the leaves are like normal foliage leaves, the inflorescence is called **frondose**.

brady- (Gr. *bradys* = slow), denoting slow.

bradycardia (Gr. *kardia* = heart), slowness of the heart rate. *Syn.* brachycardia. *Adj.* **bradycardiac**. *Cf.* tachycardia.

bradykinin (Gr. *kinein* = to move), a nonapeptide in the group of plasma → kinins in vertebrate tissues; produced from an active decapeptide, **lysylbradykinin** (kallidin, bradykininogen), which can be converted to bradykinin by aminopeptidase in the blood. Bradykinin is a potent vasodilator, acting e.g. in allergic reactions. *See* kallikrein.

bradymetabolic animals, animals with slow metabolism, i.e. the ectothermic animals, ectotherms. *See* ectothermy. *Cf.* tachymetabolic animals.

brain, (L. *cerebrum*, Gr. *enkephalos*), the enlarged anterior part of the central nervous system in most multicellular animals; receives and integrates the sensory information, and controls and coordinates the functions of different organs and organ systems. In vertebrates, the brain develops as an enlargement of the anterior neural tube of an embryo, first comprising three parts with vesicles: the → **prosencephalon** (forebrain), **mesencephalon** (midbrain), and **rhombencephalon**. Later the prosencephalon forms the → **telencephalon** and **diencephalon** (together often called cerebrum), and the rhombencephalon develops into the → **metencephalon** (cerebellum and pons, often called hindbrain), and **myelencephalon** (medulla oblongata). The brain and → spinal cord form the central nervous system; *see* also pallium.

The brain consists of **nervous tissue** with a vast number of → **nerve cells** (neurones) and a supporting tissue, → **neuroglia** (glia). There are grey and white areas in the brain tissue. **Grey matter**, chiefly consisting of a great number of cell bodies and their many dendritic branches, form the cerebral and cerebellar cortices, several brain nuclei deeper in the brain tissue, and the → reticular formation along the central brain stem. **White matter** consists of neuronal pathways formed from numerous parallel → nerve fibres with white → myelin sheaths. White matter forms laminar pathways deep in the cerebral and cerebellar areas continuing along the outer surface of the brain stem towards the spinal cord.

There is a fluid-filled cavity system inside the brain, comprising four → **brain ventricles** from which the fluid (**cerebrospinal fluid**) flows into the subarachnoid space on the brain surface between the specialized membranes, **meninges**, covering the brain (*see* meninx). Thus, the brain tissue is suspended between the two fluid beds. Many neuronal pathways leave directly from the ventral brain stem, i.e. through the → **cranial nerves** innervating mainly the head areas, the others run through the spinal cord and the → **spinal nerves** to innervate the other parts of the body, both including somatic and autonomic nerves or nerve fibres.

During the evolutionary development of species the importance of the brain as a principal control system has increased (**encephalization**); the role of the telencephalon (**cerebralization**) has become more important, and especially its cortical area (**corticalization**) has greatly developed in many species, as in man. It is estimated that the human brain (about 1.5 kg) consists of 30—100 thousand million nerve cells.

In invertebrates, the brain is usually an enlargement of anterior ganglia, often called **cerebral ganglia**; e.g. specialized nervous systems of annelids and arthropods consist of the **bilobed brain**, the highly specialized form of the cerebral ganglia. The brain of cephalopods is formed from three pairs of well developed ganglia and includes a great complexity of nerve cells (in octopus e.g. 160 million cells) with complicated neural functions, very much resembling those of vertebrates.

brain atlas, *see* stereotaxic apparatus.

brainstem, the stem areas of the vertebrate brain, including the → mesencephalon (midbrain), pons, and myelencephalon (medulla oblongata), sometimes also diencephalon; along the surface areas it contains mainly myelinated motor and sensory nerve tracts, and more centrally the → reticular formation and many special nuclei controlling autonomous functions, e.g. including the respiratory and vasomotor centres.

branchia, pl. **branchiae** (L. < Gr. *branchia* = gills), gill, i.e. the respiratory organs of aquatic animals. *Adj.* **branchial**. *See* gills.

branchial arch, → gill arch.

branchial cleft, gill cleft, gill slit; **1.** a cleft between the → gill arches (branchial arches) in fishes through which water taken into the mouth passes out bathing the gills; cyclostomes and some sharks have less differentiated → branchial pouches; **2.** a corresponding rudimentary groove in the embryo of higher vertebrates; *see* branchiogenous organs.

branchial heart, an accessory circulatory organ at the base of the gills in cephalopods; pumps blood through gill arteries to gills from which it returns to the ordinary heart through gill veins.

branchial pouch, gill pouch, pharyngeal pouch; **1.** one of the six respiratory cavities in the branchial clefts of cyclostomes and some sharks, opening out through spiracles; **2.** any of paired invaginations of embryonic pharyngeal endoderm between the branchial arches during the development of all chordates; in aquatic vertebrates they develop into branchial clefts (gill clefts/slits), in other vertebrates into → branchiogenous organs.

branching, in botany, the branching of a plant shoot. The most primitive type is → **dichotomous branching**, in which two almost or totally equal branches develop; this is caused by division of the only apical cell of the shoot, and the type occurs in lower plants, especially in mosses and clubmosses. In vascular plants the branches originate from → apical buds, and there are two types, the **monopodial** and **sympodial branching**. In the monopodial type the main trunk grows continuously and keeps its position (eg. spruce). In sympodial branching the main trunk changes at intervals, while some of the side branches take its role (e.g. linden).

branchiogenous organs (Gr. *branchia* = gills, *gennan* = to produce), organs which in tetrapod vertebrates during the embryonic development derive from the primordial branchial arches; e.g. the ear drum, auditory tube and ossicle(s), parathyroid gland, and thymus.

Branchiopoda (Gr. *pous*, gen. *podos* = foot), **branchiopods, gill-footed shrimps**; a crustacean class, mainly freshwater species with gills situated in leaf-shaped thoracic appendages. The common order is Cladocera (**water fleas**) which have a bivalved carapace and the second antennae enlarged for swimming.

branchiostegite (Gr. *stege* = roof), the ventral part of the → carapace covering the gills in some crustaceans.

Branchiura (Gr. *oura* = tail), **branchiurans, fish lice**; a crustacean class including mainly ectoparasites of fish; about 150 species are described.

breakage and reunion, the mechanism of → crossing-over where homologous chromosomes break at homologous sites and reunite in a new way.

breastbone, the → sternum of tetrapod vertebrates, especially of birds and mammals.

breed, 1. to reproduce; **2.** an arranged mating group having a common ancestor; the term is used especially for a genetically improved stock of a domesticated species.

breeding, the genetic improvement of domesticated animal species or cultivated plants; especially in → plant breeding, different strategies are used according to the types of reproduction.

breeding size, the number of individuals in a population which are actually involved in reproduction of a given generation.

breeding system, the total of the mechanisms which affect the fusion of gametes in a sexually reproducing population.

breeding value, in quantitative genetics, the central concept of → value in which individuals are estimated on the basis of their offspring. When an individual mates randomly in a population, its breeding value is calculated by subtracting the → phenotypic value of its offspring from the average phenotypic value of the population and multiplying by two. The breeding value of an individual is also the sum of the average effects of its genes, therefore also called the additive genotypic value of the individual. The symbol of the breeding value is **A**.

Broca's area/field/centre (*Pierre P. Broca*, 1824—1880), motor speech centre, corresponding the → Brodmann's area 44; the cerebral cortex area posteriorly in the inferior frontal gyrus usually on the left hemisphere of the human telencephalon; controls the motor mechanisms in articulation of speech. *Cf.* Wernicke's area.

Brodmann's areas (*K. Brodmann*, 1868—1918), in human brain anatomy, areas of the cerebral cortex mapped on the basis of the cellular structure and functional specificity. The areas are presented in the figure. *See* neocortex.

divides into two **main bronchi** (primary bronchi) and these further into secondary and tertiary bronci, i.e. **lobar**, **segmental**, and **subsegmental bronchi**. The main bronchi usually have small pieces of cartilage supporting the musculomembranous elastic walls. As generally in the respiratory pathways, a ciliated pseudostratified epithelium lines the walls of the bronchi on the inside (non-ciliated only in the smallest airways).

brood parasitism, a behavioural pattern of some animals when a reproducing individual leaves its eggs or offspring in the care of another individual or individuals, either of the same or another species; e.g. the cuckoo. *Cf.* nest parasitism.

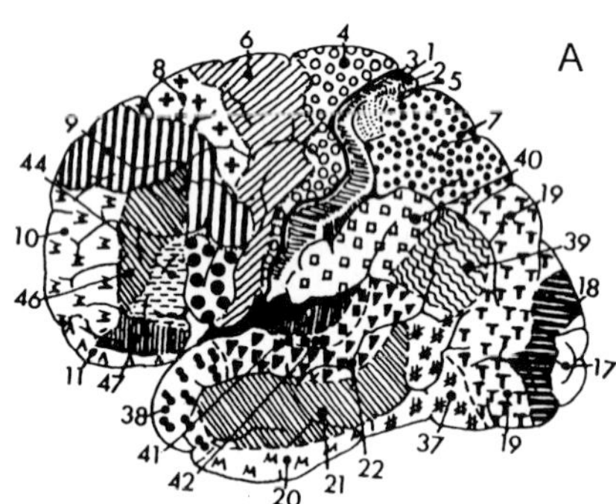

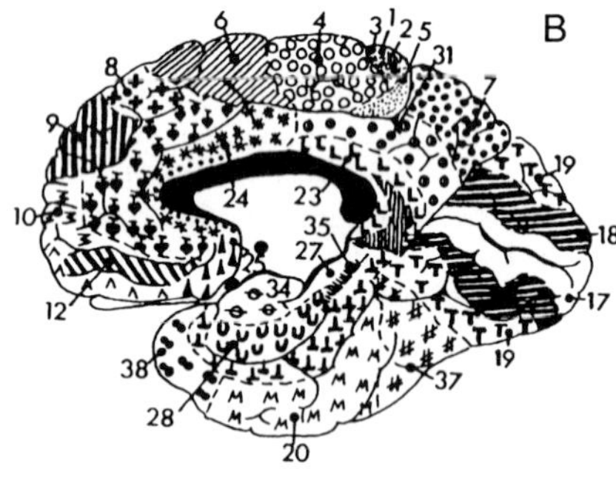

Brodmann's areas. Brodmann's areas in the human brain cortex, the left hemisphere depicted laterally (A) and the right hemisphere medially (B).

broken-winged appearance, injury-feigning; a behavioural pattern observed in many birds when a parent in a threatened situation pretends to be flightless and thus, by drawing attention to itself attempts to lure the predator away from its nest or offspring; related injury-feigning behaviour is also found in many mammals.

bronchiole (L. *bronchiolus* < Gr. *bronchion*, dim. of *bronchos* = trachea, windpipe), one of the many finer branches of the bronchi in the vertebrate lung; unlike the → trachea and bronchi, the bronchioles have no cartilages to support the wall but have a strong smooth muscle and elastic connective tissue layer, a ciliated pseudostratified epithelium lining the inner surface.

bronchus, pl. **bronchi** (L.), one of several branches of the → trachea forming the pathways to the vertebrate lung; the trachea first

brood pouch, a sac-like cavity in some invertebrates into which the parent lays its eggs or embryos; e.g. the brood pouches of isopods located at the base of the thoracic appendages.

brother-sister mating, a form of inbreeding, especially a mating system of experimental animals by which the stock reaches homozygosity in the fastest possible way.

brown algae, → Phaeophyta.

brown fat, brown adipose tissue, BAT, a highly specialized mammalian tissue that efficiently produces heat in response to cold exposure (nonshivering thermogenesis, *see* thermoregulation). BAT is found in hibernators, rodents, and many other mammals, also in humans; especially prominent it is in many newborns and cold-acclimatized mammals. In rodents the tissue also plays a role in the regulation of energy balance after eating (diet-induced thermogenesis). Brown fat is found mainly under the skin between the shoulder blades (scapulae), and around the kidney, large arteries, and trachea. The colour of BAT is typically reddish-brown. Brown fat cells contain a large population of mitochondria, and fat is located multilocularly in cells (unilocularly in white fat). The cells and surrounding arterioles and arteries are richly innervated by sympathetic nerve fibres. Heat is generated in cells through a mitochondrial proton conductance pathway, which is regulated by uncoupling protein (UCP, thermogenin), i.e. ATP is not produced and the energy is released totally as heat. Thermogenesis is evoked by noradrenaline secreted from the sympathetic nerve endings which is transmitted through β_1-receptors (*see* adrenergic receptors). Low ambient temperature results in trophic response in BAT that induces marked proliferation of mitochondria and increases uncoupling of → oxidative phosphorylation. These effects are results from selective increase in transcription of thermogenic genes in brown fat cells.

Brownian movement (*Robert Brown*, 1773—1858), the oscillatory random movement of microparticles in solutions, such as colloidal suspensions and emulsions; based on the thermal motion of molecules. *See* diffusion.

brown rot, *see* rot.

brush border, the apical surface of the absorptive epithelial cells found e.g. in the intestine and the proximal tubule of the nephron; composed of the dense covering of microvilli on the cell membrane. *See* microvillus.

Bryidae (Gr. *bryon* = moss), **true mosses**; a subclass in the class → Bryopsida (Musci) in the division → Bryophyta, in some classifications a class in the division Bryophyta; include the most generally known **mosses**. The protonema is well developed and branching, the leaves of the → gametophyte have a middle vein and the sporangium of the sporophyte has a long stalk. The number of species is ca. 13,500.

bryology (Gr. *logos* = word, discourse), a branch of science studying mosses, → Bryopsida.

Bryophyta (Gr. *phyton* = plant), **bryophytes**; a division of Plant kingdom; small nonvascular plants, have a photosynthetic → haploid gametophyte and a diploid sporophyte which develops in connection with the former and is dependent on it. The division comprises the classes → Hepaticopsida (liverworts), Anthocerotae (hornworts), and Bryopsida (Musci), i.e. mosses. In some classifications all these three classes are categorized as divisions, the class Bryopsida being named the division Bryophyta.

Bryopsida, Musci, mosses, a class in the division → Bryophyta, Plant kingdom; characteristically small, thallic, mostly terrestrial plants. The class is divided into subclasses → Sphagnidae (peat mosses), Bryidae (true mosses) and Andreaeidae. The dominating generation of mosses is a haploid, independent photosynthetic gametophyte, which has a stalk and leaves but the roots are replaced by rhizomes. The gametangia develop in the tip of the gametophyte, where also the sporophyte grows after fertilization, comprising a seta and a capsule and producing spores but being dependent on the gametophyte. A spore first grows into a protonema which further develops into a new gametophyte.

Mosses are obviously near relatives of → green algae, both having e.g. chlorophylls a and b, and starch as an energy storage. Some plants are called mosses although they belong to other plant groups; e.g. reindeer mosses are lichens, → club mosses are vascular plants, and sea mosses are algae. In some classifications Bryopsida is categorized as the division → Bryophyta.

Bryozoa (Polyzoa) (Gr. *zoon* = animal), **moss animals, bryozoans**; the former name of → Ectoprocta.

BSA, → bovine serum albumin.

BSS, → basal salt solution.

bucc(o)- (L. *bucca* = cheek), pertaining to the cheek.

buccal cavity, that portion of the mouth cavity (oral cavity) that is bounded between the teeth and cheeks.

bud, an outgrowth from the stem or body of an organism, capable of differentiating into a new individual, e.g. **1.** in anatomy of animals, a bud-like primordial part in an embryo, developing into an organ, e.g. limb bud, liver bud, tail bud, tooth bud, etc.; also similar formation in → regeneration; **2.** → taste bud; **3.** a lateral swelling on the surface of some primitive animals, such as some cnidarians, which develops to a new individual; *see* budding; **4.** in botany, a small shoot with immature leaves, protected by scale-like bracts; buds may be terminal or axillary, the latter developing into stem branches or flower shoots. The buds are also important wintering organs, which rapidly begin their growth in favourable environmental conditions in spring. *Cf.* bulbil, Raunkiaer's life forms.

budding, 1. a form of asexual reproduction occurring in yeasts, plants, and primitive animals, e.g. cnidarians; the buds formed on the surface of a parent individual are able to develop into new individuals and to stay permanently attached to the parent, or to detach from it thus developing into an independent solitary individual and beginning a new colony; **2.** departure of viruses from a host cell.

bud mutation, a → mutation in a plant, the effect of which appears in a part of the shoot developing from a mutated → bud; can be utilized in plant breeding, e.g. many cultivars of fruit trees are originally bud mutations.

buffer, anything that alleviates the effect, force, or reaction of interacting agents; e.g. **1.** pH buffer; a substance that reduces (buffers) acidity changes in a solution (→ pH). Many substances with buffering properties occur in organisms, e.g. bicarbonate (bicarbonate ion and carbonic acid), HCO_3^--H_2CO_3), phosphates (hydrogen and dihydrogen phosphates, HPO_4^{2-}-$H_2PO_4^-$), proteins, and amino acids. Organisms keep the pH of cells and interstitial fluid as constant as possible with the aid of these buffers. Many natural and artificial buffers are used in chemical and biological studies, e.g. in cell culture media, enzyme solution, and physiological salt solutions; *see* Hepes, Tris; **2.** any substance that controls the quantity of a physiologically active substance in cells and tissues; e.g. some proteins which control the amounts of free calcium, iron, copper, or hormones; a common artificial calcium buffer used in physiological and biochemical studies is → EGTA.

bufogenin (< *Bufo* = the genus of toads), a toxic steroid genin, $C_{24}H_{34}O_5$, secreted by the skin glands of a toad (*Bufo agua*); similar toxic genins secreted by toads are e.g. bufotoxin, bufotenin(e), and bufalin. They have inhibitory effects on ion transport mechanisms of the cell, especially Na^+,K^+-ATPase (*see* adenosine triphosphatases).

bulb (L. *bulbus* = onion, bulb), **1.** any fusiform or globular anatomical structure in animals, e.g. bulb of the heart (*see* bulbus arteriosus), dental bulb, bulb of the eye, hair bulb, etc; **2.** a plant shoot staying at a bud stage throughout its life. The bulb has a shoot-like part and fleshy, scale-like leaves, the internodes being very short. Bulbs are reproductive organs and energy stores at the same time, being typical of → monocotyledons such as tulips and lilies.

bulbar, 1. pertaining to a bulb or a bulb-like structure; **2.** relating to the hindbrain, i.e. → rhombencephalon of a vertebrate embryo.

bulbil, a plant bud, mostly an axillary bud, that separates from the plant and develops into a new individual, e.g. in some lilies; serves also as a wintering organ. The corresponding structure of many water plants is **hibernaculum**.

bulbourethral gland (L. *glandula bulbourethralis*), a pair of small glands near the → prostate in male mammals, producing mucoid secretion discharged through the duct into the urethra at the beginning of copulation. Also called Cowper's gland, anteroprostate, antiprostate.

bulbus, pl. **bulbi** (L. = plant bulb), → bulb; a bulb-like anatomical structure.

bulbus arteriosus, 1. bulb of the heart (*bulbus cordis*); the foremost of the three parts of the developing heart in vertebrate embryos; **2.** the related part from which the aorta begins in the heart of cyclostomes and bony fishes; the corresponding structure of cartilaginous fishes is called **conus arteriosus**.

bulla, pl. **bullae** (L.), a large vesicle or a bubble-like anatomical or pathological structure; e.g. **auditory bulla**, a bone around the middle ear of placental mammals; comprises the tympanic bone which forms the ring around the eardrum, and the cartilaginous part (endotympanic cartilage). Together with the periotic bone, the auditory bulla forms a compound temporal bone in many mammals. *Adj.* **bullous**.

bullheads, → Scorpaeniformes.

bundle scar, the scar from the vascular bundle in the area of the → leaf scar, seen after leaf abscission.

bundle sheath, a sheath around some types of → vascular bundles in plants; e.g. around the closed collateral vascular bundle common in many monocotyledons. The bundle sheaths normally protect the bundles, but they are especially important in → C_4 plants such as in many grasses, in which their large cells are active in → photosynthesis.

Bunsen burner (*Robert Bunsen*, 1811—1899), a gas burner having a non-luminous hot flame, used in laboratories.

buoyancy, the property based on the → buoyant force; in living organisms the property to float at a certain depth in various fresh, brackish, and salt waters. It includes structural (passive) properties and physiological (active) processes by which organisms can adjust their density and floating properties, e.g. they may have a large relative surface with different types of membranous and filamentous structures, or they may have → air bladders and other hydrostatic organs. Some organisms can change their tissue density by reducing or increasing heavy elements, such as salts (sodium, calcium, magnesium, etc.), or lighter substances , such as fats, oils, or water.

buoyant force, the upward thrust affecting a body immersed in a liquid; according to *Archimedes'* (287—212 B.C.) principle, the loss in weight of the body is equal to the weight of the liquid displaced.

burette, a graduated, calibrated glass tube with a tap, used in laboratories to dose solutions, particularly in → titration.

Burgess Shale fauna, a very diverse fossil fauna in the deposits of Canadian Rockies in British Columbia from the Middle Cambrian about 540 million years ago. The Burgess fauna comprises well preserved extinct invertebrates (e.g. the first cephalocordates) which are classified into many present but also into several extinct phyla. In all, the fauna shows very high diversity among the ancient animals.

bursa, pl. **bursae,** Am. also **bursas** (L. = bag, burse), a sac-like anatomical structure filled with some fluid, usually in areas subjected to friction, e.g. **1. bursa Fabricii**, the bursa of Fabricius, a lymphatic gland on the postero-dorsal wall of the cloaca in birds; produces mature B lymphocytes; *see* lymphocytes; **2.** bursae in the joints of the vertebrate skeleton, as **bursa subcutanea calcanea, b. s. infrapatellaris, b. s. olecrani**, etc.; **3. bursa copulatrix**, a sac-like structure in many insects where spermatophores are deposited during insemination; **4.** one of the ten **internal sacs** in ophiuroids, formed from invagination of the oral surface and functioning in gas exchange.

bursicon, a neurohormone (small protein) in insects secreted by neurosecretory cells in the brain and nerve cord; tans and hardens the cuticle after moulting.

bush, a perennial wooden plant branching near the ground, forming no trunk like trees. *Syn.* shrub.

butane, C_4H_{10}. *See* alkanes.

butanoic acid, → butyric acid.

butanol, butyl alcohol, a 4-carbon alcohol produced in small quantities in alcoholic fermentation; used as an organic solvent in biological and chemical studies as well as in industry. Its isomeric types are: 1-butanol i.e. n-butanol, $CH_3(CH_2)_2CH_2OH$, 2-butanol, i.e. secondary butanol, $CH_3CH_2CH(OH)CH_3$, isobutanol, i.e. 2-methyl-1-propanol, $(CH_3)_2CH$-CH_2OH, and tertiary butanol, $(CH_3)_3COH$. *See* alcohol.

butene, C_4H_8. *See* alkenes.

butterflies, → Lepidoptera.

buttress root, a flat adventitious root structure typical of big tropical trees, often reaching the height of 2-4 m above ground, giving support for the trees.

butyl alcohol, → butanol.

butyric acid, butanoic acid; a saturated four-carbon fatty acid, $CH_3CH_2CH_2COOH$, which is an end product of the carbohydrate fermentation of microorganisms and thus occurs also in the alimentary canal and in the sweat of mammals. Butyric acid is converted by epithelial cells of the intestine into β-hydroxybutyric acid which enters the blood stream; it is processed in the cellular metabolism and may have special functions in tissues, as has → gamma-aminobutyric acid (GABA) in the nervous tissue. In butter, butyric acid occurs as glycerol ester and the typical unpleasant odour of free butyric acid forms when butter turns rancid by microbial action. Salts and esters are **butyrates** (butanoates).

BX-C, → bithorax complex.

byssus, pl. **byssi,** also **byssuses** (L. < Gr. *byssos* = fine flax, linen), a silk-like tuft of threads secreted by a gland (**byssus gland**, byssal gland) in the reduced foot of some bivalve molluscs by which they attach to rocks.

C

C, symbol for **1.** carbon; **2.** cysteine; **3.** cytosine and cytidine; **4.** coulomb; **5.** complement (in immunology).

C, symbol for **1.** capacitance; **2.** heat capacity; C_p at constant pressure, C_v at constant volume.

c, centi- (L. *centum* = hundred), pertaining to hundred, or a hundredth part.

c, symbol for **1.** concentration; **2.** the velocity of light, 299,792.5 km/s.

°C, degree centigrade.

CAAT box (C < cytosine, A < adenine, T < thymine), a part of a conserved DNA sequence located upstream (about —75 base pairs) of a gene from the transcription initiation site in eukaryotic organisms. It is a part of the → promoter and recognizes RNA polymerase II and a large group of → transcription factors.

cac(o)- (Gr. *kakos* = bad), denoting bad or ill.

cachectin, a hormone-like polypeptide produced by endotoxin-activated mammalian macrophages; has an effect e.g. on fat mobilization from stores and causes lysis of bone tissue, and of tumour cells, later proved to be the same substance as the → tumour necrosis factor (TNF-α).

cachexia (Gr. *hexis* = habit), a general bad health due to a chronic disease, such as cancer or malnutrition.

cacodylic acid (Gr. *kakodes* = bad-smelling), dimethylarsinic acid; its salts, **cacodylates**, have been used as a pH buffer in electron microscopy and in medicine.

cadaver (L. *cadere* = to perish, fall), a dead body, corpse. *Adj.* **cadaveric**, pertaining to a cadaver; **cadaverous**, resembling a cadaver.

cadaverine, pentamethylenediamine, a bad-smelling nitrogenous base produced by bacterial action when lysine is decarboxylated in decaying protein material.

cadherins (c < cell, L. *adhaerere* = to cling together), cell-surface proteins belonging to **adhesion proteins**; include many different glycoproteins in various junctions between animal cells, such as **E-cadherin** (uvomorulin) e.g. in belt → desmosomes of epithelial cells, **P-cadherin** e.g. in heart, lung, and intestine, and **N-cadherin** in the nervous system, lung, and heart. The cadherin molecule comprises 720 to 750 amino acid residues and its function is calcium-dependent. Cadherins mediate the homophilic cell to cell interaction binding to a similar molecule on an adjacent cell; e.g. E-cadherin binds only to another E-cadherin. Cadherins are indirectly linked to the → cytoskeleton of the cells, hence the cytoplasmic parts of the molecules are bound to its structural protein, catenin. Cadherins are important for cell differentiation and tissue morphogenesis.

cadmium (Gr. *kadmeia* = zinc mineral), a metallic element, symbol Cd, atomic weight 112.41. Cadmium is toxic to cells and after being released into nature, has become one of the worst environmental pollutants. It is taken up by plants and then accumulated in food chains of animals, particularly in high quantities in carnivores. Cadmium is removed from soil very slowly. It occurs in minerals together with some other metals such as copper and zinc, and spreads to the environment within wastes from quarries, and metal and dye industries.

Symptoms caused by cadmium in man were first observed in Japan where an unknown disease with aching bones occurred, called **itai-itai**. Cadmium disturbs calcium metabolism resulting in softening of bones or osteomalacia, and causes e.g. anaemia. In birds, cadmium also causes the thinning of egg shells, and this allows developing embryos to dry. In many biochemical and physiological functions, primary effects result from cadmium binding to -SH groups in protein molecules.

caducous, (L. *cadere* = to fall), **1.** in zoology, pertaining to an animal subject to shedding; **2.** in botany, tending to fall; pertaining e.g. to leaves, flowers, or sepals dropping off early.

caec(i)-, caeco-, Am. **cec(i)-, ceco-** (L. *caecus* = blind), denoting relationship to the → caecum (blind gut). *Adj.* **caecal, cecal**.

caecilians, → Gymnophiona.

caecotrophy, Am. **cecotrophy** (L. *trophe* = nourishment), the property of an animal to use its own caecal microorganisms for nourishment by eating special excrements; e.g. in the digestive tract of hares and rabbits two kinds of faeces are formed, brown and hard, and green and soft. The latter are formed by the activity of microorganisms during the night and are rich in proteins, vitamins, minerals, and water. The animal uses these excrements by eating them in the morning directly from the anus. *See* coprophagy.

caecum, cecum, also **coecum,** pl. **caeca, ceca, coeca, 1.** (L. *intestinum caecum*), **blind gut**; the first part of the large intestine in mammals, and a related structure of the less differentiated intestine in many other vertebrates, also found in many invertebrates such as leeches and arthropods; forms a blind pouch with very effective function of microorganisms. The caecum is exceptionally long and large in herbivorous animals, but reduced in carnivores and omnivores; in man, it forms distally the *appendix vermiformis*; **2.** cupular sac (L. *caecum cupulare*) and vestibular sac (*caecum vestibulare*), i.e. the blind extremities of the cochlear and vestibular ducts in the inner ear of tetrapod vertebrates.

caen(o)-, Am. usually **cen(o)-, 1.** (Gr. *koinos* = shared in common), pertaining to a common characteristic; also **coen(o)-, coin(o)-, koin(o)-**; **2.** (Gr. *kainos* = new), denoting new, fresh; also **cain(o)-, kain(o)-**.

Caenorhabditis elegans, a small soil nematodean worm which is an important experimental animal in genetics and developmental biology especially because it has a constant number (959) of somatic cells. The lineage of these cells has been analysed in every detail. Only hermaphrodites and males but no females of this species exist.

Caenozoic, Cainozoic, Cenozoic, also written with initial K (Gr. *kainos* = new, *zoos* = living), the era after Mesozoic, characterized by mammals and birds; began about 66 million years ago lasting to the present. The era is divided into Tertiary and Quaternary periods.

caesium, Am. **cesium** (L. *caesius* = blue), symbol Cs, atomic weight 132.91, a highly reactive silvery-white metal that belongs to the same element group as sodium. Caesium is not biologically active, but can incorporate into organisms as well as its dangerous radioactive isotope ^{137}Cs (half-life about 30 years) which has been deposited accidentally from nuclear power stations to nature. Caesium is a dense element and its salt, CsCl, is used as a gradient substance in → density gradient centrifugation.

caffeine (L. *Coffea* = the genus of coffee tree), a white, odourless, bitter alkaloid, 1,3,7,-trimethylxanthine ($C_8H_{10}N_4O_2$), obtained from seeds of the coffee tree, *Coffea arabica*. It activates the sympathetic nervous system of vertebrates and thus e.g. stimulates the brain and cardiac functions, and increases blood pressure and the excretion of urine. In high doses caffeine enhances directly the contraction activity of muscle cells by liberating calcium from the sarcoplasmic reticulum. Tea alkaloid, **thein(e)**, obtained from dried leaves of *Thea sinensis*, is the isomer of caffeine. The properties and effects of the cacao alkaloid, **theobromine**, resemble those of caffeine.

Cainozoic, → Caenozoic.

calc(i)-, calco-, 1. (L. *calx* = lime(stone)), pertaining to calcium, or calcium salts, such as lime or chalk; **2.** (L. *calx* = heel), denoting relationship to the heel.

calcaemia, Am. **calcemia** (Gr. *haima* = blood), a hypernormal calcium concentration in the blood; hypercalcaemia.

calcaneus, pl. **calcanei** (L. = heel), **1.** the largest tarsal bone of mammals, forming the prominence of the heel; **2.** the corresponding bone of other tetrapod vertebrates, also called **calcaneum** (pl. calcanea).

calcar, pl. **calcaria** (L. = spur), a small projection from a bone or other anatomical structure.

Calcarea (L. *calcarius* = limy), **calcareous sponges**; a class in the phylum Porifera including marine species which have a calcareous skeleton, the body surface is bristly; familiar genera are e.g. *Leucosolenia* and *Scypha*.

calcareous, pertaining to lime or calcium, or a calcic material; chalky.

calcareous sponges, → Calcarea.

calcemia, → calcaemia.

calcic, pertaining to lime or calcium, rich in calcium.

calciferol (L. *ferre* = to bear), → vitamin D.

calciferous glands, glands on both sides of the oesophagus in some annelids, e.g. earthworm, secreting calcium salts to neutralize an acid food.

calcification, the deposition of calcareous material; **1.** the deposition of calcium salts within a cartilage or connective tissue in the process of ossification, normal or abnormal; **2.** the accumulation of calcium and magnesium salts in the soil; **3.** a calcified formation in anatomy or geology. *Adj.*

calcific, forming lime. *Verb* to **calcify**, to deposit calcium salts.

calciphilic (Gr. *philein* = to love), pertaining to a plant species (calciphile, calcicole) which prefers rich calcium in its environment. Calcium itself is a very common element in the ground and too scarce hardly anywhere, but its richness decreases the acidity of the soil and makes other nutrients easier available to plants. *Cf.* calciphobic.

calciphobic (Gr. *phobos* = fear), pertaining to a plant species (calciphobe, calcifuge) which avoids calcium and prefers acidity in its environment; e.g. *Rhododendron*. *Cf.* calciphilic.

calcite, calcspar; a naturally occurring mineral consisting of crystalline calcium carbonate, $CaCO_3$. Calcite is found in many forms, such as chalk, limestone, and marble. In basic calcite-rich soils, the plants suffer from phosphorus deficiency because phosphorus is bound to calcium forming nearly insoluble calcium phosphates. *See* calcium, apatite.

calcitonin (Gr. *tonos* = tone), a polypeptide hormone of vertebrates (with 32 amino acid residues), produced mainly by parafollicular cells of the thyroid gland of mammals, and from the → ultimobranchial organ in other vertebrates. Calcitonin is released into the blood as a response to hypercalcaemia. The hormone decreases calcium release from the bones and facilitates the deposition of calcium and phosphate in the bone. Thus calcitonin lowers the calcium level in the plasma. Calcitonin acts as an antagonist to the parathyroid hormone. The calcitonin released from the thyroid gland is often called **thryro-calcitonin**. *See* calcium, parathyroid gland.

calcium, an alkaline earth metal, chemical symbol Ca, atomic weight 40.08. Its salts such as → apatite accumulate in supportive tissues of animals (e.g. bone, cartilage, and other connective tissues, and chitin) making them hard. Calcium ion (Ca^{2+}) is a stabilizing factor in cellular membranes and it controls the activity of many enzymes and cellular movements. Calcium also controls the secretion of glandular cells and acts in many hormonal target cells as a → second messenger; in animals it controls the release of transmitter substances in → synapses. In the control of intracellular action the calcium-calmodulin complex (*see* calmodulin) plays an essential role. For the regulatory role of calcium in cells, it is important that the **Ca^{2+} pump** of the cell membrane keeps the cytoplasmic Ca^{2+} concentration mostly at a very low level. In cellular activation by calcium, the concentration inside the cell is rapidly increased by calcium influx through opened **Ca^{2+} channels**; this is caused e.g. by extracellular signals or physical factors. In activation, calcium ions may also be released from intracellular organelles, such as the → sarcoplastic reticulum in muscle cells (*see* muscle contraction). Plants take calcium from soil as calcium ions (Ca^{2+}), the cell walls being rich in it. Calcium is used as a fertilizer and to lower the acidity of soil. *See* cell membrane transport.

calix, pl. **calices** (Gr. *kalyx* = a cup of flower), a funnel-shaped structure, especially the → renal calix in the kidney (metanephros). *Syn.* calyx.

callose (L. *callus* = callus, hard skin), a plant cell-wall polysaccharide (β-1,3-linked glucose polymer) that plugs the pores of non-functional sieve elements and is synthesized after wounding in parenchyma tissue.

callosum, pl. **callosa,** → corpus callosum.

callous, pertaining to a hardened tissue, e.g. the skin; having → callus.

callus, pl. **calli, calluses** (L. = hard skin), **1.** a localized thickening of the horny layer of the vertebrate epidermis as a result of pressure or friction; **2.** new tissue formed at a bone fracture, i.e. fibrous tissue and cartilage, ultimately developing into hard bone; **3.** an undifferentiated plant tissue protecting a wounded area; also formed in → cell and tissue cultures in which the callus can be activated to divide and differentiate e.g. by hormonal treatment into new plant individuals. This is a rapid way to propagate plants.

calmodulin, calcium-binding protein in all eukaryotic cells, regulating calcium actions by binding 1 to 4 calcium ions (Ca^{2+}). The calcium-calmodulin complex mediates most calcium-sensitive processes by activating cellular enzymes such as → phospho-diesterase, phospholipase A_2, phosphorylase kinases, calcium-dependent protein kinases, enzymes in calcium ion transport, contractile processes, and neurotransmission. It has also an effect on cellular → microfilaments, as well as on → microtubules and thereby e.g.

on the arrangement of the → mitotic spindle in cell division.

calomel, mercurous chloride (Hg_2Cl_2); used especially in reference electrodes in electrophysiology. *See* mercury.

calorie, also **calory, cal** (L. *calor* = heat), an out-of-date unit of heat or energy, displaced by joule. *Adj.* **caloric**, pertaining to calories, or heat. *See* Appendix 1.

calorimeter (L. *metron* = measure), an apparatus for measuring the quantity of heat liberated in a system, e.g. the quantity of heat produced by an animal in a given time, or the quantity of heat liberated in a chemical reaction, e.g. in burning a nutriment; the method is called **calorimetry**.

calvaria (L. = skull), **1.** the top of the head; **2.** the bones on the roof of a skull.

Calvin cycle (*Melvin Calvin*, b. 1911), the dark reactions of → photosynthesis. *Syn.* Calvin—Benson cycle.

calx, pl. **calces** (L.), **1.** lime(stone); **2.** heel.

calyptopsis, pl. **calyptopses** (Gr. *kalyptos* = covered, hidden, *opsis* = sight), a larval stage after the metanauplius in the life cycle of krill crustaceans; has a long covering test, paired eyes, and abdominal legs.

secondary tissues; two types exist, the vascular cambium and cork cambium. The **vascular cambium** is situated in the root and stem: its **fusiform initials** form secondary → xylem inside and secondary → phloem outside, and the **ray initials** give rise to parenchymatous → ray cells. The **cork cambium** is located near the surface of roots and shoots and forms cork cells to the → **periderm**, which replaces the epidermis in older structures. *See* meristem.

Cambrian (L. *Cambria* = Wales), the oldest period of the Palaeozoic era 570 to 505 million years ago when many invertebrate phyla originated and the speciation of marine invertebrates increased quickly; e.g. trilobites, graptolites, and marine algae occurred commonly in oceans. *Adj.* **Cambrian**.

camels, → Tylopoda.

camera (L. = chamber), in anatomy, a closed structure, chamber or ventricle; e.g. *camera oculi*, the chamber of the eye.

camera eye (L. *oculus*), a type of the eye found in all vertebrates and some invertebrates, e.g. cephalopods. The eyeball is composed of three layers (coats) with several sublayers: the outermost coat, **sclera**, with the

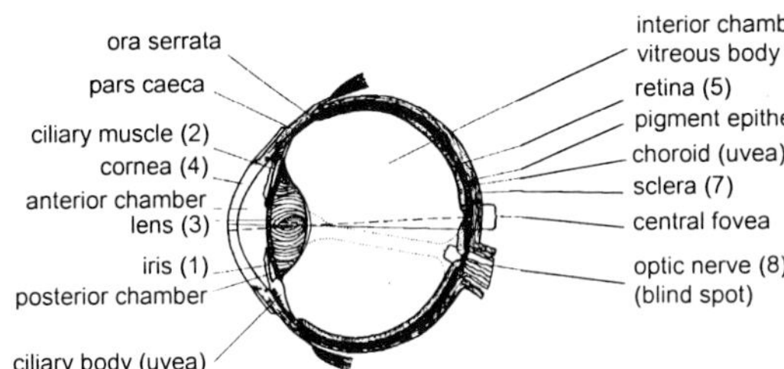

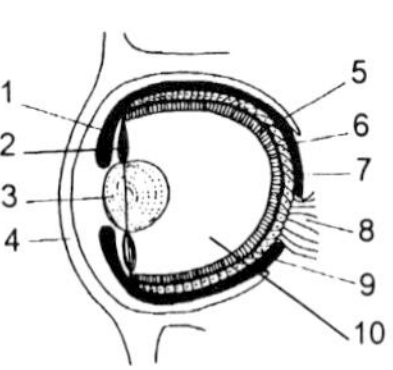

Camera eye. Main structures of the eyes of vertebrates and cephalopods.

calyptra (L. < Gr. *kalyptra* = veil), the hood protecting the developing sporophyte and often also the spore capsule in mosses, → Bryophyta.

calyptrogen (Gr. *gennan* = to produce), *see* derivative meristems.

calyx, pl. **calyces,** Am. also **calyxes** (L. < Gr. *kalyx* = a cup of flower), **1.** → calix; **2.** a funnel-shaped structure, as the outermost whorl of the flower formed by sepals; protects the flower during its development.

cambium, pl. **cambiums** or **cambia** (L. *cambiare* = to exchange), a thin cylinder-like layer of meristematic cells in vascular plants with actively dividing cells, producing

transparent area, **cornea**, in front; the middle layer is the **choroid**, composed of connective tissue with dense network of blood vessels, and the innermost coat, → **retina**, is derived from the brain and contains numerous photoreceptor cells in the photosensitive back half of the eye. The pigmented, circular membrane, → **iris**, behind the cornea regulates the diameter of the **pupil** through which light passes to the eye. The refracting media comprise the cornea, aqueous humour, **lens**, and vitreous humour. The lens is connected with zonular fibres to the folded structure of the choroid, called **ciliary body**, with ciliary muscles for accommodation of the lens. *See* photoreception. *Cf.* compound eye.

camerostome (Gr. *stoma* = mouth), a hollow anterior part of the → podosoma of Acarina (mites and ticks) to which the gnathosoma is attached.

cAMP, → cyclic AMP.

camphor (L. *camphora*), a translucent, crystalline, odorous terpenoid ketone, $C_{10}H_{16}O$, obtained from the bark and wood of the camphor tree, *Cinnamomum camphora*, or from a herbaceous camphor plant; used in the manufacture of celluloid and explosives, and in medicine.

CAM, cell adhesion molecule. *See* cell adhesion.

CAM plants (< **C**rassulacean **a**cid **m**etabolism), plants, such as *Crassula*, *Kalanchoë*, and *Agave*, which grow in dry areas and absorb carbon dioxide (CO_2) at night when their stomata are open. CO_2 is bound into phosphoenolpyruvate in the reaction catalysed by phosphoenolpyruvate carboxylase (PEP carboxylase), when carbon-4 organic acids (C-4 acids), i.e. oxaloacetate and malate, are formed. CO_2 is liberated from malate in the light and is then assimilated to carbohydrates in the Calvin cycle. *See* photosynthesis, C_4 plants.

campylotropous ovule (Gr. *kampylos* = twisted, *trope* = turning), a type of plant ovule. *See* ovule.

Canada balsam, a liquid resin, oleoresin, which is produced by some North-American spruce species (especially *Abies balsamea*). The refracting factor of Canada balsam is the same as that of glass (1.53), and therefore it is used in optical and microscopical work, e.g. as a transparent cement for mounting microscopic specimens, or in optical instruments. Also called Canada turpentine.

canaliculus, pl. **canaliculi** (L. dim. of *canalis* = canal, channel), a narrow tubular channel in intracellular or intercellular structures of cells and organs, as **apical canaliculi** in the epithelial cells of renal tubules, or **biliary canaliculi** between liver cells.

canalization, 1. the formation of canals or channels in a tissue; **2.** the property of developmental pathways to achieve a standard phenotype in spite of genetic or environmental disturbances, i.e. the genetic buffering of embryonic development against environmental changes so that the development proceeds in a certain direction, i.e. like in a canal.

canalizing selection, a form of stabilizing → selection that favours those genotypes whose development is well canalized, i.e. buffered against the changes of environmental disturbances. Like other forms of stabilizing selection, canalizing selection protects the → adaptive norm of a population.

canavanine (< *Canavalia ensiformis* = jack bean), an amino acid, $H_2NC(NH)NHOCH_2$-$CH_2CH(NH_2)COOH$, not present in proteins; forms a nitrogen store in some beans, toxic e.g. to many mammals.

cancer (L. < Gr. *karkinos* = crab, cancer), **1.** in animals, the malignant and invasive type of a cellular tumour the natural course of which is fatal. Cancer cells are neoplastic and characterized by various types of structural changes in the genetic material causing the loss of cellular differentiation and cell adhesion, usually the formation of → metastases, and unlimited growth with continuous cell divisions. The genetic process begins when a **proto-oncogene** is released from the normal genetic control by gene or chromosome mutation, and changes into an **oncogene** resulting in neoplastic cellular properties. Cancers derived from animal tissues are divided into two main categories: → carcinomas and → sarcomas. The factors inducing cancer are called → **carcinogens**; **2.** in plants, tumour formation called → crown gall, caused by the bacterium *Agrobacterium tumefaciens*.

candela (L. = candle), the unit of luminous intensity; *see* Appendix 1.

canine (L. *canis* = dog), **1.** pertaining to a dog; **2.** any animal of the Canidae (dog family); **3.** → canine tooth.

canine tooth (L. *dens caninus*), dog-tooth of mammals; primarily a long and sharp tooth, a pair in both jaws, one on each side between the incisors and premolars. Canine teeth are well-developed in carnivorous mammals; they may be enlarged to form the tusks, such as in the extinct sabre-toothed tigers, or they may be reduced or totally lacking, as e.g. in many ungulates and rodents. In human anatomy called also cuspid, cuspidate, or eyetooth.

cannabis (Gr. *kannabis* = hemp), the dried flowering tops of the pistillate plants of *Cannabis sativa* (Moraceae); its crude preparation is → marihuana.

cannibalism, a practice of feeding on conspecific individuals; **cannibal**, an animal eating individuals of its own species. Cannibalism is rather common in overdense populations, e.g. in dense gull colonies where adults eat conspecific chicks.

cannula, also **canula,** pl. **can(n)ulae** or **can(n)ulas** (L. dim. of *canna* = reed), a tube for insertion into a body cavity, or a duct, for infusing or drawing fluid; also a tube for perfusion of an isolated organ.

caoutchouc (Span. *caucho*), natural → rubber; a hydrocarbon, $(C_5H_8)_n$, obtained from the latex of many tropical plants, the most important being *Hevea* species. *Syn.* India rubber.

CAP, → catabolite activating protein.

capacitance (L. *capax* = able to contain), symbol *C*, in physics the quantity of electric charge stored in a body. *See* Appendix 1.

capacitation, the process occurring in the female genital tract of at least birds and mammals; certain inhibitive proteins are removed from the surface of spermatozoa, thus rendering them capable of fertilizing the ovum; includes the rupture of the acrosomal cap of the spermatozoon which releases enzymes functioning in penetration at fertilization.

capillaries, sing. **capillary** (L. *capillus* = hair), minute tubes; in anatomy, capillary vessels, such as **1. blood capillaries** which are the smallest blood vessels forming a capillary network between arterioles and venules. In vertebrates, they are usually 0.1 to 1 mm long and 5 to 20 μm wide vessels and walled by a single layer of flat endothelial cells with the → basement membrane under it; precapillary sphincters regulate the blood flow to the capillaries. In → capillary circulation, the wall permits exchange of water, respiratory gases, nutrients, and metabolic wastes between the blood and tissue fluid. The capillary wall may contain pores (fenestrated capillaries), as in many endocrine glands or in renal glomeruli. **Sinusoidal capillaries** found e.g. in the liver and spleen in vertebrates, are discontinuous capillaries which are wider than ordinary capillaries and lined with a more fenestrated type of endothelium;

2. lymph capillaries, the smallest of the lymphatic vessels in vertebrates; are blindly beginning capillaries, their histological structure resembling closely that of blood capillaries; however, the basement membrane is poorly developed and thus more permeable, and the lumen varies more in size;

3. bile capillaries (bile canaliculi), minute intercellular channels in the liver as a part of biliary canaliculi; about 1 μm in diameter.

capillary, 1. denoting relationship to a hair, resembling a hair; **2.** pertaining to → capillaries, capillary vessels; as a noun, a capillary vessel itself.

capillarity, capillary action, the rise of water or other liquids in a thin capillary, caused by the **capillary force** due to the surface tension and adhesion between the capillary wall and the liquid molecules. The phenomenon is an important factor in the water transport in soil and in → xylem vessels in plants. It is based on the adhesion between liquid molecules and capillary walls and on the cohesion between liquid molecules.

capillary circulation, the blood flow from arterioles to the venules through → capillaries; usually each arteriole branches into a capillary net then joining into one venule. The flow is regulated by precapillary smooth muscle sphincters, by the hydrostatic pressure of the blood, and by the resistance of the capillary net. When hydrostatic pressure (**filtration pressure**) exceeds the colloid osmotic pressure (**oncotic pressure**) due to plasma proteins and other colloids, it causes filtration of some plasma through capillary walls into the surrounding tissue. Most of the filtered plasma, however, returns immediately into the same capillary net; this occurs when the oncotic pressure exceeds the filtration pressure. Thus the **hydrostatic pressure gradient** between the capillary fluid and interstitial fluid, and the **osmotic pressure gradient** across the capillary wall determine the movement of the fluids. Some tissue fluid returns into the blood through the lymphatic system. The capillary circulation serves in the exchange of respiratory gases, nutrients, and metabolic wastes between the blood and tissue fluid; the circulation is controlled by the → autonomic nervous system, and many hormones and other local factors (e.g. oxygen). *See* capillaries, blood circulation.

capillary electrophoresis, *see* electrophoresis.

capillitium, pl. **capillitia** (L.), sterile threads among the spores in sporangia of some → Myxomycetes, or in the fruiting body of some fungi.

capitulum, pl. **capitula** (L. dim. of *caput* = head), **1.** a small anterior projection carrying the mouthparts of the mite and tick (Acari); *syn.* gnathosome; **2.** a major part of the body of a stalked barnacle (Cirripedia) which is attached to the substratum by a long stalk (peduncle); **3.** one of the two projections in the ribs of primitive tetrapods; **4.** an → inflorescence with numerous separate, small flowers (florets) on a usually flattened receptacle, often accompanied by an → involucre; the capitulum is typical of families Asteraceae and Cichoriaceae.

capping, 1. in genetics, the joining of the → cap structure to the 5' end of the → messenger RNA molecule of eukaryotes after transcription; **2.** in immunology, the grouping of → antibodies on the cell membrane to a certain region caused by binding of → antigens; in the grouping phase a cap-like appearance can be seen in immunohistochemical staining.

capric acid, n-decanoic acid (L. *capra* = she-goat), a solid fatty acid, $CH_3(CH_2)_8COOH$, occurring as glyceride in coconut and palm oils, but also in some animal fats, as e.g. in goat's and cow's milk; its esters are used for artificial fruit flavouring and perfumes. Related oily fatty acids are **caproic acid** (n-hexanoic acid), $CH_3(CH_2)_4COOH$, and **caprylic acid** (octanoic acid), $CH_3(CH_2)_6$-COOH, and its alcohol, **capryl alcohol**, i.e. → octanol.

Caprimulgiformes (L. *caprimulgus* = goatsucker), an order of nocturnal or twilight birds catching flying insects by their wide, bristler-fringed mouths. The order includes 109 species in five families, e.g. **frogmouths** (Podargidae), **potoos** (Nyctibiidae), **nightjars** and **goatsuckers** (Caprimulgidae). The birds are closely related to owls.

capsaicin (< L. *Capsicum*), $C_{18}H_{27}NO_3$, a pungent substance in fruits of many Solanaceae species, isolated from paprika and cayenne; a powerful irritant causing intense pain in the throat. In animal tissues, capsaicin probably causes the production of → substance P which has an effect on thermoregulation, which is why chili peppers cause sweating.

capsid, 1. (L. *capsa* = box), the coat protecting the nucleic acid of a virus, composed of several structural proteins, **capsomeres**; **2.** a common name for bugs in the family Miridae (*syn.* Capsidae); many species are severe pests of cultivated plants, some species also vectors of plant diseases.

capsomere, also **capsomer,** one of the protein units forming the coat, capsid, of a → virus.

cap structure, the methylated guanosine triphosphate structure at the 5' end of the → messenger RNA of eukaryotic organisms; linked to the messenger RNA after transcription. The cap structure protects RNA from the action of exonucleases and acts as a ribosome binding site. *See* capping.

capsule (L. *capsula*, dim. of *capsa* = case, box), the structure enclosing an organ or a part of an organism, e.g. **1.** in animal anatomy, a soft or hard, membranous, fibrous, or cartilaginous structure enclosing an organ or a body part, such as the **articular capsule** (*capsula articularis*) in the joint of vertebrates, or the **renal capsule** (*capsula fibrosa renis*), the fibrous capsule around the kidney; various capsules also in invertebrate anatomy; **2.** in plants, a many-seeded dehiscent → fruit formed from two or more carpels; different types, separated according to the way they open, are **septicidal capsule** (opens along the margins of the carpels), **loculized capsule** (opens along the midrib of each carpel), **poricidal capsule** (opens through holes in the structure), and **pyxis** (pyxidium), in which a part of the wall opens as a lid; **3.** in liverworts and mosses, the structure containing spores; **4.** the gelatinous envelope surrounding the cell wall in many bacteria.

captacula (L. *capere* = to capture), tentacle-like extensions with sensory receptors around the mouth of tusk shells (Scaphopoda); serves in catching food.

Captorhinida (Gr. *kaptein* = to seize, *rhis*, gen. *rhinos* = nose), **captorhinids**; an extinct order of anapsid reptiles which lived during Carboniferous and early Permian, and became extinct in Triassic; included the most primitive reptiles with some amphibian features. *Syn.* Cotylosauria.

capture-recapture methods, in animal ecology, the methods used in estimation of population densities; most of them are derived from the → **Lincoln index** which, however, excludes immigration, emigration, and selection. The technique is based on the capture, marking, release, and one or more recapture occasions (or observations) of individuals. The population density is estimated from the relation between the numbers of marked animals at different catching times and using the equation N = Mn/m, where N = estimated population size, M = number of marked and released animals in the first catch, n = number of animals caught in the second catch, and m = number of previously marked animals in the second catch. Other related methods are e.g. the **Manly—Parrin's**, and **Jolly—Seber—Comarch's methods**, and **Bailey's three-catching method**. *Cf.* removal methods.

caput, pl. **capita** (L.), the head or upper part of an organism or organ; end; e.g. *caput femoris*, the proximal end of the femur, thigh bone.

carapace (Sp. *carapacho* = covering, shield), carapax; **1.** a hard shield covering the body of several crustaceans, composed of calcareous chitin; it may cover the whole body as in ostracods, the head and thoracic segments as in crayfish, or it may be absent as in copepods; **2.** the dorsal shield of turtles and tortoises (Chelonia), composed of two layers: the outer horny layer of keratin and an inner bony layer; the carapace is fused to thoracic vertebrae and ribs; the ventral shield is called → plastron.

carb-, carbo- (L. *carbo* = charcoal), pertaining to charcoal, coal, or carbon. *Adj.* **carbonic**.

carbamate, a salt or an ester of → carbamic acid; e.g. **urethane** (an ester), used as a hypnotic; affects the cholinergic transmission in the nervous system acting as an inhibitor of acetylcholinesterase; carbamates are used in medicine and as insecticides, and some derivatives (e.g. chlorpropham) as herbicides.

carbamide, → urea.

carbamic acid, an organic acid, H_2NCOOH, known in the form of its derivatives, ie. salts, esters (→ carbamate) and amides (→ urea).

carbaminohaemoglobin, Am. **carbaminohemoglobin,** carbohaemoglobin; a chemical combination of haemoglobin and carbon dioxide ($HbCO_2$); one of the several forms in which carbon dioxide is transported in the blood.

carbamoyl, the acyl radical, NH_2CO-, of → carbamic acid; called also **carbamyl**.

carbinol, *see* methanol.

carbogen, a mixture of oxygen and carbon dioxide, usually in proportion 95/5, used as a respiratory gas for patients, or for gassing an incubation fluid for cells and tissues to provide oxygen and maintain the right pH value in a bicarbonate → buffer.

carbohydrase (Gr. *hydor* = water), any enzyme that catalyses the hydrolysis of carbohydrates; e.g. → β-galactosidase and maltase which hydrolyse disaccharides, or different types of → amylases and → cellulases hydrolysing polysaccharides.

carbohydrates, saccharides; aldehydic or ketonic derivatives of polyhydric alcohols; organic compounds primarily synthesized by plants in → photosynthesis. Sugars, i.e. low molecular weight carbohydrates contain carbon, hydrogen, and oxygen in the ratios: $C_x(H_2O)_y$, where x = the number of carbon atoms, y = x—(n—1), n = the number of monomer units. The simplest carbohydrates are **monosaccharides** (e.g. trioses, tetroses, pentoses, hexoses) having the formula $C_x(H_2O)_x$, e.g. glucose $C_6H_{12}O_6$. Sucrose, $C_{12}H_{22}O_{11}$, is one of the **disaccharides**, and raffinose (a trisaccharide), $C_{18}H_{32}O_{16}$, one of the **oligosaccharides**. Macromolecular carbohydrates, i.e. → **polysaccharides**, act as energy storage, as e.g. → starch in plants and → glycogen in animals, or they form structural elements, such as → chitin in animals and fungi, or → cellulose in plants. When linked with proteins and lipids, → proteoglycans, glycoproteins, and glycolipids are formed.

carbolic acid, → phenol.

carbon (< L. *carbo*), a chemical element, symbol C. Carbon is the basic atom in all organic compounds being the fourth commonest element in our solar system. The most common isotope is ^{12}C (C-12, atomic weight 12.01), and the other stable isotope is ^{13}C. The third isotope, ^{14}C, is unstable and radioactive, having a half-life of 5,760 years; it is produced by cosmic radiation in the upper layers of the atmosphere. Carbon isotopes are used in labelling of biological material e.g. in studying different biological syntheses and in the dating (→ carbon dating) of archaeological, palaeontological and geological samples. *See* carbon assimilation, carbon cycle.

carbon assimilation (L. *assimilare* = to make similar), the incorporation of CO_2 to an organic molecule in → photosynthesis; the end products of the photosynthetic reactions are carbohydrates.

carbonate, a salt of → carbonic acid.

carbonate dehydratase, → carbonic anhydrase.

carbon cycle, the circulation of carbon atoms between living organisms and atmosphere. Carbon in the form of carbon dioxide (CO_2) is assimilated into organic compounds by the action of autotrophic organisms, i.e. plants and some other photosynthetic organisms such as cyanobacteria, and a few chemosynthetic bacteria. Organic substances are then used by all organisms, so also by heterotrophic organisms, i.e. fungi and animals; carbon is liberated from the organisms as CO_2 into their environment in metabolic reactions, mainly involved in respiration and fermentation, in microbial decay processes of dead organic matter, or in burning. A part of cycling carbon is withdrawn into reserves of fossil fuel deposits.

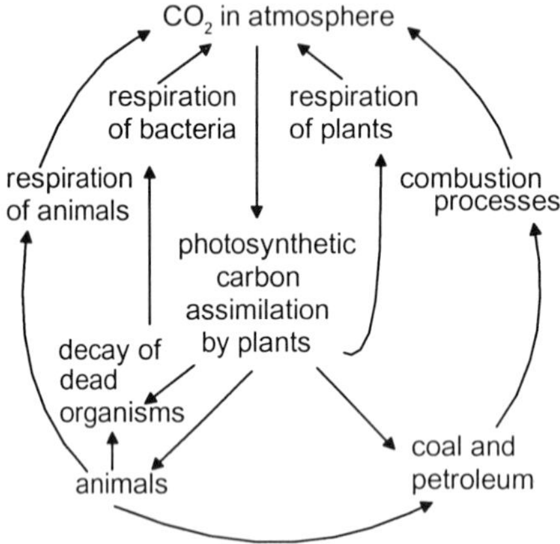

Carbon cycle. Chain of events by which carbon is circulated through the environment.

carbon dating, ^{14}C dating, carbon-14 dating, C-14 dating, radiocarbon dating; a method of dating the age of archaeological objects by estimating their ^{14}C contents (*see* carbon). Radioactive ^{14}C carbon atoms are incorporated together with ^{12}C via photosynthesis into organic molecules. By comparing a new tissue specimen, e.g. a piece of wood, with a sample of unknown age, one can estimate the age of an old specimen after calculating the decrease in the proportion of ^{14}C, based on the knowledge that the half-life of ^{14}C is 5,730 years.

carbon dioxide (Gr. *oxys* = keen, acid), an odourless, colourless gas, CO_2, produced in the oxidation of carbon-containing compounds in respiration and fermentation processes of organisms, and in burning of organic materials; present in the air (appr. 0.03 %). It dissolves slightly in water and reacts with it chemically, forming → carbonic acid, H_2CO_3. Plants absorb CO_2 and convert it to organic compounds, first to → carbohydrates. *See* photosynthesis, respiration, fermentation, greenhouse effect.

carbonic acid, the acid, H_2CO_3, formed when carbon dioxide reacts with water. Carbonic acid is a weak acid and dissociates into bicarbonate (HCO_3^-) and hydrogen (H^+) ions. It is an important agent in the carbon dioxide transport (CO_2 transport) in organisms and acts as a buffer substance against pH changes in tissue fluids. The reaction $CO_2 + H_2O \rightleftharpoons H_2CO_3 \rightleftharpoons H^+ + HCO_3^-$ occurs spontaneously very slowly in tissues but is often catalysed by → carbonic anhydrase. The salts of carbonic acid are called **carbonates**.

carbonic anhydrase, a cellular zinc-containing enzyme, EC 4.2.1.1, which regulates the balance of carbonic acid and carbon dioxide, catalysing the reaction $CO_2 + H_2O \rightleftharpoons H^+ + HCO_3^-$; the enzyme occurs in high concentration in the kidneys (proximal tubules) and red blood cells serving in carbon dioxide transport. Also called carbonate dehydratase.

Carboniferous, a period in the Palaeozoic era 360 to 285 million years ago when the climate on the Earth was warm and moist. Vegetation, especially ferns and seed ferns, were very abundant producing extensive coal deposits; radiation in amphibian development was intensive and the first reptiles appeared. In USA the period is divided to two epochs: **Mississippian** and **Pennsylvanian**. *Adj.* Carboniferous.

carbon monoxide, a colourless, almost odourless gas, CO, formed together with carbon dioxide in incomplete burning of organic material in industry, heating, and traffic; it is a poison with high affinity for haemoglobin and cytochromes thus preventing the transport of oxygen and its utilization in cells. *Syn.* coal gas.

carbon-nitrogen ratio, C/N ratio, the ratio expressing the nitrogen conditions in soil; calculated by dividing the content (%) of organic carbon by the content (%) of nitrogen. The C/N ratio is about 120 in peat which consists almost entirely of organic substances and contains very little nitrogen. In the soil of deciduous forests, decay processes rapidly liberate nitrogen from organic matter into soil and the C/N ratio is thus low, usually 10—15.

carbon tetrachloride, tetrachloromethane; a poisonous organic liquid, CCl_4, which in the presence of water produces hydrochloric acid, HCl, and phosgene, Cl_2CO. It causes nausea, vomiting and pain, and the exposure to high concentrations may cause e.g. severe damage in the nervous system, liver, and kidneys. It is used as a solvent in industry, and as a fire extinguisher and cleaning agent.

carbonyl, a chemical group, =C=O, occurring e.g. in → aldehydes and ketones.

carboxyhaemoglobin, Am. **carboxyhemoglobin,** carbomonoxyhaemoglobin, CO-haemoglobin, HbCO; a tight chemical combination of haemoglobin and carbon monoxide linked together with a covalent bond; prevents oxygen transport thus causing the symptoms of carbon monoxide poisoning in animals.

carboxyl, a chemical group, -COOH, present in many organic acids, → carboxylic acids.

carboxylase, a common name for enzymes, which in organisms catalyse the addition of carbon dioxide (CO_2) to a molecule to form an additional carboxyl group (-COOH); often contains → biotin as the prosthetic group of the enzyme. Carboxylases are grouped either into EC sub-subclass 4.1.1., e.g. Rubisco EC 4.1.1.39, or into the sub-subclass 6.4.1., e.g. pyruvate carboxylase EC 6.4.1.1. *Cf.* decarboxylase.

carboxylation, any reaction in which carbon dioxide is incorporated into a compound and a carboxyl group (-COOH) is formed; → **carboxylases** catalyse the reactions. The most important carboxylation occurs in → photosynthesis of plants when carbon dioxide is incorporated into an organic molecule, the reaction catalysed by ribulosebisphosphate carboxylase-oxygenase (Rubisco). *Cf.* decarboxylation.

carboxylic acid, an organic acid having one or more -COOH groups; e.g. acetic acid, CH_3COOH, malic acid, $HOOCCH(OH)CH_2$-COOH, and citric acid, $HOOCCH_2C(OH)$-$COOHCH_2COOH$.

carboxypeptidase, carboxypolypeptidase, an enzyme of the hydrolase group catalysing the removal of amino acids from peptides or proteins at the free carboxyl end; e.g. carboxypeptidase A (EC 3.4.17.1) removing any amino acid, and carboxypeptidase B (EC 3.4.17.2) that preferentially removes terminal arginine or lysine from a polypeptide. The enzymes act as digestive enzymes in many animals; both are found in pancreatic juice of vertebrates.

carboxysome (Gr. *soma* = body, piece), a structure in some bacteria, containing the CO_2-fixing Rubisco enzyme which in plants is involved in → photosynthesis.

carcinogen (Gr. *karkinos* = crab, cancer, *gennan* = to produce), a physical or chemical agent causing → cancer; most carcinogens are also → mutagens. Carcinogens are either physical factors, such as radioactive and UV radiation, or they are chemical compounds, such as many industrial chemicals or many compounds generated in the burning of organic materials. Several compounds used or studied in laboratories are or may be carcinogens, e.g. acrylamide monomers, aflatoxins (especially B_1), benzene, benzidine, chloroform, *o*-dianisidine, methylene chloride, 2-naphthylamine, *o*-toluidine, trichloroethylene, etc. *Adj.* **carcinogenic**.

carcinogenesis, the origin and propagation of cancer.

carcinogenicity, the ability of certain physical and chemical agents to cause cancer.

carcinoma, a type of cancer in animals derived from epithelial tissues, such as the skin epidermis or intestinal epithelium and associated glands; tends to form metastases.

cardamom, also **cardamon** (Gr. *kardamomon* < *kardamon* = cress), a spice obtained from the aromatic seeds of some zingiberaceous plants native to tropical Asia; used also in medicine.

cardi(o)- (Gr. *kardia* = heart), **1.** denoting relationship to the heart; **2.** pertaining to the stomach around the orifice (*ostium cardiacum*) of the oesophagus.

cardiac, 1. pertaining to the heart; **2.** pertaining to the vertebrate stomach at the junction of the oesophagus near the heart.

cardiac muscle, the muscle of the heart; heart muscle; in vertebrates, comprises the atrial and ventricular muscles joined together by connective tissue and the → **conduction system of the heart**. The cardiac muscle is composed of **muscle fibres** (cardiac muscle cells, cardiac myocytes) 100—400 μm in length and 10—40 μm in diameter, being thinner in lower vertebrates. Each cell has one or two nuclei in the centre, and cross striations inside the cell due to the arrangement of myofibrils like in → skeletal muscle cells.

Cardiac myocytes are connected with their cell ends and some short branches tightly to their neighbouring cells by → **intercalated discs**, and thus the whole muscle forms a net-like structure supported by connective tissue. This structure can function as a → syncytium, i.e. allowing action potentials to propagate electrically from cell to cell through special intercellular junctions, → **gap junctions**, between the two adjacent membranes of the intercalated disc. The cardiac muscle contracts rhythmically by the action of the conduction system of the heart. The frequency and contraction force are regulated by the autonomic nervous system, the sympathetic nervous system stimulating both these functions and the parasympathetic nervous system decreasing the heart rate.

The muscle of the invertebrate heart mostly consists of very small myocytes structurally resembling the smooth muscle of vertebrates (e.g. in molluscs). In some animal groups, as in insects, movements of the heart may be caused by striated muscles located outside the tubular heart.

cardiac stomach, an enlarged foregut of some crustaceans, e.g. decapods; its walls have chitinous ridges, denticles and calcareous ossicles for food crushing. *Syn.* triturating stomach.

cardiac valves, the valves between the atria and ventricles, i.e. the → **atrioventricular valves**, and at the bases of the aorta and pulmonary artery, i.e. the → **semilunar valves**.

cardinal gene (L. *cardinalis* = central, important), → interpretation gene.

cardinal veins, two pairs of venous channels in the circulatory system in adult fish and in the embryos of tetrapod vertebrates; comprise the **cranial cardinal veins** (anterior c. v.) from the dorsocranial parts of the body, and the **caudal cardinal veins** (posterior c.v.) from the kidneys and caudal parts of the body, and the **common cardinal veins** (ducts of Cuvier).

cardioinhibitory area (Gr. *kardia* = heart), the brain area in the medulla oblongata (nucleus ambiguus) of the vertebrate brain, inducing bradycardia in the heart; initiates the tonic vagal discharge to the heart, i.e. through the parasympathetic nervous system. The afferent nerve fibres from the → baroreceptors in the heart and great vessels pass directly to the cardioinhibitory area. The opposite effect (tachycardia) produced e.g. by excitement, is due to the discharge of the sympathetic nerve fibres to the heart, although there is no cardioacceleratory centre in the brain.

cardiovascular (L. *vas* = vessel), pertaining to the heart and blood vessels.

cardo, pl. **cardines** (L. = hinge), a basal segment of the first maxillae of insect mouthparts.

carinate (L. *carina* = keel), pertaining to an organism or organ which has a keel-like anatomical structure, such as the keel of the sternum; e.g. the group **Carinatae** includes birds with a keeled sternum, i.e. all the modern birds except ratites.

carmine, a red pigment (lake) obtained from **cochineal**, a substance produced by cochineal insects; the essential compound in it is **carminic acid**, $C_{22}H_{20}O_{13}$. Carmine is used as a biological stain, and as colouring in food, drugs, and cosmetics.

carnassial teeth (Fr. *carnassier* = carnivorous), **shearing teeth**; a pair of large cutting teeth in both jaws in carnivorous mammals specialized to shear flesh, i.e. the last premolars in the upper jaw and the first molars in the lower jaw.

carnitine (L. *carni-* < *caro* = flesh), an amino acid, 3-hydroxy-4-(trimethylamino)butyric acid, $C_7H_{15}NO_3$; a derivative of lysine. Carnitine occurs in animal tissues and is required in cells for mitochondrial respiration of long-chain fatty acids serving as a carrier of active fatty acids (acyl CoA) from the cytoplasm across the mitochondrial membrane to its matrix. It is especially important for liver parenchymal cells and cardiac and skeletal muscle cells of vertebrates.

Carnivora (L. *vorare* = to eat), **carnivores**; a mammalian order of terrestrial species which feed mainly on other vertebrates and have well developed → carnassial teeth. Carnivores have occurred from Palaeocene and include e.g. the families of **dogs** (Canidae), **cats** (Felidae), **weasels** (Mustelidae), **civets,**

(Viverridae), **hyenas** (Hyaenidae), and **bears** (Ursidae). In some classifications the order Carnivora is divided into two suborders: Fissipedia (terrestrial carnivores) and Pinnipedia (aquatic carnivores).

carnivore, 1. an animal feeding merely or nearly exclusively on flesh, e.g. owls, eagles, hawks, and carnivorous mammals; also some plants, such as sundews, feeding on self-caught animals,; **2.** an individual in the mammalian order Carnivora; **3.** carnivores of a consumer level (primary, secondary etc. carnivores) in the food web of a community, regardless of their systematic position; *cf.* producer, decomposer. *Adj.* **carnivorous**, feeding on flesh.

carnivorous plant, → insectivorous plant.

carotenoids (L. *carota* = carrot), yellow and red natural pigments, comprising **carotenes** and **xanthophylls**. A common carotenoid is → beta-carotene that occurs abundantly in carrot, and is a precursor for → vitamin A. Green plants contain many xanthophylls, such as lutein, and neo-, viola- and zeaxanthin. Carotenoids protect plants against high light intensities and give colours to flowers and fruits rendering them attractive to animals. Most carotenoids found in animals derive from plants but may be metabolized in animal tissues. They protect animals against UV radiation but may also have special functions, as in photoreception in the eye.

carotic (Gr. *karos* = deep sleep), **1.** pertaining to stupor; stuporous; **2.** → carotid.

carotid, pertaining to the major neck artery, **carotid artery** (*arteria carotis*) found in higher vertebrates; also carotic.

carotid body, (L. *glomus caroticum*), intercarotid body; a small neurovascular structure in carotid arteries of vertebrates, in mammals associated to the carotid wall just above the bifurcation of the common carotid artery on each side. It has a rich network of sensory nerve fibres of the glossopharyngeal nerve, serving as a chemoreceptor organ involved in the control of respiration; responds to lack of oxygen, excess of carbon dioxide, and to increased pH in the blood. A similar organ is the → aortic body.

carotid sinus/bulb (L. *sinus caroticus*), a dilatation of the internal carotid artery of vertebrates, in mammals located at the bifurcation of the common carotid artery on each side. It contains sensory nerve fibres which serve as stretch receptors (→ baroreceptors) responding to changes in blood pressure, thus being involved in the control of circulation.

carp-like fishes, → Cypriniformes.

carpal (Gr. *karpos* = wrist), **1.** pertaining to the wrist, → carpus; **2.** one of the carpal bones.

carpal bones, *see* carpus.

carpel (Gr. *karpos* = fruit), a type of flower leaves in seed plants; the carpels form the pistil or gynaecium, which is the female reproductive structure in plants containing the → ovules.

carpellate, pertaining to a flower having one or more carpels but no (functional) stamens. *Syn.* pistillate. *Cf.* staminate.

carpogonium (Gr. *gone* = birth), female → gametangium in red algae, → Rhodophyta.

carpophore (Gr. *pherein* = to bear), a structure supporting the two → mericarps of the schizocarp in the plant family Apiaceae.

carpopodite (Gr. *karpos* = wrist, *pous* = foot), *see* biramous appendage.

carpospore (Gr. *karpos* = fruit, *sporos* = seed), a spore type in red algae, → Rhodophyta. *See* spore, cystocarp.

carposporophyte (Gr. *phyton* = plant), a stage of the life cycle in red algae, → Rhodophyta.

carpus, pl. **carpi** (L. < Gr. *karpos* = wrist), **wrist**; a flexible region in the forelimb of tetrapod vertebrates between the forearm and metacarpals; in a primitive tetrapod limb it consists of 12 bones (8 in man): three proximal bones, i.e. **radiale**, **intermedium**, and **ulnare**, four **central carpals**, and five **distal carpals**. In general, there tends to be considerable reduction or fusion of the carpal bones in more specialized tetrapod groups.

carrageenan, carrageen(in) (*Carragheen*, Irish coastal town), a gel-forming polysaccharide obtained from red algae; used for ice cream, some cheese and pastries, as well as in medicine and cosmetics.

carrier, 1. genetic carrier; in human genetics, an individual heterozygous for a recessive gene allele, causing a disease in homozygous condition, i.e. an individual having a latent genetic disease; **2** an individual harbouring a pathogenic microbe which may transmit the disease to other individuals; **3.** carrier molecule, e.g. a permease protein that transfers certain molecules across cellular membranes, or a carrier protein for hormones and other active substances in tissue fluids, e.g. in blood; the carrier is usually specific to the substance to be carried; **4.** a chemical catalytic agent that transfers an element or group of atoms from one compound to another. *See* cell membrane transport.

carrying capacity, symbolized by the constant **K** (upper asymptote of the sigmoid curve); in ecology, the maximum population density of a particular species which a given habitat or locality can support without deterioration. The carrying capacity prevents the continuous growth of the population because the intraspecific competition becomes too severe, thus resulting in the maintenance of an equilibrium level in population density. Mathematically the growth can be described by the **logistic equation** (*see* logistic growth). The carrying capacity is a hypothetical conception which cannot be verified as such in nature but population densities fluctuate around the equilibrium state.

cartilage (L. *cartilago*, Gr. *chondros*), gristle, chondrus (carrageen); **1.** cartilaginous tissue; a connective tissue type in vertebrates derived from the → mesenchymal tissue. The tissue consists of cells, i.e. **chondrocytes** in mature cartilage and **chondroblasts** in developing cartilage for forming the tissue, and **chondroclasts** (tissue → macrophages) for decomposing it; abundant interstitial matrix (chiefly → collagen) produced by the cells fills the intercellular space. On the basis of the constituent of this interstitial substance, several cartilage types are distinguished: **hyaline cartilage**, the flexible, semi-transparent, bluish tissue, such as articular cartilage; **elastic cartilage** (reticular or yellow cartilage), the elastic, flexible, yellowish tissue containing elastin fibres which branch in all directions, e.g. cartilage of the external ear; **fibrocartilage** (stratified cartilage), containing abundantly compact, parallel collagen fibres in the interstitial matrix, e.g. cartilage of vertebral discs.

Embryonically cartilage is derived from the mesoderm forming most of the temporary skeleton of the embryo. Later the cartilages serve as models for the formation of most bones (cartilage bones or endochondral bones). In this process also chondroclasts are active in the cartilage. In the cartilaginous fishes this tissue type remains as the final skeleton;

2. an organ or a part composed of cartilaginous tissue, covered by the connective tissue membrane, **perichondrium**.

cartilage bone, endochondral bone, replacement bone; a bone of cartilaginous origin. *Cf.* dermal bone.

cartilaginous fishes, → Chondrichthyes.

caruncle (L. *caruncula* = a bit of flesh), **1.** an appendage containing fats or carbohydrates on the seeds of some plants; allures especially ants to collect the seeds and disperse the plant; *syn.* elaiosome; **2.** wattle, comb.

caryopsis, pl. **caryopses, caryopsides** (L. < Gr. *karyon* = nut, nucleus, *opsis* = sight), a one-seeded fruit in which the pericarp and seed coat are fused; caryopsis is a subtype of → achene and is typical of graminaceous plants.

cascade (It. *cascare* = to fall), in biology, a series of subsequent, interdependent events in living organisms or in constituent systems; e.g. a series of genes functioning in a hierarchical order during the differentiation of a cell, or a series of reactions e.g. in → blood coagulation.

casein (L. *caseus* = cheese), a mixture of phosphoproteins present e.g. in beans, nuts, and in milk as a calcium salt, **calcium caseinate**, precipitating in souring processes by lactic acid bacteria or → rennin when sour milk and cheese are produced.

Casparian strip (*Robert Caspary*, d. 1887), a secondary thickening of suberin forming a continuous belt in cell walls of the → endoderm in roots of vascular plants. The endoderm separates the → central cylinder from the cortex, and because of the Casparian strip all the nutrients taken by a plant have to pass through the cytoplasm of endoderm cells, i.e. the ions cannot pass through the endoderm cell walls.

cassette (It. *cassetta*, dim. of *cassa* = box), in genetics, an inactive DNA segment which in the process of DNA rearrangement can in the reciprocal shift replace another, active segment (cassette), thus becoming active itself. So far, cassettes are known only among yeast genes, controlling the mating type shift. Mating type switching occurs when the active cassette is replaced by information from a silent cassette which is then expressed. A cassette differs from a → transposon so that it can move only to a specific position, whereas the transposon can move to random positions.

cassowaries, *see* Casuariiformes.

cast antler, an antler type found in the deer family Cervidae; shed annually after the breeding season.

caste (L. *castus* = pure), a group of individuals of many eusocial insects which may be of the same sex but structurally and functionally different; e.g. the females of some ant species may be soldiers, workers, or queens; the males form a caste of → drones. The determination of the caste is regulated by → pheromones and/or the food quality fed during the larval stage; e.g. protein food produces a reproductive female, food containing carbohydrates a worker. *See* queen substance.

castration (L. *castratio*), the removal or destruction of gonads; **castrate**, **1.** being or pertaining to an individial treated in such a way; **2.** as a verb, to deprive the testes (to emasculate, geld), or the ovaries (to spay).

Casuariiformes (*Casuarius* = the type genus, *forma* = shape), **cassowaries** and **emus**; an order of primitive birds, from Pliocene to Recent, including four species in two genera: *Casuarius* (**cassowaries**) and *Dromaius* (**emus**), living in rain forests of New Guinea and Australia. They are large, flightless birds with three front toes on each foot; the dark feathers have two shafts. *See* ratites.

cat(a)- (Gr. *kata* = down), denoting down, under, lower, along with, against.

catabolism (Gr. *bole* = a throw, casting), pertaining to the degrading or destructive reactions in the metabolism of organisms; **catabolite**, a product of catabolism. *Adj.* **catabolic**. *Verb* to **catabolize**. *Cf.* anabolism.

catabolite activator protein, CAP, a dimeric regulatory protein in prokaryotic organisms which binds to a promoter region of cAMP; facilitates transcription by RNA polymerase of certain adjacent genes in catabolite-sensitive operons.

catabolite repression, the decreased expression of certain adjacent genes in catabolite-sensitive → operons in prokaryotic organisms by the presence of glucose or other carbohydrate metabolites. This depression of genetic transcription adapts the carbon catabolic enzymes to the energetically most advantageous use of carbon sources. It is mediated by a → catabolite activator protein.

catadromous (Gr. *dromos* = course, act of running), pertaining to an animal that migrates down the river to the sea, as e.g. female eels migrating from lakes to spawn in the sea. *Cf.* amphidromous, anadromous.

catalase (Gr. *katalyein* = to dissolve), an enzyme (EC 1.11.1.6) catalysing the degradation of hydrogen peroxide into water and oxygen; occurs both in plants and animals, in high quantities e.g. in liver cells. Catalase is involved in the elimination of → oxygen radicals from cells and tissues. In plants, catalase functions in peroxisomes (*see* photorespiration) and in glyoxysomes which are important organelles in the degradation of lipids.

catalysis, the causing or acceleration of a chemical reaction by a substance which itself does not change in the reaction; may be an **autocatalysis**, in which a new reaction product begins to act as a **catalyst** of the reaction. → Enzymes and enzyme systems act as organic catalysts, **biocatalysts**, in biological reactions. *Adj.* **catalytic**.

cataphyll (Gr. *phyllon* = leaf), a simple and small leaf in flowering plants, often providing protection, as a bud scale or a scale leaf.

cataplexy (Gr. *kataplessein* = to strike down), **1.** "animal hypnosis", a state of motionlessness and rigidity of muscles induced in many animals by certain external stimuli or by a certain posture of the body (e.g. pressing an animal on its back). It is often triggered by the presence of a predator; **2.** a condition with abrupt muscular weakness and hypotonia triggered e.g. in man by a strong emotional experience, as fear, anger, or mirth. *Adj.* **cataplectic**.

cataract (Gr. *katarassein* = to dash down), **1.** waterfall; **2.** the loss of transparency of the eye, mostly the lens.

Catarrhini (Gr. *katarrhin* = hooknosed), an infraorder in the suborder Anthropoidea, including the superfamilies Cercopithecoidea (Old World monkeys) and Hominoidea with the families Hylobatidae (gibbons), and Hominidae (gorilla, orang-utan, chimpanzee, and human species).

catastrophe, in ecology, an **ecocatastrophe**, a dramatic change in an environment causing large-scale disaster to a whole community or ecosystem. Catastrophes occur so seldom that natural selection has no possibility of affecting gene pools of organisms.

catecholamines, physiologically active monoamines of animals acting as hormones and neurotransmitters, synthesized from phenylalanine; include → adrenaline, noradrenaline, and dopamine.

category, in biology, a hierarchical term in a systematic classification of organisms including all taxa at a particular level; e.g. the category of the species comprises all the taxa of the species-level, correspondingly other categories are genus, family, order, etc. The level of a category is determined, taxa are described.

caterpillar, a larval stage of lepidopterans, some hymenopterans, and scorpion flies (Mecoptera); caterpillars have jointed legs on the anterior part of the body and various numbers of → prolegs posteriorly.

catfishes, → Siluriformes.

cathepsin, one of several enzymes in the peptide hydrolase group, catalysing the hydrolysis of peptide bonds; most cathepsins are found in → lysosomes of animal cells, where the enzymes actively dissolve cellular material in cell → apoptosis, and cause postmortal autolysis of the body; e.g. cathepsin B_2 is serine carboxypeptidase (EC 3.4.16.1), and cathepsin C is dipeptidyl aminopeptidase I (EC 3.4.14.1).

catheter, (Gr. *kathienai* = to send down), a tubular instrument employed to withdraw or introduce body fluids.

cathode (Gr. *kathodos* = way down), a negative electrode, i.e. the electrode to which positively charged ions (cations) migrate in electrolysis. *Cf.* anode.

cation (Gr. *kation* = going down), a positively charged → ion which moves to the cathode in electrolysis; e.g. proton H^+, sodium ion Na^+, potassium ion K^+, and calcium ion Ca^{2+}. *Cf.* anion.

cation exchange, *see* ion exchange.

catkin, an → inflorescence, a hanging spike whose individual flowers are simple, unisexual and apetalous, but mostly with bracts; e.g. in birch and poplar species.

caudal (L. *cauda* = tail), pertaining to the posterior end of the body; of, at, or near the tail; **caudate**, having a tail or an appendage like it.

caudal fin, tail fin; a fin in the terminal part of the fish body, differing anatomically and structurally in various fish groups; five main types are distinguished: the two-lobed **heterocercal** fin is asymmetrical both by external and internal structures, the backbone (vertebral column) extending into the larger dorsal lobe; this kind of fin is found in modern sharks and many primitive fishes, e.g. sturgeons. The **diphycercal** fin is externally and internally symmetrical, the backbone running straight to the base of the fin; is found in lungfish. The **homocercal** fin is externally symmetrical but the internal structure is asymmetrical because the backbone tilts upwards at the base of the fin; common in most bony fishes. The **protocercal** fin is an embryonal and larval type resembling the diphycercal type with external and internal symmetry; the symmetrical axis may be the notochord or backbone. The **hypocercal** fin is a reversed heterocercal type in which either the notochord or backbone extends to the ventral lobe which is also longer than the dorsal lobe.

Caudata (**Urodela**), **newts** and **salamanders**; an amphibian order in which all the species have a conspicuous tail; the order is divided into four suborders: Cryptobranchoidea, which includes e.g. "giant salamanders" in China and Japan, Sirenoidea (mud salamanders), Salamandroidea, and Ambystomatoidea; about 360 species are known in the order.

cauliflory (Gr. *kaulos* = stem, L. *flos* = flower), the formation of flowers or inflorescences from axillary buds in the stem; e.g. in *Theobroma* (cacao tree) and *Daphne*. *Adj.* **cauliflorous**.

caulimoviruses (*cauli*flower *mo*saic virus), a group of plant viruses containing double-stranded DNA; the viruses are restricted to infect a few closely related host plants, and they are transmitted by aphids. The best studied member of this group is the **cauliflower mosaic virus, CaMV**, that attacks Brassicaceae and some Solanaceae plants. The size of its circular DNA molecule is appr. 8,000 base pairs. CaMV is a possible vector for transferring foreign genetic material into plant cells.

cauline (Gr. *kaulos* = stem), pertaining to plant structures belonging or being attached to the stem.

caulonema (Gr. *nema* = thread), a richly branched part of the → protonema with few → chloroplasts in some mosses.

causality, a causal relationship; the dependence of events on each other so that one is the cause of the occurrence of the other. The determination of causal relationships is very important in natural sciences. Causality is connected with **determinism**, according to which all events are causal, proceeding deterministically, i.e. according to unfailing natural laws. *Adj.* **causal**.

cavitation (L. *cavus* = hollow, hole), in botany, the formation of air bubbles in vessels of the water transporting xylem in vascular plants; the process is not usually harmful to the plant because the bubbles are small and do not block water transport in the vessels.

cavity, a hollow space in an anatomical structure, as the abdominal cavity or body cavity.

cavum, pl. **cava** (L.), a hole, cavity; also **cavitas**.

C-banding (C = centromere), a staining method for chromosomes in which → heterochromatic regions become deeply stained and discerned from the unstained bands of → euchromatic regions.

CCC, → chlormequat.

cDNA, → complementary DNA.

Ceboidea (Gr. *kebos* = long-tailed monkey), a superfamily of primates in the infraorder Platyrrhini (suborder Anthropoidea); are New World monkeys characterized by the wide flat space between nostrils, absence of cheek pouches and by a long prehensile tail. Familiar species are e.g. the **capuchin monkey** (*Cebus*), **spider monkey** (*Ateles*), and **howler monkey** (*Alouatta*). *See* Catarrhini.

Celcius, °C (*Anders Celcius*, 1701—1744), a

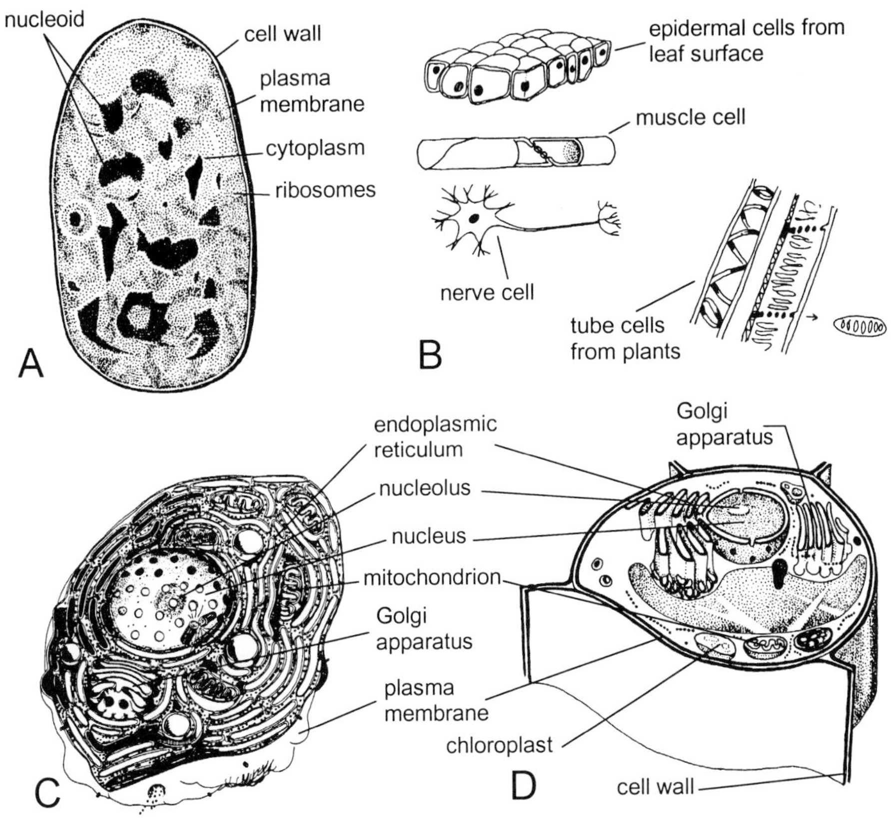

Cell. Different types of cells. (A) Prokaryotic cell (a bacterium). (B) Specialized types of cells. (C) Animall cell. (D) Plant cell. (After Postlethwait & Hopson: Nature of Life - McGraw-Hill 1992; with the permission of the authors).

temperature unit. *See* Appendix 1.

celenteron, → coelenteron.

celiac, → coeliac.

cell (L. *cella*), the smallest living unit of the structure and function of all living organisms; an organized entity capable of growth, metabolism, and reproduction. Cells are the principle units of life (e.g. bacteria and protozoa), or they are associated with other cells forming cell colonies or tissues, from which the different organs of multicellular organisms are formed. The size and shape of the cells vary according to their function (most are microscopic in size) but all have a common basic structure. The organization of the cell consists of → **cytoplasm** and various → **cell organelles** within it, including in eukaryotic organisms usually one → **nucleus**, or in prokaryotic organisms, → **nucleoid**. Some cell types are binucleate, some multinucleate, and some cells, such as mature red blood cells, secondarily lack the nucleus (*see* macronucleus, micronucleus). Each cell is surrounded by a → **plasma membrane**, and in certain groups or organisms (specifically plants and fungi), a → **cell wall**.

cell adhesion (L. *adhaerere* = to cling together), in multicellular animals, the attachment of cells to each other to form tissues, or in cell culture conditions an adhesion of cells to a substratum on which they grow; in the process several types of intercellular junctions are formed. Cell adhesion involves several junctional substances, **cell adhesion molecules, CAMs**, such as → cadherins. *See* cell junctions.

cell biology, a branch of biology dealing with the structure and function of cells; its research areas include → cytology, cytogenetics, and cell physiology.

cell body, that part of a branched cell which contains the nucleus and the surrounding cytoplasm, to be distinguished from the branches, such as dendrites and axons of nerve cells; also called perikaryon or soma.

cell capping, *see* capping.

cell centre, → centrosome.

cell culture, *see* cell, tissue and organ culture.

cell cycle, mitotic cycle; a recurrent series of events involving cell growth, division, and differentiation; the passage of a reproducing cell from one phase of life to the next. The cell cycle is divided into → interphase with G_1, S, and G_2 phases, and into → mitosis (M phase). During G_1 phase (the presynthetic gap) the cell prepares itself for DNA replication. During S phase (synthetic phase), DNA and the chromosomes are duplicated. During G_2 phase (the postsynthetic gap) the cell prepares itself for division, and in mitosis the cell divides. The cell cycle is an irreversible event, and if S phase has begun, the cell cycle will always be completed, only its rate can vary. If the cell differentiates, the "decision" for → differentiation must occur during G_1 phase, and consequently the cell turns into G_0 phase from which it in certain conditions can return to the cell cycle.

cell division, an event consisting of → mitosis or → meiosis and → cytokinesis, resulting in the reproduction of cells.

cell fractionation, the series of procedures to separate cell contents from a cell group or a tissue sample into different organelle fractions to be studied separately, and to isolate from the organelles different substances, such as carbohydrates, lipids, enzymes and other proteins, nucleic acids, hormones, pigments, etc. The first step in the fractionation is the homogenization (→ homogenize) of a tissue in a suitable buffer solution. The homogenate is then fractionated using differential → centrifugation techniques separating first the largest particles and organelles, e.g. cell wall fragments and nuclei; the second centrifugation at higher speeds separates smaller organelles such as mitochondria and chloroplasts, and the following centrifugations give spherosomes, lysosomes, glyoxysomes, peroxisomes, ribosomes of different size, etc. The purity of each fraction is tested electronmicroscopically or using specific markers, e.g. specific enzymes. The fractionation is completed using also other analytical techniques, such as → gel filtration and electrophoresis.

cell fusion, the fusion of somatic cells to produce viable somatic → cell hybrids.

cell heredity, the phenomenon in which all the daughter cells of a given cell are phenotypically identical with the mother cell and with each other, i.e. → gene expression does not change in cell divisions during the development of the organism.

cell hybrid, a cell derived from the fusion of two cells or parts of cells, consisting of material from the → cytoplasm and the → nucleus of both cells. The fusion of certain cells in the **cell hybridization** can exceptionally occur spontaneously e.g. in cell cultures, but usually the fusion is accomplished e.g by adding Sendai viruses or polyethylene glycol into the culture. The fusion of an enucleated cytoplasm with an intact nucleated cell give rise to a cytoplasm hybrid, **cybrid**. *Cf.* minicell.

cell junctions (L. *junctura* = a joining), → junctions between animal cells; patches or zones where the adjacent cell membranes with special structures join close to each other. *See* desmosome, gap junction, tight junction, neuromuscular junction, synapse.

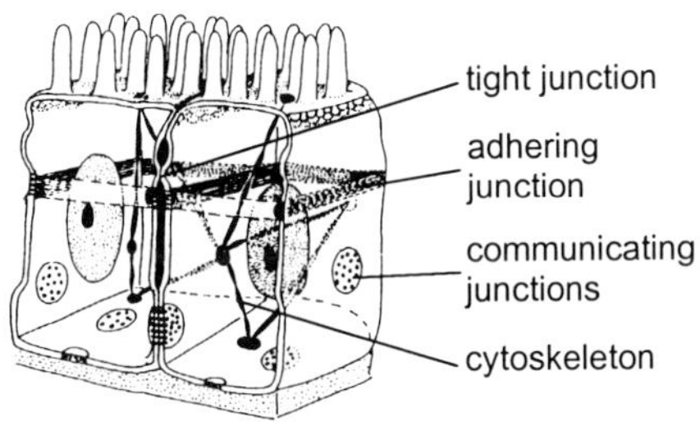

Cell junctions and links between cells. (After Postlethwait & Hopson: Nature of Life - McGraw-Hill, 1992; with the permission of the authors).

cell lethality (L. *lethalis* = causing death), the death of an individual cell due to a → mutation that causes death. Lethal mutations are changes of the genetic material which by killing individual cells cause the death of the whole individual.

cell lineage, a pedigree of cells related through asexual divisions, i.e. the pattern of daughter cells in succesive generations.

cell locomotion, the motility of cells; **1.** the locomotion of unicellular protozoans with many different locomotor cell organelles, as cilia, flagella, and pseudopodia; in amoeboid movement, the animal extends pseudopodia from the cell body to move along the substratum. The functions of the locomotor organelles of protozoans are involved not only in the motility of the animals but also in many other activities, such as handling food, reproduction, excretion, and osmoregulation; **2.** the locomotion (migration) of spermatozoa and phagocytic cells of multicellular animals; spermatozoa use flagella for their locomotion; phagocytes, which can move through many tissues with the aid of pseudopodia (amoeboid movement), are involved in physiological defence mechanisms (*see* macrophages); **3.** among plants, the locomotion of many male gametes with flagella, or of some unicellular algae, especially the → desmids, which can move by secreting mucilage through the pores in their cell wall; **4.** the locomotion of many bacteria with cilia or flagella. *See* cilium, flagellum, pseudopodium.

cell-mediated immunity, *see* immunity.

cell membrane, a component of the cell surface enclosing every living cell and representing an envelope that selectively regulates the flow of nutrients and ions between the cell's internal and external milieu. The membrane, on average 7.5 nm in thickness, consists of a **phospholipid bilayer** punctuated by globular **transmembrane**

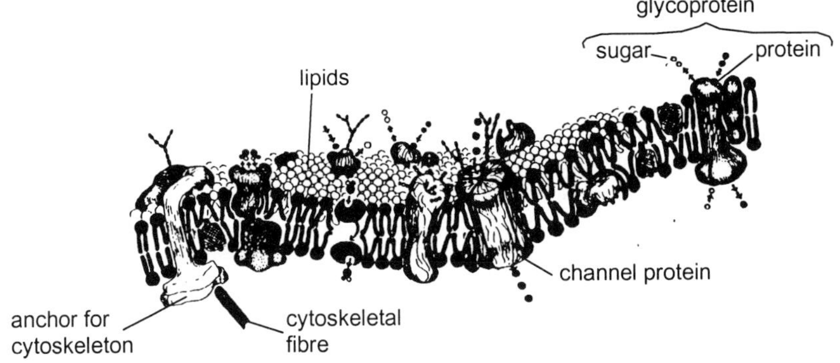

Cell membrane. Fluid mosaic model of cell membrane. (After Postlethwait & Hopson: Nature of Life - McGraw-Hill, 1992; with the permission of the authors).

proteins which as **receptor proteins** are mainly responsible for signal transduction between cells, or they create **channels** across the bilayer (*see* ion channels). Not all proteins are transmembranous but some are attached onto the inner or outer surface of the membrane.

The cell membranes of animal cells have usually → **glycolipids** and **glycoproteins** with oligosaccharide chains exposed on the external surface, often forming a thin layer called → **glycocalyx**. These oligosaccharides are involved in immunological responses, e.g. acting as blood group factors. Specialized cell membrane areas function as various types of intercellular → **cell junctions**.

The cell membrane acts as a passive diffusion barrier to ions and large molecules, and carries out facilitated diffusion, active transport, endocytosis, and exocytosis (*see* cell membrane transport). These functions produce an electrochemical gradient across the cell membrane (*see* resting potential).

The cell membrane and internal membranes of → cell organelles are essentially alike, some cell organelles having two membranes around them. *Syn.* plasma membrane, plasmalemma.

cell membrane transport, the transport of various substances across the cell membrane; comprises systems transporting ions and small organic molecules across the cell membrane, including passive diffusion, facilitated diffusion, and active transport.

In **passive diffusion** ions and molecules diffuse through membrane pores or channels passively according to the → electrochemical gradient without chemical energy. Water and some ions with a high concentration gradient are transported in this way, e.g. Ca^{2+} in both plant and animal cells and H^+ in plant cells.

In **facilitated diffusion** the transport takes place through special **channels** or **carrier proteins** integrated in the cell membrane. These systems function according to the electrochemical gradient without energy. The → ion channels allow the transport of certain ions, each having their specific channel types, e.g. Ca^{2+}, K^+ and Cl^- channels, in animal cells also Na^+. The carriers function in transporting some ions (usually against their electrochemical gradient) or organic molecules across the membrane coupled to the counter- or co-transport of Na^+, in plants of H^+. The energy to drive this form of transport is obtained from the favourable entry of Na^+ (H^+) along the chemical gradient.

The **active transport** functions against the electrochemical gradient, using chemical energy of ATP. The transporter is a special protein, for ions a special → adenosine triphosphatase (**ATPase**) also called ion pump; these are e.g. Na^+,K^+-ATPase in animal cells, Ca^{2+}-ATPase in both animal and plant cells, or H^+-ATPase in some animal and all plant cells.

In the facilitated diffusion and active transport the transporting system may be a **uniport, symport**, or an **antiport**. In the uniport system only one type of ions or molecules are transported in one direction, in the symport two types of ions or molecules are transported simultaneously in the same direction, and in the antiport two substances are transported at the same time in opposite directions.

The symport systems, driven by Na^+ gradient, are important in animal cells for active accumulation of sugars and amino acids into cells. Some bacteria are capable of accumulating sugars by coupling their transport to phosphorylation, e.g. glucose is converted into glucose 6-phosphate. The best known mechanism is the phosphotransferase system (PTS) in which phosphoenolpyruvate (PEP) is the phosphoryl donor instead of ATP:

$$\text{sugar}_{\text{outside}} + \text{PEP} \xrightarrow[\text{PTS, Mg}^{2+}]{\text{membrane}} \text{sugar-P}_{\text{inside}} + \text{pyruvate}$$

Also many other sugars or sugar derivatives are transported by a similar system through cellular membranes. Those compounds are e.g. fructose, mannose, N-acetylglucosamine, sorbitol, and lactose, but the enzymes located in membranes are specific for each sugar.

cell migration, the locomotor activity of gametes and phagocytic cells in tissues of multicellular organisms. *See* cell locomotion.

cell nucleus, *see* nucleus.

cellobiose (Gr. *bios* = life), a disaccharide consisting of two glucose molecules; produced by the hydrolysis of cellulose.

cell organelle, a specialized structure of the cytoplasm having specific functions in the cell, e.g. → endoplasmic reticulum, Golgi apparatus, lysosome, mitochondrion, chloroplast, peroxisome, cytoskeleton, nucleus, vacuole, cytopharynx, microneme, micropore.

cell plasm, → cytoplasm.

cell plate, a plate-like structure developing at first in the division plane during plant cell division; it is thought to precede the middle lamella in the → cell wall.

cell respiration, enzymatic reactions of cellular metabolism which produce energy in the form of adenosine triphosphate, **ATP**, to be used in cellular processes; includes 1) → **glycolysis**, in anaerobic reactions of which glucose or pentose is stepwise degraded and small quantities of ATP are formed, 2) → **citric acid cycle** in which protons (hydrogen ions, H^+) and carbon dioxide are liberated from smaller organic molecules, and 3) → **electron transport chain** in which the electrons originating from hydrogen, are transferred to oxygen to form water. This oxidative process is coupled to the synthesis of ATP from ADP and inorganic phosphate catalysed by F_0F_1ATPase (ATP synthase). This is called → **oxidative phosphorylation**. About 90% of cellular ATP is produced in this process. In many organisms and cell types, glycolysis can function temporarily or continuously in anaerobic conditions, i.e. → anaerobic respiration, but the whole respiratory system functions only in aerobic conditions, i.e. aerobic respiration. *See* photorespiration.

cell sap, the vacuolar solution of a plant cell containing dissolved organic substances and inorganic ions. *See* turgor, osmosis.

cell theory, one of the central theories of biology invented by *Mathias J. Schleiden* (1804—1881) and *Theodor Schwann* (1810—1882) in the 1830's. The theory states that 1) all organisms (except viruses) consist of cells; 2) the → cell is the smallest building element of multicellular organisms and as a unit is itself an elementary organism; 3) each cell in a multicellular organism has a specific function to accomplish and represents a working unit, division of labour; 4) a cell can only be produced from another cell by → cell division.

cell, tissue and organ culture, the cultivation of microorganisms, or animal or plant cells, tissues, and organs in highly controlled conditions.

The cultivation of animal material uses chiefly embryonic tissue sources beginning from separated cells, tissue samples, or whole organs (**primary cultures**), or from established, usually transformed, practically immortal **cell lines** derived from normal or tumour tissue samples. The growth of the cell culture by continuous cell divisions is optimal in conditions resembling the natural environment of the original cells and tissues, embryonic and transformed cells dividing best in the culture. Although the dedifferentiation of cells tends to occur, the differentiation of cells, tissues, and even organs can be induced experimentally to some extent in culture conditions. Hundreds of different special **culture media**, physiological buffers with necessary ions, amino acids, carbohydrates, vitamins, cofactors, hormones, and growth factors with added serum are used, and **culture dishes** and **instruments** have been developed for various cell and tissue types, most materials available in sterile packings. The cultivation provides **aseptic** conditions carried out by a special sterile technique; however, antibiotics are often added. Primary cells grow on firm substratum, as glass or plastic, forming **monolayer cultures** on it; many cell lines can also grow in **suspension cultures**. The culture of animal cells and tissues is an important aid in many fields of biology, biochemistry, medical sciences, and e.g. in drug industry.

The cultivation of plant material follows the same principles as that of animal cells and tissues. Differentiated plant cells show → totipotency and begin to divide when they, isolated from tissues, become free of the control of other cells. Tissue samples transferred on an agar culture medium begin to produce an undifferentiated tissue, **callus**. Under suitable chemical and physical conditions callus cells can differentiate and form shoots or roots depending on the hormone concentrations and ratios. Callus or other cells can be transferred also into liquid media and so suspension cultures can be obtained. When somatic cells differentiate they form **embryoids** which develop into new plants. The culture can even be started from → **protoplasts**. Using the culture technique it is possible to produce new cell mass or new plants rapidly and in great quantities, and thus the life cycle of many plants, such as trees, can be shortened by many years. Cell and tissue culture is widely used for the micropropagation of many useful and ornamental plants, and it is one way to produce virus-free seedlings. Cultured cells, proto-

plasts, and plants serve as material for many studies in physiology and molecular biology as well as for plant breeding.
The cultivation of microbes, such as bacteria and fungi, is largely used to produce material for research, e.g. to elucidate their metabolism, their genetic and pathogenic properties, etc. Microbes are cultivated in liquid media in bottles and test tubes, or on solid media in Petri dishes. A large scale cultivation requires bioreactors. The cultivation of microbes has many applications, such as for producing drugs, enzymes and other proteins, food, such as cheese, wines and beer. Cultivated bacteria have widely used as objects for genetic engineering.

cellula, pl. **cellulae** (L. dim. of *cella* = cell), **cellule**; **1.** a small cell; **2.** broadly, a small enclosed space, as a small compartment in macroscopic anatomy.

cellular, pertaining to, made up of, or characterized by cells.

cellular heredity, → cell heredity.

cellular immunity, → immunity transmitted by T lymphocytes. *See* lymphocytes.

cellular slime moulds, → Acrasiomycetes.

cellulase, endo-β-1,4-glucanase (EC 3.2.1.4); a complex enzyme system catalysing the hydrolytic decomposition of cellulose into glucose and cellobiose. Many bacteria, fungi, plants, and protozoans, as well as probably some multicellular animals, such as silverfish, earthworms, and snails, are able to synthesize cellulase and thus use glucose from cellulose in their metabolism. Ruminants and other herbivorous vertebrates have cellulose-decomposing microorganisms in their alimentary canal to catalyse the hydrolysis of cellulose. Plant cellulases increase in ageing tissues, ripening fruits, and in the abscission layer of leaves before the abscission. In the preparation of → protoplasts, cellulase is used to decompose the wall material of plant cells.

cellulate, to form into cells.

cellulose, a straight-chain polysaccharide, glucan, $(C_6H_{10}O_5)_n$, in which n is over 3,500; polymerized from glucose units bound by β-1,4-linkages; the molecular weights of cellulose vary from 200,000 to 2 million. Cellulose occurs abundantly in the secondary walls of plant cells, the primary walls contain to some degree low molecular weight cellulose. Hydrolysis catalysed by microbial enzymes, → **cellulase** and **cellobiase**, produces glucose and cellobiose (disaccharide with two glucose units). Cellulose is an important raw material in paper and textile industries. *Adj.* **cellulosic**, containing cellulose.

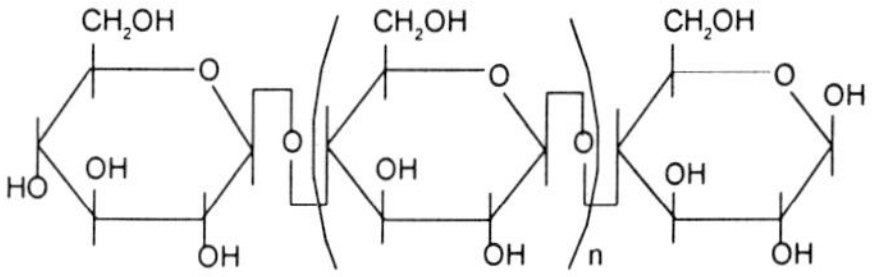

Cellulose. A glucose polymer.

cell wall, the rigid, usually multi-layered structure located outside the cell membrane surrounding most plant and fungal cells. In the course of cell division, the first developing **primary wall** is composed of pectin and hemicellulose and contains only a little cellulose, but the **secondary wall**, generated later, is chiefly composed of cellulose, often also of lignin (trees) or other carbohydrates (especially in different algae). The cell wall supports the cell and separates the cells from each other. Between adjacent cells there are, however, connecting plasma bridges, → **plasmodesmata**, across the walls. The cell wall of fungi differs from that of plants, containing e.g. chitin.
The bacterial cell wall consists of peptidoglycans, which determine the staining of the cells (→ Gram staining). Around a Gram-negative bacterium there is a membrane outside the peptidoglycan layer which may further be covered by a polysaccharide capsule.

celom, → coelom.

cementum, pl. **cementa** (L.), cement; the bone-like connective tissue (*substantia ossea dentis*) which covers the dentine at the root area of the → tooth.

cen(o)-, 1. → caeno-; **2.** (Gr. *kenos* = empty), denoting empty; **3.** (Gr. *koinos* = shared in common), denoting common, original.

cenobium, → coenobium.

cenocyte, → coenocyte.

cenotype, → coenotype.

Cenozoic, → Caenozoic.

centimorgan, cM (L. *centesimus* = hundredth, *T.H. Morgan*, 1866—1945), in genetics, a → map unit (one hundredth of → morgan), i.e. the unit of the map distance of linked genes. The map distance is determined on the basis of the appearance of → crossing-over in a chromosome, and it is a percentage figure that indicates the relative number of recombinant gametes (*see* genetic recombination).

centipedes, → Chilopoda.

central (L. *centrum*, Gr. *kentron* = centre), pertaining to, or situated at a centre. *Cf.* peripheral.

central biotope, the optimal biotope for a population (species) where the reproductive output is maximal and which becomes first populated.

central capsule (L. *capsulum centrale*), a capsule-like, porous centre of the cell body of radiolarians, dividing the protoplasm into outer and inner portions; is surrounded by a foamy **extra-capsulum**.

central cylinder, a structure in the middle of seed plant roots including conducting tissues (xylem and phloem); is a part of the → stele.

central dogma, the rule that describes the flow of → genetic information in the cell. According to it, DNA has autocatalytic and heterocatalytic properties. In **autocatalysis**, DNA promotes its own duplication or → replication; in **heterocatalysis**, DNA guides the → genetic transcription, i.e. the synthesis of → messenger RNA, which then in → genetic translation guides the synthesis of proteins.

central mother cell, a large vacuolated cell in the apical meristem of a plant shoot.

central nervous system, CNS, that part of the → nervous system which in vertebrates comprises the → **brain** and → **spinal cord**; other parts belong to the peripheral nervous system. In the most primitive groups of invertebrates, e.g. in cnidarians, no division into central and peripheral nervous system can be made. The primitive type of CNS with the anterior ganglia and two main nerve trunks appears first in flatworms. The → **brain ganglia** and the → **nerve cord** with segmental ganglia of more developed invertebrates correspond to the central nervous system of vertebrates.

centric fusion, in genetics, a → reciprocal translocation in which the long arms of two acrocentric chromosomes fuse and hence form one → metacentric chromosome. In addition, a small, usually → heterochromatic fragment is formed, but since this fragment will disappear the centric fusion usually leads to reduction of the chromosome number by one, and therefore the centric fusion plays an important role in the → evolution of → karyotypes. Centric fusion is also called Robertsonian translocation.

centrifugation (L. *fugere* = to flee), the process of separating the components of a mixture, e.g. tissue homogenate, using centrifugal force. Different particles and substances have different sedimentation rates according to their size, shape, and specific gravity. Using a → centrifuge, the sedimentation rate depends on the centrifugal field, $\omega^2 r$, in which ω is the angular velocity (rad/s $= 2\pi \cdot$ RPM/60), and r (mm) is the radius of a rotor; RPM is revolutions per minute. The relative centrifugal force (RCF) needed for sedimenting different particles varies and is calculated from the formula: $RCF = r \cdot \omega^2/g$, in which g is the standard acceleration of gravity, $1\ g = 9.80665\ m/s^2$; i.e. $RCF = 1.12 \cdot r \cdot (RPM/1000)^2$. For example, the g-value for ribosomes to be sedimented is about 100,000. The sedimentation is dependent also on the angle and k-factor ("clearing factor") of the → rotor used. The sedimentation coefficient (*see* Appendix 1) of particles, indicating their sedimentation rate, can be calculated and is based on the parameters above.

The most common technique is **differential centrifugation** with several modifications. In → **density gradient centrifugation** the components migrate in a centrifuge tube to the density area which equals their own density; two methods can be identified: **rate zonal** and **isopycnic centrifugation**. Gradient centrifugation can also be carried out using a zonal → rotor. **Continuous centrifugation** can be performed using a special rotor into which a mixture is continuously added, and the pellets (e.g. bacteria cells) are sedimented in the tubes while the liquid flows out. **Ultracentrifugation** is used to separate the smallest particles and molecules, and **analytical centrifugation** is a special technique for determining the relative molecular mass of particles (*see* ultracentrifuge).

centrifuge, an apparatus by which particles or macromolecular components of a suspension can be separated by spinning the mixture so that the centrifugal force moves the particles toward the periphery of the rotating tube (*see* centrifugation). Centrifuges of different sizes and spinning properties have been developed for laboratory use from table-top models to large analytical centrifuges.

Centrifuges can be grouped according to their revolving speed, → RPM, i.e. into **low speed centrifuges**, a few thousand RPM, **high speed (super speed) centrifuges**, max. about 30,000 RPM, and **ultracentrifuges**, over 30,000 RPM. Centrifuges are usually programmable and automatical, the rotor runs in a vacuum, and the temperature of the rotor chamber is controlled. In **analytical centrifuges** there are additional devices, such as a fluorescing plate and camera, for the visual observation and documentation of a sedimentation profile. Those centrifuges are used to determine the sedimentation coefficients of macromolecules and particles as well as the molecular weights of molecules.

centriole, a cell organelle which is a short cylinder-like particle (length 0.4 µm and width 0.2 µm) in the cytoplasm; found in the cells of animals and certain lower plants, and it has a central role in the formation of new centrioles, mitotic spindle, microtubules, and flagella of cells. In most cells, two centrioles are located near the nucleus in an area called **centrosome**; the pair of centrioles, located perpendicularly to each other, are called **diplosomes**. In electron microscopy each centriole is seen as a cylindrical structure formed by nine sets of three microtubules (triplets).

At the beginning of cell division the centrioles duplicate, separate, and move to opposite sides of the nucleus, thus forming the poles of the cell; the mitotic spindle forms between them. The formation of daughter centrioles begins when the centrioles separate in the beginning of nuclear division.

centrolecithal (Gr. *lekithos* = yolk), pertaining to a type of animal egg in which the yolk (deutoplasm) is accumulated centrally in the cytoplasm; occurs e.g. in most arthropods. *Cf.* telolecithal.

centromere (Gr. *meros* = part), that region of the chromosome where the sister chromatids are attached in → mitosis, dividing the chromosome into two arms; the fibres of the → mitotic spindle are attached to the centromere during the nuclear division. Centromeres have a division cycle of their own, i.e. they divide during the beginning of the anaphase of → mitosis, or in → meiosis during the second meiotic division. During the first meiotic division the centromeres remain undivided guiding the cellular movements, like → segregation of chromosomes. In most cases the centromere is localized on a certain site of the chromosome, but some species have unlocalized centromeres, i.e have holokinetic chromosomes. *Syn.* spindle attachment site.

centrosome (Gr. *soma* = body), cell centre; the region of the cell cytoplasm in which the → centrioles are located; divides during nuclear division forming the poles of the → mitotic spindle, thus serving as the microtubule-organizing centre.

cephal(o)- (Gr. *kephale* = head), denoting relationship to the head. *Adj.* **cephalic**.

Cephalaspidomorphi (Gr. *aspis* = shield, *morphe* = form), **lampreys**; a class in the superclass Agnatha (jawless vertebrates); have a cartilaginous skeleton and a narrow median fin extending along the back via the tail ventrally to the anus, pectoral and pelvic fins are not developed, seven gill openings on each side, and the single nostril located middorsally on the head (*cf.* Pteraspidomorphi). The class includes only one order, Petromyzoniformes. *See* ammocoete.

cephalic index, length-breadth index of the head, especially used to describe the human skull; expressed by the formula: 100 x breadth/length.

cephalins, also **kephalins,** a term formerly used of a group of phosphatidic esters containing ethanolamine or serine, now known as phosphatidylethanolamine and phosphatidylserine; widely distributed in animal tissues, especially in the nerve tissue.

cephalization, 1. the tendency of the embryonic development and growth to initiate in and concentrate to the anterior end of the body; **2.** the phylogenetic development of the head as the central structure of a body with the brain and the most important sensory organs.

Cephalochordata (Gr. *chorde* = string), **cephalochordates, lancelets (amphioxus)**; a subphylum of chordates including three genera and about 20 species (mainly in the genus *Branchiostoma*); are slender, translucent animals, about 5 to 7 cm in length, living on shallow sandy sea shores. The adults have typical structures of chordates, i.e. the notochord, nerve (spinal) cord, pharyngeal gill slits, and postanal tail. *See* amphioxus, Acrania.

Cephalopoda (Gr. *pous*, gen. *podos* = foot),

cephalopods; the most highly developed class of molluscs comprising many extinct species; at present only 650 species of all the known 10,000 species are still extant. Cephalopods are mostly large marine animals whose mouth is surrounded by several prehensile tentacles with cup-like suckers. The sensory and nervous systems are well developed, and they have a camera eye which structurally and functionally corresponds to the vertebrate eye. The class is divided into two extant subclasses: **nautiloids** (Nautiloidea) and **coleoids** (Coleoidea) including the orders of **cuttlefishes** (Sepioidea), **squids** (Teuthoidea), **octopuses** (Octopoda), and **vampire squids** (Vampyromorpha, with only one extant species). Cephalopods include the well-known fossil species, **ammonites** (the subclass Ammonoidea), which became extinct in the Cretaceous period.

cephalothorax (Gr. *thorax* = chest), the body region formed from fusion of the head and thorax in crustaceans; in arachnids the corresponding part is usually called prosoma.

ceptor (L. *capere* = to take), an element which receives; often used as a suffix, as in → receptor.

ceramides, a group of → sphingolipids in which the NH_2 group is acylated to form an N-acylsphingosine; e.g. galactosyl ceramide occurs abundantly in the plasmamembrane of nerve cells. *See* cerebrosides.

ceratin, → keratin.

ceratobranchial (Gr. *keras* = horn, *branchia* = gills), *see* gill arch.

Ceratodiformes, *see* Dipnoi.

cercaria, pl. **cercariae** (Gr. *kerkos* = tail), a tadpole-like stage in the life cycle of digenean trematodes (flukes); e.g. the cercariae of liver flukes, emerged from the → rediae reproduced asexually in the first intermediate host (a snail), penetrate into a second intermediate host (a fish). Cercariae of blood flukes have no second intermediate host but they penetrate directly into a definitive host, i.e. into a mammal, e.g. man. Cercariae may also encyst on vegetation to develop → metacercaria. Definitive hosts are often mammals, man included. *Cf.* miracidium, sporocyst. *See* Trematoda.

Cercopithecoidea (Gr. *kerkos* = tail, *pithekos* = monkey), **cercopithecoids, Old World monkeys**; a superfamily of primates in the suborder Anthropoidea; characterized by a narrow space between nostrils, usually by internal cheek pouches, but never by a prehensile tail; they have ischial tuberosities on the buttocks. The superfamily includes e.g. the **proboscis monkey** (*Nasalis*) in Borneo, the **langur** (*Presbytis*) in India, the **savage mandrill** (*Cynocephalus*), and the **rhesus monkey** (*Macaca*). *See* Catarrhini.

cercopod (L. *cercopodium*, Gr. *pous*, gen. *podos* = foot), a paired appendage in the last abdominal segment of some branchiopods. *Cf.* cercus.

cercus, pl. **cerci** (L.), a stiff hair-like anatomical structure, especially the jointed paired appendage acting as a sense organ on the 11th abdominal segment in many arthropods; the size and form are highly variable.

cerebell(o)- (L. dim. of *cerebrum* = brain), denoting relationship to the → cerebellum. *Adj.* **cerebellar**.

cerebellar hemispheres, → cerebellum.

cerebellum pl. **cerebella,** Am. also **cerebellums,** the dorsal enlargement of the metencephalon of vertebrates, concerned in the coordination of movements, in mammals comprising 10 to 20% of the total brain volume; together with the → pons forms the **hindbrain**. The cerebellum consists of a middle lobe (**vermis**) and two lateral lobes (**hemispheres**) separated by the longitudinal fissure and connected with the brain stem by three pairs of peduncles. The large surface area, **cerebellar cortex**, consists of superficial neurones in three layers, which together with several large nuclei deeper in the cerebellar tissue form **grey matter** (*see* basket cell, Purkinje cells). The surface of the highly developed cerebellum, as in primates, is full of folds extending the surface area manifold. Like the cerebrum, the cerebellum contains specialized areas which are evolutionarily of different ages: the **archaeocerebellum**, controlling mainly the body balance, the **palaeocerebellum**, controlling automatic movements, such as flight, running, and jumping, and the **neocerebellum**, functioning in the control of finer movements, e.g. those needed for using fingers. The representation of the latter areas has increased during evolution, being best developed in birds and mammals.

cerebr(o)- (L. *cerebrum* = brain), pertaining to the brain, cerebrum. *Adj.* **cerebral**.

cerebral cortex (L. *cortex* = bark, rind, shell),

→ pallium.

cerebral ganglion, pl. **cerebral ganglia,** Am. also **cerebral ganglions,** brain ganglion; also called brain; a pair of knot-like mass of nervous tissue in the head of many invertebrates, e.g. annelids, arthropods and molluscs, forming the highest category of the central nervous system in these animals. The cerebral ganglia are connected via two longitudinal nerve cords (**circumpharyngeal connectives**) to the subpharyngeal ganglia.

cerebral hemispheres, the two halves of the telencephalon in the vertebrate brain. *See* cerebrum.

cerebrosides, cerebrogalactosides; galactolipids; a class of glycosphingolipids; specifically, monoglycosylceramides (ceramide monosaccharides); contain a sphingosine and a long-chain fatty acid amide (with C_{24}, C_{18}, or C_{16} chain) to which galactose or glucose (in plants always glucose) are combined with glycosidic linkage, i.e. attached to the -CHOH group. Cerebrosides belong to the larger group of → sphingolipids. Cerebrosides, such as **kerasin** and **phrenosin**, occur in cell membranes in animals, being especially abundant in the myelin sheath of nerve fibres. Cerebrosides are also called galactocerebrosides, or glucocerebrosides.

cerebrospinal fluid, CSF (L. *liquor cerebrospinalis*), the special body fluid which fills all the four brain ventricles, the central canal of the spinal cord, and the → subarachnoid space of the vertebrate brain; the fluid is formed when blood plasma percolates through the walls of choroid plexuses (special blood vessels) on the roof of the brain ventricles. CSF flows slowly backwards, and through the foramina of Magendie and Luschka in the fourth brain ventricle it flows to the subarachnoid space; from there the fluid is filtered through the walls of the arachnoid villi into the cerebral venous sinuses and so into the blood.

cerebrum, pl. **cerebra,** Am. also **cerebrums** (L.), → telencephalon, often including → diencephalon, sometimes also mesencephalon or even the entire brain.

Ceriantipatharia, a subclass of anthozoans comprising animals with unbranched tentacles; e.g. **tube anemones** (Ceriantharia) and **black corals** (Antipatharia).

cervic(o)- (L. *cervix* = neck), pertaining to the neck or a neck-like part of an organ. *Adj.* **cervical**.

cervical nerves, *see* spinal nerves.

cervical vertebrae, *see* spinal column.

cervix, pl. **cervices** (L. = neck), **1.** a neck-like anatomical structure; **2.** *cervix uteri*, the neck of the uterus; **3.** → collum.

cesium, → caesium.

Cestoda (L. *cestus* = ring, belt), **cestodes, tapeworms**; a class in the phylum Platyhelminthes comprising parasitic, segmented worms with no digestive tract (*see* proglottid). They have a complicated life cycle with two or more host animals, the adult mature stage living e.g. in the intestine of man and many other vertebrates. Cestodes are divided into two subclasses: **Cestodaria** and **Eucestoda**. Species parasitic in humans are e.g. the broad tapeworm (*Diphyllobothrium latum*) and the beef tapeworm (*Taeniarhynchus saginata*). Intermediate hosts of the former are some fish species and those of the latter, cattle.

Cetacea (L. *cetus* = whale), **cetaceans, whales**; a marine order of mammals including 78 species in two suborders: the **toothed whales** (Odontoceti) whose food consists of large squids and some fish, and the **baleen whales** (Mysticeti), feeding mainly on krill which they filter from water with their many parallel horny plates of the whalebone on the palate. Toothed whales include e.g. dolphins, porpoises, and sperm whales; baleen whales include e.g. rorquals, right whales, grey whales, and the blue whale which is the largest animal ever lived. The forelimbs of whales are modified into broad flippers, the hindlimbs are absent but rudiments of the pelvic bones still exist. The skin is hairless and glandless, except for a few hairs on the muzzle and mammary glands; the subcutaneous fat layer is thick, serving heat insulation.

CFC, → chlorofluorocarbon.

chaeta, pl. **chaetae** (Gr. *chaite* = hair), a stiff, hair-like structure, **seta**; e.g. **1.** a chitin bristle on the body surface in annelids, serving locomotion; the number of chaetae varies depending on the animal group being rather few in oligochaetes (the earthworm has eight stiff bristles on each segment), but numerous on each body segment of polychaetes; *see* parapodium. **2.** a simple type of → sensillum in insects, located chiefly on the body wall and associated with a sensory nerve cell;

chaetae act as sense organs for different types of stimuli.

Chaetognatha (Gr. *gnathos* = jaw), **chaetognaths, arrow-worms**; an invertebrate phylum comprising species which live in the plankton of pelagic zones of oceans; their transparent, dart-like body is unsegmented and comprises the head, trunk, postanal tail, and the true coelom. Arrow-worms are predators which have eyes, sensory bristles, and a ciliary loop around the neck, probably for receiving chemical stimuli. Only about 70 species are described but the number of individuals may occasionally be very high.

chaetotaxy, chaetotaxis (Gr. *taxis* = arrangement), the arrangement (location and numbers) of bristles or → chaetae, and a pattern formed from them on the surface of the body and legs of an animal; an important feature in taxonomy of many invertebrates.

chalaza, pl. **chalazae,** or **chalazas** (L. < Gr. = grain, hail), **1.** a suspensory band attaching the vitellin sac to the membranes of a bird egg; **2.** a stalk-like part of a plant → ovule fastening the ovule into the → placenta.

chalazogamy (Gr. *gamos* = marriage), a type of plant fertilization in which the → pollen tube grows to the ovule through the → chalaza.

chalc(o)-, also **chalk(o)-** (Gr. *chalkos* = copper), pertaining to copper and its occurrence in tissues.

chalk (L. *calx* = lime, limestone), natural calcium carbonate, $CaCO_3$.

chalones (Gr. *chalan* = to slacken), a group of tissue-specific but not species-specific hormone-like proteins or glycoproteins found in animal tissues. They suppress the mitotic activity of cells and regulate → differentiation. The differentiated cells of a tissue produce a chalone whose regulatory action is targeted to the progenitor cells of the same tissue involved in the → mitotic cycle. The progenitor cells can either continue in the mitotic cycle or begin to differentiate. This determination occurs at the end of the G_1 phase of the mitotic cycle. The higher the concentration of chalones, the higher the number of cells which begin to differentiate. Thus the tissue reaches a fixed size typical of the adult animal. Nowadays the term growth-inhibiting factor is mainly used instead of chalones. Sometimes the term chalone refers to a suppressor of the action of hormones in general.

chamae(o)phyte (Gr. *chamai* = on the ground, *phyton* = plant), a plant (often twig), whose wintering buds are situated on the earth surface, protected by snow in winter. *See* Raunkiaer's life forms.

changeable cations, positive ions electrically adhered to small soil particles; the electrical charge and the size of the cations determine the strength of their affinity and they may displace each other on the particles. Many cations (e.g. Ca^{2+}, K^+, Mg^{2+}) are important plant nutrients and the process is important for plants.

chaperones (Fr. *chaperon* = hood, apron), proteins acting stoichiometrically in the folding of other proteins. The chaperones form complexes during molecular arrangement of proteins by binding reversibly to nascent peptide molecules, and prevent their incorrect folding; they also aid in their transport, particularly across intracellular membranes into different cellular components. Many → stress proteins are chaperones.

character convergence, the evolutionary similarity of a structure, behaviour, or some other characteristic in species that are unrelated, e.g. the thorns of cacti and spurges (*Euphorbia*), or the fins of fishes, ichthyosaurs, and whales. *See* convergent evolution.

character displacement/divergence, an evolutionary differentiation of a structure, behaviour, or some other characteristic of two (or more) related species which compete with each other. Thus the differences become more pronounced in sympatric areas (where the species live together), and less marked in allopatric areas. For example, the difference in beak size between the rock nuthatch (*Sitta neumayer*) and the eastern rock nuthatch (*S. tephronata*) is greater in areas where the species occur sympatrically, and smaller in allopatric areas.

character species, a plant species occurring constantly in certain plant → associations, being important for their identification. *Syn.* characteristic species. *See* dominant species.

Charadriiformes (L. *Charadrius* = genus of plovers, *forma* = form), **shore birds**; a diverse order dispersed everywhere in the world; many of them are colonial, and good flyers. The order includes 337 species in 18 families; most familiar are **oystercatchers**

(Haematopodidae), **avocets** and **stilts** (Recurvirostridae), **plovers** (Charadriidae), **sandpipers** and **snipes** (Scolopacidae), **skuas** (Stercorariidae), **gulls** (Laridae), **terns** (Sternidae), and **auks** (Alcidae).

Charophyta (L. *chara* = plant), **charophytes, stoneworts**, a division in Plant kingdom; multicellular algae with branched green thallus which is often covered by a supporting layer of calcium carbonate. Charophytes grow in fresh and brackish water, often on a soft bottom of shallow water bodies. They are related to green algae, to which group they have sometimes been connected.

chat, khat, quat; a material obtained from the leaves of *Catha edulis* (khat plant), containing alkaloids called **cathine** (norpseudoephedrine) and **cathinone** which have an amphetamine-like stimulatory effect on the nervous system, and therefore chat is used as a psychoactive recreational drug, especially in Africa.

chela, pl. **chelae** (Gr. *chele* = claw), **pincer**; a claw-like distal segment in certain legs of some arthropods; usually is located opposite to the top of the preceding segment forming a pincer-like structure, such as the thoracic appendages of crabs (chelicerates and pereiopods). Chelae are specialized e.g. for defence, handling food, and cleaning the body.

chelate, 1. *adj*., claw-like, pincer-like; e.g. the appendage of some arthropods having a → **chela** on the distal end; **2.** *noun*, a soluble complex compound formed from a metal cation and an organic molecule bound by a → coordinate bond. In plants, organic acids, proteins, and purine and pyrimidine bases can form chelates in tissues. In → haem, found in → cytochromes of all cells, the Fe^{2+} ion is chelated by the porphyrin ring. Chelates e.g. prevent essential elements, such as iron, zinc, and copper to sediment in basic soils and nutrient solutions. In homogenization and fractionation media of cells and tissues, chelating agents such as EDTA and EGTA are used to decrease the content of free calcium (Ca^{2+}), magnesium (Mg^{2+}), or heavy metals. *Verb* to **chelate**.

chelicera, pl. **chelicerae** (Gr. *keras* = horn), the first, claw-like, paired mouth appendage (preoral appendage) in chelicerates which is developed into a three-jointed prehensile appendage in spiders; e.g. the poison gland of arachnids and the silk gland of pseudoscorpions open at the tips of chelicerae. *See* pedipalps.

Chelicerata, chelicerates; a subphylum in the phylum Arthropoda including the classes of **arachnids** (Arachnida), **sea spiders** (Pygnogonida), and **aquatic chelicerates** (Merostomata). The first appendages of the animals are formed into → chelicerae (poison claws), and the second appendages developed into pedipalps. Chelicerates have four pairs of walking legs, antennae and mandibles are absent. Most of the chelicerates are terrestrial predators, some live also in fresh or salt waters. In some classifications chelicerates are considered as a phylum.

chelipeds (L. *pes* = foot), the first pair of the → pereiopods in crayfishes and lobsters; large and pincer-like chelipeds are used for defence, attack, and prey capture.

Chelonia, Testudines (Testudinata) (*Chelonia* = the type genus), **chelonians**, including **turtles**, **tortoises**, **terrapins**; an order of anapsid reptiles; sluggish, long-lived, and mainly herbivorous animals, some aquatic species feed on animals, teeth are absent. The body is covered by shields which are composed of several flattened bones enveloped by horny scales; i.e. of the convex dorsal shield (**carapace**) and of the flat ventral shield (**plastron**). About 220 species are known in the world in marine, freshwater, and terrestrial habitats.

chem(i)-, chemo-, chemico-, chemio- (Gr. *chemeia* = alchemy), pertaining to → chemistry or a chemical substance. *Adj*. **chemical**.

chemical defence, chemical defence in organisms; **1.** defence with the aid of toxic agents or scent chemicals; *see* poisonous animals, poisonous fungi, poisonous plants, scent marking, continuous defence mechanism; **2.** → detoxification.

chemical oxygen demand, COD, → biological oxygen demand.

chemiluminescence (L. *lumen* = light), *see* luminescence.

chemiosmotic theory, → Mitchell's theory.

chemistry, a branch of science dealing with compositions, properties, and reactions of the elements, various compounds of elements, and atomic relations of matter, e.g. inorganic chemistry, organic chemistry, physical chemistry. *See* biochemistry.

chemoautotroph, *see* chemotroph, autotroph.

chemocline (Gr. *klinein* = to slant), a chemical

→ cline; usually the halocline due to gradual differences of salinity in natural waters, especially in brackish waters. *See* stratification, meromictic.

chemoheterotroph, *see* chemotroph.

chemokinesis, *see* kinesis.

chemoreceptors (L. *recipere* = to receive), sensory receptors specialized for excitation by chemical substances; may be a sensory cell or a part of it, or a free nerve ending. In most animals several types of chemoreceptors are found: **olfactory receptors** functioning in → olfaction, and **gustatory receptors** functioning in → taste. In addition to olfactory receptors, many animals have also other **distance chemical receptors**, well developed e.g. in many insects. These are usually highly species-specific and are important e.g. in reproductive behaviour to attract partners by specific → pheromones and non-specific odours. Vertebrates receive pheromone signals by specialized olfactory receptors (*see* Jacobson's organ, terminal nerve system). Beside olfactory and gustatory receptors, animals have many **internal chemoreceptors** in their tissues to monitor the chemical state of the body, e.g. vertebrates have chemoreceptors in → aortic and → carotid bodies for sensing the carbon dioxide content of the blood, and special receptors in the kidney responding to salt concentration of the primary urine (*see* macula densa). Similar receptors may well exist also in invertebrates although not yet known. Many **pain receptors** (especially in the mouth) respond to chemical substances, i.e. they function as chemoreceptors (*see* pain).

chemosynthesis (Gr. *syntithenai* = to put together), the synthesis of organic compounds carried out by some bacteria (e.g. *Sulfobolus*) in which energy needed for chemosynthetic reactions is obtained from the oxidation of inorganic substances. In the synthesis the chemical energy comes e.g. from the reactions of iron and sulphur compounds and not from light as in green plants. *Adj.* **chemosynthetic**. *See* chemotroph.

chemotaxis (Gr. *taxis* = order, arrangement), movement of an organism or a cell towards a chemical stimulus (positive chemotaxis), or away from it (negative chemotaxis). *Adj.* **chemotactic**. *See* taxis.

chemotroph (Gr. *trophe* = nutrient), **chemotrophic organism**; an organism obtaining energy for its metabolism by oxidizing inorganic compounds, e.g. hydrogen sulphide (H_2S) or ammonia (NH_3) (**chemoautotrophs**), or by degrading organic material in processes which are independent of light (**chemoheterotrophs**); usually the term refers to bacteria. *Cf.* autotroph, heterotroph.

chemotropism (Gr. *trepein* = to turn), a growth movement (growth curvature) of a plant or plant organ towards a chemical stimulus (positive chemotropism), or away from it (negative chemotropism); an example of the positive chemotropism is the growth of pollen tubes. *Adj.* **chemotropic**. *See* tropism.

chiasma, pl. **chiasmata** (Gr. = a cross resembling the letter chi), a crossing of two longitudinal structures, such as tracts; e.g. **1.** the point at which non-sister chromatids of the → homologous chromosomes are joined to each other; occurs during the first meiotic division from pachytene to the first anaphase while the homologous chromosomes form a bivalent or multivalent (*see* meiosis).

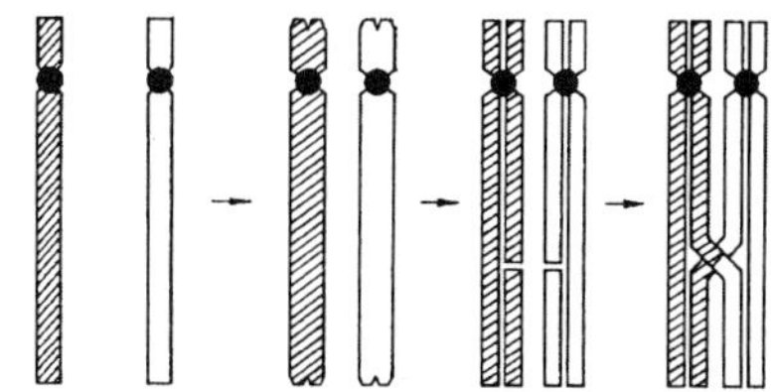

The formation of the chiasma in a chromosome pair. (After Rieger, Michaelis & Green: Glossary of Genetics - Springer Verlag, 1968; with the permission of the publisher).

Chiasmata are formed during pachytene as a consequence of → crossing-over when the homologous chromosomes break and rejoin in a new way. In normal chiasmatic meiosis usually at least one chiasma exists in each bivalent, the structure holding the members of the bivalent together and thus ensuring the → segregation of the homologous chromosomes; **2.** optic chiasma (*chiasma opticum*), i.e. the point of crossing, decussation, of optic nerve fibres on the ventral surface of the hypothalamus in the vertebrate brain.

chiastoneural, pertaining to the nervous system of prosobranch gastropods (phylum Mollusca) in which the main nerve cords (pleurovisceral commissures) are twisted into the shape of number 8, as a result of torsion of the visceral sac.

chief cells, 1. cuboidal or columnar epithelial

cells in the lower parts of the gastric glands of vertebrates; secrete an inactive enzyme, **pepsinogen**, which by the hydrolytic action of hydrochloric acid is activated into **pepsin** (a proteinase); *syn.* peptic cells, zymogenic cells; **2.** sometimes refers to several other cell types, such as pinealocytes of the pineal gland, chromaffin cells of the paraganglia, or cells of the parathyroid glands.

chill torpor, → torpor caused by cold; the condition of animals when the behavioural activity has disappeared and physiological functions have greatly diminished in cold, as occurs in most invertebrates below 0°C. *Syn.* chill coma.

Chilopoda (Gr. *cheilos* = lip, margin, *pous,* gen. *podos* = foot), **chilopods, centipedes**; a class of arthropods including worm-like, dorsoventrally flattened animals which have a body with several homonomous somites, each of which (except the first and the last two) bears a pair of seven-jointed legs. The first segment has a pair of four-jointed poison claws. Chilopods are predators feeding on earthworms and insects, large species may capture small lizards, or mice. The class contains about 2,500 species in two subclasses, Anamorpha and Epimorpha.

chimaera, Am. **chimera** (Gr. *chimaira* = a mythological monster), **1.** a plant or a plant part (seldom an animal), composed of genetically different cells or tissues originating mostly from different species; may be caused by → mutation, grafting, or fusion of embryos. If the chimaera is due to genetic differences in the layers of the → apical meristem of plants, it is called **periclinal chimaera**; such a plant may have e.g. green leaves with white spots; **2.** a protein or DNA molecule containing segments derived from different proteins or DNA segments. *Adj.* **chimaeric** (chimeric). *Cf.* mosaic.

Chimaeriformes, *see* Holocephali.

Chironomidae (*Chironomus* = the type genus, Gr. *cheironomos* = gesticulating with hands), **chironomids**; a dipteran family in Insecta including about 5,000 species in 120 genera. A mass swarming in the air is typical of adults; the larvae live mainly in fresh waters feeding on algae, diatoms, and detritus; some species are terrestrial. In some species, the blood of larvae contains haemoglobin that effectively binds oxygen, and therefore the larvae are used as indicators of oxygen deficiency in eutrophic waters where other species cannot live any more.

Chiroptera (Gr. *cheir* = hand, *pteron* = wing), **chiropterans, bats**; a cosmopolitan order of mammals including altogether about 900 species. The forelimbs of the animals are modified into wings in which the thin integumental flying membrane is supported by four elongated digits; the first digit is short and provided with a claw. Also the hindlimbs support the flying membrane. Bats are known for their ability to orientate by echolocation; in flight they emit pulses of ultrasounds (30,000—70,000 Hz), perceive echos of these with their specialized ears, and orientate accordingly. Chiroptera is divided into two suborders: **Microchiroptera** are mostly insectivorous, some species, however, have adapted to feed on fish, blood of large mammals (vampire bats), nectar, or pollen. **Megachiroptera** occur in tropical regions of Eurasia and live mainly on fruit, as e.g. the flying fox; some species have specialized to use pollen and nectar as food.

chitin (Gr. *chiton* = mantle), a polysaccharide composed of N-acetylglucosamine units. Chitin occurs in the cell walls of many fungi and is the base substance (*see* chitin cuticle) of the exoskeleton of arthropods. Chitin occurs to a lesser extent in the tissues of sponges, cnidarians, moss animals, lamp shells, molluscs, and annelids. *Adj.* **chitinous.**

chitin cuticle (L. *cuticula* = thin skin), a shell on the body wall of arthropods secreted by epidermal cells; forms the exoskeleton of arthropods and is chiefly composed of → chitin that is a very permanent substance against friction and chemicals. The chitin cuticle includes various layers: the uppermost is the **epicuticle** (*epicuticula*), under it the pigment layer, **exocuticle** (*exocuticula*), and undermost the **endocuticle** (*endocuticula*), with a differentiated upper part, **mesocuticle** (*mesocuticula*). The chitin cuticle is produced by epidermal cells which form a monocellular layer under the endocuticle. Substances supporting the exocuticle are quinones in insects, sulphur in arachnids, and calcium salts (lime) in crustaceans and some myriapods. At ecdysis or moulting, the old cuticle is discarded. *See* procuticle.

chitons, → Polyplacophora.

Chlamydiales (*Chlamydia* = the type genus, Gr. *chlamys* = cloak, mantle), an order of bacteria comprising Gram-negative coccoid microbes, which mainly live as parasites in vertebrates causing a variety of diseases also in man; includes the family Chlamydiaceae.

chlamydospore (Gr. *sporos* = seed), a spore type in → smut fungi.

chlor(o)- (Gr. *chloros* = light green), denoting green or chlorine.

chloragogen cells (Gr. *agogos* = leading), cells surrounding the intestine and dorsal blood vessel in annelid worms; the cells are derived from the **chloragogen tissue** (coelomocytes) which is derived from the peritoneum. Chloragogen cells function in the synthesis and storage of glycogen and fat, thus corresponding to the liver cells of vertebrates. Mature cells, **elaeocytes**, which are released into the coelom, transport fat to other tissues.

chloral hydrate, 2,2,2-trichloro-1,1-ethanediol, $C_2H_3Cl_3O_2$, a sedative and hypnotic compound manufactured from ethanol and chlorine; used e.g. as an anaesthetic for animals.

chloramphenicol, CAP, an antibiotic produced by a *Streptomyces* bacterium, now prepared synthetically; inhibits the translation of RNA to protein by blocking the peptidyltransferase on ribosomes in many prokaryotic organisms. It also binds to the 50S subunits of chloroplast and bacterial ribosomes and prevents the formation of peptide bonds in protein synthesis. CAP is used internally or externally against many pathogenic bacteria in medication of man and domestic animals. *Syn.* Chloromycetin (trade name).

chlorenchyma (Gr. *enchyma* = cast), a type of the plant → parenchyma tissue; is a green, photosynthetically active tissue occurring especially in leaves.

chloride cells, epithelial cells of fish gills functioning in the active transport of salts (chlorides).

chlorinated hydrocarbons, chlorocarbons, chlorohydrocarbons; chemicals prepared by chlorination of hydrocarbons or by addition of HCl or Cl_2 to olefins; used e.g. as solvents (CCl_4), chemical intermediates (C_2H_5Cl) or pesticides, particularly as insecticides (e.g. → DDT, $C_{14}H_9Cl_5$, and metoxychlor i.e. methoxy-DDT, $C_{16}H_{15}Cl_3O_2$), or for the treatment of seeds, bulbs and saplings (e.g. lindane, $C_6H_6Cl_6$). Chlorocarbons are very persistent and toxic to animals particularly affecting the nervous system, reproductive organs, and gills. Chlorinated hydrocarbons concentrate in tissues and organs, especially in fat, and gradually accumulate in organisms of various food chains (*see* **biomagnification**). Chlorinated hydrocarbon pesticides have become an environmental problem and therefore their usage is restricted or totally forbidden in many countries.

chlorine, a gaseous, greenish halogen element with a pungent odour, symbol Cl, atomic weight 35.453; reacts at its lowest oxidizing state with hydrogen and forms gaseous hydrogen chloride, HCl, i.e. **hydrochloric acid** in water. This is a strong acid dissociating into Cl^- and H^+ ions. The corresponding salts are **chlorides**, e.g. NaCl, KCl and $CaCl_2$. Cl^- ions are essential for organisms and are required e.g. for the light reactions in → photosynthesis; they are also involved in the generation of electric potentials in cells (*see* action potential).

The salts of **hypochlorous acid** (HClO) are hypochlorites and those of **chloric acid** ($HClO_3$) are chlorates, e.g. $KClO_3$. Chlorates can be converted into **perchlorates** (e.g. $KClO_4$) by electrolytic oxidation. The corresponding acid, **perchloric acid**, $HClO_4$, is a strong oxidizing agent and is used in some chemical analyses, e.g. to precipitate nucleic acids and proteins, and in the → anthrone reagent for the determination of carbohydrates. Many chlorine compounds are toxic to organisms, e.g. → chlorinated hydrocarbons.

chlormequat, CCC, chlorocholine chloride, a widely used growth retardant that inhibits the growth of the stem and therefore prevents cereal stems from being laid flat by rain; inhibits the synthesis of → gibberellins.

chlorocruorin (L. *cruor* = blood), an iron-containing blood pigment (respiratory pigment) similar to → haemoglobin, found in the haemolymph of some marine polychaetes; red in colour but in a diluted solution greenish. *Cf.* haemerythrin.

chlorofluorocarbon, CFC, a hydrocarbon containing chlorine and fluorine, Freons by trademark, e.g. dichlorodifluoromethane, CCl_2F_2, (Freon-12). CFC compounds are inert substances which have been used e.g. in the manufacture of plastics, in chemical cleaning, as aerosol propellants, in insulating materials, as a cleaning agent of electronic cards, and as a coolant in refrigeration equipment. When liberated into the air, CFSs move slowly to the upper atmosphere, and at a height of 10 to 15 km UV radiation affects these compounds liberating reactive chlorine atoms, which can gradually destroy the protecting atmospheric ozone layer. Together with many other gases, as carbon dioxide, → methane, and → nitrogen oxides, CFCs also promote the development of the global → greenhouse effect. The use of CFCs is now restricted, and safer substances have been developed, e.g. by adding hydrogen into the CFC molecule (HCFC compounds).

chloroform, trichloromethane, $CHCl_3$; a colourless, volatile organic solvent that belongs to metal halogenoids; generally inhibits neuronal functions in animals and is therefore used as an anaesthetic for animals, formerly also for humans, but because of its toxic effects on the liver and kidney, chloroform has been replaced by other, less toxic anaesthetics. It is used as a solvent in chemical industry and research laboratories.

Chlorophyceae (Gr. *phykos* = seaweed), a class in the division → Chlorophyta, Plant kingdom, containing all the best known and most important green algae.

chlorophyll (Gr. *phyllon* = leaf), the group name for the green pigments of plants, the most common being the bluish green **chlorophyll a,** $C_{55}H_{72}MgN_4O_5$, and yellowish green **chlorophyll b,** $C_{55}H_{70}MgN_4O_6$. There are also many other chlorophylls, as the **chlorophylls c, d** and **e** in plants, and the → **bacteriochlorophyll** in many bacteria. Chlorophylls are ring molecules consisting of four pyrrole rings bound to a central atom of magnesium. The long hydrocarbon chain, phytol, is bound to one pyrrole ring, and the lipid solubility of chlorophylls is due to that chain. Chlorophylls differ from each other depending on the side groups of the pyrrole rings and the sites of the double bonds. The light absorption curves are specific for different chlorophyll molecules, and the identification and determination of chlorophylls are based on those properties. The two absorption maxima of chlorophyll a in an acetone solution are at 460—480 nm and 650—670 nm. Chlorophylls play a central role in → photosynthesis where they act as → photon acceptors.

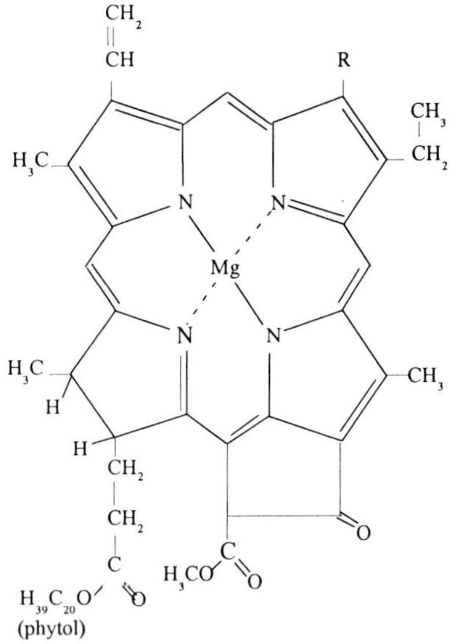

Chlorophyll. Chlorophyll a, R = CH_3; chlorophyll b, R = HCO.

Chlorophyta (Gr. *phyton* = plant), **green algae**, a division in the Plant kingdom, containing the classes Chlorophyceae and Prasinophyceae. The green algae may be unicellular, thread-like, or form green thalli, containing chlorophylls a and b, betacarotenes, and xanthophylls. Most species grow in fresh water but many occur in seas, some in soil, or e.g. on tree trunks. Green algae are evidently related to → mosses and further to vascular plants. The division consists of ca. 10,000 species.

chloroplast (Gr. *plastos* = moulded), a cellular organelle (plastid) of plants in which → photosynthesis takes place; each plant cell may contain numerous chloroplasts. Chloroplasts are disc-shaped, 3—10 µm in diameter and surrounded by a two-layered **envelope**, the outer membrane of which is very permeable to various substances, the inner membrane being more selective. The basic substance is **stroma** where the carbon assimilation and further metabolism take place; it is analogous to the matrix of the → mitochondrion. The internal membranes are called **thylakoids** which consist of **grana (grana thylakoids)**, i.e. stacks of thylakoids, and of **stroma thylakoids** between grana. This structure occurs in all seed plants, pteridophytes, mosses, and green algae, whereas the brown and red algae have chloroplasts with a more simple structure. Two kinds of chloroplasts occur in → C_4 plants; mesophyll chloroplasts with numerous large grana, and bundle sheath chloroplasts without them.

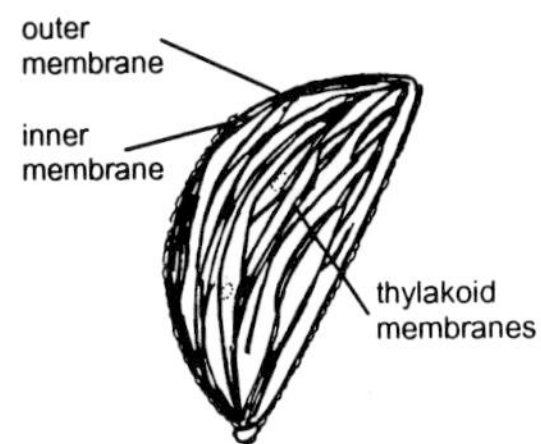

Chloroplast. Model of chloroplast. (After Postlethwait & Hopson: Nature of Life - McGraw-Hill, 1992; with the permission of the authors).

Thylakoid membranes contain → **chlorophylls** and other photosynthetic pigments, being the centre of the light reactions of photosynthesis. The **lumen** of thylakoids is an aqueous region and contains proteins, magnesium ions, and charged molecules, resulting in a potential gradient across the thylakoid membrane. The movement of ions across the membrane produces changes in pH, potential, and magnesium concentration controlling the activities of some enzymes. Stroma contains starch grains, lipid-containing spherical **plastoglobuli**, enzymes, circular DNA strands (→ chloroplast DNA), and 70S → ribosomes, indicating that the chloroplast is a very independent cell organelle. As chloroplast ribosomes are the same size as in bacteria, this is taken as evidence for evolutionary origin of chloroplasts from bacteria (*see* endosymbiosis theory).

chloroplast DNA, cpDNA, the circular double-stranded DNA which occurs in multiple copies, forming the genetic information of chloroplast. This DNA encodes ribosomal RNA (rRNA), transfer RNA (tRNA), and proteins such as Rubisco and some membrane proteins of photosystems.

chlorosis, 1. a plant disease caused by lack or degradation of chlorophyll; **2.** a rare type of anaemia in man.

choana, pl. **choanae** (Gr. *choane* = funnel), a funnel-like opening, e.g. **1.** the opening from the nasal cavity on both sides into the nasopharynx in → choanates (Choanichthyes) and tetrapod vertebrates; called also posterior naris, postnaris; **2.** a funnel-like collar in → Choanoflagellata.

choanates, a group of vertebrates which have nostrils opening into the mouth cavity; the animals have (or have had) functional lungs and paired fins with a median lobe, or paired limbs; the group includes the **lungfishes, coelacanths,** and **tetrapods.**

Choanichthyes (Gr. *ichthys* = fish), the former name for → Sarcopterygii (lungfishes and coelacanths).

choanocytes (Gr. *kytos* = cavity, cell), **collar cells**; cells occurring abundantly in sponges; with long flagella the choanocytes keep water in motion through flagellated chambers and in the spongocoel. The collar-parts of the cells filter food particles from the water which are transported to → archaeocytes for intracellular digestion.

Choanoflagellata, Choanoflagellida (L. *flagellum* = whip), **choanoflagellates**; a protozoan group of mostly stalked zooflagellates; have a long flagellum the base of which is surrounded by a funnel-like collar, choana; choanoflagellates may be ancestors of sponges.

chol(e)-, cholo- (Gr. *chole* = bile), denoting relationship to bile.

cholecalciferol, an antirachitic vitamin, D_3, $C_{27}H_{44}O$, in the group of calciferols (*see* vitamin D); produced in the human skin from 7-dehydrocholesterol by the action of UV light.

cholecystokinin, CCK (Gr. *kystis* = bag, bladder, *kinein* = to move), a hormone secreted by certain mucosal cells in the anterior part of the intestine in vertebrates; includes several short polypeptides, also found in the central nervous system where they probably act as neurotransmitters. CCK stimulates contractions of the gallbladder and thus the bile secretion to the intestine; it also stimulates the exocrine activity of the pancreas, especially the secretion of amylase. *Syn.* pancreozymin.

cholera, 1. an infectious disease in man caused by an enterotoxin (**choleragen**) produced by the bacterium *Vibrio cholerae*; it is endemic in India and southeast Asia, causing serious disturbances in the function of the intestinal mucosa; **2. fowl cholera** (chicken cholera) caused by the bacterium *Pasteurella multocida*); a haemorrhagic septicaemia occurring in domestic and many other birds all over the world.

cholesterol (Gr. *stereos* = solid), a steroid alcohol, $C_{27}H_{45}OH$, in the group of lipids; a

principal compound of cell membrane structures in animals found in all tissues, tissue fluids, and secretions such as in bile, blood, milk, egg yolk, fats and oils, etc. It is also a precursor for the biosynthesis of the steroid hormones, bile acids and vitamin D. In vertebrates, cholesterol is synthesized in the liver and secreted in bile to the intestine, from which most of it with some dietary cholesterol and other lipids is absorbed into the intestinal lymph vessels, some also into the blood vessels.
Although cholesterol is fat-soluble it stays soluble in the blood and tissue fluids with the aid of protein molecules surrounding it against water; the lipoprotein complexes thus formed may contain also other lipids. These complexes are e.g. **LDL cholesterol** (LDL = low density lipoprotein) and **HDL cholesterol** (HDL = high density lipoprotein). In LDL complexes cholesterol is transferred to tissue cells, and so also to endothelial cells of blood vessels, in which it may stay and be accumulated, causing a plaque (atheroma); therefore the name "bad cholesterol" for LDL cholesterol. HDL is capable of taking excessive cholesterol from the endothelial cells, and therefore called "good cholesterol". Obviously the oxidized form of cholesterol tends to remain more persistently in these cells. The cells have specific sites on the cell membrane, **cholesterol receptors** (e.g. LDL receptors, HDL receptors), the number of which determines the intensity of the cholesterol intake to cells. The commercial preparation of cholesterol used in pharmacy is called cholestrin.
Cholesterol has been found also in some groups of lower plants, e.g. in red algae, and even in some higher plants, e.g. potato; quantities are, however, very small.

choline, 2-hydroxyethyltrimethylammonium, $CH_2OHCH_2N^+(CH_3)_3$; present in organisms as a constituent of membrane → phospholipids such as lecithin (phosphatidylcholine). In animals, choline is also needed for the biosynthesis of → acetylcholine. Because choline has an effect on liver metabolism (inhibits accumulation of fat) it is sometimes included in the vitamin B group.

choline acetyltransferase, an enzyme (EC 2.3.1.6) which in cholinergic synapses of the nervous system catalyses the biosynthesis of → acetylcholine: acetyl-CoA + choline $\rightleftharpoons$ CoA + acetylcholine. Also called choline acetylase.

cholinergic (Gr. *ergon* = work), pertaining to structures or functions related to → acetylcholine; e.g. cholinergic nerves, cholinergic synapses, cholinergic transmission, or cholinergic drugs.

choline esterase, also **cholinesterase,** an enzyme which in different tissues of animals catalyses decomposition of → acetylcholine, i.e. the reaction: acetylcholine + H_2O $\rightleftharpoons$ choline + acetate. Two types exist: **choline esterase I** (acetylcholinesterase, true cholinesterase), EC 3.1.1.7, is specific to acetylcholine acting in the central nervous system and peripheral neuroeffector junctions; **choline esterase II** (pseudocholinesterase, serum cholinesterase), EC 3.1.1.8, is unspecific catalysing a variety of choline esters in all tissues.

chondr-, chondro- (Gr. *chondros* = grain, cartilage), denoting relationship to the cartilage or cartilaginous tissue. *Adj.* **chondral**, cartilaginous, **chondroid** (chondroitic), pertaining to or resembling a cartilage.

Chondrichthyes (Gr. *ichthys* = fish), **cartilaginous fishes**; a class of fishes with a cartilaginous skeleton and a skin covered by placoid scales; the swim bladder is absent; they reproduce through internal fertilization and are either oviparous or ovoviviparous, most of them are oceanic. The class comprises two subclasses, **sharks** and **rays** (Elasmobranchii) and **chimaeras** (Holocephali). *Cf.* Osteichthyes.

chondri(o)- (Gr. *chondrion* = granule), pertaining to a granule, e.g. chondriosome, → mitochondrion.

chondriome, the entity containing all the → mitochondria of a cell.

chondroblast (Gr. *chondros* = cartilage, *blastos* = germ), any of dividing cells forming cartilage in the body of vertebrates; derived from the embryonic mesenchyme. In the mature → cartilage these cells occupy the lacunae within the cartilage matrix, and are called **chondrocytes**.

chondroclast (Gr. *klastos* = broken in pieces), a multinucleated cell, a type of → macrophage, functioning in the resorption of → cartilage.

chondrocyte (Gr. *kytos* = cavity, cell), *see* chondroblast.

chondrocranium (Gr. *kranion* = skull), cartilaginous skull; **1.** in all vertebrates, the cartilaginous parts of the embryonic skull; i.e. the part of the skull surrounding the brain, auditory organs, and partially also eyes and olfactory organs; **2.** the skull of cartilaginous fishes.

chondroitin, a → glycosaminoglycan composed of about 60 repeating disaccharide units with glucuronic acid (or iduronic acid) and galactosamine residues; a component in the ground substance of the connective tissue, especially in cartilage and bones. Several forms exist, e.g. chondroitin sulphate B (dermatan sulphate).

Chondrostei (Gr. *osteon* = bone), **chondrosteans**; a primitive infraclass (or superorder) of ray-finned fishes (Actinopterygii), characterized by a largely cartilaginous skeleton, ganoid scales, a heterocercal tail, and unsegmented notochord. The infraclass includes e.g. **sturgeons** (Acipenseriformes), **bichirs** (Polypteriformes), and several extinct species.

chord(o)- (L. *chorda* = cord), denoting relationship to a cord, or the → notochord. *Adj.* **chordoid**, resembling the notochord.

Chordata, a phylum of animals which have four common features at least in some stage of their life cycle, i.e. a single, dorsal tubular **nerve cord**, a **notochord**, **gill slits**, and a **postanal tail**. In adults, these features may be altered or may have disappeared. The most developed is the body of vertebrates in which the anterior end of the nerve cord is enlarged to form the brain, and the notochord is developed into the vertebral column. In tetrapod vertebrates, the → branchiogenous organs are derived from the gill slits. The phylum contains three subphyla: **tunicates** (Tunicata), **lancelets** (Cephalochordata), and **vertebrates** (Vertebrata). There are about 45,000 species of chordates, nearly 44,000 of which are vertebrates.

chorioallantoic membrane, chorioallantois, allantochorion; the combined extraembryonic membrane in the Amniota group of vertebrates, formed by the partial fusion of mesodermal layers of the → chorion and → allantois. In birds and reptiles, this membrane lies immediately beneath the porous shell containing an extremely rich vascular network which provides for the oxygen-carbon dioxide exchange and transports calcium from the eggshell into the embryo for bone production. In eutherian mammals, the chorioallantoic membrane forms the foetal part of the placenta (chorioallantoic placenta).

chorioid, *see* choroid.

chorioidea, → choroidea.

chorion (Gr. = membrane), **1.** chorionic sac; the outermost of the → extraembryonic membranes enclosing the embryo and → amnion in reptiles, birds, and mammals; derives from ectodermal and mesodermal folds. In many species it partly fuses with → allantois to form the → chorioallantoic membrane (allantochorion) which in mammals forms the foetal part of the placenta. Chorion is highly vascularized and involved in respiratory gas exchange, nutrition, and waste removal; **2.** the non-cellular membrane around eggs of some animals, such as insects. *Adj.* **chorionic**.

chorionic gonadotropin, a glycoprotein hormone of mammals secreted by the placenta; especially in the beginning of pregnancy it maintains the function of the maternal corpus luteum, stimulates testosterone secretion from foetal testicles, and activates steroidogenesis in the placenta.

C-horizon, a soil layer below the upper layers, A- and B-horizons; typical e.g. of the → podsol.

choroid, (Gr. *chorion* = membrane, *eidos* = form), **1.** choroid coat; the middle vascular coat between the retina and sclera in the eye of vertebrates and some invertebrates (e.g. squids); also chorioid, choroidea, chorioidea; **2.** choroid plexus, tuft projections of blood vessels in the brain ventricles, secreting the cerebrospinal fluid; **3.** choroid rete, a countercurrent arrangement of arterioles and venules behind the retina in the eyes of teleost fish; **4.** as an adjective, resembling the → chorion or corium (*see* dermis); also choroidal or chorioidal.

choroidea, chorioidea, *see* choroid.

chrom(o)-, chromat(o)- (Gr. *chroma* = colour), denoting relationship to colour, or chromium.

chromaffin cells (L. *affinitas* = affinity), cells containing secretory granules of → catecholamines and therefore stain with chromium salts; occur in the adrenal medulla and paraganglia of the sympathetic nervous system in vertebrates.

chromatid, the longitudinal subunit (half chromosome) of duplicated chromosomes visible in the light microscope. The chromatids arise as the result of chromosome duplications, and become visible during the early prophase of → mitosis and second meiotic division, and remain visible until metaphase (*see* meiosis). After this, when the chromatids of each chromosome separate at anaphase, they are called daughter chromosomes. **Sister chromatids** are those two chromatids which during → interphase arise from the duplication of an individual chromosome and are seen during the prophase and metaphase of mitosis and meiosis.

chromatin, the material of which both interphase and mitotic chromosomes consist (*see* mitosis). In interphase, chromatin constitutes of about equal parts of DNA, → histones, and non-histone proteins by weight. All the three macromolecules contribute to form the nucleoprotein component of the nucleus. DNA and histones form complexes, called → nucleosomes, which are the structural units of the chromosomes. During mitosis, the relative amount of non-histones increases since these proteins play an important role in the condensation of chromosomes.

chromatin diminution, a process found in some nematodes, copepods, insects, and some ciliates by which portions of → chromatin are eliminated (*see* elimination chromatin) from the presumptive somatic cells, thus leading to somatic tissues containing less genetic material than the → germ line cells.

chromatogram (Gr. *gramma* = drawing, mark), the result of a chromatographic analysis expressed as a diagram. *See* chromatography.

chromatography (Gr. *graphein* = to write), a method in chemistry and biochemistry for isolating, purifying, and determining different organic compounds. The technique is based on the adsorption of compounds on the surface of a solid component and on the solubility and volatility of adsorbed compounds, or on the specific affinity of one molecule for another (affinity chromatography). Two phases are required to separate substances. The **stationary phase** is either a solid substance or a liquid adsorbed on a solid carrier. The **moving phase** is a gas or liquid. Compounds to be separated are divided into these two phases in different ratios. Practically one phase flows through the other and the molecules move continuously from one phase to another, often very rapidly. Different types of chromatography have been developed:

1) **column chromatography** is the simplest of these techniques. A column is filled with an adsorbent and an eluent flows through the column. The compounds are adsorbed as zones in the column where they can be identified and determined;

2) **thin-layer chromatography**, **TLC**, resembles column chromatography. The stationary phase (e.g. silica gel or cellulose) is a thin layer spread on a glass plate;

3) in **paper chromatography**, water is the stationary phase adsorbed by filter paper and an organic solvent mixture is the moving phase;

4) in **gas-liquid chromatography** (**GLC**) an inert gas or carrier gas serves as the moving phase and the stationary phase is on the surface of a support substance in a column (a packed glass column), in **glass capillary** columns without any support materials, the inner surface of the capillary forming the stationary phase. When a column is heated, the volatile substances of a sample are carried out with a carrier gas and they are recorded in a detector, e.g. by a flame ionization detector, and a chromatogram is obtained as a result;

5) **high performance liquid chromatography**, **HPLC**, has the advantages both of column chromatography and gas chromatography. The separation is carried out in a metal column at a high pressure (higher than 50 MPa) which considerably accelerates separation of compounds;

6) in **supercritical fluid chromatography**, compounds are separated using a supercritical fluid that is the moving phase and its properties can be controlled. The compounds dissolve into the supercritical fluid, and the technique includes the methods of partition and adsorption chromatographies;

7) in **ion exchange chromatography**, the stationary phase consists of polymers (ion exchange resins) which contain acid (-COOH, $-SO_3H$) or basic groups ($-NH_2$, $-NR_2$). H^+ ions are changed with other cations and -OH groups are replaced by other anions. Usually water serves as the moving phase. The compounds of a sample are adsorbed in a column and they are liberated in a certain order when the composition or strength of an eluent changes.

chromatophore (Gr. *phorein* = to carry along), in animals, a **pigment cell**, usually a branched cell in which a pigment can be dispersed into the whole cell, and gathered into a small granule near the cell centre, thus increasing or decreasing the colour intensity of the cell, respectively. As movements of the pigment in the skin are controlled by the nervous system, the animal can change its colour quite quickly. The chromatophores are named according to the pigment: **melanophores** (melanocytes in birds and mammals) are brown or reddish, **lipophores** red or yellow, and **guanophores** with reflecting crystals are polycoloured in reflective colours. Chromatophores protect structures of the skin and eyes from ultraviolet radiation, they also protect the eggs and embryos which, developing outside the body, are exposed to sunshine. In the eyes special pigment cells act in the direction and dimming of light.

In botany, chromatophore refers to a cell organelle containing pigments, e.g. the photosynthesizing membrane structure in → cyanobacteria. *Cf.* chromophore.

chromatosome, the core particle of a → nucleosome containing 166 base pairs of DNA, a → histone octamer, and one molecule of histone H1.

chromium, a bluish metal, symbol Cr, atomic weight 51.996. The oxidation state of chromium varies from +VI to -II, and the most stable is +III (e.g. chromic oxide, Cr_2O_3). Chromium ion, Cr^{2+}, is a micronutrient for animals acting e.g. in glucose metabolism, but is toxic in higher concentrations. Chromium is used in dye and metal industry, and has become an environmental pollutant that disturbs the cellular metabolism of organisms.

chromocentre, Am. **chromocenter,** the deeply stainable region in the cell nucleus during → interphase, consisting of tightly coiled chromosomes or their parts; contains → heterochromatin and is positively heteropycnotic (*see* allocyclic). Because the chromocentres tend to fuse, their number per nucleus may vary even within the same type of cells.

chromomere (Gr. *meros* = part), any of the bead-like thickenings of the chromatin organized along the chromosome. Chromomeres can be separated from their interval regions because of their deeper stainability. They can be seen as chromosome specific structures during the prophase of → mitosis or → meiosis. During the interphase of → giant chromosomes they are visible as transverse bands. The chromomeres are formed when in certain parts of the chromosome a stronger coiling occurs than in their interval regions. It is believed quite commonly that the chromomeres are the material counterparts of genes. When genes are activated, the coiling of the chromomere opens to form loops which are visible as specific → puffs in the → lampbrush chromosomes and → polytene chromosomes.

chromonema, pl. **chromonemata** (Gr. *nema* = string), **1.** the smallest structural fibre of a chromosome and chromatid, visible in the light microscope; consists of one double strand of DNA, and is connected to → chromomeres; **2.** a coiled substructure inside the → chromatid.

chromophore (Gr. *phorein* = to carry along), a chemical group or radical that gives colour to a compound and acts as a light absorbing pigment in organisms; has an unsaturated group, often with a double bond, as -C=C-, -C=O, or -N=N- (in azo dyes). *Cf.* chromatophore.

chromoplast (Gr. *plastos* = moulded), a pigmented plastid in plant cells, containing → carotenoids. Chromoplasts give yellow and red colours to autumn leaves, many flowers, and fruits. A chloroplast may develop into a chromoplast when the internal structure and chlorophyll disappear.

chromoprotein, a protein molecule containing a coloured atom group, → chromophore; e.g. → phytochrome, cytochrome, haemoglobin, and myoglobin.

chromosomal complement, → chromosome set.

chromosomal replication, a phenomenon based on the ability of DNA to replicate. The replication of DNA and the whole chromosome is semiconservative, i.e. the double-stranded structure of DNA opens, and both strands build a new complementary strand from free → nucleotides. This results in two identical DNA molecules which after duplication combine with → histones to form the → **chromatids** of the chromosome. In → prokaryotes, DNA forms a single replication unit, → **replicon**, while in → eukaryotes the chromosomes consist of many replicons, the number of which varies according to the developmental stage. The regulation of the chromosomal replication speed is based on the variation of the number of replicons.

chromosome (Gr. *soma* = body), any of the stainable bodies in the nuclei of eukaryotic organisms consisting of DNA and proteins; contains the → genes. Chromosomes are characterized by a linear structure, staining with alkaline dyes, and a typical behaviour during → mitosis, they are capable of duplication, and their number per cell, shape, and organization are typical of each species of organisms. Between cell divisions during interphase, → genetic transcription occurs in chromosomes transferring a part of their → genetic information to cytoplasmic → protein synthesis, and hence guide the metabolism and differentiation of the cell. During interphase the chromosomes also duplicate for nuclear division. During nuclear division the chromosomes are coiled and become visible in a light microscope as rod-like or string-like structures. The chromosome consists of two sister chromatids, and the → centromere divides the chromosome into two arms. On the basis of the stainability of chromosomal material the chromosome consists of → **euchromatin** and → **heterochromatin**.

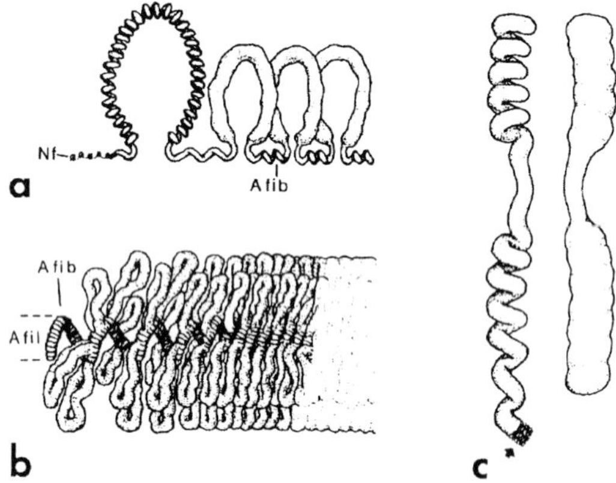

A schematic model of the hierarchical coiled structure of the metaphase choromosome. (a). The basic fibre of the chromosome, i.e. the 30 nm nucleoprotein fibre (Nf) is coiled into chromomere loops, between which an axial fibre (A fib) exists. (b). The axial fibre will further be coiled into a tube-like axial filament (A fil), from which the loops protrude. (c). The structure shown in (b) will form by further coiling the mature metaphase chromosome. (After Nokkala & Nokkala: Hereditas 104, 1986; with the permission of the publisher).

The chemical structure of the **chromosomes of eukaryotic organisms** consists of one single giant DNA molecule, → histone proteins and non-histone proteins, and a small quantity of RNA which is the transcription product of the genes activated in the chromosome. During interphase there are equal quantities of DNA, histones, and non-histones. The fine structure of the eukaryotic chromosome consists of **nucleosomes**, each of which consists of nine molecules of five different histones, and of approx. 200 base pairs of DNA. Depending on the length, one chromosome may contain tens of thousands to millions of nucleosomes. DNA is wrapped around the histone cores, and forms the **basic fibril of chromosome** (10 nm in width). The histone core comprises two molecules of histones H2a, H2b, H3 and H4. Histone H1 molecule holds the basic fibril united by binding to DNA between the histone cores. The basic fibril is further coiled forming a coil of the next order, i.e. the **solenoid coil** that is 30 nm in width. This is the basic state of the interphase chromosome. In this condition chromosomal DNA is at rest, i.e. no transcription occurs. By activation of the gene for transcription the nucleosomal structure is reversibly loosened, and the non-histone proteins remove the histones from their association with DNA.

The **chromosomes of prokaryotic organisms** are simpler. They have no nucleosome structure but DNA occurs in the cell free, i.e. bound by a few histone molecules into a loop-like structure. *See* chromosome number.

chromosome aberration, → chromosome mutation.

chromosome arm, the → centromere divides the chromosome into two parts, i.e. two arms.

chromosome jumping, a technique for isolation of genes; the chromosome is cut with the aid of → restriction enzymes distal to a known gene and a wanted gene. By circularizing this chromosome segment these two genes are brought into proximity of each other. Then the wanted gene can be isolated by using → chromosome walking from the known gene to the wanted gene.

chromosome map, a linear graphical presentation of the chromosome, the genes belonging to the same → linkage group being in a linear order; gene distances are indicated in → map units which correspond to their distances on the chromosome. There are two types of chromosome maps, i.e. the **genetical map** (relative map) and the **cytological map** (physical map). The former type is the map of linkage groups, on which the loci and distances of genes are based on frequencies of → crossing-over observed in crossing experiments. The cytological map describes the chromosome on which the physical locations of the genes has been indicated; these are based on cytological observations in which → chromosome mutations or → DNA hybridization are employed.

Genetical and cytological maps of the same chromosome are **colinear**, i.e. genes are located on them in the same order. The distances of the genes on these two maps may deviate from each other, because the cytological map represents physical distances whereas the genetical map represents relative distances which are dependent on the probability of crossing-over on different regions of the chromosome.

chromosome mutation (L. *mutare* = to change), chromosome aberration; a change in the structure of the chromosome involving the quantity or organization of the genetic material in the chromosome. Chromosome mutations may be → **deletions** in which a part of the chromosome is missing, → **duplications** in which a part of the chromosome is represented twice in the genome, → **inversions** in which a given segment of the chromosome has rotated 180°, → **translocations** in which non-homologous chromosomes have exchanged parts reciprocally, or → **transpositions** in which a given segment of a given chromosome has been removed to another location either in the same or a non-homologous chromosome.

chromosome number, the number of all the chromosomes in a given cell of the certain species, usually constant in all somatic cells of the species. In somatic cells the chromosome number is → diploid but in gametes → haploid. The diploid chromosome number between different species can vary from 2 to tens, e.g. in man it is 46 and in fruit fly 8.

chromosome painting, the staining of the whole chromosome with a specific fluorescent dye.

chromosome pairing, the chromosome

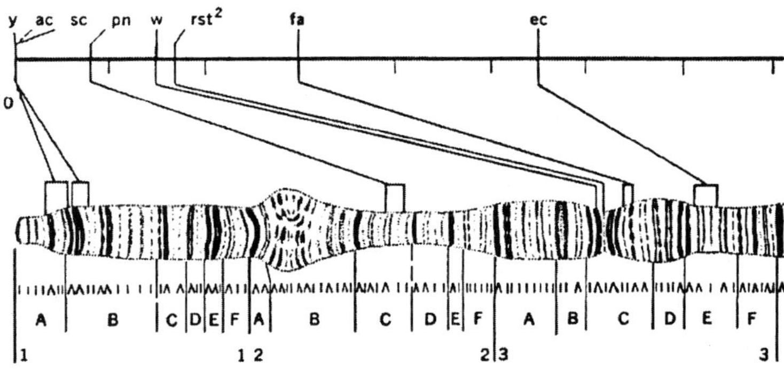

Chromosome map. Comparison of the genetic map (upper) and the cytological map (lower) of the tip of the X chromosome of the fruit fly. The lower case letters on the genetic map represent genes and their locations. The numbers below the cytological map indicate sections of the chromosome set, and the upper case letters subsections. (After Bridger: J. Heredity 29, 1938; with the permission of the publisher).

conjugation; has several forms:
1) **pairing of homologous chromosomes**, the specific association of → homologous chromosomes side by side; can be divided into four types:
a) **meiotic pairing** or **synapsis** is the tight alignment of homologous chromosomes clearly observable during the first prophase of → meiosis that leads to the formation of → bivalents in diploid and of → multivalents in → autopolyploid organisms. The bivalents and multivalents begin their formation in the zygotene of the first meiotic division and are maintained until the first metaphase. Synapsis is a necessary condition for the → segregation of homologous chromosomes occurring in the first anaphase. In electron microscopy a specific → synaptonemal complex, holding the homologous chromosomes together, is observed during → zygotene and → pachytene. Meiotic pairing begins at specific contact points of the chromosomes and proceeds in a zip-like manner until the chromosomes are fully paired at pachytene. The contact points are located at the ends of the chromosomes, at the centromere, or at both. In the multivalent formation each chromosome is paired at each point only with one homolog but usually changes its pairing partner at given points.
The mechanism in meiotic pairing is not fully understood, but it is apparent that the delay of the replication of the zygotene DNA until pachytene, is a necessary condition for pairing. Zygotene DNA consists of 0.3% of the total DNA of the chromosome and most likely forms the axis of the chromosome. At the beginning of the pairing the chromosomes are functionally single-stranded, even though most of their DNA (99.7%) has been replicated already during the proceeding interphase. Meiotic pairing is a genetically controlled phenomenon;
b) **somatic pairing** is the tight alignment of homologous chromosomes found during the prophase and metaphase phases of the → somatic cells of dipterans;
c) **salivary gland pairing** is a specific form of somatic pairing found in the salivary glands and other strongly excreting tissues of dipteran larvae; it leads to the formation of polytene → giant chromosomes. It is also found in the nurse cells of oocytes of adult dipteran females, and in the → polyploid cells of certain plants;
d) **secondary pairing of bivalents**, the non-random grouping of bivalents found in the first meiotic metaphase of certain polyploid species. Bivalents appear in pairs or groups if they are formed by genetically or evolutionarily related chromosomes;
2) **nons-pecific pairing** occurs between any chromosomes, and can be divided into three types:
a) **heterochromatic fusions** of chromosomes that lead to formation of → chromocentres, i.e. random associations of the heterochromatic parts of chromosomes;
b) **terminal non-specific association**, which is the association of the ends of univalent chromosomes during the first meiotic metaphase; these can also be consequences of heterochromatic fusions;
c) **non-specific pairing in meiotic metaphase**, the pairing of non-homologous chromosomes or chromosome regions in the case when homologous pairing has been interrupted due to structural heterozygosity or lack of the pairing partner.

chromosome polymorphism (Gr. *polys* = many, *morphe* = form), the occurrence of one or more structural forms of chromosomes in a population. The structurally changed chromosomes are consequences of → chromosome mutations. In chromosome polymorphism the mutations exist either as homozygous or heterozygous combinations in the population. The frequencies of different chromosome forms usually vary according to environmental circumstances which demonstrate the significance of chromosome polymorphism for the adaptation of organisms.

chromosome puff, *see* puff.

chromosome set, the haploid number of chromosomes per cell or individual.

chromosome theory, a theory stating that the factors of inheritance, i.e. the genes, are located on chromosomes in the cell nucleus. The theory was first presented by *Theodor Boveri* in 1902 and *Walter S. Sutton* in 1903. Sutton explained the Mendelian rules on the basis of the behaviour of chromosomes, and Boveri demonstrated the specificity and continuation of chromosomes, the necessary characters of the genetic material. The chromosome theory was confirmed when *Thomas H. Morgan* in 1911 explained the sex-linked inheritance on the basis of specific sex chromosomes. In 1915 Morgan with his group showed that the number of linkage groups was equal to the → haploid number of chromosomes which was very strong evidence in favour of the chromosome theory of inheritance.

The chromosome theory was finally proved by *Calvin B. Bridges* in 1916 when he showed that in fruit flies the → non-disjunction of the sex-linked genes was accompanied by an analogous non-disjunction of X chromosomes. Thus the exceptional behaviour of given genes was accompanied by a completely analogous exceptional behaviour of chromosomes, which must mean that the genes are located on chromosomes. More direct evidence in favour of the chromosome theory was obtained when the cytological mapping of genes was developed in the 1920's and 1930's. For the first time the exact location of given genes on the chromosome was indicated (*see* chromosome map).

chromosome walking, a method for the isolation of the genes the locations of which are known on the chromosome. The procedure occurs stepwise starting from a gene isolated earlier; an overlapping segment is isolated with it and further overlapping segments with this, etc. In this way the walking proceeds along the chromosome until the wanted gene is finally reached.

chron(o)- (Gr. *chronos* = time), denoting relationship to time.

chronaxie, *see* rheobase.

chronic toxicity (Gr. *toxikon* = arrow poison), a long-lasting toxicity; toxic effects of a poison in an animal, usually lasting for several months of continuous or repeated exposure to this poison. Chronic and acute toxicity tests must by law be performed on animals for new compounds and preparations (e.g. drugs) meant for human use.

chronobiology, a research branch of biology dealing with → biorhythms in organisms.

chronospecies, 1. a species that is represented in more than one geological time sequence; **2.** → palaeospecies.

chronotropism (Gr. *trope* = turn), a change of the rate of a periodic movement, e.g. heartbeat, through some influence; may be positive or negative chronotropism, i.e. acceleration or retardation of the rate, respectively. *Adj.* **chronotropic**.

chrysalis, pl. **chrysalises** or **chrysalides** (Gr. *chrysos* = gold), a hard-shelled → pupa of some holometabolous insects, e.g. lepidopterans.

chrysolaminarin, → Chrysophyta.

Chrysophyceae (Gr. *phykos* = seaweed), → Chrysophyta.

Chrysophyta (Gr. *phyton* = plant), **golden algae** and **diatoms**; a division in Plant kingdom, divided in subdivisions Chrysophyceae and Bacillariophyceae; unicellular algae living in plankton in waters and containing large amounts of carotenoid pigments and polysaccharide called **chrysolaminarin**. The bacillariophycean algae, i.e. **diatoms**, are important plankton algae. They have no cellulose cell wall but their walls consist of an inner pectin membrane and an outer armour of hydrated silica.

chyle (Gr. *chylos* = juice), the milky fluid found within lacteal lymph vessels in vertebrates after absorption of fats from the gut.

chylomicrons (Gr. *mikros* = small), lipoprotein complexes (emulsion particles, micelles) formed from lipid molecules and proteins when absorbed from the small intestine into the blood or lymph vessels (lacteal vessels). In vertebrates, each chylomicron contains about 80% triglycerides, 10% phospholipids, 3% cholesterol, and on the surface about 2% proteins (forming lipoproteins) necessary for maintaining the chylomicron particles in emulsion state.

chyme (Gr. *chymos* = juice), the semifluid material formed from food mass in the stomach by gastric secretion and then entering the intestine; called also **chymus**.

chymosin, → rennin.

chymotrypsin (Gr. *tryein* = to wear out), an endopeptidase, EC 3.4.21.1, found in the alimentary canal of vertebrates catalysing the hydrolysis of polypeptides preferentially at the carboxyl end of an amino acid, especially of phenylalanine, leucine, tyrosine, and tryptophan. It is secreted by the exocrine cells of the pancreas in the form of inactive precursor, **chymotrypsinogen**, which is activated when a part of the molecule is split off by another digestive enzyme, → trypsin.

Ciconiiformes (L. *ciconia* = stork, *forma* = form), an order of long-legged, long-necked birds of medium or large size, found on inland and marine shores in warm and temperate zones of the world. Six families and 117 species are known: **herons** and **bitterns** (Ardeidae), **hammerkop** (Scopidae), **whale-**

headed stork (Balaenicipitidae), **storks** (Ciconiidae), **ibises** and **spoonbills** (Threskiornithidae), and **flamingos** (Phoenicopteridae).

cili(o)- (L. *cilium* = eyelash, eyelid), denoting relationship to cilia. *See* cilium.

ciliary body (L. *corpus ciliare*), the part of the tunic of the vertebrate eye between the iris and the choroid coat; consists of the ciliary processes with the suspensory ligaments of the lens, and of the ciliary muscle accommodating the eye.

Ciliata, Ciliatea, a class in the phylum → Ciliophora.

ciliate, 1. having → cilia; **2.** a ciliophoran, i.e. any protozoan of the phylum Ciliophora.

ciliated epithelium, an epithelium, usually a columnar epithelium, the surface of which is covered with cilia; occurs e.g. on internal surfaces of respiratory pathways and in the oviduct.

ciliates, → Ciliophora, Ciliata.

Ciliophora (Gr. *phorein* = to carry along), **ciliates**; a protozoan phylum including species which live free in wet habitats or in water (e.g. *Paramecium*); some are parasitic. The cell body is binucleate with a → macronucleus and a micronucleus, and covered with cilia at least in immature stages of their life cycle. The well-developed cell organelles comprise e.g. the **cell mouth** (cytostome) and **gullet** (cytopharynx); the undigested food remains are pushed out through the **cell anus** (cytopyge). The classification of ciliates varies; some researchers divide the phylum into four subphyla Holotrichia, Spirotrichia, Suctoria, and Peritrichia; some others consider them as subclasses in the class Ciliata. The phylum is also divided into three classes: Kinetophragminophora, Oligohymenophora, and Polymenophora. Ciliophora are often called Ciliata or Infusoria.

cilium, pl. **cilia** (L.), **1.** eyelash; **2.** any of hair-like, vibratory organelles on the cell surface found in many protozoans, but also on the free surfaces of many epithelial cells (**ciliated epithelium**) in many tubular structures of multicellular animals; exist e.g. in the gills of molluscs, or in the respiratory pathways and in the oviduct of vertebrates. Often the cell surface is covered by cilia which beat rhythmically serving **cell locomotion** in an aquatic environment, or the cilia function in the transfer of material along the cell surface (*cf.* flagellum).

Structurally all cilia are similar. They are tubular extensions out of the cell membrane, having 9 pairs of → microtubules arranged around the two central microtubules (9 + 2 arrangement), all anchored onto basal bodies in the cytoplasm. The excitation state may propagate from the base of one cilium to another causing synchronous wave-like motion (**metachronal rhythm**) in the ciliary surface, or all the cilia may beat simultaneously (**isochronal rhythm**). In some protozoans, rows of beating cilia may be fused together to form an **undulating membrane**. Each single cilium has two functional phases, i.e. an **effective stroke** when the cilium stiffens and strikes forwards fast, and a **recovery stroke** when the cilium bends slowly backwards. *Cf.* stereocilium, flagellum.

cinnamon, 1. the aromatic inner bark of lauraceous trees (genus *Cinnamomum*) growing in the East Indies and Ceylon; **2.** any of such trees; **3.** a spice obtained from *Cinnamomum* trees; the genuine cinnamon originates from Sri Lanka and is obtained from the bark of young shoots.

cinnamic acid, cinnamylic acid; 3-phenyl-2-propenoic acid, $C_6H_5CH{=}CHCOOH$, obtained from strong aromatic **cinnamon oil** extracted from *Cinnamomum* trees. The oil contains 60% cinnamic aldehyde and 10% eugenol. Cinnamic acids and its derivatives are used in spices, cosmetics, and pharmaceuticals.

circadian rhythm (L. *circa* = about, *dies* = day), an internal, genetically controlled → biorhythm in organisms, the length of the rhythm being approximately one day, i.e. varying from 20 to 28 h depending on the species. The circadian rhythm is best seen in the behaviour of animals, but also found in various physiological and biochemical functions of most organisms, in plants e.g. in leaf and growth movements. In normal environment the rhythm is paced exactly to 24 h (**daily rhythm**) due to the rhythmicity of sunlight and other physical environmental factors, and appears as an endogenous rhythmicity, i.e. **free-running rhythmicity**, only in circumstances where the effects of these external factors are artificially eliminated. The activity of an organism may be high during the day and low during the night (**diurnal rhythm**), or vice versa (**nocturnal**

rhythm).

circannual rhythm (L. *annus* = year), a → biorhythm the length of which is about a year; occurs e.g. in the reproduction of organisms, moulting of animals, and blooming of plants. *Cf.* circadian rhythm.

circinotropous ovule (L. *circinatus* = roundish, *trope* = turning), a plant ovule type. *See* ovule.

circular (L. *circulus* = ring, circle), pertaining to a circle; having a form of a circle.

circular migration, loop migration; a type of the → seasonal migration which is characteristic of some migratory birds; in circular migration, the route of the return migration in the spring deviates from that of the autumnal migration; in this way migrate e.g. the golden plover in America, or the black-throated diver and the curlew sandpiper in Siberia.

circulation (L. *circulatio*), moving something in a circle or circuit, e.g. **1.** → blood circulation, including capillary circulation, pulmonary circulation, systemic circulation, etc; **2.** lymph circulation (→ lymphatic system); **3.** → enterohepatic circulation; **4.** cytoplasmic circulation, → cyclosis; **5.** air circulation; **6.** → water circulation; **7.** hydrological circulation, → hydrological cycle.

circulatory system, in physiology, the channels and spaces through which blood, lymph, or haemolymph circulates in an animal; consists of vessels and heart(s). *See* blood circulation, heart, blood vessels, lymph vessels.

circumnutation (L. *nutatio* = to sway), *see* nutation.

circumpolar (L. *polus* = end of axle), around a pole; especially denoting a location or distribution around the North Pole or the South Pole.

Cirripedia (L. *cirrus* = curl, *pes*, gen. *pedis* = foot), **barnacles**; a crustacean class in the class Maxillopoda; free-swimming and pelagic in larval stages, but as adults attached to a substratum either directly as acorn barnacles, or by a stalk as goose barnacles. Adults secrete a hard limy shell around themselves. The head and abdomen are reduced and the thorax has six pairs of legs for catching food from the opening of the shell.

cirrus, pl. **cirri** (L.), a curl or a curl-like structure; e.g. **1.** barbels around the mouth of barnacles; **2.** dorsal and ventral cirri in the parapodium of polychaetes; **3.** any of the hair-like structures in the legs of many insects; **4.** the copulation organ (penis) of some flukes and molluscs; **5.** a tuft of fused cilia serving as a sensory or locomotor organelle in some protozoans; **6.** a cloud formation.

cirrus sac, a sac-like structure of a fluke, including the **penis** (cirrus) and the opening of the sperm duct.

cis- (L. *cis* = on this side), pertaining to this side; on the same side of a structure.

cis-acting locus, a gene → locus that regulates the function of other genes located in the same DNA molecule or on the same chromosome. Usually it does not encode any protein but receives → *trans*-acting protein signals from elsewhere in the genome.

cis-acting protein, a rare regulatory protein having the capacity of regulating only the function of such genes which reside in the same DNA molecule or on the same chromosome as the gene encoding it (*see* genetic code).

cis-configuration, cis-position, to be situated on the same side of a structure; e.g. **1.** the location of two genetic factors in the same chromosome or molecule; **2.** the location of atoms or atom groups on the same side of a molecule.

cis-regulatory module, a DNA fragment containing multiple → transcription factor target sites; when activated, produces a particular supplement of the overall pattern of the gene expression.

cisterna, pl. **cisternae** (L.), **cistern**; e.g. **1.** a reservoir for lymph or other body fluids in vertebrates, as *cisterna chyli*, the dilated part of the thoracic duct receiving several lymph-collecting vessels from the lumbar and intestinal areas, or *cisternae subarachnoidales*, the subarachnoidal cisternae which are enlargements of the subarachnoid space on the cranial surface of the vertebrate brain; **2.** a sac-like vesicular cell organelle in the → endoplasmic reticulum and → Golgi complex, involved in the transport of cellular material.

cis-trans test (L. *trans* = on the other side), a genetic test for determining whether two recessive gene mutations (m_1 and m_2) are located in the same → cistron or in different cistrons of a functional gene. It is based on the following crossings: m_1/m_1 x m_2/m_2 giving the *trans*-heterozygous individuals, and m_1m_2/m_1m_2 x ++/++ giving the *cis*-

heterozygous individuals. If the *cis* heterozygote (m_1 m_2/+ +) is phenotypically different (wild) from that of the *trans* heterozygote (m_1 +/+ m_2) (which is mutant), m_1 and m_2 belong to the same cistron. If *cis* and *trans* heterozygotes are phenotypically identical (both of a wild type), the mutations belong to different cistrons. *Cf.* complementation test.

cistron, a unit of genetic function as defined by means of the → *cis-trans* test. The concept of the cistron corresponds to the concept of the → gene defined on the basis of the gene function. Materially one cistron corresponds to a DNA segment which contains information for the synthesis of one → polypeptide. *See* protein synthesis.

citric acid cycle, a series of the aerobic reactions of → cell respiration, following → glycolysis; takes place in the mitochondria of the cell in all aerobic organisms. The cycle gets new material in forms of **pyruvate** which originates from glucose processed in the glycolysis, and of **acetyl coenzyme A** which is generated in the lipid metabolism by β-oxidation. These are then oxidized in the reactions of the citric acid cycle while CO_2 and H^+ ions are liberated. The cycle consists of many intermediates and in the reactions also NADH and $FADH_2$ are produced. These act as reducing agents in further reactions of the → electron transfer chain associated with oxidative ATP production (*see* oxidative phosphorylation). *Syn.* Krebs cycle, tricarboxylic acid cycle (TCA cycle).

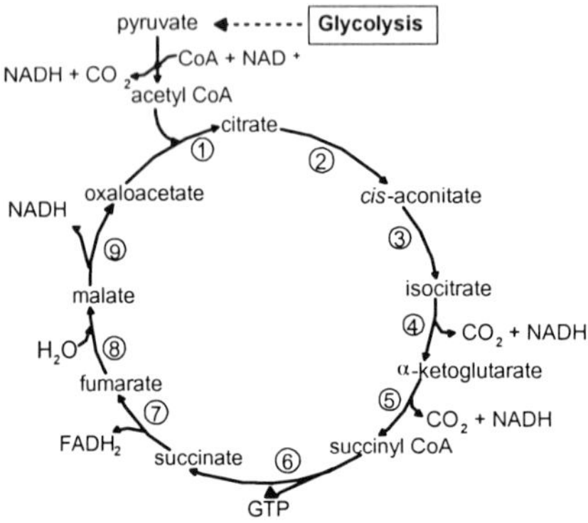

Citric acid cycle. Enzymes: 1 = citrate synthase, 2 = aconitase, 3 = aconitase, 4 = isocitrate dehydrogenase, 5 = α-ketoglutarate dehydrogenase complex, 6 = succinate thiokinase, 7 = succinate dehydrogenase, 8 = fumarase, 9 = malate dehydrogenase.

clade (Gr. *klados* = branch, sprout), **monophyletic group**; a group of species that includes the common ancestor of the group and all its descendants. In an evolutionary tree it is a branch denoting a monophyletic group of taxa formed in → cladogenesis.

cladistics, cladistic classifications; a method of classification which arranges organisms (and taxa) on the basis of their ancestral (primitive) and evolutionarily derived characteristics, and thus reflects phylogenetic relationships between different organisms. Cladistics ignores more or less the results of adaptation, such as phenetic similarities.

Cladocera (Gr. *keras* = horn), **cladocerans, water fleas**; a branchiopod order of crustaceans, important animals in fresh water plankton; they have characteristically a two-part carapace and large second antennae used for swimming. Their reproduction is unusual: in summer the parthenogenetic reproduction produces only females in a rapid cycle, but males are born under severe circumstances, especially in autumn, and then **winter eggs** are produced by sexual reproduction for the next summer.

cladode, a green, flat plant shoot resembling a leaf; is also called platyclade or phylloclade; e.g. butcher's broom.

cladogenesis (Gr. *gennan* = to produce), **1.** a type of the phyletic evolution in which a species branches into two new (daughter) species; **2.** a branching of different phylogenetic lineages constructed by cladistic methods.

cladogram (Gr. *gramma* = mark, drawing), a graphical description of branching of different phylogenetic lineages. The cladogram is not equivalent to a phylogenetic tree, the branches of which describe real lineages from the evolutionary past. *See* cladogenesis.

clamp, an instrument used for compression of a structure; e.g. used in surgery to grip, compress, or fasten parts together, or in neurophysiology to keep voltage or current at a fixed level. *See* voltage clamp, patch clamp.

clamp mycelium (Gr. *mykes* = fungus), a → mycelium, typical of → Basidiomycetes, in which the adjacent cells in dikaryotic hyphae are connected by a conjugation tube (**clamp connection**). This ensures that each cell contains two genetically different nuclei. Only clamp mycelium forms → fruiting bodies.

clams, → Bivalvia.

clasper, in zoology, an organ of some animals, such as insects, crustaceans, and cartilaginous fishes, by which the male clasps the female during copulation; e.g. a rod-shaped structure of the pelvic fin in male sharks and rays, used in copulation for the introduction of sperm. *Syn.* gonopod.

class, 1. a → taxon in classification and systematics of organisms; in botany it is below division or subdivision, e.g. the classes Monocotyledonae and Dicotyledonae in the subdivision Angiospermae, division Spermatophyta; in zoology it is below phylum or subphylum, e.g. the classes Reptilia and Mammalia in the subphylum Vertebrata. In botany and zoology the class is above order or superorder; **2.** a group arranged by classification of various structures such as macromolecules, cells, anatomical differences, etc., e.g. enzyme classes with subclasses and subsubclasses. *See* enzyme classification.

class I MHC, class II MHC antigens, *see* major histocompatibility molecules.

classical conditioning, *see* conditioning.

classical ethology, a branch of → ethology which emphasizes the classification and comparison of behavioural features of different animal species, especially on the basis of → fixed-action patterns. The central object of interest in the classical ethology is the species whose profit is mainly regulated by the behaviour of an individual. *Cf.* behavioural ecology.

classical nature conservation, the initial branch of environment conservation concentrated upon the original nature, especially by preserving species and natural landscapes with various protection procedures.

classification, a systematic arrangement into taxa and other groups; e.g. an arrangement of organisms into hierarchical systems on the basis of similarities or dissimilarities in their structures or other features. Species, populations or population groups are usually the objects of classification. At present, classification includes three major methods: numerical taxonomy, evolutionary taxonomy, and cladistics. The term is partly a synonym with the terms → taxonomy and → systematics. *See* hierarchy, class.

clastogenic (Gr. *klastos* = broken down, *genos* = origin), causing breakage of chromosomes; **clastogen**, an agent capable of causing such breakages.

clathrin (L. *clathrare* = to provide with a lattice), a fibrous protein in cells of eukaryotic organisms; a clathrin molecule consists of three heavy protein molecules (about 180 kD) and of three light chain proteins (35—40 kD), together forming a molecular triskelion structure. Clathrin molecules combine or polymerize to form a net-like structure that is associated with the function of the cell membrane in the receptor-mediated endocytosis. At a certain point of the cell membrane, **coated pit**, clathrin forms a coat around pinocytotic or phagocytotic vesicles resulting in a **chlathrin-coated vesicle** (about 50 nm in diameter). Similarly clathrin participates in the formation of coated vesicles separated from the → endoplasmic reticulum and transferred to Golgi complex.

clavicle (L. *clavicula* = a small key), **collar bone**; the paired long bone on the ventral side of the shoulder girdle (pectoral girdle) in tetrapod vertebrates. In mammals its medial end articulates to the anterior part of the sternum and the lateral end to the scapula. In birds the both clavicles have grown distally together to form the → **furcula** ("wish bone"). The clavicle is absent e.g. in actinopterygians and odd-toed ungulates (Perissodactyla).

clavus, pl. **clavi** (L. = nail), **1.** the posterior of the hard basal portions in the forewing (hemelytron) of the bugs Hemiptera; **2.** the → ergot disease in grasses.

clay, a soil material with very fine particles (below 0.002 mm in diameter) consisting of hydrated silicates of aluminum ($H_4Al_2Si_20_9$); a highly impermeable stratum.

cleaning symbiosis, *see* allogrooming.

clearance, the process of clearing, a clear space; in physiology, the removal of a substance from the blood, especially **renal clearance**, the rate (or its measurement) at which a substance is completely removed from a certain volume of plasma per time unit; usually is expressed as ml/min, i.e. ml plasma cleared in a minute.

clearing, in microscopic technique, a method used to make tissue specimens more transparent and more permeable to paraffin wax with e.g. benzene or xylene in the preparation of slides for light microscopy.

cleavage, 1. cleavage divisions; a series of successive cell divisions in the ovum of animals immediately following its fertilization; leads to the formation of the embryonal → morula stage; **2.** the scission or splitting of a molecule into two or more simpler molecules; **3.** the breaking of a crystalline susbstance so that cleavage planes often correspond to layers of atoms or molecules in the lattice.

cleid(o)-, 1. (Gr. *kleis* = clavicle), denoting relationship to the clavicle; also clid(o)-; **2.** (Gr. *kleistos* = closed), pertaining to a closed structure or state.

cleidoic (Gr. *kleidoun* = to lock in), in embryology, isolated from the surroundings; e.g. **cleidoic egg**, an egg type of terrestrial animals, such as reptiles and birds, in which the content is enclosed within a protective membrane or shell, permitting only exchange of gases and some water.

cleistocarp (Gr. *kleistos* = closed, *karpos* = fruit), → cleistothecium.

cleistogamy (Gr. *gamos* = marriage), the fertilization of a flower which does not open; the flowers are thus self-pollinated and self-fertilization takes place; occurs e.g. in some violets.

cleistothecium (Gr. *theke* = box), the closed, roundish fruiting body (→ ascocarp) of certain ascomycete fungi. *Syn.* cleistocarp.

Cleland's reagent (*W. Wallace Cleland*), → dithiothreitol.

cleptogamy, also **kleptogamy** (Gr. *kleptein* = to steal, *gamos* = marriage), a feature found in some animal groups, e.g. in birds and mammals, when a foreign male fertilizes the female which has mated with another male or belongs to the harem of another male.

cleptoparasitism, also **kleptoparasitism** (Gr. *parasitos* = parasite), a type of parasitism when an animal steals food from other animals which have caught it, e.g. the skua steals food from gulls. Cleptoparasitism may also be intraspecific, like in great herring gull colonies where the members rob each other.

climacteric (Gr. *klimakter* = rung of ladder), **1.** climacterium; the syndrome of changes occurring at the termination of the reproductive period in man, culminating in women in the menopause; **2.** any critical period in human life; **3.** the large increase in cellular respiration during ripening of some fruits, e.g. avocado.

climate zones, climatically different zones on the Earth. The best known zonation has been presented by *Wladimir Köppen* (1846—1940), describing six main zones: 1) zone of tropical climates, including tropical rain forest climate (*see* tropics, rain forest) and savanna climate (*see* savanna); 2) prairie climate (*see* prairie); 3) desert climate (*see* desert); 4) warm rain climates (e.g. deciduous forests); 5) snow and forest climate (coniferous forests), and 6) tundra climate zone (*see* tundra). *Cf.* vegetation zones.

climatic factors, abiotic environmental factors of the climate, as temperature, humidity, air currents, and atmospheric pressure, which influence physiological functions of organisms and behaviour of animals, consequently influencing population densities and distributions, especially in marginal areas.

climax (Gr. *klimax* = a ladder), **1.** the period of greatest intensity, as the orgasm in sexual excitement, or the crisis in the course of a disease; **2.** the final, stable state of a → succession, leading to a climax community.

climbing plants, plants climbing along other plants; woody climbing plants are called lianas.

climbing root, an → adventitious root in a plant, acting as a climbing organ; the → root hairs of climbing roots become wooden and fix the plant to the support; e.g. ivy and many plants in the family Araceae.

cline (Gr. *klinein* = to lean), a gradual change in certain features; e.g. a successive series of populations in which frequencies of certain genetic characteristics are changing from one population to another. If the cline is correlated with a change of some ecological factor, it is considered to be the result of an adaptation of the feature to the changing factor; e.g. many mammalian species show clines of decreasing length of ears and feet toward the colder areas of their distribution range (*see* Allen's rule).

clinogradual (L. *gradus* = step), oblique gradual changing; pertains e.g. to a gradually decreasing solubility curve of oxygen in a water body when oxygen content is lower in the hypolimnion than in the epilimnion

clinokinesis, also **klinokinesis,** *see* kinesis.

clinostate, → klinostate.

clinotaxis, → klinotaxis.

clisere, the → succession of communities as a result of climatic changes.

Clitellata (L. *clitellae* = packsaddle), a group name for oligochaetes and hirudines which have a → **clitellum**.

clitellum, pl. **clitella,** Am. also **clitellums** (L.), a glandular saddle-like swelling on the anterior body of oligochaetes (e.g. earthworm) and hirudines (e.g. leech); its cells secrete mucus to hold copulating individuals together and later to form a silk-like **cocoon** to protect developing embryos.

clitoris (Gr. *kleitoris* < *kleiein* = to shut), an elongated, erectile body at the ventral angle of the vulva in the external genitals of female mammals; homologous with the penis.

cloaca, pl. **cloacae** (L. = sewer), **1.** a common passage in the caudal end of the intestine into which the hindgut and the ducts from reproductive and excretory organs open; exists in many insects and all vertebrates except the eutherian mammals. In some invertebrates, such as sea cucumbers, the cloaca serves also as a respiratory organ; **2.** in early embryos of eutherian mammals, the terminal part of the hindgut where the hindgut and allantois empty; divides later into the rectum, bladder, and genital primordia.

clonal dispersal (Gr. *klon* = twig, sprout, bud), a scattering movement or growth of different parts (modules) in a → modular organism. In general but not always, the parts detach from each other; e.g. the fell birch disperses by clonal roots.

clonal selection, the phenomenon occurring during the ontogenetic development of an animal, when each antibody-producing → lymphocyte becomes committed to react with a particular → antigen, before being exposed to it, and then selected to produce a → clone of these lymphocytes when the animal comes into contact with this antigen.

clone, a group of organisms or cells having identical gene content, i.e. a population which is a descendant of one single individual by asexual reproduction, as e.g. corals and vegetatively reproduced plants. Hence, a clone is usually genetically homogeneous, but not necessarily because → mutations can cause genetical differences in it (*cf.* genet). Clones can be produced also in cell and tissue cultures by starting the culture from one single cell.

cloning, 1. the production of a → clone; **2.** a method by means of which DNA fragments are enriched mediated by a vector, like a → plasmid, in a bacterial cell; the desired gene is multiplied by cultivating the bacterial strain containing that gene.

closed population, a population in which neither immigration nor emigration occurs, thus forming a closed ecological system.

closed reading frame, a sequence of DNA or RNA which usually contains several → terminator codons preventing protein synthesis.

clotting, *see* blood coagulation.

cloud forest, a type of tropical rain forest, situated in mountain areas at the normal cloud height, and is thus mostly very humid; small deciduous trees and shrubs and numerous → epiphytes are characteristic.

CLSM, confocal laser scanning microscope. *See* confocal microscopy.

club mosses, → Lycopodiales.

clumped dispersion/distribution, *see* dispersion.

Clupeiformes (L. *Clupea* = the type genus, herring, *forma* = form), **herrings** and **herring-like fishes**; a large order of bony fishes mainly including small-sized, slender species living in large shoals in the pelagic zones of oceans and seas. They feed on plankton animals and play an important role in oceanic ecosystems. Herrings are common fish species and represent about one third of the global fish catch. The order includes about 300 species in four families.

clustering, in genetics, denoting genes (cluster genes) which locate consecutively as a cluster on the same chromosome.

clypeus, pl. **clypei** (L.), a broad chitinous plate between the frons and labrum in the insect head.

C mitosis (C = colchicine; Gr. *mitos* = string), a modified form of → mitosis due to partial or complete inactivation of the spindle mechanism with consequent disturbances in the movement of → chromosomes (*see* mitotic spindle). The modification is caused by → colchicine or other spindle poisons. In C mitosis, prophase proceeds normally and the → nuclear envelope disappears at the end of prophase. However, because the spindle is not formed and the prometaphase movement of the chromosomes does not occur, the chromosomes in **C metaphase** are evenly distributed throughout the whole nucleus. Chromosomes have a typical visual appearance; the sister chromatids are attached only at the centromeres, otherwise they are clearly separated. **C anaphase** follows the division of the centromeres in which stage the daughter chromosomes are aligned along their whole length. The total suppression of the spindle mechanism leads to the formation of a so-called restitution nucleus which contains a double dose of chromosomes. A partial inactivation of the spindle mechanism often leads to multipolar anaphases. The induction of C mitosis is a useful method in many cytogenetic studies, e.g. in karyosystematics.

CMP, cytidine monophosphate. *See* cytidine phosphates.

CMS, → cytoplasmic male sterility.

Cnidaria (Gr. *knide* = nettle), **cnidarians**; a primitive invertebrate phylum containing primary radially or biradially symmetrical animals; the body is composed of only two germ layers, **ectoderm** and **gastroderm**, the latter corresponding to the endoderm. A jelly-like **mesogloea** between the layers forms the gelatinous elastic skeleton. The gastrovascular cavity corresponding to the coelom of higher animals, has a single opening which serves as the mouth and anus. Around this opening, cnidarians typically have several tentacles with superficial → **nematocysts**. Cnidarians may be either solitary or colonial, the life cycle usually including both the polyp and medusa stages which are variously modified in different classes. The phylum is divided into four classes: **hydrozoans** (Hydrozoa), **jellyfishes** (Scyphozoa), **cubomedusae** (Cubozoa), and **anthozoans** (Anthozoa), such as **corals** and **sea anemones**. The phylum comprises about 10,000 aquatic species, mainly in oceans. *See* Coelenterata.

cnidoblast (Gr. *blastos* = germ), *see* cnidocyte.

cnidocil (L. *cilium* = eyelash), a small spine-like structure on the dorsal surface of the → cnidocyte; its stimulation discharges the → nematocyst.

cnidocyte (Gr. *kytos* = cavity, cell), **thread cell**; a modified interstitial cell of cnidarians, often resembling a footed cup with a small dorsal spine, **cnidocil**, and enclosing the stinging organelle, → **nematocyst**. During the development of the nematocyst, the cnidocyte is called **cnidoblast**.

CNS, → central nervous system.

co-, con- (L. *cum* = with), together, with, in association.

CoA, → coenzyme A.

coacervate (L. *coacervare* = to collect in a mass), an aggregate of colloidal particles in a solution, usually also a colloidal solution or emulsion, e.g. occurs when gum arabic is added to gelatin solution. Coacervates may have played a role in prebiotic evolution during the → origin of life on the Earth.

coadaptation, 1. the evolutionary development of two or more species dependent on each other, beneficial for both; e.g. the co-adaptation of the structure of a flower and the proboscis of the nectar collecting bee; **2.** the adaptation caused by → natural selection resulting in gene groups which act adaptively in coordination, being accumulated in the → gene pool of a population.

coagulant, 1. an agent stimulating or accelerating coagulation, especially blood coagulation; **2.** as an adjective, promoting or making possible the process of coagulation; coagulative.

coagulation (L. *coagulatio*), clotting; e.g. **1.** → blood coagulation; **2.** in chemistry, the solidification of a fluid material (sol) into a gelatinous form (gel, coagulum), as occurs in soluble proteins when a change in net charge allows molecules to aggregate; usually an irreversible process. *Adj.* **coagulable**, susceptible of being coagulated; **coagulative**, pertaining to coagulation. *Verb* to **coagulate**.

coated pit, *see* clathrin.

coated vesicle, a vesicular cell organelle with its surface membrane formed from a layer of protein coat, such as → clathrin.

cobalamin(e) (< cobalt + amine), a compound containing four pyrrole rings bound to a cobalt atom in the centre; its derivative is vitamin B_{12} (cyanocobalamin). *See* vitamin B complex.

cobalt, a metallic element of the iron group, symbol Co, atomic weight 58.93; an essential micronutrient and trace element for organisms, occurring e.g. as a constituent in vitamin B_{12}. One of its isotopes is used in the → cobalt gun.

cobalt gun, an apparatus emitting → gamma radiation, used in radiotherapy e.g. in the treatment of cancer; its radiation source is Co-60 (half-life 5.27 years) which is activated by neutron rays in a nuclear reactor.

coca alkaloids (S.Am. *kuka*), the alkaloids of the coca bush, the most important being → cocaine.

cocaine, benzoylmethylecgonine; an alkaloid

($C_{17}H_{21}NO_4$) in leaves of the coca bush (*Erythroxylon* species), or produced synthetically from ecgonine and its derivatives; has stimulating effects on the nervous system, and is used as a local anaesthetic and a narcotic drug.

cocc(o)-, cocci- (Gr. *kokkos* = grain, seed, berry), pertaining to, or resembling a berry, grain, or seed.

coccidiosis, pl. **coccidioses** (*Coccidia* = a group of parasitic protozoans), a group of intestinal diseases sometimes causing severe symptoms in vertebrates, especially in mammals, birds, and fish. Coccidiosis is caused by parasitic sporozoans in the subclass Coccidia. The term coccidiosis is especially applied to infections caused by sporozoans of the genera *Eimeria* and *Isospora*. Although coccidiosis is usually rather insignificant to humans, it may be very serious e.g. in AIDS patients.

coccolith (Gr. *lithos* = stone), a calcified scale formed on some unicellular algae; sedimented coccoliths can be utilized in studies on earlier algal flora.

cochineal, *see* carmine.

cochlea, pl. **cochleae** (Gr. *kochlos* = snail with spiral shell), **1.** any spiral-shaped structure; **2.** the spiral tube which forms the acoustic organ of the internal ear of tetrapod vertebrates; *see* spiral organ, ear.

coconut milk, also **coconut water,** a milky liquid of the coconut, the seed of the coconut palm *Cocos nucifera*; contains minerals, carbohydrates, and oil, used as such or in food; because of cytokinin-like properties its deproteinized preparations are used in culture media for plant cell and tissue cultures.

cocoon (Fr. *cocon* = shell), a capsule of many animals which is secreted around eggs or larvae for protection against disadvantageous environmental factors; e.g. the cells in the clitellum of oligochaetes secrete a protective cocoon for developing embryos; some insect larvae spin a silky capsular cocoon around themselves for the pupal stage, and some spider females spin a similar shelter around their eggs where the eggs endure to the next summer.

code, *see* genetic code.

cod fishes, *see* Gadiformes.

codein, methylmorphine; a drug obtained from opium or morphine, used as an analgesic and antitussive drug; dependence may develop.

coding strand, one of the two complementary DNA strands, i.e. that strand whose base sequence is written down when the sequence of the gene is published. Its complementary strand, the antisense strand, is used as a template in → genetic transcription. Thus the coding strand carries the same information as the → messenger RNA. *Syn.* sense strand.

codominant alleles, → alleles of a gene which do not show usual dominance and recessiveness in relationship to each other, but are both expressed in the → phenotype. E.g. the I^A and I^B factors of the human ABO blood group system are codominant; they both dominate the I factor which causes the O blood group. Hence, when they appear together in the I^AI^B genotype they cause the AB blood group, i.e. they cause the expression of both A and B blood group factors in the phenotype.

codon, the smallest sequence of nucleotides in DNA (or RNA, genetic material in some viruses) determining the position of each amino acid in the native polypeptide chain during → protein synthesis. The codons consist of groups of three consecutive → nucleotides called code triplets (→ genetic code).

The term codon is also used for the corresponding complementary group of three nucleotides in the → messenger RNA (mRNA) into which the original sequence of DNA will be transcribed in → genetic transcription. A codon can change in a → mutation so that it then codes for another, different amino acid (**missense codon**) or for no amino acid (**nonsense codon**). It is shown that nonsense codons are **terminator codons** with triplet bases UAG, UAA, and UGA; (U = uracil, A = adenine, G = guanine). **Initiator codons** are trinucleotides which, when they appear in the beginning of the genetic message, encode formylmethionine and initiate the protein synthesis; usually they are AUG, sometimes GUG. In the middle of the message the former codes for methionine and the latter for valine. **Synonymous codons** are codons which code for the same amino acid, (the regeneracy of the → genetic code). *Cf.* anticodon.

codon usage, the frequence of the usage of synonymous → codons inside the species; it includes the mode and variation of choice among synonymous codons, resulting in a distinct bias (codon bias) in a frequency with which a particular synonymous codon is used to code for a given amino acid in different genes; a natural phenomenon in all organisms.

coefficient (*co-* < L. *cum* = with, *efficio* = to accomplish), a ratio or factor that is constant in standard conditions for a given substance or process; e.g. absorption coefficient, diffusion coefficient, or → selection coefficient.

coel(o)-, Am. also **cel(o)-** (Gr. *koilos* = hollow), denoting relationship to a cavity.

Coelacanthini (Coelacanthiformes, Actinistia), (Gr. *akantha* = thorn, spine), **coelacanths**, **lobe-finned fishes**, **tassel-finned fishes**; an order (suborder in some classifications) in the subclass Sarcopterygii (class Osteichthyes); the only living species, *Latimeria chalumnae* is a relict of an ancient group which was abundant in the Devonian period. Coelacanths flourished 350 million years ago and were believed to have become extinct 70 million years ago, until this fish was found on the fish-market in Zanzibar, East-Africa, in 1938. Coelacanths are the closest living relatives to tetrapods: they have a diphycercal, tri-lobed tail fin, and the other fins (except the first dorsal fin) are supported by movable stalks. In some classifications coelacanths are placed in **crossopterygians** (Crossopterygii).

Coelenterata (Gr. *enteron* = intestine), **coelenterates**; a common name for cnidarians and ctenophores, previously considered as a phylum which included two subphyla, Cnidaria and Ctenophora; sometimes the name is also used as a synonym of → Cnidaria.

coelenteron (L.), **1.** → archenteron; **2.** a body cavity of cnidarians and ctenophores (Coelenterata) having only one opening which serves both as the mouth and anus; *syn.* enteron, gastrovascular cavity.

coeliac, Am. **celiac** (Gr. *koilia* = belly, cavity), pertaining to the abdomen or the abdominal cavity.

coelom, coeloma, Am. also **celom(a),** pl. **coeloms** or **coelomata** (Gr. *koilos* = hollow), **body cavity**, **visceral cavity**; an internal cavity of multicellular animals providing space for visceral organs; there are two kinds of coeloms, **pseudocoelom** and **true coelom**. The structure of the coelom varies among invertebrates. Primitive bilateral animals, such as mesozoans, flatworms, and nemertines, have no body cavity, and are classified into the group → **Acoelomata**. The animals of the → **Pseudocoelomata** group have a body cavity with no mesodermal peritoneum, and the most developed animals of the group → **Coelomata** have a true coelom which is lined with mesodermal peritoneum; it may be formed either from the splitting of mesodermal bands (→ **schizocoelom**) or from the pouches of the archenteron (→ **enterocoelom**). In annelids, e.g., the coelom is divided into several parts and forms the **hydrostatic skeleton** (*see* skeleton). The coelom of molluscs and arthropods serves partly as a circulatory system (→ haemocoelom).

All vertebrates have a true coelom of the enterocoelom type. During embryonal development, the mesodermic coelomic primordium develops first as uniform but splits later to form two **peritoneal membranes** and the narrow **peritoneal cavity** between them. The membranes are **parietal peritoneum** (*peritoneum parietale*) lining the walls of the body cavity, and **visceral peritoneum** (*peritoneum viscerale*) covering the internal organs. In adults of most vertebrate groups, only the membranous **false diaphragm** separates the anterior part (*cavum pericardii, cavitas pericardialis*) with the heart from the rest of the coelom. In mammals the coelom is divided into two portions, **thoracic cavity** and **abdominal cavity** which are separated by the **diaphragm**. The membranes of the cavities are called → **pleurae** (parietal and visceral pleura) which are lining walls of the thoracic cavity and lungs, → **pericardium** around the heart, and **abdominal peritoneum** (parietal and visceral abdominal peritoneum) lining walls of the abdominal cavity and internal organs. The visceral peritoneum includes the → omentum and serosa covering the digestive tract, and the perimetrium lining the uterus.

Coelomata, coelomates; comprise all the animal phyla which have a true → coelom; such are Mollusca, Annelida, Arthropoda, Phoronida, Bryozoa, Brachiopoda, Echinodermata (secondarily reduced), Chaetognatha, Hemichordata, and Chordata. *Cf.* Acoelomata, Pseudocoelomata.

coelomocytes, Am. also **celomocytes** (Gr. *kytos* = cavity, cell), a common term for cells present in body fluids of many invertebrates, as for amoebocytes or mesenchymic cells present in the body cavity of nematodes. In the body fluids of brachiopods, there are many kinds of coelomocytes some of which contain haemerythrin. The coelomocytes of echinoderms take part in excretory functions, regeneration, and different recovery processes.

coelomoduct, a duct from the coelom to the body surface found in many invertebrates; e.g. the long tubule in the excretory organ through which wastes flow from the nephrostome to the nephridiopore on the surface of an animal. Coelomoducts which are probably derived from the genital duct, are found e.g. in some annelids, molluscs, and arthropods.

coen(o)-, Am. often **cen(o)-,** also **caen(o)-** (Gr. *koinos* = common), denoting common, general.

coenenchyme, also **coenenchym** (L. *coenenchyma* < Gr. *enchyma* = infusion), a viscous mass secreted by corals (anthozoans) containing high quantities of proteins; binds polyps or zooids together to form colonies.

coenobium, pl. **coenobia** (Gr. *bios* = life), a type of cell colony formed by some unicellular organisms, e.g. some green algae (*Volvox* species); there are a definite number of cells connected by plasma bridges, and usually a certain division of labour between the cells. Thus, the coenobium is more developed than many other colony types.

coenocyte (Gr. *kytos* = cavity, cell), a multinucleate mass of protoplasm in a plant or fungal structure, consisting of only one cell with numerous nuclei; formed when the cell undergoes successive mitoses without cell divisions and cell wall formation. A coenocytic structure is typical of the order Siphonales in green algae and of slime moulds (→ Myxomycetes).

coenosarc (Gr. *sarx* = meat), a tubular gastric system in the common stem of hydroid colonies connecting the gastric cavities of different polyps or zooids. *Cf.* perisarc.

coenospecies, a group of individuals of common evolutionary origin comprising more than one taxonomic species which are capable of producing fertile hybrids, e.g. individuals of a superspecies or a supergenus.

coenotype, cenotype, the original type from which other forms have derived.

coenzyme (L. *con-* with, together, Gr. *en* = in, *zyme* = leaven, yeast), a non-protein organic compound that is bound to an enzyme protein, i.e. to → **apoenzyme**, and changed chemically in the course of the reaction, and often easily liberated from it. All coenzymes can be regenerated in association reactions binding again to the apoenzyme. The enzyme activity is based on the presence of the coenzyme; e.g. many vitamins act as coenzymes. Important coenzymes involved in cellular metabolism are e.g. → **coenzyme A (CoA)** which is converted into acetyl CoA in the reaction between → glycolysis and → citric acid cycle, and **coenzyme Q** (→ ubiquinone) in mitochondrial electron transfer. *See* prosthetic group.

coenzyme A, CoA, CoA-SH, a → coenzyme consisting of pantothenic acid that is linked to adenine mononucleotide; an intermediate involved in the transfer of an acyl group (the combined form is acetyl CoA) in living cells. It has an important role in the metabolism of fatty acids forming acetyl CoA that enters the → citric acid cycle (TCA cycle). *See* cofactor, prosthetic group.

coenzyme M, CoM, 2-mercaptoethanesulphonic acid, involved in the production of methane from CO_2 in some bacteria.

coenzyme Q, CoQ, → ubiquinone.

coevolution (L. *e-* from *ex* = out, *volvere* = to roll), the evolution of two or more species in which the phylogenetic development is interdependent, enhancing the → fitness of both partners, i.e. leads to → coadaptation; e.g. many plant/insect relationships, host/ parasite relationships, etc. *See* evolution.

cofactor, a non-protein complement of an enzyme that is essential for its catalytic activity; usually a metal ion, such as magnesium, manganese, or zinc; often also → coenzymes and → prosthetic groups are considered as cofactors.

cohesion (L. *cohaerere* = to stick together), a force between the atoms and the molecules of any substance; e.g. water strands in narrow xylem capillaries of high trees remain unbroken because of the cohesion between water molecules.

cohesive ends, 5'-termini of DNA fragments artificially made using a → restriction enzyme (endonuclease) treatment (e.g. *Eco* R1 treatment); the fragment (restriction fragment) can link, according to the base pair rule, to any other fragment made by the same enzyme, i.e to a fragment with complementary cohesive ends. *Syn.* sticky ends.

cohort (L. *cohors* = enclosure, one of the ten divisions in an ancient Roman legion), any group of individuals who have a common feature; e.g. a group of individuals in the same age-class of a population, in the same systematic taxon, or in the same caste.

coincidence, co-occurrence; **1.** in genetics, specifically the co-occurrence of the events of → crossing-over, indicated by the **coefficient of coincidence**. It demonstrates the relative proportion of double crossing-overs from the expected number of random co-occurrence of single crossing-overs. The coefficient of coincidence is the most used genetic measure for → interference. If the interference is positive, the coefficient is smaller than 1, and if the interference is negative it is bigger than 1. In the case of complete interference, when double crossing-overs are lacking, the value of the coefficient is zero; **2.** in ecology, coincidence has a great significance in → population dynamics of animal stocks when limiting factors have simultaneous parallel effects.

colchicine, an antimitotic alkaloid, $C_{22}H_{25}NO_6$, intefering cell division by preventing the formation of the → mitotic spindle, i.e. it binds to tubulin molecules inhibiting their polymerization, disturbing the formation of → microtubules. The use of colchicine in cytological experiments leads to → C mitosis, causing doubling of the number of chromosomes in a cell; therefore it is used e.g. for obtaining new agricultural plant varieties. Colchicine is prepared from the roots of the autumn crocus (*Colchicum autumnale*).

cold-blooded animals, poikilotherms. *See* poikilothermy.

cold hardening, in botany, a series of events induced in plants during the development of their resistance to low temperatures. Inducing factors are a shortened day length and lowered temperature. During the hardening process, cold-inducible genes start reactions which e.g. produce new proteins required for cold/freezing resistance. Cellular carbohydrate metabolism and water content change considerably, and changes are also found in membrane lipids. Temperatures close to 0°C (0—4°C) start the cold hardening processes in a few days, e.g. in winter rye in two days. The hardening increases when the temperature decreases further, and many plant species in northern latitudes reach the freezing resistance of -50°C to -60°C; many species can survive even at lower temperatures. The same plants can be injured or killed during the growth season at temperatures which are only a few degrees below zero. The state of cold hardening disappears in the spring; the process is called dehardening.

In zoophysiology, cold hardening means the rapid increase of **cold tolerance** of animals occurring after a short exposure to cold environment. The phenomenon is found also in separated tissues and cells and even in cell cultures, and is associated with the induction of new proteins, → stress proteins, in the same way as in → heat shock. A slower physiological adaptation process of animals to cold occurs in → acclimatization (acclimation), in which state also cold tolerance may be improved. In man, cold hardening often refers to the reduced sensation of cold, occurring after successive short exposures to cold, e.g. cold water. This is probably due to increased endorphin secretion and/or decreased sensory activity of cold receptors.

cold receptors, *see* temperature sense.

cold resistance (L. *resistentia* = opposition, counteraction), the ability of organisms to oppose and tolerate cold; the term **cold tolerance** is used to denote the limits of cold resistance.

Cold resistance of homoiothermic animals (birds and mammals) with quite a stable and high **body temperature** in varying environmental temperatures (**ambient temperatures**) is based on the thermoregulatory capacity of the body (*see* thermoregulation), depending on the size, insulation, and heat generation of the animal. If the capacity is not efficient enough for prolonged cold, the body temperature decreases and the animal dies when the body temperature drops to 10—20°C. Exceptions are the hibernating animals which tolerate body temperatures close to zero (*see* hibernation).

Cold resistance of poikilothermic animals varies greatly according to temperature changes of their natural environments, to which the animals have adapted during their evolution. If the temperature has been stable during the evolution of a species, the ability to tolerate temperature changes may not have developed. For example, poikilothermic animals living in tropical seas may die of cold at temperatures of 10—15°C. In general, poikilothermic animals living naturally at varying temperatures, as in northern countries, tolerate at least 0°C, and usually much lower temperatures. Most species prevent freezing of body fluids by producing high quantities of → **cryoprotectants**. Resting stages, as eggs and pupae, are usually more resistant to cold. For example, resting stages of *Tardigrada* species have, without chemical treatment, tolerated experimental temperatures near absolute zero. In general, germ cells, early embryos, and undifferentiated isolated cells can tolerate those very low temperatures in artificial physiological solutions with added glycerol or dimethyl sulphoxide (DMSO).

Cold resistance increases in organisms if they have time to adapt to cold for some weeks (*see* **acclimatization**). Cold tolerance much depends on cell membrane properties, especially on the viscosity of membrane lipids, which further affects permeability properties and enzyme activities (*see* homeoviscous adaptation).

Cold resistance in plants is related to → cold hardening.

Coleoidea (Gr. *koleos* = sheath, *eidos* = form), **coleoids**, **dibranchiates**; a subclass of → Cephalopoda, characterized by an internal shell and one pair of gills; includes all living cephalopods except *Nautilus*.

Coleoptera (Gr. *pteron* = wing), **coleopterans, beetles**; an order of insects, the largest order of all land animals, comprising about 300,000 species all over the world in dry, wet, and aquatic environments. Their mouthparts have adapted to biting and chewing, the forewings are thick, leathery, and veinless; the hindwings are membranous, slightly veined, and fold forward under the forewings at rest.

coleoptile (Gr. *ptilon* = feather), a sheath-like structure around the → cotyledon in the embryo of graminaceous plants (family Poaceae).

coleorhiza, pl. **coleorhizae** (Gr. *rhiza* = root), a sheath-like sructure around the → radicle of the embryo in the seeds of graminaceous plants (family Poaceae).

Cole's rule, a rule stated by *L.C.Cole* in 1954, associated with population demography and the theory of evolutionary ecology; states that the intrinsic rates of increase (r_m) are equal in semelparous and iteroparous populations when each semelparous individual produces, on average, one offspring more than an iteroparous individual.

colicin (< colibacillus), any of a class of bacterial toxins (protein antibiotics) which are encoded by extrachromosomal → plasmids (col plasmid) and can kill other bacteria.

Coliiformes (L. *Colius* = the type genus, *forma* = form), **mousebirds**; an order of small, long-tailed birds which live in savannas, south of the Sahara and in Ethiopia. The birds are grey or brown, with a stout structure, resembling small passerine birds; only six species exist.

colinearity, linear similarity; in genetics means the precise relationship between the amino acid sequence of a polypeptide chain and the → codons of the gene coding for it, i.e. the situation where the participating molecules of the code correspond place by place to the linear structure of the polypeptide chain. *See* genetic transcription, genetic translation.

collagen (Gr. *kolla* = glue, *gennan* = to produce), a fibrous protein which forms supporting intercellular structures in the connective tissues of most animal organs, such as the skin, ligaments, tendons, bones, and cartilages; the most abundant protein in the Animal kingdom, e.g. in vertebrates, representing about 1/3 of total body proteins. The quaternary structure of the collagen molecule (tropocollagen) is a triple superhelix composed of three long polypeptide chains wound around each other and linked with → hydrogen bonds; each chain is in the right-handed α-helical conformation (*see* alpha helix). Collagen is produced by fibroblasts into extracellular fluid where the protein molecules join together to form fibrils and fibres with high tensile strength. There are several collagen types in tissues, named collagen I, collagen II, etc. Collagen is rich in certain amino acids, i.e. in glycine (1/3), proline, and hydroxyproline. In boiling water, dilute acids, or alkalies, fibrous collagen is partially hydrolysed into soluble **gelatin**; for commercial use gelatin is prepared from the skin, bones, and cartilages of slaughtered animals.

collagenase, any enzyme of the sub-subclass, EC 3.4.24., catalysing the hydrolysis of peptide bonds of → collagen; different types are found in animal tissues, especially in the alimentary canal, and in microbes which use animal tissues as their substrate.

collapse (L. *collapsus*), falling in or together, as the walls of an organ, or a mental breakdown (shock) of an animal or man.

collar cells, → choanocytes.

collateral (L. *com-* = together, *latus* = side),

situated side by side, e.g. blood vessels in the limbs; also used as a noun to denote such a structure. *Cf.* contralateral, ipsilateral.

collateral bundle, the most usual type of plant → vascular bundles; may be open as in dicotyledons, or closed as in monocotyledons.

collector filterers, aquatic animals who sift their food by selecting small organic particles from surrounding water; e.g. whalebone whales and herrings. *Cf.* filter feeders, gathering collectors.

Collembola (Gr. *kolla* = glue, *embolon* = peg, wedge), **springtails**; an order of small insects having no compound eyes, Malpighian tubules, and usually no tracheae; the wings are primarily absent. On the underside of the abdomen they have a forked **springing organ** (*furcula*) by means of which the springtail can leap distances which are many times its body length. The order includes about 2,000 species which are common especially in soil.

collenchyma, pl. **collenchymata,** also **collenchymas** (Gr. *enchyma* = cast, poured in), a type of → supporting tissue of plants; consists of living cells with thickenings in their walls. Collenchyma supports especially leaves and the → cortex of shoots; it occurs often as thin layers below the → epidermis. Different types of collenchyma are **angular collenchyma** (the corners of the cell walls are thickened), **lamellar collenchyma** (the side walls are thickened), and **lagunar collenchyma** (the cell walls surrounding the intercellular spaces are thickened).

collencyte (Gr. *kytos* = hollow), a cell type in the body wall of sponges corresponding to connective tissue cells (fibroblasts, fibrocytes) in more developed animals; collencytes secrete dispersed collagen fibres.

colliculus, pl. **colliculi** (L. dim of *collis* = hill), an elevation, heap, mound; e.g. **colliculi inferior** and **superior** in the vertebrate brain, i.e. the inferior and anterior eminences of the corpora quadrigemina of the mesencephalon, participating in neural mechanisms of auditory and visual pathways.

colligative (L. *colligare* = to join together), pertaining to a chemical property of a solution that depends only on the number of dissolved molecules or ions but not on their quality; e.g. osmotic potential, or boiling and freezing points.

colloblasts (Gr. *kolla* = glue, *blastos* = bud, germ), pear-shaped cells in the tentacles of ctenophores secreting adhesive substance for catching and holding prey animals. *Syn.* glue cells, lasso cells.

collochore (Gr. *chorein* = to withdraw, go), a small non-specific heterochromatic region of a chromosome causing the chromosome association in the achiasmate → meiosis.

colloid (Gr. *kollodes* = glutinous), **1.** substances such as large molecules or aggregates of atoms (usually 1—100 nm in diameter), remaining dispersed in solution; the substance thus dispersed is called the disperse phase of the colloid system. A **colloidal solution** is jelly-like and opalizing, as e.g. the base matter of the cytoplasm in cells. There are different types of colloid solutions depending on the properties of **disperse phase** (dissolved substance) and **dispersion medium** (solvent), such as solid in liquid (suspensoid solution), or liquid in liquid (emulsion). Common colloids are e.g. starch (polysaccharide) and albumin (protein) solutions, and colloidal metals; also small soil particles, as clay particles, form soil colloids, i.e. particles together with their water mantle; **2.** the secretion stored within the follicles of the thyroid gland; **3.** as an adjective, resembling glue; glutinous, gelatinous, colloidal.

colloidal, pertaining to a colloid; glutinous, gelatinous.

colloid osmotic pressure, osmotic pressure (*see* osmosis) caused by colloidal substances in a fluid; e.g. in the blood plasma (oncotic pressure) caused by proteins, especially albumin, which cannot be filtered through capillary walls into tissue fluid, and consequently the colloid osmotic pressure is higher in the blood than in the interstitial fluid. *See* capillaries.

colocynth, colocynth apple (Gr. *kolokynthe* = bitter cucumber), a cucumber plant growing in Mediterranean areas; its peeled, dried fruits are used as laxative for purging the bowels. Also called bitter apple, bitter cucumber, bitter gourd.

colon, pl. **colons,** also **cola** (L. < Gr. *kolon*), a part of the large intestine (*intestinum crassum*) of mammals, extending from the caecum to the rectum; sometimes used to denote the entire large intestine in amniotic vertebrates. In mammals, comprises four parts: the **ascending**, **transverse**, **descending**, and **sigmoid colon**, serving in absorbing of water and some minerals, and the formation of excrement in which processes intestinal microorganisms play an essential role. The wall of the colon has strong muscular bands (*taeniae coli*) with longitudinal muscle fibres causing periodical strong contractions, mass peristalsis. *See* alimentary canal, digestion.

colonization, an entry and settlement of a new area or a new habitat by a population (species or genes).

colony (L. *colonia* = farm), a group of individuals living together, e.g. bacterial, fungal, or animal colonies; may be either intraspecific or interspecific. Among animals, interactions between the individuals vary from a mere coexistence to the highly integrated entireties with specified castes or physical unions of the members. For example, many gull species form mixed colonies during breeding without any division of labour. Different zooids (**polyps**, e.g. hydranths and gonangia) of hydroid colonies with different working roles are connected with branched stalks to a functional entirety. In the colonies of many eusocial insects (e.g. termites, ants, and bees) the organization and division of labour between the members are very complex and characterized by the caste differentation and polymorphism. *Adj.* **colonial**. *Cf.* modular organism.

colony-stimulating factors, CSFs, glycoprotein → cytokines produced by stimulated lymphocytes, monocytes, and tissue macrophages; are required for differentiation of bone-marrow stem cells into granulocyte and monocyte cell colonies and for their development into leucocytes.

colorimeter (L. *color* = colour, Gr. *metron* = measure), chromatometer, chromometer; an optical device used for measuring the intensity of the colour of liquids; e.g. an apparatus for measuring the haemoglobin content of the blood. Colorimeters are also used to measure the turbidity of a suspension, e.g. that of a bacterial culture. A more complicated but principally similar optical equipment is the → spectrophotometer. The procedure is called **colorimetry** (colorimetric analysis). *Adj.* **colorimetric**.

colostrum, pl. **colostra, colostrums** (L.), beestings, foremilk; the milk secreted from the mammary glands at the time of parturition and a few days later for feeding the offspring at the most critical period of their lives; contains more minerals and proteins, especially immunoglobulins, and less fat and carbohydrates than does ordinary milk.

colour blindness, the deficiency of colour vision in man due to genetic factors; may be complete colour blindness (tritanopia), yellow-blue blindness (deuteranopia), or red-green blindness (protanopia), depending on which cone pigments are absent from the retina. Of these, protanopia is the most common form that is characterized by absence of the red-sensitive pigment in cone cells, and it is inherited in a recessive sex-linked manner within the X chromosome (*see* sex-linked inheritance). Therefore red-green blindness occurs mainly in men (approximately 8% of all men). Colour blindness can be demonstrated with the aid of special colour tables designed for this purpose. *See* colour vision.

colour change, the rapid colour change of the skin in many animals, as in squids, fishes, frogs, and reptiles, due to changes of lightness and colour in the environment. The message is transmitted through sensory nerves from the eyes to the brain and then back through autonomic nerves and by certain neurohormones to the skin → chromatophores.

colour plant, any plant used to stain textile fibres and other materials, even drinks and food; eg. indigo, saffron.

colour vision, the ability of many animals to distinguish different wavelengths of the visible light spectrum; it is based on different types of photoreceptor pigments in the eye some of which are more sensitive to shorter and some to longer wavelengths. E.g. the human eye, like most vertebrate eyes, has three types of cones (photoreceptor cells), i.e. with **erythrolabic**, **chlorolabic**, and **cyanolabic** pigments, each having its own special sensitivity to light wavelengths. From the three-colour vision of the cones certain neuronal couplings in the retina produce the fourth colour, yellow, i.e. the vision of four main colours in numerous intensities and shades. The retinal rods have only one kind of photopigment, → rhodopsin, and are thus colour-insensitive.

Most vertebrates possess colour vision. Apparently many mammals, however, have partly lost this ability during later evolution. Also some invertebrates, such as many insects, have colour vision. Ultraviolet radiation can be seen by insects feeding on honey in flowers, and probably by some predator birds which can thus better observe a prey's urine tracks e.g. on snow.

Columbiformes (L. *columba* = dove, *forma* = form), **pigeons, doves**; a cosmopolitan order of middle-sized and short-necked birds, found world-wide except in arctic and antarctic regions; they are characterized by a well-developed crop with epithelial cells secreting **crop-milk** in both sexes during the nestling period. The order includes one family, Columbidae, which contains 303 species.

columella, pl. **columellae** (L. dim. of *columna* = column), a small column-like part; e.g. **1.** the middle column of the cochlea of the internal ear; **2.** *columella auris*, the single ossicle (stapes) of the middle ear in amphibians, reptiles, and birds; **3.** *columella cranii*, the epipterygoid bone in the skull of lizards; **4.** the central axis in the coiled shell of some gastropods; is attached to a strong muscle for pulling the animal into the shell; **5.** the middle column in the sporangium of → mosses; **6.** a dome-shaped structure in the tip of the sporangiophore inside the sporangium in the fungal class Zygomycetes (e.g. *Mucor*).

columna, pl. **columnae** (L.), a column; e.g. **1.** *columna vertebralis*, → vertebral column (spinal column); **2.** *columna anterior, posterior, lateralis*, respectively the ventral, dorsal, and lateral columns comprising the grey matter projections in the cross section of the spinal cord; **3.** a structure formed by united stamens in mallows. *Adj.* **columnar**.

coma (L. < Gr. *koma*), a deep state of unconsciousness.

com- (L. *cum* = with), denoting with or together.

comb row, → meridional ridge.

commensalism (L. *cum* = with, *mensa* = table), an interaction between two or more species in which one benefits from the relationship (or the interaction is of vital importance to it) and to the other species (host) the relationship is neutral and harmless; e.g. a tree and a bird nesting on it. *Cf.* mutualism, parasitism, symbiosis.

commissura, pl. **commissurae** (L. = a joining together), **commissure**; a joint or seam; a connecting band between corresponding structures, especially in the nervous tissue, often across the midplane.

common scab disease, a disease of potato, caused by bacteria of the genus *Streptomyces*.

communal courtship, a courtship behaviour of many animal species, in which males get together to the same, often traditional → lek to compete for the right to copulate with a female. Finally, the female selects its sexual partner from the winners. Communal courtship is common among many fowl-like birds, some waders (ruff), and deer, but also among many other vertebrates, and some invertebrates.

community, any group of organisms of several species occurring simultaneously in the same area or habitat and forming a functional entity with different trophic levels; usually characterized by certain dominant species. *Syn.* biocoenosis.

community ecology a branch of → ecology dealing with internal structures and functions of an organic share in an ecosystem.

community stability, a tendency of a community to return to its previous (original) state after a disturbance.

community structure, faunal and floral structures and abundance in a community, i.e. numbers of animal and plant species.

companion cell, a cell type in the → phloem of vascular plants; living cells, always situated beside the sieve elements of → sieve tubes. Companion cells evidently participate in the transporting activity of phloem.

compartment, a small enclosure within a larger space; e.g. **1.** fluid compartments of an organism; the term is used especially referring to different body fluids of animals; *see* fluid; **2.** a group of embryonic cells (polyclone) determined in a given way; the determination occurs stepwise so that every step creates a pair of new compartments of the previously existing compartment; **3.** any of the membrane bound units or parts within a cell, i.e. the → nucleus, mitochondria, plastids, cytosol, endoplasmic reticulum, Golgi apparatus, lysosomes, peroxisomes, ribosomes, and glycogen granules.

compass orientation, an orientation of animals with a compass method utilizing the position of the Sun or the starry sky; e.g. the orientation of bees according to the position of the Sun, and migration of many birds on the basis of positions of the stars.

compass plant, a plant having its leaf edges permanently in the north—south direction which protects the photosynthesizing surfaces from the midday heat of the sun, e.g. *Lactuca* species (lettuce). *Silphium laciniatum* (family

Asteraceae) is commonly known by the name compass plant.

compatible solutes, compounds such as sugars, which balance osmotic potential in tissues and also replace water in the hydration sphere of proteins, and thus prevent their denaturation.

compensation hypothesis, in ecology, a theory stating that an increase of reproductive effort increases the risk of death which causes the decrease of the residual reproductive value (RRV). According to the hypothesis, an organism can, at least partly, compensate this risk by increasing its food intake or by using stored reserves. *See* reproductive value.

compensation point, any stage of a physiological action where the events caused by a variable are in equilibrium; e.g. in green plants the point at which photosynthesis and respiration are in balance at a certain temperature. If a green plant is in darkness, photosynthesis is not possible and CO_2 production dominates. When light intensity increases photosynthetic reactions start, and at a certain light intensity CO_2 consumption and production are in balance, i.e. the **compensation point** of carbon dioxide is reached.

compensation zone/layer, in ecology, the depth zone of a sea or lake where the energy produced in photosynthesis and consumed in respiration of plants, are equal.

competent, in developmental physiology, describing the ability of embryonic cells to respond to an inductor or organizer and so differentiate into specialized cells and tissues. *Noun* **competence**.

competition, in ecology, a reciprocal relation in situations where two or more organisms or organism groups (populations) utilize the same limited and, from the point of success, important environmental resources. Thus the interaction is harmful for their success, e.g. by decreasing the survival or offspring numbers. Competition usually concerns food, territory, nesting place, etc.

Several kinds of competition are found: → **intraspecific competition** occurs between individuals of the same species, in **interspecific competition** the competitors belong to different species. Competition may be **resource competition** (exploitation competition, indirect competition) when individuals utilize common, scanty resources without any reciprocal physical contact, or **interference competition** (direct competition) when the individuals struggle against each other or secrete poisons towards the competitors, e.g. aggressive competition of hole-nesting birds for the nest holes, or an allelopathy of plants. When competition is equal in both participants, it is called **symmetric competition**. When one competitor is stronger than the other, it is called **asymmetric competition**. In **diffuse competition** harmful effects occur occasionally or persistently between interspecific populations because of partial overlaps of their niches.

A disputable term of **ghost of competition past**, means a situation where an interspecific competition, previously prevailed as an evolutionary factor, has ceased, but has left traces in behaviour, distribution, and morphological structures of the competitive species. *See* conventional competition, apparent competition, competitive exclusion principle.

competitive exclusion principle, a hypothesis, based on theoretical considerations and laboratory experiments, stating that two species which are identical in their ecological demands cannot exist sympatrically and simultaneously in the same unchangeable environment, but before long one has to depart from the area. *Syn.* Gause law.

competitive inhibition, a type of enzyme inhibition in which substrates compete for binding to the active site of the enzyme. In this competition some of the enzyme molecules bind a substrate which cannot take part in the enzyme reaction. This kind of binding is caused by a molecule which resembles the true substrate so that it can compete in binding to the active site of the enzyme but is not suitable for catalytic action. The same principle concerns also receptor antagonism when a drug inhibits a receptor in animal tissues. Enzyme inhibition that is not competitive is called **non-competitive**. In this type of inhibition the inhibitory compound does not compete with the natural substrate for the active site but binds elsewhere, e.g. to the allosteric site of the enzyme in the enzyme-substrate complex formed (*see* allosterism).

competitive release, *see* ecological release.

compilospecies (L. *compilatio* = gathering together), a species originated from the hybridization of two different species.

complement, *see* complement system.

complementary DNA, cDNA, a single-stranded DNA molecule experimentally copied in → reverse transcription from the → messenger RNA molecule thus being complementary to it.

complementary male, usually a reduced male who lives attached on or in the female body; e.g. some rotifers.

complementation, → genetic complementation.

complementation map, *see* genetic complementation.

complementation test, a genetic test that defines allelism of mutations (*see* allele). If the heterozygous individual m_1/m_2, resulting from the cross m_1/m_1 x m_2/m_2, has a wild phenotype, complementation is said to have occurred, and the mutants are not alleles. If, however, the heterozygous individual m_1/m_2 expresses a mutant phenotype, complementation has not occurred, and the mutants m_1 and m_2 are allelic forms of the same gene. *Cf. cis-trans* test.

complementation unit, a unit of the genetic material of a → cistron which is formed by those mutations which do not complement the action of each other, i.e. no → genetic complementation occurs between mutations which belong to different complementation units; the complementation unit corresponds to the gene. *Syn.* complon.

complement system, a part of the vertebrate immune system that comprises a cascade of over 10 distinct serum proteins (complements); these are enzymes which become activated in a proteolytic chain reaction when the first protein component (C_1) binds to an antibody—antigen complex, i.e. to an immunoglobulin that has bound to an antigen, such as a microbe. The last proteins of the activation chain render the microbial cell membrane permeable to different compounds and thus cause cytolytic processes in this microbe (immune cytolysis). The proteins of the complement system can also act as chemotactic substances in → opsonization.

complete metamorphosis, *see* metamorphosis.

complex (L. *complexus* = woven together), **1.** an anatomical entity made up of several structurally and/or functionally interrelated parts; **2.** in chemistry, the combination of two or more compounds into a larger structure without → covalent bonds.

complex compound, the relatively stable combination of two or more compounds into a large molecule by → coordinate bonds, e.g. → chelates.

complex locus (L. *locus* = place), a gene in which intragenic → genetic recombination and intragenic → genetic complementation occur.

complon (L. *completus* = full), → complementation unit.

compound eye, an eye type found in many adult arthropods, such as insects and most crustaceans; consists of cone-shaped photoreceptor units, **ommatidia**, located side by side, their number varying from 20 (in some coleopterans) to about 10,000 (in odonates). The whole compound eye is covered by a transparent **cornea**, which is divided into **facets** to form the outer surface of each ommatidium. Each ommatidium comprises two **corneagen cells** producing the **lens**, below the lens locates the **crystalline cone**, **distal pigment cells**, and the long tapering **retinula** with several photosensitive cells (retinula cells). The photosensitive parts (rhabdomeres) of the retinula cells form the → **rhabdome**. The retinula is surrounded by **basal pigment cells** and each ommatidium is connected by a **nerve fibre** to the **optic nerve** which connects the eye to the brain ganglia.

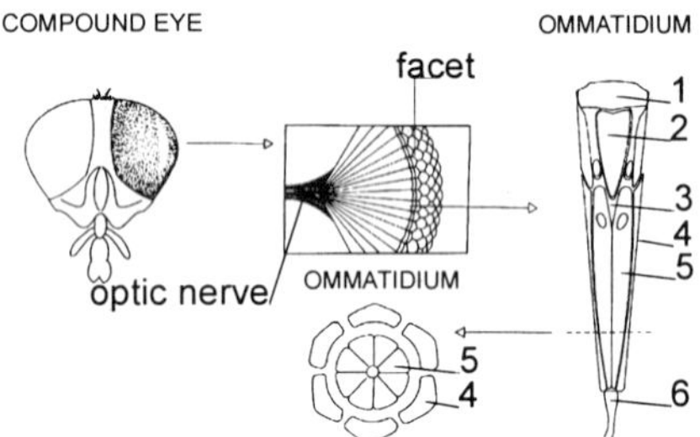

1. corneal lens 2. crystalline cone 3. rhabdome 4. pigment cells 5. retinular cells 6. nerve fibre

Functional features of the compound eye are not exactly known, but in intense light the pigment extends to prevent light passing from one ommatidium to adjacent ommatidia, and thus the eye forms an **apposition image** (mosaic image). The eye can efficiently register movements of an object. In weak light the pigment recedes towards the distal and basal parts of the ommatidia, and the light can spread laterally to adjacent ommatidia. This results in a **superposition image** (continuous image), which is less precise but more sensitive to light. The compoud eye is capable of distinguishing colours and to see ultraviolet and polarized light. Flying insects, such as honeybees and bumblebees, can probably also use the compound eye for measuring speed and distances as well as the angle of sun elevation. The most simple compound eye, such as found in some adult copepods, is called → **nauplius eye** (maxillopodan eye). *See* ocellus, eye.

compound inflorescence, a type of → inflorescence in which the basic structure is repeated in each peduncle; e.g. compound raceme, or umbel.

compound leaf, a leaf composed of several leaflets.

compression wood, → reaction wood.

concanavalin A, con A, a lectin (plant protein) obtained from the bean *Canavalia ensiformis*; acts as a haemagglutinin and mitogen in mammals, i.e. it agglutinates various animal cells, e.g. red blood cells, and stimulates the production of → lymphocytes, predominantly T cells.

concentrate, (L. *con-* < *cum* = together, *centrum* = centre), **1.** to intensify, make stronger or denser, or purer by removing inessential or foreign substances; **2.** a concentrated form of a liquid, as an extract or a solution.

concentration, 1. the quantity of a dissolved substance in a solution per unit of volume or weight; also a molar proportion of a gas in a gas mixture; **2.** increasing the amount of a solute in a solution by evaporation.

concentric bundle, a type of plant → vascular bundle in which either → xylem surrounds the → phloem, or the phloem lies around the xylem. In the former type the vascular bundle is **leptocentric**, found in many rhizomes; the other is the **hadrocentric** type, common in ferns.

Concentricycloidea (Gr. *kyklos* = circle, *eidos* = form), **sea daisies**; a class of Echinodermata including only two species living in marine waters of New Zealand; first described in 1986. They are disc-like pentaradial animals, about 10 mm in diameter, dorsally covered by scales and circular marginal plates with spines. One of the species has a stomach but the intestine is lacking; the other species has no alimentary canal. The water-vascular system includes two concentric ring canals.

conceptacle, a crater-like depression in some algae, containing → gametangia.

conception (L. *concipere* = to receive, apprehend), **1.** the act of forming a general abstract idea or notion, as a principle in a scientific system; concept; **2.** → fertilization; **conceptus**, the product of conception, i.e. an embryo or foetus.

concha, pl. **conchae** (L. < Gr. *konche* = mussel), a shell or shell-shaped anatomical structure, e.g. the **concha of auricle**, the human external ear, or the **nasal concha**, the ethmoid labyrinth in the nasal cavity.

concordance (L. *concordare* = to agree), agreement, conformity; in genetics, the similarity of the groups or pairs, especially twins, regarding a given character under comparison. *Cf.* discordance.

condensation (L. *condensare* = to pack together, condense), the act of making, or the process of becoming more solid or thick, as e.g. the conversion of gas into liquid, or liquid into solid.

condensation reaction, a reaction in which two molecules join together after the removal of a small molecule, such as water or ammonia; e.g. ethanoid anhydride is a condensation product of acetic acid (ethanoic acid).

conditional mutation, a → mutation which is expressed only in certain environmental conditions like in a given medium or at a certain temperature.

conditioned reflex, conditional reflex, a physiological response in an animal evoked gradually by repeated stimulus. Initially the stimulus (signal) is incapable of evoking the response but after having experienced it earlier (often several times) the animal associates it with a certain meaning which releases the response. The reflex is based on new interneuronal couplings in → synapses and it includes learning processes. *See* conditioning.

conditioning, Pavlovian conditioning; the process of acquiring, especially a form of **associative learning** in which → conditioned reflexes are involved. **Classical conditioning** is an experimental technique first used by *Ivan Pavlov* (1849—1936); using the technique on dogs (**Pavlov's dogs**) he noticed that a response of the animal (e.g. salivation) elicited by a natural stimulus (like food) became soon elicited by an unrelated stimulus which was connected to that natural stimulus, e.g. bell ringing preceding feeding; this happened when the animal learned to associate the bell ringing to feeding. Respectively, the stimuli are called **unconditional** and **conditional** stimuli, and the responses are called unconditional and conditional responses. *See* reflex.

conductance (L. *conducere* = to bring together), capacity for conducting; conductibility; a measure of conductivity; particularly the electric conductivity which is the inverse value of resistance. In biology, conductance measurements are used e.g. to study amounts of nutrients in waters and ion transport across cellular membranes; *see* Appendix 1.

conduction system of the heart, the system that serves for the induction and propagation of rhythmical electric impulses in the vertebrate heart. The system comprises 1) **sinus node** (**sinoatrial node**), a small cell group in the sinus venosus (in fishes and amphibians) or in the wall of the right atrium near the opening of the superior vena cava (in higher vertebrates); 2) **atrioventricular node** of similar structure situated in the wall between the right atrium and ventricle; 3) **internodal atrial pathways**, three bundles of specialized muscle fibres connecting the two nodes, so far found only in higher vertebrates; 4) **bundle of His** which is a conductive tract from the atrioventricular node along the septum branching to various portions of the ventricles, and 5) **Purkinje system**, a netlike system consisting of thinner conducting fibres branching everywhere in the ventricular muscle.

The cells of the system are specialized cardiac muscle cells which are differentiated in the nodes for rhythmic induction of cardiac impulses (**action potentials**) and along the tracts for rapid propagation of impulses from the atrioventricular node to ordinary, contracting muscle cells. Impulses from the sinus node to the atrioventricular node propagate slower than in the ventricles. The sinus node acts as a normal **pacemaker** (sinus rhythm), but if its impulses do not generate or propagate, the atrioventricular node induces a slower rhythm (nodal rhythm, nodal bradycardia) for the ventricles. The activity of the conduction system is regulated by the → autonomic nervous system; the sympathetic nervous system accelerates the rhythm (poositive chronotropism) while the parasympathetic nervous system slows it down (negative chronotropism).

condyle (L. *condylus* < Gr. *kondylos* = knuckle), **1.** a rounded protuberance of a bone, usually for articulation with another bone; e.g. condyle of humerus, *condylus humeri*; **2.** a similar process in arthropods formed from their chitinous integument. *Adj.* **condylar**, pertaining to a condyle; **condyloid**, knuckle-like, resembling a condyle.

cone (L. *conus*, < Gr. *konos*), **1.** cone cell, conus; a cone-shaped photoreceptor cell of the retina in the vertebrate eye; *see* colour vision; **2.** the female inflorescence of a coniferous plant (*see* Coniferae); in the cone the carpels are separate and arranged spirally, having the → ovules bare on them.

conexus, → connexus.

configuration (L. *configurare* = to form together), a relative arrangement or disposition of constituent elements, like body parts; in chemistry, the spatial three-dimensional organization of atoms or atom groups in a molecule around an asymmetric carbon atom. *See* conformation, isomers.

confluence (L. *confluere* = to flow together), a flowing together, meeting of streams, e.g. in embryology, the movement of cells in the process of → gastrulation. *Adj.* **confluent**.

confocal microscopy (L. *con-* < *cum* = together, *focus* = hearth, focus), a sophisticated modification of the electron → microscope. It uses lasers and special optics for "optical sectioning"; only regions within a narrow depth of focus are imagined. The picture is formed by a computer, and can be saved digitally for video research. The digital image can further be analysed in the computer.

conformation (L. *conformare* = to form, conform), uniformity, harmonization; e.g. **1.** the process of adjustment, adaptation (*see* conformers); **2.** the formation of the spatial three-dimensional structure of macromolecules like DNA and proteins which depends on ion strength or other cellular properties; **3.** in chemistry, the rotation of an atom or an atom group around single covalent bonds in the spatial structure of a molecule (conformative isomers).

conformers, in biology, animals whose physiological reactions follow environmental changes, such as the body temperature of poikilothermic animals follows the environmental temperature, or the osmotic concentration of poikilosmotic animals (osmoconformers) follows the salinity of the water they live in. However, a conformer in relation to one function may be a regulator in another function.

congenital, pertaining to a → phenotype or a characteristic present at birth; a descriptive term which neither includes all inherited characteristics, nor excludes traits which are caused by the environment only. For example, malformations which are observable immediately after birth are congenital and can be due either to genetic or external factors.

conical flask, Erlenmeyer flask (*Emil Erlenmeyer*, 1825—1909), a cone-shaped glass or plastic dish used in laboratories.

conidiophore (Gr. *konis* = dust, *phorein* = to carry along), *see* conidium.

conidium, pl. **conidia** (L. < Gr. dim. of *konis* = dust), an asexual spore of fungi, formed by constriction in tips of hyphae, or of special **conidiophores** which often have a stalk and tree-like branches; the conidia develop into new → mycelia.

conies, → Hyracoidea.

Coniferae (L. *conus* = cone, *ferre* = to carry), a class in the subdivision Gymnospermae, division Spermatophyta, Plant kingdom; comprises the families Pinaceae (e.g. pines), Taxaceae, and Cupressaceae; their leaves are needle-like and the male and female flowers are separate, the seeds developing in cones. In some systems Coniferae is classified as a separate division, named **Coniferophyta**.

coniferous, pertaining to cone-bearing plants, e.g. the trees in the group → Coniferae; all cone-bearing plants are usually called **conifers**.

coniferous forest, forests in the northern hemisphere in the → cool zone south of the arctic (tundra)zone; form the coniferous forest zone, i.e. **taiga**.

coniferyl alcohol, 4-hydroxy-3-methoxycinnamyl alcohol, $C_{10}H_{12}O_3$, an alcohol that occurs in → lignin of the cell walls of conifers; formed from amino acids containing aromatic rings.

conjugated proteins, proteins that are linked to non-protein moieties, e.g. to metal (metalloprotein) or lipids (lipoprotein).

conjugation (L. *con-* < *cum* = together, *jugare* = to join), a joining together; in biology means several phenomena in which chromosomes, nuclei, cells, or individuals join or unite; e.g. **1. chromosome conjugation,** → chromosome pairing; **2. conjugation of nuclei**, a) the union of nuclei, i.e. → karyogamy associated with fertilization, b) the formation of binucleated mycelium, i.e. → dikaryon during the life cycle of fungi; **3. conjugation of cells**, a) usually the union of gametes in fertilization or syngamy, b) the union of cells into a multinucleate → plasmodium or → syncytium;

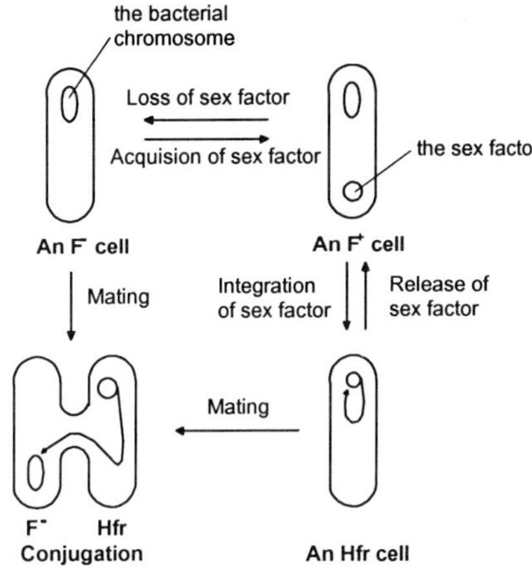

Conjugation. Schematic representation of the bacterical conjugation. (After Rieger, Michaelis & Green: Glossary of Genetics - Springer Verlag, 1968; with the permission of the publisher).

4. conjugation of individuals, the union of unicellular organisms or strings of cells of the same value, as a) the union of cells of aligned strings found in certain **algae** that leads to the fusion of nuclei, b) the temporary union of two **ciliate** (Ciliata, Protozoa) individuals during → meiosis where reciprocal fertilization occurs (*see* macronucleus), c) a substitute system of sexual reproduction found in bacteria, i.e. **bacterial conjugation** in which unidirectional transfer of genetic material occurs from a donor cell to a recipient cell through a cytoplasmic bridge. In conjugation, usually only a part of the donor DNA is transferred to the recipient. Conjugation occurs only between cells of opposite mating types. From the partial transfer of genetic material a → merozygote of the recipient cell arises. Bacterial conjugation is often associated with → genetic recombination; **5. molecular conjugation**, the association of a molecule with another, e.g. the joining of a toxic substance to a molecule synthesized by an organism, usually resulting in a non-toxic form (*see* detoxification). *Verb* to **conjugate**.

conjunctiva, pl. **conjunctivae,** Am. also **conjunctivas** (L. *tunica conjunctiva*), a mucous membrane on the vertebrate eye best developed in mammals; lines the inner surface of the eyelids (palpebral conjunctiva) and covers the exposed surface of the sclera (ocular or bulbar conjunctiva). *Adj.* **conjunctival**.

conjunctive, joining, connecting, connective.

connectins, a collective term for cytoskeletal proteins of cells, originally described in muscle cells. *See* cytoskeleton.

connective tissue, a tissue type of animals that connects, binds, supports, or surrounds other tissues and organs; usually of mesodermal origin developing from the embryonic connective tissue, mesenchyme (*mesenchyma*). Connective tissue includes 1) **loose connective tissue** (areolar connective tissue) with **collagenous**, **elastic** and **reticular** subtypes, binding various structures together as in dermis, mucous membranes, sheaths, and capsules; a specialized form of this is → adipose tissue; 2) **dense connective tissue**, including **ligaments**, **fasciae**, and **tendons** in which collagen fibres align along the direction of the tension, and in vertebrates also 3) **osseous tissue** (→ bone), and 4) **cartilaginous tissue** (→ cartilage) which form the skeleton of the body. Also blood can be considered as a type of connective tissue, having fluid intercellular matrix.

The connective tissue consists of abundant extracellular matrix with → **collagen**, → **elastin** (yellow) and → **reticulin** (argentophilic) **fibres**, and of specialized cells, **fibroblasts**, which in mature tissue are called **fibrocytes**, in the bone and cartilage correspondingly **osteoblasts** (osteocytes) and **chondroblasts** (chondrocytes). Especially loose (areolar) connective tissues have temporarily also various motile cells, such as → **leucocytes**, **mast cells**, and **histiocytes** (tissue → macrophages), which all serve as a part of the physiological defensive system.

connexon (L. *con(n)exus* = connection), *see* gap junction.

connexus, pl. **connexus** (L.), a connecting anatomical structure; e.g. *connexus intertendineus*, comprising fibrous bands between diverging tendons in the vertebrate forefoot or hand; also spelled conexus.

consciousness (L. *conscius* = aware), the state of being aware of one's self and capable of orienting in surroundings, and having clear, intact sensory functions and sensations. Limits for the conscious life in animals are often difficult to assess, especially in invertebrates, although behavioural criteria can be used. Self-consciousness is usually thought to be a property of humans and perhaps of some other mammals with a highly developed brain. In normal sleep the consciousness weakens, and in a certain stage of sleep it totally extinguishes; many drugs which affect the reticular formation of the brain stem have similar effects. Also total sensory deprivation causes unconsciousness, at least in higher vertebrates.

consensus sequence (L. *consensus* = agreement, conformity, *sequi* = to follow), a sequence of → nucleotides found in several different genes, and believed to have an identical function in all of these genes; e.g. the sequence determining the removal of → introns in → heterogeneous nucler RNA (hnRNA) encoded by the genes of eukaryotic organisms. Consequently, the proteins encoded by these genes carry consensus sequences of amino acids. The promoters of both prokaryotic and eukaryotic organisms often have consensus sequences which are not necessarily identical but possess common functions, such as ability to recognize RNA polymerases and transcription factors.

conspecific (L. *con-* = together, + species), pertaining to individuals of the same species.

constant species, a species living regularly in a certain community, present at least in 95% of random samples.

constitution (L. *constituere* = to place, establish), a composition, make-up; e.g. **1.** the structural, biochemical, or physiological characteristics of an individual; **2.** in chemistry, the composition of a molecule related to the way the atoms are linked together.

constitutive (L. *constitutivus* = a thing belonging to any structure), constituent, essential; e.g. pertaining to an enzyme (**constitutive enzyme**) or other proteins which are continuously produced by a cell in similar quantites under different circumstances not depending on requirements. In contrast to the constitutive type, many enzymes are adaptive enzymes, i.e. either inducible or repressive enzymes which are synthesized only when needed.

constitutive genes, genes expressed as a function of the interaction of RNA polymerase with the → promoter without additional regulation, i.e. these genes are functional in different conditions in all organisms. Thus the function of these genes is necessary for the basic constitutional functions of organisms; e.g. genes encoding enzymes of cell respiration. Called also housekeeping genes. *Cf.* regulated genes.

constitutive mutation, a gene mutation producing alleles which in contrast to wild alleles, are expressed without regulation, i.e. the mutant gene functions constantly in all tissûes.

constitutive phenotype, a → phenotype produced by → constitutive genes which encode products required in the maintenance of basic constitutional cellular processes or cell architecture. *Cf.* regulated genes.

constriction (L. *stringere* = to draw tight), the act or state of shrinking, compression, or pressing in, especially a constricted (shrunken) part of an organ; a stricture; e.g. **1.** → **vasoconstriction** in blood vessels; **2. duodenopyloric constriction** in the junction of the stomach and duodenum; **3.** constrictions of the chromosome (visible in the light microscope): **primary constriction**, location of the → centromere; other constrictions in the chromosome are called **secondary constrictions**, which can be e.g. the loci of → ribosomal RNA. *Verb* to **constrict**.

constrictor, anything that constricts; e.g. **1.** the muscle that constricts a hollow, vessel, tube etc. (*see* constriction); **2.** a snake that kills its prey by pressing.

consumer, in ecology, an organism feeding on other organisms; there are usually different kinds of consumers in a community, such as primary consumers (**herbivores**) feeding on producers, secondary consumers (**first carnivores**) of the third trophic level feeding on herbivores, tertiary consumers (**second carnivores**) feeding on carnivores etc., and **parasites** and **decomposers**. *Syn.* heterotroph.

consummatory act, a behavioural act of an animal immediately following the → appetitive behaviour and completing the act; e.g. a hungry animal searches for food (appetitive behaviour) and after finding it, eats it (consummatory act). The consummatory act is usually composed of → fixed-action patterns.

consumption efficiency, *see* ecological efficiency.

consumption tolerance, in environmental biology, the tolerance of the natural vegetation against mechanical erosion, caused e.g. by travellers, motor cars, camping. The consumption tolerance is very different in various vegetation types. *Cf.* carrying capacity.

contact inhibition (L. *inhibere* = to prevent), the inhibition of cell division especially in a cell culture when the cell density has become so high that the cells growing in a monolayer have contact with each other.

contamination (L. *contaminare* = to bring into contact), the act or state of becoming impure, unclean or bad, as a pure chemical by other chemicals, or an organ by pathological microbes, or some material (as soil) by radioactive compounds or other pollutants. *Verb* to **contaminate**. *Noun* **contaminant**, something that contaminates.

contest competition, *see* intraspecific competition.

contig, a group of cloned and isolated contiguous DNA segments.

continental bridge theory, a theory stating that the continents have previously been connected with each other by continental bridges and thus made possible the migration of organisms between the continents.

continental drift theory, a commonly presented theory stating that the continents have moved to their present positions by breaking off from the large ancient continent, → **Pangaea**; the drift still continues at the speed of some centimetres per year. The drift explains several, otherwise hardly explainable features of faunal compositions between the present continents, e.g. South America and Africa. *Syn.* Wegener's continental drift theory.

continental glacier, → ice sheet.

continental shelf, a submarine extension of a continent sloping to about 200 m in depth; the breadth of the shelf varies according to the coastal structure, but may be even 200 km. In the area of the continental shelf, nutrients are in the photic zone resulting in high organic productivity. The **continental slope** (rise) succeeds the shelf down to the → abyssal zone. *See* bathyal zone.

continental slope/rise, *see* continental shelf.

continuous defence mechanism, in ecology, the mechanisms by which an organism defends itself against predators. The mechanisms may be either structural or chemical. **Chemical continuous defence** is based on chemical compounds which exist permanently in organisms, as e.g. **secondary plant substances** (*see* secondary compounds) in plants. These are produced permanently regardless of animals feeding on them (*see* poisonous plants), as e.g. cardiac glycosides in milkweed (*Asclepias curassavica*) and tannins in oak trees. Also animals may have a permanent content of poisonous chemicals (*see* poisonous animals) in addition to physiological defence mechanisms which are inducible. **Structural continuous defence** involves permanent structures, as thorns, spines, hairs, teeth, etc., which have developed for protecting an organism against predation. *See* defence, chemical defence.

contour feather, *see* feather.

contra- (L. *contra* = against), denoting against, opposed.

contraception, deliberate prevention of fertilization and pregnancy.

contractile ring (L. *con-* = together, *trahere* = to pull), a cell organelle formed from a bundle of filaments which divides the animal cell into two parts during → cytokinesis.

contractile root, a root which pulls the shoot closer to the ground or, as in bulbous plants, deeper into the soil. Root contraction is rather common especially in monocotyledons and occurs in certain individual roots, being limited to parts of these roots; e.g. *Hyacinthus*, *Crocus*.

contractile vacuole (L. *vacuus* = empty), a cell organelle in many protozoans regulating the water content in the unicellular cell body. It functions frequently in a hypotonic freshwater environment where the osmotic pressure tends to bring water into the hypertonic cell. Water regulation in the cell is a rhythmic filling and emptying (out of the cell) of the contractile vacuole. The form and function of the vacuole differs considerably among the various kinds of protozoans; e.g. in the amoeba the location of the contractile vacuole varies in its endoplasm, but in the *Paramecium* it is located in a specific position beneath the cell membrane with a pore leading out. *Syn.* water expulsion vesicle.

contraction (L. *contractus* = drawn together), an act of reduction of size; a state of being drawn together or reduced; shortening; e.g. **1.** → muscle contraction; **2.** a pathologic or morbid shrinkage. *Verb* to **contract**. *Adj.* **contractile**, having the tendency or ability to contract, or capacity for shortening, i.e. **contractility**.

contracture (L. *contractura*), an abnormal constant shortening or distortion of a muscle due to spasm or fibrosis.

contralateral (L. *contra* = against, *latus* = side), pertaining to, or situated on the opposite side. *Cf.* collateral, ipsilateral.

control element, any element on the chromosomes of eukaryotic organisms that appears in addition to the actual genes and regulates the function of genes during the development of an organism. Evidence for the existence of control elements has been obtained specifically in cases where the elements move from place to place in the chromosomal complement. Therefore control elements are often called jumping genes or → **transposons**. Transposons are found in all eukaryotic species, and they are the main cause of spontaneous mutations.

Control elements suppress totally or partly the function of the genes located nearby in the chromosomal complement. Their removal restores the gene function. The function of the control elements can be autonomous or dependent on other control elements located in the same nucleus.

control group, *see* experiment.

controlled experiment, the intentional systematic arrangement of experimental material and conditions to elicit an answer to a question usually testing a hypothesis or aiming to find out the causal relations of things. *See* experiment.

conus, pl. **coni** (L.), → cone.

conus arteriosus (L.), **1.** the enlarged part between the heart ventricle and the ventral aorta in cartilaginous fishes and amphibians; *see* bulbus arteriosus; **2.** pulmonary cone, in the mammalian heart, the anterior part of the right ventricle terminating in the pulmonary artery.

conus medullaris, medullary cone; the cone-shaped terminal part of the spinal cord.

convection (L. *convehere* = to convey), the transmission of heat by streaming of gases (e.g. air) or liquids; other types of heat transmission are conduction and radiation.

conventional competition, a competition usually observed within the groups of **epideictic**

animals when individuals compete for limited environmental resources, especially food (*see* epideictic behaviour). Conventional competition may be directed to a **hierarchical position**, when the achievement of a high position ensures sufficient chances to exploit resources. Those individuals who lose in the competition usually die or emigrate to other areas.

convergence (L. *convergere* = to incline together), in biology, e.g. **1.** the structural, physiological, or ethological similarity of species which are not related; convergence is caused by similar environmental conditions; **2.** a structural or functional direction towards the same goal, as neuronal pathways from many neurones to fewer neurones, or the coordinated turning of both eyes to a near point. *Verb* to **converge**. *Adj*. **convergent**. *Cf.* divergence.

convergent evolution, the evolution of different taxa caused by similar environmental factors resulting in the formation of similar structures or functions in different, even distant evolutionary lines; e.g. the similarity of swimming organs in fishes, extinct ichthyosaurs, turtles, penguins, and whales.

conversion, in genetics, the non-reciprocal intragenic → genetic recombination. Conversion can most easily be observed in fungi in which all results of one single → meiosis are directly observable, and makes a → tetrad analysis possible. In conversion, the aberrant segregation ratios, i.e. 6:2, 5:3, 3:5 or 2:6, are observed instead of the normal 1:1 (4:4) segregation.

Conversion is correlated with reciprocal recombination, i.e. → crossing-over, and in fact they both have a common mechanism. This mechanism involves the formation of → hybrid DNA which is followed by the repair synthesis of DNA, provided that the strands forming the hybrid contain different alleles. Inevitably, in this case erroneous → base pairs are formed which are corrected by repair synthesis. In repair synthesis the DNA strand is converted to match one of the two strands of the hybrid.

convulsion (L. *con-* = together, *vellere* = to pull, tear), an involuntary, violent contraction or series of jerkings in skeletal muscles; an abnormal state found in vertebrates.

cool zone, a → vegetation zone in which the average temperature of the warmest month is over 10°C but the coldest is below -3°C. In the northern hemisphere the cool zone is situated south of the tundra zone and is characterized of coniferous forests. In the southern hemisphere there is no continuous cool zone, but the southernmost parts of South America and the southern islands in the Pacific Ocean are included in it.

cooperative breeding, 1. a breeding type in which non-breeding birds attend the nests of conspecific breeding birds and carry out various behaviour patterns associated with nesting, such as nest-building, incubation, and feeding of the nestlings (*see* helper). Cooperative breeding is found especially in passerine birds nesting in holes, but also among many colonial birds; → **kin-selection** may have played a main role in the evolution of this behaviour; **2.** a type of breeding in which all the members of a breeding group reproduce simultaneously and rear their young together, partly at least; e.g. cooperative breeding permits penguins in Antarctica to leave their young in the protection of a few adults, while the majority of the colony are fishing in the open sea. Cooperative breeding is found among many species, especially among invertebrates, and it also can be interspecific (*cf.* symbiosis).

coordinate bond, a chemical bond in which the two electrons shared by a pair of atoms originate from only one of the atoms. The compounds formed by coordinate bonds are called coordination compounds, e.g. → complex compounds. *Cf.* covalent bond.

coordinate genes, genes acting during the maturation of the oocyte in female organisms; the products of these genes form the → positional information or a kind of coordinate system in the egg.

coordination, the act or state of ordering in harmonious relation; a harmonious interaction of cells, tissues, organs, or organisms with each other; involves the propagation of information inside the system to be coordinated. In animals, the highest level of coordination occurs in the central nervous system. *Verb* to **coordinate**. *Adj*. **coordinate**, pertaining to, or involving coordination.

coparasitism (Gr. *parasitos* = parasite), a type of parasitism in which many parasite species inhabit simultaneously the same host.

Copepoda (Gr. *kope* = oar, *pous* = foot), **copepods**; a class in the subphylum Crustacea including both parasitic and free-living species; occur abundantly in planktonic, benthic and interstitial habitats. The carapace and compound eyes are lacking but a median, single → nauplius eye functions as a visual organ. The Copepoda comprise eight orders, e.g. **calanoids** (Calanoida) and **cyclops** (Cyclopoida); more than 8,400 species are described.

copepodid larvae, a series of larval stages following the nauplius stage in the development of copepods; the abdomen is already observable in copepodids.

Cope's rule (*Edward D. Cope*, 1840—1897), a hypothesis stating that the size of animals of the same evolutionary line increases in the course of time. A large animal is more advantageous in its energy economy and in its ability to compete for a mating partner. Cope's rule has been strongly criticized.

copia, a family of → transposons in *Drosophila melanogaster*, the fruit fly.

copper (L. *cuprum*, Gr. *kypros*), a metallic element, symbol Cu, atomic weight 63.55; together with gold and silver forms the copper group. Copper ores contain e.g. chalcopyrite ($CuFeS_2$) and copper sulphide (Cu_2S). Copper is a soft metal having a good electrical and heat conductivity. The cells of all organisms need copper ions (Cu^{2+}) in minute quantities (*see* micronutrients, trace elements) for enzyme reactions, e.g. phenolases and ascorbic acid oxidase in plants. Cu^{2+} is an essential element for the light reactions of photosynthesis, present as a component of → plastocyanin in plants; in animals it is found in → haemocyanin of some invertebrates.

coppice (L. *colpaticum* = cut-over area), a wood of small trees, usually cut down at certain intervals; **short rotation coppice**, a growth of small trees and bushes (energy trees), such as willows and poplars, growing abundantly close to the ground. The produced material is harvested every two to four years and used as a renewable energy source; the material is chopped and burnt to produce heat and electricity.

coprin, a substance in the mushroom *Coprinus atramentarius* (ink cap), causing harmful effects in man if ingested together with alcohol; this occurs because coprin inhibits the oxidation of acetaldehyde (alcohol metabolite) in the liver (antabuse effect) resulting in toxic concentrations of acetaldehyde in the body.

coprophagy (Gr. *kopros* = dung, *phagein* = to eat), feeding on dung, as e.g. some beetles; **coprophage**, an animal feeding on dung. *Adj.* **coprophagous**. *See* caecotrophy.

coprophilic, coprophilous (Gr. *philos* = loving), favouring faeces; e.g. a plant favouring places near excrements, growing on dung; **coprophil(e)**, a coprophilic organism. *Cf.* ornithocoprophilous.

copulation (L. *copulatus* = bound together), **1.** coupling; *adj.* **copulative**; **2.** a sexual union or intercourse between a male and female; coitus; *adj.* **copulatory**; *verb* to **copulate**; *noun* **copula** (L.), sexual union, copulation (used chiefly in law).

cor (L.), → heart.

coracidium, pl. **coracidia** (L. < Gr. *korax* = raven), the first larval stage of some cestodes, e.g. the tapeworm; free-living and covered by cilia. The larva has to get in a copepod crustacean for developing further into the next stage, → procercoid. Coracidium occurs only in the species whose development takes place in water. *Cf.* onchosphere.

Coraciiformes (L. *Coracias* = the type genus, *forma* = form), a polymorphic order of birds comprising ten families; characteristic of them are the brightness of the plumage, a straight and stout beak, and strong feet. The distribution of Coraciiformes extends from tropical to temperate districts. In total, 204 species are described, and well-known families are e.g. **kingfishers** (Alcedinidae), **bee-eaters** (Meropidae), **rollers** (Coraciidae), **hoopoes** (Upupidae), and **hornbills** (Bucerotidae).

coracoid (L. *coracoideum* < Gr. *korakoeides* = raven-like, like a raven's beak), **1.** a ventral bone in the → shoulder girdle (pectoral girdle) of the tetrapod vertebrate; it joins the **scapula** to the midventral breastbone, **sternum**, in the → glenoid cavity; in mammals, excluding monotremes, the coracoid is dwarfed into a tiny projecting "raven's beak" (**coracoid process**, *processus coracoideus*) at the lower edge of the scapula; *Adj.* resembling the coracoid; **2. coracoid plate** (*pars coracoidalis*) in the pectoral girdle in fishes; may not be homologous with the coracoid of tetrapods.

coral, an anthozoan polyp (*see* Anthozoa), either colonial or solitary, which secretes a calcareous and usually richly coloured skeleton; e.g. **soft** and **horny corals** (→ Alcyonaria), **stony corals** (zoantharian corals), and **thorny corals** (ceriantipatharian corals). *See* coral reefs.

coral reefs, usually large geographical formations of limestone in shallow tropical oceans between the latitudes of 28° N and 28° S where the water temperature is always over 20°C. Coral reefs are mainly formed by **hermatypic corals**, e.g. stony corals (the cnidarian subclass Zoantharia) and **coralline algae**. Several types of coral reefs are known: a **fringing reef** situates near the shore without any lagoon or with a narrow lagoon, a **barrier reef** surrounds the island or runs parallel to shore, being separated from it by a wider lagoon; an **atoll** (circular reef) encircles a lagoon without any island in the middle. Coral reefs are most productive aquatic ecosystems with an enormous diversity of species.

coralline, describing some species of red algae which are encrusted with lime.

Cordaitinae (*Cordaites*, the type genus), an extinct class in the subdivision Gymnospermae, division Spermatophyta, Plant kingdom; in some classifications the order Cordaitales in the division Coniferophyta. The plants were tall trees with simple, spirally arranged leaves and separate male and female flowers in cone-like inflorescences; occurred in the Carboniferous and Permian periods.

co-repressor, a metabolite, i.e. the end product of an anabolic reaction chain, which by binding to the → repressor specifically prevents the synthesis of those enzymes which catalyse the reaction chain. *See* effector, operator, operon.

Cori cycle (*Carl F. Cori*, 1896—1984, and *Gerty T. Cori*, 1896—1957), the phases of carbohydrate metabolism in animals; includes 1) → glycogenolysis in the liver, 2) the glucose-lactate reaction chain (anaerobic → glycolysis) in muscle cells during muscular activity, and 3) partial resynthesis of glucose from lactate in the liver (gluconeogenesis).

corium, pl. **coria** (L. = hide), → dermis.

cork, dead plant cells filled with → suberin; occurs mostly in external layers of roots and shoots as a part of the → periderm, being formed by the cork → cambium. The cork layer protects the plant and makes the surface impermeable to water. *Syn.* phellem.

cork cambium, *see* cambium.

cork cells, cells in the leaf epidermis of graminaceous plants; they alternate with the silica cells and long cells, all being arranged in straight rows.

corm (Gr. *kormos* = tree trunk), a short, vertical, bulb-like, underground stem, i.e. → rhizome, acting as energy store and reproductive organ. *Cf.* bulb.

cormidium, pl. **cormidia,** a group of different individuals in siphonophores, formed at the coenosarc base; after maturation they separate from the colony.

cormophyte (Gr. *phyton* = plant), a plant having differentiated roots, shoot, and leaves, comprising the divisions → Pteridophyta and → Spermatophyta.

cornea (L. *corneus* = horny), the transparent membranous structure forming the anterior surface of the eye in vertebrates and many invertebrates. In vertebrates, it comprises several tissue layers composed of connective tissue with epithelium on both surfaces, and has no blood vessels. Beside the lens the cornea acts as a light refracting structure. Sometimes called also keratoderma.

corneal reflex, the reflex closing the eyelids by an irritation of the cornea; the extinction of the reflex shows that the animal is in deep coma and near death. *Syn.* lid reflex, blink reflex, eyelid closure reflex.

cornification, keratinization; conversion of the epidermal tissue into a keratinized, horny tissue by accumulation of the fibrous protein, **keratin**, into skin cells, called **keratinocytes**; keratin finally fills the cells totally, causing the death of keratinocytes. Such tissues are e.g. horny teeth in cyclostomes, ceratothrichia in the fins of cartilaginous fishes, horny scales and shields especially in reptiles, beaks, claws, nails, and feathers and hairs of birds and mammals.

corolla, pl. **corollae** or **corollas** (L. dim. of *corona* = circle, crown), the entity formed of → petals in a flower.

corolla tube, a tube-like structure formed by the fusion of petals in a flower.

corona, pl. **coronae, coronas** (L. < Gr. *korone* = crown, circle), a crown-like structure; e.g. **1.** *corona ciliaris*, the ciliary crown on the anterior inner surface of the → ciliary body in the vertebrate eye; **2.** *corona dentis*, the dental corona, the upper part of the tooth; **3.** *corona radiata*, a) the radiating formation around the developing ovum in the Graafian follicle, or b) the crown of radiating nerve fibres ascending from the thalamus to the cerebral cortex; **4.** a ciliated disc or crown, such as the corona surrounding the mouth in rotifers; **5.** a frill-like appendage of a corolla tube in a flower, e.g. daffodil; **6.** a coloured circle around a luminous body, as around the sun or the moon. *Adj.* **coronal**.

coronary, encircling another structure, often like a crown; especially pertains to blood vessels, nerves, or ligaments resembling a crown. *See* coronary vessels.

coronary vessels, the blood vessels of the vertebrate heart; the **coronary arteries**, in the heart of tetrapod vertebrates, comprising the right and left coronary arteries conveying oxygenated blood from the base of the aorta to all parts of the heart muscle. The coronary arteries of the fish heart begin as branches from the branchial arteries behind the gills; some species have no coronary vessels at all, but a spongeous cardiac muscle gets oxygen and other substances directly from the blood flowing through the ventricle. The corresponding **coronary veins** collect blood from cardiac capillaries and end in the atrium, in tetrapods in the right atrium. In tetrapods, most veins unite to form the short trunk, **coronary sinus**, before emptying into the atrium.

coronary sulcus, pl. **coronary sulci,** atrioventricular groove; a groove on the outer surface of the vertebrate heart between the atria and ventricles; parts of it are occupied by the major → coronary vessels.

corpora allata, *see* corpus allatum.

corpora cardiaca, *see* corpus cardiacum.

corpus, pl. **corpora** (L. = body), a body, mass, anatomical part, or the entire organism, e.g. **1.** corpus gastricum, the gastric body between the fundus and the pyloric part of the stomach; **2.** → corpus luteum of the ovary; **3.** corpus vitreum, → vitreum of the vertebrate eye; **4**. corpus callosum; **5.** corpus allatum; **6.** corpus cardiacum; **7.** in plants, the innermost group of cells in → apical meristem, dividing both periclinally and anticlinally. *See* corpus allatum, c. callosum, c. cardiacum.

corpus allatum, pl. **corpora allata** (L.), a paired non-neuronal endocrine gland of insects located in the head behind the brain ganglia and posterior to another endocrine gland, → corpus cardiacum, nerve fibres connecting the glands; secretes → juvenile hormone.

corpus callosum, a flat nerve commissure connecting the cerebral hemispheres of the vertebrate brain; due to myelinated nerve fibres forms anatomically the white matter area in the median section of the telencephalon.

corpus cardiacum, pl. **corpora cardiaca** (L. < Gr. *kardia* = heart), a paired neurohaemal gland of insects, located in the head immediately behind the brain ganglia and anterior to → corpora allata; connected with nerve fibres to the neurosecretory cells of the brain. The gland stores and releases → **prothoracotropic hormone** (ecdysiotropin) which stimulates the prothoracic glands to secrete → ecdysone that further regulates the moulting of the animal. The corpus cardiacum also releases **eclosion hormone** to induce the emergence of the adult from the puparium. The secretion of the corpus cardiacum is regulated by the nervous system. The corpora cardiaca, corpora allata, prothoracic glands, and certain neurosecretory cells in the brain form the principal endocrine system of insects.

corpusculum, pl. **corpuscula** (L. dim. of *corpus* = body), **corpuscle**; a small body, or mass, as *corpuscula tactus*, tactile corpuscles (touch receptors) in the skin.

corpus luteum (L. *luteus* = saffron-yellow), **yellow body**; a yellow endocrine gland formed by a cell mass in the → ovarian follicle of mammals after ovulation. The corpus luteum secretes mainly → **progesterone** and some oestrogen. If pregnancy does not occur, it (*corpus luteum spurium*) progressively degenerates forming a small scar, *corpus albicans*. The secretion of progesterone maintains uterine endometrium until menstruation. During pregnancy, the gland (*corpus luteum verum*) grows and becomes strongly activated for maintaining the hormonal state of the pregnancy. *See* luteinizing hormone.

corpus striatum, striate body; a part of the → basal ganglia (basal nuclei) in the vertebrate brain; formed from a subcortical mass of grey and white matter in front of and lateral to the thalamus in both hemispheres, including the lentiform (lenticular) nucleus (putamen and globus pallidus) and caudate nucleus.

correlation, the interdependence of two random variables; the quantitative degree of the relation is indicated by a statistical procedure which yields the **correlation coefficient** referred to as **r** (from -1 to +1). In biology, e.g. **phenotypic**, **genetic**, and **environmental correlations** can be calculated.

corridor, 1. in biogeography, a broad continual land connection between different regions which makes possible a migration of organisms from one area to the other. Many **continental bridges** are such corridors, as has been the land bridge over the Bering Strait during Pleistocene, permitting organisms to move from Asia to America and vice versa; **2. ecological corridor**, a rather new term in the planning of nature reserves; in urban ecology means green belts and park zones in a city along which organisms can move from one urban district or nature reserve to another; the term is based on the theory of island biogeography.

cortex, pl. **cortices,** Am. also **cortexes** (L. = bark, rind, shell, husk), an external layer of an organ or tissue as distinguished from the internal structure; e.g. **1.** in animals, adrenal cortex (*see* adrenal gland), → cerebral cortex and cerebellar cortex in the brain, or renal cortex in the → kidney; **2.** in plants, a structure covering the root and shoot, formed mostly of the → parenchyma tissue. The primary cortex is often thin and covered by a unicellular layer of → epidermis; these are replaced later by a thick → periderm which gives effective protection to the inner structures; cortex is also the outermost, hard layer of → lichens. *Adj.* **cortical**.

cortical vascular bundle, a → vascular bundle in a plant shoot, seen in the cross-section exceptionally in the area of the → cortex where the bundle is turning to enter a stem branch or a leaf.

corticoliberin, a neuropeptide hormone produced by a group of hypothalamic neurones in the vertebrate brain; carried through circulation to the anterior → pituitary gland where it causes the release of → **corticotropin**. Its own release is controlled by brain activity, such as emotional functions, and by the negative feed-back action of → glucocorticoids. *Syn.* corticotropin-releasing hormone, **CRH**, corticotropin-releasing factor, **CRF**. *See* releasing hormones.

corticosteroids, corticoids; a general term for the steroid hormones (C21 steroids) synthesized from cholesterol in the cortex of the vertebrate → adrenal gland; include → mineralocorticoids, → glucocorticoids, and some other steroids with androgenic or estrogenic activity.

corticosterone, 11β,21-dihydroxy-pregn-4-ene-3,20-dione, $C_{21}H_{30}O_4$; a natural corticoid steroid with preferentially glucocorticoid activity; however, has also mineralocorticoid activity, in man much more than have cortisol or cortisone. *See* glucocorticoids, mineralocorticoids.

corticotropin, a polypeptide hormone produced by the anterior → pituitary gland of vertebrates; released in stress situations stimulating the hormone production in the adrenal cortex, especially the synthesis and liberation of → glucocorticoids. *Syn.* adrenocorticotropic hormone, **ACTH**, adrenocorticotropin, (adreno)corticotrophin. *Adj.* **corticotropic**. *See* corticoliberin.

corticotropin-releasing hormone, → corticoliberin.

cortisol, hydrocortisone, 17-hydroxycorticosterone; 11β,17α,21-trihydroxypregn-4-ene-3,20-dione, $C_{21}H_{30}O_5$; a → corticosteroid produced by the adrenal cortex of vertebrates, having strong **glucocorticoid** activity in the body. Cortisol is released by the action of → corticotropin (ACTH). Cortisol is an important stress hormone (*see* stress), and because of its anti-inflammatory and immunosuppressant effects it is also manufactured for pharmaceutical use. *See* glucocorticoids.

cortisone, 17α, 21-dihydroxypregn-4-ene-3,11,20-trione, $C_{21}H_{28}O_5$. A corticoid hormone released in stress situations from the adrenal cortex in vertebrates; it has fairly strong **glucocorticoid** activity but also some mineralocorticoid properties. Cortisol is released by the action of → corticotropin (ACTH). *See* glucocorticoids, mineralocorticoids.

corymb (Gr. *korymbos* = top, cluster of flowers or fruits), a raceme-like → inflorescence in which the flower stalks of the lower flowers are more elongated than those of the upper flowers, all the flowers thus being at the same level.

cosmid, cosmid vector (< cohesive end + plasmid), a synthetic → plasmid used as a → vector, consisting of two parts, a bacterial plasmid and the so-called **cos site** of a certain bacteriophage virus (lambda phage). A cosmid replicates like a plasmid but can be packed into the lambda or any other phage. The cosmid can be used as a vector in genetic engineering, specifically for cloning (*see* clone) of large DNA fragments. The cos site describes the cohesive closed ends of the DNA fragments of the lambda phage.

cosmine (Gr. *kosmikos* = regular), a dentine-like layer with branching tubules between the enamel and the spongy layers of the lamellar bone in → cosmoid scales of some fishes.

cosmoid scale, a scale type of crossopterygians and primitive, fossil lungfishes; is covered by a thin enamel layer with a spongy layer of porous, dentine-like cosmine underneath, and further a lamellar bony layer undermost. *Cf.* ganoid scale, placoid scale. *See* scale.

cosmopolitan (Gr. *kosmos* = order, universe, *polites* = citizen), belonging to all the world; in biology, a species or some other taxon which is distributed all over the world.

costa, pl. **costae** (L.), **1.** rib, *os costale*; **2.** a rod-like structure at the base of the → undulating membrane in some flagellates; **3.** a strong longitudinal vein in the anterior edge of the insect wing; **4.** the midrib of a leaf in mosses.

costal, relating to the ribs.

cost of meiosis, the apparent disadvantage of sexual reproduction due to the fact that, as a consequence of → meiosis, only a half of an individual's genes are transmitted to the next generation in each progeny.

cotransfection (L. *co-* = together, *trans* = on the other side, *facere* = to do), the artificial infection of a cell with isolated nucleic acid (→ transfection) in such a way that two → vectors are introduced into the cell at the same time.

cotton, a soft, downy, white matter obtained from seed hairs of the malvaceous plants, e.g. *Gossypium hirsutum* and *G. barbadense* used for making thread and cloth.

cotyledon (Gr. *kotyledon* = pit, hole), **seed leaf**, a leaf, being a part of the seed plant embryo; the monocotyledons have one, the dicotyledons two in their embryos; in → gymnosperms the number varies. The cotyledons have an important role in seed → germination. In → epigeal germination, the cotyledon rises first above the ground and begins to photosynthesize, producing energy for the growing seedling. In → hypogeal germination, the cotyledon contains energy stores and remains in the soil, giving energy for the growing parts of the plant.

Cotylosauria (Gr. *kotyle* = cup, *sauros* = lizard), **cotylosaurs**, → Captorhinida.

cotype (L. *con-* = with, *typus* = type) a term used by bacteriologists to describe any new microbe strain from the same taxon. In taxonomy, the term was earlier used as a synonym for syntype, but rarely used any more.

coumarin, cumarin, 1,2-benzopyrone ($C_9H_6O_2$); a substance that is a natural germination inhibitor occurring in many plants, e.g. in *Asperula* (woodruff) and *Melilotus* (sweet clover) species and in lavender oil. Coumarin is released on wilting (smell of hay). A derivative of coumarin, **dicoumarol**, prevents in the vertebrate liver the production of those blood clotting factors whose syntheses are dependent on vitamin K. Therefore dicoumarol and its derivatives are used as → anticoagulants. → Aflatoxins which are produced by some fungi have derived from coumarin. Coumarin has a scent of vanilla and is used in perfumery.

countercurrent, flowing in an opposite direction; a current running in an opposite direction to another current.

countercurrent system, a system in which two materials are flowing in opposite directions which makes possible the exchange of material or heat; e.g. countercurrent heat exchange between arterial and venous blood flowing in adjacent collateral vessels in the limbs (*see* rete), or countercurrent exchange of substances between the descending and ascending nephric loops of Henle (*see* nephron).

counterstaining, double staining. *See* staining.

counter tube, *see* Geiger counter.

coupling (L. *copula* = bond), e.g. **1.** in genetics, the occurrence of two genes in one of the members of a → homologous chromosome pair; **2.** in the heart, the repeated pairing of a normal sinus beat with a ventricular premature extrasystole, causing abnormal bigeminal rhythm; **3.** → excitation-contraction coupling in muscle cells; **4.** the coupling of energy metabolism to the production of ATP from ADP and P_i in → oxidative phosphorylation or → photophosphorylation.

coupling factor, *see* oxidative phosphorylation.

coupling phase, a double → heterozygote, in which the dominant → alleles (A,B) lie on one of the homologous chromosomes and the recessive alleles (a,b) on the other (AB/ab), is said to be in the coupling phase. *Cf.* repulsion phase.

courtship, a behavioural pattern preceding the

mating of animals.

courtship area, a traditional area to which males get together year after year for → communal courtship displays and fights. *Syn.* → lek.

courtship feeding, an epigamic behavioural pattern occurring especially among birds between a pair when the male presents the female with food or other objects symbolizing food. The purpose of courtship feeding is to strengthen the pair bonds and especially in late courtship and egg-laying periods to improve the nutritional state of the female; **greeting feeding** is a part of the courtship feeding where the male presents the food only during and after copulation.

COV, crossing-over value. *See* crossing-over.

covalent bond (L. *co-* = together, *valens* = strength), a bond between two atoms when the atoms share one common electron pair (single bond), two pairs (double bond), or three pairs (triple bond). The electrons of the covalent bond belong to both atoms; e.g. the covalent bond is between one hydrogen and one oxygen atom. The covalent bond is the strongest type of chemical bond. *Cf.* coordinate bond.

coverage, in botany, indicates the abundance of different plant species on the field. Coverage is determined by estimating the united vertical projection of all individuals of different species in a study area, generally 1 m^2.

covering, in agriculture, the handling of plant seeds with pesticides in order to prevent plant diseases and e.g. mould formation on germinating seeds.

cover leaf, a leaf type of some epiphytic ferns; totally pressed on the shoot (often swollen), while the later developing foliage leaves are lobed and hanging.

coxa, pl. **coxae** (L. = hip), **1.** in tetrapod vertebrates, the lateral part of the hip or → hip bone, including the hip joint; loosely used to denote the hip joint; **2.** in invertebrate anatomy, the short, proximal segment of the leg in some arthropods, such as crustaceans, insects, and arachnids; the coxa articulates the leg with the thorax.

coxal glands, excretory organs of arachnids on the sides of the → prosoma; modified nephridia opening at the coxa of the first and third walking legs. Some arachnid groups have both the excretory system of → Malpighian tubules and the coxal glands, some groups have only one of them. *Cf.* green glands.

coxopodite (Gr. *pous*, gen. *podos* = foot), the first segment of the basal protopodite (the basal segment) in the legs of crustaceans. *Cf.* basipodite.

cpDNA, → chloroplast DNA.

CpG island (C = cytosine, p = phosphorus, G = guanine), in vertebrate genomes, DNA regions with a high C + G content and a high frequency of CpG dinucleotides relative to the bulk genome. These CpG islands are usually located at the 5' ends of genes, and they serve to indicate the proximity of a gene. They probably act in post-transcriptional regulation of various → house-keeping genes.

C_3 plants (C = carbon), the C_3 group of common plants assimilating carbon in the Calvin cycle of → photosynthesis; ribulose 1,5-bisphosphate acts as the acceptor of carbon dioxide (CO_2), the first synthesized carbohydrates being 3-carbon compounds (C-3 compounds). Most plants belong to this group.

C_4 plants, plants which differ from → C_3 plants concerning the assimilation of carbon dioxide (CO_2) which is first absorbed into → phosphoenolpyruvate (PEP), and oxaloacetate and malate (4-carbon acids, C_4 acids) are formed in mesophyll cells (*see* Hatch—Slack pathway). CO_2 is then liberated in the decarboxylation reaction and assimilated via the Calvin cycle into carbohydrates in bundle sheath cells; e.g. many tropical grasses which have a great photosynthetic capacity belong to this group. *See* CAM plants.

cranio- (L. *cranium*, Gr. *kranion* = head), denoting relationship to the skull, cranium. *Adj.* **cranial.**

cranial nerves (L. *nervi craniales*, *nervi encephalici*), the encephalic nerves; the nerves of vertebrates which emerge from, or enter the brain, i.e. in fish 10 pairs of sensory, motor, or mixed nerves and in higher vertebrates 12 pairs, having their roots in the lower part of the brain stem. The first cranial nerve, **olfactory nerve** (*n. olfactorius*), comprises many separate small bundles (*fila olfactoria*) of nerve fibres. It transmits olfactory signals from the nasal mucosa to the olfactory bulb (*bulbus olfactorius*) in the most anterior part of the cerebrum (telencephalon). The second cranial nerve, **optic nerve** (*n. opticus*), transmits impulses from the retina of the eye to the thalamus in the diencephalon, the nerve pair forming the optic chiasma (*chiasma opticum*) on the ventral surface of the hypothalamus where medial nerve fibres of the nerves cross. The third cranial nerve is the **oculomotor nerve** (*n. oculomotorius*), the fourth is the **throchlear nerve** (*n. throchlearis*) and the sixth, **abducent nerve** (*n. abducens*), which all innervate the external muscles of the eyeball. The oculomotor nerve also contains parasympathetic nerve fibres. The fifth cranial nerve is the **trigeminal nerve** (*n. trigeminus*) which is divided into three main branches containing sensory and motor fibres to facial areas. The seventh is the **facial nerve** (*n. facialis*) innervating facial muscles but containing also some sensory fibres e.g. from the tongue, and parasympathetic fibres to lacrimal and salivary glands. The eighth is the **vestibulocochlear nerve** (auditory nerve, acoustic nerve, *n. vestibulocochlearis*, *n. statoacusticus*) the branches of which innervate the auditory and vestibular organs in the inner ear. The ninth is the **glossopharyngeal nerve** (*n. glossopharyngeus*) containing mainly sensory fibres from nasopharyngeal and lingual areas, and from the middle ear. It also contains parasympathetic fibres e.g. to the sublingual salivary gland. The tenth cranial nerve, **vagus nerve** (*n. vagus*), consists almost totally of parasympathetic nerve fibres innervating the inner organs in the thoracic and abdominal parts of the body. It also contains motor fibres to the muscles of the throat. The eleventh cranial nerve is the **accessory nerve** (*n. accessorius*) and the twelfth the **hypoglossal nerve** (*n. hypoglossus*), innervating neck and tongue muscles.

Recently a thin additional nerve, zero cranial nerve, **terminal nerve** (*n. terminalis*) is found in some fishes, amphibians, and mammals; it is associated with nasal olfactory mucosa. *See* terminal nerve system.

Craniata, craniates, → Vertebrata.

cranium, pl. **crania** (L. < Gr. *kranion* = skull), skull; **1.** generally denoting **neurocranium**, i.e. that part of the head in vertebrates which encloses the brain; **2.** broadly means the entire bony or cartilaginous skeleton of the head region of vertebrates, and its simplest type, in addition to neurocranium, includes also **splanchnocranium** and **dermatocranium**. In higher vertebrates these two latter parts are reduced, they have other functions (e.g. have developed into auditory ossicles), or they are closely associated with neurocranium. Most important bones of the neurocranium in mammals are: **parietal bone** (*os parietale*), **frontal bone** (*os frontale*), **temporal bone** (*os temporale*), **ethmoid bone** (*os ethmoidale*), **sphenoid bone** (*os sphenoidale*), and **maxillary bones** (*maxillae*) forming the upper jaw, and **dentary** (*os dentalis*) forming the lower jaw; *see* mandible.

On the basis of the association of the lower jaw with the skull, several different skull types can be distinguished. In the **hyostylic cranium** of some cartilaginous fishes and all bony fishes, the **hyomandibula** articulates the lower jaw to the skull, in the **autostylic cranium** of tetrapods, the bone joining the dentary is the **quadrate**, and the hyomandibulare is reduced into an auditory ossicle of the middle ear; some cartilaginous fishes have **amphistylic cranium** in which the lower jaw is connected with the skull by the posterior part of the quadrate, and hyomandibula (sometimes by the palatoquadrate).

Skulls can be classified also according to their internal flexibility, i.e opening of the mouth; e.g. mammals have an **akinetic skull** in which only the lower jaw moves; reptiles have a **kinetic skull** in which both jaws and some other facial bones move when the mouth opens. On the basis of the movable bones in the cranium several skull types are found, such as the **metakinetic**, **amphikinetic**, or **mesokinetic skull**. The mesokinetic skull of snakes is most flexible and makes it possible to swallow whole, large prey animals.

crassulacean acid metabolism, CAM, a type of carbon assimilation characteristic of many succulent plants in arid areas, e.g. in the families of Crassulaceae and Cactaceae. Carbon dioxide (CO_2) reacts with phosphoenolpyruvate in the reaction catalysed by PEP carboxylase while 4-carbon acids (C_4 acids), oxaloacetate, and malate are formed. This takes place at night when the stomata of CAM plants are open. In the light the stomata are closed and CO_2 is liberated from organic acids and metabolized in the → Calvin cycle into carbohydrates. *See* C_4 plants.

crayfish plague, a disease caused by the

fungus *Aphanomyces astaci* in the crayfish *Astacus astacus*; has killed the crayfish populations in many lakes and rivers, especially in Northern Europe.

creatine (Gr. *kreas* = flesh), N-methylguanidinoacetic acid, $C_4H_9N_3O_2$; a nitrogenous compound synthesized in animals. The phosphorylated form of it (**phosphocreatine**, creatine phosphate) is a high-energy compound serving in animals as an important energy store for immediate ATP production, especially in the muscle cells of vertebrates, **creatine kinase, CK** (EC 2.7.3.2.) catalysing the reaction, i.e. phosphocreatine + ADP $\rightleftharpoons$ creatine + ATP. *See* arginine.

creatinine, 1-methylglycocyanidine, $C_4H_7N_3O$, an anhydride of creatine; the final product of creatine metabolism of vertebrates, excreted in urine.

creationism, an unscientific doctrine stating that all life forms were created by an omnipotent Creator at one time, and thereafter only minor changes have occurred without → evolution.

Creodonta (Gr. *kreas* = flesh, *odous* = tooth), **creodonts**; an extinct order of carnivorous mammals which lived from Palaeocene to Pliocene; probably ancestral to present fissipeds (terrestrial carnivores).

cresol, tricresol, hydroxytoluene, methylphenol, $CH_3C_6H_4OH$; the mixture of three isomeric organic substances, i.e. ortho-, meta-, and paracresol, obtained from coal tar and naphtha; cresol is used as precursor for pesticides, in the plastic and dye industries, and as a disinfectant (lysol).

crest, → crista.

Cretaceous (period) (L. *creta* = chalk), the last period in the Mesozoic era between the Jurassic and Tertiary periods about 144 to 65 million years ago, characterized by thick chalk deposition. During Cretaceous a big portion of terrestrial and aquatic animals became extinct. The climax period of giant reptiles and ammonites ceased through mass extinction, probably due to the impact of a meteorite on the Earth. These were substituted by tooth-billed birds (e.g. the genera of *Ichthyornis* and *Hesperornis*). The evolution of flowering plants (Angiospermae) was dramatic; other plant taxa are relatively unchanged after Cretaceous.

cretinism (Fr. *crétin* = fool), a chronic, congenital condition in man, due to deficiency or total lack of thyroid secretion during the ontogenic development; marked by disturbances in physical and mental development and decreased basal metabolism. The condition is possible also in other vertebrate species.

Crinoidea (Gr. *krinon* = lily, *eidos* = form), **crinoids** (sea lilies and feather stars); an echinoderm class in the subphylum Crinozoa (Pelmatozoa in some classifications); the animals have a stalked, often cup-like body, and several arms bear numerous pinnules. Many characteristics are primitive. Crinoids were previously far more numerous than they are now; the class includes about 600 extant species in one subclass, Articulata.

Crinozoa, a subphylum of echinoderms including one class, → Crinoidea.

criss-cross inheritance, a phenomenon associated with the inheritance of sex-linked, specifically X-linked genes, in which the male offspring has certain characteristics (i.e. regulated by the genes on the X chromosome) resembling their mother and the female offspring their father.

crista, pl. **cristae** (L. = crest, tuft, comb), a crest, ridge, or projection in an anatomical structure, especially on a bone; e.g. **1. crista tuberculi majoris** and **minoris**, the crests of the humerus; **2. crista sterni**, the crest of the bird sternum; **3. crista ampullaris** (ampullar crest), the prominent thickening of the ampullary membrane in the semicircular duct of the inner ear; **4. mitochondrial cristae**, the transverse inner membrane infoldings in the → mitochondrion.

critical temperature, 1. the upper limit (**upper critical temperature**), or the lower limit (**lower critical temperature**) of the → thermoneutral zone of homoiothermic animals (birds and mammals); in practice means the ambient temperature in which physiological thermoregulatory mechanisms begin to work in the animal to increase heat loss in warm environment, or to increase heat production in cold. **2.** the upper limit (**critical temperature maximum**), or the lower limit (**critical temperature minimum**) of the ambient temperature (equal to body temperature) of poikilothermic animals, i.e. the temperature that endangers the survival of an animal; e.g. a criterion for the upper temperature limit of the **water hog-louse** (*Asellus aquaticus*) is considered to be that temperature at which the animal loses its ability to right itself.

Crocodylia (L. *crocodilus* = crocodile), **crocodilians**; an order of archosaurian reptiles, species of which are scale-covered predators with a long tail and snout. They live on shores of warm inland waters and river mouths preying on other vertebrates. The order includes 22 extant species, belonging to **crocodiles** (*Crocodylus*), **alligators** (*Alligator*), and **gavials** (*Gavialis*).

Cro-Magnon man (Cro-Magnon, a place in Dordogne, southwestern France), the earliest form of the modern man (*Homo sapiens*) who presumably arrived in Europe about 35,000 years ago. Individuals were more slender and taller than the Neanderthal man who previously inhabited Europe. Cro-Magnon man used stone and bony tools, and their cave art, especially in the cave of Lascaux, France, gave expression to their highly developed mental ability.

crop, 1. a thin-walled and expanded anterior portion (oesophagus) of the alimentary canal in birds and some invertebrates such as annelids and insects, mainly serving as food storage; **2.** harvest, yield; the cultivated product (growing or gathered) of the ground, i.e. the yield produced by cultivated organisms in a certain area per time and areal units; the term can be applied to a certain species or group, as barley crop, crop of grain.

crop-milk, milk-like liquid in the crop of doves and some parrots produced by the breakdown of epithelial cells of the lining tissue. Both males and females are able to produce this "bird milk" in which contents of protein, fat, and vitamin A and B are very high; its secretion is stimulated by → prolactin.

crop rotation, in agriculture, the alternation of cultivated plants in successive years; prevents the weakening of the soil, erosion, and the increase of plant diseases. Especially important is the occasional cultivation of leguminous plants because of their ability to fix nitrogen.

cross, in genetics, **1.** a process or product of cross-breeding (cross-fertilization); **2.** an experimental method bringing together genetic material from two different individuals who usually are genotypically different, and create by this means a → genetic recombination; **3.** the corresponding method used in animal and plant breeding; **4.** as a verb, to produce a hybrid organism; to interbreed, to cross-breed.

cross-breeding, the union of → gametes from two genetically different individuals in → fertilization. The advantage of cross-breeding is that it increases genetic variability in populations, and thus creates raw material for → selection. *Syn.* outbreeding, cross-fertilization, exogamy. *Verb* to **cross-breed**. *Cf.* inbreeding, self-fertilization.

cross-fertilization, → cross-breeding.

crossing, in genetics, **1.** the act of → cross-breeding (interbreeding); **2.** the union of genetic material from two different individuals.

crossing-over, crossing over, in genetics, the exchange of genetic factors, i.e. the phenomenon that leads to → genetic recombination of linked genes both in prokaryotic and eukaryotic organisms. Formally, crossing-over is the reciprocal exchange of parts at homologous sites, and is based on symmetrical breakage and crosswise reunion of → homologous chromosomes. In eukaryotes both meiotic and mitotic crossing-overs are found (*see* meiosis, mitosis).

Meiotic crossing-over is the occurrence of reciprocal exchange of parts of homologous chromosomes in meiosis, the microscopic counterpart being a → chiasma. The probability of forming any crossing-over between two linked genes is dependent on the distance between these genes on the chromosome. This is the basis of genetic mapping (*see* genetic map). Genetically the phenomenon of crossing-over is seen as the relative proportion of → recombinant offspring among all offspring of a heterozygous individual. On the basis of the number of recombinant offspring it is possible to calculate the number of the recombinant gametes, from which these offspring have arisen.

Mitotic crossing-over (somatic crossing-over) occurs either in → somatic cells or in mitotically dividing progenitor cells of the gametes, i.e. in a → germ line. Mitotic crossing-over, like meiotic crossing-over, occurs during the four-strand stage of the chromosomes, i.e. in the G_2 phase of mitosis. Mitotic crossing-over is an exceptional event, usually caused experimentally e.g. with the aid of X-rays. In spite of this, it is a regular phenomenon in the → parasexual cycle of certain fungi. The molecular mechanism of mitotic crossing-over is unknown.

crossing-over value, COV, the frequency of → crossing-over occurring between two linked genes; expressed as percentage of crossover individuals among progenies; also called cross-over value.

Crossopterygii (Gr. *krossos* = tassel, *pterygion* = fin), in some classifications a subclass of lobe-finned fishes, comprising the order → Coelacanthini.

crossover, the result (→ chiasma) of → crossing-over between homologous chromosomes.

cross-pollination, the → pollination of a flower by pollen from another plant; leads to cross-fertilization.

crown gall, a plant tumour; an undifferentiated plant tissue caused by the bacterium *Acrobacterium tumefaciens*. This kind of callus tissue is formed at a wound of a root, root collar, and stem, where infecting bacteria can penetrate the tissues.

crozier formation, hook formation; a process taking place in the tips of ascogenous hyphae in → Ascomycetes. In the process the tip of the terminal cell of a dikaryotic hypha bends backwards and forms a stalked hook, the nuclei divide, cytokinesis occurs and thus the hook separates the nuclei of different sex before the final formation of an → ascus. The process resembles the clamp formation in → Basidiomycetes (*see* clamp mycelium).

crus, pl. **crura** (L. = leg); **1.** in human anatomy, the segment of the hindleg between the knee and the ankle, i.e. the leg; **2.** a leg-like part, as **crura cerebri** (ventral pedunculi), stem-like structures in the cerebrum, each constructed of a massive bundle of corticofugal nerve fibres from the cerebral cortex to the brain stem and the spinal cord. *Adj.* **crural**.

Crustacea (L. *crusta* = shell), **crustaceans**; a subphylum in the phylum Arthropoda; the species have a segmented body in three tagmata, including the **head** (*cephalus*), **thorax**, and **abdomen**; often the first two structures are fused to form the **cephalothorax**. The body is covered with a cuticle composed of chitin, protein, and calcium. Nearly all segments have a pair of biramous appendages which are specialized for different functions, as for feeding, breathing, walking, sensing, etc. Most crustaceans are marine, some live in fresh waters or in terrestrial habitats. There are about 40,000 species of crustaceans in nine classes, the most familiar of which are → **branchiopods** (*Branchiopoda*), **ostracods** (*Ostracoda*), **copepods** (*Copepoda*), **fish lice** (*Branchiura*), **cirripeds** (*Cirripedia*), and **malacostracans** (*Malacostraca*). Some taxonomists consider crustaceans as a phylum. *See* Mandibulata, Uniramia.

cry-, cryo-, also **kryo-** (Gr. *kryos* = icy cold), pertaining to cold or freezing.

cryobiology, a branch of biology dealing with effects of low temperatures on organisms, especially freezing of cells and tissues.

cryology, the study dealing with snow and ice.

cryomicrotome (Gr. *tome* = section), an instrument used for cutting of microscopic sections (1-10 µm in thickness) from biological material for examination in the light microscope. The object is frozen in the cryomicrotome by liquid nitrogen and then cut without any chemical fixation and preparation; thus the structures of the sample are not disturbed by chemicals. **Cryoultramicrotome** is also used for cutting thinner sections (< 1 µm) to be studied as frozen in a cryoelectronmicroscope. Cryoultramicrotomy is especially used for preparing material → for X-ray microanalysis. *Cf.* microtome.

cryophytes (Gr. *phyton* = plant), organisms growing on bare snow or ice; may be algae, mosses, fungi, or bacteria. The so-called red snow is caused by the very abundant occurrence of the green alga *Chlamydomonas nivalis*.

cryoprotectants, antifreeze substances; substances which prevent freezing and freezing damage in organisms. Cells and tissues always contain many substances in their fluid compartments, e.g. glucose and other sugars, amino acids, and salts, which maintain a certain osmotic concentration in organisms, thus lowering the freezing point. In addition, overwintering plants and poikilothermic animals can accumulate some organic substances, called cryoprotectants, in their cells and tissue fluids before the cold season. Those substances may be sugars, sugar alcohols, glycerol, (poly)ethylene glycol, amino acids, and proteins, which protect cells by increasing osmotic concentration but also by preventing the formation of intracellular ice crystals. Also synthetic compounds which prevent the ice formation in cells are called cryoprotectants.

cryoscopy (Gr. *skopein* = examine), the examination of the freezing point of solutions in reference to the principle that the freezing point depends on the concentration (osmotic potential) of a substance dissolved in a solution. This method is used to determine the relative molecular mass of the substance dissolved.

cryostat (Gr. *statos* = permanent, constant), a freezing chamber, an automatic apparatus maintaining a very low constant temperature.

cryoturbation (L. *turbare* = to disturb), a process occurring in soils in spring when thawn ground refreezes; causes injurious iceburn in plants.

cryoultramicrotome, *see* cryomicrotome.

crypsis, cryptic coloration (Gr. *kryptos* = hidden), a protective coloration by breaking up the body outline so that an animal becomes difficult to be seen by predators. The cryptic coloration may be spotted or striped, or the pattern may be similar to the background of the familiar habitat of the animal, e.g. resembling a sandy bottom or green foliage.

crypt(o)-, also **krypto-,** pertaining to hidden, invisible, covered, secret, e.g. to a hidden or concealed anatomical structure, or to a → crypt.

crypt (L. *crypta*), a blind tube or pit opening onto the free surface of an organ, e.g. **crypts of Lieberkühn**, i.e. → intestinal glands, or **anal crypts**, which are the pits in the mucous membrane of the anal rectum.

cryptobiosis (Gr. *bios* = life), the state in which external symptoms of life are absent, as in a dormant organism; active life returns when the conditions change to normal. Also called anabiosis.

cryptogam (Gr. *gamos* = marriage), a plant producing no flowers or seeds, reproducing mainly by means of → spores; e.g. plants of the groups Bryophyta and Pteridophyta.

cryptogene, a gene of the → mitochondrial DNA of trypanosomes in which → RNA editing occurs in such a way that → nucleotides are removed and added in the coding regions of the primary transcription product. Thus in the function of cryptogenes, the gene does not alone determine the primary structure of → proteins to be synthesized, but information for it is also gained outside the gene. *See* guide RNA.

Cryptophyta (Gr. *phyton* = plant), **cryptophytes**, an algal division in Plant kingdom; the species are unicellular, free-living, and characterized by a special pharynx in the upper part of the cell and two flagella of different length. About 100 species are known.

cryptophyte, 1. a plant whose wintering parts (buds, rhizomes, bulbs, corms) are below ground (geophyte) or in water (helophyte); *see* Raunkiaer's life forms; **2.** a member of the group Cryptophyta.

cryptorchid, pertaining to a mammal whose testes have not descended from the abdominal cavity into the scrotum.

Cryptozoic (Gr. *zoe* = life), a geological eon in the early Precambrian era between the Archaeozoic and Proterozoic eons, over 2,400 million years ago; only very primitive organisms existed.

cryptozoite (Gr. *zoon* = animal), a pre-stage of merozoites in the life cycle of haemosporidians (malarial organisms), as of *Plasmodium*; develops in the liver of the vertebrate host from the sporozoite inoculated by an infected mosquito. *See* Haemosporidia.

crystal (Gr. *krystallos* = ice, crystal), **1.** a clear transparent mineral, or special glass, as the transparent form of crystalline quartz; **2.** in mineralogy, a solid body having a characteristic, regular internal structure with arranged plane surfaces, such as found in many chemicals and minerals; **3.** a systematically arranged solid anatomical structure or other precipitated material containing minerals, as the ear crystal, i.e. → statolith in the inner ear, or sperm crystals, crystals of precipitated spermine phosphate found in the sperm of mammals; **4.** non-living particles in plant cells, formed of calcium oxalate, either of monohydrate ($Ca(C_2O_4) \cdot H_2O$) or dihydrate ($CaC_2O_4) \cdot 2\ H_2O$); the monohydrate forms monolithic crystals, which may be solitary or may be deposited as raphides or sand-like mass. Dihydrate calcium oxalate forms mostly single tetragonal crystals. *Cf.* cystolith. *Adj.* **crystalline**, resembling a crystal, composed of crystals. *Verb* to **crystallize**.

crystalloid, 1. crystalline, resembling a crystal; **2.** a crystal-like structure in cells or tissues, e.g. crystalline formations found in the Sertoli cells and the Leydig cells of testes, or water-insoluble protein crystalloids in plant seeds; **3.** a crystallizable substance which in solution readily diffuses through biomembranes, as distinguished from a colloid that cannot do that.

CSF, 1. → cerebrospinal fluid; **2.** → colony-stimulating factor.

ctenidium, pl. **ctenidia** (Gr. *kteis* = comb, *-idion* = dim.), a feather-like or comb-shaped structure, such as the gills of gastropods and bivalves, or comb-shaped plates on the body of ctenophores.

ctenoid scale (Gr. *eidos* = form), a scale type of fish having a comb-like ridge on the exposed edge which may be an adaptation for reducing frictional drag; ctenoid scales occur e.g. in the perch. *See* scale, cycloid scale.

Ctenophora (Gr. *phorein* = to carry along), **ctenophores, sea walnuts, sea combs**, or **comb jellies**; an invertebrate phylum of biradially symmetrical animals including less than 100 species; free-living marine animals occurring in all seas, but especially in warm waters. On the surface of a fragile transparent body are eight rows of comb-like plates (ctenidia) for locomotion; many of the species emit light (*see* bioluminescence). The phylum is divided into two classes: Tentaculata whose species have several tentacles, and Nuda, species without tentacles; the former includes e.g. the familiar species **Venus' girdle** (*Cestum veneris*). *See* Coelenterata.

Cubozoa (Gr. *kybos* = cube, *zoon* = animal), **cubomedusae**; a class of cnidarians comprising animals whose medusoid stage is predominant; the polypoid stage is either inconspicuous, or unknown. Cubomedusae are predators, commonly feeding on fish. Stings of some species may be dangerous, even fatal, to humans. The class was previously regarded as an order in the class Scyphozoa.

cuckoos, *see* Cuculiformes.

cuckoo-spit, a foam secreted by the larva of a spittle-bug (Cercopidae); inside the foam the animal is protected against predators and desiccation.

Cuculiformes (L. *cuculus* = cuckoo, *forma* = form), an order of long-tailed and sharp-winged birds of different size; all the 160 species have two toes in front and two behind. Especially in the Old World, many species are nest-parasites. The order is divided into three families: **cuckoos** (Cuculidae), **turacos** (Musophagidae), and **hoatzin** (Opisthocomidae); the last one is often considered as an order.

cultivar, cv, an intraspecific plant variety; the term is used for cultivated plants only. Cultivars are formed in → breeding and they must retain their special characteristics during cultivation; e.g. *Solanum tuberosum "Bintje"* is a potato cultivar.

cultural ecology, *see* ecology.

cultural gene, → meme.

cultural evolution, the changing of behaviour of individuals as the result of learning; the culture changes and develops on the basis of the selections made by individuals, and the changes may apply to any behavioural pattern, such as social habits, working systems, feeding methods etc. In man, cultural evolution is based, partly at least, on cognitional actions and selections. → **Meme** is a unit of the cultural evolution (*cf.* gene).

culture (L. *cultura*), **1.** the cultivation of microorganisms or other cells or tissues in special media; *see* cell tissue and organ culture; **2.** manipulation of an environment by human activity; **3.** the development of the human mind through education; **4.** as a *verb*, to **cultivate**.

cumarin, → coumarin.

cuneus, pl. **cunei** (L. = wedge), a wedge-like structure; e.g. a wedge-shaped area on the elytron of a bug. *Adj.* **cuneate, cuneiform**.

cupric (L. *cuprum* = copper), containing copper in its bivalent form (Cu^{2+}), in its +2 oxidation state, as in $CuCl_2$; **cuprous**, containing copper in its monovalent form (Cu^{+}), in its +1 oxidation state, as in CuCl.

cupula, pl. **cupulae** (L. dim of *cupa* = tub), a dome-like or cup-shaped structure, e.g. in the inner ear, *cupula cochleae*, the apex of the cochlea, or *cupula cristae ampullaris*, the cap-shaped gelatinous mass that overlies the sensory cells in the ampullary crest of the semicircular ducts.

cupule, 1. a sucker cup in some invertebrates; **2.** a type of → involucre formed from bracts around the fruit or seed, found in some trees.

curare (Sp.), originally an extract from species of *Strychnos* and some other South-American plants, used as an arrow poison; later a pure alkaloid of *Chondodendron tomentosum*, called **tubocurarine**, which has been used in medicine as a muscle relaxant. Curare inhibits → nicotinic receptors in cholinergic transmission in autonomic ganglia and skeletal muscles, i.e. in → neuromuscular junctions (end-plates).

curie, Ci (*Marie*, 1867—1934, and *Pierre Curie*, 1859—1906), a former unit of radioactivity, now replaced by becquerel. *See* Appendix 1.

Cushing's disease/syndrome (*H. Cushing*, 1869—1939), a syndrome found originally in man caused by the excessive secretion of → glucocorticoids from the adrenal cortex, or excessive intake of these hormones or related synthetic substances. *Cf.* Addison's disease.

cushion plant, a grass or dwarf shrub the shoot of which forms a hemispherical, firm cushion-like structure near the ground; this kind of structure protects the inner parts of the plant from cold and drying, and is useful in insufficient conditions in high Alpine and arctic areas.

cusp (L. *cuspis* = point), one of the conical elevations (cones) of a mammalian molar tooth; the cuspid patterns are specific to a species and are taxonomically important.

cutaneous (L. *cutis* = skin), pertaining to the skin. *See* cutis.

cutaneous bone, → dermal bone.

cutaneous glands, the glands derived from the skin epithelium (originally from the ectoderm) and extended inside the corium (dermis), e.g. the sebaceous, mammary, and sweat glands, as well as several scent and poison glands.

cuticle (L. *cuticula*, dim. of *cutis* = skin), **1.** a protective, organic layer covering the surface of many invertebrates, forming the **exoskeleton** of an animal; in arthropods contains → chitin forming a hard → **chitin cuticle**, that may be hardened by accumulation of calcium salts, as in molluscs; cuticle is noncellular and secreted by the cells of epidermis; **2.** sometimes denotes just the surface of an organ, as the cuticle of a developing tooth, which will develop into enamel (**enamel cuticle**); **3.** a layer of waxy material, → cutin, on the outer surface of the → epidermis of most plants; protects the plants from drying. *Adj.* **cuticular**.

cuticularization, the formation of a → cuticle.

cutin, the mixture of esters of fatty and hydroxy fatty acids, resembling → suberin; a component of the plant cuticle. **Cutinization** is the process in which cell walls are impregnated with cutin. The cutinization of old roots decreases their water absorption capacity. **Cutinase**, the enzyme catalysing the hydrolysis of cutin, is produced in the pollen tube during its growth.

cutis, pl. **cutes** or **cutises** (L.), the vertebrate → skin; includes the dermis (corium) and epidermis.

cutting, a part of a plant, as a twig, leaf, or bud, used for → vegetative propagation of plants.

cuvette (Fr., dim. of *cuve* = tube), a glass or plastic tube or right-angled sample vessel used especially in optical analyses, e.g. in spectrophotometry.

Cuvierian duct (*Georges Cuvier*, 1769—1832), the paired, largest vein returning blood from the cardinal veins to the fish heart; in embryos of tetrapod vertebrates it develops further to form the superior vena cava.

Cuvierian tubules, Cuvier's organ, an organ of some holothurians (Echinodermata), in the genera *Actinopyga* and *Holothuria*; formed of a mass of white, pink, or red sticky tubules which are attached to the base of one or both → respiratory trees. When a sea cucumber is attacked by a predator, the tubules (containing a toxin) and the cloaca are shot out of the anus for defence.

C value (C = constant), the DNA content of the → haploid genome.

cyanelle (*cyano*bacterium + organ*elle*), an endosymbiotic cyanobacterium, often living in association with Protozoa.

cyanide, a salt of hydrocyanic acid, HCN. Cyanides such as potassium cyanide, KCN, are very poisonous inhibiting the electron transfer reactions of cell respiration. In many plants, particularly the hard endocarps (fruit stones), contain cyanogenic alkaloids which can produce cyanide e.g. in the alimentary canal of animals.

cyanide-resistant respiration, a type of cell respiration in plant mitochondria which is not inhibited (as is usual) by → cyanide. In those mitochondria there are two electron transfer chains: the usual, cyanide-sensitive electron transport system that transfers electrons through cytochromes and cytochrome oxidase to oxygen, and the cyanide-resistant chain in which the electrons are transferred from flavoprotein to a terminal oxidase and finally to oxygen. This kind of dual respiration occurs in some seeds, fruits, and flowers (e.g. in *Arum*) in which it is linked to heat production. The cyanide-resistant respiration probably protects against cyanide which is formed in many plants containing → thioglycoside.

Cyanobacteria, earlier called **blue-green algae**; a prokaryotic group (division) often classified together with Bacteria in the kingdom Monera; unicellular, filamentous or colonial photosynthetic microorganisms, which are widely distributed and living in all kinds of waters and also in terrestrial environments. The cellular structure is more developed than that of other bacteria and e.g. photosynthesis takes place in special membranes, but other cell organelles are lacking. Many species are symbiotic acting as algal parts in lichens; some species can fix atmospheric nitrogen. They also occur as endosymbionts and are thought to be the ancestors of plant → chloroplasts. Cyanobacteria are evidently the first photosynthesizing and thus oxygen-producing organisms, thus being responsible for early oxygen formation in the atmosphere. The photosynthesizing pigments in Cyanobacteria are chlorophyll a and phycocyanin (*see* chlorophyll, phycobiliproteins). Some strains of the genera *Anabaena* and *Microcystis* may produce toxins and the mass occurrence of their colonies may cause → bloom, especially in polluted waters during late summer; it may cause harmful changes in the taste and smell of the water.

cyanocobalamin, vitamin B_{12}. *See* vitamin B complex.

cyanogenic glycosides, alkyl cyanides; organic compounds containing the cyanide group -CN, the general formula $R\text{-}CH_2CN$; produced in the metabolism of many plants, e.g. nitrile and amygdalin (laetrile) found in bitter almonds and seeds of cherries, apples and plums.

cyanophycin granules, polypeptide granules occurring in cyanobacteria; usually rich in aspartic acid and arginine. Their number varies during the growth cycle of the bacteria.

cyanosis, blueness of the skin or mucous membranes due to imperfect oxygenation of the blood. *Adj.* **cyanotic**.

cyathium, pl. **cyathia** (Gr. *kyathos* = cup), the unique, reduced inflorescence of the spurges (*Euphorbia*). Each cyathium has a single, terminal, long-stalked pistillate flower without a perianth, surrounded by male flowers, each of which consists of only one stamen. The whole inflorescence is surrounded by bracts which form a cup-shaped structure. The cyathium is an inflorescence but resembles a single flower; it has also been called a false flower.

cybrid, cytoplasmic hybrid; a cell hybrid produced by the fusion of an enucleated cell (i.e. the cytoplasm of a cell) and a cell having the nucleus.

Cycadinae (L. *Cycas* = the type genus), **cycads**; a plant class in subdivision → Gymnospermae, division Spermatophyta; in some classifications the division Cycadophyta. Cycads resemble ferns but are flowering, evolutionarily very old "living fossils". The micro- and macrosporangia lie in cone-like structures; male and female sporophytes are separate plants. A special feature in the group is the formation of freely moving sperm cells. Cycads live in tropical and subtropical areas of the southern hemisphere.

Cycadophyta, *see* Cycadinae.

Cycadeoidophyta, cycadeoids; an extinct division in Plant kingdom; in some classifications the class Bennettitinae in the subdivision Gymnospermae, division Spermatophyta; tree-like plants which occurred from Trias to Cretaceous.

cycle (Gr. *kyklos* = circle, ring), a round or recurring series of events or phenomena, usually repeating in the same sequence and at the same intervals; e.g. cardiac, reproductive, oestrus, menstrual, and hormonal cycles in organisms, or many enzymatic cycles in cells. *Adj.* **cyclic**. *See* cell cycle, Calvin cycle, citric acid cycle, urea cycle.

cycle sequencing, an experimental procedure resembling the → polymerase chain reaction (PCR) using repeated cycles of thermal denaturation of DNA, → primer annealing, and extension with termination of the reaction to increase the quantity of the sequencing product. This makes it possible to sequence rather long stretches (over 0.5 kb) of a known DNA using minimal amount of a template without cloning. Cycle sequencing is based on a combination of the chain termination method of DNA sequencing and PCR.

cyclic, characterized of a cycle, occurring periodically; in chemistry, **cyclic compounds**, organic compounds containing one or more closed rings, such as the → benzene ring; homocyclic rings are composed only of carbon atoms, heterocyclic rings also of other atoms, in natural compounds usually of nitrogen and oxygen atoms. *Cf.* aliphatic.

cyclic AMP, cAMP, cyclic adenosine monophosphate, adenosine 3',5'-cyclic monophosphate (*see* adenosine phosphates); a compound acting in animal cells as an activator of phosphorylase kinase, that further activates many cellular enzyme reactions. cAMP is formed from ATP when **adenylate cyclase** (EC 4.6.1.1) is activated by certain hormones and is broken down by **phosphodiesterase** to form inactive 5'-AMP. cAMP is an important → **second messenger** produced in response to most water-soluble hormones in animal cells. It acts as an intracellular transmitter in certain reaction chains initiated by a hormone-receptor interaction (seven-spanning receptors) from the cell membrane. In some reactions it is replaced by cyclic GMP, **cGMP**, formed from guanosine triphosphate (GTP) by the enzyme guanylate cyclase (EC 4.6.1.2). cAMP is a metabolic regulator also in bacteria, some slime moulds, but probably not in plants.

cyclic nucleotides, → nucleotides in which the phosphate group forms a ring structure, e.g. → cyclic AMP (cAMP) and cyclic GMP (cGMP) which act as → second messengers in many cells.

cyclins, proteins regulating the → cell cycles and the onset of → mitosis in eukaryotic organisms; the proteins accumulate continuously throughout the cell cycle and are then destroyed by proteolysis during mitosis. *See* mitosis-promoting factor.

cyclocytic stoma, a type of plant stomatal systems. *See* stoma.

cycloheximide, an antibiotic obtained from certain strains of *Streptomyces griseus*; inhibits translation in protein synthesis of eukaryotic organisms by inactivating the peptidyl transferase enzyme; used e.g. as a fungicide.

cycloid scale, a thin and flexible scale type of advanced bony fishes; the scales are arranged in overlapping rows and exposed edges are evenly round and smooth; e.g. the scales of salmon. *Cf.* ctenoid scale.

cyclomorphosis, pl. **cyclomorphoses** (Gr. *morphe* = form), a phenomenon in which structural features of organisms change cyclically between generations; usually the changes are seasonal and always produced by environmental factors; e.g. variation in body structures of some oceanic plankton organisms is caused by changes in salinity during the reproduction and development of a new generation. Variations in many freshwater microorganisms are caused by changes in temperature or by water turbulence.

cyclopia, a congenital malformation (of man) the individual having only one eye in the middle of the forehead.

cyclosis, cytoplasmic streaming; the circulation of protoplasm in many eukaryotic cells, especially in large plant cells.

cyclosporin(e)s, cyclic non-polar decapeptides isolated from *Tolypocladium inflatum* and some other fungi; have immunosuppressive activity in vertebrates and are used clinically to prevent rejection in organ transplant recipients.

Cyclostomata, 1. a suborder in the phylum → Ectoprocta; **2.** → cyclostomes.

cyclostomes, jawless vertebrates: lampreys and hagfishes which in some classifications are included in the order Cyclostomata (class in some classifications). *See* Agnatha.

cydippid (*Kydippe*, a girl in Greek mythology), a spherical larval stage of ctenophores; superficially resembles an adult.

cyme, an → inflorescence in which the main axis forms a flower while its growth ceases, the lateral branch(es) continuing their growth and branching further. The cyme with two side branches in each node is called **dichasium**, common e.g. in the family Caryophyllaceae. If there is only one side branch, the structure is called **monochasium**, the most common type of which is **bostryx**, common in the family Boraginaceae. In this type the side branch is always on the same side. When the single side branches grow alternately on different sides, the structure is called **rhipidium** (e.g. *Gladiolus*). If there are numerous side branches at each node, the structure develops into a complicated **pleiochasium**, also common in the family Caryophyllaceae. *Adj.* **cymose**.

cyphonautes, pl. **cyphonautae** (Gr. *kyphos* = bent, *nautes* = sailor), a larval stage of bryozoans (Ectoprocta) having a triangular, compressed shell which usually is covered by a chitinous valve.

Cypriniformes (L. *Cyprinus* = the type genus, *forma* = form), **carp-like fishes**; an order of bony fishes comprising small or middle-sized freshwater species distributed to North America, Eurasia, and Africa. Carp-like fishes are important aquarium fishes, and produced in many fish farms for food and for planting in ponds and lakes. About 2,000 species are known in 24 families; most familiar are e.g. **carps, roaches, minnows**, and **electric eels.**

cypris, pl. **cyprides** (L. *Cypris* = Venus), a larval stage of cirripeds (barnacles) following the nauplius stage; the structure of the cypris is already adult-like in many respects. It fastens on a substratum using secretion of the cement glands in the first antennae and develops into a sessile adult.

cypsela, a dry fruit resembling an → achene but formed from an inferior ovary, thus being covered by some layers originating in the receptacle of the flower; typical of the plant families Asteraceae and Cichoriaceae with a → capitulum as their inflorescence; e.g. the sunflower.

cyst(i)-, cysto-, cystid(o)- (Gr. *kystis* = pouch, sac, bladder, cyst), pertaining to a sac-like structure, or a cyst.

cyst, 1. any normal or abnormal cavity in the anatomy of animals; e.g. **arachnoid cysts**, the cysts between the layers of brain meninges, or **epithelial cysts** in the skin; **2.** a protective coat or sac surrounding a resting cell or animal; e.g. a cyst surrounding the gametocytes of many sporozoans; a cyst of a parasitic amoeba, *Entamoeba histolytica* causing amoebic dysentery; an inactive developmental stage of a phytoflagellate, *Euglena*; a tough coat enclosing the metacercaria of some flukes (*Fasciola*); a cyst of some cestodes developing into a bladderworm, → cysticercus; **3.** resting spore, i.e. a thick-walled → spore of bacteria, algae, and fungi; resists poor conditions.

cysteine, abbr. Cys, cys, symbol C; an amino acid, $HSCH_2CH(NH_2)COOH$, which contains a reactive sulphydryl group (-SH); occurs as a constituent in proteins and acts as an antioxidant. *See* cystine.

cystic, cystous, 1. relating to a cyst; **2.** pertaining to the urinary bladder.

cysticercus, pl. **cysticerci** (Gr. *kystis* = bladder, *kerkos* = tail), **bladderworm**; an encysted larval stage of some cestodes, e.g. *Taenia* species, following the onchosphere; characteristic of the cysticercus is a conspicuous bladder in which the **proscolex** develops into the head segment, **scolex**. The larva usually lives in muscles of a mammalian intermediate host (e.g. in many rodents), and at this stage the larva has to pass into a definitive (final) host to develop into the adult worm. In man, the infection with cysticerci of the *Taenia solium*, i.e. **cysticercosis**, may cause severe symptoms.

Originally *Cysticercus* were described as the genus of bladderworms, and the name is still often used e.g. in veterinary books when referring to the larval encysted form of cestodes, as *Cysticercus bovis*, the cysticercus larva of *Taenia saginata* living in cattle, or *Cysticercus fasciolaris*, the larva of *Taenia taeniaeformis*, found e.g. in the liver of mice and rats. *See* hydatid cyst. *Cf.* procercoid, plerocercoid.

cystidium, pl. **cystidia,** a sterile sac-like structure in the → hymenium of many fungi in the group → Basidiomycetes; occurs together with sterile hyphae, paraphyses.

cystine, dicysteine; abbr. Cys-Cys, cys-cys; an amino acid formed from two cysteine residues; occurs as a constituent in proteins in which two -SH groups are oxidized to form a disulphide bond (-S-S-) between polypeptide chains, strengthening the protein structure; e.g. scleroproteins, such as → keratin, are rich in cystine.

cystocarp (Gr. *karpos* = fruit), a "fruit-like" structure in red algae (→ Rhodophyta) composed of gametophytic filaments; encloses the → gonimoblasts (sporogenous filaments) which are formed from the zygote after fertilization and produce asexual spores.

cystolith (Gr. *lithos* = stone), **1.** a calcium carbonate crystal inside a plant cell, **lithocyst**, which is a specialized cell, larger than its neighbour cells. Lithocysts are present mostly in → epidermis and hairs, and also in parenchyma tissues; **2.** a urinary stone (calculus) in the bladder.

cystous, → cystic.

cyt(o)- (Gr. *kytos* = cavity, cell), denoting relationship to a cell.

cytidine, cytosine ribonucleoside, symbol **Cyd** or **C**; a → nucleoside composed of cytosine and ribose, occurs e.g. in RNA; in DNA ribose is replaced by deoxyribose and the corresponding nucleoside is **deoxycytidine**.

cytidine phosphates, → nucleotides synthesized from cytosine, ribose (or deoxyribose), and one, two, or three phosphate groups, five together, depending of the site of the phosphate to the ribosyl OH groups; the corresponding nucleotides are cytidine monophosphate (CMP, cytidylic acid), diphosphate

(CDP), and triphosphate (CTP), respectively with deoxyribose dCMP, dCDP, and dCTP. Cytidine phosphates are energy-rich compounds which are involved e.g. in nucleic acid metabolism and in the synthesis of phosphatides, e.g. the activated choline is CDP-choline.

cytochalasins (Gr. *chalasis* = relaxation), a group of fungal poisons which break down contractile cellular microfilaments by inhibiting actin polymerization in eukaryotic cells, resulting in changes in the cell shape. Cytochalasins disturb the division of cytoplasm, e.g. causing the formation of multinucleate cells and the extrusion of the nucleus; they also inhibit cell movement and phagocytosis. Especially **cytochalasin B** and **D** are used in cell biological studies.

cytochromes (Gr. *chroma* = colour), a class of haemoprotein compounds formed from iron porphyrin and a protein; act as oxidoreductase enzymes in the → electron transfer chain of cellular energy metabolism (photosynthesis and oxidative cell respiration) and in many other processes. When linked with FAD they form a flavoprotein-cytochrome system. Different cytochromes are named with indexes, e.g. **cytochrome a**, **cytochrome b**, and **cytochrome c** (c_1 and c_2). **Cytochrome oxidase** (EC 1.9.3.1) is the terminal enzyme in the electron transport chain accepting electrons from cytochrome c and carrying them to molecular oxygen. **Cytochrome f** is involved in the electron transport in photosynthesis. **Cytochrome P_{450}** is a mitochondrial oxidase and is acting in hydroxylation of steroids in the adrenal cortex. Cytochrome P_{450} may also be found on smooth endoplasmic reticulum (ER) e.g. of liver cells, and is involved in the metabolism of xenobiotics.

cytogene (Gr. *gennan* = to produce), a determinant of an inherited character located in the cytoplasm. *Syn.* plasmagene.

cytogenetics, a branch of genetics, originally generated as the union of → genetics and → cytology. The task of cytogenetics is to explain the cellular basis of the phenomenon of inheritance. Hence, its basic subjects of research are the structure of → chromosomes and their behaviour during → mitosis and meiosis. At present, cytogenetics has a clear biochemical emphasis.

cytohistogenesis (Gr. *histos* = tissue, *gennan* = to produce), the development and differentiation of cells and cellular structures during the development of tissues; the corresponding research branch is called **cytohistology**.

cytoid (Gr. *eidos* = form), resembling a cell.

cytokines (Gr. *kinein* = to move), hormone-like non-antibody polypeptides and small proteins released from cells acting in the vertebrate immune system, e.g. from activated T lymphocytes on contact with a specific antigen, → macrophages, endothelial cells, and glial cells. The cytokines serve as intercellular messengers binding to surface receptors of certain cells, so causing the differentiation or proliferation of these cells. When cytokines are secreted from lymphocytes they are called → lymphokines, when from monocytes or other macrophages they are called monokines, and when from other cells, cytokines, including e.g. → interleukins, interferons, platelet-derived growth factor (PDGF) and platelet-activating factor (PAF), and tumour necrosis factors such as cachectin and lymphotoxin. Monokines are often included in lymphokines.

cytokinesis, an event in which the cytoplasm is divided between the two daughter cells in cell divisions, i.e. → mitosis or → meiosis.

cytokinins, natural or synthetic plant hormones which mainly are derived from adenine; also some urea derivatives have cytokinin activity; cytokinins are synthesized by root cells. The best known cytokinin is **kinetin** (6-furfurylaminopurine) which is not found in plant cells. Natural cytokinins are e.g. **zeatin** and **isopentenyl adenosine** (IPA); the latter can bind to tRNA and thus regulate the translation of proteins. Cytokinins promote cell division, the enlargement of cells, and the germination of seeds, induce shoot differentiation in cell and tissue cultures, control together with auxin apical dominance, and retard senescence in cells. Cytokinins are sometimes called phytokinins.

cytology (Gr. *logos* = word, discourse), the research branch in cell biology dealing with the structure, function, development, reproduction, and the life cycle of cells. Cytological methods include especially light and electron microscopy and biochemical procedures.

cytolysis (Gr. *lyein* = to dissolve), the destruction or dissolution of cells, as produced e.g. by the → complement system in immune cytolysis. *Adj.* **cytolytic**.

cytoma, a cell tumour or neoplasm in which no tissue structure has developed; e.g. sarcoma.

cytopharynx (Gr. *pharynx* = throat), *see* cytostome.

cytoplasm (Gr. *plassein* = to form, mould), the substance and the organelles inside the cell, enveloped by the → cell membrane; the cell → nucleus of eukaryotic organisms (prokaryotes have no nucleus) is excluded. In plant cells the cytoplasm is separated from the cell wall by the → cell membrane and from the nucleus by the → nuclear envelope. Numerous different → **cell organelles** and the → **cytoskeleton** are arranged in the liquid basic material of multiphasic colloidal plasm, called **cytoplasm matrix** (cytoplasmic ground substance, cytosol). The matrix consists of saline fluid containing macromolecules, such as soluble proteins with many enzymes, and a vast number of ions and smaller molecules, e.g. carbohydrates, amino acids, trace elements, and metabolites, i.e. precursors and products of enzyme processes. Many metabolic and biosynthetic cell functions occur in the cytoplasm. The nucleus and cytoplasm have mutual interaction so that neither can survive without the other. *Adj.* **cytoplasmic**.

cytoplasmic bridge, a structure formed between two bacteria during bacterial conjugation through which unidirectional transfer of genetic material can pass from a donor cell to a recipient cell.

cytoplasmic ground substance, cytoplasmic matrix, cytosol. *See* cytoplasm.

cytoplasmic inheritance, the non-Mendelian inheritance of such characteristics, factors of which are not located on the → chromosomes but mediated by the → plastids and → mitochondria in the cytoplasm. *Syn.* extrachromosomal inheritance.

cytoplasmic male sterility, CMS, the maternally inherited inability of higher plants to produce viable → pollen.

cytoplasmic matrix, *see* cytoplasm.

cytoplasmic streaming, → cyclosis.

cytoproct (Gr. *proktos* = anus), → cytopyge.

cytopyge (Gr. *pyge* = rump), the cell organelle, "anus"; is the anal orifice through which waste products are ejected from the unicellular body of some protozoans, e.g. *Paramecium*. *Syn.* cytoproct.

cytosine, symbol C or Cyt; a pyrimidine base present in → nucleosides, nucleotides, and nucleic acids.

cytoskeleton (Gr. *skeleton* = a dried body), the internal, filamentous framework of the → cytoplasm of a cell, reinforcing the cell structure and shape and generating the spatial organization in it and its extensions, like cell branches and microvilli; it is also involved in intracellular transport and movements of cell organelles, chromosomes, and the cell itself. The cytoskeleton consists of several proteins which form filaments and tubules, usually forming a net-like structure inside the cell. In eukaryotic cells the cytoskeleton comprises mostly four types of fibrous structures: 1) → **microfilaments**, actin filaments, 4 to 5 nm in thickness, 2) → **microtubules**, hollow tubes of about 24 nm in thickness formed from tubulin protein, 3) → **intermediate filaments** (diameter about 10 nm, not in plant cells), and 4) the → **terminal web** just beneath the plasma membrane. A large number of proteins, such as → dynein and kinesin (motor proteins) and connectins, are found to be associated with the cytoskeleton.

cytosol (sol < solution), cytoplasm matrix; cytoplasmic ground substance; the liquid part of the cytoplasm between cell organelles. *Adj.* **cytosolic**. *See* cytoplasm.

cytostatic (Gr. *stasis* = halt), **1.** pertaining to the suppression of multiplication and growth of cells; **2.** an agent that causes this suppression in cells.

cytostome (Gr. *stoma* = mouth), a cell organelle forming the mouth opening in some unicellular organisms, as e.g. in many ciliates and some other protozoans; opens into a nonciliated gullet-like canal, **cytopharynx**, which further passes into the endoplasm of the cell body.

cytotaxonomy (Gr. *taxis* = order, *nomos* = law), the study of natural relations of organisms by uniting → cytology and → taxonomy, i.e. the grouping of organisms according to their kindred relations by using the methods of cell research.

cytotoxic T cell/T lymphocyte (CTL), *see* killer cell.

cytotoxin (Gr. *toxikon* = arrow poison), a substance that has a specific poisonous or lethal (**cytotoxic**) effect on cells; especially denotes toxic drugs and antibodies which selectively kill dividing cells in a given organ, e.g. nephrotoxins, neurotoxins, etc. Also

→ killer cells destroy other cells cytotoxically. The degree to which a cytotoxin possesses its toxic action is called **cytotoxicity**. *Adj*. **cytotoxic**.

cytotype (L. *typos* = model, type), any variety (→ race) of a species whose chromosomal complement differs in number or structure from the common chromosomal complement of this species.

D

D, symbol for **1.** aspartic acid; **2.** deuterium (isotope of hydrogen); **3.** dioptre; **4.** dalton.

***D*,** symbol for **1.** dose of absorbed radiation; **2.** electric flux density; **3.** diffusion coefficient.

D., abbreviation for **1.** *dosis* (dose); **2.** *dexter* (right); **3.** density; **4.** duration.

D-, (small capital D), a chemical prefix that specifies the relative configuration of an optical isomer. *See* isomers.

d, symbol for **1.** day; **2.** deci- **3.** diameter; **4.** deoxyribose.

***d*-,** (L. *dextro-* = right), *see* isomers.

2,4-D, → 2,4-dichlorophenoxyacetic acid.

dactylopodite (Gr. *daktylos* = finger, toe, *pous* = foot), *see* biramous appendage.

dactylozooid (Gr. *zoon* = animal, *eidos* = form), a club-shaped polyp with long tentacles in some colonial hydroids including groups of nematocysts for catching prey and defending the colony (defensive polyp). *Cf.* gastrozooid, gonozooid.

DAG, → diacylglycerol.

dalton, D, Da (*John Dalton*, 1766—1844), the atomic mass unit, one-twelfth of a single atom of carbon 12, i.e. $1.660 \cdot 10^{-27}$ kg; the unit is used for expressing the molecular size of macromolecules, usually in kilodaltons, kD or kDa.

damping-off, a lethal plant disease caused by algal fungi, e.g. by *Pythium debaryanum*.

dark field microscope, *see* microscope.

dark reactions, the reactions of the → Calvin cycle in photosynthetic plants where CO_2 is bound to ribulose 1,5-bisphosphate and reduced to carbohydrates in the presence of NADPH and ATP; chemical reactions in which no light is needed. *See* photosynthesis.

dart sac, a small sac near the genital pore of some gastropods; includes an arrow-like or stiletto-shaped calcareous **dart** which acts as a stimulant in copulation.

Darwin's finches, Galapagos finches; an endemic subfamily, Geospizinae in the family Emberizidae, living on the Galápagos Islands in the middle of the Pacific Ocean; contains 14 species in six genera, one of them living on the Cocos Islands. It is supposed that Darwin's finches have developed by adaptive radiation from one founder species which was misplaced to the islands from the South American continent. At present, each species differs from the others in the size and shape of the beak, and in feeding habits.

Darwinism, the theory of evolution by → natural selection, presented by *Charles Darwin* (1809—1882) in 1859; the central idea is the survival of the fittest varieties of organisms in the struggle for existence (→ fitness). The basic principles of Darwinism are the **principle of variation**, the **principle of inheritance**, and the **principle of selection**, which are the necessary and sufficient conditions for → evolution. According to the principle of variation, all natural populations are variable. According to the principle of inheritance, this variation is at least partly inherited to the progenies, and according to the principle of selection, the best adapted varieties are most likely to survive in the struggle for existence, and as a consequence the mean adaptiveness of the → population increases. This sequence of events, Darwin proposed, leads gradually to the formation of new species. *See* evolution theory, neo-Darwinism.

dauermodification (Ger. *Dauer* = permanence), a change of a character usually caused by a strong environmental factor, e.g. the black colour of mill moths caused by high temperature. The change prevails during several generations in vegetative but also in sexual reproduction, even though its environmental cause no longer is present. However, dauermodification weakens gradually in the course of generations, and finally disappears. Using reciprocal crosses it has been shown that the durability of dauermodification is transmitted through the → cytoplasm, not the genes.

daughter chromosomes, the sister → chromatids of a chromosome separated from each other during cell division, i.e. in the anaphase stage of mitosis or of the second meiotic division. *See* mitosis, meiosis. *Cf.* sister chromosome.

daughter cells, cells resulting from the division of a single cell.

daughter nucleus, any of the two nuclei resulting from a nuclear division of a single cell. *See* mitosis, meiosis.

day-neutral plant, any plant which develops and flowers regardless of the day length.

d.d., degree days. *See* thermal sum.

DDT, 1,1,1-trichloro-2,2-bis(p-chlorophenyl)-

ethane, $C_{14}H_9Cl_5$; a widely used chlorinated insecticide; synthesized from trichloro-acetaldehyde, CCl_3CHO, and chlorobenzene C_6H_5Cl, and was first produced in 1874. DDT's properties as an insecticide were discovered in 1939. DDT has succesfully been used against many serious diseases such as plague, malaria, spotted fever, and yellow fever, spread by insects. DDT was a very effective pesticide but has gradually lost its efficiency because of the resistance formed in DDT-treated insects; in resistant animals the hydrochlorinase enzyme converts DDT to a less toxic form. DDT is a very serious ecological problem because it decomposes very slowly and accumulates in the food chain in organ systems, particularly in the adipose tissue. The increased DDT content of waters is a threat especially to fish-catching birds.

DDT is not a contact poison to vertebrates, but if swallowed, can cause severe damage in nerve cells. Lethal doses for man are estimated to be 0.5 to 10 g/kg body weight, but much smaller doses can cause symptoms of poisoning. The use of DDT is forbidden in most countries but is still allowed in many developing countries, chiefly because of → malaria.

de- (L. *de* = from, down, away), denoting away, from, cessation, or without; has often a negative or privative connotation.

DEAE, → diethyl-aminoethyl-.

deamination, the removal of an amino group (-NH_2) from organic compounds such as amino acids, amines, purines, and pyrimidines; the reaction is catalysed by **aminotransferases** (transaminases, EC sub-subclass 2.6.1.) transferring in cells amino groups from an amino acid usually to a 2-keto acid, or by **deaminases** (EC sub-subclass 3.5.4.) catalysing the simple hydrolysis of C-NH_2 linkages. In animals, the latter reactions take place mainly in the liver, the -NH_2 groups being eliminated in the → urea cycle. In plants, e.g. the oxidative deamination produces indole-3-pyruvic acid from tryptophan in the synthesis of auxin. The loss of amino groups from the purine and pyrimidine bases of nucleic acids results in the deamination of nucleic acids; e.g. adenine is converted to → hypoxanthine.

death, the definite and permanent cessation of life. Unicellular organisms are, however, practically immortal, i.e. in optimal circumstances they divide continuously into daughter organisms (cells). In multicellular organisms the death of many types of tissue cells is a normal, continuous process for renewal of organs and organ systems (*see* apoptosis), and the dead cells are substituted by new cells derived from certain stem cells in the tissue (in animals, e.g. the cells of the skin, bone marrow, or alimentary canal). There are also cell types like nerve cells, which do not usually divide in a mature animal, and the functions of dying cells are partly compensated by activities of other cells.

The division of cells of a multicellular animal is genetically controlled and always limited, and thus determines the life length. For example, many normal mammalian cells can produce maximally only about 50 cell generations, and the rate of the division cycle and consequently the life-time of cells seem to be related to the rate of metabolism, i.e. large animals usually live longer. Damage to and death of different cell and tissue types are differentially important for an organism. In animals, most critical is the death of nerve cells in the central nervous system, especially in the brain areas which control circulatory and respiratory functions.

Plants can be grouped according to their genetically determined life length. **Annuals** flower and produce seeds during one growing season after which they die. **Biennials** make a leaf rosette in the first year, then they overwinter and flower in the second year. The life of → **pluriens** varies from a few years to hundreds of years. Certain parts of the pluriens die every year which is controlled by the day length, such as the death of the leaves of deciduous trees.

decapitation (L. *de* = away, down, *caput* = head), beheading; the removal of the head, e.g. of an animal, foetus, bone, etc. *Verb* to **decapitate**.

Decapoda (Gr. *deka* = ten, *pous*, gen. *podos* = foot), **decapods** (crayfishes, lobsters, crabs, and shrimps); **1.** an order (a suborder in some classifications) of the crustacean class Malacostraca. The order is the largest among crustaceans representing almost one third (about 10,000 species) of the known crustaceans. It contains rather big animals whose **cephalothorax** is covered by a **carapace**; five pairs of walking legs (**pereiopods**) in the thorax, the first of which are large pincers (**chelae**). Compound eyes beneath the rostrum are stalked and movable; the jointed **abdomen** ends to a **telson**. Decapoda is divided into two suborders, Dendrobranchiata and Pleocyemata, the latter includes such groups as crabs and lobsters; **2.** a previously used name for the order Teuthoidea (**cuttlefish** and **squids**) of cephalopod molluscs (subclass Coleoidea).

decarboxylases, enzymes of the sub-subclass EC 4.1.1. of the lyase class, catalysing the removal of carbon dioxide, CO_2 from a molecule, especially from the carboxyl group, -COOH, of alpha keto acids; the reaction is named **decarboxylation.**

decay, 1. the decomposition of proteins caused by bacteria in anaerobic conditions. The proteins are first split into amino acid molecules, which are further broken in deamination (the amino group chips off as ammonia) or in decarboxylation (carbon dioxide splits from the amino acid). Decay is very important in nature because it liberates nitrogen from dead organic material for building of new organic substances in plants. The decay process can be utilized in purifying of waste waters; **2.** a stage of declined strength, as in ageing; **3.** → rot; **4.** the breaking up of radioactive materials.

decidua, pl. **deciduae** (L. *decidere* = to fall off), that part of the uterine mucosa (endometrium) of primates which is cast off in menstruation and at parturition; the shedding is called **deciduation**. *Adj.* **decidual**, pertaining to the decidua; **deciduate**, characterized by shedding.

deciduous, falling off at maturity or at the end of the growth period, or at certain stages of the development.

deciduous teeth, milk teeth; primary teeth. *See* tooth.

deciduous trees, the trees which drop their leaves yearly at a certain time after the growth period; typical in temperate and cool zones.

decimation (L. *decimare* = to choose each tenth; a punishment directed against a body of troops in the ancient Rome when every tenth soldier was doomed to death by casting lots), the destruction of a considerable part; in biology, the extinction of → taxa and the respective survival of others that seems to occur by chance alone.

decomposer (L. *de* = down, from, *componere* = to put together), any of the microorganisms and fungi feeding on and disintegrating or decomposing dead organisms thus returning the material to the natural cycle; **decomposer food web,** *see* detritus.

decussate leaf arrangement, *see* phyllotaxy.

decussation (L. *decussare* = to cross in the form of an X), intersection; an intercrossing structure; e.g. *decussatio pyramidum*, the intercrossing of corticospinal motor nerve tracts (pyramidal tracts) in the ventral part of the myelencephalon in vertebrates.

deduction (L. *deducere* = to conduct as a result), a conclusion; the conduction of a unitary truth or the proposition from general truths.

defaecation, Am. **defecation** (L. *defaecare* = to deprive of dregs), **1.** the removal of excrements from the gut through the anus; **2.** the chemical removal of impurities, or the clearing of dregs. *Verb* to **defaecate** (defecate).

defect, a fault, imperfection, failure, or absence, such as of an organ. *Adj.* **defective**.

defence, measures of self-protection; may involve physical, chemical, physiological, morphological, or behavioural mechanisms, such as hard superficial tissues, spines, claws, teeth, echinated hairs, thorns, or poisonous substances of organisms, in animals also detoxification and immunological mechanisms. *See* alkaloids, chemical defence, inducible defence, detoxification, immunity.

deficiency, 1. a defect or lack, e.g. the lack of vitamins (deficiency disease); **2.** a → chromosome mutation in which a terminal segment of the chromosome and genes located in it will be lost. Also small → deletions are called deficiencies even though they do not involve a terminal segment; symbol Df.

definitive host, final host; a host having a parasite which usually lives the main part of its life cycle and reaches the reproductive maturity in or on the host. *Syn.* primary host. *Cf.* intermediate host.

defoliation (L. *de* = off, down, *folium* = leaf), the removal of leaves by using synthetic chemicals, **defoliants**; e.g. the defoliation of cotton to increase the yield.

deformation (L. *formare* = to form), deterioration; a deviation from the normal; an abnormal alteration in structure or shape (deformity).

degeneration (L. *generare* = to generate), deterioration; e.g. **1.** a change from a higher to a lower form of an organ or organism during the course of evolution with the result that it becomes vestigial; **2.** a pathologic change in the structure and activity of cells, tissues, organ, or organism; **3.** a worsening of physical or mental qualities. *Adj.* **degenerative**.

degradation (L. *degradare* = to degrade), the act or process of disintegrating or degrading, especially the splitting of a chemical compound into a less complex compound.

degranulation (L. *granulum* = granule, grain), loss of a granular structure, e.g. the disappearance of cytoplasmic granules in a granular cell.

degree days, d.d., *see* thermal sum.

degustation (L. *(de)gustare* = to taste), the act of tasting.

dehiscence (L. *dehiscere* = to split open), the spontaneous opening of a structure; e.g. **dehiscent fruits** open along the original seams between the carpels. *Adj.* **dehiscent**.

dehydr-, dehydro-, 1. (L. *de-* = off, Gr. *hydor* = water), dehydrated; pertaining to the loss or removal of water; **2.** (< *hydrogen*), dehydrogenated; pertaining to the loss or removal of hydrogen.

dehydratases, hydratases; enzymes of the sub-subclass EC 4.2.1. of the lyase class, catalysing reversibly the removal of hydrogen and oxygen in the form of H_2O from a substance.

dehydration, removal or reduction of water content; desiccation, anhydration; e.g. **1.** removal of water from tissue samples in a successively concentrated ethanol series when preparing them for microscopical observation; **2.** removal of water from soil or any biological material e.g. for chemical analyses by heating in an oven at 90—100°C; **3.** removal of water from any molecule. *Verb* to **dehydrate**.

dehydrogenases, a class of the enzymes which catalyse the transfer of hydrogen from one compound to another, the former being oxidized and the latter reduced. Dehydrogenases function in cell metabolism; e.g. succinate dehydrogenase (EC 1.3.99.1) and isocitrate dehydrogenase (EC 1.1.1.42) in the → citric acid cycle, and lactate hydrogenase (EC 1.1.1.27-28) in anaerobic → glycolysis.

delayed implantation, embryonic diapause; the ceased development of a mammalian embryo at the blastocyst state (trophoblast) before becoming implanted to the uterine wall; may last several months. The further development is reactivated by elevated maternal progesterone and oestrogen synthesis. Delayed implantation occurs in many mammals such as seals, bears, some deer, etc., i.e. in animals who breed once a year and copulate about one year prior to the birth of the offspring.

deletion (L. *delere* = to wipe off, annihilate), in genetics, a chromosome → mutation which leads to the failure of a chromosome segment and the genes located in it. In eukaryotic organisms, they may occur at the end of the chromosome (**terminal deletions**) or elsewhere on the chromosome (**intercalary deletions**); originally the former were called deficiencies and the latter deletions. In practice, also small deletions, whether terminal or intercalary, are often called deficiencies, and only the larger changes are called deletions. Both terminal and intercalary deletion lead to the formation of an acentric or centric fragment in those eukaryotes which have a localized → centromere.

In eukaryotes, the presence of rather a long deletion can be observed cytologically on the basis of the formation of acentric and centric fragments in → mitosis, or on the basis of the lack of pairing of a given chromosome segment at the prophase of → meiosis, or also on the basis of lack of pairing in polytene → giant chromosomes.

The genetic demonstration of deletion is based on two facts. Firstly, deletions are irreversible mutations, i.e. no → back mutation can occur. Secondly, because the homologous alleles are missing, the recessive → alleles located on the corresponding region of the → homologous chromosome manifest in the phenotype in single doses, i.e. these genes are → hemizygous. The point mutations located in the region involved in a deletion do not recombine (*see* genetic recombination). Large deletions are often dominant → lethal mutations.

deletion mapping, the localization or mapping of → deletions in the chromosomal complement. In a eukaryotic organism the location of a deletion can be determined because deletions cause the expression in the phenotype of recessive point mutations occurring in the corresponding region on the homologous chromosome. In prokaryotic organisms the mapping of deletions is based on → genetic recombination. Naturally, the overlapping deletions cannot recombine but non-overlapping deletions can.

deme (Gr. *demos* = region, nation, populace), a **local population** of closely related individuals that are in panmictic (→ panmixis) reproductive relation to each other; the smallest collective unit of a plant or animal population.

demographic transition (Gr. *graphein* = to write), a hypothesis based on the progress of the human populations in western industrialized countries, especially in the course of the 19th and at the beginning of the 20th century. The progress is divided into four stages: in the first, the growth of population is slow and the birth and death rates are equal. In the second stage, the birth rate is still high but the death rate begins to decrease owing to improving living conditions, and the population increases exponentially. In the third stage, also the birth rate begins to decrease, the death rate is low, and population growth begins to moderate. In the fourth stage, both the birth and death rates are low and growth has ceased. At present, the developing countries are in the second or third stage, while the industrialized countries are in the fourth stage.

demography, the statistical study of population size, growth, density, age or sex distribution, and other features in a population; **demographic factors**, such as birth rate, mortality, and migration, regulate the changes in the structure of a population.

Demospongiae (L. *spongia* = sponge), a class of poriferans comprising about 95% of the species of living sponges; large leuconoid animals which have siliceous spicules and spongin to bind them together; the spicules may be absent in some species.

denaturation, making unnatural; **1.** a destructive change of the properties of organic molecules observed in their secondary and tertiary structures, solubility or other properties, or in their physiological and biochemical functions; initial changes may be reversible. Factors causing denaturation are e.g. pH, salt concentration, radiation, temperature, organic solvents and other chemicals. In biology, denaturation is usually applied to proteins and nucleic acids, like the denaturation and coagulaton of egg albumin (ovalbumin) at high temperatures; **2.** adulteration, such as of ethyl alcohol.

denaturing gradient gel electrophoresis, DGGE, a technique for isolating a gene based on the fact that even one base pair change of a short DNA fragment (0.5 kb or more) is often sufficient to make a perceptible difference in the stability of the double-stranded structure. If double-stranded fragments are applied to run through an electrophoretic gel within an increasing gradient of urea and/or formamide, the fragments will move faster or slower according to their size until they reach the point in the gradient where their complementary strands begin to separate, i.e. begin to denaturate. Hence, the prerequisite for DGGE is that the sequence of the gene is known. *See* electrophoresis.

dendr(o)- (Gr. *dendron* = tree), denoting relationship to a tree or a tree-like structure.

dendrites, thread-like, relatively short processes extending from the soma (cell body) of a nerve cell (neurone); dendrites have usually numerous branches which together with the soma function as the receptive surface of the neurone. In unipolar and bipolar neurones dendrites often resemble structurally the → axon.

dendritic, 1. branched like a tree; **2.** pertaining to a → dendrite.

dendrochronology (Gr. *chronos* = time, *logos* = word, discourse), the study dealing with the age of trees (living, dead, or fossil) by counting the annual rings or by studying the carbon isotopes; can be utilized in different ways, e.g. in palaeoclimate studies.

dendrogram (Gr. *graphein* = to write, record), a tree-like diagram used to describe degrees of evolutionary relationships or phenetic resemblances (phenogram) of the species of an organism group. *See* cladogram.

dendrology (Gr. *logos* = word, discourse), the study of trees.

denervation (L. *de* = away, *nervus* = nerve), resection of a nerve; the denervation of a skeletal muscle ceases the trophic action of the nerve, and therefore after a short hypersensitivity period the muscle begins to degenerate, i.e. becomes atrophic. *Verb* to **denervate**.

denitrification (L. *nitrum* = nitrate, *facere* = to do), the decomposition of nitrates to molecular nitrogen by anaerobic soil bacteria (denitrifying bacteria). Nitrate ions are electron acceptors in respiration of those bacteria instead of oxygen in aerobic organisms. The denitrifying process causes nitrogen loss from the soil to the atmosphere and is an important step in the → nitrogen cycle in nature. *Cf.* ammonification, nitrification.

denitrifying bacteria, *see* denitrification.

densitometer (L. *densitas* = density, Gr. *metrein* = to measure), an optical instrument used to measure the optical density of a chromatographic film or electrophoretic gel or plate for recording separated components in the sample. The sample is passed through a sharp light beam, and changes in light intensity are detected by a photomultiplier; also → fluorescence excited by ultraviolet light in the sample can be measured. The results are recorded on a chart and the data are collected and calculated quantitatively. The method is called **densitometry**.

density, 1. the number of any biological unit, such as individuals or species, per unit area or volume; e.g. the number of breeding birds per hectare or square kilometre, or plant individuals e.g. per m^2. Density is thus an absolute quantity and also useful in mathematical ecology; *see* absolute density, relative density; **2.** the compactness of a substance, such as a chemical element or compound, expressed often in g/ml, in the SI system kg/m^3; symbol ρ; **3.** optical density; *see* absorbance; **4.** electric flux density, symbol *D*; **5.** quantum (photon) flux density; **6.** magnetic flux density, symbol *B*.

density-dependence, in population ecology, denoting the situations in which the importance of any factor having an effect on the birth and death rates of a population changes as a function of the density. The effect is either negative (**negative density-dependence**) regulating or balancing the population density, or it is positive (**positive density-dependence**) which may result in a chaotic development (**chaos dynamics**) in the population. In the former situation, within the increasing population density the regulating factor tends to decrease birth rate and/or increase death rate. Correspondingly, within the decreasing population density the regulating factor increases birth rate and/or decreases death rate; e.g. in an increasing population, increased infections of parasites or diseases tend to increase death rates, and on the other hand, in a decreasing population, surplus of food supplies increases birth rates and survival of young.

density gradient centrifugation, a technique of → centrifugation for separating macromolecules, particles, or cell organelles in a solution gradient. The density gradient is prepared e.g. using sucrose, sorbitol, or caesium chloride solutions so that the concentration increases towards the bottom of a centrifuge tube linearly (**continuous**), stepwise (**discontinuous**), exponentially, or variably (**multiple gradient**). The gradient system can be prepared by pipetting, or using a peristaltic pump or a special gradient maker. Also it is possible to make a self-generating gradient from some special gradient media like Percoll, in which the gradient is formed during centrifugation. When using a zonal → rotor, both a gradient solution and sample suspension are pumped into the rotating rotor.

In **isopycnic centrifugation** particles form sediment in the gradient densities which correspond to the density of particles. The running time is rather long but can be shortened by using a vertical rotor. In **rate zonal centrifugation** the sedimentation velocity is chiefly dependent on the size of particles. The method is suitable for the separation of particles with the same density but of different size. When a run is completed, the sediment fractions can be taken out from the tubes using a syringe or a special fraction collector. The fractions can be monitored with an optical equipment.

dent(i)-, dento- (L. *dens*, pl. *dentes*= tooth), denoting relationship to teeth. *Adj.* **dental**.

dental formula, a formula describing the dentition of mammals; the teeth are given in order from the front backwards in the upper and lower jaws on one side: **i** = **incisors** (*dentes incisivi*), **c** = **canines** (*d. canini*), **pm** = **premolars** (*d. premolares*) and **m** = **molars** (*d. molares*). The number is expressed as a formula as follows (i, c, pm, m): e.g. for deer: 0.0.3.3/3.1.3.3, cat: 3.1.2.0/3.1.2.1, and man: 2.1.2.3./2.1.2.3.

dentary (bone), one of the bones of the lower jaw of many vertebrates, bearing the marginal tooth row in primitive vertebrates; in mammals the lower jaw comprises the dentary bone alone.

dentate(d), having teeth or tooth-like processes; toothed, notched, cogged.

denticle, → placoid scale.

denticulate(d), having small teeth; finely dentated, notched, serrated.

dentine, also **dentin, dentinum,** the hard, calcareous, bone-like tissue of the vertebrate tooth. On the crown of the tooth, dentine is usually covered by a harder substance, → enamel, and on the root by a softer tissue, → cementum. The dentine cells, **odontoblasts,** responsible for the formation of dentine, are located on the inner surface of the dentine, i.e. lying in the dental pulp and only thin projections of the cells are intruded into small **dentinal tubules** inside the dentine. The organic matrix (about 20%) is chiefly collagen with some elastin and mucopolysaccharide, and the inorganic matrix is mainly hydroxyapatite, with some magnesium and fluoride. *See* tooth.

dentition, the type, number and arrangement of the teeth of an animal. *See* dental formula, tooth, heterodont, Homodonta.

deoxyadenosine, a → nucleoside composed of adenine and deoxyribose; occurs in DNA.

deoxycorticosterone, DOC, 11-deoxycorticosterone; 21-hydroxypregn-4-ene-3,20-dione; an adrenocortical steroid acting as a precursor of → corticosterone and aldosterone; has some mineralocorticoid activity but is rarely secreted from the adrenal cortex to the blood.

deoxycytidine, a → nucleoside composed of cytosine and deoxyribose; present in DNA.

deoxyhexose, a → hexose in which one -OH group is replaced by H.

deoxyguanosine, a nucleoside formed by guanine and deoxyribose; present in DNA.

deoxyribonuclease, DNase, also **DNAse, DNAase,** a group of enzymes belonging to hydrolases catalysing the breakdown of deoxyribonucleic acid (DNA) by hydrolysing phosphodiester bonds. The enzyme can cleave DNA by starting from the end of the molecule (**exodeoxyribonucleases**) or from the middle (**endodeoxyribonucleases**). Examples of DNases are **DNase I** (EC 3.1.21.1) producing 5'-phosphodinucleotides and 5'-oligonucleotides, and **DNase II** (EC 3.1.22.1) which produces 3'-phosphomononucleotides and 3'-oligonucleotides. DNases are found in the nucleus, some of them attached to chromatin, and e.g. in the secretions of the alimentary canal.

deoxyribonucleic acid, DNA, a polymer of **deoxyribonucleotides** (*see* nucleotide) which constitutes the genetic material of all cells and some cell organelles; also the genetic material of most viruses consists of DNA (RNA viruses such as retroviruses, have another nucleic acid, RNA). DNA occurs in organisms usually as a long, unbranched chain of deoxyribonucleotides. The nucleotides of DNA consist of a phosphorous acid residue, deoxyribose sugar, and a purine or pyrimidine base; the purine may be adenine or guanine and the pyrimidine thymine or cytosine.

In all organisms the proportional quantities of

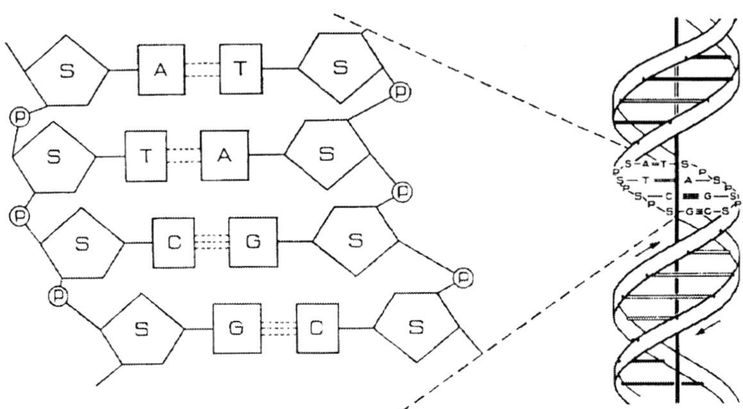

Deoxyribonucleic acid. The structure of the deoxyribonucleic acid molecule. Nucleotides are covalently linked together by phosphorous acid residues (P), and the bases (A = adenine, T = thymine, G = guanine, C = cytocine) are bound together by hydrogen bonds. (S = deoxyribose sugar). (After Rieger, Michaelis & Green: Glossary of Genetics - Springer Verlag, 1968; with the permission of the publisher).

the bases are regular, i.e. there is always **adenine** (A) and **thymine** (T) on one hand and **cytosine** (C) and **guanine** (G) on the other hand in equal quantities. Thus, the proportion of → purines to → pyrimidines always equals one (**A + G : C + T = 1**). Only the proportion **A + T : C + G** varies but has a species-specific character, i.e. is the same in all tissues and cells of the same species.
The DNA macromolecule, running from end to end at each → chromosome, comprises a long chain of nucleotides. The molecule is generated when the nucleotides link to each other by means of a diester bond formed via the phosphorous acid residue (phosphodiester bond). The phosphorous acid residue in this bond links the 5' carbon atom of one nucleotide to the 3' carbon atom of the next nucleotide. Thus the formed DNA strand has **5'-3' polarity**. The sequence of nucleotides in the strand forms the **primary structure** of DNA which can be understood as a backbone formed by the sugar moieties and the phosphorous acid residues, and the bases linked to each sugar moiety.
Usually DNA consists of two polynucleotide chains which are connected to each other by hydrogen bonds between the bases. In this double strand forming the **secondary structure** of DNA, the polarities of participating strands run in opposite directions since adenine always forms a pair with thymine, and guanine with cytosine. This **base-pair rule** explains the relative proportion of the bases presented above. The existence of the base pairs means that the sequences of the nucleotides in the two strands are **complementary**. The secondary structure of DNA is generated when the complementary strands wind around each other forming a double helix. The double helix has different conformations. The most common, the so-called B conformation (→ B-DNA), in which the double helix twists to the right, has a diameter of 2 nm and the rise of one twist is 3 nm. The length of one twist consists of ten nucleotide pairs. This structural model of DNA was first presented by *James Watson* and *Francis Crick* in 1953, and it has since been confirmed in many ways.
DNA has the properties generally required of genetic material; the ability to duplicate or replicate, a great specificity which is preserved in → replication, and the property to contain information (→ genetic information).

deoxyribose, a pentose sugar in which one OH is replaced by H; occurs in deoxynucleosides being a component of → nucleotides in DNA.

deoxythymidine, a → nucleoside formed by thymine and deoxyribose; occurs in deoxyribonucleic acid, DNA.

DEP, DEPC, → diethylpyrocarbonate.

dephosphorylation, the removal of a phosphate group from a molecule. *Cf.* phosphorylation.

deplasmolysis (L. *plasma* = shaped, *lyein* = to decompose), the reversion of → plasmolysis; a property of a living, plasmolysed plant cell to retake water and swell to its original volume in a hypotonic solution.

depolarization (Gr. *polos* = pole), the neutralization of polarization, or the change in direction of polarity; in electrophysiology, usually a localized reversal of → resting potential of a cell, i.e. a change of the negative membrane potential towards zero, occurring in the initial phase of → action potentials. Smaller depolarization occurs e.g. in → receptor potential or in the function of → synapse. *Cf.* hyperpolarization.

deposit (L. *deponere* = to put down), **1.** something precipitated, as a sediment or dregs; **2.** extraneous matter collected in the cavities or tissues of organs, or plaque on the tooth surface; **3.** as a verb, to let fall, set, or lay down.

deposit feeders, aquatic animals, such as some oligochaetes and chironomid larvae, feeding on organic matter sedimented on a bottom.

depression (L. *de* = down, away, *premere* = to press), **1.** an area pressed down; a hollow or downward displacement; **2.** a decrease in a functional activity; **3.** depression of freezing point; **4.** the mental state of general emotional dejection and sadness, despair, and withdrawal. *Adj.* **depressive**.

depressor, anything that decreases or retards functional activity, e.g. an agent, muscle, nerve, or an instrument.

deprivation (L. *privare* = eliminate), absence, dispossession, or loss, as of an organ, power, etc.; **sensory deprivation**, the absence of information through senses. *Verb* to **deprive**.

deproteinization, the removal of proteins from organic extracts and solutions using e.g. a perchloric acid or phenol treatment and the centrifugation of precipitated proteins.

derepression (L. *de* = off, *reprimere* = to press

back, smother), the release of the prevention or → repression of a function, specifically a biochemical reaction involved in an inducible enzyme system. Derepression returns the activity of an enzyme when the repressor of the coding gene is removed. The process is due to the enhancement of protein synthesis in a cell when a given specific metabolite, repressor, is combined with a specific substrate and thus released from repression allowing the activation of the operator gene (→ operon), i.e. derepression occurs.

derivative meristems, plant tissues belonging to → apical meristems, produced by meristematic initial cells (promeristem) in root and shoot tips. The cells of derivative meristems are still capable of dividing and produce new material but their development is already determined. In the root, **protoderm (dermatogen)** is the derivative meristem of the epidermis (or rhizodermis), **periblem** forms the cortex, and **plerome** gives rise to the central cylinder with its conductive tissues; **calyptrogen** forms the root cap. In the shoot, the derivative meristem of the epidermis is protoderm as in the root, **procambium** forming the structures of vascular tissues, and **ground tissue meristem** producing the cells for ground and supportive tissues.

derm(a)-, dermo-, dermat(o)- (Gr. *derma* = skin), denoting relationship to the surface or skin of an organism.

dermal, 1. in animals, pertaining to the skin, especially to the vascular layer, → dermis (corium), of the vertebrate skin; also dermic; **2.** in plants, pertaining to → dermal tissues.

dermal bone, a bone formed by the ossification of the → dermis (corium). *Cf.* cartilage bone.

dermal gills, dermal branchiae, skin gills, papulae; small, soft branchiae formed as projections from the body cavity in some echinoderms, e.g. asteroids; covered with epidermis and lined internally with peritoneum. The dermal gills function as respiratory and excretory organs.

dermal tissues, plant tissues serving as protecting or bordering layers in plants; different types are: 1) → **epidermis**, covers all parts of young plants, → hairs and stomata (*see* stoma) being epidermal appendages; 2) → **endodermis**, has the bordering role around some structures inside other tissues, e.g. around the central cylinder in roots, or the vascular bundle system in conifer needles; 3) → **periderm**, replaces the epidermis in old roots and shoots of plants. *Cf.* dermis.

Dermaptera (Gr. *pteron* = wing), **dermapterans, earwigs**; an insect order comprising animals with very short, leathery forewings and large, membranous hindwings folded beneath the forewings; wings are reduced or absent in many species, and cerci form stout horny forceps at the end of the abdomen. Dermapterans feed on other insects or on green plants; about 1,000 species are described.

dermatan sulphate, *see* chondroitin.

dermatogen (Gr. *genos* = origin), *see* derivative meristems.

dermatome (Gr. *temnein* = to cut), **1.** the mesodermal area of a vertebrate embryo developing into the dermis of the skin (*see* somites); **2.** the area of the vertebrate skin supplied with sensory nerve fibres of a single dorsal spinal root; *syn.* dermatomic area; **3.** a medical instrument for cutting skin slices.

dermatomere (Gr. *meros* = part), a segment of the integument especially of a vertebrate embryo; called also metamere.

dermatophyte (Gr. *phyton* = plant), a fungus causing some skin diseases such as ringworm (tinea) in animals, man included.

dermis (Gr. *derma* = skin), the vascular layer under the skin epithelium or → epidermis; in most multicellular animals, consists of connective tissue with blood and lymph vessels and sensory nerve endings with sensory receptors; in vertebrates may be ossified to scales or larger bones, called membrane bones (dermal bones). *Syn.* corium. *Adj.* **dermal** (dermic), pertaining to the dermis; **dermoid**, resembling the skin, also pertaining to a cystic tumour in skin or epithelium. *Cf.* dermal tissues.

Dermoptera (Gr. *pteron* = wing), **dermopterans, flying lemurs**; a mammalian order of two species (genus *Cynocephalus*) resembling flying squirrels in appearance; they are related to the true bats but cannot fly, only glide. Their distribution area is the Malayan Peninsula.

dero- (Gr. *dere* = neck), denoting relationship to the neck or a neck-like structure.

desaturase, a member of the complex of three membrane-bound enzymes, i.e. NADH-cytochrome b_5 reductase, cytochrome b_5, and desaturase; the complex catalyses in the fatty acid metabolism the reaction: stearoyl CoA + NADH + H^+ + O_2 —> oleoyl CoA + NAD^+ + 2 H_2O, thus introducing a double bond into a long-chain acyl CoA.

desaturation, the act of making something less saturated. *See* saturation.

descendence, descendance (L. *descendere* = to go down), in biology, derivation from an ancestor, → evolution.

descendence theory, → evolution theory.

desert, a → biome characterized by scanty rainfall and consequently by little or no plant cover; may be cold or hot. **Hot deserts** are mostly situated in the tropics, and the lack of rain in these areas is often caused by cold ocean currents along the coasts and the continuous high pressure over the seas. **Cold deserts** are situated in the inner regions of continents in temperate zones where mountain areas prevent humid winds from bringing moisture.

desiccation (L. *desiccare* = to dry up), the process of being dried; dehydration, exsiccation; **desiccator** (exsiccator), an apparatus with a drying agent (often silica gel) to render a material totally dry. *Verb* to **desiccate**. *Adj.* **desiccant**. *Noun* **desiccant**, a drying agent.

desmetryne, a → triazine herbicide.

desmids, a group (informal taxon) formed by two families of unicellular algae belonging to the class Chlorophyceae, division → Chlorophyta (green algae), Plant kingdom. The family Mesotaeniaceae includes the so-called **saccoderm desmids**, the cells of which do not form a new semicell during division. The cells of **placoderm desmids** (family Desmidiaceae) are formed from two semicells separated by a narrow median constriction, isthmus, where the nucleus is situated and the septum is formed; during cell division, each parent semicell forms a new semicell for itself and therefore some genera (especially *Micrasterias*) are well known objects in studies dealing with plant morphogenesis.

desmin (Gr. *desmos* = band), a protein component of muscle cells forming intermediary filaments (*see* cytoskeleton) which encircle the Z-disc, and make additional connections to neighbouring Z-discs; with longitudinal filaments they bridge between successive Z-discs in the same myofibril. They also serve to cross-link different myofibrils into bundles. Desmin filaments form a supportive structure in muscle cells and are attached to the sarcomere through interactions with myosin filaments (thick filaments). In smooth and cardiac muscles, desmin also acts as a component in different types of → desmosomes between adjacent muscle cells. Desmin filaments lie outside the sarcomere and do not actively produce contractile force.

desmodeme (Gr. *demos* = region, nation), → volvent.

desmogen (Gr. *genos* = origin), undifferentiated vascular bundle in developing plant leaves.

desmosome (Gr. *soma* = body), any of small, circular or belt-like structures between two adjacent animal cells forming intercellular junctions attaching the cells together. Desmosomes are especially found between cells of stratified epithelium, or in → intercalated discs joining cardiac muscle cells. Cytoskeletal filaments and cell membranes in adjacent cells form a discoidal, thickened structure, in which the cell membranes are closely attached to each other by means of intercellular fibrous proteins. According to the shape or structure, different types can be distinguished, i.e. **spot desmosomes** (*macula adhaerens*) with keratin filaments, found in epithelial cells and e.g. in smooth muscle, **belt desmosomes** (adherent junctions, *zonula adhaerens*) with contractile actin filaments situated as a belt near the apical end of each epithelial cell, or **hemidesmosomes** which are spot-shaped structures in membranes attaching epithelial cells to the underlying basal lamina. Bundles of intermediate filaments, associated with the → cytoskeletons of the adjacent cells, form interconnections in the desmosomes. *Cf.* tight junction, gap junction.

desmotubulus, pl. **desmotubuli** (L. *tubulus*, dim. of *tubus* = tube), **desmotubule**; the tubular structure formed from → endoplasmic reticulum, traversing a → plasmodesm between two adjacent plant cells.

detention period, in hydrobiology, a theoretical time interval indicating how long a water molecule is detained in a water system. *See* residence time.

detergent (L. *detergere* = to purify), a substance which decreases the surface tension, e.g. common dish-washing chemicals. In biochemical studies detergents are used to facilitate the separation of enzymes and cell organelles from other cell materials. Some detergents, such as → Triton X-100 and

Tween 80, are also used to promote the penetration of herbicides.

determinant (L. *determinare* = to limit, to fix), an agent or factor that establishes or determines a process or event; e.g. **antigenic determinant**, a site of an → antigen to which a single antibody binds. *Adj.* **determinable**, capable of being determined. *See* determination.

determinate growth, the limited growth typical of many plant structures, such as of leaves and flower organs; it is determined by genes.

determination, 1. a movement or course towards a certain direction or towards a given point; e.g. a sequence of events in which the daughter cells of a → totipotent fertilized egg are determined to accomplish certain parts of the developmental programme of the organism. Determination is the first step of → differentiation, and it is stable and → cell-heritable, occurring stepwise in groups of cells so that each group is guided to accomplish a certain part of the developmental programme, approaching the goal with each step. Determination is usually irreversible and proceeds most likely so that at each step the cell group has two alternative **developmental pathways**, of which one will be realized. Determination can be understood as **gene activation**, i.e. the activation of certain groups of genes or gene batteries at certain stages of development. During the developmental process, the activation of certain genes is manifested as a certain effect in the → phenotype of each cell group. The result of determination is also dependent on external factors, the effect being best seen in different plant individuals; **2.** the estimation or measurement of a quality or quantity in an experiment or study.

deterministic explanation, an explanation in which the fulfilment of a phenomenon is strictly causal without any probabilities; in ecology, **deterministic forces** affect organism groups strictly without any element of chance or probability. *Cf.* stochastic.

deterministic chaotic model, a mathematical model in which all the relationships between variables are expressed with constant coefficients, and thus the relationships are fixed; in biological systems this leads to chaos. *See* stochastic.

detoxification, also **detoxication** (L. *de* = away, Gr. *toxikon* = arrow poison), the elimination or reduction of the toxicity of a substance (poison). In the metabolism of most organisms toxic compounds are processed enzymatically to a less poisonous form. These **detoxification reactions** include several types of oxidation-reduction and conjugation reactions, such as conjugation of → glucuronic acid with a poisonous substance. The reactions occur in animals chiefly in the liver, hepatopancreas, and alimentary canal. *See* cytochromes.

detritivore (L. *vorare* = to devour), detritus feeder; an organism feeding on dead organic remains, → detritus.

detritus, pl. **detritus** (L. *deterere* = to rub away), any disintegrated material worn away from a mass; in ecology, any dead and decaying organic material settled on the ground or on the bottom in aquatic environments; supports a rich **detritus food web** (decomposer food web) which contains high quantities of decomposers, such as microorganisms and fungi. *See* soil.

deuter(o)-, deut(o)- (Gr. *deuteros* = second), **1.** denoting second, or one of the two; **2.** pertaining to → deuterium.

deuterium, heavy hydrogen, hydrogen-2, symbol D; a hydrogen isotope composed of one proton and one neutron; it is obtained from heavy water (deuterium oxide, D_2O) and used e.g. in studying fat and amino acid metabolism; **deuteron**, heavy hydrogen ion, D^+. *Cf.* tritium. *See* hydrogen.

Deuteromycetes (Gr. *mykes* = fungus), **Fungi imperfecti**, the only class in fungal division → Deuteromycota; divided into subclasses Blastomycetidae, Coelomycetidae, and Hyphomycetidae.

Deuteromycota, a fungal division including the class → Deuteromycetes (Fungi imperfecti). Deuteromycota is actually a form-division, lying outside the fungal classification, because all species are lacking the sexual reproduction which is the base of the taxonomy. Some species may be secondarily asexual, others possibly have never had a sexual stage. These fungi are found practically in every habitat; many species are saprotrophs, some are even severe pathogens causing diseases in animals, man included. Among the species are also some common → moulds, such as the genera *Penicillium* and *Aspergillus*. In some systems the species of

Deuteromycota are grouped under the name Mitosporic fungi with no taxonomic status.

Deuterostomia (Gr. *stoma* = mouth), **deuterostomes**; animals which have a proper → coelom; in the embryonic development the anus is formed from the blastopore and the mouth later at the other end of the → archenteron. The group includes the phyla → Phoronida, Ectoprocta, Brachiopoda, Echinodermata, Chaetognatha, Hemichordata, and Chordata. *Cf.* Protostomia.

deuterotoky, deuterotocia (Gr. *tokos* = offspring), *see* parthenogenesis.

deutomerite (Gr. *meros* = part), a posterior segment in the cell body of gregarines (Sporozoa). *Cf.* epimerite, protomerite.

deutonymph (Gr. *nymphe* = bride, nymph), the second larval (nymphal) stage in the life cycle of mites and ticks (Acari). *Cf.* protonymph, tritonymph.

deutoplasm, products elaborated by the cell and stored in the cytoplasm, e.g. lipid droplets, yolk bodies, pigment, and secretion granules.

Deutsch's motivation model (*J.A. Deutsch*), in ethology, a theory which emphasizes the importance of the motivation in releasing of a behavioural action of an animal by considering the negative feedback control; the strength and persistence of the motivation depends on the balance or imbalance in the internal medium of the animal; e.g. a dog finishes eating because of the feedback control from the full stomach to the central nervous system. *Cf.* Lorenz's drive model.

development, the process of differentiation and growth of an organism, such as embryonic development (*see* embryogenesis), prenatal development, postnatal development, or ontogenic development (life-span development). *Adj.* **developmental**.

developmental biology, the branch of biology dealing with the ontogenetic development of organisms, i.e. the differentiation and growth of tissues, organs, and organisms; the more concise field of it is developmental physiology.

developmental genetics, a branch of genetics concerning how genes regulate the → ontogeny of organisms. The central problems of developmental genetics are the function and regulation of genes which form the basis of the → differentiation of cells. *Syn.* **phenogenetics**.

developmental potential (L. *potentia* = power, potency), the capacity of a cell to conduct a certain part of the developmental programme of an organism. The fertilized egg is → totipotent, but the developmental potential of cells narrows stepwise during the process of cell divisions in the embryo, and each cell will be determined (→ determination) and differentiated into its own special structure and function in developing tissues and organs.

deviation (L. *deviare* = to turn aside), the act of diverging or deflecting; e.g. **1.** turning aside from the standard course; **2.** an abnormality; **3.** in statistics, the difference between an individual value and a mean value in the set of values.

Devonian (period) (*Devon* = a county in southwestern England), a geological period of the Palaeozoic era between Silurian and Carboniferous from 408 to 360 million years ago. Devonian was the age of fishes which were especially abundant in fresh waters. The development of ferns was intensive, and the first trees, insects, sharks, and amphibians appeared.

dextr(o)- (L. *dexter* = right), **1.** denoting right; **2.** a chemical prefix, symbol (+)-, *d*-, or D-; *see* isomers.

dextran, a branched, high molecular weight polysaccharide formed in microorganisms such as yeasts and bacteria; commercially produced in cultures of a bacterium (*Leuconostoc mesenteroides*). Dextran is composed of glucose units linked chiefly by α-1,6 linkages, but also α-1,3 and α-1,4 linkages occur. Dextrans are used e.g. as plasma expanders and a cross-linked form of it (Sephadex) in → gel filtration.

dextrin, an intermediate formed in the hydrolysis of starch into sugars in the reactions catalysed by α-amylase; α-dextrin contains several glucose units which are linked together by α-1,4 and α-1,6 bonds; **α-dextrinase**, EC 3.2.1.10, is the enzyme that decomposes dextrin into glucose.

dextrose, → glucose.

DGGE, → denaturing gradient gel electrophoresis.

di- (Gr. *dis* = twice), denoting two, twice, double; in chemical terms meaning diatomic (also *bi-*).

dia- (Gr. *dia* = through), denoting through, across, between, apart, completely.

diabetes, pl. **diabetes** (Gr. *bainein* = to go), originally means disorders characterized by excessive excretion of urine; **1.** *diabetes (mellitus)*, a metabolic disturbance due to deficiency of → insulin excretion or decreased efficiency of insulin resulting in symptoms of increased concentration of glucose in the blood, tissue fluids and urine. Especially the metabolism of carbodydrates and lipids is disturbed and the excretion of urine increased; the disease is well known in man but can occur also in other vertebrates; **2.** *diabetes insipidus,* a disorder in the neurohypophyseal system, i.e. in the function of the posterior lobe of the pituitary gland or the related brain nuclei, resulting in the deficient secretion of antidiuretic hormone (→ vasopressin). That causes a failure in water reabsorption in renal tubules, and thus diuresis.

diacylglycerol, DAG, the intermediate of the synthesis of triacylglycerol; in diacylglycerol two -OH groups of glycerol are esterified with fatty acids. DAG is also a → second messenger formed from PIP_2 to give IP_3 + DAG. *See* inositol.

diacytic stoma (Gr. *kytos* = cavity, cell), a type of plant stoma. *See* stoma.

diadelphous (Gr. *adelphos* = brother), pertaining to a flower with stamens in two groups due to fusion of the filaments; the stigmata are separate. *Cf.* monadelphous.

diadromy (Gr. *dromos* = course), the travel of animals, especially fish, from salt waters to fresh river and lake waters. *Adj.* **diadromous**. *Cf.* anadromous, catadromous.

diageotropism, → diagravitropism.

diagnosis (Gr. *gnosis* = knowledge), the act of identifying a disease; also the determination of a species, a family, etc.

diagravitropism (Gr. *geos* = earth, *tropos* = turn), a type of → gravitropism in which plant parts, such as twigs, grow at right angles to the direction of gravitation. Also called diageotropism.

diakinesis (Gr. *kinesis* = movement), the last phase of meiotic prophase in which the → bivalents condensate and the nuclear envelope breaks down. *See* meiosis.

diallate, *see* thiocarbamates.

diallel cross (Gr. *diallelos* = reciprocating), the set of all possible matings between several genotypes, i.e. individuals, clones, homozygous lines, species, etc.; used e.g. to determine the importance of each genotype in the genetic transmission of a certain quality to the offspring.

dialysis (Gr. *dia* = through, *lyein* = to dissolve), the separation of small molecules from large by diffusion through a semipermeable membrane, the method being used especially for purification of proteins. Dialytic treatment is used clinically for removal of urea and other substances from the blood of patients with kidney failure.

diamine oxidase, histaminase, EC 1.4.3.6; a copper-containing enzyme in animal tissues catalysing the oxidation of amines, e.g. histamine; especially active in vertebrate kidneys where it oxidizes → putrescine and some other diamines. *See* amine oxidase.

diandry, diandria (Gr. *dis* = twice, *andros* = male), a condition in which a flower has two free stamens, or a moss has two antheridia. *Adj.* **diandrous**.

diapause (Gr. *diapausis* = pause), a prolonged inactive state or arrest in the development of organisms; occurs in many invertebrate animals (e.g. eggs, embryos, and pupae), or plant seeds. In animals, e.g. the neuronal and metabolic inactivity is deeper than in torpor, and the state is characterized by inactivity of reproductive organs, decreased activity of many hormones, and fasting. Diapause helps an organism to survive through unfavourable periods (dry, cold, or hot periods, shortage of food, etc.). Although it is genetically determined, local environmental factors, such as extreme temperatures and dryness, may play a role in commencing a diapausal period. Diapause is a controlled state in which the genetic and environmental factors lead to activation of certain regulatory processes which end the arrest. Diapause lasts generally for months, but may last for several years. It may be involved in the winter, summer, or cold torpor.

diapedesis, pl. **diapedeses** (Gr. *pedan* = to leap), a leap through; especially the passing of white blood cells across intact capillary walls into different tissues. *Adj.* **diapedetic**.

diaphorase (Gr. *pherein* = to carry), **1.** an old term for any flavoprotein which in experimental conditions catalyses the oxidation of → NADH or NADPH to NAD^+ or $NADP^+$ by using a non-physiological agent, such as methylene blue, as an electron acceptor in the reaction; **2.** dihydrolipoamide dehydrogenase, EC 1.8.1.4, oxidizing dihydrolipoamide.

diaphragm (Gr. *diaphragma* = dividing wall, barrier), any enclosing or separating disc-like structure, such as a separating structure in various organs; e.g. **1.** the midriff which is the flat membranous muscle separating the abdominal and thoracic cavities in mammals, acting as the most important inspiratory muscle; called also **diaphragma** or diaphragmatic muscle; **2.** a septum between compartments of body cavities in many invertebrates, e.g. in insects a delicate transverse septum separating the shallow pericardial cavity from the coelom in the dorsal abdomen. *Adj.* **diaphragmatic**.

diaphysis, pl. **diaphyses** (L. < Gr. *diaphyesthai* = to grow between), the shaft of an elongated bone between the ends (epiphyses). *Adj.* **diaphyseal**, also diaphysial.

diapsid (Gr. *dis* = twice, *apsis* = arch), pertaining to a skull type with two pairs of openings, one in the roof and the other in the cheek (*supra-* and *infratemporal fossae*); this diapsid skull type with various modifications, is found in some reptiles, such as tuatara and crocodilians, and originally in ancestral birds but not seen any more in modern birds. *Cf.* anapsid, parapsid, synapsid.

Diapsida, diapsids; a reptilian subclass characterized by a skull type with two temporal openings; includes most of the living reptiles, such as snakes, lizards, tuatara, and crocidilians. *Cf.* Anapsida, Synapsida.

diarch (Gr. *arche* = origin), a root structure in dicotyledons with disc-like primary xylem (in cross-section) with phloem on the other sides in the central cylinder (stele).

diaspore (Gr. *diaspora* = spreading), **1.** a part of a plant, e.g. a spore, seed, or fruit, by which the plant can disperse to new areas; *syn.* propagule; **2.** a mineral, α-aluminium hydroxide, α-AlO(OH).

diastase, → amylase.

diastema, pl. **diastemata** (Gr. = interval), a space, cleft, or fissure between two adjacent anatomical structures, e.g. two adjacent teeth.

diastole (Gr. = dilatation), the relaxation of the cardiac muscle; the dilatation of the atria and ventricles due to the filling of the heart with blood; alternates rhythmically with the → systole. *Adj.* **diastolic**.

diathermy (Gr. *dia* = through, *therme* = heat), the local heating of body tissues by electric current, ultrasonic waves, or microwaves; used in surgery and thermal therapy.

diatomaceous earth, a soil type formed mainly from shells of → diatoms which have deposited in water.

diatoms, unicellular algae characterized by a double shell rich in silica. *See* Bacillariophyceae.

dicamba, a benzoic acid derivative used as a herbicide.

dichasium, pl. **dichasia** (Gr. *dichazein* = to divide into two), an inflorescence (dichasial → cyme) in which the main axis forms a flower when its growth ceases, but two lateral branches continue their growth branching further into two, etc. *Cf.* monochasium.

dichlamydeous (Gr. *dis* = twice, *chlamys* = mantle), a flower having two whorls in its perianth, i.e. separate calyx and corolla. *Cf.* monochlamydeous.

dichloromethane, methylene chloride, CH_2Cl_2; a synthetic, colourless volatile liquid used as a solvent for non-polar and polar compounds, in gas chromatography, as an anaesthetic, and a refrigerant.

2,4-dichlorophenoxyacetic acid, 2,4-D, a synthetic plant hormone with auxin-like activity; used as a herbicide.

dichogamy (Gr. *dicha* = in two, *gamos* = marriage), the ripening of stamens and gynaecium at different times; ensures cross-pollination.

dichoptic (Gr. *ops* = eye), pertaining to a type of insect compound eye in which the eyes are separate from each other. *Cf.* holoptic.

dichotomous venation, a venation type in plant leaves; in dichotomous veining the leaf veins branch into two equal parts and do not fuse with each other after the branching. The type is primitive, typical of some ferns and seed plants e.g. of *Ginkgo biloba*, which is a "living fossil".

dichotomy (Gr. *temnein* = to cut), the division into two parts, groups or classes, as of an anatomical structure or a phylogenetic developmental line into two equal branches. The structural dichotomy is common in lower plants, e.g. in algal thalli. In vascular plants structural division takes place through development of → axillary buds and there is no dichotomy, but sometimes **pseudodichotomy** in which two branches develop at the same time from two buds and the result resembles dichotomous branching. **Adj. dichotomous**, also **dichotomal**.

dicolpate (Gr. *di* = two, *kolpos* = vagina), a

pollen grain with two → germ pores; also mono- and tricolpate types are common.

Dicotyledonae, (Gr. *kotyle* = cup), **Magnoliatae, dicotydelons**; a class in subdivision Angiospermae, division Spermatophyta, Plant kingdom; in some classifications the class Dicotyledones in the division Anthophyta. These plants have two cotyledons (seed leaves) in their embryo, the leaf venation is mostly net-like, the vascular bundles in the shoot locate in a ring, secondary growth caused by the function of cambium is typical, and the numbers of flower parts are multiples of four or five. The other developmental line of Angiospermae is represented by → Monocotyledonae.

Dictyoptera (Gr. *diktyon* = net, *pteron* = wing), **dictyopterans**; an insect order including e.g. cockroaches and mantids; are animals with generalized biting mouthparts, the anterior wings are narrower and stouter than the hindwings which are more membranous and fold under the forewings; eggs are laid in an ootheca. In some classifications dictyopterans are placed in the order Orthoptera.

dictyosome (Gr. *soma* = body, bit), an organelle in plant cells, composed of stacks of flat circular cisternae, each bound by a unit membrane. The number of the cisternae is mostly 2—7, sometimes more. The cisternae pinch off small vesicles at their margins; the vesicles may contain secreted materials, which is accumulated in the → vacuoles of the cell, or they may contain material for the primary → cell wall in dividing or growing cells. Dictyosomes occur throughout the cytoplasm; the totality of them is analogous to the → Golgi complex or Golgi apparatus of animal cells.

dictyostele (Gr. *stele* = column), one of different → stele types in plants.

didactyl, didactylous (Gr. *dis* = twice, *daktylos* = finger, toe), **two-toed**; pertaining to a leg with two fingers, toes, or claws. *Cf.* pentadactyl.

diecious, → dioecious.

diel (L. *dies* = day), pertaining to a 24 hour period, i.e. a day and night; → circadian.

dieldrin, a crystalline, water-insoluble naphthaline derivative, $C_{12}H_8OCl_6$, used as an insecticide.

diencephalon (Gr. *dia* = through, *enkephalos* = brain), the second of the five brain parts in adult vertebrates, i.e. the posterior part of the forebrain (embryonic prosencephalon) located between the telencephalon (cerebrum) and mesencephalon (midbrain); comprises the epithalamus, → thalamus, subthalamus, → hypothalamus, and the third brain ventricle. Called sometimes betweenbrain or interbrain.

diestrus, → dioestrus.

diethylaminoethyl-, DEAE-, DEA-, a group, $(C_2H_5)_2NCH_2CH_2$-, that gives a positive charge to cellulose or dextran (Sephadex) which can be used as an anion exchange material in chromatography and gel filtration.

diethyl ether, *see* ether.

diethyl pyrocarbonate, DEP, DEPC, pyrocarbonic acid diethyl ester; a RNase inhibitor used to prevent the degradation of RNA.

differential centrifugation, *see* centrifugation.

differentiation, the act or process of making different, i.e. differentiating, as differentiation of cells, tissues, and organs in a developing organism; consists of those processes which distinguish the daughter cells from the mother cell during development. The processes are based on increase or decrease of specific metabolic abilities (enzymes) during the specialization of the cells.

In multicellular organisms, differentiation is the most significant feature of → development including the formation of different cells and tissues; its basis is the **regulation of gene function** in a cell that begins differentiation in a certain direction. This occurs by the activation of the group of those genes which encode the synthesis of proteins specific to specialized requirements of that particular cell type.

Differentiated cells are not only morphologically different, but they are also characterized by certain functional differences. These differences and changes are consequences of the function of certain → **morphogens** acting within and outside the cell. Differentiation always reflects changes in biochemical functions of cells, even though it can be studied in cell organizations that differ structurally. Changes in the biochemical functions of the cells are due to specific regulatory factors which affect the transcription of genes, the behaviour of the transcription products during their processing into → messenger RNA, the translation of messenger RNA, and the function of the products of translation (proteins).

Through changes in cellular structure and function, differentiation leads to the formation of different tissues, i.e. the organization of differentiated cells into entities. This process of events leads to **morphogenesis**, the formation of organs and structural entities of a certain kind. Differentiation is a two-step process. First → **determination** occurs, during which a route of differentiation is determined; the second step is **actual differentiation**, the specialization of the cell for a given function. During differentiation the daughter cells become distinct from the mother cells by losing certain characters and gaining some new. Differentiation is guided by the microenvironment of the cells (neighbouring cells and tissues) which at each stage provides an → **induction** to develop in a given direction. Thus, guided by the regulatory chain of inductive interactions, the development of cells is determined to a given developmental pathway which begins from early embryonic stages and ends in the formation of mature structures, and eventually in → regeneration of tissues.

differentiation zone, the area near the tip of a plant root in which the cells formed in the division zone of the tip differentiate, forming dissimilar tissues.

diffraction (L. *dis-* = apart, *frangere* = to break), in physics, the spreading of waves, particularly electromagnetic waves (e.g. light waves) when they pass through an aperture or by the edge of an opaque obstacle. The secondary waves thus formed interfere with the primary waves and give rise to a diffraction pattern, i.e. a series of concentric light and dark bands of constructive and destructive interference.

diffuse competition, *see* competition.

diffusion (L. *diffundere* = to spread out), the act or state of being spread out, i.e. diffused; the mixing of substances in a gas, liquid, or other media based on random movements of atoms, ions, or molecules by thermal motion (Brownian motion). In the process the concentration differences of substances disappear in the media. The diffusion events are expressed in → Fick's law. *Adj.* **diffusible**, capable of diffusing. *See* cell membrane transport.

Digenea (Gr. *dis* = twice, *genea* = race), **digenean trematodes**, e.g. liver flukes, blood flukes; a large subclass (order in some classifications) in the class → Trematoda, phylum Platyhelminthes, comprising endoparasitic worms whose intermediate host is a mollusc and the definite host (final host) a vertebrate; they diverge to inhabit nearly all organs in their hosts. *See* Schistosoma, schistosomiasis.

digestion (L. *dis-* = apart, *gerere* = to carry), the conversion of organic food material into smaller molecules which can be absorbed and used in cellular metabolism of animals; occurs mainly by the action of → digestive enzymes. The digestion of protozoans, sponges and coelenterates is **intracellular**, i.e. the animals take food phagocytotically directly into cells and digest it with intracellular lysosomal enzymes (*see* phagocytosis). Even some protozoans have special digestive organelles, such as the food vacuole, cytostome, and cytopharynx; multicellular animals have specialized cells for intracellular digestion. In some groups of primitive worms and molluscs, digestion may still be partly intracellular, but in other animals it is **extracellular** with differentiated digestive organs and fluids containing hydrolytic enzymes that sometimes are excreted outside directly on the prey, e.g. in some insects, mites, and earthworms. Usually gastrointestinal microorganisms participate in digestion, especially decomposing cellulose.
Extracellular digestion is found also in some plants, i.e. → carnivorous plants. *Adj.* **digestive**. *See* alimentary canal, alimentary glands, bile.

digestive enzymes, hydrolytic enzymes found in the alimentary canal of most multicellular animals, or intracellularly in → lysosomes of more primitive animals and certain microorganisms (*see* digestion). Digestive enzymes of multicellular animals are produced by many types of glands, such as the salivary and gastric glands, pancreas, hepatopancreas, and intestinal glands. Digestive enzymes are secreted for digestion in the alimentary canal, but similar enzymes are also found in the lysosomes of other cells in tissues. The digestive enzymes belong to → proteinases (e.g. pepsin, trypsin, and different aminopeptidases and carboxypeptidases), to → carbohydrases (amylase, saccharase, lactase etc.), and to → lipases (gastric lipase, pancreatic lipase, etc.). For digestion of cellulose most animals use symbiotic microorganisms which live in their alimentary canal, and only a few invertebrate species (e.g. silverfish) secrete their own cellulase enzymes for digesting cellulose.

digit (L. *digitus*), a finger or toe. *Adj.* **digital**, pertaining to fingers or toes, or to numerical methods; **digitate**, finger-like.

digitate veining, *see* phyllotaxy.

dihybrid (Gr. *dis-* = twice, double), an individual heterozygous for two → allele pairs. *See* heterozygote.

dihybrid cross, a cross in which the individual organisms to be crossed differ from each other by two pairs of hereditary characters to be studied. In the F_1 generation all progenies are similar and manifest both → dominant characters. If → linkage does not occur, the numerical proportions of the progenies in the F_2 generation (*see* F) are the following: nine which have both dominant characters, three which have the first dominant and the second → recessive character, three which have the first recessive and the second dominant character, and one which has both recessive characters (Mendelian ratio 9:3:3:1). Of the genes regulating the characters, the dominant and recessive → alleles appear in the cross, e.g. AABB - aabb or AAbb - aaBB.

dihydroxyacetone phosphate, an intermediate of the glycolysis of cell respiration and the Calvin cycle of → photosynthesis.

dikaryon, also **dicaryon** (Gr. *karyon* = nut, core, nucleus), a stage in the life cycle of ascomycetes and basidiomycetes which can be a binucleated cell, spore, or mycelium formed from binucleated cells. A dikaryon can be homokaryotic, i.e. the two nuclei of the cells are identical, or it may be heterokaryotic, i.e. the two nuclei of the cells are genetically different. *Adj.* **dikaryotic**.

dilatation, also **dilation** (L. *dilatare* = to expand), the conditon or act of a tubular or alveolar organ of being expanded or dilated; e.g. **1.** in animals, vasodilatation (dilatation of blood vessels), or bronchodilatation (dilatation of bronchi of the vertebrate lungs); a factor, such as a muscle which causes the dilatation, is called **dilator**; **2.** in plants, dilatation means lateral, structural enlargement; e.g. increase of cells in primary cortex through anticlinal cell divisions, or enlargement of secondary rays in the area of the secondary phloem in many trees.

diluent, 1. diluting; **2.** an agent that dilutes. *Verb* to **dilute**.

dilution (L. *diluere* = dilute), **1.** the act of diluting, i.e. reducing or weakening in concentration, strength, or quality; **2.** the state of being diluted, e.g. a diluted liquid or mixture.

dimer (Gr. *dis* = twice, *meros* = part), the combination of two subunit molecules, monomers; **homodimer**, with similar monomers, **heterodimer**, with different monomers. *Adj.* **dimeric, dimerous**.

dimethyl sulphoxide, DMSO, an organic liquid, C_2H_6OS, used in biological studies as a solvent to facilitate the absorption of different compounds into cells; artificially renders the cells tolerant to freezing and is therefore used for storing of cells and tissues in deep freeze. DMSO denatures partially DNA and is therefore slightly toxic to cells. Also called methylsulphoxide.

dimictic (Gr. *miktos* = mixed), *see* stratification.

dimorphism (Gr. *morphe* = shape), a condition in which individuals of a species exist in two different forms. Dimorphism may be associated with any characteristic, such as the anatomical or morphological structures, physiological and ethological features etc.; in plants seen e.g. as → heterophylly. In **sexual dimorphism** the sexes are different; also the individuals of the same sex may be different, like the queens and sterile workers of bees. **Dimorphism in division of labour** occurs e.g. among ants between the castes of workers and soldiers. **Seasonal dimorphism** is found in many lepidopterans with different summer and winter forms; some species have different generations (→ alternation of generations). *Cf.* polymorphism, genetic polymorphism.

2,4-dinitrophenol, DNP, Dnp, a toxic compound that in cells can act as an uncoupler of → oxidative phosphorylation (ATP production) from the respiratory chain reactions. Oxygen consumption of cells or of an organism usually increases after DNP treatment, but the cellular metabolism is in imbalance because ATP is not formed. Some DNP derivatives are used in biochemical studies of oxidative processes, as haptens in experimental immunology, and as herbicides (e.g. dinoseb) to inhibit the photosynthetic phosphorylation.

Dinoflagellata (Gr. *dinein* = to rotate, L. *flagellum* = whip), **dinoflagellates**; a protozoan class in the phylum Sarcomastigophora comprising unicellular flagellate organisms in marine and freshwater planktons.

They are either autotrophic (often classified as → Pyrrophyta or Dinophyta in botany), or heterotrophic animals; some species are both. The cell body with two flagella is covered by a firm cellulose armour formed from two to many plates. A known species is the luminescent *Noctiluca scintillans*.

Dinophyta (Gr. *phyton* = plant), *see* Dinoflagellata.

dinosaurs (Gr. *deinos* = terrible, *sauros* = lizard), extinct reptiles of two orders, Saurischia and Ornithischia; both of them became extinct during Cretaceous. Saurischia included four-footed herbivores, such as the huge *Apatosaurus* (previously called *Brontosaurus*), or bipedal carnivores such as *Tyrannosaurus*, the most ponderous flesh eater ever lived on the Earth. Ornithischia comprised four- or two-footed herbivores. Dinosaurs included the largest terrestrial animals ever lived; e.g. *Apatosaurus* living in the wet bogs of North America was about 20 m long and 4.5 m high, and weighed about 90 tons.

dioecious (Gr. *di-* < *dyo* = two, *oikos* = house), pertaining to a plant in which the unisexual flowers are on separate individuals; e.g. willow species. *Cf.* monoecious.

dioptre, diopter (Gr. *dioptron* = mirror), **1.** an optical instrument for observing distant objects and for measuring their size; **2.** a unit (symbol D) of measure of the refractive power of lenses; the refractive power in dioptres is the reciprocal of the focal length in metres (e.g. if the focal length is 0.5 m, the refractive power is 2 D).

diorama (Gr. *dioran* = to see through), a museum scene imitating native environments for plants and animals on show.

dioxin, a heterocyclic hydrocarbon, 2,3,7,8-tetrachlorodibenzo-*p*-dioxin (TCDD); one of several related compounds (dioxins), all extremely toxic to organisms, i.e. superpoisons which have caused severe environmental pollution. In animals, dioxins can disturb genetic transcription, causing e.g. damage in embryonic development and tumours in adults. Dioxins are released to the environment as a by-product in the manufacture of some pesticides, in traffic, and in some other combustion processes. *See* chlorinated hydrocarbons.

diphenylamine, N-phenylbenzeneamine, $(C_6H_5)_2NH$; a synthetic crystalline organic compound, soluble in e.g. alcohol, glacial acetic acid, and ether; insoluble in water. It is used for the detection of NO_3, Cl_2, and other oxidizing substances, and for the quantitative determination of DNA.

1,3-diphosphoglycerate, an intermediate in the → glycolysis of cell respiration and the Calvin cycle of → photosynthesis.

diphycercal (Gr. *diphyes* = twofold, *kerkos* = tail), *see* caudal fin.

diphyletic (Gr. *di-* < *dyo* = two, *phyle* = family), pertaining to a taxon which is phylogenetically developed from two founder lines.

diphyodont (Gr. *diphyes* = twofold, *odous* = tooth), pertaining to, or being a vertebrate species which has deciduous (milk) and permanent sets of teeth, as man.

diplobiont (Gr. *diploos* = double, *bioun* = to live), diplohaplontic organism, diplohaplont; an organism with → alternation of generations, i.e. has a → haploid and diploid stage; e.g. most plants.

diploblastic (Gr. *blastos* = germ, shoot), a term pertaining to the animals, e.g. coelenterates, whose body wall consists of only two cell layers, i.e. the ectoderm and endoderm.

diplohaplont, → diplobiont.

diploid (Gr. *eidos* = form), pertaining to a cell, organism, or stage in the life cycle of an organism in which the nucleus contains two homologous sets of chromosomes; of these one (paternal) is derived from the father and the other (maternal) from the mother. In a diploid cell each chromosome exists in two copies. The symbol of the diploid chromosome number is 2n. The other possibilities are → haploid (n) or → polyploid chromosome number. A diploid cell is generated when two haploid gametes unite in → fertilization (**diploidy**), and hence, such a holozygote (*see* zygote) is formed in which the whole → genome exists in two copies. In viruses and bacteria, forms of diploidy are known which involve only part of the genome; such individuals are called merozygotes. These can be formed in different ways and are usually labile.

diploid apogamy, a form of → agamospermy in which the → sporophyte is produced from a somatic, i.e. diploid, cell of the → gametophyte.

Diplomastigomycotina, *see* Mastigomycota.

diplont, 1. an organism having a life cycle in which the products of → meiosis behave directly as gametes, i.e. only the gametes are haploid, all somatic cells being diploid; the diplonts include all diploid multicellular animals and some plants, and have no → alternation of generations; **2.** → diplophase.

diplophase (Gr. *phasis* = emergence), a stage of the life cycle of a cell or an organism during which the chromosome set is double. Also called diplont. *See* diploid. *Cf.* haplophase, haplont.

Diplopoda (Gr. *pous*, gen. *podos* = foot), **diplopods, millipedes**; a class (in some classifications subclass) of arthropods comprising animals which have two pairs of appendages in each abdominal somite due to the fusion of primary somites in pairs.

diplospory (Gr. *sporos* = seed), *see* agamospermy.

diplotene (Gr. *tainia* = band), *see* meiosis.

Dipnoi, Dipneusti (Gr. *dis* = twice, *pnein* = to breathe), **lungfishes, dipnoans**; a group of primitive fishes in the subclass Sarcopterygii breathing both with gills and primitive lungs. Lungfishes (six species) are found in tropics of Australia (order Ceratodiformes), in Africa and America (order Lepidosireniformes).

dips(o)- (Gr. *dipsa* = thirst) denoting relationship to thirst; e.g. **dipsogen**, an agent inducing thirst; in the form **-dipsia** used as a word ending.

Diptera (Gr. *dis* = twice, *pteron* = wing), **true flies, two-winged flies, dipterans**; an order of holometabolous insects comprising about 120,000 species in two suborders, Nematocera and Brachycera; their hindwings are reduced to minute knobbed halteres, acting as balancing organs; the mouthparts are specialized for sucking, sponging, lapping, or piercing. The order includes e.g. **mosquitoes, flies, midges**, and **gadflies**. Many species are of medical importance acting as disease vectors, many are beneficial as pollinators, and as parasitoids in harmful invertebrates.

directional selection, a type of selection in a population that favours a single optimum in gene frequency; operates in a progressively changing environment and leads to a new state of adaptation.

disaccharide (Gr. *di-* < *dyo* = two, *sakcharon* = sugar), a sugar molecule formed from two monosaccharides; e.g. **sucrose** (saccharose) consisting of one glucose and one fructose residue; other common disaccharides are **maltose** (two glucose residues) and **lactose** (one glucose and one galactose residue).

disc flower, a tubular flower in the flower head in the middle of the capitulum; occurs in many species of the family Asteraceae, e.g. in the daisy.

discus, pl. **disci** (L. < Gr. *diskos*), **disc** (disk); a circular, flat plate in anatomical nomenclature, e.g. **1.** *discus articularis*, an articular disc composed of a dense fibrous tissue and found in some joints in vertebrates; **2.** *discus intervertebralis*, an intervertebral disc (cartilage) between the bodies of adjacent vertebrae; **3.** a disc-like → nectary in certain plants, as in *Acer* species (maple).

disc-like receptacle, a disc-like apex of the flower stalk; common in the families Asteraceae and Cichoriaceae. *Cf.* receptacle.

discordance, in genetics, the difference between groups or pairs, especially twins, concerning a given character. *Cf.* concordance.

discrete generations, successive generations which are clearly isolated from each other, i.e. a preceding generation is dead when the next one is born. However, usually the final life cycle stages of the preceding generation and the first ones of the succeeding generation occur simultaneously; e.g. in annual plants.

disinfection (L. *dis-* = apart, *inficere* = to stain, infect), the process of destroying pathological microorganisms e.g. from rooms, clothing, animal fur, etc.; **disinfectant**, an agent or factor, as a chemical used for disinfection or freeing from infection. *Verb* to **disinfect**.

disjunction (L. *disjungere* = to separate), loosening; **1.** in genetics, the separation of daughter chromosomes during cell division, i.e. at the anaphase of → mitosis or during the second meiotic division of → meiosis; also the segregation of the paired chromosomes during the anaphase of the first meiotic division; **2.** in biogeography, usually a small isolated area out of the main distribution area of an organism.

dismutases (L. *dis* = apart, *mutare* = to change), a group of enzymes catalysing the reaction (**dismutation reaction**) involving two identical molecules of which one gains what the other loses; dismutases e.g. produce different states of oxidation (e.g. → superoxide dismutase) or different states of phosphorylation (e.g. glucose 1-phosphate phosphodismutase).

dispersal (L. *dispergere* = to disperse), the

migration of individuals to new areas between two breeding seasons (**breeding dispersal**) or between the birth and the first breeding period (**natal dispersal**). Dispersal may occur either by active movement or by passive floating. Also, it may be spontaneous or forced e.g. in consequence of dominance hierarchy. Dispersal results in dispersion.

dispersal barriers, geographical structures, such as oceans, mountains, and deserts, which prevent dispersal of organisms to new areas. Also large vegetation zones may be barriers to some organisms.

dispersal polymorphism (Gr. *polys* = many, *morphe* = form), a phenomenon when two or more types of dispersal structures are found in a populaton or among the progeny of an individual; e.g. winged and wingless forms of some aphids when the former colonize new host plants.

dispersion, 1. the dispersion of a substance inside an organ system, often attached to endogenous substances, like many drugs in the body; **2.** a colloidal suspension where particles are evenly mixed or dispersed in another substance; **3.** in ecology, dispersion (**distribution**) denotes a spatial pattern of individuals, pairs, and sometimes also greater groups. Three main types are found: **even dispersion** (**regular distribution**) when the distances between individuals or groups are even, **random distribution** when the distances are random, and **aggregated distribution** (**clumped distribution**) when the individuals occur unevenly in groups.

displacement activity, a behavioural pattern of an animal, irrelevant to its normal functional significance; occurs often in conflict situations when two or more opposing drives affect simultaneously, such as attack and flight during a fight. Displacement activities are common also in situations where motivation is strong but the action, normally released by it, is somehow arrested. Thus the animal may act in a way not normally relevant to the motivation: e.g. a bird pursued by a hawk begins to sing, or fighting gulls suddenly rip grass from the ground.

displacement loop, D-loop, in covalently circular DNA, a region in which a short stretch of RNA is paired with one strand of DNA, displacing the original partner DNA strand in this region. D-loops are early intermediates in mitochondrial DNA replication (displacement replication).

display, behaviour of animals including several → fixed action patterns; associates with different communications between individuals, e.g. in courtship and territorial behaviour. Often a **displayer** exhibits conspicious structural signals, such as a richly coloured plumage, strong antlers, etc. The signals may be also olfactory, acoustic, or tactile.

disruptive colouration, a distinct colour pattern which disrupts the outline of an animal and thus makes it difficult for predators to see.

disruptive selection, a form of → natural selection in which the selection in a heterogeneous environment favours two or more → genotypes which usually represent the extreme ends of the variation, i.e. the average genotypes are not favourable. Disruptive selection leads to maintenance of balanced → polymorphism in the population.

dissection (L. *dis-* = apart, *secare* = to cut), the act or state of cutting apart or dissecting; e.g. the cutting of an organism for anatomical observation. *Adj.* **dissective**.

dissimilation (L. *dissimilare* = to make unlike), the act of making or a process of becoming dissimilar; e.g. decomposition of a substance into simpler compounds, as occurs in catabolic reactions. *Cf.* assimilation.

dissociation (L. *dis-* = apart, *sociatio* = union), disassociation, separation; e.g. **1.** the rupture of an anatomic structure; **2.** in chemistry, the decomposition of a compound into ions, smaller molecules, or radicals, e.g. by dissolving or by heating. *Cf.* association.

dissolution (L. *dissolvere* = to dissolve, loosen), **1.** the process of dissolving one substance in another, such as in liquid or gas; **2.** the splitting of a substance into its components by chemical action; **3.** the process of loosening, as death; **dissolvent**, a solvent medium. *Verb* to **dissolve**.

distal (L. *distans* = distant), outer, remote; situated farther away from a point of reference. *Cf.* proximal.

distemper (L. *dis-* = apart, *temperare* = to mix), **1.** canine distemper, a fatal disease of dogs and other canines, caused by the canine distemper virus; **2.** feline distemper (panleukopenia), a similar disease in cats and other felines caused by the feline panleucopenia virus.

distichous (Gr. *distichos* = in two rows), *see*

phyllotaxy.

distillation, a procedure in which a liquid is heated and converted into its vapour state and then condensed back into liquid (distillate); used to separate a mixture of liquids into pure liquid fractions with different boiling points, or to separate solvents from non-volatile substances.

distribution, 1. a spatial range of a species or other systematic taxa in a certain area, e.g. in Europe or the whole world; **2.** → dispersion.

distyly (Gr. *stylos* = column), a type of → heterostyly of flowers having two kinds of styles of different lengths with the stigmata on two different levels.

disulphide bond, a chemical bond (-S-S-) formed from two sulphydryl groups (-SH groups), usually of two cysteine residues; occurs in proteins between polypeptide chains, strengthening the tertiary and quaternary structures of → protein.

disymmetric flower, pertaining to a flower with two planes of symmetry. *Syn.* bilateral flower. *Cf.* asymmetric flower, dorsiventral flower, symmetric flower.

dithionite (*dithio-* = a chemical group with two atoms of sulphur, $-S_2-$), a salt of hyposulphurous acid, $H_2S_2O_4$; a strong reducing agent used generally in biological and biochemical research.

dithiothreitol, DTT, threo-1,4-dimercapto-2,3-butanediol, $C_4H_{10}O_2S_2$; an organic compound which breaks down disulphide bonds in protein molecules forming -SH groups. DTT is therefore used as an antioxidant in protein analyses and extraction media for cell organelles. *Syn.* Cleland's reagent, dierythritol.

diuresis, pl. **diureses** (Gr. *dia* = throughout, complete, *ourein* = to urinate), the excretion of urine, especially urine production in abnormally large quantities; **diuretic**, promoting the urine excretion, or an agent or factor that induces an increase in urine production.

diurnal (L. *diurnus* = daily < *dies* = day), pertaining to a day (not night); each day, daily; occurring during the day-time; **diurnal rhythm**, the day-time activity rhythm of an organism. *Cf.* nocturnal.

diuron, 3-(3,4-dichlorophenyl)-1,1-dimethylurea; a chlorinated urea derivative that interferes with electron transport in photosynthesis and inhibits → photophosphorylation; used in experimental research and as a herbicide.

divergence (L. *dis-* = apart, *vergere* = to tend), a spreading or parting, or radiating in different directions, e.g. the spreading of axon terminals of a neurone to form synapses with several other neurones. *Adj.* **divergent**. *See* divergent evolution. *Cf.* convergence.

divergent evolution, a type of → evolution in which a single population is split into two (or more) reproductively isolated populations where genetic differences gradually accumulate, and in the course of time may lead to speciation.

diversity, *see* biodiversity, species diversity, species richness.

diversity-stability hypothesis, a hypothesis describing the correlation between the stability of a community and species diversity; a community with high diversity is structurally and functionally more stable, and the fluctuations of population densities are lower than in communities with a low number of species.

diverticulum, pl. **diverticula** (L. *divertere* = to turn aside), a sac or pouch occurring normally or abnormally in the structure of an organism or organ; e.g. colonic diverticula, abnormal hernial protrusions of the mucous membrane of the colon.

diving, submersion of many air-breathing vertebrates, varying considerably in depth, duration, and exercise level; occurs e.g. in amphibians, and many reptiles, birds and mammals. Some whales can dive to depths of more than 1,500 m and stay submerged for over an hour. These animals are phylogenetically adapted to aquatic life having special arrangement of circulatory and respiratory systems; in mammals e.g. marked bradycardia and reduced cardiac output, rearrangement of blood flow selectively to the heart, brain, and some endocrine organs, high myoglobin content in muscles, large blood volume with high haemoglobin content, complete lung compression (air entering large rigid trachea and bronchi), and e.g. in whales a special substance to bind nitrogen of the breathing air. Without these adaptations, e.g. man suffers from decompression sickness, because nitrogen dissolves in body fluids resulting in formation of bubbles in rapid ascent.

diving reflex, a nerve reflex found in many aquatic reptiles, birds and mammals, causing cessation of breathing automatically at the beginning of diving. The reflex is activated by sensory receptors in the facial skin, and holding of breath further results in slowing of the heart rate. The reflex is active also in man at the early postnatal age, and occurs weakly still in adults.

division, the main group in taxonomy of plants, corresponding to the **phylum** in classification of the Animal kingdom.

division zone, the area in the plant root tip where cells are dividing actively.

dizygotic, pertaining to twins derived from two → zygotes, i.e. who arise from different ova fertilized by separate spermatozoa, in contrast to → monozygotic twins who develop from the same fertilized egg. **Dizygotic twins** are genetically as similar and as different as ordinary siblings; half of their genes are identical. The zygosity of twins can be verified by two methods: by investigating the foetal membranes and by comparing the similarities or → concordances and differences or discordances of the twins. *Syn.* fraternal twins.

D-loop, → displacement loop.

DNA, → deoxyribonucleic acid.

DNA binding protein, a protein able to bind a certain sequence of DNA, thus regulating the function of the gene(s) on the binding site in a chromosome. The majority of these are → transcription factors, but also TATA-binding factors (*see* TATA-box), → chaperones, and → tumour necrosis factors belong to these proteins. They contain a specific DNA-binding domain, such as → homeodomains in proteins encoded by → homeogenes.

DNA cloning, the artificial multiplication of a DNA sequence, usually involving the isolation of appropriate DNA fragments and their *in vitro* joining to a cloning → vector capable of replication when introduced into an appropriate host cell.

DNA fingerprint, a structural label of deoxyribonucleic acid (DNA) of a human individual, based on the VNTR polymorphism of the DNA structure; becomes visible by using gel electrophoresis. VNTR polymorphism in man is so wide that every individual (with the exception of identical twins) can be identified on the basis of their DNA fingerprints. The method of DNA fingerprinting uses DNA probes which hybridize to hypervariable → minisatellite regions of e.g. human DNA. The probe hybridizes to any fragment that contains members of the particular set of minisatellites represented by the probe. *See* variable numbers of tandem repeats, VNTR.

DNA footprint, a segment of deoxyribonucleic acid (DNA) which becomes observable when DNA is split with → deoxyribonuclease, DNase. A certain protein recognizes this segment, attaches to it, and thus shields it from DNase cleavage.

DNA fragment (L. *fragment* = piece), a part of → deoxyribonucleic acid (DNA) which has been accomplished by cleaving DNA into pieces by → restriction enzymes.

DNA gyrase, an enzyme of *Escherichia coli* that catalyses negative supercoiling of closed duplex DNA.

DNA hybridization, 1. a method used to determine the similarity of genomes from different species; preparations of single-stranded deoxyribonucleic acid (DNA) from two species are brought together (DNA hybrid), then the number of double-stranded DNA formed and their speed of formation, are measured; **2.** a method used in → Southern blotting; the existence of a given DNA sequence is demonstrated in a preparation using a → DNA probe.

DNA library, a store of cloned DNA in recombinant cloning → vectors containing chromosome-specific DNA fragments. *Cf.* gene library.

DNA ligases (L. *ligare* = to bind), → ligases catalysing the joining of breaks in the strand of deoxyribonucleic acid (DNA). Several DNA ligases, EC sub-subclass 6.5.1., are found in cells, and they all repair breaks in the DNA double helix only if the break has a 5' phosphate and a free 3' hydroxyl group. DNA ligases probably play an important role in → genetic recombination, in the repair synthesis of errors of DNA, and in DNA → replication.

DNA methylation, the addition of a methyl group onto the cytosine bases of deoxyribonucleic acid (DNA), catalysed by methylase enzymes; the methylated DNA is inactive. DNA methylation is cell-heritable, and hence a stable form of DNA inactivation; e.g. in mammals, DNA in the → inactive X chromosome is methylated, and normal embryologic development depends on cytosine methylation.

DNA polymerases, DNA pols (Gr. *poly* = many, *meros* = part), enzymes of the sub-subclass EC 2.7.7., catalysing the synthesis of a deoxyribonucleic acid strand (complementary DNA strand) from deoxyribonucleoside triphosphates, using DNA as a template (RNA in retroviruses). Eukaryotic cells have alpha, beta and gamma polymerases. Alpha polymerase is responsible for the replication of nuclear DNA, beta for repair functions, and gamma for the replication of mitochondrial DNA. DNA polymerases I, II and III are known in *Escherichia coli*, polymerase III being most important in genome replication. Polymerase I is involved in → DNA repair mechanisms, and also widely used in genetic engineering and gene manipulation, and in converting a single-stranded DNA into the double-stranded form or a single-stranded → complementary DNA (cDNA) into duplex cDNA.

DNA probe, a → probe consisting of deoxyribonucleic acid (DNA), i.e. a defined and fairly short DNA sequence which is labelled e.g. with a radioactive isotope; is used in → DNA hybridization to detect a complementary DNA sequence, usually a part of a gene.

DNA repair mechanisms, any enzyme-mediated process that removes lesions from DNA and/or restores the intact DNA molecule; usually results in a correct (**error-free**) DNA repair which does not alter the original information of the DNA molecule concerned, but sometimes is accompanied by a modification of the base sequence, resulting in a → mutation (**error-prone repair**). → DNA ligases and DNA polymerases are important enzymes in repair.

DNA replication, the direct copying of a → deoxyribonucleic acid molecule (DNA) resulting in the duplication of genetic material in cells. Replication requires the concerted action of many proteins and enzymes which include e.g. → DNA polymerase, DNA topoisomerase, and DNA ligase. In DNA replication the leading strand and the lagging strand of the double helix of DNA are distinguished. Copying of the leading strand is contiguous and occurs in 5'→3' direction while in the lagging strand the copying is discontiguous involving the synthesis of short intermediates, the → Okazaki fragments. DNA replication is always **semiconservative**, i.e. both strands of the double helix serve as templates for the synthesis of a new double-helical molecule, and it proceeds in the form of → replication fork, i.e. the two strands of DNA separate to form a fork of the leading and lagging strands.

DNase, → deoxyribonuclease.

DNase I-hypersensitive site, a short segment of → chromatin which is observed on the basis of its sensitivity to the cleaving activity of DNase I (→ deoxyribonuclease) and other nucleases; a site of chromatin from which the → nucleosomes have been removed, and thus DNA occurs naked. Active genes are in this state, accordingly being DNase I-hypersensitive. The DNase-hypersensitive sites occur in promoter regions of active genes where transcription factors are bound.

DNA sequence (L. *sequens* = the following), the order of nucleotides in deoxyribonucleic acid (DNA).

DNA sequencing, the determination of the order of nucleotides in a DNA molecule. Two methods for DNA sequencing dominate: the enzymatic dideoxynucleotide method of *Sanger et al.* and the chemical method of *Maxam* and *Gilbert*. Both consist of several common procedures, as of the selection of DNA clones and the treatment of these clones individually by base-specific reactions. The methods further include the separation of the reaction products by size on sequencing gels, and the computer-aided reading of separated bands as sequence of adenine, cytosine, guanine, and thymine bases.

DNA specificity, the order of nucleotide pairs in deoxyribonucleic acid (DNA); maintained in → replication due to the → base pair rule.

DNA theory of inheritance, a theory stating that → genes are chemically constituted of deoxyribonucleic acid (DNA). The theory is universally true with the exception of → RNA viruses.

DNA topoisomerase, an enzyme group catalysing the breaking and rejoining of DNA strands in cell nuclei, thus altering the topology of DNA. The enzymes are involved in processes such as → DNA replication, genetic replication, and genetic transcription.

dolichol, a terpenoid consisting of 13 to 24 isoprene units; acts as a transmembrane carrier, e.g. in the endoplasmic reticulum in cells, for glycosyl units in the biosynthesis of glycolipids and glycoproteins.

doliolaria (L. *dolium* = wine-cask), a last larval stage before adult metamorphosis in holothurians developing from the auricularia;

the barrel-shaped body resembles the vitellaria of crinoids.

Dollo's law (*Louis Dollo*, 1857—1931), a biological law stating that evolutionary changes are irreversible.

dolomite, a double carbonate of calcium and magnesium, $CaMg(CO_3)_2$; a common mineral, used to reduce the acidity of soil and as a calcium-magnesium fertilizer.

domain (L. *dominium* = proprietary rights), a solid, structurally distinguishable region of a macromolecule, like of polypeptide or protein; usually serves a specific function, e.g. the domain of a 110—120 amino acid unit of an antibody molecule serves in recognizing and attaching an antigen.

domestication, changing of characteristics of plants and animals by artificial selection for man's purposes.

dominance, 1. in genetics, the expression of genetically controlled characters and the corresponding gene → alleles in all → heterozygous progenies of the first generation (F_1 generation) from a cross of two → homozygous strains differing in respect to these characters; *see* dominant character; **2.** in ecology, the situation where one or more species in a community dominate other species because of their size, number, or coverage, and thus have a considerable influence upon the conditions of the community; **3.** → dominance hierarchy. *Adj.* **dominant**.

dominance deviation, symbol D; that part of the → genotypic value of an individual which is due to the dominance of a given quantitative character, i.e. dominance interactions of the genes regulating that character. D increases the value of the additive effects of these genes (*see* breeding value). *See* dominant character.

dominance hierarchy, rank order; in ethology, a hierarchic system found especially in social vertebrates; formed by → agonistic behaviour so that each member in a group has its own hierarchic position in relation to other members, and thus some individuals have better access to resources than others. When established, the dominance hierarchy is long-lasting and rather stable with scant fightings. Top dominant individuals, usually old males, are fearless and fairly strong, often characterized by conspicuous structures or colouration. In birds dominance hierarchy is also called **peck order**, but the terms are also used as synonyms.

dominance mating, a type of sexual selection in which the males compete by threatening and fighting each other to attain a dominant position in the social hierarchy and thus a chance to copulate. Dominance mating is an **intrasexual selection** in which a male can maintain a harem according to its hierarchic position in the society. *See* intersexual selection.

dominance modification, pertaining to the function of **dominant genes** (**dominance modifier genes**), i.e. genes influencing the degree of → dominance of homozygous and/or heterozygous non-alleles and their corresponding phenotypic characters.

dominant character, a genetically controlled character due to the corresponding dominant → allele (**dominant gene**), expressed in all progenies of the first generation (F_1 generation) when two pure homozygotic strains, which differ regarding this character, are crossed. In diploid, sexually reproducing organisms the dominant character is expressed in 3/4 of the progenies of the F_2 generation in a monohybrid cross. Those characters and alleles which will be masked by the dominant genes are called recessive. *See* recessive characters.

dominant species, a species in a community with a strong influence on the structure and function of the community, either numerically or in the proportion of biomass; e.g. in bird communities dominant species represent over 10% of the total number of pairs.

Donnan equilibrium (*Frederick Donnan*, 1870—1956), the ion equilibrium generated between two ionic solutions which are separated by a selectively permeable membrane through which some ions can move but some others not. In the system, ion concentrations and electrical charges influence ion movements. At equilibrium the ion concentrations can be expressed:

$$\frac{[\text{Cations inside}]}{[\text{Cations outside}]} = \frac{[\text{Anions outside}]}{[\text{Anions inside}]}$$

Donnan equilibrium determines the ion movement through the cell membrane particularly because large intracellular ions, e.g. proteins, cannot diffuse across the cell membrane, and thus an electrical gradient between the inside and outside of a cell is generated. This is one of the factors that affect the → resting potential of the cells.

donor (L. *donare* = to give as a present), **1.** an individual that supplies parts of itself, such as cells, a tissue, or an organ, to be used in another organism (recipient); **2.** an atom or a molecule which contributes electrons or radicals to another compound (acceptor).

donor splicing site, the borderline between an → intron and the preceding → exon in chromosomal DNA at which the → heterogeneous nuclear RNA is cleaved in → splicing.

dopa, DOPA, 3,4-dihydroxyphenylalanine; 3-hydroxytyrosine; an amino acid of which the L-form or **levodopa** is biologically active; an intermediate in the biosynthesis of → dopamine, noradrenaline, adrenaline, and melanin. Levodopa is used in the treatment of Parkinson's disease to improve dopaminergic neurotransmission in the brain.

dopamine, 3-hydroxytyramine, decarboxylated → dopa, C_8H_{11} NO_2; an active monoamine of the catecholamine group functioning as a hormone or neurotransmitter (**dopaminergic transmission**) in certain parts of the nervous system in vertebrates and invertebrates; the neurones are called dopaminergic neurones. Dopamine is an intermediate in the synthesis of noradrenaline, adrenaline and melanin.

dormancy (L. *dormire* = to sleep), **1.** in animals, the state of being asleep; a sleep-like (**dormant**) condition, a torpid state (*see* torpor); e.g. → aestivation, cryptobiosis, diapause, hibernation; **2.** in plants, the resting stage of seeds, bulbs, and buds, during which metabolism is very low and no growth occurs; may be winter dormancy (e.g. trees at northern latitudes) or summer dormancy (e.g. most Liliaceae plants such as tulip and lily-of-the-valley). Dormancy is controlled by both environmental factors such as the day length, temperature, and drought, as well as endogenous factors, particularly hormones of which → abscisic acid induces dormancy and → gibberellins and → cytokinins prevent or break it.

dormin, → abscisic acid.

dors-, dorsi-, dorso- (L. *dorsum* = back), denoting relationship to the back of an organism or organ; situated from a reference point towards the back surface. *Adj.* **dorsal**.

dorsal aorta, *see* aorta.

dorsal lip, the upper lip of the → blastopore; acts as an organizer in the → induction of overlying tissues in → gastrulation of an animal embryo.

dorsal placentation, *see* placenta.

dorsiventral, also **dorsoventral** (L. *venter* = belly), indicating a difference between the upper and lower surface of a (mostly flattened) structure or its part, e.g. a plant leaf.

dorsiventral flower, a flower with only one plane of symmetry. *Syn.* monosymmetric flower, zygomorphic flower. *Cf.* symmetric flower, bilateral flower, asymmetric flower.

dorsolateral, also **dorsilateral** (L. *latus* = side), pertaining to the back and the side.

dorsomedian (L. *medium* = middle), pertaining to the median line or area of the back.

dorsoventral, → dorsiventral.

dosage (L. *dosis* = act of giving), dosing; a schedule of frequency, number, size, and administration manner of doses given to an animal (or man). *Cf.* dose.

dosage compensation, in genetics, a mechanism responsible for the phenomenon that the expression of the → sex-linked genes is equal in both sexes of those organisms which have an XX-XY or XX-XO mechanism for → sex determination. In mammals, dosage compensation is achieved so that in all somatic cells of the female, one of the two X chromosomes (either the paternal or the maternal) is inactivated. Hence, in both sexes there is only one active X chromosome per somatic cell. In the cells the inactive X chromosome is heterochromatinized into a → Barr body, sex chromatin (→ heterochromatin). In insects, dosage compensation is achieved so that the two X chromosomes in females function with only half the activity of the single X chromosome present in males.

dose, 1. an amount of a substance given to an animal at one time; **2.** in radiology, the quantity of energy absorbed into an object, as e.g. organism, organ, or a unit mass of tissue.

dose equivalent (L. *aequus* = similar, equal, *valens* = worthy of), any dose which has a similar biochemical or physiological effect as the reference dose; particularly the quantity of a radiation dose. The effects of different radiation types differ from each other and therefore the effects are compared with the damaging effects caused by gamma radiation. The radiation dose equivalent is expressed as sieverts (Sv). *See* Appendix 1.

dose-response relationship, the relationship between the quantity of a substance and the response caused by it in an organism (cell, organ, or tissue). In biological systems this relationship increases linearly at certain concentration levels, when the quantity of the substance has increased exponentially.

double-blind test, a type of drug test for humans where neither the person testing nor the person taking the drug know whether a given substance contains a drug or not, i.e. whether it is an effective preparation or a placebo.

double bond, a → covalent bond in which two atoms are bound to each other by two electron pairs, e.g. C=O or C=C; double bonds are characteristic of unsaturated compounds, such as unsaturated fatty acids.

double circulation, a circulation system of tetrapods comprising **pulmonary circulation** (lung circulation, lesser circulation) and **systemic circulation** (greater circulation). *See* blood circulation, heart.

double fertilization, 1. the fertilization of the → embryo sac of a flowering plant with two sperm cells one of which fuses with the egg cell to form the zygote and the other with the diploid endosperm nucleus to form the primary triploid → endosperm; **2.** in animals, the exceptional fertilization of a single egg with two sperms; leads to a mosaic zygote.

double helix, in molecular biology, the double-stranded structure of → deoxyribonucleic acid (DNA) in which the two twisted nucleotide chains are bound together by their base parts.

double minute chromosomes, small, nearly point-like chromosomes existing in pairs in cultured tumour cells.

double recessive, pertaining to an individual, strain, or stock in which each of the two gene loci, involved in crossing or breeding work, is homozygous for → recessive alleles.

doves, pigeons, → Columbiformes.

down, *see* feather.

Down syndrome (*John Down*, 1828-1896), a form of human mental retardation caused by a → trisomy of chromosome 21; the most important form of genetic mental retardation, having a frequency of about one per 800 live births. Besides of mental retardation, it is characterized by a small stature and typical facial features with oblique eyes. The individuals suffering from Down syndrome usually live only 40 to 50 years. The risk of bearing a Down syndrome baby increases with the age of the mother. Cytologically three types are distinguished: 1) a free trisomy of chromosome 21, 2) a translocation in which the long arm of chromosome 21 is translocated onto chromosome 15, and 3) an → isochromosome 21. The types 2 and 3 are heritable because the father or the mother can be a carrier of the translocation or the isochromosome. Type 1 is mostly a new genome mutation, and it is arisen in 70% of the cases as a consequence of → non-disjunction during oogenesis, and in 30% during spermatogenesis; 95% of all cases of this syndrome are of this type. *Syn.* mongolism.

DPN, diphosphopyridine nucleotide, the former name for → nicotinamide adenine dinucleotide (NAD).

drive, a neural and hormonal tension inside an animal producing spontaneous and goal-searching behaviour. After the action the tension usually is moderated or it ends, i.e. "the drive is satisfied". The terms of drive and motivation are hardly discernable and sometimes they are considered as synonyms.

driving force, in cell biology, the force indicating the natural direction and transport affinity of ions moving across the plasma membrane. It is determined by the equilibrium potential of each ion type (*see* Nernst equation) and is calculated by subtracting the equilibrium potential from the membrane potential. Positive driving force means the removal of a positive ion type from the cell and uptake of a negative ion; negative driving force affects vice versa. Driving force is important in cellular ion metabolism and in generation of → resting potential and → action potential, and it shows whether the transport of some ion type is active or passive. *See* cell membrane transport.

drone, the male of some social insects, such as honeybees, ants, and wasps.

Drosophila, a genus of → fruit flies including more than 2,000 species, the best known of which is *Drosophila melanogaster*, the most important experimental animal in genetics, most recently also in developmental biology. In the genus *Drosophila* important studies on taxonomy, especially cytotaxonomy, and speciation are carried out.

drupe (Gr. *dryppa* = olive), a fleshy indehiscent fruit in which the endocarp around the seed is thick and hard, as in peach; in an aggregate drupe there are many separate drupelets, as in raspberry.

druse, a single → crystal formed from calcium oxalate in plant cells.

dry deposition, *see* acid deposition.

dryopithecines, also **dryopithecids** (Gr. *drys* = tree, oak, *pithekos* = ape), a group of ancestral hominoids of Miocene and Pliocene; existed in Africa and India 17—8 million years ago and probably represented the point of divergence of different hominoid lines.

duct (L. *ductus* < *ducere* = to lead), a tube-like passage in organisms, especially for the transport of excretion and exocrine secretion; also some trunk veins or other canals conducting blood or lymph, or amniotic fluid, are called ducts.

ductless gland, → endocrine gland.

ductus arteriosus (L.), duct of Botallo, arterial duct; an embryonic vascular passage between the pulmonary trunk and the aorta in reptiles, birds and mammals, functioning as a bypass from the right ventricle when the lungs are still functionless; closes and disappears when the pulmonary circulation opens, i.e. in mammals at birth.

ducts of Cuvier (*Georges Dagobert*, Baron de la *Cuvier*, 1769—1832), common cardinal veins; **1.** two half-circled vein trunks in fish, formed by anastomoses of the anterior and posterior cardinal veins; are the main venous drainage channels opening to venous sinus of the heart; **2.** the corresponding vein in vertebrate embryos, reducing in all tetrapod vertebrates during later development; it is originally paired, the right one developing into the superior vena cava.

duoden(o)-, pertaining to, or situated in the → duodenum. *Adj.* **duodenal**.

duodenal glands (L. *glandulae duodenales*), coiled, branched tubular glands in the submucous layer of the → duodenum, secreting mucus; called also Brunner's or Wepfer's glands.

duodenum, pl. **duodena** (L. *duodini* = twelve at a time, the name comes from the length of the duodenum in man, i.e. about twelve fingerbreadths, about 20 cm), the anterior region of the small intestine of mammals, extending from the pyloric valve to the jejunum; the bile and pancreatic ducts open into this part of the intestine. *See* alimentary canal, intestine.

duplex, a molecule consisting of two chains or strands bound by hydrogen bonds, e.g. duplex DNA as a synonym for double-stranded DNA.

duplication, in genetics, a → chromosome mutation in which a given segment of a chromosome is duplicated, the size of the duplication varying considerably. The existence of duplications in → eukaryotes can be observed cytologically by changes in the regional pairing of → homologous chromosomes during the first prophase of → meiosis, or in the somatic pairing of → polytene giant chromosomes. The smallest duplications that are cytologically observable are in the length of one band of the polytene chromosome. However, even whole arms of chromosomes can exist as duplications (*see* isochromosome).

Duplications can be interchromosomal or intrachromosomal. In the former case the duplicated segment has been transposed to a non-homologous chromosome; in the latter case the duplicated segment exists in the same chromosome as the original one. Often they are located consecutively forming a **tandem duplication**. The formation of duplications is an important mechanism which increases the quantity of genetic material in evolution.

dura (mater) (L. *dura* = hard, *mater* = mother), the outermost of the three membranes, meninges, covering the central nervous system of vertebrates. *See* meninx.

dwarfism, nanism, nanosomia; underdevelopment of the body.

dyad (Gr. *dyas* = a pair), a bipartite structure; e.g. **1.** a pair of cells which are formed as a consequence of the first meiotic division (*see* meiosis); **2.** a chromosome derived from a chiasmatic bivalent in the anaphase of the first meiotic division; due to → crossing-over, a dyad consisting of parts of → homologous chromosomes.

dynam(o)- (Gr. *dynamis* = power), pertaining to a force or power. *Adj.* **dynamic**.

dynamic equilibrium, the term used e.g. in chemistry and ecology to describe a condition of equilibrium where two opposing forces are equal.

dynein, a motor protein found in microtubules of those cells which have flagella and/or cilia; uses ATP as energy source for the motility of these cell organelles, i.e. ATP binding and hydrolysis in dynein bridges sequentially form and break new bonds with adjacent microtubule doublets. *Cf.* muscle contraction.

dys- (Gr. *dys-* = bad), denoting relationship to

bad, disordered, abnormal.

dysfunction, impairment of the function of a cell, tissue, organ, or organism.

dysgenesis, *see* hybrid dysgenesis.

dysmorphism (Gr. *morphe* = form), pertaining to malformation of an anatomical structure.

dysphotic, disphotic (Gr. *phos* = light), pertaining to feeble illumination, as a **dysphotic zone**, an aquatic zone between the euphotic and aphotic zones, in oceans at the depths of 80 to 600 m where light is too scant for photosynthesis.

dysplasia (Gr. *plassein* = to form), the disturbed, abnormal growth or development of a tissue or organ.

dystrophic, 1. pertaining to → dystrophy; **2.** in ecology, pertaining to a dark water body with high quantities of sterile humus, i.e. the water contains abundant humus acids but is poor in nutrients. *Cf.* eutrophic, mesotrophic, oligotrophic.

dystrophy, also **dystrophia** (Gr. *trephein* = to nourish), foul nutrition; the impairment or disorder of cells, tissues, or organs due to defective nutrition or trophic influence from other tissues; e.g. muscular dystrophy is caused by dysfunction or inactivity of innervating nerve cells.

E

E, symbol for **1.** electromotive force (redox potential); **2.** energy; **3.** glutamic acid.

***E*,** einstein. *See* Appendix 1.

E-, (Ger. *entgegen* = opposite), a stereodescriptor for compounds having double bonds.

e, symbol for **1.** base of natural logarithms (appr. 2.71828); **2.** positron (e^+); **3.** electron (e^-).

E1A, E1B, E5—E7, → oncogenes derived from different viruses.

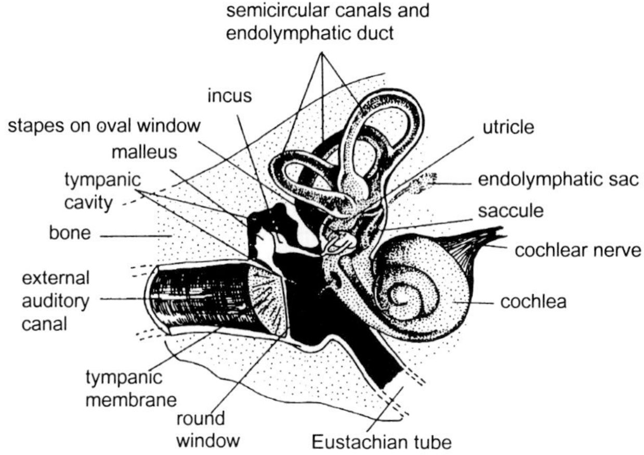

Ear. Diagram of mammalian ear.

ear, a sensory system in the vertebrate head comprising organs for hearing (auditory organ) and balance (equilibrium organ). In mammals, the ear is composed of 1) the **external ear** (outer ear) with the ear lobe (auricle), auditory canal, and eardrum (tympanic membrane), 2) the air-filled **middle ear** with the tympanic cavity and → auditory ossicle(s), communicating with the pharynx via the auditory tube, and 3) the fluid-filled **inner ear** (internal ear) with sensory receptors. The inner ear includes the → **cochlea** for hearing, and the **utricle**, **saccule**, and **semicircular canals** for balance. Between the middle and inner ear there are two membrane-covered openings, the **oval window** (*fenestra ovalis*) at the base of the vestibular duct of the cochlea and the **round window** (*fenestra rotunda*) at the base of the tympanic duct of the cochlea. Against the oval window, the auditory ossicle (three ossicles in mammals) form a bridge from the tympanic membrane. The inner ear comprises the osseous labyrinth (*labyrinthus osseus*) that is the hole inside the temporal bone; inside the osseous labyrinth lies the membranous labyrinth (*labyrinthus membranaceus*), filled with **endolymph** (fluid); the osseous labyrinth outside the membranous labyrinth is filled with **perilymph**. There are no external ears in reptiles and birds, and the tympanic membrane is situated at the level of the skin surface, or in a skin pouch (birds). In snakes and frogs the tympanic membrane and cavity are absent. Fish have no cochlea, and in the amphibian ear the cochlea is a simple, slightly bent membranous tube, *lagena*, but in higher vertebrates it is always a spiral organ. Sense receptors originate embryonically from the ectoderm, but all embryonic layers participate in the development of the ear as a whole. For invertebrates, *see* hearing.

eardrum, tympanic membrane; **tympanum**; **1.** in vertebrates, the fibrous membrane separating the middle ear from the external ear; may occur on the surface of the head as in most anurans and turtles, or at the end of the auditory canal as in most reptiles, birds, and mammals. The eardrum transmits sound vibrations to the auditory ossicle (three ossicles in mammals) of the middle ear; **2.** in invertebrates, the very thin membrane on the most developed hearing organ, **tympanal organ**, in some insects (e.g. some lepidopterans, homopterans, and orthopterans). *See* hearing.

ear ossicles, → auditory ossicles.

earth alkali metals, *see* alkali metals.

earthworms, → Lumbricidae.

earwigs, → Dermaptera.

EC, *see* enzyme classification.

EC_{50} (*effective concentration 50%*), median effective concentration, i.e. the concentration required to produce the desired effect in 50% of individuals in a population within a certain time period. EC_{50} is usually given in molar concentration or → ppm (mg/kg or mg/l). EC_{50} is also used to refer to the concentration of a substance that causes the half-maximal response in any reaction of a cell tissue, or organ, studied *in vitro*. *Cf.* ED_{50}, LD_{50}.

ecad, → ecophenotype.

eccentric (Gr. *ek* = out, *kentron* = centre), **1.** not concentric; not situated or occurring in a centre but away from it (e.g. in anatomical structures); **2.** pertaining to an odd character.

eccrine (Gr. *krinein* = to separate), pertaining to merocrine → sweat glands (small sweat glands) secreting fluid product extracellularly through → exocytosis. *See* merocrine. *Cf.* apocrine, holocrine.

ecdysiotropin, → prothorac(ic)otropic hormone.

ecdysis, pl. **ecdyses** (Gr. *ekdysis* = getting out), **moult**, **moulting**, Am. **molt**, **molting**; a periodic shedding of the cuticle of many invertebrates, or skin epidermis of some amphibians and many reptiles (lizards and snakes), due to growth. In invertebrates, it begins by softening and loosening of the epidermis below the cuticle, usually associated with growth. In the course of moulting, great changes occur in the exoskeleton of arthropods, e.g. most of the endocuticle is discharged and resorbed. Hormones, especially → **ecdysone**, regulate the timing and simultaneousness of the moulting between different parts of the body. *See* moult.

ecdysone, moulting hormone, **growth-and-differentiation hormone**; the steroid-like hormone secreted from the prothoracic gland of insects; it is synthesized from cholesterol and its release is controlled by → prothoracicotropic hormone. Ecdysone is secreted as a prohormone, called α-ecdysone, and converted to the physiologically active form, β-ecdysone, in several peripheral target tissues. It induces developmental processes leading to the differentiation and maturation of larvae, i.e. → ecdysis (moulting) and metamorphosis. In crustaceans, a corresponding hormone is secreted from the → Y-gland.

ecesis (Gr. *oikein* = to inhabit), the first stage in → succession; domestication of new, colonizing plant species.

ECG, → electrocardiography.

echin(o)- (Gr. *echinos* = sea urchin, hedgehog), pertaining to spines or other spine-shaped structures, or to sea urchins and other echinoderms.

echinococcosis (Gr. *kokkos* = berry), hydatid disease; a zoonosis caused by a small tapeworm, the hydatid worm, *Echinococcus granulosus*. The worm develops into the adult stage usually in the intestine of foxes and dogs (Canidae) which serve as definitive hosts. The eggs pass out within the host's faeces and are transferred into intermediate hosts, such as pigs, dogs, cattle, sheep, and reindeer, sometimes also into man, in whom the larval cysts form large **hydatid cysts**, e.g. in the liver, lungs, kidneys, and brain. Echinococcosis is harmful and sometimes fatal to man. *See* Stelleroidea.

echinocyte, any spiny cell, especially a crenated red blood cell.

Echinodermata (Gr. *derma* = skin), **echinoderms**; a marine invertebrate phylum including about 6,000 described species; the animals are evolutionarily rather advanced in spite of a radial symmetry. They are deuterostomes (*see* Deuterostomia) with a → coelom, and their larvae are bilaterally symmetrical. Nearly all adult echinoderms are pentaradially symmetrical, star-shaped, globular, or oval. They have a calcareous endoskeleton and a → water-vascular system, usually with many tube feet serving locomotion, food handling, and respiration. Echinodermata includes e.g. the classes (or subclasses): Asteroidea (starfish), Echinoidea (sea urchins), Holothuroidea (sea cucumbers), Ophiuroidea (brittle stars), and Crinoidea (sea lilies).

echinoderms, → Echinodermata.

Echinoidea, sea urchins, **sand-dollars**, **heart urchins**; a class in the phylum Echinodermata; about 950 species are described. The body is globular, disc-like, or heart-shaped without limbs; the skeleton is usually rigid comprising numerous movable spines, and there are numerous pedicellariae with three "jaws" in the skin. *See* Aristotle's lantern.

Echiura, (Gr. *echis* = viper, serpent, *oura* = tail), **echiurans, echiurids, spoon worms**; an invertebrate phylum comprising unsegmented worms with a long proboscis and true → coelom; some structural characteristics suggest they belong to annelids, and previously they were classified into Annelida as a class Echiuroidea. Echiurans include about 140 species in two classes: Echiurida and Sactosomatida. They are widespread marine animals in oceans, often living in littoral zones of temperate waters, some species in arctic waters. They live e.g. in mud and sand, in rocky crevices, or in empty snail shells.

echolocation, a method used by some animals for location of objects by reflecting sound, usually ultrasound; the animals can determine the delay and direction of the very high frequency sound they emit, reflected from objects in their environment. Echolocation is found in many nocturnal animals, and animals living in dark caves, as e.g. bats and oilbirds (*Steatornis caripensis*), but also in some cetaceans, insects etc., for orientation or catching prey.

eclipse plumage (Gr. *ekleipein* = to leave off), the dull plumage developed after the courtship period in many birds, especially in waterfowl's moulting in the summer. *See* nuptial.

eclosion hormone, a peptide neurohormone released from the terminals of neurosecretory cells in the → corpus cardiacum of insects; induces the emergence of the adult from the puparium.

eco-, also **oeco-, oiko-** (Gr. *oikos* = house, household, habitation), denoting relationship to environment, habitat (ecosystem), or household (economy).

ecocatastrophe, *see* catastrophe.

ecocline (Gr. *klinein* = to slant, lean), a genetical gradient seen in structural features or ecological demands of organisms (→ ecotypes), occurring along an environmental gradient or a geographical transect.

ecological counterparts, species in different geographical areas adapted to use similar environmental resources although the species may be phylogenetically far from each other, e.g. kangaroos and odd-toed ungulates, opossums and rodents, or the red deer and the wapiti. Adaptation is often seen in physiological and structural features of the counterparts. *Cf.* vicarious species.

ecological efficiency, the efficiency of energy transfer from one trophic level to the next; the **coefficient of ecological efficiency** denotes the ratio of energy assimilated at a certain trophic level to that assimilated at the preceding level; e.g. the percentage of the energy consumed by herbivores from that assimilated by plants. On average, the producers, i.e. green plants, assimilate about 1% to 5% of the energy of sunlight into organic substances; animals use about 10% of the food energy into new tissues (growth and reproduction). Ecological efficiency is usually presented as the ratio of net productivity at one trophic level to the net productivity of the preceding trophic level. *Syn.* transfer efficiency.

ecological energetics, a branch of ecology studying energy flow through communities.

ecological isolation, a situation in which a group of organisms live sympatrically with other conspecific groups but are, by their ecological demands, separated from them; thus, gene flow is lacking between the groups. Isolations of this kind are e.g. a **habitat isolation**, when the groups are living in different habitats, or a **seasonal isolation**, when reproductive times are different, as in many fish species.

ecological lifetime, *see* life time/span.

ecological niche, *see* niche.

ecological pyramid, a graphical diagram of quantitative energy relations between the different trophic levels in the food web of a community. The level of producers, i.e. green plants, is placed at the bottom, and that of carnivores on the top. An ecological pyramid can be drawn on the basis of the numbers of individuals (**individual pyramid**), of biomasses (**biomass pyramid**), or of energy relations (**energy pyramid**).

ecological race, *see* race.

ecological release, a niche expansion of a species in a situation when the number of competitive species has decreased or disappeared from the shared habitat. The ecological release is common on islands where the numbers of competitive species are smaller.

ecological segregation, a niche segregation of two or more species when competition between them has increased; e.g. competitive species try to find new habitats or new feeding resources. *Syn.* niche differentiation.

ecological species group, a group of coexistent species with similar demands on environmental factors.

ecological zone, *see* adaptive zone.

ecology (Gr. *oikos* = household, *logos* = discourse), a branch of biology studying interactions between organisms and their organic and physical environments. According to *Charles Krebs* (1978), ecology examines the factors affecting the distribution and abundance of organisms. Ecology includes e.g. **plant ecology**, **animal ecology**, and **bioecology**, the last of which concerns interactions between different organism groups, such as fungi and animals, fungi and plants, or plants and animals, e.g. pollination, feeding of insectivorous plants, herbivory, etc. **Synecology** studies interactions in plant and animal populations, or relationships of different species, while **autecology** explains innate reactions of a population in response to various environmental factors.

At present, **human ecology** has gained importance seeking to resolve problems between man and his environment. **Cultural ecology** examines distribution and abundance of animals and plants living in habitats manipulated by man. Recently, **urban ecology** has developed strongly with global human urbanization; it deals with adaptation of organisms to urban-like environments. **Reconstructive ecology** resolves the restoration of an environment, dramatically changed by man, to a natural or slightly manipulated condition. *See* community ecology, environmental ecology.

economical trait loci, ETL, gene loci which affect the genetic → variance of an economically important trait in domestic plants or animals.

ecophenotype (Gr. *phainein* = to appear, *typos* = pattern), a habitat form of an organism; a form e.g. of a plant or animal whose special features are caused by environmental factors, not genetically; typical ecophenotypes are found among unicellular algae and cladocerans. *Syn.* ecad. *Cf.* cyclomorphosis, ecotype.

ecophysiology, the research field of biology studying physiological reactions of organisms subjected to their natural environmental factors; deals especially with actions of abiotic factors and adaptation of organisms to these factors. *See* ecology, physiology.

Eco R1, a → restriction enzyme derived from *Escherichia coli*; used to cut DNA into fragments; its recognition site is G'AATTC; G= guanine, A = adenine, T = thymine, C = cytosine.

ecospecies, a group of → ecotypes able to interbreed without any decrease in the fertility or viability of the offspring.

ecosphere (Gr. *sphaira* = globe), a collection of all the ecosystems in the world including both organic (living and dead organisms) and inorganic components (energy and chemicals). *Cf.* biosphere.

ecosystem (Gr. *systema* = composite whole), a functional entity or unit formed locally by all the organisms and their physical (abiotic) environment interacting with each other; e.g. a pond, lake, or forest, but also more extensive entities like the oceans. *Syn.* biogeocoenosis. *Cf.* biosystem, community.

ecosystem model, a method used in researching the functions of ecosystems; e.g. the effects of environmental pollutants when a miniature ecosystem is built in a laboratory for examining the effects of pollutants e.g. on a food web.

ecotone (Gr. *tonos* = tone, tension), a marginal zone between two communities, e.g. an edge of a forest; the biodiversity in such an area is greater than in either of the communities alone. *See* edge effect.

ecotourism, a highly developed form of tourism concerning the nature and ecology of a particular country; it is important e.g. in many African states, such as Kenya and Tanzania, with several large nature reserves. At present, ecotourism is rapidly increasing all over the world.

ecotoxicology, the research field of → toxicology dealing with harmful environmental factors, and with toxic effects on organisms living in different ecosystems.

ecotype (Gr. *typos* = form, figure), a local population or race of a species which as a consequence of natural selection is genetically adapted to a restricted habitat; individuals of an ecotype can interbreed with other members of the species. Usually, the term is used for plant races. *Cf.* ecophenotype.

ect(o)- (Gr. *ektos* = outside), denoting outside, situated on the outside; sometimes used as a synomym for ex(o)-.

ectexine, *see* exine.

ectoblast (Gr. *blastos* = germ, shoot), **1.** →

ectoderm in an early animal embryo, or the outer layer, **epiblast**, of the gastrula, giving rise to the ectoderm; **2.** more generally, an outermost layer or membrane of an organism, organ, or cell.

ectocommensalism (L. *cum* = with, *mensa* = table), an interaction in which an organism (**commensal**) is living on the surface of the other (**host**) without producing any harm or benefit.

ectoderm (Gr. *derma* = skin), **1.** the outermost of the three germ layers of the early embryo in most metazoan animals; gives rise to the nervous system and epidermal tissues with glands, scales, hairs, feathers, nails, pigment cells, and many sensory receptors; **2.** the outer of the two body layers of some primitive animals, as cnidarians and ctenophores (sea combs). *Adj.* **ectodermal**.

Ectognatha (Gr. *gnathos* = jaw), *see* Insecta.

ectomy (Gr. *ek* = out, *-temnein* = to cut), the excision of an organ or a part of a tissue; e.g. **hypophysectomy**, excision of the hypophysis (pituitary gland), **ovariectomy**, excision of the ovary.

ectomycorrhiza (Gr. *ektos* = outside, *mykes* = fungus, *rhiza* = root), *see* mycorrhiza.

ectoparasite (Gr. *parasitos* = parasite), a parasite living on the surface of its host organism; e.g. lice. *Syn.* epiparasite. *Cf.* endoparasite.

ectophloic siphonostele (Gr. *phloos* = bark, *siphon* = tube, pipe, *stele* = column), *see* stele.

ectoplasm (Gr. *plasma* = moulded, formed), the peripheral portion of the cell cytoplasm that is less fluid and contains no cell organelles; sometimes means only the cell membrane. Also called exoplasm. *Adj.* **ectoplasmatic**, pertaining to ectoplasm.

ectoplastic, having a formative induction force on the surface of a tissue structure, e.g. in an animal embryo.

Ectoprocta (Gr. *ektos* = outside, *proktos* = anus), **ectoprocts**; an invertebrate phylum including small, marine colonial animals, some species live in fresh or brackish waters. Previously ectoprocts and entoprocts were called **bryozoans** (phylum **Bryozoa**), or **moss animals**; at present they are classified as distinct phyla (in some classifications Ectoprocta is still considered as a synonym of Bryozoa). A solitary individual, **zooid**, consists of a feeding → polypide which is enclosed by a horny, calcareous or gelatinous capsule, **zooecium**. At the top of each zooecium is a circle of tentacles (→ **lophophore**) around the mouth. The end of the U-shaped digestive tract forms the anus that opens near the mouth outside the lophophore (*cf.* Entoprocta). Ectoprocts have a true coelom but excretory and circulatory systems are absent. About 4,000 species are described; they are divided into three classes: Gymnolaemata, Phylactolaemata and Stenolaemata.

ectopy, also **ectopia** (Gr. *ek* = out of, *topos* = place), malposition of an organ. *Syn.* heterotopy. *Adj.* **ectopic**.

ectothermic (Gr. *therme* = heat), pertaining to → ectothermy; also ectothermal. *Cf.* endothermic, exothermic.

ectothermy, the dependence of the body temperature of an animal on ambient temperature only. Ectothermic animals (**ectotherms**), which include more than 99% of all animal species, are bradymetabolic animals which are not able to use their own tissue metabolism for thermoregulation. However, they usually have precise thermal sensory receptors associated with good behavioural thermoregulation. Nearly all poikilothermic animals belong to ectotherms. *Adj.* **ectothermic**. *Cf.* endothermy. *See* poikilothermy, homoiothermy.

ectotoxin, → exotoxin.

ectotrophic (Gr. *trophe* = nourishment), *see* mycorrhiza.

ED_{50} (effective dose 50%), median effective dose, i.e. the dose required to produce the desired effect in 50% of individuals in a population; ED_{50} is usually given as mg/kg body weight. ED_{50} is also used to refer to the dose that gives 50% of the total or maximal effect. *Cf.* EC_{50}.

edaphic (Gr. *edaphos* = soil, ground), pertaining to soil, living in soil; **edaphic factors** have great importance for the success of plants in different areas.

edaphology (Gr. *logos* = word, discourse), a research field of biology dealing with soil; especially studies the effects of the soil on soil organisms.

edaphon, a community of organisms living in soil.

edema, → oedema.

Edentata (L. *ex* = without, *dens* = tooth), **edentates**; an order of mammals having no incisor and canine teeth; also premolars and molars may be absent (anteaters), or are needle-shaped without enamel and grow throughout life (sloths and armadillos). All edentate species live on the American continent. The order includes one suborder Xenarthra and 29 species in three families, **anteaters** (Myrmecophagidae), **arboreal sloths** (Bradypodidae), and **armallidos** (Dasypodidae).

edge effect, a common tendency of organisms to increase the number of species and density of individuals in → ecotones, i.e. in marginal zones of communities, as compared with more central parts of a community or an ecosystem.

EDTA, ethylenediaminetetraacetic acid, edetic acid, edathamil, $(HOOCCH_2)_2N(CH_2)_2$-$N(CH_2COOH)_2$; a chelator of divalent metals; is used as a chelating agent to bind magnesium, calcium, iron, and zinc ions into complex compounds. EDTA is used to buffer free calcium or magnesium in a solution to a certain concentration, being a common protecting agent e.g. in isolation media for cell organelles. EDTA is also used for preventing the precipitation of metallic nutrient ions in culture media. Common trade names are e.g. Titriplex, Versene, Complexon. *See* EGTA.

EEG, → electroencephalography.

eels, → Anguilliformes.

EFA, → essential fatty acids.

effective population size, symbol N_e; that size of a population which determines how effectively random fluctuations of gene frequencies, such as a → genetic drift, can occur in the population; can be different from the actual number of individuals in the population. The smaller the effective size, the more efficient is the action of genetic drift.

effector, any structure or an agent such as an organ, tissue, cell, or substance, that has a specific effect on a biochemical or physiological system; e.g. **1.** in genetics, a small molecule (metabolite) which interacts with the repressor of an → operon activating or inactivating the repressor molecule. In its active form the repressor is capable of binding to the → operator, and hence inhibiting transcription in the operon; **2.** in animal physiology, a cell or tissue which becomes active in response to nerve stimuli, e.g. muscular and glandular cells; **3.** in immunology, denotes differentiated lymphocytes which are immunologically effective carrying ability for antibody production or killer function; **4.** in enzymology, an allosteric effector is an activator or inhibitor of an enzyme having its effect on a site other than the catalytic site of the enzyme.

effector plasmid (Gr. *plasma* = moulded), an additional chromosome (→ plasmid) of a bacterium onto which a gene or a regulatory region of a gene is experimentally attached in order to study its function; is transferred to the cell in a → cotransfection together with a → reporter plasmid, the latter revealing the function of the gene to be studied.

efferent (L. *ex* = out, *ferre* = to bear), conducting or conveying away from a centre, like an organ, i.e. conveying centrifugally; e.g. an efferent nerve, duct, or vessel (*vas efferens*). *Cf.* afferent.

egestion (L. *egerere* = to carry out), discharging of unabsorbed food residues from the digestive tract; **egesta**, the material cast out. *Verb* to **egest**.

EGF, → epidermal growth factor.

egg, 1. → **ovum**; female reproductive cell, female germ cell, female gamete, egg cell, sometimes denoting also its developmental stages after fertilization; *see* gametes; **2.** in common language, a roundish reproduction body laid by female animals, such as birds, reptiles, amphibians, fish, and most invertebrates, consisting of the ovum and its envelopes, i.e. vitellin, albumen, jelly, and various membranes and a shell composed e.g. of calcium or chitin.

egg cell, 1. the female germ cell of animals, i.e. the female gamete, → ovum; **2.** the female gamete of plants formed in the → embryo sac of the ovule in the gynaecium; will be fertilized after → pollination and develops into an embryo inside the seed.

egg-laying mammals, → Prototheria.

egg membranes, membranes surrounding an animal egg cell; in vertebrates usually includes a vitelline membrane (primary membrane), a secondary membrane formed from the layer of follicle cells, and tertiary membranes secreted by accessory glands in the oviduct.

egg sac, an external, sac-like structure of copepod crustaceans containing developing eggs in the breeding time; depending on the species, the number of sacs may be one or two.

egg tooth, a small tooth-like cornification in the top of the upper jaw or beak with which young birds crack the egg shell when hatching.

EGTA, ethylene glycol-bis(β-aminoethyl ether)-N,N,N',N'-tetraacetic acid; a chelator of divalent metals, especially of calcium (Ca^{2+}) and magnesium (Mg^{2+}); used e.g. in media for cell fractionations to bind free calcium and magnesium in solutions, and thus inhibit the functions which are dependent on these cations, such as phospholipase activity. EGTA also helps to break tissue and cellular structures because many fibrous proteins need calcium to retain normal structure. As a chelator EGTA is more specific for calcium than → EDTA.

eicosanoids (Gr. *eikosi* = twenty), physiologically active substances composed of twenty carbon atoms in a chain. Eicosanoids function in animals as local hormones, such as → prostaglandins, thromboxanes, and leucotrienes, and are synthesized from unsaturated fatty acids, especially from arachidonic acid (eicosatetraenoic acid) and eicosapentaenoic acid (EPA). Also called icosanoids.

Eimer's organs (*T. Eimer*), sense organs with sensitive papillae in the snout of the mole; are probably tactile organs.

einstein, E (*Albert Einstein*, 1879—1955), one mole of photons; used as a unit for illumination measurement. *See* Appendix 1.

ejaculation (L. *ejaculari* = to throw out), a sudden action, such as the discharge of semen of animals; **ejaculate**, as a verb to expel the semen, as a noun (L. *ejaculum*), the semen produced in a single ejaculation.

elaioplast (Gr. *elaion* = oil, *plastos* = moulded), a plant cell → plastid, containing high quantities of → lipids.

elaiosome (Gr. *soma* = body, particle), an oil-containing appendage in plant seeds; allures ants to carry the seeds and thus help to disperse the plant, such as the castor oil plant.

Elasmobranchii (Gr. *elasmos* = metal plate, *branchia* = gills), **elasmobranchs (sharks** and **rays**); a subclass in → Chondrichthyes; carnivorous fishes, about 700 species, which have well developed olfactory organs, from five to seven gill arches with gill slits, and a pair of small external → spiracles behind the eye connecting to the pharynx. The subclass is divided into several orders, e.g. Lamniformes and Squaliformes (**dogfishes** and **sharks**), and Rajiformes (**rays, skates**).

elastase, *see* elastin.

elastin, also **elasticin** (Gr. *elastos* = ductile, elastic), an elastic, yellow scleroprotein produced by cells in the elastic connective tissues of animals; elastin, usually together with → collagen, forms the interstitial substance e.g. of the elastic cartilage, tendons, many ligaments, and of the connective tissues of the blood vessel walls and the respiratory pathways. The enzyme catalysing the decomposition of elastin in animals are **leucocyte elastase** (lysosomal or neutrophil elastase, EC 3.4.21.37), and **pancreatic elastase** (pancreatopeptidase E, EC 3.4.21.36) which in vertebrates is produced especially by the acinar pancreas for digestion. The term elastase has formely used to denote a serine protease (EC 3.4.21.11, formerly 3.4.4.7).

elater (Gr. *elater* = driver), one of many elongated cells between spores in spore capsules of → liverworts and → horsetails; coils or uncoils depending on humidity, assisting in the spreading of the spores.

electr(o)- (Gr. *electron* = amber, that becomes electric by rubbing), denoting relationship to electricity.

electric organs, electrophores; paired organs of some fish species, producing neuronally controlled electric discharges to the surrounding water; they are composed of specialized skeletal muscle cells (sometimes epithelial cells) of the cranial or lateral regions of the body. The basic unit of the organ is a flattened cell, **electroplax** (electrocyte). Usually thousands of electroplaxes sit on top of each other to form long pillars (usually several hundreds in number) which lie side by side in the organ. The electroplax cells are electrically connected in series coupling and the pillars in parallel coupling, and thus the → action potentials of 70 to 90 mV produced in each electroplax cell are summated in the control of nerve impulses into simultaneous discharges of hundreds of volts (even greater than 500 V).

Some species, such as the electric eel, electric ray, or electric catfish, have high-voltage organs which they can use for frightening and predation of other species. Weaker electric organs (low-voltage organs) are present in many other fish species, many of which live in muddy waters. The organs are used for mutual communication or for orientation like a radar, since these animals also have a very sensitive → electric sense.

electric sense, electromagnetic sense, a sense in many lampreys, cartilaginous fishes, and in some bony fish species, based on the function of **electroreceptors** by which these animals can sense the electricity of their surroundings. The organ is always found in species which have a low-voltage → electric organ, but also in many species without this. Electroreceptors are differentiated from the lateral line organ, and two types are found: bottle-shaped, ampullary organs, called → **Lorenzini's ampullae**, and smaller, tubule-shaped organs, called **tuberous receptors**. The animals can use the electric sense for predation, mutual communication, and orientation. *Cf.* magnetic sense.

electrobiology, the research branch of biology dealing with electrical phenomena in organisms.

electrocardiography, ECG (Gr. *kardia* = heart, *graphein* = to write, record), the technique for measuring the electrical activity of the heart muscle; usually recorded by electrodes fixed on the skin of limbs and thorax; results in graphical records, **electrocardiograms**; **electrocardiograph**, an instrument used in electrocardiography.

electrochemical gradient, the gradient across a membrane in regard to an ion type, comprising both the concentration gradient of those ions and the electrical gradient on both sides of the membrane. Electrochemical gradient determines the direction to which the ions spontaneously tend to move across the membrane, and the → driving force of the ion. *See* cell membrane transport, resting potential.

electrodes, elements at the extremities of an electric circuit, e.g. in an electrical apparatus or system like batteries, transistors, and meters; the positive electrode is called **anode** and the negative **cathode**. In biologial studies, different kinds of metal electrodes (often of steal or platinum) are used e.g. in → electrocardiography and → electroencephalography to study the functions of the heart and the brain. **Microelectrodes** are thin glass capillaries filled with a salt solution (often 3 M KCl), and they can be used in studies on electrical properties of individual cells and cell membranes.

electroencephalography, EEG (Gr. *enkephalos* = brain, *graphein* = to write, record), the technique for measuring electrical activity of the brain, usually by means of metal electrodes on the scalp; results in graphical records (**electroencephalograms**) with various types of rhythmic activity or brain waves, such as alpha, beta, and delta waves or rhythms; **electroencephalograph**, a device used in electroencephalography.

electrofusion (L. *fusio* = melting together), an electric pulse technique used to help isolated animal cells or plant cell protoplasts to fuse; used as a gene transfer method in molecular biology.

electrolysis (Gr. *lyein* = to dissolve), **1.** decomposing of a material by electric current, e.g. a salt or other chemical components; **2.** the destruction of tumours by electric current. *Adj.* **electrolytic**.

electrolyte, a substance which dissociates into → ions in aqueous solutions or when melted. Electrolytes are electrically conductable and important material for carrying current in organisms. Many inorganic and organic salts, bases, and acids, are dissociated to form electrolytic solutions in cells and tissues, as well

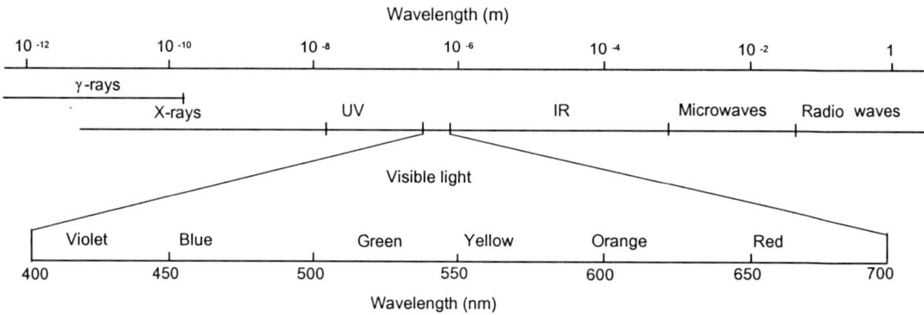

Electromagnetic radiation. The enlarged section shows the wavelength ranges of the various colours of visible radiation.

as in soil. The total ion concentration of a solution can be measured by means of its electric conductance.

electromagnetic radiation (Gr. *magnes* = magnet), radiation produced by the periodical changes of a combined electric and magnetic field. The radiation moves in space as electromagnetic waves (as energy quanta) approximately 300,000 km/s. The electromagnetic spectrum comprises radio and micro waves, infrared radiation, visible light, UV radiation, X-rays (roentgen waves), and gamma radiation. The last three types are dangerous for organisms, breaking macromolecules of cellular structures, the last two being → ionizing radiations. The shorter the wavelength of the radiation, the higher is the quantum of energy, i.e. the amount of energy is inversely proportional to the frequency of the wavelength. *See* UV radiation.

electromagnetism, magnetism produced by electric current.

electromotive chain/series, electrochemical series; the order of chemical elements arranged according to their tendency to accept or release electrons (i.e. form ions) in a solution thus producing a certain electrode potential; e.g. litium -3.02 V, potassium -2.92 V, calcium -2.76 V, sodium -2.72 V, magnesium -2.34 V, aluminium -1.67 V, zinc - 0.76 V, nickel -0.24 V, lead -0.13 V, hydrogen 0 V, copper +0.34 V, silver +0.80 V. The order in the electromotive chain determines the → redox potentials of elements.

electromotive force, EMF, the force (measured in volts, V) that causes the flow of electricity from one point to another. In a battery the force is determined by the difference of the two electrode metals in the → electromotive chain, e.g. EMF in a zinc-copper cell is 1.1 V.

electromyography, EMG (Gr. *mys* = muscle, *graphein* = to write, record), a diagnostic method for recording electrical functions of muscles, especially of skeletal muscles; electrodes are fixed on the skin or inserted into the muscle; **electromyograph**, the instrument used in electromyography; **electromyogram**, the record obtained by this method.

electron, a negatively charged subatomic particle. Electrons orbit the positive nucleus of the atom at particular distances called shells. The innermost shell usually has two electrons, the other shells contain eight electrons if possible (*see* octet rule), and the outermost shell may contain fewer. *Cf.* positron.

electron microscope, EM (Gr. *mikros* = small, *skopein* = look), a type of microscope in which the "light" source is a filament that emits electrons. The wavelength of this irradiation is very short (e.g. 0.39 nm with an accelerating potential of 100 kV) and thus the → resolution is much better than in any light microscope; the primary enlargement may be 800,000 times. The different types of electron microscopes are **transmission electron microscope** (TEM), **scanning electron microscope** (SEM), and **tunnelling electron microscope**. *See* microscope.

electron spin resonance, ESR (L. *resonantia* = echo), a phenomenon in which the energy states of an electron spin are mutilated in a magnetic field (*Zeeman* effect); the spin of a free electron may be divided between two atoms. Most electrons are in a low energy state, and an outer magnetic field drives electrons to a higher state if the frequency of this field has a resonance frequency. Resonance can be seen as a sharp peak in an oscilloscope. The location of this peak on the axis of the magnetic field depends on the electron structure and on the interaction of the electron structure in a compound. Molecular changes of a compound, such as those in the molecules of the cell membrane caused by low temperature, can be studied exposing the compound to the high frequency magnetic field in an ESR instrument; changes are seen as an altered signal when compared to the original. *Syn.* electron paramagnetic resonance, EPR.

electron transport/transfer, electron transport in cell respiration along an enzyme chain, **electron transport chain**, **ETC** (respiratory chain), from one enzyme to the next; occurs in aerobic energy metabolism of cells by means of transport compounds like → cytochromes (conjugated enzyme proteins), and is involved also in → photosynthesis. The electrons derive from hydrogen atoms (H^+ are formed) from the breakdown of substances in the → citric acid cycle and pass along ETC to molecular oxygen with the formation of water. Electron transport is based on oxidation-reduction changes (**redox reactions**) in the chain enzymes embedded in the mitochondrial inner membrane. During

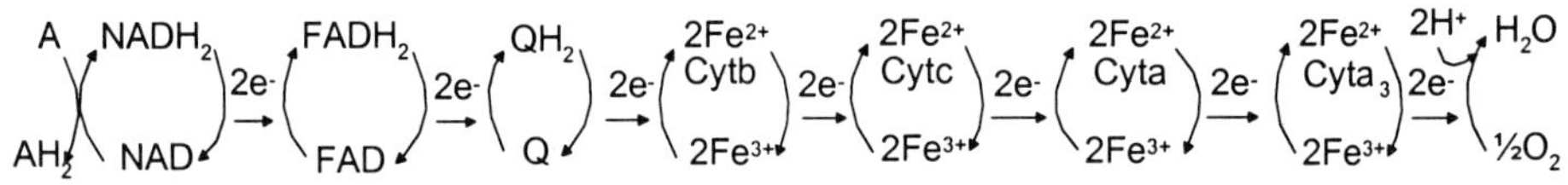

Electron transport chain. $A(H_2)$ = intermediates of citric acid cycle, $FAD(H_2)$ = flavin adenine dinucleotide, $NAD(H_2)$ = nicotinamide adenine dinucleotide, $Q(H_2)$ = quinone, cyt = cytochrome.

the transport, electrons fall from a higher to a lower energy state (**redox potentials**), and the energy so released is used for pumping protons (H^+) across the mitochondrial inner membrane. The potential energy of a proton gradient thus generated is then used for production of adenosine triphosphate (ATP) in → oxidative phosphorylation.

electronvolt, eV, *see* Appendix 1.

electrophores (Gr. *phorein* = to carry along), → electric organs.

electrophoresis, a technique for separating charged molecules or particles in an electric field generated by a current through a buffer solution. Molecules move in the field towards the positive or negative pole (electrode), their speed depending on their size and electric charge. Uncharged compounds can be separated also if electrically charged carriers are combined with them. Components to be separated are applied on starch, agarose, or polyacrylamide gels, on a cellulose acetate sheet, or on a filter paper. A gel is prepared as a plate (slab gel), or as a cylinder in a glass or plastic tube (disc gel). Capillary electrophoresis is a technique in which the separation is performed using thin glass capillaries.
Electrophoresis is either vertical or horizontal. Different compounds are separated as bands which can be detected if they are coloured, e.g. chlorophyll proteins. Many molecules are stained (e.g. proteins or polypeptides), and some compounds such as nucleic acids and their components can be detected as fluorescent bands under UV light. *See* immunoelectrophoresis, pulse-field electrophoresis.

electrophysiology, the research branch of → physiology dealing with electrical phenomena in cells, organs, and organisms involved in physiological processes.

electroplax, *see* electric organs.

electroporation (Gr. *poros* = passage, pore), the method of genetic engineering where an electric pulse is used to bore holes through the cell membrane in order to transfer DNA into a cell.

electroreceptors, *see* electric sense.

electroretinography, ERG (Gr. *graphein* = to write, record), a diagnostic method for recording electric functions of the retina after stimulation by light; **electroretinograph**, the instrument used in electroretinography; **electroretinogram**, the record obtained by this method.

electrotonic potential (Gr. *tonos* = tension, L. *potentia* = power), the term for slightly different types of small, attenuating changes in a → resting potential, i.e. electric potential prevailing in most cells; includes generator potentials, postsynaptic potentials, and end-plate potentials occurring in sensory receptors, synapses, and neuromuscular junctions, respectively.

electrotropism (Gr. *tropein* = to turn, direct), the growth movement of an organism or some of its organs as a response to an electric field. In plants it is based on the electric polarity of growing structures, supported by the distribution of Ca^{2+} and IAA^- ions. Growth is directed towards the negative electrode. Also called galvanotropism, in animals also electrotaxis or galvanotaxis.

element (L. *elementum* = first principle, rudiment, element), **1.** any of the primary constituents of a thing; **2.** in chemistry, a compound consisting only of atoms of the same atomic number; *see* Appendix 2.

elementary particle, a small basic constituent or part; e.g. **1.** any of the subatomic particles including → protons, neutrons, positrons, electrons, neutrinos, muons, tauons, mesons, hadrons, etc; **2.** elementary particle of mitochondria, any of numerous club-shaped granules (9 nm) with spherical heads attached to the inner surface of mitochondrial cristae, probably acting as the → electron transport system.

elephantiasis (Gr. *elephas* = elephant), Malabar leprosy; a tropical inflammation disease caused in humans by the filarial worms *Wuchereria bancrofti* or *Brugia malayi*, small parasitic nematodes; cause the fibrosis and hypertrophy of the skin blocking lymphatic vessels with consequent swelling of the legs, resulting in "elephant legs".

elephants, → Proboscidea.

elephant shrews, → Macroscelidea.

elicitor (L. *elicere* = to lure out), a biotic or abiotic factor which induces the production of → phytoalexins in plants or cell cultures. Biotic elicitors are e.g. proteins, glycoproteins, polysaccharides and unsaturated fatty acids; abiotic elicitors are e.g. UV light, heavy metals, and extreme temperatures.

elimination chromatin (L. *eliminare* = throw out, Gr. *chroma* = colour), Feulgen-positive chromatin material (*see* Feulgen staining) separating from the chromosomes in the early anaphase of → meiosis or → mitosis of certain species. *See* chromatin.

ELISA, → enzyme-linked immunosorbent assay.

elite plant production, a production of healthy seedlings free from plant diseases and viruses. Seedlings are produced using cell and tissue culture from apical meristems of plants cultivated in a warm environment (about 37°C) where the meristem cells are dividing rapidly.

elodeid (Gr. *helodes* = marshy), a water plant growing from the bottom and living totally under water; e.g. *Elodea*, *Ceratophyllum*.

elongation factors, EFs, proteins that regulate the prolongation of polypeptide chains during → genetic translation in → protein synthesis. In a cyclic process, the factors are involved in correct positioning of aminoacyl-tRNA on the mRNA-ribosome complex.

elongation zone, the area near the root tip in which the cells elongate but divide no more.

elution (L. *eluere* = to wash off), the removal of a substance from an adsorbent by dissolving it in a liquid; **eluent**, the liquid used in elution. *Verb* to **elute** (elutriate).

elytr(o)- (Gr. *elytron* = sheath, wing cover), denoting relationship to → elytron or a sheath-like structure such as vagina.

elytron, pl. **elytra,** wing cover; the hard or leathery forewings in some insects, e.g. in coleopterans and earwigs (Dermaptera); the elytra cover the membranous hindwings at rest.

EM, → electron microscope.

emasculation (L. *e-* = away, *masculus*, dim of *mas* = male), **1.** loss of masculine characters of a male animal, i.e. eviration; results from castration or other decrease of the efficiency of androgenic hormones; caused also by some environmental pollutants; **2.** the removal of stamens from a hermaphrodite flower in order to inhibit self-pollination; used in artificial hybridization.

Embden—Meyerhof pathway (*Gustav Embden*, 1874—1933; *Otto Meyerhof*, 1884—1951), → glycolysis.

embedding, in biology, the casting of a tissue sample in a solidifying medium like paraffin, resin, celloidin, or plastic so that the sample can be cut into thin sections for microscopic examination. Called also imbedding. *Verb* to **embed** (imbed).

embole (Gr. *embole* = insertion), the formation of the → gastrula by invagination of cells; also embolia, emboly.

embolus, pl. **emboli** (Gr. *embolos* = plug), **1.** a plug, especially a clot brought by the blood obstructing a blood vessel; **2.** a penis-like projection at the tip of pedipalps in chelicerates, serving as part of the copulatory organ.

embryo (Gr. *embryon*), the early developmental stage of an organism; **1.** in animals, the developmental stage beginning from the fertilized ovum and ending in an offspring with all major organ structures; when the structures typical of each species are developed (in man at 7—8 weeks) the mammalian embryo is called **foetus**; **2.** in plants, the embryo develops from the fertilized egg cell inside the seed and develops into a seedling and a new plant (diploid sporophyte) when the seed germinates. *See* embryogenesis.

embryoblast (Gr. *blastos* = germ, shoot), a group of cells in the early mammalian embryo; formed by inner cell mass of the early blastocyst from which the embryo develops inside the → trophoblast.

embryogenesis (Gr. *gennan* = to produce), the process during early developmental phases of a multicellular organism. In animals, embryogenesis starts by successive cell divisions beginning from a fertilized ovum, and leads to the growth and → differentiation of cells and tissues in the → embryo. Embryogenesis is preceded by the development of the ovum, i.e. → oogenesis. Several phases occur in embryogenesis; the following description refers mainly to chordates.

At **fertilization** a haploid male gamete (spermatozoon) fuses with a haploid female gamete (ovum) forming the diploid zygote and inducing its divisions. Fertilization is not always necessary and embryogenesis can exceptionally begin without it (*see* parthenogenesis). In the **cleavage phase** the ovum divides through several successive, synchronized divisions into 2, 4, 8, 16, etc. cell-stages, the cell size decreasing gradually so that the mass of the embryo does not increase. The grape-like cell mass, **morula**, is formed by these cell divisions.

In **blastulation** a fluid-filled cavity is formed inside the morula, and forms the **blastula**, a hollow, cellular ball, which in amniotic vertebrates is a flattened, two-layered disc, called **blastoderm**. The next stage, **gastrulation**, begins when cells from a certain area, i.e. the **blastopore** (**primitive streak** in birds and mammals), begin to invaginate under the external cell layer of the blastula, resulting in either two cell layers, i.e. → **ectoderm** and **endoderm** (in sponges and coelenterates), or three cell layers (in other metazoans) when the → **mesoderm** develops between the two other layers.

In the next stages the nervous system begins to develop in **neurulation** (<- neurula), and thereafter different structural features (→ **morphogenesis**), tissues (→ **histogenesis**), and organs (→ **organogenesis**) develop and the embryo grows in size.

The initial stages of development vary morphologically depending upon the type of ovum. Division of the ova (eggs), containing large amounts of yolk (vitelline) such as in birds and reptiles, is restricted to a narrow region, the **germ plate**, lying above the yolk. The embryo develops as a plate-like structure with the germ layers forming one below the other.

In mammals, the ovum with a small quantity of vitelline divides totally resulting in the → **blastocyst**. This implants on the mucous membrane of the uterus via extraembryonic tissue, called → **trophoblast**, from which the → amnion and the chorion derive later. The development of the plate-like embryo resembles that of a bird embryo. The mammalian embryo is called the **foetus** when species-specific features begin to appear (in man at the age of 7—8 weeks). In all amniotic vertebrates (→ Amniota), the → extraembryonic membranes develop to surround the embryo, which process is a major feature in the evolution of development in terrestrial vertebrates. Many animal groups involve larval stages which through → **metamorphosis** develop further into the final adult stage.

In plants, embryogenesis begins when the egg cell in the → embryo sac (in the → ovule) is fertilized by one sperm cell formed by the male gametophyte in the pollen grain and pollen tube; the other sperm cell will fuse with the nucleus of the endosperm mother cell. The egg cell is now a zygote, which divides first transversely and forms the base cell and apical cell. After this the development continues in different ways in different groups and families. In typical dicotyledons the apical cell divides further forming the → **suspensor** and the embryo with the **hypocotyle**, two **cotyledons**, and the **radicle**. The fully developed embryo is situated in the ovule so that the radicle is toward the → micropyle. The ovule develops into the seed, and during germination the radicle grows out through the micropyle forming the first root of the plant. → Monocotyledons have only one cotyledon. The embryogenesis of grasses differs from the other members of this group. The first leaf develops in a sheath-like **coleoptile**, and the radicle is also in a sheath-like **coleorhiza**. The scutellum transmits energy from the usually large → endosperm.

embryogeny, the origin and the production of the embryo. *Adj.* **embryogenic**, producing an embryo, or pertaining to → embryogenesis.

embryoid, also **embryonoid, embryoniform,** resembling an embryo; e.g. a structure formed from somatic cells in a cell culture.

embryology (Gr. *logos* = word, discourse), the branch of science studying the early ontogenic development of organs, i.e. embryos and foetuses.

embryonic, also **embryonate, embryonal,** pertaining to, or being in the state of an → embryo.

embryonic disc, the → blastoderm of amniotic vertebrates.

embryonic membranes, → extraembryonic membranes.

embryo sac, a structure in the → ovule of angiosperms; it is actually the female → gametophyte, producing the female gamete, i.e. the **egg cell**, which develops into the embryo after fertilization. The other parts of the fully developed embryo sac are: the **endosperm mother cell**, developing into the → endosperm and serving as an energy store in the seed, → **antipodal cells**, and → **synergids**.

embryo sac mother cell, a cell inside the ovary of an angiosperm, developing into the → embryo sac. *Cf.* macrospore, macrosporogenesis.

embryo transfer, the transfer of an early embryo into the uterus of a recipient female; the embryo is usually produced by *in vitro* fertilization.

emergence, the act of coming forth (emerging); e.g. **1.** the emergence of an adult insect from its pupa; **2.** an outgrowth of a plant, formed not only from the epidermis but also inner tissues, often also conducting tissues; the emergences often bear glands, as do the tentacles of *Drosera*, or they serve in defence and as attachment organs, as e.g. the thorns of roses; *cf.* hair; **3.** the appearance of new properties in the course of ontogenic development or evolution, not foreseen in an earlier stage.

emergent property, a property of a living material resulting from its organization into higher structural levels; e.g. the consciousness of higher animals, like man, is an emergent property developed during the course of organic evolution.

Emerson effect (*Ralph Emerson*, b. 1912), the special effect of different wavelengths of light on the photosynthetic rate, proving the occurrence of two different → photosystems in the photosynthetic machinery of plants. Emerson found that the photosynthetic efficiency of light longer than 680 nm in wavelength was increased by additional shorter light, i.e. shorter than 675 nm. The rate of photosynthesis in the two wavelengths together is greater than the added rates in either alone.

EMF, → electromotive force.

emigration (L. *emigrare* = to emigrate), the → migration or departure of animals from a population or area. *See* migration.

eminence (L. *eminentia*), a projection or prominence, such as an anatomical area raised above the surrounding level; e.g. the median eminence (*eminentia mediana*), the prominent lower segment of the hypothalamic infundibulum.

emission (L. *emittere* = to send out), the act of sending or discharging; e.g. **1.** emission of radiation, such as → electromagnetic radiation; **2.** emission of traffic or industrial fumes and smoke, etc.; **3.** discharge of fluid from the body, e.g. semen. *Cf.* deposition.

emmetropia (Gr. *emmetros* = in proper measure, *ops* = eye), a state of the normal refractive property of the eye, i.e. normal sight.

emotion (L. *emovere* = to disturb), a state of strong feeling, as distress, excitement, fear, or anger. *Adj.* **emotional**. *See* limbic system.

emphysema (Gr. *en* = in, *physan* = to blow), an abnormal accumulation of air in tissues, especially in the lung, called hyperlucent lung.

emulsion (L. *emulgere* = to milk or drain out), **1.** a colloidal suspension (→ dispersion) in which both phases are liquids; one of the liquids is usually water or an aqueous solution and the other a water-immiscible liquid, such as oil; e.g. milk in which fat occurs in small drops in an aqueous solution, or dietary fats emulsified by bile salts; **2.** the light-sensitive layer on a photographic film, plate or paper, consisting of silver halides in gelatin; **emulsifier**, **emulgent**, an emulsifying agent that in small quantities helps to form or stabilize an emulsion. *Adj.* **emulsive**. *Verb* to **emulsify.**

emus, *see* Casuariiformes.

enamel, any hard glassy coating; e.g. **dental enamel** (*substantia adamantina*) which covers the dentine of the crown of the vertebrate tooth; produced by **ameloblasts** (adamantoblasts) and is almost completely composed of calcium salts, especially calcium phosphate as hydroxyapatite crystals, and fluoride apatite that is the hardest substance of animal tissues.

enantiomers (Gr. *enantios* = opposite, *meros* = part), optical → isomers.

enantiomorphism (Gr. *morphe* = form, shape), the occurrence of substances in two crystalline forms which are mirror images of each other. *See* isomers.

enation (L. *enasci* = to sprout), an abnormal outgrowth on a plant (usually on a leaf), caused by tissue hyperplasia, i.e. by facilitated cell divisions in the tissue infected by a virus.

encephal(o)- (Gr. *enkephalos* = brain), denoting relationship to the brain.

encephalography (Gr. *graphein* = to write, record), the roentgenography of the brain. *Cf.* electroencephalography.

end-, endo- (Gr. *endon* = within), pertaining to an inward location, within.

endarch (Gr. *arche* = beginning), pertaining to the arrangement of the primary xylem in plants; in the endarch type the oldest xylem elements (protoxylem) are located inside the younger ones (metaxylem). Plant → stems typically have an endarch type of the primary xylem. *Cf.* exarch.

endemic (Gr. *endemos* = native), pertaining to a species or some other taxon which occurs in its original, usually restricted area; **endemic disease**, prevailing continually in an isolated population. The phenomenon is called **endemism**.

endergonic (Gr. *endon* = within, *ergon* = power), pertaining to a chemical reaction needing external energy to react. *Cf.* exergonic.

endexine, *see* exine.

endobiont (Gr. *bion* = living), → endosymbiont.

endocardium (Gr. *kardia* = heart), the thin membrane lining the cavities of the vertebrate heart; composed of the → endothelium lying on the connective tissue layer.

endocarp (Gr. *karpos* = fruit), the innermost layer of the → pericarp of a plant, i.e. the wall of a → fruit.

endocrine, also **endocrinous, endocrinic** (Gr. *krinein* = to separate), secreting internally; pertaining to hormonal secretion (**endocrine secretion**, internal secretion), i.e. the type of secretion in which a physiologically active substance is released directly from a special cell (**endocrine cell**) into body fluids such as the blood, lymph, or interstitial fluid, i.e. not released through a secretory duct like in → exocrine secretion. Endocrine secretion is called **paracrine** if the secretion acts locally on adjacent cells only, and **autocrine** if it has an effect on the secreting cell itself. *See* endocrine glands.

endocrine glands, ductless glands; glands of animals secreting physiologically active substances, → **hormones**. The glands in the **endocrine system** (*endocrinium*) are ductless and the secretion is released into the body fluids, mostly directly into the blood. In some cases, as in the thyroid gland, the secretion is stored in follicles surrounded by glandular cells. The most important endocrine glands of vertebrates are the → **pituitary gland, thyroid gland, parathyroid glands, adrenal gland, pancreatic islets, pineal body, ovaries, testes**, and **placenta**; in invertebrates e.g. the → **corpus cardiacum, corpus allatum**, and **prothoracic glands** of insects, and the **X-glands, sinus glands**, and **Y-glands** of crustaceans. In addition to the production of germ cells, the ovaries and testes secrete sex hormones. Beside the proper endocrine glands there are several types of single cells or cell groups which produce hormones in animal tissues. *Cf.* exocrine glands.

endocrinology (Gr. *logos* = word, discourse), the research field of biology and medicine studying the biochemical and physiological roles of the hormone system (endocrine system) of animals.

endocuticle (L. *cuticula* = thin skin), the elastic inner layer in the → chitin cuticle. *Cf.* exocuticle.

endocytosis (Gr. *kytos* = cavity, cell), active uptake of extracellular material in small batches inside the cell; occurs in small vesicles formed by invagination of the plasma membrane, and includes the transfer of liquids (→ **pinocytosis**) or of solid material (*see* **phagocytosis**). In most cases the process is preceded by the recognition of a material by cellular receptors (receptor-mediated endocytosis), after which the material is enclosed by the plasma membrane to form a small vesicle, **endosome**, inside the cell. The endosome then joins and fuses with a → lysosome which releases hydrolytic enzymes to digest the endocytotic material for use by the cell. *Cf.* exocytosis. *See* clathrin.

endoderm, also **entoderm** (Gr. *derma* = skin), **1.** the innermost cell layer of an early animal embryo; differentiates into the inner surface (epithelium) of the branchial and alimentary canal and of the respiratory pathways in the lung, and into the glands associated with them, such as the → thymus, thyroid, liver, pancreas, gastric glands, and intestinal glands; sometimes called hypoblast; **2.** → gastroderm.

endodermis, a type of → dermal tissue in plants; comprises a single layer of cells serving as a border around structures lying within other tissues, e.g. around the central cylinder in roots, or the vascular bundle system in conifer needles. In roots, endodermis has a special role because of the → Casparian strip in the walls of its cells. The strip forces all the ions to go through the cytoplasm of the endodermal cells to the central cylinder and conductive tissues, to be transported to other parts of the plant.

endoenzyme, an enzyme acting inside the cell, i.e. an intracellular enzyme. *Cf.* exoenzyme.

endogamy, all systems of sexual reproduction in a population in which the mating partners are more closely related than individuals picked randomly from the population. *Syn.* inbreeding.

endogenote, *see* merozygote.

endogenous, also **endogenic** (Gr. *gennan* = to produce), occurring, originating, or growing within an organism.

Endognatha, also **Entognatha** (Gr. *gnathos* = jaw), *see* Insecta.

endolymph, the special tissue fluid within the membranous labyrinth of the inner ear of vertebrates, i.e. within the cochlea and vestibular apparatus. *Adj.* **endolymphatic**. *Cf.* perilymph.

endolymphatic duct (L. *ductus endolymphaticus*), a membranous canal of the membranous labyrinth in the inner ear of vertebrates, filled with → endolymph; connects the → utricle and saccule, terminating in a dilated blind extremity (**endolymphatic sac**) on the surface of the temporal bone beneath the dura mater.

endometrium, pl. **endometria** (Gr. *metra* = womb, uterus), the mucous membrane lining the uterus of mammals; hormonally controlled cyclic growth and breakdown phases occur during sexual maturity, → oestrus cycles, and implantation of embryos. In primates with → menstrual cycles, the endometrium can be divided functionally into three layers, i.e. stratum basale, stratum spongiosum, and stratum compactum, the latter two layers forming the functional part with large changes during the menstrual cycle.

endomitosis (Gr. *mitos* = thread), a form of → somatic polyploidization (*see* polyploid) which is rather common in differentiating or differentiated tissues of some animal and plant species. Endomitosis occurs within the → nuclear envelope, and involves the duplication and multiplication and often separation of the interphase chromosomes without actual division of the cell nucleus (endopolyploidy). Examples of **endomitotic cells** are salivary gland cells of dipteran insects, differentiated cells of the root tips of plants, and several cancer cells. *Syn.* endoreduplication. *See* polyteny, mitosis.

endomycorrhiza (Gr. *mykes* = fungus, *rhiza* = root), *see* mycorrhiza.

endomysium, pl. **endomysia** (Gr. *mys* = muscle), a thin fibrous membrane composed of connective tissue surrounding each single muscle fibre (muscle cell) inside the muscle. *Cf.* perimysium, epimysium.

endoneurium, pl. **endoneuria** (Gr. *neuron* = nerve), a thin fibrous membrane composed of the connective tissue surrounding each single → nerve fibre inside the nerve. *Cf.* perineurium, epineurium.

endonucleases, → nuclease enzymes e.g. in the sub-subclass EC 3.1.30., catalysing the cleavage of nucleic acids at interior bonds, i.e. phosphodiester bonds between sugars and phosphates. The reactions produce oligo- and polynucleotide fragments. Some DNases and RNases belong to this enzyme group. *See* restriction enzymes.

endoparasite (Gr. *parasitos* = parasite), a parasite living inside its host organism. *Syn.* entoparasite. *Cf.* ectoparasite.

endopelic (Gr. *pelos* = mud), pertaining to aquatic organisms living within the bottom sediment. *Cf.* epipelic.

endopelon, a community of algae living in waters within the bottom sediments, being endopelic. *Cf.* epipelon.

endopeptidases, a group of enzymes e.g. in the sub-subclasses EC 3.4.14., 3.4.21. and 3.4.24., catalysing the hydrolytic cleavage of peptide bonds of peptides; include digestive enzymes such as → pepsin, chymotrypsin, and trypsin. *Cf.* exopeptidases. *See* peptidases.

endophyton (Gr. *phyton* = plant), a community of unicellular algae living between cells or small leaves of other plants.

endoplasm (Gr. *plasma* = moulded, formed), the central portion of the cell cytoplasm surrounded by → ectoplasm.

endoplasmic reticulum, ER, an ultramicroscopic cell organelle composed of a network of membranous tubules and cavities (cisternae) in the cytoplasm of eukaryotic cells. The membrane (unit membrane) of ER is in connection with the outer membrane of the nuclear envelope. Certain parts of the surface of the ER membrane is covered by → ribosomes forming **rough ER**, **rER**, which is also in connection with the plasma membrane; this form of ER functions in the synthesis of and processing of secretory and membrane proteins. Other parts, **smooth ER**, **sER**, lacking ribosomes are involved in the synthesis of lipids. ER is a dynamic structure changing its form depending on the activity and differentiation state of the cell; it is almost absent in metabolically inactive cells, but abundant in cells in which synthesis is active, e.g. in secretory cells. Also → glycosylation of proteins takes place in the ER lumen and is mediated by membrane-bound glycosyl transferase. In muscle cells → sarcoplasmic reticulum (SR) derives from ER. *See* cell.

endoplasmin, a calcium-binding glycoprotein (100 kD) occurring abundantly in microsomal preparations fractionated from mammalian cells; concentrated particularly to the endoplasmic reticulum.

endopodite (Gr. *pous*, gen. *podos* = foot), the inner of the two distal branches of the → biramous appendage in crustaceans. *Cf.* epipodite, exopodite.

endopolyploidy, pertaining to cells the chromosome number of which have been increased by → endomitosis.

Endoprocta (Gr. *proktos* = anus), endoprocts, → Entoprocta.

Endopterygota (Gr. *pterygotos* = winged), **endopterygotes**; an insect superorder comprising species with holometabolous metamorphosis (also called **Holometabola**); the larvae have no compound eyes and their wings develop internally. The superorder includes nine orders: Neuroptera (e.g. alderflies, snakeflies, lacewings, ant lions), Coleoptera (beetles), Strepsiptera, Mecoptera (scorpion flies, snow flies), Lepidoptera (butterflies, moths), Diptera (true flies), Trichoptera (caddis flies), Siphonaptera (fleas), and Hymenoptera (e.g. bees, ants, wasps, saw flies). *See* Exopterygota.

endoreduplication, → endomitosis.

end organ, a small structure at nerve endings, such as a sensory organ or a neuromuscular junction around the peripheral end of a nerve fibre.

endorphins (< endogenous morphines), short polypeptides produced probably by all animals, acting as → neurotransmitters or neurohormones in many parts of the brain and peripherally e.g. in the gastrointestinal tract. The length of endorphin molecules varies, e.g. β-endorphin consisting of 31 amino acid residues. Their effects are transmitted through receptors, known as **opiate receptors**, and thus their physiological influences are similar to those of opiate alkaloids; in vertebrates, endorphins e.g. inhibit synaptic transmission in pain tracts (**analgesic effect**) and affect the limbic system in the brain (**euphoric effect**). In stress conditions endorphins are secreted from the pituitary gland into the blood and have effects also on various inner organs. Pentapeptides are specifically called **enkephalins**, which together with endorphins are called **opioid peptides**.

endoskeleton, *see* skeleton.

endosome (Gr. *soma* = body), endocytotic vesicle. *See* endocytosis.

endosperm (Gr. *sperma* = seed), a tissue developing in the seeds of seed plants after → germination, functioning as an energy store for the developing → embryo and the growing seedling. Endosperm is rich in starch and sometimes even in lipids. It is formed from the diploid endosperm mother cell in the → embryo sac after fertilization caused by the germinated pollen. One of the two sperm cells in the pollen tube fertilizes the egg cell, and the nucleus of the other sperm cell fuses with the diploid nucleus of the endosperm mother cell which becomes triploid and develops into endosperm.

endosperm mother cell, *see* endosperm.

endospore (Gr. *sporos* = seed), **1.** a spore type in → bacteria and cyanobacteria; the endospores are formed inside the cell through the division of the cytoplasm into many spores; in bacteria usually one per cell; **2.** any spore formed inside any cell or a → sporangium.

endostyle (Gr. *stylos* = pillar), **1.** a band of epithelium in the oesophagus of the tornaria larva of acorn worms; **2.** a mucous secreting groove with ciliated and gland cells in the midventral pharynx of tunicates and lancelets; **3.** the hypobranchial groove; a ciliated groove on the ventral surface in the pharynx of cyclostome larvae, ammocoetes, passing food particles to the oesophagus. The endostyle binds iodine, and in adult lampreys it develops into the **thyroid gland**.

endosymbiont (Gr. *symbionai* = to live together), an organism living in → endosymbiosis. *Syn.* endobiont.

endosymbiosis, a symbiotic relation between two species (**endosymbionts**), one living inside the other without causing harm to it; e.g. some bacteria living in protozoan cells, or microorganisms living in alimentary canals of animals, being essential for the digestion of the host. According to the **endosymbiosis theory,** the chloroplasts in plant cells are originally endosymbiotic cyanobacteria, and photosynthetic bacteria are ancestors of mitochondria. *Adj.* **endosymbiotic**. *Cf.* endoparasite.

endosymbiosis theory, *see* endosymbiosis.

endothecium (Gr. *theke* = box), a cell layer below the epidermis of the anther in a stamen; causes the opening of pollen sacs in the anther and the liberation of pollen grains.

endothelin, any of 21-amino acid peptides liberated from vascular endothelial cells in vertebrates; these peptides probably act as mediators of local hormones (e.g. endothelin-1, ET-1), causing constriction of blood vessels, evoking positive inotropic and chronotropic effects on myocardium, modulating synaptic transmission in the nervous system, and acting as growth-regulating factor in various tissues.

endothelium, the thin, simple → epithelium derived from the mesoderm, lining the inner surface of the blood and lymph vessels, and of the chambers of the vertebrate heart; comprises a single layer of flattened, polygonal cells, **endothelial cells**.

endothermic (Gr. *therme* = heat), **1.** pertaining to → endothermy; also endothermal; *cf.* ectothermic; **2.** pertaining to a chemical reaction in which external heat is absorbed, i.e. external heat must be supplied for the reaction to proceed. *Cf.* exothermic.

endothermy, the ability of certain animals to use their metabolic energy to raise body temperature above environmental temperature, and often maintain and regulate the body temperature at this higher level. Endothermic animals (**endotherms**) include all homoiothermic animals (**tachymetabolic endotherms**), but also some poikilothermic animals (**bradymetabolic endotherms**) such as large reptiles and large oceanic fish species. Also some insects are temporarily endothermic as they increase their body temperature by muscle shivering before flight, or keep their body temperature high by gathering into swarms, as e.g. bees in their winter-nest. *Cf.* ectothermy. *See* thermoregulation.

endotoxin (Gr. *toxikon* = arrow poison), intracellular toxin, especially the toxic membrane fraction (lipopolysaccharide), which forms an integral part of the bacterial cell wall in certain Gram-negative bacteria, as e.g. *Salmonella typhi* and *Vibrio cholerae*; it is not secreted but liberated when the microbe dies and disintegrates. Endotoxins are relatively heat-stable substances which are toxic to animal cells, and e.g. in mammals cause fever, diarrhoea and even shock; δ-endotoxin (glycoprotein) from *Bacillus thuringiensis* kills insects.

endotrophic (Gr. *trophe* = nourishment), *see* mycorrhiza.

endozoic dispersal (Gr. *zoon* = animal), dispersal of a non-parasitic organism to new areas within an animal, usually in the alimentary canal; endozoic dispersal is possible e.g. for hard-shelled seeds and fruits, and eggs of certain crustaceans.

endozoite, also **entozoite,** a non-parasitic animal living within another animal. *Adj.* **endozoic**, **entozoic**.

end-plate, *see* neuromuscular junction.

end-product inhibition, a biological control mechanism in sequential reaction systems in which the accumulation of the final product of a sequential metabolic reaction causes the inhibition of its own formation by a negative feedback mechanism (*see* regulation). This type of inhibition occurs in enzyme systems, e.g. in regulation of hormone synthesis and gene activity.

-ene, in chemical names indicating the presence of a carbon-carbon double bond; e.g. 1-butene, $CH_3CH_2CH{=}CH_2$.

energy, *W*, dynamic force; the capacity of any system, as an organism or a part, to work. In biological phenomena different energy types can be distinguished, like kinetic, potential, radiant, thermal, electrical, and chemical energies. The primary energy source for organisms is radiation energy, i.e. light energy emitted by the Sun. This energy form is absorbed and converted into chemical energy of organic molecules by plants in → photosynthesis. In cellular metabolic processes, chiefly in aerobic → cell respiration and less in → glycolysis and some other processes, energy is conserved in the formation of **adenosine triphosphate**, **ATP**. This phosphate-bond energy serves in organism as an immediate energy source for cellular processes involved in growth, development, movements, irritability, reproduction, etc.

Energy (and work) is measured in joules, J (*see* Appendix 1). Some examples of the magnitude of energies: the rotation energy of the Earth is 10^{29} J, the annual radiation energy from the Sun reaching the surface of the Earth is 10^{24} J, the thermal energy of one ton of coal is 10^{10} J, the daily food energy for an adult man is 10^{7} J, and the work of one heart beat is 0.5 J. *See* energy flow, energy metabolism, energy stores, ecological pyramid.

energy ecology, a branch of ecology dealing with the energy flow through the different trophic levels in an ecosystem. *See* ecological pyramid.

energy flow, the transfer of energy through living organisms in an ecosystem; the amount of this energy decreases in transfer from one trophic level to another, so that only about 10% of the available chemical energy passes to the next trophic level. *See* ecological efficiency.

energy metabolism, the part of the cellular metabolism that produces adenosine triphosphate (ATP) which is used as the immediate energy source for all cell processes involved in growth, development, reproduction, irritability, etc. Energy metabolism includes → glycolysis, citric acid cycle, electron transfer system, and photosynthesis. *See* oxidative phosphorylation, photosynthetic phosphorylation.

energy pyramid (Gr. *pyramis* = pyramid), → ecological pyramid.

energy stores, in animals, the storage of → glycogen and neutral → fats in cells; in multicellular animals glycogen is found in small granules inside cells, especially abundantly in muscle and liver cells. Neutral fats are stored particularly in → adipose tissue cells, found mostly around the inner organs, in many mammals also under the skin (subcutaneous fat). Proteins are not stored in special cells or tissues but structural proteins of most tissues (especially muscle) are available as energy sources in starvation and other stress situations.

In plants, energy stores may contain dissolved sugars (in vacuoles) or solid substances, composed of carbohydrates, lipids and proteins. These materials are usually present in seeds or storage tissues of shoots, roots, tubers, and corms. The most important energy store in plants is starch.

energy trees, *see* coppice.

Engelmann's test, a test used to prove the aerobic requirements of bacteria. *Theodor W. Engelmann* (1843—1909) demonstrated how the illuminated areas of the thread-like alga, *Spirogyra*, produced oxygen, and aerobic bacteria moved towards this oxygen source. The phenomenon is called aerotaxis.

enhancer, in genetics, **1.** a → modificator gene which enhances the phenotypic expression of other genes, or increases their propensity to → mutation; **2.** *cis*-acting regulatory sites in a eukaryotic gene usually located upstream of the → promoter; enhancers recognize the regulatory proteins of the → gene, and stimulate the function of the gene under their regulation by increasing transcriptional activity. In some cases, enhancers are located downstream of the gene or inside → introns; **3.** in chemistry, a compound by which a given reaction is accelerated, e.g. in autoradiography or in chemiluminescence.

enkephalins, *see* endorphins.

enolase, phosphopyruvate phosphatase, EC 4.2.1.11; an enzyme of → glycolysis catalysing the removal of water from 2-phosphoglyceric acid for forming phosphoenolpyruvic acid (PEP).

enrichment culture, a technique used to isolate a specific type of microorganism from a natural mixed culture or inoculum. The cultivation conditions (the composition and pH of the medium, temperature, light, oxygen etc.) are adjusted to favour the faster growth of the required organism but not that of unwanted species.

ent(o)- (Gr. *entos* = inside, within), denoting inner or within.

enter(o)- (Gr. *enteron* = intestine), denoting relationship to the intestine, gut. *Adj.* **enteral, enteric.**

enterocoel, also **enteroc(o)elom** (Gr. *koilos* = hollow), **enteroc(o)elous cavity**, a type of body cavity, → **coelom**, formed by the evagination of pouches from the embryonic gut; characteristic of many deuterostomes, such as echinoderms, hemichordates, and chordates, as a group called **enterocoelomates**. *Cf.* schizocoel.

enterogastrone (Gr. *gaster* = stomach), a peptide hormone secreted by cells in the duodenum of mammals, but probably also in the small intestine of lower vertebrates. It is suggested that enterogastrone is equivalent to → **gastric inhibitory polypeptide** (GIP) described later and shown to inhibit gastric secretion and muscle movements in the stomach.

enteroglucagon, *see* glucagon.

enterohepatic circulation (Gr. *hepar* = liver), the circulation of bile substances, especially bile salts from the liver to the intestine and then through the portal vein back to the liver; in every cycle, however, a small part of bile salts leave the body in faeces (in mammals about 10%).

enterokinase (Gr. *kinein* = to move), → enteropeptidase.

enteron, 1. the gut or alimentary canal; **2.** → coelenteron.

enteropeptidase, a proteolytic enzyme, EC 3.4.21.9, secreted by special mucosal cells in the intestine of vertebrates (small intestine in mammals); converts an inactive protease enzyme, **trypsinogen**, into active → **trypsin** by cleaving off a part of the trypsinogen molecule. *Syn.* enterokinase.

Enteropneusta (Gr. *pnein* = to breathe), **enteropneusts, acorn worms, tongue worms**; a class in the phylum → Hemichordata, comprising about 70 worm-shaped species, from 25 mm to 2.5 m in length. The animals live solitarily in the bottom of oceans burrowing into U-shaped tubes of sand or other debris. The body is divided into three regions, proboscis, collar and trunk. Respiratory water comes through the mouth to the pharynx, then entering the gill chambers through gill slits in the sides of the pharynx and further through small pores to the surface behind the collar. The pores are considered to be homologous with the gill apertures of cyclostomes.

enthalpy (Gr. *en* = in, *thalpein* = to heat), the heat content of a system; symbol H; enthalpy is a thermodynamic function described by the formula: $E + pV$, where E = internal energy of a system, p = pressure, and V = volume. *See* thermodynamics.

entoderm, → endoderm.

Entognatha, also **Endognatha** (Gr. *entos* = within, *gnathos* = jaw), *see* Insecta.

entomogenous (Gr. *entomon* = insect, *genes* = born), living in or on insects.

entomology (Gr. *logos* = word, discourse), a branch of zoology studying insects.

entomophagous (Gr. *phagein* = to eat), pertaining to an animal feeding on insects. *Syn.* insectivorous.

entomophily (Gr. *philein* = to love), the pollination by insects.

Entoprocta, also **Endoprocta,** or **Kamptozoa** (Gr. *entos* = inside, *proktos* = anus), **entoprocts** (endoprocts); an invertebrate phylum comprising about 150 species of solitary or colonial animals. They have a pseudocoelom and a U-shaped digestive canal ending in the anus near the mouth. The body **(calyx)** is cup-like with a crown or circle of tentacles on the top around the mouth and anus. Most of the species are microscopic, sessile (attached to a substratum by a single stalk), ciliary filter feeders which live in the bottom of oceans. Entoprocts were previously classified with ectoprocts (→ Ectoprocta) into the phylum Bryozoa.

entozoite, → endozoite.

entropy (Gr. *entropia* = a turning inward), the magnitude of the heat energy content expressing the thermodynamic measure of the degree of disorder within a system; given usually as joule/K/mole. Entropy describes the random motion of the atoms or molecules in the system and thus that energy fraction which is not available for work. In closed reaction systems, the amount of entropy increases with each thermodynamic change when some energy is lost as heat, and at equilibrium state all energy is converted to heat. Living organisms are open systems and are capable of decreasing locally and temporarily the amount of entropy by using energy and matter from outside to increase the order in the system; primarily they use solar energy. Finally, however, entropy leads to heat death of the universe. *See* thermodynamics.

enucleation (L. *e-* = away, *nucleus*, dim. of *nux* = nut), the removal of the total core part, as **1.** removal of the nucleus from the cell; achieved e.g. by treating cells with → cytochalasin B and then separating the nuclei by centrifugation; **2.** the removal of an organ or a tumour from the surrounding tissue; e.g. *enucleatio bulbi*, removal of the eye.

environment, the total whole of the physical, chemical, and → biotic factors which affect the function and activity of a given object, such as a gene, cell, organ, organism, or population. The phenotype of an organism is determined by a complex, bidirectional interaction between the genotype and the environment. The components of the environment can be divided into genetic (*see* genetic background) and non-genetic (*see* environmental variance). The term environment is usually confined to those factors that, in addition to the genotype, affect the development of the object.

environmental chemical, *see* pollutant.

environmental deviation, that part of the → phenotypic value of the individual which is due to environmental factors. Phenotypic value (P) consists of → genotypic value (G) and environmental deviation (E) so that P = G + E. The mean environmental deviation of the population is zero.

environmental ecology, a branch of ecology dealing with effects of human activities on regional abundances and dispersals of organisms as well as on the environmental factors regulating them; thus, environmental ecology is an ecological study which concerns nature conservation.

environmental gradient, a continuum of conditions ranging between extremes, e.g. the temperature gradation from the tropics to the arctic tundra.

environmental patch, the term used in the → island theory considering the patch as a deviating habitat in a uniform environment. The patches may be e.g. islands in an ocean, or lakes, and ponds, a meadow in a uniform forest area, and tops of mountains. The colonization of patches provides the organisms with effective dispersal ability. *See* patchy habitat.

environmental resistance, a total assemblage of the limiting factors which prevent individuals of a population to reach their → biotic potential and so limit the increase of population density. In the population level, environmental resistance is determined by the difference between the maximal and actual growth rates of population densities. *See* innate capacity for increase.

environmental resource, in ecology, any environmental factor which an organism can exploit without disadvantage to other organisms; e.g. nutrient, food objects, places for reproduction, individuals of the opposite sex, etc. To acquire possibilities for exploitation of environmental resources is a main aim of competition between organisms in natural conditions.

environmental variance (L. *variantia* = variability), that part of the phenotypic variability of a population which results from differences in the environment.

enzyme (Gr. *enzymos* = leavened), an organic catalyst; a protein produced by cells and catalysing a biochemical reaction, i.e. causing chemical changes in a substance (**substrate**) but remaining itself unchanged. Most chemical reactions in organisms are **enzymatic**. An enzyme molecule consists of a protein, → **apoenzyme**, which confers specificity, and a non-protein part, → **coenzyme**, which is the smaller component necessary for the activity of the whole enzyme, **holoenzyme**.
The enzyme binds specifically to its substrate forming an enzyme-substrate complex, and lowers the activation energy of the enzyme reaction. The reaction is reversible, although in organisms often functions in one direction. The chemical structure of enzyme protein is very complicated, its primary structure being genetically determined (*see* protein synthesis). The molecular weight of most enzymes is over 20,000, each enzyme protein molecule consisting either of a single polypeptide or of two or more polypeptides. Polypeptide subunits alone have no catalytic activity but together they form a specific three-dimensional conformation into which the substrate fits with the → lock-and-key model. The binding site for the substrate is called an **active centre**.
Changes in the secondary or tertiary conformation of an enzyme affect dramatically enzyme activity; this is affected by the → allostery of the enzyme structure, temperature, pH, and the concentration of the available substrate. The enzymes which are synthesized continuously are called **constitutive enzymes**, in comparison to **inducible enzymes** which are synthesized only when needed, i.e. when their substrates are present in the cell.
Enzymes are named by adding the suffix **-ase** to the name of the substrate upon which the enzyme acts, and they are classified by the Enzyme Commission of the International Union of Biochemistry (*see* enzyme classification).

enzyme classification, according to recommendations of the Enzyme Commission, EC, of the International Union of Biochemistry, the main groups of enzymes are numbered: EC class 1. = oxidoreductases, EC 2. = transferases, EC 3. = hydrolases, EC 4. = lyases, EC 5. = isomerases, and EC 6. = synthetases (ligases); the subsidiary grouping is indicated by the second (subclass), third (sub-subclass), and fourth (enzyme) numbers; e.g. alcohol dehydrogenase EC 1.1.1.1, or alpha-amylase EC 3.2.1.1.

enzyme inhibition, *see* competitive inhibition.

enzyme kinetics (Gr. *kinesis* = movement), the study of the rates of enzymatically catalysed reactions; it is expressed e.g. using the equations and graphics of → Michaelis—Menten kinetics (substrate kinetics) and of → Arrhenius plot (temperature kinetics). The Enzyme Commission of the International Union of Biochemistry has recommended the following symbols to be used in enzyme kinetics: v = observed velocity, V = maximum reaction velocity, K_m = Michaelis constant, K_s = substrate constant, K_i = inhibitor constant.

enzyme-linked immunosorbent assay, ELISA, an analytical method in which an enzyme-labelled antigen (or an antibody) reacts with an antibody (or correspondingly with an antigen) in an insoluble phase. The colour or fluorescence formed in the enzyme reaction is in a quantitative proportion to the substance to be determined. The method is very sensitive; less than a nanogram (10^{-9} g) of a protein can be measured.

enzymology (Gr. *logos* = word, discourse), a branch of biological sciences dealing with enzymes, and reactions catalysed by enzymes.

Eocene (Gr. *eos* = dawn, *kainos* = recent), an early epoch in the Tertiary period between Palaeocene and Oligocene about 55 to 38 million years ago when modern mammalian orders and suborders developed on the Earth.

eosin(e), a red dye, tetrabromofluorescein, $C_{20}H_6Br_4Na_2O_5$; used e.g. in tissue and cell staining. Structures which can be stained with eosine or other acid substances are called **eosinophilic** or acidophilic.

eosinophil (Gr. *philein* = to love), a cell or some other histological structure which is readily stainable with → eosin, especially a type of → **granulocyte** with cytoplasmic granules containing e.g. hydrolytic enzymes (e.g. histaminase) and toxic proteins against certain parasites. Eosinophils are highly phagocytic for antigen-antibody complexes but, like → neutrophils, posses also some general activity for → phagocytosis. Eosinophilic granulocytes are also called eosinocytes, eosinophilic cells, eosinophilic leucocytes, oxyphilic leucocytes, acidocytes, acidophils.

ependyma (Gr. *ependyma* = upper garment), the membrane lining the brain ventricles and the central canal of the spinal cord in vertebrates; is formed by a single layer of **ependymal cells** which are a type of glia cell (*see* neuroglia).

ephapse (Gr. *ephapsis* = a touching), electric → synapse; a synapse with electrical transmission; a primitive type of junction between nerve cells through which nerve impulses can transmit electrically, not chemically as occurs in synapses; it is a rather uncommon junction type between nerve cells in invertebrates, occurring however also in the vertebrate brain. The ephapse contains many → gap junctions for free ion transport thus conducting electricity (impulses) without delay in both directions. The junctions between the cardiac muscle cells, and between many smooth muscle cells, principally act like ephapses.

ephedrine, an alkaloid, 2-methylamino-1-phenyl-1-propanol, $C_6H_5CHOHCH(CH_3)$-$NHCH_3$, obtained from the leaves of several *Ephedra* species, or synthetically; in animals, ephedrine influences the adrenergic functions, i.e. acts as a sympathomimetic agent in the body, and is used in medicine e.g. as a bronchodilator, topical vasoconstrictor, to decongest the nasal mucosa, and to stimulate the central nervous system.

ephemeral (Gr. *ephemeros* = lasting a day), momentary; pertaining to a short duration, such as a short-lived plant or animal species (e.g. Ephemeroptera, mayflies); also → environmental patches may be ephemeral.

Ephemeroptera (Gr. *pteron* = wing), **ephemerids, mayflies**; an order of insects whose adult stages live only a few hours or days without feeding (have no alimentary canal); the nymphs are aquatic and feed on plant material. The mouthparts of adults are chewing but vestigial, the antennae are short. The wings are held vertically above the body at rest: the forewings are much larger than the hindwings. At the end of the abdomen there are two or three jointed cerci. About 2,000 species in 20 families are described.

ephippium, pl. **ephippia** (Gr. *ephippion* = saddle, saddlecloth), **1.** a saddle-like capsule under the carapace of female daphnids transformed from the walls of the brood chamber and protecting fertilized eggs. At moulting, the ephippium with the winter eggs detaches from the female and sinks to the bottom; **2.** sella turcica; *see* sella.

ephyra, pl. **ephyrae** (Gr. *Ephyra* = sea nymph), a free-swimming larval stage of medusae (jellyfishes, Scyphozoa) developing into an adult. *Cf.* planula, scyphistoma, strobila.

epi- (Gr. *epi* = on, upon), denoting on, upon, above.

epibiotic (Gr. *bioun* = to live), living on the surface of another organism.

epiblast (Gr. *blastos* = germ, shoot), **1.** the uppermost of the two cellular layers of the embryonic disc of the early reptilian, avian, or mammalian embryo, giving rise to the ectoderm; sometimes just refers to ectoderm; **2.** a small scale located opposite to the → scutellum in the grain in graminoids. It is supposed to be a rudiment of the other → cotyledon in these monocotyledonous plants. *Adj.* **epiblastic**.

epiblem(a), the outermost cell layer of the root in plants; in a young root it comprises the → rhizodermis (epidermis), which is replaced by exodermis in an older root. *Cf.* peridermis.

epibranchial (Gr. *branchia* = gills), *see* gill arch.

epibranchial groove, → hyperbranchial groove.

epicardium, pl. **epicardia** (Gr. *kardia* = heart), the fibrous membrane structure (*lamina visceralis pericardii, pericardium internum*) lining the surface of the vertebrate heart; formed as a visceral (inner) layer of the → pericardium growing attached to the cardiac muscle. *Cf.* endocardium.

epicotyl (Gr. *kotyle* = cup), the first actual internode in plants; is the part of the shoot between the → cotyledons and the first foliage leaves. *Cf.* hypocotyl.

epicranium, pl. **epicrania** (L. < Gr. *kranion* = skull), top of the head, e.g. **1.** the skin, muscle, and aponeurosis of the scalp, i.e. structures covering the vertebrate skull (cranium); **2.** a sclerite on the insect head between and behind the eyes.

epicuticle (L. *cuticula* = thin skin), a thin, waxy membrane upon the exocuticle in a → chitin cuticle of arthropods. *Cf.* endocuticle.

epideictic behaviour (Gr. *epideiknynai* = to exhibit, display), a behavioural pattern of animals by which they are able to regulate their population densities; e.g. **epideictic displays (contests)** are a synchronized communal behaviour in **epideictic groups** serving as indications of the population density; a **clumped roosting**, known in many bird species, informs of the relation between individual numbers and environmental resources and may result in behavioural or physiological changes in the members of the group.

epideictic groups, *see* epideictic behaviour.

epidermal growth factor, EGF, any of several peptides promoting growth of epithelial and mesenchymal cells of animals, best known in vertebrates; an important factor for the development of embryonic epithelium but also affects adults, e.g. it is secreted from salivary glands, it passes into the stomach where it promotes the proliferation of epithelial cells and inhibits gastric secretion. *See* growth factors.

epidermis, pl. **epidermides** (Gr. *epi* = on, upon, *derma* = skin), **1.** in animals, the outermost epithelial layer on all parts of the body, derived from the embryonic ectoderm. In vertebrates it is a non-vascular tissue of the skin but lies upon the vascular **dermis** (corium). Epidermis consists of several sublayers which in the mammalian skin are: a) **basal layer** (*stratum basale epidermidis*), the innermost layer composed of columnar cells which by continuous cell divisions produce new epidermis, b) **prickle-cell layer (spinous layer,** *stratum spinosum e.*), composed of slightly flattened cells, c) **granular layer** (*stratum granulosum e.*) in which cornification begins in flattened granular cells, d) **clear layer** (*stratum lucidum e.*) composed of flattened transparent cells, and e) **horny layer** (*stratum corneum e.*) composed of flattened, dead cells totally cornified with → keratin, peeling off from the skin surface. A similar structure also occurs in the skin of other vertebrate groups but may be partly or totally covered by different types of → cornifications.

In invertebrates, a single-layered epidermis is the principal covering of the body. Many species secrete non-cellular **cuticle** to cover the epidermis; e.g. the epidermis of arthropods with two-layered cuticle provides both protection for the animal and a skeletal support. The molluscan epidermis is soft and contains mucous glands, which in some species secrete calcium carbonate to form a shell;

2. in plants, epidermis is the outermost cell layer covering all parts of the plant. The cells are mostly colourless, without chloroplasts, and the cells of aerial parts are covered by a wax layer, cuticle. Root epidermis is also called **rhizodermis**, and an **exodermis** may develop within it replacing the former. In aerial parts a **hypodermis** (sometimes many layers as in cacti) may develop under the epidermis. *Adj.* **epidermal, epidermic.**

epididymis, pl. **epididymides** (Gr. *didymos* = testis), **1.** the longitudinal organ on the surface of the testis in reptiles, birds and mammals. It consists of several tubules (**epididymal tubules**) beginning from the testis and joining in the caudal section of the epididymis into one long tubule which continues as the **vas deferens** (ductus deferens) to the cloaca or penis. Spermatozoa pass slowly through the epididymis mainly by the peristaltic movements (back and forth) of the smooth muscle in the tubular wall, and become mature and fertile by the influence of epididymal secretions; **2.** the term epididymis is used also for analogous structures of some invertebrates, e.g. leeches.

epidural space, a narrow cavity in the vertebral canal between the spinal dura mater (hard meningeal membrane) and the periosteal membrane. On the brain surface similar space may be formed if the dura mater becomes detached from the bone surface.

epiedaphic (Gr. *edaphos* = soil), pertaining to an organism living on the soil. *Cf.* hemiedaphic.

epigamic (Gr. *gamos* = marriage), pertaining to courtship or reproduction of an animal with structural or behavioural features which especially in the course of courtship performance attracts or stimulates individuals of the opposite sex; e.g. bright and conspicuous nuptial plumages of birds, different rich-coloured structures of reptiles, songs of birds, or courtship fights. Individuals of the same sex react to epigamic features usually in an opposite, often aggressive manner.

epigeal (Gr. *ge* = earth), living on the earth; e.g. **1.** describes an insect or other invertebrates living near or on the ground; **2.** the germination (**epigeal germination**) of seeds in which the → cotyledons rise upon the earth surface. *Cf.* hypogeal.

epigenesis (Gr. *gennan* = to produce), **1.** the formation of new structures during → embryogenesis. According to the **theory of epigenesis**, real development occurs during embryogenesis, i.e. new structures are formed from the originally undifferentiated germ cells. The theory of epigenesis has displaced the former theory of → preformation; **2.** regulation of the expression of the gene activity without alteration of the genetic structure.

epigenetic, 1. pertaining to the interaction of genetic factors during ontogenic development of an organism; **2.** denoting the regulation of realization of genetic possibilities. **3.** relating to → epigenesis.

epigenetic crisis, a period of the embryonic development during which some part of the embryo is damaged more easily than the other parts.

epigenetics, developmental genetics; a branch of biology dealing with the causal relations of the ontogenic development of an organism. It is the study of genetic processes which lead to the formation of the phenotype.

epiglottis (Gr. *glotta, glossa* = tongue), a cartilaginous flap-like structure on the floor of the mammalian pharynx descending against the entrance to the larynx during swallowing; prevents food etc. from entering the trachea. *Adj.* **epiglottidean**.

epigyne, also **epigynum** (Gr. *gyne* = woman, female), a chitinous plate in front of the genital opening of a female arachnid.

epigynous, pertaining to a plant → ovary being enclosed by the receptacle, the other flower parts arising from the receptacle above; the gynoecium is thus **inferior**. *Cf.* hypogynous, perigynous.

epilimnion, pl. **epilimnia** (Gr. *limne* = lake), a water layer found during the summer stagnation in lakes and seas above the thermocline and → hypolimnion. The temperature of the epilimnion is evenly warm and the content of oxygen is usually high.

epilithic (Gr. *lithos* = stone), pertaining to organisms living on stones; e.g. mosses and lichens, in aquatic habitats such as bryozoans on bottom stones.

epilittoral (L. *lit(t)us* = shore), *see* littoral.

epimer (Gr. *meros* = part), either of two diastereomers (molecules) differing only in the spatial configuration around one asymmetric atom, e.g. glucose and galactose with respect to carbon-4. *Adj.* **epimeric**.

epimerases, → isomerase enzymes in the subclass EC 5.1. catalysing spatial arrangement of atoms about a single carbon atom in a molecule, i.e. epimeric molecular changes in cell metabolism. An epimerase acts in sugar metabolism, catalysing the conversion of xylulose 5-phosphate into ribulose 5-phosphate; in the reaction the positions of a hydroxyl group and a hydrogen atom are changed, the epimerase reaction being reversible.

epimere, the dorsal part of a → somite developing into skeletal muscles of the vertebral embryo.

epimerite, a prolongation of the → protomerite in some gregarine protozoans, anchoring the animal to the host's tissue (**fixing organ**). *Cf.* deutomerite.

epimeron, pl. **epimera** (Gr. *meros* = thigh), the posterior part of the → pleuron in the thorax of insects. *Cf.* episternum.

epimorphosis (Gr.*morphe* = form), a form of larval development in some segmental animals, such as arthropods, in which all successive stages in the larval metamorphosis are passed in the egg and at hatching the structure of the animal is like that of the adult; e.g. in some chilopods (subclass Epimorpha) when the young before hatching have the full complement of legs and segments; **2.** regeneration of a structure of an organism. *Cf.* anamorphosis.

epimysium, pl. **epimysia** (Gr. *mys* = muscle), the fibrous sheath of connective tissue covering an entire muscle. *Cf.* perimysium, endomysium.

epinasty (Gr. *nastos* = close-pressed, firm), the increased growth of the upper side of a plant organ; in leaves causes the leaf blade to curl downwards. *Cf.* hyponasty.

epinekton (Gr. *nektos* = swimming), organisms which are incapable of moving actively against water currents but move attached onto the surface of actively swimming animals. *See* nekton.

epinephrine (Gr. *nephros* = kidney), → adrenaline.

epineurium, (Gr. *neuron* = nerve), the fibrous sheath of connective tissue covering the peripheral nerves. *Cf.* endoneurium, perineurium.

epineuston, *see* neuston.

epiparasite (Gr. *parasitos* = parasite), a parasite, such as a louse, living on the surface of its host. *Syn.* ectoparasite. *Cf.* endoparasite.

epipelagic zone (Gr. *pelagos* = sea), the uppermost pelagic zone in oceans between the surface and mesopelagic zone, i.e. from the surface to the depth of about 200 m.

epipelic (Gr. *pelos* = mud), pertaining to aquatic organisms living on the bottom sediment. *Cf.* endopelic.

epipelon, a community of (mostly microscopic) algae living on the bottom sediment, being epipelic. *Cf.* endopelon.

epipetalous, pertaining to plant stamens being more or less fused with the petals.

epipharynx, pl. **epipharynges** (L., Gr. *pharynx* = throat), **1.** an unpaired process attached to the posterior surface of the labrum in insect mouthparts; an especially large and tube-like structure (with the hypopharynx) in the species specialized for sucking blood, as e.g. in some bugs and fleas. **2.** a chitinous plate of the rostrum in arachnids. *Adj.* **epipharyngeal**.

epiphragm (Gr. *epiphragma* = covering, lid), **1.** a protective calcified membrane or plate secreted during dry weather by some pulmonate molluscs to cover the shell aperture and thus avoid desiccation; **2.** in botany, → peristomium.

epiphyll (Gr. *epi* = on, *phyllon* = leaf), a plant growing on a leaf of another plant; epiphylls are mostly mosses and lichens, common especially in tropical cloud forests. *Cf.* epiphyte.

epiphysis, pl. **epiphyses** (Gr. *phyesthai* = to grow), **1.** *apophysis ossium*, the expanded end of a long bone in which longitudinal growth occurs; *cf.* diaphysis; **2.** → pineal body. *Adj.* **epiphyseal** or **epiphysial**.

epiphyte (Gr. *phyton* = plant), a plant growing on another plant, e.g. on a twig or a branch of a tree, without being parasitic or exploiting the host plant. Epiphytes are common especially in tropical rain forests; e.g. various lichens, mosses and orchids. Typical epiphytes are also many algae living on other plants in aquatic habitats. The community formed by epiphytes is **epiphyton**. *Cf.* epiphyll.

epiphyton, *see* epiphyte.

epiplankton (Gr. *planktos* = wandering), plankton organisms living on floating objects or in the upper water zone from the depth of 200 m to the surface.

epipodite (Gr. *pous*, gen. *podos* = foot), an additional extension attached to the → protopodite of some postoral cephalic and/or thoracic legs in many crustaceans; depending on species, it is modified for different functions such as food handling and respiration. *Cf.* endopodite, exopodite, biramous appendage.

epiproct (Gr. *proktos* = anus), an unpaired plate or longitudinal appendage in some insects at the end of the abdomen above the anus; particularly long in some thysanurans and ephemeropterans.

epipsammon (Gr. *psammos* = sand), aquatic organisms living on gravel or sandy bottoms. *Adj.* **epipsammic**. *See* psammon.

episome (Gr. *soma* = body), a genetic element (e.g. plasmid) consisting of DNA which may exist as a free element or integrated as a part of the chromosome; common in bacteria.

epistasis (Gr. *epistasis* = stopping), **1.** in genetics, a form of gene interaction in which a gene masks the phenotypic expression of some other non-allelic gene (→ allele); *adj.* **epistatic**; respectively the masked gene is called hypostatic. In **dominant epistasis**, the dominant A gene masks the expression of the hypostatic B or b gene in A-B- and A-bb genotypes. In **recessive epistasis**, the recessive aa combination suppresses the expression of B and b in aaB- and aabb genotypes. In population and quantitative genetics all non-allelic, non-additive interactions of genes are called epistatic. In the example above, A and B describe dominant, a and b recessive genes; **2.** a pellicle or scum on the surface of a liquid, especially of urine.

episternum, pl. **episterna** (L. < Gr. *epi-* = on upon), *sternon* = breast), **1.** a cartilaginous appendage in the anterior part of the sternum of amphibians, reptiles, and monotremes; **2.** the anterior part of the pleuron in the thorax of insects; *cf.* epimeron.

epistropheus (Gr. *epistrephein* = to turn about), → axis, the second cervical vertebra in tetrapod vertebrates except amphibians.

epithelial-mesenchymal interaction, a tissue-interaction between the → epithelium and the → mesenchyme in the early embryonic stage of an animal; determines the direction of development in both these tissues.

epitheliomuscular cell, → musculo-epithelial cell.

epithelioma, a neoplasmic growth chiefly of epithelial origin; may be manifested in a benign form (e.g. adenoma and papilloma) or in a malignant form (carcinoma).

epithelium, pl. **epithelia** (Gr. *thele* = nipple, thin skin), the tissue covering the external surface of the body and the internal surfaces of cavities, ducts, and other pathways and vessels of animals, including most glands (**secretory epithelium**), some sensory organs or their parts (**sensory e.**), and epidermal → **cornifications**. The epithelium consists of cells joined together with → tight junctions and there may be scant cementous substance between the cells. Epithelial cells have a → basement membrane agaist connective tissue. The epithelium may comprise a single cell layer, **simple e.**, as e.g. the glandular and intestinal epithelia, or it consists of several cell layers (**stratified e., laminated e.**), e.g. the vertebrate skin, mouth, pharynx, anterior part of the oesophagus, and anal area. A **pseudostratified e.** gives an impression of being stratified because the cell nuclei locate at different levels in a simple epithelium, as e.g. often in a sensory epithelium with receptor cells and supporting cells. The urinary organs of vertebrates are lined by a distensible **transitional e.**, which changes so that in an empty ureter or bladder it displays several cell layers, whereas in a filled, stretched cavity only two layers. Epithelial cells may be **columnar**, **cuboidal**, or **squamous**; some cells may be ciliated forming a **ciliated e.**, as found in the respiratory pathways. Embryologically epithelial tissues are mostly derived from the ectoderm or endoderm. Special epithelium types are the → **endothelium** and **mesothelium**, derived from the mesoderm. The formation of the epithelium is called **epithelialization** (or epithelization). *Adj.* **epithelial**.

epithem (Gr. *epithema* = something put on), a group of cells found in plant leaves beneath a water stoma with which it forms a water excreting → hydathode.

epitonic, → dorsiventral.

epitope (Gr. *topos* = place), an antigenic determinant of a complex antigenic molecule. *See* antigen.

epitreptic behaviour (Gr. *treptos* = turned), the behaviour of an animal resulting in the approach of another conspecific animal.

epizoic dispersal (Gr. *zoon* = animal), the → dispersal of an organism to new areas by attaching itself on the surface structures of animals, such as on feathers or hairs; e.g. the burdock. *Cf.* endozoic dispersal, phoretic dispersal.

epizoite, also **epizoon, 1.** a non-parasitic animal living on the body of another animal; e.g. a barnacle living on the shell of a clam; **2.** an animal which spreads to new areas by attaching to the body surface of another animal. *Adj.* **epizoic**. *Cf.* ectozoic dispersal.

epizootic, pertaining to an epidemic disease (**epizooty**) among animals.

epoophoron (Gr. *oophoron* = ovary), a rudimentary structure associated with the ovary of amniotic vertebrates consisting of mesonephric tissue; also called parovarium or pampiniform body.

EPSP, → excitatory postsynaptic potential.

equation of state, an equation describing the relations of pressure, temperature, and volume in any system. The equation for ideal gases is derived from → Boyle—Mariotte's and Gay—Lussac's laws, also applied to solutions: $pV = RT$, where p = pressure, V = volume, R = gas constant (8.3144 J mol^{-1} K^{-1}) and T = temperature (K). The equation is valid for one mole of matter in a system but an application for n moles can be derived according to → van't Hoff's law.

equatorial plate, the equilibrium position of chromosomes of a dividing cell in the equatorial region of the → mitotic spindle during the metaphase stage of → mitosis or → meiosis. *Syn.* metaphase plate.

equilibrium (L. *aequus* = equal, *libra* = balance), a state of balance; i.e. a state when opposing forces counteract each other; **equilibration**, the achievement of this state. *See* acid-base balance, Donnan equilibrium, genetic equilibrium.

equilibrium population, 1. a population in which the frequencies of different genotypes correspond to those expected on the basis of the persistence of random mating; **2.** in ecology → stable equilibrium.

equilibrium potential, *see* resting potential.

equilibrium sense, static sense; in vertebrates, the sense based on the function of the mechanical receptors of the → **utricle** and **saccule** of membranous labyrinth in the vertebrate inner → ear, and on the brain processes integrating the sensory information. Mechanical receptor cells (**hair cells**) in the thickened epithelial spot, **macula**, of the utricle (in many vertebrates also in the saccule) react as a response to the Earth's gravity. This occurs when the calcareous crystalline particles, **otoliths** (statoliths, otoconia), floating on the top of the receptor cell group, press the long sensory hairs of these cells. Body movements for maintaining the balance are chiefly based on reflexes initiated from these receptors and controlled by the cerebellum. Also receptor cells inside the **ampullae** of the three **semicircular canals** are involved in the control of equilibrium. The cells are irritated by movements of the endolymphatic fluid inside the canals, i.e. originally by the movements of the head. The collective term **vestibular organ** (vestibular apparatus) is used for the utricle, saccule, and semicircular canals. Neuronal information from the vestibular organ propagates as nerve impulses along the **vestibulocochlear nerve** (the eighth → cranial nerve) to the **vestibular nuclei** in the medulla oblongata, and from there chiefly to the cerebellum and motor tracts.

Analogous sense organs are found also in many invertebrates (e.g. crustaceans), called → **statocysts**.

equilibrium theory, a theory in ecology concerning the organization of a community; claims that a certain equilibrium state is prevailing in a community, and this state shall return after a disturbance in the system. The conception of equilibrium is disputable. At present, it has to include at least three conditions: 1) the ability of the community to resist change, or to prevent it, 2) the stability of the community (*see* stable equilibrium), or the ability of the populations living in the community to maintain population densities within certain limits, and 3) the capability of returning to the equilibrium state after a disturbance. *See* stabilizing factor.

equipotent, having equal capacities or effects, e.g. a group of embryonic cells having the equal capability of developing into any tissue.

Equisetales, horsetails, a plant order in subdivision Sphenopsida, division Pteridophyta, in some classifications an order in the division Sphenophyta; evolutionarily an old group, the only living family is Equisetaceae which comprises the present horsetails, other families are extinct. Horsetails reproduce with spores, leaves are scale-like and green stem branches carry out photosynthesis.

equivalence, Eq (L. *aequus* = equal, *valere* = to be of some value), an equality in value, force, importance etc; in chemistry, the quality of having equal valence. *Adj.* **equivalent**, equal in any respect; in chemistry, having the same capacity to react. *See* equivalent weight and gram equivalent in Appendix 1.

equivalence group, a group of things of the same, equivalent value. The term is especially used to describe cells which have the same → developmental potential but can differentiate into different directions; e.g. in a fruit fly embryo the cells of the neurogenic region have the same developmental potential but can differentiate either into neural or epidermal cells.

era (L. *aera* = beginning of new age), a period of time marked by distinctive events or character, e.g. → geological eras, consisting of a number of periods.

erection (L. *erigere* = to raise up, erect), an act or a state of raising up or erecting, or being erected; in biology, e.g. a rigid state of an organ, as of the penis (or clitoris), which is a function of **erectile tissue** filled with blood; in vertebrates, erection is controlled by the parasympathetic nervous system. *Adj.* **erectile**, capable of being erected; **erective**, tending to erect.

erepsin, a common name for the mixture of proteolytic peptidase enzymes secreted by intestinal glands of vertebrates.

erg(o)-, 1. (Gr. *ergon* = work), denoting relationship to work or muscular activity; **2.** pertaining to → ergot.

ergastic substances, non-living passive products in the cytoplasm or vacuoles of a cell; e.g. various crystals, fat globules, fluids, or starch grains.

ergastoplasm, the term used earlier for → endoplasmic reticulum.

ergograph (Gr. *graphein* = to write, record), an instrument for recording muscular work, resulting in an **ergogram**. *Adj.* **ergographic**.

ergonomics (Gr. *nomos* = rule, law), the study dealing with human activity in the working environment; attempts to find out the anatomical, physiological, and psychological principles affecting the effective use of human energy, and use this knowledge in the design of the physical environment.

ergosterol, 7,22-didehydrocholesterol; a sterol obtained originally from → ergot but was later found to be a common substance in animal and plant tissues; in animals ergosterol is an important provitamin which by action of ultraviolet light is converted into ergocalciferol (vitamin D_2).

ergot, rye smut; a disease of rye and some other grasses caused by the fungus *Claviceps purpurea*. Ergot means both the fungus and the disease, as well as the hardened mycelial mass forming the → sclerotium of the fungus, replacing the infected seeds. Ergot contains several alkaloids, e.g. → ergotamine.

ergotamine, an alkaloid obtained from → ergot blocking the adrenergic nervous system (chiefly through alpha receptors) in animals, especially in smooth muscles. Ergotamine and its less toxic derivative, **dihydroergotamine**, are used in the treatment of migraine. A mixture of three ergot alkaloids (ergocristine, ergocornine, ergocryptine) is **ergotoxine** that have similar effects but is more toxic.

Erlenmeyer flask, conical flask (*Emil Erlenmeyer*, 1825—1909), a conical glass or plastic dish used in laboratories.

erosion (L. *erodere* = to wear away), **1.** the process of wearing and washing away surface materials of the earth by the action of water, glaciers, wind, etc. At present, global erosion is one of nature's greatest risk factors because it expands with increasing speed, resulting from the increased growth of human populations, extensive agriculture, and destruction of forests. It was estimated at the beginning of the 1990's that global erosion wore away 24,000 million tons of humus annually from the cultivated fields, and one third of the cultivated areas had problems of erosion; **2.** an injury or rubbing of a tissue, such as the mucous membrane of the digestive tract, causing a shallow ulcer.

eruciform (L. *eruca* = caterpillar, *forma* = shape), pertaining to a larval form of some insects, e.g. lepidopterans; has three pairs of jointed legs anteriorly and, depending on species, varying numbers of → **prolegs** on the abdomen.

Erysiphales (Gr. *erysi-* = red, *siphon* = tube), a fungal group in the class Ascomycetes, subdivision Ascomycotina, division Amastigomycota. The species are parasites of vascular plants and form typically a white covering of hyphae on their host. Many species are economically hazardous because they decrease crops and hinder cultivation.

erythema (Gr. *erythros* = red), flush on the skin; inflammatory redness of the skin, due to congestion of the blood capillaries.

erythr(o)- (Gr. *erythros* = red), pertaining to red, red blood cells (erythrocytes), or redness of the skin.

erythroblast (Gr. *blastos* = germ, shoot), an immature cell in the bone marrow developing into a red blood cell; has several developmental stages.

erythroblastosis, 1. erythroblastaemia, i.e. the presence of abnormally large numbers of →

erythroblasts in the peripheral blood; **2.** a disease in fowl characterized by an increased number of erythroblasts in the blood; called also erythroleucosis (erythroleukosis).

erythrocruorin (L. *cruor* = blood), a pigment found in the blood or haemolymph of many invertebrates, such as annelids and molluscs; also called invertebrate haemoglobin. *See* blood pigments.

erythrocyte (Gr. *kytos* = cavity, cell), → red blood cell.

erythromycin (Gr. *mykes* = fungus), an antibiotic obtained from *Streptomyces erythraeus*, a fungus-like Gram-positive bacterium.

erythropoietin (Gr. *poiesis* = making), erythropoietic hormone; a glycoprotein hormone in vertebrates enhancing the production of red blood cells from their stem cells, i.e. the hormone which stimulates **erythropoiesis** in the bone marrow; it also stimulates the release of → reticulocytes from the bone marrow. Erythropoietin is produced by the liver of the embryo and foetus, and chiefly by the kidneys of the adults. The usual stimulus for the secretion of erythropoietin is hypoxia. The hormone is used in medicine, and illegally as a doping drug (EPO) to increase the haemoglobin content of the blood. Also called haematopoietin, haemopoietin.

erythrose, a four-carbon monosaccharide, $C_4H_8O_4$, with two central -OH groups in the *cis* orientation (in threose in the *trans* orientation); occurs e.g. in the Calvin cycle of → photosynthesis.

Escherichia, the genus of rod-shaped Gram-negative motile or non-motile bacteria of the family Enterobacteriaceae, *Escherichia coli*, occurring in the intestine of vertebrates and thus widely distributed in nature; some strains can cause infections. These bacteria are widely used in biological, biochemical and medical research.

eserine, → physostigmine.

esophagus, → oesophagus.

ESR, → electron spin resonance.

essential amino acids, amino acids which are not usually synthesized in animal tissues but are necessary in nutrition of animals. For the adult man, like most mammals, eight amino acids are needed in the diet: isoleucine, leucine, lysine, methionine, phenylalanine, threonine, tryptophan, and valine. In addition, arginine and histidine are needed for normal growth in the immature age.

essential fatty acids, EFA, some unsaturated fatty acids which are not synthesized in animal tissues, but are necessary in nutrition of animals. Most mammals need three essential fatty acids, which are linoleic (linolic), linolenic, and arachidonic acid (some synthetic activity of arachidonic acid from linoleic acid may occur). EFAs are needed as material for cell membranes, and for → prostaglandin synthesis.

essentialism, a philosophical view stating that there is a definitive set of properties in each individual. This view has earlier been adopted as a method (typology) in → taxonomy of organisms, based on the supposition that all individuals of a taxonomic unit conform to a given structural plan without remarkable variation and evolution.

essential oils, plant products, such as pinene, terpenes, camphor, or menthone, giving characteristic odour and taste, essence, to the plant.

ESS-theory, *see* evolutionary stable strategy.

EST, → expressed sequence tag.

ester, a compound formed from an alcohol and an organic acid when their -OH and -COOH groups react with each other by eliminating water, H_2O; e.g. ethyl alcohol and acetic acid form ethyl acetate. Many esters are sweet-scented.

esterases, enzymes of the subclass EC 3.1. of the → hydrolase class; catalyses the hydrolysis of the ester bond producing an alcohol (or phenol) and an acid anion; occur e.g. in → lysosomes inside the cells, and in the alimentary canal and various tissues of animals. *See* choline esterase.

estivation, → aestivation.

estrogen, → oestrogen.

estrus, *see* oestrus cycle.

estuary (L. *aestuarium* < *aestus* = tide), a coastal water area or passage in the mouth of a river, where fresh water of the river enters the ocean (or sea) forming brackish water conditions with haloclines, especially in waters where tidal activity is weak. *Adj.* **estuarine, estuarial.**

etaerio, aggregation of fruits, e.g. of follicles in buttercups or drupes in raspberry and blackberry.

ethanal, → acetaldehyde.

ethane, bimethyl, dimethyl, C_2H_6; the second hydrocarbon in the alkane series, constituent of natural gas (about 9%); widely used in organic syntheses.

ethanol, ethyl alcohol, an organic compound, C_2H_5OH, produced from sugars in alcoholic → fermentation; in small concentrations found also in the alimentary canal of many animals being produced by intestinal microorganisms, but also in tissues of some animals as the end product of → anaerobic respiration. Anaerobic conditions prevailing e.g. under an ice cover may cause ethanol production also in plant cells. Ethanol is freely soluble in water and it is used as a solvent in biology, chemistry and industry. It is also used for the conservation of biological material, in the preparation of specimens for microscopic observation, and as a common sterilizing agent in laboratories.

Ethanol is the effective component of alcoholic beverages. It affects the properties of nerve cells, depressing brain function; in high concentrations it inhibits functions of all cell types. *See* alcohol.

ethene, → ethylene.

ether (L. *aether* < Gr. *aither* = upper air, pure air), **1.** any organic compound in which two carbon atoms are linked to a common oxygen atom, thus containing the group -C-O-C, i.e. ether bond; **2. diethyl ether,** $(CH_3CH_2)_2O$, a colourless, highly inflammable and volatile liquid, used e.g. as a solvent of lipid compounds and especially earlier as an anaesthetic; **3.** pure air.

ethereal oils, volatile and sweet-scented oily compounds produced by plants; occur in hairs of leaves and fruits possibly having a role as a lure for pollinating animals. Ethereal oils, obtained e.g. from the rose, anise, pink, eucalyptus, and many coniferous trees, are used in cosmetics, medicines and drinks.

ethidium bromide, a poisonous chemical used in molecular biology to stain nucleic acids which become fluorescent and thus can be detected under UV light. *Syn.* homidium bromide.

Ethiopian region, a zoogeographical region including Africa south of Sahara and the southern part of the Arabian peninsula. The region is divided into subregions of Africa and Malagasy. *See* zoogeographical regions.

ethmoid (Gr. *ethmos* = sieve), sieve-like, cribriform; e.g. **ethmoid bone**, the bone of the nasal cavity of vertebrates consisting of the cribriform lamina; **ethmoidal**, pertaining to the ethmoid bone.

ethogeny, also **ethogenesis** (Gr. *ethos* = custom, *genos* = descent, birth), behavioural ontogeny; a postnatal development and maturation of the innate behaviour of an animal.

ethogram (Gr. *graphein* = to write), a description and inventory of all fixed-action patterns and vocal patterns in an animal species; important in ethological studies, especially when explaining the evolution of a behavioural pattern.

ethological race, *see* race.

ethology (Gr. *logos* = word, discourse), a branch of biology studying the behaviour of animals, especially its evolution as an adaptive feature in relation to the environment. For this reason ethological studies concentrate on observations and experiments carried out in natural conditions. Important objects of studies have been e.g. signals and fixed-action patterns. Ethology is divided into several aspects with their own specialities, as **descriptive ethology**, **comparative ethology**, and **classical ethology**; **human ethology** and **cultural ethology** have developed within the comparative ethology. Especially *Konrad Lorenz* (1903—1989) and *Niko Tinbergen* (1907—1988) have creditably developed ethology. *See* behavioural ecology.

ethyl, a hydrocarbon radical, CH_3CH_2-, occurring in structures of many organic molecules, such as in ethanol.

ethyl alcohol, → ethanol.

ethyl acetate, ethyl ethanoate, acetic ether, $CH_3COOC_2H_5$; an ester composed of ethanol and acetic acid, and obtained by the slow distillation of an acidified mixture of these compounds. It is a narcotic liquid used e.g. in insect traps, pharmaceutical preparations, and artificial fruit essences.

ethylene, ethene, a gaseous hydrocarbon, CH_2CH_2; a natural plant hormone which promotes the ripening of fruits, prevents the growth of root and stem and accelerates the leaf abscission. Some bacteria produce ethylene and it is also liberated from some herbicides like ethephon. Ethylene influences the nervous system of animals producing anaesthesia. Ethylene is an explosive constituent of ordinary fuel gas. Also called olefiant gas.

ethylenediaminetetraacetic acid, → EDTA.

ethylene glycol, 1,2-ethanediol, $HOCH_2CH_2$-OH; occurs as a natural → cryoprotectant in some insects. It is used in cooling systems of different machines and in biological studies to prevent ice formation in tissues at low temperatures.

etiolation (Fr. *étioler* = to bleach), the bleaching of plants by reducing or excluding light; the reduction of light turns a plant yellow because of the lack of chlorophyll, the stem becomes long and thin, the leaves remain small, and the anthocyanin pigments are not formed.

etioplast (Gr. *plassein* = to mould), a → plastid without chlorophyll; etioplasts are yellow and formed from proplastids in plants growing in the dark or in very low illumination. In the light etioplasts change rapidly into chloroplasts.

ETL, → economical trait loci.

eu- (Gr. *eu* = good), denoting relationship to good, beneficial, normal

Eubacteria (Gr. *bakterion* = little rod), a subdivision of Bacteria. *See* bacterium.

eubiotics (Gr. *bios* = life), the study of healthy, hygienic life of man.

eucarpic (Gr. *karpos* = fruit), pertaining to a fungus having distinct vegetative and reproductive parts in its mature thallus. *Cf.* holocarpic.

euchromatin (Gr. *chromatikos* = concerning colour), less condensed portions of → chromatin, i.e. chromosomes or parts of chromosomes which maintain normal coiling (condensation cycle) and normal characteristics of stainability, and do not become → heteropycnotic; euchromatin is genetically active, in contrast to → heterochromatin.

eugenic (Gr. *eugeneia* = being of noble birth), pertaining to the improvement of the genetic abilities and constitution of the human population. *See* eugenics.

eugenics, the artificial influence on the human → gene pool, and its study aiming at improving the physical and mental genetic characteristics of future generations. Eugenics can be divided into negative (preventive) and positive (progressive) eugenics. The former tends to decrease the frequencies of harmful genetic characters like diseases, and the latter to increase the frequencies of beneficial characters. Thus, eugenics always implies a social decision or agreement on what is harmful and what beneficial to the human population.

Euglenophyta (*Euglena* = the type genus, Gr. *phyton* = plant), **euglenoids**; a division of the Plant kingdom. They are one-celled organisms living mostly in freshwater habitats but are sometimes found also in marine environments. Typical euglenoids have a pair of flagella and an eyespot. Because they frequently are photosynthetic, containing chloroplasts, they are traditionally treated as algae. However, the cells lack the cell wall typical of plant cells and many euglenoids also feed like animals. Thus, they may be considered as transitional between plants and animals and are classified also into Protozoa, in the group → Phytomastigophora.

euhaline (Gr. *eu* = good, *hals* = salt), pertaining to brackish water having the salinity between 16.5 and 22 parts per thousand, or the normal sea water in oceans, i.e. 34—35 parts per thousand. *Cf.* oligohaline, mesohaline, polyhaline.

euhemerobe (Gr. *hemeros* = tame), a plant habitat which has totally changed or is originally formed through human activity; **euhemerobic** plants are weeds which do not succeed elsewhere.

eukaryotes (Gr. *karyon* = nut, nucleus), **eukaryotic organisms**; a group of organisms including the animals, plants, fungi, and protists, i.e. all organisms except bacteria, cyanobacteria and viruses. Eukaryotes have a genuine nucleus surrounded by → nuclear envelope, and genuine chromosomes within. Hence, their cell cycles involve → mitosis and → meiosis during which the chromosomes are visible under a light microscope. *Adj.* **eukaryotic**. *Cf.* prokaryotes.

Euler—Lotka equation (*Leonhard Euler*, 1707—1783; *A.F. Lotka*, 1880—1949), an equation which explains relations between the innate capacity for increase (r_m) of a population and the probability of the productivity and mortality in different age groups in the same population during unchanged age-structure. Probabilities are e.g. → fecundity, average lifetime, average age of sexual maturity, and probability of survival in different age groups.

eulittoral (Gr. *eu* = good, L. *lit(t)us* = seashore), **1.** an intertidal zone on an oceanic coast; **2.** in lakes and inland seas, the area of the littoral zone between the lowest and highest water levels which is influenced by water breakers. *Cf.* infralittoral, littoriprofundal.

Eumetazoa (Gr. *meta* = after, between, with, *zoon* = animal), a group (subkingdom in some classifications) of multicellular animals, excluding the → sponges, placozoans, and mesozoans; the body is constructed from specialized tissues and organs whose functions are coordinated by the nervous system. *Cf.* Metazoa.

Eumycota (Gr. *mykes* = fungus), the division of fungi in the old taxonomy, in which the fungi were included in the Plant kingdom.

Euphausiacea (Gr. *phausi* = shining bright), *see* krill.

euphenics (Gr. *phainein* = to show), the study dealing with the improvement of human conditions of life, especially the environment of an individual. *Adj.* **euphenic**.

euphoria (Gr. *pherein* = to bear), a strong feeling of well-being, especially such an exaggerated feeling having no basis in reality, as e.g. caused by various drugs. *Adj.* **euphoric**.

euphotic (Gr. *phos* = light), well-illuminated; pertaining to the **euphotic zone**, i.e. the upper level of the → photic zone; it is a surface zone in a water body above the → compensation zone where light is sufficient for effective photosynthesis. *Cf.* dysphotic.

euploid, pertaining to individuals, tissues, or cells which have the normal chromosome number; i.e. one complete set of chromosomes as a single (monoploid), double (diploid), or multiple (polyploid) dosage. *Cf.* aneuploid.

eurhythmy, also **eurythmy** (Gr. *eurhythmia* = harmony), **1.** the harmonious function and development of the human body, such as the regularity of the heartbeat; **2.** rhythmical movements, especially the type of body exercise (eurhythmics) developed by *Rudolf Steiner*. *Adj.* **eurhythmic**.

eury- (Gr. *eurys* = wide), denoting wide or broad.

eurybenthic (Gr. *benthos* = depth), pertaining to a species or any other taxon living at varying depths on the bottom of oceans and lakes.

euryhaline, also **euryhalin** (Gr. *hals* = salt), pertaining to an aquatic organism which is able to live and reproduce in waters with a wide range of salinity. *Cf.* holeuryhaline, stenohaline.

euryphage (Gr. *phagein* = to eat), **euryphagous animal**; an animal species capable of using a wide variety of food. *Syn.* polyphage. *Cf.* stenophage, omnivore.

eurythermic, also **eurythermous,** or **eurythermal** (Gr. *therme* = heat), pertaining to an organism enduring a large variation of temperatures. *Cf.* stenothermic.

eurytopic (Gr. *topos* = place), pertaining to a species or any other taxon which is geographically widely distributed; such a species has a wide tolerance range to changes in environmental conditions. *Cf.* stenotopic.

eusociality (Gr. *eu* = good, L. *socius* = companion, comrade), the most developed form of sociality among animals. Evolutionarily it is preceded by **prosociality** where many patterns of social behaviour are already observable. The characteristics of eusociality are seen e.g. in individuals of a species which cooperate in caring for the young by dividing labour between the members of a community. Sterile individuals work on behalf of reproductive individuals, or when generations overlap, the young, at least during a part of their life time, help their parents. The eusociality is common especially among **termites** and **social hymenopterans**, such as ants, wasps and bees. An eusocial mammal is *Heterocephalus glaber* (naked mole-rat). **Eusocial** is the formal equivalent of the expressions "truly social" or "highly social" which are commonly used with less exact meaning in the study of social insects. *See* caste.

eusporangiate (Gr. *sporos* = seed), pertaining to a type of sporangium in pteridophytes; in this type the wall of a sporangium is formed of two or several cell layers and the spore production is more abundant than that in the **leptosporangiate** type.

Eustachian tube (*Bartolommeo Eustachio*, 1524—1574), → auditory tube.

eustele (Gr. *eu* = good, true, *stele* = column), one of the different stele types in a plant structure. *See* stele.

euthanasia (Gr. *thanatos* = death), **1.** painless, easy death; **2.** mercy killing; i.e. helping an animal (or a person) to a painless ending of life.

Eutheria (Gr. *therion* = wild animal), **eutherians**, **Placentalia**; viviparous placental mammals; an infraclass in the subclass Theria, including all mammals except monotremes and marsupials; all have a placenta by which the developing foetus is attached to the uterus. *Cf.* Prototheria, Metatheria.

euthermy (Gr. *therme* = heat), the state of normal body temperature in homoiothermic animals; the term is especially used to describe the body temperature during the active state of hibernating mammals and birds, which are heterothermic during hibernation. *Adj.* **euthermic**. *Cf.* heterothermy.

eutrophic (Gr. *trophe* = food), pertaining to waters rich in mineral nutrients; in general, the primary production of green plants and the secondary production of animals is high in eutrophic waters. The depletion of oxygen occurs in → hypolimnion during summer, in northern lakes also during winter. *Cf.* dystrophic, mesotrophic, oligotrophic.

eutrophication, an increase of mineral nutrients (especially nitrogen and phosphorus) in water bodies resulting in increased productivity in all trophic levels. This commonly results in the depletion and deficiency of oxygen in → hypolimnion, which further releases more nutrients, especially phosphorus; eutrophication has become a feedback process. *See* eutrophic.

evaporation (L. *e-* = out, *vaporare* = to steam), the conversion of a substance from a liquid state into vapour; occurs without necessarily reaching the boiling point; e.g. vaporization (evaporation) of water is an important phase of → transpiration, i.e. removing excessive water from the surface or stomata of a plant. Evaporation causes cooling of the surfaces, e.g. the skin (*see* thermoregulation). Evaporation techniques are used to remove water from solids or non-volatile substances when processing biological and other materials. The evaporation heat of water is 46.4 kJ/mol at 0°C and 40.3 kJ/mol at 100°C. *Cf.* evapotranspiration.

evapotranspiration, the sum of → evaporation (from soil) and transpiration (from plants) in an ecosystem.

even dispersion, even distribution, regular distribution. *See* dispersion.

even-toed ungulates, → Artiodactyla.

evergreen plants, plants retaining their foliage through two or more consecutive seasons. Conifers such as spruce and pine have needle-shaped leaves, but most evergreens such as most tropical trees are broad-leaved. Some shrubs and trees are both deciduous and evergreens depending on where they grow. *Cf.* deciduous trees.

evocator (L. *evocare* = to evoke), a specific substance or factor released from an → organizer or inductor initiating the differentiation and morphogenesis of the early embryo; the process is called **evocation**.

evoked response/potential, a change in the electric activity in brain areas where sensory stimuli proceed and become processed, as in the sensory cortex; produced by stimulation of sense organs, and thus may be e.g. an auditory, visual, or somatosensory evoked response.

evolution (L. *evolvere* = to unroll), the natural phenomenon in organisms where structures and ways of life have changed and are constantly changing in such a way that the progenies differ from their parents. Necessary and sufficient conditions for evolution are → **variation, heredity**, and **natural selection**.

The causes of variation are → **mutation** and **genetic recombination**. Of these, mutations are the fundamental source of all genetic variation, and recombination could not exist without mutations. Recombination on its own is important because it rapidly produces extensive variation. Due to recombination, all the progenies of a sexually reproducing organism are usually genetically different.

From the heredity follows that related individuals resemble each other more closely than do individuals in the population on average. Due to natural selection, those individuals that are best adapted to the environmental conditions produce more fertile offspring than the less adapted individuals; consequently the average → fitness of the population increases.

The **basic step of evolution** is a change of the → gene frequencies in a population, called **microevolution**. **Macroevolution**, i.e. the origin of species, can be deduced from microevolution with the help of the concept of → **isolation**. When two populations of the same species are geographically isolated from each other, i.e. are → **allopatric**, their gene frequencies evolve in different directions, due to different → **selection pressures**, giving rise to different → **races**. The **species** evolve when the races have been geographically isolated for a sufficiently long time. Thus, many genetic differences accumulate in the races resulting in the inability to interbreed, even if the populations would become contiguous, i.e. **sympatric**. Then they would be different species.

Megaevolution refers to the origin of taxa above the species level, like orders or classes. The laws of megaevolution are for the present unclear.

evolutionarily stable strategy, ESS-theory, a strategy applied from the → game theory to evolutionary studies; it is a successful behavioural pattern that leads to higher → inclusive fitness than any other alternative strategy or pattern prevailing at the same time in the population, and so tends to be established by natural selection.

evolutionary clock, the rate by which mutations accumulate in a given gene; the rate is constant and specific for every gene. According to this it is possible to determine the time when certain phylogenetic lineages have branched from each other; e.g. with the aid of the evolutionary clock it has been concluded that the lineages leading to African hominids and man have separated about 5 million years ago. *Cf.* molecular clock.

evolutionary ecology, a branch of ecology which stresses evolutionary factors and adaptations as regulators of viability and fecundity of organisms in a population.

evolutionary opportunism, an evolutionary strategy in which the adaptations regulating the direction of evolution are based on an opportunistic, preadaptive gene pool. *See* preadaptation.

evolutionary taxonomy (Gr. *taxis* = arrangement), a traditional branch of → taxonomy that classifies organisms on the basis of → phenetic and phylogenetic characteristics. *Cf.* cladistics.

evolutionary transformation series, a series of → homologous characters in an evolutionary line each derived from the previous one.

evolutionary tree, a graphical presentation that indicates phylogenetic relationships and evolutionary trends between groups of organisms. *See* phylogenetics.

evolution theory, a theory stating that all organisms presently living on the Earth have evolved from organisms which have lived earlier, but were different and in a certain respect more primitive.

The evolution theory introduced by *Charles Darwin* in 1859 is the most central of all biological theories because it creates unity in biology, and makes e.g. the adaptation of organisms to their environments scientifically comprehensive. Darwin presented ample versatile evidence in favour of → evolution, and, moreover, he gave reasons for a causal explanation of evolution, namely the mechanism of → natural selection.

By the end of the 19th century the theory became commonly accepted by biologists. However, arguments occurred regarding the causes of evolution. In the resurgence of genetics in 1900 when → Mendel's laws of inheritance were rediscovered, Mendelism and → Darwinism were in apparent conflict, because Darwinism emphasized change in hereditary factors, while Mendelism spoke in favour of the stability of these. This discrepancy was fruitfully resolved up to 1940 when both traditions were united in the birth of the **synthetic evolution theory**, i.e. **neo-Darwinism**. In a narrow sense, the synthetic evolution theory means the synthesis of genetics and Darwin's principle of natural selection. In a broader sense, the synthetic evolution theory allowed the union of all branches of biology within the theoretical framework of the evolution theory, and supported it. In the original form of the synthetic evolution theory → **population genetics** played a central role. Beginning from 1960's the significance of ecology has been stressed due to the development of **evolutionary ecology**.

The evolution theory has been confirmed theoretically and empirically in many ways. Several scientific proofs of the evolution theory have been presented. The **fossil record of organisms** shows that in the petrified remnants or fossils of the organisms, the older geological stratifications contain the more primitive organisms, and younger strata the fossils of more advanced organisms. The **distribution of organisms** indicates that the distribution of the present living forms on the Earth appear in phylogenetic lines demonstrating that all individuals of each species are descendants from a common strain. **Comparative anatomy** provided evidence that the basic structure of → homologous organs of living organisms in the same taxonomical grand group is similar in all species. The basic metabolism and structure of cells are similar in all organisms, and one of the strongest arguments in favour of the evolution theory is the universality of the → genetic code. **Comparative embryology** shows that, within a phylogenetic group, the earlier the stage of embryogenesis considered, the more closely do the embryos resemble each other. During development general characteristics common to all species of the group are generated first, and the characteristics which are specific for each species develop later. **Taxonomy** shows that the organisms can be classified into hierarchic groups that describe their kinship.

exarch (L. *ex* = out, *arche* = origin), pertaining to the arrangement of primary xylem in plants; in the exarch type the oldest xylem elements (protoxylem) are located outside the younger elements (metaxylem). *Cf.* endarch.

excision (L. *caedere* = to cut), the removal of some material, such as a tissue or an organ, from an organism by cutting.

excision repair, a repair mechanism of DNA in which a sequence of nucleotides of the damaged or faulty paired DNA is enzymatically cut out and removed from the faulty site, and replaced by the correct new sequence of nucleotides synthesized according to the → base pair rule.

excitation (L. *citare* = to call), a state of stimulation or irritation; e.g. **1.** excitation of a neurone or a group of neurones when, due to synaptic action, the membrane potential approaches the threshold potential for generation of nerve impulses; **2.** excitation of an atom or molecule by absorption of energy, such as vibration, rotation, radiation (e.g. excitation light); in excitation the state of a molecule changes to a higher energy level, due to the jumping of electrons in an atom from a shell to another. *See* fluorescence.

excitation light, the excitation of an atomic system to a higher energy level, caused e.g. by UV light. *See* fluorescence.

excitation-contraction coupling, the coupling of the → action potential of the muscle cell membrane to the contraction apparatus inside the cell. In the skeletal muscle, the action potential (muscle impulse) is transmitted into myofibrils of the muscle cell via the → T system, triggering the release of calcium ions (Ca^{2+}) from the cisterns of → sarcoplasmic reticulum. The Ca^{2+} initiates contraction. In the cardiac muscle, the action potential causes the influx of small amounts of extracellular calcium (Ca^{2+}) triggered by activation of the dihydropyridine channels in the T system. The influx of calcium further causes calcium release from the sarcoplasmic reticulum in a concentration high enough to induce the contraction of myofibrils, i.e. the action potential causes the calcium-induced calcium release. In the smooth muscle, having a poorly developed sarcoplasmic reticulum, the coupling is more dependent on the influx of extracellular calcium than in the other muscle types.

excitatory postsynaptic potential, EPSP, a weak electrotonic change (partial depolarization) of the → resting potential generated by synaptic action in the postsynaptic membrane of the dendrites and/or of the soma of the neurone. This change is induced by the transmitter substance, released from an axonal end button into the synaptic cleft, attaching to specific receptors on the postsynaptic membrane of the → synapse. EPSP may occur as a short-lasting electrotonic potential, i.e. **fast EPSP** lasting tens of milliseconds, or as a long-lasting electrotonic potential, i.e. **slow EPSP** lasting even seconds. The occurrence and size of EPSP and the summation from different synapses determine the generation of nerve impulses (→ action potentials) at the initial segment (axon hillock) of the axon of the postsynaptic neurone. In some neural junctions EPSP can propagate from one neurone to another. Similar potentials are end-plate potentials (EPP) which occur in → neuromuscular junctions.

excretion (L. *excretio* < *ex* = out, *cernere* = to separate, to sift), **1.** the process by which waste products and harmful substances are separated and removed from an organism; **2.** the substance (**excrement**) so separated and eliminated, such as urine, sweat, or certain plant products; *see* excretory and secretory tissue. *Verb* to **excrete**.

excretory, pertaining to → excretion.

excretory and secretory tissues, a type of plant tissue secreting or excreting materials, such as slime, water, oils, resins or latex; the excreting structures are e.g. glandular hairs, resin tubes, and latex tubes.

excretory organs, generally denotes only urinary organs of animals, as the → kidneys, Malpighian tubules, coxal glands, antennal glands (green glands), or nephridia.

exergonic (Gr. *ergon* = work), describes a chemical reaction in which heat is released to the surroundings. *Cf.* endergonic.

exhalation (L. *halare* = to breathe), **1.** breathing out; exhaling; **2.** emission of vapour, sound, etc. *Verb* to **exhale**.

exhaustion (L. *exhaurire* = to drain out), **1.** the act or state of being emptied or drained, as in the creation of vacuum; **2.** a state of being extremely weak and tired, e.g. an animal in heat exposure (heat exhaustion). *Verb* to **exhaust**.

exine, the outermost, chemically very durable layer in the wall of a → pollen grain, outside the → intine. The exine is divided into **ectexine** and **endexine**, the former being a rather thick outer layer, the latter a thinner layer between the ectexine and the intine.

exo- (Gr. *exo* = out, outside), denoting exterior, external, outward; sometimes used as a synonym for ecto-.

exobiology (Gr. *bios* = life, *logos* = word, discourse), the study of life beyond the Earth's atmosphere, i.e. life of other planets.

exocarp (Gr. *karpos* = fruit), the outermost, protecting layer of the → pericarp, i.e. the wall of a plant → fruit.

exocrine (Gr. *krinein* = to separate), **1.** secreting externally; pertaining to secretion of animals produced by → **exocrine glands**; **2.** denoting an exocrine gland. *Cf.* endocrine.

exocrine glands, glands of animals discharging their secretion from the gland through a duct to the external surface of the body or into body cavities, such as to the alimentary canal or respiratory pathways; e.g. sweat glands, sebaceous glands, lacrimal glands, salivary glands, intestinal glands. Also single exocrine cells occur, secreting their products directly into a cavity or onto the skin but not into interstitial tissue fluids, lymph, or blood. *Cf.* endocrine glands.

exocrine secretion, also **external secretion, 1.** the secretion (secretive process) of → exocrine glands; **2.** the product of exocrine glands.

exocuticle (L. *cuticula* = thin skin), an outer layer of the → chitin cuticle, above the endocuticle; the layer contains substances like chitin, cuticulin, and phenols which by oxidation form a dark pigment in the cuticle. The exocuticle is covered by a thin membrane, **epicuticle**.

exocytosis (Gr. *kytos* = cavity, cell), the discharge of material in membranous vesicles out of a cell. The vesicles formed in the cell in the → Golgi complex, or in → dictyosomes of plant cells, move to the cell membrane, attach to and fuse with the membrane so that the vesicles empty their contents out of the cell. Most secretion processes occur exocytotically; in animals called merocrine secretion. In plant cells the most important function of the exocytosis is to deliver material for the → cell wall. *Cf.* endocytosis.

exodermis (Gr. *derma* = skin), a cell layer with suberized walls, developing below the → epidermis (rhizodermis) of plant roots, gradually replacing it.

exoenzyme, an extracellular enzyme, i.e. functioning outside cells. *Cf.* endoenzyme.

exogamy, → cross-breeding.

exogenote, *see* merozygote.

exogenous, also **exogenic** (Gr. *gennan* = to produce), derived externally; originating outside an organism; e.g. pertains to external substances or influences, or additional cellular growth upon the surface. *Cf.* endogenous.

exon, a DNA sequence or segment encoding with the aid of messenger RNA the amino acid sequence in → protein synthesis. In interrupted or mosaic genes the exons alternate with intervening sequences, → introns. Exons are known in most genes of eukaryotic organisms and their viruses.

exon shuffling, a hypothetical reorganization of → exons by → genetic recombination in the formation of new genes during the course of evolution.

exon trapping, a molecular technique for scanning of the genome for the presence of → exons. It uses a **DNA vector** that contains a strong → promoter and a single intron interrupting two exons. When this vector is transfected into cells, its transcription produces large amounts of RNA that contains the sequences of the two exons (**RNA trap**). A restriction cloning site lies within the intron, and is used to insert genomic fragments from a DNA region of interest. If a fragment does not contain an exon, there is no change in the splicing pattern of the produced RNA (*see* splicing), and the RNA that is produced in transcription contains only the same exon sequences as the parental vector. But if the genomic fragment contains an exon flanked by two partial intron sequences, the splicing sites on either side of this exon are recognized. The sequence of the exon is inserted into the RNA trap between the two exons of the vector. This can be detected readily by reverse transcription of the cytoplasmic RNA into complementary DNA (cDNA), and using → polymerase chain reaction to amplify the sequences between the two exons of the vector. Thus the appearance in the amplified population of DNA sequences from the genomic fragment indicates that a genomic exon has been trapped.

exonucleases, → nucleases in the enzyme subclass EC 3.1. catalysing the hydrolysis of nucleic acids beginning at one end of a polynucleotide molecule, thus producing mononucleotides. Some DNases and RNases belong to this enzyme group. *Cf.* endonuclease.

exopeptidases, enzymes cleaving off the terminal amino acid residue of a peptide chain; **carboxypeptidases** (EC sub-subclass 3.4.17) acting at the carboxyl end and **aminopeptidases** (EC sub-subclass 3.4.11.) at the amino end of the peptide. *Cf.* endopeptidases. *See* peptidases.

exoplasm, → ectoplasm.

exopodite, exopod (Gr. *pous*, gen. *podos* = foot), the outer of the two distal branches of the → biramous appendage in crustaceans. *Cf.* endopodite, epipodite.

Exopterygota (Gr. *pterygotos* = winged), a superorder of insects which have an incomplete metamorphosis and are therefore also called **Hemimetabola**. The larvae (nymphs) have compound eyes and the wings develop externally. The superorder includes 14 orders: Ephemeroptera, Odonata, Orthoptera, Dermaptera, Plecoptera, Isoptera, Embioptera, Psocoptera, Zoraptera, Mallophaga, Anoplura, Thysanoptera, Hemiptera, and Homoptera. *See* Endopterygota.

exoskeleton (Gr. *skeletos* = dried up), external skeleton; an externally supporting structure of an animal secreted by the ectoderm (epidermis); most invertebrates have an exoskeleton, mostly formed from → chitin cuticle. *See* skeleton.

exosphere (Gr. *sphaira* = ball, sphere), the atmospheric layer of the Earth's atmosphere beyond the → ionosphere.

exospore (Gr. *sporos* = seed), a type of → spore in bacteria and cyanobacteria; formed from a whole cell which develops into a spore.

exothermic, also **exothermal** (Gr. *therme* = heat), pertaining to a chemical reaction that is accompanied by the liberation of heat, i.e. the enthalpic change is negative. *Cf.* endothermic.

exotic (Gr. *exotikos* = foreign), pertaining to an organism which is not native but disperses or is introduced into an area.

exotoxin (Gr. *toksikon* = arrow poison), extracellular toxin; a type of poison secreted by certain microorganisms, especially by Gram-positive bacteria, such as *Bacillus*, *Clostridium*, *Escherichia*, *Salmonella*, *Yersinia*, but also by some algae. Toxins are produced within cells but released into the surroundings, e.g. in animal tissues. Most exotoxins are heat-labile proteins which are toxic for animals in extremely small quantities; e.g. lethal doses for mice are less than 10^{-8} mg. Sometimes called also ectotoxin. *Cf.* endotoxin.

expansive growth (L. *expandere* = to expand), a vigorous growth, e.g. typical of tumours.

experiment (L. *experiri* = to try), an intentional tentative procedure in order to discover, test, or demonstrate some fact, occurrence, principle, or general truth. The experiment is usually formulated to answer the question: what is the effect of x on y, as e.g. what is the effect of temperature on the oxygen consumption of an organism. The experimental situation is reduced so that only one variable (x, free variable) is changing during the experiment while all other variables which might affect the tested phenomenon (y, dependent variable) are kept constant. If possible, the procedure includes a **control** group which gives the zero value to be compared with the experimental values.

experimental animals, 1. laboratory animals, animals kept and mostly reared in special animal houses for scientific research and testing (test animals). Usually they are different strains of certain species reared for hundreds or thousands of generations for experimental usage, such as laboratory rats, mice, rabbits, aquarium fish, cockroaches, or fruit flies; also domestic animals such as hen, sheep, and pig are used. In most countries, the use of vertebrates for experimental purposes is controlled by law and statutes stating what experiments are allowed to be performed, what experiments have to be done and what animals have to be used for testing drugs, how the animals must be reared and treated, and what kind of special education the staff dealing with experimental animals must have; *see* specific pathogen free; **2.** wild animals, who are experimentally studied in their natural environment or gathered for temporary laboratory studies.

expiration (L. *ex* = out, *spirare* = to breathe), **1.** the act of breathing out, i.e. exhaling air from the lungs; **2.** the termination of life; death. *Verb*, to **expire**. *Adj.* **expiratory**.

explant (L. *plantare* = to plant), **1.** to move cells, tissues or organs from an organism to grow in an artificial culture medium; **2.** such a sample; **explantation**, the act of explanting.

exploration (L. *explorare* = to search out, examine), the act of investigation or examination; the investigation of unknown regions or occurrences.

exploratory behaviour, goal-oriented behaviour. *See* appetitive behaviour.

exponential growth, exponential rate of increase; in population ecology, describes a growth in the size of a population (or other entity) where the logarithm of individual numbers increases linearly as a function of time; in optimal situations obeys the formula: $N_t = N_0e^{kt}$, in which N_t is the number of individuals at time t, N_0 the number of individuals at the beginning, e is 2.718 (the base of the natural logarithm), and k is a species-specific constant, (b—d, i.e. *birth-rate minus death-rate*). Growth follows the J-shaped curve using linear scales and a straight line in the logarithmic scale. Exponential growth occurs e.g. in cell cultures in optimal conditions, as well as in natural populations of various organisms when environmental resources are unrestricted. *See* geometrical growth, logistic growth.

exposition (L. *positio* = location), in botany, the effect of the direction and gradient of a slope on the vegetation; e.g. the vegetation on a slope facing the south differs from that on the north side.

exposure (L. *exponere* = to put out), the act or condition of organisms being subjected (exposed) to harmful conditions or circumstances, such as pollutants, infectious agents, radiation, or extremes of weather.

expressed sequence tag, EST, a part of DNA sequence which is known experimentally to be expressed, and hence represents a part of an → exon of a given gene. ESTs are usually stored in data banks and used for searching genes expressed in a given tissue.

expression (L. *expressio* = manifestation), **1.** expelling, squeezing out by pressure; **2.** the appearance of features on the face determined by an emotional state; **3.** in genetics, the transcription of a gene into messenger RNA (mRNA) and the translation of this into protein; in a broader sense, the manifestation of the function of the gene in enzymatic activities, in the structure and function of cells and tissues, and in the behaviour of an organism.

expression vector, a → vector (a modified plasmid or virus) efficiently expressing inserted foreign genes to be transferred into a host cell; contains all the elements, such as → enhancers and → promoters, needed for starting a gene function.

expressive behaviour, an ability to express feelings and intentions with hairs, feathers, gestures, and other expressions which may be either innate or learned. Expressive behaviour is especially well developed in man and can be compared to signals of animals; often the origins of signals and airs are common and identical. In man, the face and hands are important in expressive behaviour.

expressivity, in genetics, the degree of the strength of the → expression of a gene or genotype in the → phenotype; can be different in different individuals. Expressivity is dependent on the → genetic background and the environment of the organism. *Cf.* penetrance.

exsiccate (L. *exsiccare* = to dry up), **1.** a sample of dried plants or other organisms, used for scientific studies; **2.** evaporation residue; **3.** as a verb, to make dry, dehydrate.

exsiccator, a glass or plastic vessel for drying chemical or biological material; the material is kept together with a hygroscopic material like silica gel, and often kept under vacuum.

extended phenotype, all gene effects expressed in the → phenotype of an individual but also in the environment produced by the activity of this individual; e.g. the genotype of the beaver and a dam built by it with all its effects on the environment, forms an extended phenotype of the beaver.

extensor (L. *extendere* = to extend), any muscle that extends a joint, tending to straighten a limb. *Cf.* flexor.

exterior, externus (L.), outside; external.

external ear, *see* ear.

external fertilization, a fertilization of animals occurring outside the female genital organs, usually in water; the female may first extrude the eggs before the male discharges sperm (e.g. fish), or the female and male simultaneously discharge their gametes (e.g. amphibians). *Cf.* internal fertilization.

external genital organs (L. *genitalia* = sex organs), external genitals, external genitalia; sex organs of animals which are visible or extend outside the body; the term is usually used only to denote the organs of vertebrates. External genitals are found only in those vertebrates which have internal fertilization, i.e. in cartilaginous fishes, reptiles, birds, and mammals. In internal fertilization sperm is transferred into the female genitals. For this, male cartilaginous fishes have a **clasper** (*mixipterygium*) extending from the pelvic fins; snakes and lizards have a → **hemipenis**. Turtles, crocodilians and some birds (ostriches and geese) have an extrusible structure in the ventral wall of the cloaca, corresponding to the → **penis** of mammals. In most male mammals, external genitals also include the **scrotum**. The external genitals of female mammals include the **clitoris** which is homologous with the penis, and the **outer** and **inner lips of vulva** (*labia pudendi*).

exteroceptors (L. *receptor* = receiver), any external sense organs or sensory cells responding to stimuli from outside the body; e.g. various skin receptors, eyes, ears, and the receptors for taste and smell.

extinction (L. *extinguere* = to quench, extinguish), the act or process of extinguishing or suppression; e.g. **1.** the extinction of a species or some other taxon after the death of the last individual; may be either global or regional (*see* island theory); **2.** a progressive decrease and disappearance of a physiological function, as of a conditioned response of an animal as a result of reinforcement; **3.** the decrease or depression of any physical quantity in a time or distance, as e.g. the absorption of light in a solution (*syn.* → absorbance). *Adj.* **extinct**, ended, died out, extinguished, no longer active; **extinctive**, tending to extinguish.

extra- (L. *extra* = outside), outside of, beyond, or in addition.

extra-capsulum, *see* central capsule.

extracellular (L. *cella* = compartment, cell), outside of cell.

extracellular space, the space outside the cell; in tissues equal to intercellular space.

extrachromosomal inheritance, inheritance mediated by cell organelles other than chromosomes, such as by → mitochondria and → chloroplasts. *Syn.* cytoplasmic inheritance.

extract (L. *extrahere* = to pull out), a solution obtained from animal or plant tissues by using some solvent.

extraembryonic coelom, that part of the embryonic body cavity, → coelom, which is situated between the mesodermic layers, extending outside the body lining the outer surface of the → amnion and the inner surface of the → chorion during the embryonic development in amniotic vertebrates, i.e. reptiles, birds, and mammals.

extraembryonic membranes, the membranes outside the embryo of amniotic vertebrates, i.e. reptiles, birds, and mammals; include the → **amnion** for physical protection, → **chorion** for gas exchange, → **allantois** for waste products, → and **yolk sac** for energy store; also *see* chorioallantoic membrane. *Syn.* embryonic membranes; in late mammalian embryos called also foetal membranes.

extrapyramidal system (Gr. *pyramis* = pyramid; here the pyramid-shaped structure in the ventral medulla oblongata), the neuronal system of the vertebrate brain controlling automatic voluntary movements, such as those functioning in walking; comprises several nuclei, such as the → corpus striatum in the telencephalon, the red nucleus in the mesencephalic tegmentum, and the black nucleus from the subthalamic region to the rostral border of the pons. *Cf.* pyramidal system.

extrasystole (Gr. *systole* = contraction), a premature beat sometimes found in the vertebrate heart; arises in response to an impulse from somewhere (**ectopic focus**) outside the sinus node (sinoatrial node) normally pacing the contractions. Each extrasystole is followed by a transient interruption, slightly longer than the normal diastole.

extrinsic (L. *extrinsecus* = situated on the outside), not inherent; being outside of something, such as extrinsic influences or factors affecting an organism from outside. *Cf.* intrinsic.

extrinsic factor, vitamin B_{12}. *See* vitamin B complex. *Cf.* intrinsic factor.

extrorse (L. *extrorsus* = outward), pertaining to the way of facing of the anthers, bending away from the floral axis. *Cf.* introrse.

eye, an organ of vision or sight; the sense organ responding specifically to light. There are many different types of eyes which can be divided into three main types: → **eyespot** (*ocellus*), → **compound eye**, and → **camera eye** (*oculus*). All these occur in invertebrates but only the camera eye in vertebrates. Specific → **photoreceptors** and light-refracting and supporting auxiliary structures are found in all eye types. In the simplest forms of eyespots there are only a few photoreceptor cells and the lens may be lacking, but in more complicated forms the structures resemble those of the camera eye. The compound eye is structurally quite different.

eye muscles, ocular muscles, 1. extraocular muscles; the muscles serving the rotation of the eyeball; in vertebrates, comprise six skeletal muscles, innervated by the third, fourth, or sixth cranial nerves; **2.** intraocular muscles; the circular and radial smooth muscles in the iris constricting and dilating the pupil, and the smooth muscle of the → ciliary body in the periphery of the iris, controlling the convexity of the lens during accommodation.

eyespot (L. *ocellus*), the simplest type of eye found in most invertebrate groups. The number of photoreceptor cells varies from a few to several hundreds and generally there is a lens of non-accommodating type; in the **nauplius eye** a narrow light-refracting pore substitutes the lens. In many animal groups, e.g. in insects, eyespots (ocelli) are found beside the → compound eyes, apparently perceiving only the intensity of light. The most complex eyespots, sensitive also to the polarity of light, are found in spiders. The eyespot is sometimes called stigma.

F

F, symbol for **1.** fluorine; **2.** phenylalanine; **3.** Fahrenheit (°F); **4.** farad; **5.** F_1 or F1, (L. *filius* = son), the first filial generation arising from crossing the members of the parental generation, P, and further F_2, F_3, etc. from crosses of successive generations; **6.** coefficient for inbreeding.

***F*,** **1.** symbol for gravitation (force); **2.** Faraday's constant; *see* Appendix 1.

f, femto-, denoting one thousand million millionth, i.e. 10^{-15}.

***f*,** symbol for frequency.

Fab, (F = fragment, ab = antigen-binding), originally, either of usually two identical fragments obtained by enzymatic separation from an immunoglobulin molecule, each having an antigen-binding site; later denoting one of the arm regions (Fab region, Fab segment) of the Y-shaped immunoglobulin (antibody) monomers. *Cf.* Fc. *See* immunoglobulin.

facet, facette (Fr. *facette* = little face), **1.** a plane surface on an anatomical structure, especially on a bone; **2.** the outer surface of a single ommatidium of the compound eye; tetragonal in the eyes of crustaceans and hexagonal in those of insects.

facies, pl. **facies** (L.), **1.** face; **2.** a special surface of an anatomical structure; e.g. *facies anterior cruris*, anterior surface of the leg, *facies anterior lentis*, arterior surface of the lens of the vertebrate eye.

facialis (L.), pertaining to the face, **facial**, e.g. *nervus facialis*, facial nerve, the seventh → cranial nerve in vertebrates.

facilitated diffusion, (L. *facilis* = easy), *see* cell membrane transport.

facilitation, the promotion of any process, especially the increased responsiveness of the postsynaptic membrane to successive stimuli (nerve impulses), resulting in an increased efficacy of transmission for subsequent impulses passing the synapses. *See* social facilitation.

factor (L. = maker), any element, e.g. a substance or an energy form, that contributes to a particular result; often an unknown substance before receiving a more specific name; e.g. → biotic factors, abiotic factors.

facultative, **1.** not obligatory; **2.** capable of living in or adaptating to more than one type of conditions; e.g. a **facultative parasite** which is not totally dependent on its host but is also capable of living free, or a **facultative aerobe** (or anaerobe), a microbe which can live in both aerobic and anaerobic environments. *See* facultative apomixis.

facultative apomixis, a type of reproduction in plants in which, in addition to → apomixis, the plants also have the ability for sexual propagation; e.g. occurs in *Potentilla* and *Hieracium* species.

FAD, flavin adenine dinucleotide, *see* flavin nucleotides.

faeces, Am. **feces** (L. *faex* = refuse), **1.** waste matter or excrement of animals discharged through the anus (products of the intestine) or cloaca (products of the intestine and kidneys); contains microbes, intestinal cells, secretions, and food residues; **2.** dregs, sediment. *Adj.* **faecal**, (fecal).

Fahrenheit scale (*G.D. Fahrenheit*, 1686—1736), *see* Appendix 1.

fairy ring, an expanding circle of fruiting bodies of some mushrooms on the ground, caused by the centrifugal underground growth of the mycelium. The grass is also lighter inside than outside the sphere. A well known species forming fairy rings is e.g. *Marasmius oreades*.

Falconiformes (L. *falco* = falcon, *forma* = form), an order of diurnal birds of prey; strong fliers with a keen vision, a stout beak hooked at the top, and feet with sharp, curved claws. Falconiformes are cosmopolitans and feed mainly on other vertebrates, some small-sized species also on large insects. The order includes 290 species in five families: **condors** and **New World wultures** (Cathartidae), **secretary bird** (Sagittariidae), **hawks** and **eagles** (Accipitridae), **ospreys** (Pandionidae), and **falcons** (Falconidae).

Fallopian tube (*Gabriele Falloppio* (L. *Fallopius*), 1523—1562), the → oviduct of mammals.

false diaphragm, the membranous **transverse septum** (*septum transversum*) in the fish body cavity (→ coelom) separating the **pericardial cavity** from the **peritoneal cavity**; comprises parts of the parietal peritoneum and pericardium attached together. *See* diaphragm.

false pregnancy, *see* pseudopregnancy.

false scorpions, → Pseudoscorpiones.

false transmission, false neurotransmission; the chemical transmission of nerve impulses in a → synapse by means of a non-specific substance, not produced by the nerve ending but by other cells in the vicinity. This substance (false transmitter) is transferred into the nerve terminal (end button) by a non-specific uptake mechanism and liberated then into the synaptic cleft with a specific neurotransmitter; e.g. non-specific serotonin is liberated together with noradrenaline (specific neurotransmitter) in certain sympathetic nerve endings. Formerly, artificial substances resembling a specific neurotransmitter were also called false transmitters.

falx, pl. **falces** (L. = sickle), a sickle-shaped anatomical structure; e.g. **falx of cerebellum** (*falx cerebelli*), the fold of the dura mater in the midline of the cerebellum separating the cerebellar hemispheres in vertebrates; **falx of cerebrum** (*falx cerebri*), the fold of the dura mater separating the two cerebral hemispheres.

familial, connected with family, occurring in families; e.g. a genetic characteristic, such as a certain disease.

family (L. *familia*), a taxonomic category of organisms, consisting of one or more genera; above the family is the → order.

farad, F (*Michael Faraday*, 1791—1867), *see* Appendix 1.

Faraday cage, a cage or room surrounded by metal plates or nets, usually of copper; in biology, used to prevent the effects of external electrical disturbances on an object or equipment inside the cage in electrophysiological studies.

Faraday's constant, F, *see* Appendix 1.

farnesyl pyrophosphate, a 15-carbon compound formed from three → isoprene units; a precursor in the biosynthesis of → steroids and terpenoids in many organisms, such as → gibberellins and abscisic acid in plants.

far red radiation, FR, long-wave red light (wavelength 700—800 nm) which has specific effects on → phytochromes of plants. FR (appr. 730 nm) converts a phytochrome pigment to an inactive form (P_r) in moist light-sensitive seeds and inhibits germination. FR also controls the flowering of plants. *See* photoperiodity.

fascia, pl. **fasciae,** also **fascias,** (L. = band, bandage), a fibrous sheet-like structure or membrane in vertebrate anatomy, e.g. **muscular fasciae** enclose the muscles. *Adj.* **fascial**. *Cf.* fasciculus.

fasciation, in botany, coalescing of shoot branches of a plant to form thick bundles; especially common in lianas.

fascicle (L. *fasciculus*, dim. of *fascis* = bundle), a bundle of fibres; e.g. **1.** in animal anatomy e.g. muscle, tendon, and nerve fascicles; **2.** in botany, the vascular bundle in the needles of conifers; *see* Coniferae.

fascicular cambium (L. *cambiare* = to exchange), the → cambium tissue present in the area of the → vascular bundles of plants. *Cf.* interfascicular cambium.

fasciculus, pl. **fasciculi** (L. dim. of *fascis* = bundle), **fascicle**; a small bundle of fibres, e.g. of nerves, muscles, or tendons. *See* fascicle.

fat, 1. → adipose tissue; **2.** any of **neutral fats**, triacylglycerols (triglycerides), formed in esterification of glycerol with three fatty acids. Fats being in liquid form at room temperature are called **oils**. Most triacyl glycerols of plants are oils, rich in unsaturated fatty acids. Reserve fats of homoiothermic animals are rich in unbranched saturated fatty acids and are solid, but the saturation stage of fatty acids in poikilothermic animals varies according to the environmental temperature to which they have adapted, i.e. the lower the temperature, the more unsaturated (fluid) are the body lipids. *See* lipids, fatty acids, lipase.

fat body, 1. the fatty organ that fills partly the body cavity (haemocoel) in insects, changing its formation and size many times during the ontogenic development. The fat body regulates the synthesis, storage and transport of proteins, carbohydrates and fats, and so corresponds functionally to the liver of vertebrates; **2.** the term is also used to mean the adipose tissue in the abdomen of many amphibians and lizards, serving as an energy store for hibernation.

fate map, a map of the surface of fertilized egg or early embryo describing the developmental fate of different parts of an embryo, i.e. in which direction the embryonic tissues will develop.

fatty acid oxidation, → beta-oxidation.

fatty acids, long-chain monocarboxylic acids with an even number of carbon atoms; described chemically by R-COOH where R is an aliphatic moiety, formed from hydrocarbons. Fatty acids may be unsaturated, i.e.

Fatty acids. Examples of unsaturated and saturated fatty acids.

Common name	Systematic name	Formula
Lauric acid	Dodecanoic acid	$CH_3(CH_2)_{10}COOH$
Palmitic acid	Hexadecanoic acid	$CH_3(CH_2)_{14}COOH$
Arachidic acid	Eicosanoic acid	$CH_3(CH_2)_{18}COOH$
Oleic acid	*cis*-Δ^9-octadecenoic acid	$CH_3(CH_2)_7CH{=}CH(CH_2)_7COOH$
Linoleic acid	*cis,cis*-Δ^9,Δ^{12}-octadeca-dienoic acid	$CH_3(CH_2)_4(CH{=}CHCH_2)_2(CH_2)_6$-COOH
Arachidonic acid	all-*cis*-Δ^5,Δ^8,Δ^{11},Δ^{14}-eicosa-tetraenoic acid	$CH_3(CH_2)_4(CH{=}CHCH_2)_4(CH_2)_2$-COOH

have double bonds between adjacent carbon atoms (monounsaturated acids have one, polyunsaturated several double bonds), or they may be saturated having single bonds between all carbon atoms. If formic acid and acetic acid are not included, fatty acids (from butyric acid to serotinic acid) may contain 4 to 26 carbon atoms, which are usually numbered or marked starting at the carboxyl end, for example:

$$\underset{\omega}{H_3C}-(CH_2)_n-\underset{\beta}{\overset{3}{C}H_2}-\underset{\alpha}{\overset{2}{C}H_2}-\overset{1}{C}{\scriptstyle\begin{matrix}{/\!\!/O}\\ {\diagdown OH}\end{matrix}}$$

For example, *cis*-Δ^9 means that there is a *cis* double bond between the carbon atoms 9 and 10, as in *cis*-Δ^9-hexadecanoate (palmitoleate). The position of a double bond can alternatively be denoted counting from the distal end, with the ω carbon atom as number 1.

In tissues of organisms free fatty acids occur in low concentrations. Common fatty acids are e.g. palmitic acid (C-16, saturated) and oleic acid (C-16:1, unsaturated). The esterified forms of fatty acids are important structural components and energy storage molecules. When esterified with glycerol, fatty acids form monoacyl- or diacylglycerols, or **neutral fats** (triacylglycerols), according to how many OH groups of glycerol are esterified with fatty acid. The hydrolysis of the esters e.g. in the alimentary canal of animals produces again free fatty acids which are important energy sources for cells and structural material for → phospholipids of cellular membranes. *See* essential fatty acids, triacylglycerol, lipase, spherosome.

fauna (L. *Fauna*, sister of *Faunus* = mythical deity of herdsmen), all the animals living in a certain area and/or at a certain time; in palaeontology, also the animals during a certain geological period; e.g. the fauna of Finland, the fauna of the Mesozoic Era.

faveolus, pl. **faveoli** (L. dim. of *favus* = honeycomb), a small pit or depression in an anatomical structure; also foveola.

favouring selection, *see* intersexual selection.

Fc, (F = fragment, c = crystallizable), the shaft part of the Y-shaped structure of the immunoglobulin (antibody) monomers; does not contain an antigen-combining site but e.g. activates the → complement system. *Cf.* Fab. *See* immunoglobulin.

F-duction (F = fertility; L. *ductio* = conduction), a genetic system found in bacteria, corresponding to sexual reproduction. In F-duction a part of the bacterial chromosome is transferred from one cell to another with the aid of the → F factor. *Syn.* sexduction.

feather (L. *pluma*), a complicated, cornified structure of the epidermis covering the body of birds. It consists of a hollow, basal **quill** (*calamus*), a solid extension of the quill, called **shaft** (*rachis*), many narrow, parallel **barbs** (*rami*), joining the sides of the shaft; each barb has numerous smaller **barbules** (*radioli*) which are provided with minute **barbicels** (**hooklets**) to hold the barbules together. The plumage of birds usually includes five types of feathers: 1) **contour feathers** provide the external covering of the body, 2) long and large **flight feathers** (plumes) on the wings and tail, 3) **down feathers** (plumulae) against the skin of the body, having a short quill, reduced shaft, and long flexible barbs with short barbules but without hooklets, 4) **filoplumes** (pin feathers), situated beside other feathers, are thin small feathers with a long thread-like shaft and a few barbs and barbules at the tip, and 5) **bristles**, found e.g. around the beak of some songbirds, are hair-

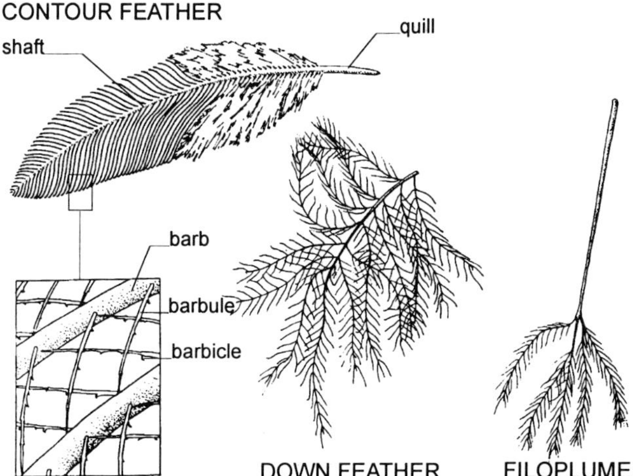

like feathers with a short quill and slender shaft, and with a few vestigial barbs at the base. Chicks are first covered only by down feathers. **Powder-down feathers**, found on some birds such as herons, hawks, and parrots, release a talc-like powder that helps to waterproof the feathers. *Cf.* apterium, pteryla.

feather follicle, a cylindric pit extending down into the dermis surrounded by an epidermal sheath to which the quill of the feather is attached. The development of the feather begins in the form of a dermal papilla including the **periderm** (feather sheath) and the developing barbs within the sheath; later the structure sinks into a follicle.

febris (L.), fever; **febrile**, pertaining to fever, feverish, pyretic.

feces, → faeces.

fecundation (L. *fecundare* = to fertilize), → fertilization, conception. *Adj.* **fecund**. *See* fecundity.

fecundity, reproductive output, i.e. the number of progenies produced by an individual or a population in a given time interval; can be presented e.g. in numbers of gametes, seeds, eggs, or offspring. *Cf.* fertility. *See* reproductive output.

feedback, the return of some of the output energy in a system, thus regulating the process. In biological systems it may inhibit the input (**negative feedback**), or facilitate it (**positive feedback**); e.g. in the control of hormone secretion negative feedback is involved when the secreted hormone inhibits (feeds back) an earlier step in the chain of hormone synthesis (**end-product inhibition**). Positive feedback occurs e.g. in the blood coagulation, or in parturition when the head of an offspring stretches the neck of the uterus and this stretch reflectively causes stronger contractions in the uterine wall muscles, leading to the birth. *See* homeostasis.

feeder layer, a layer of irradiated or otherwise treated cells (non-prolifering cells) used in cell or protoplast cultures, e.g. producing growth factors so improving cell division.

Fehling's test (*Hermann von Fehling*, 1812—1885), a quantitative colour test to determine monosaccharides, some reducing disaccharides (sugars), certain aldehydes, and other reducing agents. When these compounds are heated in an alkaline copper salt solution the reduction reaction occurring produces a red precipitate of copper(I) oxide.

fell field, the treeless vegetation zone in fell areas rich in mosses and lichens. The **fell vegetation** (regio alpina) is further divided into numerous subtypes. *See* height zones.

female (L. *femella* = young woman), the sex that produces eggs, and only eggs.

female genital organs, the organs of reproduction in female animals; in vertebrates, include **internal** female genital organs: the → ovary and oviduct, in mammals also the → uterus and vagina, and **external** female genital organs, i.e. in mammals the vulva with labia and clitoris. In invertebrates, female genital organs always include the ovary and some kind of oviduct, often also the uterus and vagina, and the spermatheca (seminal receptacle) for the storage of sperm (*see* ovotestis). Female genital organs are also called feminine genital organs.

female sex hormones, female reproductive hormones; sex hormones produced by female animals; in vertebrates, steroid hormones which are synthesized mainly by specialized cells in the ovary, smaller quantities in the adrenal cortex, and in mammals also by placental cells; probably fat tissue produces female sex hormones in some species, e.g. in humans. Male animals produce female hormones in small quantities in the adrenal cortex and testis. The hormones are divided into two groups, → **estrogens** and → **gestagens**, which are anabolic hormones resulting in the development of female characteristics and then maintain reproductive functions. Their production is controlled by → gonadotropins. Steroid female hormones are also found in many invertebrate groups, e.g. in insects.

feminization (L. *femina* = woman), **1.** the development of feminine features during sexual maturation of a female, such as the development of girls to women; **2.** development of feminine features in a male animal as a consequence of hormonal disturbances, e.g. abnormal change of masculine features to feminine in a man.

femur, pl. **femora** (L.), **1.** thigh; **2.** thigh bone; the femoral bone (*os femoris*) of tetrapod vertebrates that extends from the pelvis to the knee; it articulates with the acetabulum of the hip bone and distally with the tibia forming the knee joint with the patella; **3.** the third segment between the trochanter and tibia in the insect leg, or between the trochanter and patella in arachnids; *see* leg.

fen, a wet plant community where bog mosses (→ Sphagnidae) produce a thick but not very acid (even neutral) peat layer; often treeless, *Carex* species and dwarf shrubs are common. Fens are common especially in Scandinavia and northern Russia. *See* mire.

fenestra, pl. **fenestrae** (L. = window), a window-like aperture in an anatomical structure, often closed by a membrane; e.g. *fenestra vestibuli* and *fenestra cochleae*, i.e. the oval and round windows between the inner and middle ear of vertebrates. *See* cochlea.

feral (L. *fera* = wild animal), pertaining to domesticated animals which live in a wild state.

ferment (L. *fermentum* = yeast, leaven), **1.** → enzyme; **2.** to undergo the anaerobic enzymatic conversion, → fermentation, of organic compounds, especially decomposition of carbohydrates.

fermentation, 1. in metabolism, an anaerobic enzymatic breakdown of organic substances by living cells, especially by bacteria and yeasts; the end products are incompletely oxidized compounds such as ethanol, acetic acid, or lactic acid. Fermentation produces energy in the form of ATP, but also heat and waste gases, as CO_2. For example, in alcoholic fermentation glucose is enzymatically decomposed into ethanol and carbon dioxide: $C_6H_{12}O_6 \longrightarrow 2\ CH_3CH_2OH + 2CO_2$, and ethanol further into acetic acid, i.e. $CH_3CH_2OH + O_2 \longrightarrow CH_3COOH + H_2O$; *see* anaerobic respiration, glycolysis; **2.** the process in which microorganisms and other cells are cultured in a special vessel i.e. **fermenter** (bioreactor) in a medium to produce certain substances, e.g. medicines or cosmetics. These containers vary in size from less than one litre to thousands of litres.

ferns, common name for plants in the subclass → Polypodidae, class Pteropsida, division Pteridophyta.

ferredoxin (L. *ferrum* = iron, + → redox), an iron-sulphur protein that acts as an electron transporter e.g. in → photosynthesis and → nitrogen fixation.

ferricyanide, hexacyanoferrate(III); a salt or ion of unstable ferricyanic acid, $H_3Fe(CN)_6$, ionic form $[Fe(CN)_6]^{3-}$. **Ferrocyanide,** i.e. hexacyanoferrate(II), is a salt of an unstable ferrocyanic acid, $H_4Fe(CN)_6$, ionic form $[Fe(CN)_6]^{4-}$. Cupric ferrocyanide has semipermeable properties and is used in the osmometer (*see* Pfeffer's cell). Potassium ferri- and ferrocyanides are poisonous substances formerly used e.g. in cleaning of laboratory dishes.

ferritin, *see* apoferritin.

ferrum (L.), → iron, symbol Fe; **ferro-,** a bivalent iron atom (FeII) or its ion (Fe^{2+}); **ferri-,** a trivalent iron atom (FeIII) or its ion (Fe^{3+}). *Adj.* **ferrous** and **ferric,** respectively.

fertile (L. *fertilis* < *ferre* = to bear, produce), capable of reproduction; opposed to sterile.

fertility, the ability of an organism or population to reproduce. *Cf.* fecundity.

fertility factor, → F factor.

fertility index, the percentage of the matings resulting in fertilization, of all the matings in a population.

fertility rate, in ecology, presented as the number of offspring in relation to the number of adult females in a population.

fertility research, 1. in zoology, the study measuring the ability of animals to produce fertile gametes, zygotes, and offspring; **2.** in botany, dealing with the quantities of nutrients in soil; necessary for planning the use of fertilizers.

fertility table, in population ecology, an expanded type of the → life table in which also fecundity per female by age group has been considered.

fertilization, syngamy; the union of male and female → gametes to form a → zygote. The main event of fertilization is the union of the nuclei of the gametes, i.e. **karyogamy** where the homologous chromosomes which have separated in → meiosis come together again; may be → anisogamous or isogamous. The alternation of fertilization and meiosis forms the sexual cycle. Different fertilization types are found among organisms. **Self-fertilization** occurs when the uniting gametes are produced by one single individual which may be → haploid, diploid, or polyploid. **Cross-fertilization** occurs when the uniting gametes are produced by different haploid, diploid, or polyploid individuals. **Double fertilization** occurs in seed plants when one of the → generative cells (or sperm cells) of the pollen grain unites with the egg cell and forms the → zygote, and the other unites with the endosperm mother cell and forms the endosperm.

In animals, fertilization may be **external** as in fish and amphibians, in which the ova and spermatozoa unite in water, or it may be **internal** as in most invertebrates, cartilaginous fishes, and amniotes (i.e. reptiles, birds, and mammals), in which the ova and spermatozoa unite in the female reproductive ducts. Among invertebrates, hermaphroditism is quite common, but only some hermaphrodites have self-fertilization; most of them reproduce through cross-fertilization, producing eggs and sperm at different times. Some species have both self-fertilization and cross-fertilization.

In seed plants, fertilization is preceded by → **pollination** in which pollen is transferred to the stigma of the pistil. In → **Angiospermae**, fertilization occurs in the → embryo sac of the → ovule, in the ovary of the pistil. A **pollen tube** grows into the embryo sac. In the tip of the growing pollen tube there is a **vegetative nucleus**, and behind it are two generative cells, **sperm cells**. When they reach the embryo sac, the vegetative nucleus disappears, and one of the sperm cells fertilizes the female gamete, i.e. the → **egg cell**. The nucleus of the other sperm cell fuses with the diploid nucleus of the **endosperm mother cell** (*see* endosperm).

In female flowers of → **Gymnospermae** there are no closed ovaries, and the ovules are located on the carpels, i.e. cone scales. After pollination the cone will close and fertilization takes place inside it, in a way analogous to Angiospermae. In more primitive plants (pteridophytes, mosses, algae) fertilization is mediated by water. *Cf.* parthenogenesis, pseudogamy.

fertilizer, a substance used for adding nutrients into soil; can be natural manure, or a strong fertilizer which is an industrially produced salt mixture, usually containing nitrogen, phosphorus, and potassium.

fetal, → foetal.

fetoproteins, → foetoproteins.

fetus, → foetus.

Feulgen staining (*Robert Feulgen*, 1884—1955), a staining method based on the process in which a colourless Schiff's base (*see* Schiff reagent) forms a red complex with the aldehyde groups of biological macromolecules. This staining method is used either after an acidic hydrolysis of the tissue, as e.g. in a DNA-specific staining method of chromosomes, or after a periodate oxidation of specimens to demonstrate → polysaccharides.

fever (L. *febris*), increased body temperature of homoiothermic animals, i.e. birds and mammals, above the normal level; a special case of hyperthermia, also called **pyrexia**. Fever may be caused by several factors, such as microbes, inflammation, abnormal changes in metabolism, or cancer. In fever, **endogenic → pyrogens** are released into the blood changing the set point of the thermoregulatory centre in the hypothalamus to a higher temperature value. Then, shivering of skeletal muscles produces extra heat and increases the body temperature to this new set point value. For most mammals, fever becomes dangerous at about 41—42°C; the normal body temperature of birds is 2—3°C higher and they can correspondingly tolerate higher febrile temperatures. The importance of fever for animals is not totally understood, but it is shown that fever inhibits the growth of many pathogenic microbes, and the immune mechanisms are accelerated at higher body temperatures. Many poikilothermic animals also tend to choose higher environmental temperatures when infected by microbes (**behavioural fever**) and thus improve their survival. *See* hyperthermia.

FFA, free → fatty acid; a fatty acid occurring in free form (not esterified) in organisms.

F factor, fertility factor; an → episome the presence of which in a bacterial cell determines its "sex" and the transfer of genetic material from the donor cell to the recipient cell in bacterial → conjugation. Based on the presence of the F factor, bacterial cells are divided into four types. 1) **F^- cells** are bacteria which are lacking the F factor, functioning as recipients in conjugation; 2) **F^+ cells** are bacteria which contain an autonomous F factor located outside the chromosome; these cells function as donors in conjugation and often transfer the F character to the F^- cells; 3) **F' cells** are bacteria which contain an autonomous F factor to which, however, substantial quantities of chromosomal material are connected, thus being capable of → F-duction; 4) **Hfr cells** (Hfr = high frequency of recombination) are bacteria in which the F factor is integrated into the chromosome as a part of it; these cells act as donors in → conjugation, and as compared to F^+ cells, usually donate their genetic material more easily in conjugation. *Syn.* F plasmid, sex factor.

F generation (F < L. *filius* = son), filial generation; whichever generation that follows the parental generation (P generation). The generation which is a consequence of the crossing of the individuals of the parental generation is called the F_1 generation. The generation following self-fertilization or crossing of F_1 individuals is called the F_2 generation, and the progenies of F_2 individuals form the F_3 generation, and so on.

fiber, → fibre.

fibr(o)- (L. *fibra* = thread, fibre), denoting relationship to → fibre, or fibre-like and filamentous structures.

fibre, Am. **fiber, 1.** a slender, thread-like anatomical structure of animals, such as the → nerve fibre, muscle fibre (muscle cell), or collagen fibre; **2.** a long, pointed-ended cell type in plants, belonging to the → sclerenchyma tissues, occurring in connection with both the → xylem and → phloem tissues; fibres are lignified and give support to structures. Economically important fibres are e.g. jute, hemp, manilla hemp, and flax.

fibril (L. *fibrilla*, dim. of *fibra* = fibre), a microscopical thread-like structure, usually thinner than a → fibre (often a component of a fibre) but larger than a → filament; e.g. nerve fibrils inside nerve fibres, muscle fibrils (myofibrils) inside muscle fibres (muscle cells), or collagen fibrils as components of collagen fibres. The cellulose material of plant cells consists of → microfibrils and → macrofibrils.

fibrillar, fibrillary, 1. pertaining to a → fibril; having a fibril-like structure, **fibrous**; **2.** pertaining to → fibrillation.

fibrillation, 1. rapid, randomized contractions of muscle fibres in some part of a muscle; in the heart it is involved in cardiac arrhythmia in the atrium (atrial fibrillation) or ventricles (ventricular fibrillation); **2.** the process in a tissue of becoming fibrillar, fibril-like in structure.

fibrin, a fibrous protein formed from the soluble protein precursor, **fibrinogen**, in the final step of blood coagulation; the hydrolytic decomposition of fibrin, **fibrinolysis**.

fibrinolysin (Gr *lyein* = to dissolve), *see* plasmin.

fibrinolysis, the hydrolytic degradation of → fibrin, resulting in the dissolution of blood clots; occurs as a reaction of the **fibrinolytic system** that also balances and limits → blood coagulation *in vivo*. The active component of this system is → **plasmin** (fibrinolysin).

fibroblast (Gr. *blastos* = germ), an immature or reactivated cell of connective tissue; irregular in shape having several protrusions and a flattened oval nucleus. Fibroblasts produce proteins into interstitial fluid which become polymerized into different types of fibres (collagen fibres, elastic fibres). Fibroblasts are capable of differentiating into chondroblasts and osteoblasts in the formation of cartilage and bone; they are active also in wound healing. Less active fibroblasts of the mature tissue are conventially called **fibro-**

cytes, although many histologists consider the terms to be synonymous.

fibrocyte (Gr. *kytos* = hollow, cell), *see* fibroblast.

fibroma, a fibrous tumour derived from connective tissue, also called **fibroid**; the process of its formation is called **fibromatosis**.

fibronectin (L. *nectere* = to connect, bind), a family of extracellular multiadhesive glycoproteins in animal tissues, binding to other matrix components and cell-surface receptors (integrin family). One form is present in the blood acting as an opsonin (*see* opsonization) for phagocytic cells and in aggregation of thrombocytes. In tissues of many organs it occurs as fibrillar **cell-surface fibronectin**, which mediates adhesive interactions between cells, or between cells and the → basement membrane; it is also involved in cell migration and is important in wound healing. In embryogenesis it is related to cellular differentiation and morphogenetic movements. A similar glycoprotein in the bone tissue is called **osteonectin**.

fibrosis, pl. **fibroses,** the formation of a fibrous tissue in repair, usually associated with a disease; e.g. cystic fibrosis (a disease of the pancreas), or endomyocardial fibrosis.

fibrous, composed of, or resembling fibres.

fibrous lamina, → nuclear lamina.

fibrous roots, → adventitious roots of equal diameter, arising from the stem base, gradually replacing the tap root; common in many monocotyledons, e.g. in grasses.

fibrous tissue, dense connective tissue (a form of connective tissue) comprising large bundles of collagen fibres; it is the tissue of tendons, ligaments, and hard membranes such as the dura mater.

fibula, pl. **fibulae** also **fibulas** (L. = buckle), the thinner of the two bones between the knee and the ankle in the hindlimb of tetrapod vertebrates; in some groups, such as in amphibians, the fibula has fused with the → tibia, and in birds it is often incomplete.

fibulin, a cysteine-rich, calcium-binding glycoprotein found in the cytoplasm and the extracellular matrix of plants; based on their amino acid content, three types of fibulin are found, i.e. with 566, 601, and 683 amino acid residues.

Fick's law (*Adolf Fick*, 1829—1901), the rule describing the dependence of the diffusion on the concentration of a substance; is presented in the formula: $dQ/dt = -D \cdot dc/dx$, where D = diffusion coefficient, c = concentration, Q = the quantity of a diffusible substance, t = time, x = diffusion distance.
D is dependent on temperature, on the molecular size of any diffusible substance, and on the viscosity of the medium.

field inversion gel electrophoresis, FIGE, a special type of → gel electrophoresis used to separate DNAs in a horizontal agarose gel. The FIGE (polarity reversal) system uses an asymmetric switch time format to improve the separation of DNA types of different size. The duration of the forward switch time is typically three times longer than that of the reverse switch time, e.g. 66 and 22 milliseconds, respectively. The voltage is either constant or asymmetric.

field layer, herb layer; a vegetation layer in forests, including plants and other organisms between the ground layer on the floor and the shrub layer.

FIGE, → field inversion gel electrophoresis.

filament (L. *filamentum* < *filum* = thread), a thin fibrillary structure; e.g. **1.** in the anatomy of animals, a thin extracellular thread e.g. in a connective or nervous tissue; **2.** a thread-like organelle projecting from a cellular body of certain protozoans or spermatozoa; **3.** an ultramicroscopic, intracellular, fibrillary structure, usually a component of a → fibril; e.g. **myofilaments** (actin and myosin filaments), bundles of which form the myofibrils of the muscle cells, or **microfilaments** forming a supporting structure of the → cytoskeleton; **4.** the stalk of the → stamen in seed plants, supporting the anther; **5.** the thread-like structure of many algae and fungi; **6.** a very thin fibre in a plant cell; **7.** in electronics, a thin heating element such as the cathode in an → electron microsope. *Adj.* **filamentous**.

filarial worms, filarial nematodes (L. *Filaria* = the type genus), a group of parasitic nematodes living in many animal species, e.g. human endoparasites whose intermediate hosts are some blood-sucking insects, e.g. mosquitoes. The familiar species is the filarial worm, *Wuchereria bancrofti* causing the disease of → **elephantiasis**; another filarial worm, *Onchocerca volvulus*, causes river blindness, → **onchocerciasis**. *See* microfilaria.

filariasis, a pathologic condition when → filarial worms are present in the blood and lymph systems and other tissues.

filial generation, → F generation.

Filicales, *see* Polypodidae.

filoplume (L. *filum* = thread, *pluma* = feather), *see* feather.

filopodium, pl. **filopodia** (Gr. *pous* = foot), **filopod**; a thin, usually branched extension of a cell; e.g. a thread-like pseudopod of rhizopods (Sarcodina).

filter (L. *filtrare* = to filtrate), **1.** any material, as cloth, paper, sand, charcoal, porous porcelain, glass, or plastic, through which a liquid or gas is passed to remove some material, e.g. impurities; **microfilters**, in which the pores are equal and of exact minute size, are used for cell culturing; in animals, e.g. the walls of blood capillaries function as microfilters; **2.** a screen to absorb different types of radiation; **3.** in biogeography, a land or continental connection between isolated areas, having selective effects on the numbers and types of organisms migrating over them during different geological periods; e.g. the continental connection of the Bering Land Bridge between Asia and America in Pleistocene. Wooded zones in cities filter the dispersal of organisms between different city areas; **4.** as a *verb*, to **filter**, also **filtrate**. *See* filtration.

filter feeders, aquatic animals such as bivalves (mussels, clams) which catch food by filtering organic material (mostly microorganisms) from the surrounding water; also some large vertebrates, such as certain sharks and whalebone whales, are filter feeders; the phenomenon is called **filter feeding**. *Cf.* collector filterers, gathering collectors.

filt(e)rable, capable of passing through a filter; **filtrate**, a liquid that has passed through a filter, i.e. been filtered.

filtration, 1. the process of separating a solid from a liquid or gas by passing them through a filter; may be accomplished by pressure or vacuum (suction); in → **gel filtration** substances are separated according to their molecular size; **2.** a process of attenuating a radioactive or electromagnetic radiation when absorbed in some material. *Verb* to **filter** (filtrate).

filum, pl. **fila** (L. = thread), a thread-like anatomical structure; e.g **fila olfactoria**, thin nerve bundles (olfactory nerves) from the nasal mucosa through the cribriform plate of the ethmoid bone to the olfactory bulb of the vertebrate forebrain.

fimbria, pl. **fimbriae** (L. = fringe), a fringe or border in an anatomical structure; e.g. **fimbriae of uterine tube**, fringed processes around the ostium of the oviduct extending along the free border of the mesosalpinx.

fin, a common term for organs of aquatic vertebrates used for locomotion, steering, balancing, etc., including also paddle-like fin-limbs of seals and whales (*see* leg). Fish fins are formed from membranous extensions of the integument and are supported by cartilaginous or bony rays. They may be either soft and flexible, usually branched **soft fin rays**, or hard, calcified **spinous fin rays**. The **pectoral** and **pelvic fins** are paired structures homologous with the tetrapod limbs, while the unpaired **dorsal** and **anal fins** are derived from the skin, and the → **caudal fin** (tail fin) is related to the tail. Salmonids and catfishes have a caudo-dorsal **adipose fin** (fleshy fin).

final host, → definitive host.

fine sand, a soil type having a grain size from 0.02 to 0.2 mm in diameter. *Cf.* silt.

finger domain, a domain of → transcription factors in which a finger-like subunit of the transcription factor protein binds to certain sequences of DNA.

finger gland, a finger-like, slime secreting gland in the female reproductive system of gastropods.

fingerprint, in genetics, the specific pattern of → microsatellite DNA sequences of an individual; every individual can be recognized on the basis of its → DNA fingerprint.

Finnish disease inheritance, inheritance of genetic diseases typical of Finns. Due to the → founder effect, approximately 30 genetic diseases, rare elsewhere in the world, have been enriched in Finland. On the other hand, Finns are lacking certain genetic diseases which are common elsewhere in the world. In addition, due to the population history of Finland, each of the diseases of the Finnish disease inheritance has been enriched in a certain area of the country.

first messenger, the group name for messenger substances which specifically bring a signal to a cell, often to the cell membrane, where its information is transduced to produce a specific intracellular messenger (→ second messenger) which causes changes in the activity of the cell; e.g. many animal hormones act as first messengers.

FISH, → fluorescence in situ hybridization.

fishes, a common name for a group of poikilothermic aquatic vertebrates, especially for cartilaginous and bony fishes, often lampreys (jawless fishes) included. *See* Chondrichthyes, Osteichthyes, Petromyzoniformes.

fission (L. *fissio*), **1.** the act of splitting; **2.** a type of asexual reproduction of unicellular organisms in which the cell divides into two (binary fission) or several (multiple fission) equal daughter cells, each of which develops into a new independent organism; **3. nuclear fission**; the nuclear reaction in which an atom nucleus splits into nuclei of lighter atoms.

Fissipedia (L. *fissus* = cleft, *pes* = foot); a suborder of terrestrial carnivorous animals, as dogs, cats, bears and martens, weasels. At present, the name → Carnivora has replaced Fissipedia. *Cf.* Pinnipedia.

fissura, pl. **fissurae** (L. *findere* = to split), **fissure**, a deep groove or cleft in an anatomical structure, as e.g. a deep sulcus of the brain surface.

fistula, pl. **fistulae,** also **fistulas** (L. = pipe), an abnormal passage from a hollow internal organ to another, also from an internal organ out to the body surface.

fitness, the ability of an organism to adapt to its environment, i.e. the relative ability of an individual or a → genotype to produce fertile offspring. Thus, fitness includes the ability to survive to reproductive age, the capacity of reproduction, and the fertility of the progenies. Fitness is a relative concept which compares an individual or a genotype with the mean genotype of a population or with other individuals or genotypes. If the differences in fitness are due at least to some extent to genetic differences, this leads to → selection and to changes in the gene frequencies of the population. All phenotypic differences in fitness lead to selection, but they lead to changes in gene frequency only if the differences are genotypic, i.e. heritable. The same genotype does not necessarily lead to equal fitness in different individuals, but fitness is also dependent on environmental conditions. *See* inclusive fitness. *Syn.* selective value, adaptive value.

fixation (L. *figere* = to fix), **1.** in biology, a method to stabilize the structure of a tissue sample for microscopic examinations; it is carried out using special chemicals such as → glutaraldehyde, formaldehyde, and osmium tetroxide; whole organs or even entire small animals are sometimes treated by **perfusion fixation**, i.e. through circulation; **2.** the fixation of chemical elements in cell metabolism to form organic substances, e.g. carbon dioxide fixation in → photosynthesis, or → nitrogen fixation; **3.** in genetics, the phenomenon in which a particular gene locus becomes homozygous in a population; **4.** in medicine, the fixation of broken bones; **5.** in psychology, the turning of learned behaviour to habits.

fixed-action patterns, behavioural patterns of animals which are innate and highly stereotypic, almost without a learning component; are usually released by certain key stimuli. The fixed-action patterns of related species resemble each other, and therefore they are used as key characteristics in classification of the species, corresponding e.g. to morphological and anatomical structures.

Flagellata (L. *flagellum* = whip), **flagellates;** a highly diverse group of protozoans, sometimes considered as a synonym of Mastigophora in the phylum Sarcomastigophora; unicellular organisms with one or more flagella as locomotor organelles. The group includes **autotrophic species** with chloroplasts, capable of photosynthesis, and **heterotrophic species** without chloroplasts which are either free-living organisms in marine and fresh waters or parasites in other animals. Flagellates are divided into three subgroups (classes or superclasses in some classifications): **dinoflagellates** (Dinoflagellata) and **zooflagellates** (Zoomastigophora) comprise species which are either parasitic, commensal, or symbiotic; the genus *Trypanosoma* causes severe blood diseases in humans. The third group, **phytoflagellates** (Phytomastigophora) are organisms which contain chloroplasts and are therefore included also in the botanical classification, usually within the division Pyrrophyta.

flagellated chamber, a chamber in the body wall of leuconoid sponges (*see* leucon grade) the inner surface of which is lined by several → choanocytes. *See* Porifera.

flagellum, pl. **flagella** (L.), a long, cylindrical locomotor organelle projected from the free surface of a cell. Flagella are found in unicellular organisms, as e.g. in some protozoans (flagellates) and many unicellular algae. Also many multicellular organisms have flagellated cells, e.g. certain spores and male gametes. The motile internal structure (axoneme) of the flagellum consists of nine doublets peripherally and two centrally located microtubules (9+2 structure) which at the base of the flagellum are connected to the basal body, thus resembling the structure of cilia. Functionally flagella deviate from cilia, the flagellum propelling water parallel to its main axis. In the formation of the flagellum the → centriole plays an active role. Some prokaryotic organisms, such as *Escherichia coli*, have flagella which are structurally simpler and not homologous with those of eukaryotes.

flame cells, terminal cells in the → protonephridia of some primitive invertebrates, such as flatworms and nematodes. Each flame cell has several flagella by which the cells collect fluid from the body cavity into the collecting tubule of the protonephridia, and after absorption and secretion processes of the tubule cells, the fluid passes to the canals which release excretion ("urine") to the body surface. These cells are named flame cells because in a microscope the rhythmical beat of the tuft of flagella resembles a flickering flame. *Cf.* solenocytes.

flamingos, → Ciconiiformes.

flanking markers, in genetics, → markers which are located in the chromosome on both sides of a given gene; are used to study intragenic → recombination.

flatworms, → Platyhelminthes.

flatus (L. = act of blowing, a puff of wind), gas in the gastrointestinal tract, or expelled through the anus.

flav(o)- (L. *flavus* = yellow), denoting yellow.

flavin, also **flavine, 1.** any of yellow pigments found in animals and plants, consisting of one isoalloxazine ring bound to D-ribitol. Flavins are light-sensitive, fluorescent, water-soluble and reducible compounds. The most important of the cellular flavins is **riboflavin** (vitamin B_2), which when bound to a protein forms **flavoprotein** acting as an electron carrier in cell respiration; *see* flavin nucleotides; **2.** a yellow acridine dye, **quercetin,** $C_6H_7(OH)_5$, obtained from oak bark.

flavin nucleotides, two important cellular → nucleotides with → riboflavin moieties: **flavin mononucleotide** (oxidized form **FMN,** reduced form **$FMNH_2$**) is riboflavin 5'-phosphate, acting as a prosthetic group or coenzyme of some flavoprotein enzymes, e.g. of the enzyme that catalyses the oxidation of NADPH; **flavin adenine dinucleotide** (oxidized form **FAD,** reduced form **$FADH_2$**) is a condensation product of riboflavin and adenosine diphosphate, functioning as a prosthetic group of flavoprotein enzymes (many aerobic dehydrogenases) in cellular oxidation-reduction reactions; these enzymes act indirectly via NAD or NADP or directly on their substrates.

flavonoids, the group name for glycosidic phenol derivatives occurring in plants. The molecule has a C_6-C_3-C_6 carbon skeleton in which C_6 parts are benzene rings and the structure of C_3 varies. Flavonoids are red, purple, or yellow pigments located in the vacuoles of plant cells. Anthoxanthins, flavones, quercetins, and anthocyanidins (*see* anthocyan) are examples of flavonoids. One flavonoid activates the transcription of *Rhizobium* nodulation genes.

flavoprotein, any protein having a → flavin nucleotide as its prosthetic group; flavoproteins act e.g. as dehydrogenase enzymes in the proton and electron transport system of cell respiration.

fledging period, the interval between the hatching and the stage when a young bird is capable of flying, or when its plumage is fully developed. For nidifugous birds the fledging period is difficult to determine but among nidicolous birds the period lasts from hatching to the time when fledglings leave the nest. *Cf.* nestling period.

flexor (L. *flexio* = the act of bending), *musculus flexor*; any muscle that bends (flexes) the joint. *Adj.* **flexible**. *Cf.* extensor.

flexure (L. *flexura*), a bending in an anatomical structure; e.g. **cephalic flexure** (cranial flexure), the curve formed at the embryonic midbrain of vertebrates.

flip-flop, in cell biology, the transfer of a phospholipid molecule from one monolayer of the lipid bilayer membrane to the other.

float, *see* pneumatophore.

flora, (L. *Flora* = mythical goddess of flowers < *flos* = flower), **1.** all the plant species of an area, habitat, or period; **2.** a list (book) describing these species in systematic order; **3.** the population of microorganisms living naturally in (or on) an animal or a plant, e.g. the intestinal (micro)flora.

floral apex, the plant → apical meristem which develops into a flower or an inflorescence.

floral diagram, a diagram illustrating the

structure of a flower. *Cf.* floral formula.

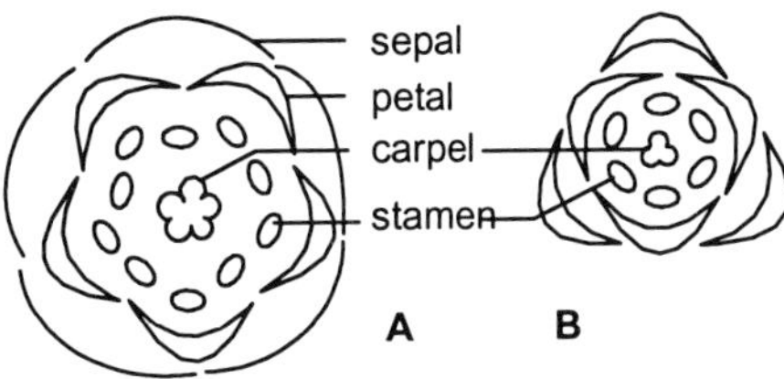

Floral diagrams. Typical dicotyledonous flower (A), typical monocotyledonous flower (B).

floral formula, a formula describing the structure of a flower in a short form: e.g. the floral formula of the species in the family Apiaceae (e.g. shepherd's purse) is K4C4A4+2*G*(2), i.e. there is a calyx (K) with four sepals, a corolla (C) with four petals, an androecium (A) formed from 4 + 2 stamens and a gynaecium (*G*), which is superior (shown by underlining) and composed of two fused (parentheses) carpels. Another way to show the structure of a flower is the → floral diagram.

floral leaf (L. *flos* = flower), a leaf occurring in a → flower; the floral leaves have developed from → foliage leaves and are often differentiated into complicated reproductive structures such as the → carpels and the → stamen; the petals and the sepals are structurally simpler, or may be totally lacking.

floral shoot, flower shoot; the part of a plant shoot which bears flowers.

floral tube, a tube-like structure formed by fusion of flower leaves, mostly both petals and sepals.

floret, one of the small flowers in the → capitulum or → spike of grasses; the flowers in the capitulum may be tube-like **disc florets** or leaf-like **ray florets**.

floridean starch, a storage polysaccharide occurring in red algae; gives a brown colour with an iodine reagent.

florigen (Gr. *gennan* = to produce), a hypothetical flowering hormone without any known structure; its existence has not been proven.

floristic kingdoms, large areas on the Earth having a more or less uniform and characteristic flora, and being mostly separated by oceans, deserts, or mountain areas. The **Holarctic kingdom** embraces the entire arctic and northern temperate zones. The **Palaeotropical kingdom** comprises tropical or some subtropical areas of the Old World, i.e. the main part of Africa, India, and oriental areas and the Pacific Ocean. The **Neotropical kingdom** consists of corresponding areas in the New World, i.e. of Mexico, Central America and South America except of Patagonia. The **Australian kingdom** comprises the Australian continent and Tasmania, the **South African kingdom** (Cape Region), i.e. the southernmost part of South Africa, and the **Antarctic kingdom**, including Patagonia and Antarctica. Sometimes called floral kingdoms. *Cf.* vegetation zones. *See* zoogeographical realm.

floristics, the study of the composition of vegetation, i.e. the plant species in an area; floristics may also denote analytical study e.g. on the origin of the flora and different geobotanical properties.

flow cytometry (Gr. *kytos* = cavity, cell, *metron* = measure), a technique used to detect and separate fluorescently activated cells, microorganisms, or cell organelles from nonfluorescent ones. In a flow cytometer → fluorescence is excited by a laser-based detector system. Positively fluorescent cells (e.g. marked by monoclonal antibodies labelled with fluorochromes) give an electric charge and can be separated from non-fluorescent uncharged cells when passing between electrically charged plates. Many parameters can be measured by this technique, such as the DNA content, cell surface markers, and karyotypes. The technique has applications in screening for chromosome abnormalities and in cancer research.

flower, the reproductive structure of → seed plants, comprising **flower leaves**, attached to the → **receptacle**. The most important flower leaves are the **stamens**, forming → pollen grains and further male gametes, and the → **carpels**, producing female gametes, seeds, and fruits. In the flowers of → Angiospermae often also **sepals** and **petals** exist, but are not essential in reproduction and are sometimes totally lacking.

flower constancy, the fidelity of an insect to visit flowers of the same plant species.

flowering plants, seed-bearing plants; plants belonging to the division Spermatophyta, devided into the subdivisions → Gymnospermae and Angiospermae. In some classifi-

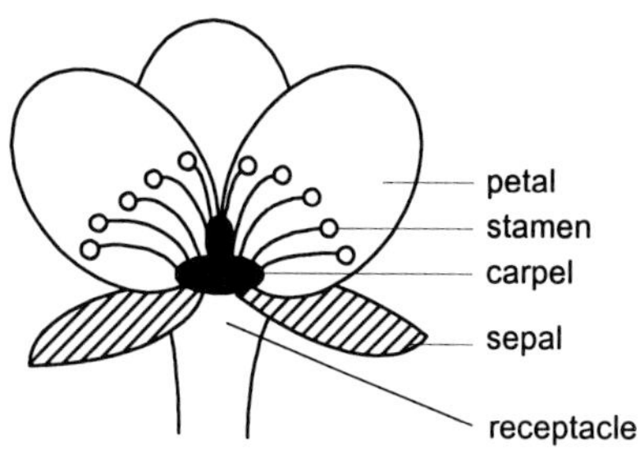

Flower. Structure of a typical single flower.

cations the groups spermatophytes, gymnosperms and angiosperms are without any taxonomical status.

flower shoot, → floral shoot.

flower stalk, → peduncle.

fluid (L. *fluidus*), a non-solid substance, as a liquid, solution, or a suspension of solid particles, capable of flowing. There are different fluids in organisms, called **fluid compartments**, i.e. **intracellular fluid** and different types of **extracellular fluids** (tissue fluids). In animals, extracellular fluids include e.g. interstitial fluid, plasma, lymph, haemolymph, and cerebrospinal, pericardial, synovial, and intraocular fluids. Allantoic and amniotic fluids are formed in the allantois and within the amniotic cavity of a foetus. Ascitic fluid is a serous liquid that abnormally accumulates in the peritoneal cavity. In plants, extracellular fluids are e.g. xylem and phloem fluids. In addition, many different artificial fluids, such as fixatives and physiological solutions, are used in biological research. *Adj.* **fluid** (also **fluidic, fluidal**), capable of flowing, changing readily. *See* Ringer fluids.

fluidity, the reciprocal of → viscosity; the unit is rhe = poise^{-1}; e.g. the fluidity of cell membranes which depends on the properties of the lipid moieties of the membranes. *See* homeoviscous adaptation, poise.

fluid mosaic model, fluid bilayer model; a generally accepted model for cellular membranes in which the matrix is formed by a bilayer of phospholipid and glycolipid molecules (bimolecular leaflet). Proteins are associated with the bilayer either peripherally (extrinsic) or they may be embedded in the bilayer (intrinsic). In many cases intrinsic proteins span the bilayer. A major feature of the model is that it predicts that the bilayer is fluid and so proteins and other associated molecules may migrate laterally in the membrane. *See* cell membrane.

flukes, → Trematoda.

fluorescein, (< fluorspar, i.e. fluorite), $C_{20}H_{12}O_5$, **resorcinolphthalein** (resorcinol phthalic anhydride); a fluorescent dyestuff; a crystalline, organic compound that in alkaline solutions produces orange colour and a green-yellow fluorescence (absorption maxima 460 and 493.5 nm) when excited by UV light of 365 nm. Fluorescein is used in colouring liquids of various instruments and for biochemical determinations, e.g. to prove the viability of cells, or to examine changes in blood vessels. The reduced form of fluorescein is called **fluorescin**.

fluorescence, the emission of one or more → photons when a molecule or atom is excited by electromagnetic radiation (**excitation light**, usually ultraviolet light). The emission is of longer radiation (**emission light,** usually visible light) than the excitation radiation, and the emission occurs immediately (within 10^{-8} seconds) when excited by light. In fluorescence, excitatory radiation causes an electron leap from one electron shell to another, and on return of the electron energy is liberated as emission light. The phenomenon is used to determine different substances in organisms by making them fluorescent or linking substances such as antibodies to fluorescent molecules, e.g. to → fluorescein or → rhodamine. The intensity of fluorescence is quantitatively proportional to the quantity of a substance to be determined. **Fluorometers** (fluorimeter) and **fluorescence microscopes** are optical instruments designed to measure fluorescence. *Cf.* phosphorescence, luminescence.

fluorescence in situ hybridization, FISH, a method for detecting the locations of genes on the chromosomal complement; a DNA or RNA molecule, complementary to the searched gene and labelled by a fluorescent stain, is hybridized to the chromosomal complement, and the fluorescent label reveals the location of the gene.

fluorescence microscopy, a special technique of light microscopy in which → fluorescence is used for localization of structures and materials in specimens. *See* microscope.

fluorescent antibody technique, a method for labelling a specific, monoclonal or polyclonal → antibody with a fluorescent compound, e.g. **fluorochrome**, binding specifically to an antigen to be detected and determined in biological materials by fluorometric methods. *See* flow cytometry.

fluorescent dye, a substance that absorbs light at one wavelength (usually UV) and emits it at a specific longer wavelength within a visible spectrum (*see* fluorescence), e.g. → fluorescein; the dye is usually linked to another molecule in fluorescent staining.

fluorine, a non-metallic element, symbol F, atomic weight 18.998; a toxic and corrosive halogen gas. It resembles chlorine but is more reactive. It occurs e.g. in the ore fluorite, CaF_2 (fluorspar), and in some silicates. **Fluorides** are salts of hydrofluoric acid (HF) which are formed in reactions by fluorine with a metal, non-metal, or organic radical. Minute concentrations of fluorides may exist in organisms, e.g. in enamel of the teeth, and in fluoridation minute quantities of fluorides are applied to the teeth for prevention of caries. In higher concentrations fluorides are toxic to organisms. *See* apatite, chlorofluorocarbon.

fluorometer, also **fluorimeter,** an apparatus with an ultraviolet light source, a monochromator for the selection of wavelength, and a light detector; it is used in quantitative analyses of substances for measuring → fluorescence e.g. in biological samples; the method is called **fluorometry** (fluorimetry).

fluoroscope, roentgenoscope; an apparatus for rendering the shadows of X-rays visible on a surface containing fluorescent material.

fluorouracil, a fluorine analogue of → uracil inhibiting the RNA synthesis; used as an anticancer drug.

flying lemurs, → Dermoptera.

FMN, flavin mononucleotide. *See* flavin nucleotides.

focused mutagenesis (L. *focus* = fireplace), the induction of a mutation in a given gene by gene manipulation. *Syn.* site-directed mutagenesis, targeted mutagenesis.

foetal, Am. **fetal,** pertaining to the → foetus.

foetal membranes, Am. **fetal membranes,** *see* extraembryonic membranes.

foetoproteins, Am. **fetoproteins,** proteins released from tissues of the foetus and the newborn mammals, especially from the liver, yolk sac, and gastrointestinal canal; the best known is **alpha-foetoprotein** (**AFP**) the amount of which is determined for diagnostic purposes especially in humans.

foetus, Am. **fetus** (L. *foetus*), the embryo of a viviparous animal, especially of a mammal, after the stage when characteristics typical of that species can be seen, i.e. in man at 7—8 weeks.

foliage leaf, a leaf which represents the "normal type" of leaves in their structure and location, as distinguished from the floral leaves, scales, and bracts. *See* leaf sequence.

foliage trees, trees belonging to → Angiospermae, having leaves and being more demanding for climate and soil than → coniferous trees. Many foliage trees are **deciduous**, losing all their leaves at a certain time of the year.

folic acid (L. *folium* = leaf), pteroylglutamic acid, often its oligoglutamic acid conjugates included; in organisms its physiologically active, reduced form, **tetrahydrofolate** (THFA), acts as a coenzyme of some cellular enzymes functioning as a carrier of single carbon groups in many reactions, e.g. in the biosynthesis and catabolism of several amino acids, and in the synthesis of creatine, choline, purines and pyrimidines, and thus of nucleotides. In vertebrates, folic acid is necessary for the normal production of red blood cells, belonging to the → vitamin B complex. Its ionic form, salts, and esters are called **folates**.

folivore (L. *folium* = leaf, *vorare* = to eat), an animal feeding on leaves. *Adj.* **folivorous**.

folliberin, follicle-stimulating hormone-releasing hormone/factor, FSH-RH, FSH-RF. *See* gonadoliberin.

follicle (L. *folliculus*), a small sac-like anatomical structure, cavity or pouch; e.g. **1.** → ovarian follicle; **2.** thyroid follicle (*see* thyroid gland); **3.** hair follicle (*see* hair); **4.** dental follicle (tooth follicle) around the developing tooth; **5.** a dry fruit formed of one carpel, opening on one side; common e.g. in the plant family Ranunculaceae.

follicle-stimulating hormone, FSH, a glycoprotein hormone in vertebrates, produced by certain cells in the anterior part of the → pituitary gland; regulates the development and maturing of female germ cells (ova) in the ovarian follicles, but also of spermatozoa in the testes. *Syn.* follitropin. *See* folliberin.

follicle-stimulating hormone-releasing hormone/factor, FSH-RH, FSH-RF, → folliberin, gonadoliberin.

follitropin, → follicle-stimulating hormone.

fontanel(le) (Fr. dim. of *fontaine* = spring), a soft spot on an anatomical structure; especially the space between the parietal cranial bones of a mammalian foetus, covered only by the skin and membranes, i.e. the anterior fontanelle (frontal f.) and the posterior fontanelle (occipital f.). Because of the extensive growth of the brain, human fontanelles are exceptionally large and close late after the birth, normally between the first and second year of age.

food chain, a hierarchical system in a community through which material and energy flow. **Primary producers** (usually green plants) absorb the radiant energy of sunlight by the photosynthetic action of chlorophyll, changing it into chemical energy, conserved mainly in carbohydrates, to be used first for many microbes, saprophytic plants, and herbivorous animals (**primary consumers**) and then for carnivorous animals (**secondary, tertiary,** etc. **consumers**). All food chains in a community form a **food web**. *Cf.* trophic level.

food chain efficiency, the efficiency of an ecosystem in which energy is transferred from one level to the other in a → food chain; the efficiency is usually presented as percentage of the energy used by organisms in one trophic level from the energy available in the preceding level. *See* ecological efficiency.

food patch, food resources occurring in patches which are available to a predator; the profitability of the patch is determined by the prey density. *See* patchy habitat.

food pyramid, *see* ecological pyramid.

food specialist, *see* specialist.

food vacuole, an intracellular vesicle containing food particles; acts as a digestion organelle of the cell body in many protozoans. *See* phagocytosis.

food web, a complex network composed by many interconnected → **food chains** in a community, exposing inherent relationships between different species in food utilization and predation. In the **grazing food web** herbivores feed on growing plants and are then in turn pray for carnivores and omnivores. *See* detritus.

foot (L. *pes*, Gr. *pous*), **1.** a distal or pedal extremity of a limb in tetrapod vertebrates, sometimes only of the hindlimb. In the hindlimb, the foot comprises the → tarsus (ankle), metatarsus, and phalanges (toes); in the forelimb, the → carpus (wrist), metacarpus, and phalanges (toes or fingers); **2.** an unpaired ventral surface of the body of chitons and gastropods (Mollusca); **3.** the unpaired muscular foot between the mantles in clams and mussels.

foot-gill, → podobranch.

footprinting, → DNA footprint.

forage plants, plants cultivated and used as food for animals, e.g. many grasses, clovers, and cabbage species.

foramen, pl. **foramina** (L. = opening, aperture), any small perforation in a bone, or other tissues; e.g. **great foramen** (*foramen magnum*), the posterior opening in the vertebrate skull through which the brain continues as the spinal cord, or **oval foramen** (*foramen ovale)*, the opening between the atria of the embryonic heart in tetrapod vertebrates.

Foraminifera (L. *ferre* = to carry), **foraminiferans**; a protozoan group in the subphylum → Sarcodina; comprises mainly oceanic animals from the Precambrian time to the present. Their hard, usually calcareous tests, comprise one to several chambers, often braid-like or spirally coiled. Long pseudopods which extend out through the test openings are thread-like and reticulate (**reticulopodia**). The tests of foraminiferans have accumulated as bottom deposits in the oceans, and formed calcareous rocks; known groups are e.g. fossil **nummulites**. At present, the **Globigerina** ooze, composed of foraminiferan tests, covers about 35 percent of the ocean bottoms.

forcing, in botany, speeding up the development of flowers or buds in cultivated plants; usually presupposes a change in cultivation temperature.

forebrain, prosencephalon; the most anterior part of the three expansions of the early embryonic brain of vertebrates; the forebrain develops further into → telencephalon and diencephalon.

foregut, (L. *stomodaeum*), **1.** in vertebrates, a region in the embryonic digestive tract anterior to the pylorus; in fish differentiating into the oesophagus, stomach, liver, and pancreas, and in tetrapod vertebrates also into the pharynx, trachea, and lungs; **2.** in invertebrates, → **stomodaeum**, the anterior part of the digestive tract e.g. in insects, comprising usually the buccal cavity, pharynx, oesophagus, crop, and proventriculus; the inner wall is lined with ectodermal epithelium, chitin included. In arthropods generally, the gut comprises anteriorly the large stomodaeal region (foregut) and posteriorly the proctodaeal region (hindgut), and the midgut between them.

forest, a plant community characterized by trees; in its natural state, a forest remains in a fixed, self-regulated condition. The tree species are determined by climate, topography, and soil; certain shrubs and herbs are characteristically associated with the trees. The vegetation affects the composition of the soil, i.e. the influence is reciprocal.

The main general types of forests are: 1) **Northern coniferous forests** form a worldwide belt in subarctic and alpine regions of the northern hemisphere. Typical trees are spruce, fir, pine, and larch species. The soil is characterized by → podzol. 2) **Deciduous forests of the temperate regions** are common in temperate areas, e.g. in Central Europe and in America; typical are deciduous foliage trees such as oak and beech species. The forests of this type at the same latitude in the northern and southern hemispheres are quite different. 3) **Deciduous monsoon forests** resemble the former ones but the areas are characterized by more abundant rains and thus the species are partly different. These forests are common throughout southeast Asia and India, and also in the Pacific coastal regions of Mexico and Central America. 4) **Temperate evergreen forests** are found in the subtropical regions of North America and the Caribbean islands; typical trees are magnolias and palms. 5) **Temperate rain forests**, with broad-leaved evergreen trees, are common on the temperate west coast of North America and widely in Australia. 6) **Tropical savanna forests** are found in regions where the forest and grassland meet, e.g. in Africa and South America. 7) **Tropical rain forests** are typical of Central Africa and the Amazon area; the forest is active around the year, there are several separate tree layers, numerous tree species, and large lianas are common. In every main general forest type numerous subtypes are found.

forestomach (L. *antrum cardiacum*), in vertebrate anatomy, the cardiac part of the stomach near the oesophagus.

forest tundra, a → tundra area with very sparsely growing trees.

form, f, (L. *forma*), **1.** in zoology, the term used as a group name without any official position in the classification; the individuals of a form deviate only by one or a few characteristics from the individuals of other forms; **2.** the lowest category in plant classification; a species can be divided in lower groups which are → subspecies, → variety and form. Forms often differ from the main type in only one characteristic, such as in the colour of the corolla: e.g. *Nymphaea candida f. rosea* (water lily).

formaldehyde, methanal, formic aldehyde, methyl aldehyde; a pungent, toxic gaseous → aldehyde, HCHO; its 37% water solution is known as **formalin** (formol) that is an important raw material in industry and used earlier as a disinfection solution. Formalin is a common conservation liquid for tissues and cells as well as for whole animals and human corpses. As a fixative for biological specimens it is used alone or together with → glutaraldehyde; formaldehyde reacts with carbohydrates and proteins, and is a tanning substance. Its use is partly restricted because it is mutagenic, probably even carcinogenic. **Paraformaldehyde** is a solid polymer of formaldehyde.

formatio, pl. **formationes** (L.), **formation**; the process creating a shape or an entity, or a structure of definite shape; e.g. **1.** *formatio reticularis*, → reticular formation of the brain; **2.** in botany, a physiognomic unit of a → plant community, in the case that the classification in the community has been done (regardless of the floristic composition) according to the predominance of particular growth forms; e.g. coniferous forest, grassland, etc.

formic acid (L. *formica* = ant), methanoic acid; the simplest of the carboxylic acids, HCOOH; a corrosive acid with a pungent smell, present in sweat and urine, and abundantly in ants and many plants, e.g. in nettles. Formic acid is a good solvent for many organic and inorganic compounds, e.g. used in textile dyeing, in leather tanning, in manufacturing of other chemicals, and in solvent systems for chromatography. Its salts and esters are called **formates** (methanoates).

formol, *see* formaldehyde.

formula, pl. **formulae,** also **formulas** (L. dim. of *forma* = form), **1.** chemical formula, an expression of the components of a compound by symbols or figures, e.g. a molecular formula or structural formula; **2.** in medicine, a recipe for a drug preparation.

fornix, pl. **fornices** (L.), an arch-like anatomical structure; especially *fornix cerebri*, the fornix of the cerebrum of vertebrates, which is the paired, arch-like neural pathway from the hippocampus to the mamillary bodies and the habenular nuclei.

fossa, pl. **fossae** (L. = ditch), a hollow or depressed area in an anatomical structure, e.g. on a bone; dim. *fossula*.

fossil (L. *fossilis* = dug up), any remains of an organism from a former geological age (at least 10,000 years old) in earth or sedimentary rock; may be an organism itself, a part of it, a mould, or a petrification. In general, permanent fossils are formed only from organisms which include hard structures, as bones, teeth, shells, or wood. The fossils have rarely been preserved complete, such as are the frozen mammoths in Siberia or the insects in amber. Sometimes hardened clay or ooze have filled up the hollows of shelled animals from which soft structures have rotted away, resulting in a **cast**; also many tree species have left casts. Plant fossils have often been preserved as carbonized remains. Fossil **footprints** of some animals have been preserved in ancient sedimentary deposits (**trace fossils**). *Cf.* index fossils. *See* living fossils.

fossorial, forrosious (L. *fodere* = to dig), pertaining to animals adapted to digging or burrowing.

founder effect, a population-genetic phenomenon in which the → gene frequencies of a new population arisen through migration, can by chance differ from the gene frequencies of the original population; occurs when the number of the immigrant individuals is small, thus the founders (immigrants) carry only a small fraction of the total genetic variation of the original population.

founder principle, a population-genetic phenomenon in which the status of the → gene frequencies of a population is decisively dependent on the gene frequencies of the founder individuals of the population.

fovea, pl. **foveae,** Am. **foveas** (L. = small pit), a small cup-shaped pit or supression in the surface of an anatomical structure; especially the **central fovea** of the retina (*fovea centralis retinae*) in the vertebrate eye. *See* macula lutea.

foveola, pl. **foveolae** (L. dim. of *fovea*), a small pit, depression, e.g. *foveolae gastricae*, gastric pits. *See* gastric glands.

fowl-like birds, → Galliformes.

F plasmid, → F factor.

FR, → far red radiation.

fractionation (L. *fractio* = act of breaking), the separation of components of a mixture; e.g. division of a chemical mixture to its components, or dividing tissue or cell homogenates into smaller parts (fractions), such as nuclei, mitochondria, chloroplasts, and ribosomes. In the fractionation of biological materials the first steps of a procedure are homogenization and → centrifugation. *See* cell fractination.

fracture (L. *fractura*), **1.** the breaking of a structure, especially of a bone; **2.** as a verb, to break.

fragile X syndrome, a form of genetic mental retardation which is inherited in a sex-linked manner, and which is cytologically characterized by the fragility at a given site of the X chromosome. The → penetrance of this genotype is incomplete. After → Down's syndrome this form is the most common genetic mental retardation in man.

frameshift mutation, a class of mutations which causes a change in the → reading frame, arising from the insertion or deletion of a nucleotide, or any number of nucleotides other than three or multiples of three, into DNA; results in misreading of → genetic information.

fraternal (L. *frater* = brother), pertaining to dizygotic twins (fraternal twins) in contrast to monozygotic twins.

free energy, a thermodynamic function representing the energy absorbed or liberated during a reversible reaction. Under constant conditions, Gibbs' free energy (*Josiah Gibbs*, 1839—1903) is defined as follows: $G = H - TS$, where H is the heat content (*see* enthalpy), T temperature and S → entropy. In chemical processes the change of free energy (ΔG) determines the direction in which a reaction proceeds. In biochemical reactions ΔG is more important than the absolute magnitude of energy, G, and is given by $\Delta G = \Delta H - T\Delta S$.

If a reaction liberates heat, ΔH is negative, as in spontaneous (exergonic) reactions. If $T\Delta S$ is not large in comparison to ΔH, ΔG is also negative, and this reaction tends to reach equilibrium where $\Delta G = 0$. If ΔG is positive, some energy supply is needed to start the reaction, as in non-spontaneous (endergonic) reactions.

freemartin, an intersexual bovine female twin foetus transformed in a masculine direction due to hormonal influence of the male co-twin through abnormal connections (anastomoses) in their placental blood circulation; results in hermaphroditism with a disturbance of gonadogenesis, and thus sterility in the female cow.

free nuclear division, a stage in the development of an organism in which many nuclei without formation of any cell membrane result from repeated divisions of the primary nucleus; causes the formation of a → syncytium.

free radical, an atom group with an unpaired valence electron existing independently in a chemical reaction for short periods (e.g. 10^{-5} s or less); under special conditions also long-lived free radicals may appear. Most free radicals are extremely reactive. *See* oxygen radicals.

freeze-drying, drying of deep-frozen biological material in a vacuum; many cellular structures and most substances remain unchangeable under those circumstances. Freeze-drying is also a preservation method in food technology.

freeze-etching, *see* freeze-fracture.

freeze-fracture, a technique used in electron microscopy to prepare cellular membranes for studies without chemical fixation. Frozen cells are fractured with the aid of a microtome e.g. in -100°C and in low pressure. In **freeze-etching** the ice on the fractured surface will be evaporated in a vacuum and the membrane surface is shadowed e.g. with platinum before examination.

freeze tolerance, *see* cold resistance, cold hardening, cryoprotectants.

freezing stress, *see* cold hardening, cold resistance.

frenulum, pl. **frenula** (L. dim. of *frenum* = bridle), a fastening fold or bridle; e.g. **1.** the frenulum of the tongue, a membranous fold of the tongue to the midline of the mouth floor, found in many vertebrates; **2.** a structure of bristles on the anterior border of the hindwing in some insects, overlying the posterior border of the forewing thus coupling the wings together.

frenum, pl. **frena** (L. = bridle), a bridle-like anatomical structure that curbs or restrains movements of an organ or part of it.

freon, → chlorofluorocarbon.

frequency, the number of occurrences per time unit of a periodic process; rate of recurrence, e.g. cycles per second (*see* hertz in Appendix 1); e.g. **1.** → **gene frequency**, the relative number of a given → allele of a given gene in a population; **2. infrasonic frequency** (subsonic f.), any frequency below the audio-frequency range, i.e. below audible sound waves; **3. ultrasonic frequency** (supersonic f.), any frequency above the audio-frequency range.

frequency-dependent selection, a form of → selection in which the → fitness of a given genotype is dependent on its frequency in the population. Usually the fitness of this genotype is inversely proportional to its frequency, i.e. the rarer the genotype is, the better selective advantage it has. Frequency-dependent selection leads in this case to balanced → genetic polymorphism.

Freund's adjuvant (*Jules Freund*, 1891—1960), *see* adjuvant.

frogs, → Anura.

frond (L. *frons* = leafy branch), a divided leaf, especially of a fern or palm; a leaf-like structure formed e.g. by the thallus of a lichen.

frondose inflorescence (L. *frondosus* = leafy), an → inflorescence inside which there are also leaves in addition to the flowers; the leaves are often like typical foliage leaves.

frontal (L. *frontalis* < *frons* = forehead), pertaining to a front; denoting the forehead or any other anterior part of the body.

frontal bone (L. *os frontale*), a pair of bones closing the front part of the cranial cavity in the vertebrate skull; in many mammals, it is united into a single bone and contains a large air cavity, **frontal sinus**, on both sides connected to the nasal cavity.

frontal suture (L. *sutura* = seam), **1.** L. *sutura frontalis*, in higher vertebrates the suture between the halves of the frontal bone; **2.** the narrow slit along the margins of the forehead between the ocelli and antennae in some dipterans; **ocellar bristles**, **orbital bristles** and **vibrissae** are attached to the suture.

frost hardening, *see* cold hardening.

frost line, a geographical boundary for frost sensitive plants.

fructification (L. *fructificare* = to bear fruit <

fructus = fruit, *facere* = to make), the reproductive organ or structure of algae and bacteria; also used as a synonym for the fruit body of fungi.

fructokinase (fructose + Gr. *kinein* = to move), a specific enzyme (EC 2.7.1.4) which catalyses the phosphorylation of fructose to fructose 1-phosphate using the energy from ATP. Fructokinase is found in animal cells, especially in liver cells, and also in plant cells. *Cf.* hexokinase.

fructosans, polymers of fructose (β-D-fructofuranose), originally synthesized from sucrose. The **inulin**-type fructosans occur as storage carbohydrates in the underground storage organs of plants, e.g. in the tubers of *Dahlia* and Jerusalem artichoke. The **levan**-type fructosans are present in the stems, leaves and roots of many monocotyledonous plants, particularly in Gramineae species.

fructose, fruit sugar, laevulose (levulose); a monosaccharide, ketohexose, $C_6H_{12}O_6$. Many plants, particularly their fruits, are rich in fructose, which when linked with → glucose forms sucrose. Animals obtain fructose from plants, or by hydrolysis of sucrose, and they can use fructose in cell metabolism by converting it first to glucose.

fructose bisphosphate aldolase, *see* aldolase.

frugivore (L. *fructus*, *frux* = fruit, *vorare* = to devour), an animal that feeds on fruit. *Adj.* **frugivorous**.

fruit, a structure which is developed from the plant → ovary and protects the seeds formed from the → ovules inside it. The fruits also help the dispersal of plants by animals who eat and transport them. The ovary wall develops into the fruit wall (**pericarp**) which consists of **exocarp** (outermost), **mesocarp**, and **endocarp** (innermost).

There are many types of fruits, such as dry or fleshy, dehiscent or indehiscent, and single or aggregate. **Dry dehiscent simple fruits** may be formed from one or more (fused) carpels; in the latter case the gynaecium is syncarpous. **Follicles** and **legumes** are fruits formed from a single carpel; from several carpels develop a **siliqua** and different types of → **capsules**. **Dry indehiscent fruits** are the **lomentum**, **lomentose siliqua**, **nut** and its special types **caryopsis** and the **achene**, and also different types of **schizocarps**, which break into **mericarps** at the maturity of the fruit, corresponding to the original, single carpels. Fleshy indehiscent fruits are the **berry** and **drupe**.

Aggregate fruits are formed from an apocarpous gynaecium, the separate carpels forming fruitlets which are combined together; e.g. the **aggregate drupe** of raspberry. Also other parts than the ovary may be involved in the formation of fruits; most common of these types is the **pome** (e.g. apples, pears, roses), developed from the inferior ovary surrounded by the receptacle, which finally forms the main part of the fruit. Even these types are classified into aggregate fruits. The **infructescence** is formed from the whole inflorescence in which all the flowers produce separate fruits combined to a fruit-like structure (e.g. pineapple).

fruit body, fruiting body, a structure producing spores in → fungi, formed of fungal hyphae, i.e. the pseudoparenchyma. The form and size of the fruit bodies vary much in different fungal groups, and in some species they are totally lacking.

fruit flies, the family **Drosophilidae** with more than 2,000 species, of which 800 occur on the Hawaiian Islands alone. Fruit flies have served as experimental organisms in genetics, e.g. in population-genetic and evolutionary studies, the most important species being *Drosophila melanogaster* which lives in numerous genetically specific laboratory stocks.

fruitlet, a part of an → aggregate fruit formed from one ovary.

frustration, a negative motivational state when the behaviour of an animal has not produced the results it expected on the basis of its earlier experiences. Frustration may result in behavioural patterns not valid to the situation, as e.g. displacement activities and aggressiveness.

frustule, the silicified bipartite wall of diatoms. *See* Bacillariophyceae.

fruticose (L. *frutex* = shrub), pertaining to a lichen which is shrub-like, clearly erect or pendulous; e.g. reindeer mosses.

fry, a newly hatched young of fish.

FSH, → follicle-stimulating hormone.

FSH-RH, FSH-RF, follicle-stimulating hormone releasing hormone (factor), folliberin. *See* gonadoliberin.

fucose (L. *fucus* = rock lichen), 6-deoxygalactose; methylpentose that in L-configuration occurs in polysaccharides and mucopolysaccharides in organisms; **fucosan**, a polysaccharide composed of fucose units found as a storage carbohydrate in cellular vesicles of brown algae.

fucoxanthin (Gr. *xanthos* = yellow), a → xanthophyll pigment found with other pigments like chlorophylls in many different groups of algae, especially in brown algae and diatoms; absorbs light wavelengths of 500—580 nm.

-fugal (L. *fugere* = to run away, flee), a word termination indicating away from a part, as e.g. centrifugal, away from the centre.

fugitive species, a species that is capable of rapid colonization of temporary habitats; reproduces and then leaves before the habitats vanish or a strong competitor occupies the habitat. Fugitive species have good dispersal methods and great reproductive capacity but poor ability for competition; usually they are r-strategists, like weeds and many insect species. *See* r-selection.

full circulation, *see* stratification.

fullerene (*R. Buckminster Fuller*, 1895—1983), one elemental form of carbon (C); the commonest type is the spherical C_{60}, consisting of regular five and six-carbon rings ("football structure"). C_{60} resembles graphite and its structure is very firm. Fullerene reacts with some other chemical elements, e.g. $C_{60}O$ is a new oxide of carbon. Fullerene has good superconductivity and light absorption properties, and can be used e.g. as membranes. Fullerene can also be used as raw material for some lubricants and as catalysts in chemical reactions.

fumarase, fumarate hydratase; a cellular enzyme (EC 4.2.1.2) which in the → citric acid cycle catalyses the conversion of → fumaric acid to malic acid, or reversibly of malic acid to fumaric acid; in the reaction, water is bound or cleft off, respectively.

fumarate reductase, a cellular enzyme (EC 1.3.1.6) which in the → citric acid cycle catalyses the reduction of fumarate to succinate. Also called succinate dehydrogenase.

fumaric acid, an organic acid, HOOCHC=CHCOOH, occurring in cells as an intermediate in the → citric acid cycle; synthesized from succinic acid via dehydrogenation, or from malic acid by cleaving off water; its ionic form, esters, and salts are called **fumarates**. *See* fumarase, fumarate reductase.

function (L. *functio* < *fungi* = to do, perform), an action or activity; e.g. **1.** the biochemical, physiological or behavioural activity of a cell, organ, or organism; **2.** the property of a chemical substance in relation to other substances; according to their functional groups, substances are grouped e.g. into acids, bases, alcohols, etc.; **3.** as a verb, to perform an action. *Adj.* **functional**.

functional individual, a self-maintaining individual or a group of individuals capable of independent life and reproduction, as e.g. a → modular organism; also the eusocial colonies of hymenopterans and termites can be considered as functional individuals. *Cf.* superorganism.

functional response, in ecology, a change in the rate of the prey consumption of an individual predator in relation to the change of the prey density, e.g. occurs in relationship between many small rodents and hawks feeding on them. *See* response.

fundamental niche, *see* niche.

fundamental number, in genetics, the number of chromosome arms in a given chromosome set (karyotype) of an organism. The fundamental number may change in chromosome mutations, as e.g. in → centric fusion or fission.

fundamental theorem of natural selection, a theorem presented by *R.A. Fisher* in 1930, stating that the rate of increase in → fitness of a given population during a given time interval is equal to the additive genetic variance of fitness at that time. The theorem is one of the most central postulates of the theory of evolution. According to the basic nature of evolution, it follows from the theorem that the average fitness of the population increases continuously as long as there exists additive genetic variance in the fitness. *See* natural selection.

fundus, pl. **fundi** (L.), the bottom or base of a organ; e.g. **gastric fundus** (*fundus gastricus/ventriculi*), the bottom part of the stomach; in man, the part of the stomach to the left and above the level of the entrance of the oesophagus.

fungal poisons, chemically diverse substances present in fungi, poisonous to animals; the most important (or dangerous) are produced in mushrooms, among which poisonous species are rather common. The substances can act as general poisons in cells, e.g. inhibiting cell division or damaging cell membranes or other structures, or they can have more specific effects, such as inhibiting the function of → synapses in the nervous system. Many poisons damage especially the liver and kidneys. For man, only a small piece of a fruit body of some *Amanita* or *Cortinarius* species may be lethal. Some invertebrates have evolutionally adapted to feed on mushrooms which are poisonous to other animal species. *See* amatoxin, phalloidin.

fungal spore, *see* spore, fungi.

Fungi (L. pl. of *fungus*), the kingdom of eukaryotic heterotrophic organisms which absorb nutrients of dead organic material or are parasitic. Fungi are unicellular or multicellular organisms; the latter are composed of the → **hyphae** forming a → **mycelium** whose cells are surrounded by a cell wall composed mostly of chitin, but may contain also cellulose and other complex organic compounds. In many fungal groups, the mycelium occasionally forms → **fruit bodies** which produce → **spores**. In addition to the spore production some fungi reproduce by asexual conidia formed by the constriction of hyphal tips. Many fungi are serious plant pathogens. The group is divided in four divisions: → Gymnomycota, Mastigomycota, Deuteromycota, and Amastigomycota which includes the best known groups and species, such as mushrooms and toadstools. The number of species in the kingdom Fungi is approx. 100,000. *Cf.* yeasts, Uredinales, Ustilaginales, slime moulds.

fungicide (L. *caedere* = to kill), a substance used to kill fungi, as for controlling plant diseases.

Fungi imperfecti, → Deuteromycetes.

fungus, sing. of → fungi.

funiculus, pl. **funiculi** (L. dim of *funis* = cord), **funicle**; **1.** in zoology, a cord-like anatomical structure, usually consisting of several fibres, ducts, or vessels, such as a group of nerve fibres in the **lateral funiculus of spinal cord** (lateral white column), or the **umbilical funiculus** (umbilical cord), a flexible structure connecting the umbilicus of the foetus with the placenta; **2.** in botany, the stalk-like structure of an → ovule; fastens the ovule to the → placenta in the → ovary.

funnel, → siphon.

furca, pl. **furcae** (L. = fork), any forked structure; e.g. the forked floating-apparatus at the top of the abdomen of copepods. *Cf.* furcula.

furcilia (L.), the last larval stage in the life cycle of krill crustaceans after the calyptopsis stage.

furcula, pl. **furculae** (L. dim. of *furca* = fork), **1.** "wishbone"; a bone in the pectoral girdle of birds formed from the paired **clavicles** which are fused at their ventral ends; pectoral muscles are attached to it; **2.** springing organ, i.e. a forked structure on the ventral surface of the fourth abdominal segment in springtails (Collembola). In the resting position, the furcula is locked by the hamulus on the third segment, but at the threat of danger it is released and by its rapid straightening the animal springs up.

furfural (L. *furfur* = bran), 2-furaldehyde, furfuraldehyde; a colourless liquid, $C_5H_4O_2$, with a pleasant rye bread odour; liberated from pentose-containing material when heated with steam under pressure. Furfural occurs in some → essential oils, and is obtained e.g. from cereal straws; it is used as a raw material for the manufacture of cellulose acetate, linseed oil, and some insecticides.

fusicoccin, a toxic compound produced by the fungus *Fusidium coccineum*. It is used in biological studies; it e.g. prevents the closure of the stomata of plant leaves by increasing the extrusion of H^+ ions and the uptake of K^+ ions; also acts as an antagonist of abscisic acid. *See* acid-growth hypothesis.

fusiform initial (L. *fusus* = spindle, *forma* = shape), a spindle-shaped cell in the vascular → cambium of plants, giving rise to cells of secondary xylem or phloem. *Cf.* ray initial.

fusion (L. *fusio* < *fundere* = to pour, melt), **1.** a process of liquefying, e.g. by melting; **2.** a union or coherence of adjacent parts, such as separate cells in **cell fusion**, or atomic nuclei in **nuclear fusion**.

fusion gene, a recombinant DNA molecule consisting of parts of two or more genes, usually the regulatory part of one gene and the structural part of one or more producer genes. Fusion genes are usually produced by gene manipulation technique for the efficient transcription and translation of these producer genes. The function of producer genes reveals the regulatory capacity of the regulatory part of the fusion gene. Fusion genes may also be generated in organisms spontaneously by mutations.

G

G, symbol for **1.** giga- (10^9); **2.** gauge; **3.** glycine; **4.** guanine, guanidine, or guanosine.

G**,** symbol for **1.** conductance; **2.** free energy; **3.** G phase (growth phase) in a cell cycle.

g, symbol for gram.

g**,** standard gravity (= 9.80665 m/s^2 at the sea level and 45° latitude).

G_1, G_2, phases of cell cycle; presynthetic gap and postsynthetic gap respectively.

GA, gibberellic acid. *See* gibberellins.

GABA, → gamma-aminobutyric acid.

GABAergic, pertaining to the chemical transmission between neurones in synapses where → gamma-aminobutyric acid (GABA) acts as the neurotransmitter; an important inhibitory transmission in the nervous system of animals.

G-actin, the globular form of → actin.

Gadiformes (L. *Gadus* = the type genus, *forma* = shape), an order including originally oceanic fishes, some of them secondarily adapted to fresh waters. Usually, they have soft-rayed fins and a sensory barbel in the lower jaw. About 500 species in 11 families are described; best known are the families **rattails** (Macrouridae), **cods** (Gadidae), **eelpouts** (Zoarcidae), and **goosefishes** (Lophiidae). Most species are distributed in northern waters and are of great commercial importance.

Gaia theory (Gr. *Gaia* = Mother Earth, goddess of Earth), a theory stating that life itself regulates and maintains on the planet Earth the conditions which are suitable for organisms. According to the theory, the Earth is a super organism maintaining the balance between land, water bodies, atmosphere, and biosphere, and is thus self-sustaining life. However, biologically it is a theory with many controversies.

gain-of-function mutation, a gene mutation in which the efficiency of gene function is increased. *Cf.* loss-of-function mutation.

galactans (Gr. *gala* = milk), polysaccharides consisting of galactose units, e.g. → agar and carrageenan.

galactokinase (Gr. *kinein* = to move), a phosphotransferase enzyme (EC. 2.7.1.6), which catalyses the phosphorylation of → galactose in cells: D-galactose + ATP —> α-D-galactose 1-phosphate + ADP.

galactolipid, any compound consisting of galactose and a lipid moiety; in animals, most important galactosides are → cerebrosides; in plants, galactolipids exist e.g. in chloroplasts.

galactomannan, a storage polysaccharide formed from galactose and mannose; found in some seeds.

galactosamine, an aminosugar present in → chondroitin sulphate; e.g. N-acetyl-D-galactosamine.

galactose, an aldohexose, $CH_2OH(CHOH)_4$-CHO, i.e. $C_6H_{12}O_6$; a monosaccharide found as a constituent in → polysaccharides (e.g. → raffinose, stachyose, galactans, hemicelluloses, and pectic substances), galactolipids, and polysaccharide-protein compounds in plants and animals. Galactose and glucose are the constituents of lactose (milk sugar). In animal physiology, galactose is also called cerebrose (brain sugar). *See* cerebroside.

β-galactosidase, β-D-galactoside galactohydrolase; **lactase**, EC 3.2.1.23; an enzyme catalysing the hydrolysis of β-D-galactosides, as e.g. → lactose into galactose and glucose; occurs abundantly in the intestine of young mammals. The enzyme is coded by the *Lac* gene, widely used as a reporter gene in studies of molecular biology. *See* lactose intolerance.

galactoside, a → glycoside that yields galactose on hydrolysis; formed when H of the OH group in carbon-1 of → galactose is replaced by an organic radical. Galactose can be linked to another sugar or a non-sugar compound.

galacturonic acid, uronic acid formed on oxidation of galactose, occurs in → pectic substances of plants; pectic acid is polygalacturonic acid.

Galápagos finches, → Darwin's finches.

galea (L. = helmet), **1.** *galea aponeurotica*, epicranial aponeurosis, a fibrous sheet forming the epicranium which connects occipital and frontal muscles of the scalp; **2.** the external pair of the lateral lobes of maxilla in insects; *syn.* lobus externus; *see* mouthparts.

gall (L. *galla* = gall, gallnut), **1.** → bile; **2.** disordered growth or swelling of a plant tissue. A gall can appear in any part of the plant and it is caused by an external stimulus, e.g. by nematodes, gall mites, or by insects, as e.g. gall wasps, gall flies, some thysanopterans, aphids, psyllids, weevils, or butterfly larvae.

gall bladder (L. *vesica fellea*), a musculomembranous bladder derived from the **bile duct** in most vertebrates; serves in concentrating and storing bile, the expulsion of which from the bladder is stimulated by an intestinal hormone, → cholecystokinin.

gallery system, passages bored by bark beetles (Scolytidae) under the bark of a tree; has usually species-specific patterns. A female bores the main passage in which she deposits her eggs. Subsequently, the larvae bore their own passages from the original one, resulting in a specific pattern. Some species, e.g. in the genus *Trypodendron*, bore a gallery system inside the wood. In mass occurrence, some bark beetles are notorious pests in forestry.

Galliformes (L. *gallus* = cock, *forma* = shape), **fowl-like birds**; an order of middle-sized and roundish birds; their feet and neck are short, the digestive tract is adapted to plant food and comprises a large crop, an anterior proventriculus, a well developed gizzard and two slender caeca. Many fowl-like birds are desirable **game birds** or **domestic birds**, as the chicken (*Gallus domesticus*). Galliformes have dispersed around the world, absent only from the Antarctic. Most of them are resident. The order includes 274 species, divided into five families: **megapodes** (Megapodiidae), **curassows** and **guans** (Cracidae), **turkeys** (Meleagrididae), **grouse** (Tetraonidae), **quails, pheasants** and **peafowl** (Phasianidae).

galvan(o)- (*Luigi Galvani*, 1737—1798), pertaining to galvanic, i.e. electrolytic current.

galvanic skin response, GSR, a change in electrical conductivity of the skin, especially on palms of mammals; it is related to changes in the activity of sweat glands due to the emotional state. In psychological tests this reaction is used to detect emotional changes in humans, e.g. in the lie detection technique.

galvanotaxis (Gr. *taxis* = order), electrotaxis; the orientation movement of cells or organisms towards an electrical stimulus, usually towards the cathode (at fields around 1 mV/mm); thought to be involved e.g. in cell guidance in morphogenetic processes of organisms. *See* taxis.

gametangial contact (Gr. *gamos* = marriage, *angeion* = vessel), → gametangiogamy.

gametangiogamy, a type of sexual reproduction in which no formation of → gametes occurs but the whole gametangia fuse. Gametangiogamy occurs in fungi, especially in → Ascomycetes. The process takes place through a hair-like elongated structure, **trichogyne**, which fuses the gametangia and through which the nuclei of the → antheridium (male gametangium) are transferred to the female → ascogonium. After this the ascogonium grows into **ascogenous hyphae** where the **spores** develop. *Syn.* gametangial contact.

gametangium, pl. **gametangia,** an organ in plants and fungi in which → gametes develop in gametogenesis. The male gametangium, **antheridium**, produces male gametes, which are called **spermatozoids** when the male gametes differ from female gametes, i.e. the male gametes are motile and smaller than the female gametes, **egg cells**, which are usually immotile (*see* oogamy). The female gametangium, **oogonium**, produces female gametes, and is called **archegonium** e.g. in liverworts, mosses, and ferns and related plants. In some groups the oogonia may have specific names, e.g. in → Ascomycetes they are called **ascogonia**. The gametes generally fuse in sexual reproduction, but in some groups, such as Ascomycetes, gametes are not produced, and in fertilization two whole gametangia of the opposite sexes fuse (*see* gametangiogamy). In principle, all the sexually reproducing plants and fungi have gametangia, but in angiosperms the → gametophyte is so reduced that the term is not used.

gamete, germ cell; a mature reproductive cell (or only a nucleus) in eukaryotic organisms, capable of fusing with a similar cell of the opposite sex to yield a zygote. Gametes are formed in gametogenesis, that in plants occurs in → gametangia; in animals it occurs in genital organs, i.e. female gametes (ova) in → oogenesis in the ovarium, and male gametes (spermatozoa) in → spermatogenesis in the testis. In → meiosis the chromosome number of the gametocytes is usually reduced to half of that of the somatic chromosome number, i.e. gametes usually have a haploid chromosome set. *See* macrogamete, fertilization, isogamy, anisogamy.

game theory, a model of linear mathematics, applied in situations in which certain functions in relationships are determined by certain rules or strategies (game); e.g. in a population of animals, behavioural phenotype (learned behaviour) gives possibilities to choose the flexible strategy to maximize profit in the interaction with other players (individuals). Game theory models are applied, when the result of the game from the player's viewpoint depends on how the other players act. Game theoretical models have been applied e.g. in sociobiology and behavioural ecology. *Cf.* optimization models.

gametocyte (Gr. *kytos* = cavity, cell), a cell which undergoes → meiosis to produce gametes; either a **spermatocyte** which produces four male gametes in → spermatogenesis, or an **oocyte** which usually produces one female gamete and three polar bodies in → oogenesis.

gametogenesis (Gr. *gennan* = to produce), the production of female or male gametes (germ cells). *See* oogenesis, spermatogenesis, gametangium.

gametophyte (Gr. *phyton* = plant), the → haploid phase in the → life cycle of plants in which the → gametangia and the → gametes are produced. The gametophyte alternates in the life cycle with the → **sporophyte** which is the diploid phase, arising by the fusion of gametes and producing haploid → **spores**. In developed plants (pteridophytes and spermatophytes) the sporophyte phase is dominant and the gametophyte is reduced; e.g. in spermatophytes (seed-producing plants) the gametophyte structures are confined entirely inside the flowers and their organs.

gametophytic apomixis, a type of → apomixis in plants. *See* agamospermy.

gamma, the third letter in the Greek alphabet, used as a prefix in connection with a biochemical or biological term, wholly written or as a letter mark (Γ, γ).

gamma-aminobutyric acid, GABA, γ-aminobutyric acid; an amino acid that functions as an inhibitory transmitter in the nervous system of most animals, i.e. in **GABAergic synapses** (GABAergic transmission); increases Cl^- conductance in → synapses. In plants, it is one of the various free amino acids. *See* glycine.

gamma globulins, immune serum globulins having γ electrophoretic mobility; a protein fraction in the vertebrate blood plasma, containing different → antibodies (immunoglobulins).

gamma motoneuron(e), a motor neurone innervating the intrafusal muscle fibres of → muscle spindles in a vertebrate skeletal muscle; controls the sensitivity of the receptor to stretch by changing the tension of the sensory area of the spindle. *See* motoneurone.

gamma radiation, γ-rays; an electromagnetic X-ray-like radiation with wavelengths shorter than 0.01 nm. Gamma radiation occurs in cosmic radiation and radioactive decay. After a radioactive decay, daughter nuclides remain at an excited stage having extra energy which is liberated as γ-quanta. Gamma radiation occurs usually together with the emission of alpha and beta particles, i.e. alpha and beta radiation. Gamma radiation is an energy-rich → ionizing radiation that penetrates thick barriers and damages cell structures; it is used in radiotherapy and examination of metals and other hard materials, also for sterilizing certain foodstuffs.

gamogony (Gr. *gamos* = marriage, *gonos* = offspring, seed), a developmental process in some protozoans, e.g. plasmodia, in which trophozoites develop into gametocytes.

gamone, a chemical compound which chemotactically promotes the copulation and fusion of the gametes of algae, e.g. fucoserraten in the brown alga, *Fucus*. *Cf.* termone.

gamont, a protozoan gametocyte, i.e. the gamete-producing generation or individual in the life cycle of some protozoans.

gamopetalous (Gr. *petalon* = a thin plate, leaf), pertaining to a flower having united petals. *Syn.* sympetalous.

gamosepalous, pertaining to a flower with united sepals. *Cf.* polysepalous.

ganglion, pl. **ganglia,** also **ganglions** (Gr. = knot), a knot-like mass of tissue, especially **1.** nerve ganglion, i.e. a group of nerve cell bodies in a peripheral nerve looking like a knot on a nerve, or sometimes a group of nuclei in the central nervous system, such as the → basal ganglia in the vertabrate brain, or segmental, suboesophageal, and brain ganglia in the ladder-like nervous system of many invertebrates; *see* nervous system; **2.** a cyst on a tendon, e.g. in the wrist. *Adj.* **ganglionic** (ganglionary), pertaining to ganglia or ganglion cells; **ganglionated**, provided with ganglia.

ganglion cell, 1. a nerve cell which has its cell body outside the central nervous system, i.e. in a peripheral ganglion; *syn.* gangliocyte; **2.** a cell type of the retina of vertebrates; it is the third neurone of the visual tract, the axons of the cells forming the optic nerve from the retina to the brain; **3.** the pyramidal ganglion cell in the cerebral cortex of vertebrates being the first neurone of the pyramidal tract; also called Betz's cell or giant pyramidal cell.

ganglioside, *see* sphingolipid.

ganoid scale (Gr. *ganos* = sheen, *eidos* = form), a scale type occurring in some primitive fishes (holosteans and chondrosteans); a large scale covered by several ganoine layers with a silvery glimmer, cosmine and successive layers of compact bone lying under the ganoine layers. Usually ganoid scales lie side by side, and not overlapping at the margins as do other kinds of scales. *Cf.* cosmoid scale. *See* scale.

gap genes, zygotic → interpretation genes of animals that read the → positional information created by the → coordinate genes. Mutations of gap genes lead to lack of segments or somites.

gap junction, a junction type between cells in some animal tissues; they are channels lined by hexagonal arrays of proteins (**connexons**) between adjacent cells allowing the passage of ions and small molecules between the cells. At the junction area the distance between the adjacent cell membranes is only 2—3 nm. Inside the area there are many minute **pores** (about 2 nm in diameter) opposite to each other in both membranes through which monosaccharides, amino acids, nucleotides, vitamins, hormones, and several ions can pass easily from one cell to another. The pores are composed of 6 or 12 polypeptide subunits which line the pore. Gap junctions are found in epithelial, cardiac and smooth muscle tissues, and between certain nerve cells; in the last three tissues gap junctions allow **electrical transmission** in which → action potentials propagate from cell to cell without the help of transmitter substances. Gap junctions are also found in electric synapses, → ephapses, between some nerve cells.

gas bladder, → swim bladder.

gas liquid chromatography, GLC, *see* chromatography.

gastr(o)-, gastri- (Gr. *gaster* = belly, stomach), pertaining to the stomach. *Adj.* **gastric**.

gaster, 1. → stomach; **2.** an enlarged caudal part of the abdomen posterior to the first segment in some hymenopterans.

gastric caecum/cecum, pl. **gastric c(a)eca,** one or more outpocketings (evaginations) of the midgut in most insects; their positions vary, but they are usually located at the anterior end of the foregut.

gastric filaments (L. *filum* = thread), internal, tentacle-like projections in the **gastric pouches** of medusae (Scyphozoa); contain batteries of → nematocysts for killing prey animals entering the coelenteron.

gastric gland, any of the tubular glands in the mucosa of the vertebrate stomach; the glands open usually in pairs or small groups into numerous small pits (**foveolae**) on the surface of the mucosa, and produce mucus, hydrochloric acid (HCl), and enzymes into the gastric fluid. The glands have several cell types: **oxyntic cells** (**parietal cells**) producing hydrochloric acid; **chief cells** secreting pepsinogen that is the precursor of the enzyme pepsin, and **mucous neck cells** secreting mucus. There are also several types of **gastric endocrine cells** producing hormones which regulate the function of the alimentary canal.

gastric inhibitory polypeptide, GIP, a peptide hormone secreted by some cells in the anterior section of the small intestine in vertebrates, in mammals in the duodenum and jejunum. The secretion of GIP is stimulated by glucose and fat in the chyme. GIP inhibits the secretion and motility of the stomach and stimulates the secretion of insulin in the pancreas.

gastric ulcer (L. *ulcus* < *ulcerare* = to make a sore), an erosion of the mucous membrane of the stomach; in many vertebrates, it is induced e.g. by some bacteria (*Helicobacter pylori* in man), irritating substances, and stress conditions, associated with hypersecretion of acid gastric juice. Peptic ulcers may occur also in the oesophagus and duodenum (duodenal ulcer).

gastrin, a peptide hormone produced by some cells in the gastric and intestinal mucosa in vertebrates, stimulating the secretion of hydrochloric acid and pepsinogen in the stomach; it also has trophic action on the mucous membrane. The secretion is induced by certain substances in the chyme, particularly by peptides and amino acids. Gastrin is also found in the brain and some peripheral nerves, as well as in the anterior and intermediate lobes of the pituitary gland.

gastrocoel, Am. also **gastrocoele** (Gr. *koilos* = hollow), → archenteron.

gastrodermis (Gr. *derma* = skin), an inner cell layer of the body of cnidarians and ctenophores lining the gastrovascular cavity (coelenteron). Also called endoderm.

gastroenteral (Gr. *enteron* = intestine), → gastrointestinal.

gastrointestinal (L. *intestinum* = gut), pertaining to the stomach and intestine. *Syn.* gastroenteral.

gastroliths (Gr. *lithos* = stone), **1.** mineral accretions in the walls of the stomach of some crustaceans in which calcium salts, withdrawn from the old cuticle to be moulted, are stored and used for hardening of the new cuticle; **2.** small stones swallowed by some birds into the gizzard.

Gastropoda (Gr. *pous* gen. *podos* = foot), **gastropods** (e.g. slugs and snails); a class of molluscs comprising species which usually have a calcareous, spirally coiled shell; the shell may be reduced to a small shield in some gastropods (slugs). Many species exhibit bilateral asymmetry due to the coiling of the visceral sac. Gastropods have a flattened large foot and a distinct head bearing tentacles and eyes; a coiled mantle covers the visceral sac. About 40,000 living and 15,000 fossil species are described and grouped in three subclasses: **prosobranchs** (Prosobranchia), which are both freshwater and marine animals, **opisthobranchs** (Opisthobranchia) are marine species only, and **pulmonates** (Pulmonata) which comprise both terrestrial and freshwater species.

Gastrotricha (Gr. *thrix* = hair), **gastrotrichs**; a small phylum including about 400 microscopic worm-like, pseudocoelomate invertebrate species which live both in marine and freshwater environments; previously considered as a class in the phylum Aschelminthes. The body is dorsally convex bearing bristles, spines, or scales, and ventrally flat and thickly covered with cilia. Reproduction is partly parthenogenetic, some species are hermaphroditic. The class is divided into two orders: Chaetonotida and Macrodasyida; the former live mainly in fresh waters, the latter are marine.

gastrovascular system (L. *vasculum* = small vessel), the branched canal system evolved from the central cavity (**gastrovascular cavity**) in cnidarians and ctenophores; has a single opening serving as both mouth and anus; functions in digestion and circulation; **2.** the highly branched intestine of turbellarians. *See* water-vascular system.

gastrozooid (Gr. *zoon* = animal, *eidos* = form), **feeding polyp**; a polyp type in hydroid colonies (Hydrozoa) specialized for food catching and digesting; the gastrozooids may be tubular, bottle-shaped, or vase-like, all of them having a terminal mouth surrounded by a circle of tentacles. *Syn.* hydranth. *Cf.* dactylozooid, gonozooid.

gastrula, pl. **gastrulae, gastrulas** (L. dim. of *gaster* = stomach), an early embryonic stage of most multicellular animals after → morula and → blastula. Typically, as in amphibians, **gastrulation** begins when a group of blastula cells move inside the early embryo via the upper lip of the blastopore (invagination), forming the third germ layer, **mesoderm** (mesodermal layer). Cells of the lower lip and the bottom of the blastocoel grow and spread up forming the **endoderm** which lines the internal cavity, now called archenteron (gastrocoel). At the same time the mesoderm spreads down pushing its cells between the endoderm and outer layer, **ectoderm**. The gastrula stage begins the differentiation of the cells, starting morphogenesis and histogenesis. The genes of the zygote control the development from this stage forward, whereas the genes of the ovum mainly conduct it before gastrulation.

gas vacuole, a structure occurring in the cells of most Cyanobacteria; the vacuole is surrounded by a protein membrane which is permeable to gases. It helps in buoyancy and helps in the shielding of light.

gated channel, *see* ion channels.

gathering collectors, a group name for small aquatic bottom animals which catch food by picking it from their environment; e.g. the larvae of ephemeropterans and chironomids in the bottom, and the water-strider (*Gerris*) on the surface film of water. *Cf.* scrapers, shredders, deposit feeders, filter feeders.

gauge, G, a standard measure; originally a plate full of holes of various sizes for measuring the outside diameter of tubes; now the numerical scale (ratio) for thickness measures of injection needles, catheters, tubes, etc.; some examples in gauges for injection needles (mm in brachets): 15 G (1.5 mm), 17 G (1.3 mm), 19 G (1.1 mm), 21 G (0.9), 23 G (0.7 mm), 25 G (0.5 mm).

Gause law/hypothesis (*G.F.Gause*, b. 1910), → competitive exclusion principle.

Gaviiformes (L. *Gavia* = the type genus, *forma* = form), **loons**, **divers**; a bird order including only one family Gaviidae, highly specialized for aquatic conditions; the beak is strong and sharp, the short legs are at the end of the shuttle-shaped body, and the toes are fully webbed. All these features render the bird an excellent diver. The order includes only five species confined to northern regions.

Gay-Lussac's law (*Joseph Louis Gay-Lussac,* 1778—1850), a law describing the temperature dependence of the volume of any gas; the volume of gases expands 1/273th of their volume at 0°C for every degree Celcius. *Syn.* Charles' law.

G-banding (G from the name *Giemsa*), a method for staining prometaphase or metaphase chromosomes using → Giemsa staining; the chromosomes are subjected briefly to mild heat or proteolysis and then stained with Giemsa reagent, a permanent DNA dye. The staining produces visible dark bands and unstained pale interbands in the chromosomes. By this method each chromosome of a mammal can be identified.

GDP, guanosine diphosphate, *see* guanosine phosphates.

Geiger counter/tube (*Hans Geiger*, 1882—1945), an apparatus for measuring radioactive radiation; detects the quantity of ionization in gas inside the tube.

geitonogamy (Gr. *geiton* = neighbour, *gamos* = marriage), a type of → self-pollination in plants.

gel (L. *gelare* = to clot, congeal), **1.** a jelly-like organic material in cells or tissues due to a solid or semisolid phase of a colloidal solution; **2.** any jelly-like colloidal solution, such as gelatinous starch or polyacrylamide. *See* agar, colloid, gelatin.

gelatin, a mixture of several proteins, chiefly denatured collagen, obtained from bone, cartilage, fasciae, and tendons from cattle and pigs by boiling tissues in water or dilute acids; used e.g. in cell culture, as a plasma substitute, in medical and glue industry, and in cookery. *Adj.* **gelatinous**, pertaining to gelatin, jelly-like.

gel electrophoresis, an electrophoretic separation method especially used to analyse proteins and nucleic acids applied to a → gel. *See* electrophoresis.

gel filtration, a chromatographical technique for fractionating a sample into its molecular components in a porous → gel, e.g. in cross-linked dextran, cross-linked polyacrylamide, or polystyrene. A liquid sample is transferred onto a gel column which is then washed with a solvent. Large molecules pass through the gel column; smaller molecules attach to the pore walls of gel and their passing is retarded depending on the properties of the gel and molecules in question, e.g. their solubility in a solvent. The molecules pass through the column in proportion to their molecule weights, the largest molecules moving fastest. The method can be used for the separation and purification of proteins, peptides and nucleic acids, and to determine the molecular weight of proteins.

gelsolin (L. *gel(osus)* = stiff, *sol(otus)* = fluid), an actin-severing protein found in animal tissues; in the presence of calcium it binds an actin filament in the cytoskeleton of cells, breaking the filament into short pieces, i.e. converts a gel form of actin into a soluble (sol) form. Gelsolin is also found in the circulating blood.

gemination (L. *geminare* = to double), doubling, duplication, e.g. an exceptional formation of two wholly or partially separated organs from one embryonic primordium; *gemellus*, dim. of *geminus* (twin).

gemma, pl. **gemmae** (L. = bud), **1.** a bud-like anatomical structure in animals, e.g. a taste bud; **2.** a structure serving reproduction in many mosses, liverworts, and in some algae; consists of a small group of cells detaching from the parent plant and developing into a new individual; resembles the fungal conidia. The corresponding structure in seed plants is the → bulbil.

gemma cup, a cup-shaped structure containing gemmae in some liverworts. *See* gemma.

gemmation, 1. budding, the formation of → gemmae; **2.** the arrangement of buds on a twig.

gemmule (L. *gemmula,* dim. of *gemma* = bud), **1.** an internal bud with several cells developing asexually in some fresh-water sponges. Covered by a shelter, the gemmule is preserved in a resting stage through unfavourable periods, such as winter, and begins to develop after the death of the parent organism; the process is called **gemmulation**; **2.** a hypothetical unit presented by *Charles Darwin* in his explanation of hereditary mechanisms in the theory of pangenesis.

-gen (Gr. *-genes* = born), as a suffix pertaining

1. to a source for a given agent, e.g. pepsinogen for pepsin; **2.** to the influence produced by a factor indicated by the word stem, e.g. allergen producing allergy, carcinogen producing cancer, gestagen a hormone with progestational activity, and pathogen producing disease.

gena, pl. **genae** (L.), cheek; a lateral part of the head of an animal, e.g. a lateral shield on the head of an insect. *Adj.* **genal.**

gene (Gr. *genos* = parentage, origin), a factor of inheritance. The gene can be defined on the basis of four genetic phenomena, i.e. → genetic transmission, → genetic recombination, → gene mutation, and → gene function. From the classical view, established in 1910—1940, the gene was the smallest unit of transmission, recombination, mutation, and function.

Gene as the unit of transmission states that when a heterozygous individual produces → gametes, the maternal and paternal factors of inheritance segregate as pure units from each other, and hence the gamete contains one piece of each gene. **Gene as the unit of recombination** requires each gene to be separated from the other genes in recombination, but it cannot be divided into parts by means of recombination. **Gene as the unit of mutation** states that each gene changes as a single unit in a mutation, and hence the different → **alleles** were mutually exclusive. **Gene as the unit of function** applies specifically to the phenomenon of → genetic complementation in such a way that allelic gene mutations cannot complement the function of each other, but non-allelic mutations can.

The concept of the gene as the unit of function became more accurate when *G.W. Beadle* and *E.L. Tatum* in 1941 formulated the hypothesis of → **one gene - one enzyme**. According to this hypothesis each gene controlled the synthesis of one enzyme.

The classical concept broke down when the phenomenon of intragenic recombination was observed. This was first noticed by *P. Oliver* in 1941 and by *E.B. Lewis* in 1945 using the fruit fly as the experimental animal. Alleles which in the complementation test appeared as alternative forms of the same functional gene, sometimes recombined. Thus, the gene defined on the basis of function and recombination was not the same unit. Alleles which recombined were called **pseudoalleles**. In the 1950's several authors observed intragenic recombination, i.e. pseudoallelism, in the genes of microorganisms which were known to control the synthesis of one enzyme, thus fulfilling the criterium of the one gene - one enzyme hypothesis. Later on, the intragenic recombination has been demonstrated to be a regular, although rare phenomenon, involving all genes.

The DNA theory of inheritance was born in 1944 when *O.T. Avery*, *C.M. Macleod* and *M. McCarthy* showed that the substance causing → genetic transformation was **deoxyribonucleic acid (DNA).**

The modern view begun to take shape in 1950's and 1960's from the work of *S. Benzer*. He studied the rII region of the T4 phage of *Escherichia coli* which regulates the reproduction of the phage, and formulated the concepts of the → **cistron**, **muton**, and **recon**. The cistron is the gene defined on the basis of function, and it is demonstrated using the → *cis-trans* test. *C. Yanofsky* and his coworkers showed in 1961 that materially a DNA segment which contains the information for the synthesis of one polypeptide, corresponded to the cistron. Thus the hypothesis of **one cistron - one polypeptide** took form. Benzer showed that inside the cistron, sites existed which mutated independently, and he proposed that the smallest unit of mutation was termed the muton. Mutations of different origin within the same cistron also recombined, and Benzer named the smallest unit of recombination the recon. Yanofsky with his group was able to show that materially one nucleotide pair in DNA corresponded to both the muton and the recon.

In modern context, the gene is defined as a cistron by means of the *cis-trans* test. In eukaryotic organisms a **structural part** and **regulatory part** can be distinguished within the cistron. The former is the → **structural gene** (**producer gene**) which contains the information for the polypeptide encoded by the gene. The latter (→ **regulator gene**) regulates the transcription of the structural gene. The regulatory part usually contains several sites, acting in *cis* position, which either enhance or silence the function of the gene, and one or more → **promoters** onto which the → RNA polymerase is attached when it starts transcription.

In the 1970's and 1980's it was observed that

prokaryotic organisms have overlapping genes and eukaryotic organisms have mosaic genes (interrupted genes). In mosaic genes → **introns** and → **exons** alternate. The exons are segments of DNA which contain the information for the polypeptide encoded by the gene, and introns are their intervening sequences. Both introns and exons are transcribed, but during the processing of the primary transcription product, which is the → heterogeneous nuclear RNA (hnRNA), into the mature → messenger RNA, the introns are removed in a specific splicing reaction (*see* splicing). Thus, in eukaryotes, the unit of transcription can be different from that of translation.

At the end of the 1980's nested genes were also found in eukaryotes, i.e. a certain gene resides at the intron of another gene. Also in the 1980's, alternative splicing was observed in many eukaryotic genes which demonstrated that exons are combined together in different ways in different tissues and/or different developmental stages during the maturation of the messenger RNA. In this way a single gene can code for several polypeptides, i.e. the gene is **genuinely pleiotropic** (*see* pleiotropy). In addition, it has been observed that if a gene has more than one promoter, it can produce more than one transcript, different promoters serving as starting points in different cases. These new observations have made the unambiguous definition of the gene difficult, and consequently the concept of the gene is again subject to reconsideration.

gene action, the expression of genes in the regulation of the biosynthetic processes of the cell. The ability of the cell to develop as an organized functional unit is dependent on the genes. Genes express themselves differentially during the different stages of the growth of an organism so that in each tissue of a given developmental stage a given → gene battery, typical of it, functions. To some extent the action of the genes is dependent on the environment, and hence the regulation of the gene action is a necessary condition for the development of an efficiently functional organism.

In molecular biology, the study of gene action often deals with the **primary gene action**, i.e. the mechanisms regulating → protein synthesis. This restriction of research is justified to a certain degree because the metabolism and the biosynthesis of the structural parts of the cell are based on enzymatically active proteins (*see* enzymes) and on structural proteins. However, also understanding of the higher levels of organization of the gene action is necessary for understanding the entirety of the organism. Primary gene action consists of those processes which lead to the expression of the genetic information residing in the genes as the primary gene products, i.e. the proteins. These processes are → genetic transcription in which RNA is synthesized according to the genetic information residing in DNA, and → genetic translation in which the information, now encoded in RNA, is translated into the primary structure of the polypeptide.

The study of higher than primary levels of the gene action is difficult because of their enormous complexity. The way from the gene to the morphological or functional → phenotype is multiphasic and indirect because it is affected also by the interaction of genes and by environmental factors. Gene action is called **cell-specific, tissue-specific,** or **stage-specific** if it is restricted to a given developmental stage. The action of a certain gene is **sex-limited** if that gene is expressed only in one of the two sexes, and **sex-controlled** if it is different in the two sexes. Sex-limited or sex-controlled genes can be located likewise in the sex chromosomes as well as in the → autosomes.

gene activator, any molecule, usually a protein or RNA, which enhances the activity of a certain gene.

gene amplification, the multiplication of a gene in the chromosomal complement, i.e. the → replication of a given gene in a cell, producing many copies of this gene without replication of other genes. Gene amplification involves genes whose products are needed in high quantities, like the genes of → ribosomal RNA in an egg cell.

gene bank, 1. a collection of living animals, plants, or fungi, or their reproductive vehicles (e.g. seeds, spores, stored embryos) striving to preserve the genetic material of the species threatened in nature; e.g. zoos and botanic gardens can act as gene banks; **2.** a selected → gene library; **3.** a database of gene structures.

gene battery, a group of → producer genes whose action is typical of cells differentiated in a certain way; e.g. different gene batteries act in cells which differentiate and organize into different tissue types.

genecology, a branch of ecology dealing with population genetics.

gene complex, a group of genes tightly linked together, usually acting under a common control; determines a phenotypic character.

gene conversion, a phenomenon in which one allele of a gene is converted into another during meiotic recombination; non-reciprocal → genetic recombination, i.e. the non-reciprocal recombinational transfer of genetic information between homologous DNA sequences without an accompanying information exchange. This recombination produces parental and recombinant gene combinations in other than 1 : 1 ratios, i.e. in gene conversion one gene allele is converted to another during meiotic recombination.

gene dosage, the number of gene → alleles in a particular genotype; depends mainly on the degree of ploidy (*see* polyploidy), but also on the existence of gene duplications in the genome.

gene expression, the manifestation of genes in the → phenotype of an organism; can be observed on several levels, i.e. from → genetic transcription to different forms of structure, function, and behaviour of an organism.

gene family, in eukaryotic organisms any set of sequence-related genes that apparently share a common ancestor, e.g. **1.** a group of phylogenetically related genes which encode similar proteins, such as the globin gene family encoding the globins (e.g. haemoglobins, myoglobins) during the different developmental stages of a vertebrate individual; **2.** a gene group formed when the same gene appears in several or many copies in the → genome, usually consecutively in the chromosome; e.g. the histone gene family.

gene flow, the spread of genes from one population to another; occurs through dispersal of seeds, pollen, or spores, or through the migration of individuals. Gene flow can lead to changes in → gene frequencies, and hence it is an effective factor in evolution. The degree of gene flow in each case depends on the size and structure of populations, both of which vary largely between different organisms.

gene frequency, the frequency of an → allele of a given gene locus in a population. The genetic constitution, i.e. the → gene pool of a population is described with the aid of its gene frequencies, which can be determined on the basis of the frequencies of different → genotypes. Both the gene and genotype frequencies remain constant in the population if no other factor than random mating is effective (*see* Hardy—Weinberg law). The analysis of gene frequencies is an important method of population genetics, and to a certain degree, makes possible the assessment of the genetic history of the population. → Mutations, selection, migration (systematic factors), and → genetic drift (random factor) cause changes in gene frequencies.

gene interaction, the interaction of allelic and/or non-allelic genes in a given → genotype during the formation of a given → phenotypic character. Between alleles the most important form of interaction is → dominance. Of the non-allelic interactions the easiest to observe are those which cause modifications in the relative amounts of progenies in the F_2 generation of a dihybrid cross. The normal relations are

9 *A-B-* : 3*A-bb* : 3 *aaB-* : 1 *aabb*

(in the formula *A* and *B* are dominant genes, *a* and *b* recessive genes). The following interactions can be distinguished:

1) recessive → epistasis:
9 *A-B-* : 3 *A-bb* : 4 (3 *aaB-* + 1 *aabb*)
2) dominant epistasis:
12 (9 *A-B-* + 3 *A-bb*) : 3 *aaB-* : 1 *aabb*
3) complementary gene interaction:
9 *A-B-* : 7 (3 *A-bb* + 3 *aaB-* + 1 *aabb*)
4) modificator gene interaction:
13 (9 *A-B-* + 3 *aaB-* + 1 *aabb*) : 3 *A-bb*
5) additive gene interaction:
9 *A-B-* : 6 (3 *A-bb* + 3 *aaB-*) : 1 *aabb*
6) duplicated gene interaction:
15 (9 *A-B-* + 3 *A-bb* + 3 *aaB-*) : 1 *aabb*

The interaction of so-called polymeric genes is additive or cumulative. Important interactions occur between genes which regulate the function of each other. Such interactions can be enhancing (positive) or silencing (negative). The regulative action of a given gene can involve → genetic transcription, genetic translation, or the function of the protein encoded by another gene. *See* genetic regulation.

gene knockout, a method of gene manipulation for the permanent inactivation or elimination of a gene in an organism.

genelet, a sequence of DNA in the gene complex of immunoglobulins in the cellular → germ line. In lymphocytes producing antibodies, the functional genes are assembled from different combinations of genelets through somatic → crossing-over, thus making the millionfold versatility of antibodies possible.

gene library, a collection of cloned DNA fragments in bacterial cultures which have been constructed as → restriction fragments of chromosomal DNA, and which represent the total genome of a given species or a part of it.

gene locus, pl. **gene loci** (L. *locus* = position, place), the position of a gene on a chromosome or, respectively, in the linkage group on a cytological or relative → chromosome map. Gene loci can be observed by cytological or genetic methods. By means of intragenic → genetic recombination the locus can be further mapped into subunits, called sites, which are located in a linear order (*see* gene map). The → alleles of a gene are situated at the same loci but on homologous chromosomes.

gene manipulation, a general term for different methods of genetic engineering; includes isolation and transfer of genes, → recombinant DNA techniques, → focused mutagenesis, and production of transgenic organisms. It is used as techniques *in vitro* in the deliberate manipulation of genes for genetic analyses and for improving gene products. Gene manipulation includes the reproduction of → restriction fragments, and their joining into an appropriate cloning → vector, which then can be introduced into a recipient cell. Also called gene technology.

gene map, a graphic representation of the internal structure of a gene. The subunits of the gene, i.e. the mutational → sites, are located on the gene map so that their distances represent recombination frequencies (genetic distances). A **physical gene map** is based on the representation of the cleavage sites of → restriction enzymes in DNA of a gene, or in its most accurate form on the sequence of nucleotides in DNA. *Cf.* chromosome map.

gene mutation (L. *mutatio* < *mutare* = to change), any hereditary change in a single → gene (**point mutation**), as compared with changes in the structure of chromosomes (→ chromosome mutation) or changes in the number of chromosomes (→ genome mutation); gene mutations may occur in germ cell lines as well as in somatic cells (*see* somatic mutation). The distinction of gene mutations from small chromosome mutations is practically and even conceptually often impossible because e.g. small intragenic → duplications, deletions, and insertions are gene mutations but merge without limit to chromosome mutations. As a consequence of gene mutations the alternative forms or → alleles of a gene are borne. Gene mutations are produced spontaneously without any definitive cause, or they can be induced experimentally by different physical or chemical → mutagens.

A gene mutation can be a consequence of any change in the normal nucleotide sequence of the gene, such as a change of nucleotides or their base parts which are called → transitions and → transversions; further the mutation can result from → insertions, deletions, inversions, or transpositions of nucleotides.

The most important cause of **spontaneous gene mutations** appears to be the addition or removal of small movable DNA segments, → transposons, which have the ability to move from one location to another in the chromosomal complement.

Most gene mutations are harmful and therefore will be eliminated by natural selection. Actions of mutation can cause death (*see* lethal factor) or they can by some other means decrease the fitness of their carriers. The effect of a gene mutation depends on the genetic background and the environmental circumstances. Specific **conditional mutations** express themselves only in a certain environment like a given culture medium or a certain temperature. *See* silent mutation.

gene pool, the genetic constitution of an interbreeding → Mendelian population at a certain time. A gene pool which remains constant for generations is said to be in equilibrium. The gene pool stays in equilibrium if no other factor than random mating is effective. Changes in the gene pool of a population can be produced by → genetic drift, mutation, migration, and selection.

general adaptation syndrome, the term given by *Hans Selye* (1907—1982) for the physiological stress response in mammals (especially man), comprising an alarm reaction, stage of resistance, and state of exhaustion. Although the stress factors vary from mental shock to injuries, cold exposure, and different types of diseases, the basic neuronal and hormonal responses elicited by them are the same. *See* stress.

generalist, an organism or a species which has a large niche, e.g. in regard to feeding ecology or to habitat selection. *Cf.* specialist.

generation (L. *generare* = to beget, create), **1.** a unit comprising all the individuals of the same age-group; **2.** in genetics, individuals of one reproductive cycle, comprising those individuals of a population who are descendants of a common ancestor; **3.** in ecology, populations can be divided into two groups on the basis of their generation types: the populations which have **discrete generations** and those with **overlapping generations**.

generation length, generation time; an average interval between two generations, i.e. the interval between the birth of parents and the birth of their young; is unambiguous only in semelparous species reproducing only once in their lifetime, the generation length corresponding to the age of sexual maturity. The generation length of the iteroparous species, which reproduce many times, is more difficult to assess; the generation length of man is estimated to be 25—30 years.

generative (L. *generativus*), **1.** pertaining to reproduction; **2.** pertaining to generation.

generative cell, the cell in a → pollen grain, i.e. the microspore of seed plants, containing the **generative nucleus**; divides mitotically and forms two **sperm cells** during the germination of the pollen grain. At this stage the male → gametophyte is completely formed and the sperm cells perform the fertilization of the plant.

generative nucleus, *see* generative cell.

generator potential, → receptor potential.

gene regulatory proteins, nuclear proteins which are not → histones, and bind to a certain site in DNA, regulating the expression of the gene located at that site; include e.g. transcriptional activators, → repressors, and → transcription factors.

generic (L. *genus* = birth, race), **1.** pertaining to a genus, presented by the first part of the scientific name; **2.** describing a class or type of a single compound that is in common use and not protected by a trademark; e.g. saccharide, alcohol, amine, and steroid are generic names.

gene surgery, → gene therapy.

genet, an entity descended asexually from a single zygote; members of the same → clone form one genet as long as they are genetically identical. Often used as a synonym of clone. *Cf.* ramet.

gene targeting, the integration of exogenous DNA into a → genome at sites where its expression will be suitably controlled; integration occurs by homologous recombination.

gene technology, → gene manipulation.

gene therapy, gene surgery; the cure of genetic diseases by substituting a defective gene with a normal gene transferred to the diseased tissue using → gene manipulation. **Somatic gene therapy** aims at curing or preventing a disease by focusing the therapy on the diseased cells of an individual, and also by the replacement or supplementation of affected cells with genetically corrected cells. **Germline gene therapy** tends to manipulate the cells of the germ line to prevent the inheritance of a disease. Gene therapy has been widely studied in animals, and the first efforts in man were performed in 1990. Research in gene therapy is in progress for curing several diseases, also including non-genetic diseases, such as cancer.

genethics, a branch of → bioethics specifically involved in the ethical implications of gene manipulation.

genetic, 1. heritable; pertaining to inheritance, genes, genetics, or genesis; **2.** relating to reproduction, origin, ontogeny, or birth.

genetic assimilation (L. *assimilare* = to make similar), the genetic fixation of acquired characters (phenotypes) by selective processes, the new characters having greater heritability than they had before. It results in the establishment of seemingly "acquired characters" in a population by means of natural selection. These characters appear first as changed responses to a given environmental stimulus, i.e. as modifications with genetically controlled variation limits. After fixation (assimilation) into the gene pool of the population these characters appear without this environmental stimulus. Genetic assimilation is the consequence of a particular threshold effect. Certain genotypes are able to respond to the environmental stimulus thus producing a modification, other genotypes are not. Due to selection, the frequency of the former genotypes increases in the population until the character which first appeared as a modification will be fixed genetically in the population.

genetic background, all genes of an organism other than the one (or ones) under study; also known as the residual genotype. Genetic background can affect the function of a gene (or genes) under consideration.

genetic balance, the coordination or balance of jointly adapted genes of a given genotype. The genetic balance of a genotype is a precondition for the gene function leading to the development of an integrated functional entirety of an organism.

genetic code (L. *codex* = book), the rule for how a gene guides the synthesis of the corresponding polypeptide. According to the ge-

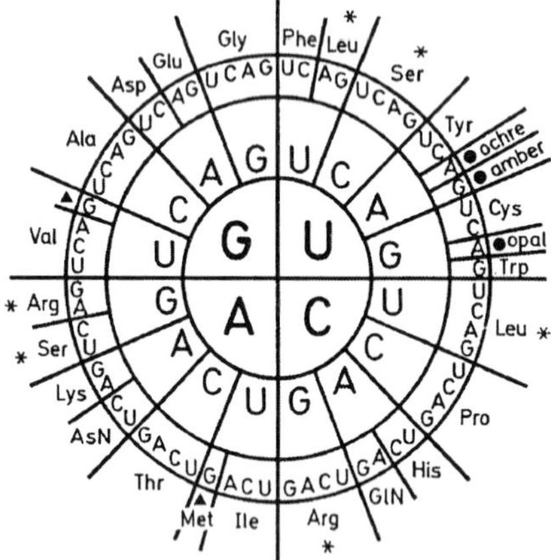

The genetic code of the messenger RNA bases' level described as a "code sun" figure. The codons can be read from inside out in the 5' → 3' direction of the messenger RNA. (A = adenine, U = uracil, G = guanine, C = cytosine). The outermost circle represents the amino acids which each codon encodes for. Asterisk represents amino acids occurring twice. Triangles represent the initiator codons, and circles the terminator codons. (After Bresch & Hausmann: Klassische und Molekulare Genetik - Springer Verlag, 1970; with the permission of the publisher).

netic code, the → genetic information residing in DNA is transcribed into → messenger RNA in → **genetic transcription**, and this code is further translated into a certain polypeptide and protein structure in → **genetic translation**. In the genetic translation the genetic code determines that the → transfer RNA molecules, loaded with amino acids, take their correct places in the right order during → protein synthesis. The following principles can be seen in the genetic code:

The genetic code is a **triplet code**, i.e. each amino acid is coded in DNA by a group of three nucleotides, i.e. a triplet. Because there are four types of nucleotides, whose base parts are adenine, thymine, guanine, and cytosine, it is possible that $4^3 = 64$ different triplets or → codons can be formed to code for 20 amino acids found in proteins. With the exception of three → terminator codons, all 64 codons indeed code for amino acids. In genetic transcription, one of the two strands of DNA in a gene is transcribed into single-stranded messenger RNA forming its copy, a → template. Thus, this strand of DNA is transcribed into a complementary messenger RNA strand according to the → base pair rule. Messenger RNA carries the genetic information from the nucleus into the cytoplasm where it is translated in protein synthesis into the amino acid sequence of the encoded polypeptide.

The genetic code is **degenerate**, i.e. there exist synonymous codons. This follows from the fact that there are 64 different codons but only 20 amino acids to be coded. The genetic code is **non-overlapping**, i.e. the codons do not overlap but are consecutive. The genetic code is **commaless**, i.e. the codons follow each other immediately, and no nucleotides form punctuation between them. The reading of the genetic code starts with specific → **initiator codons** and ends with specific → **terminator codons**.

The genetic code is **universal**, i.e. it is the same in all organisms. This fact is one of the strongest proofs of the → evolution theory and proves also the one common → monophyletic origin of organisms on the Earth. An exception from the universal constitution is the code of → mitochondrial DNA which is different from the nuclear code in five codons and also differs between species; the codes of certain ciliates are also different. These exceptions give important hints of the origin and evolution of the genetic code itself.

genetic complementation, the function of homologous genomes complementing each other; complementation occurs if the phenotype of a → heterozygote is more close to a wild phenotype than that of crossed mutant → homozygotes.

Genetic complementation is a principal phenomenon in the definition of the gene. Usually mutations that complement each other are mutations of different genes, i.e. non-allelic mutations. Allelic mutations of one single gene can hence be defined on the basis of the **complementation test**, i.e. → *cis-trans* test. If the *cis* heterozygote *a b*/+ + and the *trans* heterozygote *a* +/+ *b* are phenotypically different the former being wild and the latter mutant, *a* and *b* are mutations of the one gene, → cistron (*a* and *b* = mutant genes or their sites, + = wild type gene or its site). On the contrary, if *cis* and *trans* heterozygotes are identical by their phenotypes (usually wild), *a* and *b* are mutations of different cistrons. Thus, they complement each other and the phenomenon is called **intercistronic complementation** in which different cistrons code for different genetic functions (different proteins).

In addition to the intercistronic complementation also **intracistronic complementation** exists, in which two allelic mutations complement each other although they do not complement a possible third mutation. The intracistronic complementation differs from the intercistronic by being incomplete and temperature sensitive. **Complementation mapping** is based on genetic complementation. Mutations which fail to complement each other, form a **complon** or **complementation unit** which in the **complementation map** is described by a segment of a line.

genetic counselling, medical service aiming at preventing the birth of genetically diseased children. In genetic counselling for families already having genetically diseased children or being at risk of having one, the nature and prognosis of the disease is described in addition to its mode of inheritance; also prenatal diagnosis can be used. By analysing genes, chromosomes, or gene products of a foetus, a definition of its health status is possible.

genetic death, elimination of harmful genes from the population by means of natural selection; may be a consequence of the decreased viability or fertility of an individual carrying harmful → alleles.

genetic disease, heritable disease; any genetic disturbance which basically is dependent on biochemical metabolic defects determined by a gene, a chromosome, or genome mutations, e.g. → Down syndrome, fragile X syndrome, and haemophilia. Most diseases involve both genetic and environmental components, but proper genetic diseases are those in which the environmental component is of little importance. The relationship between inheritance and diseases is the topic of medical genetics.

Genetic diseases are not necessarily incurable; the progress of a specific disease may be stopped by therapy at different levels of the gene action, in future, possibly also by means of → gene therapy.

genetic drift, random fluctuations of → gene frequencies in a small isolated population; when reaching one or zero the genetic drift ceases. Due to the drift the gene loci in the population become gradually homozygous, i.e. the population loses its variability. This phenomenon is called **genetic decay**.

In the state of **persistent drift** the population is constantly so small that random fluctuations of gene frequencies occur. In the **intermittent drift** the population lives through a sudden and short-term decrease of size, and chance decides which alleles remain in the population, and in which frequencies. The intermittent drift is also called the **bottleneck effect**. *Syn.* Sewall Wright effect.

genetic engineering, → gene manipulation.

genetic equilibrium, a condition in a population where gene frequencies stay constant from generation to generation. A population in which random mating prevails, is in a genetic equilibrium if its gene and genotype frequencies are consistent with the → Hardy—Weinberg law, i.e. the population remains in equilibrium if no factor other than random mating is effective. Changes in the equilibrium are caused by → mutation pressure, migration, selection, and genetic drift. Of these the first three are called systematic factors, and the genetic drift is called random factor. The systematic factors drive the gene frequencies into a stable equilibrium state which prevails as long as the pressure of these factors remains the same. Due to random factors the population is driven away from the equilibrium state.

genetic flux, the horizontal transfer of genetic material from one individual to another or even from one species to another mediated by → transposons.

genetic information, the total whole of the structural and functional instructions residing in the genes of an organism; occurs in the form of the linear structure of → nucleic acids (DNA or RNA). DNA is the genetic material and the primary carrier of genetic information in all organisms, except in RNA viruses where it is RNA.

In the **transfer of genetic information** four mechanisms function: 1) → replication of DNA; 2) replication of RNA in RNA viruses; 3) → genetic transcription in which the genetic information residing in DNA is transcribed in prokaryotic organisms into → messenger RNA and in eukaryotic organisms into → heterogeneous nuclear RNA which will be processed into mature messenger RNA; the messenger RNA carries the genetic information from the nucleus into the cytoplasm; 4) → genetic translation in which the genetic information carried by the messenger RNA is translated in → protein synthesis into the primary structure of the protein or amino acid sequence.

genetic load, the average number of harmful mutations, mostly unexpressed in the genome of an organism. The → gene pool of a population always contains a certain number of harmful genes decreasing the → fitness of their carriers. All forms of → selection permit a certain quantity of genetic load, some forms of selection even maintain it. The quantity of the genetic load (L) is related to → fitness (w):

$$L = (w_{max} - w)/ w_{max}$$

where w_{max} is the fitness of the fittest genotype, and w is the mean fitness of the population.

The genetic load consists of four important components: 1) **mutational load** is a consequence of the equilibrium between → mutation pressure and normalizing selection; 2) **segregational load** is created by selection favouring the heterozygotes at the expense of the crossing homozygotes, i.e. striving for heterotic balance; 3) **substitutional load** is generated by dynamic selection at the time interval when the former → adaptive norm of the population is substituted by a new one, i.e. there are harmful genes in the population as a remnant of the former genetic constitution of the population; 4) **incompatibility load** is a consequence of genetic → incompatibility; e.g. the genotype of the foetus can be incompatible with that of the mother, causing a genetic disease in the foetus.

genetic map, either a → chromosome map or a → gene map.

genetic marker, any → allele or DNA sequence used as an experimental probe to mark a nucleus, chromosome, or gene; in practice, any difference of the genes which is employed in studies of → genetic recombination; e.g. white vs. red eye colour of the fruit fly.

genetic material, the carrier of the primary → genetic information, i.e. the matter of which the genes are formed. Usually the genetic material is double-stranded DNA, but also single-stranded DNA in some → bacteriophages and RNA in RNA viruses. The genetic material must fulfil the following three universal conditions: 1) it must be capable of conducting its own → replication and thus be responsible for reproduction typical of living nature; 2) it must be very specific, and the specificity must be maintained during replication in order to be capable of taking responsibility for the big qualitative differences between genes; 3) it must contain information, i.e. it must be an informative macromolecule having the ability to regulate the development of the structure and function of cells and individuals. This task the genetic material fulfils by conducting the synthesis of the most important structural material of organisms, i.e. the proteins.

genetic mosaic, an individual which consists of genetically different tissues and shows different characters in its different parts (the individual is called a → chimaera or a mosaic). The mosaic parts of the individual can correspond to the effects of the allelic genes of the genotype or they can be a consequence of → gene mutation, chromosome mutation, somatic crossing over, aneuploidy, or of double fertilization. A genetic mosaic is called → gynandromorph if it is partly female and partly male. Generally an organism comprising clones of cells with different genotypes is derived from one single zygote.

genetic polymorphism (Gr. *polys* = many, *morphe* = form), a condition prevailing in a population in which regularly and simultaneously exist two or more heritable → variants or → genotypes (morphs) which clearly differ from each other in such frequencies that their existence cannot (even in the case of the rarest morph) be interpreted to be due to → mutation pressure only.

Six types of the polymorphism can be distinguished. **Balanced polymorphism** is maintained in a population by balancing selection so that it favours → heterozygotes at the expence of crossing → homozygotes (so-called heterotic balance). The balanced polymorphism is characterized by a stable and optimal equilibrium between the frequencies of the different → morphs. At the equilibrium state the mean fitness of the population is at its maximum; e.g. → sickle-cell anaemia.

Transient polymorphism arises as a consequence of dynamic selection; during the substitution of the former → adaptive norm of the population by a new one, the population is polymorphic. In **neutral polymorphism** the → fitnesses of morphs do not differ from each other but are of equal value in relation to each other. In **geographic polymorphism**, different morphs inhabit different areas or habitats of the distribution area of a population. **Unisexual polymorphism** is expressed in one of the two sexes only. In **cryptic polymorphism**, the morphs cannot be distinguished from each other on the basis of their phenotypes.

In nature, all populations are polymorphic, and the amount of polymorphism is high, e.g. of the gene loci of the natural populations of the fruit flies (*Drosophila*) approximately one third are polymorphic, as assessed from enzyme polymorphism. At the DNA level polymorphism is even larger, but it is partly cryptic.

genetic recombination, a process in which two (or more) organisms unite so that in their offspring such gene combinations occur which did not exist in their parents. These offspring are called **recombinants**. Recombinants can be seen after → meiosis (meiotic recombination) or after → mitosis (mitotic recombination) or in bacteria and viruses after processes corresponding to meiosis or mitosis. Genetic recombination requires the union of genetic material from two (or more) genotypes. This may occur in different ways in the → genetic systems of different organisms.

In **eukaryotic organisms** the union is achieved by the alternation of meiosis and fertilization. In these organisms recombination can occur through four mechanisms: 1) **interchromosomal recombination** is based on the independent behaviour of → bivalents during meiosis; 2) **intrachromosomal recombination** is based on → crossing-over occurring in meiosis in which homologous chromosomes change parts reciprocally; 3) **gene conversion** is intragenic recombination; 4) **parasexual cycle** is a genetic system found in certain fungi parallel with the sexual cycle. It leads to recombination through → haploidization and mitotic crossing-over.

In **bacteria**, the union and recombination of genetic material can occur through certain mechanisms related to sexual reproduction. These are → **transformation**, **conjugation**, and **transduction**. Also in eukaryotic organisms phenomena exist which correspond to transformation; in these, recombination is mediated by → transposons. In **viruses**, the genetic material unites when two or more virus particles infect the same host cell at the same time, and recombination occurs through breakage and reunion of the viral DNA molecules. A virus recombinant can have more than two parents.

In interchromosomal recombination, the **molecular mechanisms** of genetic recombination are based on chance, but in intrachromosomal recombination and gene conversion on the breakage and reunion of DNA molecules.

genetic regulation, the general regulation of the quantity and quality of metabolic events in a cell conducted by the function of specific genes. In multicellular organisms, genetic regulation results in the formation of tissues and developmental stages due to the specific activity of the genes. In higher organisms, usually specific → regulatory genes regulate → gene batteries consisting of structural genes (producer genes), in bacteria, correspondingly gene complexes called → operons.

In bacteria, genetic regulation is almost exclusively regulation of → genetic transcription, but in eukaryotic organisms genetic regulation can occur at several functional levels. In these organisms, genetic regulation may involve transcription, the processing of the primary transcription product, → genetic translation, and function of proteins. In higher organisms such as plants and animals, genetic regulation may involve whole chromosomes or individual genes.

Genetic regulation can be **positive** or **negative**; in positive regulation, a regulatory molecule enhances the function of a gene or a group of genes under regulation, in negative regulation it suppresses the function. Both types exist in prokaryotes and in eukaryotes. Genetic regulation serves as a basic element for the differentiation of cells and tissues, and accordingly, is one of the most fundamental of biological phenomena.

genetics, a branch of science dealing with heredity and variation. Genetic research uses viruses (virus genetics), bacteria (bacterial genetics), plants (plant genetics), animals (animal genetics), and humans (human genetics). Traditionally genetics is divided into formal genetics, cytogenetics, developmental genetics, and population genetics.

Formal genetics studies inheritance from a purely formal view. Its objects are → genetic transmission, the transfer of genes from one generation to the next, → genetic recombination, and → mutation. **Cytogenetics** deals with the composition and function of the cellular material which forms the base of inheritance. Thus, its objects are the structure and function of → chromosomes and genetic material in general. **Developmental genetics** or **phenogenetics** studies the action of genes and their regulation, and thus the route from genes to the characters, phenes. Objects are → genetic transcription and → genetic translation in addition to their regulation and the problems of development in general. **Population genetics** deals with the consequences and the expression of the laws of inheritance in populations. Thus, population genetics is central to the study of evolution.

Biochemical genetics studies the action mechanisms of genes using biochemical methods, and **molecular genetics** the expression and mechanisms of all the phenomena of inheritance (including formal genetics, cytogenetics, phenogenetics, and population genetics) at the level of molecules, dealing with the structure and function of the genetic material (*see* quantitative genetics).

genetic system, a system producing → genetic recombination; includes the sexual system and its substituting systems. The sexual system consists of the alternation of → meiosis and fertilization during the flow of generations. Substituting systems are the parasexual cycle (*see* parasexuality) in some fungi, → conjugation, transformation, and transduction in bacteria, and the recombination phenomena in viruses. Also in other organisms than bacteria, asexual systems resembling transformation are found, → transposons transmitting genetic material from one individual to another and even from one species to another.

genetic toxicology, a branch of biological sciences dealing with the estimation and assessment of genetic changes caused by environmental factors.

genetic transcription (L. *trans* = across, *scribere* = to write), the enzymatic synthesis of cellular → ribonucleic acids (RNAs) on a deoxyribonucleic acid template (DNA template), i.e. **DNA-dependent RNA synthesis**. In the process, the genetic information residing in the DNA structure in the form of the → genetic code is transcribed into a complementary RNA molecule. The enzyme performing transcription is an → **RNA polymerase**, which in eukaryotes is found in three forms. RNA polymerase I catalyses the synthesis of → ribosomal RNA, RNA polymerase II the synthesis of → heterogeneous nuclear RNA, and RNA polymerase III the synthesis of → transfer RNA and other small RNA molecules. In prokaryotes, one RNA polymerase performs all these functions.

In prokaryotes, the unit of messenger RNA transcription is the → **operon**; in eukaryotes, the unit of the transcription of heterogeneous nuclear RNA is the → **gene**. A prerequisite for transcription is selective gene activity. In prokaryotes, the transcription product of the operon is a polycistronic messenger RNA. In eukaryotes, the transcription product of the gene, i.e. the heterogeneous nuclear RNA, is processed in a multistep manner into monocistronic messenger RNA (→ RNA processing).

The chemical mechanism of transcription and its regulation is dependent on the specificity of the polymerases, the initiation and ending of transcription, the cleavage of the transcription product from DNA, and for polarity and asymmetry.

The **initiation of transcription** occurs from a specific starting point and proceeds until a certain end point is reached (*see* initiator codon). Chemically, initiation involves the interaction of the polymerase enzyme and a particular site on the DNA template, resulting in the synthesis of a certain RNA. The **polarity of transcription** means that the transcription proceeds along the DNA molecule in one direction only. The transcription of messenger RNA or of heterogeneous nuclear

RNA starts from the → promoter and proceeds from the 5' end towards the 3' end. The **asymmetry of transcription** means that during transcription in the operon or gene only one strand of DNA is copied; this strand is called the → antisense strand, and can be different in different operons or genes of the chromosome. RNA polymerase recognizes the antisense strand on the basis of a certain nucleotide sequence located in the complementary strand, called → sense strand. This sequence is called the → **TATA box** (→ Hogness box in eukaryotes); it is analogous to the → Pribnow box in prokaryotes.

genetic transformation, *see* transformation.

genetic translation, the second step of the reading of genetic information in the cell following → genetic transcription, i.e. → protein synthesis on the messenger RNA template; a specific nucleotide sequence of the messenger RNA (mRNA) is translated into a specific sequence of the amino acids in a polypeptide (protein). There are three stages in the genetic message from mRNA to a protein: **initiation** with → initiator factors, **elongation** with → elongation factors, and **termination** with → termination factors. Each of the 20 amino acids present in the proteins is represented by a triplet, → **codon**, formed by three nucleotides in mRNA. In both prokaryotic and eukaryotic organisms, the unit of translation is the → cistron, and its product is a polypeptide. A colinear relationship exists between the cistron and the polypeptide encoded by it, in such a way that the sequence of the codons in the cistron determines the sequence of amino acid residues in the polypeptide.

During translation, mRNA is attached to → ribosomes which serve as the sites of polymerization of amino acids during protein synthesis. According to the **adaptor hypothesis**, the ribosomes read the mRNA so that a → transfer RNA (tRNA), which contains the respective → anticodon, is attached at the location of each mRNA codon; the tRNA molecule carries the corresponding amino acid. At the ribosome two codons and therefore two tRNAs are present simultaneously, and a peptide bond is formed between the two corresponding amino acids. After this the ribosome moves one codon step forward in its reading of the mRNA, and tRNA containing the corresponding anticodon is transferred onto the ribosome. Hence the elongation of the polypeptide chain is achieved.

Translation begins at specific → initiator codons, proceeds in 5'→ 3' direction, and stops at a → termination codon at which the ribosome detaches the mRNA. *See* protein synthesis.

genetic transmission (L. *mittere* = to send), the transfer of genes from one generation to the next.

genetic variance, → genetic variation.

genetic variation, the formation and existence of genotypically different individuals in a population, as opposed to the inheritable differences caused by environmental factors (*see* variation). The relative degree of genetic variability in a population is measured by **genetic variance** which is the relative portion of the genetic variation of the total → phenotypic variation. The causes of genetic variability are → mutations and → genetic recombination. Mutations are important because they are the primary source of all genetic variation, whereas recombination is important since it has the property of producing an enormous quantity of genetic variability, to the extent that in a cross-breeding sexual population no genetically identical individuals exist (there are certain exceptions, like monozygotic twins).

The degree of genetic diversity in populations is usually enormous. For example, in natural populations of fruit flies one third of the genes are polymorphic, as measured at the protein level, and the degree of heterozygosity is approximately 13%. In human populations, 28% of the loci are polymorphic and the degree of heterozygosity approximately 7%. As measured at the DNA level the degree of diversity is even larger. If also → introns are included, every locus of every individual is in a heterozygous state, and when any two human individuals are compared, every hundredth of the nucleotides is different. *See* genetic polymorphism.

gene transfer, the transfer of genes from one cell or one organism to another using → gene manipulation. Several → vectors can be used as vehicles of gene transfer, or genes can be injected into the recipient cell directly by using microinjections. In the technique **random** and **focused** gene transfers are used. In the former the transferred gene is integrated into a random position in the chromosomal complement of the recipient cell, in the latter into a given position of the chromosomal complement by means of homologous → genetic recombination. The organisms produced by gene transfer techniques are called transgenic organisms.

geni(o)- (Gr. *geneion* = chin), denoting relationship to the chin. *Adj.* **genial**.

-genic (Gr. *gennan* = to produce), a suffix denoting producing or forming, or pertaining to a gene.

genic selection, a hypothesis of the evolution theory regarding the gene as the unit of selection. According to this hypothesis, natural selection involves foremost the selection of alternative gene alleles. This occurs through the → phenotypes produced by the genes. However, because the phenotypes in sexual reproduction are vanishing entities but the alleles are persistent in the germ line, the target of selection is finally the gene, not the phenotype. The theory of genic selection has been an important research topic, especially in studying the evolution of behaviour.

geniculum, pl. **genicula** (L. dim of *genu* = knee), a sharp knee-like bend in an anatomical structure.

genital (L. *genitalis*), pertaining to **1.** reproduction; **2.** generation; **3.** → genital organs.

genital organs (L. *organa genitalia*), **genitalia**, **genitals**, **sex organs**; reproductive organs of animals, including the **gonads** (sex glands), either testes or ovaries in different individuals (male or female) except in → hermaphrodites which have both in the same individual, and **accessory genital organs**, such as the copulatory organs and different ducts and glands associated with the gonads. In vertebrates, the ontogenic development of the genitals occurs in close correlation with the development of the kidneys (*see* Müllerian duct). In many invertebrate species, the structures of male and female genitals are species-specific acting with the lock-and-key model, thus preventing interspecific hybridization. *See* male genital organs, female genital organs, external genital organs.

genital plate, a lamellar structure bearing a genital pore; e.g. the five plates surrounding the periproct in sea urchins; one of them is the larger, porous plate, called **madrepore**.

genome (gene + chromosome), the basic haploid chromosomal complement of eukaryotic organisms; the **haploid set** of chromosomes. It includes the number of chromosomes characteristic for the species, and the genes in them. In the diploid phase of the life cycle there are two genomes, but in the haploid phase only one. In prokaryotic organisms like bacteria, the term genome denotes the total number of genes present in their single chromosome. In common language, genome denotes the entity of the genes of an organism which usually is the diploid set of chromosomes. *See* haploid, diploid.

genome mutation, a → mutation in which the number of chromosomes has changed; genome mutations includes → polyploidy and → aneuploidy.

genomic imprinting, the phenomenon that results in functional differences in the expression of homologous gene alleles, depending on whether they are derived from the mother (maternal imprinting) or from the father (paternal imprinting); e.g. in marsupials, the paternal X chromosome is always inactivated in females, i.e. it becomes paternally imprinted inactive when it goes through → spermatogenesis.

genotoxic (Gr. *toxikon* = arrow poison), a physical or chemical agent which in genes and chromosomes causes mutation-like defects; e.g. → mutagen or → carcinogen.

genotype (Gr. *typos* = form), all the genes of an individual, i.e. the genetic constitution of an individual; also used with respect to gene combination at one special gene locus. The genotype determines the norm of reaction, i.e. how an individual reacts to the environment during its developmental and adult life. The → phenotype of the individual arises as the result of a complicated interaction between the genotype and the environment. In contrast to the phenotype, the genotype is stable, being the total complement of the possibilities residing in the genes of each individual. *Adj.* **genotypic**.

genotypic value, that part of the → phenotypic value of an individual which is due to the genes. The genotypic value of a given individual can be measured if the individual can be cloned (*see* clone). The mean of the phenotypic values of genotypically identical individuals is the genotypic value of the clone. The genotypic value (G) consists of → breeding value (A), dominance deviation (D), and → interaction deviation (I) so that $G = A + D + I$.

gens, pl. **gentes** (L. = clan, family, race), a term used in various meanings to describe a kind of a group of organisms; the group may be taxonomic, but not necessarily; e.g. in a population of the European cuckoo (*Cuculus canorus*), the gens means the group of females which lay eggs primarily in the nests of a certain host species.

genu, pl. **genua** (L.), the knee, or any anatomical structure bent like the knee. *Adj.* **genual**.

genus, pl. **genera** (L. = race), a unit used in taxonomic classification including closely related species and ranked between family and species. The genus may be either monotypic comprising only one species, or polytypic including several species. The scientific name of each species is either binomial or trinomial the first word expressing the genus, the second the species, and the third the subspecies; e.g. *Corvus corone corone*, the carrion crow.

geo- (Gr. *ge* = earth), pertaining to the earth, soil, ground, or geography.

geobotany, a branch of botany studying distribution of plant taxa, areal differences of vegetation, and history of flora and vegetation.

geographic(al) isolation, *see* isolation.

geographic(al) race, *see* race.

geographic(al) variation, genetic variation between allopatric populations. The variation may concern all the features of an organism, such as anatomical, physiological, and ethological features; e.g. → Allen's, Bergmann's and Gloger's rules describe different kinds of geographical variations among animals. *See* Rensch's laws.

geolittoral zone, *see* littoral.

geological eras (L. *aera* = the beginning of new age), the geological age of Earth (about 5,000—4,600 million years) is divided into four eras in order of age: 1) → **Precambrian**, 4,600—570 million years ago, usually divided into two eons: Archaeozoic (Archaean), 4,600—2,500 million years ago, and Proterozoic, 2,500—570 million years ago. 2) → **Palaeozoic**, 570—245 million years ago, including six periods, i.e. → Cambrian, Ordovician, Silurian, Devonian, Carboniferous, and Permian, 3) → **Mesozoic**, 245—66 million years ago, with three periods, i.e. → Triassic, Jurassic, and Cretaceous, and 4) → **Caenozoic**, from 65 million years ago to the present, including two periods, i.e. → Tertiary and Quaternary, the latter two comprising several further epochs.

geological indicator species, plant species used as indicators of ore-bearing minerals; most commonly used are indicators of copper, the best known species being red alpine catchfly (*Viscaria alpina*).

geology (Gr. *logos* = discourse), the branch of science dealing with structures, activities, and history of the Earth.

geometric growth, geometric rate of increase, the rate by which the density of a population increases over a certain time period so that at even time intervals the relation between successive density values is constant, such as in series of 2, 4, 8, 16 or 4, 6, 9, 13.5. *See* exponential growth.

geophyte (Gr. *phyton* = plant), a plant whose overwintering parts, such as the buds, bulb or rhizome, are situated totally below ground; a special type is a helophyte with the wintering parts in water. *See* Raunkiaer's life forms.

geotaxis (Gr. *taxein* = to arrange), a → taxis of organisms, especially of animals, in which the gravitational force is the directive factor for locomotion; in plants, the related reaction to gravitation is called → gravitropism, which is a growth movement (turning movement). *Adj.* **geotactic**.

geotropism, → gravitropism.

ger(o)- (Gr. *geron* = old man, the aged), pertaining to old age or aged. *Adj.* **gerontal**, senile.

germ cell (L. *germen* = germ, sprout, bud), generative cell, → **gamete**; any reproductive cell of a multicellular organism as opposed to a → somatic cell.

germ-free animals, GF animals, experimental animals not infected by any microorganisms; are reared within germ-free conditions. *See* specific pathogen free, gnotobiota.

germinal (L. *germinalis* < *germinare* = to sprout, put forth), pertaining to a seed, germ cell, or germination.

germinal epithelium, a surface cell layer, epithelium, of the peritoneum covering the vertebrate gonads from their earliest development; thought to be the source of germ cells.

germinal vesicle, the enlarged nucleus of an animal → oocyte whose development is arrested in → oogenesis during diplotene of the first meiotic division (*see* meiosis). This state of the nucleus may be prolonged especially in mammals from the foetal period to sexual maturity, i.e. in large mammals for years.

germination, the growth of a plant embryo in a seed to form a seedling (first the primary root), the formation of a prothallus or mycelia from spores, or the growth of the pollen tube out of the pollen grain. Germination requires water uptake and the activation of enzymes which hydrolyse storage compounds. Germination is controlled by several biotic and abiotic factors, such as hormones, light, temperature, etc.

germ layers, the three tissue layers of an early embryo of most multicellular animals, i.e. → **ectoderm**, **mesoderm**, and **endoderm**; in the most primitive animal groups, such as sponges and cnidarians, the mesoderm does not develop. *See* embryogenesis.

germ line, the cell line which leads from the → gametes of an earlier generation to the gametes of the next; e.g. in vertebrates and insects, the germ line is a concrete concept because the cells of the germ line separate from the rest of the cells at an early developmental stage. In contrast, in many other animal groups and especially in plants, the germ line is a more abstract concept because any cell can give rise to the cell line which produces gametes.

germ plasm, the matter of inheritance which constitutes the physical basis of inheritance and is transmitted from one generation to the next by → germ cells. The term was taken into use in biology when the nature of hereditary material was still unknown.

germ pore, a thin site in the outer wall of the pollen grain (exine), through which the pollen tube grows during germination of the grain. The number of germ pores in one pollen grain is usually 1 to 3.

gestagen (L. *gestare* = to bear, *-genes* = produced, born), any hormone with progestational activity. *See* progesterone.

gestation (L. *gestatio*), the process occurring in a viviparous female animal during the development of one or more zygotes (embryo, foetus) in the body until parturition, i.e. during a certain **gestation period**. The condition of the mother animal is called **pregnancy** (especially in humans), although the terms gestation and pregnancy are often used as synonyms. Gestation is controlled by female sex hormones, especially → **progesterone.** Some examples of gestation periods: elephant 22 months, horse 340 days, cow 282 days, goat and sheep 148 days, dog and cat about 60 days, rat 22 days, mouse 19 days.

GH, → growth hormone.

ghosts, erythrocytic ghosts; **red blood cell ghosts;** erythrocyte plasma membranes obtained when red blood cells are haemolysed.

GHRH, growth hormone-releasing hormone, → somatoliberin.

giant axons, large nerve fibres in the nerve cord of annelid worms, formed by the fusion of axons of many nerve cells; function in rapid flight reactions. *See* giant cells.

giant cells, exceptionally large animal cells; e.g. **1. megakaryocytes** which split into several pieces producing blood platelets; **2. giant neurones**, nerve cells which are much larger than neurones in general, found in cephalopod molluscs and fish. Giant cells are sometimes spontaneously formed also in cell culture. Egg cells are not usually called giant cells although most of them are exceptionally large because of storage of vitelline (yolk). *See* giant axons.

giant chromosome, a special type of polytene chromosomes (*see* polyteny) occurring as a bundle of many interphase chromosomes; formed by repeated replication of chromosomes without their separation (endomitosis). In contrast to other chromosomes, the giant chromosomes stay in permanent interphase and are thick, cable-like structures visible in the light microscope. Such chromosomes are typical especially of the secretory cells of the larvae of dipteran insects, such as the cells of the salivary glands, whose cells no longer divide. The number of chromatids in a giant chromosome is a species-specific character. The length of giant chromosomes varies between 100 to 250 μm, the thickness from 15 to 25 μm.

A clear banded structure is typical of giant chromosomes. On the basis of the bands, every segment of the chromosome can be identified which is especially important in physical or cytological gene mapping. As structural modifications of the bands, specific → **puffs** occur, which are the sites of active genes. In the puffs, the compact structure of DNA in the → chromomeres is loosened like a coil of wire, in order to make → genetic transcription possible. Giant chromosomes are also found in certain plants and protozoans. *Cf.* endomitosis.

gibberellins, GA, a group of plant hormones (diterpenes) consisting of more than 70 different GA types. The first gibberellin was isolated from the fungus *Gibberella fujikuroi* (*Fusarium moniliforme*). Each gibberellin molecule contains a gibbane skeleton, and the different types are named GA_1, GA_2, GA_3 etc. Gibberellins promote the growth and flowering of plants as well as the germination of seeds. Their primary effect is at the gene level, e.g. **gibberellic acid** (GA_3) induces certain genes to start encoding the synthesis of α-amylase, needed for starch hydrolysis in germinating seeds.

Giemsa staining (*Gustav Giemsa*, 1867—1948), a method for staining blood cells and especially chromosomes. *See* G-banding.

gigantism (Gr. *gigas* = huge, giant), excessive size, abnormal overgrowth; e.g. **pituitary gigantism** in vertebrates, is due to excess secretion of → **growth hormone**; if hypersecretion occurs in adults it induces acromegalic gigantism, i.e. the overgrowth of distal bones e.g. in the limbs and the chin. In plants, gigantism is caused by some parasites, e.g. by the fungus *Gibberella fujikuroi*, which produces → gibberellins, thus increasing cell division and elongation. In some evolutionary line the increase of size has gradually lead to normal gigantism.

gigaseal, a seal between the edge of the patch electrode and the cell membrane in the → patch clamp studies, the electrical resistance of which has the magnitude of gigaohms.

GIH, GHRIH, growth hormone (release) inhibiting hormone, → somatostatin.

gill (L. *branchia*), the respiratory organ of aquatic animals; usually a lamellary structure that widens the respiratory surface covered by a thin **gill epithelium** through which oxygen can easily diffuse from water into the capillary net of the **gill lamellae**, and carbon dioxide in the opposite direction. Animals can usually improve the effectiveness of gas exchange by active gill movements or by pumping water through the gills. The gills may be either **external** outside the body, as in anuran tadpoles, or **internal** in the gill chamber, usually on each side of the branchia, as in fish. Large crustaceans have three types of gills: **arthrobranchiae** (joint gills), **pleurobranchiae** (pleural gills), and **podobranchiae** (foot-gills) in the gill cavity enclosed by the carapace. Some aquatic insect instars, such as trichopterans and odonates, have paired movable **tracheal gills** on the abdomen. These are thin extensions of the body wall with a thick tracheal network.

gill arch, gill bar, branchial arch; one of several jointed, bar-shaped, cartilaginous or bony structures behind the jaws in fish and all larval and some adult amphibians; originally six or seven arches located in the pharynx posterior to the hyoid arch, occurring as rudimentary ridges in embryos of higher vertebrates (*see* branchiogenous organs). In most jawed fishes, five pairs of the gill bars are found in sequence along the walls of the pharynx, forming successive gill slits between them. Each gill bar of a jawed fish comprises the following structures on both sides: the dorsal element, **epibranchial**, the ventral element, **ceratobranchial**, at the top of the arch the **pharyngobranchial**, and the major ventral element, **hypobranchial**. Ventrally between the arches, the **basibranchials** connect the successive arches to each other. Usually, the gill arches bear a row of short **gill rakers** and **gill rays** which stiffen the gills. In jawed vertebrates, the jaws, hyoid arch, and gill arches together form the **pharyngeal arches** (visceral arches).

gill book, five posterior appendages modified into leaf-like respiratory organs (gills) in the abdomen of the horseshoe crabs (Merostomata).

gill fungi, fungi belonging to the class → Basidiomycetes, whose → hymenium is borne in gill-like structures on the undersurface of the cap of the fruiting body; includes most mushrooms and toadstools.

gill pouch, → branchial pouch.

gill rakers, a row of raker-like bony structures on the inner margin of the gill bars in jawed fish; prevent small food particles from being flown out through the gill slits.

gill slit/cleft, → branchial cleft.

Ginkgoinae, a class in subdivision Gymnospermae, division Spermatophyta; nowadays includes only one species, the maidenhair tree, *Ginkgo biloba*, which is evolutionarily very old, i.e. a living fossil, growing naturally in China.

GIP, → gastric inhibitory polypeptide.

gizzard, 1. L. *gigerium*, a muscular enlargement of the stomach, posterior to the **proventriculus** (glandular stomach) in many birds serving as a grinding chamber; it is well developed in herbivorous species, e.g. gallinaceous birds swallow small stones (gastroliths) to improve its function; **2.** an enlargement of the foregut in insects and some other invertebrates, such as annelids; it is lined with plates or thick, firm, muscular walls for grinding rough food.

glacial relict, a → relict species from the late Ice Age being preserved in an area; glacial relicts are common e.g. in the Baltic Sea, such as the species *Saduria (Mesidotea) entomon*, *Pontoporeia affinis*, *Mysidacea*.

gland (L. *glandula*), an organ with cells specialized for secreting substances which serve many functions in animals. The glands are usually composed of a group of epithelial cells involuted into underlying connective tissue, a tube connecting the gland to the original surface; thus the product (exocrine secretion) can be transported out onto the body integument or into different cavities of inner organs. These types of glands are called **exocrine glands**. Many glands lose their connection so that the gland forms a closed structure from which the product (endocrine secretion, incretion) is secreted to the surrounding tissue fluid or directly into the blood; these are called **endocrine glands**. Secretions of the glands serve in protection (e.g. sebaceous glands), in digestion (e.g. pancreatic and duodenal enzymes, bile), or as messenger substances between cells (hormones).

The mechanisms of secretion differ in different glands. **Merocrinic glands**, including e.g. all endocrine glands, digestional glands, and eccrine sweat glands, secrete their products through the cell membrane in a process called → exocytosis. **Apocrinic glands** secrete the apical parts of glandular cells within their excretion, as e.g. do the axillary and inguinal sweat glands and fat secreting cells of the mammary glands (protein secreting cells are merocrinic). The product of **holocrinic glands** is formed by the whole glandular cell released in the secretion process, as e.g. in sebaceous glands.

Tubular, **alveolar** and **tubuloalveolar glands** are distinguished by their form. Glands with only one glandular follicle (simple glands) are found, but also glands with several follicles (compound glands). In some tissues the glandular cells do not form an organ but occur as single cells, as e.g. the goblet cells in the epithelium of the respiratory pathways and the alimentary canal of vertebrates. In addition to ordinary glandular cells, there are many other cell types that can secrete, like white blood cells, nerve cells, and atrial muscle cells of the vertebrate heart.

In plants, glands belong to the → excretory and secretory tissues. They may be glandular hairs secreting e.g. aromatic oils, or special glands such as → salt glands. *Adj.* **glandular**.

glandular hair, a plant hair secreting continuously e.g. oils or salts out from the plant. *See* excretory and secretory tissue, hair.

glans, pl. **glandes** (L. = acorn) an acorn-shaped anatomical structure, especially the tip of the penis (*glans penis*) or a small mass of erectile tissue of the clitoris (*glans clitoridis*) in many mammals.

glass sponges, → Hexactinellida.

Glaucophyta (Gr. *glaukos* = gleaming, grey, *phyton* = plant), in some botanical systems, the group of algae, the cells of which are eukaryotic but are lacking chloroplasts, having instead endosymbiotic cyanobacteria which carry out photosynthesis. These species are considered important for studies on evolution of the chloroplasts. *See* endosymbiosis.

GLC, gas-liquid chromatography, *see* chromatography.

glenoid cavity (Gr. *glenes* = socket, pit), glenoid fossa of scapula; a socket in the shoulder girdle (pectoral girdle) of tetrapods to which the coracoid, scapula, and humerus join.

glia (Gr. = *glue*), → neuroglia.

gliacyte, → gliocyte.

gliadin, a → prolamine protein of wheat grains; quite insoluble in water but soluble in 70% ethanol. Gliadin is an important protein in food, containing high quantities of glutamine but can cause coeliac disease in some people. *See* gluten.

gliocyte (Gr. *kytos* = cavity, cell), **glial cell**; any of different cell types of the → neuroglia; also called gliacyte; **glioblast**, a developing, immature gliocyte.

global (L. *globus* = ball, globe), pertaining to the whole world; world-wide, overall; e.g. global distribution.

Globigerina ooze, a bottom ooze in oceans made up of shells of the protozoan genus *Globigerina* (*see* Foraminifera). The calcium and silica-rich ooze covers about one third of the sea bottom, being especially abundant in the Atlantic.

globin, the protein constituent of haemoglobin or myoglobin, or of similar globoid protein.

globoid, 1. globe-shaped, spheroid; **2.** a globular structure formed by protein granules in plant cells.

globulin (L. *globulus* = small ball, globule), any of the group of globular proteins which are typed to **euglobulins** (true globulins) being insoluble in water but soluble in physiological salt solutions, and **pseudoglobulins** being soluble in water but othervise resembling euglobulins. In the blood plasma of vertebrates, there are usually high concentrations of euglobulins, called → **immunoglobulins** (gammaglobulins), which serve as → antibodies in immune defence. Globulins occur also in plants, e.g. as storage proteins in many seeds.

glochidium, pl. **glochidia** (Gr. dim. of *glochis* = arrow point), a larval stage of some freshwater clams, as *Unio* and *Anodonta*; glochidia attach to fish gills and in this way can expand their distribution area.

Gloger's rule (*C.W.L. Gloger*, 1803—1859), a zoogeographical rule stating that many insects, birds and mammals are less pigmented in dry and cool areas than are species living in wet and warm regions. *Cf.* Allen's rule, Bergmann's rule.

glomerulus, pl. **glomeruli** (L. dim. of *glomus* = ball), a cluster or tuft in an anatomical structure; e.g. **1.** renal glomeruli in the vertebrate kidney; *see* nephron; **2.** a primitive excretory organ of hemichordates comprising a network of blood sinuses; **3.** a united mass of spores. *Adj.* **glomerular**.

gloss(o)- (L., Gr. *glossa*, *glotta* = tongue), denoting relationship to the tongue.

glossa, pl. **glossae** or **glossas** (L.), **1.** the tongue of vertebrates; **2.** the second distal lobes in the labium (lower lip) of the insect → mouthparts, with paraglossae forming the ligula. *Adj.* **glossal**, lingual.

glottis, pl. **glottides** (L.), the vocal organ of the → larynx of terrestrial vertebrates comprising true vocal cords (*plicae vocalis*) and the vocal opening (*rima glottidis*).

gluc(o)- (Gr. *glykys* = sweet), pertaining to sweetness or glucose; also **glyc(o)-**.

glucagon, pancreatic hyperglycaemic hormone; a polypeptide with 29 amino acid residues, produced by alpha-cells (A cells) of the → pancreatic islets in vertebrates; similar polypeptides, called **glucagon-like polypeptides** (GLP_1, GLP_2, enteroglucagons), are secreted by special cells in the mucous membrane of the upper gastrointestinal tract. Its secretion is induced e.g. by decreased blood glucose, and by amino acids present in the intestinal chyme. Glucagon stimulates → glycogenolysis in the liver by activating the phosphorylase enzyme by cyclic AMP, thus increasing the blood glucose level (*cf.* insulin). It also stimulates the secretion of growth hormone, insulin, and pancreatic somatostatin. The exact molecular structure of glucagon varies slightly depending on the vertebrate group.

glucan, a polyglucose; any of glucose polymers occurring as reserve substances (glycogen, starch) in organisms, or as structural elements (callose, cellulose) like in cell walls of plants.

glucobrassicin, a sulphur-containing → glycoside or glucosinolate in cruciferous plants (Brassicaceae) where it can act as a precursor for the biosynthesis of → auxin.

glucocorticoids (L. *cortex* = rind, bark, shell), steroid hormones (C21 corticosteroids) produced in the cortex of the → adrenal gland in vertebrates; the release is stimulated by → **corticotropin**, ACTH. They affect the intermediary metabolims of carbohydrates, proteins, and fats. They increase glucose-6-phosphatase activity in the liver releasing more glucose into the circulation, which is an important final reaction in → **gluconeogenesis**. They also stimulate the production of enzymes which act in gluconeogenesis, and thus ensure the sufficiency of glucose using non-carbohydrate sources, such as lactic acid, pyruvate and amino acids from different tissues, i.e. they stimulate **protein catabolism**. On the other hand glucocorticoids increase the formation of the active form of glycogen synthase, and in that way stimulate glycogen deposition (**hepatic glycogenesis**), and decrease glucose utilization in peripheral tissues. They also increase plasma lipid level and have anti-insulin action in peripheral tissues but not in the brain and heart.

Glucocorticoids are **stress hormones** which are secreted especially in prolonged stress situations such as in cold and starvation. They improve the resistance of an animal and thus its chance to survive over difficult situations (*see* stress). Indirectly, glucocorticoids also influence other functions, such as blood circulation and the functions of the alimentary canal (*see* permissive action) and the brain.

Glucocorticoids have anti-inflammatory actions in high concentrations, and are therefore used in medicine. Natural glucocorticoids are e.g. → **cortisol (hydrocortisone)** and **corticosterone**, synthetic glucocorticoids and their derivatives e.g. **dexamethasone** and **prednisolone**.

glucogenic (Gr. *gennan* = to produce), producing glucose.

glucogenic amino acid, an amino acid in which the carbon skeleton can at least partially be converted to glucose; e.g. alanine, serine, glycine, cysteine, and threonine. *Cf.* ketogenic amino acid.

glucokinase (Gr. *kinein* = to move), a **hexokinase** enzyme (hexotransferase, EC 2.7.1.2); occurs in most cells catalysing the conversion of glucose to glucose 6-phosphate; this phosphorylation reaction requires energy from ATP. The enzyme is very active e.g. in liver cells.

glucomannan, a hemicellulosic polysaccharide formed from glucose and mannose by β-(1,4)-glycosidic bonds; a major polysaccharide of gymnosperm softwood and a storage substance in some seeds.

gluconeogenesis (Gr. *neos* = new, *gennan* = to produce), the formation of glucose from non-carbohydrate precursors in animal tissues, as e.g. from pyruvate, lactic acid, amino acids and proteins, or fatty acids and glycerol. These sources of gluconeogenesis may be structural tissue molecules (e.g. proteins) or serve primarily as an energy source (e.g. fats). In vertebrates the gluconeogenesis is stimulated by → glucocorticoids released especially in prolonged stress situations, the liver playing the central role in the process.

gluconic acid, a crystalline organic compound, $CH_2OH(CHOH)_4COOH$, soluble in water and alcohols; hexonic acid derived from glucose by oxidation of -CHO group to -COOH. Gluconic acid is produced e.g. by oxidative action of many moulds and bacteria. *See* glucose oxidase.

glucosamine, 2-amino-2-deoxyglucose; an α-amino sugar occurring in cell membranes and as a constituent of mucus and chitin, the latter consisting of N-acetylglucosamine residues linked by β-1,4-glycosidic bonds. Called also glycosamine or chitosamine.

glucosaminoglycan, → glycosaminoglycan.

glucose, dextrose, grape sugar, an aldohexose, $C_6H_{12}O_6$; a major product of photosynthesis in plants and used as a main energy source both in plants and animals. Like all hexoses, it exists in different forms (aldehyde, pyranose and furanose forms). Glucose is the most common monosaccharide in organisms, chemically it is a reducing sugar and occurs in nature in the D-form. It is also a common component of many other sugars, such as saccharose and lactose, and of storage and structural polysaccharides including starch, glycogen, and cellulose. It is an important source for biosyntheses of several organic compounds, as organic acids (e.g. amino acids, acetic acid) and alcohols. Glucose and its phosphates play a central role as energy sources for cell respiration being gradually metabolized in → glycolysis and the → citric acid cycle into water and carbon dioxide, when ATP is produced.

glucose balance, the maintenance of the glucose level within certain limits in the blood and the interstitial fluids of animals. In most mammals, plasma glucose level tends to stay between 3 to 5 mmol/l. Most animals use glucose as their main energy source, and excess glucose is converted into → **glycogen** and stored in the liver (may be 10% of its weight) and muscles. When needed in tissues, glucose is quickly released from the liver cells for general use, but in muscle cells glucose from glycogen breakdown is used directly, and is not released into the blood.

The mechanisms controlling glucose balance are best known in vertebrates. Feeding increases slightly the blood glucose level, but soon the level is decreased by the action of **insulin**, which facilitates glucose uptake and usage in cells. Many hormones increase the glucose concentration in the blood. **Glucagon** and **adrenaline** stimulate → glycogenolysis in the liver from which glucose immediately is transferred into the blood. **Glucocorticoids** facilitate → gluconeogenesis, and together with the **growth hormone** and **thyroid hormones** increase the glucose concentration in the plasma. The **sympathetic nervous system** influences the glucose balance by regulating the secretion of adrenaline and insulin, or by having a direct effect on hepatic cells.

glucose oxidase, a flavoprotein enzyme (EC 1.1.3.4) in some fungi converting glucose to gluconolactone and hydrogen peroxide in the presence of oxygen; it is obtained from *Aspergillus* and *Penicillium* fungi for antibacterial use and for quantitative determination of glucose. Also called glucose oxyhydrase.

glucose-6-phosphatase, an intracellular enzyme (EC 3.1.3.9) in cells catalysing the hydrolysis of glucose 6-phosphate to glucose and inorganic phosphate; very active in liver cells but absent e.g. from muscles and the brain.

glucose 1-phosphate, a phosphorylated form of glucose formed from glycogen and starch by phosphorylase, initiating the metabolism of these substances in cell respiration; **glucose 1,6-bisphosphate** is the intermediate in the conversion of glucose 1- phosphate to **glucose 6-phosphate**, catalysed by phosphoglucomutase.

glucose 6-phosphate, a phosphorylated form of glucose formed by the transfer of a phosphate group from ATP in the reaction catalysed by hexokinase or formed from glucose-1-phosphate by phosphoglucomutase; an intermediate substance of → glycolysis and the → pentose phosphate pathway.

glucose-6-phosphate dehydrogenase, G-6-PDH, G6PD, an intracellular enzyme (EC 1.1.1.49) which catalyses the conversion of glucose 6-phosphate to 6-phosphogluconate in the first step of the → pentose phosphate pathway.

glucose-6-phosphate isomerase, → phosphoglucose isomerase.

glucosidases, enzymes catalysing the breaking of the glycosidic bonds (1,4- and 1,6- bonds) e.g. in carbohydrates and thioglucosides, causing the release of terminal glucose. The enzyme group is divided into two types, α-D-glucosidase (maltase, EC 3.2.1.20), and β-D-glucosidase (EC 3.2.1.21) which includes gentiobiase and cellobiase.

glucoside, a → glycoside having glucose as the sugar component.

glucosinolate, a synonym for sulphur-containing glycosides, i.e. → thioglycosides which particularly occur in cruciferous plants (Brassicaceae).

glucostats (L. *recipere* = to receive), glucose sensors, glucose receptors; according to the glucostatic hypothesis of appetite regulation, glucostats are specialized neurones in the brain tissue becoming irritated by changes in the blood glucose level. In vertebrates, glucostats are found in the feeding and satiety centres of the hypothalamus, controlling food intake.

glucosuria (Gr. *ouron* = urine), the presence of glucose in urine.

glucosylation, the transfer of glucose residue, usually from UDP-glucose to another compound.

glucuronic acid, an oxidation product of glucose that in the liver tends to conjugate with various effective endogenic or exogenic compounds (e.g. toxic substances) forming soluble **glucuronides** to be excreted in the urine. Glucuronides are usually physiologically inactive, non-toxic, and easily excreted from the body. Glucuronic acid is also a constituent in → glycosaminoglycans (mucopolysaccharides). Its salts are **glucuronates**. Many cell types have → lysosomes with **β-glucuronidase** (EC 3.2.1.31) activity, i.e. the enzyme which catalyses the hydrolysis of glucuronides. Glucuronic acid is an important constituent of hemicelluloses in plants.

glumes, the bracts enclosing the whole → spikelet of graminoids (family Poaceae).

glutamate (L. *gluten* = glue), a salt, ester, or ion of → glutamic acid, functioning in animals as an excitatory neurotransmitter. In the vertebrate brain and spinal cord it is the main excitatory transmitter. *See* glutamate receptors.

glutamate dehydrogenases, three enzymes (EC 1.4.1.2-4) which in cell metabolism catalyse the deamination of glutamic acid to α-ketoglutaric acid with reduction of NAD^+ or $NADP^+$.

glutamate receptors, neurotransmitter receptors of synaptic membranes in which glutamate serves as the specific excitatory transmitter substance; two types are found in vertebrate synapses, i.e. **metabotropic** and **ionotropic receptors**. The former are serpentine G protein-coupled large proteins which increase intracellular IP_3 and DAG levels (*see* inositol). The inotropic receptors are ligand-gated ion channel proteins which are divided into three subtypes: **kainate receptors**, **AMPA receptors** (alpha-amino-3-hydroxy-5-methyl-4-isoxazole propionate receptors), and **NMDA receptors** (N-methyl-D-aspartate receptors).

glutamic acid, abbr. Glu, glu, symbol E; an amino acid, $HOOCCH_2CH_2CH(NH_2)COOH$, occurring as a component of proteins in all organisms; also occurs as a free amino acid or as → **glutamates** with special functions (*see* glutamate receptors). The radical of glutamic acid, from which one hydroxyl group has been removed, is called **glutamyl** (Glu). Earlier called also glutaminic acid. *Cf.* glutamine.

glutamine, abbr. Gln, symbol Q; glutamic acid 5-amide; 2-aminoglutaramic acid, H_2NCO-$CH_2CH_2CH(NH_2)COOH$; an amino acid present in proteins. Cells can synthesize it from glutamic acid by binding one NH_2 group to the molecule; the energy for the reaction comes from ATP. Many animal and plant tissues are rich in glutamine and it is an important transportable form of nitrogen. **Glutaminase** (EC 3.5.1.2) catalyses the release of ammonia from glutamine e.g. in the kidneys.

glutaminic acid, → glutamic acid.

glutamyl, *see* glutamic acid.

glutaraldehyde, the dialdehyde of glutaric acid, $COH(CH_2)_3COH$; used as a common fixative in electron microscopy. Its -COH groups react with free amino groups (-NH_2) of polypeptide chains in proteins fixing the structure of cells and tissues. It is a good fixative for microtubules and nucleic acids. Glutaraldehyde, like → formaldehyde, is toxic to organisms and is used as a disinfectant. Also called glutaral or pentanedial. *See* fixation.

glutaric acid, pentanedioic acid, HOOC-$(CH_2)_3COOH$, an intermediate in the cellular metabolism of tryptophan and lysine.

glutathione (Gr. *theion* = sulphur), a tripeptide composed of glutamic acid, cysteine, and glycine; it functions as a coenzyme in some enzymes, e.g. glyoxalase, and is an important antioxidant in biochemical defence reactions of cells; **oxidized glutathione** acts in cells as a hydrogen acceptor and **reduced glutathione** as a hydrogen donor. **Glutathione peroxidase** (EC 1.11.1.9) and **glutathione reductase** (EC 1.6.4.2) are enzymes associated with these reactions.

glutelins, storage proteins present in the seeds of wheat and other cereals; dissolve in acids and alkalis.

gluten, a group of storage proteins in cereal grains; e.g. glutenin and gliadin in wheat. The baking properties of a dough are dependent on gluten. Also called wheat gum. Some people suffer from sensitivity to gluten (coeliac disease) with chronic inflammation and atrophy of the intestinal mucous membrane.

glutenin, a glutelin protein in wheat grains.

glutinant, a kind of nematocyst of cnidarians which is characterized by a long thread; on the basis of their thread structures the glutinants are divided into two types: the **streptoline glutinant** with a coiled thread and the **stereoline glutinant** with a straight thread. Glutinants produce a sticky secretion used in locomotion and food capture. *See* penetrant, volvent.

glyc(o)- (Gr. *glykys* = sweet), pertaining to sweetness, glucose; also **gluc(o)-**.

glycan, a polysaccharide. *See* glycosaminoglycan.

glyceraldehyde 3-phosphate, 3-phosphoglyceraldehyde; an intermediate of carbohydrate metabolism in → glycolysis and photosynthesis.

glyceric acid, a three-carbon acid, $CH_2(OH)$-$CH(OH)COOH$, formed by oxidation of → glycerol; its ionic form, salts, and esters are called **glycerates**; glyceric acid occurs as phosphates, i.e. as **3-phosphoglycerate** (→ phosphoglyceric acid) and **1,3-diphosphoglycerate,** in → glycolysis and the Calvin cycle (*see* photosynthesis).

glyceride, an ester composed of → glycerol and organic acids; on the basis of the number of fatty acids linked with the -OH groups of glycerol, the glycerides are called **mono-, di-,** or **triglycerides** (mono-, di-, or triacylglycerols). Neutral fats, serving as storage fats in organisms, are long-chain triglycerides.

glycerin, glycerine, → glycerol.

glycerokinase, glycerol kinase; an intracellular enzyme (EC 2.7.1.30) catalysing the conversion of glycerol and ATP to α-glycerophosphate and ADP; commercially prepared from many microorganisms, such as *Escherichia coli* and *Bacillus stearothermophilus.*

glycerol, glyceryl alcohol; 1,2,3-propanetriol; a colourless, odourless, syrup-like, sweet alcohol, $HOCH_2CH(OH)CH_2OH$. Glycerol is synthesized in cells, and by reacting with fatty acids forms **acylglycerols** (glycerides), which in the hydrolytic degradation are decomposed back into glycerol and fatty acids. Storage fats of organisms are **triacylglycerols** (triglycerides). Glycerol is manufactured by the saponification of fats and is widely used in industry. Also called glycerin(e).

glycine, *abbr.* Gly, gly, symbol G, aminoacetic acid; H_2NCH_2COOH, the simplest amino acid in proteins; acts also as an inhibiting → neurotransmitter in the nervous system of most animals, in vertebrates especially in the spinal cord, increasing Cl^- conductance in → synapses (**glycinergic synapses**). Called also glycocin, glycocoll, gelatin sugar (a major component of gelatin). *See* gamma-aminobutyric acid.

glycocalyx, also **glycocalix** (Gr. *kalyx* = rind, cup of flower), a veil-like layer seen in the electron microscope covering the cell membranes of most eukaryotic cells; it is formed of long molecules of glycoproteins and glycosaminoglycans anchored through the cell membrane onto the → cytoskeleton. With the glycocalyx the cells recognize and contact each other and different other structures and substances outside the cell. *Cf.* lectins.

glycogen (Gr. *gennan* = to produce), animal starch; a long-chain, branched glucose polymer, $(C_6H_{10}O_5)_n$, which is the most important storage polysaccharide in animals. It is synthesized and stored as small cytoplasmic granules, mainly in the liver and skeletal muscle cells, in which it is formed enzymatically from glucose in → **glycogenesis**, and correspondingly decomposed back to glucose in → **glycogenolysis**. Glycogen is also found in many fungi.

glycogenesis (Gr. *genesis* = production), **1.** the formation of → glycogen in animal and fungal cells; a synthesis catalysed by glycogen synthase (EC 2.4.1.11) producing polyglucose with α-1,4-linkages from UDPglucose, or from dextrin by dextranase (EC 2.4.1.2); **2.** more generally, the production of sugars.

glycogenolysis (Gr. *lyein* = to dissolve), the decomposition of → glycogen to glucose occurring in two types of enzymatic reactions, i.e. in **hydrolysis** by hydrolytic enzymes in extracellular digestion and in → lysosomes within cells, and in **phosphorolytic reactions** of cellular energy metabolism when glycogen phosphorylase catalyses the formation of **glucose 1-phosphate**. The reaction in vertebrate cells, stimulated by adrenaline and glucagon and inhibited by insulin, uses ATP as the phosphate and energy source. Only this phosphorylated form of glucose can be used in → glycolysis for cell respiration.

glycolipid (Gr. *lipos* = fat, lipid), a lipid molecule bound to one or more carbohydrates such as galactose, glucose, etc. Glycolipids are usually located in cellular membranes, as e.g. → cerebrosides in the nerve cells.

glycolysis (Gr. *lysis* = dissolution), the stepwise, energy-yielding decomposition of glucose to pyruvate forming the first reaction series in cell respiration, generating two ATP molecules as a net gain. The glycolytic reactions of energy metabolism in the cell cytoplasm occur without the presence of oxygen, and therefore it is also called **anaerobic glycolysis**. However, because of the accumulation of end products, glycolysis cannot usually work for a long time in anaerobic conditions, but needs aerobic reaction systems i.e. → citric acid cycle and electron transport chain, to oxidize pyruvate further. Many animals and plants can use glycolysis for → **anaerobic respiration**, and some microbes for → fermentation. *Syn.* Embden—Meyerhof pathway.

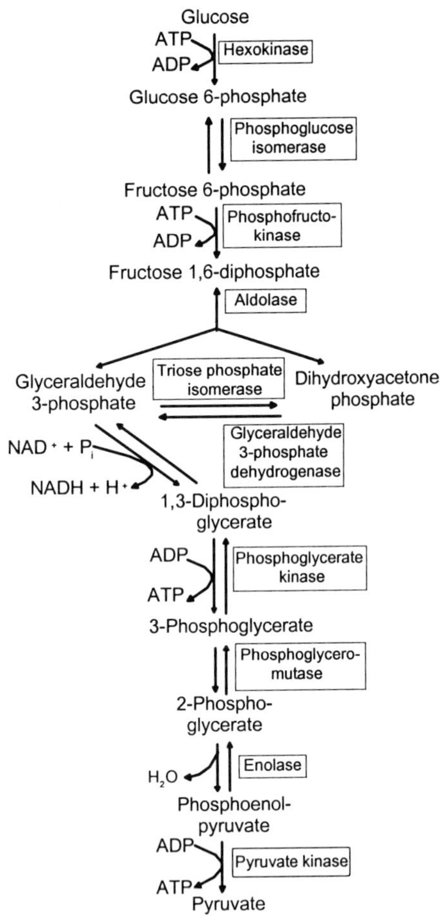

Glucose is converted stepwise into pyruvate (pyruvic acid).

glycopeptide, a compound containing sugar(s) linked covalently to amino acids or peptides, the latter forming the major part; glycopeptides occur e.g. in bacterial cell walls. *Cf.* peptidoglycan.

glycoprotein, a protein bound to a carbohydrate moiety; such conjugated proteins are e.g. mucins, mucoids, amyloid, or chondroproteins; glycoproteins which are very rich in polysaccharides are called proteoglycans.

glycosamine, → glucosamine.

glycosaminoglycan, GAG, any of those heteropolysaccharides composed of repeating disaccharide units containing a derivative of an amino sugar, either glucosamine or galactosamine (usually in sulphuric form). Compounds like chondroitin sulphate, heparan sulphate, and hyaluronic acid are glycosaminoglycans which are constituents of glycoproteins or proteoglycans, occurring e.g. in → mucins, in intercellular substance, and in → glycocalyx of animal cells. Called also glucosaminoglycan.

glycoside, a derivative of a cyclic sugar in which the hydrogen atom of a hemiacetal hydroxyl group (-OH) is replaced by an aglycone, i.e. non-carbohydrate portion (an aryl or alkyl group); *see* glycosidic bond. Usually the sugar moiety is a hexose, sometimes a pentose; if glucose is the sugar component of a glycoside it is also called glucoside. The aglycone can be replaced by a sugar and therefore also oligosaccharides are glycosides. Also oligosaccharides have glycosidic bonds, and therefore considered as glycosides. In plants, aglycone glycosides are important acting as storage substances; many are poisonous serving in defence against herbivorous animals. Many glycosides, such as digitalis glycosides and streptomycin, are used in medicine. Spices, such as vanilla and mustard oils, and many stains, contain glycosides. Glycosides are broken down enzymatically by **glycosidases** (EC subclass 3.2).

glycosidic bond/linkage, a covalent chemical linkage between a carbon-1 atom of a monosaccharide and an oxygen atom of the hemiacetal OH group of another monosaccharide residue or a non-carbohydrate component (aglycone). A glycosidic bond (-C-O-C-) occurs e.g. in di-, oligo- and polysaccharides, also in alcohol-glycosides (glycosido-alcohols). Glycoside bonds with a -C-N- linkage occur in glycosyl compounds such as nucleosides and thioglucosides.

glycosome (Gr. *soma* = body), a → microbody containing glycolytic enzymes.

glycosylation, the process in cells after protein synthesis in which a carbohydrate component (glycosylate) is bound to a protein to form glycoprotein; occurs in the → endoplasmic reticulum and Golgi complex.

glyoxylate cycle, a modified form of the → citric acid cycle occurring in algae, microorganisms and in higher plants. This cycle takes place in the glyoxisomes of plants where lipids are rapidly broken after → beta-oxidation to acetyl CoA, e.g. during the germination of seeds. Isocitrate formed in the glyoxylate cycle is converted to glyoxylate and succinate which is metabolized in the mitochondria through the → citric acid cycle.

glyoxysome, the cell organelle of plants and microbes in which fatty acids are broken in the beta-oxidation pathway and → glyoxylate cycle. *Cf.* peroxisome. *See* microbody.

glyphosate, N-(phosphonomethyl)glycine, a herbicide used to control different types of weeds, both herbaceous and woody plants; inhibits e.g. photosynthesis and the synthesis of aromatic amino acids.

GMP, guanosine monophosphate, *see* guanosine phosphates.

gnathochilarium (Gr. *gnathos* = jaw, *cheilos* = lip), one of the mouthparts of diplopods resembling the labium of insects; is formed by the fusing of paired appendages, especially those of the labial segment.

gnathopods (L. *gnathopodium*, Gr. *pous* = foot), the second and third thoracic appendages in amphipod crustaceans acting as prehensile legs in food handling and copulation.

gnathosome (L. *gnathosoma*, Gr. *soma* = body), the region around the mouth in some arachnids to which oral appendages (chelicerae and pedipalps) are attached. *Syn.* capitulum.

Gnathostomata (Gr. *stoma* = mouth), **gnathostomes**; **1.** a taxonomic group, one of the two superclasses in Vertebrata, in some classifications a subphylum in the phylum Chordata; comprises all the vertebrates except jawless fishes (Agnatha), i.e. includes jawed fishes and all tetrapods; **2.** a superorder of echinoids (sea urchins).

Gnathostomulida (Gr. *stomulus*, dim. of *stoma* = mouth), **gnathostomulids, jaw worms**; an acoelomate invertebrate phylum including about 80 species which live in the interstitial spaces of fine sandy and sulphide-silt shores. The first species was found in 1928 in the Baltic. Gnathostomulids are ciliated worm-like animals, 0.5 to 1 mm in length, and show some structural similarities with turbellarians, first included with Platyhelminthes.

Gnetinae, a class in the subdivision Gymnospermae, division Spermatophyta, Plant kingdom, in some systems classified as a separate division Gnetophyta. The group comprises three living genera, *Gnetum*, *Ephedra* and *Welwitschia*, all having structural features (e.g. real vessels and structure of the flowers), which indicate that they are evolutionally closely related to the more developed → Angiospermae.

gnotobiota (Gr. *gnotos* = known, *bios* = life), the specifically known microbe colonies or species assembled from pure isolates; **gnotobiotic** refers to certain experimental animals reared behind the microbe barrier; in practice gnotobiotic animals are often germ-free or have only a few, known microbe colonies. A microorganism of a gnotobiota is called **gnotobiote**, and the research field dealing with gnotobiota, **gnotobiotics**. *Cf.* germ-free animals.

GnRH, gonadotropin-releasing hormone, → gonadoliberin.

goblet cells, single cells (unicellular glands) in the epithelium of the mucous membrane, secreting mucus; found in vertebrates especially between epithelial cells in the alimentary canal, respiratory tracts, and genital ducts. The shape of the cells is due to mucin that is collected in the upper part of the cell. Similar mucous cells are also found in the skin of lower vertebrates, e.g. in fish.

goitre, goiter, an abnormal enlargement of the thyroid gland in vertebrates; especially in man often due to a shortage of iodine in the normal diet (endemic goitre); **goitrogen** is a factor, usually a substance, that produces goitre.

Goldman's equation (*David E. Goldman*, b. 1911), a formula used to determine the electrical potential **E** (→ membrane potential) occurring across the plasma membrane in a cell, caused together by the distribution of different ions and the permeability of the membrane to each of these ions. In animal cells, in which the most important ions are Na^+, K^+ and Cl^-, the formula is mostly written as follows:

$$E = \frac{RT}{F} \cdot \ln \frac{P_{K^+}[K^+]_o + P_{Na^+}[Na^+]_o + P_{Cl^-}[Cl^-]_i}{P_{K^+}[K^+]_i + P_{Na^+}[Na^+]_i + P_{Cl^-}[Cl^-]_o}$$

P_K^+, P_{Na}^+ and P_{Cl}^- are the **permeability coefficients** of these ions across the membrane, $[K^+]_o$ etc. are the **concentrations** of the ions outside the cell and $[K^+]_i$ etc. the concentrations inside the cell; R = gas constant, T = the absolute temperature in kelvins (K), and F = Faraday's constant. *Cf.* Nernst equation.

Golgi complex/apparatus/body (*Camillo Golgi*, 1843—1926), a complex organelle in eukaryotic cells composed of several flattened membranous sacs (cisternae) with associated vesicles. It is closely related to the → endoplasmic reticulum, and some scientists regard the Golgi complex as a differentiated part of it (Golgi area). The Golgi complex produces cellular secretions (e.g. glycoproteins, mucopolysaccharides) by adding carbohydrate chains to proteins and then packing them into secretion granules. The Golgi complex plays an important role in the synthesis of wall components in plant cells. *Cf.* dictyosome.

Golgi tendon organ, a minute, spindle-shaped mechanoreceptor associated with skeletal muscle tendons; these kinaesthetic receptors are sensitive to stretching of the tendon and function in the control of muscle function inhibiting reflexively motoneurones innervating the contracting muscle, and thus prevent too strong, destructive contractions of the muscle (autogenic inhibition). *Cf.* muscle spindle.

gonad (Gr. *gone* = seed), an animal organ producing germ cells (gametes), i.e. the → testis and → ovary.

gonadoliberin, gonadotropin-releasing hormone/factor, **GnRH**, **GnRF**; a decapeptide hormone of vertebrates produced by special neurones (neurosecretory cells) in the → hypothalamus of the brain. Gonadoliberin is transported to the anterior pituitary gland through hypothalamic portal veins, and stimulates the release of gonadotropins, i.e. → luteinizing hormone (LH) and → follicle-stimulating hormone (FSH) from the pituitary. Thus gonadoliberin acts as both **luliberin** (luteinizing hormone-releasing hormone/factor, LHRH, RHRF) and **folliberin**

(follicle-stimulating hormone-releasing hormone/factor, FSH-RH, FSH-RF), which first were thought to be two separate hormones but later considered as a single hormone.

gonadotropic (Gr. *tropos* = a turning), stimulating the gonads; especially pertaining to gonadotropic hormones (→ **gonadotropins**) of the → pituitary gland regulating the endocrine function of the gonads in vertebrates. Also called gonadotrophic.

gonadotropin, any hormone stimulating the gonads; e.g. → **follicle-stimulating hormone**, **FSH**, and → **luteinizing hormone**, **LH**, which are secreted from the anterior pituitary of vertebrates, and → **chorionic gonadotropin** from the placenta in most mammals. Gonadotropins are also called gonadotropic hormones, formerly also gonadotrophins.

gonadotropin-releasing hormone, GnRH, → gonadoliberin.

gonangium, pl. **gonangia,** Am. also **gonangiums** (L. < Gr. *angeion* = vessel), a reproductive polyp of a hydroid colony. *Syn.* gonozooid.

Gondwanaland (*Gondwana* = a highland in India), a large southern continent in the Jurassic period formed of the breakup of → Pangaea about 200 million years ago; it began to fragment about 135 million years ago into the present continents, i.e. Australia, New Zealand, India, Africa, South America, and Antarctica. *See* Laurasia, continental drift.

gonial apospory (Gr. *gone* = seed, *apo* = away, *sporos* = seed), *see* agamospermy.

gonidial layer, a layer in the structure of → lichens, formed by algal cells, **gonidia**; located mostly near the upper surface of the lichen thallus.

gonidium, pl. **gonidia, 1.** an algal cell, the algal partner in the structure of lichens; gonidia form the → gonidial layer; **2.** an asexual reproductive cell, such as a → gemma of mosses and some algae.

gonimoblast (Gr. *blastos* = germ), **sporogenous filament**; a structure in red algae (Rhodophyta), developed after fertilization from the zygote, producing asexual spores. The gonimoblasts are usually enveloped by special gametophytic filaments which form an enclosed "fruit-like" structure, **cystocarp**.

gonochorism (Gr. *chorismos* = separation), the normal differentiation of sexes into males and females, i.e. into unisexual organisms; opposed to hermaphroditism (→ hermaphrodite).

gonocyte (Gr. *kytos* = cavity, cell), any primordial or immature germ cell.

gonoduct, a tube from the gonad to the exterior, e.g. the opening in the genital plate of echinoids.

gonophore (Gr. *phorein* = to carry along), a sac-like mass of generative cells found in many modified cnidarians, such as in the stalk of colonial hydroids.

gonopod (L. *gonopodium*, Gr. *pous* = foot), → clasper.

gonopore (Gr. *poros* = opening, pore), a genital opening in many invertebrates.

gonotheca (Gr. *theke* = case, chest), a protective cup covering the gonozooid (gonangium) of colonial hydroids; includes a → blastostyle.

gonozooid (Gr. *zoon* = animal, *eidos* = form), **1.** a reproductive polyp; an individual polyp in a hydroid colony for producing generative cells (medusoids) from the blastostyle; *syn.* gonangium. *cf.* dactylozooid; **2.** a type of individual developed from a → probud of doliolids (Thaliacea).

G phase (G < growth) the growth phase of a → cell cycle; in the G_1 phase the cell prepares itself for DNA synthesis, and hence the duplication of chromosomes, and in the G_2 phase the cell prepares itself for division. In the G_0 phase the cell differentiates and stays in a permanent interphase.

G protein (G < guanine), a guanine-nucleotide binding protein; any heterotrimeric protein which mostly binds guanine triphosphate (GTP) in the cell; usually located on the inner surface of the cell membrane. G protein participates in transmembrane signalling, transmitting effects from the receptors of many hormones and neurotransmitters to various reactions inside the cell. In this pathway G protein acts as a GTP hydrolysing enzyme (GTPase activity), or stimulates or inhibits intracellular enzymes like phospholipase C (G_p protein) and adenylate cyclase (G_s or G_i protein); e.g. the retinal G protein (transducin) links the photon receptor (rhodopsin) to cyclic GMP phosphodiesterase.

Graafian follicle (*Reijnier (Regnier) de Graaf*, 1641—1673), the mature → ovarian follicle.

grade (L. *gradus* = step), a taxon or organizational level representing a distinct adaptive line or zone; e.g. penguins with their paddle-like wings and shuttle-shaped body represent a deviating adaptive grade (line) among birds. Concerning structural organization levels, the same grade may be evolved in different lineages; e.g. locomotion in water with the aid of fins found in fish, and fin-like limbs in whales and some extinct reptiles. Also different structural types of sponges (→ Porifera) represent organization grades. *See* convergent evolution.

gradient, the rate of change of pressure, temperature, concentration, or other parameters as a function of distance, time, etc.; e.g. the diastolic pressure gradient between the atrium and the ventricle of the heart, or the concentration gradient and the electric gradient (together with the → electrochemical gradient) across the cell membrane. *See* positional information.

gradient analysis, in ecology, e.g. the analysis of species' composition along a gradient of environmental conditions, such as temperature, moisture, and altitude.

gradient centrifugation, → density gradient centrifugation.

gradual metamorphosis, paurometabolic metamorphosis. *See* metamorphosis.

gradualism, *see* phyletic gradualism.

graft, 1. animal graft, zoograft; a tissue or organ used for transplantation or implantation to an animal; may be corneal graft, dermal graft, fascicular graft, fat graft, nerve graft, etc.; *see* allograft, autograft, isograft, xenograft; **2.** a part of a plant shoot which is joined to the shoot of another individual. **3.** as a verb, to place such a tissue or organ. *See* grafting.

graft hybrid, → chimaera.

grafting, the implanting or transplanting of any tissue or organ to an organism; **1.** in zoology, the implanting of an animal graft or zoograft; *see* graft; **2.** in botany, a method in which a part of a plant shoot (a graft) is joined to the shoot (basic stem) of another individual plant, after which the grafted shoot continues its growth in the basic stem. The method is useful in propagation of plant strains, especially of dicotyledonous trees and conifers. The joining of the cambium layers of two individuals is important, and therefore the method is unsure in monocotyledonous plants because they are lacking a continuous cambium.

gramicidins, antibiotic substances produced by the bacterium species *Bacillus*; form pore-channels in cellular membranes and facilitates the penetration of ions. Because of their ability to kill bacteria by changing the permeability of their cell membranes, they are used as tools (→ ionophores) in cell biological research, and as antibiotic drugs.

Gram staining (*Hans Gram*, 1853—1938), a staining procedure (crystal violet, iodine, and potassium iodide) in which one type of bacteria (**Gram-positive bacteria**) are stained purple, but the other bacteria (**Gram-negative bacteria**) do not stain at all. This occurs because the stain is easily removed by an alcohol wash from the surface of the latter bacteria. In the process Gram-negative bacteria are, however, poststained pink. Gram-positive bacteria (e.g. *Bacillus subtilis*) are more sensitive e.g. to antibiotics than Gram-negative bacteria (e.g. *Escherichia coli*), which have lipoprotein and lipopolysaccharide layers outside the thin peptidoglycan cell wall. Gram-positive bacteria have a thick cell wall composed of peptidoglycan, teichoic and lipoteichoic acids but are lacking the outer layers present in Gram-negative bacteria.

grana (L. *granum*, pl. *grana* = grain), structures formed of stacked, flat and roundish membrane bags in → chloroplasts; the photosynthetic pigments are located in granal → thylakoids and are important in → photosynthesis.

granivore (L. *vorare* = to eat), an animal feeding on grain.

granul(o)- (L. *granulum*, dim. of *granum* = grain), pertaining to a small grain or grain-like structure.

granular cell, a cell type found in the epidermis of vertebrate skin. *See* epidermis.

granule cell, a small nerve cell body seen in the granular layers of the cerebral and cerebellar cortices.

granuloblast (Gr. *blastos* = germ, shoot), an immature precursor cell of → granulocytes in the myeloid tissue of vertebrates; also called **myeloblast**.

granulocytes (Gr. *kytos* = hollow, cell), a type of white blood cell (leucocyte) in the blood of vertebrates; in these leucocytes the cell nucleus is divided into **lobes**, and the cytoplasm contains minute **granules**. According to staining properties of the granules, three different granulocyte types are distinguished: i.e. 1) **neutrophils** (neutrophilic granulocytes, about 50% of all leucocytes), bacteriocidic and phagocytic cells that penetrate from capillary vessels into other tissues; 2) **eosinophils** (acidophilic granulocytes, only a few percent of all leucocytes), the cells which can kill many microbes and are activated e.g. in allergic reactions; 3) **basophils** (basophilic granulocytes, less than 1% of all leucocytes), the cells secreting → histamine and → heparin. The formation of granulocytes is called **granulocytopoiesis** (granulopoiesis); immature precursor cells are **myeloblasts**. *Syn.* granular leucocytes, polymorphonuclear cells, multinuclear leucocytes. *Cf.* lymphocytes, monocytes.

granulocytopenia (Gr. *penia* = poverty, lack), granulopenia, agranulocytosis; the decreased number of → granulocytes in the blood.

granuloplasm (Gr. *plassein* = to form), the granular portion of the cell cytoplasm; especially denotes the **endoplasm** of some protozoans such as amoebas, to be distinguished from the more homogeneous ectoplasm, hyaloplasm.

granulosa cells, the cells of the membrana granulosa lining the → ovarian follicle; the cells produce oestrogen, and after ovulation develop into gluteal cells of the → corpus luteum.

granum, *see* grana.

Graptolita (Gr. *graptos* = painted, written, *lithos* = stone), **graptolites**; fossil invertebrates in the Palaeozoic era resembling the present hydroids. Their systematic position is unclear, earlier they were classified in Coelenterata or Bryozoa, at present they are considered as a class in Hemichordata. Graptolites were most abundant in Ordovician.

grasping reflex, a characteristic reaction of a newborn mammal, especially of those who hang in mother's fur, such as primates, man included; when the palm of a newborn baby is touched, the fingers fasten around the touching object.

grassland, a plant community growing grass, with a few or no trees. These characteristics are mostly caused by climatic factors (cold winter, low annual rainfall) in inland areas. Different types of grasslands in different geographical areas are → savanna, pampa, prairie, steppe, and veld.

gravidity (L. *gravis* = heavy), the state of a mammal, especially a human, when the embryo (later foetus) is developing in the uterus, in man lasts 280 days on average. *Adj.* **gravid**. *Syn.* pregnancy. *See* gestation.

gravitropism (Gr. *trope* = turn), a phenomenon observed in plants comprising growth movements caused by gravity. The shoot is normally **negatively** gravitropic and grows away from the stimulus, i.e. upwards. The root reacts **positively** and grows downwards. Some plant parts grow horizontally like runners, or diagonally like branches and twigs. Gravitropic reactions are based on the location of cell particles and hormone (auxin) gradients in cells leading to an uneven growth. Ca^{2+} ions play an important role in gravitropism because they control the transport of auxin in tissues. An increased concentration of auxin inhibits the growth of roots but stimulates that of stems. *Syn.* geotropism.

gravity (L. *gravitas*), the force of attraction between the Earth (or another planet) and a body within its gravitational field. *See* Appendix 1.

grazer, a herbivorous animal who does not kill its food organisms but uses only parts of them, seldom lethal but often harmful. A grazer may easily change to feed on a new object and later return again to the previous one thus exploiting several organisms during its lifetime; e.g. the insects feeding on leaves, or herbivorous mammals feeding on grass.

gray, Gy (*Louis H. Gray*, 1905—1965), the SI unit for an absorbed dose of ionizing radiation. *See* Appendix 1.

greater circulation, → systemic circulation.

grebes, → Podicipediformes.

green algae, → Chlorophyta.

green flagellates, → Phytomastigophora.

green gland, a pair of excretory glands anterior to the oesophagus in crustaceans, especially in decapods; comprises a **bladder** and an end sac with a small vesicle (**saccule**), and a **labyrinth** connected by a coiled **excretory tubule** to the bladder (urinary bladder). The green glands open either at the base of the antennae (**antennal glands**), or at the base of the second maxillae (**maxillary glands**). The green gland is analogous to the coxal gland of chelicerates.

greenhouse effect, the temperature effect caused by water vapour and the **greenhouse gases** (especially carbon dioxide) in the atmosphere. The gases absorb solar radiation, re-radiated from the earth's surface, and keep the temperature on the Earth tens of degrees (°C) warmer and more stable than it would be without this phenomenon. In a narrower sense, greenhouse effect means the **increase of greenhouse gases** caused by man, and the consequent, slow, continuous increase of temperature on the Earth. The effective gases in addition to carbon dioxide (CO_2) are → ammonia (NH_3), methane (CH_4), nitrogen oxides, and CFC compounds (→ chlorofluorocarbons).

Gregarinia (Gr. *grex* = herd, flock), **gregarines**; a protozoan subclass in the phylum Apicomplexa (class Sporozoa) including rather large, worm-like parasites, some even visible to the naked eye (more than 10 mm in length). They are endoparasites, either intracellular or extracellular, living in many invertebrates. They inhabit especially the intestine and body cavities of annelids and arthropods, producing cysts filled with spores. Some species form chains of individuals.

gregarious, living in groups or flocks.

grey matter/substance, Am. also **gray matter** (L. *substantia grisea*), the grey tissue type of the central nervous system in vertebrates; is composed of nerve cell bodies and unmyelinated nerve branches, chiefly dendrites, with supportive neuroglial cells. The grey matter areas are the cerebral and cerebellar cortices, the brain nuclei (such as basal ganglia) deeper in the brain tissue, and the longitudinal central area of the brain stem and the spinal cord. *Cf.* white matter.

grid, a grating of crossed bars; e.g. **1.** a 200—300 mesh metal net, made e.g. from nickel, diameter usually 3 mm and thickness 0.05 mm; used as a specimen frame in electron microscopy; **2.** an electrode in a vacuum tube, consisting of a coil of wires, parallel wires, or a screen, used for controlling the electron flow between an anode and cathode; **3.** a series of slits or barriers in optical equipments for controlling the formation of the light spectrum; **4.** a firm structure in a crystal formed by atoms and molecules.

gRNA, → guide RNA.

grooming, a common behavioural pattern of animals pertaining to cleaning of the skin, hairs, or feathers of mammals and birds (called **preening** in birds); similar behaviour occurs also in many other animal groups, such as in insects which use e.g. mouthparts as cleaning organs. Besides cleaning, grooming behaviour is also important in maintaining social interactions among animals.

Grooming may be **self-grooming** (autogrooming) performed by an individual upon itself. **Social grooming** is used in many different situations between individuals (*see* allogrooming), especially in supporting the **hierarchic system** in a social group when a subordinate member retards aggressions of a dominant individual by grooming (*cf.* appeasement behaviour). **Mutual grooming** is important as a placatory gesture in primates, but also in many other animal groups.

gross production, the energy used by the individuals of each trophic level of organisms per unit time, comprising the chemical energy assimilated by individuals and their offspring (*see* net productivity) and the energy consumed in respiration. The gross production of the producer level is called **gross primary production** (photosynthetic production of plants) and the gross production of the consumer levels (metabolizable production for animals), **gross secondary production**. *Cf.* net production. *See* productivity.

ground layer, the lowest layer of vegetation in forests, comprising mosses and lichens.

ground meristem (Gr. *merizein* = to divide), those parts of the → primary meristem in plants which produce ground tissues, i.e. different types of → parenchyma. *See* meristem.

ground substance, the homogeneous matrix of intercellular tissue matrix, especially that of connective tissue of animals in which fibres and cells are embedded.

ground tissues, parenchymal tissues. *See* parenchyma.

groundwater, water reserves situated in wide beds or channels beneath the surface of the ground; forms the water source in springs and wells. *Cf.* surface water. *See* hydrological cycle.

group selection, differential proliferation or extinction of groups of organisms by selection, influenced by genetic composition of particular groups. Certain characteristics may be disadvantageous to individuals bearing them, but beneficial to the group in which the characteristics increase in frequency, e.g. → altruistic behaviour in a group of relatives (*see* kin selection). Thus the groups, not the individuals, are the units of selection.

growing form, the outer appearance of a plant; growing forms are e.g. trees, shrubs, dwarf shrubs, and grasses.

growing point, in plants, the tip of a shoot and root, or of their branches with the → meristem tissue in which the division of cells and the growth are concentrated.

growth, 1. enlargement of an organism or its parts; may be enlargement of cells, increased cell number by cell divisions, and/or the accumulation of extracellular matrix; growth is controlled by genes and regulated by many growth factors and hormones, and is also strongly affected by many environmental factors; *see* hypertrophy, proliferation; **2.** an increase in the population density; *see* exponential growth, geometric growth, logistic growth.

growth factors, GF, in animals, protein or polypeptide molecules which stimulate the growth and development of cells, tissues, and organs. After being secreted from certain cells they are capable of binding to **specific receptors** of other cells, promoting their proliferation, i.e. cell division. Sometimes cells, like most tumour cells (tumour growth factors), can promote their own proliferation (autocrine secretion).

Many types of growth factors are found in vertebrates, particularly in embryonic tissues. Their effects are often directed to a special type of tissue causing multiplication and/or development of tissue cells, as e.g. **epidermal growth factor** (EGF) to epidermal cells, **fibroblast growth factor** (FGF) to connective tissue cells, and **nerve growth factor** (NGF) to nerve cells. Many hormone-like substances function as growth factors, e.g. → somatomedins such as **insulin-like growth factors** (IGF I and IGF II) in various tissues, interleukins and other → cytokines in the immune system, and → colony stimulating factors for proliferation of blood cells. Also **growth inhibitory factors** are produced in tissues, such as clonal inhibitory factor (CIF) and proliferation inhibitory factor (PIF), the two lymphokines that inhibit the growth of some vertebrate cells. The difference between the growth factors and hormones is sometimes difficult to assess.

In botany, GFs are factors like the availability of light energy, nutrients, temperature, or humidity of the environment, affecting the development, productivity, or quality of plant products. Also plant hormones are sometimes considered to be growth factors. *Cf.* growth substance.

growth hormone, GH, somatotropin; a peptide hormone of the anterior → pituitary gland of vertebrates; in humans composed of 191 amino acid residues. It is an anabolic hormone stimulating the growth of somatic tissues during the ontogenic development, and is involved in the maintenance and renewal of tissues in adult animals. Hypersecretion during development causes **gigantism** (hypersomia), and in adults **acromegaly** with abnormally enlarged extremities, such as the feet and jaws; its hyposecretion causes **nanosomia**. The effects of the growth hormone on cells are mediated by growth factors called → **somatomedins**, and the influence occurs in several types of tissues, most prominently in bones and muscles. The secretion of growth hormone from the pituitary gland is stimulated by → **somatoliberin** (growth hormone-releasing hormone) and inhibited by → **somatostatin** (growth hormone-inhibiting hormone), which both are produced in the hypothalamic area of the brain. Growth hormone is diabetogenic because it increases glucose output from the liver and has anti-insulin effect on muscle cells.

growth hormone-inhibiting hormone, GIH, → somatostatin.

growth hormone-releasing hormone, GH-RH, GRH, → somatoliberin.

growth period, the period during which plants are capable of net production; in practice, the time during which the daily mean temperature is +5°C or higher. The length of the growth period is important in temperate and cool zones because it determines the amount of the annual vegetation production and e.g. the success of cultivated species and cultivars.

growth rate of population density, *see* biotic potential, innate capacity for increase, environmental resistance.

growth retardants, synthetic substances used to retard the stem growth of cereals and other cultivated plants; one of the best known compounds is CCC, i.e. → chlormequat. Growth retardants also affect the timing of flowering, the development and ripening of fruits, and the improvement of the disease resistance of plants. Other commercial substances are e.g. B-nine, i.e. N-(dimethylamino)succinamic acid, Phosphon-D, i.e. 2,4,-dichlorobenzylphosphonium chloride, and AMO-1618, i.e. 2′-isopropyl- 4′- (trimethylammoniumchloride)-5′-methylphenyl piperidine carboxylate.

growth ring, → annual ring.

growth substance, a term used to refer to plant → hormones and artficial substances which have physiological and morphogenetic effects on plants.

Gruiformes (L. *grus* = crane, *forma* = shape), a diverse, probably polyphyletic order of birds, inhabiting open marshes or prairies (e.g. long-legged cranes) or semiaquatic marshes (e.g. short-legged and medium-sized rails). The order includes 191 species in families such as **cranes** (Gruidae), **rails**, **coots**, and **gallinules** (Rallidae), **bustards** (Otididae), and **button quails** (Turnicidae).

GSR → galvanic skin response.

GT-AG rule (G = guanine, T = thymine, A = adenine), a rule describing the fact that these dinucleotides are the first and the last dinucleotides of the → intron.

GTP, → guanosine triphosphate.

GTPase, → guanosine triphosphatase.

GTP-binding protein, *see* G-protein.

guanidine, imino urea, $NHC(NH_2)_2$; forms a part of some amino acid molecules such as arginine and creatine. Guanidine is a metabolite in some plants and lower animals, but is as such quite toxic to organisms. It is a raw material in industry e.g. for some drugs, synthetic resins, insecticides, and explosives. In molecular biology, guanidine hydrochloride is used for separating RNA from proteins.

guanine, 2-amino-6-oxypurine, $C_5H_5N_5O$, symbol G; one of the two purine bases present in all nucleic acids, i.e. DNA and RNA. It is the silvery substance in the eyes of some animals (e.g. in cats) or in the skin (e.g. in fishes); *see* guanophores; also → guano contains guanine.

guanine-nucleotide binding proteins, a protein family occurring as complexes with guanine nucleotides; become activated by guanosine triphosphate (GTP). The proteins, such as → G proteins and transducer molecules, are involved in intracellular signalling from cell-surface receptors of primary messenger substances to second messengers by activating certain enzymes.

guano (Peruvian *huanu* = dung), a natural deposit composed of excrements of seabirds and fish remains on the islands of the western coast of South America; contains large amounts of calcium phosphate, purines such as guanine, and → uric acid and its decomposition products, and is therefore commonly used as a fertilizer.

guanophores (Gr. *pherein* = to carry), cells found in the skin of many fish species, containing light-reflecting guanine granules which give the skin a metallic silver or gold luster. *Syn.* iridocytes, iridophores.

guanosine, a → nucleoside; chemically 9-β-D-ribofuranosylguanine, composed of → guanine and ribose; major constituent of RNA and guanine nucleotides.

guanosine phosphates, nucleotides composed of guanine, ribose (or deoxyribose) and one, two, or three phosphate groups. According to the number of phosphate groups they are called **guanosine mono-**, **di-**, or **triphosphates**, i.e. **GMP**, **GDP**, **GTP** (or dGMP, dGDP, dGTP). GMPs (3'-phosphate and 5'-phosphate and their mixture) are often called **guanylic acid** (guanylate). Guanosine phosphates are structural components of nucleic acids and are important for energy metabolism and protein synthesis. In some cells GTP may change into **cyclic guanosine monophosphate, cGMP**, which takes part in the transmission of some hormonal signals within cells. *See* guanosine triphosphate, G protein. *Cf.* adenosine phosphates.

guanosine triphosphatase, GTPase, one of → guanine nucleotide-binding proteins, activated by guanosine triphosphate (GTP) and inactivated when the high-energy phosphate is transferred from the molecule; important in activation of many enzymatic cellular processes; the group includes e.g. → G proteins.

guanosine triphosphate, GTP, guanosine 5'-triphosphate; a high-energy nucleotide that is a precursor in nucleic acid synthesis and is involved in protein synthesis, microtubule assembly, and signal transduction (*see* G protein) in cells. *See* guanosine phosphates.

guanylate cyclase, guanyl(yl) cyclase, EC 4.6.1.2; an enzyme mainly associated with the cell membrane in some animal tissues, producing cyclic guanosine monophosphate (**cGMP**) from guanosine triphosphate (**GTP**), but is also found free in the cytoplasm. Like → adenylate cyclase, it acts as a link in some cellular control mechanisms, such as in the action of the atriopeptide hormone released from the vertebrate heart.

guanylic acid, guanosine monophosphate. *See* guanosine phosphates.

guard cells, two specialized cells surrounding each → stoma in the plant epidermis; regulate the opening (in the light) and closing (in the dark) of the central pore of the stoma. The regulation is based on the changes of the cellular → turgor pressure and is dependent on a complicated ion transport system in the plasma membrane of the guard cells.

guide RNA, gRNA a small molecule of ribonucleic acid (RNA) from which a → cryptogene receives information for the editing of messenger RNA (mRNA). These gRNAs are encoded at some other region of the genome than the cryptogene itself. Guide RNA molecules are found in kinetoplasts (mitochondria) of trypanosome parasites. *See* RNA editing.

guild, a group of species exploiting the same environmental resources simultaneously so that they are competing with each other either occasionally or continuously.

gum arabic, gum acacia; a water-soluble, sticky substance obtained from the stems and branches of some *Acacia* species; used as a mucilage, emulsifier, food thickener, and in inks and pharmaceutical preparations (in tablets).

gustatory (L. *gustare* = to taste), pertaining to the sense of taste.

gut, → alimentary canal.

guttation (L. *gutta* = drop), the means by which plants remove excess water, forced by the action of root pressure under humid conditions in which transpiration is not possible. The water is removed from → vessels via → hydathodes on the margins of leaves. In the process, also some salts may be excreted with the water.

Gymnomycota (Gr. *gymnos* = naked, *mykes* = fungus), a division in the kingdom Fungi, including the subdivisions → Acrasiogymnomycotina and → Plasmodiogymnomycotina, the latter including the well-known group of slime moulds, → Myxomycetes. In some classifications the group is divided into the divisions Acrasiomycota, Dictyosteliomycota, and Myxomycota.

Gymnophiona (Gr. *ophis* = snake), **Apoda, caecilians**; an amphibian order comprising limbless, worm-like species which live in tropical regions, mainly burrowed in the soil; some species are aquatic. The order contains about 150 species in five families, some of which include animals more than 1 m in length.

Gymnospermae (Gr. *sperma* = seed), **gymnosperms**; a subdivision in the division Spermatophyta, seed plants; they have no ovaries but the ovules and seeds are lying on the surface of the sporophylls. Gymnospermae are divided into seven classes: Pteridospermae (seed ferns), Cycadinae (cycads), Gnetinae (e.g. *Ephedra* and *Welwitschia*), Bennettitinae, Cordaitinae, Ginkgoinae (maidenhair trees), and Coniferae, including conifers (order Pinales). In some classifications the group gymnosperms has no exact taxonomical status.

gynae(o)-, gynaec(o)-, gyne(o)-, gynec(o)- (Gr. *gyne* = woman), denoting relationship to a female or woman.

gynaecium, gynoecium, the female reproductive structure of a flower; may be composed of one or more → carpels, which may be separate or grown together. *Syn.* pistil.

gynandromorph (Gr. *aner*, gen. *andros* = man, *morphe* = form), a genetically mosaic individual whose tissues are partly female and partly male. Gynandromorphs are found in animals (e.g. insects) which have a cell autonomous mechanism of → sex determination, and they are borne from XX zygotes which lose one X chromosome in an early mitotic cleavage; hence, XX and XO cells and consequently female and male tissues, will arise respectively. In insects, which have a → haplodiploid mechanism of sex determination, gynandromorphs can arise as a consequence of double fertilization. Also other mechanisms which can lead to the formation of gynandromorphs are known. *Cf.* intersexual individual.

gynandrous, pertaining to a flower having the stamens fused with the gynaecium; e.g. many orchids.

gynandry, 1. hermaphroditism; *see* hermaphrodite; **2.** pseudohermaphroditism of the female animal, or the manlikeness of a woman.

gynobasic, pertaining to a style arising from the base of the ovary in plants.

gynodioecious, gynodiecious (Gr. *dis* = twice, *oikos* = house), pertaining to a plant having female and bisexual flowers in separate individuals. *Cf.* gynomonoecious, androdioecious.

gynoecium, → gynaecium.

gynogenesis (Gr. *gennan* = to produce), the parthenogenetic development of the → ovum or → egg cell after being penetrated but not fertilized by sperm or pollen. *See* parthenogenesis.

gynogenote (Gr. *genos* = parentage, origin), an individual which is experimentally produced by removing the male → pronucleus from the fertilized ovum, and replacing it with a female pronucleus derived from a different embryo. Thus the gynogenote has two female nuclei and two maternal sets of chromosomes.

gynomonoecious (Gr. *monos* = alone, single, *oikos* = house), pertaining to a plant having female and bisexual flowers in the same individual. *Cf.* gynodioecious.

gynophore (Gr. *phorein* = to carry along), a structure in a flower, in which the filaments of the stamens and the style of the gynaecium form a united column having the anthers and stigmata in its tip.

gynosporangium, → macrosporangium.

gynospore, macrospore. *See* macrosporangium.

gynostemium, pl. **gynostemia** (Gr. *stemon* = thread), a flower structure formed from the composed gynaecium (with two stigmata) and the stamen (bearing two club-like pollinia); typical of orchid flowers.

gyrase (L. *gyrare* = to turn, gyrate), a → topoisomerase (type II) of *Escherichia coli* with the ability to introduce negative supercoils into DNA.

gyration, revolution in a circle; rotation. *Verb* to **gyrate**, to turn, revolve. *Adj.* **gyratory**, moving in a circle or spiral; pertaining to a convoluted structure.

gyrogonite (Gr. *gonos* = offspring, seed), a fossilized oogonium with the surrounding sheath cells of the stoneworts (Charophyceae).

gyromitrin (L. *Gyromitra esculenta* = false morel), monomethylhydrazine toxin; a very poisonous compound occurring in fresh false morels; the approximate LD_{50} value for man is 60 mg per kg body weight. Gyromitrin is a water-soluble substance and can be removed from fruit bodies by boiling and rinsing them several times.

gyrus, pl. **gyri** (L. < Gr. *gyros* = circle), one of the folds of the surface of the cerebral or cerebellar cortices in many vertebrates. *Cf.* sulcus.

H

H, symbol for **1.** hydrogen (L. *hydrogenium*); **2.** histidine, histone, or histamine; **3.** humidity; **4.** henry (unit of inductance).

***H*,** symbol for **1.** enthalpy; **2.** dose equivalent.

h, symbol for **1.** hour (L., Gr. *hora*); **2.** hect(o)- (Gr. *hekaton* = 100).

***h*,** symbol for → Planck's constant.

habenula, pl. **habenulae** (L. dim. of *habena* = strap, band), a cord-like structure; e.g. **1.** a stalk of the pineal gland and the adjacent neuronic cell mass on the epithalamus (pineal habenula, pedunculus of the pineal body); **2.** the inner part (*habenula arcuata*) or outer part (*habenula pectinata*) of the basilar membrane of the cochlea in the vertebrate ear.

Haber—Bosch process (*Fritz Haber*, 1868—1934; *Carl Bosch*, 1874—1940), a technical process for producing **ammonia** for nitrogen fertilizers from hydrogen and atmospheric nitrogen gas: $3\ H_2 + N_2 \longrightarrow 2\ NH_3$. Nitrogen is bound to hydrogen at elevated temperature and high pressure.

habitat (L. *habitare* = to inhabit), a native environment of an organism, i.e. such a place in the environment which an organism requires for successful existence. Usually the habitat is characterized by a dominant vegetation (→ dominant species) and physical structures, often in different vertical levels of vegetation. The soil in a coniferous forest, a shrub layer, a foliage of a coppice, a spring, or a sandy shore are examples of habitats. A habitat may be *predictable*, such as seasonal, with regular changes of favourable and unfavourable periods, or it may be *unpredictable* when favourable periods of variable duration are interspersed with variable, unfavourable periods (*cf.* biotope). The term **habitat breadth** means a variety of different habitats inhabited by a certain species or a population. Thus there are **pretentious** and **unpretentious** species in respect of their habitat demands. *See* patchy habitat.

habituation (L. *habituare* = to accustom, to habit), becoming accustomed to something; a process of forming a habit; e.g. **1.** in zoology, especially means the gradually weakening response of an animal to a repeated harmless stimulus, and extinction of a conditioned reflex by repetition of the stimulus. The habituation is an adaptive process in the central nervous system (not sensory adaptation) and thus a process of the most primitive and common type of learning; **2.** in medicine, psychological habituation to a drug, i.e. psychic dependence on the repeated consumption of a drug to maintain a feeling of well-being; **3.** in plant physiology, habituation of cultured cells to a growth medium, i.e. cultured cells which usually require hormones to be added into the growth medium may become capable of synthesizing them. *Verb* to **habituate**.

habitus, pl. **habitus** (L. = appearance, condition, characteristic), the appearance of an organism; the body build.

hadal zone (Gr. *hades* = unseen), the deepest zone in oceanic trenches below 6,000 m.

hadrom (Gr. *hadros* = thick, bulky), the tracheary elements (tracheids or tracheae) in the → xylem tissue, excluding parenchyma cells and supporting cells (fibres and sclereids). *Cf.* leptom.

Haeckel's postulate (*Ernst Haeckel*, 1834—1919), recapitulation theory. *See* recapitulation.

haem-, haemo-, haemato-, Am. **hem-, hemo-, hemato-** (Gr. *haima* = blood), pertaining to blood. *Adj.* **haemal** (hemal).

haem, Am. **heme,** a → porphyrin, containing ferrous iron, Fe(II), as its central atom; an active part (prosthetic group) in → haemoglobin and some other blood pigments, in myoglobin of muscle cells, in some → peroxidases, and in → cytochromes of mitochondria in all cell types of organisms. In cytochromes and peroxidases the state of iron oscillates between Fe(II) and Fe(III), i.e. Fe^{2+} and Fe^{3+}, during reduction and oxidation; the oxidized form with ferric iron is called **haematin (hematin)**.

haemagglutination, Am. **hemagglutination** (L. *agglutinare* = to glue), the agglutination or clumping of blood cells, especially of erythrocytes; may be caused e.g. by antibodies, **haemagglutinins**, which are used for testing blood groups.

haemal, Am. **hemal, 1.** pertaining to the blood or blood vessels; **2.** pertaining to the ventral region of the vertebrate body where the heart and great blood vessels are located; **3.** referring to the → haemal system of echinoderms.

haemal system, Am. **hemal system,** a scattered and poorly observable vascular system of echinoderms comprising rather large **sinuses** and **perihaemal canals**. It has little or nothing to do with the circulation of blood fluids that occurs mainly by means of peritoneal cilia. The haemal system may be useful in distributing digested products but its specific function is not known. Also called haemal strands.

haematin, Am. **hematin,** methaem (metheme); a → porphyrin, containing ferric iron, Fe(III), as its central atom; produced in oxidation of → haem. *See* methaemoglobin.

haematoblast, Am. **hematoblast,** → haemocytoblast.

haematocrit, HCT, Hct, Am. **hematocrit** (Gr. *krites* = judge), **1.** the percentage of the volume of packed (centrifuged) red blood cells of the total blood volume of a sample; **2.** a centrifuge or a tube with graduated markings used for such a separation of red blood cells.

haematocyte, Am. **hematocyte,** → haemocyte.

haematogenesis, Am. **hematogenesis** (Gr. *genesis* = production, origin), the formation of blood cells, → haemopoiesis. *Adj.* **h(a)ematogenous.**

haematoma, Am. **hematoma,** a localized collection of blood in a tissue bled from a broken blood vessel.

haematopoiesis, Am. **hematopoiesis,** → haemopoiesis.

haemerythrin, haemoerythrin, Am. **hemerythrin** (Gr. *erythros* = red), a red, non-haem → blood pigment (respiratory pigment) present in the plasma or blood cells of sipunculids, priapulids, some polychaetes, and brachiopods from the genus *Lingula*; contains iron which is not bound to the haem group, as it is in most blood pigments, but directly to the protein moiety. The oxygen-carrying capacity of haemerythrin is poor. *Cf.* haemocyanin, chlorocruorin.

haemin, Am. **hemin,** a salt (usually hydrochloride) of → haematin.

haemocoel, Am. **hemocoel** (L. *haemocoeloma* < Gr. *haima* = blood, *koilos* = cavity), **1.** that part of the embryonic coelom of vertebrates which includes the developing heart; **2.** the body cavity of many invertebrates, such as molluscs, arthropods and tunicates, which takes part in the circulation; in these animals the circulation is open, the blood around the internal organs flowing outside the blood vessels in sinuses of the haemocoel; also **haemocoelom**; *cf.* coelom.

haemocyanin, Am. **hemocyanin** (Gr. *kyanos* = blue), a non-haem blood pigment in the plasma of some invertebrates such as molluscs and arthropods; the protein molecule have two copper atoms to hold one oxygen molecule. In the oxygenated form it is blue, but turns colourless when the oxygen is dissociated.

haemocyte, Am. **hemocyte** (Gr. *kytos* = cavity, cell), any blood cell; especially in vertebrates also called h(a)ematocyte.

haemocytoblast, Am. **hemocytoblast** (Gr. *blastos* = germ), any of the haematopoietic stem cells of vertebrates from which the blood cells develop in various cell series; derived from embryonic mesenchyme. *Syn.* h(a)emoblast, h(a)ematoblast.

haemocytometer, Am. **hemocytometer,** a device for counting blood cells.

haemodialysis, Am. **hemodialysis** (Gr. *dia* = through, *lysis* = dissolution), purification of blood from its waste substances in a special apparatus (**haemodialyzer**) by virtue of their diffusion difference through a semipermeable membrane.

haemodynamics, Am. **hemodynamics** (Gr. *dynamis* = power), the study of forces prevailing in the blood circulation.

haemoglobin, Hb, Am. **hemoglobin,** the respiratory pigment (→ blood pigment) of vertebrates transporting oxygen to tissues. It is packed into red blood cells, but many animals also have it in red muscle cells in addition to → myoglobin. Haemoglobin is composed of a globular protein molecule with four polypeptide chains (globin subunits) and of four → haem molecules, of which each can non-covalently bind one oxygen molecule (O_2). Oxyhaemoglobin (HbO_8) is bright scarlet but turns lilac when deoxygenated.

The **oxygen capacity** (on average 200 ml O_2/l blood in man) depends on the Hb content in the blood. **Oxygen association** onto Hb depends on the partial pressures of oxygen in alveolar air of the lungs or in water surrounding the gills, and **oxygen dissociation** from Hb depends on oxygen pressures in different peripheral tissues. These properties of Hb, however, change according to the physiological state when local concentration of carbon dioxide, acidity (pH, *see* Bohr effect), or 2,3-diphosphoglycerate produced by tissues, vary in the blood; *cf.* carbamino-haemoglobin, carboxyhaemoglobin, methaemoglobin. Erythrocruorin of some invertebrates are structurally quite similar to Hb and therefore sometimes called invertebrate haemoglobin.

Haemoglobin is found also in plant root nodules, as so-called → leghaemoglobin, binding oxygen that tends to inhibit → nitrogen fixation in the nodules. *See* thalassaemia, sickle-cell anaemia.

haemolymph, Am. **hemolymph** (L. *lympha* = springwater), the circulating fluid of invertebrates (e.g. arthropods and molluscs) whose circulatory system is open; flows through the → haemocoel and therefore has properties of blood and of a lymph-like interstitial fluid.

haemolysis, Am. **hemolysis,** pl. **h(a)emolyses** (Gr. *lyein* = to dissolve), the membrane disruption of red blood cells resulting in haemoglobin liberation from cells into an extracellular medium, or *in vivo* into the blood plasma; may be caused e.g. by bacteria (bacterial haemolysis) or by activation of certain immune reactions (immune haemolysis), or by a hypotonic solution *in vitro*; **haemolysin**, a substance that causes the haemolysis.

haemometer, Am. **hemometer,** an apparatus for measuring blood haemoglobin. *Syn.* haemoglobinometer.

haemophil(e), Am. **hemophil(e)** (Gr. *philein* = to love), **1.** a microbe that grows best in the blood or in media containing haemoglobin; *adj.* **haemophilic**; **2.** a haemophiliac person; *see* haemophilia.

haemophilia, Am. **hemophilia,** a genetic disease in man showing → sex-linked inheritance, characterized by internal bleeding that does not stop even in small wounds. Two types of haemophilia are known, type A and a milder type B. The former is caused by the deficiency of the blood clotting factor VIII in the blood plasma and the latter by the lack of factor IX. Factor VIII is antihaemophilic globulin, and factor IX is plasma thromboplastic component (PTC). Haemophilia A and B are not allelic. *Adj.* **haemophilic**, pertaining to haemophilia; **haemophiliac**, a person suffering from haemophilia.

haemopoiesis, Am. **hemopoiesis** (Gr. *poiesis* = making), blood formation, especially the formation of blood cells. In adult vertebrates the blood cells are formed from stem cells in the red bone marrow (myeloid tissue), the red cells (erythrocytes) in one stem cell line, and the white cells (leucocytes) in their own stem cell line differentiating later into several cell types. The former process is called **erythropoiesis** and is regulated by the hormone → **erythropoietin**. In the vertebrate embryos, erythropoiesis occurs first in the yolk sac and later in the liver, spleen and bone marrow. Fishes have no bone marrow and they have special myeloid tissue organs on the stomach (Leydig organ) and gonads (epigonal organ). Plasma proteins are mainly formed in the liver of all vertebrates. *Syn.* haematopoiesis.

haemorrhage, Am. **hemorrhage** (Gr. *rhegnynai* = to break, burst), a bleeding somewhere in the tissue; as a verb, to bleed. *Adj.* **haemorrhagenic**, causing haemorrhage; **haemorrhagic**, characterized by haemorrhage.

haemosiderin, Am. **hemosiderin** (Gr. *sideros* = iron), aggregated deposits of ferritin molecules in cells of vertebrate tissues, especially abundant in liver, bone marrow, and spleen cells. It is a yellowish storage form of iron in lysosomal membranes, forming intracellular granules consisting of proteins and polysaccharides with bound iron (about 50%). *See* apoferritin.

Haemosporina, Haemosporidia, (L. *sporidium*, dim. of *spora* = seed), a protozoan group in the class Coccidia, class Sporozoa, comprising unicellular animals which parasite especially in blood cells of homoiothermic vertebrates; include e.g. plasmodians (the family Plasmodidae) which cause malaria in humans.

haemostasis, Am. **hemostasis** (Gr. *stasis* = halt), the interruption of the blood flow in a vessel, or the arrest of bleeding; **haemostat**, an agent used to produce haemostasis. *Adj.* **haemostatic**.

haemovanadium, Am. **hemovanadium,** the green → blood pigment found in the **vanadocytes** (blood cells) of tunicates; contains vanadium combined with a protein. Probably it does not function as a respiratory pigment.

hagfishes, *see* Pteraspidomorphi.

hair, a general term for a slender, flexible or stiff filament on the surface of a cell or an organ of an animal or a plant.

Hairs on the skin of animals are produced by epidermal cells. The hairs (*pili*) of mammals are cornifications of the skin epidermis, formed of specialized scale-like keratinocytes which form a tubular structure, then cornify and die. Thus the hair grows in the proximal end by the activity of these living keratinocytes which form the **hair bulb** (*bulbus pili*) surrounded by the **hair follicle** (*folliculus pili*). The **hair papilla** (*papilla pili*) with nerves and blood vessels intrudes from dermis (corium) into the hair bulb. The pore of a sebaceous gland opens to the follicle, and a small smooth muscle (erector pili muscle, *musculus arrector*) is attached to the base of the hair. Pigment cells in the bulb produce → pigments (mostly melanin), which, when transferred to keratinocytes bring about a hair colour.

In mammals, there are several hair types distinguished on the basis of their functions, e.g. a dense fine **underhair** for body insulation, heavier long **guard hairs** for protection against wear. Around the nose of carnivores and rodents are long sensory **vibrissae**. Spiny anteaters, hedgehogs and porcupines are covered by sharp pointed **quills** (modified hairs). Minute cellular hairs serve as mechanical sensory receptors in → hair cells of auditory and equilibrium organs in vertebrates and invertebrates, or in hair → sensillae of vibration senses in invertebrates.

Plant hairs are mostly appendages in the plant epidermis, whose structure and functions vary greatly. The simplest hairs are formed by one cell, e.g. → **root hairs**. Other hairs are divided into **trichomes** and **emergences**. Trichomes may be non-excreting or excreting; they may be unbranched, or branched as stellate hairs, vesicles, hooks, spines, etc. The best known secretory trichomes are the glandular hairs in the family Lamiaceae, many species of which are known as herbs and used as spices, like mint. The structure of emergences is complicated; typical are the tentacles of *Drosera* with specialized cells and tissue under the epidermis.

hair cell, a type of sensory cell with hair-like processes, functioning as a mechanoreceptor in vertebrates and many invertebrates. Many stiff **stereocilia** stand on the free surface of the hair cell, often with a more flexible **kinocilium**. The bending of the cilia causes the formation of a receptor potential initiating the sensory processes in the sense organ. Hair cells are found e.g. in the cochlear and vestibular labyrinth, i.e. in the organ of Corti and in the utricle and saccule in the internal ear of vertebrates, and in acoustic and vibration receptors of invertebrates.

hairpin loop, a loop-shaped formation of a single-stranded DNA or RNA forming a two-stranded hairpin-like region when the complementary bases in the loop pair with each other.

hairworms, → Nematomorpha.

Haldane's dilemma (*John Haldane*, 1892—1964), the difficulty in explaining the high degree of → polymorphism of populations without including a heavy → genetic load caused by polymorphism; according to Haldane, the costs of natural selection in a population are heavier than the population can tolerate.

Haldane's rule, a rule stating that the hybrid individuals who die during development, or appear rarely or sterile among the crossed offspring of two species or races, represent the → heterogametic sex, i.e. males in mammals and dipteran insects, females in birds and butterflies.

half-chromatid (Gr. *chroma* = colour), the longitudinal subunit of the metaphase chromosome; according to an outdated conception, the chromatid consists of two half-chromatids. *See* mitosis.

half-life, 1. physical half-life, symbol $T_{1/2}$ or $t_{1/2}$, a period in which the → radioactivity of a substance decays to half of its original activity; **2. biological half-life**, a period in which the concentration of a radioactive substance, antibody, hormone, drug, etc. in an organism decreases to half of its original concentration; **3. effective half-life**, the sum of the inverses of physical or biological half-lives. *Cf.* half-time.

half-rosette plant, a plant which forms a leaf rosette near the ground but long internodes in the upper part of its shoot; e.g. *Thlaspi* species.

half-sibs, half-siblings, individuals who share only one parent.

half-tetrad (Gr. *tetra* = four), one half of a quadripartite structure; e.g. in genetics, two of the four chromatids of a → bivalent. *See* tetrad.

half-time, the time during which one half of a substance is converted in a chemical reaction. *Cf.* half-life.

half translocation, a chromosome mutation which is one half of a → translocation, i.e. a half of a reciprocal exchange of parts of non-homologous chromosomes. Two processes can lead to the formation of a half translocation: 1) one of the two reciprocal fragments of a chromosome does not fuse, and hence an incomplete translocation is formed; 2) a half translocation arises as a consequence of → segregation in a heterozygote for a symmetrical translocation whereby only one of the two reciprocal exchange products occurs in the gamete.

Haller's organ (*Albrecht von Haller*, 1708—1777), a bundle of stiff hairs, setae, on the most distal segments of the two anterior pairs of legs in some ticks, e.g in the family Ixodidae; contains chemoreceptors.

hallucination (Gr. *alyein* = to be distraught, to wander in the mind), a sensory experience in humans without a source in the external world, i.e. the activation of a sensory brain area, not caused by external stimuli but induced in the central nervous system; may be related to different sensations, such as **auditory**, **gustatory**, **olfactory**, **tactile**, and **visual hallucinations**. Probably hallucinations also may occur in other mammalian species with highly developed brain functions.

hallux, pl. **halluces** (L., Gr. = big toe), hallus; the innermost digit (big toe) in the hindlimb of tetrapod vertebrates; turned backwards in most birds. *Cf.* pollex.

hal(o)- (Gr. *hals* = salt), pertaining to a salt or halogen.

Halobacteriaceae, a family of aerobic bacteria which can live in environments of high salinity.

halobiont, *see* halophilic.

halocline (Gr. *klinein* = to lean), a zone of sharp vertical increase in salinity, occurring in salt or brackish waters of stratified water bodies, as in many estuaries; e.g. in the Baltic Sea, the halocline exists between the fresh water or light salt water epilimnion and the more saline hypolimnion. *Cf.* thermocline.

halogens (Gr. *gennan* = to produce), chemical elements forming the group 7A in the periodic table: include → chlorine, fluorine, bromine, iodine, and astatine (unstable). Except astatine, halogens occur in natural conditions as salts.

halon, a halogenated alkane that resembles → chlorofluorocarbons (CFC); halons are unreactive compounds used e.g. in fire extinguishers. They contain → halogens which are harmful environmental pollutants; e.g. bromine decomposes atmospheric → ozone 3 to 10 times more effectively than chlorine.

halophilic (Gr. *philein* = to love), salt-loving; pertains to an organism which can grow in, or even favour salty habitats like e.g. seashores; such an organism is called **halobiont** (halophile), e.g. many microbes and some crustaceans; halophilic plants are called **halophytes**.

halophyte (Gr. *phyton* = plant), *see* halophilic.

halothane, 2-bromo-2-chloro-1,1,1-trifluoroethane; an inhalation anaesthetic used for general anaesthesia of air-breathing vertebrates, especially mammals.

halter(e) (Gr. *halter* = leap, jump), one of the two balancers, i.e. the knob-pointed, reduced hindwings of dipterans; oscillate rapidly during flight serving as balancing organs. The knobs have sensory hair cells for "measuring" air flow.

hamathecium, pl. **hamathecia** (L. *hamatus* = hooked, Gr. *theke* = box), → pseudoparenchyma occurring in → Ascomycetes, separating the asci from each other. The formation site and direction of growth vary, the different types being **paraphyses**, **paraphysoids**, **pseudoparaphyses**, **periphyses**, and **periphysoids**.

Hamilton's rule (*W.D. Hamilton*, b. 1936), a rule associated with the theory of sociobiology, especially with the heredity of → kin selection. According to the rule, → altruistic behaviour is evolutionary profitable if an altruistic individual gains sufficient advantage to its own relatives. Profits and costs are measured as changes of fitness.

Hamilton—Zuk hypothesis, a hypothesis presented in 1982 by *William Hamilton* and *Marlene Zuk* claiming that the → sexual selection made by females of a population reduces the number of individuals which are more sensitive to parasites with long-lasting effects on a phenotype. For example, a male bird infected by Haemosporidia may sing more softly and have a more obscure plumage than that of an uninfected male, clearly diminishing his chance to get a reproductive partner.

hammer, → malleus.

hamulus, pl. **hamuli** (L. dim. of *hamus* = hook), a hook-like structure; e.g. **1.** a two-forked appendage on the 3rd abdominal segment of springtails (Collembola) to engage the → furcula (springing organ); **2.** in hymenopterans, the two wings are connected by the hamuli, which on the anterior border of the hindwing attach to a fold of the posterior border of the forewing; **3.** a minute hooklet (barbicel) on the barbule of the contour feather of birds; holds opposing rows of barbules loosely together.

handling time, a term in animal ecology associated with the → optimum foraging theory; it is the time which a predator spends in pursuing, killing, cutting, and swallowing a prey item. *See* searching time.

hapaxanthous plant (Gr. *hapax* = once, *anthos* = flower), a plant which flowers only once and dies after the flowering and ripening of the seeds. There are three types of hapaxanthous plants: 1) **annual** plants, monocyclic plants, may germinate in the spring and produce seeds in the autumn (e.g. *Capsella*, shepherd's purse, and *Thlaspi*, penny-cress), or germinate in the autumn and produce seeds during the next growing season (*Viola tricolor*, heartsease, and *Myosotis arvense*, forget-me-not); 2) **biennial** plants germinate in the spring, form a wintering leaf rosette and flower, and produce seeds during the next season (*Verbascum*, mullein); 3) **pluriens** grow many years before flowering and die after that (*Agave*).

haplobiont, an organism not exhibiting a regular alternation of haploid and diploid generations during its life cycle. *Cf.* diplobiont.

haplochlamydeous (Gr. *chlamys* = mantle), pertaining to a flower which has only one whorl in its perianth, i.e. only one calyx or one corolla. *Syn.* monochlamydeous.

haplodiploid (Gr. *haploos* = onefold, *diploos* = twofold), a species the sexes of which differ so that the male is → haploid and the female → diploid; e.g. bees and wasps.

haplodiplont, diplohaplont, → diplobiont.

haploid, pertaining to or being a cell or an individual which carries a single set of chromosomes. Viruses and bacteria, the genomes of which consist of one → linkage group only, are regularly haploid. The haploid phase of the life cycle of eukaryotic organisms is called → haplophase, the chromosome number being n, varying with → diplophases, 2n; eukaryotes may also be → polyploid.

haploid parthenogenesis (Gr. *parthenos* = virgin, *genesis* = birth, origin), in botany, the development of an embryo from a haploid, unfertilized egg cell into a haploid gametophyte; is exceptional and has only minor evolutionary impact. *Cf.* apomixis.

haploidization, the change of → diploid cells into → haploid; found in certain fungi occurring so that the number of chromosomes is reduced step by step, i.e. one chromosome pair at a time, during several cycles of cell divisions. Otherwise the haploidization occurs normally in → meiosis in which the whole chromosome number is reduced in two subsequent meiotic divisions.

Haplomastigomycotina, *see* Mastigomycota.

haplont, 1. an organism with a life cycle in which → meiosis follows immediately after → karyogamy in fertilization. Thus the organism is haploid having the diploid phase in its life cycle restricted to one cell, → meiocyte, only; **2.** → haplophase. *Adj.* **haplontic**. *Cf.* diplont.

haplophase (Gr. *phasis* = appearance), that phase of the life cycle of a cell or an organism during which the chromosomes appear only in a single, i.e. → haploid set, as e.g. in gametes. Also called haplont. *Cf.* diplophase.

Haplorhini, a suborder of → Primates in some classifications, comprising tarsiers, monkeys, apes, and humans.

haplostele (Gr. *stele* = column), one of different stele types in plant structures. *See* stele.

haplotype, 1. a haploid genotype; **2.** a group of → alleles from closely linked loci, usually inherited as a unit; **3.** a set of restriction fragments or microsatellites of a chromosome closely linked to each other and to a gene under study.

hapten (Gr. *haptein* = to fasten), a small molecule that can elicit an immune reaction in an animal when conjugated to a larger molecule, such as a protein, i.e. the conjugated hapten acts as an antigen inducing the formation of specific antibodies; e.g. the dinitrophenol group is used as a haptenic determinant in experimental immunology. The hapten is not antigenic by itself.

hapteron, pl. **haptera** (L.), the bottom part (fastening plate) of many algae, attaching the plant to a substratum.

haptonema (Gr. *nema* = thread), a long flagellum with three concentric membranes in certain algae (Prymnesiophyceae); has no locomotory function. *See* Prymnesiophyta.

Haptophyceae, Haptophyta, a class in the division Xanthophyta (yellow algae), Plant kingdom, in some systems classified as a separate division → Prymnesiophyta.

haptotropism, → thigmotropism.

hard phloem, the part of plant → phloem which consists of phloem fibres. *Cf.* smooth phloem.

hardening, 1. → cold hardening of animals and plants; **2.** an increased tolerance in man to environmental conditions, particularly extreme temperatures.

hardening of fats, *see* hydrogenation of oils.

Hardy—Weinberg law/theorem/principle, the basic law of population genetics discovered in 1908 independently by *Godfrey Harold Hardy* (1877—1947) and *Wilhelm Weinberg* (1862—1937). According to the law, the → gene frequencies in the population remain constant if no other factor than random mating is effective. If the frequency of allele A is nominated by p and that of allele a is nominated by q, the frequencies of the genotypes AA, Aa, and aa will be $(p+q)^2 = p^2 + 2pq + q^2$. Such a population is said to be in an equilibrium. A = dominant gene; a = recessive gene. *See* dominant character, recessive character.

harlequin staining (Fr. *arlequin* = clown), a staining method with the aid of which the sister chromatids of a chromosome can be stained differently. The method is used for determining the number of sister chromatid exchanges e.g. when studying the ability of a chemical compound to cause breaks in chromosomes. In harlequin staining fluorescent dyes are used resulting in a **harlequin chromosome**.

harvestmen, *see* Opiliones.

Hatch—Slack pathway, *M.D. Hatch* and *C.R. Slack* reported (1970) that malate is formed in photosynthesis of → C_4 plants. In a reaction catalysed by phosphoenolpyruvate carboxylase (PEP carboxylase), carbon dioxide (CO_2) in mesophyll cells binds to phosphoenolpyruvate (PEP). The first product is oxaloacetate which is rapidly reduced to malate or aminated to aspartate. C_4 acid, aspartate or malate, is transported to bundle sheath cells and decarboxylated yielding free CO_2 and pyruvate. This CO_2 is assimilated photosynthetically through carboxylation with Rubisco, as in C_3 plants. *See* crassulacean acid metabolism.

haustorium, pl. **haustoria** (L. *haurire* = to draw up), a specialized root or hypha of a parasitic plant or fungi, penetrating into tissues of another plant and absorbing food from the host plant.

Haversian canal/space (*Clopton Havers*, 1650—1702), any of the branching, longitudinal canals in the histological structure of the compact long bone, forming the cavities for blood vessels, lymph vessels and nerves; together with concentrically arranged lamellae around the canal, it forms the basic structure of the bone, called **Haversian system** or osteon. *See* bone.

Hb, → haemoglobin.

HDL, *see* cholesterol.

hearing (L. *auditus*), audition, auditory sense, sense of hearing; the sense for perceiving and processing sound waves, i.e. vibration of surrounding air (or water). Hearing is based on the function of mechanoreceptor cells which together with some supporting structures and neurones form the organ of hearing, i.e the **auditory organ** (acoustic organ), in vertebrates called the **ear**. The nervous system acts as a processor of the signals (receptor potential and nerve impulses) from the receptors. The **sensation of hearing** is produced by the brain, in higher vertebrates by the cerebral cortex. Vibration frequencies sensed by an animal vary: in humans from about 20 Hz to 20,000 Hz, but e.g. dogs, bats, and nocturnal moths can hear **ultrasounds** even up to 100,000 Hz. Some animals, such as big whales and elephants, can hear **infrasounds** (below 10 Hz). Very intensive sound can cause pain (pain threshold) and can even damage the ear. In invertebrates, hearing is quite rare existing only in some arthropod groups and is based on the function of a simple auditory organ formed of only a few receptor cells (*see* ear). Most invertebrates have single receptor cells in joints or in the skin associated with hair sensillae by which they can sense vibrations of ground or water, some also slow vibration frequencies of air.

heart (L. *cor*, Gr. *kardia*), a muscular internal organ found in most multicellular animals, maintaining the circulation of the blood or haemolymph by rhythmic contractions. The heart of invertebrates lies dorsally to the digestive tract. Some invertebrate groups, e.g. annelids, have no heart, but five pairs of **aortic arches** ("lateral hearts") maintain the circulation. The heart may be **tubular**, as in insects, cephalochordates, and in embryos of all vertebrates, or is formed from a **chamber** (or chambers) as in some invertebrates (molluscs) or in adult vertebrates. Some invertebrates have **accessory hearts** in addition to the ordinary heart, improving the peripheral circulation, e.g. wing hearts in insects and branchial hearts in cephalopods.

The fish heart comprises four consecutive parts: → sinus venosus, two main chambers (**atrium** and **ventricle**), and → bulbus arteriosus (conus arteriosus). The circulation is a **single-circuit system** and **complete**, i.e. it is unidirectional and the oxygenated and deoxygenated blood are not mixed; the blood from the heart passes first through the gills and then supplies the rest of the body. The heart of amphibians consists of three chambers, i.e. two atria and one ventricle in which the oxygenated and deoxygenated blood, however, keep quite well separated in both sides of the ventricle. Higher vertebrate groups (reptiles, birds, mammals) have a four-chambered heart, although in reptiles (except crocodiles) the wall between the ventricles is not complete. The circulation in three- and four-chambered hearts is **double**, consisting of the → **pulmonary circulation** (lung or respiratory circulation) and the → **systemic circulation** (major circulation). The circulation of amphibians and most reptiles is **incomplete** i.e. oxygenated and deoxygenated blood mix in the heart, but birds and mammals have complete circulation. In all animals the valves and/or septa determine the direction of the blood flow.

The wall of the heart of vertebrates consists of three layers: *epicardium* (the inner leaf of the serous → pericardium) lines the outer surface, *myocardium* forms the muscular layer, and *endocardium* lines the inner surface, and also forms the cardiac valves. The heart is enveloped by the outer leaf of the pericardium. Especially in amphibians there are large, chambered lymph vessels under the dorsal skin called **lymph hearts** which are involved in gas exchange in the lymph fluid through the skin. *See* cardiac muscle, cardiac valves, conduction system of the heart.

heartwood, the harder and darker inner part of a tree stem, consisting of dead cells, often lignified.

heat, 1. the energy of atomic and molecular vibration (thermal agitation); heat is produced e.g. in combustion, by kinetic friction, electric current, and in nuclear reactions. Heat is absorbed from the surroundings in endothermic chemical reactions and released in exothermic reactions; heat is transferred by conduction through a substance, by → convection of a substance (e.g. water, body fluids, gas), and by radiation as electromagnetic waves; *see* infrared radiation, calorie, joule; **2.** the sensation of high temperature; **3.** oestrus in animals; *see* oestrus cycle.

heat resistance, the ability of organisms to withstand high temperatures. In **homoiothermic animals** (homoiotherms) it depends on the capacity of temperature regulation and is related to the animal group, age, and physiological state, but also to other environmental factors, such as moisture. The fatal body temperature of mammals varies between 41—44°C and of birds 44—47°C. In plants and **poikilothermic animals** (poikilotherms) the resistance is related to the temperature of their natural habitats; e.g. spiders and insects living on hot deserts can resist temperatures over 50°C, but for antarctic fish species 10-15°C may be intolerable. Heat resistance is most effective in organisms living in hot springs at temperatures close to 100°C, e.g. some bacteria and cyanobacteria. Organisms living in naturally varying temperature conditions have the ability to acclimatize to seasonal variation of temperature, and with → **acclimatization** also their heat resistance changes. Many higher plants which grow in dry and hot habitats have some properties which show adaptation to extreme environmental conditions; e.g. they have thick cuticle covering the epidermis of leaves and stems, and CAM type photosynthesis (*see* CAM plants) is typical of them.

At cellular level, heat resistance is based on the stability of chemical bonds in enzymes, the properties of cellular membranes, e.g. the fluidity of cell membrane lipids and permeability properties of membranes, and on the synthesis of heat-shock proteins (*see* heat shock). The term **heat tolerance** is used to describe the limit of heat resistance. *Cf.* cold resistance.

heat shock, the exposure of an organism for a short period to temperature higher than to which it has adapted. After a heat shock, a specific group of genes are activated and the organism begins to produce **heat-shock proteins** (**HSP**) which are quite similar in different organisms. This is accompanied by decreased activity of other genes. The response has been shown in bacteria, plants and animals. After a heat shock the heat resistance of an organism increases. Also other stress situations can induce the production of the same or similar proteins, and therefore these proteins are often called → **stress proteins**.

heat stroke, a dangerous state in homoiothermic animals, especially in man, due to exposure to excessive heat so that the thermoregulatory capacity is exceeded, resulting in hyperthermia. Other symptoms are headache, thirst, nausea, muscle cramps, and dry skin. Called also **thermoplegia**.

heavy metal pollution, the environmental pollution caused by metallic elements of high atomic weight, as e.g. lead, mercury, arsenic, cadmium; often also aluminium. These are discharged to nature usually by industry, farming, traffic, and other human activities. These heavy metals are foreign elements in organisms and toxic in minute concentrations because they compete with metal cofactors of proteins and thus inhibit many enzymes, such as nitrate reductase, enolase, and the electron transport enzymes in photosynthesis and cell respiration. Also those heavy metals which are necessary for organisms, as e.g. chromium, copper, and zinc, may be harmful pollutants in high concentrations. Toxic symptoms in organisms differ but decreased uptake of necessary nutrients and reduced growth are obvious, in general. In multicellular organisms heavy metals often accumulate in tissues with high calcium concentration, in animals e.g. in bones and egg shells, but also in fat deposits. In nature, heavy metals usually accumulate in food chains, but toxic effects cannot usually be seen until in the highest levels of a food chain, as e.g. in predator birds.

hectocotylus, pl. **hectocotyli** (Gr. *hekaton* = hundred, *kotyle* = cup), a specialized tentacle of a male squid by which the animal transfers the **spermatophore** (sperm packet) into the mantle cavity of a female in copulation.

height zones, vertical vegetation zones in areas where height differences are great. In high mountain or mountain areas, the uppermost zone is called **nival zone**, characterized by permanent snow and ice. Below that is the **alpine zone** (regio alpina) and the lowest is the **subalpine zone** (regio subalpina); the former is situated above the tree line and is characterized by hay vegetation, the latter grows low forest. Between the alpine zone and tree line there may be a zone called **krummholz**, growing scattered and stunted trees and shrubs. The next zone is the **mountain zone** where trees are taller, and below it locates the **submountain zone**. The vegetation in different height zones is dependent on the latitude, the characteristic being affected by temperature, wind, and humidity.

HeLa cells, continuously cultured, transformed human cells, derived from a uterocervical carcinoma of a patient *Henrietta Lacks* in 1951; these cells are used worldwide for research in hundreds of laboratories.

helic(o)- (Gr. *helix*, gen. *helikos* = spiral), denoting relationship to a coil-like structure, e.g. a snail (*Helix*) or the → cochlea.

helicase, 1. any enzyme which can open the double helix of DNA (DNA helicase i.e. unwindase) at the replication fork when DNA polymerase can move forward during the duplication; **2.** an enzyme preparation obtained from snails, used to break down the walls of yeast cells.

helicotrema (Gr. *trema* = hole), a minute channel at the apex of the cochlea in the vertebrate inner ear; connects the scala vestibuli and scala tympani. Called also Breschet's or Scarpa's hiatus.

helio- or **heli-** (Gr. *helios* = sun), relating to the sun or sunlight.

heliotaxis (Gr. *taxis* = arrangement), the movement of an organism towards sunlight (positive heliotaxis), or away from it (negative heliotaxis); it is a special form of phototaxis and probably of thermotaxis.

heliotropism (Gr. *trope* = turning), the diurnal movement of leaves and flowers of many plants in response to sunlight. These plant organs orientate either perpendicular (**diaheliotropism**) or parallel (**paraheliotropism**) to the sunrays. *Cf.* phototropism, heliotaxis.

Heliozoa (Gr. *zoon* = animal), **heliozoans**; a class of actinopod sarcodines (superclass Actinopoda, subphylum → Sarcodina, phylum Sarcomastigophora); includes mainly protozoan species living in fresh waters. Heliozoans are radially symmetrical unicellular animals with several extensions, **axopodia**, protruding from the cell body (**central capsule**).

helix, pl. **helices** (Gr. = spiral), **1.** any coiled structure; **2.** the margin of the pinna of the human ear (auricle); **3.** → alpha helix, a secondary structure of proteins; **4.** → double helix (Watson—Crick helix).

helminthiasis (Gr. *helmis* = intestinal worm), an infection caused by parasitic worms, **helminths**, such as flatworms and roundworms.

helminthology (Gr. *logos* = discourse), a branch of biology especially dealing with flatworms and roundworms, particularly parasitic worms, helminths.

helminthosporic acid, a gibberellin-like substance, plant growth regulator, produced by the mould *Helminthosporium* that grows in cereal plants; its reduced form, **helminthosporol**, has similar properties. *Cf.* gibberellins.

helophyte (Gr. *helos* = swamp, *phyton* = plant), **1.** a perennial water plant having its stem and leaves mainly over the water surface; **2.** one of a → Raunkiaer's life forms; helophytes have their overwintering buds in mud.

helper, in ecology, a non-breeding animal such as a bird or mammal which in breeding activities helps other conspecific individuals by feeding or defending their offspring. Usually, but not always, the helper is a young individual and closely related to the breeders. Thus **kin selection** may have had an important role in the evolution of the helping behaviour. Occasionally the helper may be of different species than the breeder. *See* cooperative breeding.

helper cell, helper T cell; a type of T lymphocyte capable of activating some other cells of the lymphatic system of vertebrates. They produce interleukin-2, i.e. the growth factor which increases the synthesis and maturation of suppressor and cytotoxic T lymphocytes, and they can cooperate with B lymphocytes in the production of antibodies. *See* lymphocytes.

helper virus, a virus which offers for another virus, deficient in its genome, those metabolic functions which the latter needs to accomplish its life cycle. Helper viruses are used in gene manipulation if the → vector being used is a virus which is deficient in its genome.

hem-, hema-, hemo-, hemat(o)-, Am. (Gr. *haima* = blood), denoting relationship to blood; the British English form is → **haem-**.

hematoxylin, a dye obtained from the heartwood of logwood (*Haematoxylon campechianum*), used as a pH indicator (pH 5—6) and as a microscopic stain for animal and plant tissues. Its oxidized form is brownish red and its reduced form blue. Usually it is used together with other dyes, such as eosin, safranin, and Orange G. Hematoxylin is suitable e.g. for staining of mitochondria and chromosomes of cells.

heme, → haem.

hemelytron, pl. **hemelytra** (Gr. *hemi-* = half, *elytron* = wing cover), a forewing of heteropteran bugs (*Hemiptera*); pigmented and cuticularized at the base extending distally to form a thin membrane. At rest, the hemelytra are crossed and placed flat over the back. *Syn.* hemielytron. *Cf.* elytron.

hemeradiaphore (Gr. *hemera* = day, climax period of evolution, *dia* = through, *pherein* = to carry), a plant whose distribution has not been affected by man; may also include a species which has lost its habitat as a result of man's activity but has replaced it with a new habitat.

hemerobe (Gr. *hemeros* = tame), a culture-loving organism, especially a plant. *Adj.* **hemerophilous**.

hemerochore (Gr. *chorein* = to spread), a species which has spread to new habitats with man. *Syn.* anthropochore.

hemerophilous (Gr. *philein* = to love), pertaining to a culture-loving plant species which has gained advantage from man's activity. *See* hemerobe.

hemerophobe (Gr. *phobos* = fear), culture-avoider; a plant which has become more rare due to the activity of man.

hemerythrin, → haemerythrin.

hemi- (Gr. *hemi-* = half), pertaining to one half; in chemical names denotes structural similarity, having same characteristic groups, e.g. the structure of hemicellulose in comparison with cellulose.

hemianatropous ovule (Gr. *anatrope* = turning around), a type of plant ovule. *See* ovule.

hemibranch (Gr. *branchia* = gills), a type of → gill arch carrying gill filaments on one surface only. *Cf.* holobranch.

hemicelluloses, a large group of polysaccharides composed of various sugar residues, such as D-glucose, D-galactose, D-mannose, D-xylose, L-arabinose, D-glucuronic acid, D-galacturonic acid, L-rhamnose, and L-fucose; hemicelluloses, e.g. arabinoxylan, xylan, glucomannan, galactomannan, and arabinogalactan, are components of the primary wall of plant cells and occur as storage carbohydrates in seeds.

Hemichordata (Gr. *chorde* = string, cord), **hemichordates**; a phylum (previously considered as a subphylum of chordates) of worm-like marine animals, about 90 species in total. They have a true coelom, pharyngeal gill slits, a tubular dorsal nerve cord, and a buccal diverticulum, **stomochord** (mouth-cord), which resembles a rudimentary notochord but is not homologous with the chordate notochord. Some hemichordates have **tornaria** larvae resembling the bipinnaria and dipleurula larvae of echinoderms. The phylum is divided into two classes: *Pterobranchia* and *Enteropneusta* (**acorn worms**).

hemicryptophyte (Gr. *kryptos* = hidden, *phyton* = plant), a plant whose overwintering buds lie very close to or at the soil level. *See* Raunkiaer´s life forms.

hemidesmosome, *see* desmosome.

hemiedaphic (Gr. *edaphos* = soil), pertaining to a biota living in the detritus. *Cf.* epiedaphic.

Hemimetabola (Gr. *metabole* = change), → Exopterygota.

hemimetabolous, hemimetamorphic; pertaining to an organism which goes through an incomplete → metamorphosis; e.g. some aquatic insects (Ephemeroptera, Odonata, and Plecoptera). *Cf.* holometabolous, heterometabolous.

hemin (Gr. *haima* = blood), → haemin.

hemiparasite (Gr. *hemi-* = half, *para* = beside, *sitos* = food), an organism being partly parasitic but also capable of living without a host; e.g. a plant developing from a seed and assimilating independently while parasiting on other plants by root connections; e.g. many species of the family Scrophulariaceae. *Cf.* parasite.

hemipenis, pl. **hemipenes** (L.), a copulatory organ of snakes and lizards; a pair of cloacal pockets in the male, often containing thorn-like spines. In copulation these are extruded and inserted into the female cloaca. The hemipenis is not homologous with the penis of mammals.

hemiplacenta, a primitive type of the → placenta, found in marsupials; formed mainly from the yolk sac and chorion. Also called choriovitelline.

Hemiptera (Gr. *pteron* = wing), **hemipterans, (true) bugs** ; an order of aquatic or terrestrial insects which have a jointed beak for piercing and sucking; the food is composed of plant sap or body fluids of animals; the forewings are → hemelytra, the hindwings membranous, at rest folded under the forewings. The order comprises about 35,000 species including many notorious pests, carriers of diseases, and predators; e.g. bedbugs, plant bugs, lace bugs, water scorpions, water stinkbugs. *See* Homoptera.

hemisphere (Gr. *sphaira* = ball), one half of a ball; sphere; e.g. **1.** cerebral hemisphere, cerebellar hemisphere; i.e. the halves of the cerebrum or cerebellum; **2.** one of the two halves of an early embryo in many animals, as e.g. amphibians, called **animal** and **vegetative hemispheres**; **3.** one half of the Earth globe: the **northern hemisphere**, the **southern hemisphere**.

hemizoic (Gr. *zoon* = animal), pertaining to an organism which has chlorophyll-bearing chromatophores and the ability for photosynthesis, but which also ingests food, as green flagellates. *Cf.* holozoic.

hemizygote (Gr. *zygotos* = yoked), a cell or an individual in which certain genes appear as single, → haploid set, whereas the other genes appear in double, → diploid set; occurs e.g. in those organisms in which a XX - XY or XX - XO mechanism of sex determination prevails, the males being hemizygous for the X chromosomal genes (XY or XO). *See* sex determination. *Adj.* **hemizygous**.

hemo- (Gr. *haima* = blood), for words so beginning *see* haem-.

Henle's loop (*Friedrich Henle*, 1809—1885), in the vertebrate kidney the thinnest, U-shaped middle part of a renal tubule in the → nephron, concentrating the glomerular filtration (primary urine). Henle has also given his name to some other anatomical nomenclature, e.g. Henle's ampulla of vas deferens in mammals.

Henry's law (*William Henry*, 1774—1836), a physical law stating that the mass of a gas dissolved by a certain volume of a liquid (solution) at a constant temperature is directly proportional to the pressure of the gas.

Hensen's node/knot (*Victor Hensen*, 1835—1924), the anterior part of the → primitive streak of an early embryo of amniotes, i.e. reptiles, birds, mammals; it corresponds to the upper lip of the blastopore of amphibians, and acts as an inductor for further development.

HEPA filter, a high-efficiency particulate air filter used to purify the air taken into sterile laboratories and operating rooms.

hepar (L., Gr.), → liver.

heparan sulphate, a heparin-like heteropolysaccharide (glycosaminoglycan) in animal tissues composed of glucuronic acid and glycosamine sulphates; a constituent of many glycoproteins and proteoglycans in extracellular matrix and cell surface, being important in intercellular recognition.

heparin, heparinic acid; an anticoagulant factor (glycosaminoglycan) found in vertebrate tissues, especially in the liver and lung (in mast cells), and in certain blood cells (basophils). In conjunction with a serum factor (heparin cofactor) heparin acts as an antithrombin and antiprothrombin, thus preventing the coagulation of the blood (*see* blood coagulation). Heparin is also a cofactor for lipoprotein lipase; by activation of this enzyme heparin promotes fat transport from blood into fat cells.

hepatic, pertaining to the liver.

Hepaticopsida (Gr. *opsis* = appearing), **liverworts** (Hepaticae); a class in the division Bryophyta, Plant kingdom; in some systems classified as a separate division Hepatophyta. The plants consist of a small thin prostrate or erect thallus with leaf-like structures; in contrast to mosses, the "leaves" have no midribs. The life cycle is similar to that of mosses; the sex organs, antheridia and archegonia, vary in size and location. Most liverworts live in moist soils. The group is divided in three orders, Marchantiales, Metzgeriales, and Jungermanniales.

hepatic portal system, *see* portal vein, liver.

hepatocyte, liver cell. *See* liver.

hepatopancreas, a digestive gland found in many invertebrates, such as molluscs and crustaceans, secreting hydrolytic enzymes into the intestine for digestion of fats and proteins. The hepatopancreas corresponds to the vertebrate liver (hepar) and pancreas.

Hepatophyta, *see* Hepatocopsida.

HEPES, Hepes, 4-(2-hydroxyethyl)-1-piperazineethanesulphonic acid; a buffer substance used in biochemical and physiological experiments.

hept-, hepta- (Gr. *hepta* = seven), denoting seven.

heptane, an alkane hydrocarbon with seven carbon atoms, C_7H_{16}, found in petroleum; **heptose** and **heptulose** are the corresponding → monosaccharides.

herb (L. *herba* = grass, herb), a seed plant with no woody parts above ground.

herbaceous, pertaining to a plant having characters of a herb.

herbarium, pl. **herbaria,** Am. also **herbariums,** a systematically arranged collection of dried plants or plant parts, or the place (room or building) where the collection is stored.

herbicides (L. *caedere* = to kill), pesticides used to control weeds on plantations and associated uncultivated areas; include inorganic and organic substances. The grouping of herbicides can be made e.g. 1) on the basis of how they act in biochemical and physiological reactions, e.g. as inhibitors of mitosis, photosystem I, photosystem II, or respiration, or as growth regulators or soil herbicides; 2) on the basis of chemical structure, e.g. aliphatic acids, amides, benzoic acids, bipyridyliums, and phenoxy compounds; 3) on the basis of their selective effects on different plants; e.g. they may kill all plants (total herbicides) or only certain plants (selective herbicides). The selectivity is based on structural differences in plants (e.g. monocotyledons vs. dicotyledons) or on the different absorption, transport, and/or metabolism of herbicides in plants and plant cells. Herbicides are the largest group of the commercial pesticides.

herbivora (L. *vorare* = to devour), **1.** *Herbivora,* in old classification, a group of mammals nearly equivalent to → **Ungulata**; **2.** → herbivore.

herbivore, herbivorous animal; an animal feeding only on plants. *Syn.* phytophage.

herbivory, feeding on plants; often concentrates on certain parts of a plant only, causing injuries such as growth retardation, delay of flowering, cessation of root-growth, and increased sensitivity to plant diseases (*see* grazer). Remaining parts of the plant may partly or completely compensate the feeding damage e.g. by increasing the effectiveness of photosynthesis, improving the transportation of nutrients, and exploiting storage carbohydrates. Herbivory usually starts a chemical defence in plants (*see* inducible defence).

hereditary (L. *hereditare* = to inherit), transmitted from a parent to offspring; inherited, as e.g. a character or tendency determined by genes; descended from an ancestor. *Cf.* congenital.

heredity, the phenomenon that related individuals resemble each other more than individuals of the species on average. Heredity involves the persistence of the specificity of the genetic material during its replication and transmission to progenies. Heredity is carried out by the expression of genetic information in → genetic transcription and translation.

Hering—Breuer reflexes (*Ewald Hering*, 1834—1918; *Josef Breuer*, 1842—1925), the reflectory function of the pulmonary vagus nerve that maintains the rhythmicity of the pulmonary ventilation in vertebrates. The sensory nerve endings in lungs are irritated because of dilatation of alveoli. This irritation ceases inspiration by an **inflation reflex** that causes the relaxation of inspiration muscles through the control of the **respiratory centre** in the medulla oblongata. Then expiration starts passively, in vigorous respiration also actively, a **deflation reflex** ceasing the duration of expiration when lung alveoli empty.

heritability (L. *heres* = heir), the concept denoting the proportion of genetic variation of the whole phenotypic variation in a population (*see* variation). In the broad sense, heritability (H) is expressed by dividing the → genetic variance (V_G) with the phenotypic variance (V_P), i.e. $H = V_G/V_P$. In the narrow sense, the concept of heritability (h^2) is the proportion of the additive genetic variance (V_A) of the whole phenotypic variance (V_P) in a population, i.e. $h^2 = V_A/V_P$. The **realized heritability** measures the effect of (artificial) selection in a population, and is expressed as the proportion of selection response (R) to the → selection differential (S), or $h^2 = R/S$.

heritable, inheritable; capable of being inherited, or of inheriting.

hermaphrodite (*Hermaphroditos* = in Greek mythology the son of *Hermes* and *Aphrodite* who became merged in the virgin of the Salmakis fountain in Karias resulting in a creature, half man and half woman), a → **bisexual** individual; may be an animal which produces both male and female gametes or a flowering plant having both stamens and carpels in the same flower. *Adj.* **hermaphroditic**, also hermaphrodite.

hermaphroditic duct, a small convoluted duct in many hermaphroditic snails (gastropods) connecting the ovotestis to the albumen gland.

herpesviruses (Gr. *herpein* = to creep), double-stranded DNA viruses of the family Herpesviridae, genus *Herpesvirus*, causing skin and mucosal diseases (different types of herpes) in many mammals, in man e.g. the pathogen of herpes simplex, causing e.g. fever blisters, or of genital herpes in the genital region of both sexes; often the viruses may remain latent in tissues for years.

herpetology (Gr. *logos* = word, discourse), a branch of biology dealing with amphibians and reptiles.

herrings, → Clupeiformes.

hertz, Hz (*Heinrich Hertz*, 1857—1894), the unit of frequency. *See* Appendix 1.

heter(o)- (Gr. *heteros* = other), denoting other, different.

heteroallelic (Gr. *allelon* = of one another), pertaining to allelic genes which are mutated at different mutational sites in contrast to homoalleles which are mutated at homologous sites. *See* allele.

heteroauxin, the old name for indole-3-acetic acid. *See* auxin.

heteroblastic (Gr. *blastos* = germ), developing from more than one single tissue type.

heterocatalysis, the ability of a biological macromolecule to control the synthesis of other macromolecules, e.g. DNA that guides the synthesis of proteins in a heterocatalytic manner. *See* protein synthesis.

heterocercal (Gr. *kerkos* = tail), *see* caudal fin.

heterochlamydeous flower (Gr. *chlamys* = mantle), a flower in which sepals and petals are different, typical of most dicotyledons. *Cf.* homochlamydeous flower.

heterochromatin (Gr. *chromatikos* = concerning colour), those densely staining parts of chromosomes, or whole chromosomes, which have a tight, compact structure during → telophase, → interphase and early → prophase stages of cell division; in contrast to → euchromatin, the heterochromatin is genetically inactive. There are two types of heterochromatins. **Constitutive heterochromatin** is structural heterochromatin which contains no genes, and is found on chromosomes around the → centromeres, and in → telomeres; DNA in constitutive heterochromatin is highly repetitive. This type of heterochromatin exists in all somatic cells of eukaryotic organisms. **Facultative heterochromatin** is conditional heterochromatin which can be either in a euchromatic (*see* euchromatin) or heterochromatic state. Facultative heterochromatin is found only in some of the somatic cells.

heterochromatinization, **1.** the change of facultative → heterochromatin from the euchromatic state into the heterochromatic state; **2.** the change of → euchromatin to heterochromatin due to the → position effect in a chromosome mutation in which euchromatin is relocated in the vicinity of constitutive heterochromatin.

heterochromia, the difference of colouration in structures which should normally be of one colour; e.g. heterochromia iridis, the difference of colour in the two irides, or in different parts of the same iris.

heterochronia (Gr. *chronos* = time), **1.** the occurrence of a phenomenon at an unusual time; **2.** the rate difference between two processes.

heterochrony, also **heterochronism,** a gradual change in the order of the appearance of characteristics of organs and body parts during the ontogenic development, found between an ancestor and its evolutionary descendant. Forms of heterochrony are → paedogenesis (progenesis) and → neoteny.

heterocyclic (Gr. *kyklos* = circle, ring), in chemistry, describing an organic cyclic compound in which at least one of its ring atoms is not a carbon atom.

heterocyst (Gr. *kystis* = bladder), a colourless, thick-walled cell situated among other cells in the cell chain of some thread-like → cyanobacteria, often near → akinetes. The heterocysts lack photosynthetic apparatus but are active in biological nitrogen fixation.

heterodimer (Gr. *dis* = twice, *meros* = part), a macromolecule composed of two different kinds of subunits; especially refers to a protein composed of two polypeptides. *See* heteropolymer.

heterodont (Gr. *odous* = tooth), pertaining to a dentition with specialized tooth-types for different purposes. *See* tooth.

heteroduplex (L. *duplex* = twofold), duplex DNA containing one or more mispaired bases; it is a two-stranded DNA molecule which consists of nucleotide chains derived directly or through → replication from two parental molecules. **Mutation heteroduplex** is a double-stranded DNA molecule which bears a mutation in one strand. In replication this kind of molecule gives rise to two different daughter molecules. **Recombination heteroduplex** is a double-stranded DNA molecule in → genetic recombination, and appears as its intermediate stage during the formation of a → hybrid DNA. The concept assumes that the strands of the hybrid DNA differ at least in one nucleotide pair in such a way that no base pair is formed, but a mismatch site occurs. Recombination heteroduplex produces parental and recombinant molecules.

heteroduplex DNA, *see* heteroduplex.

heteroecious, also **heterecious** (Gr. *oikos* = house), **1.** pertaining to a parasitic animal whose life cycle includes several host species; **2.** pertaining to parasitic fungi having different spore types during their life cycle, the haploid and diploid phases developing on different hosts; e.g. rust fungus *Puccinia graminis*; *cf.* autoecious.

heterofacial (L. *facies* = face), describing a structure with clearly different upper and lower sides, as a typical plant leaf. If the upper side is more dominating, the structure is called **epitonic**, in the opposite case, **hypotonic**.

heterogametic (Gr. *gamos* = marriage), pertaining to that sex which produces two types of gametes, i.e. the male and female determining gametes (*see* sex determination). In most animals, e.g. mammals, the heterogametic sex is the **male** bearing e.g. → X and Y chromosomes, but e.g. in birds, reptiles and butterflies it is the **female** who has → W and Z chromosomes. *Cf.* homogametic.

heterogamy, **1.** the appearance of two different types of gametes, i.e. male and female; the gametes uniting in fertilization differ in size, form, and behaviour; e.g. sperm and egg; *syn.* anisogamy; **2.** the alternation of two generations one of which is genuine sexual and the other parthenogenetic; **3.** in mating, the preference of a phenotypically or genotypically dissimilar partner, e.g. a female prefers a partner who differs from herself as much as possible. *Adj.* **heterogamous**. *Cf.* homogamy. *See* mating system.

heterogeneity, heterogeneous quality or state; e.g. **1.** in ecology, the inconstancy of conditions in relation to time and space; **2.** genetic heterogeneity, i.e. genetic variance of a population.

heterogeneous (Gr. *gennan* = to produce), **1.** composed of different, dissimilar elements; **2.** in genetics, pertaining to a character that can be produced by different genes or combination of genes. *Cf.* homogeneous.

heterogeneous nuclear RNA, hnRNA, the

primary transcription product in eukaryotic organisms; a RNA molecule varying in size from 3,000 to 30,000 nucleotides, depending on the gene; hnRNA is processed in a multi-step series of events into messenger RNA. *See* RNA processing.

heterogenesis, → alternation of generations.

heterogenotes, partially → diploid bacteria which as a result of → transduction, F-duction, or conjugation in addition to their own → genome (**endogenote**) have received a part of the genome of a donor cell (**exogenote**). Thus, they are heterozygous for those genes which reside in the exogenote.

heterogenous, 1. having a dissimilar or different origin; not originating within the body, e.g. describes a → graft derived from a different species (xenograft); opposed to autogenous; **2.** consisting of dissimilar constituents; having different values.

heterogony (Gr. *gone* = seed, begetting), **1.** a cyclic → parthenogenesis where one or more parthenogenetic generations interchange with sexual generations (alternation of generations); **2.** different growth rates between the portions of two organs or organisms; **3.** → heterostyly. *Adj.* **heterogonic, heterogonous**.

heterograft, a tissue → graft taken from a donor of one species and transferred to a recipient of another species. *Syn.* xenograft. *Cf.* allograft, autograft.

heterokaryon (Gr. *karyon* = nut, nucleus), a cell with more than one functional nucleus produced by the fusion of two or more genotypically different cells; **1.** a fungal cell, a spore, or usually a mycelium, which contains genetically different nuclei within a single cell. The number of nuclei can vary depending on the species. In fungi, the phenomenon is the contrast to homokaryon in which the nuclei are genetically identical. Both heterokaryons and homokaryons can be di-, tri- or multikaryotic; **2.** a → cell hybrid produced through the artificial fusion of nuclei of two or more cells from different species. The association of genetically different nuclei in a joint cytoplasm is called **heterokaryosis**. *Adj.* **heterokaryotic**.

heteromerous (Gr. *meros* = part), consisting of parts or elements which differ in number; in biology, e.g. pertaining to tarsal segments in the legs of an insect, or to a lichen in the thallus of which the algal cells are restricted to a specific layer.

heteromeric, pertaining to another part; e.g. **1.** denoting spinal neurones with axons passing over the opposite side of the spinal cord; also heteromeral; **2.** in chemistry, a compound composed of different subunits.

heterometabolous, pertaining to animals whose developmental stages include gradual (partial or incomplete) → metamorphosis. *Cf.* ametabolous, hemimetabolous, holometabolous.

heteromorphic, heteromorphous (Gr. *morphe* = form), multiformic; differing in form, changed in form; e.g. **1.** describing homologous chromosomes which differ in size or form (*see* bivalent); **2.** pertaining to a plant with two or more morphological types within a population, such as many algae whose sporophytes and gametophytes are different. The phenomenon is called **heteromorphy** or **heteromorphism**. *Cf.* homomorphic.

heteromorphosis, a regeneration of an organ so that the new organ is dissimilar (usually smaller and less perfect) in shape and structure to the original organ.

heteromultimeric protein (L. *multus* = much, many, Gr. *meros* = part), a multimeric protein composed of two or more dissimilar polypeptide subunits.

heteronomous (Gr. *nomos* = law), involving different laws, e.g. **1.** being subject to different laws of growth or specialization; **2.** being composed of dissimilar segments; pertaining to a structure being divided into parts, such as metameres or segments which structurally deviate from each other; e.g. heteronomous segments in the body of many invertebrate animals, such as insects; **3.** in medicine, different from the type, abnormal. *Cf.* homonomous.

heterophylly (Gr. *phyllon* = leaf), the occurrence of morphologically different leaves in the same plant, e.g. pond crowfoot (*Ranunculus peltatus*) with undivided floating leaves and thin-lobed submerged leaves. *Cf.* anisophylly.

heteroplasia (Gr. *plassein* = to mould), alloplasia; **1.** the replacement of normal tissue by abnormal tissue; **2.** malposition of tissue, i.e. an abnormal locality of cells or tissues.

heteroplastic, pertaining to heteroplasia, or heteroplasty.

heteroplasty, → heterotransplantation.

heteroploid, pertaining to all chromosome numbers deviating from the normal chromosome number of a species; may be either → euploid or → aneuploid.

heteropolymer (Gr. *polys* = many, *meros* = part), a macromolecule composed of different kinds of subunits.

heteropolysaccharide (L. *saccharum* = sugar), a macromolecular carbohydrate composed of two or more different monosaccharides; e.g. glucomannan. *See* polysaccharides.

Heteroptera (Gr. *pteron* = wing), **heteropterans**; an insect group classified as a hemipteran suborder; includes e.g. plant bugs and water bugs.

heteropycnotic, also **heteropyknotic** (Gr. *pyknoun* = to condense), differing in density; especially describing chromosomes or parts of chromosomes which, as compared to other chromosomes, are in a different stage according to their condensation and stainability. Heteropycnosis is one of the properties of → heterochromatin. The heteropycnotic chromosomes or parts of chromosomes can be stained stronger (**positive heteropycnosis**) or weaker (**negative heteropycnosis**) than the rest of the chromosomal complement.

heterosexuality, the direction of sexual instinct to the opposite sex. *Adj.* **heterosexual**. *Cf.* homosexuality. *See* bisexual.

heterosis, hybrid vigour, i.e. the superiority of the heterozygous genotype (heterozygous advantage) regarding one or more characteristics (such as fertility, growth, survival), as compared to the corresponding → homozygotes. Heterosis arises as a consequence of dominance or interaction of genes in a heterozygous organism. *See* heterozygote.

heterospory (Gr. *sporos* = seed), production of two kinds of → spores by the same plant. **Macrospores** (or megaspores) give rise to the female → gametophytes and **microspores** to male gametophytes. Typical heterosporous plants are represented by the species *Selaginella* and *Isoëtes*, but also all seed plants are actually heterosporous. *Adj.* **heterosporous**. *Cf.* homospory.

heterostyly (Gr. *stylos* = column), the condition in which a plant species has two (**distyly**) or three (**tristyly**) kinds of flowers, with different types of styles differing in their length; the flower with short styles has long anthers and vice versa. Heterostyly ensures cross-pollination. *Syn.* heterogony.

heterothallism (L. *thallus* = stem), the occurrence of two different types of thalli which represent different mating types, sexual reproduction being possible only between the different types; appears in some algae and often in fungi. *Cf.* homothallism.

heterothermia, heterothermy (Gr. *therme* = heat), the state or quality of many **homoiothermic animals** capable of controlled changes of the body temperature above or below their usual body temperature. It is related to periodical or occasional changes in the physiological state of the animal, such as annual and circadian rhythms (e.g. daily torpor), or is e.g. due to exceptional cold, heat, or dryness of their environment (*see* hibernation, winter sleep). Then the body temperature is adjusted to a lower or higher set point by the thermoregulatory centre of the hypothalamus. Heterothermia is also related to immature postnatal stages of many small animals, such as many rodents, when temperature regulation is not yet fully developed and temperature changes are less controlled. Thus many homoiotherms have, in addition to their **normothermic** periods, also heterothermic periods that may appear as **hypothermia** or **hyperthermia**.

In all homoiotherms, **local heterothermia** occurs; e.g. the extremities are colder than the body trunk or inner organs (core temperature). Fever is a special case of heterothermia (hyperthermia). Sometimes the term heterotherm is also used as synonym for poikilotherms, and heterothermia for → poikilothermy. *Adj.* **heterothermic**.

heterotopy, heterotopia (Gr. *topos* = place), congenital displacement (malposition) of an organ or its part. *Syn.* ectopy. *Adj.* **heterotopic**.

heterotransplantation (L. *trans* = across, *plantare* = to plant), the → transplantation of a heterograft (xenograft), i.e. a tissue from one species to another. *Syn.* heteroplasty.

heterotrichous (Gr. *trichion*, dim of *thrix* = hair), pertaining to an organ with different types of filaments, cilia, or hairs; e.g. describes the thallus of some algae (e.g. order Chaetophorales in division Chlorophyta) consisting of the prostrate system (basal plate) of creeping, branched, mostly pseudoparenchymatous filaments, closely appressed to the substratum, and the branched erect filament system bearing the reproductive organs.

heterotroph (Gr. *trophe* = nutrient, nourishment), any organism not able to synthesize organic compounds from inorganic substances, and therefore dependent on other organisms, living or dead. Animals, fungi, most bacteria, and some parasitic plants, using complex organic compounds, are **chemoheterotrophs**. Some bacteria are **photoheterotrophic** using solar energy but also organic compounds as their energy source. *Cf.* autotroph.

heterozygote (Gr. *zygotos* = yoked), an organism which has different → alleles in one or more gene loci of its homologous chromosomes. If the differences involve two → genes it is a diheterozygote, if three, it is a triheterozygote, etc. *Adj.* **heterozygous**. The state of being heterozygous is called **heterozygosity** or **heterozygosis**. *Cf.* homozygote.

heterozygous advantage, *see* heterosis.

hex(a)- (Gr. *hex* = six), denoting six.

Hexactinellida (Gr. *aktinella*, dim. of *aktis* = ray), **glass sponges**; a poriferan class mainly including species living in deep oceanic waters; most of them are of radially symmetrical leuconoid type and their glass-like skeletons are composed of six-rayed siliceous spicules which form a kind of network. About 600 species are described.

hexad, 1. a group of six; **2.** a hexavalent atom or radical.

hexadecanoid acid, → palmitic acid.

hexane, any of five isomeric hydrocarbons of alkanes, C_6H_{14}; n-hexane, $CH_3(CH_2)_4CH_3$, a colourless and volatile lipid solvent which is the main component of petroleum ether; in biological studies used e.g. for extracting lipids from tissues.

hexanoic acid, → caproic acid.

hexaploid, a → polyploid organism which in its somatic cells carries six sets of the basic haploid number of chromosomes.

Hexapoda (Gr. *pous* = foot), **hexapods,** → Insecta.

hexokinase (Gr. *kinein* = to move), a phosphotransferase enzyme (EC 2.7.1.1.) that in cells catalyses the transfer of a phosphate group from ATP to hexoses; e.g. in phosphorylation of glucose produces glucose 6-phosphate. The hexokinase reaction requires the presence of Mg^{2+} ion.

hexosamine, an amino sugar with a hexose as a sugar component, e.g. → glucosamine.

hexosan, a polysaccharide composed of hexose units, e.g. starch, cellulose, and inulin.

hexose, a monosaccharide containing six carbon atoms in its molecule, $C_6H_{12}O_6$; e.g. → glucose (aldohexose) and → fructose (ketohexose).

hiatus, pl. **hiatus** or **hiatuses** (L.), in anatomy, an opening or orifice; e.g. *hiatus aorticus*, the opening in the diaphragm for the aorta.

hibernaculum, pl. **hibernacula** (L. = winter residence), an overwintering bud in water plants, a type of → bulbil.

hibernation (L. *hibernare* = to overwinter), a torpid state in which some small homoiothermic animals such as many bats and insectivores, some rodents, and a few birds overwinter. This state, often called true or obligatory hibernation, is brought about by a drastic but controlled decrease of the body temperature, usually close to the enviromental temperature, sometimes down to 0°C (**heterothermic state**). Then the metabolism of the animal is decreased to the level of 1—10% of that of the active state (**normothermic state**). During hibernation, the brain thermoregulatory centre remains functional, adjusting the set point for maintaining the body temperature at this new low level, and the centre ensures that the animal is not in danger of freezing. This occurs by increasing heat production if needed, especially in → brown fat, which raises the body temperature, and the animal may even become normothermic, stay awake for a while, and urinate. The ability to hibernate is genetically determined and involves complicated hormonal and neuronal regulatory mechanisms. The term hibernation is sometimes also used to mean the lighter hypothermic state, → **winter sleep** (facultative hibernation), found in some mammals, and the winter torpor of poikilothermic animals. *Cf.* aestivation.

hidr(o)- (Gr. *hidros* = sweat), denoting sweat or sweat glands.

hierarchy (Gr. *hieros* = holy, *archos* = rule), any system of individuals or things ranked one above the other. There are several hierarchical systems in biology: e.g. **1. hierarchical levels of nature** consisting of various organization levels, such as molecules, cellular organelles, cells (somatic cells and gametes), organisms, species, populations, communities, and ecosystems. The basic units of each level include various units of the lower level; **2. taxonomic hierarchy**, in which the basic unit is the species, and the whole hierarchy is composed both of intraspecific and supraspecific categories (taxa). In zoology, hierarchic main categories are kingdom, phylum, class, order, family, genus, species, and within them often several subcategories usually named with prefixes super-, supra-, sub-, or infra- (e.g. superorder, subfamily); **3.** in ethology, → **dominance hierarchy**.

high-energy phosphates, the group of phosphorylated compounds transferring chemical energy required for cellular metabolic and kinetic reactions; the most important is → adenosine triphosphate, ATP, which donates its phosphate group in a phosphorylation reaction while releasing a large amount of free energy bound in the bond between the two last phosphate groups of the molecule. ATP is continuously formed in → cell respiration. Other important high-energy phosphates in cells are guanosine triphosphate, GTP (*see* guanosine phosphates), cytidine triphosphate, CTP (*see* cytidine phosphates), uridine triphosphate, UTP, and thymidine triphosphate, TTP. All these phosphates are needed for nucleic acid synthesis, but they also have special functions in cell metabolism.

Hill reaction (*Robert Hill*, 1899—1991), the reaction which shows that isolated → chloroplasts can cause the reduction of some electron acceptors in the light, such as ferricyanide to ferrocyanide; oxygen is liberated in the reaction. The reaction is a property of → photosystem II.

hilum, pl. **hila** (L. = a small bit, trifle), **1.** in anatomy of animals, a small pit in an organ where vessels and nerves enter, as *hilum splenicum/lienis*, hilum of the spleen, *hilum pulmonis*, hilum of the lung, or *hilum renalis*, hilum of the kidney; also called porta; **2.** in plants, a scar on the seed coat on the point where the funiculus was located; water penetrates the hilum tissue more easily than the seed coat, and the hilum is thus important to the water uptake before germination; **3.** the deposition centre of a starch grain in plant cells.

hindbrain, → rhombencephalon.

hindgut, endgut; **1.** the posterior part of the intestine of vertebrates, especially during embryonic development; also used as a synonym for the large intestine, comprising the caecum, colon, rectum, and anal canal; **2.** the posterior part of the alimentary canal in arthropods and many other invertebrates; the hindgut is the major area for absorption of water and formation of faeces.

hinge, in anatomy, a structure holding dorsally the two shells (valves) of clams together; in some species, the hinge comprises **hinge ligaments** and **hinge teeth** which bind and align the valves.

hinge cells, cells in the epidermis of some graminoids (family Poaceae); absorb or lose water according to air humidity and so regulate the epidermal area and the rolling of the leaf.

hinge teeth, → hinge.

hip, 1. (L. *coxa*), the region of the vertebrate body lateral to and including the hip joint; **2.** loosely, the hip joint formed by articulation of the hip bone and femur.

hip bone (L. *os coxae*), a flat bone of the **pelvic girdle** (hip girdle), built by fusion of the ilium, ischium and pubis in adult tetrapod vertebrates; forms the lateral part of the pelvis. Also called hip, coxa(l), or innominate bone.

hippocampus, pl. **hippocampi** (L. < Gr. *hippokampos* = sea horse), in anatomy, the curved, complex folding of the cerebral cortex (chiefly of grey matter) forming the ventral cortical margin of the forebrain (telencephalon) and the floors of the temporal horns of the two lateral ventricles in mammals; its two main gyri are the Ammon's horn and dentate gyrus. The hippocampus, having the shape of a sea horse in the longitudinal section, forms a part of the → limbic system.

hirudin (L. *hirudo* = leech), a substance in the secretion (saliva) of the buccal glands in leech (*Hirudo*); prevents → blood coagulation in vertebrates by inhibiting the activity of thrombin; hirudin is used in medicine.

Hirudinea (L.), **hirudines**, **leeches**; a class of annelids including predatory or ectoparasite species mostly living in fresh waters, a few also in oceans and moist soil. The body with anterior and posterior suckers has a fixed number of segments with many annuli. The class includes about 500 species which are grouped into four orders: Acanthobdellida, Rhynchobdellida, Gnathobdellida, and Pharyngobdellida; familiar species are e.g. **medicinal leech** (*Hirudo medicinalis*) and **horseleech** (*Haemopis sanguisuga*), a predator without blood-sucking habit. *See* hirudin.

His' bundle (*Wilhelm His, Jr.*, 1863—1934), a

part of the → conduction system of the heart.

hist(o)- (Gr. *histos* = web, tissue), denoting relationship to animal and plant tissues.

histaminase, → diamine oxidase.

histamine, 1-H-imidazole-4-ethane amine, $C_5H_9N_3$; a potent vasodilator found commonly in animal tissues. It is synthesized from histidine in a decarboxylation reaction, in vertebrates produced chiefly by mast cells and basophilic granulocytes in blood, and the secretion of histamine is involved in normal regulatory mechanisms, but especially in inflammatory and allergic reactions. Histamine causes dilatation and increased permeability of blood capillaries causing hypotension, and contraction of smooth muscles in many inner organs, e.g. constriction of respiratory pathways. It also increases the secretory activity of the stomach and slightly accelerates the heart rate. The effects are transmitted by three separate receptor types (H_1, H_2, H_3). Histamine also acts as a neuronal transmitter e.g. in the central nervous system and eye. Histamine has been found also in some fungi, as in the ergot, and in some plants.

histidine, abbr. His, his, symbol H; a heterocyclic amino acid, $N{:}CHNHCHCCH_2$-$CH(NH_2)COOH$, occurring in most proteins of organisms; also serves as the precursor for → histamine. It belongs to the → essential amino acids of animals.

histioblast, also **histoblast** (Gr. *blastos* = germ), any animal cell capable of forming tissue; also used to denote a local → histiocyte.

histiocyte, also **histocyte** (Gr. *kytos* = cavity, cell), a tissue → macrophage present in loose connective tissue; **histiocytosis**, the abnormally increased number of histiocytes and their appearance in the blood.

histoblast, → histioblast.

histochemistry (Gr. *chemeia* = alchemy), the microchemistry of tissues; a research branch that deals with chemical compositions of animal and plant tissues, using chemical, biochemical, and tissue staining techniques and microscopical examination. *Adj.* **histochemical**.

histocompatibility, the quality of immunological similarity of tissues as tested by acceptance of homograft tissue transplantation, depending on how many **histocompatibility antigens** the tissues share; implies the presence of identical histocompatibility genes. *Adj.* **histocompatible**. *See* major histocompatibility molecules, HLA antigens.

histocyte, → histiocyte.

histogenesis, also **histogeny** (Gr. *gennan* = to produce), the formation and development of tissues from undifferentiated cells of the embryonic germ layers. *Adj.* **histogenetic**, pertaining to histogenesis; **histogenic**, producing tissues; **histogenous**, formed by tissues.

histology (Gr. *logos* = word, discourse), a research branch dealing with microscopical structure and composition of animal and plant tissues; is a part of microscopical anatomy. *Adj.* **histological**.

histolysis (Gr. *lyein* = to loosen, dissolve), the dissolution of tissues; a substance produced in this process is called **histolysate**. *Adj.* **histolytic**.

histomorphometry (Gr. *morphe* = form, *metron* = measure), a method using a computer for quantitative measurement and characterization of microscopical images.

histones, basic, water-soluble proteins belonging to the fundamental structure of chromosomes; exist in five qualities: H1, H2a, H2b, H3 and H4. Together with DNA, histones form the basic fibril of the → chromosome consisting of → nucleosomes. Histones are very similar in all eukaryotic organisms. Histone-like proteins are also found in some prokaryotes.

histotoxic (Gr. *toxikon* = arrow poison), poisonous or venomous to tissues; **histotoxin**, a histotoxic agent.

HIV, human immunodeficiency virus; HI virus. *See* AIDS.

HLA antigens, human leucocyte antigens; include → antigens (histocompatibility antigens) of human tissues found on the surface of most somatic cells (except red blood cells) but are best known in white blood cells, leucocytes. HLA antigens are encoded by a huge → gene complex (MHC genes) consisting of hundreds of genes (*see* major histocompatibility complex). In antigen-presenting cells, polypeptide products from the digestion of foreign antigens are coupled to HLA antigens on the cell membrane and presented there to → lymphocytes. The HLA system is involved e.g. in the rejection of transplanted tissues and organs. *See* major histocompatibility molecules.

hnRNA, → heterogeneous nuclear RNA.

hnRNP, *see* informosome.

Hoagland's (nutrient) solution (*D.R. Hoag-*

land), a nutrient solution used in the → water culture (hydroponics) of green plants; beside macronutrients it contains also many micronutrients.

hoatzins, → Opisthocomiformes.

Hogness box (*David Hogness*), → TATA box.

hol(o)- (Gr. *holos* = whole), denoting whole, entire.

Holarctic (realm) (Gr. *arktikos* = northern), a zoogeographical realm (region) and a floristic kingdom including the main part of the northern hemisphere, i.e. Europe, the middle and northern regions of Asia, and North America. In zoogeography, the Holarctic realm is divided into → Palaearctic and Nearctic regions. *See* zoogeographical regions, floristic kingdoms.

holeuryhaline (Gr. *eurys* = wide, *hals* = salt), pertaining to an aquatic animal whose distribution is not regulated by salinity but is able to live in fresh, brackish, and marine waters; e.g. eel and salmon species.

Holliday structure (*Robin Holliday*, b. 1932), a four-stranded, cross-shaped hybrid structure of DNA molecules occurring as an intermediate stage in homologous → genetic recombination. *See* chiasma.

holoblastic (Gr. *holos* = whole, *blastos* = germ), describing the ovum type (egg type) with a complete cleavage, i.e. the entire ovum divides in the beginning of embryogenesis; holoblastic egg cells are → isolecithal or moderately → telolecithal.

holobranch (Gr. *branchia* = gills), a type of → gill arch carrying gill filaments on both surfaces. *Cf.* hemibranch.

holocarpic (Gr. *karpos* = fruit), pertaining to fungi which have a fruiting body formed from the entire thallus; e.g. some parasitic fungi living in their host cells. *Cf* eucarpic.

Holocene (Gr. *kainos* = new, present), **Recent**; the latest epoch, after Pleistocene and the Ice Age; the time of the modern man beginning about 10,000 years ago.

holocentric (Gr. *kentron* = centre), pertaining to chromosomes which have a diffuse → centromere. A holocentric chromosome has no special point to which the microtubules of the → mitotic spindle attach, but the microtubules grasp the chromosome along its whole length, i.e. these chromosomes are consequently → holokinetic. Holocentric chromosomes occur e.g. in some lepidopterans.

Holocephali (Gr. *kephale* = head), a subclass of cartilaginous fishes including only one order, **Chimaeriformes** (chimaeras, rabbitfishes, ghostfishes, ratfishes) with about 25 grotesque-appearing species which live in deep ocean waters. They are characterized by jaws with large flat plates, a whip-like tail, and a scaleless body. Best known is *Chimaera monstrosa*, living in hollows of the Mediterranean Sea and Atlantic Ocean.

holocrine (Gr. *krinein* = to separate), pertaining to a type of exocrine secretion or a gland type in which the entire secreting cells with accumulated secretions die and are shed; thus the secretion is composed of these disintegrated cells, e.g. in sebaceous glands. *Cf.* apocrine, merocrine.

holoenzyme, any enzyme containing both of its structural parts, i.e. the → apoenzyme and → coenzyme, required to achieve the enzymic activity.

hologamy (Gr. *gamos* = marriage), a kind of reproduction in some unicellular animals (protozoans) in which two complete individuals, each acting as a gamete, unite to form a new large gamete.

holokinetic (Gr. *kinein* = to move), pertaining to chromosomes which are kinetically active (movable) in their whole length, carrying diffuse centromere.

Holometabola (Gr. *metabole* = change), → Endopterygota.

holometabolous, holometamorphic; pertaining to animals with complete → metamorphosis, e.g. some insects (endopterygotes) such as lepidopterans, coleopterans, or dipterans. *Cf.* ametabolous, hemimetabolous, heterometabolous.

holomicty (Gr. *miktos* = mixed), a phenomenon occurring in water bodies of cold areas implying the overturn of a water column from the surface to the bottom in a lake after the summer and winter → stagnations; in the process the oxygenation of water occurs. *See* stratification, water overturning.

holoparasite (Gr. *parasitos* = parasite), an organism which throughout its life is parasitic in (or on) a host and is unable to live independently; e.g. a plant holoparasite which has no chloroplasts. *Syn.* obligate parasite. *Cf.* meroparasite.

holopelagic (Gr. *pelagos* = sea), pertaining to an organism which is → pelagic throughout its life.

holophyte (Gr. *phyton* = plant), any organism

which is able to use light energy in → photosynthesis and needs no organic nutrients, i.e. is photoautotrophic. Holophytes include green plants and some bacteria. *Adj.* **holophytic**.

holoplankton (Gr. *planktos* = drifting, wandering), organisms which throughout their life cycles live in → plankton. *Adj.* **holoplanktonic**. *Cf.* meroplankton.

holoptic (Gr. *ops* = eye), pertaining to a type of the insect compound eye which unites or nearly unites on the vertex; e.g. the eyes of dragonflies and some dipterans. *Cf.* dichoptic.

Holostei (Gr. *osteon* = bone), **holosteans**; a group (infraclass) of primitive bony fishes in the subclass Actinopterygii (ray-finned fishes); includes two orders, **gars** (Semionotiformes) and **bowfins** (Amiiformes).

Holothuroidea (Gr. *holothourion* = sea cucumber, *eidos* = form), **holothuroids**, **sea cucumbers**; a class of marine echinoderms comprising soft-bodied, spineless animals without pedicellariae and arms; they are greatly elongated in the oral-aboral axis, skin ossicles are reduced.

Holotrichia (Gr. *trichion*, dim. of *thrix* = hair), **holotrichans**; a diverse protozoan group of ciliates, at present usually without any taxonomic status (in some classifications considered as a subphylum in the phylum Ciliophora). The group comprises unicellular animals with numerous cilia distributed evenly on the cell body; cell organelles are well developed, as e.g. in *Paramecium*. Holotrichia are common especially in lentic, eutrophic, small water bodies.

holotype (Gr. *typos* = type), *see* type specimens.

holozoic (Gr. *zoon* = animal), pertaining to organisms which ingest visible particles of food (holozoic feeders), i.e. animals. *See* hemizoic, holophyte, autotroph.

home(o)- (Gr. *homos* = same), denoting similarity or sameness; in British English also → homoi(o)-.

homeobox, a DNA segment containing 180 base pairs and found in → selector genes of animals which encode a → homeodomain. The homeodomain protein is involved in regulation of → structural genes thus controlling the development of the animal. In plants, the corresponding DNA sequence is called → MADS box.

homeodomain (L. *dominium* = ownership), a part of a protein encoded by the → homeobox; rich in basic amino acids and capable of binding to certain regions of DNA, and hence regulates the function of genes.

homeogene, any of the homeobox-containing genes which play a central role in pattern formation during the development of animals and plants. *See* homeobox.

homeosis, the change of an organ into some homologous structure; mutations which cause this type of changes are called **homeotic mutations**, e.g. the change of the antennae into leg-like structures in a fruit fly.

homeostasis (Gr. *stasis* = standing), the tendency of an organic system to maintain a certain dynamic equilibrium state so that under disturbances its control mechanisms return the system towards the initial equilibrium state; e.g. **1. physiological homeostasis** denotes the stability of the normal physiological state (internal environment) of an organism. It involves complicated biochemical, hormonal and neuronal systems which buffer an organism against environmental changes; thus the maintenance of internal homeostasis is the basis for adaptation to the environment (*see* regulation). The efficiency of the homeostatic processes are often related to heterozygosity that gives an organism better fitness than homozygosity; *see* heterozygote, homozygote; **2. collective homeostasis** describes the tendency of a population to maintain the prevailing → gene frequency; **3. epigenetic homeostasis** denotes the capacity of the developmental pathways to produce a normal phenotype in spite of developmental or environmental disturbances; also called **developmental homeostasis** and **developmental canalization**; **4. ecological homeostasis** describes the equilibrium of nature, i.e. the equilibrium between an individual and its environment. *Adj.* **homeostatic**.

homeotic, pertaining to genes that assign spatial identity to groups of cells with respect to their morphogenetic fates; i.e. they are required for proper morphogenesis of an organism.

homeoviscous adaptation (L. *viscosus* = sticky), an adaptive process for the maintenance of the physical state and function of membranes. Membrane fluidity is largely a property of the degree of saturation of component lipids. It is directly influenced by temperature; a rise in temperature increases and a fall decreases fluidity. Cells respond to long term (seasonal) changes in temperature by altering membrane lipid saturation so as to compensate for direct temperature effects on fluidity. This adaptive process, homeoviscous adaptation, results in membranes from cold acclimated cells which are more fluid than those from warm acclimated cells. Apparently also other factors, such as pressure and some chemicals (e.g. alcohols), can progressively induce homeoviscous adaptation that at least partly compensate the initial change of fluidity caused by these factors.

homeothermy, → homoiothermy.

home range, an area where an animal or animal group regularly lives. The home range is usually much larger than the → territory. In general, animals do not defend their home ranges as they do their territories. Also animals without territories may have their home ranges. Some scientists include migratory routes and winter territories of birds in the term home range.

homing, a behavioural act of an animal, usually a bird or fish, to return to an original area, as salmon to the river where they are born; e.g. **homing pigeons** were much used by the ancient Egyptians, Greeks and Romans to convey messages; still a sport in some countries.

Hominidae (L. *homo* = man), **hominids**; a family in the superfamily Hominoidea (order Primates); previously included only the genus *Homo* and → Australopithecines. Today, on the ground of molecular evidence, the chimpanzee (*Pan*) and the gorilla (*Gorilla*) are incorporated into the subfamily Homininae, and the orang-utan (*Pongo*) in the subfamily Ponginae. These subfamilies form the family Hominidae.

Hominoidea, hominoids; a superfamily in the order Primates including great apes and human species in the two extant families: **gibbons** (Hylobatidae) and **hominids** (Hominidae).

Homo (L. = man), a genus of man in the order Primates, including modern man (*Homo sapiens sapiens*) and an extinct subspecies of Neanderthal man (*H. s. neanderthalensis*), which some paleontologists consider to be the independent species *H. neanderthalensis*. In addition, the genus includes the extinct species, *H. habilis* and *H. erectus*, which preceded the modern man about 1 to 2,5 million years ago, and probably were ancient species of modern man. Two other fossil *Homo* species, *H. rudolfensis* and *H. ergaster*, have been described by some anthropologists.

homo- (Gr. *homoios* < *homos* = same), **1.** denoting the same or similar; **2.** in chemical names, describes insertion of one additional carbon atom or -CH_2- group in the chain of an otherwise similar molecular structure of some aliphatic, alicyclic, or aromatic compounds.

homoallelic (Gr. *allelon* = of one another), pertaining to genes which have identical → alleles in a gene locus, as opposed to → heteroallelic genes.

homocercal (Gr. *kerkos* = tail), *see* caudal fin.

homochlamydeous flower (Gr. *chlamys* = mantle), a flower in which the sepals and petals are similar; typical of many monocotyledons, like the tulip.

homocysteine, 2-amino-4-mercaptobutyric acid, $HSCH_2CH_2CHNH_2COOH$, an amino acid produced from methionine by the cleavage of one methyl group, i.e. by demethylation of methionine; homocysteine is a homologue of → cysteine and an intermediate in its biosynthesis.

homocystine, the synthetic disulphide of → homocysteine, $(SCH_2CH_2CHNH_2COOH)_2$; homologous with → cystine.

Homodonta (Gr. *odous* = tooth), **homodonts**; a group name for fishes, amphibians and reptiles, i.e. for all the species whose teeth are structurally similar. *Adj.* **homodont**. *Cf.* heterodont.

homogametic (Gr. *gametos* = husband), pertaining to that sex which produces only one type of gametes (*see* sex determination); e.g. in mammals, the female which produces only gametes containing the → X chromosome. *Cf.* heterogametic.

homogamy (Gr. *gamos* = marriage), **1.** in mating, the preference of a phenotypically or genotypically similar partner; *cf.* heterogamy; **2.** the property of a plant having similar flowers throughout; **3.** the maturing of the stamens and pistils at the same time.

homogeneous (Gr. *homogenes* = of the same kind), **1.** pertaining to a uniform structure or composition throughout, or to a similar kind or nature; not heterogeneous; **2.** consisting of only one phase. *Noun* **homogeneity**. *Cf.* homogenous.

homogenize (Gr. *homos* = same, *gennan* = to produce), to render a substance uniform by breaking and blending, i.e. to make it **homogeneous**. In biochemical and physiological methods the cells and tissues are homogenized in a suitable solution using a cutting or massaging blender, a **homogenizer**. The homogenized solution is called **homogenate**. To maintain the cell structure as intact as possible the temperature as well as the ionic and osmotic concentrations are controlled and pH and free calcium are buffered. **Homogenization** is usually carried out at low temperature (0 to 4°C). If it is not necessary to obtain living cells or cell organelles or to maintain enzymic activity, organic solvents such as 80% acetone or alcohols can be used as homogenization media for extracting organic materials, e.g. chlorophylls from plant tissues.

homogenous (Gr. *genos* = origin), having a structural similarity because of a common origin, due to descent from the same ancestral type. *Noun* **homogeny**. *Cf.* homogeneous.

homograft, → allograft.

homoiosmotic, also **homeosmotic** (Gr. *osmos* = push), pertaining to an animal capable of → osmoregulation, i.e. able to regulate the osmotic concentration of its body fluids. *Cf.* poikilosmotic.

homoiothermy, also **homeothermy** (Gr. *therme* = heat), the property of an animal implying the active maintenance of the body temperature at a certain, stable level by controlling heat loss and production in the body. **Homoiothermic animals** (homoiothermic animals, homeotherms) are birds with an average body temperature of 40°C and mammals with 37°C (in duckbills and marsupials some degrees lower), these values denoting the temperature of internal organs (**core temperature**). The temperatures of peripheral tissues, especially of extremities, are often much lower. The maintenance of homoiothermy is based on the controlled heat loss and production, due to body insulation, cold-induced thermogenesis in many tissues, and the control mechanisms involving complex neural and hormonal reactions. *Adj.* **homoiothermic**, **homoiothermal**, **homeothermic**, **homeothermal**. *See* thermoregulation.

homokaryon (Gr. *karyon* = nut, nucleus), a fungal cell, spore, or usually a mycelium in which each cell contains two or more genotypically identical nuclei. *Cf.* heterokaryon.

homologous (Gr. *homologos* = agreeing, corresponding), equal; resembling in structure and origin; e.g. **1.** pertaining to → homologous chromosomes; **2.** pertaining to structures or characteristics which are of the same phylogenetic origin but may have different function, such as the hand of a monkey and the wing of a bird, and the fin-like paddle of a whale. Also many of the behavioural → fixed-action patterns in different animal species are homologous. *Cf.* analogous.

homologous chromosomes, the members of a chromosome pair, i.e. similar chromosomes of which one is inherited from the mother and one from the father. *Syn.* homologues.

homology (Gr. *homologia* = uniformity, agreement), **1.** uniformity; **2.** of the same origin; **3.** the state of being → homologous.

homomorphic, homomorphous (Gr. *homos* = same, *morphe* = form), similar in form; e.g. conspecific individuals are homomorphic by appearance if they have only one form without any varieties. The phenomenon is called **homomorphy** or **homomorphism**. *Cf.* heteromorphic.

homomultimeric protein (L. *multus* = many, Gr. *meros* = part), a multimeric protein consisting of two or more identical polypeptide subunits.

homonomous, also **homonomic** (Gr. *nomos* = law), pertaining to equal (homologous) serial parts of an organism, such as segments of an annelid worm, somites of an vertebrate embryo, or fingers and toes of a primate; **homonomy**, the condition of being homonomous. *Cf.* heteronomous.

homonym (Gr. *onyma, onoma* = name), a scientific name given to two species; according to the principle of priority, the name given later has to be changed.

homoplastic (Gr. *plassein* = to mould), **1.** similar in form and structure but not in origin; e.g. pertaining to a structural or morphological similarity of organisms adapted to a similar environment without a common origin; **2.** pertaining to → homoplasty.

homoplasty, the replacement of tissues of an organism by an allograft (homograft), i.e. a → graft taken from a genetically identical individual; e.g. **homoplastic** is a graft that is transferred from an animal which belongs to the same inbred strain as the recipient. *Cf.* homoplasy.

homoplasy (Gr. *plassein* = to form), similarity in form or structure being the consequence of a convergent, parallel, or reversal evolution without a common origin.

homopolysaccharide (Gr. *polys* = many, L. *saccharum* = sugar), a macromolecular carbohydrate composed of identical → monosaccharides; e.g. cellulose that is a polymer of glucose. *See* polysaccharides.

Homoptera (Gr. *pteron* = wing), **homopterans**; an order of insects, often considered as a suborder in the order → Hemiptera. The wings are held roof-like over the body; some species are wingless. Mouthparts are adapted for piercing or sucking; all are plant feeders, many species pests. The order includes e.g. cicadas, aphids, scale insects, leaf hoppers, and tree hoppers.

homoserine, 2-amino-4-hydroxybutyric acid, an amino acid containing one carbon or -CH_2-group more than in → serine; homoserine occurs as a free acid e.g. in pea plants, not found in protein.

homosexuality, orientation of sexual intrest towards the same sex. *Adj.* **homosexual**. *Cf.* heterosexuality. *See* bisexual.

homospory (Gr. *sporos* = seed), production of → spores of only one type, giving rise to → gametophytes which all are similar; some mosses and ferns are **homosporous** plants. *Cf.* heterospory.

homostyly (L. *stilus* = stake, style), the situation when all flowers of a species have styles of equal length. *Cf.* heterostyly.

homothallism (Gr. *thallos* = stem), a situation in which plants or fungi form thalli of only one type and the gametes produced by the same thallus are able to fuse. *Cf.* heterothallism.

homozygote (Gr. *zygotos* = yoked), a diploid or polyploid organism which carries identical → alleles in the gene loci of its homologous chromosomes. *Adj.* **homozygous**. *Cf.* heterozygote.

homunculus, pl. **homunculi** (L. dim. of *homo* = man), **1.** according to a belief in the 17th century, a minute human figure was ready inside each human germ cell, especially in the spermatozoon; **2.** the figure resembling a human describing the representation of motor and sensory areas of the brain cortex.

honey-sac, a large crop in the oesophagus of the honeybee worker for storing nectar. *Syn.* honey-stomach.

hook formation, → crozier formation.

hookworms, nematodes from the family Ancylostomatidae parasiting mainly in homoiothermic vertebrates; they live e.g. in the human intestine and suck blood with sharp cutting plates and a thick, pumping pharynx. The worms cause anaemia, in young animals and children also retarded mental and physical growth. It is estimated that the species *Ancylostoma duodenale* and *Necator americanus*, causing the **hookworm disease**, kill annually millions of people in tropical and subtropical countries.

HOP, → hydroxyproline.

hordein (L. *hordeum* = barley), a storage protein that belongs to → prolamines and occurs e.g. in barley seeds.

horizon (Gr. *horizein* = to separate, part, define), **1.** skyline; **2.** horizontal zone in a sediment, soil structure (→ podzol), water, or geological layer.

horizontal cell, a nerve cell type of vertebrate retina modulating the optic information from the eye to the brain.

horizontal rhizome, a plant stem structure located horizontally below ground. *See* rhizome.

hormogonium, pl. **hormogonia** (Gr. *hormos* = chain, *gone* = seed, offspring), a part of cyanobacterial filament between two heterocysts, being able to form a new organism when breaking away from the original structure.

hormone receptors (Gr. *hormaein* = to set in motion, *recipere* = to receive), specific macromolecular structures (usually proteins or glycoproteins) in the cell membrane or inside the cell, i.e. they are cell surface receptors or intracellular receptors (usually nuclear receptors). Hormone receptors are capable of selective binding of a hormone or hormone-like ligand. Hydrophilic hormones, such as amines and peptides usually bind to cell surface receptors, and hydrophobic hormones, as e.g. steroids, diffuse across the target cell's membrane and bind to intracellular receptors. The formation of a **hormone-receptor complex** causes a chain reaction in the cell that further results in responses in cellular function, such as contraction, secretion, and synthesis of enzymes for growth or energy metabolism, or liberation of energy sources.

In animal cells, the hormone-receptor complex formed in the cell membrane usually activates → **adenylate cyclase** that is associated with the cell membrane, resulting in an increase of → **cyclic AMP** inside the cell. This further activates an enzymatic cascade of reactions changing function in the target cell. Also other → second messengers may be involved in these hormone actions. Hormone-receptor complexes formed in the cytoplasm or nucleus affect the genome, i.e. activate **gene expression** to produce a certain protein (e.g. an enzyme) in the target cells.

hormones, chemical messenger substances produced by certain cells of an organism for specific regulatory effects on the biochemical, physiological, or behavioural activity of target cells, tissues, and entire organs. Hormones are transported in the circulation to target cells, and they produce their effects in nano- and millimolar concentrations through specific molecules, → **hormone receptors**.

Animal hormones are secreted typically from the → **endocrine glands**, but also from cells differentiated principally for other functions, such as nerve cells (neurohormones), certain cardiac muscle cells, and some blood cells. Hundreds of animal hormones are known, and chemically they belong to **amines** and their derivatives (e.g. → adrenaline, noradrenaline, thyroxine), to **oligopeptides**, **polypeptides**, and **proteins** (e.g. → vasopressin, oxytocin, peptides of the alimentary canal, neuropeptides, corticotropin, insulin), or to **steroids** (→ corticosteroids, sex hormones).

Hormones are traditionally classed as **general hormones** which circulate in the blood, or locally acting **tissue hormones** (paracrinic secretion); however, this division is not possible in all cases. → Neurotransmitters are not included in hormones although the difference between these and hormones is sometimes unclear, because the same substances, e.g. noradrenaline, can act as a neurotransmitter in one cell type or tissue but as a hormone in another. Also many hormones can modulate the neural transmission. The distinction between hormones, → neurosecretion, and → growth factors is in many ways formal. The most important endocrine gland of vertebrates is the → pituitary gland which controls the function of many other endocrine glands.

Plant hormones (phytohormones) act mainly like animal hormones, their effects being mediated by specific hormone receptors located in the plasma membrane or in the cytoplasm. Plant hormones control e.g. the germination of seeds, the division and growth of cells, flowering, growth movements, the development of fruits, and leaf abscission. The main groups of plant hormones are → auxins, gibberellins, cytokinins, abscisic acid, and ethylene. Some herbicides, such as → phenoxy compounds and → polyamines, have hormone-like properties. *See* signal transduction.

horns, 1. true horns; hard projections on the head of mammals of the cattle family (Bovidae), i.e. cattle, sheep, goats, gazelles, antelopes, and some reptiles such as chameleons. They are paired, unbranched structures, growing from frontal bones, usually in both sexes, larger in males. The bony core of the horn is a spike of bone, sheathed by a hollow cone of cornified epidermis (*see* cornification). The horns are usually permanent; *cf.* antlers; **2.** horns of some other mammals, such as the unpaired **rhinoceros horn** which is a hairlike, horny mass produced by the group of dermal papillae on the snout and forehead bone, or the paired **giraffe horns**, i.e. the bony, hair-covered, knob-like projections which are not shed. The paired **pronghorns** of the American prongbuck are bony projections from the skull, covered by the skin that produces a branched horny sheath which is annually shed; **3.** any of usually several horn-like structures on the head of some fish species, as e.g. some scorpion fishes. Also some invertebrates have chitinous horn-like processes; **4.** a horn-like structure (*cornu*) in vertebrate anatomy; e.g. **Ammon's horns**, the two gyri of the → hippocampus in the brain; **uterine horns**, the two horns of the uterus, horn-like projections partially fused to form the median corpus uteri (in duplex, bipartite, or bicornuate types of uteri); **horns of the spinal cord**, dorsal, lateral, and ventral projections seen in the cross section of the grey matter (butterfly figure) of the spinal cord.

hornworts, → Anthocerotae.

horny layer (L. *stratum corneum*), *see* epidermis.

horotelic (L. *horolo* = clock, *teleos* = goal), referring to standard rates of evolution.

horsetails, → Equisetales.

horsehair worms, → Nematomorpha.

host, in biology, **1.** an object of parasitism which can be either an organism, population, or a species; **2.** a recipient organism receiving a tissue or organ transplant from another organism, the donor. *See* definite host, intermediate host.

host range, the range of host organism types that a parasite can infect, e.g. of cell types that a virus can infect.

host range mutant, any mutant of a bacteriophage virus which differs from the corresponding wild-type bacteriophage so that it can live in different bacterial host strains. *See* mutant.

hotspot, 1. a site of DNA that is sensitive to mutation; **2.** an informative name for an area where concentrations of some environmental pollutants are exceptionally high; e.g. radioactive substances, air pollutants, or water pollution loads.

housekeeping gene, any → gene the function of which is necessary in all cells of an organism; e.g. the genes encoding the enzymes of cell respiration, or the genes which encode the proteins of → endoplasmic reticulum and cell membranes. *Cf.* luxury gene.

HOX genes (so called because they contain the → homeobox), a family of → homeogenes occurring in several clusters in vertebrate cells; specify segment identity in the body structure. The individual members of this gene family are related to the genes of the complex loci → ANT-C and → BX-C in fruit flies.

HPLC, high performance liquid chromatography. *See* chromatography.

HSP, heat-shock protein. *See* heat shock.

5-HT, 5-hydroxytryptamine, → serotonin.

human (L. *humanus*), **1.** pertaining to, or having the nature of mankind; consisting of people; **2.** as a noun, human being; **humane,** pertaining to the compassion of man for other people and animals, or to humanistic studies.

human chorionic gonadotrop(h)in, hCG, HCG, a → gonadotropin secreted during pregnancy by endocrine cells of the human → chorion; a glycoprotein regulating the hormonal secretion of foetal gonads. *See* chorionic gonadotropins.

human genetics (Gr. *genesis* = origin, birth), a branch of genetics dealing with the similarities and differences of human beings controlled by genes, and the inheritance of these properties from one generation to the next. *See* medical genetics.

human immunodeficiency virus, HIV, *see* AIDS.

human pinworm, the nematode species *Enterobius vermicularis*, a parasite found in the human intestine; causes only minor symptoms, such as itching around the anus when the pinworm females migrate to the area for egg laying; the human pinworm is common all over the world, especially among children.

human placental lactogen, hPL, HPL, a polypeptide hormone secreted from the human placenta; its effect resembles that of → prolactin and growth hormone, having lactogenic, luteotropic, and growth-promoting activity. Called also e.g. placenta protein, placental growth hormone, or human chorionic somatomammotropin.

human races, races of modern man, *Homo sapiens*; during different periods of the history of anthropology, tens of human races have been distinguished, but nowadays only three main geographical races are suggested, i.e. the **caucasoid** or **white race**, the **negroid** or **black race**, and the **mongoloid** or **yellow race**. Also arguments for two main races, the white and black race, have been presented. The white race originally inhabited Europe, the black race Africa and Australia, and the yellow race Asia and its archipelago. Also the American aborigines (Indians) belong to the yellow race. The human races differ genetically only slightly from each other, i.e. about 7% of the total human genetic variation occurs between races. Consequently, some scientists have proposed that the concept of race in human biology should be discarded.

humerus, pl. **humeri** (L.), a long bone in the forelimb of tetrapods between the shoulder and the elbow.

humidity of air (L. *humidus* = humid, moist), the amount of water vapour in the air: **absolute humidity**, ρ_h, the mass of water vapour per volume unit of dry air; **maximum humidity**, $\rho_{h\,max}(t)$, the greatest possible density of water vapour in the air (depends on temperature); **relative humidity**, φ, absolute humidity divided by maximum humidity and expressed as percentage.

humidity sense, a sense of some invertebrates detecting the humidity of the air; e.g. found in some terrestrial isopods. Obviously some vertebrates which live on dry deserts have similar property related to the sense of smell.

humification (L. *humus* = earth, ground), the process of the formation of → humus.

hummock, a vertical → rhizome forming numerous sympodial branches spreading widely to all sides; a typical growth form of many grasses and sedges.

humoral (L. *humor* = liquid, fluid), pertaining to body fluids; occurring in plasma, lymph, or interstitial tissue fluids.

humoral immunity (L. *immunis* = free), → immunity mediated by soluble antibodies in the blood plasma and lymph.

humster (human + hamster) an ovum of a hamster fertilized with human sperm; the method of producing humsters is used to study the chromosome content of human spermatozoa.

humus (L. = earth, ground), a dark brown layer in the upper soil (**A_1-horizon**), composed of decomposed plant and animal remains, thus forming an organic component in the soil. Above the humus layer is detritus (**A_0**). *See* soil formation.

hunger, a need or desire for food; its physiological basis is best known in mammals. According to the **glucostatic hypothesis**, the control of feeding is associated with special areas of the hypothalamus, the **feeding centre** (hunger centre) in the bed nucleus of the medial forebrain bundle at its junction with the pallidohypothalamic fibres, and the **satiety centre** in the ventromedial nucleus. Some nerve cells (**glucostats**) found in these centres react to changes in the blood glucose concentration, inducing either the sensation of hunger or satiety. The centres also receive impulses through **sensory nerves** from the alimentary canal, especially from stretch receptors of the stomach, which influence the activity of the centres. The control is affected by the hormonal state of an animal, → **leptin** having a special role, and is associated with thermoregulation, high temperatures decreasing the appetite. Also the **lipostatic hypothesis** has been presented, stressing the role of lipids in the control of hunger.

hyalin (Gr. *hyalos* = glass), **1.** the intercellular gelatinous material of cartilage, bone, and other connective tissues of vertebrates, composed of protein and → hyaluronic acid; **2.** more generally, any of gelatinous substances (even certain tumours and hydatid cysts) of the body; **hyalinization**, the formation of hyalin in a tissue. *Adj.* **hyaline** (or **hyaloid**), glassy, nearly transparent, hyalin-like.

hyaline cartilage, a glassy type of → cartilage.

hyaloplasm (Gr. *plassein* = to form), the term earlier used to denote the non-fibrillar part of the cell cytoplasm; the term is still used to mean the agranular ectoplasm of some protozoans, e.g. amoeba.

hyaluronic acid, a natural long-chained → mucopolysaccharide formed of → glucuronic acid and N-acetylglucosamine; when joined to proteins it forms → proteoglycans (e.g. hyalins) which are important intercellular substances in animals in the connective tissue matrix, especially in cartilage, skin, blood vessels, synovial fluid, umbilical cord. Salts, esters, and the ionic form of hyaluronic acids are called **hyaluronates**.

hyaluronidase, 1. any of the three enzymes found e.g. in sperm or in bee and snake venoms, i.e. hyaluronoglucosaminidase (EC 3.2.1.35), hyaluronoglucuronidase (EC 3.2.1.36), or hyaluronate lyase (EC 4.2.2.1); the enzymes catalyse the decomposition of → hyaluronic acid and other hyaluronates; **2.** a soluble enzyme preparation with hyaluronidase activity prepared from mammalian testes.

H-Y antigen, H-W antigen, a minor histocompatibility antigen encoded by a locus on the → Y chromosome in most vertebrates or → W sex chromosome in birds and reptiles; formerly wrongly believed to be responsible for the differentiation of testes. *See* major histocompatibility molecules.

hybrid (L. *hybrida*), any offspring born from the cross between two genetically different individuals (often belonging to different species). *Cf.* cell hybrid.

hybrid DNA, a double-stranded DNA molecule or hybrid DNA region where the two strands do not have completely complementary base sequences. Hybrid DNA can arise from mutations, recombination, or by annealing single strands *in vitro*.

hybrid dysgenesis, HD (Gr. *dys-* = disturbance, *genesis* = origin, birth), a genetic phenomenon found in crossing of certain fruit fly strains; the fruit flies can be divided into two types, M and P types, of which the genome of the latter type contains → P elements, which are lacking in the genome of the M type. When an M female is crossed with a P male, a dysgenic progeny that is sterile containing many mutants, is born. This occurs because the P elements, a kind of → transposons, are mobilized in the M cytoplasm. In the reciprocal cross when a P female is crossed with a M male, a completely normal progeny will develop.

hybrid inviability, the inviability of hybrids from crosses between species; it is one form of postzygotic isolation mechanisms (*see* isolation).

hybridization, 1. the reproduction of two genetically different organisms, leading to a → hybrid progeny; occurs normally in reproduction of individuals living in a natural population, also between related plant species and rarely between some animal species; e.g. may occur between related *Anas* species, or e.g. between the black grouse and capercaillie; **2.** a concept in molecular biology and gene technology, meaning the union of two single-stranded nucleic acid molecules, opposite in their polarity, thus forming a double-stranded molecule through base pairing; **3.** the artificial union of cells being true hybrids if the cell nuclei are fused, or cytoplasmic hybrids (→ cybrids) if only their cytoplasms are fused; *see* cell hybrid.

hybridoma, a clone of somatic → cell hybrids artificially produced by fusion of a myeloma cell (tumour cell) and an antibody producing cell (lymphocyte); hybridoma cells can multiply indefinitely and produce large quantities of monoclonal antibodies in cell cultures.

hybrid sterility, the inability to produce hybrid offspring from crosses between species; one form of postzygotic isolation mechanisms (*see* isolation).

hybrid swarm, a collection of hybrids produced by complete or local breakdown of isolation barriers between two → sympatric species or subspecies, allowing their interbreeding.

hybrid vigour, the vegetative superiority of hybrids from crosses between different races or species. *Syn.* heterosis.

hybrid zone, a geographical area where two, previously allopatric populations (geographical races) hybridize with each other owing to the absence of genetic isolation.

hydathode (Gr. *hydor* = water, *hodos* = entrance), a water-secreting structure found in the margin of a leaf, excreting water even in high relative humidity of air when normal → transpiration is not possible. *See* guttation.

hydatid cyst (Gr. *hydatis* = blister), a large cyst in various internal organs of many mammals including humans, caused by larval stages of some tapeworms, such as the hydatid worm, *Echinococcus*; the larva is a special kind of cysticercus or onchosphere, and the cyst may grow larger than 20 cm in diameter, causing the **hydatid disease**.

hydr(o)- (Gr. *hydor* = water), denoting relationship to water or hydrogen.

hydranth (Gr. *anthos* = flower), → gastrozooid.

hydrarch succession (Gr. *arche* = beginning), the temporal development of vegetation during upgrowth of a pond or lake. *See* succession.

hydratases, → anhydrases.

hydrate, 1. a compound crystallizing with water; crystalline hydrates contain a constant number of water molecules at a certain temperature, e.g. $CuSO_4 \cdot 5\ H_2O$; **2.** as a *verb* to **hydrate**, i.e. to combine chemically with water. *Cf.* carbohydrate.

hydration, the act of combining, or the conditon of being combined with water.

hydrobiology (Gr. *bios* = life, *logos* = discourse), a branch of biology studying biological phenomena in water bodies. *Cf.* limnology.

hydrobiont (Gr. *bion* = living), an organism living in water.

hydrocarbon, an organic compound containing only hydrogen and carbon, the **alicyclic** and **aromatic** hydrocarbons having ring structures (e.g. cyclohexane, benzene), the **aliphatic** hydrocarbons having straight chains (e.g. alkanes, alkenes). Hydrocarbons occur in natural gas, petroleum, coal, bitumens, etc., produced by the petroleum industry; most are toxic to organisms, some are carcinogenic (e.g. benzene).

hydrocaulus, also **hydrocaulis,** pl. **hydrocauli** (Gr. *kaulos* = stalk, stem), the basal stalk developed from a hydrorhiza in colonial hydroids which, in the course of development, gives rise to new branches by lateral budding; is covered by a transparent shelter, **perisarc,** and a central tube, **coenosarc.**

hydrocephaly, hydrocephalus (Gr. *kephale* = head), a condition of disturbed development of a foetus, marked by dilated brain ventricles due to the accumulation of cerebrospinal fluid in the skull; it is chiefly due to obstruction of the pathways for cerebrospinal fluid. The condition is found e.g. in some mammalian species, man included.

hydrochloric acid, an aqueos solution of hydrogen chloride, HCl; as a concentrated solution it is a colourless or sometimes yellowish, fuming and corrosive inorganic liquid. In many animals it is formed in the stomach for hydrolysing food materials and protecting the animal against microbes. The parietal cells (oxyntic cells) of gastric glands secrete protons (H^+) and chlorine ions (Cl^-) thus producing strongly acidic juice (pH appr. 1). The commercial product of hydrochloric acid is widely used in industry but also in biological studies e.g. to hydrolyse tissues or to adjust pH of solutions.

hydrochloride, a salt formed when hydrochloric acid reacts with an organic base; also a salt which contains the $[ClHCl]^-$ anion.

hydrochore (Gr. *chorein* = to spread), an organism spreading with water; especially a plant whose → diaspores (seeds or spores) spread mainly with the aid of water. *Cf.* anemochore, anemohydrochore.

hydrocoel (Gr. *koiloma* = hollow), a coelomic vesicle of an echinoderm embryo giving rise to the → water-vascular system. *Cf.* axocoel, somatocoel.

hydrocortisone, → cortisol.

hydrogen (L. *hydrogenium* < Gr. *hydor* = water, *gennan* = to produce), the first member of the periodic table of elements, symbol H, relative atom mass 1.00794, melting point -259.14°C; occurs free as a diatomic gas, H_2, the lightest of all gases. Hydrogen is a reactive element explosively reacting with oxygen to form water, H_2O, and it is widely spread in inorganic compounds (water and minerals), but also forms an essential component in organic compounds. It has biological importance acting as a reducing agent in cellular reactions, and in its ionic form, H^+, is present in physiological fluids of organisms (*see* proton, pH, citric acid cycle, oxidative phosphorylation).

Hydrogen is produced by electrolysis of water or by catalytic reforming of petroleum and other hydrocarbons. It is used e.g. as a reducing agent, in the manufacture of synthetic ammonia (→ Haber—Bosch process) and synthetic oil, and for hydrogenation of oils. There are also two heavy isotopes of hydrogen, **deuterium** (^2H, hydrogen-2) and **tritium** (^3H, hydrogen-3) which are of importance in nuclear physics. Tritium is also widely used as a radioactive label in chemical and biological studies.

hydrogenase (Gr. *genesis* = origin, to be born), hydrogenlyase; any enzyme that removes hydrogen from a compound, especially from NADH, such as the enzymes of the → citric acid cycle; also an enzyme which adds hydrogen to a compound, as to ferricytochrome, e.g. hydrogen dehydrogenase (EC 1.12.1.2) and cytochrome C_3 hydrogenase (EC 1.12.2.1), or a bacterial hydrogenase (EC 1.18.3.1) that uses molecular hydrogen for reduction of many compounds. *Cf.* dehydrogenase.

hydrogenation, the addition of hydrogen to a compound, particularly to unsaturated fatty acid or fat.

hydrogenation of oils, the artificial solidifying (hardening) of liquid oils and fatty acids by adding hydrogen; in the process e.g. the liquid triolein, $C_{57}H_{104}O_6$, occurring abundantly in soft fats and oils, is converted into solid form i.e. tristearin, $C_{57}H_{110}O_6$ by the action of hydrogen in the presence of a nickel catalyst. The process is used in the manufacturing of margarine.

hydrogen bond/linkage, a weak chemical bond in which the hydrogen atom forms a bridge between two other atoms; it is a non-covalent association between a hydrogen atom and an electronegative atom, the bond being much weaker than → covalent bonds. Hydrogen bonds may occur between hydrogen and strongly electronegative atoms such as fluorine (F), chlorine (Cl), oxygen (O) and nitrogen (N). The bonds are important in stabilizing the threedimensional structure of biological macromolecules, especially in joining polypeptide chains to form secondary, tertiary, and quaternary protein structures, and for base pairing between complementary nucleic acid strands. Also water molecules are bound with each other by hydrogen bonds.

hydrogen carbonate, → bicarbonate; the hydrogen-containing salt or ion (HCO_3^-) of carbonic acid; important in nutrition of water plants and acts as a pH → buffer in organisms (carbon dioxide-bicarbonate system).

hydrogen cyanide (Gr. *kyanos* = dark blue), HCN, a gas forming hydrocyanic acid (prussic acid) in water, a colourless, very poisonous liquid with the smell of bitter almond; inhibits the oxidative cell respiration. Hydrogen cyanide is used for manufacturing synthetic resins. *See* cyanide.

hydrogen electrode, a glass electrode used for measuring of hydrogen ion (H^+) concentrations in liquids (pH measurements); has a H^+ selective membrane.

hydrogen fluoride, HF, a soluble gas forming hydrofluoric acid in aqueous solution which is a fuming liquid used e.g. as a solvent and for etching glass.

hydrogen ion, H^+, → proton.

hydrogen-ion concentration, *see* pH.

hydrogen peroxide, H_2O_2, a colourless, viscous liquid, usually sold as 30% solution in water; it is used as a disinfectant, bleaching agent and a powerful oxidizing agent e.g. in a wet combustion of material for determining of total nitrogen. Hydrogen peroxide is formed in small quantities in cell metabolism when → oxygen radicals are processed. It is toxic to cells, and therefore both animals and plants have → catalase enzyme that decomposes it into water and oxygen.

hydroid, 1. hydrozoan, i.e. an individual in the order → Hydroida; **2.** the polyp form of a hydrozoan as discerned from the medusa form.

Hydroida (Gr. *eidos* = form), **hydroids**; an order of hydrozoans comprising species which in their life cycle have a well developed sessile polyp stage, either solitary or colonial. *See* Hydrozoa.

hydrolases, hydrolytic enzymes, EC class 3., catalysing the decomposition of carbohydrates, proteins, and lipids by combining water at the point of cleavage; e.g. saccharase and other carbohydrases, peptidases, proteinases, lipases, phosphatases, nucleases, occurring intracellularly in → lysosomes, or extracellularly produced by glandular cells e.g. in the alimentary canal.

hydrolittoral (L. *lit(t)us* = shore), *see* littoral.

hydrological cycle/circulation, physical water cycle; includes the global water cycling on the Earth and in the atmosphere, i.e. water evaporating from water bodies (seas, lakes, rivers), moist land, and organisms to the air, rains back to the earth; on land, **surface water** runs to water bodies or is absorbed into the ground forming **ground water** that stays for longer periods. *See* water circulation.

hydrology (Gr. *logos* = word, discourse), a branch of science studying physical phenomena of water bodies, as well as occurrence, distribution, and cycling of water on the Earth. *Cf.* hydrobiology, limnology, oceanology.

hydrolysis (Gr. *lysis* = dissolution), the decomposition of a compound by adding water into the molecule at the point of cleavage. The water itself is also broken down, the hydrogen atom incorporating into one fragment of the molecule and the hydroxyl group into the other. In organisms, the reaction is catalysed by hydrolase enzymes, but the hydrolytic dissolution may occur also non-enzymatically in acid or basic solutions. *Adj.* **hydrolytic**.

hydromedusa, pl. **hydromedusae** (Gr. *Medousa* = one of the Gorgons in the Greek mythology), a → medusa stage in the life cycle of small hydrozoans, especially of *Trachylina*. The medusa has a shelf-like velum beneath the bell margin, and by contracting and extending the velum the animal transfers water into the body.

hydrometer (Gr. *metron* = measure), an instrument for measuring the specific gravity (density) of a liquid, consisting of a weighted bulb and a graduated slender stem. The apparatus floats vertically in the liquid and sinks to a certain depth based on the density of the liquid being tested.

hydronium ion, name formerly used for hydroxonium ion, H_3O^+. *See* protolysis.

hydrophilic (Gr. *philein* = to love), having an affinity to water; **hydrophilous**, pollinated by means of water.

hydrophobic (Gr. *phobein* = to be afraid), **1.** lacking affinity to water, insoluble in water, not absorbing water; **2.** pertaining to **hydrophobia**, i.e. fear to drink in → rabies.

hydrophyte (Gr. *phyton* = a plant), *see* water plants.

hydroponics (Gr. *ponein* = to toil), *see* water culture.

hydroquinone, quinol; benzene-1,4-diol, $C_6H_4(OH)_2$; a white crystalline substance used as an antioxidant and photographic developer.

hydrorhiza, pl. **hydrorhizae** (Gr. *rhiza* = root), → stolon.

hydrosere (L. *serere* = put in order, link), the temporal → succession of vegetation in a wet environment.

hydrosphere (Gr. *sphaira* = ball), a common name for all globally existing water; includes the water in oceans, lakes, rivers, ponds, and in the ground and atmosphere.

hydrostatic (Gr. *statikos* = standing), pertaining to the pressure of a liquid in the state of equilibrium.

hydrostatic organs, the organs by which animals regulate their specific gravity in water; e.g. air sacs of water insects or swim-bladders of most fishes. *See* buoyancy.

hydrostatic pressure, the pressure at a certain level on a liquid (water) due to the weight of the liquid above it; it (*p*) is directly proportional to the density (ρ) of the liquid and the distance (*h*) of that point from the liquid surface: $p = \rho gh$; hydrostatic pressure is an important factor for movement of tissue fluids in organisms.

hydrostatic skeleton, a type of skeleton found in many soft-bodied invertebrates, as e.g. in cnidarians, annelids and priapulids, in which coelomic fluid provides support for antagonistic muscle action; serves as an internal hydrostatic skeleton when circular and longitudinal muscles contract alternately against the coelomic fluids in a limited space.

hydrotaxis (Gr. *taxis* = arrangement), an orientation movement of a cell or organism as a response to water or moisture.

hydrotheca (Gr. *theke* = box), a chitinous, protective sheath surrounding the hydranth (→ gastrozooid) in hydroid colonies. *See* perisarc.

hydrotropism (Gr. *trepein* = to turn), a growth movement of a plant or fungus towards water (positive hydrotropism) or away from water (negative hydrotropism).

hydrox- (< hydrogen + oxygen), pertaining to a chemical compound containing a → hydroxyl group, -OH.

hydroxide, a base containing a potentially ionizable hydroxyl group (OH^-); formed particularly when an alkaline earth metal (e.g. calcium) or an alkali metal (e.g. sodium or potassium) reacts with water, forming the corresponding hydroxides, $Ca(OH)_2$, NaOH, and KOH.

hydroxonium ion, H_3O^+, *see* protolysis.

hydroxyapatite, a form of → apatite, $Ca_{10}(PO_4)_6(OH)_2$; occurs in rocks and as a component of bone and dentine. Its commercial product is used e.g. as a column packing material in → chromatography.

hydroxylamine, an unstable intermediate compound, NH_2OH, formed when nitrite (NO_2^-) is reduced into the ammonium form (NH_4^+) in the nitrate reduction pathway; the reaction occurs e.g. in plant leaves. Hydroxylamine is a toxic substance acting e.g. as an uncoupler of photophosphorylation. It is a highly specific mutagen reacting with cytosine to give a derivative that pairs with adenine rather than with guanine. *Syn.* oxammonium.

hydroxyl group, a univalent atom group, -OH, that occurs e.g. in → hydroxides, phenols, alcohols, and sugars. In water the hydroxides produce **hydroxyl ions** (OH^-) rendering the solution alkaline; hydroxyl ions are also formed in dissociation of water. Alcohols or sugars do not change pH, but phenols are acids because their OH groups liberate protons, i.e. hydrogen ions, H^+.

hydroxyl radical, a strongly reactive free radical, $OH^\cdot$, that is formed when hydrogen peroxide oxidizes metals such as Fe^{2+} and Cu^{2+}; e.g. $Fe^{2+} + H_2O_2 \longrightarrow Fe^{3+} + OH^\bullet + OH^-$. *See* oxygen radicals.

hydroxyproline, abbr. Hyp, HOP; 4-hydroxy-2-pyrrolidinecarboxylic acid; an imino acid present abundantly in proteins of connective tissue, especially in → collagen; → non-essential amino acid in animals; **hydroxyproline NAD$^+$ oxidoreductase** (EC 1.1.1. 104) is the enzyme that catalyses the oxidation and reduction of hydroxyproline.

8-hydroxyquinoline, oxine, "brown needles", C_9H_7ON; an organic substance forming insoluble complexes with metals and is therefore used in determination of magnesium, aluminium, zinc, and many other metals in tissue preparations. It is also used as a pretreatment substance for metaphase chromosomes to be examined in a light microscope; 8-hydroxyquinoline sulphate is antiseptic and therefore used in deodorants and as a disinfectant.

5-hydroxytryptamine, 5-HT, → serotonin.

Hydrozoa (Gr. *zoon* = animal), **hydrozoans**; a cnidarian class comprising either solitary or colonial species. Their life cycle includes both the polyp and medusa stages, or only one of them. In species having both stages, the polyp is asexual and the medusa sexual. Often the medusa stage is lacking, as in many freshwater hydrae. Most hydrozoans are marine and colonial; some marine hydroids do not have free sexual stages, medusae, which have been reduced into the gonadal tissue, called **gonophores** (e.g. the genus *Tubularia*). Hydrozoan orders are e.g. **hydroids** (Hydroida), **trachyline medusae** (Trachylina), **fire corals** (Milleporina), and **siphonophores** (Siphonophora).

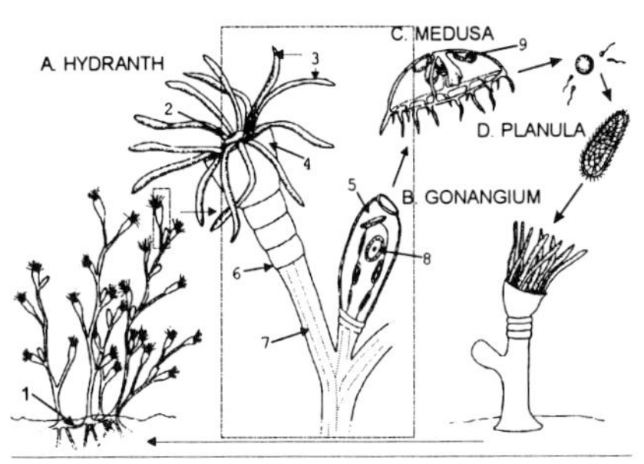

LIFE CYCLE OF COLONIAL HYDROZOANS

1. stolon 2. mouth 3. tentacles 4. hydrotheca 5. gonotheca 6. perisarc 7. coenosarc 8. medusa bud 9. ovary

Hydrozoa.

hygr(o)- (Gr. *hygros* = moist), denoting moist.

hygrokinesis, *see* kinesis.

hygrophilous, hygrophilic (Gr. *philein* = to love), an organism which favours moisture; e.g. hygrophyte.

hygrophyte (Gr. *phyton* = plant), a plant which favours moist habitats but is not a → water plant, hydrophyte.

hygroscopic (Gr. *skopein* = to view), absorbing water; pertaining to a substance which absorbs and retains readily moisture from the air, e.g. $CaCl_2$, NaOH, and many other chemicals, also some organic substances, like the hygroscopic slime secreted by frogs around their eggs. Hygroscopic substances can be used as desiccants.

hygroscopic movement, a movement of a plant part caused by unequal drying of tissues; occurs e.g. in the opening of fern sporangia. In dry weather the thick walls of the sporangium dry rapidly and the thinner walls on the opposite side are teared apart because of stretching, and dry spores spread out.

Hylobatidae, gibbons; a monophyletic family of arboreal primates in the superfamily Hominoidea; comprises six species in southeastern Asia and Indonesia. *See* Hominidae.

hymen (Gr. = membrane), a thin fold of the mucous membrane covering partly the orifice of the vagina in many virgin mammals.

hymenium, pl. **hymenia,** a superficial layer of a fungus (mushroom) producing basidia with → basidiospores in Basidiomycetes, or asci with → ascospores in Ascomycetes.

Hymenolichenes, → Basidiolichenes.

Hymenoptera (Gr. *pteron* = wing), **hymenopterans**; an insect order including about 110,000 species and two suborders, Symphyta (e.g. sawflies, horntails, woodwasps) and Apocrita (e.g. wasps, bees, and ants). Hymenopterans have chewing or chewing-lapping mouthparts and two pairs of membranous flying wings while the hindwings are clearly smaller, or are lacking in some species. The ovipositor of the female is in some species modified for sawing, piercing, or stinging. Body size and habits of the animals vary according to species. The order includes species with great economical significance, such as the species parasitic in pest insects, and many highly specified social insects.

hyoid (Gr. *hyoeides* = shaped like the letter Y, upsilon), **1.** a Y-shaped anatomical structure; **2.** pertaining to the **hyoid bone** (*os hyoideum*), lingual bone, tongue bone, i.e. a bone or a series of bones at the base of the tongue in tetrapod vertebrates; is developed from the hyoid arch.

hyoid arch, the second pharyngeal arch next behind the jaws of vertebrates, seen primarily as an arch-shaped structure in vertebrate embryos and fishes; in tetrapods it develops into hyomandibulum and → hyoid bone.

hyoscyamine, an alkaloid obtained from *Hyoscyamus niger*, *Atropa belladonna*, *Datura stramonium*, and other plants in the family Solanaceae; acts as an **anticholinergic drug** in animal tissues and is therefore used in medicine.

hyostylic cranium/skull (Gr. *stylos* = pillar), *see* cranium.

hyper- (L., Gr. *hyper* = above), denoting above, excessive, more than normal.

hyperactivity, 1. excessive acitivity, abnormally increased activity of any kind; **2.** → hyperkinesis.

hyperbranchial groove (Gr. *branchia* = gills), a groove on the dorsal wall of the pharynx of lancelets, comprising ciliated cells which carry food from the mouth cavity into the pharynx. *Syn.* epibranchial groove.

hypercalcaemia, Am. **hypercalcemia** (Gr. *kalkos* = lime, *haima* = blood), excess of calcium in the blood; causes disturbances especially in muscular and neuronal functions. *Syn.* hypercalcin(a)emia.

hyperchromatism, also **hyperchromia, hyperchromasia** (Gr. *chroma* = colour), **1.** excessive pigmentation of tissues; **2.** an increase of → chromatin in cell nuclei; **3.** the increased staining of histological tissue preparations.

hyperglycaemia, Am. **hyperglycemia** (Gr. *glykys* = sweet, *haima* = blood), abnormally increased concentration of sugar (glucose) in the blood; in vertebrates, occurs e.g. if the concentration or efficiency of insulin becomes exceptionally low, such as in → diabetes. *Cf.* hypoglycaemia.

hyperkalaemia, Am. **hyperkalemia** (L. *kalium* = potassium, Gr. *haima* = blood), excess of potassium in the blood; causes disturbances especially in neuronal and muscular functions. *Syn.* hyperkali(a)emia, hyperpotass(a)emia.

hyperkeratosis (Gr. *keras* = horn), abnormal hypertrophy of the corneous layer of the skin epidermis, or of the cornea in the eye.

hyperkinesis, hyperkinesia (Gr. *kinein* = to move), **1.** excessive muscular activity; **2.** hyperactive motility of humans due to some neural disturbances; sometimes spelled hypercinesis. *Adj.* **hyperkinetic**.

hypermorphic (Gr. *morphe* = form), pertaining to a → mutation or an → allele which acts qualitatively similarly to the corresponding wild allele but with an increased efficiency.

hypernatraemia, Am. **hypernatremia** (L. *natrium* = sodium), excessive concentration of sodium in the blood.

hyperopia (Gr. *ops* = eye), far-sightedness; a condition of the eye in which parallel rays of light are focused behind the retina, the eyeball being proportionally too short. *Syn.* hypermetropia.

hyperosmotic, pertaining to a solution with a higher osmotic concentration (i.e. lower osmotic potential) than a reference solution; e.g. the vacuolar sap of plant root cells as compared to the mineral solution in soil. *Syn.* hyperosmolar. *Cf.* hypo-osmotic. *See* hypertonic, osmosis.

hyperparasitism (Gr. *para* = beside, *sitos* = food), parasitism of a parasite by another parasite; **hyperparasite**, such an organism.

hyperplasia (Gr. *plassein* = to form, mould), abnormal increase in the number of normal cells in a tissue. *Cf.* hypertrophy.

hyperpolarization (Gr. *polos* = pole), a change in electric polarization across the plasma membrane more negative than the usual → resting potential level; most prominent in nerve and muscle cells. Hyperpolarization is caused by the outflux of positive ions or the influx of negative ions changing the negative membrane potential to a more negative value. Thus hyperpolarization decreases the discharge of → action potentials. *Cf.* depolarization.

hyperreactivity (L. *re* = back, *activus* = functioning, active), an abnormally great activity, superactivity; a hypernormal response of a cell, tissue, organ, or organism (especially behavioural) to stimuli. *Adj.* **hyperactive**.

hypersecretion, an excess secretion, especially of → gastric glands.

hypersensitivity (L. *sentire* = to feel), an exaggerated reactivity, especially an immune response to a certain antigen, e.g. to a foreign substance; may be immediate or delayed, e.g. in allergic reactions. An increased sensitivity related to the nervous system is usually called **hypersensitization**.

hypersomia (Gr. *soma* = body), gigantism. *See* growth hormone.

hypertensin (L. *tensio* = tension), the former name for → **angiotensin**.

hypertension, persistently high blood pressure. *Adj.* **hypertensive**.

hyperthermia, also **hyperthermy** (Gr. *therme* = heat), the increase of the normal body temperature of homoiothermic animals, i.e. birds and mammals. Temporary hyperthermia occurs normally in situations when the heat loss is not efficient enough, as e.g. in an excessively hot environment, especially under physical activity. For example, a hunting panther may have a hyperthermia of 43-44°C, and the body temperature of man in heavy physical training 2—3°C above normal. In these cases the thermoregulatory system of the body functions effectively to return the body temperature to normal.

In mammals living in dry desert areas, as e.g. camels, moderate hyperthermia is the normal adaptive phenomenon to save water, otherwise used for panting and sweating, during the hot day. The degree of hyperthermia is controlled by the brain thermoregulatory centre, which adjusts the set point to a higher temperature level and then back to normal when the weather cools. A special case of hyperthermia is → **fever**. *Adj.* **hyperthermic**. *Cf.* hypothermia, heterothermia. *See* homoiothermy, thermoregulation.

hyperthyroidism (Gr. *thyreoeides* = shaped like a shield, thyroid), excessive thyroid function, called also **hyperthyreosis** or **hyper-**

thyroidosis. *See* thyroid gland, thyroid hormones.

hypertonia (Gr. *tonos* = tension), **1.** excessive tension of skeletal muscles; **2.** increased blood pressure, hypertension. *Cf.* hypotonia.

hypertonic, 1. pertaining to → hypertonia; **2.** relating to → hypertonicity or → hypertonic solution. *Cf.* hypotonic.

hypertonic solution, hyperosmotic solution; any solution having a lower → osmotic potential (i.e. higher osmotic concentration) than a reference solution, usually a solution in an organism. *See* osmosis.

hypertonicity, the state of increased osmotic concentration of body fluids. *Cf.* hypotonicity.

hypertrophy, also **hypertrophia** (Gr. *trophe* = nutrition), **1.** the overgrowth of an organ resulting from the increased size of its cells, e.g. the overgrowth of muscles under training; *cf.* hyperplasia; **2.** the condition of a water system which is highly enriched with nutrients, i.e. the condition in a highly eutrophic water.

hyperventilation, overventilation, the state of excessive lung ventilation (breathing) decreasing carbon dioxide content in the lung alveoli; is a functional disturbance in the regulation of breathing in humans, resulting in respiratory alkalosis of the blood which further disturbs functions of the nervous system and muscles.

hypervitaminosis, the condition of excessive amount of fat-soluble vitamins (usually D or A) in the human body, caused by ingestion of vitamin preparations. The excess of vitamin A causes abnormal changes e.g. in the epithelium of the alimentary canal and the skin, the excess of vitamin D produces loss of weight, weakness and disturbances in the function of the liver. The condition is possible also in domestic animals fed with vitamin preparations. *Syn.* supervitaminosis.

hypervolume niche, *see* niche.

hypha, pl. **hyphae** (Gr. *hyphe* = web, fabric), any of the tubular, branching filaments which form the → **mycelium**, i.e. the basic structural unit and the vegetative growth stage of multicellular fungi. Primitive hyphae are continuous, multinucleate tubes; more developed ones have septa between separate cells. The wall of hyphae is rigid and contains chitin, sometimes also cellulose. The mycelium forms periodically fungal fruit bodies (e.g. mushrooms), consisting of **pseudoparenchyma** which resembles the related plant tissue but is composed of intertwined hyphae.

Hyphochytridiomycetes, water moulds, a fungal class in the subdivision Haplomastigomycotina, division Mastigomycota; in some systems classified into a separate division, **Hyphochytridiomycota**.

hypn(o)- (Gr. *hypnos* = sleep), pertaining to sleep or hypnosis.

hypnotoxin (Gr. *toxikon* = arrow poison), a poison secreted by certain cnidarians from their nematocyst cells, occurring e.g. in the tentacles of the *Physalia*, the Portuguese man-of-war; affects the nervous system and sense organs of other animals.

hypo- (Gr. *hypo* = under), denoting under, beneath, deficient, or below normal.

hypoblast (Gr. *blastos* = germ), → endoderm; especially refers to the endoderm of amniotes.

hypobranchial (Gr. *branchia* = gills), **1.** pertaining to a structure under the gills, e.g. the **hypobranchial space** below the gills of decapods, or the **hypobranchial groove**, i.e. → endostyle of tunicates and cephalochordates; **2.** denoting the 4th segment of a → gill arch.

hypocalcaemia, Am. **hypocalcemia** (Gr. *kalkos* = lime, *haima* = blood), deficiency of → calcium in the blood causing disturbances especially in the nervous system and muscles.

hypocercal fin (Gr. *kerkos* = tail), *see* caudal fin.

hypochlorite, a salt or ester of hypochlorous acid, HOCl; a strong oxidizer which is used as a disinfectant and bleaching substance, and in plant biotechnology as a surface-sterilizing agent for tissue pieces, e.g. sodium hypochlorite, NaOCl.

hypocotyl (Gr. *kotyle* = cup), a part of a plant embryo which during seed germination lengthens and lifts the → cotyledon(s) to the light.

hypodermis (Gr. *dermis* = skin), **1.** the cellular layer of the skin in invertebrates secreting the chitinous exoskeleton (cuticle) to cover it; **2.** the layer of loose connective tissue beneath the dermis (corium) of the vertebrate skin, also called **subcutis** (*tela subcutanea*); contains fat cells in many species; **3.** a layer or layers under the epidermal layer in some plants or certain parts of plants, as in conifer needles or in shoots of cacti.

hypogeal, hypogean (Gr. *ge* = earth), growing

or living underground.

hypogeal germination, also **hypogean germination,** a type of seed germination in which the → cotyledons of a plant embryo do not rise from the earth above the ground; e.g. the germinating of peas. *Cf.* epigeal germination.

hypoglycaemia, Am. **hypoglycemia** (Gr. *glykys* = sweet, *haima* = blood), abnormally decreased concentration of sugar (glucose) in the blood; caused e.g. by fasting or disturbed glucose regulation.

hypognathous (Gr. *gnathos* = jaw), **1.** having a downwards directed mouth (the axis is at right angles to the axis of the body) far back under the head, as e.g. in some insects; *cf.* orthognathous, prognathous; **2.** having a protruding lower jaw; especially refers to man.

hypogynous (Gr. *gyne* = woman), pertaining to a flower in which the gynoecium is situated at the apex of the receptacle, the other flower parts being below. *Cf.* epigynous, perigynous.

hypokalaemia, Am. **hypokalemia** (L. *kalium* = potassium, Gr. *haima* = blood), **hypopotass(a)emia**; abnormally decreased concentration of potassium in the blood.

hypolimnion, pl. **hypolimnia** (Gr. *limne* = lake), a water layer between the → thermocline and bottom in lakes and seas during the summer stagnation. The water overturn in the hypolimnion is slight and therefore the oxygen content decreases gradually; the temperature is usually evenly cool. *Cf.* epilimnion.

hypomere (Gr. *meros* = part), **1.** the ventrolateral part of each → myotome of a vertebrate embryo developing into skeletal muscles, innervated by the primary ventral ramus of a corresponding spinal nerve; **2.** sometimes denoting the lateral plate of the mesoderm of an embryo, developing to line the body cavities.

hypomorphic (Gr. *morphe* = form), pertaining to a → mutation or an → allele of a gene which acts qualitatively similarly to the corresponding wild allele but with a decreased efficiency.

hyponasty (Gr. *nastos* = close-pressed, firm), the growth of a flattened plant organ (e.g. leaf) which occurs more rapidly on the lower side than on the upper side; causes upward curling of the structure.

hyponatraemia, Am. **hyponatremia** (L. *natrium* = sodium, Gr. *haima* = blood), an abnormally decreased concentration of sodium in the blood.

hyponeuston, → infraneuston.

hypo-osmotic, pertaining to any solution that has a lower osmotic concentration (i.e. higher → osmotic potential) than a reference solution. *See* osmosis. *Cf.* hyperosmotic.

hypopharynx, an unpaired median projection in the mouth of insects; in the chewing → mouthparts is short and tongue-like (e.g. in grasshoppers), and in the piercing mouthparts forms a long tube (e.g. in mosquitoes).

hypoplasia (Gr. *plassein* = to mould), underdevelopment or arrested development of a tissue or an organ, mainly due to the decreased number of cells. *Cf.* hyperplasia.

hypophysis, pl. **hypophyses** (Gr. *phyein* = to grow), → pituitary gland.

hypopus, pl. **hypopi** (Gr. *pous* = foot), an intermediate stage of the deutonymph of some mites (Acari); during the stage the nymph does not eat. Dispersion of the species to new areas occurs often during this stage by → phoretic dispersal.

hypostasis, pl. **hypostases** (Gr. *stasis* = halt), the suppression of the effect of a given dominant gene by the effect of another non-allelic dominant gene or a pair of recessive genes (recessive hypostasis). *Adj.* **hypostatic**. *Cf.* epistasis.

hypostoma, pl. **hypostomata** or **hypostomas** (Gr. *hypo* = under, *stoma* = mouth), **hypostome**; a structure associated with the mouth, e.g. **1.** a hooked, serrated proboscis with a dorsal groove in the head of ticks (*Acari*); **2.** the mouth at the end of → manubrium in hydrozoan medusae; **3.** a fold in the posterior margin of the crustacean mouth.

hypotension, subnormal arterial blood pressure. *Adj.* **hypotensive**.

hypothalamus (Gr. *thalamos* = inner chamber), the ventral part of diencephalon under the → thalamus in the vertebrate brain. Important autonomic centres, such as the thermoregulatory centre, hunger and thirst centres, and the nuclei for the control of biorhythms are located in the hypothalamus. It also has nuclei for controlling the endocrine functions through the → pituitary gland. These nuclei are composed of neurosecretory cells, i.e. cell bodies of axons that terminate either on the capillary loops of the median eminence where they produce → releasing hormones (liberins) or inhibiting hormones (statins) via the portal system to the anterior → pituitary gland, or they terminate in the posterior pituitary gland (neurohypophysis), releasing → oxytocin or → vasopressin. Thus the hypothalamus exerts wide control of the endocrine system of the entire body.

hypothallus (Gr. *thallos* = stem), **1.** an undifferentiated plate-like thallus in some lichens, appearing at early stages before the formation of differentiated structures; e.g. *Cladina* species; **2.** a thin layer in the fructification (plasmodium) of slime moulds (→ Myxomycetes) under the sporangia.

hypothermia, also **hypothermy** (Gr. *therme* = heat), the temporarily decreased body temperature (core temperature) of homoiothermic animals, i.e. birds and mammals that is a normal physiological reaction (adaptive hypothermia) in many species. Many small mammals and birds are capable of saving energy in a slight hypothermia (**daily torpor**) during unfavourable periods. The initiation of hypothermia may be induced by cold, dryness, light, and shortage of food or water. Hypothermia is mostly only 5—10°C, i.e. body temperature is 5—10°C below normal, such as in small birds and rodents during cold nights, or in bigger mammals such as bear, badger, and raccoon during the → **winter sleep**, but may be tens of degrees in some animal species, such as ground squirrels and the hedgehog, in deep → **hibernation**. In all these cases, hypothermia is controlled by the thermoregulatory system (*see* thermoregulation).

In humans, especially in small children, uncontrolled hypothermia may be caused by a cold exposure, as in cold water. It is an abnormal and dangerous state, for consciousness disappears even in quite a mild hypothermia, i.e. at a body temperature of 33—34°C, although some individuals have survived in clinical restoration from profound hypothermia (below 20°C). Many drugs and alcohol increase the risk for hypothermia. In surgery, hypothermia is used in a controlled way to reduce the metabolism of tissues. *Adj.* **hypothermic**. *Cf.* heterothermia, hyperthermia. *See* homoiothermy.

hypothesis, pl. **hypotheses** (Gr. *hypotithenai* = to suppose), a scientific assumption to be tested with experiments or other studies.

hypothyroidism (Gr. *thyreoeides* = shaped like a shield, thyroid), reduced thyroid function, called also **hypothyreosis** and **hypothyroidosis**. *See* thyroid gland, thyroid hormones.

hypotonia, also **hypotony** (Gr. *hypo* = under, *tonos* = tension), **1.** decreased tension of muscles, i.e. decreased muscular tonicity; **2.** reduced tension of an organ or a part, as decreased blood pressure or decreased eyeball pressure. *Syn.* hypotension. *Cf.* hypertonia.

hypotonic, 1. pertaining to → hypotonia; **2.** pertaining to → hypotonicity. *Cf.* hypertonic.

hypotonicity, decreased osmotic concentration (pressure) of body fluids. *Cf.* hypertonicity.

hypotonic solution, any solution that has a higher → osmotic potential than a reference solution. *See* osmosis.

hypotrichia, hypotrichosis (Gr. *thrix* = hair), deficiency of hairs, especially in man. *Syn.* oligotrichia, oligotrichosis.

hypotrophy (Gr. *trophein* = to nourish), **1.** subnormal growth, i.e. the retardation of the growth of an organ or organism, in animals usually owing to deficiency of food; **2.** the growing of a plant branch or organ abnormally thick on the underside.

hypovitaminosis, deficiency of vitamins.

hypoxanthine (Gr. *ksanthos* = yellow), 6-oxypurine, $C_5H_4N_4O$; a purine base present in animal tissues as an intermediate product of uric acid; formed in purine catabolism by deamination of adenine and is converted into xanthine by xanthine oxidase, and further into uric acid. Hypoxanthine occurs also in plant tissues. A derivative of hypoxanthine, → inosine monophosphate (IMP), takes part in the biosynthesis of purine nucleotides.

hypoxia (L. *oxygenium* = oxygen), the reduction of oxygen supply in tissues below the physiological level. *Cf.* anoxia.

hypoxia inducible factor, HIF, a protein that acts as a specific → transcriptor factor at low oxygen level activating genes which improve survival in hypoxic conditions. HIF induces the synthesis of some glycolytic enzymes and thus increases the efficiency of anaerobic metabolism. In vertebrates HIF induces e.g. the synthesis of vascular endothelial growth factor, erythropoietin, and transferrin, thus improving oxygen uptake into cells.

hyps(i)-, hypso- (Gr. *hypsos* = height), denoting relationship to height, something high.

hypsodont (Gr. *odous, odon* = tooth), having long teeth; pertaining to a kind of teeth with a high crown and short roots, e.g. the molars of herbivores.

Hyracoidea (L. *hyrax* = shrew, Gr. *eidos* = form), **hyracoids, hyraxes, conies** (coneys);

a mammalian order including seven species in Africa and southwest Asia; in appearance the animals resemble rodents, but the structure of the skeleton and the nervous system are phylogenetically nearer those of the odd-toed ungulates and proboscideans.

hyster(o)- (Gr. *hystera* = uterus), denoting relationship to the uterus or womb, or to hysteria.

hysteresis (Gr. *hysterein* = to come later), a lagging behind, a time lag in the occurrence of two associated phenomena, e.g. a physical lag effect best seen in the elastic or magnetic behaviour of materials. When a material is stressed, a related strain is produced. The strain caused by a given stress is greater when the stress changes on the decreasing course than when it changes on the increasing course, i.e. there is a lag between the release of the stress and strain. On the complete removal of the stress, recidual strain may remain. Hysteresis occurs also in functions of organisms; e.g. in difference between muscle contractions with increasing and decreasing tensions, or in difference between temperature dependences of a physiological function in increasing and decreasing temperatures. This has been shown e.g. in some freezing-tolerant animals, such as insects and fish, where hysteresis proteins cause a lagging in the freezing and thawing cycles of the haemolymph.

hysterosoma (Gr. *hysteros* = after, later, *soma* = body), the posterior part of the body (idiosoma) of some acarines, including the 3rd and 4th pairs of legs, the anus, and the genital orifice. *Cf.* propodosoma.

hysterothecium (Gr. *hystera* = uterus), a fruiting body in some fungi in → Ascomycetes; actually one type of → apothecium, but opened in moist and closed in dry conditions.

I

I, symbol for **1.** iodine; **2.** isoleusine; **3.** inosine.

***I*,** symbol for **1.** radiant energy; **2.** ionic strength; **3.** electric current; **4.** sound intensity; **5.** luminous intensity.

i, intensity of selection.

-ia, (L., Gr. = a substantive-forming suffix), as a verb termination denoting a condition or state. *See* → -iasis.

IAA, indole-3-acetic acid. *See* auxins.

IAN, → indole-3-acetonitrile.

-iasis, pl. **-iases** (L. < Gr. a verb-nominalizing suffix), as a suffix, denoting a condition or state, e.g. in names of many diseases, e.g. → **helminthiasis**; sometimes **-osis**.

I-band, I-zone, isotropic band; any of the isotropic regions along a muscle fibril in striated and cardiac muscles; I-bands are seen as lighter regions between darker cross striations, i.e. anisotropic bands (**A-bands**). In the middle of each I-band is a → Z-line (Z-disc). The I-band regions are composed of thin filaments only, the A-band regions comprise thick filaments (myosin filaments) and end parts of thin filaments. *See* isotropic.

IBA, indole-3-butyric acid. *See* auxins.

ice age, a geological epoch of widespread glaciation, i.e. when large areas of the Earth were covered by ice sheets. The latest Ice Age, during which much of the northern hemisphere was covered by thick ice sheets, ended about 10,000 years ago.

ice-scorch (disease), the ice-injury which kills plants in cold areas especially in the spring; freezing of the ground during a cold spring prevents the uptake of water and may kill plants which have already started their growth.

ice sheet, a large ice cap or ice cover; one of large polar glacial areas covered by ice and snow masses. The present ice sheets of Greenland and Antarctica are relicts of massive continental glaciers from the last Ice Age. *Syn.* continental glacier.

ichthyology (Gr. *ichthys* = fish, *logos* = word, discourse), a branch of biology studying fish.

Ichthyosauria (Gr. *sauros* = lizard), **ichthyosaurs**; a marine order of fish-like reptiles which occurred in the Mesozoic era; they had four paddle-like limbs for swimming but the limb girdles were reduced.

ICSH, interstitial-cell stimulating hormone. *See* luteinizing hormone.

ideal gas, perfect gas; a theoretical gas that exactly obeys the gas laws. *See* equation of condition.

identical twins, monozygotic twins. *See* monozygotic.

ideo- (Gr. *idea* = form, notion, image), pertaining to ideas. *Cf.* idio-.

ideotype (Gr. *typos* = mark, type), a term particularly used in the breeding of forest trees; means an ideal, hypothetical model of a plant which is the goal of breeding. Sometimes means the rule for selection according to which a breeder acts.

idi(o)- (Gr. *idios* = one's own, private), pertaining to one's own, self-produced, separate, peculiar. *Cf.* ideo-.

idioblast (Gr. *blastos* = bud), a plant cell which differs in size or structure from its neighbour cells; e.g. a cell containing slime, or a → sclereid (stone cell).

idiochromosome, *see* sex chromosome.

idiogram (Gr. *gramma* = stroke, mark, letter), a schematic representation of the → chromosome set of a species.

idiosoma (Gr. *soma* = body), the body of mites and ticks which is compact and ovoid, or formed from two parts comprising the **propodosoma** and **hysterosoma**.

idiotope (Gr. *topos* = place), *see* idiotype.

idiotype (Gr. *typos* = a mark, model), **1.** the sum total of all hereditary determinants (nuclear or extranuclear) of an organism; **2.** an idiotypic antigenic determinant; a set of one or more **idiotopes**, i.e. determinants of the variable regions of an → immunoglobulin (antibody) molecule, produced by a clone of lymphocytes. Idiotypes are found outside of, within, or near to the antigen-binding site of the immunoglobin molecule, and itself may provoke the formation of other immunoglobins.

Ig, → immunoglobulin.

ileum (L. *ilia* = flanks), **1.** the last section of the small → intestine in mammals, i.e. between the jejunum and colon; **2.** the first section (small intestine) of the hindgut in insects.

ilium, pl. **ilia** (L. *os ilium*), **iliac bone**; flank bone; the paired bone in the pelvic girdle of tetrapods, articulating with one or more sacral vertebrae. *See* pelvis, hip bone.

illegitimate crossing-over, an event resembling → crossing-over but occurring between non-homologous positions.

imaginal disc (L. *imago* = picture, image, mature insect), a collective name for sheet- or pouch-like structures (**buds**) in the larvae or pupae of some insects consisting of undifferentiated but determined cells from which the legs, wings and other organs of the adult are formed. *Syn.* imaginal bud.

imago, pl. **imagoes** or **imagines** (L. = image), in zoology, the last stage in the metamorphosis of insects, i.e. the final, sexually mature stage. *Cf.* instar, larva, nymph, pupa.

imbibition (L. *imbibere* = to drink in), absorption of fluid by a solid material, such as a colloidal gel (colloidal imbibition); occurs e.g. in swelling of organs, as e.g. in seeds during germination or in absorbtion of free haemoglobin by a tissue (haemoglobin imbibition).

imbricate (L. *impricare* = to cover with imbrices or roof tiles), pertaining to overlapping structures (in plants), e.g. scales, leaves, or petals.

imidazole, a heterocyclic soluble crystalline substance, $C_3H_4N_2$; used in organic syntheses and as a buffer reagent.

imide, any compound containing the group =NH, e.g. calcium imide, CaNH; in organic compounds attached to two -CO- groups to form the imido group, -CONHCO-.

imine, a compound containing the imino group, NH=; derived from ammonia.

imino acids, heterocyclic secondary amines that occur in proteins, soluble in alcohol. Imino acids are often considered to belong to amino acids. Proline and hydroxyproline, usually considered to belong to amino acids, are the best known imino acids which are e.g. components in collagen proteins. The proline content increases in many plants during cold hardening.

immersion (L. *immergere* = to plunge into a fluid), **1.** the submergence into a liquid, e.g. during microscopical inspection by covering the microscopical objective on an object glass with liquid, usually with immersion oil or water; **2.** the disappearance of a planet behind another.

immigration (L. *immigratio* = movement into), the arrival of individuals from one population to another; immigration causes interbreeding which results in a gene flow and the increase of genetic variation. *See* emigration, migration.

immigration pressure, the rate of change of → gene frequencies in a population subject to → migration.

immobilization (L. *immobilis* = immovable), **1.** the act of making immovable; **2.** the assimilation of nutrients in the biomass. *Verb* to **immobilize.**

immune (L. *immunis* = free, exempt), **1.** being resistant against infectious diseases; **2.** pertaining to the immune response, immune system, or immunity.

immune response, a physiological response of animals consisting of cellular and molecular reactions against → antigens, such as substances or structures of various kinds of foreign organisms; **primary immune response,** occurs in the initial antigenic exposure, **secondary immune response,** occurs against the same antigen in a subsequent response, due to memory cells. *See* immunity.

immune system, a complex physiological defence system in animals consisting of cellular and molecular components. The system can distinguish between foreign and own material, and defend the animal against invading organisms and substances. It is best known in vertebrates, and it includes cellular components such as → lymphocytes and macrophages, and molecular components such as → antibodies, lymphokines, interferons, and the → complement system. *See* immunity.

immune tolerance, *see* immunological tolerance.

immunity, the physiological condition of an animal being resistant (immune) to certain → antigens, e.g. against infectious microorganisms or several kinds of foreign materials, also against transplanted foreign cells, tissues, or organs. Immune reactions maintaining the immunity of vertebrates may be **innate** (nonspecific) or **adaptive** (specific, specifically acquired); the former consist of **chemical** and **physical barriers**, as impermeable skin, mucus, lysozyme, sebum, acid in the stomach, and some intestinal factors, phagocytic cells of the reticuloendothelial system, neutrophilic leucocytes, interferons, and natural killer cells. The adaptive immune responses involve the functions of → **lymphocytes** and the → complement system. Lymphocytes either specifically kill microbes through direct cytotoxic reactions, as do the activated T lymphocytes (**cell-mediated immunity** (CMI), **cellular immunity**), or the lymphocytes secrete specific molecules, → immunoglobulins (antibodies), for processing antigenic materials, as do the activated B lymphocytes, the plasma cells (**humoral immunity**). Cellular immunity acts e.g. against fungi, various kinds of parasites, intracellular viral infections, cancer cells, and transplanted tissues; humoral immunity acts e.g. against bacteria and viruses. Usually, these two systems function simultaneously completing each other.

Immunity may be **natural** (inherited), e.g. performed by the function of → natural killer cells, or may occur through activation of specific immune reactions of B and T lymphocytes. This is called **active immunity** (acquired immunity) if induced by antigens naturally or by vaccination, or **passive immunity** if ready antibodies are transferred into an animal. This occurs during pregnancy via the placenta, or when prepared immunoglobulins or lymphocytes are administered to an animal (*see* immunization). Sometimes immune reactions are activated against one's own structures (**autoimmunity**). As far as it is known, the immune reactions of invertebrates are in principle similar to those of vertebrates. *See* immune response, antigen presenting cells.

immunization, the induction of → immunity actively or passively, i.e. rendering an animal less susceptible to certain antigens, such as pathogenic microbes or toxins; **active immunization**, the induction of immunity as a result of naturally acquired infection, or of intentional vaccination (artificial active immunization, *see* vaccine); **passive immunization**, the induction of immunity as a result of injection of appropriate → antibodies (immunoglobulins). *Verb* to **immunize**.

immunoadsorbent, immunosorbent (L. *ad* = to, *sorbere* = to suck), the preparation of insoluble → antigens, or → antibodies, used in immunoassay to bind the homologous antibodies or antigens, respectively.

immunoassay, any of the methods using the highly specific binding properties between an antigen and antibody for determination of the presence or quantity of a substance; these methods are e.g. → radioimmunoassay (RIA), enzyme immunoassay (best known is → ELISA), and fluoroimmunoassay.

immunobiology, the branch of biology that deals with immunological reactions of animals of different developmental stages and in various circumstances.

immunoblast (Gr. *blastos* = germ, shoot), *see* lymphoblast.

immunoblotting, a type of → Western blotting in which very small amounts of protein are transferred to nitrocellulose sheets and are then detected by their → antibodies.

immunocyte (Gr. *kytos* = cavity, cell), an immunologically competent cell, i.e. a cell of the lymphoid series capable (actively or potentially) of producing antibodies.

immunocytochemistry, a research branch of science using immunological techniques combined with cytochemistry.

immunodeficiency, in general, any deficiency of immune response; specifically, a disorder due to deficient immune response related to the function of T lymphocytes (cellular immunodeficiencies), or of B lymphocytes (antibody immunodeficiencies), or of phagocytes (phagocytic dysfunction disorders); best known is the acquired immunodeficiency syndrome, → AIDS.

immunoelectrophoresis, any type of → electrophoresis in which the separated molecules are detected by precipitation with an → antibody.

immunofluorescence, the use of a → fluorescence technique for immunohistochemical determinations. A fluorescent compound (fluorochrome, e.g. → fluorescein) is bound to an → antibody, and the antigen-antibody reaction occurring in an object tissue is observed using a fluorescence microscope. *See* microscope.

immunogen (Gr. *gennan* = to produce), any substance that is able to induce an immune response, i.e. is such an → antigen which after reacting with an antibody also induces the immune reaction.

immunogenetics (Gr. *genesis* = origin), a branch of genetics dealing with → antigens, → antibodies, and the reactions between them, and in general, the genetic basis of the immune response.

immunoglobulin, Ig, one of a class of structurally related, highly variable proteins in the blood and body secretions, functioning in the → immune system of vertebrates. Ig molecules are composed of two pairs of polypeptide chains, two identical **heavy chains** (H chains) and two identical **light chains** (L chains). These form a Y-shaped structure with a shaft (Fc segment) and the arms (Fab segments) with antigen binding sites. Ig molecules occur as **membrane-bound immunoglobulins** on the surface of B lymphocytes acting as antigen receptors, and as **antibodies**, secreted by activated B lymphocytes, i.e. plasma cells.

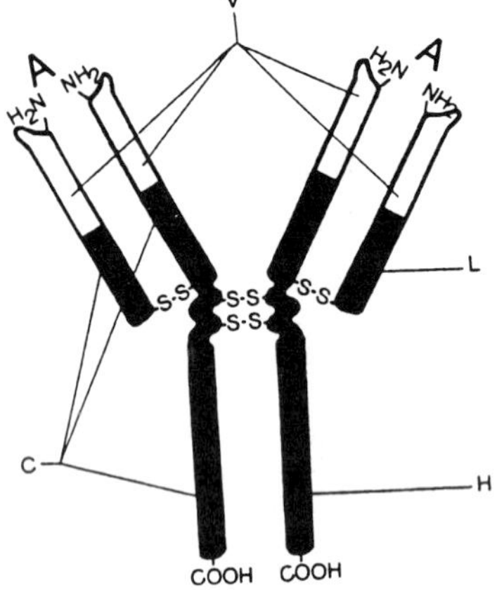

Immunoglobulin. Diagram of a basic immunoglobulin molecule; composed of four polypeptide chains, i.e. of two heavy chains (H) and two light chains (L), both with constant regions (C) and variable regions (V). A = antigen binding sites, -S-S- = disulphide bonds, NH_2 = amino group, COOH = carboxyl group.

According to the structural and antigenic properties of the H chains, immunoglobulins are classified into several groups; e.g. in humans **IgG**: 80% of all immunoglobulins, monomeric Ig, important in secondary infections and for foetal immunity because it crosses the placenta, **IgA**: 10—15%, often dimeric, abundant in intestinal secretion and milk, **IgM**: 5—10%, pentameric, abundant in the plasma, **IgD**: < 1%, attached to B lymphocytes, and **IgE**: < 0.01%, attached to basophilic granulocytes and mast cells. Many immunoglobulins act as → **opsonins** activating phagocytosis of the bound antigens, and particularly IgM and IgG activate the → **complement system**.

Both H and L chains comprise constant (C) and variable (V) regions encoded correspondingly by C and V genes. Each genome contains only a few C genes but tens of V genes, which by → recombination during maturation of → lymphocytes form millions of combinations, each lymphocyte line (clone of cells) derived from a single lymphoblast having a specific combination unit of C and V genes. *See* antibody.

immunoglobulin superfamily, Ig superfamily, a protein group with immunoglobulin-fold or related domains; includes → immunoglobulins, T lymphocyte receptors, → major histocompatibility molecules, and some cell adhesion molecules (*see* integrins).

immunologic(al) (Gr. *logos* = word, discourse), pertaining to → immunology.

immunological memory, the ability of the immune system to react quicker and more strongly against repeated antigen invasions than it did in the first exposure; the memory is based on the action of certain → lymphocytes, called **memory cells**.

immunological tolerance, immunotolerance, immune tolerance (L. *tolerantia* = ability to endure), the non-reactivity of the immune system to a given antigen; in vertebrates it generally results from a contact between the antigen and the developing immune system, i.e. usually during late prenatal or early postnatal period. Apparently those stem cells which could form antibodies against an individual's own structures are destroyed at this early stage, and thus the immune system cannot usually respond to its own tissue structures. Later, immunological tolerance may be induced e.g. by repeated administration of very large or very small doses of antigen.

immunology, a branch of science dealing with immune reactions of healthy or infected animals (chiefly vertebrates). *See* immunity.

immunoprecipitation (L. *praecipitare* = to cast down), the binding reaction between antigen and antibody molecules in a solution, resulting in their co-precipitation.

immunoradiometric assay, IRMA, an immunoradiometric analysis in which a compound to be analysed is bound to a radioactively labelled → antibody. In radioimmunoassays (**RIA**) a radioactive label is bound to an antigen.

immunosuppression (L. *supprimere* = to press down, suppress), the decrease or prevention of an immune response, caused e.g. by a disease, chemical substances, or radiation; clinically caused by procedures which suppress a patient's rejection to a transplanted tissue.

immunotoxin (Gr. *toxikon* = arrow poison), a molecule formed by combination of a toxin with an → antigen or antibody, the new molecule having the toxicity of the toxin and the specificity of the antigen or antibody; because of the specificity of the antigen-antibody reaction, immunotoxin can be targeted on destroying precisely chosen cells.

IMP, → inosine monophosphate.

impedance (L. *impedire* = to inhibit, prevent), in physics the resistance, i.e. the ratio of voltage to current in an alternating current circuit. In biology, impedance can be used as a parameter in studies of different structures; e.g. during freezing the impedances of plant structures change, and using different frequencies of electric current the type of damage can be studied.

impermeable (L. *in* = not, *permeare* = to go through), not permitting passage of substances, such as ions, gases, liquids, etc. *See* permeability.

implantation (L. *in-* = into, *plantare* = to set, plant), being planted in something; e.g. **1.** the natural attachment and embedding of the mammalian early embryo, blastocyst, to the mucous membrane (endometrium) of the uterus; *syn.* nidation; **2.** the artificial insertion of a piece of tissue, an organ, or foreign material, i.e. an **implant**, to an organism, or to a new place in an organism. *Verb* to **implant**.

impregnation (L. *praegnans* = pregnant), **1.** in engineering, impregnation of fabric, wood, etc., by filling interstices of the material with a substance; **2.** sometimes used to refer to → fertilization. *Verb* to **impregnate**.

imprinting, 1. in genetics, the inactivation of the maternal or paternal genome or chromosome, i.e. maternal or paternal imprinting, respectively; this leads to the selective expression of genes located in these regions during ontogenic development; e.g. the inactive X chromosome of marsupials is always of paternal origin; *see* X chromosome inactivation; **2.** in ethology, the associative learning of an individual soon after birth, i.e. during its **sensitive period** (critical period) with neurophysiological readiness for such a learning. As a result of imprinting, the acquired knowledge of the releasing object is retaining for life; e.g. in imprinting a duckling begins to follow any moving object that is large enough to represent its parent. In nature, however, the object is usually the parent. Because of the specific sensitive period and retaining for life, many ethologists do not consider the imprinting as an associative learning. *See* social imprinting.

improvement, in biology, a breeding programme aiming at improvement of the traits of cultivated plants and domestic animals, traditionally by means of selection and crossing. By improvement better producing animals and plants can be yielded, or their resistance to disease, cold, and other environmental stress factors, improved. Also cell and tissue culture methods and gene manipulation are used for this purpose, i.e. techniques that shorten the time needed for conventional methods.

impulse (L. *impulsus*), **1.** an impelling action or force inducing motion, such as a movement of the chest caused by a heartbeat; **2.** nerve impulse, i.e. a propagating → action potential which involves a rapid electric current across the membrane of a nerve cell; sometimes also a muscular impulse (action potential); **3.** a sudden mental inclination to perform a certain action.

in- (L. *in* = in, into), **1.** a prefix denoting in, within, inside, into; occurs as *il-* before l, *im-* before b, m, or p, and as *ir-* before r; **2.** (L. *in-* = not), a prefix meaning a sense of negation, before l, and b, m, p, and r, changes as mentioned in 1.

inbreeding, a mating system in which the pairing individuals are more related than the individuals of the population on average. Inbreeding leads to the decrease of the number of heterozygous individuals, and correspondingly to the increase of the number of homozygous individuals, as compared to a random mating. The utmost form of inbreeding is → self-fertilization which leads to a complete loss of genetic variation in the population. *Cf.* outbreeding.

inbreeding coefficient, symbol **F**; can be expressed either regarding the individual or population. The inbreeding coefficient of an individual is the probability that the individual in a certain gene → locus bears → alleles which are identical by descent, i.e. are copies of the same gene. The inbreeding coefficient of a population describes the probability that the gametes which unite in the population bear identical alleles by descent. Hence, F is the correlation coefficient of the uniting gametes.

inbreeding depression, the impairment of the fitness or viability of individuals or populations due to inbreeding; results from deleterious gene → alleles becoming homozygous.

incidence (L. *incidere* = to happen, fall into), **1.** the number of new cases of a disease in a certain population during a given time interval; broadly, the occurrence of any characteristic in the population; *cf.* prevalence; **2.** in optics, the intersection of a light ray with a surface.

incision (L. *in* = into, *caedere* = to cut), **1.** a wound made by cutting with an instrument; **2.** act of cutting.

incisor, one of the chisel-edged, cutting teeth found in most mammals, primarily three in the front on each side of the upper and lower jaws. Modified incisors are e.g. the continuously growing gnawing teeth of rodents and the tusks of elephants.

incisura, pl. **incisurae** (L. = cutting into), **incisure**, emargination; a notch or indentation at the edge of an anatomical structure, e.g. in a bone.

inciting behaviour, a mating behavioural pattern, especially typical for ducks, when a female incites its male to drive away an intruding strange male by directing her bill towards the intruder and swinging it back and forth.

inclusion (L. *includere* = to enclose, include), the act of including; the state of being included; in biology e.g. **1.** cell inclusions (inclusion granules), small intracellular granules or vesicles, such as fat droplets, or pigment and glycogen granules; **2.** a cyst inside the bone such as dental inclusion, i.e. a tooth remained unrupted in the jaw bone; **3.** a piece of material enclosed in another material, e.g. inside metal or rock.

inclusive fitness, the sum of the different components of → fitness of an individual; includes the **classical component**, i.e. reproductive success of the individual itself, and the **kin(ship) component** that expresses the effect of the individual on the success of the copies of its "own" genes located in related individuals. *See* kin selection.

incompatibility, 1. a genetic restriction of mating competence which limits or prevents fertilization between gametes of a certain type; thus, it restricts either → self-fertilization (self-incompatibility) or cross-fertilization (cross-incompatibility) in an otherwise freely breeding population; **2.** the immunological incompatibility of blood or tissue types; **3.** in plant breeding, the inability of the scion to unite successfully with the stock.

incomplete dominance, the phenomenon in which the phenotype of a heterozygous individual, Aa, represents a phenotype that is between the homozygotes AA and aa, i.e. the individual is phenotypically of intermediate type; A = dominant allele, a = recessive allele. *Syn.* semidominance, partial dominance.

incomplete flower, a reduced flower lacking one or more floral parts, i.e. sepals, petals, stamen, or carpels.

incretion (L. *in* = within, *secernere* = to separate), **1.** internal secretion, hormonal secretion; **2.** rarely, a product of an endocrine gland.

incubation (L. *incubare* = to lie upon, hatch), **1.** the act or process of brooding; sitting upon eggs; **2.** the act of maintaining something in controlled conditions in an → **incubator**, e.g. an infant, embryo, eggs, tissue sample, cultured cells. *Verb* to **incubate**.

incubation period, 1. the period required for brooding (incubating) to bring an egg ready for hatching; **2.** the period between an infection by a pathogen and the manifestation of the disease it causes.

incubator, an apparatus with controlled temperature and other conditions for maintaining cells, eggs, organs, and other living material, or premature infants in hospitals.

incubatorium, pl. **incubatoria,** Am. **incubatoriums** (L.), the skin pouch in the mammary area of the spiny anteater (echidna) where the egg is hatched; the pouch reduces after hatching.

incus, pl. **incudes** (L. *incudere* = to beat on, strike), anvil; the middle of the three auditory ossicles in the mammalian middle ear.

indehiscent fruit (L. *in-* = not, *dehiscere* = to split open), a fruit not opening spontaneously, e.g. nut and lomentum.

independent assortment, *see* Mendel's laws of inheritance.

indeterminate growth, unlimited growth of cells, as e.g. the division of cancer cells which in optimal conditions continue indefinitely.

index fossil (L. *fossilis* = dug up), a → fossil having a large regional distribution but characterizing only a narrow geological horizon representing a short time interval.

index of community diversity, *see* species diversity index.

India rubber, → caoutchouc, natural rubber.

indication (L. *indicare* = to point), anything

serving to point out or indicate; e.g. a sign that shows the cause, pathology, or treatment of a disease.

indicator, that which points out or indicates; e.g. **1.** the index finger (forefinger); **2.** any substance which e.g. by a change of colour shows a state of equivalency in a chemical reaction; **3. indicator species**, any species which is especially sensitive to some environmental factor; thus the occurrence of the species indicates this factor, e.g. some organisms have been used as **geological indicators** to reveal ore deposits, some species are found to indicate an ancient settled area or an environmental pollutant (e.g. many lichens). *See* bioindicator.

indifferent, without interest, neutral; e.g. **1.** neutral in chemical, electrical, or magnetic quality; **2.** not differentiated or specialized, such as cells or tissues. *Noun* **indifference**.

indigenous (L. *indigenus* = born in), native; pertaining to an organism which is natural to a particular geographical area or habitat where found.

indigo (Gr. *indikos* = Indic), indigo blue; a blue dye obtained from *Indigofera* plant species (Leguminosae); is used e.g. for staining biological preparations.

indirect end-labelling of DNA, a technique for studying the organization of DNA; in the procedure a break is made on a specific point of DNA, labelling it e.g. with a fluorescent marker; all the fragments located on the same side of the break are then isolated. This technique reveals the distance from one break to the next.

individual, a unit of life defined genetically, i.e. an organism which constitutes all the division products of one fertilized egg until the following → meiosis, or all the division products of one → meiospore until the following fertilization.

individual distance, in ethology, the distance within which an animal does not tolerate any other conspecific individuals.

individual selection, the term points an individual as an object of → natural selection. Intraspecific variation is a result of small changes in the genotypes of individuals, which, if positive, will become more abundant in a population. *See* selection.

indole, 2,3-benzopyrrole, C_8H_7N; a heterocyclic compound composed of one five-carbon and one six-carbon ring; a precursor for many biologically active substances, such as → auxins, tryptophan, and serotonin. Indole is formed in the intestine as a putrefaction product of proteins containing tryptophan. The radical of indole is called **indolyl**.

indole-3-acetaldehyde, an intermediate of auxin biosynthesis. *See* auxins.

indole-3-acetic acid, IAA, *see* auxins.

indole-3-acetonitrile, IAN, an intermediate of auxin biosynthesis occurring in the reactions in which IAA (indole-3-acetic acid) is synthesized from thioglucosides, as glucobrassicin in Brassicaceae plants. *See* auxins.

indolealdehyde, an intermediate of the inactivation reactions of → auxins.

indole-3-butyric acid, IBA, *see* auxins.

indolence (L. *in-* = not, *dolens* = painful), a state of being painless or slothful. *Adj.* **indolent.**

induced-fit model, a model describing the binding of an enzyme to its substrate; in the binding process the substrate induces conformational changes in the enzyme molecule altering the shape of its active site complementary to that of the substrate. *Cf.* lock-key-model.

inducer, in genetics, an anti-inhibitor (effector) that triggers transcription in specific negatively regulated genes (*see* genetic regulation) by binding to a regulator protein (*see* operon). The inducer is a molecule, such as a substrate of a specific enzyme pathway, that binds to the → repressor thus initiating the activity of the structural genes in an inducible operon.

inducible defence, any of the defence mechanisms of organisms the effectiveness of which gradually increases with use. In animals, includes **physiological defence**, such as → detoxification and → immunity, and **structural defence** in a population level, when these defence mechanisms, owing to selection, become more pronounced in the presence of a predator; e.g. the increasing growth of protective spikes on the body surface of cladoceran and bryozoan populations.

The inducible defence mechanisms in plants produce different kinds of defence chemicals against animals. Their production is induced when some parts of the plant have been damaged, e.g. eaten. The defence may be either a **rapidly inducible defence** affecting only that herbivore generation which began the feeding, or a **delayed inducible defence** affecting several generations over many years. *See* chemical defence.

inducible enzyme, an enzyme that is synthesized only when an inducing agent, usually its substrate, is present; e.g. the synthesis of β-galactosidase in *Escherichia coli. See* induction.

induction, 1. any responsive reaction in an organism or its parts caused by a stimulus; **2. enzyme induction**, the induced synthesis of so-called inducible enzymes as a result of increased concentration of a metabolite, → inducer, in a cell. In contrast to → constitutive enzymes, quantities of inducible as well as → repressible enzymes fluctuate widely depending on changes in the cellular environment; enzyme induction is a common phenomenon found in all types of organisms; **3. prophage induction**, the breakdown of the functional integration of a → prophage into the genetic system of a → lysogenic bacterium; this is followed by the vegetative replication and maturation of the phage progeny and the ultimate lysis of the bacterial cell; **4. embryonic induction**, the determination of the developmental fate of a cell or tissue by another cell or tissue; **5.** in physics, **electromagnetic induction**, the generation of an electric field by a moving magnetic field; **6.** in logic, **inductive reasoning**, i.e the derivation of a general proposition from single cases; *cf.* deduction.

inductor, any factor that causes an → induction. *Verb* to **induce**.

indumentum, pl. **indumenta** also **indumentums** (L. = garment), a hairy covering on a plant or an animal, as found on many leaves or insects.

indusium, pl. **indusia** (L. = woman's undergarment, tunic), a protective membrane covering developing sporangia on the lower surface of leaves in many → ferns, arising from the leaf tissue.

industrial melanism, the increased number of dark or melanistic forms in animal populations of polluted industrial areas, where soot and atmospheric sulphur dioxide (SO_2) contents are high. These substances darken the environment with soot and e.g. inhibit the growth of lichens. In this changed environment the dark pigmentation of animals gives better camouflage against predators, resulting in an increase in the number of melanistic forms. The phenomenon is found e.g. in many lepidopteran species in city areas.

inertia, pl. **inertiae** or **inertias** (L.), **1.** inactivity, sluggishness, lack of biochemical or physiological activity; physical state of rest; **2.** in physics, the property of a matter by which it retains its state of rest. *Adj.* **inert**.

infantile (L. *infans* = infant), **1.** pertaining to young, characteristic of an infant; **2.** being in the earliest stage of development; e.g. an infantile behaviour of some adult birds and mammals during courtship.

infarct (L. *infarcire* = to stuff full), a necrotic area of tissue mainly due to local failure of blood circulation, i.e. ischaemia; e.g. **embolic infarct** caused by an → embolus preventing the blood circulation to the infarct area, or **thrombotic infarct** caused by a thrombus; **infarction**, sudden insufficiency of blood supply; also meaning infarct.

infection (L. *inficere* = to stain, infect), **1.** invasion and multiplication of endoparasites, such as pathogenic bacteria, viruses, fungi, or protozoans, in an animal or a plant; **2.** an infectious disease. *Verb* to **infect**. *Adj.* **infective**, infectious, capable of producing infection; **infectible**, capable of being infected.

infertile, pertaining to an individual unable to produce offspring. *Noun* **infertility**.

infiltration, 1. the act of intrusion and/or accumulation of material not normal in an organ or tissue, such as foreign substances, cells, or liquids; also the accumulation of material in excess of the normal; **2.** the material deposited by infiltration; also **infiltrate**. *Adj.* **infiltrative**, pertaining to the material that is capable of being infiltrated. *Verb* to **infiltrate**, to intrude.

inflammation (L. *inflammare* = to set on fire), a localized, usually protective, non-specific response, especially known to occur in tissues of vertebrates. It is induced by physical or chemical factors, such as by the infection of microbes or by tissue damage. The inflammation serves to dilute, wall off, or destroy harmful agents and injured tissues. The reactions involve e.g. the dilatation of blood vessels, increased permeability of capillary walls, infiltration of leucocytes to the inflammatory area, and secretion of hormone-like substances. Inflammation reactions are suppressed e.g. by glucocorticoids (anti-inflammatory effect) and certain inhibitors of prostaglandin synthesis. A similar but harmful inflammation may be caused by autoimmune reactions (*see* autoimmunization).

inflation (L. *in* = into, *flare* = to blow), the act of distending, or distention of an organ with gas or fluid. *Verb* to **inflate**.

inflorescence, the arrangement of flowers on an axis. One of the two main inflorescence types is the **racemose inflorescence**, having a main axis in which the flowers develop from axillary buds. There are six different inflorescence types in this group: in the **raceme** (e.g. lupin) the stalked flowers are located laterally on the elongated main axis; the **spike** (e.g. rye) is like a raceme but the flowers are not stalked. The **corymb** (e.g. yarrow) is structurally similar to the raceme, but the stalks are shorter, the flowers being at the same level. In the **spadix** the main axis is clearly thickened and the numerous small flowers are located close to each other. In the two other types the main axis is not clearly discernible. In the **umbel** (e.g. ivy) the branches (the flower stalks) arise more or less from the same point of the axis and each flower stalk extends to the same length, all terminating in flowers; in the **capitulum** (e.g. dandelion) the end of the main axis is thickened and covered by numerous aggregated flowers. Some racemose inflorescences may be **compound**, while they undergo further branching. Compound racemes are common among plants; in the family Poaceae the type is called **panicle**. In the family Apiaceae compound umbels are common.

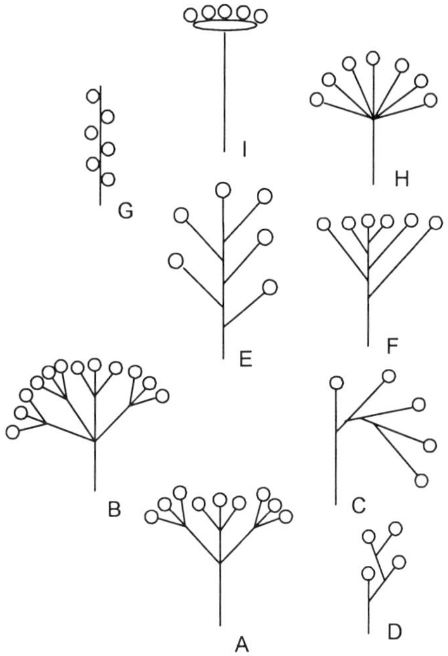

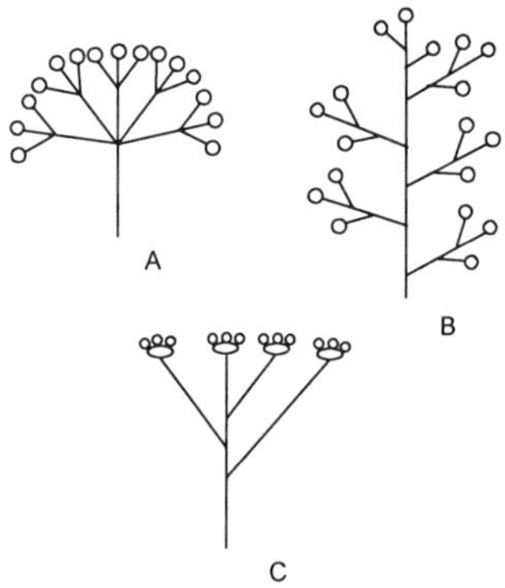

Inflorescence types of plants: cymose inflorescences: dichasium (A), pleiochasium (B), depranium (C), rhipidium (D); racemose inflorescences: raceme (E), corymb (F), spike (G), umbel (H), capitulum (I).

Compound inflorescence types of plants: compound umbel (A), compound raceme (B), capitula in corymb (C).

The other main type is the **cymose inflorescence**, having a repetitive side-branch structure: these inflorescences may be **pleiochasial**, **dichasial** or **monochasial** according to the number of side branches. In the monochasium there is always only one side branch, and its location determines the subtype: it may be **rhipidium** (e.g. iris), **cincinnus** (e.g. figwort), **depranium** (e.g. rush), or **bostryx** (e.g. day lily). In the **dichasium** the number of repetitive side branches is two, and the **pleiochasium** has three or more branches.

There are also **mixed inflorescences**, having characteristics of both the racemose and cymose types; e.g. in birch **catkins** there are small dichasia formed of three flowers grouped into a raceme, and the horse chestnut (*Hippocastanum*) has racem-like inflorescence in a corymbe-like structure.

influenza (L. *influentia* = influence), a viral infection in vertebrates, best known in mammals, often occurring in epidemics. Influenza viruses (orthomyxoviruses) spread usually with inhaled air and cause inflammation especially in the pharynx, nasal mucosa, and other respiratory pathways, giving symptoms such as fever, headache, stuffy nose, and myalgia. With a following secondary bacterial infection, like pneumonia, influenza may cause more severe symptoms. Most influenza viruses are specific for a species, as e.g. the human, equine, or feline influenzas. Also called flu, grip, grippe.

information, a message, a knowledge; according to the information theory, it is generally regarded as a form of energy which is the measure of the organization level of a system. Biological systems contain information in many forms. The most primary form is → **genetic information** which resides in the genes of each cell in the form of a linear order of → nucleotides in DNA (or RNA) molecules. Genetic information contains all the instructions for structure and function of the cell. Another important form of internal information of the individual is the **physiological information** which consists of a vast amount of different biochemical reactions in the signalling pathways within and between cells; this consists of numerous messengers, such as → hormones, growth factors, cytokines, cytokinins, and cyclic AMP. In animals, physiological signalling also involves neuronal signalling and sensory functions. Signalling maintains the cooperation of cells during different stages of development.

The third form of biological information is **signalling between individuals** in which different chemical (*see* pheromone), physical, and behavioural → signals are used like a sign language, or spoken language in man. This form of information, which is most advanced in animals, maintains the organization and coordination of the populations and ecosystems.

information centre hypothesis, in ethology, the hypothesis emphasizing the advantage of living in social groups; e.g. the members of an animal group can synchronize their activities, such as reproductive functions, feeding, guarding, etc. by signalling and observing the behaviour of other members of the group.

informosome, a complex of → heterogeneous nuclear RNA (hnRNA) and specific nuclear proteins. These complexes allow a localized accumulation of RNA until transcription is complete, acting in the condensation and stabilization of the nascent RNA and sequestering it from the transcription template. Also called heterogeneous nuclear ribonucleoprotein particle (**hnRNP**).

infra- (L. *infra* = beneath), denoting beneath or below.

infradian (L. *dies* = day), pertaining to rhythmic phenomena in organisms occurring in cycles of frequency slower than once a day (slow-frequency rhythms). *Cf.* ultradian. *See* biorhythms.

infralittoral, also **infralitoral** (L. *lit(t)us* = seashore), pertaining to a subzone of the → littoral zone situating permanently under water level in the oceans from low tide down to 100—200 m; the term is also used as a synonym for sublittoral, or denotes the upper layer of the sublittoral. *Cf.* eulittoral, littoriprofundal.

infraneuston (Gr. *neustos* = floating, swimming), planktonic organisms, such as some mosquito larvae, living suspended from the underside of the surface film of water. *Syn.* hyponeuston. *See* epineuston, neuston.

infrared (radiation), IR, thermal radiation; electromagnetic radiation of a wavelength between 0.75 and 1,000 µm; longer than visible light and shorter than radio waves.

infrared spectroscopy (L. *spectrum* = sight, appearance, Gr. *skopein* = to look), a method for chemical analyses for determining and identifying organic compounds; it is based on characteristic → absorbance of compounds in the infrared radiation, usually between wavelengths 2.5 and 16 µm.

infraspecific variation, the → variation within a species; involves the existence of → ecotypes, races, and morphs. *Cf.* polymorphism.

infructescence, an aggregated plant fruit structure formed by separate flowers of an entire → inflorescence; e.g. pineapple. *Syn.* anthocarp.

infundibulum, pl. **infundibula** (L. = funnel), a funnel-shaped anatomical structure; e.g. **1.** the infundibulum of the oviduct (tuba uterina), forming the anterior ciliated part of this organ; **2.** the outpushing prominence of the ventral hypothalamus of the vertebrate forebrain, forming a hypophyseal stalk and the posterior pituitary (neurohypophysis); **3.** the expanding part of a → calix opening into the pelvis of the kidney of metanephros type; **4.** the termination of a bronchiole in the pulmonary alveolus; **5.** the funnel or → siphon of cephalopods.

infusion (L. *in* = into, *fundere* = to pour), **1.** the introduction of a saline or other solution into a vein by gravity, or with a pump; **2.** the soaking or steeping of a crude drug preparation in water in order to extract its soluble components; also means the solution so prepared. *Adj.* **infusive**, capable of infusing.

infusoria (L. *infusus* = poured into), a name formerly used for some microscopic animals, such as protozoans, especially ciliates.

ingestion (L. *ingestio* = pouring in), **1.** the taking of food and drink into the digestive system of an animal; **2.** incorporation of material into the cytoplasm of a phagocytic cell to be digested intracellularly. *Adj.* **ingestive**. *Verb* to **ingest**.

inguinal (L. *inguen* = groin), pertaining to the groin, or inguen; e.g. **inguinal canal**, a passage in male mammals through which the testis descends into the scrotum, containing the spermatic cord; in females the passage contains the round ligament.

inhalation (L. *in* = into, *halare* = to breathe), **1.** breathing in; **2.** the drawing in of a volatile or aerosol drug (**inhalant**) with the breath; **inhalator**, the apparatus that helps one to inhale; **inhaler**, respirator, also inhalator.

inheritance, the phenomenon mediating → heredity, its mechanism comprising the transmission of genetic material from one generation to the next. **Mendelian inheritance** is mediated by genes located on chromosomes, and is characterized by regular rules which are called the → Mendel's laws of inheritance; **non-Mendelian** or **cytoplasmic inheritance** is mediated by organelles like → mitochondria and → plastids containing DNA and situated in the cytoplasm.

inhibin (L. *inhibere* = to restrain), a hormone of dimeric polypeptide structure produced by → Sertoli cells in the testes and granulosa cells in the ovaries of vertebrates (in man inhibin A and inhibin B). It inhibits secretion of → follicle-stimulating hormone (FSH) and so regulates the function of the gonads. It is found also in other tissues, such as the nervous tissue, and apparently during the ontogenic development with another similar hormone, **activin**, it regulates more widely the growth and development of tissues and organs. *See* transforming growth factors.

inhibition, the depression or arrest of a function; e.g. **1.** depression of a biochemical reaction or the action of cells or organs, caused by a chemical compound, an ion, or a physical agent, i.e. by an **inhibitor**; *see* competitive inhibition; **2.** in ethology, the prevention of a certain behavioural pattern by the influence of another behaviour; e.g. courtship behaviour arrests feeding activities; **3.** in ecology, the situation where the dispersion of a species to a certain area is prevented by the species formerly occupying the area.

inhibitor, an agent that retards, restrains, or prevents biochemical, physiological, or other reactions; e.g. an enzyme inhibitor depressing an enzyme reaction, or an inhibitor gene preventing the expression of a certain character.

inhibitory postsynaptic potential, IPSP, an electric hyperpolarization of postsynaptic membrane resulting in an inhibitory action on synaptic transmission; may be a local, **transient IPSP** lasting a few milliseconds, or may be a **slow IPSP** continuing for seconds and propagating wider. *Cf.* excitatory postsynaptic potential. *See* synapse.

initial cells (L. *initium* = beginning), small, undeveloped cells which divide continuously in plant → **apical meristems**, forming new cell material during plant growth. Initial cells occur in **promeristem** areas, forming → **derivative meristems**, the cells of which are determined to form certain tissues and structures and are also still capable of dividing. *Syn.* initials.

initials, in biology, → initial cells.

initiation complex, an obligatory intermediate of → genetic translation found both in eukaryotic and prokaryotic organisms; consists of → messenger RNA, ribosomes, initiator factors, and initiator transfer RNA.

initiator codon, initiation codon, an initiator triplet in the messenger RNA composed of three bases; the → codon which serves in the starting of → protein synthesis. The initiator codon is usually AUG (adenine-uracil-guanine) but sometimes also GUG. The initiator codon adds formylmethionine into the amino terminal of the polypeptide chain under synthesis. The amino group of formylmethionine is unable to form the peptide bond, and thus the polypeptide chain cannot grow to the left of it. Formylmethionine is removed from the amino terminus of the mature polypeptide chain.

initiator factor, initiation factor, IF, a protein which by connecting with → ribosomes catalyses the initiation of the synthesis of the polypeptide chain. Three initiator factors exist in prokaryotic organisms and at least six in eukaryotic organisms. Initiator factors are requisite for the formation of the specific → initiation complex. Ribosomes bind to messenger RNA at specific binding sites in the presence of initiator factors.

initiator transfer RNA, initiator tRNA, a unique species of methionine-transfer RNA for initiation of → genetic translation.

injection (L. *inicere* = to throw into), the act of forcing a liquid into a tissue, or the preparation so administered. Using an injection syringe and a needle, injections into animals are chiefly carried out intracutaneously (i.c.), subcutaneously (s.c.), intramuscularly (i.m.), or intravenously (i.v.). *Verb* to **inject**. *Cf.* microinjection.

injury-feigning, → broken-winged appearance.

ink sac, a sac-like organ in cephalopods (e.g. squids) containing a melanin-pigmented ink, sepia, formed by the oxidation of amino acid tyrosine. The substance is ejected through a funnel to form a "smoke cloud" when the animal is attacked.

innate (L. *innatus*), inborn; pertaining to a characteristic which is partly or totally determined by genes.

innate capacity for increase, (r_m), a maximal rate of increase in population density in an environment where all the external factors, important to the welfare of the population, are unlimited and optimal. The r_m of an organism depends e.g. on its fecundity, longevity, and speed of development. The age distribution of the maximally increasing population is stable and the growth of the density exponential. *See* biotic potential, Malthusian parameter.

innate immune response, *see* immunity.

innate releasing mechanism, IRM, according to *Niko Tinbergen*, IRM is a special neurosensory mechanism that releases a behavioural reaction in an animal to a very special combination of releasers (sign stimuli). IRM and releasers are mutually adapted to each other.

inner ear, internal ear; the innermost portion of the vertebrate → ear, containing the → cochlea as an auditory organ, and the → utricle, saccule, and semicircular canals as an equilibrium organ; the cochlea is absent in fishes but the utricle also serves as a primitive auditory organ. *See* labyrinth, equilibrium sense.

innervation (L. *in* = into, *nervus* = nerve), the supply or distribution of nerves (or nerve stimuli) to an organ.

innominate bone, → hip bone.

inoculation (L. *oculus* = bud, eye), implantation; e.g. **1.** the introduction of cells into a new culture; **2.** the implantation of infective materials, as microbes, vaccines, or serum, etc. into tissues of living animals or plants; **inoculum** (inoculant), such an inoculated substance. *Verb* to **inoculate**.

inoperculate (L. *operculum* = lid), having no → operculum; e.g. refers to an opening structure in plants (e.g. an ascus or a spore capsule) which opens in some way without an operculum or a lid. *Cf.* operculate.

inorganic (L. *in* = not, *organon* = organ), a term pertaining to compounds or reactions of inanimate nature. **Inorganic chemistry** is the study of elements other than carbon; some simple carbon compounds such as carbonates, carbides, and carbon oxides, however, are usually regarded as inorganic; the study of all other carbon compounds belongs to organic chemistry.

inosine, hypoxanthine riboside; 9-β-D-ribofuranosylhypoxanthine; a → nucleoside occurring in organisms e.g. in the → anticodon of a → transfer RNA molecule. **Inosine phosphates** are closely connected to the metabolism of → adenosine phosphates.

inosine monophosphate IMP, hypoxanthine ribonucleotide, a purine nucleotide that is the first precursor with a purine ring in biosynthesis of adenine and guanine nucleotides, i.e. adenosine monophosphate (AMP) and guanosine monophosphate (GMP) in cells. IMP is used as a flavour enhancer in many processed foods; its acid form is **inosinic acid**.

inositol, hexahydroxycyclohexane, $C_6H_{12}O_6$; a sugar-like substance found in organisms, often included in the group of vitamin B complex; has 9 isomers of which only one, meso-inositol (myo-inositol), is biologically active and widely distributed in organsims. With certain lipids inositol forms **phosphatidylinositol (PI)** that is an important membrane lipid. This is successively phosphorylated to form phosphatidylinositol 4-phosphate (**PIP**), and then phosphatidylinositol 4,5-bisphosphate (**PIP_2**). Enzymatic cleavage of PIP_2 by **phospholipase C** (specific phosphodiesterase) produces **inositol 1,4,5-tri(s)-phosphate (IP_3, $InsP_3$)** and **diacylglycerol (DAG)**. Both serve as → secondary messengers in signal-transduction pathways of many hormones, growth factors, and neurotransmitters in eukaryotic cells, such as animal and plant cells. When one of these messenger substances binds to its receptor, phospholipase C is activated to form IP_3. Then IP_3 opens calcium channels thus causing the release of calcium ions (Ca^{2+}) from intracellular stores (e.g. endoplasmic reticulum), that further evokes many responses in the cell (*see* calcium). DAG stays in the cell membrane where it activates → protein kinase C.

inquilinism (L. *inquilinus* = tenant), an interacting life form between animals when another species, e.g. an insect, lives without doing harm in the nest of the host species. Among some social insects, the workers of a host species may feed the larvae of the **inquiline**, in which case the interaction is a form of **parasitism**. *See* social parasite.

Insecta (L. plural of *insectum* = insect), **insects**, **Hexapoda**; a class in the phylum Arthropoda comprising species with a tripartite body, i.e. the **head** (*caput*), **thorax**, and **abdomen**. Three pairs of jointed legs and one or two pairs of wings (absent in some species) are attached to the thorax. The head bears one pair of antennae, paired mouthparts, and a pair of **compound eyes**. The respiratory system comprises branched tracheal tubules (*see* trachea). The life cycle includes → **metamorphosis**. About 950,000 insect species are described; they are classified into two subclasses: Apterygota (wingless insects) and Pterygota (winged insects). The latter taxon is divided into two superorders, Exopterygota and Endopterygota. In some, rarely used classification, the class Insecta is divided into the superorders Entognatha whose mouthparts are invisible below the head capsule, and Ectognatha whose mouthparts are external and clearly visible. The class Insecta includes many orders, e.g. **mayflies** (Ephemeroptera), **dragonflies** (Odonata), **stoneflies** (Plecoptera), **grasshoppers**, **crickets** (Orthoptera), **earwigs** (Dermaptera), **sucking lice** (Anoplura), **biting lice** (Mallophaga), **true bugs** (Hemiptera), **cicadas**, **aphids** (Homoptera), **dobson flies**, **ant lions** (Neuroptera), **moths**, **butterflies** (Lepidoptera), **caddis flies** (Trichoptera), **true flies** (Diptera), **wasps**, **ants**, **bees** (Hymenoptera), **fleas** (Siphonaptera), and **beetles** (Coleoptera).

insectarium, pl. **insectaria,** insectary, a cage or room for rearing insects.

insecticide (L. *caedere* = to kill), a chemical pesticide for destroying insects; e.g. DDT, parathion. *See* pesticides.

Insectivora (L. *vorare* = to devour), **insectivores**; an order of primitive insect-eating mammals from Cretaceous to the present time; in general, comprises small-sized animals with a long and tapered snout having several sharp-pointed teeth; familiar species are shrews, moles, and hedgehogs.

insectivores, 1. → Insectivora; **2.** insectivorous (insect-eating) plants or animals, e.g. many birds like passerines with narrow beaks.

insectivorous plant, a plant which catches and enzymatically digests insects for its own nourishment; e.g. *Drosera*, *Pinguicula*, *Dionea*.

insects, → Insecta.

insemination (L. *in* = into, *seminare* = to sow), the introduction of seminal fluid into the vagina or cervix; may be a natural or artificial procedure.

insertion (L. *inserere* = to join to), **1.** the act of setting in (inserting); **2.** the attachment of an organ or a part to another organ, as the attachment of a muscle to the bone or a placenta to the uterus; **3.** a form of → gene mutation.

insertion sequence, IS element, any of the distinct prokaryotic segments of DNA which are able to be transposed to numerous sites on bacterial → plasmids, chromosomes, and bacteriophages.

inside-out patch, a variant of the patch-clamp technique, in which a piece of plasma membrane (the patch) is attached on the tip of the patch electrode, with the inner side of the membrane facing outward. A patch-clamp apparatus is then used for measuring the electrical currents through the ion channels or ion pumps possibly existing in the patch. The similar measuring system uses an **outside-out patch**, in which the outer side of the membrane is facing outward.

inside-out vesicle, IOV, any of minute closed vesicles formed of cell membrane fractions by mechanical disruption of cells, used in studies of membrane functions. In inside-out vesicles the original inside of the cell membrane is facing outwards; correspondingly, in **outside-out vesicle** (OOV) the outside faces outwards.

insight learning, "aha" -experience; the highest form of learning and often a part of → **latent learning**; the ability of an animal to spontaneously combine two or more, previously separated and isolated experiences to obtain a desired result of behavioural functions without help of preprogrammed innate responses. Insight learning is common especially in hominoids but it has been shown also in many other mammals and some birds. *See* learning.

in situ hybridization, the annealing (hybridization) of labelled single-stranded DNA or RNA → probes with denatured cellular DNA or RNA on a microscopic slide, and their detection with the aid of the label.

inspiration (L. *in* = into, *spirare* = to breathe), **1.** inbreathing, the drawing of air into the lungs; **2.** an animated action. *Verb* to **inspire**.

instar (L. = form), a developmental stage between successive moults in the larval life history of an arthropod, especially an insect.

instinct (L. *instinctus* = impelled), an innate ability of an animal to produce complicated but stereotyped behavioural patterns (*see* fixed-action patterns) to various stimuli in an environment. As a concept, instinct is indefinite owing to the difficulty of determining which components in the behaviour of an animal are innate and which are learned. The problem has been studied with isolated individuals in → Kaspar Hauser experiments. *Adj.* **instinctive**.

instinctive behaviour, rather stereotyped behavioural patterns produced mainly by → instinct and determined by the genotype of an animal; thus the share of learning is diminutive. Instinctive behaviour usually comprises **tactic movements** (*see* taxis), → **fixed-action patterns**, and sometimes also the results of instinctive behaviour, thus including e.g. the nests of birds and webs of spiders.

insufficiency (L. *in-* = not, *sufficiens* = sufficient), the condition of being inadequate to some designated use or purpose; e.g. a decreased performance of an organ, such as the heart in cardiac insufficiency. *Adj.* **insufficient**.

insulin (L. *insula* = island), a hormone of vertebrates produced by the β-cells of → pancreatic islets; promotes glucose uptake and utilization by cells, protein synthesis, and formation and storage of neutral lipids. The secretion of this large polypeptide (mol. weight 5,730) increases when the concentration of blood glucose, in lesser amount also the concentrations of amino and fatty acids, increase. Several hormones (e.g. gastrin and secretin from the alimentary canal) and the autonomic nervous system, control the activity of the secretory cells. In most tissues, particularly in muscle and adipose tissues, the lack of insulin causes the cessation of glucose uptake into cells, thus resulting in an increase in the glucose level in the blood and tissue fluids. *See* diabetes.

intact (L. *in-* = not, *tangere* = to touch), not broken or impaired, remaining uninjured or whole, not changed; e.g. a laboratory animal not yet used in experiments, or cells and organs isolated sound and complete.

integral membrane protein (L. *integrare* = to make whole), a protein which is more firmly anchored in the membrane than peripheral membrane proteins. In plasma membrane the best known integral proteins are hormone receptors, ion channels, and other transport proteins. Some integral proteins do not span the membrane totally; those that do so are called **transmembrane proteins**.

integrator gene, a gene which regulates the function of different groups of → producer genes.

integrins, cell adhesion proteins in animal tissues; heterodimeric cell-surface proteins acting as receptors to which intercellular proteins, such as fibronectin, laminin, and collagen, specifically fit and fasten. They also can bind to the surface of other cells by interacting with some proteins of the → immunoglobin superfamily.

integument (L. *in* = in, *tegere* = to cover), **1.** the covering or surface of an organism; skin, especially the chitin cuticle of insects; **2.** in plants, the structure covering the nucellus of the → ovule in the → ovary; there may be one or two integuments, which form the seed coat when the ovule ripens into seed.

intellect (L. *intellectus* = understanding), the power or faculty of the mind or thinking. *Adj.* **intellectual**, intelligent, having power to understand, or pertaining to the intellect.

intensity of selection, symbol i; standardized → selection differential (S); the formula $i = S/\sigma_p$ expresses the mean deviation of the selected individuals, presented in units of phenotypic standard deviation of a metric characteristic (σ_p) of a species.

intention movement, in ethology, a preparatory movement of an animal before action; usually incomplete but predicts a future action. It is supposed that these movements have formed a basis for the development of many signals, especially in agonistic and sexual behaviour patterns.

inter- (L. *inter* = between), denoting between, among.

interaction deviation, I, a form of genetic variance caused by → epistasis and other nonallelic interactions of genes.

interactor, in genetics, an entity that can be defined as an integrated system interacting with the environment, resulting in differential replicator selections; → phenotypes are interactors. *See* replicator.

interambulacral plates (L. *ambulare* = to walk), the areas without tube feet between the zones of ambulacral plates on the body of sea urchins.

intercalary growth zone (L. *intercalare* = to insert between), *see* intercalary meristem.

intercalary meristem, a → meristem tissue situated at the base of → nodes, forming the intercalary growth zone supporting growth in node areas in some plants (e.g. in the families Poaceae and Apiaceae). This is not typical among plants; the growth normally occurs only in the tips of the shoot and shoot branches. Intercalary meristem is also found in the leaves of some monocotyledons.

intercalary vein, a longitudinal vein from the margin to the base on the wings of insects without any connections with other large veins.

intercalated disc, Am. **intercalated disk,** the cell junction between cardiac muscle cells of vertebrates; the discs can be seen as dense bands between the ends of adjacent muscle cells in a histological preparation of the heart muscle. In the disc area the cell membranes of the adjacent cells are attached close to each other having specialized membrane structures, such as → desmosomes and → gap junctions between them. Action potentials can propagate electrically without any transmitter substance from cell to cell through the → gap junctions, and so the whole heart muscle forms functionally a → syncytium.

intercellular (L. *cellula*, dim. of *cella* = cell), occurring between cells; often pertains to the ground substance (matrix) produced by the cells of a tissue, or to intercellular fluid. *Cf.* extracellular, intracellular.

intercellular junctions, specialized structures of the cell membrane at cellular margins serving in adhesion and communication between many tissue cells; e.g. → desmosome, zonula, gap junction, synapse.

intercellular matrix (L. *matrix* = womb, basic material), intercellular substance, → matrix; organic material containing glycosaminoglycans, collagen, and various adhesive proteins between the cells in animal tissues. It is produced by tissue cells forming supportive solid fibrous and/or cementitious structures in tissues between cells; can also affect the development and biochemical function of the tissue.

intercellular spaces, spaces or cavities between cells in plant tissues where vapours may move or which contain various excreted substances. Intercellular spaces may be **1. schizogenic**, formed by the division of middle lamellae in the cell walls between adjacent cells (e.g. → aerenchyma cavities and → resin ducts); **2. lysigenic**, formed after dissolving of cell walls between some cells (e.g. oil cavities of *Citrus* and *Eucalyptus* species), or **3. rhexigenic**, formed when part of the tissue has mechanically ripped to a cavity, e.g. in vascular bundles of monocotyledons.

intercostal (L. *costa* = rib), between the ribs.

intercostal muscles, intercostals; the muscles between the ribs of tetrapods, i.e. external and internal intercostals, the former elevating the ribs in inspiration, the latter depressing the ribs especially in forced expiration.

interdemic selection (Gr. *deme* = region, nation), a form of → natural selection occurring intraspecifically between populations (demes).

interfascicular cambium, parts of the → cambium formed in the area between two → vascular bundles in the plant stem; correspondingly, the cambium formed inside vascular bundles is a fascicular cambium; these two types mostly fuse later forming an unbroken cambium ring.

interference (L. *ferire* = to strike), in biology, e.g. **1. chromosome interference**: the effect of → crossing-over to another crossing-over in the vicinity, either by decreasing its probability (**positive interference**) or by increasing it (**negative interference**); **2. chromatid interference**: the non-random participation of the four chromatids of → homologous chromosomes in two subsequent crossing-over events; **3. interchromosomal interference**: the compensation of the decreased crossing-over frequency of a chromosome by an increase of crossing-over frequency in a certain segment of a non-homologous chromosome; is usually due to structural heterozygosity; **4. virus interference**: the disturbance of the growth of a certain type of virus by another virus; **5. cardiac interference**: the disturbance of the regular cardiac rhythm by two intrinsic pacemakers in the heart; **6.** in ecology, a **negative interaction** between two competitive species when the other reduces the possibilities of the competitor for coexistence (**interference competition**); **7.** in physics, the **interference of a wave motion**; e.g. the **interference of light**: when two light waves meet at a different phase of waves they enhance or damp each other; has applications in light → microscopy and holographic → interferometry; interference also occurs in the meeting of two sound waves.

interference competition, *see* interference, defin. 6.

interference (contrast) microscopy, a special technique of light microscopy. *See* microscope.

interferometry, measurement of movements or distances using equipment (**interferometer**) with two light beams in interference; acoustic interferometry respectively uses the interference of sound waves. *See* interference, defin. 7.

interferons, IFNs, glycoproteins involved in signalling between cells of the immune system in vertebrates; their production can be stimulated particularly by a viral infection. IFNs exert virus-nonspecific but host-specific antiviral activity by enhancing the activity of the genes which encode antiviral proteins, and by inhibiting the action of nucleic acid in target cells, thus preventing the multiplication of viruses. They also have some antitumour activity. IFNs also improve the efficiency of the immune system, e.g. they induce the production of immunoglobulins, activate the natural killer cells, and stimulate phagocytosis. Some leucocytes (particularly $CD4^+$ T lymphocytes) and fibroblasts begin to produce interferons after virus intrusion to tissues. Natural killer cells secrete IFN when they come in contact with tumour cells. IFNs are classified into three main groups, i.e. alpha, beta, and gamma-interferons; different animal species apparently have specific interferons. Human interferons produced by gene manipulation are in medicinal use.

interfoliar stipule (L. *folium* = leaf), a leaf-like structure formed by the fusion of → stipules of two leaves in a node; e.g. *Galium boreale*.

interglacial period (L. *glacies* = ice), a period between glacial periods.

intergradation zone (L. *gradus* = step, degree), the contact zone between the adjacent populations (usually those of subspecies) which are phenotypically different and have previously been separated. In this zone hybrid individuals are common. *Ernst Mayr* calls this the **secondary intergradation zone** (hybrid zone). The **primary intergradation zone** exists if the gradient of phenotypical features is gradual and caused by corresponding changes in environmental conditions.

interleukin, IL (< leucocyte), any of the hormone-like polypeptides secreted by lymphocytes, macrophages, fibroblasts, and some other cells in vertebrates; belong to a larger group, → **cytokines**. Interleukins are important factors in regulation of immune reactions affecting the activity of other leukocytes; e.g. **interleukin-2**, a growth factor of lymphocytes which e.g. activates T cells to change into killer cells; **interleukin-4** correspondingly activates B lymphocytes. Some interleukins act as endogenic → pyrogens which induce fever in homoiothermic animals.

intermediary (L. *medius* = middle), intermediate; being or going between, or occurring in a median stage.

intermediary metabolism, the term sometimes used for the various chemical reactions which interconvert precursor molecules such as monosaccharides, amino acids, fatty acids, and nucleotides suitable for cellular energy metabolism; in vertebrates occurs especially in the liver.

intermediate filaments, protein filaments of intermediate size (about 10 nm in diameter) of the cell skeleton (→ cytoskeleton) in eukaryotic organisms (except plants); they are thicker than actin filaments and thinner than myosin filaments. The filaments are formed from several types of subunit proteins, the protein structure varying according to the type of organism but also to the type of tissue; e.g. in epithelial cells of vertebrates the filaments consist of **cytokeratin**, in muscular cells of **desmin**, and in connective tissue of **vimentin**.

intermediate host, a host organism in which a parasite lives for a limited period of its life cycle, without reaching sexual maturity. *Syn.* secondary host. *Cf.* definite host.

intermediate leaf, a leaf inside the → inflorescence of a plant but having no flower on its base.

intermedin, *see* melanophore-stimulating hormone.

intermural (L. *murus* = wall), pertaining to an anatomical structure situated between the tissue walls.

internal (L. *internus*), occurring or situated in the interior of a structure; inner, not superficial.

internal ear, → inner ear.

internal environment, a term coined by *Claude Bernard* (1813—1878) to emphasize the importance of stability (→ **homeostasis**) of extracellular tissue fluids surrounding the cells (Fr. *milieu intérieur*) in an animal body; includes the ionic concentration, osmotic pressure, pH, temperature, etc., with complicated control systems, such as the endocrine and nervous systems, developed during evolution.

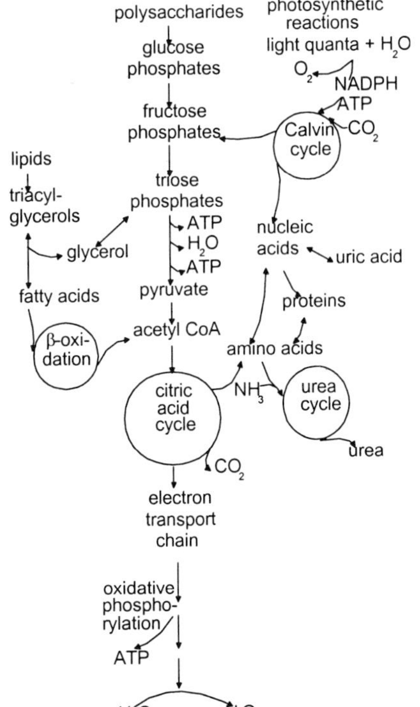

Intermediary metabolism. Main biochemical reactions in cells; some of them take place only in plant cells (e.g. photosynthetic reactions) and some only in animal cells (e.g. urea cycle).

internal fertilization, the fertilization that occurs in the female genital duct, e.g. in insects, reptiles, birds, and mammals. *Cf.* external fertilization.

interneurone, Am. **interneuron,** a relay neurone in a multineuronal tract; in vertebrates, a neurone in the grey matter of the brain stem and spinal cord, synapsing between the sensory and motor neurones in a polysynaptic reflex arc; integrative cross-connections with other neural tracts are possible through interneurones. Also called internuncial neurone.

internode, the part of the plant stem between two succesive → nodes.

interoceptor (< internal receptor), any of the sensory → receptors located inside an animal and detecting stimuli within the body; are divided into **visceroceptors**, receptors located in the viscera, and **proprioceptors**, the receptors in the muscles, tendons and the labyrinth of the inner ear. *Cf.* exteroceptors.

interphase, the period of the → cell cycle during which metabolism and synthetic events occur in the cell without visible signs of cell division. In that stage the → nucleus of the cell is called the **interphase nucleus**, and it contains the → chromosomes which replicate during interphase in such a cell that is capable of division. *See* mitosis, meiosis.

interpretation gene, any of the genes of the → zygote which read the → positional information present in the ovum. These genes are divided into **gap genes**, **pair-rule genes**, and **segment-polarity genes**. Mutations of the gap genes cause the loss of many body segments of an animal embryo, and mutations of the pair-rule genes result in the loss of either even-numbered segments or odd-numbered segments. Mutations of the segment-polarity genes cause a malformation in the body segments so that each has either two anterior parts or two posterior parts. *Syn.* cardinal gene.

interrupted gene, a gene comprising coding segments (→ exons) with non-coding segments (→ introns) between them; most of the genes of eukaryotic organisms are interrupted. *Syn.* split gene.

intersexual individual, intersex: an organism whose reproductive organs and/or secondary sex organs are partly of one sex and partly of the other, even though the individual does not consist of genetically different organs (*see* gynandromorph). Thus, intersexual individuals are clearly neither female nor male but represent a mixture of male, female, and intermediate characteristics. They are genetically sterile or produce gametes of one sex only, and therefore they can always be distinguished from → hermaphrodites which produce mature gametes of both sexes. Intersexual individuals occur rarely in the Animal kingdom and in → dioecious plants. Intersexuality is caused usually by genetic factors, such as disturbances in sex determination mechanisms, but also by hormonal factors involved e.g. in the formation of a → freemartin.

intersexual selection, the selection of a sex partner in courtship ceremonies, usually decided by the female on the basis of individual sex characters of the male (favouring selection). Sexual selection has great importance for the evolution of sex characters of males. *Cf.* dominance mating. *See* selection.

interspecific, between species.

interspecific competition, a competition in which individuals of two or more species try to use the same scarce resources in an ecosystem. *Cf.* intraspecific competition. *See* competition.

interstitial (L. *inter* = between, *sistere* = to set, place), situated between something; e.g. **1.** pertaining to a space, fluid, or cells between some anatomical or histological structures; **2.** pertaining to spaces between sand grains in soil.

interstitial biota (Gr. *bios* = life), microscopic organisms living between sand grains in soil or sandy littoral; usually smaller than 1 mm in length.

interstitial cells, 1. endocrine cells of the testes of vertebrates situated between the seminiferous tubules; secrete → testosterone; *syn.* Leydig's cells; **2.** endocrine cells of the vertebrate kidneys located between renal tubules in the medullary part, secrete → prostaglandins; also called medullary interstitial cells.

interstitial cell stimulating hormone, ICSH, *see* luteinizing hormone.

interstitial fluid, tissue fluid; the extracellular fluid between the cells in tissues of metazoan animals, but the term does not usually include the blood plasma and lymph in vessels, i.e. transcellular fluids. In vertebrates, the volume of interstitial fluid is 15 to 20% of all body fluids but varies much more between different species of invertebrates. Through the interstitial fluid, most blood substances like respiratory gases, nutrients, messenger substances, and about one third of proteins are transmitted from blood to tissue cells and vice versa. Interstitial fluid is formed continuously by filtering of blood plasma through walls of capillary vessels, and returns to the blood by force of → colloid osmotic pressure at the distal parts of the same capillaries; in vertebrates a small portion of interstitial fluid circulates through lymph vessels. In many invertebrates the interstitial fluid is mixed with blood forming haemolymph.

intertarsal joint (L. *tarsus* = ankle), a joint in the middle of the bird ankle; formed by fusion of the **tibia** with proximal **tarsals** (ankle bones), the fused bone being called **tibiotarsus**. The distal tarsals are fused with three **metatarsals** forming the "ankle-bone", **tarsometatarsus**.

interval mapping, a computer-assisted mapping method of → quantitative trait loci (QTL) making use of inbred strains of organisms marked with a sufficiently large number of mapped molecular markers. The computer programme is set up to track along each intermarker interval and to calculate the probability of the QTL located in that interval.

intervening sequence, → intron.

intestinal (L. *intestinum* = intestine, gut), pertaining to the gut or intestine.

intestinal canal, → intestine.

intestinal glands (L. *glandulae intestinales*), small tubular glands of the mucous membrane of the vertebrate intestine. The glands which almost totally cover the inner surface of the intestine are derived from the epithelial tissue involuted inside the connective tissue under it. Each gland releases its secretion into the intestinal lumen through the narrow ostium. The glands of the small intestine secrete both digestive enzymes and mucus, but those of the large intestine only mucus. *Syn.* Lieberkühn's crypts/glands/follicles. *See* duodenal glands.

intestinal juice (L. *succus entericus*), liquid secreted by the glands or single glandular cells in the wall of the intestine, the small intestine in vertebrates.

intestinal villi (L. *villi intestinales* < *villus* = tuft of hair), finger-like vascularized projections, 0.5—1 mm long, covering the surface of the mucous membrane of the anterior section of the intestine in vertebrates, the small intestine in mammals. Each villus, supported by connective tissue, contains a network of blood capillaries and a lymph vessel (lacteal), and is covered by a single layer of columnar epithelial cells with numerous → microvilli on the free cell surface. The villi (20—40 per square millimeter) together with → microvilli, increase the absorbing surface of the intestine manifold. *See* villus.

intestine (L. *intestinum*), intestinal canal; the section of the alimentary canal (digestive canal) which in vertebrates and many invertebrates lies between the stomach and anus or cloaca. In vertebrates, except fish, it comprises the **small intestine** (*intestinum tenue*) and the **large intestine** (*intestinum crassum*), the latter comprising the → **caecum**, **colon**, **rectum**, and **anal canal**. In mammals the small intestine is differentiated into three sections, → **duodenum**, **jejunum** and **ileum**. **Digestion** occurs chiefly in the stomach and in the most anterior part (duodenum and anterior part of jejunum), and **absorption** in all parts of the small intestine, water and salts also in the large intestine where the **formation of faeces** occurs.

Secretions from the liver (*see* gall bladder) and pancreas enter the anterior part of the intestine enhancing digestion; also many → intestinal glands in the mucous membrane release their secretions forming intestinal juice. Strong layers of smooth muscle function in the mixing of chyme (segmentation contractions and → peristalsis); many transversal folds and numerous → intestinal villi of the mucous membrane increase the absorptive surface manifold; for histological structure, *see* alimentary canal.

In invertebrates, the term intestine or intestinal canal usually refers to the whole alimentary canal except the mouth, often differentiated into the → oesophagus, crop, foregut, stomach, and hindgut. *See* digestion, absorption.

intestinum rectum, → rectum.

intima, pl. **intimae** or **intimas** (L. fem. of *intimus*), innermost, inmost; in vertebrate anatomy, e.g. the **tunica intima**, the innermost coat or the tunic in the wall of the blood vessels. *See* blood vessels.

intine (L. *intus* = inside, within), the inner, smooth layer of the wall of a → pollen grain. *Cf.* exine.

intolerance (L. *in-* = not, *tolerantia* = endurance), **1.** incapacity to endure or bear; **2.** lack of → tolerance or toleration. *Adj.* **intolerant.**

in toto (L. = on the whole), totally, entirely.

intoxication (Gr. *toxikon* = arrow poison), a state of being poisoned; poisoning; also drunkenness.

intra- (L. *intra* = within), denoting within, into, or during.

intra-abdominal, within or into the abdomen.

intra-arterial, within an artery or arteries; **i.a.**, given, e.g. injected into an artery.

intracellular (L. *cellula*, dim. of *cella* = cell), within a cell or cells.

intracranial, within the cranium or skull.

intracutaneous, within the skin or cutis; **i.c.**, given, e.g. injected into the skin. *Syn.* intradermal.

intradermal, i.d. (Gr. *derma* = skin), → intracutaneous.

intragastric (Gr. *gaster* = stomach), within the stomach; **i.g.**, given into the stomach. *Cf.* intra-abdominal.

intramolecular respiration, the term coined by *Louis Pasteur* (1822—1895) to mean → aerobic respiration.

intramuscular, within the muscle; **i.m.**, given, e.g. injected into a muscle.

intraoral (L. *os* = mouth), within the mouth; **i.o.**, given into the mouth.

intraperitoneal, within the peritoneal cavity; **i.p.**, given into the peritoneal cavity. *See* peritoneum.

intrasexual selection, *see* dominance mating.

intraspecific, within a species. *Cf.* interspecific.

intraspecific competition, a competition between individuals of the same species for same limited resources. If the competition occurs evenly between all the individuals, it is called **scramble competition**, in which everyone gets an insufficient share of the resources available. If the competition is uneven between the competitors, it is called a **contest competition**, in which some of the individuals get sufficiently, some are left out. Thus increasing density in a population is quickly compensated by the corresponding increase in mortality rate.

intravascular, within the blood or lymphatic vessels; injected into a vessel.

intravenous, within the veins, **i.v.**, given, e.g.

injected into a vein.

intrinsic (L. *intrinsecus* = on the inside), inherent; belonging entirely to a thing, situated within a given part, as e.g. a certain substance, muscle, or nerve. *Cf.* extrinsic.

intrinsic factor, a glycoprotein present in the gastric juice and secreted by the parietal cells of gastric glands in vertebrates. The factor is necessary for the absorption of vitamin B_{12} (cyanocobalamin, extrinsic factor), the deficiency of which results in pernicious anaemia.

intrinsic rate of natural increase, → biotic potential.

intro- (L. *intro* = inwardly, to the inside), in, into, inward, within.

introgression (L. *gressus* = course), the incorporation of genes of one species into the → gene pool of another species by hybridization and backcrossing.

intromittent organ (L. *intromittere* = to send in), an organ for the transfer of sperm into a female reproductive organ, e.g. → clasper, or penis.

intron (neologism < L. *intra* = inside), in an interrupted eukaryotic gene, a DNA segment located between the amino acid-coding segments, → exons; intron is the non-coding segment which is removed during the maturing processing of → messenger RNA. Introns are also found in genes of ribosomal RNA and transfer RNA. The function of intron is unknown. *Syn.* intervening sequence. *See* splicing.

introrse (L. *introrsus* = inward), pertaining to the way of facing of the anthers towards the inner parts of a flower (the floral axis), supporting self-pollination. *Cf.* extrorse.

intubation (L. *in* = into, *tuba* = tube), the insertion of a tube into a hollow organ or a canal, as into a trachea. *Verb* to **intubate**.

intussusception growth (L. *intus* = within, *suscipere* = to take up), a growth type of a plant cell wall in which new cellulose microfibrils insert in spaces between the older ones. *Cf.* apposition growth. *See* cell wall.

inulin, a storage polysaccharide composed of 30—35 fructose units; found in high quantities e.g. in roots or tubers of the composites as dahlia and chicory. It is used clinically as intravenous injections to determine the rate of glomerular filtration in the kidney, and instead of sugars in the diet of diabetics. *Syn.* dahlin, alantin, alant starch.

invader plant, a plant species spread to an area by human activity, being often unable to form seeds in the new area because of the climatic factors.

invagination (L. *vagina* = sheath), the process of forming a sheath; the process by which one part of a structure infolds within another part, as an inward movement of a wall of → blastula in the formation of → gastrula. *Verb* to **invaginate**.

invasion (L. *invadere* = to go in), **1.** migration or irruption of animals to new, unoccupied areas; can bring about the colonization of the area, either temporarily as many irruptive birds do, or permanently, resulting in the widening of the dirstibution range; **2.** an entrance of pathogenic microbes or metastatic cancer cells into a tissue; also an onset of a disease; **invasiveness**, ability for invasion.

invasive, intrusive; invading, tending to invade.

inversion, a turning inside out, upside down, or inward; e.g. **1.** in genetics, a chromosome mutation in which a certain segment of the chromosome has rotated 180°. There is no evidence of terminal inversions in which the tip of the chromosome would have rotated, and hence the inversion always seems to occur in regions located between the → telomeres of the chromosome. Thus, the inversion implies two break points in a chromosome, the rotation of the piece born in this way, and the reunion of the ends of the segment. In **simple inversions** two types can be distinguished: **pericentric inversions** in which the rotated segment includes the → centromere, and **paracentric inversions** in which the rotated segment does not contain the centromere but involves either arm of the chromosome (→ chromosome arm). In addition to simple inversions, several types of **complex inversions** also occur involving rotations of more than one segment of the chromosome; **2.** in chemistry, the hydrolysis of certain carbohydrates, resulting in a reversal of direction of rotatory power of the carbohydrate solution, i.e. the plane of the polarized light being bent from right to left, or vice versa; e.g. occurs in the hydrolysis of sucrose to glucose and fructose.

inversion layer, the air layer formed between warm and cold air when a warm air mass moves over colder air; the inversion layer inhibits the intermixing of the layers and, thus also the movement of e.g. smoke and exhaust fumes to the upper air layers. If the inversion layer is formed above a dense inhabited area in cold weather, it may cause a dangerous situation especially for people with respiratory insufficiency.

invertase, β-fructofuranosidase, an enzyme, EC 3.2.1.26, catalysing the hydrolysis of β-D-fructofuranosides into free fructose, or into glucose and fructose (from saccharose). The enzyme occurs e.g. in microbes, plants, and the alimentary canal of animals. *Syn.* sucrase, saccharase, β-h-fructosidase. *See* invert sugar.

invert sugar, a mixture of equal portions of fructose and glucose produced by the hydrolysis of saccharose (sucrose), in organisms catalysed by β-fructofuranosidase enzyme, → **invertase**. As the result from the reaction (**inversion**) the optical activity, i.e. the rotation of polarized light, is changed so that the dextrorotatory saccharose solution is converted with formation of laevorotatory glucose and fructose, i.e. rotating to the left. Invert sugar is the main component of honey.

Invertebrata (L. *in-* = not, *vertebra* = joint), **invertebrates**, a general term for all the animals without a vertebral column. *Cf.* Vertebrata.

inverted repeat sequence, any of a class of short DNA sequences (50 to 130 bp) which have the same nucleotide sequence but in opposite orientation to one another, e.g. at each end of an → insertion sequence (mobile element).

in vitro (L. *vitrum* = glass), within a glass; in an artificial environment, such as in a reaction vessel, bottle, Petri dish, or test tube; usually refers to a reaction or treatment carried out with a material separated from an organism, i.e. not studied → *in vivo*.

in vitro fertilization, *see* artificial fertilization.

in vivo (L. *vivus* = alive), in the living organism; usually refers to a reaction or process occurring or studied in a living organism, not → *in vitro*.

involucre (Fr. < L. *involucrum*), **1.** a protecting aggregation of bracts, usually a whorl or a rosette below the inflorescence of some plant species in the families Asteraceae and Cichoriaceae; **2.** a group of leaves around antheridia or archegonia in some mosses or liverworts.

involution (L. *involvere* = to roll inwards, wrap), the inward turning; e.g. **1.** the type of movement of a cell group in the gastrula of animals; **2.** the degeneration or retrograde change of an organ or organism, such as senile degeneration of a person. *Verb* to **involute.** *Adj.* **involutional.**

iodine (Gr. *ioeides* = violet coloured), **1.** a halogenous element, atomic weight 126.9045, symbol I; belongs to micronutrients and trace elements, and is needed by vertebrates for the synthesis of thyroid hormones (thyroxine and triiodothyronine). Iodine is also found in plants, abundantly e.g. in some algae and in roots of some graminoid plants. Iodine is taken into organisms as salts, **iodides,** which are ionized in the organ systems to I^-. The radioactive isotopes of iodine, ^{125}I and ^{131}I, are used in biological and medical studies, and in the diagnosis and treatment of disorders of the thyroid gland; **2.** a popular name for the aqueous solution of iodine and potassium iodide (or sodium iodide) which is used as a disinfectant and reagent, and in therapy. Lack of iodine in the diet is a cause of goitre. *See* Sachs' iodine test.

iodoprotein, any protein in organisms containing iodine, e,g. → thyroglobulin (iodoglobulin) in vertebrates; iodine is usually bound to a tyrosine residue.

iodopsin (Gr. *opsis* = sight), visual violet; a photopigment, 11-*cis*-retinal bound to an → opsin; most sensitive to the bluish area of the light spectrum, and is involved in → colour vision; occurs e.g. in the retinal cones of many vertebrates. *Cf.* rhodopsin.

ion (Gr. *ienai* = to go), an electrically charged atom or atom group; it has lost one or more electrons and thus is a positively charged **cation**, such as sodium Na^+, potassium K^+, and calcium Ca^{2+} ions, or it has gained one or more electrons and is a negatively charged **anion**, as e.g. chloride, Cl^-, phosphate PO_4^{3-}, and sulphate SO_4^{2-} ions. Many ions are important agents in organisms for the generation of membrane potentials across the cell membrane. *See* resting potential, action potential, receptor potential.

ion antagonism (Gr. *antagonizesthai* = to fight against), a counteraction between ions; e.g. the presence of one ion inhibits or limits the transport of another across the cell membrane. The phenomenon is based on the opposite charges of ions.

ion channels, numerous minute channels across the cell membrane, composed of protein molecules; they may be open allowing certain ions to pass through it along an electrochemical gradient, or they may be closed (gated). Opening and gating may be regulated by the level of membrane potential (**voltage-gated channels**, electrically regulated channels,) or by hormones and neurotransmitters (**ligand-gated channels**, receptor-regulated channels). The ligand may be external, as e.g. a hormone or neurotransmitter, or it may be internal, as e.g. cyclic AMP, a G protein, or Ca^{2+}. Also mechanical stimuli regulate the opening of some channels (**mechanically gated channels**), as occurs in mechanical sensory receptors. Ion types (Ca^{2+}, Na^{+}), ion concentrations, or ATP may affect the gating. Most ion channels are ion-specific, transporting e.g. only Ca^{2+}, K^{+}, Cl^{-}, or Na^{+} ions, while some channels transport several types of ions. Ion channels have been found in all groups of organisms. Their function is important in ionic metabolism and electrophysiological functions of cells. The action of many drugs is based on their effects on ion channels. *See* cell membrane transport, resting potential, action potential, patch clamp. *Cf.* ion pumps.

ion exchange, 1. reversible interchange between ions absorbed electrically by soil particles; important for the nutritive conditon of the soil and the nutrient uptake of plants. Cations are absorbed by small soil particles by electric forces of different magnitudes which depend on the charge, size, and water film (hydration stage) of the particles. Hydrogen ions (H^{+}) occurring in the soil or produced by plant roots can displace all other absorbed ions which are released into soil solution and become available to plants. Also calcium ions (Ca^{2+}) are absorbed effectively by soil particles being capable of releasing many other ions; **2.** ion exchange by resins, i.e. **ion exchangers** (cation exchangers, anion exchangers) are used in → chromatography and other chemical analyses e.g. to purify and fractionate solutions.

ionic bond, electrovalent bond; a chemical bond due to the transfer of an electron between the bonding atoms, resulting in an electrostatic attraction between the atoms; occurs between the atoms which have a big difference in their electronegativity. When sodium (Na) and chlorine (Cl) react with each other the atoms convert into ionic forms. Sodium loses one electron from its outermost shell and a Na^{+} ion is formed. When chlorine gets one additional electron, a Cl^{-} ion is formed and the → octet rule comes true. Na^{+} and Cl^{-} bind together producing sodium chloride (NaCl).

ionization chamber, a device used to measure ionizing radiation; consists of a gas-filled chamber with two electrodes. Ionizing radiation ionizes gas in the chamber and the instrument measures the ionization current produced as ions are driven by the electric field.

ionizing radiation, high energy radiation that directly or indirectly causes the excitation and ionization of atoms e.g. in organisms exposed to this radiation. In nature, ionizing radiation is a part of the cosmic electromagnetic radiation but this has practically no significance on the Earth surface because it becomes absorbed in the atmosphere and produces free electrons there, forming the → ionosphere. The main sources of ionizing radiation on the Earth are different radioactive elements which may be natural or produced by human activities. This ionizing radiation is either **particle emission** (alpha radiation, beta radiation), or **electromagnetic ionizing radiation** (the shortest wavelengths of electromagnetic radiation: X-rays, gamma rays). *See* radioactivity. In organisms, ionizing radiation can generate → free radicals, cause oxidation of substances and break cellular structures which, if not acutely killing, may lead to mutations and malignant tumours in organisms. Most wavelengths of the electromagnetic radiation (thermal radiation, visible light, and some UV) that reaches organisms are weakly ionizing or **non-ionizing radiation**, the ionizing property depending on the quantity of energy absorbed as well as the wavelength and intensity of the radiation. Throughout evolution, organisms have been adapted to a certain level of ionizing radiation by protecting themselves against it. They have developed protective tissues with pigment cells, antioxidant systems, and enzymatic repair systems for damage to nucleic acids. *See* antioxidants, DNA repair mechanisms.

ionophore (Gr. *pherein* = carry), a hydrophobic organic molecule which joins to lipid structures of cellular membranes (especially plasma membrane) increasing the penetration of ions across the membrane; often ion-specific. Ionophores either enclose the transported ions and diffuse them through the membrane (*see* valinomycin, transporting K^+ ions), or they form channel-like structures (channel former) in the membrane (*see* gramicidin). Some ionophores are produced by microorganisms, especially bacteria, which use these compounds in defence against competing species. Ionophores are used in research dealing with cell ionic metabolism.

ionosphere (Gr. *sphaira* = ball, atmosphere), the zone of the Earth's upper atmosphere between 80 and 400 km in which free electrons produced by ionizing radiation (mainly UV radiation and X-rays from the Sun) occur in high quantities. The ionosphere consists of various strata reflecting radio waves; the zone outside the ionosphere is called exosphere. *Cf.* stratosphere, troposphere, mesosphere.

ion pumps, proteins which actively transport ions across membranes of cells and cell organelles against an electrochemical gradient. They are generally called → adenosine triphosphatases, **ATPases**, because they act as enzymes using hydrolysis of energy-rich ATP as energy source for this active work. Different ion pumps are e.g. **Ca^{2+} - ATPase** in plant and animal cells, **H^+ - ATPase** (hydrogen pump) in plant cells and some animal cells, and **Na^+-K^+ - ATPase** (sodium-potassium pump) in animal cells. *See* cell membrane transport. *Cf.* ion channels.

iontophoresis, pl. **iontophoreses** (Gr. *phorein* = to carry along), a way of transporting substances in ionized form into a cell across the cell membrane using → microinjection by means of electric current.

i.p., → intraperitoneal.

IP_3, inositol 1,4,5-triphosphate. *See* inositol.

IPA, isopentenyladenosine. *See* cytokinins.

ipsilateral (L. *ipse* = self, *latus* = side), on the same side of a structure. *Cf.* collateral, contralateral.

IPSP, → inhibitory postsynaptic potential.

iridocytes, iridophores (Gr. *iris* = rainbow, *kytos* = cavity, cell), → guanophores.

iris, pl. **irides, 1.** the pigmented, circular membrane behind the cornea in the camera eye of vertebrates and cephalopods. It has circular and radial muscles which regulate the diameter of the pupil in the middle of the iris (iris reflex, light reflex); **2.** the root powder of a herb (*Iris* family) used for medical purposes.

IRMA, → immunoradiometric assay.

iron, a metallic element, symbol Fe (L. *ferrum*), atomic weight 55.847, density 7.86; occurs in natural conditions in iron compounds; common iron ores are magnetite (Fe_3O_4), haematite (Fe_2O_3), and limonite ($2\ Fe_2O_3 \cdot 3\ H_2O$). Iron exists commonly also in carbonates, silicates, sulphides, etc. Because of its oxidation-reduction property (**ferri iron**, $Fe^{3+} \rightleftharpoons$ **ferro iron**, Fe^{2+}), iron is a necessary constituent in many cellular enzymes as in → cytochromes which act in the electron transport chain in → cell respiration of all organisms; and in the light reactions of → photosynthesis in green plants. It is the central atom of → haemoglobin and → myoglobin in animals, and in plants it is essential for the synthesis of → chlorophylls. *See* haem, apoferritin.

irradiation (L. *irradiare* = to irradiate), **1.** the act of illuminating or radiating, particularly the use of some type of radiation for the treatment of an organism; e.g. irradiation with → ionizing radiation is used for the treatment of diseases, such as cancer, or to cause mutations for scientific use. Also irradiation with → infrared radiation is used for the treatment of organisms; **2.** the spreading of nervous impulses in the brain beyond the normal path from one area or tract to another.

irreversible (L. *in-* = not, *revertere* = to return), unable to change back or reverse; pertaining to any reaction or event that occurs only in one direction, e.g. some enzymatic reactions or denaturation of proteins during cooking.

irritability (L. *irritare* = to excite, irritate), the property of an organism or its cells to respond to a stimulus; a crucial characteristic of the living systems. *Adj.* **irritable**, capable of responding to a stimulus, or being abnormally sensitive to a stimulus.

irritation, 1. the state of overexcitation of the nervous system; **2.** the act of stimulation. *Verb* to **irritate**. *Adj.* **irritant**, tending to produce irritation.

irruption (L. *irrumpere* = to break in), a sudden entry or action; e.g. **1.** the act of sudden breaking through, or bursting to the surface of an organ or organism; **2.** a sudden increase in the population density as a consequence of exceptionally favourable changes in environmental conditions; it may be a reason for mass movements of the individuals in a population. Irruptions occur irregularly. *Adj.* **ir-**

ruptive, marked by irruption. *See* irruptive species.

irruptive species, an animal species, usually a bird or insect, of which an appreciable proportion migrates to new areas irregularly after a deterioration of environmental conditions or after a sudden increase in the population size (*see* irruption). Feeding conditions are important factors to bring about irruptions, and thus the irruptive species are usually food specialists; e.g. crossbills are cone specialists.

ischaemia, Am. **ischemia** (Gr. *ischein* = to suppress, *haima* = blood), local deficiency of blood in a tissue or an organ due to the constriction or any obstruction in blood vessels, usually in the artery carrying blood to the area.

ischi(o)- (Gr. *ischion* = hip joint, hip), **1.** ischium; **2.** ischial; **3.** resembling a hip joint.

ischiopodite (Gr. *pous* = foot), *see* biramous appendage.

ischium, pl. **ischia** (L. < Gr. *ischion* = hip joint) ischial bone (*os ischii*), one of the bones forming the pelvic girdle in tetrapod vertebrates. *See* pelvis.

isidium, pl. **isidia** (Gr. *Isis* = genus of gorgonians, i.e. horny corals), a type of → diaspore of lichens; a scale-like or plug-like outgrowth in the lichen thallus containing both fungal hyphae and algal cells; may separate and develop into a new thallus.

island biogeography, a branch of ecology dealing with the occurrence of species and structures of communities on islands and island-like environments. *See* island theory.

island theory, a theory presented by *R.M. MacArthur* and *E.O. Wilson* in 1963, claiming that the number of organisms living on an island are positively correlated with the size of the island and negatively correlated with the distance from a dispersion centre. Combined effects of these two factors regulate the diversity of species on the island. **Extinction** indicates the number of the species lost from the island per time unit, usually being the greater the higher the species diversity is, or the farther away the island is located from the dispersion centre (e.g. from a continent). **Immigration** presents the number of new immigrant species to the island per time unit. The number of species is in balance when extinction and immigration are equal.

The island theory may be applied also to other environmental patches (ponds, lakes, meadows, etc.) in areas where those are clearly apart from the surrounding environment. The theory is much used in planning positions and sizes of nature reserves.

islets of Langerhans (*Paul Langerhans*, 1847—1888), → pancreatic islets.

iso- (Gr. *isos* = equal), denoting alike, equal, sameness, homogeneous, uniform; e.g. **1.** in immunology, describing materials (e.g. isoantibody) which are obtained from genetically similar individuals; **2.** in chemistry, designating a structural → isomer which has one carbon branch, next to the end of the chain.

isoamyl alcohol, 3-methyl-1-butanol, an alcohol, $(CH_3)_2CHCH_2CH_2OH$, used in biochemical studies e.g. together with phenol for removing proteins from a solution, i.e. causing deproteinization. *Syn.* isopentyl alcohol.

isoantigen, → alloantigen.

isobar (Gr. *baros* = weight), a line on a weather map connecting the points of the same barometric pressure, reduced to the sea level. *Adj.* **isobaric**, pertaining to an isobar, or having equal barometric pressure.

isobutanol, isobutyl alcohol, 2-butanol. *See* butanol.

isochores (Gr. *choros* = place), very long DNA segments (longer than 200-300 kb) in vertebrate genomes; are fairly homogeneous and belong to a small number of DNA sequence classes characterized by different GC levels (guanine-cytosine levels).

isochromosome (Gr. *chroma* = colour, *soma* = body), a monocentric or dicentric → chromosome consisting of two homologous (similar and genetically identical) → chromosome arms which are mirror images of each other.

isocitrate dehydrogenase, *see* isocitric acid.

isocitric acid, an intermediate in the → citric acid cycle of cell respiration in mitochondria, $HOOCCH_2CH(COOH)CH(OH)COOH$; it is the → isomer of citric acid, and its ionic form, esters, and salts are called **isocitrates**. It is oxidized in the reaction: isocitrate + NAD^+ ($NADP^+$) —> 2-ketoglutarate + CO_2 + NADH (NADPH), catalysed by **isocitrate dehydrogenase** (EC 1.1.1.41, NAD requiring, or EC 1.1.1.42, NADP requiring).

isodeme (Gr. *demos* = district, deme, the people), in ecology, a line (isodemic line) on a map connecting areas where population densities are equal.

isodiametric, having equal diameters; e.g. describes the shape of a cell or an anatomical structure in which the length and width are approximately equal.

isodisomy (Gr. *dis* = twice, *soma* = body), a very rare and exceptional chromosomal constitution in which an individual has inherited both members of the pair of → homologous chromosomes from one parent.

isoelectric point, any pH value in which the sum of the positive and negative charges of → ampholytes, such as proteins and amino acids, is at the minimum or zero. Each protein is precipitated at its isoelectric point which property is applied to protein analyses; in **isoelectric focusing**, an electrophoretic method applied to the separation of proteins, each protein is subjected in a pH gradient to an electric field where it migrates to the pH of its isoelectric point.

isoenzyme, → isozyme.

Isoëtales (Gr. *isoetes* = equal in years), **quillworts**; a plant order in the class Lycopsida, division Pteridophyta; in some classifications an order in the division Lycophyta. The type genus is *Isoëtes*. Nearly all species in Isoëtales are submerged plants, having a bulb-like rhizome and thin, sharp-shaped leaves. The macrosporangia and microsporangia are located in expanded leaf bases. About 70 known species have distributed widely, excluding the coldest regions.

isoetid, a water plant with a leaf rosette totally submerged, growing on the bottom of a water body. *See* water plants.

isoflor (L. *flos* = flower), a curve on a map connecting areas with the same number of plant species.

isoforms, proteins encoded by the same gene, e.g. by alternative → splicing or by differential → genetic transcription; also families of functionally related proteins which are encoded by genes that are now located at different chromosomal positions but are derived from a single ancestral gene.

isogametes, *see* isogamy.

isogamy (Gr. *gamos* = marriage), a type of reproduction (**isogamous reproduction**) in which the size of the male and female gametes (**isogametes**) fusing in fertilization is quite similar. *Cf.* anisogamy.

isogeneic, (Gr. *genos* = origin), **1.** genetically identical; pertaining to any group of individuals which possess the same genotype; also called isogenic, syngenic, syngeneic; **2.** pertaining to two alleles or chromosome segments which are identical by descent.

isogenesis (Gr. *gennan* = to produce), a development through similar processes. *Adj.* **isogenous**, having the same origin, e.g. describes the development of different structures from the same cell or tissue.

isogenic, *see* isogeneic.

isograft, a tissue or organ → graft transplanted between isogeneic individuals, such as inbred rats or monozygotic twins (identical twins); between these tissue rejection does not normally occur. *Syn.* syngraft.

isohaline (Gr. *hals* = salt), a line on an oceanic map connecting the points where salinities are equal.

isokont (Gr. *kontos* = punting pole), having flagella of equal length, e.g. pertaining to a motile cell or spore of plants which have two flagella of equal length.

isolate, a group of individuals or populations, or a detached population, in which a mating partner is always chosen amongst the individuals of the group; thus the isolate is strictly separated from other groups. In human genetics, it is divided into **geographical** and **social isolates**, the latter being a consequence of economical or religious factors, or of marriage traditions.

isolation, separation from others; in ecology, the separation of a → population from the other populations of the same species, due to a geographic, ecological, climatic, physiological, or some other kind of hindrance. Sometimes isolation means the avoiding of competition between two different species in such a way that the competing species separate from each other by the usage of nutrition, selection of → habitat, periods of activity, or geographical distribution.

Due to reproductive hindrances, isolation prevents or restricts any → gene flow between the separate populations. Isolation is an important factor in evolution. Its different forms can be separated into two classes: **geographical isolation** (regional isolation) in which the populations live in different areas, i.e. they are allopatric, and **reproductive isolation** in which the interbreeding of the populations is prevented eventhough they live within the same area, i.e. are sympatric. In the reproductive isolation two types of mechanisms are functioning:

1) the **prezygotic mechanisms** prevent the

formation of hybrid zygotes. These mechanisms include **ecological isolation** in which the populations live in the same area but in different ecological niches, **seasonal isolation** in which the mating or flowering periods of the populations occur at different times, and **ethological isolation** (**sexual isolation**) in which the attraction between the sexes of different species is lacking or the → courtship is too different. It further includes **mechanical isolation** in which the external sexual organs of different species do not fit in copulation, or the parts of the flowers of different plant species do not fit in pollination or they have different pollinators, and **gametic isolation** in which the sperm of a foreign species does not survive in the genital ducts of the female;

2) the second main type of reproductive isolation comprises **postzygotic mechanisms**, which include **hybrid inviability** in which the hybrids between species are lethal, **hybrid sterility** in which the hybrids between species are sterile, and **hybrid breakdown** in which the F_1 or backcross hybrids between species are lethal or sterile.

Species are usually separated by manifold isolation mechanisms. These prevent the breakdown of the integrated → genetic systems of the species which have been adapted to certain environmental circumstances. Hence, isolation plays a central role in → speciation.

isolator, anything that isolates; e.g. a box where experimental animals or other objects are maintained behind an aseptic barrier for preventing bacterial invasion.

isolecithal (Gr. *lekithos* = egg yolk), describing an ovum type which has a moderate amount of uniformly distributed yolk.

isoleucine, abbr. Ile, ile, symbol I; 2-amino-3-methylvaleric acid, $CH_3CH_2CH(CH_3)CH(NH_2)COOH$; an aliphatic → amino acid found in almost all proteins. Isoleucine belongs to the → essential amino acids of animals.

isomerase (Gr. *meros* = part), any enzyme from the larger group (EC class 5.) that catalyses the conversion of a compound to its isomer (isomeric form). *See* isomers.

isomers, organic compounds with the same number of equal atoms in their molecules but differing in arrangement of the atoms; consequently physical and chemical properties of isomers are different. There are several types of **isomerism**:

1) **structural isomerism** in which the order of atoms in the carbon chain is different (**chain isomerism**, e.g. α-alanine and β-alanine, with an NH_2 group linked to alpha-carbon or beta-carbon). In **functional group isomerism** there are different functional groups in the molecules, as in ethyl alcohol and dimethyl ether;

2) **stereoisomerism** (space isomerism, spatial isomerism) with two subtypes, i.e. **geometric isomerism** (*cis-trans* isomerism) in which a molecule may occur in *cis* form or *trans* form. In the *cis* form the substituents are situated at the same side of the ring or double bond, but in the *trans* form at different sides; e.g. → fumaric acid and maleic acid are isomers, the former being the *trans* form:

```
     H-C-COOH
       ||
  HOOC-C-H
```

and the latter the *cis* form:

```
  H-C-COOH
    ||
  H-C-COOH
```

of butenedioic acid, $C_4H_4O_4$.

The other form of stereoisomerism is **optical isomerism**, based on the asymmetry of molecules which have atoms or atom groups at many different positions. Optical isomers, **enantiomers**, of a compound are specified by a chemical prefix D- (small capital D-from *dextro* = right) that refers to the mirror image configuration of the other form specified by L- (*laevo* = left). The two forms differ in their configuration at a chiral atom (asymmetric atom). D-glyceraldehyde is used as a reference compound. Enantiomers rotate the plane of polarization of the beam of the light to opposite directions, and are specified by prefix *d*- or (+)-, denoting that the isomers rotate polarized light in the clockwise direction (dextrorotatory isomer). The other enantiomer, specified by *l*- or (—)-, rotates the polarized light counterclockwise (laevorotatory isomer); e.g. *l*-fructose (laevulose, levulose) can be specified more precisely D-(—)-fructose. Usually only one isomer occurs in nature, e.g. amino acids in L-forms and monosaccharides in D-forms.

isometric (Gr. *metrein* = to measure), pertaining to equal length; e.g. isometric muscular contraction in which the contraction occurs under tension without changing the length of the muscle, most usual in position muscles. Isometric contractions are practised in physical training to increase the size and power of muscles. *Noun* **isometry**. *See* isotonic.

isomolar (L. *moles* = mass), pertaining to solutions with equal → molarity, i.e. solutions which contain the same number of moles in the same unit of volume.

isomorphic, isomorphous (Gr. *morphe* = form), similarity in form or appearance; e.g. **1.** pertaining to organisms (**isomorphs**) which have similar morphological appearance but different ancestry; **2.** pertaining to a type of plant → life cycle in which the sporophyte and gametophyte are separate individuals of similar shape; occurs in some primitive algae, e.g. in the genus *Ulva*; **3.** in chemistry, usually **isomorphous**, pertaining to a compound capable of crystallizing in a form similar to another compound. *Cf.* heteromorphic. *See* alternation of generations.

iso-osmotic, isosmotic (Gr. *osmos* = push, impulse), describing solutions with equal osmotic potential (concentration). *See* osmosis.

isopentenyl adenosine, IPA, *see* cytokinins.

isopentenyl pyrophosphate, an → isoprene unit with five carbon atoms; in plants e.g. → carotenoids, gibberellins, abscisic acid, and some cytokinins are synthesized from it; also an intermediate in the biosynthesis of cholesterol in animals and plants.

isopentyl alcohol, → isoamyl alcohol.

isophene (Gr. *phainein* = to show), **1.** a contour line connecting such areas on a map where frequencies of certain variant forms, as morphs, are equal; **2.** a line connecting such areas on the map where seasonal events are simultaneous, e.g. arrival dates of migratory birds.

Isopoda (Gr. *pous*, gen. *podos* = foot), **isopods**, e.g. woodlice, pill bugs; a crustacean order in the class Malacostraca, including about 4,000 species in nine suborders inhabiting waters or moist soil; some species are parasites. The isopods have a depressed body with seven pairs of unspecialized walking legs; some abdominal pleopods have developed into respiratory organs.

isoprenoid compounds, compounds containing a $(C)_2$-C-C-C chain of five carbon atoms, e.g. **isoprene**, $CH_2=C(CH_3)-CH=CH_2$; occur as a part of molecular structures of many compounds, such as → lycopene, carotenoids, phytol, rubber, vitamin A, and vitamin E.

isopropanol, isopropyl alcohol, *see* propanol.

Isoptera (Gr. *pteron* = wing), **termites**; an order of small, hemimetabolous insects whose social organization is highly complicated and the individuals are grouped in several morphologically different castes, reproductive individuals, soldiers, and workers. Termites are sometimes called "white ants" but differ from ants e.g. by having a broad union of thorax and abdomen. In many tropical areas, termites have economic significance as they destroy timber; cellulose is their principal nutrition that they, with the aid of intestinal microflora, efficiently digest. About 200 species are described.

isopycnic centrifugation (Gr. *pyknos* = dense, tight), *see* density gradient centrifugation.

isospore (Gr. *sporos* = seed), a spore type in plants which have no → sexual dimorfism, i.e. all spores (isospores) are similar. *See* homospory. *Cf.* anisospore.

isosteres (Gr. *stereos* = stiff, solid), **1.** molecules with an equal electron arrangement; **isosterism** may occur between such molecules which have the same number of atoms and electrons, e.g. CO_2 and N_2O; this leads to similarity of physical properties; **2.** in meteorology, a line on a map connecting points of equal atmospheric pressure. *Adj.* **isosteric.**

isotherm (Gr. *therme* = heat), a line on a weather map connecting points which have equal temperature at a given time, or the same mean temperature for a certain time period. *Adj.* **isothermic.**

isotocin, an nonapeptide hormone secreted from the posterior → pituitary gland of some teleost fishes; corresponds to → vasopressin in mammals.

isotomy (Gr. *temnein* = to cut), a type of dichotomical branching in which the developing branches are quite equal. *See* dichotomy.

isotones, nuclides having the same number of neutrons but a different number of protons in their nuclei, e.g. ^{132}Xe and ^{133}Cs.

isotonic (Gr. *tonos* = tension, tone), having equal tension; e.g. **1.** pertaining to solutions with equal osmotic potential (concentration), especially relating to solutions in which cells neither swell nor shrink, indicating that the solutions are osmotically equal with the intracellular fluid; the condition is called **isotonia**; *see* osmosis; **2.** pertaining to the contraction of a muscle in which the muscle shortens against a constant load and the tension state in it remains unchanged, as in lifting a weight; *cf.* isometric.

isotonicity, the state or quality of being → isotonic.

isotopes (Gr. *topos* = place), two or more atoms of the same chemical element having the same atomic number (the same number of protons) but differing in atomic weight due to the number of neutrons in the atomic nucleus, e.g. hydrogen isotopes with three different mass numbers, i.e. 1 in hydrogen, 2 in deuterium, and 3 in tritium; some isotopes are radioactive, as tritium (^{3}H), phosphorus-32 (^{32}P), or sulphur-35 (^{35}S). The isotopes of an element are identical in chemical properties, but slightly different in physical properties. Most elements found in nature are mixtures of isotopes. *Adj.* **isotopic.**

isotropic, 1. having equal physical properties in all directions, as e.g. in the → I-band of a muscle fibre; **2.** pertaining to a structure which is singly refractive in electromagnetic radiation; also isotropous. *Noun* **isotropy** (isotropism). *Cf.* anisotropic.

isotropy (Gr. *tropos* = turning), the condition or quality of being similar in all directions. *Adj.* **isotropic.**

isotype (Gr. *typos* = image, model), a similar type; e.g. **1.** in taxonomy, a duplicate of a type specimen (holotype); **2.** in immunology, an antigenic determinant (**isotypic marker**) occurring in all members of an immunoglobulin class (*see* immunoglobulin). All isotypes characteristic of the species are found in the plasma of every individual.

isozyme, isoenzyme (Gr. *en zyme* = in state of fermentation), one of different forms of an enzyme having similar catalytic activity but differing in structure due to their different subunit combinations. Isozymes have the same substrate specificity but may differ e.g. in their thermal sensitivity, and/or pH optimum, often also in their substrate affinities. Generally, the isozymes are tissue-specific and their specificity is determined by the developmental stages of cells and tissues. *Cf.* allozyme.

isthmus, pl. **isthmi,** Am. also **isthmuses** (L. < Gr. *isthmos*), a narrow structure connecting two larger parts; e.g. **1.** many isthmi in vertebrate anatomy connecting two parts of an organ or cavities, e.g. the isthmus of thyroid, connecting the two lobes of the gland; **2.** the constriction between the → semicells of placoderm → desmids; **3.** in biogeography a land bridge, i.e. a narrow terrain passage connecting two large continents or continent-like regions, e.g. the Isthmus of Suez, the Isthmus of Panama.

iteroparity (L. *iterare* = to repeat, *parere* = to give birth), a type of reproduction in which an individual produces, or is able to produce, offspring several times in the course of its life cycle. *Adj.* **iteroparous.** *Cf.* semelparity.

J

J, symbol for joule. *See* Appendix 1.

Jacob—Monod theory (*Francois Jacob,* b. 1920, *Jacques Monod,* 1910-1976), the theory of → operon.

Jacobson's organ (*Ludvig Jacobson,* 1783—1843), **vomeronasal organ**; a specialized part of the olfactory system in some tetrapod vertebrates. The organ is best developed in many reptiles, especially in snakes and lizards, in which the organ with special olfactory receptors occupies a pair of pouches opening separately to the anterior roof of the mouth cavity. The sensory epithelium of the organ is not continuously in connection with the respiratory air stream which ensures high receptor sensitivity to scent stimuli; many reptiles gather scent substances (such as → pheromones) from environmental air with their long tongue (fork-shaped in snakes), then transferring the substances to the sensory epithelium of the pouches. The organ is absent in fishes, turtles, and crocodilians. In birds and some mammals, such as higher primates, it occurs only in the embryonic stage, but many mammals, such as artiodactyls, have retained it throughout the adult age. The whole **vomeronasal system** includes also the accessory olfactory bulb located dorsally in the main olfactory bulb.

jaw-feet, → maxillipeds.

jawless (vertebrates), → Agnatha.

jaws, paired cranial bones derived from the first pharyngeal arch in vertebrates, forming a framework of the mouth cavity; the **upper jaw,** → premaxilla, maxilla; the **lower jaw,** → mandible, dentary bone; *see* cranium; **2.** → mouthparts of invertebrates.

jejunum (L. *intestinum jejunum* < *jejunus* = empty), the portion of the small intestine of mammals, between the duodenum and ileum. *See* alimentary canal.

jellyfish, → Scyphozoa.

Johnston's organ (*Christopher Johnston,* d. 1891), a pair of organs in the antennae of insects consisting of sensory cells responding to mechanical stimuli; possibly give information about flight speed and vibration of the air.

joint (L. *junctio* = connection), **1.** united; joined; **2.** junction, juncture; an anatomical structure in which two (or more) elements of the skeleton join, mostly movably together, as do the segments of → appendages in arthropods and many other invertebrates, or many bones or body segments in vertebrates (also called articulations). In the joints between the bones there is a narrow cavity (**synovial cavity**), which widens laterally forming the **synovial capsule** (*capsula articularis*) around the joint, the **synovial membrane** (**synovium**) enclosing the whole structure. Special **synovial fluid** (synovia) fills the cavity, decreasing friction in the joint. Cartilage tissue either covers the epiphyseal surfaces of the bones, or forms discs between them, as in the wrist, knee, or vertebral bones (intervertebral disc). During the ontogenic development the bones in some joints may be fused together, as in the sacral bone. There are sensory nerve endings in the synovial capsules responding to movements of the joint; **3.** a synonym of segment.

joule, J (*James Prescott Joule,* 1818—1889), the unit of energy. *See* Appendix 1.

J-shaped growth curve, the graphical description of the exponential growth of a population. *See* exponential growth.

jugular (L. *jugulum* = neck, throat), **1.** pertaining to the neck or throat; **2.** the jugular vein (*vena jugularis externa* or *interna*) carrying blood from the head in tetrapod vertebrates.

jugum, pl. **juga** (L. = yoke), **1.** a furrow or ridge of an anatomical structure in animals or plants; **2.** the posterior basal area in the forewing of some insects; in some species couples the forewings and hindwings in flight; **3.** a pair of opposite leaflets in a compound leaf.

junction (L. *junctio* or *junctura*), the act of joining; a site or point where two or more structures are joined together, e.g. where two or more cells, tissues, or organs are joined; e.g. **1.** → **cell junctions** between tissue cells; **2.** a **juncture** between anatomical structures which are points or lines of union of two parts, especially bones or cartilages; e.g. the **oesophagogastric junction** between the epithelia of the oesophagus and stomach, or the **neuroectodermal junction** in an early embryo.

junctional complex, the attachment zone between neighbouring epithelial cells where the cells are connected by various types of cell junctions, such as → tight junctions, desmosomes, and hemidesmosomes.

Jungermanniales (*Jungermannia*, the type genus < *Ludvig Jungermann*, d. 1653), a predominantly tropical order in the class Hepaticopsida (liverworts), in the division → Bryophyta, Plant kingdom; in some classifications an order in the division Hepatophyta. The number of species is ca. 9,000, the order comprising about 90% of all liverworts. Most species have a lobed thallus with a midrib and leaf-like lateral appendages (genus *Blasia*), or the thallus is totally leaf-like, broad, growing on the soil (genus *Marchantia*).

junk DNA, long segments of DNA located on all eukaryotic chromosomes between the genes; its function, if any, is unknown.

Jurassic (Jura mountains), the middle period in the Mesozoic era, between Triassic and Cretaceous about 208 to 144 million years ago. Reptiles were dominant on land, in the sea and air, e.g. dinosaurs, pterodactyls, ichthyosaurs; the first birds and archaic mammals developed.

juvenile (L. *juvenis* = young), **1.** immature, youthful; pertaining to a young, prereproductive organism, e.g. to a young bird in the stage it has fledged (if altricial) or replaced down of hatching (if precocial), or to juvenile behaviour of an animal; also pertaining to the stage of the plant development before flowering; **2.** as a noun, a young individual; **juvenility**, the juvenile state or characteristic.

juvenile hormone, a class of insect hormone produced by a paired gland, → **corpus allatum**; sustains the development of the larval stage promoting the retention of juvenile characteristics. *Syn.* neotenin. *Cf.* ecdysone.

juvenile leaf, a type of → foliage leaf in plants.

juxta- (L. *juxta* = near, nearby), denoting close by, near.

juxtaglomerular organ/apparatus, any of small organs located close to renal glomeruli in the vertebrate kidney; each is formed from specialized cells (**juxtaglomerular cells**) in the wall of the afferent nephric arterioles, and from specialized cells (**macula densa**) of the renal tubular epithelium near the juxtaglomerular cells. The cells of the macula densa are chemosensitive receptor cells which "measure" salt concentration (especially sodium chloride, Na^+, Cl^-) of the glomerular filtrate in the renal tubule. Several factors, such as blood pressure, the volume of extracellular fluid, vasopressin, prostaglandins, and the sympathetic nervous system, control the secretion of **renin** from juxtaglomerular and lacis cells in the junction between the afferent and efferent arterioles. Renin is an enzyme (EC 3.4.99.19) that is involved in the regulation of renal function in the renin-angiotensin system. *See* angiotensin.

K

K, symbol for **1.** potassium; **2.** lysine; **3.** kelvin; **4.** carrying capacity (of the environment).

***K*,** equilibrium constant, as K_d dissosiation constant, K_a acid ionization constant, K_b base ionization constant, K_m Michaelis constant.

k, symbol for kilo-.

***k*,** symbol for reaction velocity or rate constant.

Ka., abbreviation for cathode.

Kainozoic, → Caenozoic.

kak(o)- (Gr. *kakos* = bad), → cac(o)-.

kal(i)- (L. *kalium* = potassium), pertaining to → potassium.

kalium (L., Ar. *quali* = potash), → potassium.

kallidin, lysylbradykinin; a decapeptide acting as the precursor of → bradykinin; these are vasodilator agents in the blood of vertebrates. *See* kinins.

kallikrein, kininogenin; kininogenase; a proteolytic hydrolase enzyme (serine protease) found in vertebrate tissues such as blood, pancreas, salivary glands, and intestinal wall; **plasma kallikrein** (EC 3.4.21.34) breaks inactive precursors (kininogens) of → kinins (particularly bradykinin) in the blood thus activating them e.g. in inflammation reactions; **tissue kallikrein** (EC 3.4.21.35) acts in other tissues producing kallidin.

Kamptozoa (Gr. *kamptos* = flexible, *zoon* = animal), a former name for → Entoprocta.

kanamycin, an antibiotic complex (kanamycin A, B, C), produced by *Streptomyces kanamyceticus*, interfering bacterial protein synthesis by misreading the genetic code in both Gram-positive and Gram-negative bacteria; the transfer of kanamycin resistance can be used as a marker in gene transfer into cells.

kapok, also **capoc** (Malay *kapok*), a filling made from fruit fibres of a Mexican kapok tree (silk-cotton tree, Ceiba tree).

kappa, 1. a self-replicating DNA particle found in the cytoplasm of some *Paramaecium* strains, and responsible for the so-called killer phenotype, i.e. the ability to kill other (sensitive) *Paramecium* strains; **2. kappa chain**; one of the two light chains of mammalian → immunoglobulins.

kary(o)-, also **cary(o)-** (Gr. *karyon* = nut, kernel, nucleus), denoting relationship to the cell nucleus.

karyogamy (Gr. *gamos* = marriage), the fusion of two nuclei in fertilization; homologous chromosomes, separated in meiosis, fuse again and the zygote is formed as a result of the karyogamy.

karyogenesis (Gr. *genesis* = birth), the formation and development of the cell nucleus.

karyokinesis (Gr. *kinein* = to move), the division of the cell nucleus as opposed to the division of the cytoplasm, → cytokinesis. With the aid of karyokinesis the → genetic information residing in the chromosomes are divided into daughter nuclei so that two nuclei, genetically identical to each other and the mother nucleus, are born. The basic mechanisms of karyokinesis are the same in all → eukaryotic organisms. Differences exist in the

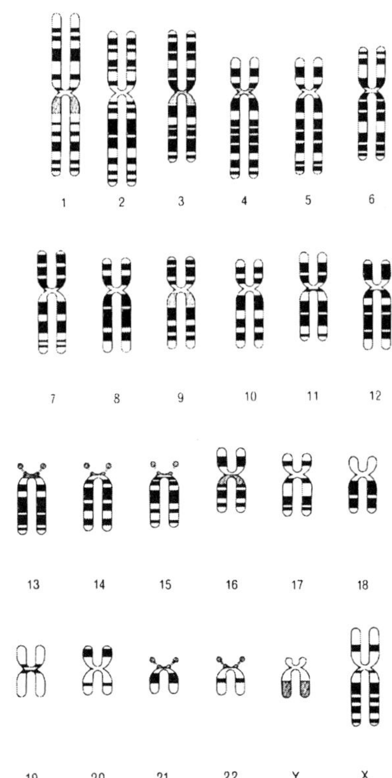

Karyotype. The normal human karyotype. The autosome chromosomes are numbered, and the sex chromosomes marked with X and Y. The dark bands are →Q bands. (After Bostock & Sumner: The Eukaryotic Chromosome - North Holland 1980; with the permission of the authors).

formation of the → mitotic spindle, and in cytokinesis which immediately follows karyokinesis. The term is often used as a synonym of → mitosis.

karyology (Gr. *logos* = word, discourse), a branch of → cytology dealing with structures and functions of cell nuclei.

karyolymph (L. *lympha* = bright water), the rarely used term for the ground substance of the cell nucleus in which the chromosomes are embedded.

karyoplasm (Gr. *plasma* = mould), nucleoplasm; the plasm of the cell nucleus.

karyotype (Gr. *typos* = form), the chromosome set of an individual or a group of related individuals, determined usually according to the number and structure of chromosomes in the mitotic metaphase (*see* mitosis). In the schematical representation of karyotypes, i.e. the **karyotypy**, the chromosomes are arranged for comparison according to their size and morphology. Karyotypy is an important method e.g. in cytotaxonomy, foetus diagnostics, and cell culture.

Kaspar Hauser experiments (*Kaspar Hauser*, 1812—1833, a youth in Nuremberg who was claimed to have lived in absolute isolation from other humans to the age of 17), a type of experiment in which an animal is reared from the birth in isolation from other individuals; in these conditions the behaviour of the animal is derivable from its genotype without influence of learning.

kb, → kilobase.

K capture, a type of radioactive decay, including the incorporation of one extranuclear electron from the inner shell (K shell) into the atom nucleus. This change takes place if the → positron emission is not possible; the atomic number decreases by one. *See* radioactivity.

K cell, → killer cell.

kD, kDa, kilodalton (= 1,000 daltons). *See* dalton.

keel, 1. → keel of sternum; **2.** a structure formed by two partially united lowest petals in the flowers of the family Fabaceae; **3.** in → diatoms (order Pennales), a ridge bearing the raphe, i.e. the longitudinal slit in the valve, where the cytoplasm streams; **4.** keel disease, a septicemic intestinal disease in ducklings caused by *Salmonella anatum*, marked by a sudden death of seemingly healthy birds.

keel of sternum, a well developed ventral keel (*crista sterni*) of the sternum of birds and bats; serves as additional attachment area for large pectoral muscles important in flight. The keel is found in all birds except the flightless ostrich-like ratites. *Syn.* carina.

kelp, a common name for marine brown algae with a large and thick thallus, e.g. *Laminaria* and *Macrocystis* used in foodstuffs (soups, Japanese sushi); kelp ash obtained from burned algae is used as a source of alkali and iodine.

kelvin, K (*William Kelvin*, 1824—1907), the SI unit of thermodynamic temperature. *See* Appendix 1.

kenenchyma (Gr. *kenos* = empty, *enchein* = to pour in), a plant tissue the cells of which have no living content; e.g. cork.

kentrogon (Gr. *kentron* = sharp point, *gonia* = angle), a larval stage of parasitic cirripeds following the cypric stage after entering the host.

keratan sulphate, keratosulphate; a sulphated → glycosaminoglycan (with D-galactose and N-acetylglucosamine residues), forming a part of proteoglycan in cartilage, bone, and other connective tissues.

keratin, also **ceratin** (Gr. *keras* = horn), a sulphur-rich, heteromeric protein (scleroprotein) synthesized by epithelial cells, → **keratinocytes** in the vertebrate skin → epidermis. Keratins are classified into five types (I—V) which occur in different epithelial cells forming a flexible cellular framework of intermediate filaments. In vertebrates, keratins form the main component in the process of keratinization (→ **cornification**) in the skin, and give rise e.g. to nails, hairs, and feathers.

keratinocyte (Gr. *kytos* = cavity, cell), an epidermal cell of the vertebrate skin synthesizing and accumulating → keratin; keratinocytes represent about 90% of all mammalian skin cells.

kernel, the inner part of a seed, as of a nut; in graminoids the whole grain or seed.

ketoglutaric acid, → alpha ketoglutaric acid.

ketones (*acetone* < L. *acetum* = vinegar), a group of organic compounds containing the carbonyl group, -C=O, the simplest being dimethylketone or → acetone, CH_3COCH_3. Ketones are synthesized in organisms in lipid metabolism, particularly when carbohydrate metabolism is retarded. In vertebrates, this process is accelerated during fasting or in deficiency of insulin, producing in liver cells ketones like acetoacetate, β-hydroxybutyrate, and acetone (**ketone bodies**, acetone bodies) from which also other ketones may be formed; the condition is called **ketosis**. In high quantities in the blood (ketonaemia or acetonaemia) ketones cause disturbances in cell metabolism, associated with the increased acidity (→ acidosis) of body fluids.

ketose, a monosaccharide or its derivative containing a ketone group (carbonyl group), -C=O, e.g. → fructose (ketohexose) or dihydroxyacetone; ketoses are called ketopentoses, ketohexoses, etc., according to the number of carbon atoms they contain.

key-factor analysis, a mathematical method for analysing population data which deals with factors affecting the population size; the analysis is based on the data of → life and fertility tables and aims at explaining mortality rates in different stages of the life cycle. *Syn.* K-factor analysis.

key species, keystone species, a species having a great effect on the structure and action of ecosystems; the key-species usually has numerous individuals, but may also be a scarce species with large individuals.

key stimulus, a stimulus or a group of stimuli which particularly activates an innate releasing mechanism in the behaviour of an animal. In general, key stimuli are species-specific, as e.g. colour patterns of nuptial plumages. *See* lock-and-key model.

K-factor analysis, → key-factor analysis.

kidney (Gr. *nephros*, L. *ren*), an animal organ excreting chiefly waste products, especially nitrogen compounds produced in the metabolism of amino acids, purines, and pyrimidines; participates also in pH, ion, and osmotic regulation. In vertebrates, the kidney is a paired organ situated dorsally in the coelom. There are different stages in the embryonic development of the kidney, **pronephros** (primary kidney) is the first stage in all vertebrate embryos, forming the functional kidney also in amphibian larvae. It is a long, segmental organ containing segmentally situated → nephrons, the functional units of the kidney, and like the two later kidney stages, it develops from the mesodermal tissue of the body cavity. Waste products are filtered from the blood through the walls of arterial glomeruli into the body cavity (coelom), and then the filtrate flows through open nephrostomes into the nephron tubules (kidney tubules). Proximally several tubules coalesce to form collecting ducts which end to the primary urinary duct opening to the cloaca.

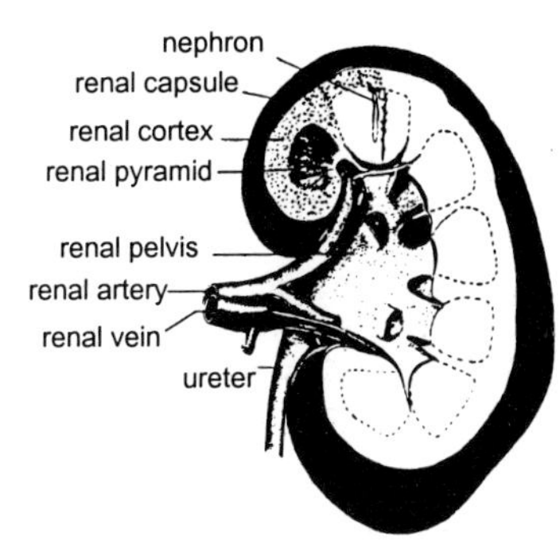

Cross-section of the mammalian kidney (metanephros).

The second stage, **mesonephros**, develops later beside the pronephros. Here the nephrostomes are closed, forming **Bowman's capsules**, each of them containing one arterial **glomerulus** enveloped by the capsule. In this kidney type, glomelurus filtrate is formed from the glomerular blood directly to the Bowman's capsule; the filtrate then flows into the nephron tubule (renal tubule).

The mesonephros remains functional in adult fish and amphibians, but in higher vertebrates it is a temporary stage which degenerates when the third stage, **metanephros**, develops caudally to it. The metanephros derives from a tissue of the ureteric bud (part of the mesonephric duct or Wolffian duct) and of nephrogenic tissue around it. It is a roundish, often bean-shaped organ containing millions of nephrons (in man more than one million per kidney), which structurally and functionally are very similar to those of the mesonephros. Urine processed in the nephrons by reabsorption and secretion is collected into the renal pelvis, from which it passes through the ureter to the urinary bladder. The function of the kidney is hormonally controlled by → vasopressin, mineralocorticoids, and the renin-angiotensin system; *see* angiotensin, renal calix, renal cortex, renal medulla.

In invertebrates, the functions of the kidney are performed by different types of organs, such as the → nephridia, Malpighi's tubules, green glands, and gills, in protozoans by a → contractile vacuole.

killer cell, 1. cytotoxic T cell (effector T cell, T_8 cell); a specifically activated T lymphocyte of the vertebrate immune system capable of killing cytotoxically foreign cells, such as bacteria; *see* lymphocytes; **2.** → natural killer cell.

kilo- (Gr. *chilioi* = one thousand), in the metric system signifying one thousand (10^3).

kilobase, kb, a length unit of polynucleotide and nucleic acid molecules, i.e. 1,000 bases or base pairs.

kilodalton, kD, kDa, 1000 → daltons.

kin(e)-, kin(o)- (Gr. *kinein* = to move), denoting relationship to movements; also **cin(e)-**.

kinaesthetic sense, Am. **kinesthetic sense** (Gr. *aisthetikos* = concerning sensory perception), "muscle sense"; the sense by which movements, position, and strain are perceived in the body, includes muscle spindles, tendon organs, and the sensory receptors of joints. *Syn.* kin(a)esthesia.

kinanaesthesia, Am. **kinanesthesia** (Gr. *an-* = not, *aisthesis* = sensation, perception), a disturbance of the → kinaesthetic sense, found e.g. in man resulting in the inability to perceive extent, position, and direction of movements. Also spelled **cinan(a)nesthesia**.

kinases, EC 2.7.1-4.; a sub-subclass of transferase enzymes catalysing the transfer of a phosphate (phosphoryl) group from high-energy phosphate, usually from ATP, to a second substrate; kinases phosphorylate e.g. hexoses (→ hexokinase), proteins (→ protein kinase), or amino acids (e.g. tyrosine kinase, serine/threonine kinase).

kinesi(o)- (Gr. *kinesis* = motion), denoting relationship to motion.

kinesin, a complex protein (→ motor protein) found in eukaryotic organisms involved in the microtubular transport process inside the cell (e.g. in neuronal axon transport); has the ability to bind to microtubules and certain cellular vesicles and transport the vesicles along the microtubule, usually in centrifugal direction, i.e. towards the distal (+) end of microtubules, using ATP as energy source. *Cf.* dynein, myosin.

kinesis, pl. **kineses** (L., Gr.), motion; movement; a motor activation or movement of a cell, organ, or organism (an animal), especially the movement induced by a stimulus; e.g. **1.** plasmic movements in a cell during → cytokinesis; **2.** a motor reaction in which a stimulus affects the enhancement or retardation of the behaviour of an animal; the movement is random and depending on the intensity, not the direction of the stimulus as occurs in a → taxis. Animals may change their locomotor speed (**orthokinesis**) or their turning frequency (**klinokinesis**). Different types of kineses are distinguished depending on the stimulus, e.g. **photokinesis** (light stimulus), **chemokinesis** (chemical stimulus), or **hygrokinesis** (water stimulus).

kinesthetic sense, → kinaesthetic sense.

kinetics (Gr. *kinetos* = moving), a branch of physical chemistry dealing with forces involved in producing the motion of masses; e.g. turnover rates of molecules in organic processes.

kinetin, a synthetic plant hormone (an adenine derivative, 6-furfurylaminopurine) with → cytokinin activity.

kinetochore (Gr. *choros* = place), the locomotive organ of the chromosome; a three-layer protein structure located in the centromere region of a chromosome attaching to the microtubules (kinetochore fibres) of the → mitotic spindle; kinetochores play an active role in the movement of chromosomes towards the poles during cell division. *See* centromere.

kinetoplast, a highly specialized mitochondrion associated with the → basal body of some flagellate protozoans (Kinetoplastida, e.g. trypanosomes); its DNA is exceptional forming a network consising of interlocked circles.

kinetosome (Gr. *soma* = body), → basal body.

kingdom, in biology, the highest taxon of living organisms; at present usually five kingdoms are classified: **monerans** (Monera), comprising → prokaryotes, **protists** (Protista) comprising protozoans, unicellular fungi, and algae, **fungi** (Fungi), **plants** (Plantae), and **animals** (Animalia). Previously only three or four kingdoms were distinguished in which classification Plantae included Fungi, and unicellular organisms were divided partly into Plantae, including Bacteria, Algae, and photosynthetic protozoans, and partly into Animalia, including protozoans. Even now the use of the five kingdoms is not categorial; e.g. green algae are mostly classified also into the division → Chlorophyta (kingdom Plantae), especially by botanists.

kinins (Gr. *kinein* = to move), **1.** in animals,

polypeptides produced in vertebrate tissues when certain proteins, **kininogens**, are cleft by the enzymatic action of → **kallikrein**. Effects of kinins resemble those of histamine, e.g. causing vasodilatation and increased permeability of capillary walls; they also irritate pain receptors causing pain (*see* bradykinin, kallidin); **2.** in plants, → cytokinins.

kinocilium, pl. **kinocilia,** a motile → cilium found on free surfaces of certain cells; especially refers to a type of sensory cilium on the surface of mechanoreceptor cells of animals, usually appearing together with many immotile → stereocilia; the internal structure of the kinocilium with 9+2 microtubules is similar to that of motile cilia, although usually it does not exhibit motility (*see* hair cell).

Kinorhyncha (Gr. *rhynchos* = snout, beak), **kinorhynchs**; a phylum comprising marine worms living all over the world on mud bottoms from intertidal shore areas to 6,000 m in depth. The spined body, not more than 1 mm long, is divided into several segments and covered by cuticular plates; the retractile head has a circlet of spines. About 120 species are known. *See* Aschelminthes.

kin selection, a type of natural selection in which the altruistic behaviour of an animal increases the reproductive value of its relatives and thus confirms the transfer of its own genes into future generations. *See* Hamilton's rule, inclusive fitness.

kiwis, → Apterygiformes.

Kjeldahl method (*Johan Kjeldahl*, 1849—1900), an analytical method for the quantitative determination of nitrogen particularly in organic materials. Nitrogen is converted to NH_4^+ with concentrated sulphuric acid and catalysts, e.g. hydrogen peroxide and copper or potassium sulphates (wet combustion). Ammonia is liberated from the solution when heated with NaOH in a distillation apparatus, then bound into boric acid and determined by titration.

Klenow fragment (*Hans Klenow*, b. 1923), the large proteolytic protein fragment (67 kD) split from the bacterial DNA polymerase I after treatment with a protease. The fragment retains 5'→ 3' polymerase and 3'→ 5' exonuclease activities and can be used in the → Sanger method of DNA sequencing.

kleptoparasitism, → cleptoparasitism.

Klinefelter's syndrome (*Harry Klinefelter*, b. 1912), a form of human developmental disturbance due to the aneuploidy of sex chromosomes caused by their non-disjunction; characterized by a tall stature, long and slender limbs, sterility, and mental retardation. The chromosomal constitution is usually 47,XXY, but may also be e.g. 48,XXYY, 48,XXXY, 49,XXXXY; these individuals are males. *See* X chromosome, Y chromosome.

klinokinesis (Gr. *klinein* = to slope), *see* kinesis.

klinostat, also **clinostat** (Gr. *statos* = arranged, placed), a device for rotating a plant to neutralize the directional effect of gravity. *See* gravitropism.

klinotaxis, clinotaxis (Gr. *taxis* = arrangement), a tactic movement (*see* taxis) of an animal in which it orients itself in relation to a stimulus by turning the head or anterior part of the body sidewise, and so tests the position and intensity of the stimulus on each side.

K_m, Michaelis constant. *See* Michaelis—Menten kinetics.

knockout organism, an individual or a certain line of a species from which a given gene has been technically removed or inactivated in order to investigate the effect of this gene on the phenotype.

Knop's solution (*J.A. Knop*, d. 1891), a simple nutrient solution for liquid culture of plants; contains $Ca(NO_3)_2$, KH_2PO_4, KNO_3, $MgSO_4$ and iron as sulphate or chloride. The original Knop's solution does not contain any → micronutrients. *Cf.* Hoagland's solution.

Kornberg enzyme (*Arthur Kornberg*, b. 1918), DNA polymerase I functioning in DNA repair.

Kranz anatomy (Ger. *Kranz* = wreath), a typical tissue structure of leaves of → C_4 plants; formed from vascular bundles which are surrounded by large, chloroplast-containing bundle sheath cells, in the cross-section resembling a wreath. Typical mesophyll cells with chloroplasts form the surrounding → mesophyll.

Krebs cycle (*Hans Krebs*, 1900—1981), → citric acid cycle.

krill, the crustacean order Euphausiacea; a significant group of relatively large (about 5—7 cm, some species up to 15 cm long), shrimp-like malacostracans living in the southern oceans; an important constituent of marine food webs, being e.g. the main food of many filter-feeding baleen whales.

Krummholz, a region between the alpine zone

and the tree line, growing shrubs and dwarfed trees. *See* height zones.

K-selection (K = from the carrying capacity, K, in the logistic equation), the type of natural selection which occurs under conditions of stable and predictable environments where populations are controlled primarily by density-dependent factors (**K-strategic environments**). The species living in such environments (**K-strategists**) have usually a slow individual development, delayed reproduction, iteroparity, a small reproductive effort but high survival rate of offspring, large body size, and good competitive ability. The population densities of the species are usually near the carrying capacity of the environment. *Cf.* r-selection. *See* density-dependence.

K-strategist, K-selected species. *See* K-selection.

Kupffer cell (*Karl W. von Kupffer*, 1829—1902), a cell type present in the walls of sinusoidal capillaries in the liver of vertebrates; the cells belong to the → reticulo-endothelial system being tissue → macrophages that phagocytose noxious particles which enter the liver from the intestine via the hepatic portal vein.

kymograph (Gr. *kyma* = wave, *graphein* = to write), an instrument with a large revolving cylinder, on which undulative movements of an organ can be mechanically recorded; replaced by electrical recorders.

kyno- (Gr. *kyon* = dog), denoting relationship to a dog.

kyto- (Gr. *kytos* = cavity, cell), pertaining to a cell; usually → **cyto-**.

L

L, symbol for **1.** litre (SI symbol is l); **2.** lambert; **3.** leucine; **4.** abbreviation e.g. for lumbar (L1—L5) or lethal (lethal dose).

***L*,** symbol for inductance.

L., abbreviation e.g. for *Lactobacillus*, left, length, or latin.

L-, (a small capital L < *laevo* = left), a chemical prefix that specifies the relative configuration of an optical isomer. *See* isomers.

l, SI symbol for litre.

***l*,** symbol for length.

***l*-,** *see* isomers.

label, 1. a marker group, an identifying mark, tag, etc.; e.g. in **radioactive labelling**, a stable atom in a molecule is replaced by a radioactive isotope to identify, localize, or determine the molecule e.g. in cells, tissues, or cell extracts. Also a great number of non-radioactive labels have been developed, e.g. using **immunochemical** and **fluorometric methods**; **2.** as a verb, to add such a marker atom or group to a molecule.

labellum, pl. **labella** (L. dim. of *labium* = lip), a pair of enlarged, pointed parts of the labium in many insects, e.g. in the housefly; functions in sponging liquid food. *See* mouthparts.

labi(o)- (L. *labium*= lip), denoting relationship to lips or lip-like structures. *Adj.* **labial**.

labial palps (L. *palpi labiales*), **1.** a pair of the outermost, jointed sensory appendages in the lower lip (*labium*) of insects; **2.** two large flap-like folds associated with tentacles on each side of the mouth of some molluscs, such as clams. *See* mouthparts.

labium, pl. **labia** (L.), a lip; an edge or a border in the anatomical structure of animals and plants; e.g. **1.** lips (*labia oris*), around the mouth in many animals, often accompanied by labial palps; **2.** the lower lip of an insect mouth; *see* labial palps, mouthparts; **3.** the edge of the shell in gastropods from which the shell grows; **4.** the pudendal lips (*labia majora* and *minora*), skin folds (two parts) of the external genitalia in female mammals; **5.** the lower lip of a flower in the family Lamiaceae.

labrum, pl. **labra** (L.), a lip, lip-shaped structure, or margin; e.g. **1.** acetabular lip (*labrum acetabulare*), a cartilaginous edge attached to a margin of the → acetabulum of the hip bone in tetrapod vertebrates; **2.** a lobe-like upper lip in front of the mouth in many arthropods, as e.g. in insects and some arachnids; the labrum is hinged with the → clypeus. *See* mouthparts.

labyrinth (Gr. *labyrinthos*), a system of intercommunicating canals or cavities; e.g. **1.** the labyrinth in the internal ear of vertebrates comprising the osseous (bony) labyrinth and membranous labyrinth; **2.** auxiliary canals or cavities of the ethmoid bone forming a part of the lateral wall of the nasal cavity of higher vertebrates; **3.** an auxiliary respiratory organ found in some fishes (labyrinth fish, family Belontiidae). It is a maze-like organ above the gills developed from folds of the oral mucous membrane through which the fish can breathe air on the water surface; **4.** a spongy structure of the → green gland in crustaceans.

Labyrinthodontia (Gr. *odous* = tooth), **labyrinthodonts**; an extinct primitive amphibian group comprising species of variable size, some attained crocodile proportions. The enamel layer on their teeth had a complicated pattern which gave the tooth a labyrinthine appearance in cross section. In the late Palaeozoic and early Mesozoic, labyrinthodonts were abundant but became extinct at the end of Triassic.

Labyrinthulomycota, in some systems a fungal division comprising the orders Labyrinthulales and Thraustochytriales.

Lacertilia (L. *lacerta* = lizard), lacertids, lizards; → Sauria.

lacinia, pl. **laciniae,** also **lacinias** (L. = flap, fringe), a fringe-like anatomical structure, e.g. **1.** *laciniae/fimbriae tubae*, numerous fringe-like processes around the distal infundibulum of the uterine tube; **2.** the inner lobe of the maxilla of insect → mouthparts.

Lack's principle (*David Lack*, 1910—1973), a rule stating that the number of offspring produced and the number of offspring reaching sexual maturity are not in complete correlation. Thus e.g. a large clutch size in birds does not ensure a large number of adult individuals, and it does not increase the maximal proportion of the parental genotypes in the next generation. The maximal clutch size increases offspring mortality during the fledging period and often brings down the fitness of the parents, and consequently the final reproductive effort remains smaller. Therefore the average clutch size is under control of natural selection so that the maximal number of offspring will reach the reproductive stage.

lac operon, a bacterial → operon specifying the synthesis of proteins involved in lactose metabolism; the encoded enzyme proteins are β-galactosidase, galactoside permease, and *trans*-acetylase.

lacrimal, also **lachrymal** (L. *lacrima* = tear), pertaining to tears; e.g. lacrimal duct, gland, sac, or bone; **lacrimation**, secretion of tears; **lacrimiform**, tear-shaped.

lacrimal gland, a gland above the eye producing fluid secretion (tears) in terrestrial vertebrates; the secretion contains salts and some organic materials, such as proteins, and moistens and cleans the eye; e.g. by means of lysozyme enzyme which inhibits bacterial growth. The secretion is discharged onto the eye surface through many small excretory ducts and further drained from the lacrimal point of the eye corner into the nasal cavity through the lacrimal canaliculi (in mammals two lacrimal ducts, a lacrimal sac and nasolacrimal duct). In many amphibians, the gland is not compact but forms a row of small glands along the inside of the lower lid. In some marine birds and reptiles the glands are differentiated into → salt glands.

lact(o)-, lacti- (L. *lac* = milk), pertaining to milk, lactic acid, or lactose.

lactalbumin (L. *album* = white), a protein found in milk, i.e. milk → albumin.

lactase, → β-galactosidase.

lactate, 1. a salt, ester, or the ionic form of → lactic acid; **2.** to secrete milk.

lactate dehydrogenase, LDH, the group of NAD^+ oxidoreductases (e.g. EC 1.1.1.27 and 28) which catalyse the reversible reaction: pyruvate + NADH + $H^+ \rightleftharpoons$ lactate + NAD^+, occurring in anaerobic → glycolysis of cell metabolism. Also called lactic (acid) dehydrogenase. *See* anaerobic respiration.

lactation (L. *lactare* = to suckle), **1.** secretion of milk; **2.** the period of the secretion of milk. *See* prolactin.

lacteal (L. *lacteus* = milky), **1.** pertaining to milk or latex; **2.** pertaining to lymphatic chyle of intestinal lymph vessels, i.e. white milky fluid containing dietary fat absorbed from the small intestine into lymph capillaries and then collected into larger lymphatic vessels, **chyle cisterns** (*cisterna chyli*), from which the chyle through the thoracic duct enters the circulation; **3.** as noun, any of the lymph vessels draining the lymphatic chyle.

lactic acid, 2-hydroxypropionic acid, CH_3CH-OHCOOH, produced in microbial fermentation and in → anaerobic respiration of animals and plants when pyruvate is reduced by NADH into lactate in → glycolysis instead of participating in reactions of the aerobic → citric acid cycle. In aerobic conditions, lactic acid can be reoxidized to pyruvate for aerobic energy production, or converted back to glucose in → gluconeogenesis. The ionic form, esters, and salts are called **lactates**. *See* lactate dehydrogenase.

lactic acid bacteria, *see* lactobacillus.

lactic acid fermentation, the enzymatic conversion of carbohydrates, especially → lactose, into lactic acid by the function of lactic acid bacteria; an important process in producing acid food products, especially in the dairy industry. *See* lactase, lactobacillus.

lactiferous, pertaining to an anatomical structure which produces milk or latex.

lactobacillus, pl. **lactobacilli** (L. *bacillus* = small rod), a group of Gram-positive, anaerobic or microaerophilic bacteria from the genus *Lactobacillus*, family *Lactobacillaceae*, producing lactic acid. Lactobacilli occur e.g. in soil participating in natural and artificial fermentation processes, and in animals as important microorganisms in the normal flora of the mouth, intestine, vagina, and skin. These bacteria are used in the dairy industry.

lactogenic hormone, → prolactin. *See* also human placental lactogen.

lactose, milk sugar; 4-(β-D-galactosido)-D-glucose, $C_{12}H_{22}O_{11}$, a disaccharide composed of → galactose and → glucose with a β-1-4 linkage; exists naturally as α- and β-lactose, abundant in milk. Lactose is relatively rare in plants, present e.g. in the anthers of *Forsythia* flowers. It is an important substance e.g. in the production of cheese and yoghurt and many pharmaceutical preparations.

lactose intolerance (L. *intolerantia* = inability), inability of humans to digest lactose, caused by the deficiency or lack of the intestinal enzyme, → **β-galactosidase** (lactase); characterized by abdominal cramps and diarrhoea after consuming lactose-containing food. After childhood, most people become unable to digest lactose thus being lactose intolerant, but among cattle-rearing people, such as the Nordic people, the expression of the genes for lactase synthesis in adulthood has persisted (or this trait has redeveloped) in most individuals, and consequently the intolerance is less common (the incidence 10—20%) than elsewhere.

lacuna, pl. **lacunae** (L. = cavity, pit), a small space or cavity in an anatomical structure; e.g. **1.** in animals, may be normal, such as lacunae in the osseous or cartilaginous tissue containing bone or cartilage cells, or an abnormal structure in a tissue; **2.** in plants, an air-filled space.

lacunar collenchyma, a type of collenchymatous tissue in plants. *See* collenchyma, parenchyma.

ladder-like nervous system, a type of nervous system found in some invertebrate groups, as e.g. flatworms. *See* nervous system.

Laemmli gels, polyacrylamide gels (PAG) used in → electrophoresis. *See* acrylamide, gel.

laesion, Am. **lesion** (L. *laedere* = to hurt), an injury or wound; a traumatic or pathological change or loss of a body tissue.

laev(o)-, Am. **lev(o)-** (L. *laevus* = left), **1.** denoting left; **2.** a chemical prefix, symbol (–)-, *l*-, or L-; *see* isomers.

laevan, Am. **levan,** any polysaccharide synthesized from fructose units.

laevulin, Am. **levulin,** a starch-like compound occurring in tubers of many plants.

laevuli(ni)c acid, Am. **levuli(ni)c acid,** 4-oxopentanoic acid, $CH_3COCH_2CH_2COOH$; a crystalline and freely water-soluble compound; obtained from starch, cane sugar, or cellulose with hydrochloric acid, and used in organic syntheses and e.g. in the manufacture of nylon, synthetic rubber, and drugs. *Cf.* aminolaevulinic acid.

laevulose, Am. **levulose,** laevorotatory D-fructose.

lagena, pl. **lagenae,** or **lagenas** (L. = flask), a rather straight type of the cochlea in the inner ear of amphibians and birds.

lagging chromosome, a chromosome that during a nuclear division remains at the equatorial plate, and does not move like the other chromosomes towards the cell poles during → anaphase.

Lagomorpha (Gr. *lagos* = hare, *morphe* = form), **lagomorphs**; an order of mammals including **pikas**, **hares**, and **rabbits**. Lagomorphs are herbivorous mammals with a rodent-like skull and constantly growing incisors; in the upper jaw, there is an additional pair of small, peg-like incisors behind the first pair (*cf.* Rodentia). → Caecotrophy is common among lagomorphs. The hindlegs are specialized for fast jumping motion, and are clearly stronger than the forelegs. The lagomorphs have evolutionarily diverged into a separate group already in Eocene. About 60 species are classified as two families, Ochotonidae (pikas) and Leporidae (hares and rabbits). In early classifications, hares and rabbits were located into Rodentia as a suborder Duplicidentata.

lag phase, the first phase of the growth of cell cultures when no increase in cell number occurs; typical e.g. of the growth of bacteria and unicellular algae.

LAI, → leaf area index.

Lamarckism (*Jean-Baptiste de Lamarck*, 1744—1829), a theory of evolution stating that the origin of species is based on the inheritance of acquired characters. The use of an organ increases the size and functional ability, while the lack of use causes degeneration. The characters acquired this way are also inherited by the following generations. The theory was discarded in the 19th century but Lamarck's significance as a pioneer of evolutionary thinking has been appreciated in recent years.

lambda (phage), a double-stranded DNA virus (bacteriophage) that infects *Escherichia coli* and can enter the lysogenic or lytic cycle of its own replication in a cell; widely used as a vector (lambda cloning vector) in DNA studies.

Lambert—Beer's law (*Johann Lambert*, 1728—1777; *Georg Beer*, 1763—1821), the rule concerning absorption of light in any homogeneous matter; i.e. every layer of the same thickness of any pure homogeneous matter absorbs an equal amount of light. The absorption equation is: $I = I_0^{-kcl}$ or $I = I_0 \cdot 10^{-\varepsilon cl}$, in which I is the intensity of the transmitted light, I_0 is the intensity of the incident light, k is a specific extinction coefficient, l is the thickness of the matter (usually a liquid layer), c is concentration (e.g. mg/ml), and ε is a molar extinction coefficient. The equation has important applications in photometry. *See* absorbance, spectrophotometer.

lamella, pl. **lamellae** (L. dim. of *lamina* = plate), any thin plate-like or scale-like structure, e.g. a bone lamella, periosteal lamella, or gill lamella. *Adj.* **lamellar**.

lamellar collenchyma, a type of collenchymatous tissue in plants. *See* collenchyma, parenchyma.

Lamellibranchia (Gr. *branchia* = gills), lamellibranchs, → Bivalvia.

lamellipodium, pl. **lamellipodia,** *see* pseudopodium.

lamina, pl. **laminae,** also **laminas** (L. = thin plate), a thin scale, plate, or layer; **1.** in vertebrate anatomy e.g. a sublayer of the mucous membrane (*tunica mucosa*) with *lamina epithelialis*, *lamina propria*, and *lamina muscularis mucosae*; also e.g. the basilar lamina (membrane) of the cochlea; **2.** in plants, the leaf blade that is mostly broad and thin, specialized for photosynthesis, gas exchange, and water evaporation; also the flattened part of algal thallus.

laminal/laminar placentation, a type of placentation in plants (*see* placenta), the ovules attaching to the inner surface of separate carpels.

laminar-flow cabinet, a chamber used for cell culture or other sterile work; laminar air at the inflow is sterilized by filtration thus preventing contamination of the inside air. Also called safety chamber.

laminarin (*Laminaria*, the genus of a seaweed), a storage polysaccharide in the → brown algae (Phaeophyta), consisting mainly of glucose units; stored in cells in minute sacs surrounded by a membrane.

laminin, a heterotrimeric, multiadhesive protein produced by epithelial cells of animals; a component of extracellular matrix in the → basal lamina, having binding sites for → collagen, heparan sulphate proteoglycans, and certain cell-surface receptors; e.g. plays a role in neuronal outgrowth.

lamins, a group of intermediate filament proteins found in the cell nucleus associated with the inner nuclear membrane, participating in the attachment of chromatin and nuclear matrix to the → nuclear envelope.

Lamniformes (L. < *Lamna* = the type genus, *forma* = shape), an order of cartilaginous fishes (subclass Elasmobranchii) including most **sharks**; large-sized, almost purely predaceous animals which feed on oceanic fish, marine reptiles, mammals, and invertebrates. The large mouth contains numerous small sharp teeth, and reproduction is → ovoviviparous. The order comprises seven families, world-wide distributed, and about 190 species; e.g. **mackerel sharks** (Lamnidae), **catsharks** (Scyliorhinidae), and **carpet sharks** (Orectolobidae).

LAMP-1, LAMP-2, → lysosome-associated membrane glycoproteins.

lampbrush chromosome, a special type of chromosome found in → bivalents of primary oocytes of some vertebrate and invertebrate groups; appears during the prolonged diplotene phase of the first meiotic division (*see* meiosis) and in the nuclei of the spermatocytes of fruit flies (*Drosophila*). Lampbrush chromosomes have been studied especially on amphibians. The lampbrush stage of the chromosomes is reversible and dependent on the phase of cell division. Lampbrush chromosomes may be as big as the polytene → giant chromosomes in dipterans.

The microscopical appearance of the lampbrush chromosomes resembles an oil lamp brush. The trunk is formed from the **basic fibril** of the chromosome, and from special **loops** protruding from the trunk; also the core is formed from the basic fibril of the chromosome. The loops are active genes whose transcription activity can be seen as the transcription of ribonucleoprotein matrix proceeds along the DNA strand. Lampbrush chromosomes are found in maturing oocytes in which high gene activity is needed. In this stage the → messenger RNA molecules are synthesized for the production of reserve nutrients and proteins in the early development of the embryo. During the lampbrush stage the volume of the amphibian egg increases a millionfold.

lampreys, → Petromyzoniformes.

lamp shells, → Brachiopoda.

lancelets, → Cephalochordata.

Langerhans' islets, islets of Langerhans, (*Paul Langerhans*, 1847—1888), → pancreatic islets.

langmuir circulation, a water current straight towards the wind on a lake; arises during strong winds and is seen as foaming paths on the water surface.

lanolin (L. *lana* = wool, *oleum* = oil), a purified fat mixture extracted from wool grease of sheep; absorbs ca. 30% water forming an emulsion with it. Lanolin is readily absorbed by the skin and is used in ointments and cosmetic creams.

lanugo (L. = down), lanugo hair; the fine hair growing on the skin of a human foetus.

LAP, → leucine aminopeptidase.

lapar(o)- (Gr. *lapara* = flank), pertaining to the flank or loin, sometimes also to the abdomen.

large intestine (*intestinum crassum*), that part of the intestine of amphibians, reptiles, birds, and mammals which extends from the small intestine to the anus; comprises the → caecum, colon, rectum, and the anal canal or → cloaca, the ileocaecal valve separates the small and large intestines. In amphibians, the large intestine is less differentiated.

larva, pl. **larvae,** Am. also **larvas** (L. = ghost, mask), an early immature stage after hatching from the egg in the life cycle of many animals, preceding the pupal (or next larval, or adult stage). A larva deviates structurally and functionally from the adult; it is independent from the parents but is mostly incapable of sexual reproduction. Usually the larva undergoes metamorphosis before the adult stage. Each animal group has its own, special larval type: e.g. → **planula** of cnidarians, **amphiblastula** of sponges, **pilidium** of nemertines, **trochophore** of polychaetes and marine clams, **veliger** of marine gastropods, **glochidium** of fresh-water clams, **nauplius** of crustaceans, **caterpillar** of holometabolous insects, **auricularia** of sea cucumbers, **bipinnaria** of sea stars, **pluteus** of sea urchins, **ophiopluteus** of ophiuroids, **actinotrocha** of phoronids, **tornaria** of lower chordates, **ammocoete** of cyclostomes, **leptocephalus** of eels, and **tadpole** of amphibians. *See* metamorphosis.

Larvacea, larvaceans; a class in the subphylum Tunicata in the phylum Chordata; small, free-swimming planktonic animals whose larval structures still remain in the adult stage. *Syn.* Appendicularia. *See* paedogenesis.

larvicide (L. *caedere* = to kill), any pesticide used for killing larvae.

larviparous (L. *parere* = to produce, bear), pertaining to an animal which reproduces by releasing larvae.

larvivorous (L. *vorare* = to eat), pertaining to an animal which feeds on larvae.

laryng(o)-, denoting relationship to the → larynx.

larynx, pl. **larynges** (L.), the upper end of the trachea of tetrapod vertebrates; a musculocartilaginous structure, lined by the mucous membrane at the junction of the pharynx and trachea. The larynx consists of several cartilages connected by ligaments, i.e. one epiglottis, thyroid, and cricoid, and two arytenoid, corniculate and cuneiform cartilages. Primarily the larynx guards the entrance into the trachea, secondarily serves as the vocal organ in mammals with a special structure in the opening (*see* glottis). In humans, the laryngeal prominence seen in the front of the neck is also called Adam's apple. *Adj.* **laryngeal**.

laser (< *l*ight *a*mplification by *s*timulated *e*mission of *r*adiation), a device that concentrates the energy of monochromatic electromagnetic radiation of the visible wavelength region into an extremely intense, narrow beam with all the waves in the same phase; laser is used e.g. in surgery, diagnosis, and research.

lasso cell, → colloblasts.

late leaf, a type of foliage → leaf in plants.

latency (L. *latens* = hidden), **1.** a state of apparent inactivity or dormancy; a latent state; **2.** a latent period (latency period), i.e. reaction time, the period between the introduction of stimulus and a response elicited by it.

latent learning, a type of → learning in animals in which an association is formed between a stimulus and the performance of a behavioural pattern without any obvious reward or punishment. Through latent learning the ability develops in the animal to connect a stimulus to circumstances which the animal could not associate earlier. Thus a learnt behavioural pattern remains latent in memory for later use; e.g. a thirsty bird returns to drink at the pond where it earlier was bathing. Latent learning often occurs during exploratory behaviour. Latent learning may include also → **insight learning** that occurs only in the most developed animals.

latent period, *see* latency.

lateral (L. *lateralis* < *latus* = side), situated at a side; denoting a position aside from the midline of the body.

lateral bulb, a new → bulb formed laterally from an → axillary bud in an old bulb in plants.

lateral flower, a flower which develops laterally in the → inflorescence, the flower in the tip being the **terminal flower**.

lateral hearts, aortic arches; five pairs of enlarged muscular transverse vessels in the circulatory system of earthworms (*Oligochaeta*) between the dorsal and ventral vessels. The lateral hearts pump blood through the ventral vessel.

lateral inhibition, 1. a mechanism of the regulation of cell differentiation in which a cell, differentiated in a certain way, inhibits its neighbouring cells from differentiating in the same way. Hence, usually a regular pattern emerges by the process, such as the distribution of bristles on the insect body; **2.** the mechanism found in both the camera eye and compound eye which helps to sharpen the edges of the stimulus and improve discrimination of boundaries between darker and lighter parts of the visual field, i.e. illumination of one visual area inhibits the response of receptors in the neighbouring area. In the vertebrate eye, lateral inhibition is probably mediated via horizontal cells in the retina.

lateral leaves, dorsal leaves of some liverworts (*Hepaticae*), situated in two rows; if there is a third leaf row, it consists of smaller, **ventral leaves**.

lateral line (L. *linea lateralis*), **1.** a dotted line along either side of the body of cyclostomes, fishes, amphibian larvae, and aquatic amphibians; often the line continues and branches onto the head region. The lateral line is a visible part of the cutaneous sense organs, → **neuromasts**, which together with the primitive internal ear form the **acousticolateralis system**, serving in the perception of sound waves and movements of water; **2.** longitudinal lines of lateral excretory canals along the epidermal ridges of nematodes.

lateral meristems (Gr. *meristos* = divided), a type of the plant → meristem; lateral meristems comprise the → **cambium** which causes secondary thickening of vascular plant stems and roots, and the cork cambium (→ **phellogen**) which forms the → **periderm** in those structures.

lateral plate, the unsegmented, thickened cell mass on both sides of the mesoderm (mesoblast), giving rise to somatic and splanchnic mesoderm in an early vertebrate embryo (embryonic disc). Also called lateral mesoblastic plate.

lateral tendrils, plant → tendrils situated on both sides of the leaf base.

laterite soil (L. *later* = brick), a reddish, ferruginous soil formed in tropical regions. *See* soil formation.

late wood, → summer wood.

latex, pl. **latices** or **latexes** (L. = fluid), a chemically variable milky liquid in latex tubes of some angiosperm plants; contains e.g. proteins, sugars, fats, mineral salts, and alkaloids. In some cases latex serves in the defence against herbivores, but may also form a store of different substances or improve wound healing in plants. An economically important latex is obtained from the caoutchouc tree (*Hevea brasiliensis*) which is raw material for natural → rubber; some latices are used in the drug industry (e.g. poppy).

latex tube/duct, a usually richly branched tube in certain plants containing → latex.

lathyrism, a chronic state of intoxication in man caused by an alkaloid, β-aminopropionitrile, $H_2NCH_2CH_2CN$, present in seeds of leguminous plants in the genus *Lathyrus*. The alkaloid inhibits the function of lysyl oxidase and thus lysine metabolism, resulting in spastic paraplegia and pain in people in countries where peas of these plants are regularly used as food.

Latimeria (*M. Courtenay-Latimer*), *see* Coelacanthini.

Laurasia (*Saint Lawrence* in North America), the large northern ancient continent in Jurassic which was formed by the breakup of → Pangaea about 200 million years ago and began to fragment into the present North America, Greenland, Europe, and Asia (excluding India) about 135 million years ago. *See* Gondwanaland, continental drift theory.

Laurer's canal, a genital duct of some trematodes from the junction of the oviducts and yolk gland to the pore on the dorsal surface. According to some assumptions it could be a reduced vagina. *Syn.* Laurer—Stieda canal.

laver (L. = water plant), any of edible seaweeds, e.g. some species of *Porphyra*.

law of heterogeneous stimuli summation, in ethology, describes the additive effect of several → key stimuli in releasing a species-specific instinctive movement. The greater the number of key stimuli operating simultaneously, the easier the release of a certain behavioural activity.

law of the minimum, → Liebig's law of the minimum.

laxation (L. *laxare* = loosen), defecation; bowel movements (normal or induced) for defecation.

laxative, 1. purgative, aperient; **2.** an agent promoting evacuation of the bowel.

layer, 1. in anatomy a stratum or lamina; sheet-like tissue structures lying upon one another, e.g. the cornified and granular layers of the epidermis; **2.** any of different layers of soil, ground, water, atmosphere, etc. *See* layering.

layering, 1. a layered structure of vegetation in plant communities; e.g. in woodlands the common layers are trees, shrubs, herbs, and bryophytes; *syn.* stratification; **2.** a vegetative propagation of economic and garden plants (mostly shrubs or trees) by bending a branch of a twig into soil and covering it partly, after which the covered part begins to form roots and develops into a new individual.

LC_{50}, *see* LD_{50}.

L chain, *see* immunoglobulin.

LD_{50}, median lethal dose, L. *dosis lethalis*; an average measure for the toxicity of chemicals, i.e. a dose at which half (50%) of the test organisms die. It is usually expressed as mg/kg of body weight, for aquatic organisms as the concentration of a chemical present in surrounding water (mg/l, µg/l, or in molar concentration); the dose is expressed as **LC_{50}** (lethal concentration). LD_{50} is always determined or otherwise estimated for drugs, pesticides, etc. which will be brought into usage. *Cf.* LT_{50}, EC_{50}, ED_{50}.

LDH, → lactate dehydrogenase.

LDL, low density lipoprotein, *see* cholesterol.

lead (L. *plumbum*), a soft metallic element, symbol Pb, atomic weight 207.2, density 11.34; occurs naturally e.g. as sulphide (galena, lead glance), PbS. Lead is abundantly used in industry, earlier also in water and gas supplies of households. Lead compounds are toxic to cells inhibiting many enzyme functions, as e.g. those acting in the synthesis of haemoglobin. In many places lead has become a serious pollutant as the result of the activity of man. Because high quantities of lead have been spread from petrol in combustion engines, lead is totally substituted by other petrol additives in many countries.

leader sequence, a rather long sequence in the beginning of the messenger RNA molecule of eukaryotic organisms; does not take part in translation but contains regulatory elements for translation.

leader sequence peptide, a sequence of 16 to 20 amino acids at the N-terminus of some eukaryotic proteins that guides the transport of the protein across the cell membrane or into cell organelles.

leaf, a plant organ formed as an outgrowth from the plant stem; mostly green, thin and broad. The leaves are specialized for photosynthesis, gas exchange and water evaporation. The leaf consists typically of a **leaf blade** and a **leaf stalk** (petiole), and is attached to the stem by the → **leaf base**. The form and appearance of leaves vary greatly; e.g. → bracts and → scale leaves, often even thorns are modified leafs.
Many of the characteristics used to identify plants are external structural features of leaves: the basic types are the **simple leaf** with an undivided blade (e.g. oak), and the **compound leaf**, with a blade composed of several leaflets (e.g. clover). The compound leaves may be **pinnate**, bearing leaflets in two rows arising from a central midrib (e.g. ash), or the leaflets may arise from a single point (e.g. horse chestnut), the leaves being **palmately** compound.
Different types of arrangement of the leaf veins (*see* venation types) are the simplest **dichotomous** type (*Ginkgo biloba*, some ferns), the **pinnate venation**, having a clear middle vein with smaller lateral veins on both sides (typical of dicotyledons), and the **parallel venation**, in which the large veins are running parallelly through the leaf (most monocotyledons). The arrangement of leaves in plants varies also greatly (see phyllotaxy) and is a species-specific feature.

leaf abscission (L. *scindere* = cut), a process caused by the separation of cells of the → **abscission layer**; it is controlled by plant hormones, especially auxins, abscisic acid and ethylene. Cell walls in the abscission layer are hydrolysed by pectinases and cellulases leading to the shedding of leaves. *See* defoliation.

leaf area index, LAI, the ratio of a photosynthetic leaf area of a certain vegetation to a given ground area.

leaf area ratio, LAR, the ratio of a plant leaf surface area (photosynthetic area) to its dry weight.

leaf arrangement, → phyllotaxy.

leaf base, the structure which joins the → leaf (mostly petiole) to the stem; its structure varies and may be modified e.g. into → tendrils or sheath. *Syn.* phyllopodium.

leaf blade, lamina, mostly a thin, broad part of a leaf, specialized for photosynthesis, gas exchange, and water evaporation.

leaf gap, a region in the vascular cylinder of the stem of a vascular plant immediately above the **leaf trace**, i.e. the vascular bundle(s) reaching the leaf from the cylinder; the leaf gap is filled with parenchyma, and although it is not empty, it is a gap in the vascular system.

leaflet, a part of a compound leaf; joins to the leaf structure with its own petiole.

leaf scar, a scar on the stem of a (woody) plant showing the former position of a shed leaf.

leaf sequence, the order in which leaves develop in a seed plant; the → **cotyledons** grow first, then the → **scale leaves** (i.e. **bracts**) and after them the normal → **foliage leaves**. In many flowering plants there are bracts (also called **upper leaves**) growing in the area of inflorescence, the **flower leaves** being the latest to grow.

leaf sheath, the modified leaf base forming a sheath around the stem; occurs e.g. in grasses (family Poaceae).

leaf stalk, → petiole.

leaf trace, a vascular bundle extending from the vascular cylinder of the stem to the leaf. *See* leaf gap.

leaf vein, the structure of the leaf consisting of a vascular bundle and the bundle sheath around it. The leaf veins are connected to the → stele of a plant and they transport water and organic compounds in the leaf. According to the order and branching of the veins, there are different → venation types in different plants.

learning, a developmental process during the life cycle of an animal when its behaviour changes as a result of experience; is based on → memory. The learning process differs from the more stereotyped inherited behaviour (*cf.* instinctive behaviour), usually forming individual modifications of behaviour through experience. Learning includes several types: → **habituation**, **conditioning**, **trial-and-error conditioning** (**learning**), **latent learning**, **insight learning**, **associate learning**; some ethologists also include → imprinting in learning.

learning index, an index used in ethological studies to describe the differences in learning capacity between different test populations. The index is calculated by subtracting the number of unlearned animals from the number of learned ones and the difference is divided by the number of learned animals.

Le Chatelier's principle/law (*Henri-Louis Le Châtelier*, 1850—1936), the principle that if a system in equilibrium is subjected to a constraint, e.g. heat or pressure, the system tends to react in such a way that the effect of the disturbing factor is reduced to a minimum thus neutralizing the effect of the constraint. The principle is functioning also in regulatory systems of organisms.

lecithin (Gr. *lekithos* = yolk), → phosphatidylcholine.

lecithinase, → phospholipase.

lectins (L. *legere* = to select), a group of proteins found in plants (especilly legumes), fish, and many invertebrates (e.g. molluscs), having a high binding affinity and specificity for certain carbohydrates; e.g. → **concanavalin A (ConA)** from the jackbean binds to the non-reducing terminals of α-mannosyl residues, **soybean agglutinin (SBA)** recognizes galactosyl units and **wheat-germ agglutinin (WGA)** binds to disaccharide or oligosaccharide units. All lectins contain two or more binding sites for carbohydrate units, and they have the capability of attaching to certain animal cells, e.g. agglutinating red blood cells, i.e. acting as phytoagglutinins, thus resembling the action of antibodies. Lectins are useful probes for examination of cell surfaces because they recognize specific oligosaccharide patterns. Their role in plant seeds or in other parts is uncertain, but it is found that a lectin participates in binding a nitrogen-fixing bacterium (*Rhizobium trifolii*) to the surface of the root of clover. This lectin links receptors on root hairs to bacterial capsular polysaccharides and lipopolysaccharides. Also some bacteria contain lectins; e.g. *Escherichia coli* bacteria adhere to epithelial cells of the gastrointestinal tract by means of bacterial lectins. Some lectins are known to have a mitogenic activity e.g. stimulating the proliferation of lymphocytes.

lectotype (Gr. *lektos* = chosen, *typos* = pattern), *see* type specimens.

leg, one of the appendages of an animal serving mainly in locomotor activity in walking or swimming; e.g. **1.** in tetrapod vertebrates, → limb; in human anatomy, the lower limb from the knee to the ankle, in common language, denoting the entire lower limb; **2.** one of the paired, segmented walking or swimming appendages of arthropods. Insects have usually three and arachnids four pairs of walking legs. Crustaceans have more legs, many of them functioning as gills; the basic structure is a → **biramous appendage.** The leg of insects includes five segments: **coxa, trochanter, femur, tibia**, and jointed **tarsus**. In addition to these, arachnids have **patella** and **metatarsus**; *see* appendage.

leghaemoglobin, Am. **leghemoglobin** (leg < leguminous plants), a haemoglobin-like protein in root nodules of nitrogen fixing plants; binds oxygen, thus protecting → nitrogenase which is sensitive to oxygen. *See* → nitrogen fixation.

legume (L. *legumen* < *legere* = to gather), a fruit characteristic of most species in the family Fabaceae (e.g. peas and beans); formed by a single carpel opening on two sides.

legumin, a globulin (protein) in the seeds of leguminous plants.

leishmaniasis, also **leishmaniosis** (*Sir William B. Leishman*, 1865—1926), an infection found in many mammals and caused by certain flagellate protozoans from the genus *Leishmania* (suborder Trypanosomatina); often occurs as secondary infection in humans, transmitted e.g. by rodents. The disease is divided into Old and New World forms, some being visceral and some cutaneous, manifested by ulcers.

lek (Sw. = play, game), → courtship area.

lemma, (Gr. = rind, shell), **1.** a thin membrane or sheath, especially the membrane around an egg cell, ovum (oolemma); *see* neurilemma; **2.** the lower of the two bracts enclosing a single flower and the inner bract (palea) in a → spikelet of graminids (family Poaceae).

lemnid(e), a plant floating on water without roots; e.g. duckweeds. *See* water plants.

Lemuria (L. *lemures* = ghosts), a suggested continental bridge in Cretaceous about 100 million years ago between the present Africa, Madagascar, and India.

lens, pl. **lenses** (L. = lentil), **1.** a light-refracting device causing convergence or divergence of light rays, made from transparent material, mostly glass; also an electron lens, an electrostatic or magnetic device, used e.g. to control the electron beam in the → electron microscope; **2.** in biology, the crystalline lens of the → eye.

lentic (L. *lentus* = slow, calm), pertaining to, or living in still waters, such as in ponds, springs, lakes, or swamps. *Cf.* lotic.

lenticel (L. *lenticella*, dim. of *lens* = lentil), a pore in the periderm and cork tissue in the stem of a tree or shrub; lenticels allow water transpiration and air uptake through the periderm.

lenticular, denoting relationship to a lens or lens-like structure, e.g. the crystalline lens of the eye.

lentil, a lens-shaped seed of the leguminous plant *Lens culinaris*, or the plant itself.

lentivirus, any virus of the subfamily *Lentivirinae*, including slow-acting viruses in mammals, e.g. maedi and visna viruses in sheep, or HIV (human immunodeficiency virus) in man; the disease caused by them is manifested years after infection. Lentiviruses are retroviruses, having reverse transcriptase, and resemble certain tumour RNA viruses (Oncovirinae).

Lepidodendrales (Gr. *lepis* = scale, *dendron* = tree), a fossil order in the class Lycopsida, division Pteridophyta (in some systems division Lycophyta), Plant kingdom; the species were tall (40 m) and very thick trees, and reproduced by spores. They grew commonly during the Carboniferous and Permian periods.

Lepidoptera (Gr. *pteron* = wing), **lepidopterans**; an insect order comprising animals which have two pairs of scale-covered, often brightly coloured wings, a sucking proboscis, and complete → metamorphosis; their herbivorous larvae (caterpillars) have chewing mouthparts. The order is highly diverse comprising 28 superfamilies, e.g. Papilionoidea (**butterflies**), Geometroidea (**carpet moths**), Bombycoidea (e.g. **silkworms**) and Noctuoidea (**nocturnal moths**). In all, the order includes 110,000—120,000 described species.

Lepidosauria (Gr. *sauros* = lizard), **lepidosaurs**; a subclass (or superorder) of reptiles; have a diapsid skull, and the limbs and limb girdle are unspecialized, reduced, or absent. The taxon includes the orders Squamata (**snakes**, **lizards** and **worm lizards**), Rhynchocephalia (**tuatara**), and Ichthyosauria (extinct **ichthyosaurs**).

Lepidosireniformes (L. *lepidus* = pretty, *siren* = mythical mermaid), → Dipnoi.

leptin (Gr. *leptos* = thin, slender), a peptide hormone found in mammals; affects the hypothalamic satiety centre in the brain thus decreasing the feeling of → hunger; leptin is apparently secreted by fat cells.

leptocentric bundle (Gr. *kentron* = point, middle point), a concentric → vascular bundle with external xylem; is one type of concentric vascular bundle in plants, the other type being the **hadrocentric bundle** which has an external phloem.

leptocephalus, pl. **leptocephali** (Gr. *leptos* = slender, thin, *kephale* = head), the translucent, leaf-shaped larva of eels; lives in oceans before the elver stage.

leptoid (Gr. *eidos* = form), a living, long, thin-walled cell type in well developed mosses; transports nutrients and forms a primitive conducting structure.

leptome, also **leptom,** the entirety formed by the conducting parts of the → phloem (sieve cells or sieve tubes and compagnion cells); in addition, the phloem consists of parenchyma cells and supporting fibres.

lepton, a class of elementary particles of the → atom; includes the electron muon, tauon, and neutrino.

leptospirosis (L. *spira* = coil), the infection with aerobic bacteria from the genus *Leptospira*, occurring in worldwide distribution in many mammals, such as in dog, swine, raccoon, skunk, fox, opossum; is transmitted also to man chiefly in urine of an infected animal. Infection syndromes vary from the mild state to a fatal disease; high fever is usually manifested.

leptotene (Gr. *tainia* = band), the first stage of prophase in → meiosis.

Lesch—Nyhan syndrome (*Michael Lesch*, b. 1939, *William Nyhan*, b. 1926), a rare, human sex-linked defect of purine metabolism due to deficient hypoxanthine-guanine phosphoribosyltransferase; is manifested e.g. by marked increase in the rate of purine biosynthesis by the *de novo* pathway, and by an overproduction of urate, causing symptoms such as atherosis, hyperreflexia, and compulsive self-destructive behaviour.

Leslie—matrix, a method by which the effects of the age structure on birth and death rates are taken into account in the description of population fluctuations. Leslie—matrix is commonly used in the description of population growth, especially in age-structured populations. The method is used to study e.g. effects of human activities on populations.

lesion, → laesion.

lestobiosis (Gr. *leistes* = plunderer, *biosis* = manner of living), the relationship between two species in which individuals of one species steal food by furtive thievery from those of another species, e.g. some ants. *Cf.* cleptoparasitism.

lethal (L. *lethum*, *letum* = death), fatal, killing; lethal dose (*dosis lethalis*). *See* LD_{50}, LT_{50}.

lethal factor, a gene or chromosome mutation which causes the death of an individual before the age of reproduction. Lethal factors are usually recessive, but they may also be dominant (→ recessive, dominant character). According to the → penetrance, lethal factors are classified into **actual lethal factors** which in all conditions cause untimely death of the individuals in at least 90% of cases, **semilethal factors** which cause death of their carriers in 50—90% of cases, and **subvital factors** which cause death in less than 50% of cases. Further, **conditional lethal factors** can be distinguished; they cause death only in certain conditions like in a given temperature or growth medium. It is estimated that every human being carries in his or her genes approximately twenty recessive lethal factors which manifest themselves only if they come from both parents.

leuc(o)-, leuk(o)- (Gr. *leukos* = white), pertaining to white.

leucine, an amino acid, abbr. Leu, leu, symbol L, 2-amino-4-methylvaleric acid, $(CH_3)_2CH$-$CH_2CH(NH_2)COOH$; a constituent of proteins, belonging to the → essential amino acids of animals.

leucine aminopeptidase, LAP, an exopeptidase enzyme, EC 3.4.11.1, which liberates amino acids from the N-terminal end of proteins and polypeptides, reacting most rapidly on leucine residues; also hydrolyses many aliphatic amides and thiolesters. High LAP activity is found e.g. in the anterior part of the vertebrate intestine, liver, and kidney. LAP is extensively used in sequence analysis of proteins and peptides.

leucine zipper, a region in DNA-binding proteins of about 30 amino acids in which leucine forms every seventh amino acid residue, forming an alpha helix; this tends to form dimers of two such proteins which occur as transcriptional regulators, i.e. → transcription factors.

leucism, the genetic characteristic in which white fur or plumage exists in an animal with pigmented eyes and skin. *Cf.* albinism.

leucocytes, Am. **leukocytes,** → white blood cells.

leucon grade, leuconoid type, the most complex structural type of Porifera (sponges); its thickened body walls include enormous numbers of microscopic, rounded **flagellated chambers** connected with branched canals to each other and to the body surface. *Cf.* ascon grade, sycon grade.

leucoplast (Gr. *plassein* = to form), a colourless → plastid in a plant cell; may e.g. contain starch and is in that case called amyloplast.

leucosin, a granular polysaccharide found in some brown algae (Chrysophyta). *Syn.* chrysolaminarin.

leucotriene, leukotriene, LT (*leucocyte* + *triene* indicating three conjugated double bonds), any of the group of physiologically active compounds (→ **eicosanoids**) of animals comprising a straight-chain 20 carbon carboxylic acid with four double bonds; are identified by letters A, B, C, D, E, and F with a number as subscription that indicates the number of double bonds, e.g. LTD_4. Leucotrienes are products of arachidonic acid metabolism, and they act as mediators (local hormones) in inflammatory and allergic reactions. They can produce bronchoconstriction, the constriction of arterioles, and an increase in vascular permeability; they also attract certain leucocytes, such as neutrophilic and eosinophilic → granulocytes, to inflammatory sites. *Cf.* prostaglandins, thromboxanes. *See* eicosanoids.

leuk(o)-, → leuc(o)-.

leukaemia, Am. **leukemia** (Gr. *haima* = blood), leukocytic sarcoma; a type of cancer including a series of malignant diseases of the blood-forming myeloid tissue in vertebrates; characterized by distorted development and overproduction of white blood cells (leucocytes), usually resulting in the reduced production of red blood cells and platelets.

leukocytes, leucocytes; → white blood cells.

lev(o)-, *see* entries beginning laev(o)-.

levodopa, L-dopa; 3-hydroxy-L-tyrosine; (-)-3-(3,4-dihydroxyphenyl)-L-alanine; the immediate precursor of → dopamine (and further other catecholamine neurotransmitters) in its biosynthesis from phenylalanine in animals; used for the medical treatment of Parkinson's disease to increase the decreased synthesis of dopamine in the brain.

Leydig's cell, Leydig cell (*Franz von Leydig*, 1821—1908), a type of **interstitial cell** in the testicular tissue of vertebrates; produces androgenic hormones, → androgens, chiefly testosterone.

LH, → luteinizing hormone.

LHRH, → luliberin.

liana, a climbing plant with a woody stem; lianas are most common in tropical rain forests.

liberating hormones, liberins, → releasing hormones.

libido, pl. **libidines** (L. *libere* = to please), desire, especially desire derived from primitive impulses, e.g. sexual desire; *ad libitum* (*ad lib.*), at pleasure.

libriform (L. *liber* = inner bark of a tree), resembling the phloem fibres.

libriform cell/fibre, a fibre type in the → xylem of plants, having thick walls and simple pits, while **fibre tracheids** have thinner walls and more numerous pits.

lice, sing. **louse,** the common name for different invertebrates, e.g. for many insects, as **book lice** (Psocoptera), **biting lice** (Mallophaga), **sucking lice** (Anoplura), or some crustaceans, such as **fish lice** (Branchiura).

lichenology (Gr. *leichen* = lichen, *logos* = word, discourse), the branch of biology dealing with → lichens.

lichens, lichen-forming fungi, the common name for fungi which symbiotically form shared thalli with some algae. The lichen-forming fungi belong mostly to the group → Ascomycetes, seldom to → Basidiomycetes. Taxonomically the lichens are fungi. The thallus of a lichen consists chiefly of → mycelium, algal cells occurring as thin layers near the surface.

Lichens grow very slowly, but they are successful even on bare ground, on rocks or tree trunks, and are therefore usually the pioneer plants on a new, bare ground, e.g. on small rocky islands rising from the sea. The main form types are **crustose**, **foliose** and **fruticose** lichens. The crustose species form a thin, flat crust on the ground, the foliose species are lobed or leaf-like and are attached to the ground by numerous rhizoidal hyphae, the fruticose species being clearly erect or pendulous (e.g. reindeer mosses). Many lichens (especially fruticose ones) are very sensitive to polluted air (particularly to sulphur dioxide) and are used as bioindicators.

Lieberkühn's glands/crypts (*Johann Lieberkühn*, 1711—1756), → intestinal glands.

Liebig's law of the minimum (*Justus von Liebig*, 1803—1873), the theory presented by von Liebig (1840) explaining that the development and growth of a plant are limited by any essential mineral element which is available in an insufficient quantity. The law can be applied also to animals.

life, a term for vital phenomena (Gr. *bios*, *zoe*, L. *vita*) involved in physico-chemical processes of organisms distinguishing them from inorganic objects and dead organisms. The basic unit maintaining life is a cell that energetically is an open system, through which material and energy (also in the form of information) flow and change from one form to another. The presence of numerous basic elements such as carbon, hydrogen, oxygen, sulphur, nitrogen, phosphorus, potassium, sodium, calcium, magnesium, and iron, are constituent elements in large organic molecules such as → proteins, carbohydrates, lipids, and nucleic acids, produced in biochemical reactions by the action of numerous → enzymes in cells and tissues. Water functions as a solvent comprising 70—90% of the content of protoplasm of an active cell.

In the cellular processes, life is manifested in **organization**, **metabolism**, **growth**, **reproduction**, **development**, **irritability**, **movement**, **heredity**, and **adaptation** with complex regulatory systems (*see* regulation). These characteristics exist already in unicellular organisms, one single cell of which is the entity of their life. The life of a multicellular organism is based on the functions of several differentiated cell types, strictly organized into functionally integrated tissues. The life of a virus begins only after having entered a cellular organism.

Life demands certain environmental circumstances, such as a permissive temperature, light, pressure, source of necessary chemical elements, and the absence of poisonous substances. Living processes follow the laws of chemistry and physics without being necessarily reduced to these phenomena, but e.g. inheritance and development are characteristics displayed solely by living organisms. *See* origin of life.

life cycle, the continuous sequence of all the different stages which an organism undergoes from the fusion of gametes to the same stage in the next generation. Important are the species-specific features which regulate reproductive value and survival of individuals (*see* life cycle trait). All life cycles are based on genetic control and many include → alternation of generations; environmental factors are involved in realization and duration of a life cycle. The life cycle of an individual is linear but it is cyclic at the population level.

life cycle strategy, a life strategy of an organism, includes many → life cycle tactics being conceptually larger and more complete than the life cycle tactics. Life cycle strategies are e.g. → K-selection, r-selection.

life cycle tactics, an entity composed of various life cycle characteristics to resolve particular ecological demands; e.g. the determining of a clutch size so that the clutch produces the maximal number of offspring to their reproductive stage, or e.g. defence mechanisms developing to maximal effectiveness so that the probability of becoming a prey is as small as possible. The characteristics are evolved by → coadaptation.

life cycle trait, life history trait, any quantitative property of the life cycle of an organism; e.g. the initial age of reproduction, the number of reproductive times, the number of offspring per reproductive time, the body sizes and sex ratio of offspring.

life expectancy, the time that a particular individual can be expected to live, calculated from life table statistics. *See* life span/time.

life form, a characteristic structure and function of an adult animal or plant, or a population, produced by certain environmental factors; e.g. marine animals living in the Baltic Sea, or → **Raunkiaer's life forms** of plants which are groups in vegetational classification based on the position of perennating → buds of plants in relation to soil level; indicate how the plants pass an unfavourable season in their annual life cycles.

life form spectrum, describes the distribution of plant species of an area in → Raunkiaer's life forms; different climate zones have characteristic spectra according to the length and type of the unfavourable season.

life history, distribution of significant events and traits in the life time of an organism from the fertilization to the death, particularly those features influencing reproduction and survival. Life history of an individual is usually divided into different phases: prereproductive (growth and development), reproductive, and postproductive phases. Often life history and life cycle are considered to be synonyms.

life span/time, the duration of the life of an organism or its part; the **physiological** and **eco-**

logical life times are distinguished, the first expressing the median age of individuals in a population when all the environmental factors are optimal, the latter expressing the median age in prevailing conditions. The physiological age is usually noticeably longer. → **Life expectancy** is the statistically determined remaining age of an individual which can be calculated from the life table information.

life table, data describing the mortality schedule of a population by age. Life tables are especially suited for studying species whose individuals display ageing. Life tables can be constructed after gathering data in either of the two different ways. The **static life table** (stationary, time-specific, current, or vertical life table) is drawn up on the basis of the age distribution of the population at a specific time. The **cohort life table** (generation, horizontal life table) is calculated on the basis of a cohort (e.g. the age group of individuals born in a certain time period) the individuals of which are followed throughout their lives.

ligament (L. *ligare* = to tie, bind), a band or sheet of the fibrous connective tissue which connects bones or cartilages, and supports joints, tendons, and fasciae of muscles, and holds organs in place; some ligaments develop as folds of fasciae, some are relics of embryonic organs. *Adj.* **ligamentous**.

ligand, 1. an organic molecule that binds to another molecule, especially to a protein, as e.g. a hormone or its analogue to a → receptor molecule, or an antigen to an antibody; **2.** any molecule or ion acting as a donor for a pair of electrons to form a coordination compound (*see* coordinate bond) with a central metal atom, e.g. the porphyrin portion of → haem attached to the nucleus of vitamin B_{12}.

ligase, EC class 6; one of the six main classes of enzymes catalysing the joining of two molecules. The reaction energy is supplied by breakdown of a pyrophosphate bond in an ATP or similar energy donor (triphosphates). Ligases are e.g. involved in the repair mechanisms of DNA fragments (DNA ligases).

ligature (L. *ligatura*), a string used to tie a vessel or duct, or the act of such a binding.

light, light waves, → electromagnetic radiation that has a wavelength approx. between 390 nm and 700 nm (from blue to red) and is visible to the human eye. Many animals are capable of seeing also a part of infrared (IR) or ultraviolet (UV) wavelengths of the radiation spectrum. The speed of light is approximately 300,000 km/s in a vacuum. → Photons (quanta of light energy) of short wavelengths contain more energy than those of longer wavelengths. Red light (650—700 nm) is the most important in → photosynthesis that converts light energy to chemical energy.

light chain, L chain, 1. one of the two identical polypeptide chains in an → immunoglobulin molecule; **2.** a polypeptide chain that forms the shaft part of the molecule of → myosin.

light reactions, light-driven reactions of photosynthesis taking place in chloroplast thylakoids; the reactions produce ATP and NADPH. *See* chloroplast, photosynthesis. *Cf.* dark reactions, Calvin cycle.

light trap, an apparatus used for catching nocturnal insects particularly; it is constructed of a plastic shelter, a mercury vapour lamp (or a blended lamp), and underneath a funnel for gathering the insects into a bottle containing poison, such as slowly evaporating tetrachloroethane or trichloroethylene. *Cf.* bait trap.

lignicole, lignicolous (L. *lignum* = wood, *colere* = to inhabit), pertaining to an organism (lignicole) living in or on wood, e.g. many fungi.

lignin, a complex phenylpropene polymer formed from → coniferyl, sinapyl, and hydroxycinnamyl alcohols; occurs in stiffened (lignified) secondary walls of plant cells, particularly in woody plants. Lignin binds cellulose fibres, thus increasing the strength of the wood. In cellulose manufacture, lignin is removed and burned or used as raw material e.g. for phenol and other aromatic chemicals, vanillin, and as a rubber additive.

ligula, pl. **ligulae,** also **ligulas** (L. = a small tongue, spoon), **1.** a structure between labial palps in the lower lip (*labium*) of the insect mouthparts; **2. ligule**; a membranous or scale-like structure formed of the plant leaf base; common in graminaceous plants (family Poaceae), in which it is situated in the opening of the leaf sheath.

Liliatae (Monocotyledonae), a class in the subdivision Angiospermae, division Spermatophyta, Plant kingdom; in some classifications a class in the division → Anthophyta.

limb, leg, wing, **1.** in vertebrates, one of the two paired appendages especially in tetrapods serving locomotion in walking, flying and swimming; the basic structure of the tetrapod limb is the pentadactyl (five toed) jointed structure which is attached to the body trunk by the → shoulder and pelvic girdles. The bones from proximal to distal are in the forelimb: *humerus* (upper arm), *radius* and *ulna* (forearm), *carpals* (wrist), *metacarpals* (palm), *phalanges* (digits, fingers), and in the hindlimb: *femur* (thigh), *tibia* and *fibula* (shank or lower leg), *tarsals* (ankle), *metatarsals* (sole of foot), and *phalanges* (digits or toes). The limb bones are highly modified in various tetrapod groups in adaptation to special modes of locomotion. Reduction in the number of phalanges, and fusion of many bones are common in many tetrapod groups. Also complete loss of limbs have occurred in various tetrapods, such as snakes and some amphibians; *see* fin; **2.** for invertebrates, *see* leg.

limb bud, a protuberance on the body trunk of an embryo from which a limb develops.

limbic system (L. *limbus* = border), "emotion brain"; brain structures of higher vertebrates (especially mammals) consisting of certain ventral parts of the telencephalon (forebrain), such as the hippocampus with its archicortex, the dentate and cingulate gyri, septal areas, and the amygdala, functioning in close correlation with certain hypothalamic nuclei. The limbic system is associated with olfaction, autonomic functions, sexual behaviour, motivation, and emotions such as rage and fear. Especially the hippocampus and amygdala are also important in fixation of long-term memory.

limbus, pl. **limbi** (L.), a fringe or border in an anatomical structure, e.g. *limbus corneae*, the junctional region between the cornea and sclera in the vertebrate eye.

lime, calcium oxide, or calcium hydroxide, or any of calcium salts.

limestone, natural calcium carbonate, $CaCO_3$; occurs in rocks of sedimentary origin containing precipitated calcium carbonate and remains of marine organisms.

limestone steppe/grassland, a treeless moor the soil of which is calcium-rich and alkaline; is common in Gotland (Sweden) and in Estonia with its islands. The vegetation has steppe-like features. *Syn.* alvar.

limiting factor, any single factor that limits a whole process; e.g. **1.** in biochemistry and physiology, the slowest step in a reaction chain limiting the function of the whole chain; in the entire organism often seen as inhibition of growth; **2.** in ecology, a single environmental factor that either alone or together with other factors limits a population density by preventing the individuals from reaching their → biotic potential. The limiting factor is usually some minimum factor, as a nutrient or trace element. *See* Liebig's law of the minimum.

limiting resource, a resource that is scarce in relation to the demand for it. *See* limiting factor.

limiting similarity, such a high degree of similarity between competing species that it limits their possibilities for long-term coexistence. May occur e.g. when to animal species utilize very similar food resources.

limn(i)-, limno- (Gr. *limne* = pool, lake), pertaining to fresh water.

limnetic, pertaining to, or living in open water of a fresh water body, such as a lake, in upper layers above the compensation zone and outside the → littoral zone.

limnobiology (Gr. *bios* = life, *logos* = word, discourse), the biological study of life in standing fresh waters.

limnology, a branch of science dealing with physical, chemical, and biological conditions in inland fresh water bodies, e.g. lakes and ponds.

limonene (L. *Citrus limon* = lemon), citrene, carvene; a lemon-scented → terpene, $C_{10}H_{16}$, occurring e.g. in sour orange and lemon oils; used e.g. in surface-active agents and in the manufacture of resins. *dl*-form is called dipentene.

limonite (Gr. *leimon* = mire), ferric hydroxide; bog iron ore; a hydrated form of iron (III) oxide, i.e. $FeO(OH) \cdot n\ H_2O$.

Lincoln index (*F.C. Lincoln*), a basic index in ecological → capture-recapture methods describing population densities in areas of random sampling. *Syn.* Petersen estimate, Lincoln—Petersen index.

lindane, 1,2,3,4,5,6-hexachlorocyclohexane; a → chlorinated hydrocarbon used as a pesticide.

Linde process (*Carl von Linde*, 1842—1934), a high-pressure process developed for producing liquid air and also other liquefied gases, e.g. helium; in the process air is compressed to about 20 MPa (200 bar) followed by refrigeration and fractionation.

LINE, → long interspersed element.

linea, pl. **lineae** (L.), a line, stripe, or narrow ridge; e.g. *linea alba*, the tendinous white median line between the two rectus muscles of the abdominal wall; *linea lateralis*, → lateral line in fish.

lineage, a line of cells or organisms of common descent, e.g. → cell lineage.

line bud, one of many → axillary buds which develop in a row on the leaf axis, i.e. in the point where usually only one bud grows, e.g. in the banana. *Syn.* row bud.

Lineweaver—Burk plot (*Hans Lineweaver*, b. 1907, *Dean Burk*, b. 1904), *see* Michaelis—Menten kinetics.

lingual (L. *lingua* = tongue), pertaining to the tongue.

lingula, pl. **lingulae** (L. = dim. of *lingua* = tongue), **1.** a tongue-like structure in vertebrate anatomy, e.g. the lingula of mandible (*lingula mandibulae*), a tongue-shaped part of the mandible giving attachment to the sphenomandibular ligament; **2.** any brachiopod of the genus *Lingula*. *Adj.* **lingular**.

linkage, the act of linking, or the state of being linked together; e.g. **1.** a → covalent bond; **2.** in genetics, the inheritance of given genes or phenotypic characters together. Linkage is due to the fact that the genes involved reside in the same chromosome. Linkage appears as an exception of Mendel's law of independent assortment of genes so that in F_2 generation the number of parental combinations is higher, and the amount of → recombinants lower than what could be expected on the basis of independent assortment.

Genes which are located in → homologous chromosomes constitute a → linkage group. The number of linkage groups of a given species is the same as the → haploid chromosome number of that species. The linkage of genes is very rarely complete. Usually it is broken down (*see* genetic recombination) when homologous chromosomes reciprocally exchange parts in → crossing-over. Genetic mapping is based on the phenomena of linkage and crossing-over (*see* chromosome map).

linkage disequilibrium, the non-random association of gene alleles at two linked loci in frequencies deviating from those expected from the random combinations in the population, i.e. the phenomenon where the two possible linkage phases of genes, A b and a B on one hand, and A B and a b on the other hand, do not occur in equal frequencies in the population. A and B are dominant, a and b recessive genes.

linkage group, consists of gene loci that show → linkage. The chromosome is the material counterpart of the linkage group (linkage group is a formal concept, chromosome a concrete concept). Thus the number of linkage groups of a given species is the same as the → haploid chromosome number of that species. By means of crossing experiments the gene loci of a given linkage group can be arranged in a linear order. The genes belonging to different linkage groups, i.e. located in different chromosomes, show independent assortment (combination).

linkage map, a → chromosome map that consists of the linear order of genes belonging to the same → linkage group and hence located on the same chromosome. The distances and linear order of genes on a linkage map are based on → crossing-over frequencies shown in crossing experiments. *See* gene map.

linker, → linker fragment.

linker DNA, that part of the DNA of a chromosome which, in addition to the DNA twisted around the histone core, belongs to a → nucleosome linking nucleosomes together.

linker fragment, a short, synthetic double strand of → nucleotides that has one restriction site of a given → restriction enzyme, and that can be coupled to the end of some other DNA fragment when a → recombinant DNA is constructed synthetically.

linking number, the number of turns of the two strands of a closed DNA molecule crossing each other.

Linnean, Linnaean pertaining to the system of binomial nomenclature, originated by the Swedish naturalist *Carolus Linnaeus* (*Carl von Linné*, 1707—1788); in the Linnean natural system the organisms are classified according to their outer characteristics, such as the number of the stamens in seed plants.

linoleic acid, (L. *linum* = flax, *oleum* = oil), 9,12-octadecadienoic acid; an unsaturated fatty acid, $CH_3(CH_2)_4CH{=}CHCH_2CH{=}CH$-$(CH_2)_7COOH$, which occurs in glycerides of many fats and oils and is one of the → essential fatty acids in the diet of many animals.

linolenic acid, 9,12,15-octadecatrienoic acid; an unsaturated fatty acid, α-isomer, CH_3-$(CH_2CH{=}CH)_3(CH_2)_7COOH$, found in many seeds; occurs also as γ-isomer, $CH_3(CH_2)_4$-$(CH{=}CHCH_2)_3(CH_2)_3COOH$, which is a principal compound of cell membranes and one of the → essential fatty acids in the diet of many animals.

lip(o)- (Gr. *lipos* = fat), denoting relationship to fats, lipids.

lipase, a lipolytic enzyme, i.e. any enzyme that catalyses the hydrolysis of lipids; occurs abundantly in spherosomes of plant seeds containing storage lipids (triacylglycerols), and in animal cells, such as adipose, muscle, and liver cells, and as digestive enzyme in the alimentary canal of animals; e.g. triacylglycerol lipase (pancreatic lipase, EC 3.1.1.3), and lipoprotein lipase (3.1.1.34).

lipid A, the lipid in → Gram-negative bacteria; is associated with lipopolysaccharides in the cell wall and is responsible for endotoxic activity. *See* Gram staining.

lipid bilayer, a bimolecular layer formed by amphipathic phospholipid molecules in an aqueous environment, each molecule having its hydrophilic group oriented outwards and the hydrophobic group to the interior. This is the basic sructure of different membranes in the cell, e.g. → cell membrane.

lipids, a common name for fats and fat-like substances containing an aliphatic hydrocarbon moiety in the molecule; the formation of lipids, i.e. lipogenesis, occurs in cells from other organic compounds, such as saccharides, carboxylic acids, and amino acids, in the last step from → acetyl CoA. Lipids do not form a uniform group, being all water-insoluble but soluble in organic solvents such as alcohol, ether, benzene, etc. Lipids are grouped into → fatty acids, monoacylglycerols, diacylglycerols, neutral fats (→ triacylglycerols or triglycerides, and waxes), and polar lipids (→ phospholipids and glycolipids). According to their solubility properties, also → steroids, chlorophylls and carotenoids are considered to belong to lipids. Some lipids, such as phospholipids and cholesterol, are essential structural molecules of cellular membranes, others, like triacylglycerols and fatty acids, serve as energy stores for cell metabolism, and some, like → eicosanoids, act as messenger substances.

lipochromes (Gr. *chroma* = colour), lipochrome pigments; yellow or yellowish fat-soluble pigments in organisms; e.g. **1.** chromolipids, e.g. carotenoids such as betacarotene, xanthophylls, occurring in both animals and plants. To animals, they or their precursors are transferred from plants with food; some, e.g. → lipofuscin, are formed by oxidation and polymerization of membrane lipids; **2.** yellow bacterial pigments.

lipocyte (Gr. *kytos* = cavity, cell), adipocyte; a fat cell of an animal.

lipofuscin (L. *fuscus* = brown), a yellow-brown pigment formed in animals when lipid components of cell membranes oxidize and polymerize in → lysosomes of many cell types. In vertebrates, the quantity of lipofuscin (age pigment) increases particularly with ageing, e.g. in liver, muscle, and nerve tissues.

lipogenesis (Gr. *gennan* = to produce), the biosynthesis of → fatty acids or → lipids in organisms; in animals also called adipogenesis. In cells, various types of metabolites, derived from carbohydrates, proteins, fatty acids, and cholesterol, may be used for lipogenesis. The biosynthesis of phospholipids occurs in all cells of all organisms. The synthesis of storage lipids in animals occurs especially in → adipose tissue, fat body, and liver cells, and cholesterol synthesis chiefly in liver cells. In plant cells, lipogenesis occurs in cytoplasmic → spherosomes, especially in seeds.

lipoic acid, thioctic acid; 1,2,-dithiolane-3-valeric acid, $C_8H_{14}O_2S_2$, formed in organims from valeric acid (a fatty acid) and disulphide; is a bacterial growth factor, and a precursor for **lipoamide** which acts as an enzyme cofactor in carbohydrate metabolism (decarboxylation) of higher organisms. Its ionic form, salts, and esters are called **lipoates**.

lipolysis (Gr. *lyein* = to dissolve), hydrolysis of lipids in organ systems. *Adj.* **lipolytic**. *See* lipase.

lipopolysaccharide, a complex compound formed of a lipid joined to a carbohydrate; bacterial lipopolysaccharides (bacterial → endotoxins) are major constituents of the cell wall components of Gram-negative bacteria, responsible for the antigenic specificity of bacteria.

lipoprotein (Gr. *protos* = first), a protein molecule that contains a lipid moiety. Lipoproteins act in animals in transport of lipids, such as triacylglycerols and cholesterol, which without the protein moiety are insoluble in plasma or other interstitial fluids. The protein constituents of the lipoproteins are called apolipoproteins. → Chylomicrons are large lipoprotein complexes. Lipoproteins of plants are found to associate with cellular membranes, often having binding sites for cations. *See* cholesterol.

liposome (L. *soma* = body, piece), an artificial vesicle enclosed by a lipid monolayer, bilayer, or multilayer; liposomes are used as models when studying transport properties of cellular membranes, and as → vectors in gene manipulation.

lipotropin, LPH (Gr. *trepein* = to turn), a hormone-like polypeptide in vertebrates; in mammals, it is **β-lipotropin** (91-amino acid peptide), which is produced by certain cells of the anterior → pituitary gland. Although the physiological role of LPH is uncertain, there is evidence that it controls the lipid metabolism by increasing lipolytic action in cells, and apparently has some effects resembling those of → corticotropin. The LPH molecule contains amino acid sequences of many peptide hormones, such as melanocyte-stimulating hormone (MSH), enkephalins, and endorphins, and probably produces active peptides by hydrolytic cleavage of the molecule.

lipoxygenases (L. *oxygenium* = oxygen, Gr. *gennan* = to produce), enzymes catalysing the linking of an oxygen molecule to an unsaturated fatty acid in organisms; e.g. arachidonate 5-lipoxygenase (EC 1.13.11.34) converts arachidonic acid to 5-hydroperoxyarachidonate in the synthesis of → leucotrienes, or linoleate lipoxygenase (EC 1.13.11.12) catalyses the reaction: linoleate + O_2 —> 13-hydroperoxyoctadeca-9,11-dienoate. *Syn.* lipoxidases.

liquid chromatography, *see* chromatography.

liquid nitrogen, a liquid state of nitrogen (boiling point -195.8°C) produced by pressurizing and cooling gaseous nitrogen; used e.g. as a coolant in some equipment, for deep-freezing of biological material, e.g. to store tissues, cells, and cell organelles.

liquor (L.), liquid, fluid; e.g. *liquor amnii*, amniotic fluid, *liquor cerebrospinalis*, → cerebrospinal fluid.

Lissamphibia (Gr. *lissos* = smooth), **lissamphibians**; in some classifications, a subclass in the class → Amphibia, comprising all extant amphibian species.

listeriosis (*Joseph Lister*, 1827—1912), an infectious disease of many mammals, as e.g. man, rat, cat, dog, cow, and of many birds, caused by Gram-positive bacteria of the genus *Listeria*. The bacteria do not necessarily cause symptoms in animals, but e.g. in man, sheep, and cattle the disease is manifested by cardiorespiratory distress, meningitis, encephalitis, and abortion. The infection is transmitted through faeces and undercooked meat. Because diseased animals tend to move in circles it is also called **circling disease**.

lithium (Gr. *lithos* = stone), the lightest alkali metal, symbol Li, atomic weight 6.94. Lithium naturally occurs in some silica minerals but not free because of its high reactivity. Lithium does not normally participate in the metabolism of organisms, but in some reactions Li^+ ions can replace Na^+ ions and thus have pharmacological effects especially on the nervous system; lithium salts are used in neurophysiological studies and as sedative and hypnotic drugs.

lithocyst (Gr. *kystis* = bladder), → cystolith.

lithophyte (Gr. *phyton* = plant), a plant growing on the surface of rocks, attached directly to the stone. Lithophytes are lichens or mosses; lichens are mostly of the crustose type.

lithosere, (L. *serere* = to put in order), the temporal → succession of vegetation on exposed rock surfaces.

lithosphere (Gr. *sphaira* = globe), the stony crust of the Earth.

lithotroph (Gr. *trophe* = nutriment), an autotrophic bacterium, e.g. nitrifying and sulphur bacteria which obtain their energy from oxidation of inorganic compounds (e.g. sulphur).

litmus, a pigment obtained from some lichens; used as a pH indicator, being red in acid and blue in alkaline solutions.

litter, 1. the offspring brought forth at one birth by a multiparous animal; **2.** the bedding (hay, straw, etc.) for animals; **3.** the upper, partially decomposed layer of the forest floor.

littoral, also **litoral** (L. *li(t)tus* = seashore), pertaining to the shore of a lake, sea, or ocean. The **littoral zone** is defined as the zone from the maximum water line to the lowest depth where rooted aquatic plants grow. In practice, it is the region where light penetrates to the bottom. In oceanic ecosystems, littoral is sometimes defined as the region between high and low tide lines, or as the continental shelf to the depth of 200 m; in lakes varies from a few metres to some tens of metres. The littoral zone is divided into → **supralittoral, eulittoral**, and **infralittoral zones**. On shores of lakes and rivers, the litto-

havioural factors, especially motivation, and external behavioural factors, such as key stimuli and releasers, in outbursts of instinct behaviour. *Cf.* Deutsch's motivation model.

Lorenzini's ampullae, ampulla or ampullary organ of Lorenzini; any of several sensory organs in the snout skin of sharks, rays, and other selachians; each ampulla is a bladder-like structure with a thin, short duct to the body surface. The walls of the bladders are covered by electroreceptor cells responding to weak electrotonic stimuli (*see* electric sense), probably also to changes in temperature, water pressure, and salinity.

lorica, pl. **loricae** (L. = corslet), an external protective case or shell of loriciferans, rotifers, and many ciliates.

Loricifera (L. *ferre* = to carry), **loriciferans**, an invertebrate phylum described in 1983; comprises only some microscopic species living on the sandy bottoms of oceans. The body with the mouth cone and spiny head is caudally covered by a vase-shaped **lorica**, which is constructed of cuticular plates. The nervous system includes cerebral ganglia, a nerve ring around the mouth, and some ventrocaudal ganglia. Loriciferans are gonochoristic and their life cycle includes at least two larval stages.

Loschmidt's number (*Joseph (Johann) Loschmidt*, 1821—1895), → Avogadro number.

loss-of-function mutation, a gene mutation in which gene function is totally lost.

loss of heterozygocity, LOH, the changing of a cell (usually a cancer cell) from a heterozygous condition into homozygous or hemizygous conditions. The change into a homozygous condition can occur by mutation or somatic → crossing-over, and into a hemizygous condition by a → deletion.

lotic (L. *lotus* = washed), pertaining to, or living in running waters; e.g. denotes organisms living in streams, rivers, or straits. *Cf.* lentic.

Lotka—Verhulst equation (*A.J. Lotka*, 1880—1949, *P.F. Verhulst*), a mathematical model in population ecology used to describe an increase of population densities in situations where the growth is density-dependent. The equation produces logistic growthcurves. *See* logistic growth.

Lotka—Volterra equations (*A.J Lotka,* 1880-1949, *V. Volterra*, 1860-1940), mathematical equations of population growth which are expansions of the logistic model (*see* logistic growth). As the logistic model considers only intraspecific competition, the Lotka—Volterra equations include also interspesific competition (competitive exclusion principle) as well as mutualistic interactions and the cyclic fluctuations in population densities caused by predator-prey relations.

LTR, → long terminal repeat.

LT_{50}, (L. *lethalis* = killing), median lethal temperature; the temperature at which 50% of test organisms or cells die. The term is used to express the temperature stress or temperature tolerance of organisms. LT_{50} value is time-dependent and comparable with → LD_{50}.

Lubrol, polidocanol; polyethylene glycol monododecyl ether, a non-ionic detergent used as a lipid solvent (emulsifier) in biochemical analyses.

luciferase (L. *lucifer* = carrier of light), a group of flavoprotein enzymes occurring in luminescent organisms with light-producing organs, → **photophores**, found in some animals, e.g. many insects such as glow-worm and firefly, deep-sea fishes, crustaceans, coelenterates, and in a few plants. In the presence of O_2 and ATP, luciferase catalyses the oxidation of a substrate, called **luciferin** (e.g. a heterocyclic phenol); energy in the reaction is emitted as cold light. In research, luciferase is used for biochemical determinations of certain substances, e.g. ATP. *See* luminescence, bioluminescence.

Ludox, a silica sol used as a gradient material in → centrifugation. Percoll, a self-generating gradient substance, is produced by coating Ludox with polyvinylpyrrolidone.

luliberin, luteinizing hormone-releasing hormone (factor), LHRH. *See* gonadoliberin.

lumb(o)- (L. *lumbus* = loin), pertaining to the loins. *Adj.* **lumbar**.

lumbar nerves (L. *nervi lumbales*), five pairs of spinal nerves arising from the lumbar area of the spinal cord of tetrapod vertebrates; ventral branches of the second to fifth lumbar nerves are partly joined together to form the **lumbar plexus**. The nerves include sensory, motor, and sympathetic nerve fibres.

lumbar vertebrae (L. *vertebrae lumbales*), the vertebrae of the caudal vertebral column between the thoracic vertebrae and the sacrum in tetrapods excluding amphibians and some reptiles; in birds, the lumbar vertebrae fuse to form the **synsarcrum** bone.

lumen, pl. **lumina,** Am. also **lumens** (L. = light, window), **1.** the cavity of a tubular structure, as blood vessel or intestine; **2.** the narrow cavity in phloem fibres of plants; **3.** the unit of luminous flux; *see* Appendix 1.

lumi- (L. *lumen* = light, window), pertaining to light, white, colourless; in chemical names refers to substances which emit light or are formed by irradiation, usually by light, e.g. → lumirhodopsin from rhodopsin in photoreceptors.

luminescence, the emission of light without liberating thermal energy, i.e. cold light; produced when electrons jump from one electron shell to another. **Chemiluminescence** is produced in some chemical reactions, and → **bioluminescence** is chemiluminescence produced by biological systems. When this kind of light is produced by radiation (visible light, UV, or X-rays), the phenomenon is called **photoluminescence**, and when by radioactive radiation, it is called **radioluminescence**. *See* fluorescence, phosphorescence, photophore.

luminophore (Gr. *phorein* = to carry along), a chemical compound or an atom group producing → luminescence, e.g. → luminol.

luminol, 3-aminophthalic hydrazide, a compound acting as a → luminophore; emits light when oxidized. Luminol is used as a substrate e.g. in assaying the production of free → oxygen radicals in cells (e.g. phagocytic leucocytes) by chemiluminescence technique. *See* luminescence.

lumirhodopsin, the colourless intermediate in the light reaction of the retinal red pigment, → rhodopsin; formed in photoreceptors during bleaching of rhodopsin by light.

lunar (L. *luna* = moon), pertaining to the moon or a moon-shaped structure.

lunar rhythm, the rhythm occurring in many organisms depending on the rhythmicity of the moon's orbiting. *See* biorhythms.

lung (Gr. *pneumon*, L. *pulmo*), a respiratory organ for gas exchange in many groups of air-breathing animals. In most tetrapod vertebrates, it is a paired (unpaired in many snakes), sac-like, alveolar organ, in which the respiratory gases, O_2 and CO_2, are ventilated and exchanged between the blood and alveolar air.

In higher vertebrates the lung is a membranous, **sac-like organ** that is highly folded forming a botryoid alveolar structure with wide surface area (60—80 m^2 in the lungs of an adult man). As an **air-pump**, the lungs follow movements of the thorax having only a microscopic narrow cavity (**pleural cavity**) between the visceral pleura covering the lung and the parietal pleura lining the thoracic cavity. The lungs are in connection to external air through the respiratory passages, i.e. the → **trachea** opening to the pharynx through the → larynx, and the **bronchi** which are the branches of the trachea. These are divided successively several times into smaller passages forming pulmonary and respiratory **bronchioles**, which end in **alveolar ducts**, each of which opens into several **pulmonar alveoli** (alveolar sacs). The alveolar walls are lined by a microscopic layer of special fluid containing phospholipids as a → **surfactant**. Diffusion of respiratory gases occurs through two single-celled layers, i.e. the **respiratory epithelium** of the alveolar sacs and the **endothelium** of pulmonary blood capillaries, both pressed against each other.

Bird lungs are more complicated and efficient in many ways. Large bronchi are divided into hundreds of small parallel passages, **parabronchi**. These are open at both ends and air flows through them always in the same direction, i.e. towards the bill, both in inspiration and expiration. Every parabronchus is connected to numerous transverse finger-shaped tubes (air capillaries), the surfaces of which serve as the respiratory epithelium. In inspiration a part of inhaled air passes the lungs flowing into nine airbags, and afterwards in expiration into parabronchi. Thus, the lungs get fresh air in both stages of the ventilation. Some bird groups have a slightly different, phylogenetically younger structure, **neopulmo**. In other vertebrate groups the lungs are simpler, in reptiles resembling those of birds, and in amphibians resembling those of mammals, but having only a few large alveoli.

Only some groups of invertebrates have lungs; in **pulmonates** the vascularized mantle wall forms a lung, arachnids have → **book lungs**, and holothurians → **respiratory trees**. *See* respiration.

lung book, → book lung.

lungfishes, → Dipnoi.

lunularic acid (*Lunularia*, the genus of liverworts), 6-(*p*-hydroxyphenethyl)salicylic acid; a hormone-like phenol derivative (dihydrostilbene), $C_{15}H_{14}O_4$, that replaces → abscisic acid in algae and liverworts.

lutein (L. *luteus* = yellow), **1.** a yellow pigment, $C_{40}H_{56}O_2$, one of the → xanthophylls in organisms; a carotenoid alcohol that occurs in plants e.g. in green leaves, blossoms, algae, and in animals e.g. in the → corpus luteum, adipose tissue, egg yolk, and feathers of birds (*see* lipochrome); **2.** the dried preparation of the pig corpus luteum, earlier used in medicine as a progesterone source.

luteinization, development of the ovarian follicle after ovulation into the → corpus luteum. *Verb* to **luteinize**, to form luteal tissue.

luteinizing hormone, LH, lutropin; a glycoprotein hormone secreted from the anterior → pituitary gland of vertebrates; one of the gonadotropins, i.e. the hormones which control the development and function of the gonads. In females, LH e.g. causes the ovulation and the development of the ovarian follicle and corpus luteum, also stimulating the hormone secretion (progesterone and oestrogens) from their cells. In males, LH stimulates the interstitial cells of testes to produce testosterone. The secretion of LH is controlled by the hypothalamic hormone, → **gonadoliberin** (gonadotropin-releasing hormone, GnRH). In males, LH is also called **interstitial cell-stimulating hormone, ICSH**.

luteinizing hormone-releasing hormone (factor), LHRH, luliberin; → gonadoliberin.

luteotropin, luteotropic hormone, LTH, → prolactin.

lutropin, → luteinizing hormone.

lux, pl. **lux** or **luxes** (L. = light), abbr. lx, a unit of illumination. *See* Appendix 1.

luxury gene, a tissue-specific gene whose action is typical only of cells differentiated in a certain way; e.g. the globin genes in the erythrocyte mother cells regulating the synthesis of the protein portion of haemoglobin.

lyases (Gr. *lyein* = dissolve), enzymes of the EC class 4, one of the six main classes; catalyse the cleavage of chemical bonds such as C-C, C-O, C-N, removing groups from a substrate without hydrolysis, oxidation, or reduction. In certain conditions they can also act as synthetases in reverse reactions. Lyases are e.g. aldolases, decarboxylases, and dehydratases. *Cf.* ligases.

lycopene, ψ, ψ-carotene; a red-yellow pigment that occurs in ripe fruits, abundantly e.g. in tomato and grape fruit; synthesized from → isoprene. The → carotenoids of plants are derived from lycopene.

Lycophyta, *see* Lycopsida.

Lycopodiales, clubmosses, an order in the class Lycopsida, division Pteridophyta (in some classifications division Lycophyta), Plant kingdom; evolutionarily a very old plant group, the most important genus of which is *Lycopodium*. Club mosses are rather small, evergreen, and grow along the ground. They reproduce by spores which develop in special spore spikes in the tips of the shoots. The order includes ca. 400 species.

Lycopsida, a class in the division Pteridophyta, Plant kingdom including the orders → Lycopodiales, Selaginellales, Isoëtales, and Lepidodendrales. The number of living species in Lycopsida is about 1,200. In some systems Lycopsida is classified as a separate division Lycophyta (lycophytes).

lymph (L. *lympha* = clear water, lymph), yellowish fluid found in the → lymph vessels of vertebrates, containing much the same inorganic salts as the blood plasma. The major part of interstitial tissue fluid turns back into the blood across blood capillary walls, but some of it (extra fluid) enters lymphatic vessels and drains through them into the blood. Before flowing back to the blood the lymph flows through → lymph nodes where → lymphocytes recognize and bind microbes and other foreign particles, while some lymphocytes are released to the lymph flow. Because of the lipids absorbed into the lymphatic vessels through the intestinal wall, the lymph passing from the gut region is called **chyle**. *See* lymphatic tissue.

lymphatic, 1. pertaining to the lymph, or organs involved in it; also called **lymphoid**; **2.** denoting a sluggish temperament; **3.** → lymph vessel.

lymphatic system, lymphoid system, → immune system; the system serving in physiological defence of vertebrates, comprising the → lymph vessels, lymph nodes and other organs with lymphatic tissue, and the lymph with circulating lymphocytes.

lymphatic tissue, a tissue system of the vertebrate body comprising tissues of the immune system where → lymphocytes develop; its histological structure comprises a lattice of reticular connective tissue with interspaces containing lymphocytes. It includes **primary lymphatic tissue**, i.e. bone marrow, thymus, bursa of Fabricius (in birds), and embryonic liver, from which the lymphocytes migrate to the **secondary lymphatic tissue**, i.e. the → lymph nodes and spleen, and from there further to circulate in the blood. In these organs, the tissue may be formed from nodular, diffuse, or loose lymphatic tissue. *Syn.* lymphoid tissue, lymph tissue, adenoid tissue.

lymphatic vessels, *see* lymph vessels.

lymph(atic) follicle, lymphoid follicle (L. *folliculus lymphaticus*), *see* lymph node.

lymph heart, the enlarged region of a lymph vessel with a strong pulsating wall musculature, pumping lymph; found in lower vertebrates, but not in birds and mammals. *See* lymph vessels.

lymph node, also **lymphonodus,** pl. **lymphonodi** (L. *nodus lymphaticus*), any of numerous bean-shaped organs of vertebrates, in man from 1 mm to 25 mm in size, situated along the course of lymphatic vessels. Lymph nodes are most abundant in intestinal and neck areas, in tetrapods also in the axillae and groins.

Several afferent lymphatic vessels enter each node through the fibrous capsule of the concave side and open into the **cortex** that consists of loose, reticular connective tissue with many dense cell groups, i.e. **lymphatic follicles** (lymph follicles/nodules), with a large number of → **lymphocytes**. After flowing slowly through the tissue into the **medulla**, the lymph, now containing many new lymphocytes released from the follicles, flows into one efferent lymphatic vessel coming out through the convex side of the node. The tissue of cortex and medulla contains mainly B lymphocytes, and between them there is a layer, **paracortex**, that contains mainly T lymphocytes.

The lymph nodes are the part of the reticuloendothelial system having phagocytic cells, → **macrophages**, lining the sinuses; these together with lymphocytes and → **antigen-presenting cells** ensure the maintenance of → immunity.

lymph nodule, lymphatic nodule (L. *nodulus lymphaticus*), **1.** any of small groups of lymphatic tissue usually beneath the epithelium of the alimentary canal (**solitary lymphatic nodules** and **aggregated lymphatic nodules**); **2.** lymph follicle, any of small dense groups of lymphocytes in the cortex of a → lymph node; **3.** occasionally used to refer to a small lymph node.

lymphoblast (Gr. *blastos* = germ, shoot), rarely called lymphocytoblast; **1.** a premature state of a lymphocyte; **2.** sometimes used to denote an activated lymphocyte.

lymphocystis (Gr. *kystis* = bladder, cyst), a viral disease found in fish, manifested by cystic skin swellings filled with lymph; may be fatal to an animal because of secondary infections.

lymphocytes (Gr. *kytos* = cavity, cell), a mononuclear, non-phagocytic white blood cell, serving adaptive → immunity and found in the blood and → lymphatic tissue of vertebrates. In birds, during the ontogenic development certain precursor cells, **lymphoblasts**, of the lymphatic cell line in the bone marrow move to the → bursa of Fabricius, but in mammals probably to the liver, to be matured into **B lymphocytes** (B cells). In all vertebrates, some other precursor cells move to the thymus to mature into **T lymphocytes** (T cells). The lymphocytes proliferate in the lymph nodes and circulate in the lymph and blood, at this stage called small lymphocytes (7—12 μm). When coming in contact with antigens (e.g. microbes) the cells are activated to develop into large lymphocytes, often also called lymphoblasts (10—30 μm). The third type of lymphocytes are → **natural killer cells**.

B lymphocytes are responsible for humoral immunity; when activated, they are differentiated into **plasma cells** which produce soluble → antibodies in the blood. Some of the B cells differentiate into **memory cells** which can cause a stronger response for subsequent infections (acquired immunity). Each lymphocyte recognizes generally only one → antigen or its determinant, and then produces a specific antibody against it. The whole lymphatic system can react to thousands of different antigens, and for that purpose a vast number of slightly different lymphocyte cell lines (clones) are produced in the body. B cells are characterized by the presence of surface immunoglobulin, monomeric IgM or IgD which constitutes the **B cell antigen receptors**.

T lymphocytes are responsible for cellular immunity, and when activated by antigens they are differentiated into **killer cells** (cytotoxic T cells, effector T cells, CD_8, T_8 cells), which can destroy cytotoxically foreign cells; some T cells differentiate into **helper T cells** (CD_4, T_4 cells) or into **suppressor T cells**, the cell types controlling

immune reactions (→ cytokines), and some differentiate into **memory cells**. T lymphocytes bear antigen-specific receptors, **T cell antigen receptors**, on their surface and react to foreign antigen that is presented to them e.g. by macrophages.
The functions of different lymphocyte types as well as other white blood cells and macrophages are coordinated by a complex system of chemical messengers.

lymphoid (Gr. *eidos* = appearance), **1.** → lymphatic; **2.** → adenoid.

lymphoid tissue, → lymphatic tissue.

lymphokines, soluble polypeptides or proteins found in vertebrate tissues, released from sensitized lymphocytes when contacted with specific → antigens. They help cellular immunity by stimulating the activity of → macrophages; include e.g. mitogenic, chemotactic, and transfer factors, e.g. some → interleukins (IL-1 and IL-2); sometimes also → monokines are included. Lymphokines belong to a larger group, → **cytokines**.

lymphoma, a tumour derived from lymphatic tissue.

lymphonodus, → lymph node.

lymphous, pertaining to → lymph.

lymph tissue, → lymphatic tissue.

lymph vessels, lymphatic vessels, lymphatics, a system of channels carrying → lymph in the vertebrate body. The vessels begin as finger-like, small lymph vessels, **lymph capillaries**, with fairly permeable walls through which some of the interstitial tissue fluid filters in; there is very little if any → basal lamina under the wall epithelium, and no tight intercellular connections exist. Lymph capillaries unite together to form larger vessels, which then join to two main trunks, i.e. the **thoracic duct** (*ductus thoracicus*) and the **right lymphatic duct** (*ductus lymphaticus dexter*), which open to the blood circulation via two subclavian veins.
The lymph vessels serve in transporting various agents from tissues into the circulation, e.g. proteins and cells, especially lymphocytes from the lymph nodes, and nutrients such as fats from the gut. The lymph flow is slowly pushed forward by kinetic energy of gravity, by contractions of skeletal muscles, by pressure pulses of adjacent arteries, and by contractions of smooth muscle in the wall of vessels. In lower vertebrates, such as in frogs, there are large cavities (lymphatic sacs, lymph hearts) with pulsating smooth muscles, connected with the vessels. Many **valves** in the vessels maintain the direction of the flow. The histological structure of the lymph vessels is much like that of the veins. Oedema caused by lymphatic obstruction is called lymphoedema; in filariasis (*see* elephantiasis), small parasitic worms migrate into lymph vessels and obstruct them. *See* blood vessels.

lyo- (Gr. *lyein* = to dissolve), pertaining to something dispersed or dissolved.

lyonization (*Mary F. Lyon*, b. 1925), → X chromosome inactivation.

lyophilization (Gr. *lyein* = dissolve, *philein* = to love), a freeze-drying technique to preserve biological material, such as serum and enzymes or other proteins. Water is evaporated under a vacuum from frozen material by means of a special device, **lyophilizer**.

lyriform organs, minute organs (slit organs) on the body and appendages of spiders; the function is uncertain, they may act e.g. in chemoreception (smell sense) or in recognizing vibrations in the web.

lys(i)-, lyso- (Gr. *lysis* = dissolution), pertaining to decomposition, dissociation, dissolution.

lysate, material produced by the lysis of cells; e.g. obtained from tissues of living organisms by means of an artificial procedure, such as *in vitro* digestion.

Lysenkoism (*Trofim D. Lysenko*, 1898—1976), an erroneous doctrine of inheritance that was first presented by the plant breeder *Ivan Mitšurin* (1860—1935), prevailing in Soviet Union during Stalin's regime (from 1948 it was the only officially permitted doctrine of genetics). For example, according to Lysenkoism the acquired characters are heritable, whereas → Mendelism was considered to be erroneous. Lysenko destroyed the whole of classical genetics in the Soviet Union because he dominated the field dictatorially. Gradually after 1965, genetics could experience a new development also in the Soviet Union.

lysergic acid (Gr. *lyein* = to dissolve + ergot), an organic substance, $C_{15}H_{15}N_2COOH$, obtained by hydrolysis from ergot alkaloids; its derivative, **lysergic acid diethylamide** (**LSD**), used as a hallusinogenic drug, affects the adrenergic and serotoninergic brain systems causing e.g. sensory distortions and changes in psychic functions, often resulting in a persistent psychotic state.

lysigenic, lysigenous (Gr. *gennan* = to produce), pertaining to formation of tissue cavities in plants by dissolution of the cell walls; occurs in formation of secretory cavities (e.g. in orange peel) or of lysigenous → intercellular spaces. *Cf.* lysogenic, schizogenic.

lysin, 1. any substance that causes the lysis of cells; **2.** specifically an antibody that causes complement-dependent lysis of cells and tissues in vertebrates; *see* complement system.

lysine, α,ε-diaminocaproic acid, abbr. Lys, lys, symbol K; $H_2N(CH_2)_4CH(NH_2)COOH$, an amino acid that occurs in proteins of all organisms. Lysine belongs to the → essential amino acids, and is thus important for nutrition of man and animals. There are continuous efforts using selective breeding for increasing lysine content in cereal.

lysis, pl. **lyses** (L. < Gr.), dissolution, destruction, decomposition; e.g. **1.** the lytic decomposition of a cell after the cell membrane is destroyed e.g. by a → bacteriophage; **2.** decomposition of a chemical compound e.g. by a specific agent.

lysogen (Gr. *gennan* = to produce), **1.** any agent causing dissolving or decomposition; **2.** a bacterium in the state of lysogeny; *see* lysogenic.

lysogenesis, the production of lysins.

lysogenic, 1. lytic, or producing lytic substances; **2.** specifically pertaining to a bacterium that on its chromosome carries a → prophage, i.e. the DNA of a bacteriophage (virus). A lysogenic bacterium (lysogen) has a tendency to be lysed; the phenomenon is called **lysogeny**. In certain circumstances the prophage is decoupled from the genetic control of the lysogenic bacterium and begins to reproduce like an autonomous virus inside the bacterial cell, causing the decomposition (lysis) of the bacterium. During lysis the progeny of the bacteriophage is released. *Cf.* lysigenic.

lysogenic intercellular space, *see* intercellular spaces.

lysogenization, the formation of lysogeny and its establishment when a temperate bacteriophage infects a bacterium and integrates itself as a part of the bacterial chromosome. *See* lysogenic.

lysogeny, *see* lysogenic.

lysol, a soapy mixture of → cresols, used as a disinfectant.

lysosome-associated membrane glycoproteins, LAMP-1, LAMP-2, lysosome-specific proteins of the cell membrane having a long epimembranic domain, and short transmembrane and cytoplasmic domains; their participation in the cellular metabolism is unknown.

lysosomes (L. *soma* = body), cell organelles which are minute cytoplasmic vesicles of half a micrometre to several micrometres in diameter. Each lysosome is enveloped by a unit membrane, containing many **hydrolytic enzymes**, especially acidic hydrolases which catalyse the breakdown of ester, glycoside, and peptide bonds of organic substances; pH in lysosomes is about 4-5. Lysosomes are found in all animal cells, except in mature red blood cells, and also in cells of fungi and some algae. They function in processes in which organic materials are decomposed, as e.g. in → phagocytosis, pinocytosis, and programmed cell death. In those processes a **primary lysosome** joins with that material to form the **secondary lysosome** (phagolysosome), wherein the digestion of engulfed material then occurs enzymatically.

lysozyme, muramidase, mucopeptide glycohydrolase; a lytic enzyme, EC 3.2.1.17, catalysing the decomposition of peptidoglycan wall material of many bacteria; occurs e.g. in lacrimal fluid, saliva, egg white, and milk of vertebrates, and similar enzymes are also found in the haemolymph of many insects. Lysozyme protects animals against bacteria.

lysylbradykinin, → kallidin.

lytic, 1. pertaining to, or producing a → lysis, i.e. decomposing or breaking down a compound or an effect; **2.** pertaining to a → lysin. *See* lytic bacteriophage.

lytic bacteriophage, a virus of a bacterium, i.e. a kind of → bacteriophage (virus) that always when it infects the bacterium also causes its lysis, and hence cannot be integrated as a → prophage into a part of the bacterial chromosome. *See* lysogenic. *Cf.* temperate bacteriophage.

M

M, symbol for **1.** molar concentration (moles per litre); **2.** mega- (millionfold); **3.** methionine.

m, symbol for **1.** metre; **2.** milli- (one thousandth part).

***m*,** **1.** *m*-, meta-; **2.** molal concentration (moles per kilogramme); **3.** mass or mass defect.

maceration (L. *macerare* = to soften), softening by soaking; e.g. the maceration of a tissue sample, i.e. softening or partly degrading the tissue for microscopical observation. *Verb* to **macerate**.

macerozyme, a pectinase which is used for the dispersion of plant cell walls. *See* pectic substances.

macr(o)- (Gr. *makros* = large), denoting large or long.

macrobenthos (Gr. *benthos* = depth), the composition of those bottom organisms which remain on a sieve with 0.5 mm mesh in fresh waters, and 1.0 mm in marine waters. *Cf.* meiobenthos, microbenthos.

macrobiosis (Gr. *bios* = life), longevity; lengthening of the life span. *Adj.* **macrobiotic**, long-lived; pertaining to a factor which lengthens life, e.g. lengthening a dormant state of seeds for years.

macrobiota, a set of living organisms in soil consisting of organisms, **macrobionts**, 4-5 cm or more in length; includes e.g. large insects, oligochaetes, and roots of plants. *Cf.* mesobiota, microbiota.

macroclimate (Gr. *klima* = inclination, slope), the climate of a large geographic area. *Cf.* microclimate. *See* climate zones.

macrocyst (Gr. *kystis* = bladder, pouch), the → resting spore of the Myxomycetes.

macrocyte (Gr. *kytos* = cavity, cell), an abnormally large red blood cell (10 to 12 µm in diameter), smaller than → megalocyte.

macroevolution (L. *evolvere* = to unroll), → evolution that leads to the formation of new species; evolution within a species with changing gene frequences is called **microevolution**. Macroevolution can be deduced from microevolution, both using substantially same evolutionary mechanisms, i.e. macroevolution can be deduced from microevolution if the concept of → isolation is included. *Cf.* megaevolution.

macrofauna (L. *Fauna*, sister of *Faunus* = mythical deity of herdsmen), the → fauna consisting of large animals living in the soil and the bottom of waters (*see* macrobiota). In biological studies concerning the inland water bodies, the macrofauna of the bottom animals comprises all those individuals which remain on a sieve with 0.5 mm mesh. The corresponding mesh size in the studies concerning oceans and e.g. the Baltic Sea is 1.0 mm. *Cf.* meiofauna, microfauna, macrobenthos.

macrofibril, the structural unit of the cellulose wall of a plant cell, consisting of ca. 250 cellulose → microfibrils.

macroflora, pl. **macroflorae,** or **macrofloras** (L. *Flora* = mythical goddess of flowers), macroscopic plants and fungi in an area; in practice other than unicellular plants. *Cf.* microflora.

macrogamete (Gr. *gamos* = marriage), megagamete; the larger of the two different conjugant → gametes in → anisogamy, usually the non-motile ovum or egg. *Cf.* microgamete.

macromere (Gr. *meros* = part), **1.** one of the large → blastomeres arising by the unequal division of a → telolecithal ovum in the lower part (vegetal hemisphere) of an early animal embryo (morula), as in amphibians; **2.** a cell type of the → amphiblastula. *Cf.* micromere.

macromolecule (L. *molecula*, dim. of *moles* = mass), a large molecule, generally a polymer, with a molecular weight more than 10,000, often having colloidal properties; e.g. → nucleic acids, proteins, large polypeptides and polysaccharides, and lignin.

macronucleus, pl. **macronuclei** (L. *nucleus*, dim. of *nux* = nut), the larger of the two different cell nuclei existing in many protozoans, such as in ciliates and some foraminiferans; the other is → **micronucleus**. During the life cycle of ciliates periods of **asexual** (vegetative) reproduction occur that alternate with periods of **sexual** reproduction. In the non-reproductive cell, a → diploid or more often an → endopolyploid macronucleus functionally dominates the regulation of metabolic functions of the cell. During asexual reproduction (**binary fission**) the micronucleus divides mitotically (*see* mitosis) and the macronucleus elongates and divides amitotically. During sexual reproduction (**conjugation** of two individuals) the macronucleus disintegrates and the micronucleus of each individual undergoes → meiosis. After the exchange of one haploid pronucleus between conjugants the exchanged pronucleus fuses with the resident one to form synkaryon pronucleus (diploid) which divides three times to form eight micronuclei. Four micronuclei of the original 8 become macronuclei. When sexual reproduction has terminated, each of the four resulting daughter cells receives one micronucleus and one macronucleus, and the macronucleus again begins to regulate most synthetic activities of the cell.

macronutrients, 1. in plant physiology, the main essential elements required by plants for normal growth and development; concentrations of the salts in complete nutrient solutions range usually between 1—5 mM. Macronutrients include nitrogen, phosphorus, potassium, calcium, magnesium, sodium, and sulphur. Sometimes also iron is included in macronutrients, but more usually in → micronutrients; *syn.* macroelements, major elements; **2.** in zoology, the term macronutrients is sometimes used to denote those nutrients required by animals in greatest quantities, as carbohydrates, proteins, and fats.

macrophage (Gr. *phagein* = to eat), a large, active phagocyte involved in the physiological defence of animals (*see* **phagocytosis**). In vertebrates, the macrophages include the → **monocytes** of blood which, when moved to other tissues, are called **tissue macrophages**, involved in the local defence of tissues and organs. Both cell types together constitute the **mononuclear phagocyte system (MPS)**, also called **reticulo-endothelial system**. The tissue macrophages include e.g. the → histiocytes of connective tissue, → Kupffer cells of the liver, macrophages of the pulmonary alveoli, → osteoclasts of the bone tissue, and microglial cells of the nervous system; especially when activated (e.g. by lymphokines from T lymphocytes) they all are called macrophages. Macrophages can actively move using → **pseudopodia** (amoeboid movement) and can penetrate into tissues. Each macrophage is able to engulf invading microbes, microbial remnants, and tissue debris formed in the organism itself. Macrophages secrete about 100 different substances, that affect lymphocytes and many other cell types. They secrete e.g. many enzymes (e.g. lysozyme, collagenase) and hormone-like substances such as → prostaglandins (of E series), → interferons, interleukins, and clot-promoting factors. They also serve as → antigen-presenting cells, i.e. present substances engulfed by them to lymphocytes (T cells) thereby activating these cells to immunological defence. *Cf.* granulocytes.

macrophyll (Gr. *phyllon* = leaf), *see* megaphyll.

macrophyte (Gr. *phyton* = plant), a macroscopic plant, e.g. a vascular plant, moss, or a big alga. *Cf.* microphyte.

macroplankton (Gr. *planktos* = wandering), macroscopic planktonic organisms varying in size between 2 and 20 mm (remain on a sifter with 2 mm mesh). In the oceans the macroplankton is limited to substantially larger organisms, 20—200 mm in length; e.g. some crustaceans. *Cf.* megaplankton, microplankton. *See* plankton.

macroplasia (Gr. *plassein* = to form), an excessive (gigantic) growth of tissues and organs.

Macroscelidea (Gr. *skelos* = leg), **elephant shrews**; a suborder of mammals in the order Insectivora (in some systems classified as an order); comprises 19 species in Africa; they are characterized by long hindlegs and an elongated snout adapted to feed on insects. The elephant shrews separated from primitive insectivores during the early Oligocene.

macrosclereid (Gr. *skleros* = hard), *see* sclereid.

macroscopic(al) (Gr. *skopein* = to watch), visible to the naked eye; visible without magnification, i.e. not microscopic.

macrosmatic (Gr. *osme* = smell), pertaining to an animal with a good sense of smell; e.g. most carnivorous mammals. *Cf.* microsmatic.

macrosomia (Gr. *soma* = body), gigantic growth of the body. *See* gigantism.

macrospecies, pl. **macrospecies,** a large polymorphic species. *Cf.* microspecies.

macrosporangium, pl. **macrosporangia** (Gr. *sporos* = seed, *angeion* = vessel), a → sporangium present in heterosporous plants (*see* heterospory) producing → **macrospores** (megaspores); in seed plants it is the nucellus of an → ovule. *Syn.* megasporangium.

macrospore, the larger of two types of spores in heterosporous plants (→ heterospory), formed by meiotic divisions (→ meiosis) from the **macrospore mother cells** in a process called **macrosporogenesis**. The macrospores develop into female → gametophytes, producing female gametes. In seed plants, the macrospores are formed in the ovule of the ovary; one macrospore mother cell forms four macrospores, three of which become stunted and the fourth develops into the embryo sac mother cell and further into the → embryo sac, i.e. the female gametophyte which contains the egg cell. *Syn.* megaspore, gynospore. *Cf.* microspore.

macrospore mother cell, macrosporocyte; a diploid cell from which the → **macrospores** (megaspores) develop through meiosis in → **macrosporogenesis** in heterosporous plants (*see* heterospory). *Syn.* megaspore mother cell, megasporocyte.

macrosporogenesis, megasporogenesis (Gr. *gennan* = to produce), the formation of → **macrospores** from a macrospore mother cell through meiosis in heterosporous plants (*see* heterospory).

macrosporophyll (Gr. *phyllon* = leaf), a leaf, or a modified leaf, bearing the → macrosporangia of heterosporous plants (*see* heterospory); in seed plants it is the carpel. *Syn.* megasporophyll.

macrostrobilus (Gr. *strobilos* = cone), *see* strobilus.

macrothermophilic, macrothermophilous (Gr. *therme* = heat, *philein* = to love), pertaining to an organism (macrotherm, megatherm), particularly plants, living in warm → habitats, especially in the tropics.

macula, pl. **maculae** (L. = spot), **macule**; a small spot that differs from the surrounding tissue, especially in colour; e.g. **1.** → macula lutea; **2.** maculae acusticae, comprising the saccular spot (macula sacculi) and the utricular spot (macula utriculi), forming the neuroepithelial sensory apparatus that contains sensory hair cells, supporting cells, and otoliths in the walls of the saccule and utricle of the inner ear in tetrapods; *see* equilibrium sense; **3.** macula densa; *see* juxtaglomerular apparatus; **4.** macula adhaerens, i.e. spot → desmosome.

macula lutea (L. *luteus* = yellow), macula retinae, yellow spot; a yellowish oval area temporal to the optic disc in the → retina of the vertebrate eye (in man 3 by 5 mm); at its centre is the fovea (*fovea centralis retinae*), the region of the most detailed, clearest vision that in most vertebrate species is associated with colour vision using cones as photoreceptors.

mad cow disease, bovine spongiform encephalopathy (BSE); a disease of the nervous system in cows caused by a → prion, with symptoms such as disturbance of coordination of movements, and ultimately death. The disease is presumed to be infectious also to other species, also to humans in whom it probably takes the form of Creutzfeldt—Jacob disease.

madreporite (Ital. *madre* = mother, Gr. *poros* = passage, pore), a porous plate associated with the → water-vascular system of echinoderms, located at the end of the → stone canal; may be totally invisible inside the animal, such as in holothurians, or visible on the → aboral surface, as e.g. in the starfish.

MADS-box (M < MCM1, a mating type factor of yeast, A < AGAMOUS, a transcription factor of *Arabidopsis thaliana*, D < DEFA, a transcription factor of snapdragon, S < SERUM transcription factor), a DNA sequence common to several → transcription factors of → selector genes in plants consisting of many zinc-finger motifs (→ zinc-finger protein); analogous but not homologous to the → homeobox in animals.

magnesium, an element belonging to the alkaline earth metals, symbol Mg, atomic weight 24.305. Because of its great reactivity, magnesium occurs naturally only in the form of compounds, such as in dolomite ($CaCO_3$.-$MgCO_3$), magnesite ($MgCO_3$), magnesium chloride ($MgCl_2$, especially in sea water), and magnesium hydroxide ($Mg(OH)_2$). Magnesium ions (Mg^{2+}) are essential for all organisms, acting in cells e.g. as activators of many enzymes, such as kinases and other enzymes involved in phosphate transfer, or transketolase in the metabolism of carbohydrates. Magnesium ions are also essential for the activity of ribosomes. In plants, magnesium is the central atom of the → chlorophyll molecule and exists also in → pectic substances of the cell wall in plants.

magnetic sense, a sense by which some animals are able to observe the magnetism of a certain region, using it for orientation and navigation by the magnetic field of the Earth. Researchers have demonstrated a magnetic sense e.g. in insects, fishes, and birds. Receptor cells (**magnetoreceptors**), found e.g in the head in bees and pigeons, contain ferromagnetic crystals attached to the plasmamembranes of receptor cells. Probably the crystals react by turning in the magnetic field which results in irritation, i.e. the formation of a → receptor potential in the receptor membrane. Similar crystals occur also in some bacteria. *See* magnetosome, magnetite.

magnetite, a magnetic black iron ore, iron oxide, Fe_3O_4, i.e. a mixture of iron (II) and iron (III) oxides, $FeO \cdot Fe_2O_3$. Crystals of magnetite are found in magnetoreceptors of some animals and bacteria. *See* magnetic sense.

magnetosome (Gr. *soma* = body), an enveloped structure in some bacteria containing magnetite crystals; bacteria living in aqueous environments detect the Earth's magnetic field and move towards the bottom.

Magnoliatae, → Dicotyledonae.

Magnoliophyta, → Anthophyta.

major circulation, systemic circulation, greater circulation. *See* blood circulation.

major elements, → macronutrients.

major gene (Gr. *genos* = origin, birth), any → gene whose function regulates the development of a clear, qualitative phenotypic trait, and accordingly the transmission of the gene can be followed by simple crossing experiments (*see* genetic transmission). In addition to major genes, also → **minor genes** function in organisms, each of which has a minor effect on the polygenic quantitative traits. The difference between major and minor genes is artificial and based on methodological distinctions; major and minor genes do not differ qualitatively.

major histocompatibility complex, MHC, a mammalian gene complex of several highly polymorphic linked genes encoding glycoproteins (→ major histocompatibility molecules) involved in cell-surface → histocompatibility and thus in many aspects of immunological recognition; the gene loci are called H-2 complex in mice, located on the seventh chromosome pair, and HLA complex in humans on the sixth chromosome pair.

major histocompatibility molecules, MHC molecules, cell-surface → glycoproteins in vertebrates involved in antigen recognition in immunological reactions both between different lymphoid cells and between lymphocytes and antigen-presenting cells, e.g. H-2 antigens in mice and HLA antigens in man. These glycoproteins are also called **MHC antigens** because they act as → antigens in transplanted tissues and organs, causing immune response in the recipient individual by binding peptides from foreign proteins and presenting them to T lymphocytes (T cells); MHC molecules were originally identified as being responsible for rapid graft rejection between individuals. There are two types of MHC molecules, i.e. **class I** and **class II** molecules; the former bind peptides derived from foreign intracellular proteins, the latter bind peptides from extracellular proteins. MHC molecules are encoded by the genes of the → major histocompatibility complex, while related glycoproteins encoded outside this complex are called **minor histocompatibility molecules** (antigens). *See* histocompatibility.

mal- (L. *malus* = bad, ill, abnormal), denoting relationship to illness, disease, or impaired function.

malacology (Gr. *malakos* = soft, *logos* = word, discourse), a branch of biology dealing with molluscs.

Malacostraca (Gr. *ostrakon* = shell), **malacostracans**; a class of crustaceans with a constant number of body segments, i.e. five segments in the head, eight in the thorax, and six in the abdomen; the head and thorax are covered with the → carapace. The class includes about 23,000 species in several orders, such as **euphausiaceans** (Euphausiacea, e.g. krill), **mysidaceans** (Mysidacea, opossum shrimps), **isopods** (Isopoda, e.g. woodlice), **amphipods** (Amphipoda, e.g. prawns, shrimps), and **decapods** (Decapoda, e.g. crayfishes, crabs, lobsters).

malaria (It. = bad air), an infectious human disease endemic in parts of Africa, Asia, Oceania, and South and Central America. Malaria is caused by intracellular protozoans of the genus *Plasmodium* transmitted to humans by the bite of an infected female mosquito of the genus *Anopheles*. The life cycle of the parasite includes asexual multiplicative stages in infected humans first in liver parenchyma cells and then in red blood cells (erythrocytic schizogonous cycles). The production of gametocytes in human red blood cells, and further of gametes in a mosquito, makes new infections transferred by mosquitoes possible. The disease is characterized by paroxysms of high fever, shaking chills, sweating, and anaemia. Called also jungle fever, swamp fever, marsh fever, or paludism. *See* merozoite, sickle-cell anaemia.

malate (L. *malum* = apple), a salt, ester, or the ionic form of → malic acid.

malate dehydrogenases, enzymes (EC 1.1.1.37-40) functioning in the energy metabolism of cells; in the → citric acid cycle of cells they catalyse the reaction: malate + NAD^+ —> oxaloacetate + NADH + H^+; in the photosynthetic carbon fixation of C_4 plants the enzyme is called **malic enzyme** (EC 1.1.1.40) catalysing the reaction: malate + $NADP^+$ —> pyruvate + CO_2 + NADPH + H^+.

malathion, an organophosphorous → insecticide, $C_{10}H_{19}O_6PS_2$, for ordinary use in gardens; blocks cholinesterase activity thus inhibiting strongly the function of the nervous system in most animals, and thus is effective e.g. against mites and insects. Malathion is highly toxic also to vertebrates, such as man.

male, 1. an individual, as an animal or staminate plant, whose reproduction organs can produce only sperm, spermatozoa, or pollen; *see* gamete; **2.** masculine, manlike.

maleate, a salt, an ester, or the ionic form of → maleic acid.

male genital organs, male genitals; reproduction organs in male animals; in vertebrates, they include **internal** male genital organs, i.e. the → testis, epididymis, vas deferens, seminal vesicle, prostate, and bulbourethral glands, and **external** male genital organs, i.e. the → penis (or hemipenis) with the ejaculatory duct, in mammals also the scrotum; in invertebrates, usually comprise the penis and some accessory organs (*see* dart sac).

male haploidy, → arrhenotoky.

maleic acid, butenedioic acid, HOOCHCCH-COOH, the *cis* isomer of → fumaric acid; used in biological buffers and as a precursor for organic syntheses, e.g. in the manufacture of antihistaminic drugs; its ionic form, esters, and salts are called **maleates**. Also called toxilic acid.

maleic hydrazide, MH, 1,2-dihydropyridazine-3,6-dion; a herbicide used for controlling grasses. It is a potent repressor of growth by interfering with cell division; it also causes the obstruction of the sieve tubes by inducing callose formation in them, and inhibits the function of many enzymes.

male sex hormones, male reproductive hormones, androgens, androgenic hormones: hormones which regulate the sexual characteristics of male animals; in vertebrates, include e.g. → testosterone, dihydrotestosterone, and androsterone, the principal and most potent being testosterone. Male sex hormones are → steroids which are secreted by the testes, in vertebrates mainly **testosterone**, but small quantities are secreted also from the adrenal cortex, chiefly **dehydroepiandrosterone**. Male sex hormones regulate primary and secondary sexual characters, as well as the growth of bones and muscles and the function of the brain. They enhance masculine behaviour, and in many animals the development of aggressive behaviour. Luteinizing hormone, LH, secreted from the → pituitary gland of the both sexes, regulates the synthesis and secretion of sex hormones in the testes, in most species varying substantially according to the season. The so-called **anabolic steroids**, synthetic derivatives of male sex hormones, are more potent than testosterone to stimulate the growth of bones and muscles; to a lesser extent they also affect sexual characteristics and the function of the nervous system. Steroid male hormones are also found in many invertebrate groups, e.g. insects. *See* female sex hormones.

male sterility, a mutant trait of higher plants in which pollen production is inhibited during development, and the plants are therefore self-sterile; it is usually caused by cytoplasmic factors like the mitochondrial → plasmid gene or mtDNA gene. The mutant is used in plant breeding to prevent self-fertilization.

malfunction (L. *malus* = ill), dysfunction; the disturbance in a function; an error function.

malic acid (*Malus* = the genus of the apple tree), a dicarboxylic acid, $HOOCCH_2CH$-(OH)COOH, one of the intermediates of the → citric acid cycle in cell metabolism; occurs in higher concentrations in many fruits, such as apples, grapefruits and gooseberries. Malic acid is formed from fumaric acid and becomes oxidized into oxalacetic acid in the next reaction of the cycle. Its ionic form, salts, and esters, are called **malates**.

malic enzyme, *see* malate dehydrogenases.

malignant, very harmful in influence or effect, becoming progressively worse, deadly; e.g. describes tumours which are metastatic and have destructive growth. *Noun* **malignancy**. *Cf.* benign.

malign transformation, the transformation of a cell to a cancer cell (transformed cell). *Syn.* malignization.

malleus, pl. **mallei** (L.), hammer; the largest of the three → auditory ossicles (ear ossicles) in the middle ear of mammals; its shaft leans against the ear drum and the head articulates with the incus.

Mallophaga (Gr. *mallos* = lock of wool, *phagein* = to eat), **biting lice**; an order comprising wingless insects which live on the skin of birds and mammals; they have highly modified mouthparts for chewing and legs adapted for clinging to the host.

malonic acid, propanedioic acid, $HOOCCH_2$-COOH, its molecular structure resembles that of → succinic acid, and therefore can be used as a competitive inhibitor of succinate dehydrogenase. In cellular metabolism malonate is associated with the coenzyme A (**malonyl coenzyme A**) which transfers 2-carbon groups to the growing carbon chain in the synthesis of fatty acids. Malonic acid and its esters are used in many organic syntheses as e.g. for manufacturing barbiturates; its ionic form, salts, and esters are called **malonates.**

Malpighian body/corpuscle (*Marcello Malpighi*, 1628—1694), an essential part of the → nephron in the vertebrate → kidney of the mesonephros and metanephros types; consists of one arterial glomerulus (Malpighian glomerulus, renal glomerulus) and → Bowman's capsule (Malpighian capsule) enveloping it.

Malpighian layer, germinative layer (*stratum germinativum*); the innermost, mitotically active layer of the → epidermis of the vertebrate skin, comprising the basal layer and prickle cell layer; its cells, keratinocytes, often contain melanin pigment originated from adjacent pigment cells.

Malpighian tubules/tubes, long, blind excretory tubules which open into the posterior part of the alimentary canal, i.e. at the borderline of the midgut and hindgut in terrestrial insects, arachnids, and myriapods. The walls of the tubules are composed of a single layer of cells for removing uric acid and salts from haemolymph. Besides the Malpighian tubules, also the hindgut takes part in the excretory function.

malt, a mixture of hydrolytic products of starch, containing e.g. maltose; usually produced from barley seeds which are germinated under controlled conditions and then heated, dried and ground, used in brewery and as a food additive.

maltase, α-glucosidase; α-D-glucoside glucohydrolase, EC 3.2.1.20; an enzyme which catalyses the hydrolytic cleavage of → maltose into two glucose molecules; occurs in the digestive system of animals, in yeast, and in starchy seeds. *See* glucosidases.

Malthusian parameter, r_m (*Thomas Malthus*, 1766—1834; the symbol r_m comes from the words rate and Malthus), in population genetics, the rate of increase of a gene allele at a locus that has age-specific effects on birth and death rates; is one measure of fitness that describes the → adaptation of a population by measuring the intrinsic rate of increase of the population. If the population increases in a time interval with a factor of z, then $r_m = log_e z$.

Malthusian theory, a theory proposed by *Thomas Malthus* stating that populations tend to increase at the rate of geometric series but the amount of food and other environmental resources only at the rate of arithmetic series, ultimately leading to the shortage of these resources. In human populations this leads, according to Malthus, to the poverty and degradation of lower social classes. The Malthusian theory had a significant influence on Darwin and Wallace when they developed the theory of natural selection and the theory of struggle for existence which underlies it.

maltose, malt sugar, maltobiose; a hard crystalline disaccharide, $C_{12}H_{22}O_{11}$, composed of two glucose molecules with an α-1,4-bond; formed e.g. in hydrolysis of starch by → amylase which occurs in the alimentary canal of animals and in germinating seeds. Maltose occurs in many plants, but usually in small quantities, mainly produced in hydrolysis of starch although some is synthesized *de novo*. Maltose is the fermentable intermediate in brewing. Also called malt sugar or maltobiose.

mamilla, pl. **mamillae** (L. dim. of *mamma* = breast), the nipple (*papilla mammae*), or any nipple-like anatomical structure; also mammilla.

Mammalia (L. *mamma* = breast), **mammals**; a vertebrate class from Triassic to present, comprising homoiothermic, hair-covered animals, although the pelage may secondarily be very scant, as e.g. in whales and man. Common features are e.g. mammary glands of females, a complete muscular diaphragm, three auditory ossicles in the middle ear, the left aortic arch, and the lower jaw consisting of the dentary bone (*dentale*) alone. Mammals are divided into two subclasses: *Prototheria* (**egg-laying mammals**) including e.g.

monotremes and several extinct orders, and *Theria* (**therians**) including marsupials and placental mammals, i.e. most of the 4,100 mammalian species described.

mammal-like reptiles, → Therapsida.

mammalogy (Gr. *logos* = word, discourse), a branch of biology dealing with mammals.

mammals, → Mammalia.

mammary glands (L. *glandulae mammariae*), paired, tubuloalveolar exocrine glands of female mammals secreting milk for offspring; usually comprises several lobes each containing several lobules. Each lobule has one secreting glandular alveole opening to the alveolar duct. The alveolar ducts of each lobe join together forming the lactiferous duct (*ductus lactiferus*) which then opens through the nipple (*papilla mammae*); the nipples are lacking in prototherians. During the embryogenesis of both sexes, the glands derive ventrally from ectodermal tissue (**mammary ridge**, "milk line"), of which a typical number of mammary glands begin to differentiate; the number corresponds approximately to the offspring number typical for a species but may vary even between individuals of the same species. When sexual maturation begins, the secretion of female sex hormones (oestrogens and progesterone) induces the growth of the mammary glands. → Prolactin induces milk secretion, and in many species → oxytonin causes contraction of ductal myoepithelial cells with consequent ejection of the milk through the nipple.

mammotropin (Gr. *trepein* = to turn), → prolactin.

mandelic acid (Ger. *Mandel* = almond), hydroxytoluic acid, phenylglycolic acid, C_6H_5-CHOHCOOH; a white, crystalline organic acid with three optically isomeric forms, appearing combined in the glycoside **amygdalin**, found e.g. in seeds of *Rosaceae* species, such as the almond and plum; mandelic acid is an antibacterial agent used e.g. in treatment of urinary tract infections.

mandible (L. *mandibula* = jaw), **1.** the lower jaw in vertebrates; *see* mandibular arch; **2.** one of paired mouth appendages of arthropods, often covering the mouth ventrally; in insects the mandibles are composed of one part only, but in crustaceans of the basal and distal parts. *Adj.* **mandibular**, pertaining to the lower jaw; submaxillary.

mandibular arch, in vertebrate embryos, the first → pharyngeal arch, in fishes developing into the jaws; in tetrapod vertebrates, the ventral part degenerates into the → Merkel's cartilage (mandibular cartilage), and the other parts into some → branchiogenous organs, such as the auditory ossicle, in mammals into malleolar ligaments, malleus, and incus.

Mandibulata, the former name for the subphylum of arthropods which have mandibles and antennae; comprised crustaceans, insects, pauropods, symphylans, and myriapods (i.e. millipedes and centipedes). In present classifications, the four last-mentioned orders are sometimes grouped in the subphylum Uniramia (uniramians) or Atelocerata.

manganese (L. *manganesium*, *manganum*), a metallic element, symbol Mn, atomic weight 54.938; occurs naturally e.g. as pyrolusite (MnO_2) and higher oxides, and as manganese carbonate ($MnCO_3$). The manganese ion (Mn^{2+}) belongs to the essential micronutrients of organisms and acts as an activator in several enzymes, e.g. liver arginase, and PEP carboxylase (*see* phosphoenolpyruvic acid) in C_4 plants; in some cases can be replaced by Mg^{2+}. The light reactions of → photosynthesis (photolysis of water) are dependent on the presence of manganese ions.

mangrove vegetation, a vegetation type occurring on subtropical and tropical tidal shores, especially on delta coasts of rivers. Typical mangrove plants are trees growing in water and having supporting roots and → pneumatophores growing upwards through the water surface.

manipulation (L. *manus* = hand), skilfull handling; treatment by hand, as in therapy.

manipulatory experiment, an experiment in which a sample of experimental organisms has been somehow manipulated to resolve the factors under study; another sample of organisms comprises the intact control group. The term is commonly used in experimental ecology.

manna (L. < Hebr. *man*), **1.** the hardened saccharine exudation from the European flowering ash (*Fraxinus ornus*) or some related plants containing high quantities of mannitol; **2.** manna lichen, *Lecanora esculenta*, probably the manna mentioned in the Bible, still gathered by arabs and mixed with meal in order to make bread; may be transported by wind, producing a local rain of food; **3.** a waxy production of some scale insects; e.g. a production of the manna insect (*Trabutina mannibara*) on the tamarisk.

mannan, a polysaccharide formed from → mannose, occurring in plants and fungi.

mannitol, a sugar alcohol, $HOCH_2(CHOH)_4$-CH_2OH, corresponding to → mannose; occurs widely in fungi and plants, such as in seaweeds, but also e.g. in tissue fluids of insects. Beside saccharose and sorbitol, mannitol is used in experimental biology e.g. in isolation media of cells, protoplasts, and cell organelles. Mannitol or its derivatives are also used as a diuretic drug in medicine.

mannoglycerate, a storage substance in red algae, derived from → mannose.

mannose, an aldohexose, $HOC(CHOH)_4$-CH_2OH, with the molecular structure close to glucose; occurs as a storage polymer (mannan) e.g. in some seeds.

manometer (Gr. *manos* = thin, loose, *metron* = measure), an apparatus used for measuring pressures of gases and liquids; the simplest type comprises a thin scaled capillary tube and a closed reaction vessel connected with it. The capillary U-tube is partially filled with stained liquid or mercury moving in the tube according to pressure changes in the vessel. The instrument is used e.g. for measuring oxygen consumption of cells, tissues, mitochondria, or small organisms.

mantle (L. *mantellum* = cloak, mantle), a covering layer in an anatomical structure; a soft integument or fold of the body wall in many invertebrates; e.g. **1. brain mantle,** → pallium; **2. myoepicardial mantle**, a layer of visceral mesoderm in the early vertebrate embryo developing to form the epicardium and myocardium; **3.** a soft epidermic fold inside brachiopods; **4.** the mantle in molluscs, in most cases secreting a shell; in bivalved molluscs it comprises thin tissue sheets adhering to the inner surface of the clam valves, forming the **mantle cavity** with respiratory epithelium; in gastropods it is a sac-like organ; **5.** the **tunica** of tunicates; a tough, translucent test covering the body of the animal; **6.** a **sheath** formed by fungal hyphae around the root of a plant.

mantle cavity, *see* mantle.

manubrium, pl. **manubria** (L. = handle), an anatomical structure resembling a handle; e.g. **1.** *manubrium sterni*, episternum (presternum), i.e. the most anterior segment of the sternum in most tetrapod vertebrates; **2.** *manubrium mallei*, the handle of the malleus (hammer) of the mammalian ear; **3.** the suspending tubular structure on the underside of a medusa leading to the → gastrovascular cavity; the mouth opens at its end.

MAO, → monoamine oxidase.

map distance, *see* mapping.

mapping, in genetics, the determination of the order and distance of genes; → chromosome map, gene map; **map distance**, the distance of two linked genes on the genetic map expressed in percentages of → **crossing-over** obtained from crossing experiments or pedigree analyses. Because a map distance is based on the total frequency of crossing-overs on a certain region of the chromosome, it can be directly translated into map distance units, → map units; **mapping function** is the relationship between the true map distance and the → recombination frequency. A specific type of gene mapping makes use of DNA markers, such as → restriction enzyme cleaving sites; using the method, both relative and physical genetic maps can be constructed. *See* restriction map.

map unit, in genetics, the unit of map distance (*see* mapping) between linked loci on the → chromosome map, i.e. crossing-over unit; is presented as one hundredth of → morgan, i.e. centimorgan (cM). The map unit is used for measuring the length of the chromosome, and it corresponds to the percentual frequency of → crossing-overs between the given loci, excluding the effects of double crossing-overs and → interference. A corrected crossing-over frequency of 1% corresponds to one map unit and represents the map distance between the given loci.

Marchantiales (*Marchantia*, the type genus < *Nicolas Marchant,* d. 1678), an order in the class → Hepaticopsida (liverworts) in the division Bryophyta (in some classifications division Hepatophyta), Plant kingdom; the thallus is leafless but with highly differentiated tissues, the sporogonium is roundish, without columella, and opens often with numerous teeth.

marginal meristem (L. *margo* = edge, border), a → meristem tissue located in marginal areas of growing leaves, forming the leaf blade.

marginal placenta, *see* placenta.

marginal population, a reproductive population inhabiting an edge of the distribution area of a certain species.

marginal value theorem, a rule derived from theoretical exploration, first formulated by *E.L. Charnov* in 1976, dealing with the optimal foraging time for a predator in each patch of → patchy habitats.

marihuana (Mexican Sp.), a crude **cannabis** preparation obtained from the flowers and leaves of the Indian hemp, *Cannabis sativa*; is smoked for its euphoric properties. Its most effective constituents are tetrahydrocannabinols. *Syn.* marijuana, Mary Jane, bhang, charas, hasach, and grass.. Hashish is a more concentrated cannabis drug.

marine (L. *mare* = sea), pertaining to sea; existing in, or produced by the sea.

maritime, living by the sea.

marker, a characteristic factor used to identify a biological structure or biochemical reaction; e.g. **1.** → genetic marker; **2.** in immunology, a specific antigen used to identify a particular class of lymphocytes; **3.** in cell biology, a specific characteristic of a cell or cell line (e.g. metabolic deficiency or specific staining property) that enables the identification of these cells in mixed cell cultures or tissues; **4.** in electrophoretic or chromatographic analyses, a molecule of known size and quality running beside unknown molecules on an electrophoretic gel.

marker-assisted selection, MAS, a method of artificial selection for detecting desirable genotypes; is based on heterozygous marker genes which are linked to → economical trait loci.

marker gene, *see* genetic marker.

marsh (Old E. *mersc* = mere, marsh), → marshland.

marshland, a low-lying treeless shoreland which is occasionally covered by seawater; the soil does not form peat. The plant community is called **marsh**.

Marsileales, *see* Marsileidae.

Marsileidae (*Marsilea*, the type genus < *Luigi F. Marsigli,* d. 1730), **water ferns**; a subclass in the class Pteropsida, division Pteridophyta, Plant kingdom; in some classifications the order Marsileales in the division Pterophyta. The group includes the genera *Marsilea* and *Pilularia* in the family Marsileaceae; e.g. *Pilularia globulifera* (European pillwort) grows commonly in waterlogged areas.

Marsupialia (L. *marsupium* = pouch), **marsupials**, **pouched mammals**; a superorder (some consider as an order in the infraclass Metatheria) of viviparous mammals (subclass Theria), comprising the non-placental animals whose embryonic development begins in the uterus, but after a short time, usually a few days, one or more premature offspring ("larvae"), are born, then crawling to the pouch of the mother's abdominal skin, **marsupium**, where each is attached closely with the mouth to a mammary nipple to get milk for further development. Marsupials live mainly in Australia, but one species (**opossum**) is found in South America where marsupials inhabit the ecological niches occupied by eutherians elsewhere. Familiar species are e.g. **kangaroos, wombats, bandicoots**. *See* Metatheria.

martelism (L. *martes* = marten), a habit of some solitarily living mammals to share space and food resources without protecting their territories against conspecific individuals, but allow the home ranges to overlap; this occurs only at times when the animals are not mating. The behaviour is common in some *Mustela* species, such as marten (*Mustela erminea*) and wolverine (*Gulo gulo*).

MAS, → marker-assisted selection.

masculinization (L. *masculinisatio*), **1.** the induction of normal development of male sexual characters in a male animal; *syn.* virilization; **2.** the induction or development of male sexual characters in a female animal, as e.g. in the case of → freemartin; **masculinity**, possession of male characteristics. *Adj.* **masculine**. *Verb* to **masculinize**.

mask, a prehensile organ of a dragonfly nymph formed from the labium and comprising basal and terminal joints, the latter containing spikes or bristles. The nymph can capture its prey by suddenly extending the mask and catching the prey with the spikes.

mass defect, symbol Δm, the difference between the actual atomic mass and the total mass of the protons and neutrons constituting the nucleus; i.e. the mass of an atom is not the true sum of the elementary particles of the atom, since a part of the mass, *m*, is converted to energy, *E*, as expressed in the mass-energy equation by *Albert Einstein* (1879—1955): $E = m \cdot c^2$; in which *c* is the velocity of light (approx. $3 \cdot 10^{10}$ cm s^{-1}).

mass number, the total number of nucleons (protons and neutrons) in the nucleus of an atom.

mass spectrometer, an instrument that, usually

together with a chromatograph, is used for identifying chemical compounds and determining relative atomic masses and the abundance of isotopes in a sample. In a mass spectrometer the sample is bombarded with a stream of high-energy electrons resulting in the ionization of molecules. Positive ions are accelerated in both electric and magnetic fields, and then collected onto a detector; ions are separated according to the ratio of their mass and charge (*m/e*). The fragments of a compound to be analysed form a line spectrum that serves for identification of the compound. The results obtained from a mass spectrometer are called **mass spectra**.

mast(o)- (Gr. *mastos* = breast), a breast-like extension in an anatomical or histological structure.

mast cell (Ger. *Mastzelle* < *Mast* = feeding, *Zelle* = cell), a cell type of the interstitial connective tissue of vertebrates, found especially around the capillaries; wandering cells which have small cytoplasmic granules containing → heparin, histamine, and proteases, in some animal species also → serotonin; the content of the granules is secreted exocytotically into the extracellular fluid and blood e.g. in inflammatory and allergic responses. The cells have IgE receptors on their cell membranes, and they become granular when IgE coated antigens bind to their surface. Mast cells are functionally analogous to basophilic → granulocytes. *Syn.* mastocyte, heparinocyte.

mastication (L. *masticare* = to chew), the act of chewing. *Adj.* **masticatory**.

Mastigomycota (Gr. *mastix* = whip, *mykes* = fungus), a division in the kingdom Fungi, including the subdivisions Haplomastigomycotina (classes Hyphochytridiomycetes, Chytridiomycetes, Plasmodiophoromycetes) and Diplomastigomycotina (class Oomycetes). The species are simple, often microscopic and mostly aquatic fungi having motile zoospores. Some species cause plant and animal diseases, as e.g. *Phytophthora infestans* (Oomycetes) causing → blight, and *Aphanomyces astaci* causing crayfish plague. In some systems the group Mastigomycota is divided into the divisions Hyphochytridiomycota, Chythridiomycota, Plasmodiophoromycota, and Oomycota.

Mastigophora (Gr. *pherein* = to bear), *see* Flagellata.

masturbation (L. *manus* = hand, *stuprare* = to rape), sexual self-stimulation.

mater (L. = mother), dura mater, pia mater. *See* meninx.

maternal (L. *maternus*), pertaining to the mother; on the mother side, inherited from the mother; **maternity**, motherhood. *Cf.* paternal.

maternal effect, any specific effect of the maternal genotype or phenotype on the first generation offspring.

maternal effect gene, a gene which affects the egg cells of the female but has no action in the male; the genes produce the gradient of → morphogenes, i.e. positional information for the development of early animal embryos.

maternal inheritance, inheritance mediated through an egg cell; is transmitted by extrachromosomal genetic elements which are located in the cytoplasm of the egg cell; e.g. the inheritance of mitochondrial genes is a typical example of maternal inheritance. *Cf.* Mendelian inheritance.

mating, pairing for breeding; the act of pairing, especially sexually; **1.** in genetics, denotes the paired union of eukaryotic unisexual individuals representing opposite sexes. Its aim is sexual reproduction which leads to the formation of the → zygote. In prokaryotic organisms the surrogate systems of sexual reproduction mechanisms resemble mating, like bacterial → conjugation; **2.** in ethology, mating means the formation of a couple for the purpose of reproduction; *see* mating system; *cf.* copulation.

mating structure, social behaviour between sexes which in different ways leads to → assortative mating. The mating structure consists of several alternative ways of behaviour according to how many individuals of the opposite sex an animal mates with, and how durable the pairing is. The mating structure may include → monogamy, polygamy, promiscuity, or dominance mating. *Cf.* mating system.

mating system, the process by which the reproductive cells, → gametes, of a species unite at the population level. Mating strategies in a population include e.g. the number of coexistent mating partners, the stability of the bond of a couple, and the degree of inbreeding (→ monogamy, polygamy, polyandry, polygyny). Three different main types of mating systems are distinguished among organisms reproducing sexually:

1) in **random mating** (panmixis) each individual has equal chances to mate with any

individual of the opposite sex; 2) in **genotypic assortative mating**, mating is dependent on the degree of kinship between the male and female. If the mating individuals are more closely related than individuals in the population on average, **positive genetic assortative mating** occurs, but if less, **negative genetic assortative mating** is in question; 3) in **phenotypic assortative mating**, mating is dependent on the phenotypic resemblance of the male and the female. If individuals of the same appearance tend to mate, it is **positive phenotypic assortative mating**, but if individuals of different shape tend to mate, it is **negative phenotypic assortative mating**.

mating type, a mating characteristic primarily of unicellular organisms; is genotypically regulated and often controlled by only one or a few allele pairs, which determine the mating capacity of an organism. Individuals of the same mating type cannot mate but the mating individuals must represent opposite mating types; e.g. in many fungi two opposite mating types, α and ***a***, can be distinguished. They cannot, however, be regarded as sexes since reciprocal → karyogamy occurs between the mating cells. *See* mating, mating system.

matrix, pl. **matrices** (L. *matr-* < *mater* = mother), the ground material of a cell organelle, cell, or tissue; e.g. **1. cell matrix** (cytoplasmic matrix, cytomatrix), the fluid, non-structural base material of the cytoplasm between the organelles and the cytoskeleton of a cell; contains salts (chiefly K^+) and many organic substances, such as soluble enzymes; **2. nuclear matrix**, i.e. the fluid base substance of the cell nucleus consisting chiefly of proteins from which chromosomes emerge at a certain stage of the nuclear cycle; **3. mitochondrial matrix**, a fine, granular proteinaceous material enclosed by the inner membrane of a mitochondrion; contains e.g. ribosomes, filaments of DNA, enzymes, and many enzyme substrates; **4. tissue matrix** of an animal tissue, such as the matrix of connective tissue (e.g. cartilage matrix, bone matrix), formed from fibres embedded in an amorphous ground substance between cells; **5.** hair matrix, the root of the hair follicle; **6.** nail matrix (*matrix unguis*), nail bed, the tissue upon which the nail rests; also demotes the proximal portion of the nail bed from which the nail grows; **7. wall matrix** of plant cells, the base material of plant cell walls consisting of pectin and hemicellulose; wall matrix is abundant in the primary walls with a few cellulose fibres, the fibrillar material which almost solely forms the → secondary wall.

matrix potential (L. *potentia* = power), *see* osmosis.

matroclinous, matroclinal, also **matriclinous, matriclinal** (L. *mater* = mother, Gr. *klinein* = to lean), derived or inherited from the mother or maternal line; refers to inheritance in which the progenies resemble the mother more than the father. *Cf.* patroclinous.

maturation (L. *maturus* = ripe), the process of becoming fully developed, as the maturation of a germ cell, organ, or organism (often sexual maturation); **maturity**, the period of achieving maturation. *Adj.* **mature**. *Verb* to **mature**.

maturation division, *see* meiosis.

maturation-promoting factor, MPF, → mitosis-promoting factor.

maxilla, pl. **maxillae,** Am. also **maxillas** (L. dim of *mala* = jaw, cheek), **1.** one of a pair of bones close behind the premaxilla in the upper jaw of most vertebrates; **2.** the upper jaw of mammals in which several bones found in lower vertebrates are fused; **3.** one of two pairs of appendages in the mouth of arthropods, located behind the mandibles. The first pair of maxillae of crustaceans is called **maxillule**. The structures of maxillae vary greatly depending on their functions: e.g. the sponging mouthparts of insects contain the sponge-like labium, but the maxillae and mandibles are reduced. In chewing and sucking mouthparts, such as in bees, the maxillae are elongated together with the labium to form a long sucking tube; the long stylet of hemipterans and homopterans is partly composed of the maxillae. *Adj.* **maxillary**. *See* mouthparts.

maxillary gland, *see* green gland.

maxillary palp (L. *palpus maxillaris*), a pair of segmented palps attached to the → maxillae of insects. *See* mouthparts.

maxillipeds (L. *pes*, gen. *pedis* = foot), usually one, two, or three first pairs of thoracic appendages immediately behind the → maxillae in some arthropods, e.g. crustaceans and uniramians; act in food handling, e.g. in filtrating food particles by crustaceans. In chilopods, the maxillipeds form the poison claws capable of injecting toxic fluid.

Maxillopoda (Gr. *pous* = foot), in some classifications, a crustacean class comprising several subclasses, such as **mussel shrimps** (Ostracoda), **mystacocarids** (Mystacocarida), **copepods** (Copepoda), **fish lice** (Branchiura), **barnacles** (Cirripedia), which are commonly regarded as classes.

maxillule (L. *maxillula*, dim. of *maxilla* = jaw), *see* maxilla.

maximum likelihood estimation, MLE, a method in human genetics by which the probability and intensity of → linkage of genes is calculated based on pedigree analyses (*see* pedigree); as the result of these calculations a → lod score is obtained.

maximum velocity, V_{max}, *see* Michaelis—Menten kinetics.

MCPA, 2-methyl-4-chlorophenoxyacetic acid; a widely used hormone-like selective herbicide with effects resembling those of → auxins (IAA and 2,4-D).

meadow, a type of vegetation on open, fresh, or moist areas; is treeless but rich in grasses. Meadows mostly develop from the destruction of the original forests by human activities and the prevention of the regeneration of trees by regular mowing or grazing. Original meadows are found e.g. in the mountains or fells above the tree line, on exposed coasts, and on marshlands and saltmarshes.

meatus, pl. **meatus** (L. = a passage), a channel or passage in an anatomical structure, especially the external opening of a channel or canal, e.g. external auditory meatus, the passage of the external ear.

mechanics (Gr. *mechane* = machine), a branch of physical science dealing with energy, motion, and actions of forces on solid, liquid or gaseous bodies; consists of kinematics, kinetics, dynamics, and statics. The mechanics of biological systems is often called **biomechanics**, in animals **body mechanics**.

mechanoreceptors, sensory receptors which specifically (adequately) respond to mechanical stimuli, e.g. auditory and equilibrium receptors in the → ear, kinaesthetic receptors in muscles and tendons (*see* muscle spindle, Golgi tendon organ), and tactile receptors in the skin (*see* tactile hairs, Meissner's corpuscle, Pacinian corpuscle).

Meckel's cartilage (*Johann Meckel*, 1781—1833), **mandibular cartilage** (tympanomandibular cartilage); the cartilaginous bar in the mandibular arch of the vertebrate embryo, supporting the lower jaw; forms the **lower jaw** in adult elasmobranchs but in other vertebrates with bony skeleton it is largely replaced by a series of dermal bones. In most of these vertebrates, the cartilage develops into a small bony structure, **articular bone**, bearing an articular surface for the quadrate bone of the skull. In mammals develops into some → branchiogenous organs such as malleus and incus.

meconium (Gr. *mekonion,* dim. of *mekon* = poppy), **1.** the content of the foetal intestine of mammals, consisting of mucus, bile and other glandular products, epithelial cells, and material from swallowing of amniotic fluid; **2.** poppy juice, opium.

mecoprop, 2-(2-methyl-4-chlorophenoxy)propionic acid; a hormone herbicide resembling 2,4-D and MCPA.

medi-, medio- (L. *medius* = middle), denoting middle or median.

media, 1. plural of medium; **2.** the middle part of a stratified structure, e.g. *tunica media* in the wall of a blood or lymph vessel; **3.** the median vein of an insect's wing.

medial (L. *medialis*), **1.** situated in the middle, as an anatomical structure near the midline or median plane; **2. medial lemniscus**, a bundle of nerve fibres in the vertebrate central nervous system transmitting proprioceptive sensory impulses from the spinal cord to the thalamus; **3.** → median.

median (L. *medianus*), medial; e.g. **1.** being in the middle; situating in the intermediate position; **2.** pertaining to a plane dividing a structure into two equal parts, as an animal into the right and left halves; **3.** the middle value such as a mean in a set of measurements.

median lethal dose, → LD_{50}.

median lethal temperature, → LT_{50}.

mediastinum, pl. **mediastina** (L. *medius* = middle), a median part, especially a septum between certain structures; e.g. the space in the body of higher vertebrates separating the right and left thoracic cavities, containing the heart with its pericardium, bases of the great vessels, trachea, main bronchi, oesophagus, thymus, and lymph nodes.

medical genetics, the branch of → human genetics dealing with relationship between heredity and human diseases; the goal is the diagnosis, treatment, and prevention of hereditary diseases.

medicinal plant, officinal plant; any plant

containing pharmacologically effective substances which can also be used as raw materials for drugs. Medicinal plants are most common in the group of seed plants, and as such they are often poisonous for most animals. The medical compounds obtained from plants are usually grouped in the following way: 1) → alkaloids, as e.g. opioid alkaloids obtained from opium poppy; 2) → glycosides, such as digoxin obtained from *Digitalis lanata* (foxglove); 3) → antibiotics, obtained from many bacteria and fungi (e.g. penicillin); 4) mucilage materials, as e.g. flaxseed oil (laxative effect); 5) evaporating oils, such as anise oil; 6) raw materials of semisynthetic medicines, such as derivatives of morphine and atropine; 7) additive materials used in the production of medicines; e.g. starch used in tablets.

medico- (L. *medicus* = physician), denoting relationship to medicine; medical.

medicobiologic(al), pertaining to biological aspects of medicine.

medium, pl. **media,** Am. also **mediums,** (L. *medius* = middle), **1.** a middle state; **2.** a means by which an action is performed; **3.** a condition or environment which is natural and optimal for an organism to grow and flourish; **4.** a mixture of substances (fluid or solid) used for the isolation, cultivation, or storage of living cells, tissue, or organs; **5.** an intervening matter transmitting something, such as air.

medulla, pl. **medullae** (L.), the innermost portion of a structure, as e.g. of an organ; marrow; **1.** in anatomy of animals, e.g. the → *medulla oblongata*, i.e. myelencephalon, *medulla spinalis*, the → spinal cord, *medulla renis*, the medulla of the → kidney, or *medulla ossium*, the → bone marrow; **2.** in botany, the central part of a plant stem in which the vascular tissue forms a cylinder; may also be found in roots if the xylem does not fill the central part of the root but there is parenchyma tissue instead; the innermost region of the thallus in lichens. *Adj.* **medullary**.

medulla oblongata, the most caudal part of the vertebrate brain between the pons and the spinal cord, forming the main part of the brain stem and derived from the posterior part of the embryonic rhombencephalon; contains the **fourth brain ventricle** enclosed dorsally by the cerebellum. The ventrolateral superficial area of the medulla oblongata comprises **white matter**, which contains mainly myelinated ascending sensory and descending motor tracts of nerve fibres. The inner part consists of **grey matter**, which contains many brain centres, **nuclei**, with numerous nerve cell bodies which control important vital functions, such as blood circulation and respiration; there are also relay nuclei for synaptic transmission of sensory and motor tracts. Most → **cranial nerves** have their roots in the medulla oblongata. *Syn.* medulla, bulb, bulbus, or myelencephalon (used especially in comparative vertebrate anatomy).

medullary plate, → neural plate.

medullary ray, a type of → ray in plants, formed from parenchyma tissue transporting materials radially in the plant; the medullary ray is a special type of ray with connection to the medulla (pith), found especially in the primary structure of the stem before secondary growth.

medullary tube, → neural tube.

medullated nerve fibre, myelinated nerve fibre. *See* nerve fibre.

medusa, pl. **medusae** (Gr. *Medusa* = in Greek mythology one of the three Gorgons), **medusome**, **medusoid stage**; a free-swimming, sexual stage in the life cycle of most cnidarians (the asexual stage is → polyp). The medusa resembles an umbrella or a bell and is characterized by a water-filled → gastrovascular system. The mouth with four long tentacles is located on the lower surface. The medusoid stage is a dominant reproductive generation in the life cycle of scyphozoans, but is small and less visible or lacking in hydrozoans, and lacking in all anthozoans.

meg(a)- (Gr. *megas* = big), **1.** denoting big, great, large; of abnormally large size; sometimes a synonym for macro-; **2.** in connection with measurement units, indicates one million (10^6), symbol M, e.g. MV (megavolt) one million volts, or MPa (megapascal) one million pascals.

megaevolution, → evolution of higher taxonomic categories, like genera, families, orders, classes, and phyla. *Cf.* macro- and microevolution.

megafauna (Gr. *Fauna*, sister of *Faunus* = mythical deity of herdsmen), **1.** the term used chiefly in palaeontology, including large, extinct mammals and birds from Pleistocene, as e.g. the mammoth, cave bear, sabre-toothed tiger, and Irish elk; the extinction of the megafauna is partly caused by man hunting them; **2.** especially in hydrobiology, large animals visible to the naked eye, such as medusae, molluscs, crabs, and vertebrates.

Megagaea, also **Megagea** (Gr. *ge* = earth), → Arctogaea.

megagamete (Gr. *gamos* = marriage), the bigger of the two gametes fusing in → anisogamy. *Syn.* macrogamete.

megagametophyte (Gr. *phyton* = plant), the female → gametophyte in heterosporous plants; in seed plants, it is located in the ovule (in the ovary) and developed from the macrospore (megaspore).

megakaryocyte (Gr. *karyon* = nut, nucleus), a cell type found in the bone marrow; an exceptionally large cell with a lobulated nucleus. Blood platelets are formed as cytoplasmic fragments from this giant cell. The immature developmental stages are **megakaryoblast** and **promegakaryocyte**.

megalecithal (Gr. *lekithos* = yolk), pertaining to a vertebrate ovum which contains a large yolk so that the actual cytoplasm including the nucleus is located as germinal plasm on the animal pole of the egg; megalecithal eggs are found e.g. in birds, and reptiles, and in many fish species. *Cf.* mesolesithal, oligolesithal.

megalocyte (Gr. *megal-* < *mega* = large, *kytos* = cavity, cell), an exceptionally large (12—25 μm in diameter) red blood cell in the blood; its immature stage is called **megaloblast**.

megalops (Gr. *ops* = eye), a postlarval stage of malacostracans, e.g. crabs; the animals have stalked eyes, a crab-like cephalothorax and unflexed abdomen with one or more pairs of → pleopods.

meganucleus, → macronucleus.

megaphyll (Gr. *phyllon* = leaf), a leaf representing the normal type, present in most vascular plants; is relatively large with leaf veins and associates with the stele in the stem. *Syn.* macrophyll. *Cf.* microphyll.

megaplankton (Gr. *planktos* = wandering), the largest organisms in the oceanic → plankton, the individuals varying from 0.2 to 2 m in size; e.g. jellyfishes. *Cf.* macroplankton, microplankton.

megasporangium (Gr. *sporos* = seed), → macrosporangium.

megaspore, → macrospore.

megaspore mother cell, → macrospore mother cell.

megasporocyte, → macrospore mother cell.

megasporogenesis, → macrosporogenesis.

megasporophyll, → macrosporophyll.

Mehlis' gland, a collective name for unicellular gland cells surrounding the ootype of trematodes; takes part in the formation of eggshells.

meio-, mio-, mi- (Gr. *meion* = less), less, smaller, fewer, slightly.

meiobenthos (Gr. *benthos* = depth), bottom organisms varying in size between 0.1 and 1 mm; the term is commonly used as a synonym of meiofauna which, however, denotes animals only. *Cf.* macrobenthos, microbenthos.

meiocyte (Gr. *meiosis* = reduction, diminution, *kytos* = cavity, cell), any cell whose nucleus is undergoing → meiosis and gives rise to gametes or sexual spores. In animals, meiocytes include the primary oocytes and the primary spermatocytes (*see* oogenesis, spermatogenesis), in higher plants, they include the macrosporocytes (megasporocytes) and microsporocytes (*see* macrosporogenesis, microsporogenesis).

meiofauna (Gr. *meion* = less, L. *Fauna*, sister of *Faunus* = mythical deity of herdsmen), animals including middle-sized aquatic invertebrates which live on the bottom and remain on a sieve with 0,1 mm mesh; sizes of individuals range between 0,1 and 1,0 mm. *Syn.* mesofauna. *Cf.* macrofauna, microfauna.

meiosis, pl. **meioses** (Gr. = reduction, diminution), **reduction division**; the process of cell division which is involved in sexual reproduction. The results of meiosis are the reduction of the chromosome number from → diploid to → haploid, and → genetic recombination. Meiosis includes two nuclear divisions, the **first meiotic division** and **second meiotic division**, but only one → replication (duplication) cycle of the chromosomes. As a result of the meiotic divisions, four haploid nuclei are always produced.

In animals, meiosis is associated with the formation of → gametes, and in plants and fungi with the production of sexual spores (*see* sporogenesis). Meiosis may begin immediately after fertilization in which case the diploid phase is limited to a single cell; this kind of **zygotic meiosis (initial meiosis)** exists in lower plants and fungi. In the second type, meiosis immediately precedes fertilization; this type of **gametic meiosis (terminal**

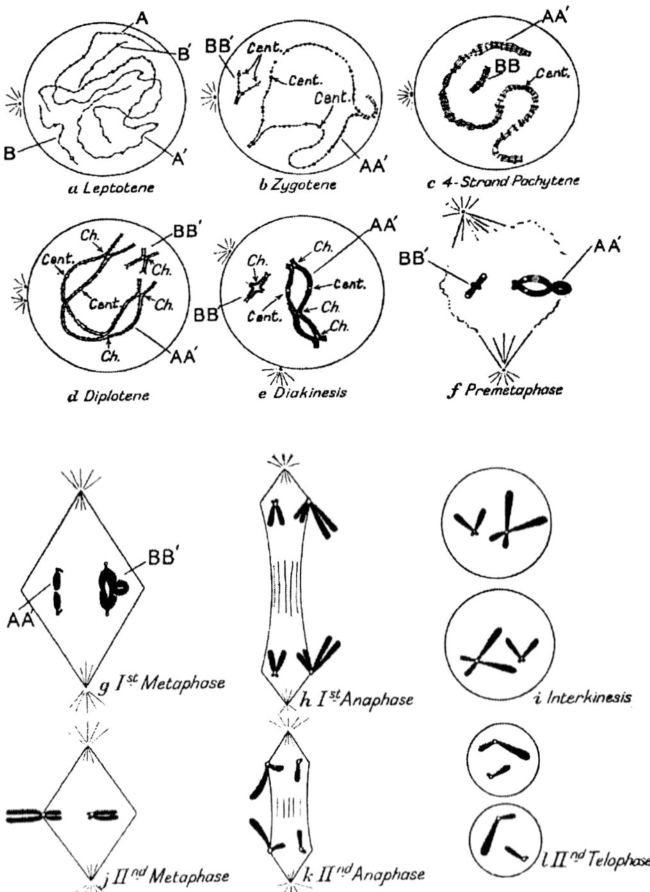

Meiosis. A schematic representation of the stages of meiosis. (Cent. = → centromere; Ch. = → chiasma). (After White: The Chromosomes - Methuen, 1961; with the permission of the publisher).

meiosis) exists in animals. The third type is **indeterminate meiosis (sporic meiosis)** which represents the intermediate form of the two previous ones and exists in higher plants. In this last type mitotic cell generations normally occur between fertilization and meiosis, and between meiosis and the next fertilization. Meiosis always has several main stages:

1) The **first meiotic division** is a modified mitosis. One of its major features is → **conjugation**, i.e. the pairing of homologous chromosomes, which is followed by the segregation of bivalents (bivalent chromosomes) formed in the pairing. This further includes several stages, i.e. prophase, metaphase, and anaphase.

The **prophase of the first meiotic division** lasts for a long time as compared to the mitotic prophase, taking usually days or even years. The substages of prophase are leptotene, zygotene, pachytene, diplotene, diakinesis, and prometaphase.

At the start of **leptotene** the chromosomes become visible as single strands, even though the largest part of the chromosomal material has been replicated during the preceding interphase, but some replicons (late replicons) are not, being responsible for the formation of the chromosome axis; thus the chromosomes are effectively single-stranded in the beginning of leptotene.

In **zygotene** the homologous chromosomes take an abreast position, starting usually at the telomeres, and a → synaptonemal complex will be formed between the homologous chromosomes. The chromosomes conjugate, i.e. the homologous chromosomes pair, and immediately after conjugation the late replicons are duplicated.

In the beginning of **pachytene**, conjugation has ended. The **bivalents**, i.e. the two-stranded pairs of homologous chromosomes, lie abreast and are observable under a microscope. Because both chromosomes of the bivalent consist of two sister chromatids, the bivalents are actually four-stranded tetrads. At the same time → **chiasmata** are formed so that either of the two chromatids in both homologous chromosomes break, and the broken ends join in a new way forming a chiasma. Any chromatid of both chromosomes may take part in a chiasma. The chiasma provides within the chromosome the recombination of paternal and maternal genetic material. Chiasma is the cytological counterpart of → **crossing-over**.

Diplotene begins when the synaptonemal complex disappears, and thus conjugation is dissolved. The bivalents are, however, held together due to chiasmata. When the condensation of the chromosomes proceeds, the bivalents become clearly visible as typical structures in which the chiasmata are visible as crosslike formations.

During **diakinesis** the chromosomes condensate further and reach a complete condensation at the end of the stage. Bivalents are evenly distributed within the nucleus.

During **prometaphase** the → mitotic spindle is formed and the nuclear envelope disappears. Microtubules are bound to the undivided centromeres of the chromosomes, and pull the bivalents into the equatorial plate.

2) The second main stage is **metaphase of the first meiotic division**; there the bivalents are located on the equatorial plate in such a way that the centromeres lie on the opposite sides of the equatorial plate, i.e. the centromeres are **co-oriented**. Every bivalent behaves in this respect independent of the others, and consequently the free combination of the genes occurs. The chiasmata lie on the equatorial plate.

3) In the third main stage, **anaphase of the first meiotic division**, the homologous centromeres move, as drawn by the spindle, into opposite poles, i.e. the segregation of the bivalents occurs. The separating chromosomes are called → **dyads** to emphasize that they are different in their genetic constitution from the pairing chromosomes in the beginning of meiosis (*cf.* anaphase of mitosis).

Between the first and second meiotic divisions, there is a short interval, **interkinesis**, which resembles interphase but is not connected with chromosome replication.

Second meiotic division follows immediately after the first one. There the dyads divide in a usual mitotic way (*see* mitosis). Both of the nuclei formed in the first meiotic division divide, and thus the final result of meiosis is four haploid nuclei in which the paternal and maternal genes exist in new combinations.

Several modifications of meiosis are known; e.g. **achiasmatic meiosis** in which no chiasmata are formed.

meiosporangium (Gr. *sporos* = seed, *angeion* = vessel), a → sporangium in which → spores are produced after meiosis.

meiospore, a → spore which results from → meiosis.

meiotic, pertaining to → meiosis.

meiotic drive, a genetically caused disturbance of → meiosis leading to offspring ratios which deviate from those expected according to → Mendel's laws. In meiotic drive the segregation of chromosomes is non-random, one member of a particular → bivalent being inherited more frequently than the other.

Meissner's corpuscle (*Georg Meissner*, 1829—1905), touch corpuscle (*corpusculum tactus*); a type of tactile receptor found in the dermal papillae of the vertebrate skin; consists of a sensory nerve end surrounded by the capsule of connective tissue with epithelioid cells. The corpuscles respond to tactile stimuli; in man, most numerous in the lips, fingers and toes.

Meissner's plexus (L. *plexus* = braid), submucous plexus; a network of thin autonomic nerve fibres in the submucosal layer of the alimentary canal in vertebrates; controls the secretion of exocrine and endocrine cells in the gastrointestinal tract.

mel-, 1. (Gr. *melas* = black), denoting black, dark, or melanin; also **mela-**, **melo-**, **melano-**;

2. (L. *mel* = honey), pertaining to honey, or sweet; also **mell(i)-**; **3.** (Gr. *melos* = limb), pertaining to extremities or limbs; **4.** (Gr. *mela* = cheeks, lit. apples, pl. of *melon* = apple), pertaining to cheeks or apples.

melancholia, melancholy (Gr. *melas* = black, *chole* = bile), originally meant a lowered temperament believed to be due to an excess of black bile in the human body; now means a gloomy state of mind or a mental depression.

melanin, a pigment found in animal tissues, synthesized by pigment cells derived embryonically from nervous tissue. The biosynthesis of melanin derives from tyrosine in oxidation and polymerization reactions. The molecular structures of different melanin types vary slightly, but can be categorized into two groups: **eumelanins**, varying from black to brown being polymers of 5,6-dihydroxyindoles, and **phaeomelanins** which contain more sulphur, the colour varying from red to yellow. Melanin occurs especially in the skin, hair, and feathers, in the pigment cells of the eye, and in the nervous tissue (especially in *substantia nigra* of the vertebrate brain). *See* melanocyte, melanophore.

melanism, excessive pigmentation of an animal, generally of genetic origin. *See* industrial melanism. *Cf.* melanosis.

melanocyte (Gr. *kytos* = cavity, cell), a type of pigment cell found particularly in tissues of birds and mammals, chiefly in the skin; a dendritic cell which produces → melanin pigment synthesized and packed in small granules, **melanosomes**. In the epidermis melanosomes are transferred to → keratinocytes, colouring the skin, hairs, and feathers. The immature or active stage of a melanocyte is also called **melanoblast**. *See* chromatophores.

melanocyte-stimulating hormone, *see* melanophore-stimulating hormone.

melanoma, a cancer derived from → melanocytes.

melanophore (Gr. *phorein* = to carry along), a → chromatophore (pigment cell) containing → melanin.

melanophore-stimulating hormone, α-MSH, a peptide hormone released from the intermediate lobe (*pars intermedia*) of the → pituitary gland of fish, amphibians, and reptiles. It stimulates the dispersion of pigment granules in → melanophores and enhances melanin synthesis in these cells. The corresponding hormone in birds and mammals, released chiefly from the anterior lobe of the pituitary gland, is called **melanocyte-stimulating hormone** (β-MSH, melanotropin), that stimulates the synthesis of the melanin pigment in → melanocytes, and has some effects also on the nervous system; its physiologic function in man is not clear. α-MSH is made up of 1—13 amino acid residues of corticotropin (ACTH), and β-MSH of 18 amino acid residues at the C-terminal end of the γ-lipotropin (γ-LPH). Consequently, ACTH has considerable MSH activity. The third type of MSH, called γ-MSH, is found in the intermediate lobe of the pituitary gland in mammals (e.g. in man). It stimulates melanocytes and has a slight corticotropic effect on the adrenal cortex. Collectively MSHs are called **intermedin**.

melanosis, an abnormally strong pigmentation of tissues and organs of animals due to a changed or disturbed metabolism of → melanin or other related pigments; may occur as a result of → melanism, various metabolic diseases, infections, neoplasms (e.g. melanoma), and in humans during pregnancy.

melanosome, *see* melanocyte.

melatonin, N-acetyl-5-methoxytryptamine; a vertebrate hormone whose biosynthesis derives via serotonin (5-hydroxytryptamine) from tryptophan by N-acetylation and O-methylation. It is secreted from the pineal body chiefly in the dark, i.e. during night, some melatonin also in peripheral nerves. Its function is associated with maintaining different types of → biorhythms in the body, linked to light rhythmicity in the environment. Melatonin has an effect on the brain activity inducing sleep, and it inhibits the activity of the ovaries and testes. It also affects the moulting period of animals, and so melatonin injections given to fur animals in the late summer induce earlier growth of the winter fur. Melatonin may also have a role as an antioxidant, and at least in amphibians, it lightens the skin colour. During the winter the secretion of melatonin is exceptionally high in animals living in polar areas, such as the reindeer. The seasonal depression syndrome found in humans living in high latitudes, is probably involved in disturbed melatonin secretion and daily rhythmicity during the winter.

melibiose (L. *mel* = honey), a galactopyranosyl-glucopyranose disaccharide, $C_{12}H_{22}O_{11}$;

obtained from raffinose by fermentation and used e.g. in culture media for plant tissues.

meliphage (Gr. *phagein* = to eat), an animal which feeds on honey. *Adj.* **meliphagous.**

melitose, melitriose, → raffinose.

MEM, → minimum essential medium.

membrane (L. *membrana*), **1.** a **thin tissue sheet** in the anatomical structure of animals, composed usually of epithelium and interstitial connective tissues; covers surfaces of inner organs, or lines body cavities; **2.** → **cell membrane** (plasma membrane, plasmalemma), surrounding the cell; **3.** an **intracellular membrane**, such as the membrane of → endoplasmic reticulum, → nuclear membrane, or membranes of some other → cell organelles. Cellular membranes are not static barriers but are effectors of → membrane traffic and other active functions, such as enzyme activity and ion and molecule transport. *See* cell membrane transport systems. *Cf.* nuclear envelope, tonoplast, unit membrane.

membrane bone, → dermal bone.

membranelle (L. *membranella*, dim. of *membrana* = thin sheet, membrane), an undulating, fan-shaped membranous organelle of some ciliate protozoans or → tornaria larvae; is located usually in the peristomial area of the cell body, and is formed by fusion of the bases of certain cilia. The membranelle serves in gathering food towards the cytostome, and is also used in locomotor activity of the animal.

membrane potential, transmembrane potential, i.e. the electric potential difference across the cell membrane; the inside of the cell is negative with respect to the outside during resting conditions (*see* resting potential), changing to positive (is depolarized) when the cell is electrically activated (*see* action potential, receptor potential). Often the membrane potential refers only to resting potential.

membrane traffic, the movement of cellular membrane material from one site to another; occurs principally in two different ways: in the **membrane cycle** (membrane recycling) which means a movement of membrane material from intracellular membranes into the cell membrane and vice versa, occurring in → exocytosis and endocytosis. The membrane cycle may also include movements of this material between the intracellular membranous organelles, such as occurs between the → endoplasmic reticulum and → Golgi complex. In the **lateral membrane traffic** the movement occurs along the cell membrane from one site to another, as occurs in → capping. *See* flip-flop.

membranous labyrinth (L. *labyrinthus membranaceus*), the membranous system of communicating sacs and ducts inside the → osseous labyrinth of the inner ear of vertebrates; includes the → utricle, saccule, semicircular canals, endolymphatic duct, and cochlear duct. The fluid inside it is called **endolymph**; outside, between the membranous labyrinth and osseous labyrinth, it is called perilymph. The membranous labyrinth contains the mechanoreceptors for equilibrium and auditory senses.

meme (*mem*ory and gen*e*), the unit of selection in cultural evolution, i.e. the "gene" of cultural evolution, proposed by *Richard Dawkins* (b. 1941). The meme is comparable to the gene concept of biological evolution. In cultural evolution the memes compete like genes in the population. When memes are copied, they usually change, i.e. "mutation" occurs. A meme can e.g. be a part of a religious ritual which occurs in different forms in different religions, or a tune which occurs in different music in various but recognizable forms. *Syn.* cultural gene.

memory, the property of storage of information in biological systems; **1. genetic memory**, includes the information stored in genes from generation to generation; *see* gene; **2. neuronal memory**, includes the information stored in the nervous system existing in all multicellular animals except sponges. This memory store provides a synthesis of new proteins and structures in the neuronal junctions, → synapses. In humans, three stages are involved in the recall of events: the **immediate memory** lasts only for a few seconds without consolidation of the memory trace, the **temporary memory** lasts for several hours and consolidation occurs in the synapses; the **long-lasting memory** remains for years. The last two memory types are based on synthesis of the synaptic material, which improves the synaptic transmission in certain neuronal circuits. Neuronal memory is the basis for → learning; **3. immunological memory**, appears as an improved response of the immune system to microbes and other antigens during the second, third, etc. infections; it is due to the maintenance of certain lymphocytes, i.e. the memory cells (memory B cells and memory T cells) in the body; *see* immunity.

Mendelian character (*Gregor Mendel*, 1822—1884), a separate character formed by the regulation of genes showing → Mendelian inheritance, i.e. its occurrence in subsequent generations follows → Mendel's laws of inheritance.

Mendelian inheritance, the mechanism of the inheritance of chromosomal genes which follows Mendel's laws. *Syn.* Mendelian heredity, Mendelism. *See* Mendel's laws of inheritance.

Mendelian population, a group of individuals which are in more or less close reproductive relationship and which share a common → gene pool. Mendelian populations can be hierarchically organized from the smallest to the largest as follows: → panmictic unit, deme, race, and species.

Mendelism, the inheritance of chromosomal genes according to the chromosome theory of inheritance, and the theory concerned. *Verb* to **mendelize**. *See* Mendel's laws of inheritance.

Mendel's laws (of inheritance), the rules of inheritance discovered by *Gregor Mendel* which constitute the basis of genetics. According to **Mendel's first law** (segregation rule) the allelic genes (→ allele) segregate as pure entities from each other during the formation of gametes so that each gamete will consist of only one copy of each gene. **Mendel's second law** (rule of independent assortment) presents that when segregation occurs, each pair of genes behaves independently of the others. Mendel's first law is universal in sexually reproducing organisms. However, the → linkage of genes constitutes a regular exception of Mendel's second law, and the law is true only for genes which are located on different chromosomes or on the same chromosome so far apart that in practice a → chiasma always occurs between them.

mening(i)-, meningo- (Gr. *meninx* = membrane), denoting meninges. *See* meninx.

meninx, pl. **meninges** (Gr.), a membrane, especially one of the three membranes, lying on top of each other enveloping the brain and spinal cord of vertebrates: **dura mater**, the hardest and thickest and located outermost, then a thinner membrane, **arachnoid**, and tightly against the nervous tissue the softest membrane, **pia mater**. Between the last two membranes lies a narrow cavity, the **subarachnoid space**, which is supported by arachnoid trabeculae and filled with → cerebrospinal fluid. The dura forms the → falx. *Adj.* **menigeal.**

meniscus, pl. **menisci** (Gr. *meniskos* = crescent), a crescent-shaped anatomical structure; e.g. **articular meniscus** (articular crescent), a crescent-shaped intra-articular cartilage found in some joints of vertebrates.

menopause (Gr. *men* = month, *pausis* = cessation), a period of time when → menstruation ceases in aged females of humans and great apes; occurs in humans usually around the age of 50. *See* climacteric.

menotaxis (Gr. *menein* = to remain, *taxis* = arrangement), a tactic movement in which an animal does not orientate straight to or away from a stimulus but the direction of the movement is at an angle to the stimulus. *Cf.* compass orientation. *See* taxis.

menstrual cycle (Gr. *men* = month), the period of the regularly recurring changes in the ovaries and uterus in humans, apes, and Old World monkeys; corresponds to the → oestrous cycle of other mammals. It includes the maturation of the ovum in the → ovarian follicle, ovulation, and physiologic changes in the mucous membrane (endometrium) of the uterus, i.e. the proliferation of the endometrial tissue for implantation of an early embryo and partial shedding with some bleeding (**menstruation**) of the non-pregnant uterus. In humans, menstruation occurs approximately at four week intervals. The menstrual cycle is regulated hormonally by female sex hormones (→ oestrogens, progesterone) secreted from the ovaries under the control of → gonadotropins from the anterior pituitary gland and further by the cyclic function of the hypothalamus. Called also **menses** (L. *mensis* = month). *Adj.* **menstrual** (menstruous). *Verb* to **menstruate**.

mental, 1. (L. *mens* = mind), pertaining to the mind; psychic; **2.** (L. *mentum* = chin), pertaining to the chin or the → mentum; genial, genian.

menthol (L. *Mentha* = mint), $C_{10}H_{19}OH$, a terpene alcohol, a compound of the camphor group, occurs in natural plant oils; used e.g. in sweets and pharmaceutical preparations.

mentum (L. = chin), **1.** one of the basal joints in the labium of insect → mouthparts; is a distal part of the postmentum, between the submentum and prementum; **2.** the medial part in the → gnathochilarium of diplopods.

mercaptans (L. *corpus mercurio aptum* = a substance linked to quicksilver), thiols; organic sulphuric compounds with a strong unpleasant odour, formed from alcohols and phenols in which -OH groups are replaced by -SH groups, and sulphur is directly linked to carbon. Mercaptans, such as methyl mercaptan (methanethiol), are produced in protein hydrolysis by bacteria in the intestine, faeces and carcases; they are also produced in manufacturing sulphate cellulose. Mercaptans are used e.g. as solvents and in drug industry.

2-mercaptoethanol, β-mercaptoethanol, thioglycol, C_2H_6OS; a strongly reducing, sulphur-containing liquid with an unpleasant odour; used e.g. in homogenizing media for tissues to protect sulphydryl groups from oxidation. It inhibits cell division.

mercury (L. *Mercurius* = ancient Roman god, Mercury), **quicksilver,** L. *hydrargyrum*; a metallic element, symbol Hg, atomic weight 200.59; is liquid at room temperature and very poisonous to all organisms. Mercury occurs in compounds as a univalent mercurous (e.g. Hg_2Cl_2, mercurous chloride, calomel) or bivalent mercuric form (e.g. $HgCl_2$, mercury bichloride). Common mercury compounds are also mercury(II) sulphide (HgS) that occurs naturally as cinnabar, and mercury oxide (HgO).

Mercury has become an environmental problem as a waste product from industry and combustion gases of coal power plants, and large quantities of mercury have earlier been released from seed dressings used in agriculture. Different treatments of soil, such as digging and forest ploughing, liberate mercury from its natural mineral sources and acidification of soil by acid rains increases its release into water bodies. Mercury occurs in nature as inorganic or organic compounds, and especially organic methyl mercury, C_2H_6Hg, causes chronic toxicity in animals because it accumulates in → food chains.

Mercury has cytotoxic effects in cells, binding to -SH groups of principal organic compounds; consequently it disturbs cellular processes, especially reproduction processes in animals and other organisms. Mercury and its chemical compounds are used in barometers, manometers, thermometers, lamps, and as tooth fillings.

mericarp (Gr. *meris* = part, *karpos* = fruit), a one-seeded part of the → schizocarp (compound fruit) that breaks off at the maturity of the fruit; occurs e.g. in the family Apiaceae.

meridian (L. *meridies* = midday), a geographical circle on the Earth running through the poles, or a half of such a circle between the poles; also used as an adjective.

meridional, 1. pertaining to a meridian; **2.** from pole to pole of a structure; **3.** following a north-south direction, southern, southernly.

meridional ridge, one of the eight ridge-like structures on the surface of ctenophores, associated with transverse comb-like plates (**comb plates**) formed by fused cilia. The animal uses rhythmical movements of the ridges in its locomotor activity. *Syn.* comb row.

meristele, leaf trace; a branch of a → stele (an individual vascular bundle in dictyostele) supplying a leaf.

meristem (Gr. *meristos* = divided < *merizein* = to divide), a plant tissue the cells of which are capable of dividing and thus producing new cell material for growth and morphogenesis of a plant. Meristem tissues can be divided into **primary meristems**, which develop directly from embryonic cells, and **secondary tissues**, which are formed secondarily from non-dividing tissues. The primary meristems include the → **apical meristems, resting meristems** and **lateral meristems**. The apical meristems occur especially in the tip areas of shoots and roots, comprising **promeristem** and → **derivative meristems**. Resting meristems consist of meristematic cells which have been in a resting state within other tissues and form later → **intercalary meristems**. These are situated at the base of plant → nodes, forming the intercalary growth zone, supporting growth in node areas in some plants, e.g. in families Poaceae and Apiaceae. Those parts of primary meristems which produce ground tissues, i.e. different → parenchyma tissues, are called **ground meristems**. The secondary meristems include the → **cambium** and **cork cambium** (phellogen) in plant roots and shoots, and the → **pericycle** in roots. *Adj.* **meristematic, meristemic.**

meristemoids (Gr. *eidos* = form), small cell groups or single cells in plants; have preserved their ability to divide and form some special structures such as stomata, root hairs, and ray cells.

meristic, segmented, divided into parts; e.g. **meristic structure**, a discretely varying, countable structure, such as digits; **meristic variation**, a variation in numbers of parts or segments.

meristoderm (Gr. *derma* = skin), the outer meristematic cell layer(s) in some → brown algae.

mero-, mer-, 1. (Gr. *meros* = part), pertaining to a part; partial; **2.** (Gr. *meros* = thigh), denoting relationship to the thigh.

meroblastic (Gr, *meros* = part, *blastos* = germ), partially dividing; pertains to a type of cell division in which during the → cleavage of an egg cell only the non-yolky portion of the egg participates in cell divisions; occurs in → megalecithal eggs with a large yolk.

merocrine (Gr. *krinein* = to separate), pertaining to such a type of exocrine secretion or a gland in which secretion passes from the cell by diffusion or by → exocytosis so that no cytoplasmic or cell membrane material is included in the secretion and thus the cell remains intact in the process; e.g. goblet cells, salivary glands, exocrine cells of the pancreas, and small → sweat glands (eccrine glands). *Cf.* apocrine, holocrine.

meroistic (Gr. *oion* = egg), describes a type of ovary, found e.g. in many insects, in which the → nurse cells serve the production of reserve nutrients and different → morphogens constitute the → positional information of a maturing ovum. *Cf.* panoistic.

merolimnic (Gr. *limne* = lake), pertaining to an animal which for a part of its life cycle lives in fresh water; e.g. ephemeropterans, dragonflies.

meromictic (Gr. *mixis* = act of mixing), pertaining to a lake in which the water layer near the bottom, **hypolimnion**, remains unmixed during autumn overturn; the phenomenon is called **meromicty**. In brackish water bodies **chemical meromicty** prevails, a condition in which heavy salt water lies permanently on the bottom, and thus a → **halocline** prevents the full circulation of water, which results in an oxygen deficiency in the hypolimnion. *Cf.* holomictic, mixolimnion, monimolimnion. *See* stratification.

meromyosin, *see* myosin.

meroparasite (Gr. *parasitos* = parasite), → facultative parasite.

meropelagic (Gr. *pelagos* = sea), pertaining to an organism which is pelagic only during a certain period of its life cycle. *Cf.* holopelagic.

meroplanktonic (Gr. *planktos* = wandering), pertaining to an organism which is planktonic only during a certain period of its life cycle, e.g. in larval stages. *See* plankton.

meropodite (Gr. *pous*, gen. *podos* = foot), *see* biramous appendage.

merosporangium (Gr. *sporos* = seed), a structure rarely formed in the apex of a → sporangiophore, producing a row of **merospores**.

Merostomata (Gr. *stoma* = mouth), **merostomatans**, **king crabs**, **horseshoe crabs**; an ancient, marine chelicerate class in which only four species of one order or subclass (Xiphosura) live; large animals with a thick, horseshoe-shaped carapace covering the large prosoma, a small, hinged opisthosoma, and the long postanal, caudal spine. Giant water scorpions (order/subclass Eurypterida), which lived from Ordovician to Permian, were the largest of all fossil arthropods, some species more than 3 m in length. The most common extant species is the horseshoe crab (*Limulus polyphenus*).

merozoite (Gr. *zoon* = animal), **schizozoite**; a motile stage in the life cycle of sporozoans, as plasmodians, developing in the liver and red blood cells of a host animal by mitotic divisions of trophozoites; is responsible for the vast reproductive capacity of sporozoan parasites. The merozoites are discharged from these cells into blood plasma and each of them penetrates a new red cell. This asexual reproduction is called **schizogony** and in the course of this process, attacks of fever (chill syndrome) occur in homoiothermic host animals. *See* malaria.

merozygote (Gr. *zygotos* = yoked), a → zygote formed from an incomplete union of two bacteria; i.e. is diploid only in a certain part of the genetic material. In the merozygote, that part of the genome of the recipient cell which is homologous to the genome fragment transferred from the donor cell, is called **endogenote**, and the genome part of the donor cell is called **exogenote**.

mes(o)- (Gr. *mesos* = middle), denoting middle, intermediate, or mean; in chemistry, signifies an optically inactive compound.

mesad, towards the median line or plane of a body; also mesiad.

mesangiocyte (Gr. *angeion* = blood vessel, *kytos* = cavity, cell), mesangial cells; a cell type in the glomeruli of the vertebrate kidney located between the endothelial cells of glomerular capillaries. These stellate cells are contractile and play a role in the regulation of glomerular filtration; they are also phagocytic taking up immune complexes and secrete various substances. The tissue formed by the cells and their surrounding matrix is called **mesangium**.

mesarch (Gr. *arche* = origin), pertaining to a structure in which the metaxylem is formed all around the former protoxylem in a plant stem; is possible if the stele structure is primitive, as in some ferns. *See* xylem.

mescaline, also **mescal, mezcaline,** 3,4,5-trimethoxyphenylethylamine, $C_{11}H_{17}NO_3$, an alkaloid obtained from flower buttons of the Mexican cactus *Lophophora williamsii* (Sp. *mezcal*); affects the central nervous system of vertebrates producing in humans psychomimetic effects (e.g. sensory distortions) similar to those produced by LSD.

mesencephalon, pl. **mesencephala,** Am. also **mesencephalons** (Gr. *mesos* = middle, *enkephalos* = brain), **midbrain**; the third main portion of the vertebrate brain developed from the middle of the three primary vesicles of the embryonic neural tube. The ventral part comprises ascending and descending tracts (**cerebral peduncles**, *crus cerebri*), and nuclei in which the third and fourth cranial nerves have their origins. The dorsal area, **tectum**, is formed of four colliculi, i.e. the paired **superior colliculi** (cranial or rostral c.) and **inferior colliculi** (caudal c.), functioning in the processing of optic and acoustic information; these are especially well developed in lower vertebrates whose cerebral cortex areas are phylogenetically less developed. In the anterior part of the mesencephalon there are two nuclei, the **red nucleus** and **substantia nigra**, involved in the control of motor functions. Longitudinally lies the narrow central canal, **aqueductus cerebri** (aqueductus mesencephali, aqueduct of Sylvius, mesocoelia), connecting the third and fourth brain ventricles.

mesenchyme, mesenchyma (Gr. *enchyma* = infusion), **1.** the meshwork of primordial embryonic connective tissue derived from the → mesoderm; consists of stellate **mesenchymal cells** and intercellular jelly-like matrix; **2.** the term mesenchyma was previously used wrongly to mean the middle layer of the body wall of sponges, i.e. the → mesohyl.

mesenteron, pl. **mesentera** (Gr. *enteron* = gut), **1.** midgut, especially the embryonic midgut before final differentiation of the alimentary canal; derived from the archenteron; **2.** the median part of the → gastrovascular cavity in anthozoans.

mesentery, also **mesenterium, 1.** a double-layered fold of the → peritoneum membrane of vertebrates, attaching certain parts of the abdominal viscera to the abdominal wall and conveying the blood and lymph vessels and nerves to them; that part of the mesentery which encircles the greater part of the small intestine is called **mesostenium**. The ligament of the mesonephros (*see* kidney) is called **urogenital mesentery**; **2.** six pairs of tissue septae dividing the → gastrovascular cavity into six radial chambers in the body of sea anemones. *Adj.* **mesenteric**.

mesobiota (Gr. *biote* = life), composed of intermediate soil organisms ranging in size from those just visible with a hand lens to those of 40 to 50 mm in length. *Cf.* macrobiota, microbiota.

mesobiotic, 1. pertaining to → mesobiota; **2.** describing seeds which survive in a dormant state from 3 to 15 years.

mesoblast (Gr. *blastos* = germ), the undifferentiated → mesoderm of an early embryo, or a synonym for the mesoderm. *Adj.* **mesoblastic**, pertaining to, or derived from the mesoblast.

mesoblastema, all the cells comprising the → mesoblast. *Adj.* **mesoblastemic**.

mesocarp (Gr. *karpos* = fruit), the middle layer of the → fruit wall, between exocarp and endocarp; is often fleshy and thick, as in the cherry.

mesocoel(e), Am. also **mesocele** (Gr. *koilos* = hollow), **1.** in molluscs, the middle section of the tripartite → coel between anterior section, procoel, and the posterior section, metacoel; **2.** in phoronids, the anterior section of the body extending to the → lophophore; the procoel is reduced but the metacoel exists. *Adj.* **mesocoelic, mesocoelian**.

mesocoelia, aqueductus cerebri. *See* mesencephalon.

mesocotyl (Gr. *kotyledon* = cup-shaped, hollow), the internode between the coleoptile and the scutellum in grass seedlings.

mesoderm (Gr. *derma* = skin), the middle layer of the three germ layers found in all animal embryos except sponges; develops into the interstitial connective tissue, cartilage, bone, notochord, blood and lymphatic organs, main portion of the muscle tissue, kidneys, gonads, and the → coelom (body cavity with its membranes). *Cf.* endoderm, ectoderm. *See* embryogenesis.

mesofauna, → meiofauna.

mesogloea, Am. **mesoglea** (Gr. *gloia* = glue), **1.** an acellular, supporting elastic lamella between the epidermis and gastrodermis in the body wall of cnidarians and ctenophores; **2.** → mesohyl.

mesohaline (Gr. *hals* = salt), pertaining to brackish water having salinity between 3.0—10.0 per thousand. *Cf.* oligohaline, polyhaline.

mesohumous (L. *humus* = ground, soil), pertaining to water or a water body containing moderate amounts of → humus. *Cf.* oligohumous, polyhumous.

mesohyl (Gr. *hyle* = wood, matter), the supporting middle layer of the body wall of sponges including numerous amoebocytes and spicules; does not correspond to the → mesoderm or mesenchyma of more developed animals. *Syn.* mesogloea.

mesokaryote (Gr. *karyon* = nut, nucleus), the nucleus of dinoflagellates and euglenophytes in which the chromosomes remain condensed throughout the mitotic cycle and thus does not undergo a convential → mitosis. The term is specifically used to denote the evolutionary position of dinoflagellates and euglenoids between prokaryotes and eukaryotes.

mesolecithal (Gr. *lekithos* = yolk), pertaining to an ovum (egg) which contains a moderate amount of yolk; i.e. more than in the oligolecithal ovum, and less than in the megalecithal ovum.

mesomere (Gr. *meros* = part), **1.** mesoblastic somite; a primordial body segment of an animal embryo; **2.** a → blastomere between a macromere and micromere in size.

mesomerism, the property of some molecules or ions having more than one form (**canonical** or **mesomeric** forms) with the same chemical formula and the same spatial structure, but with different electron arrangement in binding of the atoms of the same element; e.g. the mesomerism of carbonate anion, CO_3^{2-}, one of the three oxygen atoms is bound with a double bond, and the other two have negative charges forming only single bonds. Depending on which oxygen atoms are negatively charged, three mesomeric forms of the ion exist. It is possible that these states of bonds vary between the three atoms, and therefore the actual state is said to be a **resonance hybrid** of these three equivalent structures.

mesonephros, pl. **mesonephroi** (Gr. *nephros* = kidney), a kidney type found in vertebrate embryos developing after the pronephros stage, but in adults occurs only in fish and amphibians. In embryos of higher vertebrates (reptiles, birds and mammals) the third type, metanephros, develops while the mesonephros degenerates. The mesonephros is also called Wolffian body (*corpus Wolffi*), or mesonephron. *Adj.* **mesonephric**. *See* kidney.

mesonotum, pl. **mesonota** (L., Gr. *noton* = back), the dorsal sclerite (plate) on the middle thoracic segment (mesothorax) in insects. *Cf.* metanotum, pronotum.

mesons, a group of elementary particles, i.e. subatomic particles with either positive, negative, or zero charge. Their mass is between that of an electron and proton, about 200 times more than the mass of the electron; the life-time is only 10^{-7} seconds. They are found in cosmic radiation and are emitted from atomic nuclei when bombarded by high energy particles. Called also mesions or mesotrons.

mesopelagic zone (Gr. *pelagos* = sea), a subzone of the oceanic pelagic zone lying between the epipelagic and bathypelagic zones; ranges from 200 to 1,000 m (according to some researchers, to the depth of 2,000 m).

mesophyll (Gr. *phyllon* = leaf), parenchyma tissue between epidermal layers of a plant leaf; composed of **mesophyll cells** and structures of leaf veins, containing living cells, rich in chloroplasts, thus being a photosynthetically active tissue. The most common, **dorsiventral** type of mesophyll consists chiefly of **palisade parenchyma**, with long and narrow cells located on the upper side of the leaf, and of some **spongy parenchyma** with large intercellular spaces below the former. In the **isolateral** type there are palisade parenchyma layers on both sides of the leaf, or the whole mesophyll consists of palisade parenchyma. In the **centric** type, palisade parenchyma cells surround the leaf veins, and in the **homogeneous** type all the mesophyll cells are undifferentiated spongy parenchyma cells.

mesophyte, a plant growing in average conditions of moisture. *Cf.* hygrophyte, xerophyte.

mesoplankton (Gr. *planktos* = wandering), plankton organisms in oceans varying from 0.2 to 20 mm in size, in inland waters may be smaller. *Cf.* microplankton, macroplankton, megaplankton.

mesopleurum, pl. **mesopleura** (Gr. *pleuron* = side), the lateral sclerite (plate) of the second thoracic segment (mesothorax) in insects. *Syn.* mesopleuron. *Cf.* metapleurum, propleurum.

mesosoma, pl. **mesosomata** (Gr. *soma* = body), the middle part of the body of some invertebrates; e.g. the collar of acorn worms, or the anterior part of the opisthosoma in arachnids. *Cf.* prosoma, metasoma.

mesosome, a membrane-covered particle in the cytoplasm of certain bacteria, formed from the plasma membrane via invagination and constriction; mesosomes are found during the duplication of DNA, and in secretion.

mesosphere (Gr. *sphaira* = ball, atmosphere), the zone in the Earth's → atmosphere between stratosphere and ionosphere, from 50 to 80 km, in which temperature decreases from about 0° to —90°C.

mesosternum, pl. **mesosterna,** also **mesosternums** (L. *sternum* = breast bone), **1.** the ventral sclerite (plate) of the second thoracic segment (mesothorax) in insects; *cf.* metasternum, prosternum; **2.** the middle part of the sternum in vertebrates.

mesothelium, a thin, single epithelium-like layer formed from flattened cells derived from the embryonic mesoderm; lines the serous membranes of the → coelom of vertebrates, i.e. → peritoneum, pleura, and pericardium. *Adj.* **mesothelial**.

mesothorax, pl. **mesothoraxes** or **mesothoraces** (Gr. *thorax* = chest), the middle somite of the thorax of an insect; a pair of legs (in winged insects also the forewings) are attached to it. *Adj.* **mesothoracic**. *Cf.* metathorax, prothorax.

mesothorium, the name used for actinium and radium isotopes occurring in natural decay series.

mesotocin, a peptide hormone secreted from the posterior part of the → pituitary gland in lungfish, amphibians, reptiles, and birds.

mesotrophic (Gr. *trophe* = nourishment), **1.** pertaining to an aquatic or terrestrial environment with the average nutrition level, i.e. between eutrophic and oligotrophic; *cf.* hypertrophic, dystrophic; **2.** pertaining to an organism which has partly autotrophic and partly saprophytic nutrition; *syn.* mixotrophic; *cf.* metatrophic.

mesotropic (Gr. *trope* = a turning), turning toward; situated in the median plane of an organ or a body, e.g. in the middle of the body, such as the abdomen.

mesovarium, pl. **mesovaria** (L. *ovarium* = ovary), the mesentery of the vertebrate ovary, i.e. the peritoneal fold of the ligament of the ovary, holding the ovary in its place.

Mesozoa (Gr. *zoon* = animal), **mesozoans**; a small phylum of minute worm-like animals (about 50 species), parasitic in marine invertebrates. Their systematic position has long been unsolved, they are suggested to represent a phylum, subphylum, or even an animal subkingdom. The body comprises only a few tens of cells in two layers which are not homologous to the germ layers of metazoans. Owing to their parasitic life it is also possible that mesozoans are degenerated forms of Platyhelminthes, but more plausibly, they are intermediate forms between protozoans and metazoans. The mesozoans are grouped into two classes, Rhombozoa and Orthonectida.

Mesozoic (era) (Gr. *zoe* = life), a geological era about 245 to 66 million years ago; is called the Middle age of the Earth, or the Age of reptiles because these animals dominated the life on land. The era is divided into the → Triassic, Jurassic, and Cretaceous periods.

messenger RNA, mRNA, a type of single-stranded → ribonucleic acid (RNA) molecule of heterogeneous size arising from → genetic transcription of DNA and acting as the model (template) for polypeptides in → protein synthesis. The → nucleotide sequence of mRNA is complementary to the DNA strand of the gene, and is translated into the sequence of amino acid residues in protein synthesis (*see* genetic translation).

The mRNAs of prokaryotic organisms are **polycistronic** as the molecule contains the information of all the structural genes of the whole → operon. The mRNAs of eukaryotic organisms are → **monocistronic**, i.e. each molecule contains the information of a single gene only.

In prokaryotes, mRNA is synthesized directly in its active form, and its translation begins during transcription. In eukaryotes, the primary transcription product is the → **heterogeneous nuclear RNA**, from which the mature mRNA is produced in a complicated process (RNA processing) in the nucleus, involving → capping, splicing, and polyadenylation. A mature eukaryotic mRNA molecule has a cap structure in the 5' end protecting the molecule, followed by an untranslated → leader sequence, translation → initiation codon, actual message involved in the mRNA translation, and → terminator codon. Thereafter follows a long, untranslated → trailer sequence and the poly-A tail that consists of ca. 200 adenosine nucleotides added to the mRNA during polyadenylation. The poly-A tail protects the 3' end of mRNA, and is most likely necessary in the transport of mRNA from the nucleus to the cytoplasm where translation occurs.

In the cytoplasm the mRNA molecule which acts in the transmission of genetic information, is connected to → ribosomes where translation occurs. The half-life of prokaryotic mRNA molecules is rather short, but in eukaryotes it can vary thus constituting one possibility for the regulation of gene function.

messengers, a term for different substances which carry information from cell to cell inside an organism, inside cells, or between organisms; include → hormones, neurotransmitters, growth factors, second messengers, and pheromones.

meta- (Gr. *meta* = after, on the other side, between, with), **1.** subsequent to, behind, hindmost, or together with something else; **2.** a chemical symbol, *m*-, denoting an atom or atom group in the 1- and 3-position in a carbon ring or in another related structure; e.g. *m*-xylene is 1,3-dimethylbenzene; **3.** pertaining to a polymeric acid anhydrine such as metaphosphoric acid (always written in full).

metabasidium, pl. **metabasidia** (Gr. *basis* = base), a cell of a fungus of the group → Basidiomycetes in which → meiosis occurs in the formation of basidia and basidiospores.

Metabola, → Pterygota.

metabolism (Gr. *metabole* = change), the integrated network of biochemical reactions proceeding in a number of steps (**metabolic pathways**) in organisms (*see* glycolysis, citric acid cycle, Calvin cycle, electron transport). Metabolism consists of decomposition reactions (**catabolism**) and synthesizing reactions (**anabolism**) which are needed to maintain all vital functions, such as growth, development, reproduction, production and usage of ATP, photosynthetic reactions, etc. Metabolism is based on the function of several enzymatic reactions controlled by many activators as micronutrients, vitamins, growth factors, hormones, and the nervous system in animals. **Metabolic rate** is dependent e.g. on enzyme and substrate concentrations, hydrogen ion concentration (acidity), and temperature. The uptake of oxygen, production of carbon dioxide, and photosynthesis in plants, reflect the metabolic activity level of a whole organism. The cell metabolism expresses the activity of biochemical enzyme reactions in cells and cell organelles. *Adj.* **metabolic**.

metabolite, a product of metabolic action; a biochemical substance taking part in → metabolism, an intermediate or an end product in a serial enzymatic process.

metacarpal (Gr. *karpos* = wrist), **1.** pertaining to the → metacarpus; **2.** a bone (hand bone) of the metacarpus; primarily five metacarpals (*ossa metacarpalia*) per tetrapod forelimb.

metacarpus, pl. **metacarpi** (L.), the bony structure in the tetrapod forelimb between the carpus (wrist) and phalanges (fingers).

metacentric (Gr. *kentron* = centre, central point), pertaining to a chromosome in which the → centromere is located in the middle thus dividing the chromosome into two equally long arms.

metacercaria, pl. **metacercariae** (Gr. *kerkos* = tail), a cystic stage following the → cercaria in the life cycle of flukes (digenic trematodes); develops into the adult stage when eaten by the definite host. *See* miracidium, redia, sporocyst.

metacoel(e), Am. also **metacele** (Gr. *koilos* = hollow), **1.** the anterior extension of the fourth ventricle in the vertebrate brain; **2.** the posterior part of the tripartite coelom of molluscs; **3.** in phoronids, the part of the coelom that is located in the trunk region of an animal. *Cf.* mesocoel(e).

metacommunication, a behavioural pattern including signals to impart information about how subsequent signals should be interpreted; it is known e.g. in parent-offspring relation among primates and large carnivores when the playing parent informs offspring that subsequent aggressive behaviour is not hostile.

metafemale, an exceptional female individual which, according to the balance theory of →

sex determination, has an abnormally large number of factors which determine female development; e.g. a 3X 2A individual among fruit flies is a metafemale (X = → X chromosome, A = dosage of → autosomes).

metagenesis (Gr. *gennan* = to produce), alternation of generations; denotes the production of generations characterized by the regular alternation of sexual and asexual generations in the same species; found e.g. in cnidarians and fungi.

metal (L. *metallum*, Gr. *metallon* = mine, metal), a group of electropositive elements having a lustrous appearance and a good conductivity of heat and electricity, and the tendency to lose electrons in chemical reactions; many metal cations are important → macro and micronutrients (trace elements) in organisms. *See* alkali metals.

metalimnion, pl. **metalimnia** (Gr. *limne* = lake), a rather thin layer in a water body between epilimnion and hypolimnion; in this layer temperature suddenly cools (**thermocline**), and/or the salinity increases (**halocline**) from the surface to the bottom. *See* stratification.

metalloid, 1. semimetal; an element with some properties characteristic of metals and other properties of non-metallic elements, e.g. arsenic, bismuth, germanium, and antimony; formerly denoting an alkaline metal (e.g. sodium) or an alkali earth metal (e.g. calcium); **2.** as an adjective, resembling a metal, relating to a metalloid; *syn.* metalloidal.

metalloprotein, a protein, usually an enzyme, having a metal as its prosthetic group.

metamale, an exceptional male individual which, according to the balance theory of → sex determination, has an abnormally large number of factors which determine male development; e.g. an 1X 3A individual among fruit flies is a metamale (X = X chromosome, A = dosage of autosomes).

metamerism (Gr. *meros* = part), **1.** segmentation; having serially homologous segments, as in the body of many animals divided into segments, **metameres**, which correspond externally and internally to each other, e.g. in annelids, tapeworms, insects, and in vertebrate embryos; **2.** in chemistry, pertaining to a type of isomerism occurring in organic compounds of the same chemical type, e.g. different ethers; designates the attachment of different radicals to the same central atom; *see* isomers. *Adj.* **metameric**.

metamorphosis, pl. **metamorphoses** (Gr. *morphosis* = action of forming, shaping < *morphe* = form), a change or transition of a form or shape, especially from one developmental stage into another, as in the course of the life cycle of an animal from an embryonic or larval stage to the adult stage. Familiar examples are metamorphoses of insects which are commonly divided into three different developmental patterns: 1) in **ametabolous** insects the body never undergoes striking changes but the animals develop gradually through a series of moults, as in springtails and silverfish. 2) In **heterometabolous** insects, incomplete metamorphosis occurs which may be either **paurometabolous**, i.e. the small adult-like nymphs have external wing buds and the metamorphosis is gradual but not extreme (e.g. grasshoppers and locusts), or it may be **hemimetabolous** with only one moult, the nymph (naiad) being clearly different from the imago (e.g. dragonflies). 3) In **holometabolous** insects the life cycles include larval and pupal stages before the adult stage; this is a **complete metamorphosis** found e.g. in lepidopterans, coleopterans and dipterans.

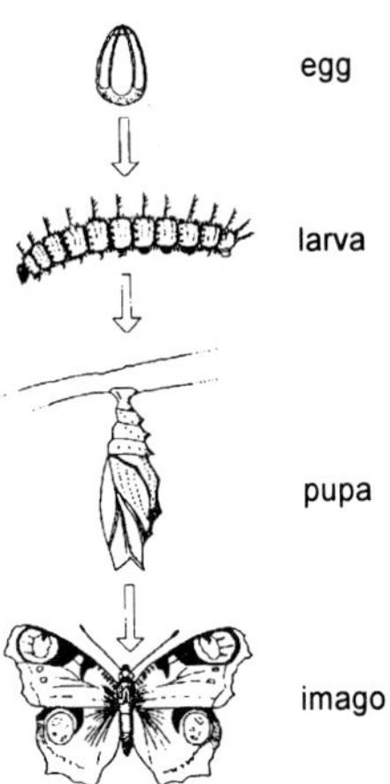

Holometabolous metamorphosis of butterfly.

The metamorphosis of insects is regulated by the balance of juvenile hormone and ecdysone, the former maintaining larval stages and the latter inducing the development of the adult. Among vertebrates, metamorphosis is found only in lampreys (Cephalaspidomorphi), in some fish species such as eel, and in

azathionium chloride; a greenish blue crystalline powder, used e.g. as a biological dye and redox indicator. *Syn.* methylthioninium chloride, tetramethylthionine chloride. *Cf.* methyl blue.

methylene chloride, → dichloromethane.

methylene group, a divalent atom group, $=CH_2$, occurring in many organic compounds.

methyl group, an atom group, $-CH_3$, found in many organic compounds, e.g. in methanol, CH_3OH.

methyl mercury, dimethylmercury, $(CH_3)_2Hg$; the most common mercury compound, found naturally in soil produced from inorganic mercury mainly by organisms; it (together with monomethylmercury, CH_3Hg) tends to accumulate in food chains becoming toxic to organisms, e.g. to birds feeding on fish. *See* mercury.

methylphenols, *see* cresols.

2-methyl-1-propanol, *see* butanol.

methyl red, $C_{15}H_{15}N_3O_2$, an indicator used in acid-base titrations; yellow above pH 6.0 and red below pH 4.4.

methyl tert-butyl ether, MTBE, 2-metoxy-2-methylpropane; an organic compound in a group of ethers, used e.g. as a chromatographic eluent and as an additive in petrol for improving combustion. When spread into the environment MTBE evaporates quite easily from surfaces but deeper in the ground may act as a pollutant which cannot be decomposed by microbes. Its toxicity is similar to that of petrol: e.g. for rats the lethal dose is about 4,000 mg/kg; 250 ppm in inhaled air causes nervous symptoms, such as drowsiness.

Metzgeriales, an order in the class → Hepaticopsida, division Bryophyta, Plant kingdom. The species are strap-shaped and dichotomous liverworts, living mostly on tree-trunks and rocks. In some classifications Metzgeriales is an order in the division Hepatophyta.

mevalonic acid, 3,5-dihydroxy-3-methylpentanoic acid; an intermediate in the biosynthesis of → terpenoids and → sterols; e.g. a precursor of → cholesterol, gibberellins and abscisic acid.

MGDG, → monogalactosyl diglyceride.

MHC, → major histocompatibility complex.

mho, reciprocal ohm, the unit of → conductance, i.e. siemens. *See* Appendix 1.

micelle, also **micella,** pl. **micelles, micellae** (L. *mica* = piece, grain), **1.** an electronmicroscopic aggregation of macromolecules seen in a colloidal suspension; **2.** a minute particle in an aqueous solution, composed of a group of non-soluble lipid molecules with polar lipid molecules on the surface of the micelle; molecules have hydrophobic ends inwards and hydrophilic ends outwards forming a membranous structure against the surrounding water; e.g. in the intestine, bile salts with non-soluble fatty acids, cholesterol, and monoglycerides form **lipid micelles** before absorption through the intestinal mucosa; **3.** the base fibre (→ microfibril) of the cellulose cell wall of plants; **4.** a base unit of the colloidal soil particle having two mantles formed from → ions, the inner and outer mantle having opposite electric charges. A micelle is therefore electrically neutral which is an important property in ion exchange and plant nutrition.

Michaelis—Menten kinetics (*Leonor Michaelis*, 1875—1949; *Maud L. Menten*, 1879—1960), the relationship between the reaction velocity (V) and substrate concentration [S] in an enzyme reaction, expressed with the parameters K_m, V_{max} and V (= $V_{max}/2$); where K_m is the substrate concentration at which half the maximum velocity (V_{max}) of a reaction is achieved. K_m is called the **Michaelis constant**. The **Michaelis—Menten equation** is expressed:

$$V = V_{max}\frac{[S]}{[S] + K_m}$$

This can also be expressed graphically and to get a straight line plot one can take the reciprocal of both sides of the equation:

$$\frac{1}{V} = \frac{1}{V_{max}} + \frac{K_m}{V_{max}} \cdot \frac{1}{[S]}$$

A plot of $1/V$ versus 1/[S] is called **Lineweaver—Burk plot** that gives a straight line.

micr(o)- (Gr. *mikros* = small), denoting small; in units of measurement indicates one millionth part (10^{-6}), symbol μ.

microbe (L. *microbium* < Gr. *bios* = living), a microscopic organism, a germ; the common name for small organisms including → bacteria, spirochaetes, rickettsiae, and viruses. Sometimes the term is used as a synonym for **microorganism** including also protozoans, algae, and many fungi, such as moulds, rust fungi, smut fungi, or slime moulds. *Adj.* **microbial** or **microbic**.

microbenthos (Gr. *benthos* = depth), bottom organisms smaller than 0.1 mm in size. *Cf.* macrobenthos, meiobenthos.

microbicide (L. *caedere* = to kill), a chemical used for killing microorganisms.

microbiology, a branch of science dealing with microorganisms; called also protistology.

microbiota (L. < Gr. *biote* = way of life), microscopic organisms (microorganisms) of a region, especially in soil. *Cf.* macrobiota, mesobiota.

microbiotic, 1. pertaining to → microbiota; **2.** short-lived; e.g. describing seeds surviving in the dormant state for a relatively brief period, usually less than 3 years.

microbivore (L. *vorare* = to devour), an organism feeding on microorganisms.

microbody, any of spherical cell organelles surrounded by a single membrane, 0.3 to 1.5 µm in diameter. They can be divided into → **peroxisomes** (in animal and plant cells) and → **glyoxysomes** (in plant cells). Peroxisomes contain enzymes, such as catalase that reduces hydrogen peroxide and oxidizes ethyl alcohol, phenols, and formaldehyde, and they also take part in β-oxidation of lipids. In plant cells, → photorespiration takes place in peroxisomes. Glyoxysomes of germinating seeds contain malate hydrogenase and isocitrate dehydrogenase which produce glucose from lipids via acetyl coenzyme A (acetyl CoA).

microcell, an experimentally produced particle of a cell nucleus surrounded by a thin layer of cytoplasm and cell membrane; can be produced by breaking the nucleus into pieces by colchicine treatment and thereafter by isolating the nuclear particles from the cell, i.e. by **enucleation** in which each nuclear particle becomes enveloped by a fragment of the plasma membrane. Microcells can be used e.g. for preparing → cell hybrids. *Cf.* minicell.

microcirculation, blood circulation through arterioles, capillaries, and venules.

microclimate (Gr. *klima* = inclination, slope), the local climate of a given site or habitat, usually in a small area often in vegetation near the ground (at the most to 2 m); differs usually from the prevailing → macroclimate.

microcyte (Gr. *kytos* = cavity, cell), any of the exceptionally small red blood cells produced e.g. in certain diseases of vertebrates, such as in some types of anaemia.

microdissection (L. *dissecare* = to cut apart), the dissection of cells, tissues, or small organisms under the microscope.

microendemic (Gr. *endemos* = native), pertaining to a taxon occurring in a very small native area. *Cf.* endemic.

microelectrode, an → electrode with the tip less than 1 µm in diameter; usually a glass electrode that allows measurement of electrical potentials and pH levels in a single cell.

microevolution, the → evolution which occurs within a species; e.g. involves changes in → gene frequencies of a population, resulting from selective accumulation of minute variations of genetic material of the species. *Cf.* macroevolution, megaevolution.

microfauna (L. *Fauna*, sister of *Faunus* = mythical deity of herdsmen), **1.** the animals of microbiota; microscopic animals, chiefly protozoans, usually smaller than 0.1 mm in length; **2.** small → fauna; the fauna of a → microhabitat. *Cf.* meiofauna, macrofauna.

microfibril (L. *fibrilla*, dim. of *fibra* = fibre), **1.** in animals, the common name of microscopic and electronmicroscopic fibrils (average diameter about 15 nm), usually inside the cell; are composed of protein (e.g. myofibrils, neurofibrils), often formed of a bundle of smaller structures such as → microfilaments; **2.** the base unit of the plant cell wall, especially of the → secondary wall, consisting of cellulose; *see* micelle; *cf.* microfilament.

microfilament (L. *filamentum* = fibre), any of the electronmicroscopic fibrous protein elements (average diameter 4—5 nm) which partly form the → cytoskeleton of a cell, e.g. actin filaments. *Cf.* microfibril.

microfilaria, pl. **microfilariae** (L. *filum* = thread), a minute larval stage of parasitic filarial worms living in the blood and other tissues of a host animal. *See* Filarioidea, onchocerciasis.

microflora (L. *Flora* = the mythical goddess of flowers), **1.** microscopic plants; **2.** flora (the plants) of a microhabitat; the entire population of microorganisms (many bacteria, unicellular algae, and fungi) living in a special location, as e.g. in or on a larger organism; *Cf.* macroflora.

microfossil (L. *fossilis* = dug up), a microscopic → fossil, as e.g. bacteria, diatoms, protozoans, pollen grains, and minute bone remains.

microgamete (Gr. *gamos* = marriage), the smaller of two different conjugant → gametes in → anisogamy, usually the motile male gamete. *Cf.* macrogamete.

microgametophyte (Gr. *phyton* = plant), the male → gametophyte in heterosporous plants (*see* heterospory), formed in seed plants within the (germinating) pollen grain. *See* microspore.

micrograph (Gr. *graphein* = to write), **1.** an instrument for measuring microscopic movement of a thin diaphragm (membrane), used for recording sound waves or other pulses which are mediated through air to the diaphragm; **2.** photomicrograph, electron micrograph; the permanent image of a specimen obtained on a photographic film or an electron-sensitive plate by means of a microscope or electron microscope.

microhabitat (L. *habitare* = to inhabit), the immediate environment of an organism; a part of a larger habitat where the individuals of a species are living (especially small organisms); e.g. a stump of a tree for certain insects, or a meadow patch in the middle of a wood area. Sometimes the term is used as a synonym for **biotope** but usually the microhabitat is considered as a more limited entirety than the biotope.

microinjection, a micromanipulation technique in which material (e.g. DNA, dyes, etc.) is injected into a cell through the cell membrane by means of a thin micropipette.

microlecithal (Gr. *lekithos* = yolk), pertaining to an egg type with little yolk. *Cf.* megalecithal, mesolecithal, telolecithal.

micromere (Gr. *meros* = part), **1.** any of the small → blastomere cells produced in the upper part of an early animal embryo by unequal cleavage of a fertilized → telolecithal ovum; **2.** a cell type in an → amphiblastula. *Cf.* macromere.

micron (Gr. neuter of *mikros* = small), μ; formerly used to mean micrometre, μm, i.e. one-thousandth of a millimetre.

microneme (Gr. *nema* = thread), a slender and elongated structure in the apical part of many sporozoans; probably aids in penetrating host's cells or tissues.

micronucleus, pl. **micronuclei, 1.** the sexual or generative cell nucleus of many protozoans, as e.g. some foraminiferans (*Foraminifera*) and most ciliates (*Ciliata*); present with another nucleus, the → **macronucleus**; **2.** an additional nucleus, separate from the actual nucleus; formed during the telophase of mitosis or meiosis from → lagging chromosomes or parts of chromosomes borne in spontaneous or artificially induced structural changes of chromosomes; **3.** the term is sometimes used to mean a small nucleus in a large cell.

micronutrients, nutrients needed by organisms in minute quantities. In animals, these essential food factors required by cells include → **trace elements** and → **vitamins**. Micronutrients needed by plants for special reactions are e.g. manganese, chlorine, and copper in the light reactions of photosynthesis, molybdenum in nitrate reduction (*see* nitrate reductase), zinc in the biosynthesis of → auxin, boron in carbohydrate metabolism, and iron in the synthesis of → chlorophyll. *Cf.* macronutrients. *See* minerals, nutrients, Hoagland's solution.

microorganism (L. *organismus* = organism), *see* microbe.

microparasite (Gr. *para* = beside, *sitos* = food), a parasitic microorganism having usually a short life cycle and a high reproductive rate; e.g. bacteria, viruses, many protozoans, and fungi.

microphage (Gr. *phagein* = to eat), **1.** an organism feeding on microscopic organic particles or minute organisms; *see* microphagy; **2.** a small phagocyte, especially a neutrophilic → granulocyte; *cf.* macrophage. *Adj.* **microphagous.**

microphagy, the property of certain animals (**microphages, microphagous feeders**) which feed on small particles or organisms; e.g. baleen whales (Mysticeti) sieving planktonic organisms through their → baleens, tunicates catching small organic particles and/or minute organisms by their pharyngeal gills, or bivalve molluscs filtering food by cilia on their lamellary gill surfaces (*see* filter feeder). Also some protozoans, like paramecia and heliozoans, and many crustaceans are microphagous feeders.

microphyll (Gr. *phyllon* = leaf), a simple small leaf such as the leaves of the horsetail or club mosses, having only a single strand of vascular tissue, i.e. a single leaf vein.

microphyte (Gr. *phyton* = plant), a minute plant, nearly or totally invisible to the naked eye; e.g. many unicellular algae in a plankton. *Cf.* macrophyte.

microplankton (Gr. *planktos* = wandering), planktonic organisms which are 20 to 200 μm in length. *Cf.* nanoplankton, picoplankton.

micropore (Gr. *poros* = passage, pore), any minute pore or channel, usually microscopic in size; in organisms e.g. **1.** a cell organelle formed by the cell membrane in some unicellular organisms, e.g. in some sporozoans; it is composed of two concentric rings and acts as a feeding organelle (cytostome, "mouth"); **2.** any of small pores including a sense organ in the shells of some chitons. *Adj.* **microporous**.

micropyle (Gr. *pyle* = gate), **1.** a pore in the cell membrane of an egg through which sperm enter, e.g. in insects; **2.** a pore in a plant → ovule, through which the pollen tube usually grows after pollination. *Adj.* **micropylar**.

microsatellites (L. *satelles* = attendant), in genetics, very short DNA segments (di- or trinucleotides) repeated in tandem; occur in the genome of all eukaryotic organisms. The number of repetition varies prominently within the species, and the number often changes. Microsatellites are used in DNA fingerprinting and genetic mapping. *Cf.* minisatellite.

microsclereid (Gr. *skleros* = hard), a very small, branching → sclereid in plants.

microscope (Gr. *skopein* = to view, examine), an optical (light microscope) or electron-optical (electron microscope) apparatus, which forms an enlarged image of an object. The light microscope contains two lens systems, the objective and the ocular, mounted at opposite ends of a closed tube. The lenses are set up so that the real image formed by the objective lies at the focal point of the ocular, and the observer looking through the ocular sees an enlarged virtual image of the real image. The total magnification is determined by the focal lengths of the two lens systems, the ability of a microscope to distinguish fine details depending on the wavelength of the used radiation. The wavelength of visible light is in average approximately 500 nm and the best → resolution is about 450 nm; i.e. objects closer together will not be resolved separately. The greatest magnification of a typical **light microscope** is ca. 1,500 times, but the magnification ability can be improved by adding immersion oil between the specimen and the lens, by shortening wavelengths, and with some special techniques.

The **ultraviolet microscope** uses the ultraviolet region of the spectrum rather than the visible region, and the resolution is improved due to the shorter wavelength of the radiation. Because glass does not transmit shorter UV, the lenses must be of quartz. UV-radiation is also invisible, and the image must be made visible by → fluorescence.

The function of a **phase-contrast microscope** and an **interference microscope** (interference-contrast microscope) is based on the fact, that the speed of light is slower in the object than in the air. In these techniques, the entering light is split into two beams one of which is directed through and the other by the specimen and then the phase differences formed in the beams are utilized to amplify the magnification and to cause better effectivity than in the conventional system. These microsope types are also suitable for examination of unstained and living material because of their ability to produce better contrasts than an ordinary light microscope. Also the **dark-field microscope** is suitable for studies of transparent, unstained biological materials and of objects that cannot be seen in normal illumination under the microscope. The dark-field microscope employs illumination in the form of a hollow cone, the objective lying in the dark portion of the light-cone. The objective thus gets only scattered light and glows brightly against the dark field.

A **polarization microscope** uses polarized light, and if there are crystalline materials in an object, they change the vibration level of penetrating light, producing illustration especially of the crystal structure of the object.

A **fluorescence microscope** is constructed for identification and localization of various chemical substances in an object. The object has to be treated with a special compound, which includes a fluorescent atom group and which specifically attaches to a substance under study. When the object is illuminated with UV light under the microscope, the intensity of fluorescence shows the location and quantity of the substance in the object, and the view can be photographed for analysis. The specific attaching of the fluorescent compound can be based e.g. on a specific immunological reaction, and the attached compound can be localized with the aid of an enzyme reaction, which causes e.g. the formation of a certain colour.

An **electron microscope** may magnify an object even some hundred thousand times.

The radiation that forms the picture is an electron beam, which is modulated with electron-optical lenses, and the picture can be seen on a fluorescent screen. The preparation of samples is much slower and more difficult than in the light microscopy. There are several different techniques: in **transmission electron microscopy (TEM)** the electron radiation penetrates the object, and in **scanning electron microscopy (SEM)** the radiation scans the object and forms a picture of the surface of the sample. In **STEM** (scanning transmission e.) these both techniques can be used in the same microscope.

Developed modifications of microscopes are a **confocal microscope** (confocal laser scanning microscope, CLSM) and a **tunnelling electron microscope**. The confocal microscope uses lasers and special optics for "optical sectioning"; only those regions within a narrow depth of focus are imagined. Regions above and below the selected plane of view appear black rather than blurry. The picture is formed by a computer, and is saved digitally. The digital-formed information can be analysed by the computer. The tunnelling electron microscope is a type of the scanning electron microscope, which can give an atomic level resolution of the structure of surfaces, or e.g. of biological macromolecules.

microsmatic (Gr. *osme* = odour, smell), pertaining to an animal with a weakly developed sense of smell, as e.g. in humans. *Cf.* macrosmatic. *See* olfaction.

microsoma, pl. **microsomata** (L., Gr. *soma* = body), **1.** a very short but not dwarfish stature; *adj.* **microsomatous**, having a small body; **2.** → microsome.

microsomal enzymes, several enzymes associated with → endoplasmic reticulum in cells; can be isolated from the microsomal fraction after homogenization of a tissue (*see* microsomes). The enzymes are important in catalysing biotransformation of foreign substances, such as environmental toxic substances. In animals they are usually oxidative enzymes with cytochrome P_{450} acting as → coenzyme; very active in the liver, kidneys, intestinal wall, and skin.

microsomes, the term originally used to mean different spherical cytoplasmic vesicles which are only just visible in the light microscope when fixed and stained; nowadays used as the group name for some cell organelles and their fragments (**microsomal fraction**) obtained in a cell fractionation and separation by centrifugation of a tissue homogenate with broken cells. Mitochondria, lysosomes, and other larger cell organelles are separated in a low-speed centrifugation, and the microsomes are then obtained by centrifugation of the supernatant in a high-speed or ultracentrifuge (e.g. for 1 h at 100,000 *g*). The microsomal fraction includes fragments of the → endoplasmic reticulum often with attached → ribosomes, and fragments of the Golgi complex, mitochondrial cristae, and plasma membranes. *Adj.* **microsomal**, pertaining to microsomes.

microspecies, pl. **microspecies,** infrasubspecific group; a small population slightly but effectively differentiated from other, related populations or forms. *Cf.* macrospecies.

Microspora, microsporidians; a phylum of unicellular animals, protozoans, living as parasites in invertebrates, especially in arthropods, and lower vertebrates; are characterized by minute infective spores with a coiled tubular polar filament. The class includes e.g. the genus *Nosema*, the species of which cause diseases in silkworms and honeybees.

microsporangium, pl. **microsporangia** (Gr. *sporos* = seed, *angeion* = vessel), a structure producing → **microspores** through meiosis in → **microsporogenesis** in heterosporous plants (*see* heterospory).

microspore (Gr. *sporos* = seed), the smaller of the two types of spores in heterosporous plants (*see* heterospory), formed in a **microsporangium** through meiosis from the **microspore mother cells (microsporocytes)** in → **microsporogenesis**. The microspores develop into male → gametophytes (microgametophytes) which then produce male gametes. In seed plants the pollen sac (in a stamen) is the microsporangium and the pollen grains are actually microspores; the male gametophyte is formed inside the (germinated) pollen grain producing the gametes (sperm cells) fertilizing the egg cell in the → ovule. *Cf.* macrospore.

microspore mother cell, microsporocyte. *See* microspore, microsporogenesis.

microsporocyte, (Gr. *kytos* = cavity, cell), microspore mother cell. *See* microspore, microsporogenesis.

microsporogenesis (Gr. *genesis* = birth), the formation of → **microspores** from a microspore mother cell (microsporocyte) through meiosis in heterosporous plants (*see* heterospory). In seed plants, microsporogenesis takes place in anthers, where the microspore mother cells divide meiotically forming microspores, i.e. → pollen grains. These are often first attached in tetrads, but soon separate and develop into independent pollen grains. *Cf.* macrosporogenesis.

microsporophyll (Gr. *phyllon* = leaf), the structure bearing the → microsporangia of heterosporous plants; may be leaf-like, or modified into a stamen in seed plants.

microstrobilus (Gr. *strobilos* = cone), *see* strobilus.

microsurgery (L. *chirurgia* < Gr. *cheir* = hand, *ergon* = work), **1.** dissection of cells or minute organisms under a microscope; **2.** in medicine, the surgery of small structures performed using a microscope; e.g. the microsurgery of the ear.

microsymbiote (Gr. *symbioun* = to live together), **microsymbiont**; the smaller of the two symbiotic species, usually microscopic in size.

microtaxonomy (Gr. *taxis* = arrangement, *nomos* = law), a classification of organisms into infraspecific taxa, such as subspecies and varieties. *See* taxonomy.

microtherm (Gr. *therme* = heat), a plant favouring relatively cold environment with a temperature about 5°C in the coldest month and between 10°C and 20°C in the warmest month.

microtome (Gr. *tome* = section), an instrument used for preparation of microscopic sections (some micrometres thick) from biological material for examination under the light microscope; an **ultramicrotome** is used for cutting even thinner sections for electron microscopy; **microtomy** (histotomy), a technique of using the microtome. *Adj.* **microtomic**. *See* cryomicrotome, cryoultramicrotome.

microtubule (L. *tubulus*, dim. of *tubus* = tube), an intracellular tubular structure occurring in the cytoplasm of plant, fungal, and animal cells. Microtubules are cylindrical structures 20 to 30 nm in diameter with varying lengths, the wall (4 nm thick) containing globular monomeric proteins, α- and **β-tubulins**; these form heterodimer subunits which are aligned end to end into long filamentous elements, **protofilaments**. The columns of protofilaments are packed side by side to form the wall of the microtubule. The microtubules have an important function as a part of the → cytoskeleton controlling the shape and strength of the cells. Vesicles and protein particles often are transported along microtubules (*see* kinesin). Microtubules also take part in the movements of cell organelles, forming fibrillar structures of → cilia, flagella, and the → mitotic spindle. *Adj.* **microtubular**. *See* neurofibrils.

microtubule organizing centre, MTOC, a region in the cell cytoplasm from which the → microtubules radiate and in which their pattern and number are determined. The major MTOCs in an animal cell are the **centrosome** with a pair of → centrioles organizing the microtubules in an interphase cell but also the → mitotic spindle in dividing cells, and the **basal body** of a → cilium or flagellum.

microvillus, pl. **microvilli** (L. *villus* = shaggy hair, tuft of hair), any of finger-like processes of the cell membrane, about 0.5 to 5 µm in length, occurring especially in epithelial cells of animals to increase the absorptive surface area, as in the mucosal epithelium of intestine, or the epithelium of renal tubules. Inside a single microvillus are **actin** filaments bound together by other proteins, e.g. **fascin, fimbrin,** and **villin**. Intestinal microvilli seen in the light microscope, are called **brush border**.

microwaves, electromagnetic radiation between the infrared and radio waves with wavelengths from about 1 mm to 1 m, frequency about 10^{10} to 10^{12} Hz; absorption of microwaves produces rotation of molecules causing heat production, and are used e.g. in microwave ovens and for inducing some polymerization reactions.

microzooid (Gr. *zoon* = animal, *eidos* = form), a free-swimming, ciliated stage of some protozoans, especially ciliates.

microzoon, pl. **microzoa,** a microscopic animal, especially a protozoan.

miction, micturition (L. *micturire* = to urinate), urination.

midbrain, 1. the middle part (vesicle) of the embryonic vertebrate brain later developing into the → mesencephalon; **2.** mesencephalon.

middle ear (L. *auris media*), the air-filled part of the ear of most tetrapod vertebrates inside the temporal bone; comprises the **tympanic cavity** (*cavum tympani*, *tympanum*), containing the **auditory ossicle** (*ossicula auditus*), the **auditory tube** (Eustachian tube, *tuba auditiva*) connecting the middle ear to the pharynx. The auditory ossicle (three in mammals) forms a bone bridge through the tympanic cavity from the **tympanic membrane** (eardrum) to the **oval window** against the inner ear; the bridge contains small muscles which with a nerve reflex regulate the tenseness of the bridge according to the strength of sounds. *See* ear.

middle lamella, a lamella between the → primary walls of adjacent plant cells, composed of pectin compounds.

middle repetitive DNA, groups of those DNA segments 100—500 base pairs in length which are repeated 100—10,000 times in the genome. This class of DNA includes some → gene families, like the family of → ribosomal RNA genes in eukaryotes, and rather short, repeated DNA segments which are located between the segments of → unique copy DNA also in eukaryotes, and which probably have regulatory functions. Called also intermediate-repeat DNA, moderately repetitive DNA. *See* sequence complexity of DNA.

midges, *see* Chironomidae.

midgut, mesenteron; **1.** the middle part of the embryonic alimentary canal of vertebrates, derived from the archenteron; **2.** the middle portion of the intestine in some invertebrates, e.g. arthropods.

midrib, the largest vein in the middle of a plant leaf, reaching the leaf tip.

migration (L. *migrare* = to migrate), the act of moving or drifting from one place or region to another; in biology, the moving of individuals of a certain population between different home ranges due to changes in environmental factors, such as season, climate, feeding conditions, etc. Migration may be either regular movements between two regions, as in **seasonal migration**, including departure from a district (**emigration**) or settling to a new district (**immigration**). In genetics, emigration and immigration mean the transfer of some genetic material with a migrating individual (**migrator**) from one population to another. Thus, migration may affect evolutionary processes by changing gene frequencies in populations. *Adj.* **migratory**. *Verb* to **migrate**.

migratory restlessness, the restlessness observed in birds before the seasonal migration; occurs as an increased motility of animals, and is seen also in birds reared in cages at the time when conspecific birds are ready for migration.

mildew, 1. a visible growth of fungi and bacteria on food, wet clothes, etc; **2.** two families of parasitic fungi causing plant diseases, growth occurring as a white powder on the leaves of certain plants. The fungi in the family Peronosporaceae (division Mastigomycota) cause **downy mildews**, and the species in the family Erysiphaceae (division Amastigomycota, subdivision Ascomycetes) cause **powdery mildews**.

milk, white liquid emulsion secreted by the → mammary glands of female mammals, serving for the nourishment of the offspring. The quantity of dry substances in the milk of most mammals is about 1—5 per cent, but may be as high as 20—30 per cent, e.g. in seals and the polar bear. Milk contains proteins (chiefly casein and albumin), fats, and carbohydrates (chiefly lactose), many trace elements and minerals (rich in calcium), immunoglobulins, endorphins, melatonin, and vitamins (particularly A and E); it usually also contains environmental pollutants. *See* prolactin, oxytocin.

milk sugar, → lactose.

milk teeth, → deciduous teeth.

millipedes, → Diplopoda.

Millon's test (*Auguste Millon*, 1812—1867), a qualitative test for proteins and amino acids; the sample is heated with the **Millon's reagent** (mercuric nitrate in nitric acid), which gives a red colour with proteins and other substances, such as tyrosine and phenol.

mimicry (Gr. *mimikos* = imitative), **protective resemblance**; the resemblance of an organism (**mimic**) to either another organism or some non-living object (**model**) for avoiding predation; may concern e.g. habits, colour, or structure. Some insects, as syrphids (flower flies) and some clear-winged moths (Sesiidae) mimic species of inedible wasps, bees, or bumblebees.

In animals, three types of mimicry are found: **Batesian mimicry** in which an edible species mimics an inedible species, to the benefit of the mimic, **Müllerian mimicry** (imitative similarity), in which a mimic has a morphological similarity with two or more harmful and disagreeable species thus being of mutual advantage, and **aggressive mimicry** in which a harmless species mimics a predator or parasite.

Among plants, some species mimic harmful species in order to avoid herbivores, e.g. some *Lamium* species resemble the nettle. The flowers of some orchids mimic insects and become pollinated by insects trying to copulate with them. *See* automimicry.

Minamata disease (*Minamata Bay* in Japan), the chronic mercury intoxication found first in the inhabitants of Minamata Bay. The disease developed in people who regularly ate fish and mussels caught from water polluted by a mercury mine. The intoxication is caused by the accumulation of organic mercury in tissues, and symptoms occur especially in the nervous system and senses, such as ataxia, dysarthria, the loss of peripheral vision, mental disability, and even death; the intoxication may occur also in animals.

mineral (L. *minera* = mine, ore), **1.** any inorganic material (atoms, atom groups, ions, or inorganic compounds) occurring as a constituent of the Earth's crust; many mineral elements are essential for the life processes of organisms. Minerals needed by animals in moderate quantities are sodium, potassium, calcium, magnesium, phosphorus, chlorine, and sulphur, all in ionic form. Animals also need some other minerals in smaller quantities (*see* micronutrients). Minerals used by plants are called → macronutrients (nitrogen, phosphorus, potassium, calcium, and magnesium). Plants also use micronutrients, which are needed in much smaller amounts; **2.** a mixture of homogeneous substances obtained from the ground, such as ore, coal, petroleum, and natural gas.

mineralization, 1. the convertion of substances into a mineralized form, such as in impregnation of minerals into bones or plant cell walls; **2.** the breakdown of organic material in nature into its inorganic components and elements by decomposers, i.e. by microorganisms; **3.** in geology, the transformation of a metallic material into an ore. *Verb* to **mineralize**.

mineralocorticoids (L. *corticalis* = pertaining to the cortex), steroid hormones of the adrenal cortex of vertebrates serving in regulation of salt balance in tissues. They control the excretion of sodium, potassium, and protons (H^+) from the body into urine, faeces, and sweat by increasing the activity of the sodium pump (Na^+, K^+-ATPase) in excretory cell membranes, especially in distal renal tubules of → nephrons. The activity increases the reabsorption of sodium into the blood and excretion of potassium from the body. The action of the hormones is transmitted by special proteins (**mineralocorticoid receptors**) in the cytoplasm, the hormone-receptor complex then stimulates the expression of a certain gene in the cell nucleus. Mineralocorticoids are vital for the life of an animal, the most important of them being **aldosterone**.

minerotrophic (Gr. *trophe* = nourishment), pertaining to a bog type (**minerotrophic bog**) rich in vegetation due to nutrient-rich water running from its surroundings. Minerotrophic bogs are common in northern parts of the coniferous forest zone. *Cf.* ombrotrophic.

minicell, an experimentally isolated cell nucleus surrounded by a small fraction of the original cytoplasm and cell membrane. Minicells can be produced e.g. from cultured cells by centrifugation in a medium with added cytochalasin B. The treatment produces enucleated cells and minicells. The minicell is formed from the cell nucleus with some cytoplasm enveloped by a fragment of the plasma membrane. Minicells can be used e.g. for the production of cell hybrids. *Cf.* microcell.

minichromosome, a nucleosomal form of SV40 DNA (simian virus 40 DNA), or of polyomavirus ring-shaped DNA. *See* nucleosome.

minigene, a shortened version of a natural gene artificially constructed by deleting internal nucleotide sequences.

minimal medium, → minimum medium.

minimum essential medium, MEM, a culture medium (physiological solution) used in the cultivation of animal cells and tissues; consists of necessary salts, vitamins, amino acids, carbohydrates, and other nutrients for these cells, but does not include hormones and growth factors which are added separately e.g. in serum. Modifications of MEM have been developed originally for mammalian cells, but later also for cells from other animal groups. *Cf.* minimum medium.

minimum factor, any physical or chemical factor in an environment which is necessary for the growth and reproduction of an organism, the deficiency of which causes the organism to suffer. Thus the minimum factor restricts the occurrence and distribution of the organism. The minimum factor may be e.g. temperature, nutrient (often micronutrient), light, or humidity. *See* Liebig's law of the minimum.

minimum medium, minimal medium, a medium used in the cultivation of cells, especially bacteria and microscopical fungi; consists only of necessary nutrients, and thus is able to maintain the growth of → prototrophic, wild-type strains, but not of → auxotrophic strains, such as mutant strains. *Cf.* minimum essential medium.

minisatellites (L. *satelles* = companion), in genetics, a class of dispersed arrays of rather short (15—100 base pairs) tandemly repeated DNA sequences in the genomes of vertebrates; exhibiting a high degree of length variation, probably due to changes in the copy number of tandem repeats. Minisatellites are much used in DNA fingerprinting. *Cf.* microsatellites.

minor genes (L. *minor* = smaller), → polygenes and → modificator genes which regulate the development of the quantitative characters of organisms. The inheritance of minor genes cannot be studied by simple Mendelian methods (*see* Mendel's laws) but rather by methods of quantitative genetics, thus contrasting the study of inheritance of → major genes which regulate the development of qualitative characters of organisms. Principal qualitative differences between the minor and major genes are artificial, based only on research methods used.

Miocene (Gr. *meion* = less, *kainos* = recent), a geological epoch in the Tertiary period, between Oligocene and Pliocene, from about 24 to 5 million years ago. In the course of the epoch, the first apes appeared, primitive elephants occupied Europe, and most bird genera developed; grassy plains spread greatly. The climate was mainly moderate in the northern hemisphere, while glaciation occurred in the southern hemisphere. *Adj.* **Miocene** or **Miocenic**.

miosis, pl. **mioses** (Gr. *myein* = to close); **1.** the contraction of the pupil of the eye; called also myosis; *cf.* mydriasis; **2.** a term incorrectly used earlier for → meiosis.

miracidium, pl. **miracidia** (Gr. *meirakidion* = stripling), the ciliated, free-swimming first-stage larva of parasitic flukes (Digenea in Trematoda), emerging from the egg. The barely visible multicellular miracidium with a nerve ganglion, two eye-spots, and two nephridia, penetrates the epidermis of the intermediate host, a snail, inhabits the host's haemocoel, and develops into a sac-like → **sporocyst**. The structure and function of the miracidium are different in different fluke species. *Cf.* cercaria, metacercaria, redia.

mire, peatland; an ecosystem in which organic material is produced faster than it is decomposed, leading to the accumulation of partially decomposed vegetative material, **peat**. In some mires the peat cover is rather thin and the ecosystem is wet, the peat cover being in contact with the surrounding soil. This kind of mire grows peat mosses (*Sphagnum* species) but is dominated especially by grass-like sedges, being called → **fen**. In some mires, → **bogs**, peat becomes so thick that the surface vegetation loses its contact with the mineral soil, and is therefore dependent on the precipitation of water and nutrients, thus, the bog is very poor and acidic, being dominated by *Sphagnum* species.

missense mutation (L. *mutatio* = a change), a gene mutation in which a → codon coding for a certain amino acid has changed to code for a different amino acid. *Cf.* nonsense mutation.

Mitchell's theory (*Peter Mitchell*, 1920—1992), a chemi-osmotic theory that explains the energy transduction in cells and cellular membranes.

mites, → Acari.

mitochondrial DNA, mtDNA, the circular deoxyribonucleic acid molecule occurring in mitochondria of all eukaryotic cells; forms the genome of mitochondria which replicates and functions autonomously when mitochondria are reproduced within the cell cytoplasm. The mitochondria contain several identical circular molecules (about 14—100 kilobase pairs, in plants even 2,500) of their own DNA (mitochondrial DNA, mtDNA), less than 1% of the total cellular DNA. It contains the information to code for certain ribosomal and transfer RNAs and for some subunits of multienzyme complexes of the respiratory chain, such as cytochrome *c* oxidase. The mitochondria carry and transport genetic information, in higher organisms being inherited maternally, i.e. the mitochondria originate from a female gamete, and they are capable of dividing under control of their own DNA (e.g. in yeasts biparental inheritance). *See* mitochondrial ribosomes, mitochondrion.

mitochondrial Eve hypothesis, a hypothesis proposed by *Rebecca Cann* based on studies of polymorphism of mitochondrial DNA in man, stating that the present human population is derived from a single woman (Eve) who lived in Africa ca. 300,000 years ago. Mitochondrial DNA polymorphism is highest in Africa and derived from the variety of remarkably uniform populations. Thus the hypothesis is in principle necessarily true because the prevailing population of any species is derived from the minority of the earlier generation, and gradually from the decreasing minority of every earlier generation until one comes down to one individual only. However, the time when this individual lived, is uncertain.

mitochondrial inheritance, inheritance mediated by → mitochondrial DNA; in almost all cases mitochondrial inheritance is maternal.

mitochondrial ribosome, a → ribosome type occurring in mitochondria; its size is 70S with 50S and 30S subunits, mitochondrial ribosomes being smaller than cytoplasmic ribosomes. The mitochondrial ribosomal RNA (rRNA) is encoded by → mitochondrial DNA, mtDNA. Mitochondrial ribosomes resemble prokaryotic ribosomes in their RNA and protein compositions, size, and sensitivity to certain antibiotics, probably reflecting the bacterial ancestry of mitochondria.

mitochondrion, pl. **mitochondria** (Gr. *mitos* = thread, *chondrion*, dim. of *chondros* = granule), any of several elongated organelles or a tubular network in the cell cytoplasm, usually 1—5 μm in length; the principal energy "factories" producing ATP in cell respiration. Any cell which has a nucleus also has mitochondria, each composed of two sets of → unit membranes; the inner membrane forms folds, **cristae**, and the smooth outer membrane envelops the whole organelle. The fluid material between the cristae is the mitochondrial **matrix** containing ions and soluble organic substances. The enzymes of the citric acid cycle are located in the mitochondrial matrix, and those of → electron transport and → oxidative phosphorylation systems are embedded on the cristae, functioning in oxidative cell respiration to produce ATP. Each mitochondrion contains its own DNA, → mitochondrial DNA, mtDNA. Mitochondria are also called chondriosomes.

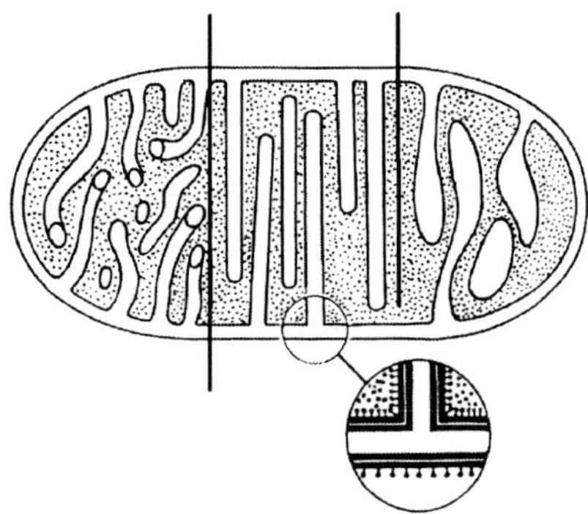

Mitochondrion. A schematic representation of the fine structure of the mitochondrion showing three structural types in combination (left: the tubular type; middle: the cristae type; right: the intermediate type frequently found in plants). Bottom: A detail section showing the arrangement of the tennis rocket-like structures being the carriers of the oxidative enzymes. (After Sitte: Bau und Feinbau der Pflanzenzellen - Fischer, 1965; with the permission of the publisher).

mitosis, pl. **mitoses, karyokinesis**; the division of the cell nucleus, resulting in two nuclei which are genetically identical with each other and with the mother nucleus. The stages of mitosis are prophase, prometaphase, metaphase, anaphase, and telophase.

In **prophase**, the chromosomes which have replicated during the preceding → interphase become visible in the form of two-stranded strings under the light microscope. During **prometaphase** the → mitotic spindle is formed by the action of → centrioles, and its microtubules attach to the → centromere of each chromosome, causing these to move towards the equator of the spindle (congression movement of chromosomes) to form the metaphase plate. At the same time the chromosomes become gradually more condensed and thus more visible and the nuclear envelope disappears.

In **metaphase**, the condensation of the chromosomes reaches its maximum and they lie at the equatorial plane (metaphase plate) in a position in which their centromeres are oriented to the equator of the cell. At the beginning of **anaphase** each chromosome divides, and the sister chromatids separate into independent chromosomes and begin to move towards the opposite poles of the nucleus by the

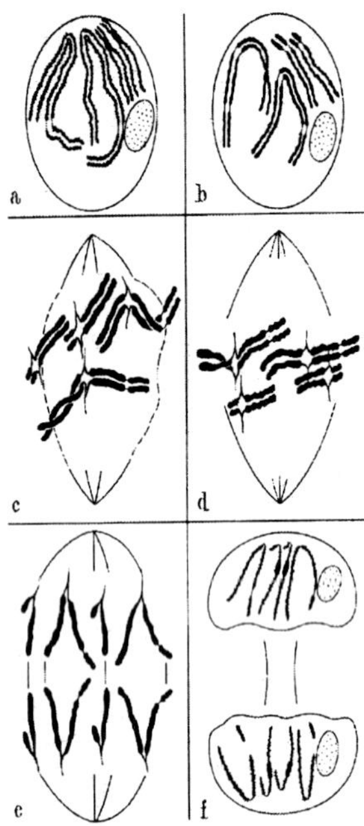

A schematic representation of mitosis. (a). early prophase, (b). late prophase, (c). prometaphase, (d). metaphase, (e). anaphase, (f). telophase. (After Rieger, Michaelis & Green: Glossary of Genetics - Springer Verlag, 1968; with the permission of the publisher).

action of the mitotic spindle, i.e. the chromosomes are pulled by the contracting microtubules attached to the centromeres. In **telophase** the movements of the chromosomes are completed, the chromosomes begin to lose their condensation, i.e. decondensation occurs, and separate nuclear envelopes are formed around the two groups of chromosomes.

Usually the telophase is followed by the division of the cell itself when the plasma membrane (in plants also the cell wall) is generated between the daughter nuclei. *Cf.* meiosis.

mitosis-promoting factor, MPF, a protein in eukaryotic cells inducing the initiation of cell division, i.e. → mitosis and → meiosis; MPF is a heterodimer that has the activity of → **protein kinase**, phosphorylating proteins which mediate mitotic processes. MPF is composed of a catalytic enzyme protein (kinase subunit) and a regulatory protein (regulatory subunit), the latter controlling kinase activity by determining which proteins are phosphorylated by this enzyme. It has been found that in dividing embryonic animal cells the regulatory unit (called **cyclin**) is synthesized continuously up to anaphase, at which phase its abrupt degradation occurs; this periodic synthesis and degradation is required for the rapid mitotic cycles of these cells. MPF was originally called **maturation-promoting factor** because of its capability of promoting the maturity of *Xenopus* oocytes.

mitosporangium (Gr. *sporos* = seed, *angeion* = vessel), a → sporangium producing spores through mitosis.

mitospore, any → spore formed through mitosis.

mitotic apparatus, a structure observable in each cell during → mitosis; is visible from prometaphase to the end of anaphase. The apparatus affects the movement of chromosomes during cell division (congression and anaphase movements). The mitotic apparatus forms in both poles of the nucleus a star-like structure (aster formation) consisting of → microtubules which further extend to form the → mitotic spindle connecting the poles. In animal cells the → centriole is in the middle of the aster formation.

mitotic cycle, → cell cycle.

mitotic index, a ratio which indicates the proportion of cells within a tissue which are presently in → mitosis.

mitotic spindle, an oval, bipolar filamentous complex which can be seen in cells during cell division; acts in the control of chromosomal movements and the division of the cytoplasm. The → centromeres of the chromosomes are connected with the **chromosomal microtubules** (kinetochore microtubules) of the spindle (*see* microtubules) whose contractions cause the anaphase movement of the chromosomes towards the cell poles. Also the **continuous microtubules** reaching from pole to pole extend during anaphase and thus play a crucial role in the chromosomal movements (*see* mitotic apparatus). The pair of → centrioles function in the organization of the mitotic spindle. Also called metaphase spindle, nuclear spindle, or spindle.

mitral cells (Gr. *mitra* = mitre), large nerve cells in the olfactory lobe of the vertebrate brain; axons of the olfactory receptor cells passing from the nasal mucosa form → synapses with dendrites of the mitral cells whose axons form the olfactory tract to the olfactory cortex.

mitral valve, → bicuspid valve.

mixed inflorescence, an → inflorescence which has characteristics of different basic inflorescence types; e.g. the inflorescence of *Hippocastanea* with small cymes in a raceme.

mixolimnion (Gr. *mixis* = act of mingling, *limne* = lake), the upper zone above the monimolimnion in a → meromictic lake where water circulation and oxygenation usually occur. *See* stratification.

mixoploid (Gr. *-ploos* = -fold, *eidos* = form), a population of cells which contain different numbers of chromosomes, e.g. in → chimaera; the condition is called **mixoploidy**.

mixotrophic (Gr. *trophe* = nourishment, food), pertaining to an organism (**mixotroph**) which has partly autotrophic and partly heterotrophic nutrition, e.g. many saprophytic or parasitic organisms. Also called mesotrophic.

MLE, → maximum likelihood estimation.

mnemotaxis (Gr. *mneme* = memory, *tassein* = to arrange), a → taxis of an animal in which the stimulus directing a movement is acquired by memory; occurs e.g. in birds' orientation to feeding or nesting places. *Syn.* pharotaxis.

mobbing, a joint harassment and assault directed usually at a nest predator such as a hawk, performed by a social group or mated pair of animals; occurs especially among many bird species, including song birds, gulls, and terns. When mobbing, the defenders swarm around and pursue the predator with the purpose of driving it away. In a slighter form, mobbing may be only a loud warning uttered near the predator, but especially in the late breeding period mobbing is active, often including determined attacks towards the predator.

mobile DNA element, one of transposable elements (*see* transposon, retrotransposon); any DNA sequence the chromosomal location of which varies between individuals of a species.

model, something that simulates or represents something else; e.g. **1.** models for describing, demonstrating, and teaching biological structures and functions; **2.** an animal model for human condition; **3.** in ethology, an artificial and simple object simulating a key-stimulus of particular fixed-action patterns; is commonly used in behavioural experiments. Usually a suitable model represents only a part of the unit, such as a beak, eyes, colour pattern, etc. *See* mimicry.

modification, in biology, e.g. **1.** any non-heritable change in the → phenotype of an individual caused e.g. by the environment; **2.** in bacteria, the addition of a methyl group to one or two bases of DNA by a modification enzyme, usually within the restriction site, protecting the bacterial DNA from cleavage by its own restriction endonuclease while it destroys foreign DNA; **3.** behavioural modification, a changed pattern of the behaviour of an animal due to conditioning or learning experiences; may teach new skills, or extinguish undesirable behaviour or attitudes.

modificator gene, a → gene which in interaction with other genes causes a change in their expression. According to their effects the modificator genes are divided into → **enhancers** and **reducers**. *Syn.* modifier (gene).

modified organ, an organ or part of an organism whose structure has changed in the course of evolution to such a degree that its origin is difficult or impossible to determine; e.g. flattened platyclades or leaves modified into thorns.

modifier gene, → modificator gene.

modular organism, an organism composed of several, iterative structural units (**modules**), organically connected with each other; typical modular organisms are e.g. colonial hydroids, bryozoans, and trees.

module (L. *modulus*, dim. of *modus* = measure, quantity), a unit of measurement; a structural unit; in biology, e.g. **1.** modules of the brain cortex, functional units formed from small columnar groups of neurones, each group (module) controlling a certain function, such as the same motor tract; **2.** a structural unit in a colonial organism or a plant clone (→ modular organism) that is organically connected with the colony and usually capable of reproducing sexually. The connection may be broken after which the separated modules develop into independent organisms. *See* genet.

molality (L. *moles* = piece, lump), molal concentration, *m*; the concentration of any solution expressed as → moles of a solute per one kilogram of a solvent; 1 molal solution = 1 mole of solute/1 kg of solvent. *Adj.* **molal**.

molarity, M, molar concentration; the concentration of any solution expressed as → moles of a solute per one litre of the final solution; 1 M solution = 1 mole of solute/1 l of solution, abbr. **mol/l** or **mol/dm^3**. *Adj.* **molar**.

molasses (L. *mel* = honey), **1.** the brown, viscid mixture (syrup) obtained from raw sugar in processes of sugar manufacture; is used in husbandry, as cattle feed, and in the production of alcohol; **2.** a similar syrup obtained by boiling fruits, fruit sap, or sweet vegetables.

mold, → mould.

mole, 1. (< molecular weight), abbr. **mol**, the SI unit of the amount of a chemical substance; the amount of a chemical compound whose mass in grams equals its relative molecular mass (molecular weight); also called gram-molecule; *see* Appendix 1; **2.** pigmented nevus (L. *naevus*), a congenital mark or spot (old English *mal*) on a human skin; **3.** (L. *moles* = mass, lump), in pathology, a fleshy mass, as e.g. formed in the uterus by the abortive development of an ovum; **4.** an insectivorous mammal (Talpidae).

molecular biology, a branch of biology dealing with the physicochemical structure and function of living matter; the biological phenomena are studied by biochemical methods at molecular level, thus being closely related to biochemistry of nucleic acids and proteins. *See* molecular genetics.

molecular clock, a concept stating that at the molecular level the rate of → evolution is constant in different evolutionary lines, and thus the differences in biological macromolecules (nucleic acids and proteins) indicate the time at which different lines have separated from each other. According to this hypothesis, each protein has a typical and constant evolutionary rate of its own. *See* evolutionary clock.

molecular genetics, a branch of genetics dealing with genetic phenomena in terms of physics and chemistry, i.e. molecular biology applied to genetics. It involves the molecular study of genetic mechanisms, such as the molecular structure of nucleic acids and the biochemical basis and regulation of gene expression. In practice, molecular genetics is the study of biochemical reactions of nucleic acids and proteins, concerned primarily with the chemical reactions relating genotype and phenotype, e.g. replication of DNA, transcription, and translation into proteins.

molecular weight, MW, mol wt, → relative molecular mass.

molecular weight marker, *see* marker.

molecule (L. *molecula*, dim. of *moles* = mass), the smallest structural unit of a substance formed by chemical bonds from two or more → atoms, i.e. a di-, tri- or polyatomic substance. It is the smallest unit to which any substance can be divided without changing its chemical properties, and is the smallest particle of matter that can exist in a free state; e.g. water, H_2O, sodium chloride, NaCl, or oxygen molecule, O_2. *See* molecular weight.

molecule filter, an ultrafilter made of synthetic aluminium silicates and used for separating macromolecules, such as proteins and nucleic acids.

Mollusca (L. < *mollis* = soft), **molluscs**, Am. **mollusks**; a phylum comprising invertebrate animals characteristically covered by a calcareous shell which may be reduced as in slugs, squids, and octopuses. In general, the soft, usually unsegmented body comprises a head, an unpaired muscular foot, and a visceral sac surrounded by mantles (the head is reduced in bivalves). The coelom is rather small, and the open → haemocoel forms a part of the circulatory system. The nervous and sensory systems are well developed especially in squids and octopuses. The molluscs are a very polymorphic taxon including about 35,000 extinct species and more than 100,000 living species in eight classes; i.e. **gastropods** (Gastropoda), **chitons** (Polyplacophora), **bivalved molluscs** (Bivalvia), **caudofoveates** (Caudofoveata), **solenogasters** (Solenogastres), **tooth shells** (Scaphopoda), **squids** and **octopuses** (Cephalopoda). Most primitive and now almost extinct are **monoplacophorans** (Monoplacophora) which first appeared in Cambrian. *Adj.* **molluscan**, Am. **molluskan**, relating to Mollusca; **molluscoid**, resembling a mollusc.

molt, → moult.

molybdenum (Gr. *molybdaina* = a piece of lead), a hard metal of the chrome group resembling iron, symbol Mo, atomic weight 95.94, density 10.220; in nature, found in various compounds, as in molybdenite (MoS_2) and molybdates (best known wulfenite, $PbMoO_4$). The ionic form of molybdenum (Mo^{2+}) is an essential trace element in organisms acting as the → cofactor of xanthine oxidase and → nitrate reductase and is a structural component of the larger subunit of bacterial → nitrogenases. Metallic molybdenum is used in alloys and special steels, and its sulphide as a solid lubricant.

mon(o)- (Gr. *monos* = single, alone), pertaining to a single element or part; same as *uni-* (L.).

monadelphous, (Gr. *adelphos* = brother), pertaining to a flower structure in which the stamens have united into one group, the anthers being separate. *Syn.* monodelphous.

monarch (Gr. *arche* = beginning), pertaining to a primitive structure of vascular tissues in plant roots in which the tissues form only one strand without any side protrusions in the central cylinder; occurs e.g. in ferns. *Cf.* triarch, tetrarch.

Monera (L., pl. of *moneron* < Gr. *moneres* = solitary), → prokaryotes.

mongolism, → Down's syndrome.

monimolimnion (Gr. *monimos* = stable, steady, *limne* = lake), the bottom zone of the lower water layer of a → meromictic lake in which the water never mixes with the upper layer.

monoacylglycerol, monoglyceride. *See* triacylglycerol.

monoamine oxidase, MAO, amine oxidase; an oxidoreductase enzyme, EC 1.4.3.4, oxidizing amines into aldehydes or ketones using oxygen; in animals this mitochondrial enzyme occurs in the nervous system regulating the impulse transmission in → synapses by inactivating amine neurotransmitters, such as → catecholamines and → serotonin. Earlier called tyramine oxidase, tyraminase.

monocarpic (Gr. *monos* = single, *karpos* = fruit), pertaining to a plant which dies after producing fruits once; also used as a synonym of monocarpous. *Cf.* semelparity.

monocarpous, a plant having only one carpel per flower, e.g. pea and bean. *Cf.* polycarpic.

monochasium (L. < Gr. *dichasis* = division), an → inflorescence in which the main axis forms only one branch, this again one branch, etc. *Cf.* dichasium.

monochlamydeous, → haplochlamydeous.

monocistronic (Gr. *cis* = on this side, *trans* = on the other side), pertaining to a → messenger RNA molecule which contains information from one → cistron only, and thus encodes only one polypeptide; typical of eukaryotic organisms.

monoclimax theory (Gr. *klimax* = ladder), a theory stating that all the succession series occurring in a certain area will result in only one climax community. *See* succession. *Cf.* polyclimax theory.

monoclonal antibody (Gr. *klon* = twig), an → antibody specific to one antigen determinant; produced by a genetically identical → clone of B lymphocytes originated from a single cell, or experimentally using a hybridoma cell line. Monoclonal antibodies have become important tools in diagnostics, therapy, and research (e.g. in immunoassay).

monocolpate (Gr. *kolpos* = fold, vagina), pertaining to a pollen grain with one → germ pore; also dicolpate and tricolpate types are commom.

Monocotyledonae, monocotyledons; a class in the subdivision Angiospermae, division Spermatophyta (Anthophyta, seed plants), Plant kingdom. Typical monocotyledons have only one cotyledon (seed leaf) in their embryo, the base number of the flowers is three, the venation of the leaves is mostly parallel, and the vascular bundles are scattered in the stem. Typical monocotyledons are lilies, orchids, and grasses like wheat, rice, barley, and maize. Another class and developmental line in Angiospermae is → Dicotyledonae.

monoculture, a planted area covered only by a single plant species, such as a crop cultivar or a tree species; may be very sensitive to plant diseases or the effects of herbivores.

monocyclic (Gr. *kyklos* = circle), having a single cycle, e.g. **1.** describing a hapaxanthous plant which flowers and after ripening of the seeds, dies; **2.** pertaining to a flower with a single whorl, i.e. having only petals or sepals; **3.** in chemistry, pertaining to a molecular structure that contains only one ring, such as a 5-carbon ring.

monocyte (Gr. *kytos* = cavity, cell), the largest (13—25 μm) of various white blood cells of vertebrates (in mammals 3—8% of all leucocytes); a spherical cell with one big, kidney-shaped nucleus. The monocytes circulate in blood for quite a short time migrating to other tissues, such as the liver (Kupffer cells), spleen (white pulp macrophages), brain (microglia), bone (osteoclasts), and lung (alveolar macrophages). After migration the cells undergo further differentiation developing into the **tissue macrophages** capable of acting as phagocytes which engulf and destroy foreign material, such as microbes and cell fragments. They also take up antigens, process them e.g. by partial degradation, and present them to specific T lymphocytes. All these cells together (including monocytes and often also lymphocytes) are called **mononuclear cells**. *Adj.* **monocytic**. *See* macrophages.

monodactylous, monodactyle (Gr. *daktylos* = finger, toe), having one digit or claw, such as in horses and some extinct ungulates (some species of the order Litopterna); **monodactyly, monodactylism,** also **monodactylia**, a condition of being monodactylous; may also be a developmental anomaly sometimes found in tetrapod vertebrates characterized by the presence of one digit on the foot (or hand).

monodelphous, → monadelphous.

monoecious (Gr. *oikos* = house), **1.** having both female and male reproductive organs in the same individual; producing female and male gametes in the same individual; *syn.* → hermaphrodite, hermaphroditic; **2.** in botany, pertaining to a plant having stamens and a pistil in the same flower, or female and male flowers in the same plant. *Cf.* dioecious.

monogalactosyl diglyceride/diacylglycerol, MGDG, a glycolipid in which galactose is bound to the non-esterified -OH group of glycerol; is an important glycolipid of → thylakoids in chloroplasts of plants. *Cf.* digalactosyl diglyceride.

monogamy (Gr. *gamos* = marriage), the mating system in which a male and a female form a pair; the relationship may last only one breeding season or throughout life. Monogamy occurs rarely among mammals but is common among birds of which about 80% are monogamous. *Adj.* **monogamous** or **monogamic**. *Cf.* polygamy, polyandrous, polygynous.

Monogenea (Gr. *genea* = race, descent), **monogeneans**; **monogenean flatworms**; a subclass (order) in the class Trematoda, phylum Platyhelminthes; many live parasitic in vertebrates, e.g. ectoparasites on the gills and skin of fish, but also in higher vertebrates, mammals included. The life cycle of monogeneans is simple, having only one generation with a free-living ciliated larva and without a change of host.

monogenic (Gr. *genos* = birth, offspring, kindred), pertaining to a character which is regulated by the → alleles of a single (known) gene.

monogenous, asexually produced, as by gemmation, sporulation, or fission.

monoglyceride, monoacylglycerol. *See* triacylglycerols.

monogynous (Gr. *gyne* = woman), **1.** pertaining to a male who has only one female mating partner at the time; in social insects, having only one functional female (the queen) in the colony; **2.** having one pistil in a flower. *Noun* **monogyny**.

monohybrid (L. *hybrida* = an organism of mixed breed), pertaining to, or being an individual which is heterozygous only for one known pair of → alleles; also the progeny of a cross in which the parents differ in the alleles of a single gene → locus, as e.g. *AA* x *aa* → *Aa* (A = → dominant gene, a = → recessive gene).

monohydrate (Gr. *hydor* = water), a compound containing one molecule of water.

monokaryon, also **monocaryon** (Gr. *karyon* = nut, nucleus), a spore, tissue, or hypha which consists of uninucleate cells. *Cf.* dikaryon.

monokines (Gr. *kinein* = to move), the common name for → cytokines released from mononuclear phagocytes such as → monocytes; are regulatory messenger polypeptides and proteins which act in communication between the cells of the immune system of vertebrates.

monolayer, 1. a single cell layer, growing as a sheet of cells on a substratum in cell culture conditions; **2.** a single layer of molecules, a molecular film formed on the water surface by certain substances, such as polar lipids, e.g phospholipids. *Cf.* bilayer.

monomer (Gr. *meros* = part), a single unit which may unite to form a larger structure, → polymer; especially a molecule of low molecular weight which reacts with similar molecules by repetition (polymerization) resulting in a dimer, trimer, tetramer, etc., or a polymer, in organisms e.g. single **nucleotides** forming DNA or RNA, **amino acids** forming a polypeptide, or **monosaccharides** forming polysaccharides. Also a single polypeptide chain which together with other polypeptides form a large protein is called monomer; the protein is then called multimer.

monomeric, 1. pertaining to a → monomer; **2.** → monomerous.

monomerous, made up, or derived from one part or segment; also monomeric.

monomictic (Gr. *mixis* = act of mingling), *see* stratification.

monomorphic, also **monomorphous** (Gr. *morphe* = shape, form), uniform; unchangeable in shape, exhibiting the same or similar structure; e.g. **1.** pertaining to a → population which comprises individuals of one type only, or ametabolous insects retaining a similar form throughout different developmental stages; **2.** pertaing to plants and fungi which produce spores of one type only. *Noun* **monomorphism**. *Cf.* dimorphism, polymorphism.

mononuclear, mononucleate, uninucleated; pertaining to a cell having a single nucleus.

mononuclear cell, 1. any uninucleated cell; **2.** in vertebrate physiology, especially refers to any cell of the mononuclear phagocyte system; *see* → monocyte.

mononuclear phagocyte system, MPS, *see* macrophage.

mononucleotide, → nucleotide. *Cf.* polynucleotide.

monophage (Gr. *phagein* = to eat), an animal feeding on one kind of food only, e.g. an insect feeding on plants of a certain genus, or preying only on one animal species. *Adj.* **monophagous**. *Cf.* oligophage, stenophage, polyphage.

monophyletic (Gr. *phyle* = tribe), **1.** in taxonomy, pertaining to a group of organisms or a taxon which contain the most recent common ancestor of all members of the group, all descending from that ancestor; *cf.* paraphyletic, polyphyletic; **2.** in animal physiology, pertaining to the theory that all blood cells derive from one common stem cell (haemocytoblast).

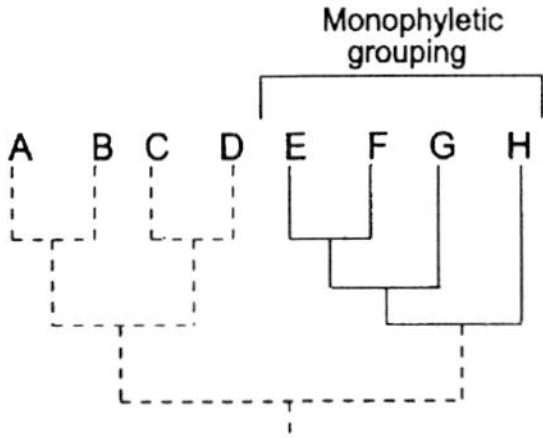

Monophyletic relationships between eight taxa (A - H).

Monoplacophora (Gr. *plax* = flat plate, *phorein* = to carry along), **monoplacophorans**; a small class of primitive molluscs, comprising mainly extinct species; the present species are living in abyssal depths of oceans. Some structural features of monoplacophorans suggest that they are possible ancestors of clams and squids. *See* Polyplacophora.

monoplanetic (Gr. *planos* = wandering), pertaining to an organism which has one motile stage in its developmental cycle, as in some fungi.

monoploid, → haploid.

monopodial branching (Gr. *pous*, gen. *podos* = foot), **1.** a type of plant branching in which the main, primary axis (trunk) remains and none of the developing younger branches gets a dominating role in the structure; the result is thus a straight axis, **monopodium**; **2.** a similar branching growth in colonial hydroids. *Cf.* sympodial branching.

monopodium, pl. **monopodia,** *see* monopodial branching.

monosaccate (L. *saccus* = sac), pertaining to a pollen grain which has only one air bladder.

monosaccharide (Gr. *sakcharon* = sugar), a simple carbohydrate with the formula $(CH_2O)_n$, where n = 3—7; e.g. → trioses (as dihydroxy acetone), pentoses (as ribose and deoxyribose), hexoses (as glucose, fructose, and galactose), and septuloses. Monosaccharides are synthesized and decomposed in the cellular metabolism but are also produced in cells by hydrolysis of di-, oligo- and polysaccharides, or of glycosides; this hydrolysis occurs also in the digestive tracts of animals.

monosiphonic, monosiphonous (Gr. *siphon* = tube), having a single tube, siphon; e.g. **1.** pertaining to a hydrocaulus of some hydrozoans; **2.** pertaining to a type of germination of a → pollen grain in which the grain forms only one pollen tube; *see* fertilization; **3.** pertaining to algae which have a single central tube in filaments.

monosomy (Gr. *soma* = body), a form of aneuploidy in which, in an otherwise diploid chromosome set, one homologous chromosome appears single, or the X chromosome is unpaired (X0). Monosomy usually results from a mitotic or meiotic disturbance, called → non-disjunction. When two or more chromosomes appear singly, the condition is called **twofold monosomy**, **threefold monosomy**, etc. *Adj.* **monosomic**. *See* Turner's syndrome.

monospore, *see* spore.

monostromatic (Gr. *stroma* = mattress, bed), pertaining to a plant with only one cell layer in its thallus; e.g. some green algae such as *Monostroma* species.

monosymmetric flower, pertaining to a flower with only one plane of symmetry. *Syn.* dorsiventral flower, zygomorphic flower. *Cf.* asymmetric flower, bilateral flower, symmetric flower.

monoterpene, *see* terpenes.

Monotremata (Gr. *trema* = hole), **monotremes, egg-laying mammals**; a mammalian order in the subclass → Prototheria comprising only six living species in two families, i.e. Ornithorhynchidae, (1 species, duckbilled platypus) and Tachyglossidae (5 species, e.g. spiny anteater). Monotremes live in Australia and Tasmania, the spiny anteater also in New Guinea. Contrary to other mammals, monotremes have a cloaca and a horny beak; the teeth are absent in adults. Females are oviparous, laying eggs, and have no uterus and vagina. In a recent classification, monotremes are classified with other mammals as an order in the infraclass Ornithodelphia, subclass → Theria.

monotypic (Gr. *typos* = type), pertaining to a taxon including only one subtaxon; e.g. a genus with only one species, or a species with no subspecies. *Cf.* polytypic.

monozygotic (Gr. *zygon* = yoke), pertaining to mammalian individuals, especially to human monozygotic twins (identical twins), developed from a single fertilized ovum splitting at the 2-cell or early embryonic stage into two separate parts. The zygocity of twins can be verified with some certainty on the basis of their foetal membranes, with higher certainty on the basis of inherited characters like blood groups, and with an absolute certainty by using DNA analysis. Even more than two genetically identical monozygotic individuals may develop if the early embryo splits into more than two parts. The armadillo (*Dasypus*) gives birth normally to four young at one time, all of the same sex (either male or female) and all derived from one zygote. *Cf.* dizygotic.

monuron (Gr. *ouron* = urea), N'-(4-chlorophenyl)-N,N-dimethylurea; a herbicide which inhibits the light reactions of photosynthesis.

moor, a treeless plant community, rich in peat and small woody plants such as heather. Moors are common in areas of humid and cool climate and on acid soils, found e.g. on shore areas of the Atlantic Ocean and the Nordic Sea.

MOPS, 3-(N-morpholino)-propanesulphonic acid; an organic → buffer, active in the pH range of 6.5—8, used in biochemical and physiological studies.

morgan, M, (*Thomas H. Morgan*, 1866—1945), a unit of gene map distance, i.e. the length of a → chromatid in which on average one → chiasma is formed per → meiosis; in practice, **centimorgan, cM** (0.01 M) is used. The distance in morgan units between two gene loci in a chromatid equals the number of chiasmata formed between these loci in a → bivalent, divided by two.

morph(o)- (Gr. *morphe* = shape, form), denoting relationship to a form or shape, often to an external structure of an organism.

morph, one of the genetic forms (individual variety) occurring in a polymorphic population; different forms are e.g. the dark and pale peppered moths. *See* polymorphism, industrial melanism.

morphactins, derivatives of fluorene-9-carboxylic acid; affect plant growth and development acting e.g. as inhibitors of auxin transport in plant tissues, thus inhibiting the geotropic response of plant parts. Morphactins also stimulate the growth of primary roots, the elongation of lateral roots being inhibited.

morphallaxis, pl. **morphallaxes** (L. < Gr. *allaxis* = exchange), a type of the tissue regeneration in which a part of an organism's structure is renewed by reorganization of present cells without proliferating new cells; e.g. the regeneration of a whole hydra from a part of the body. The new individual formed is smaller.

morphine (*Morpheus* = god of dreams in ancient Greece and Rome), a narcotic alkaloid of opium; 7,8-didehydro-4,5-epoxy-17-methylmorphinan-3,6-diol. Like all opiates, morphine acts in animal tissues through special receptors, **opiate receptors**, e.g in vertebrates causing depression of some synaptic functions and excitation of some other functions of the nervous system and some peripheral tissues; in humans, this is manifested as **analgetic** (pain-relieving) and **euphoric** effects. Endogenous ligands in tissues for these receptors are → endorphins and enkephalins which are synthesized especially in the nervous system. Repeated administration of morphine and related substances (also endorphins) leads to tolerance and physical and psychological dependence (addiction). Morphine and its derivatives are misused as narcotic drugs.

morphogen (Gr. *morphe* = shape, form, *gen-*

nan = to produce), any of the diffusible substances (polypeptides, proteins, ribonucleoproteins) of embryonic tissues, the concentration gradient of which contains → positional information and thus determines the morphogenetic differentiation of cells in a developing embryo or organ; e.g. the **bicoid** and **dorsal proteins** of the fruit fly. *See* morphogenetic field.

morphogenesis, the development of form; the differentiation of cells and tissues leading to the development of the shape of an organ or part in an early animal or plant embryo, also during → metamorphosis. *Adj.* **morphogenetic**. *See* organogenesis.

morphogenetic field, a field containing gradients of certain → morphogens in an embryo or a part of it; determines the formation of certain polar structures like the proximal and distal parts of a limb. For example, in the formation of dentition of mammals the incisors form one, canines a second, intermolars a third, and molars a fourth structure determined by different morphogenetic fields.

morphogenetic hormones, hormones affecting → morphogenesis during the development of an organism; in animals e.g. → thyroxine and → juvenile hormone.

morphogenetic map, fate map; a map describing the fertilized egg or an embryo expressing the final fate of developing embryonic regions.

morphogenetic movements, movements of cells and tissues during the early development of an animal embryo, i.e. during → morphogenesis; extensive morphogenetic movements occur especially in the formation of blastula and gastrula.

morphological race, *see* race.

morphological species, → morphospecies.

morphology (Gr. *logos* = word, discourse), a branch of biology dealing with structural features and forms of organisms. Morphological structures are examined on the basis of their evolutionary backgrounds, functional factors, ontogenic stages, and environmental factors, such as food and predation. Traditionally, the study of the macroscopic morphology of organs is called anatomy, whilst the microscopic morphology of cell organelles, cells, and tissues is called histology. Sometimes the term is used to denote also the form or structure itself. *Adj.* **morphologic(al)**. *Cf.* physiology.

morphometry (Gr. *metrein* = to measure), a method dealing with the measurement of shape, e.g. the data obtained from an object (e.g. a tissue sample) in a series of two-dimensional sections are added and analysed by a computer to predict the properties of this object in three dimensions.

morphospecies, morphological species; a → species described only on the basis of individuals' morphological structures.

mortality (L. *mors* = death), **1.** the state or quality of being subject to death, i.e. of being **mortal**; **2.** → mortality rate; **3.** a fatal outcome; a heavy loss of life, such as by disease or war.

mortality rate, death rate; a rate expressing the number of deaths in a population, i.e. the proportion of organisms which died in a certain time interval as a function of the number of all individuals which lived in a population (e.g. at the beginning of the interval); e.g. in a human population, the mortality is expressed as the number of deaths per 1,000 (also per 10,000 or 100,000) individuals per unit of time, usually a year. Called also lethality rate.

morula, pl. **morulae,** Am. also **morulas** (L. *morum* = mulberry), the first embryonic stage of most multicellular animals comprising a solid mass of blastomere cells (**blastomeres**) resulting from the cleavage divisions of a fertilized ovum (zygote); followed by the → blastula and gastrula stages.

mosaic (L. *musaicum* < *musivum* = artistic), a pattern made of numerous small pieces fitted together; in biology, e.g. **1.** an individual having genotypically different cells; *see* mosaicism; **2.** a viral disease characterized by coloured mottling of the foliage.

mosaic egg, an egg cell type of animals in which the differentiation has occurred to the extent that it determines the daughter cells, borne during cleavage divisions, to develop in a certain direction; e.g. egg cells of tunicates. *Cf.* regulative egg.

mosaicism, cellular mosaicism; the existence of genotypically different cells in a single individual, **mosaic**. In contrast to → chimaeras, the genetically different cells of a mosaic are always cells of the same species. Mosaicism may be produced artificially but also occurs in natural conditions, as e.g. very rare forms of butterflies which consist of both female and male tissues (gynandromorphic mosaicism).

mosses, the common name for plants of the class Bryopsida (Musci) in the division → Bryophyta, Plant kingdom; are characterized by a leafy gametophyte, a sporophyte growing in the tip of the gametophyte, and a filamentous young gametophyte (a protonema).

moss animals, bryozoans. *See* Bryozoa.

moths, *see* Lepidoptera

motif, 1. a unit exhibiting a particular three-dimensional architecture in proteins, usually associated with a given function; e.g. many DNA-binding proteins contain one of a small number of DNA-binding motifs, such as → homeodomain and → zinc finger; **2.** a given DNA sequence found in a given gene in one particular population only; e.g. the Saami motif of human mitochondrial DNA in the population of Saami.

motility (L. *movere* = to move), the capability of spontaneous movements. *Adj.* **motile**.

motivation (L. *motivus* = moving), a temporarily changing readiness for a goal-directed behaviour of an animal; activated by innate forces generated in the brain. In higher vertebrates, motivations with associated emotional conditions are regulated by the → limbic system, the activity of which is dependent on the nutritional and physiological (especially hormonal) state, previous experiences, and sensory stimuli. *Verb* to **motivate**. *Adj.* **motivational** (motivative). *See* Deutsch's motivation model, Lorenz's drive model.

motive, something that incites an individual to a goal-directed behaviour, as hunger to predatory behaviour.

motoneurone, motoneuron, motor neuron(e) (L. *motio* = movement, Gr. *neuron* = nerve), **1. somatic motoneurone**; any nerve cell (neurone) that conveys impulses from the central nervous system to skeletal muscles. In vertebrates, the simplest form of a motor system consists of two successive motoneurones; the first is the **upper motoneurone** in the cerebral motor cortex sending its axon to the second neurone, the **lower motoneurone**, situated in the brain stem or spinal cord from which it sends its axon to the skeletal muscle. The latter neurone is also called peripheral motoneurone, **alpha motoneurone**. A special type of motoneurone is the **gamma motoneurone** which innervates the → muscle spindle; *see* pyramidal system, extrapyramidal system; **2. autonomic motoneurone** (visceral motoneurone); nerve cells of the → autonomic nervous system innervating smooth muscles.

motor, a source of motion; in biology, pertaining to muscles or nerves, or other elements producing movements in cells, tissues, and organs, or to locomotor activity of an organism.

motor cortex, pl. **motor cortices,** the posterior area of the frontal lobe of the cerebral cortex in the brain of mammals; contains the cell bodies of the first → motoneurones which through the → pyramidal system control precise motor functions of voluntary muscles. The motor cortex corresponds approximately to → Brodmann's areas 4 and 6 of the precentral gyrus.

motor end-plate, *see* neuromuscular junction.

motor nerve (L. *nervus motorius*), an efferent somatic nerve stimulating muscle contraction, i.e. any of the nerves (or nerve branches) comprising nerve fibres to skeletal muscles; in vertebrates, e.g. the oculomotor nerve of the eye (*nervus oculomotorius*), or the motor nerve of the tongue (*nervus hypoglossus*). The functional units of motor nerves are the peripheral or lower → **motoneurones**; many nerves are mixed nerves consisting of motor, sensory, and autonomic neurones.

motor neurone, → motoneurone.

motor proteins, proteins involved in movements of cells or cell organelles, including e.g. → **myosin**, **dynein**, and **kinesin**; large proteins inside the cells with enzyme activity (ATPase activity), used for sliding movements along cytoplasmic → microfilaments and → microtubules.

motor unit, the neuromuscular unit including all skeletal muscle fibres which are innervated by the same peripheral → motoneurone; in vertebrates, one unit may contain only a few muscle fibres, as in an external ocular muscle, or several hundreds of muscle fibres depending on how precise movements a given muscle serves.

moulds, Am. **molds**, the common name for fungi with similar growing patterns forming dense layers of hyphae, mycelium, on many different objects or soil. Best known moulds are **Penicillium** and **Aspergillus** which belong to the division Deuteromycota, but moulds are known also in the groups of Ascomycetes (Ascomycota) and Myxomycetes (Myxomycota). Many moulds are important decomposers in ecosystems.

moult, Am. **molt** (L. *mutare* = to change), **1.** to cast off or shed the outgrown skin or exoskeleton, mainly in the process of growth, as in insects, or crustaceans and some other arthropods, as well as horny scales of lizards and snakes, or feathers of birds and hairs of mammals; **2.** the process of such a shedding, called also **moulting** (molting); *see* ecdysis.

moulting hormone, → ecdysone.

mountain sickness, altitude sickness, puna, soroche; a syndrome caused especially in man when exposed to high altitude with decreased atmospheric pressure (oxygen pressure). In the acute form (Acosta's disease), low pressure causes lowering of arterial oxygen content and respiratory alkalosis with breathlessness, fatigue, headache, nausea, dizziness, vomiting, pulmonary oedema, insomnia, and impairment of mental functions. In prolonged exposure (acclimatization), polycythaemia with increased haemoglobin content relieves acute symptoms.

mountain zone, *see* height zones.

mousebirds, → Coliiformes.

mouth, 1. (L. *os*, pl. *ora*), the anterior opening of the alimentary canal; in cnidarians, comb jellies, and flatworms, the mouth serves also as an osculum ("anus"), but in other animals the embryonic blastopore develops either into the final mouth (in → Protostomia), or into the anus (in → Deuterostomia), while the final mouth develops in the other end of the embryonic alimentary canal; **2.** (L. *orificium*, *ostium*), orifice, ostium; any aperture or mouth-like opening of a cavity or canal. *See* oral cavity.

mouthparts, various appendages on the head of different invertebrates, especially arthropods. In insects, the first two pairs of mouthparts, **mandibles** and **first maxillae**, lie at both sides of the mouth, while both **second maxillae** joined posteriorly to the mouth, are fused together to form the **labium**. This primitive and generalized structure is found in insects feeding on solid food, as in cockroach; includes the **clypeus** and **labrum** which lie above the median hypopharynx, and the jointed mandibles and maxillae with the **maxillary palps** (*palpi maxillares*). Each first maxilla is articulated to the head by four basal segments, the **cardo** and **stipes** as well as the two distal lobes, **lacinia** and **galea** (with *lobus internus* and *lobus externus*). In the labium the basal plates are fused to form the

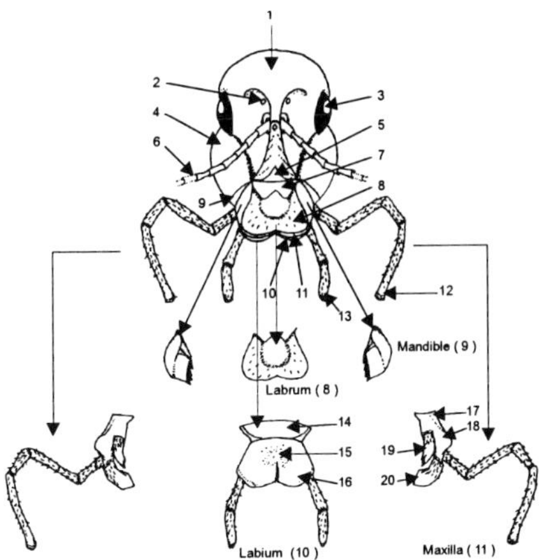

MOUTHPARTS OF GRASSHOPPER

1. vertex 2. ocellus 3. compound eye 4. gena 5. frons 6. antenna 7. clypeus 8. labrum 9. mandible 10. labium. 11. maxilla 12. maxillary palp 13. labial palp 14. submentum 15. mentum 16. ligula 17. cardo 18. stipes 19. lacinia 20. galea.

submentum and **mentum**, and the more distal segment is known as **prementum**. The prementum bears paired plates of the **ligula**, including two median **glossae** and two lateral **paraglossae**. At either side of the prementum is a **labial palp** (*palpus labialis*).

The mouthparts of insects have been modified in the course of evolution for various feeding types. Most primitive are the **chewing mouthparts** found e.g. in grasshoppers (Orthoptera), dragon-flies (Odonata), mayflies (Ephemeroptera), and beetles (Coleoptera). In **sucking mouthparts**, occurring e.g. in moths and butterflies (Lepidoptera), the galea forms a long tube through which food is sucked; the other mouthparts are absent or vestigial. In **chewing and sucking mouthparts**, found in bees and wasps, the labrum and mandibles have retained the chewing form but nectar is gathered by the elongated maxillae and the tube-like glossa. **Piercing mouthparts** are characteristic of insects (e.g. aphids and leafhoppers) utilizing either plant juices or body fluids of other animals, as do e.g. mosquitoes. Such mouthparts are elongated and are organized in various ways to form a stilus. In hemipterans, e.g. the stylet is composed of the mandibles and maxillae which lie in a groove of the labium covered by the labrum. **Sponging mouthparts** (e.g. in the housefly) have a spongeous labium for obtaining food, and the mandibles and maxillae are reduced.

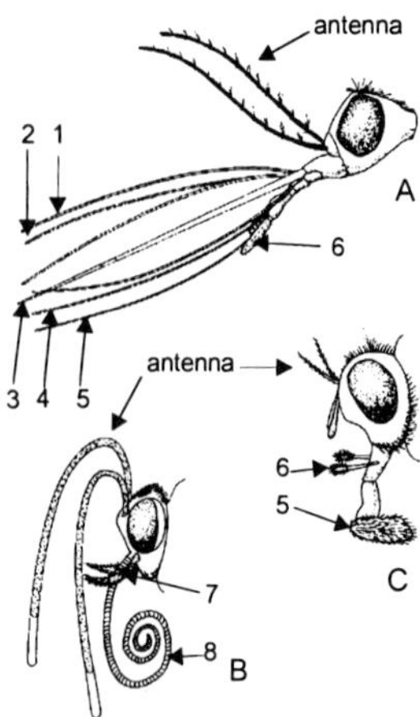

Piercing (A), sucking (B), and sponging (C) mouthparts of insects.
1. labrum 2. mandible 3. hypopharynx 4. maxilla 5. labium 6. maxillary palp 7. labial palp 8. maxillary proboscis

Structurally, the mouthparts of arachnids are simpler. The mandibles are lacking and the mouth appendages are constructed of two pairs: **chelicerae** and **pedipalps**, and their structures vary according to feeding type, i.e. sucking, biting, or piercing.

MPF, → mitosis-promoting factor.

MPS, mononuclear phagocyte system. *See* macrophage.

mRNA, → messenger RNA.

MS, 1. → mass spectrometer; **2.** multiple sclerosis, a neurological disease in man, characterized e.g. by speech disturbances, weakness, and muscular incoordination; caused by sclerotic patches of demyelination throughout the white matter of the central nervous system.

MS-222, tricaine; 3-aminobenzoic acid ethyl ester methanesulphonate, $C_{10}H_{15}NO_5S$; used for anaesthesia of aquatic animals, such as fish.

MSH, → melanophore-stimulating hormone, melanocyte-stimulating hormone.

MS medium, a synthetic medium developed by *T. Murashige* and *F. Skoog* for plant cell and tissue cultures.

MTBE, → methyl tert-butyl ether.

mtDNA, → mitochondrial DNA.

MTOC, → microtubule organizing centre.

muc(i)-, muco- (L. *mucus* = nasal mucus), denoting relationship to mucus, mucin, or slime. *See* myx(o)-.

mucilage (L. *mucilago* = musty juice), **1.** a substance containing various types of polysaccharides, viscous and slimy when wet; secreted by various plants and bacteria; **2.** an artificial viscid preparation containing vegetable substances such as gum or dextrin; used as a vehicle in pharmacy, and as a demulcent application to the mucous membranes in therapy.

mucin, any of jelly-like carbohydrate-rich → glycoproteins (mucoproteins) in mucus secreted by glandular cells of animals; a complex of → glycosaminoglycan and protein; occurs also in the ground substance of the interstitial connective tissue in most organs. *Adj.* **mucinous**, pertaining to, or containing mucin. *Cf.* mucinoid.

mucinoid, resembling mucin, mucoid.

mucoid, 1. → mucinoid; **2.** any of the group of mucus-like conjugated proteins in animals, e.g. → mucin, mucoprotein, glycoprotein.

mucolipid, any → sphingolipid containing sialic acid; e.g. ganglioside.

mucopeptide, a term formerly used for → peptidoglycans and their degradation products.

mucopolysaccharide, → glycosaminoglycan.

mucoprotein, the general term for protein-polysaccharide complex found in animal tissues and mucus, emphasizing that the protein component is a major part of the complex. Often used as a synonym for → proteoglycans or → peptidoglycans which usually contain more polysaccharides.

mucosa, pl. **mucosae,** → mucous membrane. *Adj.* **mucosal**.

mucous, 1. pertaining to mucus; **2.** covered with mucus, containing mucus, secreting mucus.

mucous gland, **1.** L. *glandula mucosa*, a gland secreting mucus in the mucous membrane or skin of vertebrates; called also muciparous gland; **2.** a slime gland of invertebrates, such as a gland in the anterior edge of the foot of gastropods whose slimy secretion helps locomotion.

mucous membrane/tunic (L. *tunica mucosa*), a membrane or coat lining the inner surface of various tubular structures, such as the alimentary canal, respiratory pathways, and genital or urinal cavities and tubes in vertebrates. Typically it comprises an epithelium (*lamina* *epithelialis*) with a → basement membrane, a layer of loose connective tissue (*lamina propria mucosae*), and a muscular layer (*lamina muscularis mucosae*). The oral, oesophageal, vaginal, and anal areas are lined by a stratified epithelium, elsewhere the epithelium is simple. There are many single gland cells (→ goblet cells) in the epithelium, and glands, e.g. gastric and intestinal glands, derived from the epithelium but located deeper in the connective tissue.

mucus (L.), **1.** viscid secretion of glands of the vertebrate mucous membrane, containing inorganic salts, mucin, loosened epithelial cells, and leucocytes; **2.** broader, any slimy secretion produced by organisms. *Adj.* **mucous**.

mud, organic material precipitated in water and composed of dead remains of plants, animals, and other organisms, usually containing also clay. *Cf.* peat.

Müllerian duct, Müller's duct (*Johannes Müller*, 1801—1858), **paramesonephric duct**; (L. *ductus paramesonephricus*), a tube developing parallel with the → Wolffian ducts from the coelomic embryonic mesothelium in jawed vertebrates; its development is associated with the formation of the mesonephros, and it is paired except in birds. The ducts arise as peritoneal pockets which extend caudally to join the urogenital sinuses, and then develop into the → oviducts (in mammals the uterine tubes, Fallopian tubes), in mammals also into the uterus and vagina. In amniotic vertebrates, the fertilization of eggs occurs in the oviduct. In males the Müllerian duct generally degenerates into vestigial structures (vagina masculina).

Müllerian mimicry (*Fritz Müller*, 1831—1897), *see* mimicry.

Müllerian regression factor, MRF, a polypeptide hormone secreted by → Sertoli cells of the embryonic testes in tetrapod vertebrates; inhibits the development of → Müllerian ducts, and together with testosterone stimulates the development of vas deferens and related structures. Also called Müllerian inhibiting substance (MIS) or Müllerian duct inhibitory factor.

Muller's ratchet (*Hermann Joseph Muller*, 1890—1967), a model by which Muller aimed to explain the evolution of sexual reproduction. According to the model, an asexual line of a population incorporates a kind of ratchet mechanism, such that the population does not get rid of mutant genes, i.e. it can never have a load of mutation (*see* genetic load) smaller than that already existing. On the contrary, in a sexual line recombination leads to elimination of harmful mutations.

multi- (L. *multus* = much, many), denoting many, numerous.

multiaxial (L. *axis* < Gr. *axon* = axis, axle), having more than one axis; e.g. pertaining to a plant structure having an axis with several apical cells which form parallel filaments; e.g. some red algae.

multicellular (L. *cellula* = small cell), composed of many cells. *Noun* **multicellularity**.

multicellular organisms, eukaryotic organisms composed of more than one cell, as distinguished from unicellular organisms, → Protista. The cells of a multicellular organism are accompanied by intercellular adhesion and intercellular communication. The organisms are usually structurally and functionally differentiated into tissues and organs with specialized functions.

multienzyme complex, an enzyme system containing several or many polypeptide monomers with different enzyme activities; e.g. the **fatty acid synthase** in the cytoplasm for the synthesis of saturated long-chain fatty acids with 7 enzyme activities, such as acetyl CoA carboxylase, acetyl transacylase, and malonyl transacylase, or the → **pyruvate dehydrogenase complex** catalysing the formation of acetyl CoA.

multiform (L. *forma* = shape, form), consisting of many forms, having many forms or shapes. *Noun* **multiformity**.

multimer (Gr. *meros* = part), a union of two or more organic macromolecules, like a union of several → polypeptides forming a **multimeric** protein, or a protein complex composed of several protein units; if the participating molecules are of different types, it is a **heteromultimer**, if of the same type, a **homomultimer**. *Adj.* **multimeric**. *Cf.* polymer.

multi-net theory, a theory about the growth of the plant cell wall, stating that the wall is formed from separate layers of → microfibrils, the directions of which turn during the enlargement of the cell to form a multi-net structure.

multineuronal, pertaining to a muscle cell which is innervated by more than one neurone, usually one or more excitatory neurones and one inhibitory neurone; multineuronal innervation is found in some invertebrate muscles. *Cf.* multiterminal.

multinuclear, multinucleate (L. *nucleus* = core, nucleus), containing several nuclei. *Syn.* polynuclear, polynucleate.

multiple alleles (L. *multiplex* = manifold), the members of a group of gene → alleles (more than two). Multiple alleles follow the same rules of inheritance as do loci with two alleles. **Multiple allelism** (genetic polymorphism) is very common in organisms, in principle concerning all genes.

multiplication index, *see* reproductive rate.

multiserial, multiseriate (L. *series* = row), arranged in many rows or series; e.g. **multiserial ray**, a plant → ray with many parallel rows of parenchyma cells. *Cf.* uniserial ray.

multiterminal, in neuroanatomy, pertaining to a neurone which ends polysynaptically, i.e. its → axon branches and ends with two or more → synapses; e.g. multiterminal innervation of a single muscle cell contains several neuromuscular junctions (end-plates) from the same neurone, found e.g. in some invertebrate muscles. *Cf.* multineuronal.

multivalent (L. *valentia* = power, capacity), **1.** having many values or meanings, as the multiple value of chemical bonds, or binding sites in an → antibody; *syn.* polyvalent; **2.** in genetics, the union of more than two homologous → chromosomes in which the chromosomes of polyploid cells or organisms are joined by chromosome pairing and chiasmata during → meiosis.

muramidase, → lysozyme.

mureins (L. *murus* = wall), → peptidoglycans composing the bacterial cell wall; consist of linear polysaccharides with alternating *N*-acetylglucosamine and *N*-acetylmuramic acid units, to which oligopeptides are linked.

muscarine, a toxic alkaloid found in some mushrooms, as e.g. in *Amanita muscaria*; a → cholinergic substance stimulating → muscarinic receptors in animals. This fungus contains also some related alkaloids which cause more severe toxic effects in animals.

muscarinic receptor (L. *recipere* = to receive), one of the two types of cholinergic receptors (acetylcholine receptors) occurring in various smooth muscles, cardiac muscle, and exocrine glands; are called muscarinic receptors because → muscarine has actions similar to those of → acetylcholine in these organs. Muscarinic receptors are also found in the central nervous system of vertebrates. These receptors mediate effects, such as cardiac inhibition, decrease in blood pressure, bronchoconstriction, salivation, and lacrimation. The muscarinic receptors are competitively blocked by → atropine and related drugs. *Cf.*

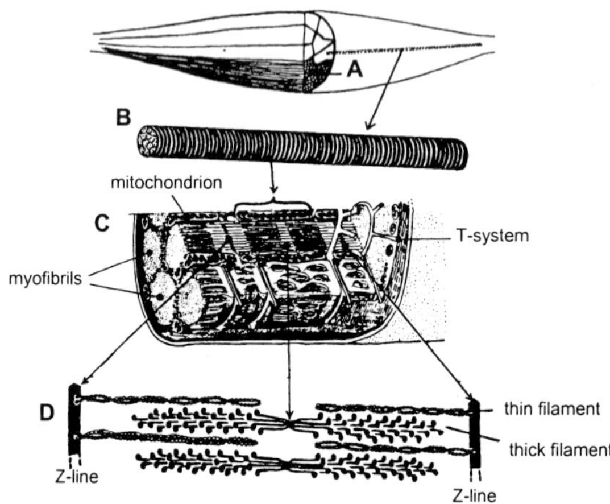

Muscle. Structural units of the skeletal muscle: A: a bundle of muscle fibres. B: a muscle fibre, enlarged in C where mitochondria and T system can be seen between the muscle fibrils. Muscle filaments (D) are thick filaments with club-shaped myosin molecules, and thin filaments with globular actin and troponin molecules and fibrous tropomyosin molecules. The thin filaments are anchored to Z-line (Z-discs). Part C is redrawn after Krstic R.V.: Ultrastructure of the Mammalian Cell. Springer-Verlag, 1979.

nicotinic receptor.

muscle (L. *musculus*, dim. of *mus* < Gr. *mys* = mouse), a contractile organ of animals consisting of → muscle tissue (muscle cells) with some supportive connective tissue; serves in motility of animals and their organs.

The muscles of vertebrates are anatomically classified into three different categories: **skeletal muscles** (striated muscles), **smooth muscles** (unstriated muscles), and **cardiac muscles** (also striated). Functionally the muscles are divided into **voluntary muscles** comprising the skeletal muscles, and into **involuntary muscles** which are the smooth and cardiac muscles. Skeletal muscles are also called **neurogenic muscles**, because the impulses (→ action potentials) inducing contractions are generated in nerve cells. The cardiac and most smooth muscles (some are neurogenic) are called **myogenic muscles** in which impulses are generated in the muscle cells themselves.

Most muscles in invertebrates are unstriated resembling the smooth muscles of vertebrates, although the size, structure, and function vary according to the group and species. Leg and wing muscles are striated resembling the skeletal muscles of vertebrates, but striation may be diagonal, as in some worms. Innervation of a single muscle cell may be → multineuronal and → multiterminal (in vertebrates unineuronal and uniterminal). Like in vertebrates, the muscles are either voluntary (serving motility of the animal) or involuntary (serving motility of inner organs), being → neurogenic or → myogenic, respectively. *See* muscle cell, muscle contraction, motor unit, red fibres.

muscle cell, a contractile cell type of a muscle serving motility of animals and their organs. Regarding their structural differences three main types are distinguished: **skeletal** (striated), **smooth** (unstriated), and **cardiac muscle cells**, which form three muscle types, respectively. Common structures in all muscle cells are contractile protein elements, the **myofibrils**. *Syn.* myocyte. *Cf.* muscle fibre. *See* skeletal muscles, smooth muscle, cardiac muscle, muscle tissue, myoepithelial cell.

muscle contraction, an active contraction of the muscle based on simultaneous or partly successive contractions of many muscle cells. The contraction mechanism is best known in skeletal muscles in which the contraction units of the cells are numerous → sarcomeres with linearly arranged protein filaments. The contraction is caused by active shortening of the intracellular protein elements, **myofibrils**, in which **thin filaments** (containing **actin**) and **thick filaments** (**myosin** filaments) react in the presence of **Mg-ATP** and **Ca^{2+}**, sliding along each other (*see* actin, myosin). The relaxation follows when the Ca^{2+} concentration decreases and ceases the reaction between the filaments. Calcium is released into the cytoplasm mainly from terminal cisternae of the **sarcoplasmic reticulum** in the cell, and contraction is regulated by four accessory proteins: **troponins C**, **I**, and **T**, and **tropomyosin**. The whole complex is called the **tropomyosin-troponin system**. When Ca^{2+} is bound to troponin C, troponin I and T move aside from the reaction site of tropomyosin, making possible the interaction of actin and myosin and thus the contraction. After the contraction, calcium is actively removed back to the → sarcoplasmic reticulum by **calcium pumps** (certain Ca^{2+}-ATPases) and the rearrangement of the four proteins inhibits the sliding reaction and consequently the relaxation of the muscle fibres. Energy needed for contractions comes from → adenosine triphosphate (ATP), → creatine phosphate acting as an immediate energy store. Skeletal muscles usually generate rapid movements (**isotonic contraction**) or maintain a contractive state without shortening of the muscle (**isometric contraction**); *see* motor unit, tetanus.

In principle, the contraction of other muscle types is quite similar, differing only in detail. The sarcomeric structure occurs also in the cardiac muscle, but the filaments of smooth muscle cells are not arranged in regular arrays of sarcomeres. The sarcoplasmic reticulum of the smooth muscle and of the cardiac muscle in poikilothermic vertebrates is poorly developed and sparse, and thus the contraction is slower and more dependent on extracellular calcium. The smooth muscle contains tropomyosin but is lacking troponin; the thin filaments contain a calmodulin-binding protein, **caldesmon**, which controls the calcium availability for contraction (*see* calmodulin).

muscle energetics, energy sources and usage of energy for muscle contraction. Energy stores in muscle cells are chiefly **glycogen** and **neutral fats** which are metabolized into glucose, and fatty acids and glycerol, respectively. When energy stores in muscle cells decrease below a critical level, the cells begin to use glucose via blood from other sources, especially from the liver (glycogen) and fat tissues. In cell energy metabolism, anaerobic → **glycolysis** may produce sufficient energy (i.e. ATP) for muscular work for a short period (glycolytic muscle fibres), but because of accumulation of lactic acid (oxygen debt), **aerobic energy metabolism** is needed almost always for constant muscular work, and in many muscle fibres (aerobic muscle fibres) constantly; → **myoglobin** acts as an oxygen store for aerobic metabolism. The energy for ATP synthesis can be stored in muscle cells in high-energy compounds, → **creatine phosphate** (in vertebrates) or → **arginine phosphate** (in most invertebrates): e.g. ATP + creatine ⇌ creatine phosphate + ADP. This immediate energy source is sufficient only for a sudden muscular action.

muscle fibre, Am. **muscle fiber,** muscle cell; especially skeletal muscle cells (skeletal muscle fibres) and cardiac muscle cells (cardiac muscle fibres), seldom used for smaller smooth muscle cells. The skeletal muscle fibres are giant multinuclear cell fusions (from millimetres to centimetres in length and 0.01 to 0.1 mm wide), every fibre being formed of many primary cells, myoblasts. Cardiac muscle fibres are quite large, single longitudinal cells, maximally about 0.5 mm in length.

muscle plate, *see* myotome.

muscle reflex, *see* reflex movements.

muscle sense, muscular sense, *see* kinaesthetic sense.

muscle spindle, any of specific stretch receptors found in skeletal muscles of vertebrates serving in the control of muscle length, i.e. the muscle tonus or tension, position → reflexes, and in homoiothermic animals also of muscle shivering in cold. This spindle-shaped organ, 0.1 to 0.2 mm in length, contains 2 to 10 small, specialized skeletal muscle fibres (**intrafusal fibres**), which are contractile only at the ends of the fibre whose slow tonic contractions are controlled by gamma motoneurones. The spindle lies parallel with the **extrafusal fibres** (ordinary muscle fibres). Sensory nerve endings are associated with the middle parts of the intrafusal fibres responding to the stretch of the muscle spindle due to the stretching of the whole muscle. Because the spindle can contract neurogenically at both ends of the intrafusal fibres, the central nervous system can control the sensitivity of the spindle to stretches (*see* reflex movement). Called also spindle organ or neuromuscular spindle.

muscle sugar, meat sugar, → inositol.

muscle tissue, also **muscular tissue,** the tissue of the animal body responsible for the contractions of muscles in multicellular animals except sponges; is usually of mesodermal (mesenchymal) origin. Three main types are distinguished: **skeletal** (striated), **smooth** (unstriated), and **cardiac muscle tissues** (the cardiac tissue is structurally also striated but physiologically differs from the others). Muscle tissue is composed of muscle cells packed closely alongside each other. The interstitial material between the cells is produced by connective tissue cells between muscle cells, forming a very thin sheath (**endomysium**) around each cell, a thicker sheath (**perimysium**) around the groups of muscle cells, and the thickest sheath (**epimysium**) around the whole muscle.

In invertebrates, the muscle tissue is quite different in various animal groups, but two main types can be distinguished: striated muscle tissue (striation may be also diagonal) resembling that of vertebrates, and unstriated muscle tissue resembling the smooth muscle tissue of vertebrates. *See* skeletal muscles, smooth muscle, cardiac muscle, muscle contraction.

muscle twitch, a quick contraction of a single skeletal muscle elicited by a single volley of impulses from the motor nerve; the most typical twitch can be elicited experimentally in a single muscle fibre or a → motor unit of skeletal muscle cells induced by one single nerve impulse. Also some smooth muscles (neurogenic smooth muscles) respond normally by muscle twitches, as well as single isolated cardiac and smooth muscle cells in experimental conditions.

muscular, pertaining to muscles.

muscular atrophy (Gr. *atrophia* < *a-* = not, without, *trophe* = nourishment), reduction and wasting of muscle tissue which may occur locally or generally in an animal; usually caused by lack of use, i.e. by decreased nerve activity in the muscle. A more severe degeneration of muscles is called **muscular dys-**

trophy.

musculature, the whole muscular system of an animal or a certain organ in which the muscles function together, such as in the legs. In the body of primitive multicellular animals, such as cnidarians, only single contractile cells exist, called musculo-epithelial cells, but more developed animals have a specialized → muscle tissue organized into muscles as a part of the musculature.

musculo-epithelial cell, epitheliomuscular cell, a slightly specialized cell type with long contractile processes found in the gastrodermis of cnidarians and ctenophores; the cells form a covering structure on the body surface but are also capable of contracting, thus resembling muscle cells. *Cf.* myoepithelial cell.

mushroom, common name for the edible members of Fungi with gills; colloquial term for any of the larger fleshy or woody fungi. *Cf.* toadstool.

musk (Sansk. *muska* = scrotum, vulva), a substance secreted by special glands under the abdominal skin near the scrotum of male musk deer, muskrats, civets, etc.; has a strong odour serving as a chemical messenger between individuals in → scent marking. Purified musks, containing e.g. a macrocyclic ketone, **muscone** (**muskone**, 3-methylcyclopentadecanone), as well as synthetic musks, are used in perfumery.

mussel, any of bivalve molluscs, especially marine bivalves of the family Mytilidae and freshwater clams of the family Unionidae.

mustard (L. *mustum* = must), **1.** (L. *sinapis*), a spice made from powdered seeds of white and black mustard species from which → mustard oils are removed; **2.** any of various brassicaceous plants producing mustard.

mustard gas, dichlorodiethyl sulphide; a highly toxic liquid used as a war gas. Also called sulphur mustard.

mustard oils, esters of isothiocyanic acid or isothiocyanates with a pungent odour; occur in some plant families, particularly in Brassicaceae and Resedaceae. **Allyl** (propenyl) **isothiocyanate**, CH_2=$CHCH_2NCS$, is rather common and present e.g. in sinigrin, the best known of mustard glycosides, obtained from mustard seeds. Some isothiocyanates are sweet-scented, such as those in the mignonette (*Reseda*).

mutability (L. *mutabilis* = variable < *mutare* = to change), the state or quality of being mutable; the property of a gene or → genotype to produce certain changes, → **mutations**, in itself. Mutability offers the ultimate basis for genetic variation and thus for the adaptation of the genetic material and the whole organism to changes in environmental conditions.

mutable gene, a labile → gene which during the development of an individual changes more often than the majority of genes which remain stable. *See* mutation.

mutagen (Gr. *gennan* = to produce), any physical or chemical agent (mutagenic agent) which increases the number of → mutations above the spontaneous background level. Mutations which are induced by treatment with mutagens are called **induced mutations**, in contrast to naturally occurring **spontaneous mutations**. Mutagens are e.g. radioactive, roentgen (X-ray), and ultraviolet irradiations and certain chemicals; several DNA base analogues and alkylating agents are experimentally used to produce both gene and chromosomal mutations. Many mutagens are → carcinogens.

mutagenesis, production of a → mutation. *Adj.* **mutagenic**.

mutagenicity, the ability of certain physical or chemical agents to cause mutations.

mutant, 1. undergoing → mutation, resulting from mutation; **2.** a cell or individual that carries a gene, chromosome, or genome mutation; **3.** a gene in which a mutation has occurred (mutant gene, mutant allele). *See* mutate.

mutase, any enzyme of the subclass EC 5.4. in the isomerase class, catalysing the transfer of a chemical group, such as a phosphoryl, amino, or acyl group, usually within a molecule from one position to another; e.g. phosphoglucomutase (EC 5.4.2.2).

mutate, 1. to undergo → mutation; **2.** as a *noun*, → mutant (defin. 2.).

mutation, a change of the genetic material which does not occur in consequence of → segregation or → genetic recombination. Mutations are transmitted to daughter cells and through the germline cells to following generations thus causing mutant cells or individuals to arise. Mutations are → **genome mutations** in which the number of chromosomes is changed, → **chromosome mutations** in which the structure of chromosomes is changed, and → **gene mutations** in which individual genes have mutated. Mutations can arise spontaneously or they can be induced artificially with → mutagens. *See* mutability.

mutational load, that part of the → genetic load which is due to → mutation pressure; consequently, deleterious genes continuously exist in the population decreasing its → fitness.

mutational site, a place in a gene or chromosome in which → mutation can occur; it is a structural unit of DNA, in which nucleotides are located within a gene in a linear order. In typical gene mutations the size of the mutational site is one nucleotide pair of DNA.

mutation frequency, the number of new mutations per cell (usually one gamete), and usually per gene locus in a chromosome.

mutation pressure, changes in gene frequences in a population produced by mutational changes alone, i.e. the tendency of the repeated occurrence of the same mutation in a → population; it is the main cause for the maintenance of mutant alleles in a population.

mutation rate, the number of new mutations per time unit, usually per one generation and per one gene locus.

mutator gene, a gene which causes a considerable increase in the → mutation rate of other genes.

muton (< mutation), the smallest unit of the genetic material a change of which is sufficient to cause a change in the phenotype of the organism; is the smallest unit of mutation. Materially a muton corresponds to one nucleotide of DNA (or RNA in RNA viruses).

mutual antagonism (L. *mutuus* = reciprocal, Gr. *antagonizesthai* = to fight against), the negative reciprocal effect of two organisms or species on each other; e.g. a competition between two species.

mutual interference, in ecology, a reciprocal interaction with negative effects, but without antagonistic actions of two animal species; e.g. interference between two predator species so that the feeding effectiveness of each individual declines. The mutual interference becomes stronger when population density increases.

mutualism, a symbiotic interaction of organisms in which different populations or individuals of two different species benefit from the association, e.g. the individuals of two populations develop and grow better and produce more offspring; may be **facultative** or **obligatory mutualism**. Occasionally the term mutualism is used as a synonym for → symbiosis; generally it is regarded as a special case of symbiosis. *Adj.* **mutualistic**.

myc(o)-, mycet(o)- (Gr. *mykes* = fungus), pertaining to fungus.

mycelium, pl. **mycelia** (Gr. *helos* = nail, wart), a branching network of fungal hyphae which is the dominating, vegetative phase of multicellular fungi. Mycelium grows in soil, on rotting wood etc., forming in certain favourable conditions → fruit bodies of tight pseudoparenchyma. The fruit bodies produce fungal spores which further form new hyphae.

mycetogenic, mycetogenetic, mycetogenous (Gr. *gennan* = to produce), caused or produced by fungi.

mycetophage (Gr. *phagein* = to eat), an organism feeding on fungi. *Adj.* **mycetophagous** (fungivorous).

mycobacterium, pl. **mycobacteria,** bacteria in the class Thallobacteria; the type genus *Mycobacterium* comprises aerobic, non-motile Gram-positive bacteria. Mycobacteria are slightly curved slender rods, some forming filaments; they are distinguished by acid-fast staining. The most important species is *M. tuberculosis*, tubercle bacillus of which *M. avium* and *M. bovis* are the avian and bovine subspecies. *M. leprae* is a bacterium causing human leprosy. *See* Actinomycetales.

mycobiont (Gr. *bios* = life), the fungal component of a → lichen.

mycology (Gr. *logos* = word, discourse), the study of → fungi.

mycoplasmas (Gr. *plassein* = to form, mould), a group of prokaryotic organisms, generally classified within → Bacteria of the class Mollicutes, being between bacteria and viruses in size; mycoplasmas have a triple-layered cell membrane, and as obligate intracellular parasites have no rigid cell wall typical of ordinary bacteria. Mycoplasmas have nucleic acids in the cytoplasm and the cellular structure is very simple and primitive. Most important is the family *Mycoplasmataceae* with the genus *Mycoplasma*, many species of which cause plant and animal diseases. **Mycoplasma-like organisms** (MLOs) have been found infecting many plants in which they cause numerous plant diseases. Mycoplasmas were earlier called **pleuropneumonia-like organisms** (PPLOs).

mycor(r)hiza, pl. **mycor(r)hizae** or **mycor(r)hizas** (Gr. *rhiza* = root), a symbiotic structure formed by a seed plant root and hyphae (mycelium) of some fungi. In **ectomycorhizae** (ectotrophic mycorhizae) the fungal hyphae enclose small roots and grow between cells in the root cortex, but do not penetrate the cells. In **endomycorhizae** (endotrophic mycorhizae) the fungal hyphae are inside the root, growing between and also into the cells. Endomycorhizae are divided into two types, **vesicular-arbuscular** type (VAM) and **orchid** type. The first is formed mainly by algal fungi: the hyphae grow in the cells, where they form vesicular structures of arbuscular, tree-like structures. In the orchid type the hyphae penetrate the cells forming tight balls, which are used as nutrition by the host; this type occurs in the families Orchidaceae and Ericaceae. In general, the mycorhizae help the host plant in taking nutrients and water from the soil, while the fungi obtain organic material for their nutrition from the host plant. Mycorhizae are essential for the optimal growth of many trees, shrubs, and even herbaceous plants.

mycosis, pl. **mycoses,** any animal disease caused by a fungus.

mycotrophy (Gr. *trephein* = to nourish), a type of nutrition in which seed plants and fungi live in → symbiosis; the fungi supply nourishment from soil for their host plants and obtain e.g. carbohydrates and some essential compounds from the plants. Together the plants and fungi form a saprophytic system. **Mycotrophic** plants are common especially in the family Orchidaceae.

mydriasis (Gr.), dilation of the pupil of the eye; may be caused by a physiological regulation, or be a result from a disease or a drug. *Adj.* **mydriatic**.

myel(o)- (Gr. *myelos* = marrow), denoting relationship to the bone marrow, spinal cord, medulla oblongata (myelencephalon), or to → myelin.

myelencephalon (Gr. *enkephalos* = brain), → medulla oblongata.

myelin, 1. the substance of the → myelin sheath around the axon of a nerve cell; it is composed of proteins and fats, such as phospholipids, neutral fats, and cholesterol, and gives white colour to myelinated regions of the brain (white matter); **2.** more generally, any of lipid substances accumulated in a tissue; occurs e.g. in certain tumours.

myelin sheath, the sheath surrounding the axons of certain neurones (**myelinated nerve fibres**) found generally in vertebrates and most developed invertebrates; serves as an electrical insulator increasing the conduction velocity of the nerve fibre. In the central nervous system of vertebrates, the myelin sheaths are formed by the action of **oligodendrocytes**, in peripheral nerves by the action of **Schwann cells**. The membranes of these cells are wrapped many times round the axon, leaving unmyelinated intervals (→ nodes of Ranvier) between adjacent Schwann cells (or oligodendrocytes). *See* myelin, nerve fibre, saltatory conduction.

myeloblast (Gr. *blastos* = germ), a stem cell of white blood cells in the bone marrow; develops into a promyelocyte, then myelocyte, and finally into a → granulocyte.

myeloma, a type of malignant tumour the cells of which are derived from a lymphocytic cell line, as e.g. from a lymphoblast in the bone marrow.

myenteric plexus (Gr. *mys* = muscle, *enteron* = intestine), → Auerbach's plexus.

myo-, my- (Gr. *mys*, gen. *myos* = mouse, muscle), denoting relationship to mouse or muscle.

myoblast (Gr. *blastos* = germ), **1.** an immature muscle cell developing into the mature muscle cell or myocyte; **2.** in adult animals a → satellite cell. *Syn.* sarcoblast, sarcogenic cell.

myocyte (Gr. *kytos* = cavity, cell), → muscle cell.

myoepithelial cell, the special term for slowly contracting stellate cells found in the epithelium (**myoepithelium**) of many exocrine glands of vertebrates; the cells are derived from the ectodermal tissue, while the ordinary muscle cells are mesodermal. With cytoplasmic processes the cellular structure resembles that of smooth muscle cells. The cells are located around glands and secretory alveoli e.g. in sweat, lacrimal, and mammary glands serving in the transport of secretion. *Cf.* musculo-epithelial cell.

myofibril (L. *fibrilla*, dim. of *fibra* = fibre), **muscular fibril**; any of contracting protein threads inside the muscle cell; myofibrils are composed of smaller units, **myofilaments**, the two types of which are **thin filaments** containing mainly actin, and **thick filaments** formed from myosin. Also called myofibrilla (pl. myofibrillae). *See* muscle contraction.

myofilament (L. *philum* = thread), *see* myofibril, muscle contraction.

myogenesis (Gr. *gennan* = to produce), the development of muscle tissue; occurs mainly during embryogenesis.

myogenetic, also myogenic, pertaining to → myogenesis.

myogenic, (Gr. *genes* = born), **1.** originating in the muscle, muscle tissue, or muscle cells; e.g. **myogenic contraction**, a contraction type of smooth and cardiac muscles in which the contractions are induced by impulses originating in the muscle itself, i.e. not by neuronal impulses; also **myogenous**; *cf.* neurogenic; **2.** → myogenetic.

myoglobin, myohaemoglobin; the red pigment of vertebrate muscle cells functioning as oxygen storage and transport inside the cell, delivering stored oxygen to muscle cell mitochondria; a conjugated protein like a subunit of haemoglobin, composed of a polypeptide chain, **globin**, and of → **haem** (heme). Myoglobin has greater affinity for oxygen than has haemoglobin, releasing oxygen only in very hypoxic conditions.

myoma, pl. **myomata,** also **myomas,** a benign tumour derived from the muscular tissue.

myometrium (Gr. *metra* = uterus), L. *tunica muscularis uteri*, the smooth muscle forming a thick layer in the wall of the uterus.

myoneme (Gr. *nema* = thread), any of several contractile fibrils in protozoans, found especially in the cytoplasm of stalked, ciliated protozoans, such as *Vorticella*. *Syn.* myonema.

myoneural junction, → neuromuscular junction.

myopia (Gr. *myein* = to shut, *ops* = eye), shortsightedness, nearsightedness; a property of such an eye that focuses light rays parallel to the optic axis in front of the retina (the eyeball is too long).

myosin (Gr. *mys* = muscle), a large → motor protein participating in contractions of different cellular elements in all eukaryotic cells; especially abundant in muscle cells. Each myosin monomer consists of a globular **head** domain composed of four polypeptides, and a long **tail** domain composed of fibrillar peptide, and between them a flexible **neck** region with certain proteins (e.g. calcium-binding calmodulin light chains) wrapped around it. In the cell the molecules polymerize together forming long **myosin filaments** (thick myofilaments in muscle cells), and the heads are situated spirally along the surface of a filament. In contraction the head acts as an → adenosine triphosphatase (ATPase); the heads bind repeatedly to special sites on the adjacent thin filament (actin filament) pulling the myosin filament along it (*see* muscle contraction). The light chains of the neck regulate myosin activity, and the tail domain controls the special function of each myosin. Fractionating of myosin e.g. by trypsin enzyme produces the **heavy meromyosin fragment** composed of the head and neck domains, and the **light meromyosin fragment** composed of tail domains of myosin.

More than ten classes of the myosin gene family are known, three major protein types being monomeric myosin I, dimeric myosin II and myosin V; types I and II are the most abundant proteins commonly present in organisms. Type II act in muscle contraction and → cytokinesis, type I and V are involved in cytoskeleton membrane interactions, such as the transport of membrane vesicles.

myotome (Gr. *tomein* = to cut), **1.** muscle plate; the middle portion of each mesodermal segment (somite) of a vertebrate embryo, developing into the skeletal muscle; **2.** muscle segment of the fish body; *cf.* dermatome; **3.** the muscles of the body segment (metamere) in a segmented invertebrate; **4.** a knife for dividing muscles in myotomy.

Myriapoda (Gr. *myrios* = countless, *pous*, gen. *podos* = foot), **myriapods**; a superclass (in some classifications class) in the phylum Arthropoda comprising the classes (subclasses) Chilopoda (**centipedes**), Diplopoda (**millipedes**), Symphyla, and Pauropoda. Some scientists regard myriapods only as an informal group without any systematic status. Myriapods have a long, slender body comprising the head and the segmental trunk with one or two pairs of legs per somite.

myristic acid (*Myristica* = a genus of tropical trees), tetradecanoic acid, a saturated fatty acid, $CH_3(CH_2)_{12}COOH$; one of the fatty acid residues in phospholipids of cellular membranes in organisms, occurring abundantly e.g. in certain vegetable oils and glycerides of milk.

myrmec(o)- (Gr. *myrmex* = ant), pertaining to ants.

myrmecochorous (Gr. *chorein* = to advance, spread), dispersed by ants; pertaining e.g. to seeds with an oily appendix (elaiosome) attracting ants. *Noun* **myrmecochory**.

myrmecology (Gr. *logos* = word, discourse), a branch of biology studying ants.

myrmecophage (Gr. *phagein* = to eat), any animal feeding on ants. *Adj.* **myrmecophagous**.

myrmecophil(e) (Gr. *philein* = to love), an organism which lives in association with ants; e.g. an insect that shares the nest of ants. It may also be a plant, e.g. *Myrmecodia* with thorny, hollow roots, with ants living there and bringing organic material utilized also by the plant. *Adj.* **myrmecophilous**. *Noun* **myrmecophily**.

myrosinase, an enzyme, EC 3.2.3.1, catalysing the hydrolysis of organic sulphur compounds (thioglucosides) to sulphate and isothiocyanates; found most abundantly in brassicaceous plants, such as mustard, cabbage, and rape. *Syn.* thioglucosidase, sinigrinase, sinigrase. *See* mustard oils.

Mysidacea (*Mysis*, the type genus), **mysidaceans** (opossum shrimps); an order of crustaceans in the class Malacostraca (subclass in some classifications); about 450 species which mainly live in pelagic zones of oceans. The first thoracic appendages are forked and the carapace covers only partly the thorax.

mysis, a larval stage after the zoea larva found in some malacostracans, as e.g. shrimps. *Syn.* schizopod.

Mystacocarida (Gr. *mystax* = mustache, *karis* = shrimp), **mystacocarids**; a class of small, primitive crustaceans living interstitially between sand grains in marine beaches; about 10 species are described.

Mysticeti (*Mysticetus*, the type genus), **whalebone whales, baleen whales**; a suborder in the order Cetacea (whales), feeding chiefly on planktonic crustaceans, such as krill, sieving them with whalebones (baleens); the teeth are totally lacking in adults.

myx(o)- (Gr. *myxa* = mucus, slime), denoting relationship to slime or mucus; → muci-.

myxamoeba, *see* Myxomycetes.

Myxini, Myxiniformes, → Pteraspidomorphi.

myxobacteria, a group of Gram-negative bacteria in the class Scotophobia in the group → Bacteria; small, aerobic rods with flexible cell walls and motile flagella, being able to creep leaving a layer of slime. They can form fruiting bodies in which the cells differentiate into resting spores. In artificial cultures some myxobacteria can feed on other living organisms, e.g. other bacteria.

myxoedema, Am. **myxedema** (Gr. *oidema* = swelling), a waxy, abnormal type of swelling of the skin associated with → hypothyroidism.

myxoid, mucoid, resembling mucus or slime.

myxomatosis, 1. a conditon characterized by the development of myxomas, i.e. tumours derived from the primitive connective tissue; **2.** a fatal viral infection in rabbits and hares characterized by oedematous swellings in the skin and mucous membranes; infection by the virus is used for controlling the number of rabbits in Australia; more specifically called **infectious myxomatosis** (*myxomatosis cuniculi*).

Myxomycetes (Gr. *mykes* = fungus), **slime moulds**, a class in the subdivision Plasmodiogymnomycotina, division → Gymnomycota, kingdom Fungi. A slime mould forms typically a naked, large (even tens of centimetres) acellular protoplasm mass, the **plasmodium**, with numerous diploid nuclei which moves on the ground while bacteria and other small material of plants and animals are taken into cells by → endocytosis and digested. When the food supply is inadequate, the plasmodium forms sporangia to produce spores (myxospores); these develop first into zoospores and then further into naked amoeboid cells, **myxamoebae**, which fuse and form a new diploid plasmodium. Slime moulds are common in damp environments. In some systems the group Myxomycetes is classified as a separate division, **Myxomycota** (**plasmoidal slime moulds**), comprising the classes Myxomycetes and Protosteliomycetes.

myxophycean starch (Gr. *phykos* = seaweed), a storage polysaccharide in Cyanobacteria resembling glycogen; occurs as granules and elongate bodies in the cytoplasm.

Myxophyta (Gr. *phyton* = plant), → Cyanobacteria.

myxospore (Gr. *sporos* = seed, spore), a spore of slime moulds, formed in a **myxosporangium**. *See* Myxomycetes.

N

N, symbol for **1.** nitrogen; **2.** asparagine; **3.** normality (the concentration unit of a solution); **4.** newton (the unit of force, *see* Appendix 1).

n, symbol for **1.** nano (Gr. *nanos* = dwarf), as a prefix of a unit of measure denoting a thousand millionth part (Am. billionth), 10^{-9}; **2.** symbol for → haploid (n) or diploid (2n) chromosome number, whereas the state of polyploidy is designated with the symbol x (instead of n) indicating the basic chromosome number as 3x = triploid, 4x = tetraploid etc.; *see* polyploid; **3.** abbreviation of nerve (*nervus*), pl. nn. (*nervi*).

NA, abbreviation of numerical aperture. *See* resolution.

NAA, → naphthaleneacetic acid.

nacreous layer, also **nacre layer** (Fr. *nacre* = mother-of-pearl), the innermost layer of the mollusc shell, next to the mantle and secreted by mantle cells; it is formed of coloured and iridescent, very thin parallel lamellae of crystalline calcium carbonate (aragonite), maximally 400—5000 layers per cm. Many bivalves such as **pearl oysters**, i.e. the marine molluscs of the family Pteriidae, and some freshwater species, e.g. *Margaritana margaritifera*, may secrete nacreous layers also around a foreign body, such as a grain of sand, in the mantle cavity thus forming a pearl. Also some gastropods such as abalones (*Haliotis*), and a cephalopod, the chambered nautilus (*Nautilus*), have the ability to form pearls. *See* periostracum, prismatic layer.

NAD, → nicotinamide adenine dinucleotide (formerly DPN); the oxidized form is **NAD^+**, the reduced form **NADH**.

NADase, → NAD^+ nucleosidase.

NADH dehydrogenase, cytochrome c reductase, EC 1.6.99.3; an iron-containing flavoprotein enzyme that acts in cell respiration oxidizing NADH to NAD^+.

NADH-FMN oxidoreductase, NADH dehydrogenase (quinone); an enzyme, EC 1.6. 99.5, reducing flavin mononucleotide (FMN to $FMNH_2$) and oxidizing β-NADH with a quinone as an electron acceptor; occurs commonly in organisms.

NAD^+ kinase, an enzyme, EC 2.7.1.23, phosphorylating β-NAD into β-NADP in the presence of ATP; occurs commonly in organisms.

NAD^+ nucleosidase, NADase, an enzyme, EC 3.2.2.5, occurring commonly in organisms; obtained e.g. from *Neurospora crassa*. NADase catalyses the hydrolysis of β-NAD to nicotinamide and adenosinodiphosphoribose (ADP-ribose). *Syn.* DPN nucleosidase (DPNase), NAD^+ glycohydrolase.

NADP, → nicotinamide adenine dinucleotide phosphate (formerly TPN); the oxidized form is **$NADP^+$**, the reduced form **NADPH**.

NADPH-FMN oxidoreductase, NADPH dehydrogenase (quinone); an enzyme, EC 1.6.99.1, reducing FMN to $FMNH_2$ and oxidizing NADPH with a quinone as an electron acceptor; occurs commonly in organisms.

nagana (Zulu *u-nakane* = weak), an infectious disease caused by several species of trypanosomes transmitted by tsetse flies to many mammals, such as cattle, horse, pig, sheep, goat, camel, and dog, especially in Central Africa. *See* Trypanosoma.

naiad, pl. **naiads** or **naiades** (Gr. *Naias* = water nymph), the aquatic nymph stage of some hemimetabolous insects, such as dragonflies. *See* nymph.

nail (L. *unguis*), **1.** a horny plate of thickened stratum lucidum developed from the skin epithelium on the tips of toes (and fingers) in some mammalian species, e.g. primates; the convex unguis lies on top of the softer **subunguis**, the shape and relative size varying according to the function; **2.** a homologous structure of other tetrapods, as the **hoof**, **claw**, or **talon**.

Na^+,K^+-ATPase, sodium-potassium ATPase. *See* adenosinetriphosphatase, ion pumps.

nalorphine, N-allylnormorphine; a drug chemically related to morphine acting in animals as an antagonist of morphine and other opiates; **naloxone** and **naltrexone** are similar antagonists of morphine and → endorphins.

nano- (Gr. *nanos* = dwarf), **1.** denoting relationship to small size; **2.** as a prefix in the units of measurement indicates one thousand millionth part (Am. one billionth), 10^{-9}, symbol n; e.g. one **nanometre** = 10^{-9} metres; one **nanolitre** = 10^{-9} litres.

nanoplankton, → plankton comprising organisms with a body size from 2 to 20 μm, filtering through the standard plankton net; includes microscopic unicellular plants, fungi, and animals. *Cf.* microplankton, picoplankton.

nanosomia, nanism (Gr. *soma* = body), dwarfism, usually genetically determined; in vertebrates, caused by decreased secretion of → growth hormone or its weak effect on cells and tissues.

naphthacene, a yellowish tetracyclic hydrocarbon consisting of four benzene rings; occurs in small quantities in coal tar.

naphthalene, $C_{10}H_8$, a silvery crystalline hydrocarbon with two benzene rings; a toxic, carcinogenic compound obtained from coal tar and used in chemistry, earlier also as an antiseptic and moth repellent. Also called naphthalin or tar camphor.

naphthaleneacetic acid, naphthylacetic acid, an organic acid, derivative of → naphthalene; it is a synthetic → auxin used to prevent premature fruit drop and to induce root formation in cuttings. Being chemically stable it is used to replace indoleacetic acid (IAA) in plant tissue cultures, and also as a substrate for → esterases, e.g. in identifying these enzymes in electrophoretograms.

naphthol, $C_{10}H_7OH$, a phenol of naphthalene occurring in two isomeric forms: 1-naphthol (α-naphthol) and 2-naphthol (β-naphthol) which are used in the manufacture of dyes, 2-naphthol also as an antiseptic.

Naphthol Blue Black, one of the two acid dyes used for staining proteins e.g. in electrophoresis. *Syn.* Amido Schwarz.

naphthyl, the radical of → naphthalene, $C_{10}C_7$–.

narc(o)- (Gr. *narkoun* = to benumb), pertaining to a stuporous state, narcosis. *Adj.* **narcotic**.

naris, pl. **nares** (L.), nostril, opening on either side of the → nasal cavity; anterior nares (**external nares**) open onto the head surface, posterior nares (**internal nares**) form the → choanae of vertebrates; in fish, each naris comprises only a closed sac.

nas(o)-, also **nasi-** (L. *nasus* = nose), denoting relationship to the nose. *Adj.* **nasal**.

nasal cavity (L. *cavitas/cavum nasi*), the cranial portion of the passage to the respiratory system of tetrapod vertebrates; a complex, membranous cavity supported by bones and cartilages, i.e. by the **nasal conchae**. The anterior portion of the cavity just inside the nares is called the **vestibule of nose** (*vestibulum nasi*). The epithelium of the mucous membrane is partly formed from **ciliated columnar epithelium** typical of all respiratory passages, partly from the **olfactory epithelium** with sensory receptors (*see* olfaction). The respiratory mucous membrane functions in moistening, cleaning, warming, and in a hot environment in cooling the inhaled air, and the olfactory mucous membrane acts in olfaction. Mammals have mucosa-lined air cavities, the **paranasal sinuses**, in the cranial bones associated with the nasal cavity.

nasal concha, pl. **nasal conchae** (L. *concha nasalis* < Gr. *konche* = shell), one of the three bony plates (inferior, middle, superior conchae) covered by the mucous membrane in the nasal cavity of tetrapod vertebrates. The conchae extend the surface of the cavity and are exceptionally large in macrosmatic animals, as e.g. in dog, with a highly developed sense of smell. Also called crest (*crista ethmoidalis*).

nascent (L. *nasci* = to be born), **1.** being born, beginning to exist, e.g. nascent polypeptide in → protein synthesis; **2.** in chemistry, pertaining to an atom or an atom group at the moment when it is set free from a compound, i.e. is in a nascent state, *in statu nascendi*; in this condition the element is highly reactive.

nastic movement, nasty (Gr. *nastos* = close-pressed), a plant movement caused by a non-directional stimulus, always occurring in the same way. In **thermonasty**, plants response to temperature changes, e.g. flowers close in cold and open in warmth. The thermonasty is a typical **growth movement** as well as **photonasty**, in which diffuse light stimuli cause leaflets to press together, or flowers to close during dark periods. **Thigmonasty** is triggered by a touch resulting in rapid changes in the turgor pressure of certain cells, e.g. of → pulvinus in the nodes of *Mimosa*, the changes of which cause rapid movements of the leaflets of the plant. *See* epinasty, hyponasty, nyctinasty. *Cf.* tropism.

natal, 1. (L. *natalis* < *natus* = birth), pertaining to birth; **2.** (L. *nates* = buttocks), pygal; pertaining to buttocks.

natality, in population ecology, usually denoting the mean number of female offspring per adult female per time unit in the population. The factors affecting natality are: 1) the initiation age and the duration of the reproductive period of a female, 2) the mean number of female offspring per female per parturition, 3) the mean number of parturitions in the lifetime of a female, and 4) the age structure and sex ratio in the population. The natality rate is nearly equivalent to the term **birth rate**, but has a broader sense including the production of new individuals by birth, hatching, germination, and fission.

national park, *see* nature reserves.

native (L. *nativus*), original, innate, being born, arising; e.g. pertains to organisms which originate in the district where they live.

natrium, Na (L.), → sodium.

natriuretic factor (Gr. *ouron* = urine), a glycoside hormone (similar to → ouabain) found in the blood of vertebrates, but apparently occurs also in other animals. Like other natriuretic hormones it acts in the kidneys inhibiting the Na^+,K^+-ATPase and so increases the excretion of sodium from the body.

natriuretic hormones, vertebrate hormones which increase sodium excretion from blood to urine in the kidneys, including the → natriuretic factor and → natriuretic peptides.

natriuretic peptides, peptide hormones of vertebrates decreasing blood pressure by increasing sodium excretion and consequently water from the body; the peptides have an inhibitive effect on vascular smooth muscles thus causing vasodilatation in blood vessels. Natriuretic peptides include **brain natriuretic peptide** (BNP) and → **atrial natriuretic peptide** (ANP).

natural cultivation, cultivation of plants by natural methods, i.e. only with natural fertilizers (compost, ash, natural minerals) without artificial fertilizers, herbicides, etc. *Syn.* biological cultivation.

natural immunity, *see* immunity.

natural killer cell, NK cell, a type of → lymphocyte in the immune defence of vertebrates by lysing target cells without involvement of antibodies or complements, thus being a component of the innate immune system. NK cells, which in mammals comprise about 10% of all lymphocytes, are closely related to T lymphocytes acting e.g. in killing of tumour cells or cells infected by viruses. NK cells are activated by → interferons.

natural selection, a type of → selection tending to secure the survival of those individuals or groups who are best adapted to the conditions under which they live. Natural selection results from the varying → fitness of different genotypes, and it preserves favourable characters in natural populations and eliminates unfavourable ones. The theory of natural selection was developed by *Charles Darwin* and *Alfred Wallace*.

Three types of natural selection are presented.

1) **Stabilizing selection** acts in circumstances which have been stable for a long period, and it strives for preservation of the → adaptive norm of the population. This type of selection is further divided into **normalizing selection** and **canalizing selection**. The former eliminates harmful mutations and the latter favours the genotypes whose developments are well canalized (*see* canalization).

2) **Balancing selection** strives for the maintenance of → polymorphism in a population. It includes firstly **heterotic balance selection**, in which the selection favours → heterozygotes at the cost of both homozygotes, and secondly **frequency dependent selection** in which the fitness of the genotypes is in inverse relation to frequencies of the genotypes, and thirdly **disruptive selection**, favouring extreme ends of the distribution of genotypes in the population at the expense of the mean types.

3) **Dynamic selection** begins when the life conditions of the population change so that the previous adaptive norm no longer offers the best possible adaptiveness. Consequently, dynamic selection creates a new adaptive norm. *See* sexual selection.

nature conservation, the careful and organized use and management of natural resources, such as soil, water, animals, plants, and minerals. From the aesthetic viewpoint, nature conservation also includes the maintenance of nature reserves and national parks, wildlife, and historic sites.

nature philosophy, natural philosophy, a paradigm of ancient Grecian and Renaissance philosophy to explain phenomena by natural causes without mythical elements.

nature reserve, a land area protected from human activities to preserve the physical features, fauna and flora in their natural conditions. There are **national parks** which are open to the general public or **strict nature reserves** which are open only to scientists.

nauplius, pl. **nauplii** (L. < Gr. *nauplios* = shellfish), the first, free-swimming, larval stage in the life cycles of some crustaceans, such as ostracods, copepods, and barnacles. The unsegmented body has a single median eye (→ nauplius eye), three pairs of appendages, i.e. antennules, antennae, and mandibles, the last two with richly branched swimming setae. *See* megalops, metanauplius, protozoea, zoea.

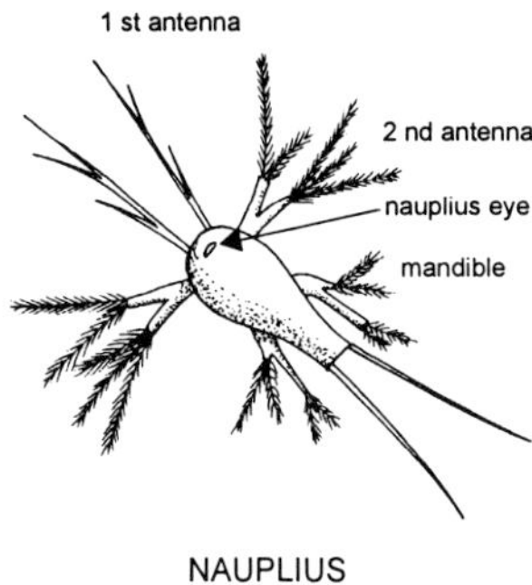

NAUPLIUS

nauplius eye, a single median eye with a few ocelli of the nauplius larva and of some adult crustaceans, e.g. copepods; the nauplius eye is able to observe the intensity and direction of light.

Nautiloidea (Gr. *nautilos* = sailor, nautilus, *eidos* = form), **nautiloids**; a subclass of Cephalopoda including one extant species, the **pearly nautilus**, which lives on the bottoms of the Indian Ocean and eastern Pacific. Its spiral shell is divided by septa into several chambers. Nautiloids were common through Palaeozoic periods.

navigation (L. *navigare* = to sail, navigate), in biology, the ability of certain animals to navigate over long distances to regions, often unfamiliar. Several sensory systems are utilized in navigating species: **clock-compass** navigation (celestial navigation), in which animals, such as bees, take and keep direction by the Sun's position (and by polarized light); correspondingly, nocturnally migrating birds navigate with respect to certain patterns of stars. Another type is **geometric navigation** in which animals (e.g. many birds and some fish species) utilize the magnetic field for navigation and orientation. This ability is based on the function of the → magnetic sense.

N-band (N < nucleolus), a band-like structure of a chromosome becoming visible in → Giemsa staining after pretreatment in a hot, acid phosphate solution. In plant and mammalian chromosomes the dye stains the region of the → nucleolar organizer, and also a part of constitutive → heterochromatin. The target substances in staining are acidic → phosphoproteins.

ne-, neo- (Gr. *neos* = new), denoting new, young, or recent.

Neanderthal man (*Neanderthal* = a valley in Germany), a subspecies, *Homo sapiens neanderthalensis*, of the species of modern man, *Homo sapiens*; settled in Central Europe and Western Asia about 230,000—30,000 years ago. Some anthropologists classify Neanderthal man into a separate species, *Homo neanderthalensis*. They were robust hunters whose brain size was equal to that of modern man, and they were able to speak; probably they buried their dead. Neanderthal man was displaced by → Cro-Magnon man for unknown reasons.

Nearctic region (Gr. *neos* = new, *arktos* = bear), a zoogeographical region including Greenland, North America and the northern parts of Mexico; together with the Palaearctic it forms the **Holarctic region**.

necatorin (L. *necare* = to kill), a poisonous substance occurring in the milk cap of the mushroom *Lactarius necator*; has mutagenic properties.

necessary and sufficient condition, a concept belonging to modal logic, describing the causal relations of different phenomena; it is defined as follows: if always and only when A occurs also B occurs, A is a necessary and sufficient condition for B, and respectively B for A. For example, the pulsation of the heart is a necessary and sufficient condition for blood circulation in mammals.

necessary condition, a concept belonging to modal logic, describing the causal relations of different phenomena; it is defined as follows: if always when A occurs also B occurs, B is a necessary condition for A. All necessary conditions of a phenomenon together, i.e. their conjunction (N_1 and N_2 ... N_n), are consequently a → sufficient condition for this phenomenon. For example, a breakage in DNA is a necessary condition for a → mutation.

neck, a constricted portion between the head and trunk of the body, or a constricted part of an organ; corresponds to the latin terms *cervix* and *collum*.

necr(o)- (Gr. *nekros* = dead body, corpse), denoting relationship to death or to dead organic material, as e.g. dead cells, tissues, or body.

necridium, a cell in the filament of → cyanobacteria the death of which results in the formation of a → hormogonium, i.e. a part separated from the parent filament giving rise to a new filament.

necromass, the weight of dead organisms, dead organic material; usually presented as a mass unit per unit of land area or water volume. Sometimes the term is used to describe dead structures or tissues in a living organism, such as horns and hairs of animals, bark of trees, etc.

necroparasite, a parasite which kills its host organism or parts of it and uses the dead material for growth and development.

necrophage (Gr. *phagein* = to eat), an organism feeding on dead animals; also called necrophile. *Adj.* **necrophagous**.

necrophile (Gr. *philein* = to love), **1.** necrophage; **2.** an organism having a preference for dead tissues of animals. *Adj.* **necrophilous**.

necrosis, pl. **necroses** (Gr. *nekrosis* = deadness), death of a circumscribed area in an organ or tissue due to local cell death with the degradative action of cell organelles and liberation of hydrolytic enzymes; may be caused by several reasons, as by the deficiency of oxygen supply, bacterial infection, etc.

necrotrophic (Gr. *trephein* = to nourish), pertaining to an organism which gets its nourishment from dead cells and tissues.

nectar, a liquid rich in sugars, amino acids, and other nutrients secreted by certain cells of many seed plants, luring insects (or birds) to pollinate the plants. **Nectaries** (**nectar glands**), i.e. the organs which secrete nectar, are found mostly in flowers, especially in petals or sepals, but they may occur even in leaves or stipules.

nectar guide, a pattern in a flower which guides an insect to the nectary; usually reflects ultraviolet radiation and thus is not visible to the human eye but to many insects with ultraviolet vision.

nectary, *see* nectar.

necton, → nekton.

negative complementation, a condition appearing if a heterozygous individual expresses a stronger mutant phenotype than either of the mutant parental individuals. *See* genetic complementation.

negative control, in genetics, includes control systems of a gene function in which a regulatory molecule (usually a protein), produced by the gene, inhibits in its active form the function of this or some other gene. In principle, it is the same phenomenon as **negative feedback** in physiology. *See* regulation.

negative feedback, *see* regulation.

negative reinforcement, *see* reinforcement.

negative staining, a staining method for light and electron microscopy by which only the background is stained, and the unstained specimen can be seen against it.

nekton, also **necton** (Gr. *nektos* = swimming), free-swimming animals of pelagic zones which, contrary to the → plankton, are able to move independently of waves and currents; e.g. large invertebrates, fish, and whales. *See* benthos.

nematoblast (Gr. *nema* = thread, *blastos* = bud), *see* nematocyst.

nematocyst (Gr. *kystis* = sac, bladder), a stinging organelle in the → **cnidocyte** of cnidarians for catching prey; includes a long coiled filament. Mechanical stimulation of the spine-like → **cnidocil** causes the nematocyst to expand rapidly, and the hydrostatic pressure pushes the filament out to catch the prey, e.g. by injecting poison into it. Over 20 different types of nematocysts have been described, the three main types being → **glutinants, penetrants**, and **volvents**. The nematocysts develop from **nematoblasts**.

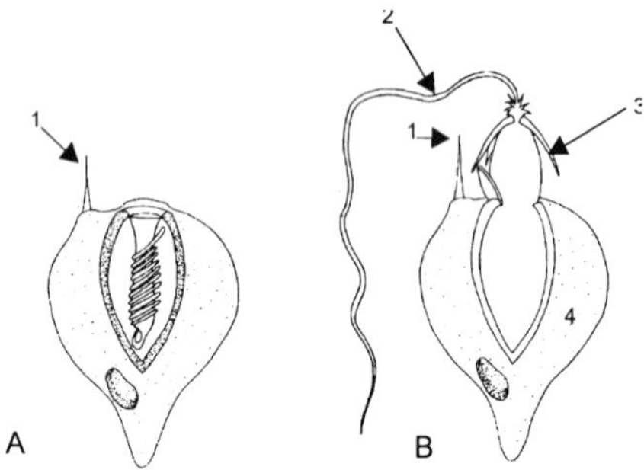

STRUCTURE OF NEMATOCYST
A. undischarged B. discharged.
1. cnidocil 2. filament 3. operculum 4. cnidocyte

Nematoda, also **Nemata** (Gr. *nema* = thread, *eidos* = form), **roundworms, nematodes**; an invertebrate phylum (in some classification a class in the phylum Aschelminthes) comprising unsegmented, slender, and often shuttle-shaped worms which have a strong longitudinal musculature, intracellular excretory system, → pseudocoelom, and thick elastic cuticle. The nervous system consists of simple nerve cords with many ganglia, and a nerve ring around the oesophagus. The lengths of nematodes vary from less than one millimetre to tens of centimetres. The phylum includes probably tens of thousands of species (about 12,000 are described) which occur almost everywhere either free-living or parasitic. Their taxonomy is not yet established. Usually they are divided into two classes: **Phasmidia** (Secernentea) with caudal sensory organs, → phasmids, and **Aphasmidia** (Adenophorea) without phasmids; the first includes e.g. **hookworms**, **pinworms**, **filarial worms**, and **ascaris**, the latter including e.g. **trichinae**. *See* human pinworm, ascariasis.

nematology (Gr. *logos* = word, discourse), a branch of biology dealing with nematodes.

Nematomorpha (Gr. *morphe* = form), **nematomorphs**, **horsehair worms**, **hairworms**; an invertebrate phylum with ca. 250 hair-like, unsegmented species which as adults live freely in soil or fresh water; their larval stages are parasitic in arthropods. The alimentary canals of adult nematomorphs are reduced. Previously nematomorphs were classified as a class in the phylum Aschelminthes.

Nemertea, Nemertinea, Nemertina, Nemertini, Rhynchocoela (Gr. *Nemertes* = a nereid), **nemerteans**, **nemertineans**, **nemertines**, **ribbon worms**; an invertebrate phylum including mainly marine species which are long, slender, acoelomate worms with an aperture above the mouth from which a muscular, tubular **proboscis** is capable of protruding to catch a prey; their larval stage is called **pilidium**. The phylum comprises about 650 species in the classes *Anopla* (proboscis worms) and *Enopla*; most of them live on shores around the tide zone.

neo-, ne- (Gr. *neos* = new), denoting new, young, or recent.

neocentric activity (Gr. *kentron* = centre, central point), a diffuse kinetic activity of → centromeres or → telomeres in a cell, usually observed in → meiosis of abnormal genotypes, like inbred lines of certain plants.

neocentromere (Gr. *meros* = part), a region outside the → centromere in a cell which in certain conditions during → mitosis or meiosis is transformed into a kinetically active element.

neocerebellum, phylogenetically the most recent part of the → cerebellum of tetrapod vertebrates, consisting of the large lateral portions of the cerebellar hemispheres receiving nerve fibres from the main motor and sensory pathways in the brain stem via the pontine nuclei and peduncles. The neocerebellar cortex consists of three layers, i.e. an external **molecular layer**, a **Purkinje cell layer**, and an internal **granular layer**. The middle layer is only one cell thick, formed from large → Purkinje cells with extensive dendritic arbours extending throughout the molecular layer and with axons passing to deep nuclei. The molecular layer contains a vast number of cell branches but also the cell bodies of inhibitory **stellate** and **basket cells**. The granular layer contains **granule cells** and inhibitory **Golgi cells**, both projecting into the molecular layer. The neocerebellum is highly developed e.g. in primates controlling automatic voluntary movements. *Syn.* corticocerebellum.

neocortex, pl. **neocortices** (L. *cortex* = bark), **neopallium**, **isocortex**; the most highly evolved part of the cerebral cortex of mammals and advanced birds and reptiles, characterized by a six-layered organization in its cellular structure; i.e. inwards from the surface: 1) molecular layer (plexiform layer), 2) outer granular layer, 3) pyramidal cell layer, 4) inner granular layer, 5) inner pyramidal layer (ganglionic layer), and 6) multiform layer. This organization is most complex in the neocortex of intelligent mammals such as primates, controlling exact voluntary motor and sensory functions and their interpreting and integration, and complex associative brain functions such as thinking. *See* pallium, Brodmann's areas.

neo-Darwinism (*Charles Darwin*, 1809—1882), a modern form of the theory of evolution developed between 1920's and 1940's. The basic principle of → Darwinism, i.e. the principle of → natural selection, has been unified with the theories of genetics. Population genetics plays a central role in neo-Darwinism. *Syn.* synthetic theory of evolution. *See* evolution theory.

Neogaea, also **Neogea** (Gr. *ge* = earth), a zoogeographical area, previously defined to comprise both the → Nearctic and Neotropical regions, at present only the Neotropical region, i.e. southern Mexico, Central America, South America, and the West Indies. *Cf.* Arctogaea, Notogaea, Palaeogaea. *See* Zoogeographical regions.

Neogene (Gr. *genos* = age), a collective term for the epochs of the later Tertiary (i.e. Miocene and Pliocene). *Cf.* Palaeogene.

Neognathae (Gr. *gnathos* = jaw), a superorder of the subclass Neornithes in the class → Aves; includes all birds other than ratites and tinamous (the other superorder → Palaeognathae).

Neolithic (Gr. *lithos* = stone), New Stone Age; a phase in human history beginning about 12,000 years ago when cultivation of plants and domestication of animals began, and stone and decorated horn or ivory tools were used; the wheel was invented.

neomorph (Gr. *morphe* = form), **1.** a structure, such as an organ of higher organisms, recently developed in the course of evolution and having only slight or no traces in lower orders; the phenomenon is called **neomorphism**; **2.** a mutant allele in which the function of a gene gives rise to a characteristic with new qualities. *Adj.* **neomorphic**.

neonatal (L. *natus* = birth), pertaining to the time period immediately after birth, e.g. in cats and dogs to the period from birth until the eyes are opened, and in children to the first four weeks after birth. *Cf.* perinatal.

neopallium (L. *pallium* = cloak), → neocortex.

neoplasm (Gr. *plassein* = to form), an abnormal, usually uncontrolled new growth of tissue; tumour; **neoplasia**, the formation of a neoplasm. *Adj.* **neoplastic**.

Neopterygii (Gr. *pteryx* = fin), **neopterygians, modern bony fishes**; in some classifications an infraclass or superorder in the subclass Actinopterygii (ray-finned fishes) comprising **holosteans** and **teleosts**; neoptertygians represent one of the two major lineage groups among the ray-finned fishes (the other is → Chondrostei).

Neornithes (Gr. *ornis* = bird), a subclass in the class Aves, characterized by the sternum with high keel and reduced spinal tail. *Cf.* Archaeornithes.

neostigmine, a synthetic organic compound, $[C_{12}H_{19}N_2O_2]^+$; a reversible cholinesterase inhibitor in cholinergic nerve terminals, preventing the breakdown of acetylcholine. Neostigmine, as well as the related substance **physostigmine** (eserine), increases the activity of cholinergic synapses and is used in pharmacological research and medication.

neotenin, → juvenile hormone.

neoteny (Gr. *teinein* = to extend), the persistence or prolonged development of the larval stage, or the occurrence of an underdeveloped form which may gain sexual maturity; examples of neoteny are the axolotl (*Ambystoma*), the olm (*Proteus*), or certain insect castes held in the larval stage as future replacement for the queen. Neoteny may also occur in newt and toad larvae after cold treatment, due to incomplete activity of the thyroid gland. Many mammals, e.g. whales, secondarily adapted to aquatic environment, possess **neotenic characteristics**. Also the slow ontogenic development with increased brain size and some anatomical and physiological properties of humans are suggested to be neotenic (*see* aquatic ape theory). *Cf.* paedogenesis.

Neotropical kingdom, *see* Neotropical region.

Neotropical region (Gr. *tropikos* = pertaining to a turn), a zoogeographical region (in botany **Neotropical kingdom**) including southern Mexico, Central America, South America, and the Archipelago of the West Indies; in geobotany, however, the southernmost part of South America belongs in the Antarctic kingdom.

neotype (Gr. *typos* = pattern), *see* type specimen.

neoxanthin (Gr. *ksanthos* = yellow), a carotenoid pigment of plants, especially algae, one of the → xanthophylls.

nephelometry (Gr. *nephele* = cloud, *metrein* = to measure), a quantitative method for measuring the turbidity of a solution; is based on the spectrophotometric estimation of the scattering of light by particles in a suspension.

nephr(o)- (Gr. *nephros* = kidney), denoting relationship to the kidney. *Adj.* **nephric**.

nephridiopore (Gr. *poros* = passage, pore), *see* nephridium.

nephridium, pl. **nephridia** (Gr. *nephridios* = pertaining to the kidney), **1.** an excretory tubule in a vertebrate embryo before the development of the kidney; **2.** a duct-like excretory organ found in many invertebrates, such as platyhelminths, nemerteans, rotifers, annelids, and lancelets; may begin with an open ciliated funnel (**nephrostome**) in the coelom, and ends in the gut or the exterior via an excretory pore (**nephridiopore**) on the surface of an animal. In fully segmented worms, e.g. oligochaetes, a pair of nephridia are repeated in each body segment, i.e. are segmented organs. Nephridia are divided into various types on the basis of their developmental stage; *see*

protonephridium, metanephridium.

nephrocyte (Gr. *kytos* = cavity, cell), an excretory cell type found in many arthropods (e.g. insects), and in ascidians; collects and destroys waste products from tissues by → phagocytosis. In insects, nephrocytes are situated around the intestine wall and → Malpighian tubules.

nephromixium, pl. **nephromixia** (L. < Gr. *mixis* = act of mixing), a tubular organ (**urogenital duct**) of some polychaetes through which the eggs, sperm, and excretory wastes pass from the → nephridium out of the body.

nephron, the anatomical and functional unit of the → kidney of vertebrates, in number from tens of thousands to millions per kidney (about one million in one human kidney). The most primitive nephrons, located segmentally in the body, exist in the **pronephros** (primitive kidney); each nephron consists only of a thin tubule that begins as a funnel through which fluid from the body cavity is filtered into the tubule.

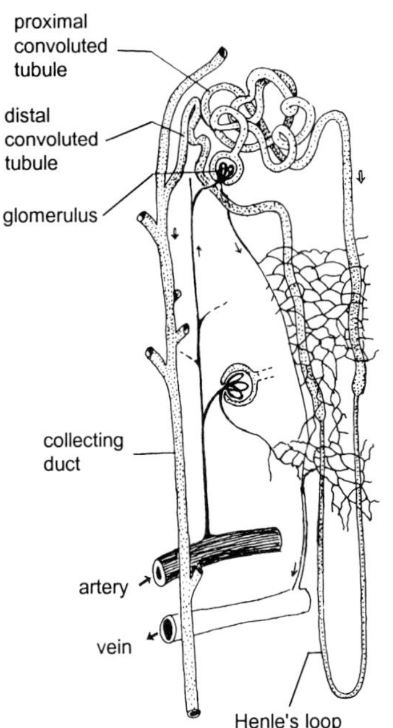

A nephron in the metanephros type of the kidney.

The nephron of more developed kidney types, the **mesonephros** and **metanephros**, consists of the **renal tubule** (nephron tubule) and of the **renal corpuscle** (Malpighi's corpuscle). The latter comprises the **glomerular capsule** (Bowman's capsule) and the → **glomerulus**, i.e. an arterial capillary enveloped by the capsulous end of the renal tubule. Blood plasma (excluding proteins) is filtered by hydrostatic pressure through the thin capillary and capsule walls into the capsule, and the glomerular filtrate flows from the capsule freely to the tubule which consists of several parts, i.e. the **proximal convoluted tubule**, **Henle's loop**, and **distal convoluted tubule**. The glomerular filtrate is processed into urine by the specialized cells of the tubule walls by means of active and passive absorption, resorption, and specialized secretion of certain substances; e.g. water and nutrients are mainly resorbed into blood capillaries around the tubules. Several nephron tubules join to form the **collecting ducts** (collecting tubules), which all open to the **renal basin** (renal pelvis). The nephronal functions are regulated by several hormonal and neural systems, such as the renin-angiotensin system (*see* renin), mineralocorticoids, vasopressin (antidiuretic hormone), natriuretic hormones, and autonomic nervous system.

nephrostome (Gr. *stoma* = mouth), *see* nephridium.

nephrotome (Gr. *tomein* = to cut), the modified part of a → somite of a vertebrate embryo which develops into a segmental excretory tubule of the pronephros (primitive kidney). *See* kidney.

neritic (Gr. *nereites* = clam), **1.** pertaining to an organism living only in the shallow littoral zone of a sea or ocean; **2.** pertaining to the sea zone over a continental shelf, specifically from the sublittoral zone to the depth of 200 metres; includes the photic zone and consequently is a habitat of benthic organisms in the oceans. *Cf.* oceanic, pelagic.

Nernst equation (*Walther Nernst*, 1864—1941), describes an electrochemical equilibrium between the diffusion force resulting from a concentration (activity) gradient of a certain permanent ion and the resulting electrical gradient across a boundary between two systems, e.g. between the cytoplasmic and extracellular compartments separated by the plasma membrane. When the ion concentrations (activities) in these compartments are known, the direction of ion transport and the equilibrium potential can be calculated (*see* resting potential). The potential difference at equilibrium is:

$$\varepsilon = RT/FZ \cdot \ln (c_i/c_o),$$

where R = gas constant (8.3144 J mol^{-1}), T = absolute temperature, F = Faraday constant (96,484 coulombs · mol^{-1}), Z = valence of an ion, ln = natural logarithm, c_i/c_o = the ratio of the ion concentrations (inside/outside of a cell).

nerve (L. *nervus*), any of the threadlike structures in most multicellular animals forming the → **peripheral nervous system** together with peripheral ganglia; comprises bundles of → nerve fibres having their roots in the central nervous system. Usually the nerves are branched. Each nerve is surrounded by a thick sheath, **epineurium**, each bundle of several nerve fibres within the nerve has its own thinner sheath, **perineurium**, and each nerve fibre is further enveloped by an **endoneurium**, all composed of fibrous connective tissue. Nerves may be **sensory**, **motor**, or **mixed nerves**, containing sensory or motor nerve fibres only, or both.

nerve cell, neurone, neuron; the structural and functional unit of the nervous system consisting of a **cell body** (soma) and its processes, i.e. typically one or more **dendrites** and one **axon**, with specialized junctional endings with other cells. The dendrites receive and conduct information in the form of → electrotonic potentials spreading towards the cell body, in which the initial segment (axon hillock) of the axon generates **nerve impulses** (*see* action potentials) which propagate along the axon to its termination to form → **synapses**, in which impulses are transmitted to another neurone, or an effector, i.e. muscles and glands.

According to the number of processes there are **unipolar**, **bipolar** and **multipolar** nerve cells in most animals. A special type is a sensory neurone of vertebrates, the nerve cell body having one process which branches immediately forming two branches (**pseudounipolar neurone**), and both branches are structurally and functionally → nerve fibres. Cell organelles in a nerve cell body are typically one large round → **nucleus**, large **endoplasmic reticulum**, **ribosomes** (Nissl bodies), large **mitochondria**, and **neurofibrils** which are bundles of → microtubules and neurofilaments passing also into the branches.

The neurones are differentiated to integrate and propagate information in electric form, but some neurones are specialized for hormonal secretion (→ **neuroendocrine cells**). In some animal groups there are giant neurones (in molluscs and fish) or giant axons (in annelids), functioning in quick flight reactions. *See* resting potential.

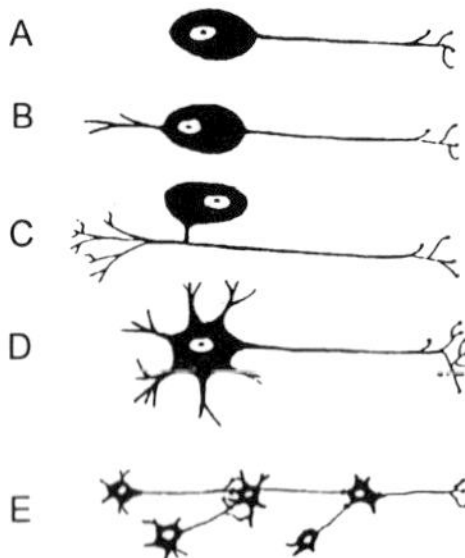

Nerve cell types; unipolar (A), bipolar (B), pseudounipolar (C), and multipolar (D) nerve cells. Nerve cells send axon branches to form synapses with dendrites and cell bodies of other nerve cells, forming neural pathways (E).

nerve cord, in invertebrates, the longitudinal **nerve chain** from the brain ganglia to the caudal end of the body, composed of nerve pathways with segmental or regional ganglia; in annelids and arthropods, usually a paired, solid, ventral nerve chain (ventral nerve cord) with segmental ganglia. *See* nervous system.

nerve ending, the termination of the peripheral → nerve fibre; may be a **sensory nerve ending** serving as a sensory receptor, which may be structurally differentiated as a part of a sense organ, or a less differentiated free nerve ending like a pain receptor. Further the termination may be a **motor nerve ending**, i.e. a nerve fibre having a specialized junctional structure with an effector cell, i.e. the → neuromuscular junction on the surface of a muscle cell. **Autonomic nerve endings** form homologous terminations to cardiac and smooth muscle cells, and to many glandular cells.

nerve fibre, Am. **nerve fiber,** the → axon of the nerve cell with a sheath around it; only exceptionally, as in somatosensory neurones of vertebrates, embryonic → dendrites of bipolar cells form nerve fibres. The length of nerve fibres varies greatly according to the tissue and the size of the animal; in peripheral

nerves of large animals, such as whales, some nerve fibres may be metres in length. Usually nerve fibres are thin, i.e. from 1 to 20 μm in diameter but may be much wider (*see* giant axons).
In nerve fibres, the axon is supported by glial cells and their action determines the type of sheath surrounding the axon. In **unmyelinated nerve fibres**, the axons are surrounded by glial cells, which support and nourish the axon. Unmyelinated fibres occur in invertebrates but also in vertebrates, especially in the central nervous system but also e.g. in peripheral autonomic nerves. In **myelinated nerve fibres** the axons are surrounded by a sheath which is formed when the membranes of glial cells wrap many times tightly around the axon (axis cylinder). In the periphery, this **myelin sheath** is produced by special glial cells, called **Schwann cells**, but in the central nervous system it is formed by glial cells, called **oligodendrocytes**. The myelin sheath is formed from elaboration of the plasma membrane of the glial cells producing a multilayered structure, the cytoplasm and nuclei of the glial cells are confined to the surface, and in the light microscope can be seen as an additional sheath, called the **Schwann sheath** (neurilemma, neurolemma). Adjacent glial cells along the length of the axon do not meet but leave short, naked (unmyelinated) areas of the axon, → **nodes of Ranvier**. At these nodes the tissue fluid is in direct contact with the axonal cell membrane (plasma membrane), and thus the transport of various substances is possible between the tissue fluid and the axonal cytoplasm. The myelin sheath functions as an electrical insulator, the impulses propagating along the fibre by jumping from one node to another (saltatory conduction). This results in a manifold increase in conduction velocity of nerve impulses as compared with those in a corresponding unmyelinated nerve fibre. In vertebrates, the fastest nerve fibres, such as sensory and motor fibres of the central and peripheral nervous systems, are usually myelinated, and the velocity of nerve impulses may be more than 100 m/s.

nerve gas, any poisonous compound that strongly depresses the function of the nervous system, usually inhibiting the cholinergic transmission between the cells, acting e.g. as an anticholinesterase. These nerve poisons have been used in gaseous form in war; most are derivatives of phosphoric acids, as sarin or tabun (dimethylphosphoramidocyanidic acid, ethyl ester). In aquous solutions nerve gases are used as insecticides (e.g. → parathion, malathion).

nerve growth factor, NGF, a protein (a neurotrophin) necessary for the growth and maintenace of sympathetic neurones and some sensory neurones in many vertebrate species; found in many different tissues. *See* growth factors.

nerve impulse, an → action potential occurring in nerve cells (neurones) serving as an electric message in the nervous system. The impulse is initiated by stimulation of sensory receptors or at the initial segment of an axon as a result of synaptic action (*see* synapse). The impulse is generated as a quick electric discharge (about 100 mV) of the → resting potential due to rapid, transient ion currents (influx of Na^+, efflux of K^+) across the axonal membrane. The impulse is initiated locally but propagates immediately along the axon membrane to the terminal button. The impulses are usually transmitted chemically to other cells at the → synapses.

nerve net, 1. the most primitive type of the nervous system found between the ectoderm and gastroderm in cnidarians, branches of the nerve net connecting to the sensory cells and to epithelial cells with contractile properties. It is a network of bipolar nerve cells which do not form proper → synapses between themselves but information is propagated electrically in both directions through intercellular junctions; a similar nerve net is found also in echinoderms and hemichordates as a part of their nervous system; **2.** the term is also used for a non-synaptic network of free autonomic nerve fibres terminating in the connecting tissue of an internal organ, such as in the walls of the alimentary canal of many arthropods, molluscs, and vertebrates.

nerve plexus (L. *plexus* = braid, network, tangle), a network of parallel nerves, such as in tetrapod vertebrates the **brachial plexus** and **lumbar plexus** consisting of several spinal nerves, or → **Auerbach's plexus** (myenteric plexus) and → **Meissner's plexus** (submucous plexus) of the autonomic nervous system in the wall of the vertebrate alimentary canal.

nerve poisons, substances which strongly inhibit the action of the nervous system, mostly affect functions of → synapses, thus causing highly toxic effects on animals; may be produced by organisms (→ neurotoxins), or by human technology, as e.g. → pesticides and nerve gases.

nerve root (L. *radix nervi*), the bundles of nerve fibres which emerge from the central nervous system and form a peripheral nerve; in vertebrates, each cranial nerve (except the fifth and eighth) have only one root, but in tetrapod vertebrates each spinal nerve has two roots, i.e. the **ventral root** (motor root) and the **dorsal root** (sensory root) which outside the vertebra join to form a single segmental spinal nerve.

nerve tissue, nervous tissue, a tissue type of most multicellular animals consisting of → nerve cells (neurones), supporting glial cells, and fluid extracellular matrix; embryonically develops from the ectodermal layer in the process of neurulation. *See* neuroglia.

nervous system (L. *systema nervosum*), an organ system of multicellular animals acting as the uppermost controller and integrator of body functions. It is composed of highly specialized electrically active cells, → **nerve cells (neurones)**, with supporting glial cells, communicating with each other through → synapses. The nervous system receives and interprets information from the environment and inner organs through sensory receptors and transmits impulses to the effector organs, such as muscles and glands.

Mesozoans and sponges have no nervous system, and the simplest form is found in cnidarians which at the body surface have a primitive → nerve net with bipolar **neurones**; these form electrical junctions with other cells, but a more developed neuronal system with specialized processes of neurones, and proper synapses between the nerve cells is found elsewhere in the body.

During the phylogenetic development of the nervous system, a certain organization of the neurones develops: cell bodies of the neurones are grouped together forming **ganglia**, and nerve fibres together to form **nerve cords** (containing nerve tracts) and peripheral **nerves**. The cranial ganglia associated with special senses in the head area combine together to form the **brain** or **brain ganglia**.

The basic type of anatomical structure in many invertebrates is the **ladder-like nervous system**. It consists of the paired anterior ganglia (**brain ganglia**) from which two main ventral nerve trunks run posteriorly and are linked together by transverse commissures at each segment (ladder structure). Peripheral nerves emerge symmetrically from the ganglia to peripheral tissues. In more developed invertebrates, such as annelids and arthropods, the ventral trunks are joined to form the double **ventral nerve cord** with ganglia and distinctive sensory and motor neurons. The **suboesophageal ganglia**, posteriorly next to the brain, are well developed. In nonsegmented molluscs there are three large ganglia, called **brain ganglia**, **pedal ganglia**, and **visceral ganglia**, large connectives uniting the last two to the first. The radially symmetrical echinoderms have a **nerve ring** with several ganglia around the pharynx, the peripheral nerves emerging radially from them.

Anatomically the nervous system of vertebrates comprises the **central nervous system** (brain and spinal cord) and the **peripheral nervous system** (nerves and their ganglia). Functionally the nervous system can be devided into 1) the **voluntary nervous system** and 2) the → **autonomic nervous system**, the former involving voluntary locomotor actions with associated sensory functions, the latter controlling autonomic (involuntary) functions of internal organs and associated sensory activity. 3) The **associative nervous system** comprises integrative brain activity (as thinking, motivation, emotions) of the cerebrum, especially the cerebral cortex. 4) The **endocrine nervous system** consists of hormone secreting neurones (neurosecretory cells), especially in the → hypothalamus.

nervus, pl. **nervi** (L.), → nerve.

Nessler's reagent (*Julius Nessler*, 1827—1905), an alkaline solution of potassium mercury(II) iodide ($KHgI_3$) in dilute potassium hydroxide; the reagent is used for detecting and determining ammonia, with which it forms a yellow compound.

nestling, a young bird not yet able to leave the nest.

nestling period, the time from the hatching of a nestling until it leaves the nest; short among nidifugous species.

nest parasitism (Gr. *parasitos* = parasite), the situation when parasites live in the nest of a host animal (common in many invertebrates), and also e.g. in the skin of bird nestlings and young mammal pups; are usually → ectoparasites. *Cf.* brood parasitism.

net photosynthesis, a term describing the net amount of carbon assimilated into carbohydrates in → photosynthesis; is obtained by subtracting CO_2 produced in dark respiration from the total CO_2 uptaken in photosynthesis.

net production, in ecology, the chemical energy bound to organic substances at a certain → trophic level, being available for organisms at the next level. The net production of producers is called **net primary production**, i.e. the total energy accumulated into organic substances by plants during photosynthesis. The net production of consumers is called **net secondary production**. **Net production efficiency** refers to the proportion of the assimilated energy that is incorporated into growth and reproduction as a percentage of **gross production**. In terrestrial ecosystems, net production efficiency is 20—50% at the producer level, and about 10% at the other levels. *See* productivity.

net reproductive rate, *see* reproductive rate.

neur(o)- (Gr. *neuron* = nerve), denoting relationship to the nervous sytem or its elements, as nerves, nerve cells, etc. *Adj.* **neural**.

neural canal, spinal canal. *See* vertebra.

neural crest (L. *crista* = ridge), an embryonic neural structure of vertebrates formed from neuroectodermal cells as paired cellular bands which during neurulation becomes separated from the dorsolateral portion of the → neural plate, later located between the neural tubc and the epidermis. The neural crest gives origin to the cranial and spinal ganglion cells with their sensory nerve fibres, autonomic ganglion cells, chromaffin cells of the adrenal medulla, Schwann cells, and epidermal pigment cells. Also a part of connective tissue (ectomesenchyme) derives from the head region of the neural crest. *Syn.* ganglionic crest, ganglion ridge.

neural gland, a gland connected to the neural ganglion (trunk ganglion) of an ascidian; apparently homologous to the neurohypohysis, i.e. the posterior lobe of the → pituitary gland in vertebrates.

neural induction, the induction of the neural plate from the ectoderm by the underlying mesoderm in an early vertebrate embryo.

neural lobe, neurohypophysis, the posterior lobe of the → pituitary gland.

neural network, the network of interconnected nerve cells in the cerebral cortex of vertebrates in which activity patterns are propagated during higher brain functions, as in perception, integration, learning, and memory.

neural plate, 1. a plate-like structure of the dorsal ectoderm (neuroectoderm, neuroepithelium) in an early vertebrate embryo, developing into the **neural tube** and the paired **neural crest** which further develop into the nervous system. An induction for its development is generated by the mesodermal tissue of the dorsal archenteron underlying the neuroepithelium. The differentiation of the plate ends the gastrula stage and starts neurulation (*see* neurula); *syn.* medullary plate; **2.** the lateral member of a vertebral arch (neural arch of vertebra).

neural tube, the dorsal epithelial tube developed from the → neural plate of an early vertebrate embryo; develops further into the central nervous system. *Syn.* medullary tube.

neuraminidase, an enzyme, EC 3.2.1.18, that cleaves the terminal acylneuraminic residues from oligosaccharides, mucopolysaccharides, glycoproteins, or glycolipids; present as a surface antigen in myxoviruses. The enzyme is used in histochemistry to remove sialomucins. *Syn.* sialidase.

neurilemma, also **neurolemma** (Gr. *lemma* = rind, peel, husk), Schwann sheath. *See* nerve fibre.

neurine, vinyltrimethylammonium hydroxide, CH_2=CH-N$(CH_3)_3$OH; a neurotoxic amine formcd in ccll dcath and decaying organisms from lecithin and choline, found also in some poisonous mushrooms.

neurite, an obsolete term for → axon; also used for neuronal branches in cell cultures.

neurobiology, the branch of science dealing with the development, structure, and function of the nervous system, includes e.g. neuroanatomy, neurohistology, neurophysiology, and neurochemistry (neurobiochemistry). *See* neurosciences, neurology.

neuroblast (Gr. *blastos* = germ), an embryonic → nerve cell, an immature nerve cell developing into a neurone.

neurocranium (L. *cranium* < Gr. *kranion* = skull), the portion of the vertebrate skull (cranium) which encloses the brain; composed of cartilaginous or bony tissue.

neuroectoderm, that region of the early embryonic → ectoderm in most animals which gives rise to the nervous system; sometimes called neuroepithelium.

neuroendocrine (Gr. *krinein* = to separate), **1.** pertaining to anatomical and functional relationships between the hormonal (endocrine) and nervous systems; **2.** describing neurosecretory cells; *see* neurosecretion.

neuroendocrinology, a research area dealing with interactions of the nervous and hormonal (endocrine) systems; a branch of → endocrinology.

neuroepithelium, pl. **neuroepithelia**, **1.** → neuroectoderm; **2.** sensory epithelium with sensory cells developed from nerve cells (neurones), such as the olfactory epithelium in the nose of vertebrates. *See* epithelium.

neurofibril (L. *fibrilla,* dim. of *fibra* = fibre), any of the very thin, thread-like intracellular structures forming the → cytoskeleton of a nerve cell; are seen when specially stained under a light microscope. Neurofibrils reach from the cell body to its processes, i.e. the dendrites and the axon and sometimes to synaptic endings. Neurofibrils are composed of smaller units, → microtubules and → microfilaments, which can be seen in an electron microscope, in nerve cells called **neurotubules** and **neurofilaments**.

neurogenesis (Gr. *gennan* = to produce), the ontogenetic (embryonic) development of the nervous system.

neurogenic, 1. forming nervous tissue; pertaining to neurogenesis; **2. neurogenous**, i.e. caused by, or originating in the nervous system.

neurogenic contraction, a muscular contraction triggered by a nerve impulse; occurs normally in skeletal muscles and in some rapidly contracting smooth muscles (e.g. intraocular muscles). *Cf.* myogenic.

neuroglia (Gr. *glia* = glue), non-neuronal cells of the → nervous tissue; dendritic cells around and between the nerve cells, also between nerve cells (neurones) and blood capillaries, supplying the nervous system. Neuroglial cells support neurones, function in the formation of sheath around the → nerve fibres, have trophic effects on neurones, and participate in physiological defence. In the central nervous system they include → oligodendrocytes (oligodendroglial cells), astrocytes, ependymal cells, and microglial cells. Schwann cells (neurolemmal cells) around the peripheral nerve fibres correspond to oligodendrocytes.

neurohaemal organ, Am. **neurohemal organ,** an organ containing axon endings of specialized neurones, **neurosecretory cells** (*see* neurosecretion) in which hormones are stored and released directly into the blood or haemolymph; e.g. the → corpus cardiacum in insects and the → sinus gland in decapod crustaceans.

neurohormones, 1. hormones produced by specialized neurones, **neurosecretory cells**, and liberated by nerve impulses (*see* neurosecretion), e.g. → releasing hormones of the hypothalamus, the hormones of the posterior lobe of the → pituitary gland, or adrenaline and noradrenaline released from the adrenal medulla; **2.** broadly, the term includes also other hormones which stimulate neural mechanisms. *Adj.* **neurohormonal**.

neurohumor (L. *humor* = liquid), a term formerly used to denote any chemical substance liberated from nerve endings into tissue fluids to activate or modify the action of other neurones or effector cells (muscles, glands); includes the → neurotransmitters and neurohormones. *Adj.* **neurohumoral**.

neurohypophysis, the posterior pituitary, i.e. the posterior lobe (neural lobe) of the → pituitary gland (hypophysis), i.e. the nervous lobe and infundibulum (neural stalk).

neurolemma, *see* nerve fibre.

neurology (Gr. *logos* = word, discourse), a branch of medical science dealing with the nervous system and its disorders, including also clinical neurology.

neurolysis (Gr. *lyein* = to dissolve), **1.** dissolution of the nervous tissue; **neurolysin**, a cytolytic substance (e.g. antibody) that specifically destroys nerve cells; **2.** the surgical release of the nerve.

neuromast, neuromast organ; a mechanosensory receptor organ of the → lateral line system of cyclostomes, fish, and totally aquatic and larval amphibians; probably developed as a structure homologous to the vestibular apparatus of the internal ear. Each neuromast is composed of a group of → **hair cells** surrounded by **supporting cells**, and covered by a gelatinous mass, **cupula**, with the cilia of the hair cells responding to vibrations of water. Typically the neuromasts are arranged in rows located either uncovered in small skin pits or in subcutaneous canals with membrane-covered orifices on the surface of the head and lateral regions (lateral lines). The neuromasts with their sensory neurones form a sensing part of the **acoustico-lateralis system** of these animals responding to low frequency vibrations.

neuromere (Gr. *meros* = part), a metameric segment containing a paired ganglion of a nerve cord, e.g. in annelids and arthropods; also a segment of the spinal cord of vertebrates at each spinal nerve, especially referring to an embryonic spinal cord.

neuromuscular, denoting relationship to nerves and muscles. *See* neuromuscular junction, neuromuscular transmission.

neuromuscular junction, myoneural junction, endplate; a specialized → **synapse** between a neurone and skeletal muscle cell. In skeletal muscles, the branched endings of a motor nerve fibre form plate-like junctions, **endplates** (motor endplates), one at the tip of each branch. Together, the neurone and the muscle cells it innervates, are called → **motor unit**. The structure and function of the endplate resemble those of the synapses between neurones. Each endplate consists of a flat, discoid expansion of an axon terminal pressed against the specialized, folded site of the muscle cell membrane. Across the cleft (**synaptic cleft**) between the neuronal and muscular membranes, the action potentials triggering muscle contraction are transmitted chemically to the muscle cell. This occurs by the action of **acetylcholine** liberated from the axon terminals principally in the same way as in synapses.

In the skeletal muscles of vertebrates, each muscle cell (muscle fibre) has only one endplate that is always excitatory (uniterminal innervation), but in many invertebrates there may be two or more endplates (**multiterminal innervation**) from the same neurone, or endplates from different neurones (**multineuronal innervation**); often comprising also inhibitory endplates. *See* neuromuscular transmission.

neuromuscular transmission, 1. synaptic transmission in → neuromuscular junctions of skeletal muscles; **2.** synaptic transmission between neurones of the autonomic nervous system and the smooth muscle cells or cardiac muscle cells (*see* synapse). In the smooth muscle, neuromuscular contacts are located in nerve endings at successive **varicosities**, which are enlargements along the ending, the same nerve ending innervating many muscle cells. This type of neuromuscular transmission is called **synapse en passant**. The neurotransmitter (**acetylcholine** or **noradrenaline**) liberated from the varicosities binds to specific receptors on the muscle cell membrane and changes the → resting potential of muscle cells. The → action potentials which trigger the smooth muscle contraction are usually induced in muscle cells at a certain area, called **pacing area**. The neurotransmitters liberated from the junction, however, enhance or inhibit the induction of these myogenic action potentials, and consequently contraction of the muscle.

In the heart ventricle the contacts between the noradrenergic fibres and cardiac muscle cells resemble those found in smooth muscle. The exact nature of neuromuscular transmission in specialized muscle cells of the → conduction system of the heart is not known.

neuronal, pertaining to a nerve cell, neurone.

neurone, also **neuron,** → nerve cell.

neurone oscillators (L. *oscillare* = to swing), groups or nets of nerve cells (neuronal network) which spontaneously function rhythmically and thus generate and maintain different types of nervous activity in the brain.

neuropeptides, small peptides functioning as hormones, or in → synapses either as proper → neurotransmitters or functional modulators, often spread in tissue fluids from other cells; include e.g. → endorphins, enkephalins, and releasing and inhibiting hormones (liberins and statins) secreted by the hypothalamus, as well as many hormones regulating the function of the alimentary canal. Neuropeptides are chiefly produced by neurones.

neurophysins (Gr. *phyein* = to grow), proteins found in neurosecretory granules in a certain hypothalamic region of the vertebrate brain; function as carriers during the transport and storage of neurohypophyseal hormones. Neurophysin is synthesized as a part of a large precursor protein in the cell body of the neurone and packaged into secretory granules in the → Golgi apparatus. The precursor molecule contains three oligo- or polypeptide sequences, i.e. a neurophysin, a hormone (→ vasopressin or oxytocin), and a signal peptide, which split from each other during the transport to the axon ending in the neurohypophysis, i.e. the posterior lobe of the → pituitary gland.

neurophysiology, the → physiology of the nervous system.

neuropil(e) (Gr. *pilos* = felt), a felt-like net comprising an enormous number of processes of nerve cells (dendrites and axons) and glial cells in which the cell bodies lie embedded; forms the bulk of the grey matter of the central nervous system of vertebrates and of the relating structures of nerve ganglia in invertebrates.

neuropodium, pl. **neuropodia** (Gr. *pous*, gen. *podos* = foot), **neuropod**; the ventral lobe of the → parapodium of polychaetes. *Cf.* notopodium.

neurosciences, the sciences dealing with the nervous system; include → neurobiology with its specialities such as neurophysiology and neurochemistry, as well as neuropharmacology, neurotoxicology, and → neurology.

neurosecretion (L. *secernere* = to secrete), **1.** the secretory process of certain neurones (**neurosecretory cells/neurones**), differentiated to produce and release hormones, **neurohormones**, from the axon terminals into the blood; in vertebrates e.g. the secretion of → releasing hormones from the hypothalamus, or neurohypophyseal hormones from the posterior → pituitary gland (*see* neuropeptides), or the secretion of the hormones of the adrenal medulla. In invertebrates, most hormones are produced by neurosecretory cells, such as the hormones of → corpus cardiacum and corpus allatum in insects; *see* neurohaemal organ; **2.** a secretory product produced by neurosecretory cells; **3.** sometimes the term is used also to include the secretion of → neurotransmitters.

neurotensin (L. *tensio* = stretching, pressure), a tridecapeptide found in the central nervous system of mammals, apparently acting as a neurotransmitter or modulator of neuronal functions; plays a role in pain perception, and its analgesic effect is not blocked by opiate antagonists. Neurotensin also increases vascular permeability, produces vasodilatation thus decreasing blood pressure, and increases plasma concentrations of some hormones, such as glucagon and growth hormone.

neurotoxin (Gr. *toxikon* = arrow poison), a poisonous substance destroying nerve cells, i.e. acting as a **neurolysin**; usually a large organic molecule produced by microbes, rendering the cell membrane leaky.

neurotransmitter (L. *transmittere* = to send across), **synaptic transmitter**; a specific substance which, when released in response to nerve impulses from the axon terminal at synapses, transmits information from one nerve cell to another, or to an effector cell, such as muscle and glandular cells. Transmitters may stimulate or inhibit the postsynaptic cell. The process is called **neurotransmission**. Tens of different neurotransmitters are known, most important of which are → **acetylcholine**, **noradrenaline**, **adrenaline**, **dopamine**, **serotonin**, **gamma-amino butyric acid** (GABA), **glycine glutamate**, and many peptides such as **endorphins**. One transmitter, as e.g. noradrenaline and many peptides, can act in several types of synapses and neuromuscular junctions, due to slightly different types of **receptor molecules** on the cell membrane of a target cell. More than one neurotransmitter may act in the same synapse, although many synapses contain only one transmitter.

The synapses are named according to their main neurotransmitter; e.g. → acetylcholine acts as the transmitter in cholinergic synapses, → serotonin in serotoninergic synapses and → noradrenaline or adrenaline in adrenergic synapses. Some neurotransmitters act also as a hormone (e.g. noradrenaline), or a hormone may act as a neurotransmitter (e.g. vasopressin). The inactivation of neurotransmitters may occur in several processes, such as in enzymatic degradation of the transmitter or in returning the transmitter into the axon terminal. *See* synapse.

neurotrophins (Gr. *trephein* = to nourish), proteins found in the vertebrate brain where they promote the growth and survival of nerve cells; e.g. prevent the → apoptosis of the cells.

neurula, pl. **neurulae,** also **neurulas** (L.), an early developmental stage of a chordate embryo after → gastrula; begins by formation of the → **neural plate** and ends when the → **neural tube** is completed.

neuston (Gr. *neustos* = floating), small or medium-sized organisms living either on the upper side (**epineuston**) or on the underside (**hyponeuston**) of the surface of the water film; epineustonic are e.g. water bugs (*Gerris*) and hyponeustonic many mosquito larvae.

neuter (L. *neuter* = neither), in biology, pertaining to, or being an organism with no reproductive organs but otherwise normal, or with no functional sex organs; e.g. a castrated animal, or non-fertile females of ants and honey-bees (workers).

neutral fat, → triacylglycerol.

neutral gene theory, a theory about the nature of → genetic polymorphism, suggesting that most cases of the polymorphism observed at the molecular level are neutral, i.e. the → alleles are equivalent with regard to → natural selection.

neutrality, in chemistry, the neutral state (pH 7.0) of a solution, i.e. neither acid nor alkaline, having equal numbers of hydrogen (H^+) and hydroxyl (OH^-) ions.

neutralization, any act or process rendering something neutral; e.g. **1.** in chemistry, the addition of acid to base or vice versa in equivalent quantities until the solution turns neutral (pH 7.0); chemical → titrations are based on the neutralization reaction; **2.** in physics, any system reaching a state of neither positive nor negative net electric charge; **3. viral neutralization**, inactivation of an infective virus by forming a complex with an antibody, or antibody plus a complement (*see* complement system).

neutrino, a practically massless elementary particle of an atom which has no electric charge and with spin 1/2; three different forms of neutrinos exist, e.g. one form emitted in certain radioactive decay, and one associated with beta decay. Neutrinos travel at the speed of light and are classified as → leptons.

neutron, an elementary particle of all atomic nuclei except that of normal hydrogen. Its mass is slightly greater than that of → proton and has no electric charge. A neutron is unstable outside the atomic nucleus (half-life 12 minutes) being decayed into an → electron, proton, and antineutrino.

neutrophil, (Gr. *philein* = to love), **1.** having affinity to neutral dyes (also neutrophilic); **2.** neutrophilic → granulocyte, a polymorphonuclear leucocyte called also neutrophile; **neutrophilia**, neutrophilic leucocytosis, i.e. an increased number of neutrophils in the blood.

neutrophilic, 1. having affinity to neutral dyes; stainable by dyes which are not acid or alkaline, such as aniline stains; *syn.* neutrophil or neutrophilous; **2.** pertaining to neutrophilic → granulocytes, neutrophils; **3.** neutrophilic mosquitoes, not zoophilous or anthropophilic but feeding on plants.

nexin (L. *nexus* = bond), a protein (165 kD) linking the outer pairs of microtubules in flagella and cilia.

nexine, the inner part of exine, which is the outer layer of the wall of a → pollen grain.

nexus, pl. **nexus,** also **nexuses** (L. = bond), connection, tie; e.g. a connection between cells, especially a → gap junction.

niacin, → nicotinic acid.

niacinamide, nicotinamide. *See* nicotinic acid.

niche, **ecological niche**; according to an idea first presented by *Roswell H. Johnson* in 1910, a niche comprises the range of all the environmental features which individuals of a population (species) need for their survival, growth, and reproduction in an ecosystem. The niche includes not only the physical space occupied by an organism, but also the functional role of the organism in relation to the other organisms (e.g. prey and especially competitors) in a community; further the niche includes the position of an organism in environmental gradients of temperature, moisture, pH, soil, etc.

The niche is described graphically by **resource utilization curves** based on as many various environmental resources (variables) as possible. In total, the variables comprise the **multidimensional niche** (n-dimensional hypervolume), within which a species or a population is able to survive. Ecological niches can be designated as the **fundamental niche** which is the maximal, abstractly inhabited hypervolume niche where the species is not constrained by competitors. The fundamental niche is reduced by increasing competition leading to the **realized niche**, whose extreme stage is the **central niche**, i.e. the smallest niche where a species can survive.

The term **niche overlap** describes niche-associations of two species. The **complementarity of niches** describes the situation in which two competing species utilize the same resource (e.g. the same feeding height in a forest), but differ from each other in respect to some other resource (e.g. food composition). The **complete niche overlap** can be estimated by considering all the environmental variables which influence the competition, which in practice is very difficult to determine.

niche differentiation, → ecological segregation.

nick, in molecular biology, a hydrolytic breakage of the phosphodiester bond between two → nucleotides in one strand of a double-stranded DNA.

nick translation, the ability of → DNA polymerase I of the bacterium *Escherichia coli* to use the → nick as an initiating point for enzymatic degradation, in which one DNA strand (the broken one) begins to disperse and synthesize a new strand. In molecular biology, nick translations are used to add radioactively labelled nucleotides to DNA.

nicotinamide, nicotinic acid amide, niacinamide; 3-pyridinecarboxamide; a component of → nicotinamide adenine dinucleotide, NAD, and its phosphate form, NADP; used in the prevention and treatment of → pellagra. *See* vitamin B complex.

nicotinamide adenine dinucleotide, NAD, (oxidized form, NAD$^+$, reduced form, NADH), an organic compound composed of ribosylnicotinamide 5'-phosphate (→ NMN) and adenosine 5'-phosphate (AMP); a derivative of nicotinic acid acting as an important coenzyme of many respiratory enzymes in → cell respiration; acts as a primary hydrogen acceptor and donor (electron carrier): NAD$^+$ —> NADH. Formerly called diphosphopyridine nucleotide, DPN, or coenzyme I. *See* nicotinic acid.

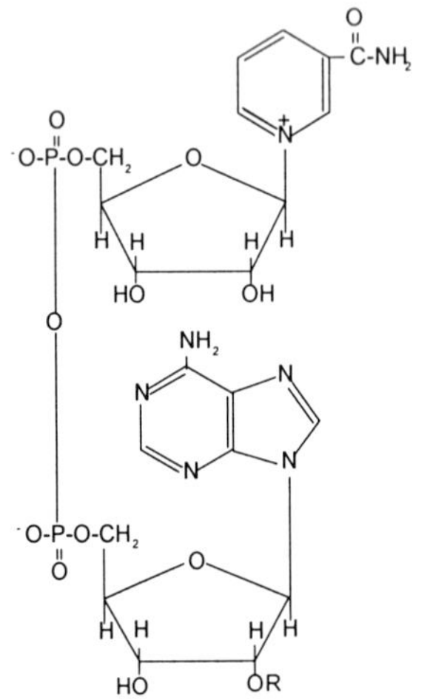

Nicotinamide adenine dinucleotides. Structure of oxidized form of NAD (NAD$^+$) and of NADP (NADP$^+$); in NAD R = H and in NADP, R = PO_3^{2-}.

nicotinamide adenine dinucleotide phosphate, NADP, (oxidized form NADP$^+$, reduced form NADPH), an organic compound, a derivative of NAD in which one phosphate group is bound to the 2' site of adenine; serves as a coenzyme of many oxidases (dehydrogenases) acting as a hydrogen acceptor and donor (electron carrier) in many biosynthetic reactions in cell metabolism: NADP$^+$ $\rightleftharpoons$ NADPH. NADPH is also the end product of the light reactions in → photosynthesis, and then used in reduction of CO_2 in the Calvin cycle. Formerly called triphosphopyridine nucleotide, TPN, or coenzyme II.

nicotinamide mononucleotide, NMN, ribosylnicotinamide 5'-phosphate; a nucleotide which is a constituent of NAD and NADP. *See* nicotinamide adenine dinucleotide.

nicotine (L. *nicotiana* = tobacco < *Jean Nicot de Villemain*, 1530—1600), 3-(1-methyl-2-pyrrolidinyl)pyridine; an oily volatile alkaloid of tobacco, formed from pyrroline and nicotinic acid. Nicotine is toxic to most animals influencing the nervous system through → nicotinic receptors; in vertebrates, nicotine first (in small doses) stimulates and then depresses synapses at autonomic ganglia and myoneural junctions, and has sedative effect on certain brain areas, therefore used as a relaxant by man. Nicotine is also used as an insecticide.

nicotinic acid, niacin, a water-soluble compound belonging to the → vitamin B complex; is a precursor for **nicotinamide** (nicotinic acid amide, niacinamide) from which are formed the coenzymes (NAD and NADP) of many enzyme proteins acting in cellular oxidation-reduction reactions in cells. *See* nicotinamide adenine dinucleotide. *Cf.* nicotine.

nicotinic receptor (L. *recipere* = to receive), one of the two types of cholinergic receptors in animals; the receptors are stimulated by low concentrations but inhibited by high concentrations of → nicotine. In vertebrates, the nicotinic receptors transmit the synaptic action of → acetylcholine in ganglia of the autonomic nervous system and in myoneural junctions of skeletal muscles; these are also found in the central nervous system, probably involved in sedative effect of nicotine. The nicotinic receptors in the autonomic ganglia are blocked by hexamethonium, and those at myoneural junctions by → curare. *Cf.* muscarinic receptor.

nictitating membrane (L. *nictare* = to wink), third eyelid; a thin membrane beneath the lower lid or in the inner corner of the eye in most vertebrates. The membrane is formed as an embryonic skinfold and is reduced in most mammals but conspicuous and very mobile in birds, many reptiles, and some cartilaginous fish such as sharks. The membrane sweeps rapidly and independently over the cornea.

nidal (L. *nidus* = nest), pertaining to a → nidus, nest, or a nest-like anatomical structure.

nidamental gland, a gland of a female cephalopod secreting protective covering around the eggs; found e.g. in squid.

nidation, the implantation of an early mammalian embryo in the uterine mucosa.

nidicolous (L. *colere* = to dwell), pertaining to a nestling reared in its nest, e.g. a bird nestling who is hatched naked, eyeless, and unable to leave the nest for a certain time. *Cf.* nidifugous.

nidifugous (L. *fugere* = to flee), pertaining to a nestling, especially a bird hatched downy and ready to leave the nest soon after the hatching, e.g. a chicken. *Cf.* nidicolous.

nidus, pl. **nidi,** also **niduses** (L. = nest), **1.** a nest or a nest-like structure where eggs or spores develop; **2.** a nucleus in the central nervous tissue, especially a group of neurones from which a nerve originates; **3.** a focus of pathogenic organisms in a tissue; **4.** the nucleus of a crystal.

Nieuwkoop centre (*P.D. Nieuwkoop*), a small area in the dorsal part of the vegetal hemisphere of an early vertebrate embryo; its chemical signals direct the development of the overlaying cells towards a more dorsal differentiation. The developmental area induced by the Nieuwkoop centre is known as → Spemann's organizer.

***nif* genes,** the gene complex of nitrogen-fixing bacteria encoding the proteins required for → nitrogen fixation, especially nitrogenase.

nigericin, an → ionophore, $C_{40}H_{68}O_{11}$, acting as a carrier for K^+ and Rb^+ and as an exchange carrier for H^+ with K^+; produced by *Streptomyces hygroscopicus* and used in studies on cellular transport systems, originally used as an antibiotic (antibiotic K 178).

night hawks, → Caprimulgiformes.

ninhydrin, 2,2-dihydroxy-1,3-indanedione; a pale yellow, crystalline compound that is freely soluble in water forming the **ninhydrin reagent** which is used for the detection of free α-amino and carboxyl groups in amino acids, proteins, and peptides; yields a blue colour when heated with these substances.

nipple, in anatomy, the breast papilla, *papilla mammae*, a protuberance of each mammary gland to which several milk ducts open. The nipples are lacking in monotremates whose mammary glands open onto the ventral skin surface spreading milk to be licked by young. In most mammals each nipple is surrounded by the areola with sebaceous glands greasing it. *Syn.* mam(m)illa, teat.

Nissl substance/bodies/granules (*Franz Nissl*, 1860—1919), basophilic material in the cytoplasm of the cell body and dendrites of a neurone; composed of granular → endoplasmic reticulum and → ribosomes.

nit, 1. a unit of luminance equal to one candela per square metre; **2.** the egg of a louse.

nitrate (Gr. *nitron* = base), a salt or ion (NO_3^-) of nitric acid (HNO_3); nitrates are the main nitrogen source for plants but not used by animals. In plants, they are reduced by → nitrate reductase into nitrite (NO_2^-) and finally by → nitrite reductase into ammonium form (NH_4^+) which is usable for metabolic reactions. Organic nitrates are used in fertilizers, explosives, and in medication (nitroglycerin). *See* nitrogen oxides, nitrite.

nitrate reductase (L. *reducere* = to lead back), an enzyme, EC sub-subclass 1.6.6., found in plants and microbes catalysing the reduction of NO_3^- to NO_2^-, i.e. nitrate to nitrite; the reaction requires FAD and NADPH. Nitrate reductase contains molybdenum in the ionic form Mo^{5+} or Mo^{6+} depending on the oxidation-reduction state.

nitric acid, HNO_3, a strong inorganic acid which is a powerful oxidizing agent, highly toxic and corrosive to tissues; its salts are → **nitrates.**

nitric oxide, NO. *See* nitrogen oxides.

nitrification (L. *facere* = to do), **1.** the reaction chain in which ammonium ion (NH_4^+) is oxidized first to nitrite (NO_2^-) and finally to nitrate (NO_3^-) by → **nitrogen bacteria**. The first reaction is carried out by bacteria of the genus *Nitrosomonas*, and the latter of the genus *Nitrobacter*. The reactions take place in soil and are components of the natural → nitrogen cycle; *cf.* ammonification, denitrification; **2.** the treatment of a substance with nitric acid.

nitrilase, an enzyme found in some plants; converts indole-3-acetonitrile into indole-3-acetic acid, i.e. into → auxin. The reaction takes place e.g. in cruciferous plants (Brassicaceae), in which the primary precursor for auxin is a thioglucoside, → glucobrassicin.

nitrile, an alkyl cyanide, i.e. any organic compound containing the cyanide group -CN; e.g. a reactive intermediate, R-CH_2-CN, of the nitrogen metabolism in plants, produced by reactions in which → cyanogenic glycosides are synthesized from some amino acids. In cruciferous plants (Brassicaceae) → glucosinolates are converted into indole-3-acetonitrile in the reactions of auxin biosynthesis. *See* glycosides.

nitrite, a salt or ion (NO_2^-) of nitrous acid (HNO_2); nitrite ions participate in → nitrification and nitrate reduction. Alkali nitrites are used as preservatives in food products.

nitrite reductase (L. *reducere* = to lead back), an enzyme (EC 1.6.6.4) in plants reducing nitrite (an intermediate of nitrate reduction reactions) into ammonium: $NO_2^- \longrightarrow NH_4^+$. *See* nitrate.

nitrogen, a chemical element, symbol N, atomic weight 14.007; molecular nitrogen is diatomic, N_2, appearing as a colourless, insoluble, inert gas at normal pressure and temperature, comprising approximately 78% of the atmosphere by volume. All → amino acids, proteins, and nucleic acids contain nitrogen which is therefore an essential element for all organisms. The oxidation state varies from +5 (nitrogen pentoxide, N_2O_5, and nitric acid, HNO_3) to -3 (ammonia, NH_3, and ammonium ion, NH_4^+). Plant roots take up nitrogen as nitrate, ammonium, or an organic form such as urea; animals use only organic nitrogen sources.

Ammonia produced in the metabolism of amino acids is rather toxic for organisms and must therefore be metabolized into other compounds or excreted from tissues. In animals, nitrogen end products derived from the metabolism of amino acids are → ammonia (ammonotelic animals), urea (ureotelic animals), uric acid and its salts uricates (uricotelic animals); → allantoin is the end product of purine metabolism. Most nitrogen end products are excreted through the kidneys into urine, but in aquatic animals ammonia is also excreted through the gills or skin, i.e. directly from the interstitial fluid or blood into surrounding water.

Nitrogen is a constituent e.g. in fertilizers and dyes. Liquid nitrogen, -196°C, is used as a refrigerant. *See* nitrogen cycle.

nitrogenase, a reducing enzyme complex occurring in nitrogen-fixing microorganisms; consists of two iron-containing subunits, the larger containing molybdenum. Oxygen is an inhibitor of nitrogenase function. Nitrogenase catalyses the reduction of molecular nitrogen to ammonia: $N_2 + 8\ H^+ + 8\ e^- \longrightarrow 2\ NH_3 + H_2$. Electron transfer is accompanied by the hydrolysis of two ATP molecules. Nitrogenase also catalyses the reduction of other compounds with a triple bond, e.g. acetylene to ethylene: $C_2H_2 + 2\ H^+ + 2\ e^- \longrightarrow C_2H_4$, the reaction used to determine nitrogenase activity. *See* nitrogen fixation.

nitrogen bacteria, 1. bacteria acting in → nitrification; **2.** nitrogen-fixing bacteria; *see* nitrogen fixation.

nitrogen balance/equilibrium, nutritive balance/equilibrium; the relation of nitrogen intake in food to nitrogen excretion in an animal; a greater intake causes a positive nitrogen balance and greater excretion results in a negative balance; when the intake and excretion are equal it is in balance.

nitrogen cycle, the cycling of nitrogen between organisms and soil, water, and the atmosphere.

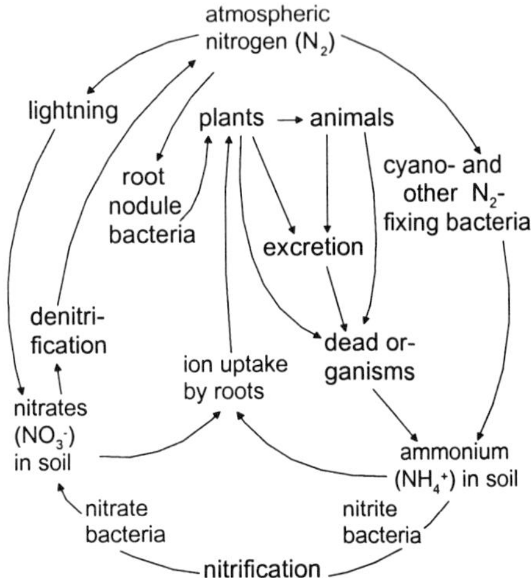

Nitrogen cycle. Chain of events by which nitrogen is circulated through environment and organisms.

Nitrogen occurs usually in different nitrogen compounds but some microbes can also fix atmospheric molecular nitrogen to the cycle. Atmospheric nitrogen compounds, e.g. nitrogen oxides, are also produced in different burning and combustion processes and by lightning. *See* nitrogen, nitrogen fixation.

nitrogen dioxide, NO_2. *See* nitrogen oxides.

nitrogen equilibrium, → nitrogen balance.

nitrogen fixation, 1. the fixation and reduction of atmospheric molecular nitrogen into ammonia, carried out by nitrogen-fixing microorganisms. Ammonia is used for syntheses of amino acids, proteins, nucleic acids, and other organic nitrogen-containing compounds. The fixation process is very important particularly in soils poor in nitrogen sources.

Green plants can use molecular nitrogen only if they live in symbiosis with nitrogen-fixing microbes, *Rhizobium* bacteria being the most common group of those microbes. The bacteria live in symbiosis with the roots of leguminous plants. Other common nitrogen-fixing organisms are e.g. *Frankia* bacteria which live in symbiosis with the roots of the alder and buckthorn. Some microorganisms can also live in association with their hosts, as e.g. *Azotobacter* species together with *Paspalus* grass in South America, and cyanobacteria in *Stereocaulon* lichens. Some of the nitrogen-fixing microorganisms are free-living bacteria (e.g. *Azotobacter*, *Clostridium*), or cyanobacteria. In all nitrogen-fixing organisms the reduction of molecular nitrogen (N_2) is catalysed by the enzyme → nitrogenase; nitrogen reduction is expressed: $N_2 + 3\ H_2 \longrightarrow 2\ NH_3$;

2. nitrogen fixation in an industrial process for production of ammonia from nitrogen (N_2) and hydrogen (H_2) (*see* Haber—Bosch reaction). Ammonia is used in the manufacture of fertilizers and other commercially important nitrogen compounds.

nitrogen-fixing bacteria, *see* nitrogen fixation.

nitrogen oxides, compounds of nitrogen and oxygen, the most common being nitrous oxide, nitric oxide, and nitrogen dioxide, all gases.

Nitric oxide, NO, (nitrogen oxide, nitrogen monoxide) is found to be an important regulatory molecule in animal tissues. It is produced in tissues by the enzymatic action of **nitric oxide synthase**, and is involved e.g. in the control of blood circulation and inflammation processes. It is released from L-arginine in endothelial cells of blood vessels by the action of nitric oxide synthase, an enzyme activated by acetylcholine (cholinergic transmission) or bradykinin. The released nitric oxide then relaxes the smooth muscle of blood vessels causing vasodilatation; it also inhibits the aggregation of blood plateles. Effects can be mimicked by nitrogen-containing drugs, e.g. → nitroglycerin. Nitric oxidase synthase is also found in nerve cells, and in some leucocytes activated by inflammation factors such as bacteriotoxins and cytokines.

Nitrous oxide, N_2O, (dinitrogen oxide, dinitrogen monoxide) is liberated by microbes in → **nitrification** and **denitrification** reactions, and its production is enhanced when nitrogen fertilizers are added to the soil. Nitrous oxide has narcotic and analgesic effects on animals and is therefore used in medication (laughing gas).

Nitrogen dioxide, NO_2, in dimeric form N_2O_4, (nitrogen peroxide) together with nitrous oxide, the end product of **combustion** processes; formed in great quantities from atmospheric molecular nitrogen (N_2) in burning reactions, e.g. in car engines. Nitrogen oxides have therefore become a serious environmental problem. *See* greenhouse effect, smog.

nitroglycerin, also **nitroglycerine,** glyceryl trinitrate, $C_3H_5(NO_3)_3$, a highly explosive liquid formed of nitric acid and glycerol; has vasodilative effects in vertebrates, also on cardiac vessels, and therefore it is used in the treatment of angina pectoris. *See* nitrogen oxides.

nitrosamines, organic nitrogen compounds with a group R_2N-NO; are formed in small quantities from nitrates, nitrites, and amines in the stomach of vertebrates, ascorbic acid inhibiting their formation; some nitrosamines are found to be mutagenic and carcinogenic.

nitrous acid, a weak unstable inorganic acid, HNO_2; its salts → **nitrites** are stable.

nitrous oxide, N_2O. *See* nitrogen oxides.

nival (L. *nix* = snow), pertaining to snow.

nival zone, the area north of the → tundra zone, or on mountains above alpine zones; is characterized by permanent snow and ice.

NK cell, → natural killer cell.

NMDA-receptor, N-methyl-D-aspartate receptor, *see* glutamate receptors.

NMN, → nicotinamide mononucleotide.

NMR, → nuclear magnetic resonance.

noble gases, rare gases, inert gases, a group of gaseous elements (the group 0) including **helium** (He), **neon** (Ne), **argon** (Ar), **krypton** (Kr), **xenon** (Xe), and **radon** (Rn); are chemically inactive elements and do not usually form chemical compounds, although some of them can form compounds (e.g. XeO_3 and KrF_2) which are rather unstable. All exist as minor components in the air, about 0.8% argon and 0.2% the others altogether. Noble gases are used e.g. for maintaining an inert atmosphere in electric light bulbs and discharge tubes (neon), and liquid helium for achieving low experimental temperatures (-268.9°C, i.e. 4.215 K).

nociceptor (L. *nocere* = to damage, *capere* = to receive), **pain receptor**; nociceptors are usually free nerve endings which are specifically stimulated by agents causing tissue injuries, as e.g. by heat, strong chemicals, mechanical strain, and many specific substances released in inflammation. *Adj.* **nociceptive**, related to pain receptors, receiving pain sensations. *See* pain.

nocturnal (L. *nox* = night), pertaining to night; in biology, pertaining to an organism which is most active during the night; e.g. an animal being active at night and resting in daytime. *Cf.* diurnal.

nodal bract, a small leaf-like nodal appendage, i.e. a → bract on a plant node.

node (L. *nodus* = knot), a swelling or knot (either normal or pathological) in the structure of an organ or organism; e.g. **1.** a → lymph node of vertebrates; **2.** the site on a plant stem where one or more leaves arise; the part of the stem between two nodes is called internode.

nodes of Ranvier (*Louis Antoine Ranvier*, 1835—1922), structural constrictions along the myelinated nerve fibres at intervals of about 0.5 to 1 mm; they are unmyelinated areas between successive segments of the → myelin sheath. *See* nerve fibre.

nodule (L. *nodulus*, dim. of *nodus* = knot), a small → node. *Adj.* **nodular**. *See* lymph nodules, root nodules.

nodulins, nodule-specific proteins produced in legumes infected by nitrogen-fixing bacteria and involved in symbiotic → nitrogen fixation in → root nodules.

nomadism (Gr. *nomades* = pasturing flocks), **1.** in ecology, an irregular wandering of animals, usually for seeking food, often associated to the maintenance of a territory when stronger competitors expel surplus individuals; nomadism is common e.g. among many owl species and crossbills; **2.** pastoral (nomadic) wandering of people.

nomen nudum, pl. **nomina nuda** (L. = naked name), in taxonomy, an invalid name owing to insufficient description and information to satisfy the criteria of availability or valid publication.

nominalism, the theory stating that there are no universal essence, and that only individuals exist; in classification of organisms, considers that the categories are not real but artificial, made for naturalistic reasons.

non-Darwinian evolution → evolution caused by other forces than → natural selection or sexual selection, i.e. caused mainly by → genetic drift. *See* Darwinism.

non-competitive inhibition, *see* competitive inhibition.

non-cyclic photophosphorylation, reactions by which light energy, absorbed by photosystems in chloroplasts, is used to generate → adenosine triphosphate, ATP, in → photosynthesis.

non-disjunction (L. *disjungere* = separate), a disturbance of a cell division in which one or more pairs of homologous chromosomes (in → meiosis) or sister chromatids (in → mitosis) pass to the same pole of the cell instead of being separated. These phenomena are called meiotic or mitotic non-disjunction, respectively, Down's syndrome being an example of meiotic non-disjunction.

non-electrolytes, nonpolar substances, i.e. which do not dissociate into ions in solutions, therefore having low electric conductivity; e.g. sugars.

non-homologous chromosomes (Gr. *homologos* = agreeing), chromosomes which do not belong to the same chromosome pair, i.e. which are not → homologous chromosomes.

non-reciprocal, *see* reciprocal.

non-renewable resources, natural resources which are not renewable but are limited and in danger of becoming totally expended; e.g. fossil fuels, many minerals, space for living, etc.

nonsense codon, termination codon, stop codon; a → codon in → messenger RNA, mRNA, which in → protein synthesis does not code for the joining of an amino acid into the polypeptide chain, but instead encodes the termination of this polypeptide synthesis. Three nonsense codons are known: UAA (ochre), UAG (amber) and UGA (opal); U = uracil, A = adenine, G = guanine.

nonsense mutation, a change (mutation) of a gene → codon, normally encoding amino acids in the polypeptide synthesis, into a → nonsense codon which leads to the precocious termination of → protein synthesis.

non-shivering thermogenesis, *see* thermoregulation.

non-striated muscle, → smooth muscle.

non-synaptic neurotransmission, volume neurotransmission, diffusion neurotransmission; chemical transmission of nerve impulses outside a → synapse; a neurotransmitter released from the synapse spreads to the surrounding tissue affecting ion currents of cells which have specific receptors for this neurotransmitter.

NOR, → nucleolar organizing region.

nor-, **1.** (short for normal), in chemical names, referring to a normal molecular structure, i.e. an unbranched chain of carbon atoms in aliphatic compounds (especially amino acids), as opposed to a branched chain with the same number of carbon atoms, e.g. norvaline as compared to valine; **2.** (Ger. *nor* = Nitrogen ohne Radikal = nitrogen without radical), pertaining to a compound of normal structure without radicals, such as the lack of a methyl ($-CH_3$) or methylene ($=CH_2$) group; e.g. adrenaline molecule contains the methyl group (a radical) which is absent in noradrenaline.

noradrenaline, NA, Am. **norepinephrine, NE** (L. *ad* = to, *ren* = kidney, Gr. *epi* = on, upon, *nephros* = kidney), α-(aminomethyl)-3,4-dihydroxybenzyl alcohol; a → catecholamine serving as a **neurotransmitter** and **hormone** (neurohormone) in animals. Noradrenaline acts in **adrenergic → synapses** in the central nervous system and peripheral sympathetic nerve endings. In vertebrates, the action is transmitted chiefly by alpha receptors, less by beta receptors (*see* adrenergic receptors).

In vertebrates, noradrenaline is secreted into the blood from the **adrenal medulla** and some other chromaffin tissues (*see* adrenal gland). It acts as a hormone e.g. stimulating functions of the respiratory and circulatory systems and decreasing the activity of the alimentary canal. Its secretion together with → adrenaline occurs especially in stress situations improving the ability of an animal to survive in difficult situations. The hormonal secretion of noradrenaline and adrenaline is controlled by the sympathetic nervous system. *Cf.* adrenaline.

norepinephrine, NE, → noradrenaline.

norm(o)- (L. *norma* = rule), denoting normal, usual; conforming to the rule.

normal, **1.** typical, approximately average; conforming to the standard; **2.** untreated experimental animal; **3.** in chemistry (symbol N), pertaining to a solution that contains one equivalent weight of an active substance in one litre of solution; *see* equivalence, titration. *Noun* **normality**.

normoblast (Gr. *blastos* = germ), the nucleated precursor cell of the erythrocyte series with four developmental stages in the bone marrow of vertebrates.

norm of reaction, the way, determined by genes, by which an organism reacts to its environment. The → genotype of an organism responds to different environmental factors during ontogenic development according to its norm of reaction.

normocyte, normoerythrocyte (Gr. *kytos* = cavity, cell, *erythros* = red), any structurally normal → red blood cell (erythrocyte) circulating in the blood.

normothermy (Gr. *therme* = heat), the normal body temperature of a homoiothermic animal, i.e. a bird or mammal; especially denotes the body temperature of a hibernating animal in its active state, as a contrast to the hypothermic state during hibernation.

Northern blotting (Northern, as distinguished from → Southern blotting), a method for isolation and identification of → ribonucleic acid, RNA. In the technique, denaturated RNA fragments are transferred from an electrophoretic gel to a nylon membrane or some other membrane, and hybridized on it with a DNA or RNA probe labelled with a radioactive or another suitable label (*see* probe). Thus, different RNA fragments can be identified.

nos(o)- (Gr. *nosos* = disease), denoting relationship to a disease.

not(o)-, **1.** (Gr. *noton* = back), denoting relationship to the back part of an animal; *adj.* **notal**; **2.** (Gr. *notos* = south, south wind), pertaining to south; southern.

notochord (L. *chorda dorsalis*; Gr. *noton* =

back, *chorde* = cord, string), the primitive rod-shaped axis of the embryonic body of all chordates consisting of mesodermal cells from the mesoblast below the primitive groove of the embryo. The notochord forms the centre for the development of the vertebral column with segmental vertebrae, disappearing during the embryonic development in vertebrates except in some groups, such as lampreys and lungfish, with an underdeveloped vertebral column. The notochord is also found in adult lancelets and the larval stages of hemichordates disappearing during metamorphosis.

Notogaea, Am. **Notogea** (Gr. *notos* = south, *ge* = earth), a zoogeographical area which previously included Australia and the Neotropical region, but at present only Australia, New Zealand, New Guinea, Tasmania, the Pacific Ocean Islands, and the Indonesian islands south and east of → Wallace's Line. *Adj.* **Notog(a)ean**. *Cf.* Arctogaea, Neogaea, Palaeogaea. *See* zoogeographical regions.

notopleuron (Gr. *noton* = back, *pleuron* = side), a triangular plate on the dorsal thorax of some dipterans.

notopodium, pl. **notopodia** (L. < Gr. *pous*, gen. *podos* = foot), **notopod**; the dorsal lobe of the → parapodium of polychaetes; in some species may serve as a gill. *Cf.* neuropodium.

notum (L. < Gr. *noton* = back), a dorsal shield on the thorax in some arthropods, e.g. insects; formed from several hard plates, sclerites. *Cf.* pleuron, sternum.

N.T.P., also **NTP, ntp,** normal temperature and pressure, → s.t.p.

nucellus, pl. **nucelli** (L. *nucella* = small nut), the inner part of a plant → ovule which finally contains the female gametophyte (embryo sac) with an egg cell.

nuclear (L. *nucleus*, dim. of *nux* = nut), pertaining to the nucleus of a cell or atom.

nuclear envelope, the double membrane surrounding the chromosomal complement in the cells of eukaryotic organisms, thus separating the content of the cell nucleus from the cytoplasm. The nuclear envelope is formed in the beginning of interphase and can be still seen in prophase, disappearing when the cell begins to divide (*see* mitosis). The nuclear envelope has a double membrane structure which consists of two **nuclear membranes**, both being → unit membranes. The narrow gap between the nuclear membranes is called the perinuclear space. The transport of substances between the nucleus and cytoplasm occurs through numerous → **nuclear pores** of the nuclear envelope.

nuclear lamina, the laminar protein layer comprising a filamentous network underlying the inner membrane of the → nuclear envelope; can connect the chromosomes to the nuclear envelope during interphase. *Syn.* fibrous lamina.

nuclear membrane, one of the two unit membranes of the → nuclear envelope; sometimes used as a synonym for the nuclear envelope.

nuclear matrix, a structural framework of the cell nucleus consisting of fibrillar proteins and some RNA, i.e. → small nuclear ribonucleoprotein particles. The matrix remains when isolated nuclei are treated to remove membranes, chromatin, and soluble molecules with detergents and DNase I, and by extraction with high concentrations of salt. The nuclear matrix is involved in many nuclear functions, such as → DNA replication, → genetic transcription, and processing of transcripts, as well as in binding of hormone receptors. It also gives support to nuclear structures. *Syn.* nuclear skeleton, karyoskeleton, nuclear scaffold, nuclear cage.

nuclear pores, pores in the → nuclear envelope allowing passage of many substances, also macromolecules, such as RNA and many proteins. Each pore is a symmetrical barrel-like structure with protein subunits forming the pore walls, spanning the inner and outer membranes of the nuclear envelope. Proteins surrounding each pore form the **nuclear pore complex**.

nuclear reaction, any reaction involving a change in an atomic nucleus (in ordinary chemical reactions changes concern only orbital electrons). Energy (E) produced by nuclear reactions is expressed: $E = \Delta m \cdot c^2$, where c = velocity of light (appr. $3 \cdot 10^{10}$ cm s^{-1}), Δm = → mass defect. *See* radioactivity.

nuclear receptors/proteins, any of the family of cytoplasmic and nuclear proteins which bind to specific chromatin regions in the cell nucleus thus acting as → transcription factors regulating → gene expression. In organisms nuclear proteins coordinate complex events controlling → morphogenesis and → homeostasis in response to the binding of their cognate ligands.

nuclear spindle, → mitotic spindle.

nucleases, hydrolytic enzymes which catalyse the cleavage of the phosphodiester bonds of → nucleic acids producing oligonucleotides or nucleotides; they are either **DNases** (→ deoxyribonucleases) hydrolysing DNA, or **RNases** (→ ribonucleases) hydrolysing RNA. Depending upon whether the cleavage begins at the end of the substrate molecule, or in the middle of it, the nucleases are classified into the sub-subclasses of **exonucleases** (EC 3.1.11-16.) and **endonucleases** (EC 3.1.21-31.), respectively. Nucleases are present generally in all organisms, e.g. in → lysosomes of the cells, in animals also in the alimentary canal. Nucleases belong to the phosphodiesterase subclass, EC 3.1.

nucleic acids, polynucleotides; information macromolecules of the nucleus and cytoplasm formed from smaller units, → **nucleotides**; two natural forms exist: → deoxyribonucleic acid (**DNA**) and → ribonucleic acid (**RNA**). They are macromolecule chains which control the development of each individual organism. DNA (except in RNA viruses) is the principal component of the → chromosomes carrying the **genetic material**, i.e. the store of the molecular information of the genes (*see* genetic transcription). RNA is involved in deciphering the genetic code of DNA in → genetic translation and → protein synthesis. *See* hybridization.

nucleohistones, complexes of DNA and → histones; the histones are bound to the large groove of DNA thus forming the basic structure of chromosomes.

nucleoid (L. *nux* = nut, Gr. *eidos* = form), **1.** the cell region in bacteria and other prokaryotes, constituting the genome; corresponds to the cell nucleus of eukaryotic organisms; **2.** as an adj., resembling a nucleus.

nucleolar organizer, a → secondary constriction of a chromosome having the genes for → ribosomal RNA. The RNA molecules, produced in the transcription of these genes, form the → nucleolus which is connected to the chromosome and serves as the storage material for cytoplasmic ribosomal RNA (*see* ribosomes). There may be one or several nucleolar organizers per genome of a cell, e.g. five in human cells.

nucleolar organizing region, NOR, the → secondary constriction on a chromosome in which the → nucleolar organizer is located.

nucleolus, pl. **nucleoli,** usually a spherical colloidal eukaryotic cell organelle inside the cell nucleus which is formed during telophase of cell division (*see* mitosis). The nucleolus is connected to specific regions on the chromosomes called → nucleolar organizers. The nucleolus contains large quantities of ribosomal RNA which moves from the nucleolus to the cytoplasm, and together with certain proteins constitutes the → ribosomes. The number of nucleoli per nucleus varies according to species, and may be constant or variable.

nucleomorph, a plastid genome located inside the endoplasmic reticulum of the plastid in members of Cryptophyta; it is a vestigial nucleus of an algal endosymbiont.

nucleoplasm, the protoplasm of the cell nucleus, i.e. the non-nucleolar ground substance. *See* nucleus.

nucleoplasmin, a pentameric protein present in the soluble ground substance of the cell nucleus but not bound to DNA or chromatin; found especially in oocytes.

nucleoproteins, conjugated compounds composed of → nucleic acids and proteins abundant in the cell nucleus but also found in the cytoplasm, particularly in → ribosomes. Nucleoproteins, composed of DNA and → histones, are essential building units of the chromosomes.

nucleoside, an organic compound composed of pentose sugar and a → purine or → pyrimidine base; phosphorylated nucleosides are called → nucleotides, which are structural units (monomers) of nucleic acids (DNA and RNA). In RNA nucleosides the sugar component is **ribose**, in DNA nucleosides **deoxyribose**. The base components in DNA are adenine, thymine, guanine, and cytosine, forming with deoxyribose corresponding nucleosides: **adenosine**, or deoxyadenosine with deoxyribose, **thymidine** (also called deoxythymidine), **guanosine** (deoxyguanosine with deoxyribose), and **cytidine** (deoxycytidine with deoxyribose); in RNA thymine is replaced by uracil which with ribose forms **uridine**. Particularly in → transfer RNA (tRNA) also other, relatively rare bases exist.

nucleosome (Gr. *soma* = body), the basic unit of the structure of an eukaryotic chromosome consisting of about 146 base pairs of DNA. In the nucleosome, DNA is wrapped around the core of histone proteins. In addition, each nucleosome associates with one → histone

H1 molecule which is located outside DNA. At low salt concentration the nucleosomes of isolated chromatin are repeated as beaded structures connected by a DNA string in the constitution of the basic fibril of a chromosome; in physiologic concentration the structure appears in a more condensed form.

nucleotide, mononucleotide; a phosphate ester of a → nucleoside; a compound comprising a nucleoside with one, two, or three phosphate groups, i.e. mono-, di- or triphosphates; are either **purine nucleotides** (adenine or guanine) or **pyrimidine nucleotides** (cytosine, thymine, or uracil as the base). Nucleotides are e.g. adenosine monophosphate (**AMP**, dAMP with deoxyribose) and its cyclic form, cyclic AMP (**cAMP**), adenosine diphosphate (**ADP**, dADP), and adenosine triphosphate (**ATP**, dATP), and corresponding phosphates of other nucleosides. Nucleotides are the structural units (monomers) of → nucleic acids (DNA, RNA) formed by the action of **polymerase** enzymes, but are essential compounds also in cellular energy metabolism (especially ATP) and as → second messengers inside the cell (especially cAMP). Nucleotides are also components of the coenzymes FAD, NAD, and NADP. The enzymes which catalyse the cleavage of nucleotides are called **nucleotidases**. *Cf.* polynucleotide.

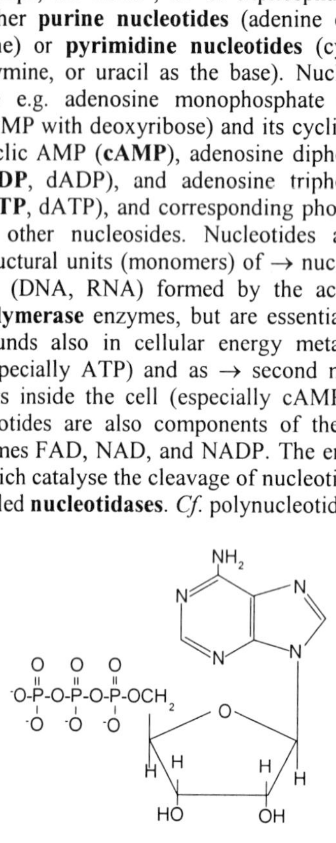

Nucleotide. ATP is a nucleotide that consists of adenine (purine base), ribose and three phosphate groups.

nucleotide sequence (L. *sequi* = to follow), the continuum of nucleotides in a DNA or RNA molecule. *See* nucleic acids.

nucleus, pl. **nuclei** (L. dim. of *nux* = nut), **1.** cell nucleus; a cell organelle of eukaryotic organisms which contains the **chromosomes**, and thus constitutes the regulatory centre of the cell. The nucleus is surrounded by the → **nuclear envelope** which separates it from the cytoplasm. The sap inside the nucleus is called the **nucleoplasm**, and the supporting framework the → **nuclear matrix**. The nuclei of somatic cells and tissues divide in → mitosis, and during the formation of gametes and sexual spores in → meiosis.

A typical nucleus, surrounded by the nuclear envelope, can be seen only during the → **interphase** of the cell cycle. When the nucleus divides the nuclear envelope disappears but will be rebuilt at the end of the division. A cell usually contains only one nucleus but bi-, poly-, and anucleate cells also exist. Examples of binucleate cells can be found e.g. in vertebrate cardiac muscle, in dikaryons of fungi, and in some ciliate protozoans. Ciliates contain two nuclei of different size, i.e. the → **macronucleus** and **micronucleus**. Polynucleate cells are e.g. the cells of striated muscles and of polykaryons of fungi. The red blood cells of mammals are secondarily anucleate. The interphase nucleus in mitotically active tissues is usually spherical but in differentiated cells its shape can vary considerably.

Chemically the nucleus consists of four main components, i.e. → deoxyribonucleic acid (DNA), ribonucleic acid (RNA), proteins, and lipids. The DNA content of the nucleus is dependent on its ploidy, in diploid cells being species-specific and constant. RNA is mainly located in the → nucleolus. Proteins exist as structural parts of the chromosomes and in the → matrix of the nucleus, and the amount of proteins varies according to the physiological stage of the cell. Most of the nuclear lipids are connected with the proteins; **2.** the nucleus as a structure of the nervous system is formed from a unified group of nerve cells (grey matter in vertebrates) within the central nervous system; is often structurally and functionally related to a certain peripheral nerve; **3.** atomic nucleus, the central element of an → atom; **4.** any core element, as a foreign particle or some precipitated material in a tissue, or a nucleus of a molecule.

nude (L. *nudus* = unclothed, naked), in biology, pertaining to an animal who is lacking the usual body covering, pelage or plumage; may be caused by a mutation, as in a **nude mouse** who lacks hair, and because of the hypoplastic thymus the cellular immune system (T cell system) is not developed.

null allele (L. *nullus* = none), → amorphic allele.

null hypothesis, a hypothesis stating that the results observed in a study, experiment, or test, are not different from those that might have occurred by chance alone. If in a statistical analysis, the experimental results differ from those predicted by the null hypothesis, the results are acceptable.

nullisomic (L., Gr. *soma* = body), pertaining to an → aneuploid chromosome mutant in which one of the chromosome pairs is totally lacking; the chromosome number is accordingly 2n-2 (*see* n). The phenomenon is called **nullisomy**.

numerical aperture, *see* resolution.

numerical phyletics, → numerical taxonomy.

numerical taxonomy, a branch of biology that classifies organisms to taxa by quantitative estimates of their phenotypic similarities. All the characteristics used in classification are in principle considered to be of the same value, thus resulting in the description of similarities, i.e. a **phenogram**. *Syn.* taxometrics, numerical phyletics. *See* phenetics.

nummulation (L. *nummulus*, dim of *nummus* = coin), the tendency of red blood cells to line up into rouleaux formations in the blood, resembling stacks of coins.

nummulites, fossil foraminiferans (the extinct genus *Nummulites*) whose coin-shaped shells formed limestone deposits during Tertiary.

nunatak (Inuit), a mountain top which penetrates up through a continental glacier. *See* refugium.

nuptial (L. *nuptiae* = marriage), in ecology, **1.** capable of breeding; **2.** pertaining to courtship or breeding; e.g. the **nuptial plumage** (breeding plumage) of a bird, or the **nuptial flight** of bees and other social insects. *Cf.* postnuptial.

nurse cell, any cell nourishing other cells; e.g. **1.** → Sertoli cell; **2.** any of the cells in a → meroistic ovarium which synthesizes the energy store and → morphogens for the ovum to which the nurse cells are connected with cytoplasmic canals.

nut, a dry, indehiscent → fruit with a hard, woody shell.

nutation (L. *nutare* = to nod repeatedly), **1.** the act of nodding, especially an involuntary nodding of the head; **2.** spontaneous movements of a plant or plant organs due to irregular growth; for example, the spiral movement of the apex of a shoot during growth (*syn.* **circumnutation**); **3.** an oscillation of the Earth's poles about the mean position due to the joint effect of the Sun and Moon.

nutlet, a small nut, or the stone of a stone fruit (drupe), such as cherry and plum.

nutrient (L. *nutrire* = to nourish), **1.** nourishing; **2.** food or a component of food, a nutritious material, nutriment; e.g. **essential nutrients**, the nutrients required for growth of animals, i.e. proteins, carbohydrates, fats, vitamins, minerals, and → trace elements, the first three called **macronutriens** and the others **micronutrients**; **3.** any substance required and used by plants; **macronutrients for plants** are nitrogen (N), phosphorus (P), potassium (K), calcium (Ca), sulphur (S), and magnesium (Mg); **micronutrients (trace elements)** are e.g. iron (Fe), copper (Cu), zinc (Zn), boron (B), sodium (Na), manganese (Mn), chlorine (Cl), molybdenum (Mo), and vanadium (V) (→ minerals). Sometimes also water is considered to be a nutrient.

nutrient cycling, the cycling of nutritive elements through an ecosystem from inorganic materials to organic compounds in the structures of organisms, and recycling of the material by the action of decomposers back into inorganic nutrients in soil and into carbon dioxide in air.

nutriment (L. *nutrimentum*), nourishment, food. *Adj.* **nutrimental**.

nutrition, the sum of processes in organisms involved in taking in and metabolizing chemical elements and food materials. In animals, nutrition includes the processes of eating, digesting, and utilizing → nutrients in cell metabolism. Fungi obtain their nutrition and energy from the environment, i.e. from dead organic material. Green plants need and use different elements, → macronutrients, and → micronutrients.

nutritional, affecting nutrition, functioning in nutrition.

nutritious, providing nourishment, nourishing; also nutritive.

nutritive, 1. pertaining to nutrition; **2.** capable of nourishing; also nutritious.

nutritive balance, *see* nitrogen balance.

nyctalopia (Gr. *nyx* = night, *alaos* = obscure, blind, *ops* = eye), night blindness; decreased ability to see in dim light; usually resulting from vitamin A deficiency.

nyctinasty (Gr. *nastos* = close-pressed), a type of → nastic movement in plants, "sleep movements"; associated with periodic alternation of day and night, e.g. the opening and closing of flowers, or rhythmic movements of leaves or leaflets. Nyctinastic movements are automatic, caused by light and temperature, which affect growth processes and changes in cellular turgor pressure. *Adj.* **nyctinastic**.

nymph (Gr. *nymphe* = bride, chrysalis), an immature stage of some invertebrates, following hatching; e.g. **1.** a wingless or incompletely winged juvenile stage of heterometabolous insects, e.g. crickets and grasshoppers; *cf.* naiad; **2.** an eight-legged juvenile stage after a six-legged larval stage of ticks and mites (Acari) with several substages: **protonymph**, **deutonymph**, and **tritonymph**. *Adj.* **nymphal**.

nympheid, a water plant with leaves floating on the surface of water; e.g. water lily. *See* water plants.

O

O, chemical symbol for oxygen; diatomic form $\mathbf{O_2}$.

***o*-,** chemical symbol for → ortho-.

O antigen (Ger. *ohne Hauch* = without breath), a lipopolysaccharide component in the cell wall of Gram-negative bacteria.

ob- (L. = to, toward, over, against), denoting inward, incompletely, in reverse order.

obduction (L. *obducere* = to draw over, obduce), a legally performed autopsy, i.e. a postmortem examination of the human body.

obesity (L. *obesus* = fat), excessive fatness, corpulence. *Adj.* **obese**.

objective (L. *obicere* = to throw before, object), in optics, an **object lens** (object glass) in a microscope, telescope, etc., i.e. the lens or a combination of lenses which receive and focus the rays coming from an object. *See* microscope.

oblanceolate (L. = *lancea* = lance), pertaining to a leaf that is inversely → lanceolate.

obligate (L. *obligatus*), necessary, obligatory, not facultative; without an alternative pathway; in biology, pertaining to a restriction in the life of an organism, especially microbes and parasites, e.g. an obligate aerobic bacterium which cannot live without oxygen. *Cf.* facultative.

obligate parasite, → holoparasite.

obtect (L. *obtegere* = to cover up), pertaining to a → pupa having the antennae, legs, and wings fixed close to the body surface by means of a hardened larval secretion.

Occam's razor, Ockham's razor (*William of Ockham or Occam*, ca. 1285—1349), a principle of the philosophy of science according to which unnecessary hypotheses should be avoided.

occipital (L. *occipitalis*), pertaining to the occiput, i.e. the back part of the head or skull, e.g. the occipital lobe of the brain. *See* occipital bone.

occipital bone (L. *os occipitale*), a bone in the vertebrate skull (cranium), situated at the posterior part of the head. The **occipital condyle(s)** on the undersurface of the bone articulates the head with the first vertebra, atlas. The condyles are paired in mammals and amphibians, single in reptiles and birds, and absent in most fish.

occiput, pl. **occipita** or **occiputs** (L.), **1.** the back part of the vertebrate head or skull; also called the occipital region or occiput of cranium (*occiput cranii*); **2.** a plate of exoskeleton forming the back part of the insect head.

occlusion (L. *occludere* = to shut up), the act or state of closure, e.g. an obstruction of material within the body cavities; in dentistry, the fitting of upper and lower teeth together when the jaws are closed.

oceanic (Gr. *okeanos* = ocean), pertaining or belonging to the ocean; e.g. oceanic currents or organisms inhabiting the open sea where it is deeper than 200 metres. *Cf.* pelagic.

oceanography (Gr. *graphein* = to write), a branch of science dealing with geographical, geological, chemical, physical, and biological phenomena occurring in the oceans. *Syn.* oceanology, defin. 1. *Cf.* hydrology, limnology.

oceanology (Gr. *logos* = word, discourse), **1.** a branch of science examining oceans from the point of natural sciences; *syn.* oceanography; **2.** a study of oceans from the human point of view, with regard to the importance of economy, politics, and military aspects.

ocellus, pl. **ocelli** (L.), the common term for different types of **simple photoreceptors**; e.g. **1.** a light-sensitive → **eye-spot** in cnidarians, ctenophores, and turbellarians; **2.** a **simple eye** of some arthropods; e.g. the eyes of arachnids, lateral ocelli of larval insects, and dorsal ocelli on the vertex of the head in some adult insects. The ocellus usually includes a chitinous lens without an ability to accommodate, and beneath the lens is a group of photosensitive cells associated with several → rhabdomes. Simple ocelli, especially when occurring in an animal in addition to the compound eyes, perceive changes in light intensity rather than form images. The ocelli of arachnids are more complicated, sensing e.g. the polarization of light, as does the compound eye of many insects; **3.** an eye-like spot e.g. in many butterflies and other insects.

ochratoxins (L. *Aspergillus ochraceus*, Gr. *toxikon* = arrow poison), mycotoxins produced by some fungi, e.g. *A. ochraceus* contaminating coffee and cocoa in warm climates and *Penicillium verrucosum* damaging cereals in temperate areas of the world. Ochratoxin A is fairly toxic to man and other vertebrates acting as a nephrotoxin and being probably carcinogenic. It inhibits e.g. the formation of phenylalanine tRNA complex in protein synthesis.

ochre, Am. **ocher** (Gr. *ochros* = pale yellow), a natural, iron-containing earthy ore (hydrated Fe_2O_3), used as a yellow or brownish red pigment.

ochrea, → ocrea.

ochre codon, ochre termination codon; one of the → termination codons, UAA, found in all organisms.

ochre mutation, "stop mutation"; a → mutation of a gene changing an amino acid coding triplet into a UAA codon which is the **ochre termination codon** in the → messenger RNA (U = uracil, A = adenine). *See* genetic code.

ochre suppressor, a mutation of the → transfer RNA molecule which reads the nonsense → ochre mutation codon as a sense codon. *See* genetic code.

Ockham's razor, → Occam's razor.

ocrea, ochrea, pl. **ocreae, ochreae** (L. = greave, legging), **1.** a sheath, such as a sheath-like structure formed of stipules at the base of the petiole in plants, e.g. in the genus *Raphanus*; **2.** partial covering on the fruiting body of some mushrooms, formed from the remains of the disintegrated → veil.

oct(a)-, octo- (Gr. *okto* = eight), denoting eight, eightfold.

octadecanoic acid, → stearic acid.

octane, any 8-carbon hydrocarbon, C_8H_{18}, of the alkane series, altogether eighteen isomers ocurring in petroleum; e.g. 2,2,4-trimethyl pentane; iso-octane, $(CH_3)_3CCH_2CH(CH_3)_2$, used e.g. as a solvent.

octanoic acid, → caprylic acid.

octanol, octyl alcohol, a group of isomeric alcohols, $CH_3(CH_2)_7OH$, used as a solvent e.g. in biochemistry and physiology.

octet, in chemistry, a stable group of 8 valence electrons (4 electron pairs) in the outer electron shell of an atom; all atoms tend to attain this stage. The octet rule concerns only atoms which have the L-shell as their outer electron shell; → noble gases (except helium) have a stable group of 8 electrons. When the atoms combine to form compounds, they either donate or share electrons so that each combined atom will have a stable octet in its outer shell. For example, in a water molecule one oxygen atom, with six electrons in its outer shell, shares the single electrons from each of the two hydrogen atoms, and the covalent bonds are formed between the oxygen and hydrogen atoms.

octopamine, an amine produced by disturbed tyrosine metabolism in the liver of vertebrates; it can replace noradrenaline in adrenergic synapses and thus has a slight adrenergic effect in animals. In many invertebrates it probably functions as a normal neurotransmitter. *Syn.* norsynephrine.

octopine, *see* opine.

octoploid (Gr. *-ploos* = -fold), pertaining to, or being an → autoploid or → alloploid cell, tissue, or individual that carries eight times the normal haploid chromosome set, symbol 8x.

octyl alcohol, → octanol.

ocul(o)- (L. *oculus* = eye), denoting relationship to the eye. *Adj.* **ocular.**

ocular plates, five small plates between the → genital plates surrounding the aboral periproct of an echinoid, each bearing a small opening connected with a → radial canal.

oculomotor nerve, *see* cranial nerves.

oculus, pl. **oculi** (L. = eye), **1.** → eye; **2.** a leaf bud in a tuber, e.g. in potato.

OD, optical density, *see* absorbance.

odd-toed ungulates, → Perissodactyla.

odont(o)- (Gr. *odous* = tooth), denoting relationship to teeth. *Adj.* **odontic.**

odontoblast (Gr. *blastos* = germ), a cell type originated from the mesenchymal connective tissue of a developing tooth in vertebrates; the cells form dentine around their long processes, the cell bodies lying in the dental pulp adjacent to the dentine layer. *See* tooth.

Odontoceti (Gr. *ketos* = whale), **toothed whales**; a suborder of marine mammals in the order Cetacea; includes seven families and about 65 predatory species mainly feeding on fish. Known families are e.g. **dolphins** (Delphinidae), **porpoises** (Phocoenidae), and **sperm whales** (Physeteridae). *See* Mysticeti.

odontoid process, a tooth-like peg (L. *dens axis*) of the second cervical vertebra (axis) of reptiles, birds, and mammals around which the first vertebra (atlas) rotates; derives from the atlas but becomes fused to the axis.

oedema, Am. **edema,** pl. **(o)edemata,** also **(o)edemas** (Gr. *oidema*), a swelling; the effusion of abnormally large quantities of tissue fluid in intercellular spaces; occurs e.g. during inflammation by decreased lymph flow from the tissue area. *Adj.* **(o)edematous.**

oesophag(o)-, Am. **esophag(o)-** (Gr. *oisein* = to carry, *phagein* = to eat), denoting relationship to the → oesophagus.

oesophagus, Am. **esophagus,** pl. **(o)esophagi** (Gr. *oisophagos*), the anterior part of the alimentary canal between the pharynx and the stomach in vertebrates and many groups of invertebrates; a part of it may form a → crop. *Adj*. **oesophageal**.

oestradiol, Am. **estradiol,** $C_{18}H_{24}O_2$; a natural → oestrogen.

oestriol, Am. **estriol,** $C_{18}H_{24}O_3$; a natural → oestrogen.

oestrogen, Am. **estrogen** (Gr. *oistros* = gadfly, vehement desire, *gennan* = to produce), any of several female sex hormones and their artificial homologues, structurally steroids. Oestrogens are produced by the theca and granulosa cells of maturing → **ovarian follicles** in vertebrate ovaries, by cells in the adrenal cortex of both sexes, and in mammals also by the → placenta, and in some species in smaller quantities by fat cells. In vertebrate testes, oestrogens occur as intermediate products in the synthesis of testosterone and are also secreted in small quantities. The secretion of oestrogens is stimulated by → **follicle stimulating hormone** (FSH) released from the anterior pituitary gland which is under the control of → **gonadoliberin** (gonadotropin-releasing hormone, GnRH) secreted by certain hypothalamic cells. Through negative feedback control, oestrogens decrease FSH secretion. In many mammals the most potent natural oestrogen is **oestradiol**, but two others, **oestrone** and **oestriol**, also possess physiological activity.

The physiological effects of oestrogens, like all steroid hormones, are produced by specific gene activation induced by the oestrogen receptor complex in the target cell nucleus. Oestrogens stimulate the development and maintenance of the female **genital organs** and **secondary sex characters**, have an effect on the **oestrus cycle**, also stimulating oestrus behaviour, and promote **growth spurt** prior to maturity. In many species, oestrogens are also involved in the maintenance of maternal behaviour. In these functions also other anabolic hormones are involved, especially → gestagens and growth hormone. Oestrogens are principal components of hormonal contraceptives. Some plant steroids (phytosteroids) have oestrogenic properties which may have some physiological effects in herbivorous animals.

oestrone, Am. **estrone,** $C_{18}H_{22}O_2$; a natural → oestrogen.

oestrous cycle, Am. **estrous cycle,** the sexual cycle of mammalian females; appears in the development and maturation of egg cells in the ovary, and as rhythmic changes in genital organs, in endocrine and mammary glands, and in some other features, and in the behaviour of animals. Following phases occur: **pro-oestrus** before **oestrus** (heat), during which the animal is most willing to permit coitus, and **diestrus** which is the sexually inactive period before the next oestrous cycle begins. The period immediately after oestrus is called **metoestrus** (luteal period) when the → corpus luteum is most active in hormone secretion. In small mammals, such as mice and rats, the cycle lasts only 4 to 5 days, in large animals about a month; in cold areas oestrous occurs usually only once a year. The → menstrual cycles of humans, apes, and Old World monkeys correspond to the oestrous cycle. In animals living in natural circumstances with large climatic seasonal variations, as in nordic areas, oestrous periods are dependent on the season and a long period without oestrus (**anoestrus**) may occur. *See* menstruation.

oestrus, Am. **estrus,** a period of sexual receptivity with intense sexual urge in female mammals; not found in humans. *See* oestrous cycle.

ohm (*Georg Ohm*, 1787—1854), the SI unit of → resistance, symbol Ω. *See* Appendix 1.

Ohm's law, a rule describing the dependence of the electrical quantities on each other; expressed as the equation: $U = IR$, in which U = the voltage in volts (potential difference), I = the current in amperes, R = the resistance in ohms.

oil (L. *oleum*, Gr. *elaion* = originally olive oil), a simple → lipid (glyceride of fatty acids) which is liquid at 20°C, and the molecule contains one or more unsaturated fatty acids; oils are usually grouped into **animal**, and **plant oils**, e.g. storage oils of plants and ectothermic animals, such as terpenes or cod-liver oil. Petroleum and many of its processed products (mineral oils) are hydrocarbons. *See* fat.

oil body, any of globular particles in plant cells containing oily substances; occur e.g. in liverwort cells.

oil gland, a gland that secretes oil, particularly the **uropygial gland** (preen gland) of birds which is a paired, adipose gland in the rump at the base of the tail. The glandular tubules open to a common papilla in the skin from which the bird preens the feathers with the oily secretion, rendering the feathers impervious to water.

oil plants, plants cultivated for oil production; the oil, obtained mostly from their seeds, is used as nutrients or raw-material of industry; e.g. soy, corn, *Brassica* species, many palms, and olive.

Okazaki fragment (*T.* and *R. Okazaki*), any of newly synthesized single-stranded DNA fragments, approximately 2,000 nucleotides in length, which is covalently joined to other such fragments during the proceeding of the lacking strand in DNA replication.

OLA, → oligonucleotide ligation assay.

-ol, pertaining to a chemical compound with a hydroxyl group, such as an alcohol or phenol.

ole-, olei-, oleo- (L. *oleum* = oil), pertaining to oil, or oleic acid and its derivatives. *Adj.* **oleic**.

-ole, pertaining to a chemical compound having a five-membered, usually heterocyclic ring, e.g. **pyrrole**, or to a compound not containing a hydroxyl group, e.g. several ethers such as **anisole**.

olecranon (Gr. *olekranon* = point of elbow), the proximal bony process of the ulna at the elbow in tetrapod vertebrates, serving for attachment of the brachial muscle straightening the forelimb.

oleic acid, *cis*-9-octadecenoic acid; a liquid unsaturated fatty acid, $H_3C(CH_2)_7CH{=}CH{-}(CH_2)_7COOH$, which occurs in glycerides in most natural lipids, i.e. in vegetable oils and animal fats and oils, particularly as a storage lipid. On hydrogenation, oleic acid is converted into stearic acid. Purified oleic acid is e.g used in manufacturing soft soap, polishing materials, lubricants, and cosmetics.

olein, triolein, glyceryl trioleate, $(C_{17}H_{33}COO)_3C_3H_5$; a glyceride of → oleic acid; a liquid neutral fat that occurs in many natural oils and fats.

oleosome, → spherosome.

olfaction (L. *olfacere* = to smell), the act of smelling; the sense of smell; a chemical sense of animals, based on the activity of the **olfactory receptors** which in vertebrates are located in a specialized area of the nasal mucosa (**olfactory epithelium**); the receptor cells, which are primarily nerve cells, have specialized cilia which are sensitive to different types of chemical compounds. The axons of the receptor cells form bundles (*fila olfactoria*) which together are called the **olfactory nerve**, proceeding to the **olfactory bulb** of the brain. There the first neurones form → synapses with the second neurones which run directly to the specific cerebral area, comprising an evolutionarily old cerebral cortex (**olfactory cortex**, "smell brain"). There are animals in all vertebrate groups with a highly developed sense of smell (**macrosmatic animals**, such as dog), but also with an feebly developed olfactory system (**microsmatic animals**, such as man). Many vertebrate species also have additional olfactory areas in the brain, such as the → terminal nerve system and → Jacobson's organ.

In invertebrates, olfactory receptors are more difficult to separate from taste and other chemical receptors. The chemically sensitive receptors in the antennae of many animal groups are usually classified into olfactory receptors. The effects of → pheromones are transmitted through olfactory receptors. *See* rhinencephalon.

olfactory nerve, *see* cranial nerves, olfaction.

olfactory receptors, olfactory cells, *see* olfaction.

olig(o)- (Gr. *oligos* = little, a few), denoting little, scanty, a few, deficient.

oligo-(dT)-cellulose, oligodeoxythymidylic acid-cellulose; cellulose used in chromatography to separate → messenger RNA (mRNA) from other RNAs; contains deoxythymine (= dT) that specifically binds the adenine chain of mRNA or poly$(A)^+$ RNA.

oligocardia, → bradycardia.

Oligocene (Gr. *kainos* = new), the middle geological epoch in the Tertiary period between the Miocene and Eocene epochs, about 37 to 24 million years ago, during which e.g. the Alps and Himalayas rose. Coniferous forests displaced deciduous forests in Europe as the result of cooling of climate; modern families of mammals, such as apes, monkeys, and whales, appeared.

Oligochaeta (Gr. *chaite* = hair), **oligochaetes** (oligochaets, oligochetes); a class in the phylum Annelida (sometimes classified as an order) comprising segmented worms living in fresh water (e.g. sludge worms, Tubificidae), or in moist soil (e.g. pot worms, Enchytraeidae), and earthworms (Lumbricidae); only a few species are marine. All species are hermaphrodite, some reproduce asexually. The animals have scarce, segmental **chaetae** serving in locomotion, and a → **clitellum** with glandular cells secreting a **cocoon** around the laid eggs. Altogether about 3,100 species are described, well known are the families Lumbricidae and Tubificidae. *Cf.* Polychaeta.

oligocythemia (Gr. *kytos* = cavity, cell, *haima* = blood), the abnormally decreased number of blood cells in the blood of vertebrates. *Syn.* oligocytosis or globular anaemia. *Adj.* **oligocythemic**.

oligodendrocyte (Gr. *dendron* = tree, *kytos* = cavity, cell), a cell type of → neuroglia (oligodendroglia), forming the myelin sheath of nerve fibres in the central nervous system. *Cf.* Schwann cell.

oligodendroglia, 1. oligodendrocyte; a cell type of the → neuroglia; **2.** the neural tissue composed of such cells.

oligohaline (Gr. *hals* = salt), pertaining to brackish water having salinity from 0.5—3.0 per thousand. *Cf.* mesohaline, polyhaline.

oligohemerobe (Gr. *hemeros* = tame), pertaining to an environment where nature is changed by human activities but its essential quality is unchanged; e.g. cultivated forests.

oligohumous water (L. *humus* = earth), water or a water body containg only little → humus. *Cf.* mesohumous water, polyhumous water.

oligolecithal (Gr. *lekithos* = yolk), pertaining to an egg cell type that contains only little yolk. *Cf.* mesolecithal, megalecithal.

oligomer (Gr. *meros* = part), a → polymer having comparatively few monomers (usually less than 20), which are bound to each other by covalent or non-covalent bonds.

oligomictic (Gr. *mixis* = intercourse, mixing), *see* stratification.

oligonucleotide, a compound composed of only a small number of → nucleotides, usually from 3 to 10.

oligonucleotide ligation assay, OLA, a method for detection of mutations in which two oligonucleotides, capable of hybridizing at adjacent positions with a target nucleic acid sequence, are joined by ligation and converted to a single nucleotide. This ligation is used as a measure of the number and precise sequence of the target molecules.

oligopeptide, a → peptide composed of a small number of amino acids, i.e. from about 3 to 10. *See* polypeptide.

oligophage (Gr. *phagein* = to eat), an animal eating only a few kinds of food, e.g. a few prey species or genera; the condition is called **oligophagy**, and it is common among animals, especially insects. *Adj.* **oligophagous**. *Cf.* monophage, stenophage, polyphage.

oligophyletic (Gr. *phyle* = family, genus), pertaining to a taxon developed from a few different ancestral lines. *Cf.* monophyletic, polyphyletic.

oligosaccharide (Gr. *sakcharon* = sugar), any carbohydrate (sugar) molecule composed of only a few monosaccharide units, usually from 3 to 10; e.g. raffinose (melitose) and stachyose. *Cf.* disaccharide, polysaccharide.

oligothermic (Gr. *thermos* = warmth), pertaining to a species which endures or demands relatively low temperatures; e.g. in aquatic ecosystems oligothermic species live at low temperatures and do not usually tolerate temperatures above 10°C, i.e. are cold stenothermic. *Cf.* polythermic.

oligotrophic (Gr. *trophein* = to nourish), pertaining to deficient nutrition; e.g. **1.** referring to a substratum for plant growth, i.e. either a water body or soil that is nutrient-poor and low in productivity, and therefore contains only few organisms; an organic matter which is only partly dissolved, accumulates in an oligotrophic ecosystem owing to low production and decomposition, such as in brown-coloured waters containing humus in temperate districts; **2.** pertaining to plant species or microorganisms succeeding in areas which contain little nutrient. *Noun* **oligotrophy**. *Cf.* dystrophic, eutrophic, mesotrophic, hypertrophy.

omasum, pl. **omasa** (L. = bullock's tripe), psalterium; the third portion of the stomach in ruminants, derived from the oesophagus. *See* rumination.

ombrotrophic (Gr. *ombros* = rain, *trophe* = nourishment), pertaining to a bog type (**ombrotrophic raised bog**), which is raised from its surroundings and gets water and nutrients only from rain. *See* mire.

omega (Greek letter), **1.** Ω, symbol for → ohm; **2.** ω, symbol for angular velocity; **3.** ω-, the carbon atom farthest away from the principal functional group of a molecule; *see* fatty acids.

omentum, pl. **omenta,** also **omentums** (L. = "fat skin"), a fold of the → peritoneum in the abdominal cavity of mammals extending from the stomach to support adjacent organs; the **lesser omentum** connects the stomach with the liver, and the **greater omentum** hangs in front of the intestines. *Adj.* **omental**.

ommatidium, pl. **ommatidia** (Gr. *omma* = eye), one of the long conical units of the → compound eye.

omn(i)- (L. *omnis* = all), denoting all, universal.

omnivore (L. *vorare* = to swallow up, devour), an animal eating all kinds of food, i.e. both plant and animal food; e.g. man and pig. *Adj.* **omnivorous**.

onc(i)-, onco-, onch(o)- 1. (Gr. *onkos* = barbed hook), denoting relationship to a hook, barb, or hook-shaped structure; **2. onc(h)o-** (Gr. *onkos* = bulk, mass, tumour), pertaining to a mass, swelling, or tumour.

onchocerciasis, also **oncocerciasis,** pl. **onc(h)ocerciases** (Gr. *onkos* = barbed hook, *kerkos* = tail), the state of being infected with nematode parasites of the genus *Onchocerca*, formerly *Filaria* (superfamily Filarioidea, filarial worms), native to Africa but now common in wide tropical areas, especially in Africa and Central and South America. Infections in mammals are caused by different species, in man by *Onchocerca volvulus*, which lives in streams and is transmitted by bites of several species of fly, e.g. blackflies (*Simulium spp.*). In man, the nematode breeds in the subcutaneous and other tissues, producing fibrous nodules (onchocercomas), and giving rise to microfilariae, which move freely out of the nodules into tissues. The adult worms cause dermatological changes and oedema under the skin by obstructing the lymph circulation; sometimes microfilariae progress into the eyes causing blindness. Called also onchocercosis, river blindness, blinding filarial disease, mal morado.

onchosphere, oncosphere (Gr. *sphaira* = globe, sphere), an oval, six- or eight-hooked embryo of cestodes; it has to enter the digestive tract of an intermediate host (e.g. a cow) and then through blood and lymph vessels finally to penetrate a skeletal muscle in order to develop further either into the → coracidium, procercoid, or cysticercus.

oncogene, a → gene that can initiate the formation of a cancerous tumour and sustain its development (*see* cancer). There are **viral oncogenes** found in viruses, causing cancer in higher organisms. Most of these oncogenes derive from normal cellular genes (**cellular oncogenes, proto-oncogenes**) which are incorporated in the genome of a virus during a viral infection. Cellular oncogenes are present normally in the genomes of cellular organisms such as plants and animals, where they regulate the growth, division, and differentiation of cells, but may change into a carcinogenic form producing a cancer cell, i.e. the neoplastic transformation of a cell. Cellular oncogenes are called proto-oncogenes as long as they act normally.

oncogenesis, the origin and growth of a tumour.

oncogenic, oncogenous, giving rise to a tumour, capable of causing a tumour; especially refers to tumour-inducing viruses.

oncology (Gr. *logos* = word, discourse), the study of cancer.

oncosphere, → onchosphere.

oncovirus, a virus of the subfamily Oncovirinae; retroviruses (RNA viruses) which can cause tumours. Called also oncornavirus.

one gene—one enzyme hypothesis, a hypothesis formulated in the early 1940s on the basis of experimental work in biochemical genetics; states that each gene regulates the synthesis of one enzyme protein. After its formulation the hypothesis has been in principle confirmed though it is replaced by the **one cistron—one polypeptide hypothesis**, presenting that each → cistron regulates the synthesis of one polypeptide.

ontogeny, ontogenesis (Gr. *on* = existing, *gennan* = to produce), the development of an individual organism from a fertilized egg to adult, as distinguished from the evolutionary development of a species, i.e. → phylogeny. *Adj.* **ontogenetic, ontogenic**.

Onychophora (Gr. *onyx* = claw, nail, *pherein* = to bear), **onychophorans**, **velvet worms**, **"walking worms"**; a phylum of animals including about 70 species, mainly living in wet detritus of tropical forests. They have many pairs of unjointed prolegs with two claws probably evolved from parapodial appendages; the eyes, similar to those of annelids, are located at the base of the antennae. On the basis of these and some other structural and functional properties they are phylogenetically placed between annelids and arthropods. In some classifications, onychophorans are considered as a class (or a subphylum) in the phylum Arthropoda; a known genus is *Peri-*

patus.

oo-, o- (Gr. *oon* = egg), denoting relationship to an egg or ovum; also → **ov(o)-**.

oocyst (Gr. *kystis* = bladder), a cyst forming a sheltering capsule around the zygote of → sporozoans, e.g. in → plasmodians. Sporogonic multiplication (*see* sporogony) produces sporozoites, the infectious agents, which are released from the oocyst and migrate to the salivary glands of the definite host (e.g. a mosquito), and are introduced into the blood of an intermediate host (some homoiothermic animal, e.g. man) by the bite of the infected mosquito.

oocyte (Gr. *kytos* = cavity, cell), the immature ovum of animals which in → oogenesis develops into the mature ovum; **primary oocyte**, the stage beginning the first maturation (meiotic) division; **secondary oocyte**, the stage between the first and second meiotic divisions. *See* meiosis.

ooecium, pl. **ooecia** (Gr. *oikos* = house), a brood pouch of some bryozoans in which embryos develop. *Syn*. ovicell.

oogamy (Gr. *gamos* = marriage), the conjugation of two dissimilar gametes, the female gamete being usually large and non-motile and the male gamete small and motile; occurs in multicellular animals and in some plants, such as in many algae and regularly in bryophytes and ferns. *Cf.* isogamy.

oogenesis, pl. **oogeneses** (Gr. *gennan* = to produce), the formation and maturation of female gametes (eggs, ova) in animals; the → ovary contains primary germ cells which in the beginning of oogenesis are called **primary oogonia**. They divide rapidly in → mitosis (goniomitosis) to form **secondary oogonia**, which then without cell division further develop into → **oocytes**. This maturation process occurs during the first and second meiotic divisions, when the oogonia give rise via **primary** and **secondary oocytes** to four **ootids**. One of these grows to an **ovum** and the other three degenerate to form → **polar bodies** (polar cells). During meiosis the ovum usually also grows substantially. Synthesis of RNA and proteins occurs during oogenesis producing reserve nutrients for early stages of → embryogenesis.

In plants, the development of the female gamete (egg cell) occurs in the female → gametangium.

oogonium, pl. **oogonia** (Gr. *gone* = birth), **1.** the primary female germ cell (egg mother cell) of animals from which the → ovum develops during → oogenesis; **2.** the female → gametangium of certain algae and fungi, containing one or several egg cells (oospheres).

ookinesis, pl. **ookineses** (Gr. *kinein* = to move), chromosomal movements of ova (egg cells) during maturation and fertilization.

ookinete (Gr. *kinetos* = moving), a motile, worm-shaped zygote of sporozoans; e.g. the ookinetes of malarial plasmodians penetrate the stomach wall of the mosquito developing there further into → oocysts.

oolemma (Gr. *lemma* = sheath), the layer surrounding the developing ovum, oocyte, especially the pellucid zone (*zona pellucida*) enveloping the mammalian egg.

Oomycetes, a class of saprophytic and parasitic fungi in the subdivision Diplomastigomycotina, division Mastigomycota; in some systems classified as a separate division, **Oomycota**. Oomycetes are primitive species which live free in natural waters, some are parasites of algae and small animals, the more developed species are parasites in higher plants. Unlike other fungi, the cell walls of oomycetes contain cellulose. The best known and most important are the species *Phytophthora infestans* which causes the serious disease, potato blight, and the genus *Saprolegnia* causing fish diseases. The group includes ca. 700 species.

oophor(o)- (Gr. *oophoros* = bearing eggs), pertaining to the → ovary.

oosphere (Gr. *sphaira* = ball), a large, spherical female gamete (macrogamete, egg cell) formed in an oogonium of algae and fungi in the class Oomycetes.

oospore, (Gr. *sporos* = seed), a thick-walled zygote which is formed from the fertilized female gamete (oosphere) in some algae and fungi in the class Oomycetes.

oostegite (Gr. *stege* = roof), one of the chitinous plates in the base of thoracic appendages of malacostracans (Crustacea) forming the floor of a marsupium for sheltering developing embryos and larvae.

ootid (Gr. *ootidion,* dim. of *oon* = egg), one of the four cells produced as a consequence of meiosis in → oogenesis. One of these develops into the ovum and the other three into polar bodies (polar cells).

ootype (Gr. *typos* = mould, type), an enlarged

distal part of the oviduct in many platyhelminths (flatworms) to which the ducts open from the Mehlis' gland (shell gland), vitelline (yolk) gland, and spermatheca, and in which the eggs are fertilized, provided with yolk, and supplied with a shell.

oozoid, oozooid (Gr. *zoon* = animal, *eidos* = form), any animal developed from an egg cell (ovum). Although most animals develop this way, the term is usually used only to denote primitive animals, such as some cnidarians and tunicates, in contrast to individuals borne by budding.

opal codon, UGA (uracil, guanine, adenine), one of the three stop codons in DNA causing the termination of → protein synthesis. *See* amber codon, ochre codon.

opal mutation, a → mutation in which a codon coding for an amino acid is changed into an → opal codon.

open bundle, a → vascular bundle in which a vascular cambium develops in a plant. The **collateral open type** is common in dicotyledons, which may also have a **bicollateral open type** with two layers of cambium.

open inflorescence, an → inflorescence having no terminal flower, i.e. the inflorescence grows continuously; represents the monopodial racemose type.

open reading frame, ORF, in genetics, the part of the DNA or RNA molecule that contains first an → initiator codon, next many codons encoding amino acids, and then a → termination codon. The nucleotide sequence of such an open reading frame encodes → protein synthesis.

open system, a system in which energy and matter are continuously exchanged between the system and its environment, such as in living organisms. *See* entropy.

operational taxonomic unit, OTU, in numerical taxonomy, a unit evaluated using taxometric methods; at a basic level, the unit is an individual but may also be a population, species, or some other taxon.

operator (L. = worker), in genetics, a short DNA sequence of an → **operon** in a bacterial or viral genome controlling the transcription of adjacent structural genes of the same operon. The operator binds to the product (**repressor protein**) of the regulator gene and hence controls the → genetic transcription of the structural genes. Typical of constitutive mutations in the operator is, that they are *cis* dominant (*see cis*-configuration), i.e. that the mutations have an effect on the structural genes located in the same DNA strand only.

operculum, pl. **opercula,** also **operculums** (L. = cover, lid), a cover or lid in organ structures; e.g. **1.** the gill cover in fishes and some amphibians; the skin fold supported by bony plates (opercular bones: preopercle, subopercle, interopercle, and opercle); **2.** the lid covering the nostrils or ears of some birds; **3.** one of the small plates in spiders covering the orifice of the trachea or lung sac; **4.** one of the two or more movable plates of the shell of a barnacle; **5.** a horny or calcareous plate on the dorsal surface of the foot of many gastropod molluscs (Prosobranchia); closes the aperture when the animal is retracted; **6.** the lid of the sporangium (spore capsule) of mosses.

operon (L. *operari* = to work), the unit of gene regulation and → genetic transcription in bacteria; consists of a → **promoter**, an → **operator**, and one or several **structural genes** which lie consecutively on the chromosome. The function of the operon is regulated by a → regulator gene.

Operons are usually divided into inducible and repressible, and negatively and positively controlled operons. **Inducible operons** guide the synthesis of catabolic enzymes (*see* catabolism) and **repressible operons** the synthesis of anabolic enzymes (*see* anabolism). In **negative regulation** the product of the regulator gene in its active form is a **repressor** (repressor protein) of gene function, and in **positive regulation** it correspondingly is an **activator** (activator protein). Both the repressor and activator are allosteric proteins.

The regulation of the function of the operon occurs at the level of transcription. As a regulatory molecule acts an **effector** which in catabolic reactions is the substrate molecule and in anabolic reactions the end product of a certain reaction chain. The former is an → **inducer** and the latter a → **co-repressor**.

In a negatively regulated, inducible operon the repressor protein attaches to the operator thus preventing transcription in structural genes if there is no inducer in the cell. If the inducer is present, it changes the structure of the repressor protein in such a way that this no longer can recognize the operator, and consequently transcription in the structural genes begins. Transcription is followed by the synthesis of those enzymes which catabolize

the inducer.
The operon is the unit of transcription and produces a **polycistronic messenger RNA** molecule (*see* polycistronic). In eukaryotes, operons are not found but certain → gene complexes of unicellular fungi are putative operons, and in → *Caenorhabditis elegans* a total of 2/3 of the genome is organized as operons. In this latter case it is not yet known whether they are units of gene regulation or simply adaptations to a small genome size. The theory dealing with the operon is called **Jacob—Monod theory.**

ophi(o)- (Gr. *ophis* = snake), denoting relationship to snakes and their venoms, or snake-like structures. *Adj.* **ophic**.

Ophidia, snakes, → Serpentes.

Ophioglossidae (Gr. *glossa* = tongue), a subclass in the class Pteropsida, division Pteridophyta, Plant kingdom; in some classifications the order Ophioglossales in the division Pterophyta. The genera *Ophioglossum* (adder's tongues) and *Botrychium* (moonworts) grow on meadows and pastures; both are

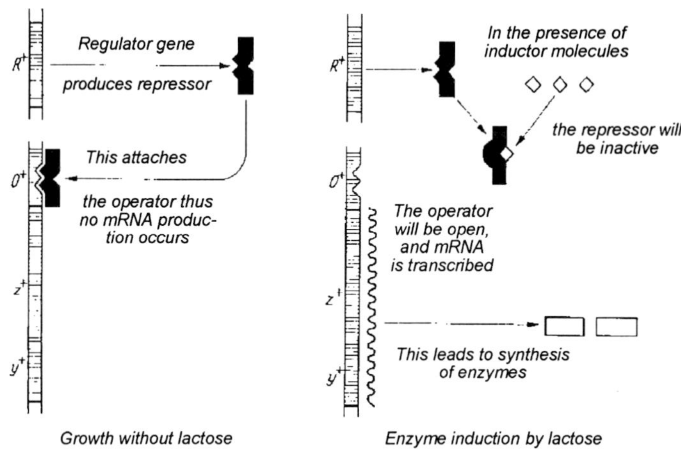

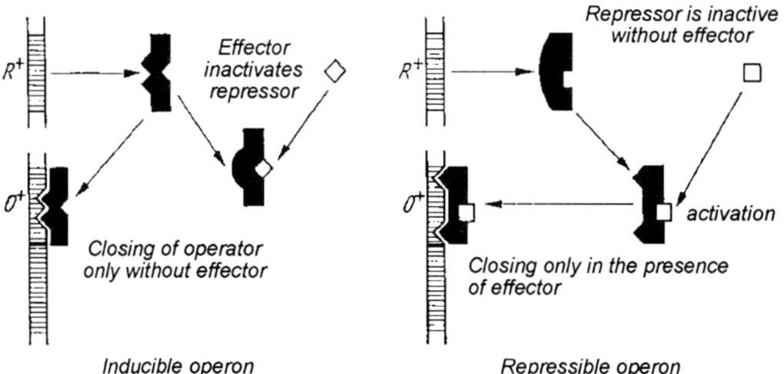

Operon. A schematic representation of the function of the → lac-operon (upper), and comparison of the principle of the function of an inducible and a repressible operon (lower). (After Bresch & Hausmann: Klassische und Molekulare Genetik - Springer Verlag, 1970; with the permission of the publisher).

small, have an underground stem, one leaf, and spike-producing spores.

ophiopluteus (L. *pluteus* = shed), a free-swimming planktonic larva of ophiuroids. *See* pluteus.

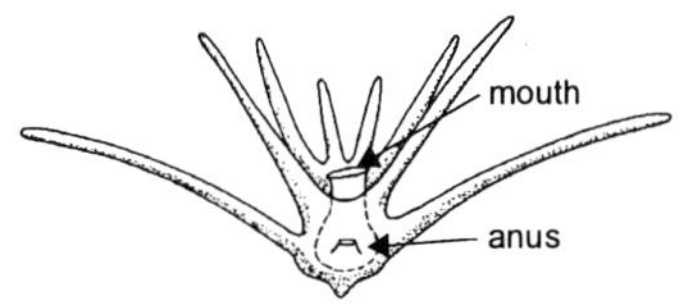

OPHIOPLUTEUS

Ophiuroidea (Gr. *oura* = tail, *eidos* = form), **ophiuroids**, **ophiurans**, **brittle stars**, **serpent stars**; a class (subclass) in the phylum Echinodermata, the animals having long and roundish branches; the → pedicellariae and anus are lacking, and the tube feet without suckers are not used for locomotion but are sensory organs which are used also in respiration. The interior of the long arms is almost filled with axial skeleton formed from solid cylindrical ossicles.

ophthalm(o)- (Gr. *ophthalmos* = eye), denoting relationship to the eye. *Adj.* **ophthalmic**. *Cf.* ocul(o)-.

-opia (Gr. *ops* = eye), denoting a condition of the eye or vision, e.g. hyperopia; also **-opy**.

opiate receptors, *see* opioid peptides.

opiates (Gr. *opion* = poppy-juice), a group name for natural and synthetic drugs which have opiate-like (morphine-like) effects on animals; often any narcotic drug that induces sleep. *See* opioid peptides.

Opiliones, Phalangida (*Opilia* = the type genus), **harvestmen, phalangids**, "daddy long-legs"; an order in the class Arachnida; the body is short and the cephalothorax (prosoma) broadly joined to the abdomen (opisthosoma) which shows weak external segmentatation. The legs are extremely long; in the centre of the cephalothorax is a tubercle with a pair of eyes. Poisonous and spinning glands are lacking. In general, harvestmen are predatory but scavenging is more important than among other arachnids.

opine, an amino acid derivative produced in the → crown gall tissue of plants induced by the Ti plasmid of the bacterium *Agrobacterium tumefaciens*; the most common opines are **octopine** and **nopaline**. The agrobacterium uses opine as a carbon and nitrogen source but the host plant cannot utilize it. The transfer of genetic material into cells can be studied by analysing the opine formation.

opioid, pertaining to → opiates.

opioid peptides, small peptides and polypeptides produced chiefly in the nervous tissue of animals, having opiate-like (morphine-like) physiological actions; in vertebrates two groups are known, → **endorphins** and **enkephalins**. Actions of both opioid peptides and → opiates (as morphine) are transmitted in tissues through specific protein molecules, **opiate receptors**, present in cell membranes, especially in postsynaptic membranes of certain → synapses e.g. in the brain. Sometimes the term opioids includes also opioid antagonists such as nalorphine.

Opisthobranchia(ta) (Gr. *opisthen* = behind, *branchia* = gills), **opisthobranchs**; a subclass of marine molluscs in the class Gastropoda including e.g. **sea hares** and **nudibranchs**. The body is asymmetrical or secondarily bilateral, one gill is usually located behind the heart. Most are hermaphrodite; often the shell has become internal or is totally lacking. About 2,000 species are known in nine orders.

opisthocoelous, Am. also **opisthocelous** (Gr. *koilos* = hollow), pertaining to a vertebra with the anterior end of the centrum flat or convex and the posterior end concave, occurs e.g. in some fish groups (Lepisosteidae, Anguilliformes) and caudate amphibians (Caudata). *Cf.* procoelous, amphicoelous.

Opisthocomiformes (Gr. *kome* = hair, L. *forma* = form), a bird order including only one species, **hoatzin** (*Opisthocomus hoazin*); a middle-sized, crested bird with a short, stout bill, long neck, long tail, and small wings, living in the tropcical forests of South America. Its young have claws on the first two digits of the wing used for climbing on trees. Called also stinkbird.

opisthosoma, pl. **opisthosomata** (Gr. *soma* = body), the posterior region in the body of some arachnids, merostomes, and pogonophores.

opium (Gr. *opion*, dim. of *opos* = latex, sap), dried latex obtained from unripe capsules of poppy (*Papaver somniferum*); contains about 20 **opium alkaloids** which are processed into drugs, e.g. → morphine, noscapine, codein, papaverine, which have analgesic and hypnotic effects on animals. *See* opiates, opioid peptides.

opportunistic (L. *opportunus* = suitable, opportune), **1.** in ecology, pertaining to a species adapted to the effective exploitation of temporary and local conditions, i.e. to a species having an ability to reproduce and disperse in an unpredictably changing environment, e.g. r-strategists (*see* r-selection); **2.** pertaining to a pathogenic microorganism, living normally in a host organism but turning pathogenic only under certain conditions.

opsin (Gr. *ops* = eye), a protein acting in → photoreception in animals. The protein is a part of visual pigments, e.g. → **rhodopsin**, from which the chromophore component, **retinal**, is split off by the action of light. In the vertebrate retina, the protein of the rods is often called **scotopsin** and that of the cones **photopsin** (actually usually three different opsins).

opsonin (Gr. *opsonein* = to buy victuals), a substance such as an antibody, enhancing → phagocytosis. *See* opsonization.

opsonization (Gr. *opsonion* = victuals), making a microbe or other antigenic material attractive to phagocytes (*see* phagocytosis) in an animal; occurs when the antigen is bound to an **opsonin**, which is a protein or polypeptide molecule usually produced by the immune system. Any immunoglobulin (e.g. IgM, IgG) or certain fragments of the → complement system (e.g. C3b, C4b) may act as opsonins in vertebrates.

opt(o)- (Gr. *optos* = seen < *ops* = eye), denoting relationship to eyes or vision. *Adj.* **optic**.

optical density, OD, *see* absorbance.

optical isomerism, *see* isomerism.

optic chiasm(a), optic decussation; the crossing point of the fibres of the optic nerve seen in front of the hypothalamic infundibulum in the vertebrate brain.

optic disc/disk, *see* blind spot.

optic nerve, the second → cranial nerve in vertebrates.

optimal biotope, a biotope in which the individuals of a population preferentially settle and where the reproductive output of the population is maximal.

optimal foraging theory, → optimum foraging theory.

optimal yield, → optimum yield.

optimization models, mathematical models used in evolutionary ecology to resolve the best solution in behavioural strategy of an animal for achieving certain aims. The basic idea in optimization models is that natural selection favours the genes producing those phenotypical features which minimize the costs of an action and maximize the profits. The risk of dying and an increase in the use of energy are e.g. used as measures of costs, while the profits are measured by an increase of reproductive output or savings in energy use. An optimum situation is achieved when the costs of an action are as minimal as possible and the profits maximal. *Cf.* game theory, optimum foraging theory.

optimum, pl. **optima** (L. = best), in biology, the best or most favourable condition for reproduction, development, growth, etc. of organisms; e.g. the most favourable conditions for activities of biochemical and physiological reactions, e.g. the optimum of temperature (temperature optimum), illumination, pH, mineral, and substrate concentration. *Adj.* **optimal**.

optimum/optimal foraging theory, a theory that natural selection favours such foraging strategies which produce the best profit for a forager per time and energy units. *Cf.* searching time, handling time.

optimum/optimal yield, the maximal amount of biomass (or individuals) that can be removed without depletion of a population. It is about half the → carrying capacity. *Syn.* maximum sustained yield, MSY.

oral (L. *os* = mouth), pertaining to the mouth; e.g. **1.** oral temperature, or oral administration, i.e. given to or through the mouth; **2.** located on the same side as the mouth in radially symmetric animals; opposed to aboral.

oral cavity (L. *cavitas/cavum oris*), mouth cavity; the first part of the digestive canal of vertebrates, including the tongue, teeth, and palate; begins from the mouth aperture and ends at the pharynx. The oral cavity is covered by mucous membrane with a stratified epithelium. Many invertebrates have a similar type of cavity between the mouth-opening and the foregut. *Syn.* buccal cavity.

oral groove, a shallow furrow on the surface of the unicellular body of some ciliate protozoans, e.g. Paramecium; ends at the cell mouth, **cytostome**.

orbit (L. *orbita* = track, route < *orbis* = circle), **1.** orbital; a round or elliptical path described by a planet, satellite, or an electron around the nucleus of an atom, or by a charged particle on an electric or magnetic field; **2.** in anatomy, the orbital cavity of the skull containing the eye, called also eye socket or **orbita**; also the skin surrounding the eye of a bird or a circle around the eye of an insect; **3.** as a verb, to move in or put into an orbit.

orbital, pertaining to → orbit.

orbital bristles, *see* frontal suture.

orcein, a purple histochemical dye derived from → orcinol and used in microscopy to stain cells and tissues, as e.g. aceto-orcein for nuclei and chromosomes.

orchid(o)-, orchi(o)- (Gr. *orchidion*, dim. of *orchis* = testis), denoting relationship to testes or orchids. *Adj.* **orchidic** or **orchic**.

orcinol, 5-methylresorcinol; 3,5-dihydroxytoluene, $CH_3C_6H_3(OH)_2 \cdot H_2O$; a substance found in certain lichens from which the histological dye, **orcein**, is prepared by treatment with air and ammonia. Orcinol is used in the quantitative determinations of sugars, glycosides, sulpholipids, and as an external antiseptic in medicine. *Syn.* orcin.

order, the systematic taxon of organisms between the family and class (or subclass). The order includes families of the same kind. When necessary, the order is divided into suborders, or several orders are united to a superorder.

Ordovician (L. *Ordovices* = ancient people in Wales), the second geological period in the Palaeozoic era, between Cambrian and Silurian, about 505 to 438 million years ago; graptolites were dominant and the number of many other invertebrate species, such as molluscs, brachiopods, and bryozoans increased greatly, trilobites decreased, and the first vertebrates (ostracoderms) appeared. Algae were common among plants whereas the development of the higher plants had not yet begun.

ORF, → open reading frame.

organ (L. *organum*, Gr. *organon*), **1.** a partly independent body part of metazoan animals, having a specialized structure formed from differentiated cells and tissues, and performing a specialized function; e.g. the heart, liver, kidney, or a sense organ. Several organs together may form a larger entity called **organ system**, as the circulatory system, including the heart and blood vessels; **2.** in → vascular plants (e.g. ferns and seed plants), a structure having a certain differentiated form and function, the main organs being the root, shoot, and leaves. The other so-called lower plants are thallophytes having no differentiated structure or organs except the mosses which have structures resembling a shoot and leaves.

organ culture, the culture of usually embryonic organs under laboratory conditions; an important method for the study of the differentiation of tissues and organs. *See* cell tissue and organ culture.

organelle (L. *organella,* dim. of *organum* = organ), cell organelle; any of minute specialized subcellular structures most of which are characteristic for all cell types; e.g. → centriole, cytoskeleton, endoplasmic reticulum, Golgi complex, dictyosomes, lysosomes, mitochondria, nucleus, peroxisomes, plasma membrane. In protozoans there are special organelles like → cytostome, cytopharynx, and contractile vacuole. The cell vacuole and the plastids (e.g. chloroplasts) are typical of plant cells.

organic (Gr. *organikos*), **1.** pertaining to an organ; **2.** living or existing in, or originating from organisms; **3.** structural, organized; **4.** pertaining to carbon compounds, e.g. organic chemistry.

organism, a living individual, i.e. an animal, plant, fungus, or microorganism; an organic unit of continuous lineage with an individual evolutionary history; comprises one or more living → cells, or in the case of → viruses which are intermediate forms between living and non-living material, of one virus particle with nucleic acids.

organizer, any group of cells in a developing embryo which through an → induction organizes and regulates the differentiation of tissues; originally referred to the **primary organizer**, i.e. the dorsal lip of the blastopore of the early vertebrate embryo inducing the development of the ectodermic → neural plate.

organ of Corti → spiral organ.

organogenesis (Gr. *gennan* = to produce), the formation of organs during embryonic development, i.e. the → differentiation and early development of the anlagen (primordia) of organs into functional organs. *See* morphogenesis.

organogeny, 1. → organogenesis; **2.** the study of organogenesis.

organology (Gr. *logos* = word, discourse), the study of the organs of plants and animals; the term is used especially in botany dealing with the outer structure of organs, i.e. their forms and mutual relationships.

organotrophic, *see* heterotroph.

organ system, *see* organ.

organum, pl. **organa** (L.), → organ; e.g. *organum auditus*, hearing organ; *organum visus*, visual organ; *organa genitalia*, genital organs.

orgasm (Gr. *orgasmos* = swelling), the culmination of sexual excitement, especially that emotionally experienced by man and higher animals. *Adj.* **orgasmic, orgastic**.

Oriental region (L. *orientalis* = eastern), a zoogeographical region including southern Asia south of the Himalayas, Southeast Asia and the Malaya Archipelago to the → Wallace's Line. *See* zoogeographical regions.

orientation (L. *oriens* = sunrise), a change of position of an organism in relation to an external stimulus, or the direction of a stimulus, also the ability to locate oneself in an environment. The **navigating orientation** (→ navigation) is common among animals, as e.g. fishes, birds, and some mammals, who make seasonal long-distant migrations. Orientation is based on several different methods in which the Sun, the magnetism of the Earth, and recalling ground topography are important. The ability is based on sensory and neuronal mechanisms, also on some special senses, such as → magnetic sense, still insufficiently known. Many insects orientate according to the sunlight and on cloudy days they sense the light direction from the polarization of light in clouds. *See* migration, taxis.

origin of life, the development of the earliest stages of life on the Earth; according to present theories, life originated most likely in the sea approximately 4,000 million years ago; the actual beginning of life was preceded by chemical evolution during which simple organic molecules, such as lipids, amino acids, small peptides, nucleotides and nucleic acids (presumably RNA first, *see* ribozymes) were formed from inorganic material, the UV radiation of the Sun, and ligthning from heavy thunderstorms providing the energy for these reactions. Subsequently polypeptides and RNA molecules formed symbiotic macromolecules each part catalysing the replication of the other, and → evolution started, due to the differential rate of replication of these macromolecules. During this early stage of biological evolution, cell membranes were formed from lipids to surround the self-replicating macromolecules, and consequently the first primitive cells evolved. RNA as the carrier of the genetic information was replaced by the more stable DNA, the RNA molecules acting as templates for DNA synthesis, and complex genetic systems evolved. Due to competition between the primitive cells for restricted resources, multicellular organisms gradually evolved, and this created an ever increasing number of new ecological → niches which became gradually filled with more complex forms of life. Thus, step by step the whole biological diversity of living material arose.

ornith(o)- (Gr. *ornis* = bird), denoting relationship to birds. *Adj.* **ornithic**.

ornithine, Orn, 2,5-diaminovaleric acid; NH_2-$(CH_2)_3CH(NH_2)COOH$; an amino acid which is an intermediate in the → **urea cycle** (**ornithine cycle**) of animals, but not a constituent in proteins; its derivative, ornithuric acid, is secreted in the urine of birds and reptiles.

Ornithischia (Gr. *ischion* = hip), an extinct order of bipedal or quadrupedal, herbivorous → dinosaurs who lived during Mesozoic.

ornithocoprophilous (Gr. *kopros* = dung, *philein* = to love), favouring the dung of birds; e.g. pertaining to a plant or fungus which prefers to grow in bird dung, as e.g. many lichens which live on small rocky islands inhabited by birds.

Ornithodelphia (Gr. *delphys* = womb), in some recent classifications, an infraclass in the mammalian subclass → Theria, including one order, → Monotremata.

ornithology (Gr. *logos* = word, discourse), a branch of zoology studying birds.

ornithophilous (Gr. *philein* = to love), favouring birds; pertaining e.g. to a plant which is pollinated by birds.

ornithosis, a disease of birds caused by the bacterium *Chlamydia psittaci*; it is transmissible to man causing a disease called → psittacosis. Originally the term ornithosis referred to the disease of non-psittacine birds, and psittacosis to that of psittacine birds (parrot disease or fever) and man. Avian infections are mostly latent but may be transferred to humans, often causing mild disease (severe in old people) with symptoms of bronchopneumonia.

oro-, 1. (L. *os*, gen. *oris* = mouth), denoting relationship to the mouth; **2.** (Gr. *oros* = mountain), pertaining to a mountain or elevation, e.g. orophyte, a subalpine plant.

oroanal (L. *os* + *anus*), pertaining to the mouth and anus; e.g. denotes an opening that functions both as the mouth and anus in some invertebrate animals, e.g. turbellarians.

orotic acid, abbr. oro (Gr. *oros* = whey), a precursor in the synthesis of → pyrimidine nucleotides, first found in milk; $C_5H_4N_2O_4$; is a growth factor for many microbes. Its salts, esters, and the ionic form are called **orotates**.

orth(o)- (Gr. *orthos* = straight), **1.** denoting straight, right, or normal; **2.** in chemistry, *o*-, pertaining to an organic compound which has two substitutions on adjacent carbon atoms of a hexagonal ring, particularly of a benzene ring; *cf.* meta-, para-; **3.** pertaining to an inorganic acid or its salt which has a higher degree of hydration, e.g. orthophosphoric acid, H_3PO_4, as distinguished from metaphosphoric acid, $(HPO_3)_n$.

orthodromic (Gr. *dromos* = run), proceeding or moving in the right (normal) direction; e.g. describes a nerve impulse propagating in the normal direction along a nerve fibre. *Cf.* antidromic.

orthogenesis, the evolution of phyletic lines (*see* phyletic evolution) which in a long process of time proceeds in one direction governed by intrinsic factors and not by natural selection; the theory is not accepted in a modern concept of evolution.

orthognathous, orthognathic (Gr. *gnathos* = jaw), **1.** having straight jaws; having a face without a projecting jaw, like most developed hominids (man); **2.** in some invertebrates, describing the head, the axis of which is at a right angle to the body axis. *Cf.* hypognathous, prognathous.

orthograde (L. *gradi* = to walk), characterized by walking or standing erect, with the body upright, as man.

orthograde axoplasmic transport, the transport of proteins from the nerve cell body along → microtubules to the axon terminal.

orthokinesis (Gr. *kinein* = to move), a random movement of an animal in response to a stimulus, changing the speed of its motility according to the strength of the stimulus. *See* kinesis.

orthologous (L. *locus* = place), pertaining to the genes which diverged in parallel with the species harbouring them, i.e. pertaining to genes of different species that are homologous when derived from a common ancestral gene, e.g. alpha-globin genes from humans and mice.

orthophosphate, the inorganic phosphate (P_i), the ions (PO_4^{3-}, HPO_4^{2-}, $H_2PO_4^-$), esters or salts of orthophosphoric acid (H_3PO_4), e.g. inorganic phosphate which in cell respiration is involved in the metabolism of nucleoside phosphates (ATP, UTP, etc.). *Cf.* pyrophosphate, metaphosphate.

Orthoptera (Gr. *pteron* = wing), **orthopterans**; an order of insects whose forewings are narrow and thickened, hindwings broad, membranous and folded like a fan beneath the forewings; they have chewing mouthparts, many species have stridulating organs (→ stridulation). The order includes several families: e.g. **cockroaches** (Blattidae), **praying mantises** (Mantidae), **stick insects** (walking sticks, Phasmatidae), **grasshoppers** (Tettigoniidae and Acrididae), **crickets** (Gryllidae). In some classifications cockroaches and mantises are located in the order Dictyoptera.

orthoselection, → natural selection which acts in the same direction for a long period of time and gives rise to what is called → orthogenesis.

orthostatic (Gr. *statikos* = standing), pertaining to, or caused by an erect posture or vertical position.

orthotropism (Gr. *trope* = turning), **1.** orientation or growth in a straight line, e.g. in a vertical direction such as a tap root of a tree; **2.** the tendency of an organism to be oriented in the direction of a stimulus. *Adj.* **orthotropic**.

oryctology (Gr. *oryktos* = formed by digging, *logos* = word, discourse), the study of fossils, → palaeontology.

os (L. gen. *oris*, pl. *ora*), → mouth.

os (L. gen *ossis*, pl. *ossa*), → bone.

oscillation (L. *oscillare* = to swing), a pendular motion; fluctuation above and below a certain value; occurs in biological regulatory systems in which a certain quantity (concentration of a substance, body temperature, etc.) to be regulated, oscillates above and below the set point. Oscillation contains increasing, decreasing, and static phases, and attenuates gradually in stabilizing conditions. *Verb* to **oscillate**. *Adj.* **oscillatory**.

oscilloscope, cathode-ray oscilloscope, CRO,

an electronic instrument which records changes in the voltage of an electric circuit by a trace of light on the fluorescent screen of a cathode-ray tube. It can be used in biology, especially in electrophysiological studies, e.g. in recording action potentials or functions of ion channels in cells. *See* membrane potential.

Oscines (L. *oscen* = singing bird used in divination), the largest suborder in the order → Passeriformes, including 57 families of the total 72 in the order.

osculum, pl. **oscula** (L. = dim. of *os* = mouth), a small aperture, opening, or pore, especially the large excurrent opening on the top of sponges (Porifera) for water currents from the spongocoel.

osm(o)-, 1. (Gr. *osmos* = push, impulsion), denoting relationship to → osmosis; **2.** (Gr. *osme* = smell), pertaining to smell, odour, or osmium.

osmiophilic (Gr. *philein* = to love), pertaining to a structure which can be stained with → osmium dyes, e.g. for electron microscopy.

osmium (Gr. *osme* = smell), symbol Os, a poisonous crystalline metal element of the platinum group, atomic weight 190.2. It is used in alloys with iridium and platinum. **Osmium tetroxide** (OsO_4) is commonly used as a stain for fatty substances (e.g. cellular membranes) in light microscopy, and as a postfixative "stain" (*see* fixation) in electron microscopy to increase the electron opacity of cell components, thus improving the contrast of specimens.

osmolality (Gr. *osmos* = push), the total osmotic concentration of a solution expressed as the number of → osmoles per one kilogram of solvent (water); 1 osmolal (osm) solution contains one osmole of a solute per one kilogram of solvent. *See* osmosis.

osmolarity, the osmotic concentration of a solution expressed as → osmoles per one litre of a solution.

osmole, also **osmol, Osm,** the molecular weight (*see* mole) of a substance (solute), in grams, divided by the number of ions or particles into which it dissociates in solution. For example, one mole of glucose (non-ionic) forms one osmole in a solution, but one mole of sodium chloride (dissociated to Na^+ and Cl^- ions) forms two osmoles.

osmometer (Gr. *metron* = measure), equipment used for measuring the concentration of a solution; its function is usually based on the → depression of the freezing point of a solution. *See* osmosis.

osmoreceptor, 1. (Gr. *osmos* = push, L. *recipere* = to receive), a neurone or a part of a neurone that responds to changes in osmotic concentration of tissues. In vertebrates, osmoreceptors are found in the thirst centre of the hypothalamus, but similar receptors probably exist in the nasal epithelium of desert species (specialized smell receptors). Related osmoreceptors function in some invertebrate species, e.g. the receptors of humidity; **2.** (Gr. *osme* = smell), any of the olfactory receptors; *see* olfaction.

osmoregulation, the regulation of osmotic concentration of cellular and extracellular fluids of an organism at a certain, more or less stable level; includes the maintenance of salt and water balance (**ionic regulation**), but also other osmotically effective substances, such as sugars, sugar alcohols, and proteins, are involved (*see* osmosis).

Among animals, all vertebrate and many invertebrate species are **osmoregulators** (homoiosmotic), but some invertebrates are **osmoconformers** (poikilosmotic) tolerating large fluctuations of osmotic concentration in their tissue fluids. The strategies of osmotic regulation vary according to the environment. Animals living in fresh or brackish water tend to prevent the swelling of tissues by restraining water intrusion from the hypotonic surroundings into hypertonic tissues. These animals have an impermeable skin, but they can actively take ions through their **gill epithelium**, some also through the **skin**.

Marine animals tend to lose water from their hypotonic tissues to the surroundings, and take up salts from seawater into the body. To prevent this they possess an impermeable skin and excrete salts from the body through the gill epithelium, **kidneys**, or special **salt glands**, the gills being the major osmoregulatory organs in fishes and many aquatic invertebrates. In some marine animals the osmotic concentration of tissue fluids is exceptionally high, even isosmotic to the seawater, as in most cartilaginous fishes. They maintain this with high concentrations of urea and trimethylamine oxide. Many invertebrates are osmoconformers allowing their osmotic concentration to follow that of the environment.

In terrestrial animals, osmotic regulation is based on the maintenance of a balance between the gain of water and salts in food and drink, and loss of water in respiration and evaporation, and water and salts by excretion. Important organs in regulation are the kidneys, skin, alimentary canal, lungs, and in some species specialized salt glands, and neural and hormonal control of their functions. In most terrestrial animals, osmotic concentration of body fluids is maintained at the level of about 300 mOsm/l.

In plants, the water balance and water transport in tissues and cells play an important role in osmotic regulation; e.g. the function of the guard cells in the opening-closing mechanism of stomata is based on osmotic regulation and further on the turgor pressure in the cells. The most important osmotically active substance in the cells is potassium, the amount of which is regulated especially by the K^+ channels. Potassium regulation is involved in rapid changes (as in guard cells), while sugars and other organic compounds are important in long-term regulation. For example, before an unfavourable season when freezing is expected, water is removed from cold-hardening plant cells in which (especially in the cell vacuoles) the concentration of the organic compounds is increased and the osmotic potential decreases (i.e. becomes more negative); these changes cause the depression of the freezing point of the cell contents which is favourable for the wintering plant. *See* water balance, cold hardening, osmosis.

osmosis, the → diffusion of solvent (water) from a dilute solution or pure water to a more concentrated solution through a semipermeable membrane which is permeable to the solvent but not to the solute. Water moves towards the lower (more negative) → **osmotic potential** and the transport velocity is dependent on the concentration difference ($\Delta\psi$) between the systems. Osmosis is expressed as **water potential**:

$$\psi = \psi_\pi + \psi_p + \psi_m + \psi_g + \psi_c$$

where ψ = water potential, zero for pure water and negative for any solution; ψ_π = osmotic potential (negative); ψ_p = pressure potential (positive or zero); ψ_m = matric potential; ψ_g = gravitational component; ψ_c = component depending on electrical charges. In practice only the first three addends are important.

When cells lose water their water potential decreases and the cell sap becomes more concentrated. When water is again available it moves into cells towards the low water potential. The water taken in causes a **hydrostatic pressure (pressure potential)** and cells become enlarged. Experimentally, red blood cells and plant protoplasts swell and break rapidly in distilled water or in dilute solutions (→ haemolysis, plasmolysis) because these cells have no cell wall. Pressure potential determines the equilibrium point and the water uptake ceases.

Osmosis is an important factor of **water transport** in organisms, particularly in plants; e.g. plant roots take water osmotically from the soil. This is possible only if the osmotic potential of root cells is lower than that of the solution between soil particles. Sometimes water may be bound very tightly on soil colloids and the matric potential (ψ_m) is often a more important factor in some soils than the osmotic potential.

In animals, osmosis is an important factor in the passive intake and removal of water. Osmosis affects the transport of cellular and interstitial fluids between fluid compartments, e.g. from blood through the walls of capillaries into surrounding tissues and back. Most animals can effectively regulate the osmotic concentration of interstitial fluids. *See* osmoregulation.

osmotic potential, symbol ψ; a property of a solution due to the presence of osmotically active solute particles: water moves between two systems of different osmotic concentrations towards a lower (more negative) osmotic potential. The osmotic potential of solutions is calculated using the osmotic potential of pure water as a reference (zero value). Osmotic potential is given as pressure units, usually kPa or MPa. The potential of solutions is negative, e.g. between –1.0 and –0.5 MPa in potato tissues. *See* osmosis, osmotic pressure, van't Hoff's law.

osmotic pressure, the pressure (symbol π) needed to prevent the water transport in → osmosis; depends on the number rather than the type of particles in a solution. The greater the number of particles, the higher the osmotic pressure. Instead of the term osmotic pressure, → osmotic potential, based on the water potential, is used nowadays in botany. *See* van't Hoff's law.

osmotic shock, the intrusion of a solvent (water) by osmosis into a cell or other membrane-bound structures causing the rupture of the membrane; used to lyse cells and cell organelles.

osphradium, pl. **osphradia** (L. < Gr. *osphradion,* dim. of *osphra* = smell), a chemical sense organ found in most aquatic molluscs, located near the gill having connections with visceral ganglia, and "smells" the quality of water flowing to the gills.

osse(o)-, ossi- (L. *os* = bone), denoting relationship to bone. *See* ost(e)-.

ossein, → collagen of the bone tissue. *Syn.* ostein.

osseous (L. *osseus*), denoting relationship to bone; composed of, or resembling bone; also **osteal**.

osseous labyrinth, the bony cavity of the inner ear in the petrous part of the temporal bone in vertebrates; the → membranous labyrinth with sensory structures is situated inside the cavity. The membranous labyrinth is filled with endolymph and surrounded by perilymph. *Syn.* bony labyrinth.

osseous tissue, → bone tissue.

ossicle (L. *ossiculum*, dim. of *os* = bone), **1.** a small bone or calcareous structure; e.g. auditory ossicles; **2.** a small calcareous plate with a projecting spine and tubercle under the skin of many echinoderms, forming the mesodermal endoskeleton of the body.

ossification, the formation of a bony structure or bone tissue. *Adj.* **ossific**. *See* osteogenesis.

ost(e)-, osteo- (Gr. *osteon* = bone), denoting relationship to bone.

osteal, → osseous.

Osteichthyes (Gr. *ichthys* = fish), **bony fishes**; a class of aquatic vertebrates in which the skeleton is completely ossified. They have either a swim bladder or lungs (lungfishes); the gills are covered by the **operculum**. The body is often covered by scales or bony plates. Most of the present fishes belong to the bony fishes (about 21,000 described species) and the class can be divided into two subclasses: **fleshy-finned fishes** (Sarcopterygii) and **ray-finned fishes** (Actinopterygii). *Cf.* Chondrichthyes. *See* Teleostei.

osteoblast (Gr. *blastos* = germ), bone-forming cell; the cells derived from → fibroblasts when osteogenesis begins in the cartilage or connective tissue, and gradually during the maturation of the bone tissue the osteoblasts become enclosed by calcified bone matrix; the cells are then called **osteocytes**. *See* bone.

osteoclast (Gr. *klastos* = broken), **1.** osteophage; a phagocytic cell associated with dissolution and removal of bone during bone formation and remodelling; a large multinuclear cell differentiated in the bone marrow from the stem cell of the monocyte series, as do all → macrophages. The cells are activated by the parathyroid hormone causing e.g. the release of calcium from the bone into the blood; **2.** an instrument used in bone surgery. *Adj.* **osteoclastic**, denoting osteoclasts, destructive to bone.

osteocyte (Gr. *kytos* = cavity, cell), bone cell, osseous cell; any of the cells derived from → osteoblasts, which in osteogenesis have become enclosed within the calcified bone matrix but with long cytoplasmic processes through small canaliculi keep contact with other osteocytes. *See* bone.

osteogen (Gr. *gennan* = to produce), the tissue layer from which bone is formed. *Adj.* **osteogenetic**, **osteogenic**, **osteogenous**, relating to osteogenesis, forming bone.

osteogenesis, osteogeny, formation of bone; development of the bone tissue. The bones are formed from various soft connective tissues (direct osteogenesis), or from the cartilaginous tissue (indirect osteogenesis); the activities of → osteoblasts and → osteoclasts are needed for the process. *See* bone.

osteoid, 1. pertaining to, or resembling bone; **2.** the organic matrix of young bone before calcification.

osteolysis (Gr. *lyein* = to dissolve), dissolution of the bone tissue caused by the activity of → osteoclasts. *Adj.* **osteolytic**.

osteomalacia (Gr. *malakia* = softness), an abnormal softening of bones caused by impaired calcification due to the deficiency of vitamin D or calcium and phosphates. Also called adult rickets.

osteomere (Gr. *meros* = part), any of a series of similar bone segments, such as a vertebra of the vertebral column.

osteon, also **osteone,** the Haversian system; a long structural unit of the compact bone comprising the central canal (→ Haversian canal) and concentric osseous lamellae arranged around it. *See* bone.

osteonectin, *see* fibronectin.

osteopenia (Gr. *penia* = poverty), abnormally reduced bone mass due to inadequate osteoid synthesis or decreased density or mineralization of the bone.

osteophage (Gr. *phagein* = to eat), → osteoclast.

osteoporosis (Gr. *poros* = passage, pore), the condition of reduced quantity of bone substance, leading to tendency to fracture, usually due to deficiency in oestrogen secretion, increased glucocorticoid secretion, or reduced motility. The condition is found in humans, especially in postmenopausal women, but may occur also in other vertebrate species.

osteosclereid (Gr. *skleros* = hard, *eidos* = form), a type of plant → sclereid.

ostium, pl. **ostia** (L. = entrance, mouth), any small opening or orifice between two cavities or leading to a tubular organ or canal; e.g. in vertebrate anatomy, **1. aortic ostium**, *ostium aortae*, the opening from the ventricle to the aorta, in a four-chambered heart from the left ventricle to the aorta; 2. **cardiac opening**, *ostium cardiacum*, the oesophagogastric orifice from the oesophagus to the stomach; **3.** *ostium pyloricum*, pyloric orifice from the stomach to the intestine; **4.** *ostium uteri*, the vaginal opening of the uterus. In invertebrate anatomy, **5.** any of several tiny openings for incoming water in the body of a sponge; **6.** one of several openings in the heart of arthropods through which blood flows from the pericardial sinus into the heart; **7.** any of numerous minute pores in the body surface of sponges for water incurrent.

Ostracoda (Gr. *ostrakon* = shell, *eidos* = form), **ostracods** (**mussel shrimps**, **seed shrimps**); a class (or a subclass of the class Maxillopoda) in the subphylum Crustacea; the animals have a mollusc-like, bipartite shell covering both the head and the body, the trunk is highly reduced in size and number of segments (considerably fused); only 2 or 3 pairs of legs are associated with the thorax. Ostracoda is a polymorphic group including ca. 2,000 species in five orders. They are widespread in both marine and freshwater habitats living on the bottom, or on plants; they are carnivorous, herbivorous, filter feeders, parasitic, or scavengers; some live in → plankton.

Ostracodermi (Gr. *derma* = skin), **ostracoderms**; an extinct group of jawless vertebrates (Agnatha) which occurred mainly in fresh waters from Ordovician to Devonian. They were the earliest vertebrates and perhaps the ancestral group to all jawless vertebrates. Some structures suggest relationship to the phylogeny of Gnathostomata. All ostracoderms were covered by a well-developed bony armour, and some earliest species lacked paired fins.

ostracods, → Ostracoda.

ostriches, → Struthioniformes.

ot(o)- (Gr. *ous*, gen. *otos* = ear), denoting relationship to the ear. *Adj.* **otic**.

otoconium, pl. **otoconia** (Gr. *konis*, *konia* = ashes, dust), → otolith.

otocyst (Gr. *kystis* = bladder, sac), **1.** the → statocyst of the vertebrate ear; **2.** the embryonic auditory sac of vertebrates.

otogenous, otogenic (Gr. *gennan* = to produce), originating in the ear.

otolith, also **otolite** (Gr. *lithos* = stone), earstone; a crystal particle of calcium carbonate located on the sensory receptors in the vesicles *utriculus* and *sacculus* of vertebrates; forms a principal part of the equilibrium organ of the internal ear. Otoliths are especially well developed in ray-finned fishes, being massive structures which fill almost the entire cavities of the two vesicles, in fishes serving also as the auditory organ. Annual rings can be seen clearly in a larger otolith, and the number of rings is used in the determination of the fish's age. *Syn.* statolith, otoconium, statoconium, otoconite. *See* equilibrium sense.

OTU, → operational taxonomic unit.

ouabain (Fr. *ouabaio* < Somalian *waba yo* = an African tree), a glycoside, $C_{29}H_{44}O_{12}$, obtained from the wood of trees of the genus *Acocanthera* or seeds of the shrub *Strophanthus gratus*, used as an arrow poison. Ouabain has a strong effect on animal cells by inhibiting specifically the function of the sodium-potassium pump (Na^+,K^+-ATPase) in the cell membrane. Especially in the vertebrate heart, calcium concentration is increased inside the muscle cells thus strengthening the contraction force. Ouabain is used in physiological and pharmacological research, and as a drug in cardiac insufficiency. Also called G-strophanthin (strophanthin G).

outbreeding, the crossing of unrelated individuals of a species; i.e. the union of gametes from unrelated individuals in → fertilization. *Syn.* **cross-breeding**, **cross-fertilization**. *Cf.* → inbreeding.

ov(o)- (L. *ovum* = egg), pertaining to an egg; egg-shaped; also **oo-**. *Adj.* **oval**.

ovalbumin (L. *album* = white), a major protein component (44 kD) of the egg white of birds, resembling serum albumin; its synthesis is stimulated by oestrogen. *Syn.* albumen.

oval window, vestibular window, fenestra of the vestibule (*fenestra ovalis*, *fenestra vestibuli*); one of the two membrane-covered openings between the middle ear (tympanic cavity) and the inner ear, leading into the vestibule; the auditory ossicle (stapes in mammals) is leaning against it. *See* ear.

ovar(i)-, ovario- (L. *ovarium* = ovary), denoting relationship to the ovary. *See* oophor(o)-.

ovarian follicle (L. *folliculus*, dim. of *follis* = bag), any of the cell aggregations in the vertebrate ovary containing the developing ovum with follicular cells around it. There are different developmental stages: the **primordial follicle**, consisting of an oocyte surrounded by a single layer of flattened follicular cells, the **primary follicle**, with one to several layers of roundish or columnar follicular cells, and the **secondary follicle (Graafian follicle)**, the mature ovarian follicle with accumulated fluid between the follicular cells, and therefore also called vesicular ovarian follicle. The accumulation of fluid finally forms a single cavity, antrum, pushing the mature ovum to the periphery of the follicle. During maturation the whole follicle pushes towards the surface of the ovary where → ovulation can occur.

The follicular cells produce hormones, mainly → oestrogens, and after ovulation the cells develop into a luteal gland, → corpus luteum. The maturation of the follicle is controlled by gonadotropins, chiefly → follicle stimulating hormone (FSH), and the ovulation by → luteinizing hormone (LH) and a prostaglandin; the number of simultaneously maturing ova and consequent ovulations varies from a few (in most mammals and birds) to hundreds and even thousands (in fishes and amphibians).

ovariole, any of the tube-like structures in the ovary of insects; the ovarioles are tapering egg tubes resembling a string of beads when filled with developing eggs.

ovarium, pl. **ovaria** (L.), → ovary.

ovary (L. *ovarium* < *ovum* = egg), **1.** the female reproductive organ (gonad) of animals; typically a paired organ (unpaired in birds and many invertebrates) which produces **ova** (eggs) and **female sex hormones**. During vertebrate embryogenesis, germ cells move from the the **germinal epithelium** of the primordial gland deeper into the stroma (connective tissue), where during sexual maturation some ova begin to develop into → ovarian follicles which also produce female sex hormones. The maturing follicles move to the surface of the ovary and one or two (e.g. primates), several (e.g. rodents), or hundreds or thousands of ova are ovulated at the same time (e.g. fishes and frogs); *see* oogenesis; **2.** in plants, the enlarged part of the carpel(s), containing the → ovule(s), developing after fertilization into a → fruit.

overdominance, a condition in which the genotypic value of the heterozygote *Aa* individual is better or bigger and the phenotype more fit than that of either homozygote *AA*, *aa* (A = dominant gene, a = recessive gene).

overexploitation, 1. an exploitation of a natural population or a community by other organisms (e.g. by man) at a rate greater than the population or the community is able to compensate with their own recruitment; the result of continuous overexploitation is extinction; **2.** the exploitation of environmental resources by an excessively grown animal population, exceeding the → carrying capacity.

overwintering, wintering, survival of an organism over the winter. The homoiothermic birds and mammals living in cold areas can overwinter either in **active** state, or in **inactive** or **resting** state such as some species in → **hibernation**. Active wintering includes several strategies, as do a well-adapted rhythmicity of long rest and short active periods, building of winter-nests, and the effective exploitation of snow as a heat-insulator, gathering of food supplies, and increasing of energy stores in the adipose tissue. Structural changes in the skin increase insulation of the fur and plumage, and in many species decreased basal metabolism saves energy. In a **group wintering** (winter aggregation) the individuals gather together, forming dense groups which better endure cold by means of social thermoregulation, e.g. the musk ox and some bird species, such as the long-tailed tit and penguins. Poikilothermic animals living in cold areas overwinter mainly in active state if the surroundings do not freeze, but otherwise in passive state, either in cold dormancy, → diapause, or torpor, or e.g. as an ovum or pupa.

The wintering type in plants is determined according to their life forms, i.e. according to the location of the buds of biennial or perennial plants; annual plants overwinter as seeds; *see* Raunkiaer's life forms.

ovicell, → ooecium.

oviduct (L. *ovum* = egg, *ductus* = tube), a tube of the female reproductive organs through which ova are transported from the ovary to the → uterus, or from the coelom to the exterior of the animal; often contains glands which secrete vitelline, albumin, and a covering or a shell around the eggs. In amniote vertebrates, the oviduct is embryonically developed from the → Müllerian duct. The mammalian oviduct is usually called **uterine tube** (*tuba uterina*) or Fallopian tube.

oviparous (L. *parere* = to bring forth), egg-laying; pertaining to an animal which lays either unfertilized eggs, or fertilized eggs whose embryonic development has not yet begun. The first group includes e.g. most invertebrates, fish, and amphibians which have external fertilization, the latter group comprises e.g birds and monotremes. *Cf.* ovoviviparous, viviparous.

ovipositor (L. *ponere* = to place), an organ in the terminal segment of the abdomen of female insects and some harvestmen, specialized for laying or positing eggs. The sting of honeybees is a modified ovipositor, present only in workers and the queen.

ovotestis (L. *testis* = testicle), the reproductive organ (gonad) of some hermaphroditic gastropods which produces both eggs and sperm.

ovoviviparous (L. *vivus* = alive, *parere* = to bring forth), pertaining to an animal which has internal fertilization and whose eggs, enveloped by membranes, hatch within the mother's body or immediately after extrusion from the parent; consequently the embryos or young usually get no nourishment from the mother. Ovoviviparous reproduction occurs in egg-laying reptiles, in cartilaginous and a few bony fishes, and in some invertebrates, such as many insects and snails. *Cf.* oviparous, viviparous.

ovulation (L. *ovulatio*), the release of an ovum from the ovary; in vertebrates, it is controlled by → gonadotropins (especially by luteinizing hormone, LH) and → prostaglandins. *See* ovarian follicle.

ovule (L. *ovulum,* dim. of *ovum* = egg), **1.** any egg-shaped structure; **2.** sometimes used for the ovum within the → ovarian follicle of the mammalian ovary; **3.** a structure in seed plants containing the female gametophyte and the egg cell, developing into a seed after fertilization. In → Gymnospermae the ovule is structurally simple and is situated naked on the carpel (cone scale). In → Angiospermae the ovules are in the → ovary (which develops into a fruit) and their structure is more complicated.

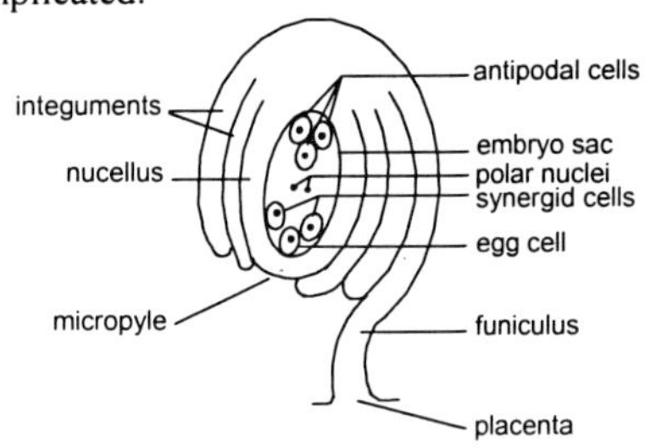

Typical plant ovule (anatropous type).

The main part is the → **nucellus**, surrounded by one or two **integuments** with an opening, → **micropyle**, on the tip. The stalk-like **funiculus** attaches the ovule to the → **placenta**. The most common type of ovule is **anatropous** in which the micropyle is turned downwards. In the **atropous** type the stalk is straight and the micropyle is directed upwards, and in the **campylotropous** type the micropyle is directed sidewards. Other ovule types, turned in different ways are **hemianatropous**, **amphitropous**, and **circinotropous** types.

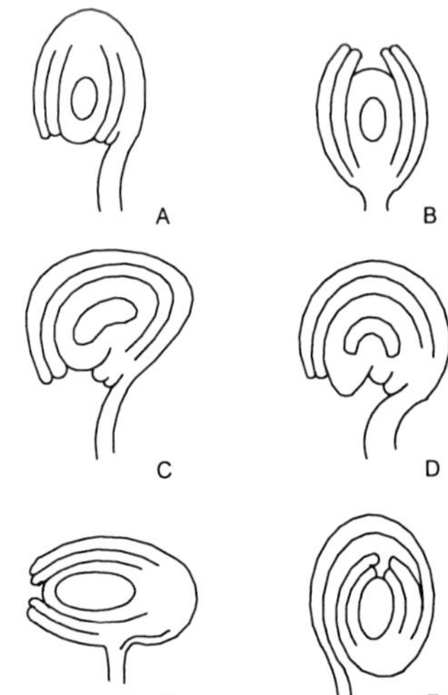

Ovule types in plants: anatropous (A), atropous (B), campylotropous (C), amphitropous

(D), hemianatropous (E), circinotropous (F).

ovum, pl. **ova** (L. = egg), **egg cell**; the female gamete of an organism when the gametes of the opposite sexes are different (*see* oogamy). The ovum is a specialized cell containing nutrients for the developing embryo whereas the male gamete (sperm cell, spermatozoon) is a motile cell fertilizing the ovum in sexual reproduction. The chromosome content of the ovum is usually → haploid, except in certain forms of → parthenogenesis, and the organization of the cytoplasm is very complicated containing e.g. → positional information. Sometimes the ovum is called macrogamete or megagamete. *See* egg, egg cell.

owls, → Strigiformes.

ox(o)- (L. *oxygenium* = oxygen < Gr. *oxys* = keen, quick, sour, acid), **1.** denoting addition of oxygen, containing oxygen; **2.** used to replace keto- in chemical nomenclature, usually in the form **oxo-**. *See* hydrox(y)-.

oxalic acid (L. *Oxalis acetosella* = wood sorrel), ethanedioic acid, an organic dicarboxylic acid, HOOC-COOH, found abundantly e.g. in rhubarb, spinach, and wood sorrel; its salts, esters, and ionic form are **oxalates**. Calcium reacts with oxalic acid forming oxalate crystals which accumulate in vacuoles of plant cells. The use of potassium fertilizers in large amounts increases the concentration of oxalic acid in fodder plants. High quantities of oxalic acid are harmful to animals because the acid easily crystallizes as oxalates in the kidneys.

oxaloacetic acid, a ketodicarboxylic acid, $COOHCH_2COCOOH$, which in the → citric acid cycle of cellular metabolism reacts with acetylcoenzyme A to form citric acid; plays also an important role in the carbon assimilation of C_4 plants and in gluconeogenesis.

oxidant, *see* oxidation.

oxidase, also **oxydase,** formerly, any enzyme that catalyses the oxidation of a substrate, i.e. belonging to enzyme class 1 (EC 1.) now named oxidoreductases; later used for enzymes which catalyse the reaction in which an oxygen molecule acts as an acceptor of hydrogen or electrons, while those oxidizing enzymes which remove hydrogen from a substrate are now called dehydrogenases. *See* oxidation.

oxidation, the act or process of oxidizing, a state of being oxidized; a chemical reaction in which a valence (oxidation state) of an atom increases by addition of oxygen, or by removal of hydrogen or one or more electrons, thus rendering the atom more electropositive, e.g. ferrous iron (II, Fe^{2+}) is oxidized into ferric state (III, Fe^{3+}). The electron acceptor reducing in the reaction is called **oxidant**. Oxidation processes take place in cells with the help of special enzymes or → **oxidases**, e.g. cytochrome oxidase or peroxidase. In the last step of cell respiration, hydrogen is oxidized when cytochrome oxidase tranfers hydrogen to molecular oxygen.

oxidation-reduction reaction, → redox reaction.

oxidative, pertaining to → oxidation; being capable of oxidizing.

oxidative burst, the marked increase of oxidative metabolism in cells, producing reactive oxygen intermediates (ROI), such as → oxygen radicals; in animals, occurs in → phagocytes (e.g. some leucocytes) following ingestion of particles. In plants, it is found under stress conditions triggering the expression of antioxidants in neighbouring cells, and limits the size of lesions.

oxidative phosphorylation (Gr. *phosphoros* = light-carrying), the synthesis of adenosine triphopsphate (ATP), in cell aerobic respiration in the presence of oxygen. ATP is a high energy phosphate produced in mitochondria in association with the stepwise reactions of the → electron transfer chain in which molecular oxygen serves as a final electron acceptor. In phosphorylation, the free oxidation energy is bound as chemical energy into ATP when inorganic ortophosphate radical (P_i) reacts with adenosine diphosphate (ADP). The energy required in the reaction comes from the proton (H^+) gradient across the inner membrane of the mitochondria. The gradient is produced in the reactions of the electron transfer chain generating protons. The principal enzyme is **ATP synthase** in the mitochondrial inner membrane. It is a bipartite enzyme composed of two oligomeric complexes (**coupling factors**, F), i.e. a transmembrane protein F_0 acting as a proton channel, and an epimembrane protein, F_1, acting as ATP synthesizing unit. ATP synthase, also called F_0F_1-ATPase or H^+-ATPase, couples proton movement down its electrochemical gradient with the synthesis of ATP from ADP and P_i. In aerobic cells, oxidative phosphorylation produces the principal part (about 90%) of the total ATP energy. *See* citric acid cycle.

oxidative stress, a state in which abnormally high concentrations of free oxygen radicals are formed due to increased oxygen content in tissues, and may cause a condition of **oxygen poisoning** (oxygen toxicity). In man, this may occur when pure oxygen or compressed air is respired, as e.g. in diving, in deficiency of cellular antioxidants, or in reoxygenation of tissues which have suffered from oxygen deficiency. The poisoning effect is worst in the lung epithelium and in the nervous system as a whole.

oxide, a chemical compound of oxygen formed when oxygen reacts with another element or a radical; e.g. inorganic oxides, such as → carbon monoxide (CO), carbon dioxide (CO_2), nitrous oxide (N_2O), nitric oxide (NO), sulphur dioxide (SO_2), and sulphur trioxide (SO_3), or organic oxides, as e.g. dimethyl sulphoxide ($(CH_3)_2SO$), ethylene oxide (CH_2-CH_2O), and propylene oxide (CH_3CHCH_2O). Oxides form acids in aqueous solutions.

oxidize, to cause an element or radical to combine with oxygen or to lose electrons.

oxidoreductases, a class of enzymes (EC 1.) catalysing oxidation-reduction reactions in cells and tissues; the enzymes transfer electrons from one compound (electron or hydrogen donor) which is oxidized, to another (electron or hydrogen acceptor) which is reduced; the reactions are reversible. The class includes e.g. → dehydrogenases, hydroxylases, oxidases, oxygenases, peroxidases and reductases, and transhydrogenases.

oxime, an organic compound formed from hydroxylamine (H_2NOH) and an aldehyde or ketone; contains the group =NOH.

oxine, → 8-hydroxyquinoline.

oxonium ion, an ion containing tetravalent basic oxygen, R_3O^+, where R is an organic group or hydrogen; e.g. hydroxonium ion, H_3O^+, or trimethyloxonium, $(CH_3)_3O^+$. *See* protolysis.

oxy- (L. < Gr. *oxys* = keen, quick), **1.** sharp, keen, quick; **2.** acid; **3.** (L. *oxygenium* = oxygen), pertaining to the presence of oxygen, either added or substituted; also oxi-, oxo-.

oxygen (L. *oxygenium*), a gaseous chemical element, symbol O, atomic weight 15.9994, boiling point -183°C. An oxygen molecule (O_2) is very reactive due to its free electrons on the outermost ring. Common oxygen, i.e. molecular oxygen, O_2, is converted into **ozone** (O_3) by electrical discharges and UV radiation. *Antoine Lavoisier* (1743—1794) proved that air contains 21% oxygen which maintains burning, and he named the gas oxygenium.

Oxygen is the most common element of the Earth found both in organic and inorganic substances, e.g. in water and rocks. It is necessary for the respiration and metabolism of nearly all organisms. Photosynthesis carried out by plants produces oxygen into air by cleaving water which is used in the light reactions of photosynthesis. This oxygen is then used by most organisms, such as animals, producing water in aerobic respiration when oxygen reacts with hydrogen. Some organisms can permanently or temporarily survive without oxygen by means of → anaerobic respiration. *See* oxygen radicals.

oxygenases, a group of oxidizing enzymes (direct oxidases, EC subclass 1.13.) of the → oxidoreductase class (EC 1.) catalysing direct incorporation of oxygen to an organic substrate; e.g. an oxygenase catalyses the breaking of double bonds of an aromatic ring (e.g. in aromatic amino acids) and binding of oxygen to the ring. **Monoxygenases** catalyse incorporation of one oxygen atom, the other becomes reduced into water; **dioxygenases** catalyse incorporation of both oxygen atoms of its molecule. For example, phenylalanine hydroxylase catalyses the hydroxylation of phenylalanine into tyrosine, and → ribulose 1,5-bisphosphate carboxylase-oxygenase acts in → photosynthesis.

oxygenation, binding of oxygen loosely and temporarily to a substance so that the valence of a reacting atom (or atoms) remains unchanged in contrast to oxidation. *See* oxyhaemoglobin.

oxygen debt, the state of oxygen deficiency in animals in which the concentration of lactic acid is increased in tissues capable of → anaerobic respiration. In vertebrates, the condition is generated especially in skeletal muscles, resulting in weakening of the muscular work. At rest, oxygen debt is paid back quite quickly by effective breathing when some of the lactic acid is oxidized in → aerobic respiration and some converted in the liver into glucose (→ gluconeogenesis).

oxygen poisoning, *see* oxidative stress.

oxygen radicals, an oxygen atom, oxygen molecule, or an atom group which contains an oxygen atom with odd electrons on its outermost electron shell. These → free radicals are e.g. → **singlet oxygen** ($^1O_2$), **superoxide radical** ($O_2^{\cdot-}$), and **hydroxide radical** ($OH^{\cdot}$), e.g. forming → hydrogen peroxide (H_2O_2). They originate e.g. from → ozone (O_3) or are produced from two-atomic oxygen (O_2) by ionizing radiation and combustion processes. In organisms some oxygen radicals are normally generated in reactions of the → electron transfer chain, and the quantity of radicals in cells and tissues increases in parallel with the increase of partial pressure of oxygen, by ionizing and UV radiation, or by the increase of certain substances, like iron. Oxygen radicals are very short-lived but strong oxidizers and harmful to organisms. Cells can defend themselves against these radicals by means of **antioxidants** which destroy them immediately. Some cells, such as white blood cells, can use oxygen radicals to destroy invasive microbes.

oxyhaemoglobin, HbO_8, Am. **oxyhemoglobin,** the oxygenated form of → haemoglobin having oxygen (O_2) covalently bound in the → haem of the molecule, but without the change of the ferrous state into ferric state of iron which occurs in oxidation. Oxyhaemoglobin is involved in oxygen transport in red blood cells giving the arterial blood its bright red colour.

oxymyoglobin, the oxygenated form of → myoglobin occurring in muscle cells; analogous to → oxyhaemoglobin.

oxyntic (Gr. *oxynein* = to make acid, sharpen), forming acid; e.g. denoting the parietal cells (oxyntic cells) of the vertebrate gastric glands.

oxyphilic, oxyphil(e), oxyphilous (Gr. *philein* = to love), acidophilic, → acidophile; **1.** having affinity for acid dyes; **2.** pertaining to certain cells, as e.g. eosinophilic leucocytes (→ granulocytes), which are stained by acid dyes.

oxytocin, OXT (Gr. *oxys* = quick, *tokos* = birth), a nonapeptide hormone secreted from the posterior lobe of the → pituitary gland (neurohypophysis) of mammals; stimulates myometrial contractions of the uterine wall at the end of pregnancy, and enhances milk ejection from the mammary gland.

oxyuriasis, oxyuriosis, infection caused by parasitic nematodes of the family Oxyuridae; the fairly harmless parasites live in the intestine of many vertebrate and invertebrate species, e.g. *Oxyuris equi* (the horse pinworm), a common parasite in the large intestine of horses all over the world, or *Enterobius vermicularis* in humans.

ozone (Gr. *ozein* = to smell), a triatomic allotropic form of oxygen, O_3, boiling point —112°C; is chemically very active being a powerful oxidizing agent. Ozone is generated from the common two-atomic oxygen (O_2) by electrical discharges and UV radiation.

Ozone forms an important gas zone (**ozone layer**, **upper ozone**, **ozonosphere**) in the → stratosphere absorbing effectively → UV radiation and thus preventing its excessive access onto the surface of the Earth. → Chlorofluorocarbons (Freons) released into the atmosphere by human activity, destroy the ozone layer especially in polar areas. With some other pollutants (e.g. nitrogen oxides) chlorofluorocarbons cause a serious threat to organisms, due to increased UV radiation. The quantity of ozone in the stratosphere can be indicated as the thickness of the layer which ozone would form above sea level at normal atmospheric pressure: the unit in use is **Dobson**, one Dobson being 10^{-3} cm. Normal values are 300 to 400 Dobsons but during most critical periods (in the spring) the value may decrease especially over the polar regions to 140—150 Dobsons, the decrease of ozone thus being more than 50% from the normal level.

Ozone is also generated by incomplete combustion processes, like in car engines, and the quantity of ozone (**earth ozone, lower ozone**) may reach harmful levels in polluted areas on sunny days with high UV radiation. Ozone is destructive to organisms because it produces reactive → free radicals which damage cell structures; the respiratory organs are particularly sensitive to ozone. Excessive earth ozone may cause the death of coniferous trees, and decreases the yields of many cultivated plants. Ozone is used e.g. as an oxidant in industry.

P

P, symbol for **1.** phosphorus, or phosphate group; **2.** (partial) pressure; **3.** proline; **4.** protein; **5.** panmictic index; **6.** peta- (10^{15}); **7.** parental generation.

p, pico-, 10^{-12}, one billionth part (Am. one trillionth).

p, **1.** chemical symbol for *par(a)-*; **2.** symbol of the short arm of a (human) chromosome.

P_{660}, P_r, *see* phytochromes.

P680, the reaction centre (the primary electron donor) of → photosystem II. *See* photosynthesis.

P700, the reaction centre (the primary electron donor) of photosystem I. *See* photosynthesis.

P_{730}, P_{fr}, *see* phytochromes.

PABA, → para-aminobenzoic acid.

pacemaker, anything that sets the pace; e.g. **1.** a cell or group of cells inducing the rate of activity in other cells or in an organ, e.g. the → sinus node in the vertebrate heart, or the → pacing area in the anterior end of the vertebrate stomach; **2.** an apparatus inplanted beneath the skin for providing the normal heartbeat by electrical stimulation.

pachytene (Gr. *pachy-* < *pachys* = thick, *tainia* = strand, string), a stage of → meiosis.

pacing area, a pacemaker area in a cardiac or smooth muscle inducing impulses (action potentials) which then are propagated to surrounding muscle cells and result in the contraction of the entire muscle, or a part of it. The cells of the pacing area are specialized muscle cells, in which sodium ions (Na^+) slowly leak into the cells causing the resting potential to change to the threshold level, and thus cause the induction of an action potential. This process is repeated either regularly as in the heart, or irregularly as is usual in the smooth muscle. *See* conduction system of the heart.

Pacinian corpuscle, Pacini's corpuscle, Vater—Pacini corpuscle, Vater's corpuscle (*Abraham Vater*, 1684—1751; *Filippo Pacini*, 1812—1883), sensory bodies (*corpuscula lamellosa*) concerned in perception of pressure in the skin, periostea, and synovial capsules throughout the body of vertebrates. The corpuscle consists of concentric layers of connective tissue encapsulating a nerve ending, which reacts shortly to the bending of lamellae being thus sensitive to pressure and vibrations.

packing ratio, the ratio of the length of a DNA molecule to the length of the chromosomal DNA strand containing it.

paed(o)-, ped(o)-, also **paid(o)-** (Gr. *paid-* < *pais* = child), denoting relationship to a child or an offspring.

paedogamy (Gr. *gamos* = marriage), a form of → automixis (obligatory self-fertilization) characterized by the fusion of isogametes or anisogametes produced by one single individual. Either a parental meiocyte (gamont, gametangium) produces directly copulating gametes in → meiosis, or the individual carries male and female sexual organs (antheridia and oogonia), producing gametes which fuse in self-fertilization.

paedogenesis, also **pedogenesis** (Gr. *gennan* = to produce), a form of → heterochrony leading to paedomorphosis; the evolutionary juvenilization process in which the reproductive organs (but not somatic tissues) undergo accelerated development becoming mature at the precocious age, i.e. in a larval or pupal stage as occurs e.g. in many insects (some mites, gall midges, wingless aphids, etc). Paedogenesis has been an important factor in evolution of new taxa. *Syn.* progenesis. *Cf.* neoteny.

paedomorphosis, Am. **pedomorphosis** (Gr. *morphe* = form), a phenomenon found in the development of some species in which the features of the juvenile (infantile) stage still persist in adulthood; e.g. the evolutionary retention of juvenile characters in adult animals, produced by processes called → **paedogenesis (progenesis)** and → **neoteny**. According to the **paedomorphosis theory**, the evolution of some groups of animals has so occurred that larval stages have directly differentiated into adults; e.g. in insects or chordates. *Adj.* **p(a)edomorphic**.

PAGE, polyacrylamide gel electrophoresis. *See* electrophoresis, acrylamide.

pain (sense) (Gr. *poine* = penalty), a sense producing an avoidance reaction in animals, and in higher animals the feeling of discomfort, distress, or agony. Pain receptors (**nociceptors**) are usually only slightly differentiated nerve endings becoming irritated by many types of stimuli, which may be mechanical, chemical, or thermal (heat or cold). They are most abundant in the skin, fewer in the membranes surrounding the organs, with only a few inside the internal organs and none in the brain tissue.

Although the quality of the pain sensation in different animal groups (especially in invertebrates) is difficult to assess, the pain sense shows primitive features and is apparently evolutionarily old and involved in the avoidance behaviour of animals. In vertebrates, different pain types are grouped according to the location of the receptors: **superficial pain** (from skin), **deep pain** (from muscles, bones and tendons) and **visceral pain** (from internal organs), and in man according to the quality of the pain, e.g. ache (dull or slight), smart, burning pain, etc. Many higher vertebrates like humans can suffer also from **psychic** (psychogenic) and **psychosomatic** pain.

pain pathways, the nerve pathways from the pain receptor (nociceptor) to the brain. In vertebrates it is formed from two successive neurones, the first of which leads from the receptor to the spinal cord or the medulla oblongata, and the second to the thalamus. Some pain tracts have yet a third neurone continuing to the cerebral cortex. It gives more precise sensations from certain skin areas, e.g. from the face.

pair-rule genes, the → interpretation genes of a zygote which read positional information created by the → coordinate genes downstream of the → gap genes. The pair-rule genes regulate the formation of segments and somites, and in insects their mutations cause the loss of either odd or even-numbered segments.

palae(o)-, pale(o)-, also **palaio-** (Gr. *palaios* = old, ancient), denoting old, ancient, early, primitive, archaic.

Palaearctic region, Palearctic region (Gr. *arktikos* = northern), a zoogeographical region including Europe, North Africa, western Asia, Siberia, northern China and Japan; the Palaearctic region is also considered as a subarea of the Holarctic region. *Cf.* Nearctic region.

palaeobotany, paleobotany, the study of fossil plants and plant remains.

Palaeocene, Paleocene (Gr. *kainos* = new), the initial epoch of the Tertiary period preceding Eocene, 66 to 58 million years ago, during which the first placental mammals appeared on the Earth.

palaeoclimatology, paleoclimatology (Gr. *klima* = climate, *logos* = discourse), a branch of science studying the climate of ancient ages.

palaeocortex, paleocortex (L. *cortex* = park, rind), the evolutionarily old → cerebral cortex of vertebrates, comprising chiefly the olfactory cortex. *Cf.* neocortex.

palaeoethnobotany, paleoethnobotany (Gr. *ethnos* = people), the study of the ancient relationship between man and plants.

Palaeogaea, Paleogea (Gr. *ge* = earth), a zoogeographical area comprising Palaearctic, Ethiopian, Oriental, and Australian regions. *Cf.* Arctogaea, Neogaea, Notogaea. *Adj.* **Palaeogaean, Paleogean**.

Palaeogene, Paleogene (Gr. *genesthai* = to be born), the geological time period in the early Tertiary 66 to 24 million years ago, including the epochs → Palaeocene, Eocene, and Oligocene. *Adj.* **Palaeogene**, pertaining to this period. *Cf.* Neogene.

palaeogeography, paleogeography (Gr. *graphein* = to write), a branch of geography examining structures of the earth in ancient times.

Palaeognathae (Gr. *gnathos* = jaw), **palaeognathous birds**; a superorder of birds with primitive features in the structure of the bony palate of the skull resembling those of archosaurians. The extant species are classified as several orders: Tinamiformes (**tinamous**) with a keeled sternum, and Struthioniformes (**ostriches**), Rheiformes (**rheas**), Casuariiformes (**cassowaries** and **emus**), and Dinornithiformes (**kiwis** and extinct **moas**) which have secondarily lost the keel of sternum and therefore are often called by the group name **ratites**. *Cf.* neognathae.

palaeontology, paleontology (Gr. *onta* = existing things, *logos* = word, discourse), a branch of science that studies the life of ancient times, dealing with fossils and fossil traces; gives information e.g. about phylogeny and relationship of modern animals. Also called oryctology.

palaeospecies, paleospecies, an extinct species only known as fossils..

Palaeotropical kingdom, Paleotropical kingdom (Gr. *tropikos* = tropics), a → floristic kingdom comprising the tropical areas of the Old World.

Palaeozoic, Paleozoic (Gr. *zoe* = life), the era about 570 to 245 million years ago, succeeding Precambrian. In the course of the Palaeozoic era the vegetation developed from algae into great tree ferns, and also some primitive seed plants appeared. Owing to the vegetation, the quantity of oxygen in the atmosphere increased and made it possible for other organisms to move from the sea to a terrestrial environment. Due to the increase of hard-shelled species many fossils are found from this era showing that invertebrates, later also vertebrates, were common. The Palaeozoic era is divided into → Cambrian, Ordovician, Silurian, Devonian, Carboniferous, and Permian periods. *Adj.* **Palaeozoic**, pertaining to the Palaeozoic era.

palaeozoology, paleozoology (Gr. *zoon* = animal, *logos* = word, discourse), a branch of science dealing with ancient and fossil animals and their traces.

palatal arches, palatine arches (L. *palatum* = palate), the oral arches of higher vertebrates formed by the anterior and posterior folds, i.e. palatoglossal and palatopharyngeal arches at the back roof of the mouth, where the palatine tonsils situate between the arches. In air-breathing vertebrates, the palatine velum (*velum palatinum*) is located in the middle of the anterior arch, the musculous palatine uvula (*uvula palatinum*) hangs in the middle. The palatal arch is also called the maxillary arch or the palatomaxillary arch.

palate (L. *palatum*), the partition that separates the oral and nasal cavities of vertebrates, consisting of the hard palate locating anteriorly, and of the soft palate posterior to it.

palatine tonsil (L. *tonsilla palatina*), either of the two small organs between the folds of the → palatal arch of higher vertebrates, consisting of lymphoid tissue.

pale(o)-, → palae(o)-.

palea, pl. **paleae** (L. = chaff, straw), the upper of the two bracts enclosing an individual flower (floret) in the spikelet of grasses.

palindrome (Gr. *palindromos* = running backwards), in genetics, a DNA sequence in which the → nucleotide sequence is the same when read in either direction.

palisade parenchyma (French *palissade* = pole fence < L. *palus* = stake), a cell layer or layers of the leaf mesophyll; the cells are long and thin, situated vertically under the upper epidermis, being specialized for photosynthesis.

pallium, pl. **pallia** or **palliums** (L. = mantle), **1.** the cerebral cortex (*cortex cerebri*) of the vertebrate brain; especially refers to evolutionary stages of the → telencephalon; **2.** the mantle of brachiopods and molluscs.

palmitic acid, hexadecanoic acid; a saturated C-16 fatty acid, $CH_3(CH_2)_{14}COOH$, found in most plant and animal lipids, such as palm oil, butter fat, and tallow; solid at room temperature. The **palmityl residue** is a common acyl residue in phospholipids of cellular membranes of many organisms. Its ionic form, esters, and salts are **palmitates**, and the corresponding fat, **palmitin** (tripalmitin), is a triacylglycerol. *See* stearic acid.

palp (L. *palpus* = soft palm of the hand, stroking), a sensory appendage; a thread-like sensory organ found in many invertebrates, e.g. the two-jointed palps on the sides of prostomium in polychaetes, the paired labial palps (*palpi labiales*) in the mouth of insects, or the maxillary palps (*palpi maxillares*) in the → pedipalps of arachnids. *See* mouthparts.

palpation (L. *palpare* = to touch, stroke), the examination by touch with fingers. *Verb* to **palpate**. *Adj.* **palpable**, perceptible by touch.

palsa (Finn.), a peat-coated mound with a permanent ice core, found in subarctic mires. The palsas may be large and several metres high.

palynology (Gr. *palynein* = to scatter, *logos* = word, discourse), the study of → pollen and spores whether living or fossil.

pampa (Sp.), Argentinean → steppe. *Cf.* grassland.

pan(o)- (Gr. *pan* < *pas* = all, every), denoting all, entire, completely, general.

panchromia (Gr. *chroma* = colour), the condition of staining with various dyes.

pancreas, pl. **pancreata** (L.< Gr. *pankreas* < *kreas* = flesh), an elongated compound gland of vertebrates located caudally from the stomach. The greater part of the tissue is **exocrine** (*pars exocrina*) consisting of ducts and **secretory alveoles** (pancreatic acini) with some loose connective tissue around them. The acinar cells secrete pancreatic juice containing different types of digestive enzymes (→ amylase, lipase, trypsinogen, etc.) and sodium hydrogen carbonate ($NaHCO_3$) which is important in neutralizing gastric acid in the intestine. The excretion passes through the converging duct system and finally through the pancreatic ducts into the initial section of the small intestine, i.e. the duodenum in mammals. A small portion of the pancreatic tissue forms the **endocrine part** (*pars endocrina*), i.e. small endocrine cell groups, called → **pancreatic islets**, which secrete hormones. *Cf.* hepatopancreas.

pancreatic islets, islets of Langerhans; any of the **endocrine** cell groups scattered throughout the pancreatic tissue of vertebrates; in primates, the islets comprise about 2% of the total pancreatic tissue mass. There are three types of hormone-secreting cells: beta cells (B cells) producing → **insulin**, alpha cells (A cells) secreting → **glucagon**, and delta cells (D cells) producing → **somatostatin**. Insulin and glucagon participate in the regulation of → glucose balance; somatostatin has various effects, but in the pancreas itself it inhibits the secretion of insulin and glucagon. Also a fourth hormone, called **pancreatic polypeptide**, is found to be secreted by pancreatic cells (at least in some mammals). Its secretion is also inhibited by somatostatin, but its physiological function is unknown.

pancreozymin (Gr. *zymosis* = fermentation), → cholecystokinin.

Paneth's cell (*Josef Paneth*, 1857—1890), a cell type in the intestinal glands (crypts of Lieberkühn) of vertebrates; these exocrine, columnar epithelial cells, containing large secretory granules, secrete lysosome (an antibacterial enzyme) and probably a peptidase.

Pangaea (Gr. *pan* < *pas* = all, *ge* = earth), an ancient supercontinent formed about 340 million years ago, comprising all the present continents which began to separate about 200 million years ago forming first Gondwanaland and Laurasia.

pangenesis theory (Gr. *genos* = birth, origin), a now discarded theory stating that each part of the body produces its representative factors (gemmules, pangenes) which are transferred to germ cells to mediate inheritance to following generations.

pangolins, → Pholidota.

panicle (L. *panicula,* dim. of *panus* = tuft), a compound → raceme, common in the family Poaceae. *See* inflorescence.

panmictic index (Gr. *pan* = all, *miktos* = mixed), an index complementing the → inbreeding coefficient *F*, indicating the number of random pairing in a population; its symbol is *P*, and it obeys the position of dependence, $P = 1—F$.

panmictic unit, a population in which random mating (panmixis) completely prevails; the smallest unit by the individual number among → Mendelian populations. *Cf.* deme, race, species.

panmixis, panmixia, pl. **panmixes, panmixias** (Gr. *mixis* = act of mingling), random mating among individuals in an interbreeding population. *Adj.* **panmictic**.

panoistic (Gr. *oion* = egg), pertaining to ovaries in which nutritive cells are absent and the nutrients and → morphogens are produced by the action of the genome of the ovum itself; e.g. amphibian ovaries. *Cf.* meroistic.

Pantopoda, pantopods, → Pycnogonida.

pantothenic acid, a water-soluble vitamin (B_5), the constituent of → coenzyme A which is a crucial substance in the metabolism of all cells; its ionic form, esters, and salts are called **pantothenates**. *See* vitamin B complex.

pantropic (Gr. *tropos* = a turning), **1.** affecting many types of tissues; used chiefly to refer to viruses; **2.** → pantropical.

pantropical, also **pantropic,** distributed or existing throughout tropical regions; e.g. pantropical distribution, pantropical plants and animals.

papain, a proteolytic enzyme, endopeptidase (EC 3.4.22.2), occurring in the fruits and leaves of the papaya tree, papaw (*Carica papaya*); used commercially for tenderizing meat and in medicine as a digestant.

papaverine (the plant genus *Papaver* = poppy), a water-insoluble, non-narcotic alkaloid, $C_{20}H_{21}NO_4$, obtained from → opium, not a morphine derivative. It is usually used as hydrochloride in medicine as a smooth muscle relaxant.

paper chromatography, *see* chromatography.

papilla, pl. **papillae** (L. = nipple), a small nodule, node, or projection; e.g. **1.** the nipple of the breast (*papilla mammae*); **2.** the root papilla of a hair (*papilla pili*) or a feather (*papilla ptili*); **3.** the papillae on the surface of the vertebrate tongue (*papillae linguales*); **4.** the bud of a developing tooth (*papilla dentis*); **4.** in botany, a small outgrowth of epidermal cells, common especially in petals. *Adj.* **papillary**.

papillary muscles (L. *musculi papillares*), the finger-like muscular projections from the inner walls of the heart ventricles to the cusps of the atrioventricular valves attached to them with tendinous cords (*chordae tendineae*); the muscles regulate the closing of the valves. There are anterior, posterior, and septal papillary muscles in the ventricles.

pappus, pl. **pappi** (L. < Gr. *pappos* = grandfather, down of the chin), a parachute-like ring of bristles or hairs formed from the calyx in flowers during seed ripening, common in the family Cichoriaceae, e.g. in dandelion; helps the distribution of seeds by wind.

papula, pl. **papulae** (L.) **1.** a small, conical, solid elevation of the skin, pimple or **papule**; **2.** a soft projection of the coelomic cavity of some echinoderms, such as sea stars, acting as external gills (dermal branchia). *Adj.* **papular**.

PAR, → photosynthetically active radiation.

par(a)- (Gr. *para* = beyond, beside), **1.** denoting adjacent, beside, by, outside of, parallel, pseudo-, sham; **2.** in a chemical compound designating two substituents linked to opposite carbon atoms in the benzene ring, abbr. *p*.

***para*-aminobenzoic acid, *p*-aminobenzoic acid, PABA,** an organic acid which is essential for the synthesis of folic acid in many organisms; is regarded as a member of the vitamin B group necessary e.g. for rats and chicks, and is essential for bacterial cells. It absorbs UV light and is therefore used for protection of the skin against solar radiation. Also called aminobenzoic acid.

***para*-aminosalicylic acid, *p*-aminosalicylic acid, PAS,** an analogue for → *para*-aminobenzoic acid (PABA) preventing the growth of some bacteria (e.g. tubercle bacillus) by inhibiting their folic acid synthesis. Also called aminosalicylic acid.

parabiont (Gr. *bioun* = to live), **1.** any organism living in close relationship with another organism, i.e. in **parabiosis**; e.g. a member of two or more species living in close relationship with each other, generally maintaining separate colonies, e.g. some ant species living together, or a fungus growing on a lichen; parabionts are often parasites but may also be symbionts; **2.** an individual formed from the fusion of whole eggs or embryos, or one of the two individuals conjoined with each other allowing mixing of their body fluids, such as Siamese twins or experimental animals joined through the vascular systems by a surgical operation. *Adj.* **parabiotic**.

parabiosis, *see* parabiont.

parabronchus, *see* lung.

paracentral, paracentric (Gr. *kentron* = sharp point, centre), located near the centre, e.g the area around the fovea centralis in the retina; in genetics, paracentric describes an → inversion that does not involve the → centromere region of the chromosome. *Cf.* pericentral.

paracrine (Gr. *krinein* = to separate), pertaining to a type of secretion of certain endocrine cells in which a messenger substance produced acts only locally in the vicinity of a secreting cell (**paracrine cell**); in animals such substances are e.g. → prostaglandins and growth factors.

paracytic stoma, pl. **paracytic stomata** (Gr. *kytos* = cavity, cell), one of the stoma types in a plant. *See* stoma.

paradidymis, pl. **paradidymides** (Gr. *didymos* = testis), a group of convoluted tubules in the cranial end of the spermatic cord of amniotic male vertebrates; thought to be a remnant of the mesonephros.

paradigm (Gr. *paradeigma* = prototype, model, example), a set of forms; the set of inflectional forms of a word; in science, the framework within which the observations, hypotheses, and theories become comprehensible.

paraffin (Gr. < L. *parum* = little, *affinis* = related), **1.** one of the series of acyclic hydrocarbons, → **alkanes**; **2. paraffin wax**, hard paraffin; a mixture of saturated hydrocarbons, i.e. of higher alkanes; a white, odourless, tasteless, water-insoluble substance processed from long chain hydrocarbons of petroleum. Paraffin wax is a chemically inert substance, solid at room temperature and liquid in the range 50°—60°C; used e.g. as an embedding material for biological specimens in microscopy.

paraformaldehyde, a solid polymer of → formaldehyde, $(HCOH)_n$; used as a fixative in microscopy.

paraganglion, pl. **paraganglia,** any of several groups of nervous chromaffin cells (chromaffin bodies) along the aorta and its large branches near the sympathetic nerve trunk in vertebrates. Like the adrenal medulla, most paraganglia secrete adrenaline and noradrenaline. The abdominal paraganglia near the bifurcation of the aorta are called **Zuckerkandl's bodies**.

paraglossa, pl. **paraglossae** (L. *glossa* = tongue), one of the pair of lateral appendages in the labium of insects; together with the →

glossa it forms the *ligula*. *See* mouthparts.

paralectotype (Gr. *lektos* = chosen, *typos* = type), *see* type specimen.

parallel evolution, the evolution of two or more species or groups that are closely related and are developing in the same direction usually in a similar environment. *Cf.* convergent evolution, divergent evolution.

parallel venation, a type of plant leaf venation with the large veins running parallel through the leaf; occurs in most monocotyledons, e.g. in grasses. *See* venation types.

paralogous, 1. pertaining to an anatomical similarity having no phylogenetic or functional implication; *noun* **paralogy**; **2.** pertaining to the genes which are located in parallel loci of the homologous gene complexes in different species, and also to genes within the genome that have arisen from duplication processes; *cf.* homologous. *See* gene complex.

paralysis, pl. **paralyses** (Gr. *lyein* = to loosen), an impairment or loss of the functions of skeletal muscles (motor paralysis) and/or of sensory functions (sensory paralysis), caused e.g. by traumatic, microbic, or toxic factors. *Verb* to **paralyse** (Am. paralyze). *Adj.* **paralytic**.

paramylum, paramylon (L. *amylum*, Gr. *amylon* = starch), a starch-like carbohydrate assimilated by some organisms, such as some algae and green flagellates.

parapatric (Gr. *patra* = fatherland), pertaining to species or other taxa whose distribution areas meet each other but are not overlapping, or overlap only to a minor degree; e.g. the populations of the hooded crow (*Corvus corone cornix*) and the carrion crow (*C. c. corone*) in Central Europe. *Noun* **parapatry**. *Cf.* allopatric, peripatric, sympatric.

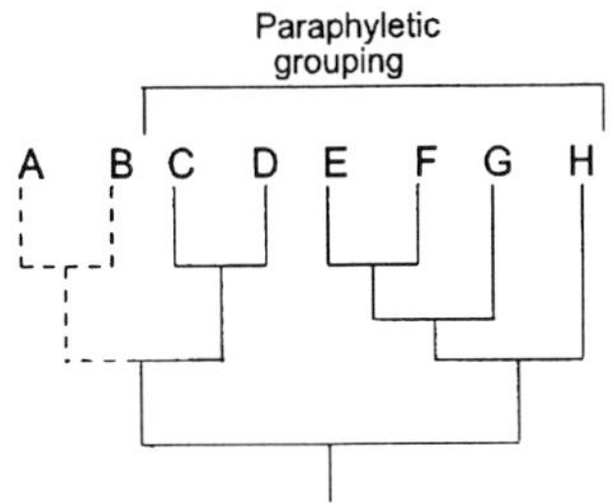

Paraphyletic relationships between eight taxa (A-H).

paraphyletic (Gr. *phyle* = tribe), pertaining to a group of organisms or a taxon which contains the most recent common ancestor of all members of the group (or taxon) and some but not all of its descendants; e.g. the class Reptilia is a paraphyletic taxon because it contains a theoretical ancestor of reptiles and most of its descendants except birds and mammals. *Cf.* monophyletic, polyphyletic.

paraphysis, pl. **paraphyses** (Gr. *physis* = growth, origin), **1.** median evagination of the roof tissue of the telencephalon of the embryonic vertebrate brain, located anteriorly to the pineal body. The developing paraphysis degenerates in higher vertebrates during embryonic or foetal development but remains in certain lower vertebrates; its physiological purpose is not known; *syn.* paraphyseal body; **2.** in botany, one of the sterile threads occurring between gametangia in reproductive organs of many algae, mosses and fungi.

parapodium, pl. **parapodia** (L. < Gr. *pous* gen. *podos* = foot), **parapod**; **1.** a paired muscular projection located segmentally in many polychaetes serving locomotion; comprises two lobes, a dorsal **notopodium** and a ventral **neuropodium**, both provided with bristle-like **setae** and a **cirrus**, as well as a needle-like chitinous **aciculum** supporting the parapodium; **2.** a lateral projection on both sides of the foot forming a swimming organ in opisthobranch molluscs.

paraproct (Gr. *proktos* = anus), a paired lobe or plate beside the anus in some diplopods and several insects.

parapsid (Gr. *apsis* = arch), describing a skull type of some extinct reptiles, e.g. plesiosaurs, having one dorsal temporal opening behind the postorbital bone on each side. *See* anapsid, diapsid, synapsid.

parasexual cycle, *see* parasexuality.

parasexuality, 1. a life cycle in fungi (parasexual cycle) occurring side by side with the sexual system producing → genetic recombination otherwise than by → meiosis and fertilization; includes the fusion of two haploid heterokaryotic nuclei forming a diploid heterozygous nucleus. This nucleus divides in → mitosis, during which → crossing-over exceptionally occurs, and haploidization (*see* haploid) through mitotic non-disjunction or chromosome extrusion takes place. The process produces non-sexual spores which are genetically different from the original mycelium; **2.** physical mapping of genes by means of genetic study of somatic cell hybrids *in vitro*. *Adj.* **parasexual**.

parasite (Gr. *sitos* = bread, food), an organism which lives at the expense of another living organism, either a plant or an animal (the host), over an extended part of its life cycle. A parasite may live upon a host (→ **ectoparasite**), or within it (→ **endoparasite**), and may harm the host to various degrees, but seldom kills it. A parasitic species may have one, a few, or many definitive host species; in the last case the interrelations between the parasite and the hosts are rather loose and less specialized. A parasitic individual can also have several → intermediate hosts in the course of its life cycle. *See* parasitism, parasitoid, microparasite, meroparasite.

parasitic castration, a situation where a parasite inhibits the reproduction of its host by changing physiological processes or the structure of reproductive organs; e.g. many parasites in plants, or some barnacles parasitic in male crabs.

parasitism, a special form of → symbiosis; a relationship between two species one of which (**parasite**) exploits the other (**host**). The effects of parasitism on the host may range from slight harm to lethal damage. Several types of parasitism occur: → **ecto-**, **endo-**, **hemi-**, **hyper-**, **clepto-**, **nest-**, or **broodparasitism**. In **facultative parasitism** other modes of life, as predation, may occur, but in **obligate parasitism** a parasite cannot live independent of its host. In **semiparasitism** (partial parasitism) an organism, a plant such as mistletoe, parasitizes a host but has also some photosynthetic activity. Sometimes parasitism occurs between the individuals of the same species, such as a parasitic male of a deep-sea fish which lives attached to a female. *See* parasitoid.

parasitoid (Gr. *eidos* = form), an animal living part of its life cycle as a parasite and another part freely; parasitoids are e.g. free-living insects whose larvae live as → parasites in individuals of another species, first being rather harmless but at last usually killing them, as e.g. ichneumon wasps and some dipterans. About 25% of the animal species (chiefly insects) in the world are parasitoids. *See* parasitism.

parasitology (Gr. *logos* = word, discourse), a branch of biology studying parasites and parasitism.

parasorbic acid (L. *Sorbus* = mountain ash), an unsaturated lactone compound, $C_6H_8O_2$, a form of sorbic acid; inhibits the germination of seeds in the same way as → coumarin; occurs abundantly e.g. in rowanberries (*Sorbus aucuparia*). *Syn.* sorbic oil.

parasympathetic (Gr. *sympathes* = having common feelings), denoting relationship to the → parasympathetic nervous system or its actions.

parasympathetic nervous system, that division of the → autonomic nervous system of vertebrates that is most active in the state of rest. Parasympathetic nerve fibres in peripheral nerves innervate smooth muscles, heart, and various glands. The pathway of the system comprises two successive neurones which are linked with cholinergic synapses in the parasympathetic ganglia near or in the organs they supply. In tetrapod vertebrates, the axons of parasympathetic neurones pass within the 3rd, 7th, 9th, and 10th (**vagus nerve**) cranial nerves and the most caudal spinal nerves, i.e. the system has **craniosacral output**. In both the parasympathetic ganglia and nerve endings the → neurotransmitter is **acetylcholine**.

The parasympathetic nervous system is activated at rest, i.e. the parasympathotonia increases and consequently the activity of many functions is adjusted to the resting level; blood pressure and heart rate decrease, respiration and many brain functions slow down, but secretion and motility of the alimentary canal increase, resulting in the activation of digestion. In many organ systems the parasympathetic nervous system acts antagonistically to the → sympathetic nervous system.

parasympatholytic (Gr. *lytikos* = dissolving), pertaining to the inactivation of the → parasympathetic nervous system, or denoting an agent that opposes the functions of this nervous system or effects of → acetylcholine, i.e. acts anticholinergically.

parasympathomimetic (Gr. *mimetikos* = imitative), producing effects similar to the parasympathetic nervous system, or pertaining to an agent that activates this nervous system or has effects of acetylcholine, i.e. acts cholinergically.

parasympathotonia, parasympathicotonia (Gr. *tonos* = tension), the domination or hyperactivity of the → parasympathetic nervous system. *Syn.* vagotonia.

parathion, diethyl-*p*-nitrophenyl thiophosphate; a synthetic compound that inhibits the function of the nervous system acting as an acetylcholinesterase inhibitor, being highly toxic to animals; used as an insecticide.

parathyroid, 1. located beside the thyroid gland; **2.** denoting relationship to the → parathyroid gland or its products; **3.** parathyroid gland.

parathyroid gland (L. *glandula parathyreoidea*), an endocrine gland of vertebrates situated in the vicinity of the thyreoidea. In mammals it is formed from 4 or 5 separate portions (glands) located on the dorsal surface of the thyroid gland; in man it weighs about 100 mg altogether. Endocrine cells, scattered quite evenly in the gland, secrete a peptide hormone of vital importance, called **parathyroid hormone** (parathormone, parathyrin). It participates in the regulation of calcium and phosphate metabolism in the body; when calcium concentration decreases in body fluids, more hormone is secreted, and this increases the availability of calcium (Ca^{2+}) by promoting the release of calcium from bones and by increasing the intestinal and renal absorption of Ca^{2+}. The effects of the hormone on the phosphate concentration are mainly opposite; the concentration in the blood decreases and that of the urine increases; → calcitonin opposes these actions.

parathyroid hormone, *see* parathyroid gland.

paratype, *see* type specimen.

Parazoa (Gr. *zoon* = animal), **parazoans**; a subkingdom of animals comprising sponges (Porifera), and in some classifications also placozoans (Placozoa); are primitive multicellular animals whose cells do not form tissues or organs. *Cf.* Protozoa, Metazoa.

parenchyma (Gr. *enchein* = to pour in), **1.** the essential functional ground tissue of an animal organ, especially that of glands, distinguished from its framework of connective tissue, the stroma; e.g. the parenchyma of the liver; **2.** the loose, jelly-like connective tissue between the internal organs of platyhelminths and some other invertebrates; **3.** the ground tissue of plants composed of thin-walled cells of varying size; different types of parenchyma tissues are the → storage parenchyma, aerenchyma, chlorenchyma, and water storing parenchyma. *Adj.* **parenchymatous.**

parenchymula (Gr. dim. of *parenchyma*), a free-swimming, flagellated larva of most sponges whose cavity contains solid gelatinous cell mass. *Cf.* amphiblastula.

parental (L. *parentalis* < *parens* = parent), pertaining to a parent; originating from parents.

parenteral (Gr. *para* = beside, *enteron* = gut, intestine), not intestinal; pertaining e.g. to an injection which is not given through the alimentary canal but via other routes, e.g. subcutaneously, intraperitoneally, or intravenously. *Cf.* enteral.

paresis, pl. **pareses** (Gr. = relaxation), a slight → paralysis.

parietal (L. *parietalis* < *paries* = wall of an organ), **1.** relating to the wall of a cavity; **2.** pertaining to the upper part of the head, as the parietal bone of the skull or the parietal lobe of the brain; **3.** in botany, pertaining to a peripheral location, as parietal placenta.

parietal cell, 1. a large spheroidal cell type scattered along the walls of the gastric glands of vertebrates, producing hydrochloric acid and → intrinsic factor; *syn.* oxyntic cell; **2.** a cell formed in the division of an archespore cell in plant reproductive organs, producing material for anther walls or → nucellus; the other cells formed are sporogenic cells; *see* archesporium.

parietal eye/organ, → pineal eye.

parotic (Gr. *para* = beside, *ous* = ear), **1.** located near the ear; **2.** pertaining to a bone projection in the skull of some reptiles.

parotid, pertaining to, produced by, or located near the parotid gland.

parotid gland (L. *glandula parotis*, *glandula parotidea*), the largest of the salivary glands in mammals; an exocrine gland located below and in front of the ear. Its secretion flows from numerous small glandular alveoli through the duct into the mouth, containing α-amylase that catalyses the digestion of polysaccharides, e.g. starch.

parotin, a hormone-like protein (globulin) obtained from mammalian parotid glands; it has effects on mesenchymal tissues, e.g. promoting calcification of dentin.

paroxysmal (Gr. *paroxysmos* = irritation, excitement), pertaining to a sudden onset of symptoms. *Noun* **paroxysm.**

parrots, → Psittaciformes.

pars, pl. **partes** (L.), a part; e.g. *pars tuberalis*, *pars distalis*, *pars intermedia*, are parts of the pituitary gland (hypophysis).

parthenocarpy (Gr. *parthenos* = virgin, *kar-*

pos = fruit), the development of a plant fruit without formation of seeds; is caused by an inhibition of pollination, fertilization, or development of ovules.

parthenogamy (Gr. *gamos* = marriage), an extreme form of → automixis (obligatory self-fertilization) in which two haploid nuclei, produced in → meiosis, fuse within a single undivided cell forming a zygote and thus starting embryogenesis.

parthenogenesis, pl. **parthenogeneses** (Gr. *genesis* = birth, origin), the reproduction of a new individual from an unfertilized female gamete without any influence of the male gamete. Parthenogenesis can be divided into several types which further include several systems. The grouping may be based on the mechanism of reproduction, the mechanism of sex determination, or on differences observed in gametes.

Occasional parthenogenesis, in which an unfertilized ovum develops into an adult by chance, is distinguished from **normal parthenogenesis**. The latter is further divided into **obligatory** and **facultative parthenogenesis**. In the facultative type the development of the unfertilized ovum into an adult is possible only in certain environmental conditions. In obligatory parthenogenesis all generations can be parthenogenetic (complete parthenogenesis), or sexual and parthenogenetic generations can alternate (cyclical parthenogenesis).

According to sex determination the following types are found: **arrhenotoky** in which unfertilized eggs develop parthenogenetically into males and fertilized eggs into females, **thelytoky** in which unfertilized eggs develop into females, and **deuterotoky** in which unfertilized eggs can develop into either sex.

On the basis of the differences observed in gametes, two types are distinguishable: **generative** (haploid) and **somatic parthenogenesis**. In the former the parthenogenetic individuals develop from eggs that have undergone → meiosis and hence are haploid. In the latter the parthenogenetically developing individuals have a diploid chromosome number. For example, some plant lice reproduce parthenogenetically in summer, but turn to sexual reproduction in autumn, and hence are examples of cyclical parthenogenesis.

When classified according to the cytological data, **generative (haploid)** parthenogenesis, or **somatic (diploid)** parthenogenesis and its subtypes **automictic** and **apomictic** parthenogenesis are distinguished. In the first case, parthenogenetically produced individuals develop from an egg cell in which the meiotic reduction of the chromosome number has occurred. In the second case the parthenogenetically produced individuals have the zygotic, i.e. diploid chromosome number. In automictic parthenogenesis this is achieved through mechanisms in which regular chromosome pairing and reduction of the chromosome number occur, but the zygotic number of the chromosomes is restored by fusion of two haploid nuclei. In apomictic parthenogenesis neither the chromosome reduction nor the fusion of nuclei takes place.

Parthenogenesis can also be accomplished experimentally; e.g. an unfertilized amphibian egg can be induced to divide using certain chemicals or mechanical stimuli. *Adj.* **parthenogenetic**. *Cf.* apomixis, agamospermy.

partheno(gen)ote, an individual that is produced by → parthenogenesis, often experimentally.

partial migrants, animal species with both migrating and resident individuals; thus the populations are dimorphic in migrating behaviour. *See* migration.

partial refuge (L. *refugere* = to flee away), *see* refuge.

partial veil, *see* veil.

particle radiation/emission, an → ionizing radiation originated from radioactive nuclides including → alpha particles and → beta emission.

partition coefficient, distribution coefficient, the ratio in which a substance is distributed between two or more immiscible phases at equilibrium; e.g. the ratio of lipid solubility to water solubility of any substance. An increasing coefficient describes greater lipid solubility. The molecules with high coefficient values can easily penetrate cellular membranes which contain high quantities of phospholipids.

parturition (L. *partus* = childbirth), the act or process of giving birth to offspring from the uterus at the end of pregnancy in placental mammals. At parturition also the placenta and extraembryonic membranes (afterbirth) are expelled. Parturition is controlled by the nervous and hormonal system involving stimulatory effects of oxytocin and prostaglandins on the smooth muscle of the uterine wall (causing labour). *Adj.* **parturient**. *Cf.* abortion.

PAS, 1. → *para*-aminosalisylic acid; **2.** PAS reagent, the periodic acid-Schiff reagent used for the quantitative determination of carbohydrates and mucopolysaccharides, and for staining of histological samples.

pascal (*Blaise Pascal*, 1623—1662), the derived SI unit of pressure; symbol Pa. *See* Appendix 1.

Passeriformes (L. *Passer* = the type genus of sparrows, *forma* = form, shape), **passerine birds (passerines)**, **perching (song)birds**; the largest order of birds including four suborders (Eurylaimi, Menurae, Tyranni, and Oscines) which comprise 72 families and 5,414 species, i.e. 60% of the species in the avian class. Passerine birds are found all over the world and in all habitats from tundra to tropical rain forests. The size and structure of the birds are various (e.g. crows, larks, and kinglets); a common characteristic is a **perching foot** with three toes pointing forward and one backward. Newly hatched young are naked and blind, → altricial and nidicolous.

passive dispersal, the dispersal of spores, seeds, and animals by means of external factors, such as winds, oceanic currents, and other animals. *See* anemochore, anemohydrochore, anthropochore, hydrochore, zoochore.

passive transport, the transport of different substances across cellular membranes without using metabolic energy. The molecules are transported by → diffusion or facilitated diffusion in the direction of the electrochemical gradient; the transport is dependent on concentrations and chemical and electrical properties of substances to be transported. *See* cell membrane transport.

Pasteur effect (*Louis Pasteur*, 1822—1895), the inhibiting action of oxygen on fermentation or glycolysis, i.e. the glucose consumption in cells is much lower under aerobic than under anaerobic conditions. The phenomenon was discovered by Louis Pasteur in studying fermentation by yeast.

pasteurization, the process of partial sterilization by rapid heating of a liquid, such as milk, wine, beer, etc., below its boiling point (e.g. milk to 80°C) or by holding it for a longer period at slightly lower temperature (e.g. at 72°C for 15 s, or at about 65°C for 30 min), followed by rapid cooling. The idea of this type of sterilization was discovered by *Louis Pasteur*. The technique destroys most living bacteria (except thermoduric microbes, such as lactic acid bacteria) from liquids but preserves flavour and consistency. Using similar techniques, different types of solid food products are also pasteurized.

patch clamp, an electrophysiological method for measuring ion currents through a small patch of the cell membrane, allowing the examination of single ion channels; the method is a special type of the → voltage clamp.

patchy habitat, a habitat within which there are significant spatial variations in the suitability for a given species; e.g. small forest islets for the forest birds on large cultivated fields where favourable and unfavourable patches are interspersed.

patella, pl. **patellae** or **patellas** (L. dim. of *patina* = pan), **1.** the knee cap; a triangular, flattened bone in front of the knee of tetrapod vertebrates; **2.** the fourth segment in the leg or pedipalpus of an arachnid; **3.** *Patella*, a gastropod genus (limpet). *Adj.* **patellar**.

paternal (L. *pater* = father), pertaining to, or derived from the male parent, father. *Cf.* maternal.

pathogenesis (Gr. *pathos* = emotion, state of suffering, *gennan* = to produce), the origination and development of a disease, especially including processes of biochemical and cellular events in the development of a disease; **pathogen**, a microorganism or other material producing disease; **pathogenetic**, pertaining to the pathogenesis; **pathogenic**, causing a disease; **pathology**, a branch of medical sciences studying diseases.

patroclinous, patroclinal, patriclinous, patriclinal (L. *pater* = father, Gr. *klinein* = to lean), pertaining to inheritance found in some animals in which the offspring resemble the male more than the female parent; e.g. X0 males (*see* X chromosome) of the fruit fly arising as a consequence of → non-disjunction are patroclinal. *Cf.* matroclinous.

patulin (< *Penicillium patulum*), 4-hydroxy-4H-furo[3,2-c]-pyran-2(6H)one, an antibiotic obtained from metabolites of many species of *Penicillium*, *Aspergillus*, and *Byssochlamys*. Important producers are e.g. *P. expansum*, the mould rot of apples and other fruits, and *A. clavatus* growing on cereals. Patulin has carcinogenic activity.

paurometabolous metamorphosis (Gr.

pauros = small, *metabole* = change), *see* metamorphosis.

Pauropoda (Gr. *pous,* gen. *podos* = foot), **pauropods**; an arthropod class of myriapods comprising small, segmented animals, mainly 0.5 to 1.5 millimetre long; they live in tropical and temperate regions, often in forest soil, mosses, and detritus.

PCB, → polychlorinated biphenyls.

PCR, 1. PCR cycle; photosynthetic carbon reduction cycle; *see* Calvin cycle, photosynthesis; **2.** → polymerase chain reaction.

peat, organic material in soil, especially bogs, formed in anaerobic conditions in water from dead, partly decomposed remains of vascular plants and mosses, especially peat mosses (the genus *Sphagnum*); peat is an early stage in formation of coal. It is used as fuel and as cultivation substratum for plants.

peat mosses, → Sphagnidae.

peck order, pecking order, *see* dominance hierarchy.

pecten, pl. **pectens,** also **pectines** (L. = comb), a comb-like anatomical structure; e.g. **1.** *pecten ossis pubis*, pectineal line of the pubis; **2.** *pecten analis,* the zone in the caudal part of the anal canal of mammals; **3.** a triangular membrane inside the retina of the eye of birds and many reptiles; this heavily capillarized membrane is protruded from the choroid membrane into the vitreum, and several assumptions of its physiological meaning are suggested, such as it being a special sense for navigation, a heater, or an organ producing more nutrients or oxygen inside the eye; **4.** an organ with long hairs in the hindleg of a worker honeybee for removing pollen from the opposite leg; **5.** any of bivalve marine molluscs, scallops, of the genus *Pecten.*

pectic substances, macromolecular carbohydrates in fruits, leaves, and roots of plants, particularly in the → cell wall, middle lamella, and primary wall. Pectic substances are synthesized via polymerization from simple carbohydrates or their derivatives, like **galacturonic acid**. First it polymerizes into **pectic acid** which then esterifies with an alcohol producing **pectin**. Several pectin chains are bound by Ca^{2+} and Mg^{2+} ions, forming water-insoluble **protopectin**. In ripening fruits protopectins are enzymatically converted into more water-soluble compounds rendering the fruits softer. The enzymes catalysing the cleavage of pectic substances are called **pectinases**, which increase strongly during the → abscission of leaves and fruits. Pectinases are used in experimental studies, e.g. when single cells are separated from plant tissues.

pectus, pl. **pectora** (L.), the breast, chest, or thorax. *Adj.* **pectoral**.

pectoral girdle, → shoulder girdle.

ped(o)-, pedi- (L. *pes,* gen. *pedis* = foot), denoting relationship to the foot. *Adj.* **pedal**. *Cf.* paed(o)-.

pedal connectives (L. *connectere* = to bind together), the nerve cords connecting the cerebral and pedal ganglia in molluscs.

pedal ganglia, a pair of nerve ganglia in the foot of molluscs; nerve fibres extend from the ganglia throughout the foot.

pedal glands, attachment organs in the foot of rotifers providing a sticky secretion by which the animal may attach temporarily to its substratum.

pedicel (L. *pedicellus* < *pediculus,* dim. of *pes* = foot), pedicle; e.g. **1.** a foot-like or stem-like structure in vertebrate anatomy, e.g. cone pedicle, a club-like ending of the retinal cone cell forming synapses with the horizontal and bipolar neurones; **2.** the thin and short segment between the → prosoma and opisthosoma in spiders; **3.** the second, narrow segment in the abdomen of some hymenopterans; *syn.* peduncle; **4.** a tubular appendage in the eggs of some insects; **5.** a short stalk of fruit or sporangium, or a stalk of an individual flower in an inflorescence.

pedicellaria, pl. **pedicellariae** (L.), any of the pincer-like structures around spines of some echinoderms, e.g. asteroids, for cleaning the body surface.

pedicle, → pedicel.

pedigree (Fr. *pie de grue* = crane's foot), a catalogue of ancestors, or a register of genealogy. In human genetics, pedigrees are described schematically using certain symbols for analysing the inheritance of characteristics in families. Such a study is called the **pedigree analysis**.

pedipalp (L. *pedipalpus*, *pes* = foot, *palpare* = to stroke), one of the second pair of appendages lying on each side of the mouth of arachnids, pantopods, and merostomatans; for instance in spiders the pedipalps include an enlarged basal segment and a six-jointed **palpus** ending in a claw. The pedipalps are used to squeeze and chew food, and in mature males the club-like tip of the palpus is specialized for transferring sperm into a female.

pedogenesis, → paedogenesis.

pedomorphosis, → paedomorphosis.

peduncle (L. *pedunculus,* dim. of *pes* = foot), a stalk, or a stem-like structure in anatomy; e.g. **1.** a large bundle of nerve fibres in the central nervous system in vertebrates; **2.** a thinned stalk in the tail of a stickleback (*Gasterosteus*); **3.** the stalk by which an invertebrate is attached to a substratum, e.g. in sedentary crinoids, brachiopods, and goose barnacles; **4.** a slender "waist" between the thorax and abdomen of arthropods; *syn.* pedicel; **5.** flower stalk, a stalk-like structure connecting a flower with the flower shoot.

PEG, → polyethylene glycol.

Peking man, a fossil hominid found in China, classified into *Homo erectus*. *See* Homo.

pelagic (Gr. *pelagos* = sea), pertaining to, or living in the open sea; oceanic.

pelagic zone, the aquatic zone in upper and middle levels of oceans, seas, or great lakes outside the littoral zone to which light penetrates but where seed plants are unable to grow.

pelagosphaera (Gr. *sphaira* = ball), the pelagic larval stage following the trochophore larva in some sipunculans.

Pelecaniformes (Gr. *pelekan* = pelican, L. *forma* = form, shape), an order of middle- or large-sized birds which are cosmopolitan, especially in the tropics; live mainly in marine coastal waters except in the Baltic Sea, some inhabit the shores of lakes and rivers. The order includes 62 species in six families: **frigate birds** (Fregatidae), **tropic birds** (Phaethontidae), **pelicans** (Pelecanidae), **gannets** and **boobies** (Sulidae), **anhingas** (Anhingidae), and **cormorants** (Phalacrocoracidae).

Pelecypoda (Gr. *pelekys* = ax, *pous* = foot), → Bivalvia.

P element (P < L. *pater* = father), a → transposon found in fruit flies; consists of repetitive circular DNA molecules and causes → hybrid dysgenesis. P elements are used as → vectors in genetic engineering to produce transgenic insects.

pellagra (It. *pelle* = skin, *agra* = rough), a syndrome due to the deficiency of niacin found especially in man; is characterized by disturbances in the skin, but also in gastrointestinal and psychic functions. *See* vitamin B complex.

pellet (L. *pila* = ball), **1.** a sedimented material at the bottom of a centrifuge tube; **2.** a pressed nutritive matter, or a certain hormone preparation to be implanted under the skin; **3.** a briquette made from peat, wood, etc.; **4.** a → vomit ball.

pellicle (L. *pellicula,* dim. of *pellis* = skin), a thin film or skin; e.g. **1.** a scum or film on the surface of a liquid; **2.** a thin, flexible and translucent covering membrane of many protozoans, such as some sporozoans, flagellates, and ciliates.

peltate leaf (Gr. *pelte* = small shield), a shield-shaped leaf with the stalk attached in the middle of the lower surface.

pelvic girdle, *see* pelvis.

pelvis, pl. **pelves** (L.), **1. pelvic girdle**; a cartilaginous or bony structure of the body trunk connecting the hindlegs or pelvic fins to the vertebral column in jawed vertebrates; in tetrapod vertebrates, the pelvic girdle (hip girdle, pelvic belt, pelvic arch) consists of the **hip bones** (ischial, iliac, and pubic bones) and the sacral vertebrae, often joined together to form the sacral bone, *sacrum*. In fish, it is a small, primitive, usually cartilaginous structure supporting the pelvic fin. The pelvis is partly degenerated in giant snakes (boa and python) or totally in smaller snakes. There are only rudiments of the pelvic bones in whales. In the pelvis of other tetrapods (except birds) both ischial and pubic bones grow caudally together forming the **pubic symphysis** (*symphysis ossium pubis*), in primates this is formed only by pubic bones. In mammals the pelvic bones have grown together forming the paired compact hip bone (coxal bone, *os coxae*); **2.** the posterior part of the vertebrate body comprising the cavity inside the pelvic girdle; **3.** → renal pelvis.

Pelycosauria (Gr. *pelyx* = wooden bowl, *sauros* = lizard), **pelycosaurs**; an extinct order of synapsid reptiles who lived in Carboniferous and Permian; they were carnivorous or herbivorous animals characterized by several anatomical structures which suggest their ancestral position in the lineage of evolution to terapsids and mammals.

penetrance (L. *penetrare* = to penetrate), in genetics, the frequency of gene expression in a population; indicates how large a portion of the individuals representing a given → genotype expresses the examined character also in their → phenotype. If the penetrance is **complete** (100%), the gene expresses itself in the phenotypes of all individuals; in **incomplete** penetrance the gene is expressed in some but not all individuals.

penetrant, one which penetrates; in zoology, the largest type of the → nematocysts in cnidarians whose spherical, footed structure has a long, coiled thread tube with long spines and small barbs. The penetrant is specialized for capturing chitin-covered animals by discharging the thread tube into their body and injecting toxic fluid. *Cf.* glutinant, volvent.

penetration, the act of piercing or entering deeply (penetrating), e.g. the penetration of light through some matter (e.g. focal depth in microscopy), or the entry of microbes into cells and tissues.

penguins, → Sphenisciformes.

penicillin (L. *penicillus* = brush), **1.** originally an antibiotic discovered by *Alexander Fleming* (1881—1955) from *Penicillium* fungi (1928). *Howard Florey* and *Ernst Boris Chain* isolated and crystallized penicillin 10 years later and soon it was taken into clinical use. The penicillin molecule consists of thiazolidine and β-lactam rings bound by peptide bonds to different side groups (a benzyl ring in benzylpenicillin). The synthesis of the wall material (peptidoglycan) of penicillin-sensitive bacteria is inhibited by penicillin, causing bacteria to burst because of the intracellular pressure. Penicillin mimics the normal substrate of glycopeptide transpeptidase, which acts in the wall formation, but the penicillin-enzyme complex is inactive and the synthesis of the cell wall is prevented; **2.** any of various natural and synthetic derivatives of penicillic acid, such as ampicillin, benzyl penicillin, and piperacillin, having activity particularly against Gram-positive bacteria and low toxic action on animal cells.

penis, pl. **penes,** also **penises** (L.), the male copulatory organ for transferring sperm into the vagina of a female, and in mammals, also urine out of the body. The penis of mammals comprises erectile tissue with blood vessel cavities, and in many species, as e.g. carnivores, also a supporting copulatory bone (*baculum*). The penis is not developed in other vertebrates, although an analogous structure, → hemipenis, has developed in some reptiles. In invertebrates, the term penis denotes usually the caudal end of the ejaculatory duct.

pentadactyl (Gr. *pente* = five, *daktylos* = finger), having five digits; five-fingered, five-toed. *Cf.* didactyl.

pentane, a hydrocarbon, C_5H_{12}, the fifth of the alkane series (*see* alkanes); the most important of the three → isomers is normal pentane (*n*-pentane), used as a solvent.

pentaploid (Gr. *haploos* = onefold), pertaining to or being a polyploid cell or individual that carries a certain basic (haploid) chromosome number in five sets; *see* polyploid.

pentasomic (Gr. *soma* = body), pertaining to → aneuploid genome mutations in which a given chromosome occurs in five sets. Thus the chromosome number is 2n + 3; *see* n.

Pentastomida, Pentastoma (Gr. *stoma* = mouth), **Linguatulida, pentastomids, tongue worms**; an invertebrate phylum, in some classifications subphylum in the phylum Arthropoda; small, flat, worm-like animals in which eyes, respiratory, excretory, and circulatory organs are lacking. Tongue worms have three larval stages, and they live parasitic in the respiratory passage or body cavity of reptiles, some species also in birds and mammals (also in man). There are five short papillose appendages in the front of the body, and therefore the name pentastomid "five-mouthed". One appendage is provided with the mouth and the four others are legs with a hook on their tips. About 90 species are described.

pentosan, any of polysaccharides, such as xylan or araban, yielding pentoses on hydrolysis; pentosansoccur in many plants, especially in stems and leaves of cereals.

pentose, a monosaccharide having five carbon atoms in its molecule; general formula $C_5H_{10}O_5$; e.g. → ribose, deoxyribose, ribulose, arabinose, xylose.

pentose phosphate pathway, the pathway of the carbohydrate metabolism in cell respiration, known since 1930's; it is alternative to → glycolysis (the decomposition of glucose to triose phosphates), and does not contain an aldolase enzyme. **Glucose 6-phosphate** is oxidized via this pathway to **ribose 5-phosphate** by the enzyme glucose-6-phosphate dehydrogenase and NADP is reduced: glucose 6-phosphate + 2 $NADP^+$ + H_2O —> ribose 5-phosphate + 2 NADPH + 2 H^+ + CO_2. Ribose phosphate (pentose) is a precursor for many important biomolecules, e.g. ATP, coenzyme A (CoA), and nucleic acids. The reducing power (**NADPH**) is used in cells for biosyntheses, e.g. in lipid biosynthesis. In photosynthesis the pentose phosphate pathway takes part in the production of hexoses from CO_2. In animal cells, the pathway is active e.g. in the adrenal cortex, adipose tissue, and mammary glands, i.e. the tissues with active lipid synthesis; the pathway lacks in skeletal muscle cells. *Syn.* pentose shunt, hexose monophosphate pathway, phosphogluconate oxidative pathway.

PEP, → phosphoenolpyruvic acid.

pepsin (Gr. *pepsis* = digestion), the common name for a digestive enzyme in vertebrates; a proteolytic enzyme group (EC sub-subclass 3.4.23.) found in vertebrates in the gastric juice, where it catalyses the hydrolysis of proteins to smaller molecules, such as → peptones, proteoses, at acidic pH, optimally at pH 2. It is secreted as an inactive precursor, **pepsinogen**, by the **chief cells** of the gastric glands. This precursor is activated by removing amino acid residues in the presence of hydrochloric acid or autocatalytically by pepsin itself. The structure of pepsin varies among different species and groups of vertebrates, and similar enzymes are also found in many invertebrates.

peptic (Gr. *peptein* = to cook, digest), pertaining to pepsin, digestion, or the stomach.

peptidases, peptide hydrolases of the enzyme sub-subclasses EC 3.4.11-23.; hydrolytic enzymes catalysing the cleavage of → peptide bonds. Peptidases which are important enzymes in digestion are grouped to **endo-** and **exopeptidases** (proteinases). Endopeptidases break peptide bonds inside a peptide chain; exopeptidases are either **aminopeptidases** which begin to cleave molecule chains from the amino end, or **carboxypeptidases** which act at the carboxyl end. Enzymes catalysing the hydrolysis of dipeptides are **dipeptidases**, and of polypeptides, **polypeptidases**. Sometimes the term peptidase is limited to mean only exopeptidases.

peptide, a compound formed from two or more amino acid residues. According to the number of amino acids, peptides are called di-, tri-, tetra- etc. peptides; oligopeptides consist of 2 to 10 amino acid residues, and larger compounds are called → polypeptides. Often the term peptide is limited only to the shorter peptides, e.g. → endorphins, vasopressin, or oxytocin. *See* peptide bond, protein, protein synthesis.

peptide bond, a → covalent bond, -CO-NH-, formed between two amino acids from one α-amino group ($-NH_2$) and one carboxylic group (-COOH) while one molecule of water is removed: R-COOH + NH_2-R —> R-CO-NH-R + H_2O. Peptide bonds are formed in → protein synthesis in the formation of → peptides.

peptide hormones, a large group of animal hormones of oligopeptide or polypeptide structure functioning as general or local hormones, some also as neuronal modulators and neurotransmitters; oligopeptides or small polypeptides are e.g. → vasopressin, oxytocin, and endorphins; larger polypeptides e.g. corticotropin, parathyroid hormone, insulin, and many hormones secreted in the alimentary canal. *See* neuropeptides.

peptide nucleic acid, PNA, a nucleic acid derivative consisting of a peptide (polyamide) backbone composed of N-(2-aminoethyl) glycine units to which nucleobases (purines or pyrimidines) are attached by carbonyl methylene linkers; presumed to have played a role in prebiotic evolution.

peptidoglycan (Gr. *glykys* = sweet), a compound containing amino acids or peptides linked covalently to sugars the latter forming the major part; especially found in the inner cell wall of all bacteria (10% of the walls in Gram-positive and 50% in Gram-negative bacteria), consisting of linear polysaccharide chains which are cross-linked by short peptides. *See* Gram staining. *Cf.* glycopeptide, proteoglycan.

peptidyl site, → P-site.

peptone, any of the water-soluble → polypeptides formed when proteins are partially hydrolysed; e.g. produced by pepsin and gastric acid in the initial stages of protein digestion.

per (L.), through, by means of, for each; e.g. *per rectum*, through the rectum, *per os*, by mouth, orally, or *per capita*, for each one.

per-, 1. denoting throughout, thoroughly, utterly, completely; **2.** in chemistry, containing the largest or a relatively large proportion of an element, usually oxygen, as in hydrogen peroxide (H_2O_2), perchloric acid ($HClO_4$), persulphuric acid ($H_2S_2O_5$), potassium persulphate ($K_2S_2O_5$).

perception (L. *percipere* = to take in, perceive), the act of mental, conscious registration of sensory stimuli; e.g. **stereognostic perception**, the recognition of objects by touch. *Adj.* **perceptive**.

Perciformes (L. *perca* = perch, *forma* = shape); **perch-like fish**, a large and multiform order of bony fishes (Osteichthyes) spread all over the world both in fresh water and marine habitats. In general, they possess a spine-rayed first dorsal fin and soft-rayed second dorsal fin; the pelvic fins are located close behind or in front of the pectoral fins. The skin is covered by rugged ctenoid scales (some species are scaleless); the duct of swim bladder is lacking. The order includes about 7,000 species and about 150 families, e.g. **fresh-water perches** (Percidae), **sea basses** (Serranidae), **goatfishes** (Mullidae), **cichlids** (Cichlidae), **mullets** (Mugilidae), **barracudas** (Sphyraenidae), **wrasses** (Labridae), **prickle-backs** (Stichaeidae), **gunnels** (Pholididae), **sand lances** (Ammodytidae), and **gobies** (Gobiidae).

perching birds, → Passeriformes.

percutaneous (L. *per* = through, *cutis* = skin), given or performed through the skin, e.g. an injection given to, or a tissue sample taken from an animal. *Syn.* transcutaneous, transdermic.

pereiopod, pereopod (L. *pereiopodium* < Gr. *peraioun* = to convey, *pous*, gen. *podos* = foot), a walking limb in the thorax of some crustaceans, e.g. crayfish and lobsters.

perennial plant, perennial (L. *perennis* = surviving from year to year), a plant persisting for several years. *Cf.* ephemeral, annual, biennial.

perforation (L. *perforare* = to pierce through), the act of piercing through, as e.g. through the wall of an organ; such a hole itself; **perforans** (L.), penetrating, e.g. nerves and blood vessels through other organs. *Verb* to **perforate**.

perforation plate, the contact area between xylem elements; may be perforated (scalariform perforation plate), or quite open (simple perforation plate) as in more developed vascular plants. *See* xylem.

perfusion (L. *perfundere* = to drench, pour over), the process of pouring through or over; especially passing a fluid, such as a physiological solution or fixative through the vessels or the cavity of an organ or over an organ; **perfusate**, a liquid used for perfusion. *Verb* to **perfuse**.

peri- (Gr. *peri* = around), denoting around, about, beyond.

perianth (Gr. *anthos* = flower), **1.** collectively the calyx and corolla in a plant flower; if the flower is **heterochlamydeous**, the calyx and corolla are formed from different leaves; in the **homochlamydeous** flower the calyx and corolla are alike (e.g. many Liliaceae species); **2.** the sheath around the archegonia in some mosses; **3.** the tubular sheath covering the developing sporophyte in leafy liverworts.

periblem (Gr. *blema* = cover), *see* derivative meristems.

peribranchial cavity (L. *cavum peribranchialis*, < Gr. *branchion* = gill), a cavity around the gills to which water passes through the gill slits in ascidians (tunicates) and lancelets. *Syn.* atrial cavity. *See* mantle cavity.

pericardium, pl. **pericardia** (L.< Gr. *perikardion* = around the heart), heart sac; the fibrous membrane enclosing the heart and the roots of the great blood vessels in vertebrates, consisting of two layers, the inner of which (visceral layer, **epicardium**) is tightly attached onto the cardiac muscle, the outer (parietal layer) forming the sac around the heart and epicardium. Between the layers is a very narrow space (**pericardial cavity**, sometimes itself called the pericardium) filled with a special fluid reducing friction around the heart. Similar pericardia and pericardial cavities around the heart are found also in many invertebrates, such as molluscs and arthropods. *Adj.* **pericardial**, also **pericardiac**.

pericarp (Gr. *peri* = around, *karpos* = fruit), the wall of a → fruit; consists of three layers, the outermost being the **exocarp** (epicarp, outer skin), the middle the **mesocarp**, and the innermost the **endocarp**.

pericellular (L. *cellula* = cell), surrounding a cell; around a cell.

pericentral, pericentric (L. *centrum* = centre), surrounding a centre; in genetics, pericentric refers to the type of chromosome inversion which contains the centromere. *Cf.* paracentral.

perichondrium, pl. **perichondria** (Gr. *chondros* = cartilage), the fibrous connective tissue membrane around the cartilage.

periclinal (Gr. *perikline* = sloping on all sides), parallel to the surface of an organ; e.g. describing the division plane of cells parallel to the surface. *Cf.* anticlinal.

periclinal chimaera, also **periclinal chimera,** *see* chimaera.

pericycle (Gr. *peri* = around, *kyklos* = ring,

circle), the outermost layer of the central cylinder (→ stele) between the → endodermis and the conducting tissue in a vascular plant root; forms the root branches and the → phellogen in the root.

pericyte (Gr. *kytos* = cavity, cell), one of the elongate, contractile cells found around capillaries and postcapillary venules in vertebrates. The pericytes regulate the flow through the junctions between the endothelial cells and release vasoactive agents. They are apparently less differentiated than proper smooth muscle cells, for in damaged tissues they are able to dedifferentiate into myoblasts or even mesenchymal cells. Also called pericyte of Zimmermann or Rouget cell.

periderm (Gr. *derma* = skin), **1.** the outer layer of the skin epidermis, especially in an embryo; **2.** the tissue that replaces the epidermis and → primary cortex in secondarily growing plants; protects the root and stem. It is formed from **cork cambium** (**phellogen**), which produces non-living cork tissue (**phellem**) outside and parenchyma tissue called **phelloderm** inside. In roots the periderm is formed inside the primary cortex, in stems outside it.

peridium, pl. **peridia,** also **peridiums** (Gr. *peridion,* dim of *pera* = leather bag, wallet), **1.** the outer layer around a fruiting body or sporangium of fungi; **2.** a cortex-like layer surrounding the sporangium in → Myxomycetes.

perigynium (Gr. *gyne* = woman), a bottle-like, protecting leaf around the ovary of a female flower in sedges (genus *Carex*); a membranous pouch around an archegonium in liverworts, the → involucre in mosses.

perigynous, pertaining to a flower in which the gynoecium is in the middle of a concave receptacle and the other flower parts on its margin. *Cf.* epigynous, hypogynous.

perihaemal system, Am. **perihemal system** (Gr. *haima* = blood), the part of the coelom of some echinoderms, e.g. asteroids, comprising tubes which surround and parallel those of the → haemal system.

perikaryon, pl. **perikarya** (L. < Gr. *karyon* = nut, nucleus), **1.** the cytoplasm around the cell nucleus; **2.** the cell body as distinguished from cell branches, especially that of the nerve cell; also pericaryon.

perilymph (L. *lympha* = clear water, lymph), a special fluid between the → bony labyrinth and the membranous labyrinth of the inner ear in vertebrates. *Cf.* endolymph.

perimetrium, pl. **perimetria** (Gr. *metra* = uterus), the peritoneal serous coat (*tunica serosa uteri*) lining the outer surface of the uterus of mammals. It is formed from the inner leaf of the embryonic → **peritoneum**, and is composed of connective tissue with mesothelium on its surface. *Adj.* **perimetric**. *Cf.* endometrium, myometrium.

perimetry (Gr. *metron* = measure), determination of the extent of the visual field in humans using an instrument, called **perimeter**.

perimysium, pl. **perimysia** (Gr. *mys* = muscle), the fibrous sheath of connective tissue surrounding separate skeletal muscle bundles (groups of muscle fibres) inside a muscle. *Cf.* endomysium, epimysium.

perinatal (L. *perinatalis* < *natus* = birth), pertaining to the period shortly before, during, and after birth, e.g. in man from about 28 weeks of gestation to 1—2 weeks after birth; **perinatal mortality**, the mortality rate at that time.

perineum, pl. **perinea** (L. < Gr. *perinaion*), the area between the anus and the external genitalia.

perineurium, pl. **perineuria** (Gr. *neuron* = nerve), the connective tissue sheath surrounding a bundle of nerve fibres within a → nerve.

perinuclear (Gr. *nucleus* = nut), occurring or situated around a nucleus.

periodic law, in chemistry, a law by *Dimitri Mendeléeff* (1834—1905) stating that chemical elements can be arranged periodically on the basis of their atomic weight (atomic number), forming the groups of elements with similar chemical properties. In the **periodic table** the elements fall into distinct periods of 2, 8, 8, 18, 18, and 32 elements. Also called Mendeléeff's (Mendeleev's) law.

periodic table, an arrangement of chemical elements (see Appendix 2) in order of their atomic numbers to demonstrate the → periodic law.

periosteum, pl. **periostea** (Gr. *peri* = around, *osteon* = bone), periost; a layer of fibrous connective tissue covering all bones.

periostracum, pl. **periostraca** (L. < Gr. *ostrakon* = shell), a hardened protein layer on the outer surface of the shell in mussels, gastropods and brachiopods; protects limy inner layers from acids and other corrosive agents. *Cf.* nacreous layer, prismatic layer.

peripatric (Gr. *patra* = fatherland), pertaining to a species living far from the majority of conspecific populations; **peripatric speciation,** an evolutionary speciation process in such populations, usually giving rise to → ring species.

peripheral (Gr. *peripherein* = to carry around), situated at or near the periphery, i.e. away from a centre or central structure, such as peripheral nerves from the brain and spinal cord, or peripheral blood vessels from the heart.

peripheral membrane proteins, *see* integral membrane proteins.

peripheral nervous system, that part of the nervous system which comprises the nerves running from the central nervous system (brain and spinal cord in vertebrates) to different peripheral tissues. In vertebrates, includes the → **cranial nerves** and → **spinal nerves** in which motor, sensory, and autonomic nerve fibres may exist in the same nerve (mixed nerve) or form special motor, sensory, sympathetic, or parasympathetic nerves. In invertebrates with a well developed nervous system, like in annelids and arthropods, the peripheral nervous system includes the nerves excluding the brain ganglia and the ventral nerve cord with their ganglia.

periphysis, pl. **periphyses** (Gr. *phyein* = to grow), *see* hamathecium.

periphyton (Gr. *phyton* = plant), a material composed of organisms and their remains (→ detritus) attached to the surfaces of aquatic plants, or underwater stones and other structures.

periplast (Gr. *plassein* = to mould), **1.** intercellular substance, stroma of a tissue; **2.** the outer covering of algal cells in the Cryptophyta.

periproct (Gr. *proktos* = anus), the spineless area around the anus of echinoderms. *Adj.* **periproctic**, circumanal; pertaining to a structure around the anus.

perisarc (Gr. *sark-*, *sarx* = flesh), a transparent chitinous integument surrounding the common stem of colonial hydrozoans. *Cf.* hydrotheca, gonotheca.

perisperm (Gr. *sperma* = seed), a storage tissue in some seeds, formed exceptionally from the → nucellus and not from the endosperm mother cell. *Cf.* endosperm.

Perissodactyla (Gr. *perissos* = uneven, extraordinary, *daktylos* = finger, toe), **perissodactyls**, **odd-toed hoofed mammals**; a mammalian order, including herbivorous species having an odd number (one or three) of toes, each with a cornified hoof, the weight being on the third toe. In some classifications, the order is divided into three suborders: **Hippomorpha**, horse-like perissodactyls, **Tapiromorpha** (Ceratomorpha), tapirs and rhinoceros, and **Ancylopoda,** extinct forms. Both Perissodactyla and → Artiodactyla are often referred to **ungulates**, hoofed mammals.

peristalsis, pl. **peristalses** (L., Gr. *stalsis* = contraction), **peristaltic movement**; a slowly propagating wave-like contraction around the wall of a tubular organ, such as the alimentary canal, propelling the content in the lumen. In vertebrates, the contractions of the smooth musculature (circular and longitudinal muscle layers) begin in certain **pacemaker areas**, and the contraction may remain local (rhythmic segmentation), or it may propagate forward along the wall as a peristaltic wave. The wave may also propagate backwards (**reversed** or **retrograve peristalsis**), as in the large intestine. The peristalsis is controlled by the nervous system and hormones; strong bursts of peristaltic movements, occurring especially in the caudal end of the intestine, are called **mass peristalsis**.

peristomium, pl. **peristomia** (L., Gr. *stoma* = mouth), **peristome**; **1.** in invertebrates, an area around the mouth, as in some protozoans, echinoderms, insects, or annelids; e.g in polychaetes, the posterior part of the head formed by two somites surrounding the mouth; bears four pairs of **peristomial tentacles** which are sensory organs of touch, smell, and sight; *cf.* prostomium; **2.** in botany, a ring of pointed teeth protecting the mouth of the opened sporangium of mosses; the teeth may turn inwards and outwards according to the moisture of the air, controlling the dispersal of the spores.

perithecium, pl. **perithecia** (L. < Gr. *perithekion*, *theke* = box, case), a flask-shaped → fruiting body of ascomycete fungi containing the asci. *See* Ascomycetes.

peritoneal cavity (L. *cavitas peritonealis*), *see* peritoneum.

peritoneum, also **peritonaeum,** pl. **peritonea, peritonaea** (Gr. *teinein* = to stretch), **1.** the serous membrane lining the wall of the abdominal cavity (**parietal peritoneum**), or covering the gut and other visceral organs (**visceral peritoneum**) in vertebrates; the narrow space between the peritoneal membranes is called **peritoneal cavity**; **2.** in a broader sense, the membranes of the body cavity in all animals which have a real → coelom. *See* mesentery, omentum.

peritreme (L. *peritrema*, Gr. *trema* = opening, hole), **1.** a margin of an opening in the shell of an animal; **2.** a small plate surrounding the tracheal spiracle of insects, ticks, and mites.

Peritrichia (Gr. *thrix* = hair), **peritrichians**; a group (subclass) of microscopic ciliates (phylum Ciliophora), usually stalked, sessile, aquatic animals; cilia are usually lacking except in the disc-like oral region.

perivascular (L. *vasculum,* dim. of *vas* = vessel), **1.** located around a blood or lymph vessel; **2.** surrounding the vascular cylinder in plants.

perivisceral (L. *viscera* = internal organs), located around visceral organs, e.g. the perivisceral cavity (peritoneal cavity).

perlite (Fr. *perle* = pearl), volcanic glass, usually appearing as pearl-like grains, e.g. used as a ground material for plants growing in a water-culture. *Syn.* pearlite.

permanent wilting point, a soil water content so low that plants cannot take water into their cells, and begin to wilt; the agreed standard for the → water potential of the soil is –1.5 MPa or lower. *See* osmotic potential.

permeability (L. *permeabilis* = permeable < *per* = through, *meare* = to move), the property of a membrane or other barrier of being passable or penetrable. *See* semipermeability, osmosis.

permease, the common name for enzymes (particularly for certain bacterial proteins) which act as carriers for substances across the semipermeable cell membrane.

Permian (*Perm* = an area in Russia), the Permian period; the last major period of → Palaeozoic from 286 to 245 million years ago succeeding the Carboniferous period. Global climate turned clearly dryer and extensive glaciation occurred in the southern hemisphere. Many marine invertebrates, e.g. trilobites, became extinct, amphibians were abundant and the number of insect species increased. Vertebrates began to settle in terrestrial habitats, and radiation of reptiles progressed gradually. Among the plants, gymnosperms developed while ferns decreased.

pernicious anaemia, Am. **pernicious anemia** (L. *perniciosus* = destructive, Gr. *an-* = no, *haima* = blood), → anaemia caused by the deficiency of vitamin B_{12} in tissues, due to the failure of the gastric mucosa to secrete an **intrinsic factor**, i.e. the glycoprotein required for the absorption of vitamin B_{12}, which is necessary for the production of red blood cells. *Syn.* Addisonian anaemia, cytogenic anaemia, malignant anaemia.

peroxidase, a group of enzymes (EC sub-subclass 1.11.1.) of the oxidoreductase class having a haem protein structure; catalyses the oxidation of various substances in a reaction in which → hydrogen peroxide (or an organic peroxide) is reduced to water by a reductant (AH_2): $H_2O_2 + AH_2 \rightarrow 2\ H_2O + A$. The enzyme is common in plant cells and is found also in many animal cells, as e.g. white blood cells, and in the cells of the liver, lung, spleen, and kidney. **Iodide peroxidase** (thyroid peroxidase, EC 1.11.1.8) is involved in iodination of tyrosine. *Cf.* catalase.

peroxide, the oxide of an element containing more oxygen than the normal oxide, and the oxygen atoms form the linkage -O-O-, as in → hydrogen peroxide, H_2O_2; other typical peroxides are e.g. barium peroxide, BaO_2, sodium peroxide, Na_2O_2; **peroxide radical** (O_2^{2-}) is the reduction product of molecular oxygen in which the oxygen molecule has two extra electrons.

peroxisome (Gr. *soma* = body), a membrane-bound cell organelle found in nearly all eukaryotic cells; the diameter of peroxisomes varies from 0.15 to 1.5 µm. They are usually divided into two groups: **microperoxisomes** (0.15—0.25 µm) and larger **microbodies** (microgranules). Peroxisomes are enclosed by a unit membrane (6.5 nm in thickness) and they contain many different enzymes, such as → catalase, amino acid oxidase, and peroxidases. The enzyme content varies according to cell type, but all cells contain enzymes which reduce oxygen to hydrogen peroxide. All animal peroxisomes (e.g. in liver and kidney cells) contain catalase which decomposes hydrogen peroxide and also oxidizes different compounds which are toxic to cells. In many peroxisomes lipids are broken via β-oxidation and acetyl CoA is produced. Animal peroxisomes contain similar enzymes also in → **lysosomes**. Peroxisomes of plant leaf cells act in → photorespiration.

per rectum (L.), through the rectum; pertaining to a method of medication, by way of the rectum.

perspiration (L. *perspirare* = to breathe through), **1.** excretion of sweat, sweating; **2.** sweat.

perturbation method (L. *turbare* = to throw into disorder, disturb), a method used in community ecology when interrelationships of different species are studied by artificial disturbance experiments; e.g. by removing one or more dominant species from the community under study.

pes, pl. **pedes** (L.), **1.** foot; **2.** a foot-like or basal anatomical structure; e.g. *pes hippocampi*, the rostral end of the ventricular wall of the → hippocampus in the vertebrate brain.

pest (L. *pestis*), **1.** an animal or plant species threatening in some way, directly or indirectly, the health, welfare, or success of man; such species may still be important in balancing ecosystems; **2.** an epidemic disease, especially → plague.

pesticide (L. *caedere* = to kill), a chemical agent used to kill pest animals, fungi, and plants; some pesticides are also used as growth retardants for manipulation of cultivated plants. Pesticides are divided into different categories on the basis of their target species or groups, e.g. **herbicides** (to kill unwanted plants such as weeds), **fungicides** (to prevent rotting), **insecticides** (to destroy pest insects), **rodenticides** (to kill rats, mice and other rodents), **acaricides** (to kill ticks and mites), **molluscicides** (to control damage caused by molluscs), and **larvicides** (to kill the larvae of various pest animals).

pest pressure, harmful effects on animal or plant populations caused by pests or parasites.

petal (Gr. *petalon* = a thin plate, leaf), a sterile leaf type of a flower forming the corolla, often brightly coloured; the petals are often totally lacking.

-petal (L. *petere* = to seek, go toward), directed towards or moving towards, the word stem indicating the point of reference, as **centripetal**, moving towards a centre, or **corticopetal**, towards a cortex. *Cf.* -fugal.

petiole (L. *petiolus* = small foot), any slender, stalk-like structure, such as stem, stalk, pedicel (pedicle); e.g. **1.** epiglottic petiole, the lower end of the cartilage of the → epiglottis; **2.** a slender segment, usually characterized by one or two nodes between the thorax and abdomen in some hymenopterans; **3.** the stalk of a leaf, connecting the leaf blade with the leaf base and further with the shoot. *Adj.* **petiolar**.

petite yeast mutation (Fr. *petit* = small), a yeast cell → mutation resulting in disturbances of the respiratory metabolism due to lack or deficiency of → mitochondrial DNA; the petite yeast mutants (also called petites) grow more slowly than wild-type yeasts and form small colonies.

Petri dish/plate (*Julius Petri*, 1852—1921), a shallow, circular glass or plastic dish with a cover over the top and sides; used in laboratories e.g. for culturing of bacteria, cells and tissues.

petrification, also **petrifaction** (L. *petra* = stone, rock, *facere* = to make), the process of conversion of organic matter into stone or stony substance, as e.g. in fossilization of organic tissues through saturation with minerals from the surrounding water.

petrol, Am. **gasoline,** a complex mixture consisting mainly of hydrocarbons (5 to 8 carbon atoms in a chain) in the boiling range 40—180°C, distilled from crude petroleum. Petrol is used as motor fuel which contains different additives, an earlier additive being lead which became a serious pollutant and is therefore mainly substituted by ethers, such as → methyl tertiary butyl ether (MTBE).

petroleum, crude oil, mineral oil; a mixture of hydrocarbons and other organic compounds formed from remnants of animals and plants in underground reservoirs under pressure at high temperatures; the composition of petroleum from different sources varies widely. By fractional distillation petroleum is separated into gases, different liquids, and solid → paraffin wax.

petroleum ether, petroleum benzin, a low-boiling mixture of hydrocarbons containing mainly pentane and hexane; a very volatile and flammable liquid and a good lipid solvent used e.g. for chemical fractionation of biological material.

Petromyzoniformes (Gr. *petros* = rock, *myzon* = sucking), **lampreys**, the only extant order in the class → Cephalaspidomorphi.

petrophyte (Gr. *phyton* = plant), a rock plant, living in depressions of rocks in very dry conditions.

Peyer's patches/glands (*Johann Peyer*, 1653—1712), lymphatic follicles (nodules) closely packed together in the wall of the vertebrate small intestine.

Pfeffer's cell (*Wilhelm Pfeffer*, 1845—1920), a cell model to show osmotic processes in which colloidal cupric ferrocyanide in the minute pores of a clay dish is used as a selectively permeable membrane. The membrane is made by immersing the dish, containing a potassium ferrocyanide solution, into a copper sulphate solution. The solutions react and form cupric ferrocyanide that has semipermeable properties resembling the cell membrane.

PG, → prostaglandin.

PGA, 1. → phosphoglyceric acid; **2.** pteroylglutamic acid, → folic acid; **3.** prostaglandin A.

P generation (L. *parens* = parent), in genetics, the parental generation; P_1, parents; P_2, grandparents, etc.

pH, the negative logarithm of the hydrogen ion (H^+) concentration which expresses the acidity, i.e. pH = -log[H^+]; e.g. in the 0.1 M (10^{-1} M) hydrogen ion concentration [H^+] pH is 1, and in the 0.01 M (10^{-2} M) concentration pH is 2, etc. Acidic solutions have pH values of less than 7, alkaline more than 7, and a neutral solution has a pH of 7.0, on the scale of 0—14.

Phaeophyceae (Gr. *phaios* = grey, *phykos* = seaweed), the only class of the division → Phaeophyta (brown algae).

Phaeophyta (Gr. *phyton* = plant), **brown algae**; a division in the Plant kingdom; includes multicellular, mostly large-sized algae containing chlorophylls a and c, beta-carotene, and fucoxanthin, together causing the brown colour of the species. Almost all species of brown algae live in sea or brackish water. The reserve carbohydrate is laminarin.

phage (Gr. *phagein* = to eat), → bacteriophage.

phagocyte (Gr. *kytos* = cavity, cell), a cell specialized for → phagocytosis; **1.** in animals, e.g. → macrophages, monocytes, and neutrophilic → granulocytes; **2.** in plants, a phagocytic cell able to digest endotrophic fungal filaments in the roots.

Phagocytellozoa, *see* Placozoa.

phagocytosis, a type of → endocytosis in which solid substances, such as other cells, microbes, dead tissue, foreign particles, are engulfed by a cell (**phagocyte**) and digested inside the cell. The material is taken up within invaginations of the plasma membrane forming membrane-bounded vesicles, **phagosomes** (phagocytic vacuoles), inside the cells. These then fuse with → lysosomes and form **phagolysosomes** (heterophagosomes), where digestion takes place. In protozoans and sponges, phagocytosis serves normal food intake (intracellular digestion in food vacuoles). In more developed animals, phagocytosis acts as a physiological defence mechanism in which the specialized phagocytes remove microbes and cell fragments. Phagocytic cells of vertebrates include most forms of leucocytes as e.g. neutrophils (neutrophilic → granulocytes, microcytes) and monocytes, and various cell types of the → mononuclear phagocyte system located in the walls of the vascular system (sinusoids) of many internal organs, such as the liver, spleen, lymph nodes, and bone marrow. Especially when activated for phagocytosis these cells are called **macrophages**. The microglia (*see* neuroglia) and → osteoclasts also belong to tissue macrophages.

phagosome (Gr. *soma* = body), *see* phagocytosis.

phalange (Gr. *phalanx* = line of soldiers), → phalanx.

Phalangida, → Opiliones.

phalanx, pl. **phalanges** (L.), **1.** any of the bones (phalanges) of the fingers or toes; **2.** a plate-like structure made up of rows of supporting cells (phalangeal cells) of the spiral organ in the vertebrate ear; **3.** a bundle of stamens joined together by their filaments.

phalloidin(e) (Gr. *phallos* = penis, *eidos* = form), a bicyclic heptapeptide, $C_{35}H_{48}N_8O_{11}S$; the best known and most poisonous of the fungal toxins (phallotoxins) found in the death cup mushroom, *Amanita phalloides*; binds to actin filaments and prevents cell movements. Lethal values e.g. for mice are 2 to 3 mg/kg, in man phalloidin causes vomiting, diarrhoea, convulsions, and death. Fluorescent derivatives of phalloidin are used to identify filamentous actin. *Cf.* amatoxin.

phallus, pl. **phalli,** Am. also **phalluses, 1.** penis; **2.** the undifferentiated embryonic organ that develops into the penis or clitoris; **3.** an external genital structure of a male invertebrate, such as of an insect; **4.** the fruiting body of the stinkhorn fungi; **5.** an image of the penis symbolizing the generative power in certain religious systems; also phallos.

phanerogams (Gr. *phaneros* = apparent, *gamos* = marriage), → seed plants.

phanerophyte (Gr. *phyton* = plant), a perennial plant having its overwintering buds clearly above the ground, e.g. many trees and shrubs. *See* Raunkiaer's life forms.

Phanerozoic (Gr. *zoon* = animal), a geological eon succeding Proterozoic from about 570 million years to the present, including the → Palaeozoic, Mesozoic, and Caenozoic eras.

pharmac(o)- (Gr. *pharmakon* = medicine, drug, poison), denoting relationship to a drug or medicine.

pharmacogenetics (Gr. *genesis* = birth, origin), a branch of → genetics dealing with genetically determined variations in responses of animals to drugs and with genetic mechanisms involved in the actions of drugs.

pharmacology (Gr. *logos* = word, speech), a branch of medicine dealing with drugs, their origin, chemistry, nature, actions, and use. The most important research fields are **pharmacodynamics** dealing with effects of drugs, and **pharmacokinetics** with absorption, distribution, localization in tissues, metabolism, and excretion of drugs; **pharmacognosia** is the study of the origin and processing of natural drugs, **pharmacochemistry** deals with biochemical and chemical reactions involving drug development and usage, and **toxicology** with poisonous actions of drugs and other chemicals on animals.

pharmacy (Gr. *pharmakeia* = usage of natural drugs), **1.** a field of health sciences dealing with manufacturing and utilization of drug preparations; **2.** a place where drugs are prepared and dispensed; a drugstore. *Adj.* **pharmacal, pharmaceutic.**

pharotaxis, pl. **pharotaxes** (Gr. *pharos* = lighthouse, *taxis* = arrangement), → mnemotaxis.

pharyngeal arch (L., Gr. *pharynx* = throat), *see* gill arch.

pharyngeal tonsil (L. *tonsilla pharyngealis*, *tonsilla* = stake), the lymphoid tissue in the roof and posterior wall of the nasopharynx.

pharyngobranchial (Gr. *branchia* = gills), pertaining to the throat and gills, or their embryonal primordia. *See* gill arch.

pharynx, pl. **pharynges** (L., Gr.), **1.** in vertebrates, the passage from the posterior part of the nasal cavity and mouth to the oesophagus and larynx. Its wall consists of mucous membrane with a stratified squamous epithelium and loose connective tissue, and of muscular tissue comprising mainly skeletal muscle fibres in similar layers as the smooth muscle more caudally in the → alimentary canal. In tetrapod vertebrates, three sections of the pharynx are separately called **nasopharynx, oropharynx** and **hypopharynx**, the soft palate separating the two first parts from each other; the auditory tube opens to the nasopharynx; **2.** in invertebrates, the part of the digestive tract between the mouth and oesophagus, its structure varying greatly between animal groups; in many liquid sucking species, the pharynx is a well developed muscular organ, as e.g. in nematodes. *Adj.* **pharyngeal.**

phase contrast microscopy, *see* microscope.

phasmid (Gr. *phasma* = apparition), **1.** one of the paired papillae or pouches in the posterior end of some nematodes (class Phasmidia) including receptors for chemical stimuli (chemoreceptors); **2.** a member of stick insects (walking sticks) and leaf insects (walking leaves) in the family Phasmida in the order of Orthoptera. **3.** *Adj.* pertaining to the Phasmida.

phellem (Gr. *phellos* = cork, + → phloem), layers of cork in the outermost part of the → periderm in secondarily growing plant roots and shoots; formed from the cork cambium.

phelloderm (Gr. *derma* = skin), *see* periderm.

phellogen, cork cambium; *see* periderm.

phen(o)- (Gr. *phainein* = to show, appear), **1.** appearing, showing, displaying; **2.** in chemistry, denoting a derivation from → benzene (phenyl-).

phene, a genotypically regulated phenotypic character; e.g. the human eye colour. *See* genotype, phenotype.

phenetic, the term based on structural similarities of phenotypes between different organisms without consideration of a genetic or evolutionary background; e.g. **phenetic species, phenetic taxonomy** (→ **phenetics**), etc. In evolutionary ecology, **phenetic variation** is an important object of study. *See* phenogram. *Cf.* phylogenetics.

phenetics, phenetic taxonomy, a method of classification based only on measurable similarities and differences in phenotypic characteristics (e.g. morphological, physiological, and ethological) without any consideration of homology, analogy, or phylogeny. *Cf.* cladistics.

phenocopy, a change in a → phenotype of an organism due to environmental factors mimicking a mutation phenotype but is not inheritable.

phenocritical period, the period during the ontogenic development of an organism in which the development of the mutant genotype for the first time deviates from that of the normal genotype.

phenogenetics, → developmental genetics.

phenogram (Gr. *gramma* = letter), a graphical presentation used in the taxonomy of organisms, based on the phenetic classification. *See* phenetics.

phenol, 1. a toxic aromatic alcohol (C_6H_5OH) containing the benzene ring with one hydroxyl group, known also by the names carbolic acid, phenic acid, oxybenzene, or phenyl alcohol; phenol is used in chemistry e.g. as a deproteinizing agent, formely also as a pesticide and disinfectant; **2.** any of the class of aromatic organic compounds containing one or more hydroxyl groups attached directly to the benzene ring; many plants produce phenols, especially after being damaged by herbivores. *See* chemical defence.

phenology (Gr. *logos* = word, discourse), a branch of biology dealing with relations between periodic biological phenomena, as e.g. flowering, reproduction, migration, irruption, etc., and environmental, especially climatic factors, such as temperature and light; thus, it explains seasonal and annual variations in these functions.

phenolphthalein, a colourless crystalline solid, $C_{20}H_{14}O_4$; its alcoholic solution is used as a pH indicator which turns deep purple-red in alkaline solutions.

phenomenalism (Gr. *phainomenon* = phenomenon), a philosophical concept stating that observed phenomena are true, and the only objects of knowledge, the only form of reality. Thus the meaning of science is only to recognize and classify phenomena.

phenon (Gr. *phainein* = to appear, show), a phenotypically uniform group of individuals verified by methods of numerical taxonomy.

phenotype (Gr. *typos* = model, type), the entity of the visible or otherwise detectable characters of an individual determined by the interaction of its → genotype and environment. *Adj.* **phenotypic**.

phenotypic value, a measure used in quantitative genetics to describe a certain character of an individual, e.g. the length of an individual; it is indicated as a certain unit of measure for this character, and usually includes the deviation from the mean of the population. Thus, the phenotypic value describes the individual as observed. The phenotypic value (P) consists of the → genotypic value (G) and the → environmental deviation (E) so that $P = G + E$.

phenoxy compounds, chlorinated organic compounds containing the phenoxy radical C_6H_5O-, derived from 2,4-dichlorophenol and acetic or propionic acid; e.g. synthetic → auxins used as herbicides, e.g. → 2,4-D, MCPA, dichloroprop, and mecoprop. *See* herbicides.

phenylalanine, abbr. Phe, phe, symbol F, an aromatic α-amino acid, $C_6H_5CH_2CH(NH_2)$-COOH, commonly present in proteins; an essential amino acid for animals serving also as a precursor for tyrosine and → catecholamines. *See* phenylketonuria.

phenylketonuria, PKU, a rare, inborn error of the amino acid metabolism in man. It is the deficiency of phenylalanine 4-monooxygenase enzyme causing inadequate oxidation of phenylalanine to tyrosine, resulting in disturbances in the function of the nervous system and the impairment of pigment synthesis.

phenylmethylsulphonyl fluoride, PMSF, α-toluenesulphonyl fluoride; a → protease inhibitor used in homogenization and isolation media for cell organelles and membranes.

pheophytin, phaeophytin (Gr. *phaios* = grey, *phyton* = plant), an electron transport pigment of the photosystem II; is derived from → chlorophylls. *See* photosyntesis.

pheromone (Gr. *pherein* = to bear, *horman* = to set in motion, to excite), a messenger substance of animals transmitting a hormone-like signal to other individuals of the same species which have special sensory chemoreceptors for this substance; e.g. **sexual pheromones** released by some insect females allure males or activate a sexual state and reproductive urge in the opposite sex (*see* primer signal). **Alarm pheromones** (alarm substances) are secreted at the threat of danger. Many pheromones of social animals, e.g. several insects, regulate also physiological processes, such as the growth and development of individuals. Pheromones are effective in minute quantities being dispersed either by contact (**tactile pheromones**) or by water and air flows. In vertebrates, the pheromones are scent signals (*see* scent marking) which are important especially in social behaviour. *Cf.* allomone.

pheromone trap, a poison trap used especially

for catching pest insects which are attracted to the trap by their natural → pheromones.

philopatry (Gr. *philein* = to love, *patra* = fatherland), → place fidelity.

phlegm (Gr. *phlegma*), slime, mucus; in ancient antique medicine of *Hippokrates* the phlegm was one of the four basic substances (humours) of the human body believed to affect the character and health of a person. *Adj.* **phlegmatic**, characterized by an excess of phlegm, therefore calm, dull, and apathetic.

phloem (Gr. *phloos* = bark), a complex plant tissue conducting organic compounds such as sugars, amino acids, and some inorganic mineral ions; belongs to the → vascular tissues. The phloem consists of different cell types, in → Angiospermae of **sieve tubes** with associated **companion cells**, **phloem parenchyma cells**, and **phloem fibres**. More primitive → Gymnospermae have separate **sieve cells**, instead of continuous sieve tubes. *See* conducting tissues.

phloem fibre, a long, thick-walled cell belonging to the → supporting tissues, occurring in the structure of → phloem.

phloem parenchyma (Gr. *parenchein* = to pour in beside), the → parenchyma that occurs in the → phloem tissue of a plant, storing organic compounds.

phloem protein, → P protein.

phloem ray, *see* ray.

phlogiston (Gr. *phlogistos* = inflammable), an imaginary substance which was supposed to be liberated in combustion. The phlogiston theory was accepted in the 17th and 18th centuries when the burning process was not understood.

Pholidota (Gr. *pholis* = scale), **pangolins** (**scaly ant-eaters**); a mammalian order the species of which were previously classified to Edentata. The proportionally long body is extensively covered by horny scales, the head is small, and the teeth are lacking. They catch ants and termites by their long and slimy tongue. The order includes only one genus, *Manis*, found in tropical regions, four species in Africa and three in Asia.

phon(o)- (Gr. *phone* = sound, voice), denoting relationship to sound, voice, tone, or speech.

phonocardiography (Gr. *kardia* = heart, *graphein* = to write), the graphic registration of heart sounds by the **phonocardiograph**; **phonocardiogram**, the graphic record obtained. *Adj.* **phonocardiographic**.

phonogram (Gr. *gramma* = letter), a graphic curve depicting the intensity and duration of a sound.

phorbol esters, diesters of phorbol; polycyclic organic compounds obtained from croton oil. They act in animal tissues as tumour promoters, i.e. cocarcinogens which facilitate growth of a tumour induced by a carcinogen. They activate → protein kinase C and increase the concentration of various growth factors. Phorbol esters are used e.g. in cell research; a common compound is phorbol myristyl acetate.

phoretic dispersal, phoresy (L. *phoresia* < Gr. *phorein* = to carry along), the dispersal of nonparasitic animals by the aid of other, larger animals; occurs especially among wingless invertebrates, such as mites, which tend to attach to the body of flying insects. *Cf.* endozoic dispersal.

Phoronida (L. *Phoronis* = Gr. *Io*, a mythical priestess), **phoronids**, **horseshoe worms**; a small invertebrate phylum including two genera and about ten sessile, unsegmented, worm-like species inhabiting chitinous tubes either in the bottom mud of oceans, or attached to rocks or clam shells, etc. Phoronids have blood vessels and the coelom is divided into various parts by membranes (**mesocoel** and **metacoel**); the mouth includes a horseshoe-shaped filter-feeding organ, **lophophore**.

phorozooid (Gr. *phorein* = to carry along, *zoon* = animal), *see* probud.

phos- (Gr. *phos* = light), denoting relationship to light.

phosph(o)-, phosphor(o)- (Gr. *phosphoros* = light-bearing), denoting relationship to phosphorus, or phosphoric compounds.

phosphagen (Gr. *gennan* = to produce), any organic compound which liberates inorganic phosphorus; particularly an organic high-energy phosphate like **creatine phosphate** (phosphocreatine) or **arginine phosphate** (phosphoarginine).

phosphatases, enzymes of the sub-subclass EC 3.1.3. of the hydrolase class (EC 3.), catalysing the release of inorganic phosphate from phosphoric esters; e.g. **acid phosphatase** (EC 3.1.3.2) which occurs in → lysosomes of animal cells and is most active below pH 6, and **alkaline phosphatase** (EC 3.1.3.1) which has an optimum pH of about 9; these can be localized cytochemically in cell membrane fractions. Phosphatase determinations from blood are performed for diagnostic purposes to indicate cell damages. Plant cells also contain phosphatases, e.g. in the nucleus, spherosomes, membranes, and cell wall.

phosphate, a salt of phosphoric acid, H_3PO_4, as phosphate ion PO_4^{3-}, hydrogen phosphate ions HPO_4^{2-} and $H_2PO_4^-$; phosphates are used as fertilizers.

phosphatide, the former name for → phosphatidic acid and phosphatidate.

phosphatidic acid, diacylglycerol 3-phosphate, the simplest → phosphoglyceride; any of the organic acids formed from glycerophosphoric acid when the remaining two hydroxyl groups of the glycerol are esterified with fatty acids; is the key intermediate in the biosynthesis of the other phosphoglycerides, i.e. many → phospholipids. The radical of a phosphatidic acid is called **phosphatidyl (Ptd)**, salts and esters are called **phosphatidates**. The former name for phosphatidic acid was **phosphatide**.

phosphatidylcholine, PtdCho, a → phosphoglyceride formed from phosphatidic acid and choline by an ester linkage, having many compositions depending on the structures of the two component fatty acids; the principal → phospholipid in most membranes of higher organisms. *Syn.* lecithin.

phosphatidylethanolamine, PtdEth, a → phosphoglyceride formed from phosphatidic acid and ethanolamine by an ester linkage; a common phospholipid found in cellular membranes of plants and animals, especially in neurones. *See* cephalin.

phosphatidylinositol, PtdIns, a → phosphoglyceride formed from phosphatidic acid and inositol by an ester linkage; a common → phospholipid in cellular membranes. *Syn.* phosphoinositide.

phosphatidylinositol phosphate, PIP, a phosphate derivative of → inositol (phosphatidylinositol 4,5-bisphosphate) found in cellular membranes; in animal tissues a hormone activates a phospholipase which hydrolyses PIP into two intracellular messengers, **diacylglycerol** (DAG) and **inositol triphosphate** (IP_3, inositol 1,4,5-triphosphate). IP_3 e.g. stimulates the release of Ca^{2+} from intracellular stores and DAG activates protein kinase C.

phosphatidylserine, a → phosphoglyceride formed from phosphatidic acid and serine by an ester linkage; a common → phospholipid in cellular membranes. *See* cephalin.

phosphite, a salt of phosphorous acid, H_3PO_3. *Syn.* phosphonate.

phosphodiesterases, enzymes of the sub-subclass EC 3.1.4. hydrolysing phosphodiester bonds such as those between nucleotides in nucleic acids. **Phosphodiesterase I** (5'-exonuclease, EC 3.1.4.1) is present e.g. in snake venom, **phosphodiesterase II** (3'-exonuclease, EC 3.1.16.1) in snake venom and bovine spleen, and **3', 5'-cyclic nucleotide 5'-nucleotidehydrolase** (EC 3.1.4.17) in many animal tissues hydrolysing → cyclic AMP (cAMP) into adenosine monophosphate.

phosphoenolpyruvic acid, PEP, $CH_2C(OPO_3H_2)COOH$; an intermediate in cellular carbohydrate metabolism, → glycolysis; formed from 3-phosphoglyceric acid, and converted further to → pyruvic acid in the reaction which produces ATP. PEP is an intermediary metabolite in the biosynthesis of aromatic amino acids, such as tyrosine and phenylalanine. In photosynthesis PEP serves as a carbon dioxide acceptor in the carbon assimilation of → C_4 plants, i.e. the mesophyll cells contain PEP to which CO_2 is bound in the reaction catalysed by **PEP carboxylase**. The first reaction product is oxalic acetic acid which is further converted into malic acid. PEP serves as the CO_2 acceptor also in → CAM plants which take CO_2 during the night. The ionic form, esters, and salts of PEP are called **phosphoenolpyruvates**.

phosphoglucose isomerase, D-glucose-6-phosphate ketoisomerase, EC 5.3.1.9; an enzyme catalysing in cellular metabolism the reverse reaction between glucose 6-phosphate and fructose 6-phosphate acting in → glycolysis.

phosphoglyceric acid, PGA, an intermediate in the metabolic pathways of carbohydrates like in → photosynthesis, glycolysis, and the pentose phosphate pathway; its ionic form, esters, and salts are called **phosphoglycerates**. When CO_2 is bound in the dark reactions of photosynthesis, **3-phosphoglyceric acid** is formed and then converted in phosphorylation into **1,3-diphosphoglyceric acid**.

phosphoglyceride, a → lipid consisting of a glycerol backbone, two fatty acid chains, and a phosphate group or phosphorylated alcohol; the simplest phosphoglyceride is → phosphatidic acid (diacylglycerol 3-phosphate).

phospholipases (Gr. *lipos* = fat, lipid), → lipases hydrolysing the ester bonds of → phospholipids; several types occur in organisms: A_1, A_2, and B release fatty acids, and C and D release diacyl glycerol or → phosphatidic acid from phospholipids. Enzymes A_1, (EC 3.1.1.32), A_2 (EC 3.1.1.4), and C (EC 3.1.4.3) occur in all mammalian tissues; A_2 is common in venoms, and C is also found in bacteria. Type B (EC 3.1.1.5) is present in snake venom, and type D (EC 3.1.4.4) in plants and some bacteria. Phospholipases are also called lecithinases.

phospholipids, → lipids containing phosphorus, derived either from → glycerol (a three-carbon alcohol) or → sphingosine, (a more complex alcohol). They are the main components of membrane lipids; e.g. phosphoglycerols and plasmalogens with a glycerol backbone, or sphingomyelins with a sphingosine backbone. *See* phosphatidylcholine, phosphatidylethanolamine, phosphatidylinositol, phosphatidylserine.

Phosphon-D, *see* growth retardants.

phosphonium ion, PH_4^+, a univalent ion or radical that forms compounds analogous to those of the ammonium ion, NH_4^+.

phosphoproteins, proteins containing a covalently bound phosphate group in one or more amino acid residues (serine, threonine, and seldom tyrosine). The liberation of inorganic phosphate (orthophosphate) from phosphoproteins is catalysed by phosphoprotein phosphatase.

phosphorescence, a form of → luminescence; a light emission followed by the absorption of excited radiation; it is generated when electrons, which are transferred by radiation to the outer shell of an atom, move back to the original shell. Emitted light is of longer wavelength (visible light) than the excitation radiation. Light emission resembles → fluorescence but continues still after the source of irradiation is removed (may last from seconds to hours).

phosphoric acids, inorganic acids: **1.** orthophosphoric acid, H_3PO_4; **2.** phosphorous acid, H_3PO_3; **3.** metaphosphoric acid, glacial phosphoric acid, $(HPO_3)_n$, a deliquescent solid obtained from phosphorus pentoxide; **4.** pyrophosphoric acid, $H_4P_2O_7$, a crystalline solid formed from phosphorus pentoxide and two molecules of water. Organic phosphoric acid compounds are important biological molecules in cell metabolism, as e.g. adenosine triphosphate, ATP.

phosphor imaging, an alternative technique to the traditional → autoradiography; capable of imaging and quantifying both radioisotopic (gamma and beta emitters) and chemiluminescent samples, e.g. on electrophoretic gels and membranes, with accuracy which exeeds that of X-ray films. A sample is exposed to the storage phosphor screen which is then scanned using a rapid laser scanner. The data are collected for display and analysis.

phosphorous acid, H_3PO_3, a colourless, deliquescent crystalline substance; used as a reducing agent; its salts are **phosphites** (phosphonates). *Syn.* phosphonic acid.

phosphorus, a non-metallic element, symbol P, atomic weight 30.974; appears in several allotropic forms of which white and red phosphorus are the commonest. White phosphorus is waxy, poisonous, and very flammable; red phosphorus is a non-poisonous dark red, inflammable powder. Phosphorus is found widely in nature, always in combinations, such as phosphates and phosphites. All organisms and cells need phosphoric compounds and ions, because phosphorus is a structural element in many essential molecules, such as nucleic acids, phospholipids, and energy rich compounds like ATP. Bones are rich in calcium phosphate, → apatite. Plants take phosphorus either as HPO_4^{2-} or $H_2PO_4^-$ ions; the ion form depends on the pH of the soil.

phosphorylase kinase (Gr. *kinein* = to move), a transferase enzyme (EC 2.7.1.38) which enhances the activity of another enzyme, → phosphorylase; catalyses the reaction: 2 ATP + phosphorylase b —> 2 ADP + phosphorylase a (the b form is inactive and the a form is active).

phosphorylase phosphatase, an intracellular hydrolase enzyme (EC 3.1.3.17) which decreases the activity of another enzyme, → phosphorylase; it catalyses the reaction: phosphorylase a + 2 H_2O —> phosphorylase b + 2 orthophosphates (the a form is active and b form is inactive).

phosphorylases, transferase enzymes of the sub-subclass EC 2.4.1. catalysing the transfer of an inorganic phosphate to some organic acceptor; often catalyse cleavages of glycoside bonds of carbohydrates by adding an inorganic phosphate and forming from a sugar molecule the corresponding sugar phosphate, such as glucose 1-phosphate; e.g. **nucleoside phosphorylase** that catalyses the phosphorolysis of nucleosides in all cells, **glycogen phosphorylase** which in animal cells catalyses the formation of glucose 1-phosphate from glycogen, and **starch phosphorylase** correspondingly in plant cells; the latter also catalyses the synthesis of starch, e.g. in potato.

phosphorylation, a metabolic reaction in which inorganic phosphate is bound to an organic molecule; e.g. **1.** → **oxidative phosphorylation** in which phosphate is bound to ADP and high-energy ATP is formed; **2. photophosphorylation (photosynthetic phosphorylation)**, the energy-producing (ATP) reaction of → photosynthesis; **3. phosphorylation of proteins**, the addition of phosphate groups to hydroxyl groups of proteins; the activity of enzyme proteins (*see* kinases) are often regulated by phosphorylation. *See* adenosine triphosphate.

phosphotransacetylase, acetyltransferase, phosphate acetyltransferase; an enzyme (EC 2.3.1.8) that in cells transfers an acetyl group (CH_3CO-) from acetyl lipoamide to coenzyme A forming → acetyl CoA.

photic (Gr. *phos* = light), pertaining to light, illuminated.

photic receptors, *see* photoreception.

photic zone/region, the upper layer of a water body into which light penetrates thus supporting photosynthesis in plants. *Cf.* euphotic, dysphotic, aphotic zone.

photoactive, reacting chemically to (sun)light or ultraviolet radiation.

photoautotroph, *see* autotroph.

photoblastic (Gr. *blastos* = sprout, germ), sensitive to light; e.g. a seed needing light for germination.

photochemical smog, *see* smog.

photodynesis, photodinesis (Gr. *dynamis* = force), the acceleration of cytoplasmic streamings in plant cells caused by intensive light; the mechanism is not known.

photoheterotroph, *see* heterotroph.

photoinhibition, inhibition by light; occurs e.g. as the decrease of CO_2 uptake and O_2 production in → photosynthesis under illumination which exceeds the level of saturable illumination; it is caused by damage in photosystem II. Stress factors such as unsuitable temperature, decrease of CO_2, or lack of water, increase the sensitivity of plants to photoinhibition.

photokinesis, *see* kinesis.

photolysis, the breakdown of any material caused by light or ultraviolet radiation. The most important photolytic reaction is the photolysis of water which takes place in → photosynthesis. The reaction liberates oxygen and electrons which fill the electron deficit of photosystem II (PSII).

photon, a quantum of light; its energy is expressed: $E = h \cdot c/\lambda$ in which h = the Planck constant ($6.6254 \cdot 10^{-34}$ Js), c = the velocity of light (appr. $3 \cdot 10^{10}$ cm/s), and λ = the wavelength of light. The energy content of one quantum increases at shorter wavelengths.

photonasty (Gr. *nastos* = pressed together, firm), a type of → nastic movement by which plants or plant parts respond to changes in light intensity; e.g. closing of flowers and leaflets in the evening. *Cf.* phototropism.

photooxidation (Gr. *oxys* = acid, sour), oxidation caused by light energy.

photoperiodism, photoperiodicity (Gr. *periodos* = period of time), the ability of an organism to respond to the relative lengths of lightness and darkness; appears e.g. in growth and development periods of organisms. Photoperiodism plays an important role in the germination of seeds as well as in the development and flowering of plants. Photoperiodic responses in plants depend not only on the duration of light and dark cycles but also on the quality (wavelength) of light, e.g. red and far red light. In plants, the substances reacting to the changes of light are → **phytochromes**. Also → biorhythms, such as breeding and migration periods of animals, are influenced by periodical changes of light; photoperiodism is shown to be the most important factor controlling and pacing the → life cycles of animals.

photophore (Gr. *phorein* = to carry along), a light-emitting organ occurring in many multicellular animals, such as in deep-sea fishes, molluscs, crustaceans, and in some insects. Light (*see* bioluminescence) may be produced by bacteria living in the photophores, or is produced by glandular cells of an animal itself. Different colours and the focusing of light are generated by the aid of a lens and various reflecting pigments in the photophore; the rhythmic activity of light emission is controlled by the nervous system.

photophosphorylation, the production of ATP using light energy in → photosynthesis.

photopigments, light-sensitive pigments. *See* photoreception.

photoreception (L. *recipere* = to receive), the process by which certain cells detect (sense) light energy, usually visible radiation energy, i.e. of wavelengths approx. from 400 to 700 nm. Photoreception in multicellular animals occurs by the action of specialized cells, or parts of cells, called → **photoreceptors** (light receptors) which have special **photopigments** (e.g. → rhodopsin) reacting chemically to light. The photoreceptors of metazoans are usually arranged to form a special light organ, the → **eye**.

There are several pigment systems in plants which absorb light. The most important are photosynthetic pigments, i.e. chlorophylls and carotenoids, which act as photoreceptors (*see* photosynthesis). Other photopigments are e.g. → phytochromes, which absorb red light and regulate photoperiodic phenomena, and yellow pigments (e.g. flavins and carotenoids) in → phototropism. *Adj.* **photoreceptive**, responding to light, sensitive to light.

photoreceptor, 1. a sense organ responding to light, such as the → eye; **2.** a cell or parts of a cell responding to light, e.g. cones and rods in the vertebrate eye; **3.** a light-sensitive molecule reacting chemically to light, e.g. rhodopsin, iodopsin, phytochrome, or chlorophyll. *See* photoreception.

photorespiration (L. *respirare* = to breathe), the reactions based on the oxidation properties of an enzyme, **ribulosebisphosphate carboxylase-oxygenase** (**Rubisco**). This enzyme which has a carboxylating role in → photosynthesis also oxidizes ribulose 1,5-bisphosphate partially into phosphoglycolate, thereby diminishing the yield of the photosynthetic carbon assimilation and the amount of sugar phosphates produced in the carboxylation process.

photosynthesis (Gr. *synthesis* = incorporation), the carbon assimilation carried out by plants and some bacteria by means of light energy. Usually photosynthesis is divided into light and dark reactions, i.e. Calvin cycle.

The **light reactions** take place in the thylakoid membranes of **chloroplasts** in which the absorption of light quanta or → photons is carried out by → **chlorophylls** and other

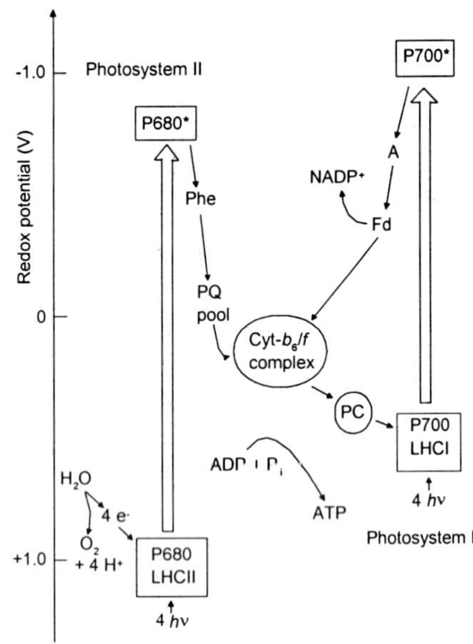

Simplified Z-scheme describing the transport of electrons from water to $NADP^+$ by two photosystems, and during this ATP is formed. A = primary electron acceptor, Cyt = cytochrome, Fd = ferredoxin, LHCI and LHCII = light harvesting chlorophyll proteins, PC = plastocyanin, Ph = pheophytin, PQ = plastoquinone, P680* and P700* = reaction centre chlorophylls excited by photons (*hν*)

pigments. Absorbed energy is transferred from those light-harvesting chlorophyll-protein complexes to the reaction centres of the photosystems (PS). Photosystem I (PSI) is rich in chlorophyll a and photosystem II (PSII) in chlorophyll b. Chlorophyll in the reaction centre of PS I absorbs effectively light of wavelength ca. 700 nm and is therefore called P700. The absorption maximum of chlorophyll in the reaction centre of PSII is close to 680 nm (P680). Chlorophyll molecules of PS are raised into an excited stage by light, thus **liberating electrons**.

The electron deficit of PSII is compensated by electrons liberated in the photolysis of water, and the electrons are then transferred from PSII to PSI through the chain of various compounds, such as plastoquinone, cytochrome *b/f*, and plastocyanin. The active electron transport chain and the photolysis of water produce **hydrogen ions** (H^+, protons) which are transferred into the lumens of thylakoids. The electrons are further transported via proton channels into the stroma of chloroplasts where high-energy **ATP** is produced by ATP synthase.

ATP is one of the end products of the light reactions and is then used for CO_2 assimilation (carbon dioxide assimilation) in the Calvin cycle that is located in the chloroplast stroma. The other end product, **NADPH**, is produced by means of the reduction energy of PSI: NADP —> NADPH + H^+. NADPH acts as a reducing agent for CO_2 in dark reactions. The light reactions of photosynthesis produce oxygen from water essential for all aerobic organisms. Light reactions require the presence of some ions such as chloride (Cl^-), calcium (Ca^{2+}) and manganese (Mn^{2+}).

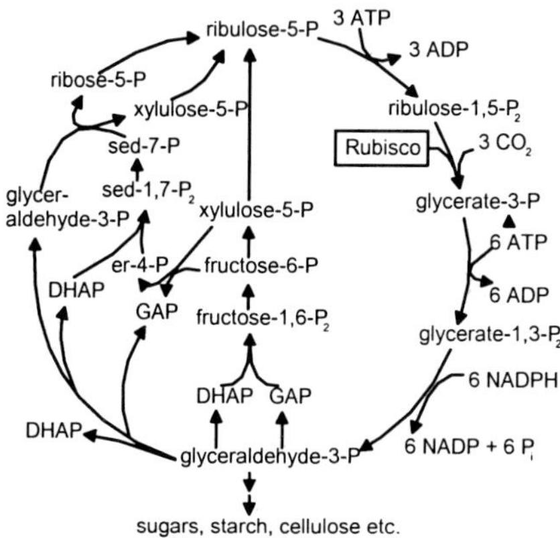

Reactions in Calvin cycle (dark reactions of photosynthesis). ADP = adenosine diphosphate, ATP = adenosine triphosphate, DHAP = dihydroxyacetone phosphate, er-4-P = erythrose 4-phosphate, GAP = glyceraldehyde 3-phosphate, sed-7-P = sedoheptulose 7-phosphate, sed-1,7-P_2 = sedoheptulose 1,7-diphosphate.

CO_2 enters the leaves through stomata and is reduced to carbohydrates in chloroplasts of the mesophyll cells (in C_4 plants only in bundle sheath cells; → phosphoenolpyruvic acid). CO_2 is incorporated into **ribulose 1,5-bisphosphate** in the carboxylation reaction catalysed by **Rubisco** (ribulose 1,5-bisphosphate carboxylase-oxygenase), and two 3-phosphoglyceric acid molecules are formed. Phosphorylation with ATP and reduction with NADPH lead to the synthesis of **glyceraldehyde 3-phosphate**. In chloroplasts, when **starch** is synthesized, many sugar phosphates acting as intermediates in the formation of ribulose 1,5-bisphosphate. Different sugars and cell wall components are synthesized outside the chloroplasts.

photosynthetically active radiation, PAR, electromagnetic radiation approx. from 400 to 700 nm, particularly the red area of the spectrum, which is most effective in photosynthetic reactions. PAR corresponds approximately to the wavelengths of visible light.

photosystem, PS, the multimolecular chlorophyll-protein complexes (PSI and PSII) forming the centres of the light reactions of → photosynthesis.

phototaxis, pl. **phototaxes** (Gr. *taxis* = order, arrangement), a free movement of an organism in response to light, either positive (towards a light source) or negative (away from light), e.g. the movement of some bacteria and microalgae to optimum light intensity; also the orientation of → chloroplasts to light. *Adj.* **phototactic**. *Cf.* heliotaxis, phototropism, heliotropism.

phototropism (Gr. *tropos* = turn, turning), a → tropism in which light is the orienting stimulus; e.g. **1.** a growth movement of plants in response to the stimulus of light; the shoot of a plant is usually positively phototropic, i.e. turns towards light. The shoots of many herbaceous plants are negatively phototropic in full sun, but positively phototropic in shade; the roots are usually negatively phototropic. A **phototropic** curvature is caused at least partially by the raised concentration of → auxin on the dark side of a shoot; **2.** an orientation of sedentary animals in response to light, found e.g. in corals. *Cf.* heliotropism, phototaxis, heliotaxis.

phragmoplast (Gr. *phragmos* = fence, enclosure, *plastos* = moulded, formed), a plate composed of vesicles and microtubules in the division plane of a plant cell in the beginning of cell division; precedes the formation of the actual cell wall.

phren(i)-, phreno-, phrenico- (Gr. *phren* = diaphragm, mind), **1.** denoting relationship to the diaphragm; **2.** pertaining to mind. *Adj.* **phrenic**.

phycobilins (Gr. *phykos* = alga, *bilis* = bile), plant pigments, phycocyanin and phycoerythrin, acting as accessory photosynthetic pigments. *See* phycobiliproteins.

phycobiliproteins, chromoproteins found especially in → phycobilisomes of cyanobacteria and red algae, having **phycobilins** as pigment parts; act as accessory pigments in → photosynthesis by transferring light energy to chlorophyll a. The phycobilin in red algae is **phycoerythrin**, which gives the algae their red colour; it is also found in some cyanobacteria and species of Cryptophyceae. Characteristic of cyanobacteria is **phycocyanin** giving the organisms the blue-green colour. *Syn.* biliproteins.

phycobilisome (Gr. *soma* = piece, body), a hemispherical structure harvesting light energy on the surface of the photosynthetic lamellae of → cyanobacteria and on lamellae of rhodoplasts (corresponding chloroplasts of green plants) in red algae. Phycobilisomes contain → **phycobiliproteins** acting as accessory pigments in → photosynthesis, transferring light energy to chlorophyll a.

phycocyanin (Gr. *kyanos* = blue), *see* phycobiliproteins.

phycoerythrin (Gr. *erythros* = red), *see* phycobiliproteins.

phycology (Gr. *logos* = word, discourse), the study of algae.

phyletic evolution (Gr. *phyle* = tribe), an evolutionary change that appears in the sequence of generations in such a way that no branching in the evolutionary line occurs (*see* evolution). This is typical of the more or less linear evolution of a → phylum.

phyletic gradualism (L. *gradus* = step, grade, value), a doctrine of a school of → evolution theory arguing that speciation occurs gradually in small steps without periods of rapid evolution. This appears in fossil records as a continuous or **gradual change**. *Cf.* punctuated equilibrium model.

phylloclade (Gr. *phyllon* = leaf, *klados* = shoot, branch), **1.** a green flat photosynthetic plant stem (e.g. in cacti); called also **cladode** if resembling a foliage leaf (e.g. butcher's broom); **2.** a flat, leaf-like thallus in some lichens. *Syn.* platyclade, platycladium.

phyllode (Gr. *eidos* = form), a flat and broad, leaf-like and photosynthetic petiole in plants, e.g. *Acacia*.

phyllopodium (Gr. *pous* = foot), *see* biramous appendage.

phyllosoma (Gr. *soma* = body), a larval stage of some lobsters (e.g. *Panulirus*) comprising a flattened, transparent body and long thoracic appendages.

phyllotaxy, phyllotaxis, pl. **phyllotaxies, phyllotaxes** (Gr. *taxis* = order), the arrangement of leaves on the plant stem. The main types are **spiral** and **whorled arrangements**. In the spiral type there is only one leaf in every node, while in the latter there are two or more leaves in each node. In the spiral type the intervals between the successive leaves are usually equal, so that the leaves are distributed uniformly about the shoot. In the most simple spiral type (called **truly alternate** or **distichous**) leaves are 180° apart, so that the passage from one leaf to the leaf precisely above it involves one whole circuit around the stem. In this the phyllotaxy is of the type 1/2 (e.g. in beech); the numerator of this fraction is the number of revolutions of the stem which are necessary to reach the next leaf of the same **orthostichy** (a vertical row of leaves), and the divisor shows the number of leaves encountered to reach this leaf. Other known possibilities are 1/3, 2/5, 3/8, etc. In the whorled arrangement the number of leaves in every node may vary from 2 to 20 or more, a special type being the **decussate** arrangement where there are always two leaves on different sides of the stem, the leaves being arranged in four orthostichies. *Syn.* leaf arrangement.

phylo- (Gr. *phyle* = tribe), denoting relationship to tribe, race, or phylum.

phylogenesis (Gr. *genesis* = descent), → phylogeny.

phylogenetic taxonomy, phylogenetics, phylogenetic systematics; a branch of biology dealing with biological classification concerned with reconstructing **phylogenies**, i.e. evolutionary relationship between different species or other taxa. With schematic presentations (**phylogenetic trees**) it describes the evolutionary relationships between different taxa, i.e. the historical course of speciation. The present method of phylogenetics is → **cladistics** which makes use of comparisons of different species or groups at different levels of biological organization, including molecular studies.

phylogeny, phylogenesis, the evolutionary history of a species or other taxa. *Adj.* **phylogenetic** or **phylogenic**. *Cf.* ontogeny.

phylotype, a common, genetically determined ground plan for the embryonic development of a structure in animals belonging to the same → phylum. *Cf.* zootype.

phylum, pl. **phyla,** one of the chief taxa in

animal classification between the kingdom and class; comprises subphyla, superclasses, and all the lower taxa. The basic structures of all animals in the same phylum are of the same kind and based on a common evolutionary background. The Animal kingdom comprises more than 30 phyla; e.g. Cnidaria, Platyhelminthes, Annelida, Mollusca, Arthropoda, Echinodermata, and Chordata. The corresponding taxon in botany is → division.

physi(o)- (Gr. *physis* = nature), **1.** denoting relationship to nature, natural; **2.** relating to the body, physical, as distinguished from the mind; **3.** denoting relationship to → physiology, physiological.

physic (Gr. *physikos* = natural, physical), **1.** the art or profession of medicine and therapeutics; **2.** a medicine (drug), especially a cathartic.

physical, 1. pertaining to material things in general; **2.** relating to the body, as distinguished from the mind; **3.** pertaining to physics (physical sciences).

physics, the branch of science dealing with the phenomena of matter and energy and the changes of the matter without losing its chemical structure and properties; includes mechanics, electricity, magnetism, heat, radiation, sound, optics, and nuclear phenomena.

physiognomy (Gr. *gnomon* = judge), **1.** determination of the characteristics (especially of man) from facial features or countenance (habitus); **2.** in botany, classification of vegetation and plants according to their outer features.

physiologic, pertaining to normal, as opposed to pathologic.

physiological, pertaining to the function of an organism or a part of it, or to → physiology.

physiological age, the age of an animal associated with its developmental stage in the life cycle; it is not equal to a chronological age but dependent on various environmental factors, such as temperature, food, light rhythm, etc. Especially in poikilothermic animals, the physiological age may greatly differ from the actual chronological age.

physiological race, *see* race.

physiological solution, any of the solutions corresponding to body fluids of animals and used for the maintenance or culture of cells, tissues and organs. For a short incubation of tissues different types of **physiological salt solutions** (physiological salines, physiological buffer solutions, such as → Ringer's solutions) with necessary ions in isotonic concentrations and pH buffers, are used, but especially for cell and tissue cultures many additional substances, such as glucose and other energy sources, amino acids, fatty acids, vitamins, and trace elements are needed in the **culture media**.

physiology (Gr. *logos* = word, discourse), a branch of science dealing with the functions of living organisms and their parts; according to the research objects different branches are distinguishable, e.g. animal physiology, plant physiology, cell physiology, or ecophysiology. More concise subjects are e.g. insect physiology, avian physiology, mammalian physiology, etc. Medical physiology is an example of applied physiology.

physis, pl. **physes** (Gr. *phyein* = to generate, produce), epiphysial cartilage; the segmental area in the tubular bone where growth in length mainly occurs; locates between the head (**epiphysis**) and the shaft (**diaphysis**). *Cf.* hypophysis.

physo- (Gr. *physa* = bellows), pertaining to air or gas.

physoclistous fishes (Gr. *kleiein* = to close), *see* swim bladder.

physostigmine, an alkaloid obtained from the seeds of Calabar bean (*Physostigma venenosum*); stimulates cholinergic action in animals by inhibiting the activity of cholinesterase enzyme and thus the decompositon of → acetylcholine. Physostigmine as well as its derivative **neostigmine** are used in physiological research and medicine. *Syn.* eserine.

physostomous fishes (Gr. *stoma* = mouth), *see* swim bladder.

phyt(o)- (Gr. *phyton* = plant), denoting relationship to a plant; associated with plants.

phytal (zone), an ecosystem in shallow waters of shores; the vegetation consists of phanerogames, large algae, and microscopic epiphytes.

phytin, the calcium magnesium salt of **phytic acid** (phosphoric acid ester of → inositol); functions as a phosphate reserve in some seeds.

phytoalexins (Gr. *alexein* = to protect), substances produced by plants in response to tissue damage or infection by fungi or bacteria, their synthesis being induced by abiotic and biotic → elicitors; include e.g. coumarin, phenanthrene, terpenes, and flavonoids.

phytobiotic (Gr. *bios* = life), pertaining to an organism living within plants; e.g. some protozoans.

phytochromes (Gr. *chroma* = colour), coloured protein pigments in plants acting as acceptors of red light and controlling physiological processes, such as the germination of seeds and flowering of plants (*see* photoperiodism); exist in two interconvertible forms: phytochrome 660 (P_{660}) absorbs red light (R, maximally at wavelength 660 nm), which converts P_{660} into phytochrome 730 (P_{730}), i.e. the form absorbing far-red light (FR, maximally at 730 nm) that is the physiologically active form. P_{730} is converted back into P_{660} by far-red light; this change can also occur slowly in darkness. All changes of phytochromes are reversible, and their physiological response is determined by the radiation type, e.g. germination is induced by red light.

phytogenesis (Gr. *genesis* = birth, origin), the evolution and developmental history of plants.

phytohaemagglutinin, PHA, Am. **phytohemagglutinin** (Gr. *haima* = blood, L. *agglutinare* = to glue together), phytolectin; a → lectin obtained from beans (*Phaseolus sp.*) and used in immunology to activate the division of T lymphocytes; also causes agglutination of red blood cells.

phytol, $C_{20}H_{39}OH$, a long-chain alcohol forming the tail of the chlorophyll molecule, attached to the porphyrin ring by an ester linkage. Lipid solubility of chlorophylls is dependent on the hydrocarbon chain of the phytol. Phytol is a constituent of vitamins E and K and used in the manufacture of these vitamins.

Phytomastigophora (Gr. *mastix* = whip, *phorein* = to carry along), **green flagellates**; a flagellate class of protozoans in the phylum Sarcomastigophora; possess chloroplasts thus being capable of photosynthesis, producing paramylum in the light. In the dark they feed heterotrophically as true animals. Known genera are e.g. *Euglena* and *Volvox.* These organisms are often also regarded as unicellular algae. *Cf.* Zoomastigophora, Flagellata.

phytophage (Gr. *phagein* = to eat), an animal feeding on plants; especially refers to invertebrates, e.g. insects. *Cf.* herbivore. *Adj.* **phytophagous**.

phytoplankton (Gr. *planktos* = wandering), **plant plankton**; a component of → plankton comprising bacteria, cyanobacteria and microscopic algae, including non-motile and motile cells and colonies, living in open water. Phytoplankton is important in the primary production of aquatic ecosystems carrying on photosynthesis near the surface, providing food for grazing zooplankton, and giving rise to complex food chains.

phytotron, a culture cabin for plants where conditions, such as light, humidity, and temperature, are accurately controlled and programmed.

pia mater (L. = tender mother), *see* meninx.

PIC, → polymorphism information content.

Piciformes (L. *picus* = woodpecker, *forma* = shape), an order of birds which have **climbing feet** (i.e. two of the four toes are directed forward and two backward) and a specialized, chisel-like bill; the long tongue can be protruded for capturing insects from inside wood. All species nest in holes. The order comprises six families: **woodpeckers** (Picidae) including 200 species of all the 381 in the order, **toucans** (Ramphastidae), **honeyguides** (Indicatoridae), **jacamars** (Galbulidae), **puffbirds** (Bucconidae), and **barbets** (Capitonidae).

picloram, 4-amino-3,5,6-trichloropicolinic acid, used as a → herbicide.

pico- (It. *piccolo* = small), denoting small; in connection with units of measurement indicates one trillionth part (10^{-12}).

picoplankton (Gr. *planktos* = wandering), a → plankton comprising organisms with a body size from 0.2 to 2.0 μm; commonly comprising → phytoplankton, chiefly bacteria (bacterioplankton). *Cf.* microplankton, nanoplankton.

Picornaviridae (*pico-* + RNA + virus), a family of small (20 to 30 nm) non-enveloped viruses having a single-stranded RNA; includes e.g. polioviruses, rhinoviruses, and 2,4,6-aphthoviruses.

picric acid (Gr. *pikros* = bitter), trinitrophenol; a yellow poisonous substance which is explosive, when dry; used as a fixative for tissue specimens.

picrotoxin (Gr. *toxikon* = arrow poison), an alkaloid obtained from seeds of *Anamirta cocculus*; stimulates the central nervous system of vertebrates and many invertebrates by blocking chloride channels in GABAergic neurones; picrotoxin is used in neurophysiological research and as an antidote for poisoning caused by central nervous system depressants. *Syn.* cocculin.

piezoelectricity (Gr. *piezein* = to press), electricity or electrical polarity produced by mechanical stretch or pressure on non-conducting crystals, such as quartz. The phenomenon is utilized e.g. in sensors of some electrophysiological equipment, e.g. in a **vibrating probe**, which can measure ionic currents occurring in living organisms, e.g. in growing plant embryos, by means of the mechanical effect of the currents on the probe.

PIF, prolactin inhibiting factor, → prolactostatin.

pigeons, *see* Columbiformes.

pigment (L. *pigmentum* = paint), any of colouring chemical elements or compounds. Different types of organic pigments, colouring tissues and cells, exist in animals and abundantly in fungi and plants; the pigment molecule contains at least one double bond. Animal pigments are e.g. → melanins, photosensitive pigments of photoreceptors (e.g. rhodopsin), and many porphyrins, such as → haemoglobin and myoglobin. Plant pigments are e.g. → anthocyans, chlorophylls, and phytochromes. Such pigments as carotenoids and porphyrins (coenzymes of cytochromes) appear generally in all organisms.

pigment cell, a cell type of animals in which a → pigment is located in cellular granules colouring the cell; often the granules are movable changing the intensity of the colour, as in the skin of lower vertebrates. Pigment cells are e.g. → **chromatophores** and **melanocytes** protecting the skin from ultraviolet radiation, and pigment cells of the eye (in the iris and tapetum nigrum) changing light access and reflexion. There are also pigment cells in the central nervous system. *Syn.* colour cell.

PIH, prolactin inhibiting hormone, → prolactostatin.

pila, pl. **pilae** (L.), a pillar or pillar-like anatomical structure; e.g. pilae (trabeculae) of spongy bone.

pileus, pl. **pilei** (L. = felt cap), an umbrella-shaped cap of mushrooms and toadstools. *Adj.* **pileate**.

pilidium, pl. **pilidia** (Gr. *pilidion,* dim. of *pilos* = felt cap), a helmet-shaped, ciliated and free-living larva of nemerteans which has a ventral mouth but no anus. *See* larva.

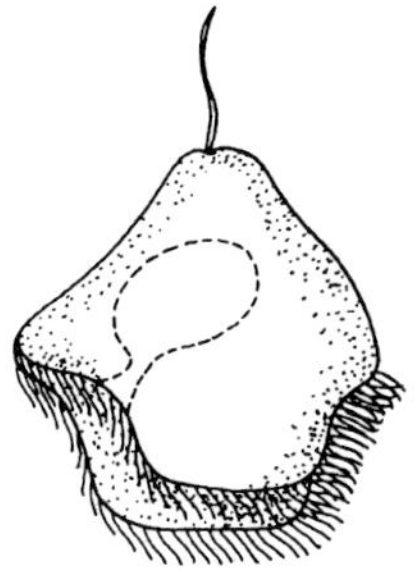

PILIDIUM

piliferous layer (L. *ferre* = to carry), the area in the plant root epidermis giving rise to root hairs.

pilocarpine, an alkaloid found e.g. in the plant genus *Pilocarpus*; it has cholinergic effect on animals transmitted by muscarinic receptors, i.e. it is a parasympathomimetic agent.

piloerection (L. *pilus* = hair, *erigere* = to rise up), erection of hairs (hair bristling) in mammals, due to sympathetic stimulation usually triggered by cold or fear. *Cf.* ptiloerection.

Piltdown man, a skull of a fossil hominid that was reportedly found in Sussex, England in 1912; in 1953 it was revealed to be falsified using the lower jaw of an orang-utan, skilfully joined to the skull of modern man.

pilus, pl. **pili** (L.), **1.** a hair or hair-like structure on the surface of an animal or plant; **2.** one of filamentous appendages of certain bacteria.

pinacocytes (Gr. *pinax* = board, plate, picture, *kytos* = cavity, cell), plate-like cells on the body surface (**pinacoderm**) of sponges.

Pinales, an order in the class → Coniferae, division Gymnospermae, Plant kingdom; contains the families Cupressaceae, Pinaceae and Taxaceae, i.e. the most important, very widely distributed coniferous trees.

pineal body/gland/organ (L. *corpus pineale* or *epiphysis*), the cone-shaped gland derived from the roof of the third brain ventricle in vertebrates. It consists of endocrine parenchyma cells (pinealocytes), glial cells, and of neuronal axons from the sympathetic nerve cord through which the pineal body is linked to retinohypothalamic neurones and suprachiasmatic nuclei in the brain. In lower vertebrates, the pineal body has also direct neural connections from the brain. In many mammals the gland begins to degenerate in adulthood. The pineal body secretes a hormone, → **melatonin,** that is synthesized from → serotonin. The concentration of melatonin in the blood increases strongly in the dark and decreases in the light, and melatonin is involved in the maintenance of circadian and annual rhythms of animals, e.g. by inactivating the sex organs during the dark seasons.

The pineal body is exceptionally active producing melatonin in arctic mammals (as in reindeer) in the autumn and winter, stimulating e.g. the growth of fur and the activity of → brown fat, and consequently energy metabolism. In lower vertebrates, the pineal body has direct connections to the epithalamus of the brain forming a part of the → pineal eye, and is also involved in the control of changes of skin colour. For many people living in nordic areas, the daylight in winter is not intensive enough for the maintenance of melatonin induced rhythmicity, and people may suffer from disturbances in their circadian rhythms.

pineal eye, an unpaired photosensitive organ located on the top of the head of the tuatara (*Sphenodon*), some lizards and lampreys. It develops from the roof of the diencephalon and from the parietal epithelium, and in many reptiles it degenerates before adulthood. It is best developed in the tuatara (Sphenodontidae) that lives on a few small islands of New Zealand. Its pineal eye has the structure of a simple eye, such as transparent skin, the lens, and retina. Its function is connected with that of the → **pineal body** (pineal gland), i.e. to the control of → biorhythms of an animal. *Syn.* parietal eye, parietal organ, epiphysial eye, median eye.

pingo (Inuit), a palsa-like hummock in the soil in the areas of permanent ice; contains mineral soil and ice, while → palsas contain frozen peat.

pinna, pl. **pinnae** (L.), **1.** feather, wing, fin, or a part resembling these; **2.** the auricle of the ear; **3.** a leaflet of a compound leaf.

pinnate, resembling a feather; having similar parts, such as leaflets (in a pinnate leaf), on opposite sides of an axis.

pinnate venation, a type of leaf venation with a clear middle vein, smaller lateral veins running on both sides; typical of dicotyledons. *See* venation types.

Pinnipedia (L. *pes* = foot), **pinnipeds,** i.e. seals, sea lions, and walruses; a mammalian order which is considered also to be a suborder in the order Carnivora. Pinnipeds are aquatic animals with fin-like limbs, a spindle-shaped body and a thick layer of adipose tissue beneath the skin. The order includes 34 species in three families: **eared seals** (Otariidae), **walruses** (Odobenidae) and **earless seals** (or hair seals, Phocidae).

pinnule (L. *pinnula*, dim. of *pinna* = feather, wing), an anatomical structure in animals resembling a barb of a feather, e.g. **1.** a part of a feeding filament, i.e. reduced → parapodium, in some polychaetes; **2.** a branchlet in radial arms of a crinoid; **3.** a secondary → pinna in a compound leaf.

pinocytosis (Gr. *pinein* = to drink, *kytos* = cavity, cell), the uptake of liquid by a cell; a type of → endocytosis in which extracellular fluid is ingested into the cell. It is the active transport process for macromolecules, e.g. hormones, beginning with a small invagination of the cell membrane which then closes and pinches off forming a fluid-filled vesicle (**pinosome**) inside the cell.

pioneer community, *see* pioneer species.

pioneer species, species which first inhabit empty spaces and thus start a primary → succession; e.g. lichens on rocky shores rising from the sea; pioneer organisms form a **pioneer community**.

PIP_2, phosphatidyl inositol 4,5-bisphosphate. *See* inositol.

PIPES, piperazine N,N'-bis-(2-ethanesulphonic acid), a pH buffer used in solutions for biological and biochemical experiments.

pipette (Fr.) **pipet, 1.** a slender, graduated glass or plastic tube for transferring precise amounts of liquid (or gas); at present automatic pipettes with disposable tips are in common use; also a dropper (Pasteur pipette) used e.g. in cell-biological laboratory work and for the dosage of medicines; **2.** as a verb, to transfer a liquid (or gas) with a pipette.

piscivore (L. *piscis* = fish, *vorare* = to eat), an animal feeding on fish. *Cf.* herbivore, carnivore. *Adj.* **piscivorous**.

pistil (L. *pistillum* = pestle), the female reproductive organ in a flower; consists of one or more carpels, which may be separate or fused together. *Syn.* gynaecium.

pistillate, pertaining to a flower with one or more carpels (a pistil) but no functional stamens. *Syn.* carpellate. *Cf.* staminate.

pit, a hole, cavity, or depression in an anatomical structure; **1.** in animals, e.g. the arm pit, nasal pit, or gastric pits (*foveolae gastricae*) in the gastric mucous membrane, or coated pits in the plasma membrane of many cells associated in receptor-mediated → endocytosis; **2.** → pit organ; **3.** in plants, an unthickened area in the cell wall between adjacent plant cells. In the pit is a pit membrane which is perforated by plasmodesmata (*see* plasmodesm), through which the adjacent cells are in contact. Pits are common in the walls of most plant cells. They may also be structurally more complicated **bordered pits**, which are common in water-conducting elements of higher plants. In these the middle part of the pit membrane is thickened, known as **torus**, which may be pressed by water pressure against the pit borders, the pit functioning thus as a valve regulating water conduction.

pitfall trap, a pot-like trap dug into the soil so that animals falling in cannot escape; the trap contains ethylene glycol solution to kill the animals and some detergent to remove surface tension. The pitfall trap is used especially to catch coleopterans and arachnids.

pith, 1. the soft inner part of an organ, especially of a feather or hair; **2.** the central part of roots or shoots of → vascular plants, formed from → parenchyma tissue. Pith cells are mostly large and thin-walled storing water and different compounds. In some plants, the cells in the pith area break up and a cavity is formed in the middle of the stem. In shoots of most monocotyledons, the vascular bundles are scattered filling the whole shoot and thus no pith is formed; **3.** as a verb, to pierce the spinal cord or brain.

pit organ, a paired, temperature-sensitive organ of some snakes, e.g. rattlesnakes (Crotalidae), located in the head between the nostrils and the eyes. The organ comprises a pit in the skin with numerous warm receptors on the lower epithelium, sensitive for infrared radiation. The pit organ acts like an infrared camera forming some sort of "warm picture" of an object, e.g. a prey in darkness.

pituicyte (pituitary + Gr. *kytos* = cavity, cell), a neuroglial cell type in the posterior → pituitary gland (neurohypophysis) in vertebrates; the fibro-, micro- and reticulopituicytes are different subtypes. The pituicytes serve for supporting and nourishing the axon endings of nerve cells.

pituitary (L. *pituita* = phlegm), **1.** denoting relationship to the → pituitary gland; **2.** the pituitary gland; **3.** a therapeutical preparation obtained from the pituitary gland.

pituitary (gland) (L. *hypophysis*), the most essential endocrine gland in vertebrates secreting about ten different hormones, most of which regulate the functions of other endocrine glands. The gland is located in the → sella turcica under the floor of the third brain ventricle maintaining the connection (stalk) to the brain. It is structurally and functionally dual, composed of two main lobes: the **anterior pituitary (adenohypophysis)** and the **posterior pituitary (neurohypophysis)**.

The anterior pituitary (anterior lobe) derives from the epithelial pouch (Rathke's pouch) of the roof of the embryonic buccal cavity. When fully developed, the organ comprises three parts: stalk (*pars tuberalis*), anterior part (*pars distalis*), and intermediary part (*pars intermedia*).

The anterior pituitary contains many different types of secretory cells which produce several peptide or protein hormones most of them regulating functions of other endocrine glands; these are → **growth hormone**, **corticotropin**, **thyrotropin**, **gonadotropins**, and **prolactin**. The secretion of the hormones is controlled by certain neurohormones, **releasing and inhibiting hormones** (liberins and statins), which via the hypophyseal portal veins are carried from the hypothalamus, where they are produced, to the pituitary gland. The anterior pituitary is also the source of → **pro-opiomelanocortin** that is the prohormone of → **endorphins** and → **lipotropin**.

Pars intermedia secretes a hormone that regulates the function of pigment cells, and thus has an effect e.g. on the skin colour. In lower vertebrates it is called → **melanophore-stimulating hormone**, **MSH**, in birds and mammals **melanocyte-stimulating hormone**. In birds and many mammals (e.g. man), pars intermedia is reduced but some MSH is secreted also by cells of the anterior lobe.

The posterior pituitary has derived from the floor of the third brain ventricle and consequently consists of nervous tissue. During embryonic development the axons (nerve fibres) of certain neurones intrude from the hypothalamus along the pituitary stalk posterior to the adenohypophysis, enlarged axon endings with some glial cells forming the tissue of the posterior pituitary. The pituitary hormones, → **oxytocin**, and → **vasopressin** (antidiuretic hormone) are nonapeptides which are produced in the cell bodies of certain neurosecretory cells in the hypothalamus, and are transported along the axons to the posterior pituitary (*see* **neurosecretion**) and stored there until secreted into the blood. The corresponding hormones in lower vertebrates may have slightly different structures, and therefore are called by different names, such as **vasotocin** (chiefly corresponding vasopressin) and **mesotocin** (corresponding oxytocin) in amphibians, reptiles, and birds, or **isotocin** in fish.

PKU, → phenylketonuria.

place fidelity, a tendency of an animal to stay permanently in an area, or return constantly to the same place; e.g. **birth place fidelity** when the animal is born in the place, and **nest site fidelity** when the animal returns repeatedly to the same place for reproduction, as do migratory birds. *Syn.* philopatry.

placebo (L. *placere* = to please), a medicinal preparation which has no pharmacological activity but is given for psychophysiological effects or in a clinical trial to a control group; broadly, any dummy medical treatment.

placenta, pl. **placentae** or **placentas** (L. = a flat cake), **1.** in zoology, an organ that during embryogenesis develops to join the foetus to the maternal tissues within the uterus of most mammals (**eutherians**), except monotremes and marsupials (**metatherians**) which have a → hemiplacenta. The placenta develops when the chorion and allantois grow together to form the allantochorion which pushes small villi into the uteral wall, thus increasing the metabolic surface area between the placenta and the uterus. The **umbilical cord** connects the placenta to the foetus.

Several morphological placenta types are recognized: **diffuse placenta** (e.g. pig, horse) in which the whole surface of the allantochorion participates in formation of the placenta, **multiplex placenta** (most ruminants) with several partial placentas, **zonary placenta** (carnivores) in which the placental surface surrounds the foetus like a belt, and **discoid placenta** (man and other primates) in which the placental tissue resembles the form of a flat cake. Histological placenta types are characterized according to the cellular layers between the foetal and maternal blood vessels, as e.g. **epitheliochorial placenta** (pig, horse), **endotheliochorial placenta** (carnivores), and **haemochorial placenta** (primates).

The placenta maintains the separation of the foetal blood from the maternal blood, but allows the diffusion of respiratory gases and nutritive substances. The foetus also receives selectively some specific substances (e.g. antibodies) from the maternal blood through the placenta. Foetal waste products are transferred through the placenta into the maternal blood, and the placenta forms a barrier, → blood-placenta barrier, against many harmful substances circulating in the maternal blood. It also acts as an endocrine gland secreting e.g. → gonadotropins and gestagens.

2. In botany, the placenta is the site in the → ovary to which the → ovule or ovules are attached. Placenta types vary according to the structure of the gynaecium and the ovary.

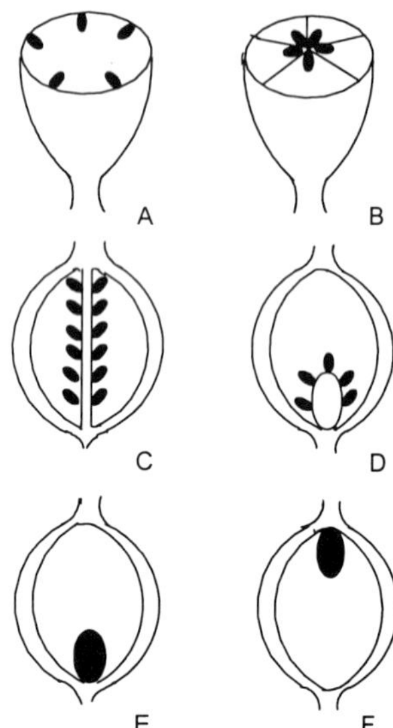

Types of placentas in plants; the black ovals are ovules: parietal marginal (A), central-axile (B), central (C), free central (D), basal (E), apical (F).

Parietal placenta is found in a gynaecium which is constructed of more than one carpel and is unilocular; in this the ovules are formed upon the walls of the ovary in the middle of the carpels (**parietal laminal placenta**), or in the joints between the carpels (**parietal marginal placenta**). If there is a central column in the unilocular ovary, the ovules being formed upon the column, it is called **free central placenta**. In a special type (**basal placenta**) only one ovule is situated in the bottom of an open ovary and only one seed will develop (e.g. in cherry).

The ovary may be partitioned with septa in separate locules. In this case the ovules are formed upon the central column, and the placenta is **axile** (**central-axile**). In a gynaecium composed of several separate carpels, there is a placenta in each gynaecium; it may be **marginal**, the ovule(s) located in the joint of every (closed) carpel, or **laminal**, the ovules being formed in the middle of the carpels.

Placentalia, → Eutheria.

placental lactogen, PL (L. *lac* = milk, Gr. *gennan* = to produce), a mammalian hormone (polypeptide) produced by the placenta; activates milk production, stimulates the growth of various tissues, and has a luteotropic effect on the ovaries. *Syn.* **chorionic somatomammotropin**. In man it is called human placental lactogen (hPL, HPL).

placental mammals, → Eutheria.

placode (Gr. *plax* = plate, *eidos* = form), a plate-formed anatomical structure, especially an embryonic plate of the → ectoderm from which a certain organ develops; e.g. lens placode, auditory placode, etc.

Placodermi (Gr. *derma* = skin), **placoderms**; an extinct, primitive class of fish whose head was covered by a thick bony armour; they were the first jawed vertebrates (Gnathostomata), living from the Silurian to the Permian, most abundant in the Devonian period.

placoid scale, a dermal scale type covering the skin of some cartilaginous fishes, e.g. sharks; it is a backward pointing spine, histologically resembling the vertebrate tooth with enamel and dentine layers, and a pulp. *Syn.* denticle.

Placozoa (Gr. *zoon* = animal), **placozoans**; a primitive invertebrate phylum including only one described species living in the littoral zone of warm oceans; has a plate-like body, some millimetres in length, which includes some thousands of cells whereas true tissues and organs as well as the coelom are not developed. The simple epithelium is composed of flagellated cover cells; the phylogenetic position is uncertain, probably nearest to sponges (→ Porifera). Placozoans are sometimes placed into a separate subkingdom, Phagocytellozoa. *See* Parazoa.

plagiotropism (Gr. *plagios* = inclined, *tropos* = turn), a type of growth of plant parts which is more or less divergent from the vertical or the direction of gravity, e.g. branches and twigs of trees are plagiotropic, more precisely plagiogravitropic.

plague (L. *plaga* = pestis < Gr. *plege* = stroke), **1.** any severe disease of wide prevalence and high mortality; **2.** pestilence (pest, pestis), a severe infection caused by the bacterium *Yersinia pestis*, primarily in rodents (especially rat) but when transmitted to humans may cause fatal worldwide epidemics; transmits by the bite of infected fleas or ingestion of infected animals; **3.** as a verb, to infect by pest bacteria, or more broadly, to cause worry or distress.

Planck's constant, Planck constant (*Max Planck*, 1858—1947), the constant $6.6254 \cdot 10^{-34}$ J · s; e.g. expresses the energy of one quantum of electromagnetic radiation: $E = h\nu$, in which E = quantum of energy, h = Planck's constant, and ν = frequency/s.

plank root, a flat, enlarged supporting root structure above ground in big trees especially in rain forests.

plankter (Gr. *planktos* = wandering), an organism in the plankton; the smallest plankters are phytoplankters, protozoans and rotifers, the biggest e.g. water fleas and copepods.

plankton, usually microscopic algae (**phytoplankton**) or animals (**zooplankton**) drifting passively or moving slowly in fresh or marine waters. The active moving of planktonic organisms, plankters, is regulated by food, light, temperature, predation and floating capacities of the organisms. On the basis of size the plankters are divided into → **pico-**, **nano-**, **micro-**, **meso-**, and **macroplankton**.

planospore (Gr. *planos* = wandering, *sporos* = seed), a motile → spore moving by means of a flagellum; occurs among many aquatic protozoans, algae, and fungi. *See* zoospore.

plant (L. *planta*) any of mostly autotrophic organisms, characterized by a cellulose wall surrounding the cells and by the lack of a nervous system and usually of locomotive movements. Most plants remain rooted to one position. Most plants are autotrophic, green due to chlorophyll and obtain energy through → photosynthesis from sunlight. Typical of plants is a high plasticity in form, largely determined by the environment during the ontogenic development. A few plants are heterotrophic, i.e. parasitic or saprophagic. The reproduction of plants is primarily sexual, with alternating generations of haploid gametophytes (producing → gametes) and diploid sporophytes (producing → spores), the former being reduced during evolution. During evolutionary development the structure of plants has been differentiated into tissues and organs (root, shoot, leaves). *See* Plant kingdom.

planta, pl. **plantae** (L.), **1.** *Plantae*, plants; **2.** something flat, such as a foot sole (*planta pedis*), or the → sclerite on the insect pretarsus. *Adj.* **plantar**.

Plantae, → Plant kingdom.

plant anatomy (Gr. *ana* = apart, *temnein* = to cut), a branch of plant morphology comprising the microscopic examination of the inner structures of plants. *Cf.* plant morphology, plant organology.

plantation (L. *plantare* = to plant), **1.** the insertion of cells, tissues, or an artificial material onto or into the body (implantation); **2.** a stand of planted plants, e.g. trees (chiefly Brit.); **3.** a farm in a tropical country.

plant breeding, the purposeful improvement of characteristics of cultivated plants, such as quantity and quality of crop, the size and appearance of flowers, etc.; essential features in plant breeding are: the increase of inheritable variation (or utilization of the existing variation), crossing of different strains, selection of the best adapted individuals and their separation, and the preservation of the new features. This kind of breeding produces new → cultivars of the plants. The breeding strategy depends on the type of reproduction. There are four main plant groups with different strategies: annual self-pollinators (wheat, barley, oat), annual cross-pollinators (corn, rye), perennial cross-pollinators (timothy, clovers, *Medicago* species), and perennial plants propagated vegetatively (potato, many fruit trees and berry plants). Cell and tissue culture, biotechnology, and gene manipulation, include several additional breeding methods.

plant community, in nature, an often repeated, regular structure of vegetation, the formation of which is determined by environmental factors such as climate and soil, and also by competition between species; e.g. forests which can further be divided into many types and subtypes.

plant disease, an abnormal condition in a plant, caused by different kinds of organisms, or by an unsuitable environment; e.g. fungi (smut and rust fungi, algal fungi) are important pathological agents, especially due to their effective distribution by spores. Also viruses, bacteria and some algae are able to cause plant diseases; some animals (mites, insects) cause → gall formation in plants. Unsuitable environments, such as wet, dry, poor as well as too rich nutritive conditions, may cause physiological non-infectious diseases.

plant ecology (Gr. *oikos* = household, *logos* = word, discourse), the study dealing with relationships of plants and their natural environments, especially the effects of ecological factors on the success, reproduction, and distribution of plants. *See* ecology.

Plant kingdom, one of different kingdoms of organisms including → plants; formerly included also the prokaryotic groups → **Bacteria** and → **Cyanobacteria** which are now classified separately (→ Monera). The classification of eukaryotic plants varies, the traditional divisions being **Chlorophyta** (green algae), **Charophyta** (stoneworts), **Phaeophyta** (brown algae), **Rhodophyta** (red algae), **Chrysophyta** (golden-brown algae), **Xanthophyta** (yellow-green algae), **Euglenophyta** (euglenoids), **Pyrrophyta** (dinoflagellates), **Cryptophyta** (cryptophytes), **Bryophyta** (mosses), **Pteridophyta** (pteridophytes, i.e. seedless vascular plants), and **Spermatophyta** (seed plants). The divisions are further divided into classes, orders and families. For another classification, *see* Appendix 3.

In conventional systems, organisms are divided into the → **Animal kingdom** (Animalia), **Plant kingdom** (Plantae) and **Fungi**. Another system divides organisms into five kingdoms, in which the group **Monera** contains prokaryotic organisms and → **Protista** the unicellular, eukaryotic protozoans, algae, and fungi.

plant morphology (Gr. *morphe* = form, *logos* = word, discourse), the study of the form and structure, both internal and external, of plants and plant organs. *Cf.* plant anatomy, plant organology.

plant organology (Gr. *organon* = organ), a branch of plant morphology, studying the outer structure of plants and their organs. *Cf.* plant anatomy, plant morphology.

plant pathology (Gr. *pathos* = suffering, sensation), a branch of botany dealing with → plant diseases.

plant physiology (Gr. *physis* = nature), a branch of botany dealing with life functions in plants, such as physicochemical reactions and their control mechanisms in cells and tissues. Modern plant physiology is directed to molecular biology.

plant quarantine (It. *quarantina* = forty) a way to prevent the spreading of plant diseases to new areas by keeping an imported plant for a certain time (originally 40 days) in controlled conditions, i.e. quarantine time and quarantine place. The quarantine rules control or totally forbid the transportation of plants from polluted areas to those not infected.

plant science, → botany.

plant sociology, the study dealing with vegetation, structural features of vegetation and species, and ecological factors affecting them. Classical plant sociology examines plant communities. The basic unit of the hierarchical plant sociology is **association**, which is often equated with the species in systematics.

plant substances, *see* secondary compounds.

planula, pl. **planulae,** Am. also *planulas* (L. dim. of *planus* = something flat, plane), a free-living, egg-shaped and ciliated larva of cnidarians consisting of two primary germ layers only, the ectoderm and endoderm; **invaginate planula**, the gastrula stage of an animal embryo.

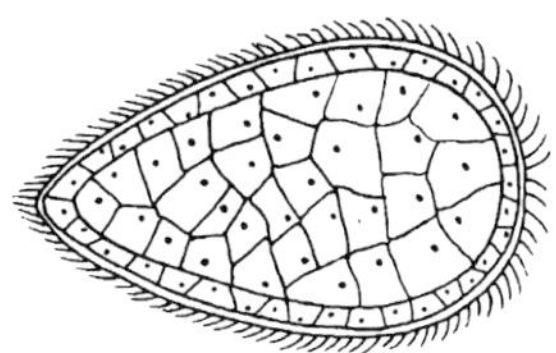

PLANULA

plaque, a divergent flat area or a patch (normal or pathological) on the surface of a structure; e.g. **1.** a clear area in a bacteria culture, when e.g. viruses have destroyed bacteria on that area; **2.** the minute plaques on the muscle cell membrane to which myofilaments attach (seen in electron micrographs); **3.** cholesterol plaques inside blood vessels, found especially in man; **4.** dental plaque, a film of debris deposited on the teeth; **5.** senile plaques (neuritic plaques), groups of amyloid fibrils in the brain tissue.

plasma, plasm (Gr. *plassein* = to mould, form), **1.** the fluid portion of the blood or lymph, *see* blood plasma; **2.** → cytoplasm; **3.** in geology, the green form of calcedon (silicate compound); **4.** in physics, the ionized gas, in which positive and negative charges are equal; also neutral atomes and molecules may be present.

plasma cell, the activated B lymphocyte producing soluble → antibodies in the blood (*see* lymphocytes); serves in humoral → immunity of vertebrates; *syn.* plasmocyte (plasmacyte); **plasmocytosis** (plasmacytosis), excess of plasma cells in the blood.

plasmagene, → cytogen.

plasmalemma (Gr. *lemma* = membrane), a special term for the → cell membrane of plants and some protozoans, such as amoeba.

plasmalogens, phosphoglyceracetals; glycerophospholipids in which the glycerol moiety has a 1-alkenyl ether group. *See* phospholipids.

plasma membrane, → cell membrane.

plasma proteins, proteins in the blood plasma. *See* blood.

plasmatic inheritance, the inheritance of those characters of organisms the genetical factors of which are not located on → chromosomes but in the → cytoplasm; mediated by → mitochondria or plastids.

plasmid (Gr. *eides* = like, resembling), a hereditary factor in an autonomous state, i.e. a piece of symbiotic DNA in bacteria which does not belong to the chromosomes; thus, as opposed to an → episome, the inheritance of a plasmid is independent of any chromosome. The DNA plasmids of bacteria and certain other unicellular organisms consist of 1,000 to 30,000 base pairs. In molecular biology plasmids are used in gene transfer technique.

plasmin, a hydrolase enzyme (EC 3.4.21.7) found in vertebrates catalysing the breaking of peptide bonds at the carbonyl end of lysine and arginine residues; catalyses the conversion of fibrin to soluble products, and is therefore also called **fibrinolysin** (or fibrinase). Plasmin is present in the blood plasma in its inactive form, **plasminogen**, which is physiologically activated e.g. by kallikrein.

plasmodesma, pl. **plasmodesmata** (Gr. *desmos* = ribbon), a cytoplasmic thread running across the plant cell wall through a minute pit, connecting the cytoplasm of adjacent cells. Plasmodesmata are common in plant cells and their density may be as much as 10^6 per mm^2. They markedly facilitate transport between separate cells.

Plasmodiogymnomycotina, in some systems a subdivision in the primitive fungal division → Gymnomycota; divided into two classes, Protosteliomycetes and → Myxomycetes (slime moulds).

Plasmodiophoromycetes, a class in fungal subdivision Haplomastigomycotina, division Mastigomycota (algal fungi). The species are endoparasites in vascular plants, algae and fungi, growing as plasmodia inside a host, and reproduced by spores and gametes. A well known species is *Plasmodiophora brassicae*, which causes clubroot in cabbage. In some systems the group Plasmodiophoromycetes is classified as a separate division, **Plasmodiophoromycota**.

plasmodium, pl. **plasmodia** (L. *plasmodium* = small plasma), **1.** a multinucleate mass (without cell walls) of streaming protoplasm in slime moulds, → Myxomycetes, forming the non-reproductive stage of the organisms; **2.** *Plasmodium*, the genus of parasitic protozoans, i.e. the sporozoan family Plasmodiidae (*see* Sporozoa), the causal agent of malaria. *Adj.* **plasmodial**.

plasmolysis (Gr. *lyein* = to loosen), the shrinking of a plant cell resulting in the loosening of the cytoplasm with its plasma membrane from the rigid cell wall; occurs in a hypertonic environment due to loss of water from the hypotonic cytoplasm. Plasmolysis can be caused by osmotically active compounds, such as sugars and salts, using them in hyperosmotic concentrations (*see* osmosis). Plasmolysis is a reversible phenomenon; in a hypotonic environment the cell can take in water again and reach the normal turgor; this phenomenon is called **deplasmolysis**. Prolonged exposure of cells to plasmolysis reduces the ability for deplasmolysis, and some plasmolytic substances can prevent deplasmolysis by destroying cellular structures like plasmodesmata and cell organelles. *Adj.* **plasmolytic**, pertaining to plasmolysis; **plasmolyzable**, capable of undergoing plasmolysis.

-plast (Gr. *plastos* = formed, moulded), denoting any small organized structure in a cell, tissue, or organ, e.g. a → chloroplast, chromoplast, or protoplast.

plastid (Gr. *eidos* = shape, form), an organelle of a plant cell; e.g. **1.** → **chloroplast**, containing chlorophylls and other pigments acting as the photosynthetic centre; **2.** → **chromoplast**, containing pigments; **3.** → **amyloplast**, accumulating starch; **4.** → **elaioplast**, containing lipids; **5.** → **leucoplast** (colourless). The primordia of plastids are called **proplastids**.

plastochron (Gr. *chronos* = time), a time interval between periodically repeated phenomena in the development of an organism, e.g. in the development of leaf primordia in a spiral leaf arrangement in plants.

plastocyanin (Gr. *kyanos* = blue), a copper-containing protein pigment which participates in electron transport in the light reactions of → photosynthesis.

plastoglobulus (L. *globulus*, dim. of *globus* = ball), an oil-containing droplet in → chloroplasts.

plastoquinone, a compound with oxidation-reduction ability participating in → electron transport in the light reactions of → photosynthesis.

plastron (Fr. *plastron* = breastplate), **1.** a flat, ventral part of the shell of tortoises and turtles, formed by plate-like bones and covered by epidermal plates; **2.** a region of the body surface in some aquatic insects (e.g. water scorpions) that is covered with short hydrophobic chitin hairs bent at the tip so that air is trapped permanently between the hairs. The plastron acts as a physical gill exchanging respiratory gases between the surrounding water and tracheal tubules opening to the skin of the plastron region. The gas exchange is called **plastron respiration**.

platelet, blood platelet, thrombocyte; any of numerous plate-like, anuclear cell fragments found in the vertebrate blood, from 2 to 4 μm in diameter; they are formed by fractionation of large cells, → megakaryocytes, in the bone marrow. Platelets function in → blood coagulation.

platinum (Sp. *plata* = silver), a precious, heavy metallic element, symbol Pt, atomic weight 195.09, resistant to most chemicals; used in the manufacture of chemical tools and apparatus, as a catalytic agent, and in jewellery.

platyclade, platycladium, pl. **platyclades, platycladia** (Gr. *platys* = flat, *klados* = shoot), → phylloclade.

Platyhelminthes (Gr. *helmis* = worm), **flatworms**; a phylum comprising bilateral, acoelomate worms without respiratory and circulatory organs. Also the digestive tract is either absent (in cestodes) or incompletely developed (e.g. the anus is lacking). Flatworms are free-living such as planarians, or endoparasites such as digeneans and cestodes. The phylum comprises about 12,500 described species divided into three classes: **turbellarians** (Turbellaria), **trematodes** (Trematoda), and **cestodes** (Cestoda). Trematodes are divided into subclasses (or orders) Monogenea and Digenea.

Platyrrhini, an infraorder of anthropoid primates comprising the superfamily → Ceboidea (New World monkeys).

plectenchyma (Gr. *plektos* = twisted, *enchyma* = poured in), a type of → pseudoparenchyma composed of tightly intertwined fungal hyphae forming thalli of lichens.

plectostele (Gr. *stele* = column), one of different stele types in plants. *See* stele.

pleio-, pleo-, plio- (Gr. *pleion*, *pleon* = more), denoting more, excessive.

pleiochasium (Gr. *chasis* = division), one of → inflorescence types. *See* cymose.

pleiomorphism, → pleomorphism.

pleiotropy (Gr. *trope* = change), a genetical phenomenon in which one → gene regulates the development of many seemingly independent characters; may be relational or genuine. In **relational pleiotropy** one gene causes one primary character which then regulates the development of several secondary characters; e.g. an enzyme that regulates the synthesis of a multifunctional hormone. In **genuine pleiotropy** one gene has several primary effects, i.e. encodes the synthesis of more than one polypeptide.

Pleistocene (Gr. *pleistos* = most, *kainos* = recent), the earlier of the two epochs of the Quaternary period, estimated conventionally to have lasted from 1.6 million to 10,000 years ago, including the last glacial periods. In the course of Pleistocene, man (*Homo sapiens*) appeared, and present arctic animals such as the arctic fox, musk ox, and rock ptarmigan appeared in Central Europe; at the end of the epoch, large mammals and birds (→ megafauna) became extinct.

pleo-, pleio- (Gr. *pleon*, *pleion* = more), denoting more, excessive, multiple; also **plio-**.

pleomorphism, pleomorphy, pleiomorphism (Gr. *morphe* = form), **1.** generally the same as → polymorphism; **2.** in fungi, a type of polymorphism with several different forms of the life cycle, such as successive spore forms in rust fungi. *Adj.* **pleomorphic**.

pleophage (Gr. *phagein* = to eat), an animal feeding on various kinds of food, or on several prey species; *syn.* polyphage. *Adj.* **pleophagous**; *syn.* pleotrophic, polyphagous. *Cf.* omnivore.

pleopods (L. *pleopodium* < Gr. *plein* = to swim, *pous,* gen. *podos* = foot), paired leg appendages on the segments of the crustacean abdomen.

pleotrophic (Gr. *trophe* = nourishment), pleophagous. *See* pleophage.

plerocercoid (L. *plerocercaria* < Gr. *pleres* = full, *kerkos* = tail, *eidos* = form), the third larval stage (the second "bladder stage") of some cestodes (e.g. *Diphyllobothrium*, fish tapeworms) which encysts especially in skeletal muscles (muscle larva). The larva has to be transferred from the fish to the digestive tract of a definite host (e.g. man) to develop into the adult form. *Syn.* plerocestoid. *Cf.* metacercaria.

plerome, also **plerom** (Gr. *pleroma* = that which fills), *see* derivative meristems.

plesiomorphic (Gr. *plesios* = near, *morphe* = form), in phylogenetics, pertaining to an ancestral (basic) character from which more developed characters have arisen; e.g. the scale is plesiomorphic in relation to the feather. *Noun* **plesiomorphy**. *Cf.* apomorphic.

plesiomorphous, similar in form. *Noun* **plesiomorphism**.

Plesiosauria (Gr. *sauros* = lizard), **plesiosaurs**; an order of extinct marine reptiles, widely distributed in Mesozoic; typically large animals (even 15 m) with a long neck, small head, and paddle-shaped limbs.

plethysmograph (Gr. *plethysmos* = multiplication, *graphein* = to write), an instrument for recording volume changes in an organ or limb; indicates the amount of blood present or passing through the organ.

pleura, pl. **pleurae** (L., Gr. = side, rib), the fibrous membrane enveloping the lungs (**visceral pleura**) and lining the thoracic cavity (**parietal pleura**) in tetrapod vertebrates; the potential space (**pleural cavity**) between the two membranes contains a thin layer of friction-reducing serous fluid.

pleurobranch, pleurobranchia pl. **pleuro-**

branchs, pleurobranchiae** (Gr. *branchia* = gills), any of the gills attached to the lateral walls of the thorax in some arthropods, especially in crustaceans. *Cf.* podobranch, arthrobranch.

pleurocarpic, also **pleurocarpous** (Gr. *karpos* = fruit), pertaining to a plant having its male and female gametangia on the tip of a lateral branch; e.g. some mosses. *Cf.* acrocarpic.

pleuron, pl. **pleura** (L., Gr. = side, rib), a lateral shield on the thorax in some arthropods, e.g. insects; formed from several hard plates, **sclerites**; called also **pleurum**. *Cf.* notum, sternum, tergum.

Pleuronectiformes (Gr. *nektos* = swimming, L. *forma* = shape), **flatfishes**; an order of asymmetrical bony fishes with both eyes on one side of the head and a strongly compressed body; the adult fish lies on its side on the bottom of marine coastal waters, the undersurface is usually unpigmented. Early stages are bilaterally symmetrical. Flatfishes are commercially important food fish, especially plaice, sole, and many of the true flounders.

pleuroperitoneum, the membrane (the → pleura combined with the → peritoneum) lining the body cavity and covering the surface of the viscera of vertebrates with no diaphragm, i.e in groups other than mammals.

pleuropneumonia-like organism, PPLO, *see* mycoplasma.

pleuston (Gr. *pleusis* = sailing), macroscopic organisms floating freely on or near the water surface; the species often have gas-filled bladders or floats. Pleuston includes floating algae, spermatophytes, and associated small animals. *Cf.* neuston.

plexus, pl. **plexus** or **plexuses** (L. = network, braid), a network formed of vessels, veins, nerves, or nerve fibres; e.g. abdominal aortic plexus, lymphatic plexus, or lumbar plexus (of lumbar nerves).

plica, pl. **plicae** (L.), a fold or ridge; e.g. *plicae circulares*, circular folds in the small intestine, *plica vestibularis,* false vocal cord in the larynx, or *plica vocalis*, true vocal cord.

Pliocene (Gr. *pleion* = more, *kainos* = recent), the last epoch in the Tertiary period between Miocene and Pleistocene, about 5 to 1.6 million years ago. Many present species of mammals were living and the Australopithecines appeared.

ploidy, the number of haploid chromosome sets in a cell, e.g. the → haploid, diploid, triploid, etc. chromosome number.

plumage (L. *pluma* = soft feather), the assemblage of → **feathers** (plumes), i.e. the cornified outermost layer on the skin of birds insulating the body against heat loss; includes four types of feathers: **contour feathers**, **down feathers** (plumules), **filoplumes**, and **bristles** (vibrissae). Feathers develop only on certain areas of the skin (**pterylae**, feather tracts) between which there are featherless areas (**apteria**, **apterylae**).

plume, → feather.

plumule (L. *plumula*, dim. of *pluma* = soft feather), **1.** down → feather; **2.** a developing shoot above the cotyledon in a plant embryo, comprising the epicotyl and young leaves.

plurien (L. *pluris*, gen. of *plus* = more), a plant that grows for many years and dies after flowering.

pluripotentiality (L. *potentia* = power), power of acting in several possible ways; e.g. **1.** the ability of early embryonic cells to develop into several types of tissue, or of haemocytoblastic stem cells in the bone marrow to develop into different types of blood cells; **2.** the pharmaceutical power of a drug to act in various ways. *Adj.* **pluripotent(ial)**.

pluteus, pl. **plutei** (L. = shed, movable shelter), a free-swimming larval stage of certain echinoderms (echinoids, ophiuroids); e.g. the larva of the sea urchin is **echinopluteus**, and that of the brittle star, **ophiopluteus**.

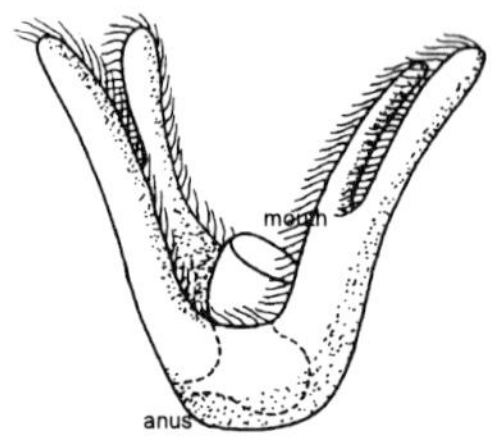

PLUTEUS

plutonium (< the planet *Pluto* = the god of harvest and wealth in Greek mythology), a transuranic element (actinoid), symbol Pu; appears in natural conditions in small quantities within uranic minerals, being formed from uranium by neutron capture. Plutonium is produced in nuclear reactions, e.g. ^{239}Pu is produced by neutron irradiation of ^{238}U, an intermediate product being neptunium. Altogether 15 different plutonium isotopes (atomic weights 232—246) are known, the isotopes ^{235}Pu and ^{239}Pu being natural.

Plutonium isotopes are radioactive; the half-lives vary from 20 minutes (^{233}Pu) to 75 million years (^{244}Pu). ^{239}Pu can be produced in large quantities in nuclear reactors and it undergoes nuclear fission when bombarded by slow neutrons. This isotope (half-life 24,400 years) is also used in nuclear weapons, e.g. in the bomb dropped in Nagasaki in 1945. ^{238}Pu is used as a nuclear power source of space probes and satellites.

Plutonium isotopes, particularly ^{239}Pu, are liberated into nature e.g. from nuclear power plants and these radioactive substances are very dangerous, damaging cellular structures in organisms. The oxides and halogens of plutonium accumulate easily in organisms, especially in the bones of vertebrates, and may e.g. induce cancer.

PMSF, → phenylmethylsulphonyl fluoride.

PNA, → peptide nucleic acid.

pneum(o)-, pneumat(o)- (Gr. *pneuma* = breath, air, wind), denoting relationship to lungs, air, gas, or respiration. *Adj.* **pneumatic.**

pneumat(h)ode, an aerial or respiratory root of a plant. *See* pneumatophore.

pneumatocyst (Gr. *kystis* = sac, bladder), **1.** in zoology, an air cavity serving as a float, e.g. the swim bladder of fish; **2.** in botany, a gas-containing cavity in some brown algae.

pneumatophore (Gr. *phorein* = to carry along), **1.** a gas-filled bladder (float) of siphonophores serving as a float; **2.** in botany, the respiratory root; a branch of a plant root rising above the ground or water surface taking oxygen to the root system, common especially in → mangrove vegetation; **3.** an apparatus containing oxygen to be inhaled by people working in conditions deficient in oxygen, e.g. in mines.

pneumatot(h)ode, a lenticell-like opening in the aerial or respiratory roots (pneumatophores) being in contact with the inner aerenchyma; facilitates the change of gases.

p.o., per os (L. *per* = through, *os* = mouth), by mouth, orally; e.g. administration of a drug to an animal. *See* oral.

pod(o)-, -pod (Gr. *podos*, gen. of *pous* = foot), denoting relationship to the foot or a foot-like structure.

podeon, pl. **podeones** (Gr. = mouth of a wine-skin, neck), the slender, second segment of the abdomen between the propodeon and metapodeon in some hymenopterans. *Syn.* podeum.

podetium, pl. **podetia,** the erect outgrowth of the → thallus in some lichens, bearing apothecium; e.g. in *Cladina* species.

Podicipediformes (L. *podex* = rump, *pes* = foot, *forma* = shape, form), **grebes**; an order of diving birds feeding on fish and other aquatic animals. The caudal legs are short and positioned at the rear of the body, and the toes are lobate-webbed. The order includes one family only, *Podicipedidae*, with 21 species in five genera.

podium, pl. **podia**, Am. also **podiums** (L. < Gr. *podion,* dim. of *pous* = foot), **1.** a foot or foot-like structure in some invertebrates, e.g. the tube feet of echinoderms; **2.** a stem axis of a plant; *see* monopodial branching, sympodial branching.

podobranch, podobranchia, pl. **podobranchs, podobranchiae** (Gr. *branchia* = gills), **foot-gill**; any of the gills attached to the basal segments (epipodites) of the maxillipeds or thoracic appendages in some crustaceans, such as crayfish. *Cf.* arthrobranch, pleurobranch.

podocarpic acid, the main acidic constituent ($C_{17}H_{22}O_3$) of the resin of coniferous *Podocarpus* trees.

podocyte (Gr. *kytos* = cavity, cell), a cell type derived from the epithelial cells of the glomerular capsule in the vertebrate kidney; the cells have foot-like processes which are pressed against the basal lamina of the wall of blood capillaries thus regulating glomerular filtration.

podosoma, pl. **podosomata** (L., Gr. *soma* = body), the portion of the body of mites and ticks having four pairs of legs. *Cf.* camerostome.

podsol, podzol (Russ. *pod* = under, *zola* = ashes), a poor soil type typical of the cool coniferous forest zone on the northern hemisphere in areas where rainfall is more abundant than evaporation. The podsol is characterized by clear layers, which are formed from acidic humus compounds by the effect of water. The uppermost layer is the slowly decomposing humus layer, underneath that a grey-white A-horizon, a red-brown B-horizon (rich in iron and aluminium compounds), and an unchanged C-horizon. *See* soil formation.

Pogonophora (Gr. *pogon* = beard, *phoros* =

bearing), **pogonophorans**, **pogonophores**, **beard worms**; an invertebrate phylum comprising slender and longish worms inhabiting chitinous tubes in the bottom of deep oceans. The body is divided into four parts: **head**, **mesosoma**, **metasoma**, and **opisthosoma**. The mouth, intestine, and anus are lacking, food is probably absorbed from the surrounding water through the surface of the tentacles. The phylum includes about 100 species which are divided into two orders, Athecanephria and Thecanephria, in some classifications into two classes, Frenulata and Afrenulata.

poietin (Gr. *poiein* = to make), a group of hormones (large polypeptides) involved in the control of blood cell production in vertebrates; e.g. **erythropoietin** induces the production of red blood cells from haematocytoblastic stem cells, and **granulopoietin** the production of granulocytes from similar stem cells. Sometimes called poetin.

poikil(o)- (Gr. *poikilos* = varied, spotted), denoting variable or irregular.

poikilosmotic, pertaining to an animal whose osmotic concentration of body fluids varies in accordance with the osmotic concentration of the environment; the animals lack osmotic regulation mechanisms and thus are isoosmotic with their media, i.e. are **osmotic conformers**, as e.g. many marine invertebrates. *See* osmoregulation, osmosis.

poikilothermy (Gr. *therme* = heat), variation of the body temperature of animals (**poikilotherms**, cold-blooded animals) in accordance with changes in environmental temperatures. More than 99% of all animal species are **poikilothermic**. *Cf.* homoiothermy, endothermy, ectothermy.

point loading, discharges of pollutants to an area from identifiable sources, such as industry, sewerages, piggeries, etc.; the control of point loading is easier and cheaper than that of scattered loading. *Syn.* point pollution.

point mutation (L. *mutatio* = change), a change in one nucleotide pair in a DNA sequence; broadly means the same as → gene mutation, i.e. a mutation that is depicted as a point on the → chromosome map. Because gene mutations may be small → deficiencies, duplications, or insertions, a more narrow meaning has been given to the concept of point mutation.

poise (*J.L.M. Poiseuille*, 1799—1869), symbol P, the unit of dynamic viscosity. *See* Appendix 1.

poison (L. *potare* = to drink), any inorganic or organic liquid, solid, or gaseous substance, which in relatively small quantities causes disturbances or damage in structures and/or functions of organisms. Many substances and elements which in very small concentrations are necessary for organisms, may be poisonous in higher concentrations, e.g. many trace elements. Special types of poisons are e.g. → toxins, venoms, insecticides, and herbicides. *Adj.* **poisonous**.

poisonous animals, animals who produce poisons and use them for defence or predation; some of them are poisonous only if they are ingested by other animals. There are poisonous animals in several groups, such as in cnidarians, echinoderms, arthropods (e.g. scorpions, some spiders, and insects), fishes, amphibians, and especially in reptiles (some lizards and many snakes). Also some birds in the genus *Pitohui* are poisonous.

Poisons vary in their chemical structures and mechanisms of action. They may change permeability of the cell membrane by acting on ion channels (e.g. → tetrodotoxin), or result in pores in the cell membrane (*see* ionophores). Some are enzymes, such as proteases and phospholipases, which damage cell and tissue structures and disturb immune reactions; many are nerve poisons e.g. inhibiting functions of → synapses.

poisonous fungi, fungi containing poisonous compounds in quantities which are toxic when ingested by animals; poisonous species are common, and the types and effects of poisons vary much more than in → poisonous plants. Only small pieces of the most dangerous species, as some species of the genera *Amanita*, *Cortinarius*, and *Gyromitra*, cause death if eaten by man. The effective compounds present in these species are cell poisons destroying cell structures and preventing cell divisions. Another main group is formed by nerve poisons (e.g. *Amanita muscaria*, some *Clitocybe* and *Inocybe* species), which affect the function of → synapses between the nerve cells or between the nerve cells and their effector cells. Some invertebrate species, as certain fly larvae, are evolutionarily adapted to feed on fungi which are poisonous to other animals.

poisonous plants, plants containing poisonous

compounds in quantities which are toxic when eaten by animals. Poisons usually act in cells, or as e.g. → nerve poisons disturb functions of intercellular junctions, → synapses. Plant poisons may be → **alkaloids** (e.g. in tobacco, poppies), **saponins** (e.g. in horse chestnut), **glycosides** (in common foxglove), or **diterpenic alcohols** (in *Poinsettia*). The poisons are obviously useful for the plants because they protect against animals. Many poisonous plants are important in medicine. *Cf.* poisonous fungi. *See* chemical defence, continuous defence mechanism.

polar body, three degenerative daughter cells separating from the oocyte (egg cell), during its maturation after the second meiotic division (*see* meiosis); if the first polar body does not undergo the second meiotic division, the number of polar bodies is two. Called also polar cell.

polarimeter (Gr. *metrein* = to measure), an instrument for measuring the rotation of the vibration plane of polarized light caused by optically active substances, such as sugars.

polarity, the property of having two different characteristics, e.g. electric, magnetic, functional, or structural properties between different points, **poles**, of the system; e.g. **1.** the structural or functional difference along the axis of cells, tissues, organs, or organisms, as e.g. head and tail, or shoot and root; **2.** the polar organization of the cytoplasm of an egg cell, resulting in polar development of most organisms (*see* morphogenesis), i.e. the first cell division takes place perpendicularly against the polarity axis between the poles. The division is unequal and produces a larger and a smaller half of a cell and thus polarity of the embryo. The direction of the polarity axis in plants can be determined e.g. by unilateral light; **3.** polarity of molecules, e.g. many macromolecules or water molecules, having a positive charge on hydrogen and negative charge on oxygen; **4.** electrical polarity; *see* resting potential.

polarization, 1. the action of polarizing; development of → polarity; **2.** the production of polarized light, i.e. the process of affecting light or other radiation so that vibrations of the wave are confined to one plane only; e.g. occurs as a result of the reflection or transmission of radiation through a certain medium.

polarization microscope, *see* microscope.

polarized light, *see* polarization.

polar nuclei, two haploid nuclei in the central cell of the plant → embryo sac; the nuclei fuse before fertilization and form the diploid secondary endosperm nucleus. In fertilization the nucleus becomes triploid and the cell develops into the → endosperm.

polar plasm, the cytoplasm in the posterior pole of some invertebrate egg cells, e.g. in insects or nematodes.

pole (L. *polus*, Gr. *polos*), **1.** either of the two extremities of the axis of a sphere, as e.g. the axis of the Earth, cell, organ, or organisms; *see* animal pole; **2.** either of two points with opposite physical qualities, e.g. electric poles (negative and positive).

polemochore (Gr. *polemos* = war, *chorein* = to spread), a plant carried to a country by troops and traffic during war.

poli- (Gr. *polis* = city), pertaining to a city area. *See* polio-.

Polian vesicles (*S. Poli*), five swellings in the ring canal of sea stars (asteroids), sea urchins (echinoids), and sea cucumbers (holothurians) regulating water flow in the → water-vascular system.

polio-, poli- (L. *polios* = grey), pertaining to the → grey matter (substance) of the vertebrate nervous system.

pollakanthic (Gr. *pollakis* = many times, *anthos* = flower), pertaining to a plant growing for many years and flowering many times; such herbaceous plants are mostly called **perennial**.

pollen (L. = fine flour), fine powder produced by anthers of seed plant flowers, composed of → pollen grains.

pollen analysis, a study, **palynology**, dealing with former vegetation and climates by analysing pollen that is preserved in the smooth layers and deposits of soil, such as peat; used for dating fossils and other remains.

pollen basket, corbiculum, *see* pollen comb.

pollen comb, a comb-like structure on the inner surface of the tarsus in the honeybee hindlegs, brushing pollen into the **pollen basket** on the external surface of the tibia; the comb is formed from downward-pointing spines in ten rows.

pollen grain, the haploid microspore of seed plants in which the male → gametophyte develops; pollinates the flower and causes fertilization by transporting the male nuclei (*see* sperm cell) of the male gametophyte to the

mation for the synthesis of more than one polypeptide; the mRNA molecules produced by → operons are usually polycistronic. *Cf.* monocistronic. *See* → cistron.

polyclimax theory (Gr. *klimax* = ladder), a theory presenting that a balanced condition between local communities is achieved by the concurrence of several environmental factors, such as climatic factors, soil, and altitude. Thus succession series in the area will result in several climax communities, **polyclimax**. *See* climax. *Cf.* monoclimax theory.

polyclonal (Gr. *klon* = twig, bud), derived from different cells. *Cf.* monoclonal. *See* polyclonal antibodies.

polyclonal antibodies, a mixture of antibodies produced against different epitopes of an → antigen; produced by lymphocyte strains derived from several mother cells. The serum obtained from animals always contains polyclonal antibodies. *Cf.* monoclonal antibodies. *See* antibody.

polycyclic (Gr. *kyklos* = ring), having many whorls, rings, or cycles (frequencies), e.g. pertaining to **1.** a flower with many whorls of petals and sepals; **2.** a plant which grows many years but dies after flowering; **3.** the vascular system with several concentric rings of vascular bundles in a plant stem, e.g. in some ferns; **4.** a chemical compound with many ring structures.

polycythaemia, Am. **polycythemia** (Gr. *kytos* = cavity, cell, *haima* = blood), an increased number of red blood cells in the blood.

polydactyly (Gr. *daktylos* = finger, toe), an anomaly caused by → mutation and characterized by more than the normal number of toes or fingers; found in many vertebrate species, also in man. Called also polydactylism, polydactylia, hyperdactylism, hyperdactylia. *Cf.* polymely.

polyderm (Gr. *derma* = skin), a plant → periderm that exceptionally contains living parenchyma cells in addition to dead, suberinized cells.

polyembryony, 1. the development of two or more animal embryos from a single fertilized ovum, e.g. identical twins; **2.** a condition in which more than one embryo develop in the → ovule of a plant. The phenomenon is not usual among plants but typical of coniferous trees in which generally four embryos are initially formed, but only one of them continues its development. In → Angiospermae polyembryony occurs in connection with → apomixis in the genus of *Citrus*. In this case a normal embryo, derived from the ovum, is generated in the normal way (*see* embryogenesis), but in addition, one or more auxiliary embryos develop asexually in the ovule from a completely vegetative tissue. *Adj.* **polyembryonic**.

polyethism (Gr. *ethos* = manner), a specialization of individuals in an animal community to bring about the division of labour. The functions may be determined by morphological structures (e.g. **caste polyethism** in social insects; *see* caste) or by age (**age polyethism**).

polyethylene glycol, PEG, any polymerization product, $HO(CH_2CH_2O)_nH$, of ethylene glycol or ethylene oxide; an odourless liquid or wax-like substance soluble in water and alcohol. PEG is used in industrial grease, wetting and washing agents, and in emulsifiers. In biotechnology it is used to improve the fusion of isolated → protoplasts.

polygalacturonan, a polysaccharide consisting chiefly of galacturonic acid residues, but may also contain monosaccharides such as arabinose, galactose, and rhamnose; a component of plant cell walls. It is hydrolysed by the enzyme **polygalacturonase** (pectinase). *See* pectic substances.

polygamy (Gr. *gamos* = marriage), **1.** polygamous mating, i.e. mating and producing offspring with more than one partner; among animals pertains to the sexual behaviour of an individual (usually a male) having several mating partners who usually have no other partner. Polygamy may be simultaneous or successive. Polygamy is called **polyandry** when a female has several males, and **polygyny** when a male has several females; **2.** in botany, the existence of different combinations of → unisexual and → bisexual flowers in the same plant species either in the same or in different individuals.

polygen (Gr. *gennan* = to produce), a complex → antigen composed of several kinds of antigenic determinants, i.e. stimulates the production of several antibodies.

polygene (Gr. *genos* = descent, offspring, family), a → gene that alone has only a minor effect on the → phenotype but together with several other non-allelic genes (polygenes) regulates the development of a certain phenotypic character, producing a cumulative effect on it. *Adj.* **polygenic**. *See* polygeny.

polygenic character, a character of an organism the development of which is regulated by several interacting → genes (polygenes). Polygenic characters are typically quantitative characters of organisms, e.g. the height or weight of man or the productivity of domestic animals and plants. Polygenic characters are inherited through mechanisms called **polygenic inheritance**.

polygeny, 1. the development of a seemingly single → phenotypic difference or character due to the function of several interacting genes, → polygenes; **2.** the descent of man from two or more independent ancestors. *Adj.* **polygenic**.

polygraph (Gr. *graphein* = to write), an instrument for simultaneous recordings of several physiological activities in an organism, organ, or cell, as represented by mechanical or electrical impulses; used e.g. in recordings of blood pressure, heart contractions, respiratory movements, etc.

polygynous (Gr. *gyne* = woman), **1.** pertaining to a mating type in animals when a male has more than one female mating partner at a time; **2.** pertaining to a union of two or more female pronuclei with a male pronucleus, resulting in polyploidy of the zygote; **3.** pertaining to a flower with numerous styles. *Noun* **polygyny**. *Cf.* polyandrous. *See* polygamy.

polyhaline (Gr. *hals* = salt), pertaining to brackish water having salinity between 10.0—16.5 per thousand. *Cf.* mesohaline, oligohaline.

polyhumous water (L. *humus* = soil), water or water body containing high quantities of → humus. *Cf.* mesohumous, oligohumous water.

polykaryotic (Gr. *karyon* = nut, core, nucleus), pertaining to a cell (polykaryon) containing many nuclei. *Syn.* multinuclear, multinucleate, polynuclear, polynucleate. *See* coenocyte.

polyketides, cyclic aromatic compounds synthesized from acetyl coenzyme A and malonyl coenzyme A in the acetate/malonate pathway of the phenolic metabolism. Polyketides are precursors of some plant → quinones and flavonoids, and they exist as intermediates in the syntheses of some mycotoxins, such as → aflatoxins and patulin.

polymely, polymelia (Gr. *melos* = limb), a rare developmental anomaly in animals characterized by the presence of more than the normal number of limbs. *Adj.* **polymelous**, an individual exhibiting polymely.

polymer (Gr. *meros* = part), a compound formed by the joining of smaller, repeating molecules (monomers) arranged in a long chain or network. Biological polymers are e.g. DNA and RNA molecules synthesized from many → nucleotides, or the polypeptides formed from many amino acids. Polymers of the largest biological macromolecules are often called **multimers**, such as the proteins synthesized from several polypeptides. *Adj.* **polymeric**. *See* polymerization.

polymerases, enzymes of the transferase class (EC 2.) catalysing → polymerization, e.g. that of nucleotides to polynucleotides.

polymerase chain reaction, PCR, a method used for gene manipulation in molecular biology in which a certain segment of DNA is amplified exponentially. A preparation of DNA is denatured, the single-stranded preparation is annealed with two short → primer sequences which are complementary to sites on the opposite strands on both sides of the target region to be amplified. DNA polymerase (normally heat-stable → *Taq* polymerase) is used to synthesize a single strand from the 3' end of each primer. The entire cycle can then be repeated by denaturing the preparation and starting again. By employing this method, high quantities of DNA can be isolated for analyses.

polymerization, a chemical reaction in which relatively small molecules (monomers) join to form larger compounds; two units of the same compound form **dimer**, three units **trimer**, four units **tetramer**, etc., and many units → **polymer**. Usually the molecules consisting of 2 to 10 units are called **oligomers**, and molecules containing more than 50 units **high polymers**. A polymerization reaction requires some catalyst like a metal, acid, peroxide, radiation, pressure, and in cells a → polymerase enzyme. Such compounds which contain double or triple bonds are easily polymerized, e.g. ethene, aldehydes and nitriles.

polymictic (Gr. *miktos* = mixed), *see* stratification.

polymorph (Gr. *morphe* = shape, form), polymorphonuclear granulocyte. *See* granulocytes.

polymorphism, appearance in more than one form; e.g. **1.** a condition in a population when at the same time two or more forms (→ **morphs**) exist together in such proportions that the rarest cannot be maintained merely by

→ mutation pressure; *see* genetic polymorphism; **2.** balanced polymorphism; the appearance of different morphs in balanced proportions maintained by selection in a population; **3.** DNA polymorphism, a condition of a DNA structure having different but normal sequencies at a certain site of DNA. *Adj.* **polymorphic**.

polymorphism information content, PIC, an estimate to observe whether a genetic → marker is sufficiently polymorphic to detect the → linkage of a given gene locus. The estimate is made on the basis of the number of alleles found in that marker locus.

polymorphonuclear (L. *nucleus,* dim. of *nux* = nut), **1.** having a cell nucleus which is lobed or so divided that it looks to be multiple; **2.** pertaining to polymorphonuclear leucocytes, → granulocytes (polymorphonuclear cells).

polymyxins, a group of basic peptide antibiotics produced by bacilli, e.g. *Bacillus polymyxa*; act against Gram-negative bacteria by increasing the permeability of their cell membranes.

polynuclear (L. *nucleus,* dim. of *nux* = nut), **1.** pertaining to a cell that contains several nuclei; *syn.* polynucleate, polykaryotic, multinuclear, multinucleate; **2.** pertaining to → polymorphonuclear.

polynucleolar, pertaining to a cell nucleus containing several nucleoli. *See* nucleolus.

polynucleotide, a → polymer that consists of many → nucleotides; e.g. DNA and RNA.

polyp (Gr. *polypous*, *pous* = foot), **1.** an individual (**zooid**) of a colonial hydrozoan; may be either a feeding **gastrozooid**, reproductive **gonozooid**, or **dactylozooid** for defence and catching prey; a branched tubular system, **coenosarc**, connects the gastrovascular cavities of various polyps; **2.** a solitary, sessile hydrozoan, such as a hydra, having several tentacles around the mouth on the top of the body; **3.** a protuberance on the mucous membrane or skin, caused e.g. by an inflammation.

polypeptide (Gr. *peptos* = cooked), a → peptide usually composed of more than ten amino acid residues. A short peptide is **oligopeptide**, and a polypeptide containing more than 100 amino acid residues is usually called **protein**, but often proteins are composed of several polypeptide chains. The **primary structure** of a long polypeptide is formed from different amino acids by → peptide bonds (*see* protein synthesis). The **secondary structure** of a polypeptide adopts the form of → alpha-helix or → beta-pleated sheet, or a mixed structure of these both, or of random coiling with hydrogen bonds. The **tertiary structure** results from the folding of polypeptide with hydrogen and disulphide bonds (*see* protein).

polyphage (Gr. *polyphagos* = eating too much), an organism feeding on many kinds of food and several prey species, e.g. rats, or insects feeding on several plant species. *Syn.* euryphage, pleophage. *Adj.* **polyphagous**. *Cf.* monophage, oligophage, stenophage.

polyphagy, polyphagia, 1. gluttony, excessive eating; **2.** use of food of many kinds; *see* polyphage.

polyphenols, polymerization products of → phenol, e.g. → tannins which cause browning of plant tissues. Normally many plants contain polyphenols which are synthesized in higher quantities when tissues are wounded; the **polyphenol oxidases** (EC sub-subclass 1.14.18.) liberated from damaged tissues turn polyphenols brown, e.g. occurs on cut surfaces of a potato or apple. Polyphenols may be important in defence against herbivores and also have antimicrobial properties.

polypheny (Gr. *phainein* = to show, to express), → pleiotropy.

polyphyletic (Gr. *phyle* = tribe), pertaining to a group of taxa which does not include the most recent common ancestor of all members of the group, but the members are derived from two or more different ancestors. *Cf.* monophyletic, paraphyletic.

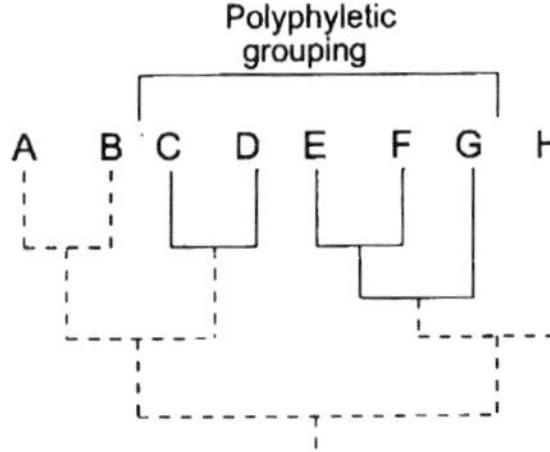

Polyphyletic relationships between eight taxa (A-H).

polypide, polypite (Gr. dim. of *polypous* = polyp), a soft, living part, e.g. a lophophore, intestine, or a muscle of a solitary individual, **zooid**, of a colonial bryozoan; the polypide is enveloped by a tubular, chitinous housing, **zooecium**.

Polyplacophora (Gr. *plax* = plate, *phoros* =

bearing), **polyplacophorans**, **chitons**; a class in the phylum Mollusca including aquatic species which are adapted to living on the rocks in the intertidal zones of oceanic coasts. They are primitive molluscs with a bilateral, segmented body covered by a calcareous shell with eight plates; the head is often reduced, and a large ventral muscular foot is used for crawling. In some classifications, chitons are considered as an order (or subclass) in the class *Amphineura*.

polyploid (Gr. *polyploos* = manifold), a somatic cell, tissue, individual, or species in which the basic chromosome number (haploid chromosome number, n) exists in more than two complete sets (tri-, tetra-, pentaploid, etc., i.e. 3n, 4n, 5n, etc.). The polyploid state, **polyploidy**, can arise spontaneously or it can be produced experimentally. Polyploids that carry the non-homologous, basic chromosomal complement in two or more sets, and hence are usually generated as a consequence of a crossing between species, are called → **allopolyploids**. On the other hand, polyploids in which the same basic number of homologous chromosomes is present in more than two sets are called → **autopolyploids**. Polyploid is also used as an adjective.

Polypodidae, spleen-worts, a subclass in the class Pteropsida, division Pteridophyta, Plant kingdom; includes ca. 9,000 species, e.g. the best known common ferns such as the lady fern and the ostrich fern. In some systems spleen-worts are classified as the order Filicales in the division Pterophyta.

polypores (Gr. *poros* = pore), a group of fungi in the subdivision Basidiomycotina, division Amastigomycota, including the bracket fungi, coral fungi, and chanterelles. The fructifications are mostly leathery or woody and perennial, and many cause diseases of trees.

polyprenols, monohydroxylic alcohols biosynthesized from mevalonic acid; their carbon skeleton is composed of → isoprene units; e.g solanesols in plant leaves and → dolichols in bacteria, also found in many animals. Polyprenol phosphates act as carriers in the sugar transport from nucleotide diphosphate sugars to different acceptors, e.g. polysaccharides or glycolipids, bound to the cell membrane and cell wall.

polyp stage, a stage in the life cycle of hydrozoans when the animal is sessile; the polyp stage may alternate with the → medusoid stage. *See* polyp. *Cf.* medusa.

Polypteriformes (Gr. *pteron* = wing), **bichirs** and **reedfishes**; an order in the subclass Sarcopterygii (class Osteichthyes); found in some lakes in Africa. They have a dorsal fin of 8 or more spinous finlets, lobate pectoral fins, ganoid scales, and a pair of lung-like air bladders. In some classifications bichirs and reedfishes are placed in the subclass Brachiopterygii.

polyribosome, → polysome.

polysaccharides (Gr. *sakcharon* = sugar), glycans; linear or branched high molecular weight → carbohydrates synthesized in cells by linking many → monosaccharides (10 or more) together with → glycosidic bonds. The **homopolysaccharides** are composed of one type of monosaccharide whereas **heteropolysaccharides** contain different monosaccharides. Important polysaccharides are e.g. → fructosans, glycogen, and starch as energy storage, and → chitin, glycosaminoglycans and cellulose as structural substances.

polysepalous, pertaining to a flower having numerous → sepals separate from each other. *Cf.* gamosepalous.

polysiphonic (Gr. *siphon* = tube), **1.** pertaining to an exceptional type of germination of a → pollen grain in which several → pollen tubes develop instead of only one as normally; **2.** a plant structure consisting of several cell rows, e.g. in thalli of some brown and red algae.

polysome (Gr. *soma* = body), **polyribosome**; a cytoplasmic organelle consisting of → messenger RNA and several → ribosomes together; it is the site of → protein synthesis. Polysomes are found abundantly in → Nissl substance of a nerve cell.

polysomia, a foetal malformation rarely found in a vertebrate embryo, having doubling or tripling of the imperfect body, usually partially fused together.

polysomic, 1. a → diploid cell or individual carrying some of the chromosomes in more than two sets. Polysomics are e.g. trisomic, double trisomic, or tetrasomic cells and individuals in which the chromosome numbers are 2n + 1 (→ Down's syndrome), 2n + 1 + 1, and 2n + 2, respectively (*see* n); the phenomenon is called **polysomy**; **2.** *adj.* pertaining to a cell or an individual exhibiting polysomy.

polyspermia, polyspermism (Gr. *sperma* =

sperm), **1.** → polyspermy; **2.** abnormal, excessive secretion of sperm.

polyspermy, the penetration of more than one spermatozoon into an egg cell regardless of their participation in the fertilization; often occurs in eggs with a large yolk, e.g. in birds, or may be produced artificially. *Syn.* polyspermia.

polysymmetric flower, pertaining to a flower with more than two planes of symmetry. *Syn.* actinomorphic flower, radial flower.

polytene chromosome (L. *taenia* < Gr. *tainia* = band), *see* giant chromosome.

polyteny, the amplification of chromosomes as a result of replication of chromosomes without separation and with somatic conjugation. *Cf.* endomitosis. *See* giant chromosome.

polythermic (Gr. *therme* = heat), pertaining to an organism which endures relatively high temperatures; e.g. the optimal temperature of polythermic organisms in northern waters may be 15°C and they can tolerate environmental temperature changes up to 25°C. *Cf.* oligothermic. *See* eurythermic, stenothermic.

polythetic (Gr. *thetikos* = fit for placing), pertaining to a classification based on many characteristics common to the members of a certain taxon. *Cf.* monothetic.

polytopic (Gr. *topos* = place), pertaining to an organism or a species living in, or originating from several areas.

polytypic (Gr. *typos* = type), having many types; pertaining to a → taxon that includes several subtaxa; e.g. a species comprising several subspecies, a genus with several species, etc. *Cf.* monotypic.

polyuria (Gr. *ouron* = urine), excessive excretion of urine.

polyvinylpyrrolidone, PVP, a synthetic, chemically inert → polymer (molecular weight ranges from 10,000 to 700,000), used as a dispersing and suspending agent, or e.g. to bind phenols in plant tissue homogenates and to protect other molecules, particularly enzymes, from inactivation. PVP is also used in gradient → centrifugation media. *Syn.* polyvidone, povidone.

Polyzoa (Gr. *zoon* = animal), → Bryozoa.

pome (Fr. *pomme* = apple), a → fruit formed by an inferior ovary, the fleshy part being largely developed from the surrounding receptacle; e.g. the apple, pear, rose pome.

Pongidae (Congo *mpungu* = ape), a family in the superfamily Hominoidea comprising the great apes, i.e. orang-utan (*Pongo*), chimpanzees, and gorilla; in modern classifications placed in the family → Hominidae.

pons, pl. **pontes** (L. = bridge), a bridge-like anatomical structure; specifically the ventral part of the metencephalon in the vertebrate brain stem, situated caudally from the mesencephalon under the cerebellum and comprising nerve tracts to and from the cerebellum; called also *pons cerebelli* or *commissura cerebelli*. *Adj.* **pontine, pontil(e)**.

pool, a small area of standing water; specifically in biology, e.g. **1.** → gene pool; **2.** a pool of cells and their parts, or tissues (as plasma, blood, etc.) obtained from different sources; **3.** the collecton or concentration of a dispersed fluid into one location; **4.** as a verb, to add together.

population (L. *populus* = people), **1.** a group of individuals living at the same time in the same area being capable of reproduction among themselves, i.e. belonging to the same species. A population can be **open** or **closed**, according to whether it exchanges individuals with other populations by → migration, or not; *see* metapopulation; **2.** a coherent group of cells, macromolecules, etc.

population cycles, fairly regular and marked changes in the number of individuals in a population. Such **cyclic oscillations** in population densities are typical especially of small mammalian and some bird populations on open tundra (e.g. lemmings, voles, and willow ptarmigans), as well as predators preying on them (e.g. rough-legged buzzard and long-tailed skua). Intervals of cycles vary according to the species; the cycles of small microtine rodents have a typical periodicity of 3 or 4 years, whereas hares and lynx undergo 10-year cycles. Population cycles are also found among some plants, e.g. in aquatic populations.

Reasons for the cycles may be different. Among animals, the most important factors are probably the changes in quantity and quality of food (*see* chemical defence) as well as the effects of predation, but in overcrowded populations also → stress may decrease the productivity of the animals.

population density, the number of individuals in a population per certain environmental unit, e.g. per unit area, unit volume, sometimes also per leaf, tree, or host.

population dynamics, changes in a population

density, size, and internal structure (e.g. age structure, sex ratio) with respect to time and space, and to the factors affecting them. These factors may be either dependent on the population density and other properties, such as food, or independent of them, as e.g. climatic factors and catastrophes.

population ecology, a branch of ecology dealing with changes in population size and density, internal qualitative changes of a population, and factors affecting them.

population fluctuation, variations in the size and density of a population in respect to time.

population genetics, a branch of genetics examining the consequences of → Mendelian inheritance in populations, often presented in a mathematical form. Population genetics deals with gene frequencies and interactions in → Mendelian populations, and studies the effects of → mutation, migration, genetic drift, and selection on gene frequencies and hence microevolution. Population genetics is divided into **theoretical** and **experimental** population genetics.

population regulation, the regulation of changes in population size caused by density-dependent factors. When the density is high, an increase of death rate, emigration, and/or a decrease of birth rate, tend to reduce it. Correspondingly, when the density is low, an increase of birth rate, immigration, and a decrease of death rate, tend to raise it. Factors affecting population densities are studied intensively, but no unambiguous explanations have been found. Factors affecting populations are divided into two groups: **intrinsic regulation factors** and **extrinsic regulation factors**.

Intrinsic factors in mammals are considered in the **stress theory** emphasizing that hormonal activities which control physiological reproductive functions of animals are disturbed when the population density has become too high, leading to reduced reproductive capacity. **Chitty's theory** emphasizes the importance of genotypic behavioural changes (caused by changed densities) in reproductive effort and ultimately in population densities. According to the **theory of Charnof and Finerty**, new aggressive immigrants whose number correlate positively with the population density, disturb social balance in the population causing decreasing reproduction.

Extrinsic factors include e.g. direct **climatic effects** on population densities and the importance of **feeding factors** which may be regulated by climatic conditions or changes in food quality (*see* chemical defence). Also **predators, parasites**, and **diseases** may effectively regulate population densities. *See* population cycles, self-regulation hypothesis.

pore (L. *porus*, Gr. *poros*), a small opening; e.g. a gustatory pore, sweat pore, or pores in the cell wall of unicellular algae.

Porifera (L. *ferre* = to carry), **poriferans, sponges**; a phylum of primitive animals which are either radially symmetrical or asymmetrical, and are formed only from two embryonic layers, ectoderm and endoderm (the mesoderm is lacking). Adult sponges are sessile marine animals, only one family existing in fresh waters.

On the basis of the structure of the canal system, sponges are divided into three different types. The simplest structural organization is found in the **ascon type (asconoid)**, having a tubular body with many microscopic dermal pores, a large cavity, **spongocoel**, and dorsally a single large **osculum**. Water flows through pores (ostia) into the spongocoel and is expelled through the osculum. Each pore is surrounded by contractile cells, **porocytes**, and the walls of the spongocoel are lined by food-collecting → **choanocytes**. In the **sycon type (syconoid)**, choanocytes are located in the **radial canals** of the thickened body wall. The **leucon type (leuconoid)** is the most complicated with choanocytes lining the walls of several **flagellated chambers** connected by numerous channels.

The **sclerocytes** secrete needle-like calcareous or siliceous **spicules** to form an endoskeleton, and **spongocytes** secrete spongin fibres. The phylum includes about 5,000 species which are divided into four classes: **calcareous sponges** (Calcarea), **glass sponges** (Hexactinellida), **coralline sponges** (Sclerospongiae), and Demospongiae which includes about 95% of living sponge species.

porins, transmembrane proteins found in the outer lipopolysaccharide membrane of Gram-negative bacteria; as trimers they form 1 nm wide channels through which small hydrophilic molecules (e.g. disaccharides) can pass; porins are also found in mitochondrial membranes of eukaryotic organisms.

porocyte (Gr. *kytos* = cavity, cell), a cell type in the body wall of sponges, capable of contraction; the cell has a median hole through which water flows into the spongocoel. *See* Porifera.

porogamy (Gr. *gamos* = marriage), growth of the pollen tube through the → micropyle to the → ovule after pollination; the normal type in most angiosperms. *Cf.* aporogamy, chalazogamy.

porometer (Gr. *metron* = measure), an apparatus for measuring the opening stage of leaf stomata (*see* stomatal resistance) and respiration of plants; measures the time taken until a certain level of relative humidity (e.g. 55%, selected on a porometer scale) is reached.

porphyrin (Gr. *porphyra* = purple), a compound formed from four pyrrole rings which easily binds a metal atom in the centre of the molecule (*see* pyrrole); the compound is called **metalloporphyrin**. It is present in the → haem of cytochromes (the enzymes of cell respiration), many → blood pigments, and myoglobin of muscle cells, as well as in → chlorophylls of plants. The first stage of the porphyrin synthesis is the linking of glycine and succinyl coenzyme A to form δ-aminolaevulinic acid. Two δ-aminolaevulinic acid molecules form porphobilinogen, and the linear-chain tetrapyrrole is synthesized by combination of four porphobilinogen molecules. The linear molecule is then converted to the ring form and porphyrin is synthesized via some intermediate steps.

portal (L. *porta* = entrance, gate), **1.** an entrance or gateway, such as the point of entry of a pathogenic microbe into an animal body; **2.** → portal vein; **3.** as and adjective, pertaining to a gate, e.g. a gateway or entrance in an anatomical structure.

portal vein, a vein connecting two capillary beds, i.e. any of short veins passing from one organ to another where it successively divides into smaller branches and capillaries, but does not pass directly to the heart as do other veins; e.g. **1. portal vein of liver** (*vena portae hepatis*) forms the **hepatic portal system**, where the large vein trunk passes from the intestine to the liver where the vessels form a capillary-like system of **sinusoids** (into which also hepatic arterioles are united). Single sinusoids join together forming small intralobular central veins and finally the larger hepatic veins (*venae hepaticae*) draining from the liver to the inferior vena cava and to the heart. The blood in the portal vein transport substances absorbed from the gut first to the liver to be at least partly metabolized there; **2. pituitary portal system**, exists in the head, with the portal veins passing from the hypothalamus to the anterior pituitary; **3. renal portal system** in the mesonephros of lower vertebrates.

ports, porters, cell membrane proteins transporting substances across the membrane. *See* → cell membrane transport.

porus, pl. **pori** (L.), → pore.

positional cloning, the isolation and cloning of genes according to their positions on the → genetic map defined by their mutations.

positional information, an information system based on differences in the concentrations of → morphogens or other regulatory substances in different parts of a developing organ or in the whole embryo; according to these differences the cells receive information on their position and are determined (*see* determination) to differentiate by this information.

position effect, the effect of the location of a → gene of the chromosome on its function; can be stable, variegated, or a *cis-trans* position effect. In **stable position effect** (S type p.e.), the influence of the location of a gene is the same in all cells of an organism. In **position effect variegation** (PEV), the function of the gene varies from cell to cell, thus resulting in phenotypic mosaicism. It appears when the gene is removed in a chromosome mutation and transferred to the vicinity of → heterochromatin, which inactivates the gene in some but not in all cells. This occurs because the inactivating effect of the heterochromatin propagates polarly into the → euchromatin area.

The *cis-trans* position effect means that the *cis* heterozygote a b/+ + is phenotypically different from the *trans* heterozygote a +/+ b. The → *cis-trans* test and hence the definition of the → cistron is based on this phenomenon (a and b = → mutant alleles, + = → wild allele).

position pseudoalleles, → pseudoalleles which show a *cis-trans* → position effect.

positive control, a form of the regulation of a gene function in which the regulatory molecule (usually a protein) in its active form enhances the gene function; positive control corresponds to positive → feedback in regulation in general. *Cf.* negative control. *See* regulation.

positive reinforcement, *see* reinforcement.

positron, the positive → electron of an atom, i.e. the antiparticle of the electron, β^+, having the same mass as the electron and the electric charge of equal magnitude but opposite sign. Positrons are formed e.g. when a → proton changes into a neutron in radioactive decay processes (beta decay); when positrons collide with negative electrons, both particles are annihilated and only electromagnetic radiation remains.

post- (L. *post* = after), denoting after or behind.

postabdomen (L. *abdomen* = belly), the posterior differentiated part of the → abdomen in some arthropods; specifically the flexible caudal portion of the abdomen of a scorpion constructed of knot-shaped segments and ending with a **stinging apparatus**. *Cf.* preabdomen.

postembryonic (Gr. *embryon* = embryo), occurring after the embryonic stage.

postero- (L. *posterus* = behind, following), pertaining to a situation behind or posteriorly.

postganglionic, pertaining to neural structures distal to the ganglion in the nerve tract, or to the functions occurring in this part. *Cf.* preganglionic.

postmentum (L. *mentum* = cheek), the proximal basal plate in the labium of insect mouthparts; divided into **submentum** and **mentum**. *See* prementum, mouthparts.

postmortal (L. *mors* = death), pertaining to, or occurring after death; **postmortem**, pertaining to a period after death.

postnatal (L. *natus* = birth), occurring after birth. *Cf.* perinatal, neonatal, prenatal.

postnuptial (L. *nuptiae* = wedding), occurring after mating or the reproductive period; e.g. **postnuptial plumage** in birds.

postsynaptic, pertaining to neuronal structures distal to the synaptic cleft in a → synapse, or to the functions occurring in this part. *Cf.* presynaptic.

postural (L. *postura* = position, posture), pertaining to a position or posture.

potassium (L. *potassa* = potash), symbol K (kalium), atomic weight 39.098; an alkaline metallic element that exists in natural conditions only in combination with other elements, e.g. in minerals and sea water. Potassium ion (K^+) is an essential element in cells and tissues of all organisms involved in the regulation of ion-water balance. K^+ is an important intracellular cation and its transport into cells across the cell membrane is controlled by an active transport mechanism (*see* ion pumps) and → ion channels. Potassium plays a central role in electrical properties of cells, i.e. in the formation of the → resting potential and → action potentials across the cell membrane.
Potassium is present in many inorganic compounds, e.g. chloride (KCl), chlorate ($KClO_3$), cyanide (KCN), phosphates (KH_2PO_4, K_2HPO_4, K_3PO_4), sulphate (K_2SO_4), and hydroxide (KOH), but also in organic compounds, e.g. acetate (CH_3COOK), oxalate ($K_2C_2O_4$), and tartrate ($C_4H_4K_2O_6$).

potassium-argon dating, K/Ar-dating, an archaeological method of dating specimens, especially from Pleistocene, by determining the ratio of argon to potassium; it is based on the natural radioactive disintegration of potassium-40 (^{40}K) into argon-40 (^{40}Ar). The half-time of radioactive ^{40}K-isotope is 1,300 million years, and the ratio of these two nuclides in the specimen indicates its age. The method is useful especially for determining the age of volcanic minerals several thousand million years old.

potassium channel, an → ion channel in a cell membrane selective for potassium ion, K^+.

potassium permanganate, a deep purple, crystalline water-soluble salt, $KMnO_4$; a strong oxidant used for different volumetric analyses and as a fixative for biological material; particularly maintains the structures of cellular membranes. It is also used for determination of oxidizing substances in water; a certain volume of $KMnO_4$ consumed (reduced) in titration is proportional to the quantity of organic matter in a water sample. Some compounds, such as methane, hydrogen sulphide, or chloride ions can interfere with the permanganate reaction.

potency (L. *potentia* = power), power, efficacy, strength, influence; e.g. **1.** male sexual ability; **2.** the capability of an embryonic tissue of developing into its completed form; **3.** the power of a medicinal agent to produce a desired effect. *Adj.* **potent**.

potential, *adj.* **1.** possible, latent; capable of being or doing; ready for action; *noun* **2.** possibility, potentiality; **3.** in physiology, an electric potential, i.e. the difference in electricity between two points of a tissue structure, e.g. outside and inside the cell membrane; *see* resting potential, action potential.

potometer (Gr. *poton* = drink, *metrein* = to measure), a simple device for measuring the transpiration of plants. A plant or a twig is placed into a water container connected with a capillary tube where the amount of transpired water can be observed.

pouched mammals, → Marsupialia.

POU domain (P = pituitary, O = octamer, U = *unc*-86, a gene of → Caenorhabditis elegans), a DNA binding region of some → transcription factors.

PPB, → preprophase band.

ppb, parts per billion (Am.), a billionth part (10^{-9}).

PPi, → pyrophosphate.

PPLO, pleuropneumonia-like organisms. *See* mycoplasmas.

ppm, parts per million, a millionth part (10^{-6}).

P protein, phloem protein, a protein in the → sieve tubes of vascular plants; it has a protective function during an unfavourable season being able to block transport through the pores of the sieve plates, but may also support the transport.

prairie, North-American → grassland area, covering the central regions of the continent.

pre- (L. *prae* = before), denoting before, in front of something.

preabdomen, also **praeabdomen** (L. *abdomen* = belly), the broader, anterior portion in the abdomen of a scorpion. *Cf.* postabdomen.

preadaptation (L. *adaptare* = to fit), the preceding occurrence of those adaptive characteristics which favour the survival of an organism in new environmental conditions; e.g. the plumage of birds has evolved as an adaptation for thermoregulation but at the same time it was a preadaptation for the evolution of the ability to fly. *See* adaptation.

preadult, 1. a developmental stage preceding the adult stage; e.g. the postlarval stage of ephemeropterans with wings and capable of flying, but has to pass through one moult before sexual maturity; **2.** as and adjective, pertaining to such a stage.

Precambrian (era) (*Cambria* = Wales), the oldest era of the Earth preceding Cambrian, 4,600—570 million years ago; first prokaryotes and then primitive eukaryotes appeared and photosynthesis developed; evidences of sponges and worm burrows are found. The era is divided into **Archaean** (**Archaeozoic**) and **Proterozoic** eons.

precious gases, → noble gases.

precipitation (L. *praecipitare* = to cast down), **1.** in chemistry, the separation (settling down) of a substance in a solid form from a solution; or a deposit (**precipitate**) of this substance; **2.** in immunological tests, the formation of an antigen-antibody complex; **3.** in meteorology, the condensation of moisture from a vaporous state into rain, snow, etc. *Verb* to **precipitate**. *Adj.* **precipitant**, a substance that causes precipitation; **precipitable**, capable of precipitation.

precocial (L. *praecox* = premature), pertaining to a newhatched bird or newborn mammal with open eyes and a capability of walking and running almost immediately after birth, e.g. chickens, colts, calves, and lambs. *Cf.* altricial, nidicolous, nidifugous.

precocious, developed unusually early or rapidly, e.g. refers to a person with exceptional mental or physical characteristics at an early age.

precursor (L. *praecursor* = forerunner), something preceding; predecessor; e.g. a substance from which another, usually a more active substance, is synthesized. *Adj.* **precursory**, preliminary, introductory.

predation (L. *praedari* = to plunder), an interaction between species where one species, the → **predator**, feeds on individuals or parts of individuals of other species, i.e. **prey**; predation may be directed to plants (**herbivory**) or to animals (**carnivory**). Also some plants can be carnivorous, as sundews. Broadly, the parasiting and grazing may be included in predation.

predator, an organism which exploits other organisms by eating, grazing, parasitizing, or in some other way. The predator may be either a specialist or a generalist depending on whether the predation is applied to a narrow or wide scale of prey groups. Predators are divided into → true predators, grazers, parasites, and parasitoids.

predetermination, the → determination of a genetically controlled character by the → maternal genotype, thus existing in an ovum before fertilization. Consequently, the characters of the hybrid progeny derived from such a predetermined → zygote are → matroclinous with respect to those characters showing predetermination.

prednisolone, a synthetic → glucocorticoid, the dehydrogenated analogue of cortisol; the dehydrogenated analogue of cortisone is

prednisone; both are used in medicine.

preening, *see* grooming.

preformation theory, a theory prevailing in the 18th century according to which a ready formed miniature adult organism exists in the egg, spermatozoon, or early zygote. Hence, the development only occurred as the growth of this miniature individual. The opposite idea is presented in the theory of → epigenesis, stating that new structural and functional organisations appear during the development of an organism.

preganglionic, pertaining to neuronal structures proximal to the ganglion in a nerve tract, or to the functions occurring in this part. *Cf.* postganglionic.

pregnancy (L. *gnatus* = born), → gestation.

pregnanediol (L. *praegnans* = pregnant), an inactive metabolite of → progesterone (female sex hormone in vertebrates) metabolized in the liver and excreted in urine.

pregnenolone, an intermediate in the biosynthesis of steroid hormones.

prehensile leg/limb (L. *prehendere* = to grasp), a leg adapted for grasping and holding; as the hand of primates, the foot of the common swift (*Apus apus*), or the gnathopodium of some crustaceans; in addition, many monkeys have a prehensile tail.

preliminary cross, the crossing of organisms performed to produce wanted genotypes, which are then used in an actual experimental cross.

premaxilla, pl. **premaxillae,** also **premaxillas** (L. *prae* = before, *maxilla* = jaw), a paired dermal bone anterior to the → maxilla in the upper jaw of most vertebrates; in tetrapods, becomes incorporated into other jaw bones in the course of embryonic development. In birds, the premaxillae form the largest part of the upper beak, and in mammals bear the incisor teeth.

prementum, pl. **prementa** (L. *mentum* = cheek), the tip of the → labium in the mouthparts of insects, bearing the palps, and ligula. *Cf.* postmentum.

premolar (L. *moles* = mass), **1.** a premolar tooth (biscuspid tooth); one of the cheek teeth of mammals (in man four in both jaws), posterior to the canines (or to the incisors if the canines are lacking) and anterior to the molars; premolars are permanent teeth but are preceded by milk teeth; **2.** pertaining to the premolar teeth; situated in front of the molar teeth.

premutation (L. *mutatio* = change), damage of the gene which, if not repaired, leads to → mutation but can still be reverted in the → repair replication of a cell. If premutation has already been replicated, it can no longer be repaired.

prenatal (L. *natus* = birth), preceding birth; with reference to the foetus, denoting the period before birth, especially refers to mammals. *Cf.* perinatal, neonatal, postnatal.

prepattern, a theoretical model of animal development; the descriptive term for the general organization of a developing system of an organism before a specific and observable pattern of organization is reached; e.g. the prepattern of the wing in the → imaginal disc of developing insect larvae.

preprophase band, PPB, a band formed at the division plane of a plant cell from → microtubules and F-actin filaments (*see* actin) immediately before the prophase (*see* mitosis), i.e. at preprophase. The cortical microtubules first concentrate in the PPB together with the actin filaments; at metaphase the microtubules depolymerize but the actin filaments remain and determine the division plane, i.e. the site of the forming middle-wall in the cell.

prepupa, pl. **prepupae,** also **prepupas** (L. *pupa* = girl, doll), a developmental stage preceding the pupa in many insect larvae.

presby(o)- (Gr. *presbys* = old man), denoting relationship to old age.

presbyopia (Gr. *ops* = eye), the physiological loss of accommodation of the eyes that in man normally occurs after the age of 45. Distant objects are seen clearly, but the eye is unable to accommodate to see near objects distinctly due to loss of elasticity of the lens.

pressoreceptor (L. *pressura* = strain, stress), → baroreceptor.

pressure potential, an intracellular positive pressure formed by water uptake in plants; in flaccid cells the pressure potential is zero. *See* osmosis.

presynaptic, pertaining to the neuronal structures and functions proximal to the synaptic cleft of a → synapse. *Cf.* postsynaptic.

pretarsus, pl. **pretarsi** (Gr. *tarsos* = sole of the foot), the most distal segment in the leg of insects, myriapods, and spiders.

prevalence (L. *praevalere* = to prevail), in ecology, the relative number of individuals in a population suffering from a certain disease at a given time. *Cf.* incidence.

prey, an organism consumed by another organism, predator. *See* predation.

Priapulida (Gr. *priapos* = phallus), **priapulids, priapus worms**; a small invertebrate phylum comprising about 15 species which live in the mud and sand to depths of several thousand metres in cold marine water bodies in both hemispheres; have a cucumber-shaped body, a eversible proboscis with curved spines surrounding the mouth, one or two caudal appendages, and → pseudocoelom.

Pribnow box (*David Pribnow*), in prokaryotic → promoters a highly conserved sequence element, i.e. a short consensus sequence of DNA (TATAAT) which is analogous with the → TATA box of eukaryotes. The box is located at position −10 upstream of the transcription initiation site in the promoter and recognized by a common transcription factor, thus being essential for the initiation of → genetic transcription. Called also −10 sequence.

prickle cell, a cell type forming a cell layer (*stratum spinosum*) in the → epidermis of the vertebrate skin; in histological preparations small artificial tufts (prickles) project from these cells.

primaquine, 8-aminoquinoline; the most effective drug to prevent spread of human malaria; disturbs the mitochondrial function of the malaria *Plasmodium*, possessing a gametocidal effect on all forms of malaria; used in its diphosphate form.

primaries (L. *primarius* = of the first rank, principal < *primus* = first), the flight feathers of the bird wing being attached to the hand bones, i.e. carpometacarpals and digits. *See* secondaries.

primary cell wall, → primary wall.

primary consumer, an organism at the lowest consumer level of a food web, i.e. feeding on plants; herbivore.

primary cortex, the primary cortical layers of plant roots and shoots consisting of → epidermis and some layers of parenchyma and sometimes collenchyma. The primary cortex is replaced by the → periderm formed by the phellogen during secondary growth.

primary endosperm nucleus, the cell nucleus resulting from fusion of one generative nucleus with the two polar nuclei in the endosperm mother cell of the embryo sac in flowering plants; is usually triploid. *See* endosperm.

primary growth, growth occurring in roots and shoots from beginning of the growth of the embryonic organs in seeds to their complete differentiation before → secondary growth.

primary host, → definite host.

primary meristem (Gr. *merizein* = to divide), the → meristem tissue in plants, including the → apical meristem, intercalary meristem and lateral meristems.

primary oocyte, *see* oocyte.

primary phloem, the → phloem formed during the differentiation of → conductive tissues in the → apical meristem of a plant; transports organic compounds and products of photosynthesis during the first growth period. The primary phloem that develops first is **protophloem**, which is soon replaced by **metaphloem**. If → secondary growth occurs, the **secondary phloem** replaces the primary phloem, the primary tissues still remaining in the structure. *Cf.* primary xylem.

primary producer, → autotroph.

primary production, 1. the production of organic substances from inorganic materials (carbon, nitrogen, etc.) by → autotrophs, chiefly in → photosynthesis; **2.** the quantity of organic material produced by green plants and other → autotrophs in an ecosystem in a certain time, as within a year, a period of growth, etc. *Cf.* gross production, net production. *See* productivity.

primary ray, *see* ray.

primary sex ratio, *see* sex ratio.

primary sex(ual) characters, genital organs with the gametes. *See* sex characters.

primary signal, 1. in ethology, a → primer signal; **2** in botany, an environmental factor such as illumination, temperature, or gravity, which may affect directly the mechanisms transporting ions across the cell membrane (*see* cell membrane transport), and further the eletrical properties of the membrane, in this way controlling cellular functions in plants.

primary spermatocyte, *see* spermatogenesis.

primary structure, *see* polypeptide, protein.

primary succession, *see* succession.

primary thickening, a type of plant thickening caused only by cell divisions in the → apical meristem, e.g. in palms and fern trees. If the thickening is supported by other meristems (secondary meristems such as → cambium), the thickening is secondary.

primary transcript, the RNA transcript that occurs in the nucleus before → splicing and polyadenylation.

primary wall, the cell wall in plants formed first to envelope a newly divided cell; thin and elastic, being composed of pectin, hemicellulose, and small amounts of cellulose fibrils. The layers of the → secondary wall are normally formed soon on both sides of the fully developed primary wall, the latter remaining as a layer in the middle of the secondary wall.

primary xylem, a type of → xylem formed during the differentiation of → conductive tissues in an → apical meristem in a plant; transports water during the first growth period. The primary xylem developing first is **protoxylem**, which is soon replaced by a firmer type, **metaxylem**. If → secondary growth occurs, **secondary xylem** replaces the primary xylem, but the primary tissues still remain inside the secondary xylem. *Cf.* primary phloem, phloem.

Primates (L. *primas* = one of the first), a mammalian order whose species have existed in the world from Eocene (prosimians) and Oligocene (monkeys and apes) to the present time. Most primates are arboreal, having binocular and colour vision, prehensile limbs specialized for climbing, and a reduced snout with a less developed sense of smell. The brain is well developed, including especially large cerebral hemispheres. Many species exhibit strong social organization. The order with about 180 species is divided into two suborders: Prosimii (**prosimians**), comprising e.g. **lemurs** (Lemuridae), **lorises** (Loridae), **tarsiers** (Tarsiidae), and **aye-aye** (Daubentoniidae); the other suborder Anthropoidea (Haplorhini, Simiiformes) includes all other primates in two infraorders: Platyrrhini (**New World monkeys**) and Catarrhini (**Old World monkeys**, **apes**, and **humans**).

primatology, a branch of biology studying primates, especially other than the recent man.

primed *in situ*-labelling of nucleic acids, PRINS, the detection of a specific nucleic acid target by hybridizing *in situ* an unlabelled synthetic oligonucleotide or a short DNA fragment (→ probe) with the complementary nucleic acid sequence. Labelled nucleotides are then incorporated by a suitable polymerase using the probe as the → primer and the target nucleic acid as the template, thus labelling the site of hybridization.

primer, in genetics, a DNA or RNA segment necessary for the initiation of DNA replication; a short nucleic acid sequence containing a free 3' hydroxyl group, that forms base pairs with the complementary template strand and acts as the starting point for addition of nucleotides to copy the template strand.

primer signal, in ethology, a signal preparing an animal to change its own behavioural patterns or activity urges; it may be any signal, e.g. a courtship display, or an olfactory signal such as a → pheromone. Primer signals may stimulate or inhibit the receiver by affecting its hormonal balance. *Syn.* primary signal.

primitive (L. *primitus* = firstly), **1.** of earliest origin; **2.** unspecialized, → primordial; **3.** pertaining to a characteristic which was already present in an early ancestral form of a plant or an animal and has remained relatively unmodified in the stem line of a taxon up to present forms.

primitive streak, a thin, white tissue streak at the caudal end of the embryonic disc of amniotic vertebrates. It is the area from which the mesoderm arises when cells move (invaginate) inward and laterally under the ectoderm in the beginning of gastrulation. The anterior end of the streak is called **primitive node/knot** (Hensen's node/knot) and it is homologous to the dorsal lip of the blastopore in amphibian embryos.

primordial (L. *primordium* = origin), first formed, initial, primitive; e.g. **primordial fauna**, Cambrian fauna, **primordial skull**, a primitive, cartilaginous skull of cyclostomes and cartilaginous fishes.

primordial germ cell, the most primitive, undifferentiated → germ cell (gonocyte) in an early embryo; in animals, found initially outside the gonads.

primordial meristem, promeristem. *See* meristem.

primordium, pl. **primordia,** original form; e.g. **1. anlage**; the rudiment or commencement of an organ or a part of an animal embryo; **2.** a cell group developing into a specialized structure; e.g. the leaf primordia in the shoot tip of a plant.

primosome (Gr. *soma* = body), a complex structure of proteins in the cell nucleus involved in the priming action that initiates synthesis of each → Okazaki fragment during discontinuous DNA replication.

principle of priority, a principle presupposing that the name first given to a taxon and acceptably published, is valid. In zoology, the first accepted publication in this respect is the 10th edition of **Systema Naturae** by Carl von Linné from the year 1758. In botany, an original priority is dependent on the plant group; e.g. concerning vascular plants and most algae, the oldest publication is the first edition of **Species Plantarum** by Carl von Linné from 1753.

PRINS, → primed *in situ*-labelling of nucleic acids.

prions, 1. proteinaceous infectious particles; proteins found in animals encoded by their own genomes but which for a reason yet unknown are changed into a pathological form, i.e. have lost their tridimensional tertiary structure and thus can no longer function normally. This form is very resistant to heat and chemical treatment and can infect other individuals, apparently even other species, causing the appearance of the disease after months or even years. For example, the brain inflammation (*scrapie*) of goat and sheep, → mad cow disease, as well as certain slow inflammations of the central nervous system in man (Creutzfeldt-Jakob disease), are caused by prions; **2.** marine birds of the genus *Pachyptila*, order Procellariiformes.

prisere, primary → sere; complete natural → succession of plants from bare earth or water to a climax.

prismatic layer, the middle layer between the periostracum and the nacreous layer (nacre layer) in the clam shell, formed from calcium carbonate arranged in prisms.

PRL, → prolactin.

Pro, pro, → proline.

pro- (L., Gr. *pro* = before), denoting before, in front of, anterior to; in chemistry indicating a precursor of a substance.

proband (L. *probare* = to try, test), the index case; in genetics an individual, e.g. affected by a genetic disorder, of the family or the pedigree under study. *Syn.* propositus.

probasidium (Gr. *pro* = before, *basis* = base), a cell in → Basidiomycetes in which karyogamy occurs before the formation of the → basidium.

probe, 1. a molecule experimentally joined to a cell or to a part of a cell which is identifiable in experimental conditions, such as a labelled antibody for identifying proteins, or a segment of nucleic acid that has been labelled e.g. by a radioactive, coloured, or fluorescent substance. Using a labelled probe, a complementary nucleotide sequence is identified in → Southern or Northern blotting, or → *in situ* hybridization; *see* DNA probe; **2.** a flexible instrument introduced into an organ cavity.

probe DNA, → DNA probe.

Proboscidea (L. *proboscis* = snout), **proboscideans, elephants**; an order of mammals comprising large herbivorous animals which have massive legs, a long and flexible trunk (→ **proboscis**), and two long upper incisors elongated as tusks. At present, only three species are extant: the African elephant (*Loxodonta africana*), Indian elephant (*Elephas maximus*), and pigmy elephant (*Elephas cyclotis*) in West Africa. Well known are also the huge, extinct mammoths (*Mammuthus*) and mastodons (*Mammut*) in ancient tundra habitats.

proboscis, pl. **proboscides,** also **proboscises** (L. = snout, trunk), a tube-like enlargement of the mouth, snout, or pharynx; e.g. **1.** the muscular and tubular pharynx of a free-living flatworm (Turbellaria) which the animal can extend for capturing food; **2.** the extendible spiny proboscis of thorny-headed worms (Acanthocephala); **3.** a tube-like sucking snout in the mouthparts of insects formed by first maxillae; **4.** a ciliated and plate-like forepart of the body in tongue worms (Hemichordata); **5.** the trunk of elephants formed by the nose and upper lip.

proboscis worms, → Nemertea.

probud, a larval bud broken off the → stolon of doliolids (Thaliacea); divides several times to form buds which grow into individuals of three kinds: **gastrozooids**, **phorozooids**, and **gonozooids**.

procaine, 2-diethylaminoethyl *p*-aminobenzoate; used as hydrochloride for local anaesthesia.

procambium, *see* derivative meristems.

Procellariiformes (L. *procella* = tempest, *forma* = form), **tubenoses**; an order of pelagic ocean birds, feeding on fish, crustaceans and other invertebrates, and nesting on isolated islands; they have tubular nostrils for secreting extra salt (*see* salt gland). The order comprises 110 species in several families: e.g. Diomedeidae, **albatrosses**, Procellariidae including **prions**, **petrels**, **fulmars** and **shearwaters**, and Hydrobatidae, **storm petrels**.

procercoid, procercaria (L. < Gr. *pro* = before, *kerkos* = tail), an early larval stage (the first bladder-stage) in the life cycle of some cestodes, e.g. fish tapeworms. The procercoid lives in the first intermediate host (a copepod), and further development into the following stage, **plerocercoid**, occurs in a fish. *Cf.* cysticercus.

process (L. *processus*), **1.** in anatomy, a projection or processus in an organ, especially in a bone, e.g. the articular process on the surface of the arch of a vertebra (the inferior and superior articular processes); **2.** a series of events, reactions, or operations, achieving a certain definite result; **3.** as a verb, to prepare or treat by series of events.

Prochlorophyta (Gr. *chloros* = yellowish green, *phyton* = plant), **prochlorophytes**; photosynthetic → prokaryotes being structurally similar to the → cyanobacteria but having both chlorophylls a and b, and lacking → phycobiliproteins. They are free-living in plankton in fresh-water lakes, and live as symbionts in colonial ascidians. It is suggested that prochlorophytes have arisen from cyanobacteria by acquisition of chlorophyll b and they are suggested to be ancestral to the chloroplasts of eukaryotic cells. *See* endosymbiosis theory.

procoel, *see* mesocoel.

procoelous, Am. **procelous** (Gr. *koilos* = cavity), pertaining to a structure with a concave anterior face; e.g. **procoelous vertebrae** of amphibians and reptiles, the centra of which are concave in front and convex behind. *Cf.* amphicoelous, opisthocoelous.

Proconsul (*Consul* = the name of a chimpanzee in the London Zoo), one of the most ancient hominids whose fossils have been found in Africa; a well known species is *Proconsul africanus*. The family Proconsulidae lived in Miocene about 22—17 million years ago. The Proconsul species had no tail and they used all four feet for moving on the ground; they probably climbed trees as do chimpanzees at present.

proct(o)- (Gr. *proktos* = anus), denoting relationship to the anus, or rectum.

proctodaeum, Am. **proctodeum,** pl. **proctod(a)ea,** also **proctod(a)eums** (Gr. *hodaios* = on the way), **1.** anal pit; the depression at the caudal end of the body of an animal embryo, i.e. at the terminal region of the embryonic hindgut. In vertebrates, the proctodeal endoderm and anal ectoderm form first the anal plate (cloacal plate), which then ruptures and forms external urogenital and cloacal (anal) orifices; **2.** the terminal part of the alimentary canal of arthropods (hindgut).

procuticle (L. *pro* = before, *cuticula,* dim. of < *cutis* = skin), a layer under the epicuticle of moulting arthropods, composed of chitin and protein; in the mature → cuticle it is differentiated into exocuticle and endocuticle.

producer (L. *producere* = to produce), an organism synthesizing complicated organic materials from simple inorganic nutrients; in general, producers are green plants but also some types of bacteria. *See* autotroph.

producer gene, a gene that encodes protein synthesis in eukaryotic organisms; corresponds to → structural genes in prokaryotes.

productivity, in ecology, the potential rate of incorporation of energy in organisms, or the efficiency of production of biological material by organisms; it is given as the total quantity of organic material and energy produced by an individual, population, or community in a unit time per unit area. Often the term is used as a synonym of **production,** that usually means the act of producing, or product. *See* primary production, secondary production, gross production, net production.

production efficiency, the proportion of energy which is fixed into the biomass of individuals, populations, or communities; calculated in percentages of a → gross production.

proembryo (Gr. *pro* = before, *embryon* = embryo), a structure in a seed formed from the zygote, differentiating further into the suspensor and the → embryo.

profitability, in ecology, the term used in connection with the → optimum foraging theory as a standard for values of food patches; e.g. describes the relation between the energy used in predation and that gained from it. The profitability is high, if a great amount of food is provided with minimum use of energy.

profundal zone (L. *profundus* = deep), a deep water zone at the bottom of a lake, situated under the → compensation zone where light does not penetrate. *Cf.* littoral zone.

progenesis, pl. **progeneses,** a form of → heterochrony. *Syn.* → paedogenesis.

progenitor (L. *progignere* = to beget), one that precedes; precursor; an ancestor in the direct line. *See* progenitor cell.

progenitor cell, a cell from which the cells of a

given tissue are derived so that the daughter cells can differentiate into such tissue cells. *Cf.* stem cell.

progeny, offspring.

progestational (L. *pro* = before, *gestation* = pregnancy), **1.** pertaining to the oestrus phase of an → oestrus cycle of mammals, i.e. to the preparation of reproductive organs for pregnancy, the state when the → corpus luteum is activated and the endometrium of the uterus is in the secretion phase; **2.** pertaining to → progesterone, or pharmaceutical preparations with progesterone-like effects.

progesterone, a steroid hormone in vertebrates; 4-pregnene-3,20-dione, $C_{21}H_{30}O_2$; the most important progestational hormone in mammals. In females it is secreted mainly by the cells of the **corpus luteum** in ovaries, in mammals also from the **placenta**, and in both sexes in lesser quantities from the **adrenal cortex**. Progesterone stimulates the development of the uterus and other female genital organs and the mammary glands during sexual maturation, and prepares them for pregnancy, maintains pregnancy, and stimulates the growth of mammary glands for lactation; has also an antioestrogenic effect. Progesterone has effects also on brain functions involved in female behaviour. It is used as a component in contraceptive pills. *See* progestin, gestational.

progestin, 1. an old name for a crude hormone preparation obtained from the corpora lutea of mammals, mainly containing → progesterone; **2.** the generic term for natural or synthetic **progestogens** or **progestational agents**, i.e. the agents having the capability of preparing mammalian organs, especially the uterus, for pregnancy.

progestogen (Gr. *gennan* = to produce), any agent producing progestational effects on female mammals, e.g. a synthetic derivative of → progesterone.

proglottid (L. *proglottis*, Gr. *glotta* = tongue), one of the segments forming the body (**strobila**) of a tapeworm, each containing separate reproductive organs; new proglottids are formed from the "neck" (budding zone) behind the head (**scolex**), and each individual proglottid thus moves posteriorly in the strobila, during which its gonads, both male and female, become mature. The most posterior proglottids with mature eggs are shed and discharged in the host's faeces, where they disintegrate and may then enter an intermediate host to continue the life cycle.

proglottis, pl. **proglottides** (L.), → proglottid.

prognathous (Gr. *gnathos* = jaw), **1.** having projecting jaws, i.e. jaws protruding in front of the skull; in primates denoting an ape-like feature; **2.** in some insects, pertaining to the head axis being in line with the body. *Cf.* hypognathous, orthognathous.

programmed cell death, apoptosis; active death of individual cells associated with the development of tissues and organs, the function of the immune system, ageing, and other biological phenomena; occurs according to a genetically determined programme triggered by certain stimuli. E.g. in the differentiation of the hand, a plate-like structure develops first where the interspaces of the fingers are differentiated through the programmed cell death.

progymnosperms (Gr. *gymnos* = naked, *sperma* = seed), a group of fossil, spore-bearing, woody plants and trees, thought to be the ancestors of gymnosperms. In some systems classified as the division Progymnophyta, including e.g. the genus *Archaeopteris*.

prohormone, an inactive precursor form of a polypeptide or protein hormone which is enzymatically cleaved to form an active hormone; e.g. → pro-opiomelanocortin.

projection (L. *jacere* = to throw), the state or act of protruding out or apart; e.g. **1.** the projection of sensory pathways through the brain stem diverging to a certain area of the brain cortex, such as the sensory projection from the thalamus (thalamocortical projection); **2.** the projection of an optical image, e.g. on the retina of the eye; **3.** in psychology, an unconscious defence mechanism of a person having a tendency to attribute (project) unacceptable feelings and thoughts to someone else.

prokaryotes (Gr. *karyon* = nut, nucleus), **prokaryotic organisms**; unicellular organisms including → **bacteria**, **archaebacteria**, **mycoplasmas**, **cyanobacteria** and **prochlorophytes**, often considered to constitute the kingdom Monera. Prokaryotes have no genuine nucleus, the nuclear material (DNA) is located naked in the cytoplasm as a → **nucleoid**. Prokaryotes neither undergo → mitosis nor → meiosis as stages of the cell division cycle, and they have no actual cellular organelles, but the processes take place in the cytoplasm and cell membrane. They divide by simple cell fission during which the replicated genetic material is divided to the daughter cells as the sister DNA molecules are attached to the cell membrane and remain in the daughter cells. *Cf.* eukaryotes.

prolactin, PRL (L. *lac* = milk), a protein hormone (about 200 amino acid residues in a molecule) secreted by specialized cells of the anterior lobe of the → pituitary gland in vertebrates. In mammals, prolactin stimulates milk secretion in the mammary glands matured by the action of other hormones (oestrogens, progesterone, growth hormone etc.), and inhibits the secretion of → gonadoliberin and the effects of follicle-stimulating hormone (FSH) on the gonads during lactation. In some mammals prolactin assists in maintaining the corpus luteum (therefore called also **luteotropic hormone (LTH)**). In mammals, birds, and many other vertebrates, prolactin has effects on sexual and maternal behaviour (e.g. egg-hatching in both sexes), and in some birds, such as pigeons, it induces the secretion of "crop-milk" for feeding nestlings. In frogs it is involved in metamorphosis by activating the thyroid gland. Substances which have stimulatory effects on milk secretion from the mammary glands are called **lactogenic hormones** (substances). *See* prolactostatin.

prolactin inhibiting hormone/factor, PIH, PIF, → prolactostatin.

prolactostatin, prolactin inhibiting hormone/factor (PIH, PIF); a peptide hormone produced by certain neurosecretory cells of the hypothalamus in vertebrates; inhibits prolactin secretion from the anterior pituitary gland. *See* statins.

prolamellar body (L. *lamella* = thin disc), a tubular structure in a colourless etioplast formed in the dark from a proplastid instead of the normal, green → chloroplast. In the light the etioplast develops rapidly (within hours) into a chloroplast, while the prolamellar body develops into thylakoid lamellae capable of photosynthesis.

prolamines, also **prolamins** (proline + ammonia), a group of plant proteins soluble in alcohol, dilute acids, and alkalies, but not in pure water or neutral salt solutions; e.g. the most important reserve proteins of maize (→ zein) and barley (hordein).

prolarva, pl. **prolarvae** (L.), a newly hatched larva which is still supported by the energy from the yolk sac; e.g. the prolarva of a dragonfly not yet having legs, or a prolarva of a fish.

prolegs, "false legs"; unjointed, fleshy, abdominal appendages in the larvae of lepidopterans and hymenopterans, e.g. wood wasps.

proleptic development (Gr. *prolepsis* = anticipation, preconception), unfinished development of some structures of plants, such as → buds, which may develop only partially but may be activated and completed later, functioning then in the normal way.

proliferation (L. *proles* = offspring, *ferre* = to bear), the growth or reproduction of a tissue by active cell division; e.g. the proliferation of the mucous membrane of the uterus after menstruation. *Adj.* **proliferative (proliferous)**, capable of proliferation; marked by proliferation.

proline, abbr. Pro, pro, symbol P; 2-pyrrolidinecarboxylic acid, $C_5H_9NO_2$; an amino acid (actually imino acid) with an aliphatic side chain. Proline differs from the other 20 amino acids of proteins; its side chain is bound to both a nitrogen and the α-carbon atom and therefore has a cyclic structure. Proline occurs in most proteins, abundantly in collagen of the connective tissue of animals. It occurs also in plant tissues and its increase often correlates with the development of cold resistance in plants.

promeristem (Gr. *merizein* = to divide), the tip of the → apical meristem in plants, consisting of actively dividing, totally undifferentiated cells. *Cf.* derivative meristems.

prometaphase (Gr. *meta* = between, *phasis* = phase), a stage between prophase and metaphase of → mitosis and → meiosis in cell division.

prometryne, 4,6,-bisisopropylamino-2-methylthio-1,3,5-triazine, used as a herbicide. *See* triazines.

prominence (L. *prominentia*), a projection or protrusion in an anatomical structure, such as the spiral prominence of the cochlea in the vertebrate ear.

promiscuity (L. *promiscuus* = diverse), a brief, ephemeral mating relationship, either polyandrous or polygynous, in which an individual copulates with two or more partners of the opposite sex without any enduring bond. *Cf.* polygamy, monogamy.

promoter (L. *promotor* < *promovere* = to move forward), **1.** the binding site of RNA →

polymerase and consequently the initiating point of → genetic transcription in DNA. Upstream of the RNA polymerase binding site in the promoter, usually several *cis*-acting → enhancers and → silencers exist, i.e. recognition sites of → transcription factors and other regulatory proteins; **2.** in chemistry, a substance that in very small quantities is capable of increasing the activity of a catalyst.

pronase, a proteolytic enzyme (EC 3.4.24.4) obtained from *Streptomyces griseus*, used e.g. for isolating cells from a tissue; has a high collagenase activity.

pronation (L. *pronatio*), **1.** rotation of the forelimb or hand to a position where the palm is facing downwards or backwards (the radius and ulna are crossed) which is a natural position in many tetrapod animals; **2.** a state of being prone, i.e. laying face downward. *Cf.* supination.

pronephros, pl. **pronephroi** (Gr. *pro* = before, *nephros* = kidney), *see* kidney.

pronotum, pl. **pronota** (Gr. *noton* = back), the dorsal sclerite (plate) on the first thoracic segment (prothorax) in insects. *Cf.* mesonotum, metanotum.

pronucleus, pl. **pronuclei,** the haploid → nucleus of the ovum (female pronucleus) or the spermatozoon (male pronucleus); exists from maturation to → karyogamy, when both pronuclei unite to produce the nucleus of the → zygote.

proofreading, in genetics, the cellular process that removes incorrect nucleotides during DNA replication; this correcting mechanism functions also in the activation of amino acids and their joining to → transfer RNA and polypeptides in → protein synthesis.

pro-opiomelanocortin, POMC, a polypeptide produced in the anterior and intermediate lobes of the → pituitary gland, and in the nervous system, lung, and gastrointestinal tract of vertebrates; it is the prohormone which proteolytically cleaves to form β-endorphin (*see* endorphins), melanocyte-stimulating hormone (MSH), corticotropin (ACTH), lipotropin (LPH), and corticotropin-like intermediate-lobe peptide (CLIP).

propachlor, 2-chloro-N-isopropylacetanilide; an amide used as a herbicide which selectively inhibits protein synthesis in the root meristem and acts as an auxin and gibberellin inhibitor.

propagation (L. *propagare* = to increase), **1.** reproduction, multiplication; **2.** spreading, transmission, or moving, e.g. propagation of nerve impulses along a nerve cell. *Verb* to **propagate.** *Adj.* **propagative.**

propagule, a part of a plant (e.g. spore, seed, or fruit) by means of which the plant can disperse to new areas. *Syn.* diaspore.

propane, $CH_3CH_2CH_3$; an inflammable gas, the third hydrocarbon of the alkane series; used as fuel and refrigerant, e.g. in freezing biological specimens for → freeze-fracture.

propanoic acid, → propionic acid.

propanol, propyl alcohol, a short-chain alcohol which occurs as two → isomers: n-propanol, $CH_3CH_2CH_2OH$, that is an end product of alcoholic fermentation, and iso-propanol, i.e. 2-propanol, $CH_3CHOHCH_3$, that is an artificial product manufactured from propene. Propanol is a water-soluble and lipid-soluble liquid which is widely used as a solvent in biochemical and biological studies. In organisms, propanol acts like ethanol but is more toxic.

propanone, → acetone.

propene, propylene. *See* alkenes.

prophage (Gr. *pro* = before, *phagein* = to eat), a temperate, latent stage of a → bacteriophage (virus) in which the viral genome is an integrated part in the chromosome of the → lysogenic bacterium. In the initiation of the lysis of the bacterium, the prophage is separated from the coordination of the bacterial chromosome and begins to reproduce autonomously, resulting in the formation of active bacteriophages and the complete lysis of the bacterial cell.

prophase (Gr. *phasis* = appearance), the first stage of → mitosis and → meiosis in cell division.

propionic acid, propanoic acid, methylacetic acid, CH_3CH_2COOH; an organic acid produced in the metabolism of many bacteria, e.g. cheese bacteria. It is also found in sweat, produced by normal skin bacteria. The ionic form, salts, and esters of propionic acid are called **propionates**. Propionic acid is used for the manufacture of lactic acid, and both the acid and its salts as preservatives in bakery products.

proplastid (Gr. *plastos* = moulded), an undifferentiated preliminary stage of plastids, e.g. → chloroplast, in meristematic cells.

propleurum, propleuron, pl. **propleura** (Gr. *pleuron* = side), the lateral sclerite (plate) of the first thoracic segment (prothorax) in insects. *Cf.* mesopleurum, metapleurum.

propodeon, pl. **propodeones** (Gr. *podeon* = neck), the first segment, between the thorax and → podeon, in the abdomen of some hymenopterans. *Syn.* **propodeum**, pl. **propodea**, also **propodeums**. *Cf.* metapodeon.

propodite (Gr. *pous*, gen. *podos* = foot), *see* biramous appendage.

propodium, pl. **propodia,** the anterior part of the foot of a mollusc.

propodosoma, pl. **propodosomata** (Gr. *soma* = body), the anterior part of the body (idiosoma) of mites and ticks (Acari) bearing two anterior pairs of legs. *Cf.* hysterosoma.

proportional counter, a counter tube for measuring ionizing radiation; the output pulse is proportional to the number of ions produced by the initial ionization.

propositus, → proband.

proprioceptors (L. *proprius* = own, *capare* = to take), sensory receptors in muscles, tendons, and the vestibular labyrinth of the inner ear, giving information about movements and position of the body. Called also proprioreceptors.

propyl, a univalent alkyl radical, $-C_3H_7$, which has two isomeric forms: normal propyl (1-propyl), $CH_3CH_2CH_2$-, and *iso* propyl (2-propyl), $(CH_3)_2CH$-; occurs in many biological compounds, e.g. in → propanol.

propyl alcohol, → propanol.

propylene, propene. *See* alkenes.

propylene oxide, 1,2-epoxypropane, CH_3-$CHOCH_2$; colourless liquid used e.g. in → embedding tissues for electron microscopy.

proscolex, pl. **proscoleces,** also **proscolexes** (L. *pro* = before, Gr. *skolex* = worm), *see* cysticercus.

prosencephalon, pl. **prosencephala** (Gr. *proso* = in front, forward, *enkephalos* = brain), the **forebrain** of a vertebrate embryo; the other two portions are the **midbrain** (mesencephalon) and **hindbrain** (rhombencephalon). During further development, prosencephalon develops into the telencephalon and diencephalon.

prosenchyma (Gr. *pros* = near, *enchyma* = poured in), **1.** a type of → parenchyma tissue in plants, the cells of which are long and thin and not short and roundish as usual; prosenchyma cells are found e.g. in the woody and bast portions of plants; **2.** loose fungal tissue in fruiting bodies, formed of separate hyphae.

Prosimii (L. *simia* = ape), **prosimians**; a suborder of chiefly nocturnal and arboreal primates, such as **lemurs**, **lorises**, **tarsiers**, **bush-babies**, and **pottos**, in the forests of Madagascar, Africa, Malay Peninsula and the Philippines. Prosimians have a long non-prehensile tail, and their second toe is provided with a claw, the others with nails. In general, prosimians are omnivorous, eating plants and small animals. In some classifications the suborder Prosimii (except the tarsiers) is known also by the name Strepsirhini.

Prosobranchia, also **Prosobranchiata** (Gr. *proso* = ahead, in front, *branchia* = gills), **prosobranchs**; a subclass of gastropods including mostly marine species having a spirally coiled body, gills located in the front of the heart, and a foot provided with an operculum for closing the shell opening. *Cf.* Opisthobranchia.

prosociality, *see* eusociality.

prosodus, pl. **prosodi,** Am. also **prosoduses** (L. < Gr. *prosodos* = procession), a short duct in the body wall of some sponges through which water flows from an incurrent canal to the spongocoel.

prosoma, pl. **prosomata, prosomas** (Gr. *pro* = before, *soma* = body), the anterior portion of the body of many invertebrates, especially arachnids. *See* cephalothorax. *Cf.* opisthosoma.

prosome, a ribonucleoprotein particle (19S) found in both the cell nucleus and cytoplasm; composed of small cytoplasmic RNA and heat-shock proteins (*see* stress proteins), and is probably involved in the repression of mRNA translation.

prosopyle (Gr. *pyle* = gate), an opening between the incurrent channel and the flagellated chamber in the body of leuconoid sponges.

prostacyclin, *see* prostaglandin.

prostaglandin, PG (< *prostate gland*), any local hormone in animals derived from 20-carbon fatty acids, → eicosanoids. The first prostaglandin was found in mammalian sperm, and was supposed to be produced in the prostate gland. Later different types of prostaglandins were found to be synthesized generally in animal tissues from unsaturated lipids of cell membranes. There are several main classes: PGA, PGB, PGE, PGF, with different subclasses, as $PGF_{2\alpha}$. Prostaglandins act locally as tissue hormones, although they may be released into the blood. Depending on the tissue and the type of prostaglandin, their physiological effects are variable: they can modify the effects of other hormones, they are involved in the regulation of digestion, circulation, body temperature, secretion, inflammation, and in reproduction processes, such as the motility of spermatozoa, ovulation, and the movements of the uterine muscle. Their derivatives, **prostacyclin** (named first **PGI_2**) and **thromboxanes**, control blood coagulation, the former inhibiting, and the latter accelerating the process; prostacyclin also dilates blood vessels. Salicylates and their derivatives attenuate actions of prostaglandins by inhibiting their synthesis in tissues.

prostate (gland) (L. *prostata* < Gr. *prostates* = standing in the front), **1.** a gland in male mammals situated around the neck of the urethra and the bladder; comprises many small glandules, which produce slightly alkaline secretion (one third of the semen volume) containing e.g. acidic phosphatases, proteases, citric acid, and prostaglandins, the secretion having an effect on the motility of spermatozoa and sperm coagulation; **2.** in oligochaetes, glandular tissues (prostate glands) which are often, but not always, associated with the male gonoducts (vas deferens).

prosternum, pl. **prosterna,** also **prosternums** (L.), the ventral sclerite (plate) in the first thoracic segment (prothorax) in insects. *Cf.* mesosternum, metasternum.

prosthetic group (Gr. *prosthesis* = addition to something), a non-protein substitutive chemical group in a protein, often containing a metal cofactor. The prosthetic group is attached e.g. to → enzymes acting as a → coenzyme or its part, but also are linked to many other proteins, such as blood pigments and photopigments. The prosthetic group may be e.g. a → vitamin, flavin, or haem and other porphyrins which in the protein form its active site.

prostomium, pl. **prostomia** (Gr. *pro* = before, *stoma* = mouth), a fleshy lobe above the mouth of some oligochaetes, e.g. earthworm and molluscs; **prostomial tentacles**, short tentacles between the palps in the head of polychaetes.

prot-, proto- (Gr. *protos* = first), denoting relationship to the first in a series or the highest in rank.

protamines, small, highly basic, arginine-rich proteins bound to DNA, found abundantly e.g. in chromosomes of spermatozoa, replacing → histones in DNA packing.

protandry (Gr. *aner* = male), **1.** a condition of sequential hermaphrodites in which the production of sperm occurs before the maturation of eggs; **2.** in plants, the ripening of the stamens before the gynaecium; prevents self-pollination. *Cf.* protogyny.

protanopia (Gr. *an-* = not, *ops* = eye), a form of red-green colour blindness characterized by absence of the red-sensitive pigment in the cones; **protanomaly**, the condition in which the quantity of the red-sensitive pigment is decreased. *See* colour blindness.

proteases, the descriptive name for proteolytic enzymes; include → exopeptidases and → endopeptidases. *See* proteinases.

proteasomes (Gr. *soma* = body), complexes of proteolytic enzymes involved in controlled protein degradation in the cell cytoplasm and nucleus; serve in the hydrolysis of the proteins which are linked to → ubiquitins.

protein, any of the enormous number of different nitrogen-containing (about 15%) organic compounds which are essential to all organisms and produced by their cells in → protein synthesis. Proteins are macromolecules with a molecular weight ranging from 5 to 6,000 kD. They consist of one or more → **polypeptide** chains formed from → **amino acids** which are bound to each other by → covalent bonds (peptide bonds). Twenty different amino acids are found in natural proteins, and one protein molecule usually contains hundreds of these amino acids alternating in a strict order determined by the → genetic code in protein synthesis.

The **primary structure** of a protein is determined by the amino acid sequence in its polypeptide structure coded by the nucleotide sequence of a gene (*see* cistron) in → genetic transcription and translation. Also the disulphide bonds between two cystine residues are included in the primary structure. The **secondary structure** of a polypeptide or protein is formed by bending of an amino acid chain (the spatial arrangement of amino acid residues), depending on the amino acid sequence of a polypeptide chain and on the physiological state of the cell. This occurs through the spatial configuration of → **alpha helix** as found in fibrous proteins, and/or → **beta-**

pleated sheet, or a mixed structure of these both, or of random coiling, as found in globular proteins. Hydrogen bonds between different amino acids are essential to the formation of secondary structure.

The **tertiary structure** is the overall spatial conformation of a polypeptide chain in a protein, brought about by complex folding of the polypeptide chain by hydrogen and disulphide bonds. It includes elements of the primary and secondary structures, and therefore the difference between the secondary and tertiary structure is often indecisive.

The **quaternary structure** is formed when two or more polypeptide chains are combined together in a spatial arrangement by hydrophobic and electrostatic forces, and by disulphide bonds, to form large proteins, forming dimers, trimers, tetramers, etc. The polypeptides of a protein may be identical or different. In plants e.g. peroxidase enzyme consists of only one 40 kD polypeptide chain, whereas ribulose 1,5-bisphosphate carboxylase-oxygenase (Rubisco, 568 kD) is composed of large (each 55 kD) and small subunits (each 16 kD), eight together. Most enzymes of animal cells consist of only one polypeptide but proteins forming ion channels and receptors contain several subunits. The primary and secondary structure of a protein together with the physiologial state of the cell will determine the final form and functional capability of the protein.

Proteins are divided into **structural proteins** (e.g. collagen, keratin), **enzymes** (catalytic proteins), **regulatory** (e.g. calmodulin, troponin), **transport** (e.g. serum albumin), **contractile** (e.g. actin, myosin), and **protective proteins** (e.g. antibodies), **hormones** (e.g. insulin), and **storage proteins** (e.g. casein, ovalbumin, and many proteins in plant seeds). Some structural proteins can also act as enzymes, and some enzymes can be structural elements e.g in cellular membranes. Proteins can also be divided into **globular proteins** which are simple, soluble proteins, common in cells, **fibrous proteins** which occur in contractile structures of cells and as supporting extracellular structures in tissues, and **conjugated proteins** which contain → prosthetic groups or carbohydrate or lipid moieties, e.g. → glycoproteins and lipoproteins.

proteinases, enzymes of the sub-subclass EC 3.4.21.—3.4.24., of the subclass endopeptidases which split peptide bonds of proteins and polypeptides initiating the hydrolysis in the middle of a molecule chain; they are divided into serine-proteinases, cysteine-proteinases, and metalloproteinases; e.g. → pepsin, trypsin, rennin (chymosin), and papain. *Cf.* exopeptidases.

protein engineering, the production and analysis of non-natural proteins, often enzymes, with desired new modifications of the primary amino acid sequences by means of the recombinant DNA technology. Aims of protein engineering are e.g. 1) improving the properties of proteins, like the stability of enzymes (e.g. thermal) which can be improved by adding a number of disulphide bridges into the molecular structure; 2) altering the substrate specificity of an enzyme; because enzymes usually catalyse a very narrow range of reaction this is carried out by changing amino acids around the active site of the enzyme; 3) changing the pharmacological specificity of proteins, or reducing harmful side effects of drugs.

protein kinases (Gr. *kinein* = to move), enzymes of the sub-subclass EC 2.7.37. in the transferase class, binding a phosphate group to a protein: ATP + protein —> ADP + phosphoprotein. Protein kinases phosphorylate specific amino acid residues, e.g. protein-serine kinase and protein-tyrosine kinase, and they are important factors which control the cellular metabolism by catalysing the conversion of proteins from an inactive form to an active enzyme. Protein kinases control the metabolism of glycogen, cholesterol, and amino acids.

proteinoplast, → proteoplast.

protein synthesis, the formation of proteins in the cell directed by gene activity. The direction results from a specific order of → nucleotides in the gene which, according to the rules of the → genetic code, produces the primary structure of the polypeptide chain or protein. Protein synthesis involves → genetic transcription and → genetic translation.

In **transcription**, the information located in the antisense strand of DNA is copied according to the → base pair rule, in eukaryotic organisms first into → heterogeneous nuclear RNA which is then processed into → messenger RNA (mRNA). In prokaryotic organisms like bacteria, the copying produces mRNA directly, which is then transferred to

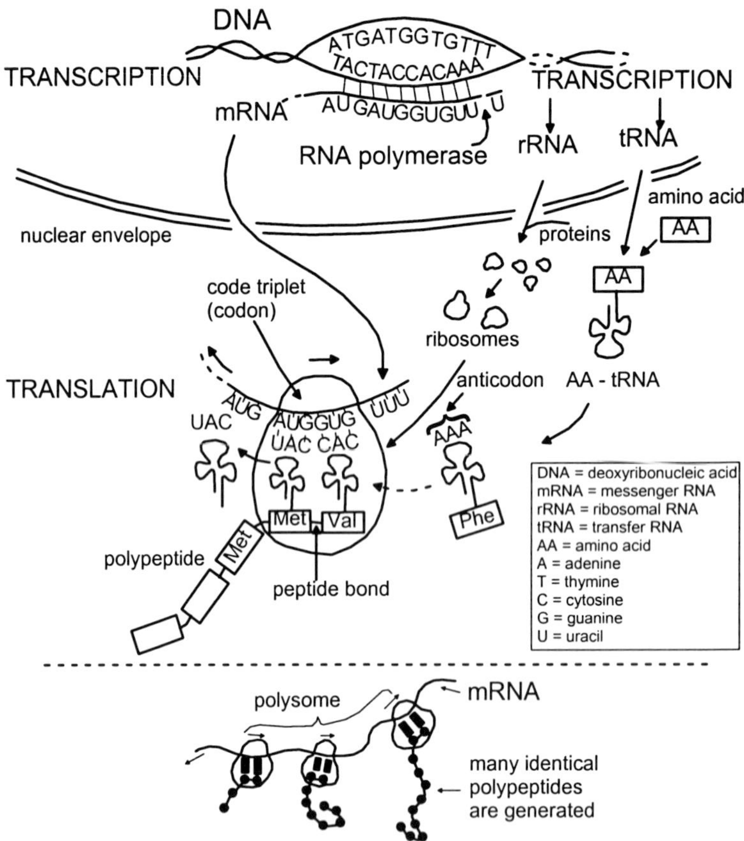

Protein synthesis. A schematic representation of the mechanism of protein synthesis.

→ ribosomes, thus forming → polysomes.

In **translation**, the ribosomes read the genetic information of mRNA and transform it into the primary structure of protein. In translation two → codons (nucleotide triplets) of mRNA are attached simultaneously to the ribosome, the first one on the **peptide site**, and the other on the **amino site**. At both sites a → transfer RNA molecule (tRNA), previously linked to a specific amino acid, is bound to each of the two codons with the aid of its → anticodon (according to the base pair rule). The peptide bond is then formed between the two amino acids now located side by side on the ribosome. After this the ribosome moves one step forward along the strand of mRNA, and thus the codon located on the amino site will move onto the peptide site, and the following codon to the amino site. The next tRNA molecule with an amino acid attaches to the site of the codon, and a peptide bond is formed again between the new and the former amino acid, and hence the polypeptide chain is elongated until the ribosome reaches a → nonsense codon that encodes the termination of the chain. Then the ribosome becomes detached from mRNA and the polypeptide chain is released. Polypeptide chains then arrange into secondary and tertiary structures, longer chains forming protein molecules. Two or more polypeptide chains may link together to form a quaternary protein structure. *See* protein.

proteoglycan (Gr. *glykys* = sweet), any substance composed of a protein and carbohydrates; contains much less protein than carbohydrate (*cf.* glycoprotein). Proteoglycans are present e.g. in various tissue fluids of animals, mucous secretions, and in the matrix of the connective tissue. In the molecule many

→ glycosaminoglycan chains (e.g. hyaluronic acid) are covalently bound to a protein stem projecting like bristles of a bottle brush. The structure binds large amounts of water and is a good viscous lubricant. *Cf.* peptidoglycan.

proteolipid, a macromolecule composed of a protein and lipid, or of a peptide and lipid, having the solubility properties of lipids; occurs e.g. in cell membranes of the brain tissue. *Cf.* lipoprotein.

proteolysis (Gr. *lyein* = to dissolve), the hydrolytic decomposition of proteins and polypeptides; breakdown of the → peptide bonds occurs in cells and tissues chiefly by the catalytic activity of **proteolytic enzymes**. *See* proteinase.

proteome, a complete set of all proteins which a living cell is able to synthesize.

proteoplast (Gr. *plastos* = moulded), a storage → plastid of a plant containing proteins. *Syn.* proteinoplast.

proteose, a mixture of intermediate products of proteolysis; structurally between protein and peptone.

Proterozoic (Gr. *proteros* = earlier, *zoon* = animal), the eon 2,500 to 570 million years ago, succeeding Archaean. Only a few fossils are found from this period containing e.g. cyanobacteria and soft-shelled invertebrates, e.g. annelids.

prothallus, pl. **prothalli** (Gr. *pro* = before, *thallos* = young shoot), a small, haploid → gametophyte developing from a spore of pteridophytes; bears the gametangia, i.e. antheridia or archegonia, or both.

prothorac(ic)otropic hormone, PTTH (Gr. *trope* = turning), an insect hormone synthesized by the neurosecretory cells of the pars intercerebralis and released from the → corpus cardiacum, controlling the secretion of → ecdysone from the **prothoracic gland**. Also called thoracotropic hormone, TTH, ecdysiotropic hormone, brain hormone.

prothorax (Gr. *thorax* = chest), the first segment in the → thorax of an insect bearing a pair of legs but no wings. *Cf.* mesothorax, metathorax.

prothrombin (Gr. *thrombos* = clot), the inactive precursor (factor II) of the enzyme **thrombin** acting in → blood coagulation; a glycoprotein produced and stored by liver cells. Also called plasmozyme, thrombinogen, or zerozyme.

Protista (Gr. *protistos* = primary), **protists**; in some classifications, a kingdom of unicellular eukaryotic organisms including fungi, algae, and protozoans. Protists are highly diverged comprising free-living autotrophs and consumers in all waters and soils, or they may be parasites; some are ectoparasites, some live as symbionts with other organisms. Protista includes 27 phyla, of which 13 form protozoans; altogether about 65,000 species are described, more than half of these are fossils. The group Protista was first proposed by Ernst Haeckel because of the difficulty of dividing one-celled organisms into plants and animals; e.g. unicellular green algae are mostly classified also in the division → Chlorophyta (kingdom Plantae), especially by botanists. The group is also called **Protoctista**. *Cf.* Monera.

protistology (Gr. *logos* = word, discourse), a branch of biology studying protists; called also microbiology.

proto- (Gr. *protos* = first), denoting first, or the highest in rank.

Protoarticulatales, an extinct order in the class Sphenopsida, division → Pteridophyta (in some systems division Sphenophyta), Plant kingdom. The best known fossil species (from Lower Devonian) belong to the genus *Hyenia*; they were small, dichotomously branched shrubs with irregular whorls of forked leaves.

protocercal (Gr. *kerkos* = tail), *see* caudal fin.

protochlorophyll (Gr. *chloros* = yellowish green), a precursor of → chlorophyll which is synthesized from protoporphyrin and magnesium (*see* porphyrin); occurs in plants grown in the dark, and turns green under the influence of light, first into chlorophyllide a and then into chlorophyll. *Syn.* protochlorophyllide.

Protochordata (L. *chorda* = cord), → Acrania.

protocoel (Gr. *koilos* = hollow), the anterior coelomic vesicle in some deuterostomes, corresponding to the axocoel in echinoderms.

protocooperation, an interaction between different species that is mutually beneficial but is not necessary (e.g. physiologically) to the survival of either. *Cf.* symbiosis.

Protoctista, protoctists, → Protista.

protoderm (Gr. *derma* = skin), the embryonic epidermis in plant embryos; one type of → derivative meristem.

protogyny (Gr. *gyne* = woman), **1.** the condition of a hermaphrodite animal in which the development of eggs occurs before the maturation of male gametes, spermatozoa; **2.** in hermaphrodite plants, the maturing and withering of the stigmata before the stamens and production of pollen; ensures cross-pollination. *Cf.* protandry.

protolysis (Gr. *lysis* = decomposition, degradation), the reaction involving the transfer of protons (hydrogen ions); the theory named by *T.M. Lowry* and *J. Brønsted* concerns acid-base reactions which are expressed: A (acid) $\rightleftharpoons$ B (base) + H^+ (proton); the acid is a proton donor that is protolysed liberating a proton to the base, and the base is a proton acceptor that is protonated in the reaction. Water can act as a base: $HCl + H_2O \rightleftharpoons H_3O^+ + Cl^-$ or as an acid: $H_2O + NH_3 \rightleftharpoons NH_4^+ + OH^-$. Oxonium ion ($H_3O^+$) is an acid corresponding to the base H_2O, and in the latter the hydroxide ion OH^- is a base corresponding to the acid H_2O. Even pure water has a weak electrolytic conductivity meaning that it contains some ions; these are produced by autoprotolysis of water: $H_2O + H_2O \rightleftharpoons H_3O^+ + OH^-$. *Adj.* **protolytic**, pertaining to the transfer of a proton.

protomerite (Gr. *meros* = part), the anterior part of the cell body in gregarine protozoans; it is separated from the larger, nucleus-containing posterior part (**deutomerite**) by a septum. In some parasitic species, a prolongation of the protomerite, called **epimerite**, exists for anchoring the parasite onto the host's cell.

proton, the positively charged elementary particle in the nucleus of an atom; protons together with → neutrons are constituents of all atomic nuclei, and one proton alone forms the whole nucleus of the hydrogen atom, forming the **hydrogen ion**, H^+, in dissociation. Each atom, which has an equal number of protons (atomic number of an element) and → electrons, is electrically neutral. The proton mass, $1.672 \cdot 10^{-27}$ kg (1.00794 atom weight units), is 1,836 times greater than the mass of the electron. Acids are electrolytes which liberate protons in water solutions (*see* protolysis).

proton ATPase, → proton pump.

protonema, pl. **protonemata** (Gr. *nema* = thread), **1.** a filamentous haploid structure formed from a spore in mosses; **2.** a filamentous stage in the development of some algae and of prothalli of ferns.

protonephridium, pl. **protonephridia** (Gr. *nephros* = kidney), a unit of a primitive excretory organ, → **nephridium**, comprising a branched blind-ended tubule with specialized **flame cells** (flagellated cells) or → **solenocytes** (with one flagellum) for collecting excretory products from coelomic fluid; occurs e.g. in turbellarians, polychaetes, priapulids, gastrotrichs, and rotifers. *Cf.* metanephridium.

proton motive force, PMF, the factor which determines the activity of proton (H^+) transport by the action of ATP synthases in energized membranes of → mitochondria and thylakoid membranes in → chloroplasts. PMF is dependent on the concentration gradient of H^+ ions and the electrical gradient (membrane potential) across the membrane, supported by protons but also other ions. *See* oxidative phosphorylation.

proton pump, proton ATPase, H^+-ATPase; an adenosine triphosphatase enzyme, ATPase, which actively transports protons, hydrogen ions (H^+), across cellular membranes using energy of the reaction: ATP —> ADP + P_i, in which ATP decomposes into adenoside diphosphate (ADP) and inorganic phosphate (P_i). In plant cells the proton pump supports membrane potential in the plasma membrane producing the electrochemical gradient used in the uptake of plant nutrients. The H^+/K^+ exchange in plant cells corresponds to the exchange of Na^+ and K^+, and thus the function of the → sodium-potassium pump in animal cells; *see* ion pumps.

In → chloroplasts, protons are transported actively across the thylakoid membrane into the thylakoid compartment, creating a proton (H^+, pH) gradient across the membrane. Protons are transported back from the compartment to the stroma through → ATP synthase, while photosynthetic ATP is produced. In the → mitochondria of all cells, protons are transported actively from the mitochondrial matrix into the intermembrane space where they create a proton gradient across the inner membrane, and when they leak back to the matrix by ATP synthase, mitochondrial ATP synthesis takes place (*see* oxidative phosphorylation).

Many animal cells, as e.g. red blood cells, parietal cells (oxyntic cells) in gastric glands, and gill epithelial cells, can exchange potassium ions with protons by the action of another enzyme, **H^+,K^+-ATPase**.

protonymph (Gr. *protos* = first, *nymphe* =

bride, nymph), the first of the three larval (nymphal) stages in the life cycle of mites and ticks (Acari); possesses four pairs of legs. *Cf.* deutonymph, tritonymph.

proto-oncogene, a gene regulating the normal growth and division of a cell but as a result of → mutation, translocation, or gene amplification is released from its normal control and hence changes into a tumour causing → oncogene.

protopectin, *see* pectic substances.

protophloem, *see* primary phloem.

protoplasm (Gr. *plassein* = to mould), a term especially earlier used to mean the "living substance" of the cell. *See* cytoplasm, nucleoplasm.

protoplast, 1. the living material of a single cell; **2.** an isolated bacterial, fungal, or plant cell without cell wall, i.e. a cell having its cell membrane exposed; used e.g. as material for physiological and gene transfer studies.

protopodite (L. *protopodium* < Gr. *pous*), the basal segment of the → biramous appendage (leg) in crustaceans. *Syn.* protopod. *Cf.* epipodite, endopodite, exopodite.

protosoma, pl. **protosomata** (Gr. *soma* = body), **protosome**; the anterior part of the body in pogonophorans and hemichordates. *Cf.* mesosoma, metasoma.

protostele (Gr. *stele* = column), a stele type in plants. *See* stele.

protostome (Gr. *stoma* = mouth), → blastopore.

Protostomia, protostomes; a group of invertebrates including e.g. **annelids**, **molluscs**, and **arthropods**, linked by some similarities in their ontogenetic development. The most important features are that the first embryonic opening, **blastopore**, develops to form a mouth, or in some species a common mouth-anal opening, and that all protostomes have a true → coelom. *Cf.* Deuterostomia.

Prototheria (Gr. *therion* = wild animal), **prototherians, egg-laying mammals**; a subclass of primitive mammals (order → Monotremata, e.g. platypus and echidnas) which have milk glands without nipples, and no external ears. *See* Theria.

prototrophic (Gr. *trephein* = to feed), **1.** pertaining to the ability of an organism (bacterium) to obtain nourishment from a single inorganic source, such as sulphur or nitrogen, as do nitrifying bacteria; **2.** describes a microorganism, e.g. a strain of a bacterium or a fungus, that has the same minimum requirements for growth factors as a certain wild-type strain; such an organism is called **prototroph**.

prototype (Gr. *prototypos* = original), in biology, **1.** a type species or example; **2.** an ancestral form of an organism from which the later forms or groups are derived. *Syn.* archetype.

protoxylem, *see* primary xylem.

Protozoa (Gr. *protos* = first, *zoon* = animal), **protozoans**; a phylum of unicellular animals; at present, considered as a subkingdom in the kingdom → Protista. The subkingdom is divided into seven phyla: Sarcomastigophora, Labyrinthomorpha, Apicomplexa, Myxozoa, Microspora, Ascetospora, and Ciliophora. Protozoans include about 35,000 species found world-wide in all possible habitats. Most of them are solitary, some species live in colonies. Several species are parasitic causing severe diseases in animals, in man e.g. malaria, trypanosomiasis, kala-azar, and amoebic dysentery. Some parasites are also beneficial symbionts, such as intestinal protozoans. Most protozoans are aquatic, and if the water body dries out the animals produce resistant cysts or spores to survive unfavourable conditions. Parasitic protozoans have usually a complicated life cycle.

protozoea (Gr. *zoe* = life), the larval stage preceding the zoea- stage in the life cycle of some crustaceans.

protrusion (Gr. *pro* = before, L. *trudere* = to push), act or state of being pushed forward; e.g. an outstanding projection of an organ, especially of a bone. *Verb* to **protrude**.

provenience (L. *proveniere* = to come out), the original growth area of a plant, especially a tree.

proventriculus (Gr. *pro* = before, L. *ventriculus*, dim. of *venter* = belly), **1.** the anterior portion of the stomach in birds in which the food coming from the crop is mixed with peptic fluid and enzymes, and then passed to the → gizzard; **2.** the portion (gizzard) of the foregut in some insects, such as grasshoppers, situated between the crop and ventriculus (stomach), lined by chitinous plates; **3.** the anterior part (gizzard) of the stomach in crustaceans including a triturating apparatus ("gastric mill") with chitinous ridges, denticles, and calcareous ossicles; **4.** in annelids, the enlargement of the oesophagus anterior to the gizzard, for storing food. *Cf.* crop.

provirus, a virus → genome attached to the genome of a eukaryotic cell and replicating in a coordinated fashion with the latter; can be released from the coordination, and starts to replicate autonomously thus producing active virus particles.

provitamin, a precursor of a → vitamin; e.g. provitamin A is beta-carotene.

proximal (L. *proximus* = next), nearest to the base, trunk, origin, or any point of reference. *Cf.* distal.

proximate factor, any of the immediate factors which is the cause (**proximate cause**, the term coined by *Ernst Mayr*) of a certain biological phenomenon; e.g. changes in hormonal secretion due to changing photoperiod are proximate factors for the coat colour change of the hare. The length of the day and the temperature conditions in the early spring are proximate factors for birds which begin to breed at the optimum time. Thus, an animal can use proximal factors as cues of future environmental changes and adapt its own behaviour to them. *Cf.* ultimate factor.

Prymnesiophyta, a group of unicellular, mostly marine algae, having two equal flagella and a → haptonema, i.e. a long flagellum with three concentric membranes, arising close to the pair of flagella. This group is often included in the division Chrysophyta (as the family Prymnesiophyceae), but is sometimes separated into a division in Plant kingdom. *Syn.* Haptophyta, Haptophyceae.

psammon (Gr. *psammos* = sand), a → biota of organisms living on sandy shores between sand grains. *Cf.* epipsammon. *See* interstitial biota.

psammosere (L. *serere* = to put in order), a → succession of vegetation in sandy coast areas. *See* sere.

pseud(o)- (Gr. *pseudes* = false), denoting false or spurious.

pseudanthium (Gr. *anthos* = flower), a very reduced inflorescence without any petals, resembling one single flower; e.g. the **cyathium** in the family Euphorbiaceae.

pseudoalleles, → mutations of a gene that are → alleles according to the functional but not to the structural classical criterion, i.e. they are alleles between which → crossing-over can occur, i.e. they can recombine. Pseudoallelism demonstrates that the → **gene** cannot be defined as the smallest unit of gene function or → genetic recombination, but the smallest unit of the gene function is the → **cistron** and that of recombination, the → **recon**.

pseudoautosomal region (Gr. *autos* = self, *soma* = body), a region of sex chromosomes (common in both the X and Y chromosomes) by means of which the conjugation of sex chromosomes occurs in → meiosis. The genes located in this region show a kind of inheritance as if they were located in autosomal chromosomes (*see* autosomes).

pseudobulb (L. *bulbus* = bulb), a thickened internode formed in the base of the plant shoot above ground, functioning as energy and water reserve; common in the family *Orchidaceae*. *Cf.* bulb.

pseudocoel(om), also **pseudoc(o)ele** (Gr. *koiloma* = cavity), the body cavity between the body wall and the intestine in certain primitive invertebrates, embryonically derived from the blastocoel between the mesoderm and endoderm; the peritoneal lining derived from the mesodermal epithelium is lacking; therefore it is not a true → coelom, and internal organs lie free in the pseudocoelom. *See* Pseudocoelomata.

Pseudocoelomata, pseudocoelomates; a polyphyletic group of several primitive invertebrate animal phyla, comprising species with → pseudocoelom; it is a heterogeneous assemblage of animals which belong to → Protostomia; the group includes the following nine phyla: Rotifera, Gastrotricha, Kinorhyncha, Loricifera, Priapulida, Nematoda, Nematomorpha, Acanthocephala, and Entoprocta. *Cf.* Coelomata.

pseudocopulation, an attempt by a male insect to copulate with a flower resembling a conspecific female; serves the pollination of the flower (e.g. in orchids).

pseudocyphella (Gr. *kyphellon* = bowl), a small pore in lichen thalli, possibly aerating the inner tissue. *Cf.* cyphella.

pseudodichotomy, *see* dichotomy.

pseudogamy (Gr. *gamos* = marriage), **1.** an exceptional development of eggs of some invertebrates; occurs without fertilization after a spermatozoon has penetrated and been absorbed by the egg cell; **2.** an atypical type of reproduction in plants, in which → fertilization is needed only for the development of the → endosperm, while the embryo is formed independently without fertilization, the union of hyphae belonging to different thalli.

pseudogene (Gr. *genos* = birth, family), a nonfunctional duplication or a derivative of a functional → gene. Pseudogenes are mutated ancestral genes which have lost their significance although they are found in all eukaryotic organisms.

pseudohermaphroditism, → androgyny.

pseudomycelium, pl. **pseudomycelia** (Gr. *mykes* = fungus, mushroom), chains of single cells, e.g. yeast cells or bacteria, resembling fungal hyphae and true mycelium.

pseudoparaphysis (Gr. *para* = beside, *physis* = growth), → hamathecium.

pseudoparenchyma (Gr. *enchyma* = poured in), **1.** the fungal tissue constructed of aggregated hyphae, resembling → parenchyma of plants; forms the → fruiting body in fungi; **2.** the tissue of thalli of some algae. *Cf.* plectenchyma.

pseudopodium, pl. **pseudopodia** (Gr. *pous*, gen. *podos* = foot), **pseudopod**; a temporary, local cytoplasmic extension of the cell body of many protozoans, especially sarcodines and many flagellate protozoans, and of amoeboid cells found generally in multicellular animals; pseudopods serve the locomotion of cells (**amoeboid movement**). In protozoans, pseudopods exist in several forms: **rhizopodia** (**lobopodia**) are blunt extensions containing both endoplasm and ectoplasm, **filipodia** are thin extensions containing only ectoplasm, **lamellipodia** are plate-shaped, **reticulopodia** form a net-like mesh, and **axopodia** are long, thin pseudopods supported by axial rods of → microtubules along which cytoplasm can flow.

According to the direction of movement, several pseudopods can be extended and retracted simultaneously in various parts of the cell. In all types the locomotor activity appears to be basically similar and related to functions of certain → microfilaments, especially actin filaments beneath the cell membrane, but also membrane proteins are involved. During movement a part of the cytoplasm changes from → sol to gel in the border of the endoplasm and ectoplasm. This is caused by actin-bound proteins, such as **filamin** and **gelsolin**, which bind actin to form a filamentous network which is then quickly dissolved.

pseudopregnancy, a condition sometimes found in female mammals, resembling pregnancy but appearing without fertilization; shows as changes in accessory sexual organs and behaviour. Its development may be induced e.g. by physical and psychic stimulation resulting from sterile mating, also by a certain hormonal treatment. Women sometimes imagine themselves pregnant with some signs of pregnancy; this condition is called pseudocyesis or false pregnancy.

Pseudoscorpiones, Pseudoscorpionida (L. *scorpio* = scorpion), **pseudoscorpions, false scorpions**; an order of arachnids whose pedipalps have developed into catching organs. They live under stones, moss, or bark, feeding on minute insects; the book scorpion lives in old books. Pseudoscorpions, comprising about 2,000 species, resemble scorpions but the tip of the abdomen is round and without any stinging apparatus.

pseudostigma, pl. **pseudostigmata,** also **pseudostigmas** (Gr. *stigma* = mark), pseudostigmatic organ; a small, paired, cup-shaped sense organ on the dorsal prosoma of the oribatid mites; has a long sensory hair (seta) sensitive to air currents.

pseudotagma, pl. **pseudotagmata** (Gr. *tagma* = corps), a part of the animal body which by its origin is not comparable to a → tagma; e.g. → gnathosoma, idiosoma, and hysterosoma in acarines are pseudotagmata.

pseudouridine, 5-ribofuranosyluracil, symbol Ψ; a → nucleoside having a structure similar to that of → uridine but the linkage is between 1-carbon of ribose and 5-carbon of the uracil ring, instead of the nitrogen atom in normal uridine; present in → transfer RNAs like many other rare nucleosides. The pyrimidine base in the pseudouridine is called **pseudouracil**.

Psilophyta, *see* Psilotopsida.

Psilotopsida (Gr. *psilos* = bare), **whisk ferns**; a class in the division Pteridophyta, Plant kingdom; in some systems classified as a separate division, Psilophyta. The species were numerous during Silurian and Devonian and at present the group is represented only by two genera (*Psilotum*, *Tmesipteris*), which are tropical plants with simple rhizomes and aerial branches with scale-shaped or bract-like outgrowths. Also called Psilopsida.

P-site, peptidyl site; the binding site for transfer RNA (tRNA) linked to the growing end of the polypeptide chain during → protein synthesis. *Cf.* A-site.

Psittaciformes (L. *psittacus* = parrot, *forma* =

form), **psittacines**; an order of birds including e.g. **lories, cockatoos, parrots**; the animals have a strong beak and often polycoloured feathers, and they live mainly in tropical forests of South America, Africa and Australia. Parrots are small or middle-sized, noisy, and social birds; 342 species are described.

psittacosis, parrot disease (fever); an infectious disease found in psittacine birds, but can be transmitted to humans; caused by the bacterium *Chlamydia psittaci*. Avian infections are mostly latent but in humans the infection may cause a mild to severe disease, especially in old age with symptoms of bronchopneumonia.

psych(o)- (Gr. *psyche* = spirit, mind, soul), denoting relationship to mind, i.e. the mental functions, especially in humans. *Adj.* **psychic**.

psychr(o)- (Gr. *psychros* = cold), denoting relationship to cold.

psychrometer (Gr. *metron* = measure), an apparatus used for measuring the humidity of air by means of temperature differences of dry and moist air. The equipment has two thermometers, one consisting of a dry, and one of a wet bulb. The dry thermometer shows the actual air temperature. The quicksilver bulb of the other thermometer, covered by a moistened cloth, shows a lower temperature; the difference between the thermometers is proportional to the quantity of evaporated water from the cloth. This evaporation is dependent on the → relative humidity of the air.

psychrophile (Gr. *philein* = to love), an organism, especially a microbe, which grows best at low temperatures (15—20 °C). *Adj.* **psychrophilic**. *Cf.* thermophile.

Pteraspidomorphi, Myxini (Gr. *pteron* = wing, *aspis* = shield, *morphe* = shape), **hagfishes**; a class of jawless vertebrates (Agnatha), closely related to lampreys; they have a long body and paired nostrils (*cf.* Cephalaspidomorphi), and paired fins are lacking. Hagfishes are marine scavengers feeding on dead or dying animals. The class includes the order Myxiniformes.

Pteridophyta (Gr. *phyton* = plant), **pteridophytes**; a division in the Plant kingdom; the most developed group of spore plants having root, stem, leaves, and conducting tissues, but no flowers. The dominant stage in the life cycle is the → sporophyte, the gametophyte being a separate, small, reduced → prothallus. The division Pteridophyta is divided into four classes: → Psilotopsida (**whisk ferns**), Lycopsida (**clubmosses**), Sphenopsida (**horsetails**), and Pteropsida (**ferns**). In some systems pteridophytes is a group name without any taxonomical status and the division Pteridophyta is abandoned.

Pteridospermae (Gr. *sperma* = seed), **seed ferns**; a fossil plant class in the subdivision → Gymnospermae, division Spermatophyta; in some systems a division called Pteridospermophyta. Seed ferns lived from the late Devonian to the Cretaceous period, they were large-sized and their leaves resembled fern leaves; therefore they were classified as ferns before their seeds were found.

Pterobranchia(ta) (Gr. *branchia* = gills), **pterobranchs**; a class in the phylum Hemichordata comprising deep-sea invertebrate animals which have a small (1—7 mm) tripartite body consisting of the **proboscis**, the large **collar** with several tentacles, and the **trunk**; most species are colonial living together in gelatinous tubes. The class includes only three genera.

Pteroclidiformes (L. *Pterocles* = the type genus, *forma* = shape), **sandgrouses**; an order of dove-like birds with a short bill, long, sharp-pointed wings, and quite a long tail; they live in large groups on open steppes and sandy or stony plains in southern Europe, Africa, and Asia, nesting on the ground; 16 species are described.

Pterophyta, *see* Pteropsida.

pteropods (Gr. *pous*, gen. *podos* = foot), **sea butterflies**; a group (without taxonomic status) of small, swimming, pelagic opisthobranchs (Gastropoda) characterized by wing-shaped extensions in the foot. The group (earlier considered as superorder Pteropoda) has two orders: **shelled pteropods** (order Thecosomata), and **naked pteropods** (order Gymnosomata) without a shell. Most pteropods have no gills but gas exchange occurs across body surface; naked forms have no mantle cavity.

Pteropsida, a class in the division Pteridophyta, Plant kingdom; includes the subclasses Archaeopteridae (extinct), Ophioglossidae (adder's tongues), → Polypodidae (spleen worts), Marsileidae, and Salvidae. In some systems Pteropsida is classified as a separate division, Pterophyta (ferns).

Pterosauria (Gr. *saura* = lizard), **pterosaurs**; "pterodactyls"; an extinct order of gliding reptiles with membranous wings; lived during Mesozoic and likely fed on marine surface fish.

pterostigma, pl. **pterostigmata** (Gr. *stigma* = mark), an opaque spot on the anterior margin of the wing of an insect.

pterygium, pl. **pterygia**, Am. also **pterygiums** (Gr. *pterygion*, dim. of *pteron* = wing), a wing or a wing-shaped structure. *Adj.* **pterygoid**, wing-like.

Pterygota, (Gr. *pterygotos* = winged), **pterygote insects**; a subclass of insects; usually they have wings which may secondarily be lacking, as e.g. in fleas and lice or in other insects living as parasites or within the ground. Pterygota includes most of the extant insects, the wingless insects belonging to the subclass Apterygota. On the basis of the → metamorphosis, the group is also called **Metabola**.

pteryla, pl. **pterylae** (Gr. *hyle* = wood, forest), feather tract; one of the areas of bird skin on which feathers grow; **pterylosis**, the arrangement of feathers in definite tracts of growth. *Cf.* apterium.

ptilinum, pl. **ptilina** (Gr. *ptilon* = feather), a bladder-shaped structure on the frontal head of some dipterans; during the emergence of the insect it expands and breaks the walls of the puparium.

ptiloerection (L. *erectus* = raised up), elevation or bristling of feathers, serving as threat signals and thermoregulation in cold.

PTTH, → prothorac(ic)otropic hormone.

ptyalin (Gr. *ptyalon* = saliva), an old name for α-amylase in saliva. *See* amylase.

pubescent (L. *pubescens* = becoming hairy), **1.** covered with soft, fine hair; downy; **2.** developing to the age of puberty.

pubis, pl. **pubes** (L. *os pubis*), *see* pelvis.

puerulus, pl. **pueruli** (L. dim. of *puer* = child), the last larval stage of some lobsters (*Palinura*, Crustacea), having long antennae, and after many moults develops into an adult crab.

puff, in genetics, a swelling at a certain location of a → giant chromosome. In the puff the tangle of DNA situated in a transverse band of the giant chromosome opens up for → genetic transcription. Transcription in puffs can be demonstrated by → autoradiography, RNA specific stains, and by inhibiting the transcription chemically; these indicate that the puffs represent active genes. Puffs are divided into non-specific, stage-specific, and tissue-specific; 80% of the puffs of the fruit fly are stage or tissue-specific. The formation of puffs is called **puffing**. *Cf.* Balbiani ring.

pulling roots, adventitious roots of plant bulbs, shortening during growth and preventing the structures from rising up from the ground; found in some bulb-bearing plants, e.g. *Crocus*.

pulmo, pl. **pulmones** (L.), → lung; *pulmo dexter*, right lung; *pulmo sinister*, left lung. *Adj.* **pulmonary**, pertaining to the lung.

pulmonary circulation, lung circulation, lesser circulation; a type of blood circulation found in tetrapod vertebrates who have a double circulation; comprises the circulation of deoxygenated blood from the right ventricle to the lungs and oxygenated blood from the lungs to the left atrium; in frogs from the right side of the single ventricle. *Cf.* systemic circulation. *See* blood circulation.

pulmonary ventilation, *see* ventilation.

Pulmonata, pulmonates; a subclass in the class Gastropoda, phylum Mollusca; comprises air-breathing, hermaphroditic species with one lung, i.e. a highly vascularized area of the mantle. The shell is spirally coiled and usually without operculum. Pulmonates include both terrestrial (order Stylommatophora) and freshwater species (mainly in the order Basommatophora); only few of the species are marine (the small order Archaeopulmonata). Aquatic pulmonates have to rise to the surface for air.

pulp (L. *pulpa* = flesh), a fleshy, soft part of a tissue; e.g. **1.** the inner substance of a tooth (dental pulp, *pulpa dentis*); **2.** the succulent part of a fruit; **3.** the pith of a plant stem.

pulsation (L. *pulsare* = to beat), a rhythmical beat, such as the heartbeat.

pulse (L. *pulsus*), a stroke, a beat, a single pulsation; especially the regular expansion of the arteries.

pulse-chase technique, a biochemical method in which cells are incubated very briefly with a radioactive enzyme substrate, the non-radioactive substrate (cold chase) is then added to replace the labelled molecules. The fate of the radioactive substance incorporated in the pulse is used to study the metabolism of the substrates in cells and tissues.

pulse-field electrophoresis, a technique of → electrophoresis used for a high-resolution separation of DNA fragments; also suitable for large (several megabases) DNA fragments. The direction of electric field can be regulated using electric pulses during the run at an angle of 90 to 120°. The alternating electric field causes DNA molecules to re-orient and move at different speeds in an agarose gel. The pulse-field technique speeds up separation of DNAs and improves resolution.

pulvillus, pl. **pulvilli** (L. dim. of *pulvinus* = cushion), a small pad or lobe; e.g. a thin lobe-like structure between the claws of insects keeping e.g. a fly on a wall or cealing by adhesion.

pulvinus, pl. **pulvini,** a swelling at the junction of the axis and the leaf stalk in plants, playing a part in the leaf movement; especially K^+ concentrations affect the turgor pressure of pulvini cells and consequently the position of the leaf.

puna (AmerInd. = high Andean plateau), → mountain sickness.

punctualism, *see* punctuated equilibrium.

punctuated equilibrium, a conception of the evolution theory assuming that a species remains unchanged for long periods (**stasis**), which are punctuated by stages of rapid changes and speciation, followed by **geographic dispersal**. Because the speciation occurs in small **peripheral populations**, practically no fossils from these populations have remained. *Syn.* punctualism. *Cf.* phyletic gradualism.

punctum, pl. **puncta** (L.), in anatomy, a small spot, area, or point of projection; e.g. *punctum lacrimale*, lacrimal point, the opening of the lacrimal duct in the papilla of an eyelid.

puncture (L. *punctura*), **1.** a small point-like depression in a morphological structure of an animal; **2.** the act of piercing or perforating the body surface or the wall of an organ with a pointed instrument; also a hole made by piercing.

Punnett square (*Reginald C. Punnett*, 1875—1967), a graphic representation of all the possible unions of gametes which bear different combinations of → alleles in a given cross.

pupa, pl. **pupae,** also **pupas** (L. = puppet), a developmental stage after the egg and larval stages in the life cycle of a holometabolous insect (*see* metamorphosis). In the course of the pupal stage, the insect does not feed but undergoes great physiological and structural changes, when many larval organs dedifferentiate and the new adult structures develop. Usually, the pupa is immobile, but in some species the head of the pupa may move suddenly, startling predators. Many insects, as e.g. lepidopterans, have a → **cocoon** previously prepared by the larva for protecting the pupa; some insects have a harder shell, → **chrysalis**. Two pupal forms are found: in the **obtect type** the appendages are glued to the body by a secretion, e.g. in lepidopterans and dipterans. In the **exarate type**, characteristic of the groups Hymenoptera, Coleoptera, Mecoptera, and Neuroptera, the antennae, wings, and legs are free. *Verb* to **pupate**, to become a pupa; the process is called **pupation.** *See* puparium.

puparium, pl. **puparia** (L.), a barrel-shaped protective case surrounding the → pupa of some dipterans; formed from the exoskeleton of the final larval instar. Such a protected pupa is called **coarctate pupa**.

pupil (L. *pupilla*, dim. of *pupa* = puppet), **1.** the opening at the centre of the iris of the camera eye in vertebrates and cephalopods; the size varies according to the contraction or dilatation of the iris by its smooth muscles; **2.** the central spot of an → ocellus. *Adj.* **pupillary**.

pupillary, pertaining to the pupil; **pupillary reflex**, the change of diameter of the pupil triggered by a stimulus, especially the constriction of the pupil by light. *Cf.* corneal reflex.

pupiparous (L. *parere* = to beget), pertaining to an insect which brings forth pupae or larvae ready to pupate at birth; e.g. tsetse flies whose larvae develop inside the body of the female fly and after birth are ready to pupate.

pure culture, a cell culture containing only one microbe strain, usually descendants of a single cell, not contaminated by other bacteria or viruses.

pure line, a series of progenies and generations derived from one completely → homozygous individual through → self-fertilization; the individuals are thus genetically identical and homozygous for all their genes. Called also isogenic strain.

purine (L. *purum* = pure, + *urine*), $C_5H_4N_4$, an organic compound consisting of two heterocyclic rings, not known to appear free in nature but as **purine bases**, i.e. **adenine** and **guanine**, which are structural components in → nucleosides and nucleotides (e.g. ATP), and consequently also in → nucleic acids

(DNA, RNA). *Cf.* pyrimidine.

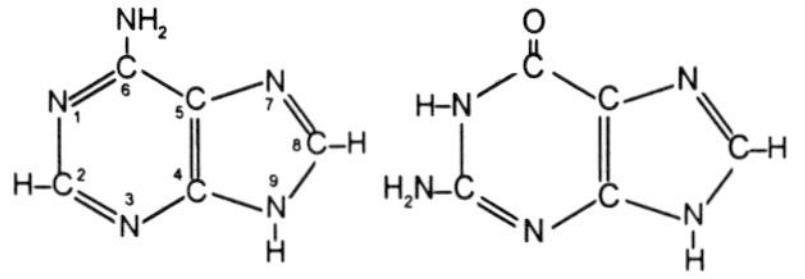

Purines. Two common purine bases: left, adenine (A) and right, guanine (G).

Purkinje('s) cell (*Johannes von Purkinje*, 1787—1869), a cell type found in the cerebellar cortex of the vertebrate brain; a large, pear-shaped nerve cell with numerous short dendrites. The cells participate in the control of motor functions of the skeletal muscles.

Purkinje('s) fibres, Am. **Purkinje('s) fibers,** bundles of specialized cardiac myocytes in the → conduction system of the heart of vertebrates; in the subendothelial tissue of the walls of the ventricles, they form a netlike structure (**Purkinje('s) network**) along which the action potentials (impulses) rapidly propagate everywhere in the ventricular muscle.

puromycin, an antibiotic obtained from *Streptomyces* fungi, used as a drug and in physiological studies; resembles aminoacyl tRNA of bacteria and can link to an amino acid site on ribosomes of a bacterial cell thus inhibiting the lengthening of a peptide chain. Puromycin inhibits protein synthesis also in many eukaryotic cells.

puszta (Hung.), Hungarian → steppe.

putamen, pl. **putamina** (L. = shell), the outer segment of the lentiform nucleus in the telencephalon (cerebrum) of vertebrates; belongs to the larger grey area, corpus striatum, which is involved in motor functions of the → extrapyramidal system.

putrefaction, the decomposition of organic matter by bacteria and fungi often resulting in foul-smelling products, such as mercaptans, in anaerobic splitting of proteins.

putrescine, 1,4-diaminobutane, $H_2N(CH_2)_4$-NH_2; a diamine formed abundantly in decay processes of organic material (putrefying tissue) by decarboxylation of arginine. It is found in small quantities in living animal tissues having e.g. hormone-like effects on cell growth. Also in plants putrescine has some hormone-like actions.

PVP, → polyvinylpyrrolidone.

pycn(o)-, pykn(o)- (Gr. *pyknos* = dense, thick), denoting dense, thick, compact, close, bulky, frequent.

pycnidiospore (Gr. *sporos* = seed), a conidium formed in the → pycnidium.

pycnidium, pl. **pycnidia** (L.), a flask-shaped structure opening to the surface of the thallus of many lichens or in the fruiting body of some fungi, such as rust and smut fungi; produces conidiospores called pycnidiospores. *Syn.* pycnium.

Pycnogonida, Pantopoda (Gr. *gony* = knee), **pycnogonids, pantopods, sea spiders**; a class in the subphylum Chelicerata, phylum Arthropoda, including about 1,000 species; sea spiders have a short body with very long legs (Pantopoda = completely legs), many species bear chelicerae; respiratory and excretory systems are absent. They are small marine animals living on oceanic bottoms to the depth of 6,000 m.

pycnospore, a spore type in → Uredinales (rust fungi).

pyel(o)- (Gr. *pyelos* = pelvis, tub), pertaining to the pelvis, especially to the renal pelvis.

pyg-, pygo- (Gr. *pyge* = rump, tail), denoting relationship to the rump, buttocks, or more generally to the caudal or caudodorsal part of the trunk of an animal.

pygostyle (L. *os pygostylis* < Gr. *stylos* = column), a bone formed by fusion of the 5 or 6 last caudal vertebrae in birds; supports the skeletal muscles which move the tail feathers.

pyknometer, pycnometer (Gr. *pyknos* = dense, thick, *metros* = measuring), an apparatus for measuring the density of a liquid.

pyl(e)-, pylo- (Gr. *pyle* = gate), pertaining to a → portal vein. *Adj.* **pylic**.

pylorus, pl. **pylori** (L. < Gr. *pyloros* = gatekeeper), **1.** a muscular structure for closing (muscular sphincter) or opening (muscular dilator) the lumen or orifice of an organ; **2.** the passage from the stomach to the intestine; in vertebrates, comprises a band of circular muscle (pyloric sphincter), pyloric antrum, pyloric canal, and pyloric opening (orifice).

py(o)- (Gr. *pyon* = pus), denoting relationship to pus, i.e. an inflammation product of vertebrates, consisting chiefly of white blood cells in tissue fluid.

pyramidal cell, a large, cone-shaped nerve cell with many branching dendrites in the cerebral cortex; the axons of the cells form the upper part of the pyramidal tracts innervating skeletal muscles. *See* pyramidal system.

pyramidal system/tract, the system of motor nerve fibres of mammals beginning from the → pyramidal cells in the cerebral cortex, their axons passing to the brain stem (**corticobulbar tract**) and/or the spinal cord (**corticospinal tract**), forming → synapses with peripheral motoneurones which innervate skeletal muscles. The pathways have been referred to as the pyramidal system because the fibres of the lateral corticospinal tract form a pyramid-shaped structure (*pyramis*) in the midline of the medulla oblongata. The pyramidal system is phylogenetically rather new, well developed in man, providing for direct cortical control of skilled movements, as those related to speech and movements of fingers.

pyramid of biomass, biomass pyramid, *see* ecological pyramid.

pyramid of numbers, *see* ecological pyramid.

pyramid of energy, *see* ecological pyramid.

pyramis, pl. **pyramides** (L.), **pyramid**; in vertebrate anatomy, e.g. **1.** the pyramid-shaped structure in the ventral medulla oblongata (*pyramis medullae oblongatae*); *see* pyramidal system; **2.** renal pyramid (pyramid of Malpighi, *pyramis renalis*), one of the conical structures in the medulla of the → kidney constructed of distal tubules and collecting duets of the → nephrons; **3.** pyramid of vermis (*pyramis vermis*), the pyramid-shaped part on the surface of the cerebellum.

pyranose, a → monosaccharide consisting of a 6-membered ring of 5 carbons and one oxygen.

pyrenoid (Gr. *pyren* = fruit "stone", *eidos* = form), a proteinaceous structure in the chloroplasts of algae and certain liverworts; mostly surrounded by numerous starch grains and probably involved in starch synthesis.

pyrethrin (Gr. *pyrethron* = feverfew), an alkaloid in → pyrethrum plants; a viscous water-insoluble liquid, a terpenoid compound such as pyrethrin I ($C_{21}H_{28}O_3$) and pyrethrin II ($C_{22}H_{28}O_5$), which are extracted from pyrethrum flowers and used as insecticides and as a drug against scabies.

pyrethrum, 1. the mixture of substances obtained by grinding and extracting dried flower heads of chrysanthemums; **2.** Spanish camomile, obtained from roots of *Anacyclus pyrethrum*; its flowers are used as a source of → pyrethrin; **3.** any of several chrysanthemums.

pyrexia (Gr. *pyressein* = to be feverish < *pyr* = fire), feverishness; the feverish (febrile) condition. *Adj.* **pyrexial**. *See* fever.

pyridine, a heterocyclic liquid, C_5H_5N, extracted and purified from coal tar; used as a solvent e.g. in chromatography; pyridine derivatives are e.g. **pyridine bases** and **pyridine nucleotides** (NAD, NADP).

pyridoxal, the 4-aldehyde of → pyridoxine, one of the forms of vitamin B_6 group; the active form is pyridoxal 5'-phosphate.

pyridoxal kinase, *see* pyridoxine.

pyridoxamine, the amine of → **pyridoxine**, one of the forms of vitamin B_6 group; the active coenzyme is pyridoxamine phosphate.

pyridoxine, 5-hydroxy-6-methyl-3,4-pyridinedimethanol, $C_8H_{11}NO_3$; the original vitamin of the B_6 group. In tissues it is converted to **pyridoxal 5'-phosphate** by pyridoxal kinase (EC 2.7.1.35), in that form acting as a coenzyme in enzymes which catalyse fatty acid and amino acid metabolism in cells of all organisms (e.g. in transaminase, decarboxylase enzymes). Also the amine form, **pyridoxamine**, and the aldehyde form, **pyridoxal**, can be phosphorylated and used as coenzymes. Pyridoxine is synthesized by plants, but especially animals require it from other sources. In vertebrates, the deficiency of pyridoxine results in e.g. a decreased utilization of unsaturated fatty acids in cells, and disturbances appear first in the skin and neuronal functions. *See* vitamin B complex.

pyrimidine, an organic compound consisting of one 6-membered heterocyclic ring, $C_4H_4N_2$; found in **pyrimidine bases**, which are structural units in → nucleosides, nucleotides, and nucleic acids. Pyrimidine bases in DNA are **cytosine** and **thymine**, in RNA cytosine and **uracil**, in → transfer RNA also **pseudouracil**. → Thiamine (vitamin B_1) is a derivative of pyrimidine. *Cf.* purine.

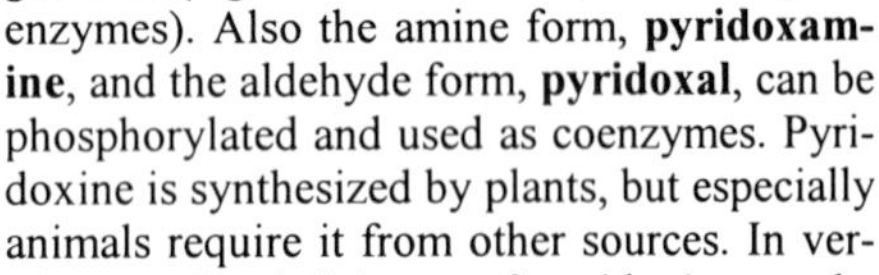

Pyrimidines. Common pyrimidine bases, from left: cytosine (C), uracil (U) and thymine (T).

pyrites (Gr. *pyr* = fire), natural sulphides of metals; e.g. iron pyrites, i.e. iron disulphide, FeS_2, the most common sulphur mineral; copper pyrites, $CuFeS_2$, "fool's gold".

pyrogen (Gr. *gennan* = to produce), a substance producing fever in homoiothermic animals; may be **exogenous pyrogens**, toxic products (endotoxins) of bacteria, moulds, and yeasts, and whole viruses or other antigens, or they may be **endogenous pyrogens** which are polypeptides (cytokines such as interleukin I) produced by phagocytic white blood cells or tissue → macrophages, and secreted mostly in response to exogenous pyrogens. The endogenous pyrogens elevate the set point of the hypothalamic thermoregulatory centre so that shivering is induced in muscles, resulting in an increase of the body temperature, i.e. causing → fever. In poikilothermic vertebrates, pyrogens may cause behavioural fever, i.e. after an infection the animal selects a warmer environmental temperature to occupy. Materials used for pharmaceutical purpose, are always pyrogen-tested, usually on rabbits in which minimal quantities of exogenous pyrogens cause fever.

pyrophosphate, PP_i, a salt or the ionic form ($HP_2O_7^{3-}$) of pyrophosphoric acid ($H_4P_2O_7$); an inorganic phosphate which occurs e.g. in the metabolism of → nucleotides in cells, e.g. when AMP or ADP is bound to inorganic phosphate (i.e. to PP_i or P_i) it is converted to high-energy ATP. The transfer of pyrophosphate is catalysed by **pyrophosphokinases** of the enzyme sub-subclass EC 2.7.6.; the breakdown of the ester bond between two phosphorus atoms in pyrophosphate is catalysed by **pyrophosphatase**, EC 3.6.1.1, while orthophosphate, P_i, is formed. *Cf.* orthophosphate.

pyrrole, azole, imidole, C_4H_5N; a heterocyclic compound with 5-membered ring present in many important biological molecules; forms → porphyrin structure e.g. in → chlorophyll and → haem.

Pyrrophyta, a division in Plant kingdom, unicellular biflagellated, planktonic organisms. The group includes both photosynthetic and heterotrophic forms and is also classified in the Animal kingdom (or Protista) as the group → Dinoflagellata. A characteristic feature of many species is the sculptured theca, formed from cellulose plates under the plasma membrane.

pyruvate, a salt, ester, or the ionic form of → pyruvic acid.

pyruvate carboxylase, a ligase enzyme, EC 6.4.1.1, catalysing the carboxylation of pyruvate to form oxaloacetate; pyruvate + ATP + CO_2 —> oxaloacetate + ADP + PP_i.

pyruvate dehydrogenase complex, a multienzyme complex of three enzymes (in the bacterium *Escherichia coli* with 60 peptide chains), catalysing the oxidative decarboxylation of pyruvate to acetyl CoA to be metabolized further in the → citric acid cycle: pyruvate + CoA + NAD $\rightleftharpoons$ acetyl CoA + CO_2 + NADH. The complex consists of three enzymes: **pyruvate dehydrogenase component** (EC 1.2.4.1), **dihydrolipoyl transacetylase** (dihydrolipoamide acetyltransferase, EC 2.3.1.12), and **dihydrolipoyl dehydrogenase** (EC 1.8.1.4). In → eukaryotes, this enzyme complex is found in the mitochondria of all cells.

pyruvate kinase, an enzyme, EC 2.7.1.40, in cellular metabolism catalysing the transfer of phosphate in a reversible reaction from phosphoenolpyruvate (PEP) to ADP, forming ATP and pyruvate e.g. in glycolysis.

pyruvic acid, a tricarbonic acid, CH_3CO-COOH; the end product of → glycolysis in cell respiration. From pyruvic acid the reaction pathway leads via acetyl CoA to the → citric acid cycle, or into lactic acid, ethanol, or some other compounds in → anaerobic respiration of cells, or in fermentation reactions by microbes. Pyruvic acid is an important intermediate in many metabolic processes. Its salts, esters and ionic form are called **pyruvates**.

pythogenesis (Gr. *pythein* = to rot, *gennan* = to produce), origination from decay, decomposition, or decaying matter. *Adj.* **pythogenic** (pythogenous).

pyxidium, pyxis, a type of fruit → capsule in which a part of the wall lifts off as a lid, opening the fruit; e.g. in *Anagallis arvensis*.

Q

Q, symbol for **1.** quantity of electric charge (also q); **2.** quantity of heat (also q); **3.** glutamine; **4.** ubiquinone (coenzyme Q).

q, symbol for the long arm of a (human) chromosome.

Q_{10}, temperature coefficient; the quotient which expresses an increase in the velocity of any reaction when temperature rises 10°C. If the reaction velocity is measured at two different temperatures, Q_{10} can be calculated according to the following equation:

$$Q_{10} = (R_2/R_1)^{10/(T2-T1)}$$

T_1 and T_2 = temperatures, R_1 = the reaction velocity at T_1, and R_2 = the reaction velocity at T_2.

Enzymatic (physiological) reactions of organisms measured in physiological circumstances usually give Q_{10} values between 1.5 and 2.5.

Q_{CO2} (Q < L. *quantum* = how much), the carbon dioxide volume in microlitres at normal temperature and pressure expressed usually per tissue milligram per hour.

Q_{O2}, Q_O, symbols for oxygen consumption of an organism.

Q-banding (Q = quinacrine stain), a method for staining chromosomes; quinacrine or other fluorescent dyes are used to stain certain areas of the metaphase chromosomes strongly fluorescent (**Q bands**), the regions between them remaining unstained. Quinacrine is intercalated in DNA helices.

Q-enzyme, an enzyme (branching enzyme) that catalyses the formation of α-1,6-bonds from α-1,4-bonds between glucose molecules in the synthesis of → amylopectin in plants; the branched structure of amylopectin is dependent on these bonds.

QTL, → quantitative trait loci.

quadr(i)-, quadru- (L. *quattuor* = four), denoting four or square; tetra-.

quadrat (L. *quadratus* = square), a delineated area of vegetation (usually 1 m^2) for the study of vegetation and animals; the quadrats may be located in the field randomly (*see* random sampling) or according to certain patterns e.g. in linear sampling. *See* systematic sampling.

quadrate, 1. having four equal sides; **2. quadrate bone**, the posterior cartilage or bone of the upper jaw (hyomandibular bone) to which the lower jaw is articulated in gnathostome vertebrates; in mammals it forms the incus of the middle ear.

quadrivalent (L. *valens* = being of some value), **1.** tetravalent, four-valued; **2.** a multivalent chromosome group that consists of four chromosomes which may be completely → homologous, as e.g. in autotetraploids (*see* polyploid), or partially homologous, as e.g. in translocation heterozygotes.

quadrupedal (L. *pes* = foot), having four feet or walking on four feet. *See* Tetrapoda.

qualitative defence, in botany, a chemical defence of plants and fungi based on the presence of specific poisonous substances in these organisms; e.g. poisonous mushrooms. It is common also in short-lived herbaceous plants, especially in those growing in the first stages of → succession. The qualitative defence comprises mainly → continuous defence mechanims. *Cf.* quantitative defence.

qualitative inheritance, the inheritance of characters showing alternative → variation and thus being inherited according to → Mendel's laws. *Cf.* quantitative inheritance.

quantasome (L. *quantitas* = quantity, *soma* = body), a granule-like structure seen under the electron microscope on the thylakoid membranes of a → chloroplast; quantasomes are thought to be structural units involved in → photosynthesis.

quantitative defence, in botany, a chemical defence of plants having inducible mechanisms by which the nutritive value of a plant for herbivores is reduced due to harmful → secondary compounds (chemicals) produced in plant tissues. The production of the chemicals begins to increase after damage caused by herbivores. The quantitative defence is typical of long-lived and large plants, such as trees. *See* inducible defence. *Cf.* qualitative defence.

quantitative genetics, a branch of → genetics studying the inheritance of quantitative or metric characteristics in organisms. The basic idea of quantitative genetics is the partition of a genetic variation into different components according to causes of the variation.

quantitative inheritance, the inheritance of metrical characteristics showing fluctuating → variation as opposed to alternative (qualitative) variation. *Cf.* qualitative inheritance.

quantitative trait loci, QTL, genetic determinants of quantitative or metric traits in (domestic) plants and animals.

quantum, pl. **quanta** (L. = how much), **1.** a certain definite amount or quantity; **2.** a unit of radiant energy, $h\nu$, in which h is Planck's constant $6.6254 \cdot 10^{-34}$ J · s, and ν is the frequency of vibration; *see* photon.

quartz, a crystalline form of silicon dioxide, SiO_2; used in optical and electric instruments, its crystals exhibiting → piezoelectricity.

Quaternary (L. *quaternarius* = containing four), a period after Tertiary in Caenozoic, comprising the Pleistocene and Holocene epochs. Quaternary began about 1.6 million years ago and continues to the present time; it has included several glacial periods. In the beginning of Quaternary, human evolution diverged from other hominids.

quaternary structure, *see* protein.

queen, a reproducing female in colonies of social insects, such as hymenopterans (e.g. bees and ants) and isopterans (termites).

queen substance, a group of chemical substances, → **pheromones**, secreted by a queen bee's mandibular glands and passed throughout the colony by → **trophallaxis**. Pheromones inhibit the initiation of development of other females to queens by controlling the behaviour of workers in reproductive actions. When the secretion of the queen-bee substance begins to decrease, e.g. owing to aging of the queen, the workers start to build a new cell (**royal cell**) and feed its larva with **royal jelly** which induces the development of a new queen. The most active of the pheromones is *trans*-9-keto-2-decenoic acid.

quenching, diminishing of radiation energy, e.g. in → scintillation counting liquids caused by chemical substances (coloured substances, halogens, etc.) and physical factors (e.g. self absorption, paper, dilution, phase formation).

quicksilver, *see* mercury.

quiescent centre, a cell group near the tip of a growing plant root between the root cap and the root meristem; in the quiescent centre the frequency of cell division (mitosis) is very low, unlike in the root tip in general.

quillworts, *see* **Isoëtales**.

quinacrine, a fluorescent dye for staining DNA (*see* Q-banding); used as an antimalarial agent, and as an anthelmintic drug against intestinal tapeworms.

quinidine, the dextrorotatory stereoisomer of → quinine.

quinine, the most important of the ca. 20 different cinchona alkaloids; a white, bitter, slightly water-soluble compound, $C_{20}H_{24}N_2O_2 \cdot 3\ H_2O$, obtained from the bark of cinchona (*Cinchona officinalis*). It is a protoplasmic poison, reducing metabolism and enzyme actions and inhibiting internal membrane fusion processes. Quinine like other cinchona alkaloids (e.g. **quinidine**) is used e.g. as an antipyretic and analgesic drug, and in the treatment of malaria and cardiac arrhythmia (atrial fibrillation).

quinone, 1. the general name for aromatic compounds bearing two oxygens in place of two hydrogens, usually in *para* position; **2.** the oxidation product of → hydroquinone; **3.** the specific name for 1.4.-benzoquinone and other aromatic dicarbonyl compounds, diketones. **Benzoquinones**, found especially in plants, are oxidation products of → phenols, i.e. in their molecules two hydrogen atoms in the same benzene nucleus are replaced by oxygen atoms. Quinones are highly coloured compounds and many of them are present in organisms, e.g. **phylloquinone**, **menaquinone** (vitamin K). According to their oxidation-reduction property benzoquinones act in cellular electron transport reactions, as e.g. **plastoquinone** in → photosynthesis and **ubiquinone** in mitochondrial cell respiration. They are used as oxidizing agents in photography and in dye manufacture.

R

R, symbol for **1.** an organic → radical; **2.** arginine; **3.** roentgen; **4.** Réaumur; **5.** → ring chromosome.

***R*,** symbol for **1.** resistance; **2.** gas constant, $R = 8.314\ \text{J mol}^{-1}\text{K}^{-1}$.

r, symbol for **1.** radius of a circle; **2.** revolution, e.g. rpm (RPM); **3.** the former symbol for roentgen, replaced by R.

RA, → retinoid acid.

rabies, pl. **rabies** (L. *rabere* = to rage), a deadly infectious disease caused by a rhabdovirus in many mammals, including man. The virus causing damage to the central nervous system, is usually transmitted in saliva by bites of a rabid animal, but also through the respiratory route or by the ingestion of infected tissues. Important vectors are e.g. fox, raccoon, raccoon dog, skunk, wolf, mongoose, bat, but also dog and cat. Symptoms are exhibited especially as disturbed function of the nervous system, i.e. a bizarre behaviour, delirium, fearfulness, convulsions, salivation, paralysis, and coma. The disease is pandemic and it kills annually about a million people. *Syn.* lyssa, hydrophobia.

Rabl-orientation (*Carl Rabl*, 1853—1917), the maintenance of the anaphasic orientation of chromosomes to the next mitotic prophase. *See* mitosis.

race, a group of interbreeding individuals which have some characteristics differing from all other individuals of the same species. The characteristics may be biochemical (**biochemical race**), physiological (**physiological race**), ethological (**ethological race**), ecological (**ecological race**), etc. The **geographical race** means the populations characterized often by morphological features which occur in specified geographical districts. In genetics, race is defined as an allopatric population diverging by its gene frequencies from other populations. The term is also used in human biology. *Cf.* subspecies.

raceme (L. *racemus* = cluster of grapes), a type of → racemose inflorescence in plants.

racemic mixture, an equimolecular mixture of the two optically active → isomers of a compound, consequently it is optically inactive; designated by the letters *dl*, sometimes by ±.

racemose (L. *racemosus*), relating to, having the form of, or resembling a raceme or cluster of grapes, especially when describing the structure of some glands in animals or a → racemose inflorescence in plants.

racemose inflorescence, an → inflorescence with flowers along a continuously growing main axis. An inflorescence with stalked flowers is called **raceme** (e.g. *Convallaria majalis*); **spike** is otherwise similar but the flowers are not stalked. A spike with a clearly thickened main axis is termed **spadix** (the female inflorescence of maize). A compound raceme is formed, if some of the branches of the first order undergo further branching; such a structure is found e.g. in **panicle** (e.g. *Syringa*, and paniculate species of Poaceae).

rachis, pl. **rachises** or **rachides** (Gr. *rhachis* = spine, ridge), a shaft or ridge-like central anatomical structure; e.g. **1.** the vertebral column, spine; **2.** the solid stalk (shaft) in the bird contour → feather, being a continuation of the hollow quill and bearing numerous barbs; **3.** the main axis of an → inflorescence; **4.** the axis of a pinnately compound leaf; shaft.

rachitis (Gr. *rhachitis* = disease of spine), **1.** → rickets; **2.** an inflammatory disease of the spinal column.

rad, **1.** abbr. of radian; *see* Appendix 1; **2.** the former unit of absorbed dose of ionizing radiation, later replaced by gray (Gy); *see* Appendix 1.

radial (L. *radius* = ray, beam, spoke), **1.** radiating or spreading from a certain point; **2.** pertaining to the → radius (bone).

radial branch, a branch of a plant stem, situated radially with other, similar branches; e.g. in horsetails.

radial bundle, a type of → vascular bundle in which the xylem and phloem lie on alternate radii of the axis usually separated by nonvascular tissue. The → stele of the roots of seed plants is representative of this type.

radial canal, **1.** a duct in the → gastrovascular system of some cnidarians such as jellyfishes, and in the → water-vascular system of some echinoderms, such as sea stars; in the jellyfish, the radial canals are directed from the gastric cavity to the ring canal at the edge of the umbrella, and in echinoderms from the ring canal around the mouth into each arm, the radial canal running along the underside of the arm and ending at the tip of a terminal tentacle; **2.** small canals lined with choanocytes opening to the spongocoel in the body wall of syconoid sponges.

radial flower, pertaining to a flower with more than two planes of symmetry. *Syn.* actinomorphic flower, polysymmetric flower.

radial symmetry, a type of symmetry in which the parts of an organism are arranged symmetrically around the central axis (oral-aboral axis in animals), e.g. in cnidarians and echinoderms; in botany, in actinomorphic (radial)flowers and some algae (centric diatoms). *See* bilateral.

radian, abbr. rad, the SI unit of plane angle. *See* Appendix 1.

radiation (L. *radiatio*), **1.** divergence from a centre; **2.** a structure formed from diverging elements, as the afferent nerve tracts to the cerebrum; **3.** adaptive radiation, involving an evolution from a primitive species to diverse, specialized species; **4.** → electromagnetic radiation, i.e. emission of energy as electromagnetic waves; **5.** particle radiation, i.e. particle emission (→ alpha particles, beta emission). *See* ionizing radiation, light, Planck's constant.

radiation dose, the energy dose of → ionizing radiation (radioactive radiation) that is absorbed in a medium, such as in cells and tissues; its unit is gray, Gy (1 Gy = 1 J/kg). Approx. a 3 Gy dose of gamma radiation is fatal to man.

radiation hybrid mapping, RH-mapping, a method of physical mapping of human genes in which fragments of human chromosomes produced by radiation are introduced into hamster cells.

radical, R (L. *radicalis* = thorough), **1.** a basic unit that serves as an element for a bigger and more complex structure; **2.** in chemistry, a group of atoms in a compound remaining unchanged though chemical changes occur in other parts of the molecule; e.g. the general formula for amino acids is designated HOOC-R-NH_2, in which R is usually a carbon chain of different length. Radicals are usually incapable of independent existence. A **free radical** is a radical in its transient uncombined state, having an unpaired electron on the outermost shell of an atom. Free radicals are usually short-lived and very reactive, and thus toxic in organisms. *See* oxygen radicals.

radicle (L. *radicula*, dim. of *radix* = root), **1.** a structure resembling a rootlet, e.g. the radicle of a vein or nerve; **2.** a part of the → embryo in seed plants. During seed germination the radicle grows first downwards out from the seed forming the taproot in dicotyledons; in monocotyledons the radicle soon dies and the root system is formed by → adventitious roots developed from the tissue at the base of the stem. *See* root.

radio- (L. *radius* = spoke, ray, beam), **1.** pertaining to the radio, radium, radiant rays, radiant energy or radioactive isotopes; **2.** relating to the → radius.

radioactive series, a series of radionuclides, i.e. a radioactive family, in which each except the first one is the decay product of the previous one, the final member being an isotope of lead that is stable. *See* radioactivity.

radioactive tracing, a technique of tracing the course of a radioactively labelled element through a biological or chemical system. *See* label.

radioactivity, the quality of atoms emitting particles (alpha and beta radiation) or electromagnetic waves (gamma radiation) in disintegration of atomic nuclei. Radioactive radiation was first found by *A.H. Becquerel* (1852—1908) when he studied uranium salts. Radioactivity is a natural property of all chemical elements having atomic number above 83, generating in three naturally occurring series, i.e. in decay series of thorium-232 (^{232}Th), uranium-235 (^{235}U), and uranium-238 (^{238}U). Radioactivity is induced by cosmic radiation of some lighter elements, such as carbon (^{13}C, ^{14}C) and potassium (^{40}K), and artificially in nuclear reactors also of other elements. There are 81 stable and 11 radioactive elements in nature, and these can produce more than 1,000 radioactive isotopes (**radioisotopes** i.e. **radionuclides**). The radioactivity of an element disappears in the course of time. The time during which radioactivity decays to half is called the **radioactive half-life** (**$T_{1/2}$**); known half-lives vary from 10^{-7} seconds to 10^{10} years (e.g. ^{232}Th 1.4 · 10^{10} years). The half-lives of some common radionuclides used in laboratories are: ^{3}H 12.26 y, ^{14}C 5,730 y, ^{32}P 14.2 d, ^{33}P 25 d, ^{35}S 87 d, ^{86}Rb 18.7 d, ^{125}I 60 d, ^{131}I 8.04 d.

In organisms, all forms of radioactive radiation disturb biochemical reactions damaging and transforming cellular structures by ionization of substances (called also **ionizing radiation**), e.g. resulting in mutations and cancer. The **biological half-life**, which is the time required for an organism to eliminate one half of the radioactive substance, is short since the radioactive substances may leave tissues through transpiration, respiration, excretion, etc.; e.g. the radioactive half-life for caesium-137 (^{137}Cs) is 30 years, but its biological half-life in man only from 50 to 200 days. Radioactive substances spread into the natural environment by man, are a serious problem. *See* radium.

radioautography, → autoradiography.

radiocarbon dating, *see* carbon dating.

radioimmunoassay, RIA, a quantitative immunological technique based on the competition between radioactively labelled and unlabelled compounds in antigen-antibody reactions. It is a very sensitive method for detecting and measuring substances, especially different kinds of proteins, using radioactively labelled specific antigens or antibodies.

Radiolaria (L. *radiolus* = small sunbeam), **radiolarians**; a diverse group of protozoans in the superclass Actinopoda, phylum Sarcomastigophora. These animals live in oceanic plankton; cell bodies are spherical and radially surrounded by siliceous needles (spicules). In oceans dead and fossil radiolarians form a prominent part of the bottom mud; important in flint formation. *See* Sarcodina.

radioulna (L. *radius* = ray), a bone in the forearm of some tetrapods, such as frogs, formed by fusion of the radius and ulna.

radium, a radioactive element, symbol Ra, belonging to the scarce alkaline earth metals in the nature, found e.g. in pitchblende (uraninite); 16 isotopes are known, of which the isotopes 223, 224, 225, and 226 are naturally existing, mainly in uranium metals. The most stable form, radium-226 (^{226}Ra), atomic mass 226.025, half-life 1,620 years, emits alpha particles and gamma rays forming radon. It is one element in the decay series in which uranium-238 (^{238}U) is stepwise converted into stable lead-206 (^{206}Pb). In this series, ^{226}Ra is converted to radon-222 (^{222}Rn) in which process alpha particles and gamma rays are emitted. Radium is used e.g. in tumour therapy. *See* radioactive series, radioactivity, ionizing radiation.

radius, pl. **radii** (L.), a line radiating from a centre; e.g. **1.** one, usually the thinner, of the two bones of the forearm in tetrapod vertebrates; *cf.* ulna, radioulna; **2.** a main longitudinal vein in the forewing of insects; **3.** radius of a circle.

radix, pl. **radices** (L.), root, base; e.g. *radix dentis*, the root of a tooth.

radon, an unstable, radioactive element, the heaviest of the → noble gases, symbol Rn; the most long-lived of its 20 known isotopes is radon-222, ^{222}Rn (half-life 3.823 days) formed in disintegration of radium-226, ^{226}Ra. Radon is formed in the bedrock when uranium decays in the series from uranium-238 (^{238}U) to lead-206 (^{206}Pb), and it is transferred with moist air into houses and thereby also to human tissues; it is the main radiation source in many geographical areas. The proportion of radon (approx. 4 mSv/year) is about 65% of the annual radiation dose to which the populations of the Nordic countries are exposed; regionally the dose varies much.

radula, pl. **radulae** (L.), a rasping tongue; a lamellar organ with several chitinous teeth in the mouth of molluscs, except bivalves and most solenogasters; its function is to rasp off fine particles of food material.

raffinose, a slightly sweet trisaccharide of plants, $C_{18}H_{32}O_{16}$, synthesized from galactose, glucose, and fructose; occurs abundantly in cottonseed meal (about 8%), in many cereals and many trees, as in the Australian eucalyptus. Its concentration increases in many plants during cold acclimation. *Syn.* melitose, melitriose.

rain forest, a type of evergreen forest typical of the tropical vegetation zone with high annual rainfall (*see* tropics). Approx. 90% of the vegetation mass is composed of trees. The vegetation consists of many layers, but the earth surface is mostly bare due to the lack of light. The soil is effectively weathered and the decomposition of organic material is very rapid. The largest rain forests exist in the areas of the Amazon and Congo rivers but are found also in Indonesia, Australia, Polynesia, and the Caribbean area. The plant and animal diversity is enormous in rain forests; it has been suggested that some 90% of all species live there, and thus, they are irreplaceable. Logging diminishes rain forests continuously; the loss in the late 1980's was estimated to be about 200,000 km^2 a year, in the late 1990's still about half of that. The devastation of forests also makes the → greenhouse effect more effective. Although some or some parts of these forests are capable of renewal, their situation is considered to be endangered.

Rajiformes (L. *Raja* = the type genus of the ray, *forma* = form), **rays, skates**; an order of cartilaginous fishes (subclass Elasmobranchii), comprising oceanic species feeding on bottom animals, such as molluscs, crustaceans, and small fish. The body of rays is flattened, with greatly enlarged pectoral fins, the tail is long and whip-like; dorsal and pelvic fins are small or absent. The order includes about 315 species.

Ramapithecus (*Rama* = Hindu epic hero, Gr. *pithekos* = ape), a genus of a primitive fossil hominid which inhabited open forest areas in Asia and south-eastern Europe in Miocene, about ten million years ago; may be ancestral to the present hominids.

ramet (L. *ramus* = branch), a potentially independent member of a → clone, e.g. modules of an asexual → modular organism which is formed by vegetative growth; usually refers to tillers and shoot modules of plant clones. *Cf.* genet.

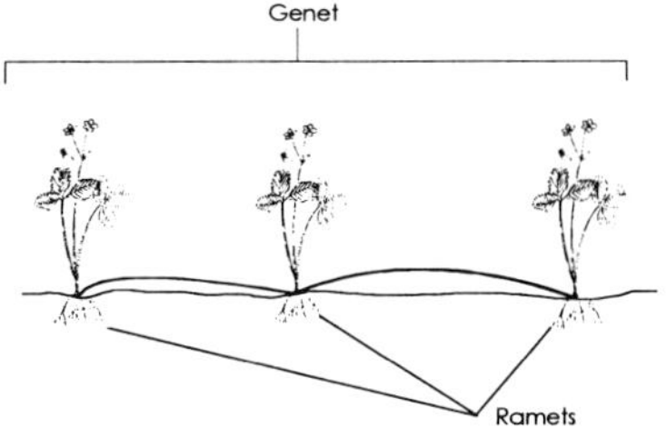

RAMETS OF STRAWBERRY

ramulus, pl. **ramuli** (L. dim. of *ramus* = branch), a small branch, branchlet. *Adj.* **ramulous**, **ramulose**, having many small branches.

ramus, pl. **rami** (L.), a branch of a plant, vein, nerve, bone, etc. *Adj.* **ramose**, consisting of, or having branches; **ramous**, ramose, or relating to branches.

random amplified polymorphic DNA, RAPD, a method used for detecting → genetic polymorphism revealed by using randomly chosen → primers to amplify DNA in the → polymerase chain reaction; makes use of non-specific PCR reaction to uncover genetic variability. In this method short → primer sequences, 8 to 12 base pairs long, are annealed to generate a ladder of bands resembling a → DNA fingerprint. Profiles compared between individuals or species, often show differences. Polymorphism is thought to be due largely to the gain or loss of primer binding sites. The method is useful because it does not require any previous knowledge of the genome of the organism under study (shorthand RAPD).

random dispersion/distribution, *see* dispersion.

random mating, a mating system in which each individual of a population has an equal probability of mating with any individual of the opposite sex.

random sampling, a → sampling method used e.g. in ecological field studies where sample areas or points on the study range are chosen randomly, e.g. using random-number tables, so that the results are statistically valid. *Cf.* systematic sampling.

Ranvier's nodes, *see* nodes of Ranvier.

RAPD, → random amplified polymorphic DNA.

raphe, also **rhaphe** (Gr. *rhaphe* = seam), **1.** a seam, gap, or border line in an anatomical structure of animals; e.g. *raphe medullae oblongatae* in the brain; **2.** a longitudinal ridge in → seeds formed from anatropous → ovules, marking the position of adherent → funiculus; **3.** in → diatoms (order Pennales), a median longitudinal slit in the valve, where the cytoplasm streams; is associated with the movement of these organisms.

raphide (Gr. *rhaphis* = needle) a needle-shaped crystal formed from calcium oxalate occurring in some plant cells. *See* crystals.

rare gases, → noble gases.

RAS, → reticular activating system.

rate-zonal centrifugation, *see* centrifugation.

Rathke's pouch/pocket (*Martin Rathke*, 1793—1860), an outgrowth of the stomodeal ectoderm of the vertebrate embryo, developing into the anterior lobe of the → pituitary gland. Called also craniobuccal pouch, neurobuccal pouch, or pituitary diverticulum.

ratites, a group of flightless birds with small or rudimentary wings and no keel on the sternum; at present the group includes **ostriches**, **rheas**, **cassowaries**, **emus**, and **kiwis**; extinct species are e.g. moas and elephant birds. *See* Palaeognathae.

Raunkiaer's life forms (*Christer Raunkiaer* 1860—1938), groups in vegetational classification based on the position of perennating → buds in relation to soil level, showing how the plants pass the unfavourable season.

Phanerophytes are woody plants with perennating buds clearly above soil level (more than 25 cm), and usually above snow level. **Chamaephytes** are woody or herbaceous plants with overwintering buds near but above soil level (less than 25 cm), utilizing thus the shelter given by snow. **Hemicryptophytes** are herbs with overwintering buds at soil level, and **geophytes** have their buds below ground; the buds of **hydrophytes** are correspondingly situated in water and **helophytes** have them in mud. **Therophytes** are annual herbs and survive the unfavourable season in the form of seeds. *See* life form spectrum.

ray (L. *radius* = spoke, beam), **1.** a beam of light, radiation, or a ray-like line, e.g. beta, gamma, or ultraviolet rays; **2.** one of the long structures at the distal end of the limb of the vertebrate embryo, developing into the metatarsal bones and phalanges; **3.** one of the cartilaginous (in sharks and rays) or bony spines supporting the fins of fishes; **4.** any cartilaginous fish in the order → Rajiformes; **5.** a plant stem structure, functioning in the transport of materials between the pith and cortex; found in → Gymnospermae and → Dicotyledonae. They are **primary rays** in a young stem, i.e. in non-differentiated areas between separate → vascular bundles. After the beginning of → secondary growth they are **secondary rays**, formed by → parenchyma cells in rows. If the secondary rays are situated in the area of the → secondary xylem, they are called **xylem rays**, and in the area of the → secondary phloem they are called **phloem rays**.

ray-finned fishes, → Actinopterygii.

ray initial, a cell in the vascular → cambium of plants, giving rise to horizontally oriented parenchymatous ray cells. *See* ray. *Cf.* fusiform initial.

rays, → Rajiformes.

R-banding (*R* < reverse), **R-banding stain**; a differential staining of chromosomes to bring out certain chromosome bands, called R bands, which become visible with Giemsa stain in hot alkali (pH 9.0) as intervening regions of coloured G bands. The target of the Giemsa stain in this method is not known. DNA in the regions of R bands replicates at the beginning of the synthetic phase, and contains more guanine and cytosine base pairs than DNA in the regions of → G-banding. The method is used in identification of specific parts of the chromosome and in the location of the sites of chromosomal breaks and rearrangements.

RCF, relative centrifugal force. *See* centrifuge.

rDNA, → ribosomal DNA.

re- (L.), **1.** again, anew; **2.** back, backward.

reaction (L. *reagere* = to react), **1.** reverse action, counter action; **2.** in physiology, a response of a cell or its part, or tissue, organ, or organism to stimulation; **3.** in psychology, an emotional state elicited in reponse to a stimulus; **4.** in chemistry, the interaction of substances resulting in chemical changes in them; i.e. the process in which the linking and detaching of atoms, radicals, and molecules are based on interaction between the outermost rings of electron sheaths in formation of chemical bonds; **5.** in physics, a) a force equal in magnitude but opposite to the direction of some other force; b) nuclear reaction, i.e. the process involving changes in atomic nuclei.

reaction wood, the structurally anomalous → secondary xylem, which is formed in bent stems and twigs of trees. The reaction wood of foliage trees, called **tension wood**, is formed on the upper side of the structures, i.e. on the side of tension. This type of wood is tight and contains more cellulose and less lignin than normal wood. The reaction wood of coniferous trees (**compression wood**) is formed on the lower side of bent structures, i.e. on the side of pressure. This kind of wood is heavy and hard, and contains, in contrast to tension wood, more lignin and less cellulose than the typical xylem.

reading frame, the sequence of the → nucleotides of DNA or RNA read as triplets; one of three possible ways of reading a nucleotide sequence as a series of triplets. The reading frame is **open** (ORF, open reading frame) when it starts with an → initiator codon, which is followed by a sequence of codons coding for amino acids. An open reading frame ends with a → nonsense codon. The reading frame is **closed** if it contains several → nonsense codons thus being unable to code for a polypeptide of full length.

reading frame shift mutation, a gene mutation due to insertion or → deletion of one or more (but not three or a multiple of three) nucleotides in the coding DNA sequence. This type of mutation results in a complete alteration of genetic information because in genetic translation, all the triplet codons from the mutation point onwards are erroneous. *Syn.* frameshift mutation.

reagent (L. *re-* = back, again, against, *agere* = to act), a common name for any substance used to produce a chemical reaction in a chemical or biochemical process; reagents are used e.g. for detecting, measuring, and examining other substances.

realized niche, *see* niche.

reassociation kinetics, → renaturation kinetics.

reaumur, R (*René Réaumur*, 1683—1757), a thermometric scale with the freezing point of water at zero (0°R) and the boiling point at 80°R, under standard atmospheric pressure.

recapitulation (L. *recapitulatio* < *capitulum* = number of heads), an old theory stating that the ontogenic development of an individual passes through stages which represent adult phases of ancestral species from which the individual has descended, i.e. the → ontogeny is a recapitulation of the phylogenetic development of a species. The theory as such is rejected but the idea is presented in the → biogenetic law. *Syn.* Haeckel's postulate.

Recent, → Holocene.

receptacle (L. *receptaculum* = store), **1.** one that receives and contains something, e.g. → spermatheca; **2.** the apex of a → flower stalk bearing the flower parts. *Syn.* torus.

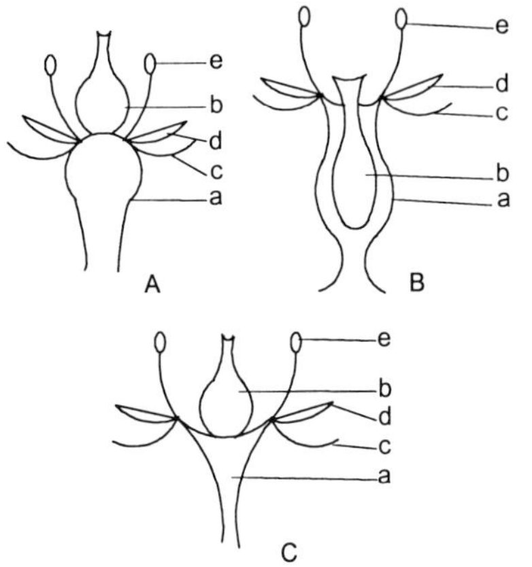

Receptacle types in plants: hypogynous (A), epigynous (B), perigynous (C); receptacle (a), ovary (b), sepal (c), petal (d), stamen (e).

receptaculum (L.), container, e.g. **1.** receptaculum chyli; → cisterna chyli; **2.** receptaculum ovarum (ovisac, egg sac), an internal bag associated with the ovary to store ova in many oligochaetes; **3.** receptaculum seminis, seminal receptacle, → spermatheca.

receptor (L. *receptio* = reception), **1.** a molecular structure, usually a protein, within or on the surface of a cell, inside the cell, or in interstitial structures, characterized by selective binding ability to a specific substance (*see* lock-and-key model); e.g. receptors for hormones, transmitters, antigens, and immunoglobulins. The binding of the specific substance is followed by a specific reaction. The receptor may be occupied also by many synthetic substances with selective binding ability either eliciting the reaction (agonistic action) or inhibiting it (antagonistic action); *see* adrenergic receptors, cholinergic receptors; **2.** sensory receptor, a receiving structure for a certain sensory stimulus; a receptor cell (a special neurone or epithelial cell), or certain differentiated parts of the cells; *see* receptor potential, sense organ.

receptor potential, generator potential; a graded initial depolarization of → membrane potential observed in a sensory receptor (cell) as a response to a stimulus; produces an → action potential in the sensory neurone when the threshold depolarization level is reached. The receptor potential is relatively low (1 to 30 mV), and results from the change in ion permeability of the cell membrane exposed to a stimulus. *Cf.* excitatory postsynaptic potential.

recess (L. *recessus* < *recedere* = to draw back), an empty space, cavity, or hollow in an anatomical structure; it may be normal, such as the cochlear recess of the vestibule of the vertebrate ear, or it may be abnormal; **recession**, the process in which a tissue draws away from its original position.

recessive, 1. drawing away, receding; **2.** in genetics, pertaining to → a recessive character.

recessive character, any of genetically regulated characters and the gene → alleles responsible for their production, in the case when this character is expressed in the → phenotype only if these alleles are in homozygous condition in the genotype (*see* homozygote). The opposite to recessive charcter is → dominant character. If two pure forms are crossed, one representing the recessive and the other the dominant character, the recessive character does not appear at all in the first (F_1) generation, but in the second (F_2) generation in one quarter of the progeny.

recipient (L. *recipere* = to receive), one who receives; receiver; e.g. **1.** an organism to which a tissue is transferred from another organism (donor); **2.** in ecology, e.g. a water body into which sewage is released.

reciprocal (L. *reciprocus*), returning the same way, alternating; in genetics, pertaining to **1. reciprocal crosses**, complementary to each other, e.g. two crosses of fruitflies, e.g. red-eyed female x white-eyed male (first cross) and white-eyed female x red-eyed male (reciprocal cross); also pertaining to a cross between two hermaphrodite individuals in which the male and female sources of the gametes, used for fertilization, are reversed, i.e. reciprocal; **2.** two **reciprocal chiasmata** of two → chromatids (two-strand double crossing-over); **3. reciprocal** → **recombinants** arising from the same or similar → crossing-over; reciprocal recombination is the production of new genotypes with the reverse arrangements of alleles according to maternal and paternal origin; **4. reciprocal** → **translocation**, involving an exchange of parts between non-homologous chromosomes.

reciprocal ohm, mho, *see* conductance, Appendix 1.

reciprocal predation, an interaction between two species (or individuals) in which both prey on each other and thus compete with each other; e.g. among many insects, reciprocal predation is seen between two species when both adults and larvae eat both eggs and pupae of each other. *See* competition.

recombinant (L. *re-* = back, again, *combinatio* = combination), an individual or cell arising from the → genetic recombination between chromosomes or within a chromosome; the individual has a genotype different from that of either parent.

recombinant DNA, a DNA molecule constructed in gene manipulation by joining experimentally DNA from more than one source; occurs naturally as a result of crossing-over during → genetic recombination.

recombinant DNA technology, the methods applied to replicate (clone) recombinant DNA molecules in appropriate host cells in order to produce large amounts of DNA or their products, or genetically modified cells and organisms.

recombinant protein, an artificially produced protein that consists of parts of two or more natural proteins encoded by different genes.

recombination, *see* genetic recombination.

recombination frequency, recombinant frequency, the number of the offspring (**recombinants**) that arise as a consequence of → genetic recombination divided by the total number of progeny, i.e. the relative proportion of recombinant progeny. Such frequencies indicate the relative distances between loci on a → chromosome map.

recombination nodules/nodes (L. *nodulus*, dim. of *nodus* = knot), tight spherical structures observable in the → synaptonemal complex of chromosomes, determining most likely the site of → crossing-over.

recombination-repair, the repairing of damage (gap) in DNA in a process resembling → genetic recombination.

recon (< recombination), the smallest unit of DNA capable of → genetic recombination; structurally equivalent to one nucleotide of DNA.

recruit (L. *recrescere* = to grow again), a number of new individuals settled in a population by birth or immigration; **recruitment**, the process of recruiting.

rectal (L. *rectus* = straight), pertaining to the → rectum.

rectal gills, *see* rectum.

rectal glands, glands attached to the rectum; e.g. paired glands in the hindgut of some starfishes or excretory salt glands in the hindgut of some fishes, such as sharks and other elasmobranchs; serve in water and ion regulation.

rectum, pl. **recta,** Am. **rectums** (L. *intestinum rectum*), the straight terminal part of the intestine of vertebrates, opening via the anus or → cloaca; the term is used also for the analogous part of the invertebrate alimentary canal. The rectum commonly stores faeces and absorbs mainly water and some salts. Aquatic turtles and some invertebrates use the rectum (cloaca) for respiration; some insect larvae, such as the nymphs of some dragonflies, have **rectal gills** in the rectum. The gills serve in respiration but are also important in ion regulation.

red algae, → Rhodophyta.

red blood cell, erythrocyte, a cell type in the blood of vertebrates containing → **haemoglobin** and transporting respiratory gases. The mature form of red blood cells in mammals is mostly a round, non-nucleated, biconcave disc, 6—8 μm in size, but in animals adapted genetically to high altitudes (like llamas) the cells are much smaller and oval in shape. The red cells of other vertebrates are mostly oval and nucleated, and in poikilothermic animals generally larger than in birds and mammals. Red blood cells are produced in the red bone marrow, and during embryonic and foetal development also in the liver and spleen; originally the stem cells come from the yolk sac. The production of these cells is hormonally controlled by **erythropoietin**, produced especially by the kidneys. In human blood, the number of red blood cells is about 5 million per mm^3 (= 5×10^{12} per litre) and they live on average 4 months. Normally there are also immature red cells (**reticulocytes**) in the blood. Old, worn cells are decomposed in the spleen, and the haemoglobin is metabolized in the liver. The glycoproteins on the surface of the red blood cells act as surface antigens specifying the → blood group.

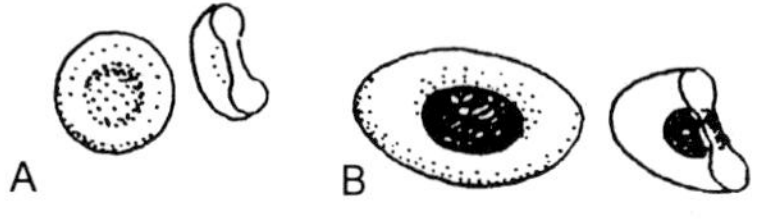

Examples of mature red blood cells; mammalian cells (A) lack the nucleus; in other vertebrates, the nucleus lies in the middle of the red blood cell (B).

Red blood cells contain **carbonic anhydrase**, i.e. the enzyme which catalyses the reaction $H_2O + CO_2 \rightleftharpoons H^+ + HCO_3^-$, enabling the exchange of carbon dioxide (CO_2) in the lungs and body tissues. Red blood cells are also called **red blood corpuscles**. *See* haemolysis.

red fibres, Am. **red fibers,** highly aerobic skeletal muscle fibres of vertebrates with numerous large mitochondria, high myoglobin content, and slower contractions than the more anaerobic → white fibres. Muscles built chiefly of red fibres are called **red muscles**; e.g. most posture muscles with a dense net of blood capillaries.

red-green colour blindness, *see* colour blindness.

redia, pl. **rediae** (*Francesco Redi*, 1626—1698), the third larval stage of most fluke species (digenean trematodes parasiting commonly in snails); the larva has a pharynx, saclike intestinum and foot-like posterior appendages, and it develops parthenogenetically into the → cercaria stage. *Cf.* metacercaria, miracidium, sporocyst.

redirected behaviour, occurs in situations when normal behavioural reactions of an animal are pent up as a result of conflict, and the animal directs its behaviour at substitute objects; e.g. a bird furiously pecks anything, instead of a competitor.

red muscles, *see* red fibres.

red nucleus (L. *nucleus ruber*), an oval pink area (nucleus) of → grey matter in the upper midbrain of vertebrates (the colour comes from an iron-containing pigment); as a part of the → extrapyramidal system it functions in the control of locomotor activity, having connections e.g. with the cerebellum, brain stem, and spinal cord.

redox potential, reduction-oxidation potential; an electrochemical measure in volts for oxidizing/reducing power of a chemical element, i.e. a power to release or accept electrons, referred to the potential of a hydrogen electrode at absolute temperature; redox potential is calculated from the equation:

$E_h = E_0 + RT/nF \ln$ (oxidant)/(reductant), where R = gas constant, 8.31434 J/K · mol; T = absolute temperature (in kelvins); n = number of electrons transferred; F = Faraday constant (*see* Appendix 1); E_0 = symbol for the potential of the system at pH 0.

redox reaction, an reduction-oxidation reaction, in biological systems usually catalysed by enzymes. In the reaction an oxidizing agent is reduced and a reducing agent is oxidized, involving the transfer of electrons from an electron donor (reducing agent) to an electron acceptor (oxidizing agent); correspondingly, sometimes hydrogen atoms are transferred.

redox zone, a zone in a water body beneath the oxygen-rich upper zone; rich in hydrogen sulphide participating in → redox reactions; common in eutrophic waters such as many lakes and the Baltic Sea.

Red Queen hypothesis, a principle concerning an evolutionary process in which a species must evolve as fast as possible to avoid extinction. If its competitors can adapt to exploiting environmental resources more efficiently than earlier, this increase in the effectiveness reduces the resources available to other species. The increase in the → fitness of one species thus means the impairment of the environment regarding the other species, causing continuous competition between the species. *Syn.* Red Queen effect.

The hypothesis was named after a character in the novel Alice Through the Looking-Glass, by *Lewis Carroll* (1832—1898). In the story the Red Queen, because of the peculiarity of the world, has to run constantly as fast as possible simply to stay in the same place.

reduction (L. *reducere* = to reduce, bring back), the act of reducing or repositioning; e.g. **1.** the development of any complex structure or function to a less complex form, e.g. the halving of the number of chromosomes in → meiosis; **2.** in chemistry, the addition of electron(s) or hydrogen to a molecule, ion, or atom, or correspondingly removal of oxygen from a substance; e.g. when ferric iron (Fe^{3+}) changes into the ferrous state (Fe^{2+}), or hydrogen is added to a double bond; *cf.* oxidation; **3.** in philosophy, the returning of a doctrine to its basis, such as a biological process to chemistry and physics.

reduction division, usually the first division of → meiosis; a phase of cell division in which the chromosome number is reduced from → diploid to → haploid, i.e. to half that of the parent cell. In reduction division, whole chromosomes separate in the first anaphase, and divide in the second meiotic division. In a few cases the chromosomes divide in the first meiotic division, and the → chromatids derived from the chromosomes, separate in the second meiotic division. In this case the second meiotic division is the reduction division.

reflectance (L. *reflectere* = to bend back), the ability to reflect back; determined as the ratio of the reflected radiation to the total radiation incident on a surface; e.g. sound or light; called also reflection coefficient; **reflection**, a bending back, or a state of being reflected.

reflex, 1. reflected or directed back, as e.g. sound, light, heat, etc.; **2.** in animals with a proper nervous system, an involuntary neuromuscular or neuroglandular action in response to a stimulus. The reflexes function in a single organ, such as in secretion of a gland or involuntary movement of a single muscle, or at the level of an organ system or whole animal, as e.g. in actions of alimentary, respiratory, or circulatory systems, or behaviour. The neuronal basis is similar in various reflexes having the **reflex arc** as a principal unit. The stimulation of a peripheral sensory neurone induces the action of an autonomic or motor neurone in the central nervous system, thus causing an action in an effector organ, i.e. in a gland or a smooth muscle (**autonomic reflex**) or in a skeletal muscle (**somatic reflex,** *see* reflex movement); e.g. excretion of a salivary gland, or a patellar reflex, respectively. Most reflexes are **innate** but some reflexes can be modified by the central nervous system; learning usually alters the response (**conditioned reflexes**). *See* fixed-action patterns.

reflex movement, motor reflex; the quick neuromuscular response to an external or internal stimulus, based on the function of a **reflex arc** (reflex circuit) which in its simplest form has a peripheral sensory neurone connected via the central nervous system to a motor neurone (**monosynaptic reflex arc**). Usually there are regulatory interneurones between the sensory and motor neurones (**polysynaptic reflex arc**). An example of reflex movements of skeletal muscles (**somatic reflexes**) in vertebrates is the **stretch reflex** appearing e.g. as the patellar reflex which is a monosynaptic reflex arc induced by a stretch stimulus in → muscle spindles in the quadriceps muscle of the thigh. The stretch reflexes function in all skeletal muscles for the maintenance of muscle tonus, and play a role in the control of muscle → shivering. Nociceptive **flexor reflexes** and **postural reflexes** are polysynaptic and associated with the avoidance of harmful stimuli and the function of the equilibrium sense in the inner ear.

In the broader sense, reflex movements include also reflexes of smooth muscles (**autonomic reflexes**), as e.g. the **pupillary reflex** controlling movements of the pupil in response to light. The **defecation reflex** and **vomiting reflex** include both somatic and autonomic reflexes.

reforestation, the establishing of forests in treeless areas by natural seeding or artificial planting; is carried out for wood production, and/or to replace vegetation lost e.g. through erosion, flood, or desertation.

refraction (L. *refringere* = to break apart, to refract), the deviation of radiation in passing obliquely from one medium to another of different density, e.g. the refraction of light by the eye. The rays coming to the refracting surface are called **incident rays** and after penetrating the surface, **refracted rays**. *Adj.* **refractive**, pertaining to refraction, or ability to refract.

refractory period, in a cell, organ, or organism, the time of reduced irritability after which it is able to become stimulated again; e.g. **1.** in the neurone, a period (often about 1 millisecond) of repolarization of the → membrane potential comprising **absolute** and **relative** refractory periods, the former lasting about 2/3 and the latter 1/3 of the whole period. During the relative period an exceptionally strong stimulus can induce a new action. Similar refractory periods occur in muscle fibres, motor units, or in a whole muscle which e.g. in the human cardiac muscle may be several hundred milliseconds; **2.** the resting period associated with decreased sexual activity and regressed gonads in many animals after intensive reproduction.

refuge (L. *refugere* = to flee away), in models of population ecology, a limited area (or time) where a prey individual or species may avoid a predator; in practice the refuge is always partial (**partial refuge**) which means that predation pressure, on average, is slighter in the refuge than in its surroundings.

refugium, pl. **refugia,** usually a limited, isolated area where a species, genus, family, or community, previously widely distributed, have endured environmental changes, such as glaciation; typical refugia in Northern Europe have been **nunataks** during the Ice Age.

regeneration (L. *re* = again, *generare* = to produce, generate), the renewal of the structure of an animal by regrowth, e.g. the renewal of a lost tissue or the whole organ or limb. Many invertebrates, such as cnidarians and many worms (e.g. planarians), are capable of regenerating the entire animal from a small part of the body, while e.g. in adult insects the regeneration ability is minimal. In vertebrates, regeneration is mostly limited to the skin (healing of wounds), although amphibians and reptiles have the ability to replace whole organs; e.g. some lizards can remove the tail by **autotomy**, but the new organ is not quite the similar to the old one (**heteromorphosis**). In some mammals, as e.g. in rat, the liver regenerates quite well after injury, growing usually to normal size (**compensatory hypertrophy**), and to some extent muscular tissue is also capable of regenerating. The ability to regenerate is usually based on the existence of certain undifferentiated cells in the tissue (such as the satellite cells in the skeletal muscles of vertebrates) which begin to divide (proliferate) and differentiate in a damaged tissue (**epimorphosis**). Some renewal of lost tissue may also occur by reorganization of the remaining tissue involving extensive cell movements, but without cell proliferation (**morphallaxis**).

In plants, regeneration is very common. In vascular plants, it usually begins by the formation of secondary meristem or the production of adventitious buds; new individual plants can regenerate e.g. from an isolated cell, a piece of tissue, or from a → protoplast. Regeneration is used in plant propagation.

regio alpina (L. *regio* = district, region), alpine mountain zone. *See* height zones.

region, 1. a sub-area in biogeographical realms (called also **subregion** when the realm and region are considered as synonyms); the regions are characterized by flora and fauna different from those in other areas; e.g. arctic, subarctic regions, Amazonian region; **2.** anatomical region; e.g. orbital region, *regio orbitalis*, the region around the eye.

regression (L. *regressio* < *regredi* = to go back), the act or process of going back; a shift towards an average or to a lower or less perfect state of function, differentiation, or development; e.g. **1.** a gradual loss of specific differentiation in a body part, such as menopausal regression of ovaries; **2.** the reversion of an organism in successive generations to a general or less advanced form; **3.** in psychology, the mental reversion to infantile patterns of thought or behaviour; **4.** a functional relationship between a dependent or random variable and another, independent variable (or several variables); graphically expressed by a regression curve, i.e. the regression of Y on X. *Verb* to **regress**. *Adj.* **regressive**.

regular dispersion/distribution, *see* dispersion.

regular sampling, → systematic sampling.

regulated genes, genes that act only under certain developmental stages and special conditions prevailing in a cell or tissue, in contrast to constitutive genes which act continuously. *Cf.* constitutive genes. *See* genetic regulation.

regulation (L. *regula* = rule), the act or state of adjusting; in organisms involves the adjustment and coordination of biochemical, physiological, genetic, and behavioural reactions serving the adaptation of different functions to environmental demands. All organisms, being **open systems**, have the tendency for **autoregulation** even in their simplest biochemical reactions, because concentrations of compounds and energy relations determine the equilibrium state of reactions. In more complex systems the qualitative and quantitative regulation of the activity of a reaction pathway involves feedback control. A product of the pathway affects its overall activity by inhibiting (**negative feedback**) or stimulating (**positive feedback**) at a particular step in the pathway. In organs and organ systems regulation is based on the actions of various messenger substances, such as hormones and → second messengers, in animals also on the function of the nervous system. *Adj.* **regulative**, **regulatory**. *See* homeostasis, genetic regulation, population regulation.

regulative egg, an egg cell (ovum) in which → differentiation has not yet begun; if the daughter cells, derived from the first cell cleavage, are separated, each will still develop into a complete individual although they normally regulate each other. The eggs of sea urchins, amphibians, and mammals are examples of regulative eggs. *Cf.* mosaic egg.

regulator, one that regulates; e.g. **1.** a hormone or growth factor; **2.** an animal able to regulate a certain physiological function in varying environmental conditions, e.g. homoiothermic animals, regulating their body temperature in varying ambient temperatures, or homoiosmotic animals (osmoregulators), regulating their osmotic concentration in different salinities of natural water bodies; *see* conformers.

regulator(y) gene, a → gene whose product (RNA or a protein molecule) regulates the expression, usually transcription of other genes (*see* genetic transcription). Regulator genes can affect their target genes either negatively (suppression of transcription) or positively (enhancement of transcription).

regulatory sequence, a DNA sequence in the → promoter to which → transcription factors (regulatory molecules) bind; regulates the expression of an adjacent gene.

regulon, 1. in bacterial gene expression a global regulatory system consisting of several → operons which involves the interplay of pleiotropic regulatory domains (*see* pleiotropy); **2.** in eukaryotic organisms, a group of different genes regulated and coordinated by one or more → regulatory genes that constitute a regulatory cascade network.

reinforcement, an increase of strength or force; in behavioural science, an event altering the response of an animal to a stimulus by reward or punishment; in **positive reinforcement** the stimulus is pleasant and increases the intensity of response, and in **negative reinforcement** the stimulus is painful or disagreeable decreasing the intensity of response.

rejection (L. *rejectio* = throwing back), refusal; e.g. graft rejection, i.e. the refusal of an individual to accept a → graft from another individual, due to the immunological incompatibility of a transplanted tissue or organ. *Verb* to **reject**.

relatedness, kinship, the degree of **symmetric** genetic similarity between individuals in a population; the number of → alleles identical by descent, i.e. the number of alleles, which are copies of the same progenitor gene, is used as its measure. In nature, **asymmetric** relatedness between conspecific individuals also exists, meaning that the relatedness of individual A to individual B is different from the relatedness of B to A; this type occurs e.g. in → haplodiploid hymenopterans.

relative atomic mass, atomic weight. *See* Appendix 1.

relative abundance, *see* abundance.

relative density, in ecology, a measure of population density at a particular time as compared to some other time or to another population; e.g. seasonal or annual fluctuations in population densities. The values are presented as indices showing the relative changes of densities, but not absolute densities.

relative humidity, RH, the ratio of absolute air humidity to the actual saturation humidity at the same temperature, usually expressed as a percentage.

relative molecular mass, molecular weight. *See* Appendix 1.

relaxant (L. *re-* = back, *laxare* = to loosen), **1.** reducing tension, especially muscle tension; relaxing; **2.** an agent that reduces tension, e.g. a muscle relaxant.

relaxation, the relief from effort, work, muscle tension, or pain; **isometric relaxation**, the relaxation of a muscle without any change in its length.

relaxin, a polypeptide hormone of mammals produced in the corpus luteum especially during pregnancy, causing relaxation of the pubic symphysis as well as softening and dilatation of the uterine cervix; found also in males (at least in man) secreted from the prostate.

release factors, → termination factors.

releaser, sign stimulus; in ethology, a → key stimulus or a part of it which is sufficiently effective to elicit a stereotype response, usually a behavioural action pattern, in a receiver (*see* innate releasing mechanism). Releasers may be either intraspecific, such as courtship stimuli from one sexual partner to another, or interspecific, as e.g. stimuli from predators or prey. Releasers may be colour patterns, various body structures, voices, motions, etc. Animals often communicate using releasers (**social signals**). In humans, intraspecific releasers are e.g. smiles and many other facial expressions. *See* supranormal releaser.

releasing hormones, RH, also **releasing factors, RF,** peptide neurohormones of vertebrates secreted by certain neurosecretory cells in the hypothalamic area of the vertebrate brain. The hormones are transferred in blood to the anterior lobe of the → pituitary gland, where they stimulate the release of anterior pituitary hormones (tropic hormones) into blood. Releasing hormones, five altogether, are also called liberins; they are **corticotropin-releasing hormone**, **CRH** (corticoliberin), **thyrotropin-releasing hormone**, **TRH** (thyroliberin), **growth hormone-releasing hormone**, **GRH** (somatoliberin), **gonadotropin-releasing hormone**, **GnRH** (gonadoliberin, luteinizing hormone-releasing hormone, LHRH, luliberin), and **prolactin-releasing hormone**, **PRH** (prolactoliberin). PRH may, however, actually be the thyrotropin-releasing hormone or some other peptide known earlier. *See* statins.

relic (L. *relictus* < *relinquere* = to leave behind), **1.** something that is left behind after disintegration decay, e.g. a part of a body structure; **2.** → relict.

relict, 1. a species or population inhabiting an area from a previously much wider distribution range; **glacial relicts** are e.g. the fourhorn sculpin and seals, known in the Baltic Sea from the period of late Ice Age; **2.** as an adjective, pertaining to such a relict species or population.

REM sleep, rapid-eye movement sleep, *see* sleep.

remex, pl. **remiges** (L. *remex* = oarsman), one of the primary or secondary wing feathers of a bird. *See* primaries, secondaries.

removal method, a method to estimate a population density by removing a known number of animals from the population by successive captures; e.g. used in studies on fish and small mammals. *See* capture-recapture methods.

ren(i)-, reno- (L. *ren* = kidney), denoting relationship to the kidney. *Adj.* **renal**.

renal calix, pl. **renal calices** (L. *calix renalis*), any of the recesses (branches) of the renal pelvis in the metanephros type of the vertebrate kidney, enclosing the renal pyramid; also called the infundibulum of kidney.

renal cortex (L. *cortex renis*), the outer part of the metanephros type of the vertebrate kidney composed of glomeruli and convoluted renal tubules; called also the cortex of kidney, or cortical substance of kidney.

renal medulla (L. *medulla renis*), the inner part of the metanephros type of the vertebrate kidney, consisting of Henle's loops and collecting renal tubules, organized into cone-shaped structures, **renal pyramids**; called also medulla of kidney.

renal pelvis (L. *pelvis renalis*), a funnel-shaped expansion of the anterior end of the ureter into which the renal calices open in the metanephros type of the vertebrate kidney.

renal pyramid (L. *pyramis renalis*), *see* renal medulla.

renal tubule (L. *tubulus renalis*), the long tubular part of a → nephron of the vertebrate kidney composed of different parts, i.e. Bowman's capsule, the proximal convoluted tubule, Henle's loop, and the distal convoluted tubule.

renaturation (L. *re-* = anew, *natura* = origin, nature), the experimental reassociating of complementary single strands of a denatured DNA double helix, or the restoration of the structure of a slightly denaturated protein.

renaturation kinetics (Gr. *kinesis* = movement), the organization of single-stranded DNA molecules experimentally into double-stranded molecules; the process can be used to determine the structure of DNA. *Syn.* reassociation kinetics.

renette (L. *ren* = kidney), a large specialized gland cell with excretory functions in some nematode worms; in general, a worm has only one or two renette cells located ventrally in the pseudocoel near the pharynx.

renewable resources, the resources which an environment replenishes by natural ecological processes at a certain rate. If the replacement and the use of resources remain in balance, the resources exist in principle infinitely; e.g. forests, fish resources, wild animals.

renin (L. *ren* = kidney), a hydrolytic enzyme (EC 3.4.99.19) produced in vertebrate kidneys, converting plasma angiotensinogen into → angiotensin; a factor in the **renin-angiotensin system**, which plays a role in the control of blood pressure. *Syn.* angiotensinogenase. *See* juxtaglomerular organ.

rennin, a hydrolytic enzyme (EC 3.4.23.4) found in the gastric juice of suckling mammals catalysing the cleavage of milk casein (soluble casein) to insoluble casein (paracasein), which then reacts with calcium to form an insoluble curd. The enzyme obtained from the abomasum of calves is widely used in the dairy industry in making cheese. Rennin is found also in some insectivorous plants. *Syn.* chymosin, chymase, pexin, rennase.

Rensch's laws (*Bernhard Rensch*, 1900—1990), stating that: **1.** mammal and bird races have larger litters and clutches in cold climatic zones than the races of the same species in warm zones; **2.** birds have shorter wings (called the **wing rule**) and mammals thinner fur and shorter hair in warm climatic areas; **3.** the shells of terrestrial gastropods are brown in cold areas and pale in warm areas; **4.** the thickness of a gastropod shell (snail) is positively correlated with the intensity of solar radiation and humidity conditions of the environment. *Cf.* Allen's rule, Bergmann's rule, Gloger's rule. *See* geographical variation.

reoxygenation (L. *re-* = back, *oxygenium* = oxygen), returning of oxygen to a tissue e.g. after reopening of blood vessels; generates large quantities of free radicals that may damage the tissue.

repair, in genetics, the restoration of the biological activity of damaged DNA by means of enzymatic mechanisms. *See* repair replication.

repair replication (L. *replicatio* = folding anew, repetition), a DNA repair mechanism in which a cell is able to recognize and repair damage in a double-stranded DNA. In the repair system a damaged region of one strand is enzymatically removed and replaced by a new one which is synthesized using the undamaged complementary strand as a template; e.g. repairs premutations. Repair replication is catalysed by several DNA polymerases.

repellent (L. *re-* = back, *pellere* = to drive), **1.** causing aversion; repulsive; **2.** something driving away, such as a → pesticide used to repel but not necessarily to kill pests; e.g. an **insect repellent**; **3.** able to reduce swellings or tumours, or an agent causing this effect.

repetitive DNA, deoxyribonucleic acid (DNA) existing in many copies in the genome; **highly repetitive DNA** exists in 10,000 to 1 million copies of very short DNA segments and constitutes the satellite DNA, and it is located in the constitutive → heterochromatin of chromosomes; → **middle repetitive DNA** exists in 1,000—10,000 copies of fairly long DNA segments, comprising some gene families like the family of the ribosome RNA genes and histone genes, as well as gene regulatory elements. Highly repetitive DNA is presumably involved in the recognition of homologous chromosomes in chromosome pairing. Gene families consist of genes, the products of which are required in large quantities, and repetitive gene regulatory elements are required for the regulation of many genes. *See* sequence complexity of DNA.

replicase (L. *replicare* = to fold back), any enzyme which is capable of catalysing the replication of nucleic acids, such as the RNA dependent polymerase (EC 2.7.7.48) catalysing the → replication of RNA in RNA viruses.

replicate, **1.** one of several parallel experiments or observations; **2.** folded back, replicated, repeated; **3.** as a verb, to duplicate; to copy itself, as does DNA.

replication (L. *replicatio*), **1.** a reply, a copy; **2.** the duplication of genetic material (nucleic acids: DNA, RNA) such as a copy of the parent molecule, which acts as a template; involves the reading and restoring of genetic information; in replication a new carrier of genetic information is generated; *see* DNA replication; **3.** the duplication of cell organelles or cells; **4.** the repeating of an experiment.

replication fork, the point at which the strands

of the parental DNA double helix are separated so that → replication of new DNA can proceed; is seen as a Y-shaped structure in electron micrographs.

replication slippage, a putative mechanism by which the number of trinucleotide repeats increases or decreases in the genome.

replicator, an entity such as a gene, which can make copies of itself. According to the inventor of this term, *Richard Dawkins*, a replicator must have longevity, fidelity, and → fecundity. *See* interactor.

replicon, a unit of → replication (duplication) of the genetic material; a part of the DNA molecule. Usually the single chromosome of a prokaryotic organism consists of only one replicon, but each chromosome of eukaryotic organisms contains many consecutive replicons, the number and length of which vary according to the physiological state of the cell. Replication starts from the initiation point located in the middle of the replicon and proceeds in both directions towards the end points. *See* DNA replication.

replisome (Gr. *soma* = body), a structure consisting of several protein molecules adhered to the → replication fork of bacteria; performs the → replication of DNA. The replisome consists of DNA polymerase and other enzymes (*see* polymerase).

repolarization (L. *re-* = back, *polus* = pole), the reestablishment of polarity; e.g. in nerve physiology, denotes the return of cell membrane potential to the level of → resting potential after depolarization during → action potential.

reporter gene, a → gene which is joined to the regulative sequence of DNA by using gene technology. Reporter genes encode proteins which are easy to detect, and thus their expression in the cells can be determined. These genes may be associated with any → promoter so that the expression of the gene can be used to assay promoter function.

reporter plasmid, a → plasmid into which a → reporter gene is joined.

repressible enzyme (L. *repressus* = kept back, restrained), an enzyme produced continuously in a cell until its synthesis is repressed by excess of a product (corepressor), that is usually a metabolite or the end product of the reaction or reaction chain in which this enzyme acts. *See* operon.

repression, 1. the act of suppressing, restraining, inhibiting; **2.** the specific suppression of gene expression resulting in the inhibition of protein synthesis (*see* operon); more generally, refers to the inhibition of transcription or translation by binding of a repressor protein to a specific site on DNA or mRNA, respectively (*see* repressor); **3.** an unconscious mental mechanism for keeping unacceptable impulses out of consciousness.

repressor, a regulatory protein which binds to an operator on DNA or RNA thus exerting direct negative control over → gene expression at or near a promoter site; produced by a regulatory gene specific for the operon. *Cf.* activator.

reproduction, generation; the production of a new organism; **1. sexual reproduction** involves regular alternation of → meiosis and → fertilization during the → life cycle of organisms; **2. asexual reproduction (agamic reproduction)**: generation of a new individual without the fusion of gametes, i.e. the development of a new individual either from a single egg or from a group of cells (vegetative reproduction) by mitotic cell fission or budding without meiosis and fertilization. *See* parthenogenesis.

reproduction cost, in ecology, means that an organism by allocating its resources for reproduction (e.g. by increasing clutch size or reproduction frequency) simultaneously brings about a decrease in survivorship and/or rate of growth. Consequently, this decreases potential reproduction in the future. *See* allocation hypothesis.

reproductive allocation, (L. *ad* = to, *locus* = place), the proportion of resource input available to an organism allocated for reproduction in a certain frequency interval. In practice, the allocation is usually measured as the weight or volume of reproductive organs or their products, because the resources used in reproductive behaviour are difficult to measure.

reproductive effort, the resources which an organism or a population invests in reproduction. *Cf.* allocation hypothesis.

reproductive isolation, an isolation regulated by various genetic or environmental factors which prevent successful interbreeding with individuals of different populations or species. Geographical isolation is the main factor in separating populations from each other and thus an important evolutionary factor in speciation. Reproductive isolation mechanisms may be either **premating mechanisms**, such as seasonal and habitat isolation or ethological or mechanical isolation, or **postmating mechanisms**, as e.g. gametic mortality, zygote mortality, hybrid inviability, and hybrid sterility. The isolation mechanisms maintain differences in gene frequencies which have previously arisen by evolution in the course of the geographical isolation of the populations.

reproductive organs, 1. → genital organs (sex organs) in animals; **2.** for plants, *see* gametangium, sporangium, flower.

reproductive output, the production of offspring by an organism or population; e.g. the number of gametes, seeds, eggs, or offspring produced by an individual during a time interval. *Cf.* fecundity, fertility.

reproductive potential, → biotic potential.

reproductive rate, the number of offspring in relation to adult individuals reaching the reproductive age in a population per time unit, usually a generation. In general, the reproductive rate is presented as the number of female offspring per adult female (multiplication index). This is called **net reproductive rate (R_0)** (*cf.* life table). The reproductive rate presents the change in the size of the population during a certain time. If R_0 is permanently < 1, the population is decreasing exponentially, if it is > 1, it is increasing exponentially, and in the case of $R_0 = 1$, there is no change in the population density.

reproductive value, RV, expected total → reproductive output (e.g. number of female offspring) of an organism in a particular age class; consists of **current reproduction** and the **residual reproductive value**, **RRV**, the latter representing the expected reproductive output of an organism during its residual life time. RV is expressed in relation to the total output of the population.

Reptilia (L. *repere* = to creep, crawl), **reptiles**; a large class of poikilothermic tetrapod vertebrates having a body covered by horny scales, plates or scutes; they breathe by lungs, have metanephric kidneys and an incomplete four-chambered heart (complete in crocodilians), and most of them lay amniotic eggs, but some species produce young developed in the oviduct. The class includes the orders: **turtles** and **tortoises** (Chelonia or Testudines), **tuatara** (Rhynchocephalia or Sphenodonta), **lizards** and **snakes** (Squamata, includes three suborders, **snakes**, **lizards**, and **worm lizards**), and **alligators** and **crocodiles** (Crocodylia). About 6,300 species are found on all continents (except Antarctica), forming a small relict from the ancient, rich vertebrate class dominant in Mesozoic. *See* dinosaurs.

repugnatorial glands (L. *repugnare* = to resist), the glands of some arthropods, such as harvestmen and millipedes, which produce offensive secretions, often quinones and phenols, for defence by squirting them at a predator. Called also defensive glands.

repulsion phase (L. *repellere* = to repel), a phase during the alternation of products of → crossing-over in which the dominant → alleles of a double → heterozygote locate in different homologous chromosomes, i.e. *A b/a B*, where *A* and *B* are dominant, *a* and *b* recessive genes. *Cf.* coupling phase.

RER, rough endoplasmic reticulum, *see* endoplasmic reticulum.

resident bird, a bird species staying in the same place throughout the year.

residual reproductive value, RRV, *see* reproductive value.

resilin, an amorphous non-fibrous protein in insect cuticle; very elastic and thus responsible for various spring-like actions of the wings. Resilin is similar to → elastin but does not form fibres.

resin (L. *resina* < Gr. *rhetine* = resin of the pine), **natural resin** is a solid or semisolid, brittle substance, excreted by resin canals in plants, composed of resin acids, esters and → terpenes. These resins are insoluble in water, but soluble in many organic solvents. Resins, which are common in coniferous trees, cover wounds and obviously protect the plant (e.g. rosin in pines). Resins are also found in animals, e.g. shellac in some insects. **Synthetic resins**, used as synthetic plastic material, are produced by polymerization from different materials, e.g. from cellulose or casein. Originally the name synthetic resins was used to describe synthetic substances whose physical properties resembled those of natural resins. *See* excretory and secretory tissues, ion exchange.

resin canal/duct, a tubular intercellular space in the wood or bark of conifers, lined with glandular epithelial cells secreting → resin.

resistance (L. *resistentia*), the act or power of opposing; counteracting force or state; e.g. **1.** the ability of organisms to resist environmental factors such as microbes or chemicals (e.g. drug resistance), or of microbes to resist antibiotics; **2.** vascular resistance (peripheral resistance) opposing the blood flow, or the expiratory resistance opposing the flow of gas from the lungs; **3.** airway resistance to air flow in respiratory pathways; **4.** → stomatal resistance in plants, one of the main factors affecting transpiration and the uptake of carbon dioxide; **5.** electrical or ohmic resistance, *R*, (*see* Appendix 1); **6.** in psychology, opposition to the rise of repressed material into consciousness. *Verb* to **resist**. *Adj.* **resistant**.

resistance factor, R factor; any bacterial → plasmid which carries genes for resistance to many antibiotics; the factors are often transmissible to other bacteria of the same or related species, rendering them resistant as well. Recombination between plasmids can give rise to multiple-resistance plasmids. *Syn.* resistance plasmid, R plasmid.

resistance genes, plant genes which confer resistance to pathogens carrying the complementary avirulence gene.

resolution (L. *resolvere* = to resolve), **1.** a complete return to a normal structure and function, e.g. the regeneration of DNA; **2.** the ability of a microscope to distinguish fine details, depending on the wavelength of the radiation used. The wavelength of visible light is on average ca. 500 nm and the best resolution is about 450 nm; i.e. objects closer to each other will not be resolved separately. Resolution, r = k(λ/NA), in which λ is the wavelength of the light used, k the diffraction constant of light (0,61), and NA the numerical aperture. NA = n x sin(α/2), showing the wideness of the illuminated area available for the objective lens, n being the refraction constant of the medium between the specimen and the objective, and alpha the opening angle of the lens. The resolution will thus improve by shortening the wavelength and increasing n; immersion oil, instead of air, between the specimen and the lens can be used to improve NA and resolution. *Syn.* resolving power.

resonance (L. *resonare* = to echo, resound), **1.** in physics, the vibration state of a system in response to an external stimulus, occurring when the frequency of the stimulus is the same (or nearly the same) as the frequency properties of the system; e.g. the bony resonance cavities amplify sounds produced by animals; **2.** electrical resonance known in atom and nuclear physics. *Adj.* **resonant**. *See* nuclear magnetic resonance, electron spin resonance, mesomerism.

resorption (L. *resorbere* = to swallow, to suck up again), **1.** reabsorption, such as the absorption of filtered glucose or salts back to circulation in the renal tubules; **2.** the loss or lysis of tissue substance in physiological or pathological processes, e.g. the resorption of bone, tooth sementum, dentine, or early foetus in the uterus. *Adj.* **resorptive**. *Verb* to **resorb**. *See* absorption.

resource (L. *resurgere* = to rise again, to lift), a reserve or a new source of supply or support; in ecology, natural resources, → environmental resources.

resource competition, in ecology, competition for → environmental resources. *See* competition.

resource depletion zone, RDZ, in ecology, a zone around a → consumer where resources are reduced; e.g. a prey depletion zone, or nutrient and water depletion zones around the plant roots.

respiration (L. *spirare* = to blow), the metabolic process of living organisms by which molecular oxygen (O_2) is taken in and used for energy release in reactions in which carbon dioxide (CO_2) and water are produced. Broadly, respiration includes three main processes: 1) external respiration, i.e. the exchange of oxygen and carbon dioxide between environment and body fluids, occurring via lungs, gills, tracheae, and through the skin (in many animals is equivalent to **ventilation** or **breathing**); 2) transportation of respiratory gases in body fluids, and 3) internal respiration, the enzyme reactions of energy metabolism in the cell, i.e. → **cell respiration**; includes the glycolytic part of reaction pathways, called → **glycolysis** (anaerobic reactions, *see* **anaerobic respiration**), and reactions of the → citric acid cycle and respiratory chain, in which molecular oxygen is consumed and carbon dioxide and water are produced, and which therefore is called **aerobic respiration**. Often respiration refers only to the exchange of gases between an organism and the environment (the first explanation). → **Photorespiration** decreases the yield of the photosynthetic carbon assimilation.

respiratory centre, Am. **respiratory center,** a brain area in the medulla oblongata of vertebrates controlling breathing; includes two groups of respiratory neurones, i.e. the **dorsal group** that is the source of rhythmic drive via contralateral phrenic neurones, and the **ventral group** which e.g. innervates accessory and intercostal respiratory muscles mainly via the vagus nerve. The presence of separate inspiratory and expiratory centres, as suggested earlier, has not been confirmed experimentally. Also some areas in the pons participate in the regulation of respiration. *See* aortic body.

respiratory chain, the enzyme chain functioning in → electron transfer of cell respiration.

respiratory enzymes, cellular enzymes functioning in energy metabolism. *See* cell respiration.

respiratory metabolism, the enzymatic functions involved in energy metabolism of organisms including the breakdown of glucose and other energy substrates, the use of oxygen (O_2), and the production of carbon dioxide (CO_2), water, and heat. Respiratory metabolism is studied using entire organisms, tissue sections, isolated cells, and mitochondria, or enzymes purified from these structures.

respiratory organs, organs in animals specialized for the exchange of respiratory gases, i.e. oxygen (O_2) and carbon dioxide (CO_2). Some primitive animals, as protozoans, sponges, coelenterates, and most worm groups have no special respiratory organs, but they respire by diffusion through the surface of the organism or through the skin (**cuticular respiration**). Most groups of invertebrates have specialized organs for respiration: aquatic animals respire through the → **gills, gill books, tracheal gills, respiratory trees**, or **water lungs**, and terrestrial invertebrates, such as insects, through the **tracheal system**, spiders through **book lungs** (**lung books**), and lung snails through the **lungs**. Vertebrates adapted primarily to aquatic environments respire through the gills (fish and larval amphibians), or through the lungs (lung fish and adult amphibians), whereas some species, especially frogs, also use the skin respiration. Other tetrapod vertebrates have the lungs for respiration. *See* lung.

respiratory pigments, → blood pigments.

respiratory quotient, RQ, respiratory exchange ratio; the ratio of the volume of carbon dioxide produced to the volume of oxygen consumed (CO_2/O_2) by an animal in a given period of time. When an animal uses pure carbohydrates for its energy metabolism the theoretical RQ value is 1.0, for proteins 0.7 to 0.8, and for fats 0.7, but may exceed 1.0 when carbohydrates convert into storage fats. Usually animals consume a mixed diet, e.g. the RQ in people with a normal diet is about 0.8.

respiratory tree, an organ serving respiration and excretion in sea cucumbers (Holothuroidea); composed of two tubes with numerous branches, opening to the cloaca which pumps sea water in and out.

response (L. *respondere* = to answer, reply), any reaction of a cell, tissue, organ, or organism to a stimulus, such as to a change in the environment. The response may be a change in enzyme activity, hormone secretion, behavioural pattern, or reproduction activity. In ecology, **numerical response**, meaning a response of population density to environmental factors, and **functional response**, seen as a change in behaviour and physiological functions.

resting bud, a plant → bud that is in a resting stage and does not grow. This stage may last tens of years and the buds may still retain their growing capacity. An example is the sprouting of foliage trees from resting buds in stumps after felling of the trees.

resting egg, a thick-shelled egg of many invertebrates which endures drying and freezing for a long time. Exposure to cold or heat may be necessary for further development of the resting egg; e.g. the winter eggs of branchiopods which are laid in the autumn and develop further in the spring.

resting meristem, an intercalary meristem with dividing cells, situated e.g. in nodes of graminoids; causes intercalary growth.

resting nucleus/cell, a cell nucleus or cell not undergoing any dividing phase, such as → mitosis or → meiosis.

resting potential, the electric potential across the cell membrane in the inactive phase of a cell, i.e. when not generating an → action potential (impulse). Resting potential can be measured as a potential difference between the outside and inside of a cell. It is a general property of cells of all organisms but varies from a few tens of millivolts to over 200 millivolts, the inside being negative as compared to the outside. In animal cells the highest potentials are found in nerve and muscle cells (-70 mV to -90 mV). Resting potential is due to **chemical** (ion concentration) and **electrical gradients** (charge of ions) generated across the cell membrane, which is selectively permeable to ions (*see* ion channels). The **sodium-potassium pump** (→ Na^+,K^+-ATPase) transports actively Na^+ out of the cell and K^+ into the cell, the **calcium pump** transfers Ca^{2+} out of the cell or into some cell organelles. Thus the Na^+,K^+-ATPase establishes a sodium gradient across the membrane with the outside containing 10 times more Na^+ and Cl^- than the inside compartment. The Ca^{2+}-ATPase similarly maintains a low calcium concentration inside by pumping Ca^{2+} out where Ca^{2+} is 10^4 times more concentrated. The intracellular compartments contain about 20—30 times more K^+ than the extracellular medium. Unevenly distributed ions tend to move along the concentration and electrical gradients across the cell membrane, and the resting potential is principally due to selective diffusion of the ions, mainly potassium (K^+). This diffusion potential can be calculated using the **Goldman constant-field equation**. The electrical gradient across the cell membrane affects the transport of ions, because a negative potential attracts positive ions and vice versa. This means that at a certain potential, i.e. **equilibrium potential**, the passage of an ion has the same magnitude in both directions. The equilibrium potential can be calculated by → **Nernst equation** if the ion concentration on both sides of the membrane are known. The direction and the **driving force** of a given ion can be calculated by subtracting the equilibrium potential from the membrane potential. If the result is positive and the ion negative, the direction is out of the cell. The activation of the cell changes the permeability of the ions and, in consequence the potential across the membrane will change; this occurs in the generation of an → action potential (impulse) important in nerve or muscle cells, or of a → receptor potential in sensory receptors, or of postsynaptic potentials in → synapses.

The facts above are mainly true also for plant cells. However, the term resting potential is seldom used, because the electric functions in plant cells are slow and rapid changes (such as action potentials) are rare. In plant cells the K^+, Ca^{2+} and Cl^- ions are significant, but the sodium ions mainly not. The H^+ ions are especially important; plant cells typically have an H^+ pump (H^+-ATPase) in their plasma membrane pumping H^+ ions out of the cell, and the gradient formed in this way is used in the uptake of negative ions whose penetration to the negative inside is normally difficult. Many of these ions are important plant nutrients (e.g. $-SO_4^{2-}$, $-NO_3^-$).

resting spore/cyst, a thick-walled → spore able to survive in unfavourable conditions, e.g. zygospore, formed from the zygote in fungi; resting spores exist also in bacteria and algae.

restriction cleavage site, the site in the DNA molecule at which a given → restriction enzyme cleaves the DNA chain, usually a palindrome.

restriction enzymes/endonucleases, a group of → endonucleases found in bacteria which restrict the host range of → bacteriophages, i.e. the ability of the phages to replicate in bacterial cells, by cutting the double-stranded DNA chains of the phage at interior bonds, producing oligo- or polynucleotide fragments of varying size. Hundreds of varieties of these enzymes are known, each recognizing a specific sequence of DNA; they are important in genetic engineering.

restriction fragment length polymorphism, RFLP, a form of → polymorphism in which the individuals of a species differ from each other with respect to the length of certain → restriction fragments, i.e. in some individuals, certain cutting points for a particular → restriction enzyme in the DNA chain of their genome are lacking. The lengths of DNA fragments can be revealed by → Southern blotting, and thus RLFP can be used for genetic mapping.

restriction fragments, cut fragments of double-stranded DNA formed by the action of → restriction enzymes; usually have → cohesive ends, and can thus experimentally be joined to any other DNA fragment produced by the action of the same enzyme.

restriction map, a map describing the internal structure of a → gene or genome; constructed by organizing the → restriction fragments produced by the action of → restriction enzymes into a linear order. The cutting points in DNA for the restriction enzymes occur as map points, and their true distances (map distances) as numbers of nucleotide pairs (base pairs). The restriction fragment mapping may produce complete nucleotide sequences of entire chromosomes.

resupinate (L. *resupinare* = to turn back or up), **1.** inverted in position; pertaining to a structure twisted upside down, as e.g. flowers of many orchids; **2.** denoting the structure of a type of the fruit body in polyporacean fungi, having no protruding part but growing along a substratum, e.g. tree stem or stump.

rete, pl. **retia** (L.), a net, meshwork, especially a network of blood vessels; e.g. **1. rete cutaneum**, the arterial network in the skin of vertebrates between the corium and the tela subcutanea; **2. rete mirabile**, a network formed by branching of an artery or a vein into a net-like formation of smaller vessels, often capillaries, which again reunite into a single vessel, present e.g. in the glomeruli of the kidney. In the limbs of many homoiothermic vertebrates such retia may exist in collateral blood vessels, functioning particularly as a heat exchanger (**countercurrent system**). A system of rete mirabile is found also in the subcutaneous tissue in some poikilotherms, e.g. in tuna fish who can maintain a body temperature about 10°C above the environmental temperature. Many mammalian species, such as artiodactyls and carnivores, have a **carotid rete**, which allows the brain to be selectively cooled. The warm arterial blood is carried to the brain via an extensive rete, a capillary network enclosed within a venous sinus. The sinus receives cool blood from the nasal cavities and cools the carotid blood within the rete before it enters the brain. Many birds have a similar rete (*rete mirabile ophthalmicum*) near the eye. Some invertebrates, such as holothurians, possess corresponding rete-like networks of blood vessels; **3.** *rete testis* (*rete Halleri*), the network of channels at the termination of the straight tubules in the vertebrate testis; called also rete of Haller.

retention (L. *retentio* < *retinere* = to retain), the act or process of retaining; e.g. **1.** the persistency or capacity to retain substances such as excretions within the body; **2.** an ability to keep things in mind. *Adj.* **retentive**, tending to retain.

reticul(i)-, reticulo- (L. *reticulum,* dim. of *rete* = net), pertaining to a reticular structure or resembling a structure of a network. *Adj.* **reticular**.

reticular activating system, RAS, the nonspecific neuronal system of the vertebrate brain ascending from the brainstem → reticular formation and the thalamus to the cerebral cortex (ascending RAS) and descending from the cerebral cortex back to lower brain parts (descending RAS); regulates the alertness, reflexes, and muscle tonus.

reticular cell, reticulum cell, a slightly differentiated, branched cell of the reticular connective tissue of vertebrates; in contact with each other the cells form a cellular network especially in the bone marrow and lymphatic tissue, and produce an intercellular substance, **reticulin**. The cells are part of the → reticuloendothelial system and probably have the capability of differentiating into → macrophages.

reticular fibres, Am. **reticular fibers,** the fibres in the reticular connective tissue.

reticular formation/substance (L. *formatio/substantia reticularis*), the grey substance along the central part of the brain stem of vertebrates (r. f. of mesencephalon and r. f. of medulla oblongata); a phylogenetically old area with diffuse aggregations of nerve cells and numerous short nerve branches, having numerous synaptic connections especially to collaterals of ascending sensory neurones (*see* reticular activating system). It controls the sleep-arousal state, skeletomuscular tonus, and many autonomic and endocrine functions of the body. The grey matter of the spinal cord is a corresponding structure and sometimes called reticular formation of the spinal cord.

reticular veining, *see* venation types.

reticulin, the protein produced by reticular cells forming **reticular** or **reticulin fibres** in the → connective tissue of animals; once thought to be distinct from → collagen but now regarded as the type III collagen that is associated with glycoproteins and proteoglycans.

reticulocyte (Gr. *kytos* = cavity, cell), the last immature stage in the development of → red blood cells in the bone marrow in mammals; still contains ribosomal material seen as a basophilic reticulum in staining; some reticulocytes enter the circulation (5% of all red blood cells in human blood).

reticulocyte lysate, an extract obtained from → reticulocytes of mammals, usually rabbits, used in studies of protein synthesis *in vitro*.

reticulo-endothelial system, RES, the system of tissue → macrophages in vertebrates consisting of various types of phagocytic, or putative phagocytic cells in many organs removing foreign particles from the blood and lymph. The cells originate from the bone marrow and enter the blood as **monocytes** which after some days enter different tissues, becoming **tissue macrophages**. These cells are found in connective tissue, and among endothelial tissue cells in the blood capillaries (sinusoids) and other ducts and cavities. The tissue macrophage system includes phagocytic cells e.g. in the spleen and other lymphoid organs, in the liver (→ Kupffer cells), lungs (pulmonary alveolar macrophages), bone (osteoclasts, reticular cells), and nervous tissue (microglia). The term **mononuclear phagocyte system, MPS**, includes both the monocytes and tissue macrophages.

reticulum, pl. **reticula** (L.), a network; any reticulated system or structure, as of intracellular, cellular, or tissue structure; e.g. **1.** → endoplasmic reticulum; **2.** → sarcoplasmic reticulum; **3.** stellate reticulum, the soft tissue in the enamel organ of a developing tooth; **4.** the second chamber of the stomach in ruminants (sometimes called honeycomb stomach because of the structure of the mucous membrane), derived from the oesophagus (*see* rumination).

reticulum cell, → reticular cell.

retina, pl. **retinae** or **retinas** (L.), the innermost of the three tunics in the wall of the camera eye, derived from embryonic brain tissue; in vertebrates it is divided into three parts: *pars optica*, *pars ciliaris*, and *pars iridica*. In the vertebrate eye, pars optica is composed of an outer, pigmented layer (*stratum pigmenti*, *pars pigmentosa*) and of an inner, transparent layer (*stratum cerebrale*, *pars nervosa*), the latter comprising nine histological layers: 1) layer of rods and cones (outermost layer), 2) external limiting membrane, 3) outer nuclear layer, 4) outer reticular (plexiform) layer, 5) inner nuclear layer, 6) inner reticular (plexiform) layer, 7) ganglion cell layer, 8) nerve fibre layer, 9) internal limiting membrane. The layer of rods and cones responds to light stimuli, the nuclear layers are composed of somas of the neurones, and the reticular layers consist of branches of neurones. The limiting membranes are formed chiefly from supporting cells (glial cells), known as Müller cells (Müller fibres).

The most sensitive area of the retina, the yellow spot, (*macula (flava) retinae*, *macula lutea*) is at the point of the optic axis, with a small depression (*fovea centralis*) in the centre consisting only of cones. The optic disc (the blind spot of the eye), comprising blood vessels, the optic nerve and no photoreceptor cells, is located quite close to the macula, medially of it. The number and types of photoreceptor cells (cones and rods), and consequently vision, vary greatly in different animal species and systematic groups; e.g. most mammals studied have rods as the major photoreceptor type in the retina.

Also the eyes of many invertebrates have a retina, e.g. in the ocelli of annelids and many molluscs, especially in the camera eyes of cephalopods. *See* eye, cone, rod, photoreception.

retinal, retinaldehyde, retinene, vitamin A_1 aldehyde; the aldehyde form of → retinol.

retinoic acid, RA, vitamin A acid produced in animals by oxidation of retinal (vitamin A_1 aldehyde); a putative → morphogen in vertebrates having similar but much stronger effects than vitamin A, especially on the growth and differentiation of epithelial cells. Retinoic acid binds to specific proteins, i.e. RA receptors (RAR), and then the formed complex binds to nuclear DNA. Synthetic retinoic acid is used e.g. for the treatment of acne. *See* retinol.

retinoid, vitamin A and its derivatives, analogues and metabolites, such as → retinol, retinal, and retinoic acid.

retinol, vitamin A_1, vitamin A alcohol, $C_{20}H_{30}O$; an organic substance metabolized in animal cells enzymatically into → **retinal**; is involved in the synthesis of the visual pigments in the photoreceptors of the retina. Retinol as well as its synthetic derivatives and → retinoic acid affect strongly the differentiation of vertebrate epithelial cells. *See* vitamin A, photoreceptors.

retinula, pl. **retinulae** (L. dim. of *retina*), **retinula(r) cell**; a sensory cell in the ommatidium of the compound eye; the photosensitive parts of several retinula cells form the → **rhabdome**.

retro- (L. *retro* = backward), denoting backward, behind, or past.

retrograde (L. *gradi* = to step), **1.** contrary to

the usual or natural course, such as degenerating or deteriorating; **2.** moving backward, retreating, reversed.

retroplasia (Gr. *plassein* = to mould, form), **1.** degeneration of a cell or tissue into a more primitive stage; **2.** the state of a cell when its activity is below normal.

retroposon, retrotransposon (L. *ponere* = to place), a → transposon which moves via an RNA form; the transposon DNA element is first copied into RNA from which a new DNA molecule is formed in → reverse transcription and is then attached to a new site in the genome.

Retroviridae, retroviruses, a family of single-stranded → RNA viruses in which a reverse transcriptase, i.e. RNA-dependent DNA polymerase, converts RNA to DNA; thus the virus is able to integrate with the DNA of its host cell. Retroviruses include e.g. RNA tumour viruses and HIV (human immunodeficiency virus).

reverse genetics, a branch of genetics which, in contrast to classical genetics, works from genes to characters by isolating and sequencing genes, thus inferring which proteins they encode and what characters they regulate.

reverse polymerase chain reaction, R-PCR, a method for analysing messenger RNA (mRNA) molecules; mRNA is first converted into complementary DNA (cDNA) in → reverse transcription, and this cDNA population is then amplified in → polymerase chain reaction (PCR) and analysed by → Southern blotting.

reverse transcriptase, RNA-dependent DNA polymerase found in → retroviruses; synthesizes complementary single-stranded DNA (cDNA) on a single-stranded viral RNA template.

reverse transcriptase polymerase chain reaction, RT-PCR, a method in which first a complementary DNA (cDNA) is synthesized on a messenger RNA template (mRNA template) by the reverse transcriptase enzyme and then amplified in the → polymerase chain reaction. The method is similar to R-PCR.

reverse transcription (L. *trans* = across, trough, *scribere* = to write), the synthesis of DNA on a RNA template catalysed by the **reverse transcriptase** enzyme; present in retroviruses, and used in genetic engineering.

reverse translation, a technique for isolation of a gene by hybridizing it with a short nucleotide sequence produced synthetically by deducing the nucleic acid sequence from the amino acid sequence of a known protein.

reversible, capable of being reversed or of reversing, such as the restoration of a chemical or physical stage; e.g. most biochemical reactions of cell metabolism can occur in both directions.

reversion, in genetics, **back mutation, reverse mutation,** the return of a mutant phenotype back to a wild phenotype either genetically (genotypic reversion) or non-genetically (phenotypic reversion). In **true reversion** the mutant allele at the original → mutation site is reverted to the wild-type. In **second site reversion** the function of the original mutant allele is reversed to the wild type function by a second mutation at another site in the same gene. In **suppression reversion** the function of the original mutant allele is reversed to the wild-type function by a second mutation in another gene (suppressor mutation); usually these are mutations of → transfer RNA coding for molecules which suppress → nonsense mutations.

R factor, → resistance factor.

RFLP, → restriction fragment length polymorphism.

R_f-value, retardation factor, the ratio that expresses the mobility of a compound in a chromatographic system, particularly in paper and thin-layer chromatography. R_f (£ 1) is determined by dividing the distance of a compound from its original site by the distance of the solvent front from its original site. In gel electrophoresis the mobility of a component is often expressed as R_s-value (£ 1); it is related to the anion front.

Rh, *see* rhesus factor.

rhabd(o)- (Gr. *rhabdos* = rod), denoting a rod or rod-shaped structures.

rhabdite, any of the minute rod-shaped structures in the epidermis of many turbellarians and some nemerteans. Rhabdites are packets of mucoid material which swell in water producing protective mucus on the body surface.

rhabdome, or **rhabdom** (Gr. *rhabdoma* = bundle of rods), a light-sensitive rod-shaped structure in the middle of several sensory cells (**retinula cells**) in each ommatidium of the → compound eye in arthropods; it is formed from membranous elements, **rhabdomeres,** differentiated from microvilli of each sensory cell. Rhabdomeres contain visual pigment for → photoreception.

rhabdomere, *see* rhabdome.

Rhabdoviridae, rhabdoviruses, a family of bullet-shaped viruses with a single-stranded (minus strand) RNA genome; contain virus-specific RNA polymerase. The rhabdovirus group includes plant and animal viruses, e.g. rabies virus (*see* rabies) and vesicular stomatitis virus of cattle. *See* RNA viruses.

rhamnose (L. *Rhamnus* = the genus of buckthorn), a hexose found in plant glycosides and cell wall polysaccharides, and in lipopolysaccharides in the walls of Gram-negative bacteria.

rhe (Gr. *rheos* = stream, current), the unit of fluidity. *See* Appendix 1.

Rheiformes (L. *Rhea* = the type genus; in Greek mythology *Rhea* was the mother of Zeus and other gods, *forma* = shape), **rheas,** American ostriches; an order of ratites which includes large flightless birds on South American pampas; two species are extant, the greater rhea and the lesser rhea (*Rhea americana* and *Rhea pennata*). *See* ratites.

rheobase (Gr. *rheos* = stream, *basis* = base), an electrophysiological quantity; the threshold voltage, i.e. the smallest possible electrical stimulus required to cause a response in a cell, tissue, or organ (e.g. in a muscle or nerve). The sensitivity of an organ is described also by **chronaxia** (chronaxie, chronaxy) which is the shortest duration of electric stimulus required to elicit the response, when the strength of the stimulus is twice the threshold voltage, i.e. twice the rheobase value.

rheotaxis (Gr. *taxis* = arrangement), a → taxis caused by the stimulus of a current, mostly water current.

rheotropism, a → tropism, i.e. growth curvature stimulated by an air or water current.

rhesus factor, Rh factor (Rh = rhesus monkey, *Macaca*), a blood group antigen in man; first found in the rhesus monkey, later shown to differ slightly from the human red cell Rh antigen and renamed LW. More than 30 different Rh antigens (Rh factors) are found on the surface of human red blood cells, the most important being **antigen D** present in about 85% of people (Rh positive, **Rh$^+$**), the rest being Rhesus negative (**Rh$^-$**). Antibodies against Rh antigens do not naturally exist in the blood, but they can be produced in the Rh negative person if Rh positive red blood cells somehow get into his or her blood. This may happen e.g. in a wrongly performed blood transfusion, or in the Rh negative mother during pregnancy and delivery when the baby is Rh positive, resulting in the haemolytic disease of the newborn.

rhexigenous (Gr. *rhexis* = a bursting, act of breaking, *gennan* = to produce), pertaining to a structure resulting from rupturing or tearing; e.g. rhexigenous intercellular space in plant tissues.

rhin(o)- (Gr. *rhis* = nose), denoting relationship to the nose or a nose-like structure.

rhinencephalon, pl. **rhinencephala** (Gr. *enkephalos* = brain), **olfactory brain**; **smell brain**; the parts of the vertebrate forebrain involved in olfactory mechanisms; it forms large areas in the brain hemispheres of fishes, amphibians and reptiles; in mammals it includes proportionally smaller areas, comprising the olfactory nerves, bulbs, and tracts, olfactory cortex, and originally also parts of the hippocampus and amygdala.

rhinoviruses, a group of RNA viruses (family Picornaviridae) causing the common cold and other respiratory ailments in man and foot-and-mouth disease in cattle.

rhiz(o)-, -rhiza, -rrhiza (Gr. *rhiza* = root), denoting relationship to a root.

Rhizobium (L.), the genus of Gram-negative rod-shaped bacteria fixing molecular nitrogen (N_2) from air. *See* root nodules, nitrogen fixation.

rhizodermis (Gr. *derma* = skin), the outermost cell layer, → epidermis, of a young root; in an older root it is replaced first by the **exodermis**, and finally by the → **periderm** formed of cork. A synonym for rhizodermis and exodermis is epiblem(a).

rhizoid (Gr. *eidos* = form), a root-like structure, or resembling a → root; e.g. **1.** a hair-like unicellular or multicellular structure serving as a root. Present in moss stems, liverworts, fern prothalli, and in some algae and fungi; **2.** sometimes root-like structures of sessile invertebrates are called rhizoids, such as the attaching "roots" of rhizocephalans (→ Cirripedia).

rhizome (L. *rhizoma*), an underground, perennial, often juicy plant stem growing either horizontally (e.g. European wood anemone, *Anemone nemorosa*) or vertically (e.g. *Plantago* species); bears roots, scale-like leaves and buds, and serves also vegetative propagation. *Syn*. rootstalk. *See* stolon.

Rhizopoda (Gr. *pous*, gen. *podos* = foot), **rhizopods**; a class or superclass in the subphylum → Sarcodina; protozoans which move using → pseudopodia of various forms, such as rhizopodia, filopodia, or reticulopodia. The class includes e.g. the genera *Amoeba*, *Entamoeba*, *Arcella*, and *Globigerina*.

rhizopodium (Gr. *pous* = foot), *see* pseudopodium.

rhizosphere (Gr. *sphaira* = ball), a soil area that immediately surrounds and is influenced by the roots of a plant. The **rhizosphere effect** means the enhancement of the growth of soil microorganisms caused by the physical and chemical effects of the roots within a rhizosphere.

RH-mapping, → radiation hybrid mapping.

rhod(o)- (Gr. *rhodon* = rose), denoting relationship to a rose, or to a red or reddish colour.

rhodamine, any of reddish, highly fluorescent, triphenylmethane derived dyes which can be linked to cellular proteins by isocyanate; it is used to visualize membrane and cell organelle structures e.g. by immunofluorescent methods. Some rhodamine dyes are also used in colouring paper and laquers.

Rhodophyta (Gr. *phyton* = a plant), **red algae**; a division of the → Plant kingdom; a group of chiefly marine, multicellular algae, some species living in fresh water. Their biochemical properties make them different from all other algae; the photosynthetic pigments (chlorophyll a and d, α- and β-carotenes, phycoerythrin) give them a typical dark red colour. Also the life cycle differs from that of other algae indicating that they are not near relatives to them. The → gametophyte and → sporophyte of red algae are separate. The oogonium is differentiated into a **carpogonium** with a threadlike tip, **trichogyne**. Non-motile → **spermatia** are formed in the **spermatangia**; they will attach to the trichogyne and invade the carpogonium. The zygote, formed after fertilization and called **carposporophyte**, stays attached to the mother plant, and forms diploid **carpospores**, which germinate and develop into **tetrasporophytes**. These form **tetrasporangia** and **tetraspores**, and the tetraspores again form new gametophytes. Red algae are used as a source of → agar, and species of the genus *Porphyra* are used as food in the Far East, especially in Japan. The total number of species is more than 4,000; about 200 are found in flowing fresh waters.

rhodopsin (Gr. *opsis* = sight), **visual purple**; the photosensitive pigment of the rods in the vertebrate eye and of related sensory cells in invertebrates; it is composed of **retinene** (11-cis-retinal, *see* retinol) and a protein called **opsin** (scotopsin) which are detached from each other by light energy forming *trans*-retinal moiety of retinene, and opsin. This activation of rhodopsin produces formation of a series of intermediates, e.g. **metarhodopsin II**, that initiates the closure of Na^+ cannels of the receptor membrane in → photoreception. Rhodopsin is one of the many → serpentine receptors coupled to G proteins.

rho factor, a protein factor which regulates the termination of transcription of some bacterial genes; probably acts by pulling the transcript away from the RNA polymerase.

rhombencephalon, pl. **rhombenencephala** (L. < Gr. *rhombos* = rhomb, *enkephalos* = brain), **hindbrain**; the most posterior part of the three primary brain vesicles of an early vertebrate embryo; during further development it divides into metencephalon (pons and cerebellum) and myelencephalon (medulla oblongata).

rhopalium (Gr. *rhopalon* = club), a marginal, club-shaped sense organ of scyphozoans (jellyfishes) containing a statocyst for maintaining equilibrium, a group of neurones to control pulsation, and sometimes an ocellus. The ocellus may be either a simple pit with pigment and photoreceptor cells or a more complex eye with a lens and retina-like arrangement of sensory cells, as in cubomedusans.

Rhynchocephalia (Gr. *rhynchos* = snout, *kephale* = head), **rhynchocephalians**; a primitive order of reptiles, species of which have lived from Trias to the present time. Only one species is extant, the **tuatara** (*Sphenodon punctatum*), living in the archipelago of New Zealand. *Syn*. Sphenodonta.

rhynchocoel (Gr. *koilos* = hollow), the fluid-filled tubular cavity in nemertean worms containing the introverted proboscis; the force for eversion of the proboscis out of the anterior tip of the worm is provided by muscular pressure. *See* Nemertea.

Rhynchocoela, an earlier name for → Nemertea.

rhythm (Gr. *rhythmos* = measured motion,

rhythm), *see* biorhythms, electroencephalography.

rib (L. *costa*), **1.** one of the paired, curved, thin, osseous or partly cartilaginous bones of the thorax in tetrapod vertebrates, or related structures supporting the abdominal cavity in lower vertebrates; in many fish species two pairs at each vertebra, i.e. haemal and lateral ribs; **2.** any elongated elevation running the length of an anatomical structure, such as the central vein of a leaf.

rib cage, *see* thorax.

riboflavin, also **riboflavine** (L. *ribo-* = ribose, *flavus* = yellow), vitamin B_2, 7,8-dimethyl-10-ribitylisoalloxazine; it is synthesized in all green plants and many microbes and is present in all cells; the richest natural source is yeast. Riboflavin occurs as a component of the coenzymes FMN and FAD which act in cellular oxidation-reduction reactions of all organisms. *See* vitamin B complex.

ribonuclease, RNase, RNAase, a group of hydrolytic enzymes, EC sub-subclass 3.1.27.; (transferases or phosphodiesterases) catalysing the hydrolysis of RNA by breaking phosphodiester bonds. In cells there are many types of RNases having different pH and temperature optima. Ribonuclease I (RNase A, EC 3.1.27.5) is a pancreatic enzyme which is one of the best known ribonucleases and used much in cytochemistry.

ribonucleic acid, RNA, a polymer of ribonucleotides. RNA is a long-chain unbranched molecule consisting of numerous → nucleotides of four different types; a ribonucleotide consists of a phosphorous acid residue, ribose (the sugar component), and a purine or pyrimidine base which in RNA are usually adenine (A), guanine (G), uracil (U), and cytosine (C). Apart from the usual → nucleosides, transfer RNA molecules contain also rare nucleosides, such as inosine, pseudouridine, ribothymidine and dimethylguanosine. Six different RNA types are found in the cells: → **heterogeneous nuclear RNA (hnRNA)**, the primary transcription product of the genes from which → **messenger RNA (mRNA)** is processed, and further → **transfer RNA (tRNA), ribosomal RNA (rRNA), small nuclear RNA (snRNA)**, and **small cytoplasmic RNA (scRNA)**.

ribonucleoprotein particle, RNP, any of the ribonucleoprotein complexes involved in → gene expression of eukaryotes. *See* small nuclear ribonucleoprotein particle.

ribonucleoprotein, RNP, a compound consisting of ribonucleic acid and protein.

ribonucleoside, a → nucleoside having ribose as the sugar component.

ribonucleotide, a → nucleotide having ribose as the sugar component. *See* ribonucleic acid.

ribose, a pentose sugar (monosaccharide), $CH_2OH(CHOH)_3COH$, occurring e.g. in RNA and ATP, present in all cells.

ribosomal DNA, rDNa, a long segment of DNA which is located in the nucleus in the → nucleolar organizing region, and constitutes the genes encoding → ribosomal RNAs in many consecutive copies. *See* deoxyribonucleic acid.

ribosomal RNA, rRNA, the → ribonucleic acid (RNA) of ribosomes and polyribosomes in all cells; different types are known and denoted by their sedimentation coefficients, e.g. in eukaryotes 28S, 18S, 5.8S, and 5S (S = → Svedberg unit). *See* ribosome.

ribosome (Gr. *soma* = body), a cell organelle composed of RNA (ribosomal RNA, rRNA) and protein, and being 10 to 40 nm in diameter, is visible only in an electron microscope. The ribosomes are principal structures of all cells acting as the sites of → protein synthesis; they have binding sites for messenger RNA (mRNA) and nascent peptide chain. The ribosomes of prokaryotes (70S, → Svedberg unit) are freely distributed in the cytoplasm, but most ribosomes of eukaryotic cells (80S) are bound to the → endoplasmic reticulum.

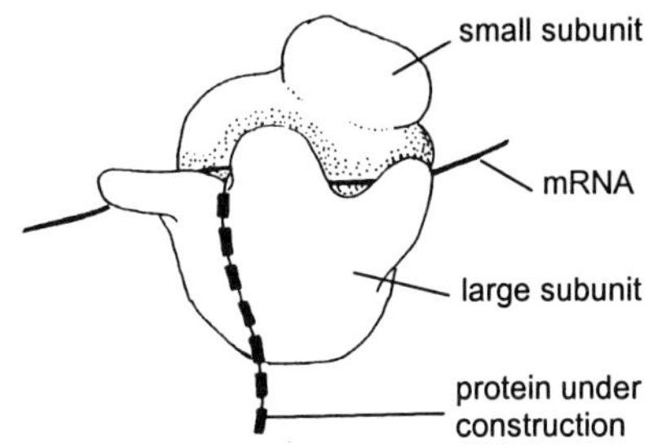

Ribosome presently active in protein synthesis. (After Postlethwait & Hopson: Nature of Life - McGraw-Hill, 1992; with the permission of the authors).

Some cell organelles, such as mitochondria and chloroplasts, have their own ribosomes. All ribosomes are composed of two dissociable subunits containing four different ribosomal RNA chains and tens of different proteins; the sizes of the subunits in bacteria are 50S and 30S and in eukaryotes 60S and 40S. Several ribosomes attach along the length of a single molecule of mRNA and form a → polysome, i.e. polyribosome that is a functional unit of protein synthesis.

ribozyme, any RNA molecule with catalytic activity, such as the self-splicing introns of some ribonucleic acids (*see* splicing). Ribozymes are presumed to have a role in the origin of life on Earth; e.g. they have been observed to undergo molecular evolution in artificial conditions, and moreover, this evolution has been demonstrated to be continuous.

ribulose, a pentose (sugar), the 2-keto isomer of ribose; its phosphorylated derivative, → **ribulose 1,5-bisphosphate**, is the carbon dioxide (CO_2) acceptor in the dark reactions of → photosynthesis.

ribulose 1,5-bisphosphate carboxylase-oxygenase, Rubisco, RUBISCO, EC 4.1.1.39, the carboxylating enzyme present in all green plants and photosynthetic bacteria; catalyses the fixation of carbon dioxide, CO_2, in → photosynthesis, also has oxidizing activity, which is involved in → photorespiration that diminishes the yield of photosynthesis. The enzyme is probably the most abundant single protein on the Earth.

ricin, a compound, 66 kD → lectin, obtained from the seeds (castor beans) of the castor oil plant, *Ricinus communis*; toxic to animals, e.g. inhibits protein synthesis and acts as an haemagglutin, i.e. causes agglutination of red blood cells.

rickets, infantile or juvenile osteomalacia; a disturbed condition of ossification in bones caused by deficiency of vitamin D, especially during infancy in man. The condition is manifested e.g. by distortion and bending of bones and delayed closure of the fontanelles. Adult rickets is called osteomalacia. *Syn.* rachitis.

rickettsia, pl. **rickettsiae** (*Howard Ricketts*, 1871—1910), any of small intracellular, parasitic Gram-negative bacteria (order Rickettsiales) found in the cells of many animals; transmitted to man and other mammals e.g. by many arthropods (lice, fleas, ticks, mites, etc.), causing **rickettsiosis**, such as epidemic and murine typhus and similar fevers, e.g. Rocky Mountain spotted fever.

rifamycin, an antibiotic produced by *Streptomyces mediterranei*; inhibits prokaryotic DNA-dependent RNA synthesis by blocking the initiation of transcription; **rifampicin** (**rifampin**) is a semi-synthetic compound of the rifamycin group. Both antibiotics are used in the treatment of pulmonary tuberculosis.

rigor (L. *rigere* = to be stiff), stiffness; the rigid state; e.g. **rigor mortis**, the temporary stiffening of the body muscles after death due to the lack of ATP in muscle cells; disappears after the autolytic processes have softened the muscle cells in tissues.

ring canal, 1. a duct in the gastrovascular system of medusoid cnidarians, e.g. jellyfish, or in the → water-vascular system of some echinoderms, such as sea stars; it is situated tightly at the margin of the umbrella of a jellyfish and connected to the gastric cavity by several → radial canals. In echinoderms, the ring canal is located around the oesophagus below the madreporite; *syn.* circular canal; *cf.* radial canal, stone canal; **2.** in genetics, cytoplasmic bridges connecting → nurse cells to the → oocyte in → meroistic ovaries.

ring chromosome, circular chromosome; the normal form of the chromosome in → prokaryotes; exists only exceptionally in the nucleus of eukaryotic organisms, e.g. a mutant chromosome, but circular DNA exist normally in → mitochondria.

Ringer's solution (*Sydney Ringer*, 1835—1910), Ringer solution, Ringer fluid. *See* physiological solution.

ringing, 1. a method of killing a tree by cutting away a ring of bark at the base of the trunk so that the continuous phloem is cut thereby preventing the nutrition of the roots; the tree dies and regrowth or sprouting is prevented; **2. bird ringing**, marking birds individually with a small metal ring around the leg; a common method for studying e.g. migration routes, population dynamics, and lifetime.

ring species, two species developed from the same original species by ring-shaped geographical distribution; a series of the intraspecific populations which have diverged to races or subspecies in different parts of their common, ring-shaped distribution area. Interbreeding is possible between adjacent populations on the distribution range, but when the terminal populations meet in the ring-closing area, no interbreeding occurs. They behave as two different species, e.g. in northern Europe the herring gull (*Larus argentatus*) and the lesser black-backed gull (*L. fuscus*) which are considered to be the terminal links of a chain of subspecies circling the north temperate region.

risk factor, a factor which disposes an organism to a certain disease, e.g. to genetic disorder or cancer.

R-loop (R = RNA), usually an experimentally accomplished structure which is generated when a RNA strand is hybridized with its complementary strand in the DNA → double helix, displacing the other DNA strand from its original connection. The loop is formed by the displaced single-stranded DNA and the RNA-DNA duplex.

RNA, → ribonucleic acid.

RNA abundance, the quantity of → ribonucleic acid (RNA) per cell.

RNA editing, a process occurring in the cells of eukaryotic organisms in which certain → nucleotides are added to → messenger RNA (mRNA), or nucleotides (other than introns) are removed from it. The RNA editing includes the → polyadenylation of mammalian mitochondrial RNA (*see* mitochondrial ribosome), and certain rare mechanisms which are mainly found in the RNA products of mitochondrial genes in parasitic organisms. These are e.g. additions and deletions of nucleotides in trypanosome mitochondria, and additions of guanosine nucleotides in paramyxoviruses. Also the replacement of → cytosine by → uracil in plant and mammalian mitochondrial messenger RNA are included in RNA editing.

RNA polymerases (Gr. *poly* = many, *meros* = a part), enzymes of the sub-subclass EC 2.7.7. catalysing the synthesis of RNA using DNA as a template in → genetic transcription. Prokaryotic organisms have only one RNA polymerase, but eukaryotes have three: **RNA polymerase I** catalyses the synthesis of → ribosomal RNA (except the 5S component), **RNA polymerase II** the synthesis of → messenger RNA and → heterogeneous nuclear RNA, and **RNA polymerase III** the synthesis of → transfer RNA and other small RNA molecules such as the 5S component of rRNA.

RNA processing, the ripening of a → heterogeneous nuclear RNA into → messenger RNA; the process involves → capping, splicing, and polyadenylation.

RNase, RNAase, → ribonuclease.

RNA virus, a virus having RNA as its genetic material.

RNP, 1. ribonucleoprotein; **2.** → ribonucleoprotein particle.

Robertsonian translocation, whole arm translocation (*J.D. Robertson*, b. 1922), a → translocation or replacement involving whole → chromosome arms; it is an abnormality in eukaryotic organisms in which non-homologous acrocentric chromosomes are involved in chromosomal fusion. *Syn.* centric fusion, Robertsonian fusion.

rod (cell), one of the two types of photoreceptor cells in the sensory retina of the vertebrate → eye. The photoreactive part (outer segment) of the cell is long and rod-shaped, containing **rhodopsin** as the visual pigment. Rods are much more sensitive to light than the other receptor type, the cone. In primates, rods serve particularly night vision (scotopic vision, non-colour vision), but in many species rods are the major type of photoreceptor cells. In humans, about 120 million rods are scattered widely in the retina, and about 6 million cones concentrated in a small retinal area, → fovea. *See* photoreception.

Rodentia (L. *rodere* = to gnaw), **rodents**; a polymorphic mammalian order which includes 33 families and over 1,650 species. A typical characteristic is a dentition with a pair of open-rooted and continually growing → incisors in both the upper and lower jaws. Only the foreside of the chisel-like incisors is covered by enamel and thus they remain sharp through the entire lifetime. Rodents are mainly herbivorous, some species also omnivorous, and found in almost all terrestrial habitats. The order is the most widespread and largest of all mammalian orders including e.g. the families: **squirrels** (Sciuridae), **beavers** (Castoridae), **voles** (Cricetidae), **rats** and **mice** (Muridae), **dormice** (Gliridae), **jumping mice** (Zapodidae), and **Old World porcupines** (Hystricidae).

rodenticide (L. *caedere* = to kill), a chemical agent (pesticide) used for killing rodents; most rodenticides are → anticoagulants causing bleeding by preventing blood coagulation.

roentgen, R (*Wilhelm Röntgen*, 1845—1923), the old unit for X radiation. *See* Appendix 1.

roentgen microanalysis, → X-ray microanalysis.

roentgen rays, → X-rays; written also röntgen

rays or Röntgen rays, discovered by *Röntgen* in 1895.

rolling circle, a mode of DNA multiplication based on a circular template of the DNA molecule; found in viruses and bacteria, but also in the amplification of the eukaryotic ribosomal DNA during the development of ova or egg cells. An extrachromosomal circular copy of the DNA sequence is produced which in turn produces many copies containing tandem repeats of the sequence. *See* gene amplification.

root, 1. an organ typical of → vascular plants, taking water and nutrients into the plant and transporting them into the plant shoot, anchoring the plant in the soil, and serving as a storage organ. The roots are mainly underground (in some plants wholly above the ground), and unlike the shoot, roots never have leaves or nodes. The main root types are the **taproot** and **adventitious roots**. The taproot develops from the → radicle, grows vertically downwards into the soil, forms branches (lateral roots); it is typical in → dicotyledons. All other roots are adventitious roots. In → monocotyledons, the tap root vanishes early and the root system is formed from adventitious roots grown from the base of the stem; *see* root bud, root cap;
2. the lowermost part (*radix*) by which something is attached in anatomical structures of animals, as the nerve root or the tooth root.

root bud, a → bud formed in a root of a plant; e.g. numerous sprouts of aspen and poplar are grown from root buds.

root cap, a cell layer covering the apex of a → root; the cells of the root cap are short-lived, break easily and lubricate the passage of the root through the soil; the root cap also protects the → meristem tissue (apical meristem) in the root tip and is important in the regulation of the gravitropic reaction of the roots. *See* gravitropism.

root hair, pilus, a hair-like swelling of a cell in the plant root → epidermis (rhizodermis). Root hairs are numerous in the → root hair zone (piliferous layer) and function in increasing the surface area of the root and thus the absorption of water and nutrients.

root hair zone, the epidermal zone near the tip of a plant root, rich in → root hairs. *Syn.* piliferous layer.

root neck, the uppermost part of a → taproot.

root nematodes, plant-parasitic nematodes, may be a few millimetres long; harmful worms causing plant diseases, such as **root galls**, abnormal enlargements on roots. *Syn.* root-knot nematodes.

root nodules, small swellings caused by symbiotic nitrogen-fixing bacteria on the roots of many plant species, such as leguminous plants (e.g. pea and clover), the most important bacteria belonging to the genus *Rhizobium*. *See* nitrogen fixation.

root pressure, a positive hydrostatic pressure in plants, created by ion uptake and movements and consequent osmotic water uptake in root cells; forces water from the roots up into the xylem of the stem. *See* guttation.

root rot, a destructive plant disease characterized by decay of roots, caused by fungi, e.g. the genera *Fusarium*, *Armillaria*, *Oozonium*, *Rhizoctonia*, or *Sphaerostilbe*.

root sprout, a new shoot originating from a → root bud of a plant.

rosette (Fr. = small rose), any formation or structure resembling a rose; e.g. **1.** a cluster of plant leaves (**rosette plant**), situated on the ground (base rosette, as in dandelion, *Taraxacum*) or on the top of the shoot (aerial rosette, as e.g. chickweed wintergreen, *Trientalis europaea*). The **rosette tree** is a tree or tree-like rosette plant having a leaf rosette in the top of the stem; e.g. many tropical palms, *Yucca*, *Dracaena*. The **half-rosette plant** forms a leaf cluster near the ground but forms long internodes and some leaves even in the upper part of its shoot; e.g. *Thlaspi* species; **2.** a small group of cells in an animal tissue, e.g. certain neoplasms of neuroblastic origin, or of leucocytes; **malarial rosette**, grouping of malaria parasites around a red blood cell.

rosette disease, any of several plant diseases characterized by rosette formation of leaves; caused e.g. by fungi, viruses, or nutritional disturbances, e.g. the deficiency of zinc and boron.

rosette theory, a model which declares the synthesis and deposition of cellulose microfibrils in the → secondary wall of plant cells. The rosettes are protein structures embedded in the plasma membrane of the cell, arranged in hexagonal arrays some tens together, each forming a microfibril on the outside surface of the membrane, between the primary wall and the cell membrane. The rosettes move in the membrane and the typical structure of the secondary wall is formed with separate layers, each with equally directed microfibrils.

rosin, the solid resin obtained from pine trees, used in plasters and ointments; formerly called colophony.

rostellum (L. dim. of *rostrum* = beak, snout), a small beak-like structure; e.g. **1.** a projecting structure, often with hooks or suckers, found e.g. on the scolex of the tapeworm; **2.** the median stigma in orchids, an organ for attachment of pollinia.

rostral (L. *rostralis*), **1.** pertaining to a → rostrum, having a beak or snout; **2.** situated towards a rostrum, in vertebrate anatomy, denoting towards the oral and nasal region; **3.** as a noun, a plate or shield on the rostrum of some reptiles.

rostrum, pl. **rostra** or **rostrums** (L.), a beak-like structure; e.g. **1.** rostrum of corpus callosum, i.e. the anterior end of corpus callosum in the vertebrate brain; **2.** sphenoidal rostrum, the ridge of the sphenoid bone; **3.** a projection of the snout, e.g. of a fish; **4.** a median prolongation at the anterior end of the crayfish carapace; **5.** an anterior tip of the epimerite of some gregarines (Gregarinia); **6.** the beak or snout of any of various insects or arachnids, as the prolonged snout of scorpionflies and snowflies (Mecoptera), or of snout beetles (Curculionidae), or the piercing and sucking mouthparts of hemipterans (Hemiptera).

Rot, R_0t (concentration of renaturated RNA per time), the product of initial RNA concentration (R_0) and time (t) of incubation in an RNA-driven hybridization reaction. Correspondingly, the concentration of renaturated DNA per time is called Cot.

rot, 1. decay caused by fungi and bacteria, especially the decomposing of plant material by heterotrophic organisms, chiefly by → saprophages. Most important decomposers are various fungi, because they are able to decompose the main materials of plants, → cellulose and → lignin; bacteria decompose less effectively these materials. In **brown rot**, caused by many *Polyporus* fungi, cellulose is dissolved and brown lignin remains, in **white rot,** caused by e.g. *Clitocybe* and *Marasmius* species, lignin is decomposed; **2. liver rot**, a disease of sheep caused by the liver fluke *Fasciola hepatica*; may also occur in other herbivorous mammals, and in man; **3. foot rot**, a disease of the feet of sheep, goats, and cattle, caused by *Bacteroides nodusus* and *Fusobacterium necrophorum*.

rotation (L. *rotare* = to turn, rotate), the act of turning around an axis; in biology; e.g. **1.** movement of a body or limb about its axis; **2.** the slow movement of cytoplasm that is observable under the microscope, such as the movement of cell organelles and the flow of cytoplasm; it is caused by the function of microfilaments; in plant cells it is important during cell division in the transport of vesicles containing cell wall material; **3.** optical rotation; *see* optical isomerism; **4.** revolving of the rotating part of an engine, such as the rotor of a → centrifuge. *Adj.* **rotatory** (rotary, rotative, rotational), pertaining to rotation, occurring in rotation, caused by rotation.

rotation culture, a cell culture technique in which a culture tube or vessel is kept rolling to prevent cells from adhering to the bottom or each other.

Rotatoria, → Rotifera.

Rotifera (L. *rota* = wheel, *ferre* = to carry), formerly known as **Rotatoria**; **rotifers**, "wheel animalcules"; an animal phylum (in some classifications a class in the phylum Aschelminthes) comprising small, mainly microscopic free-moving animals living solitarily usually in fresh water; some live in colonies, a few species are → sessile. The animals have a → pseudocoelom and ciliated crown (→ corona) around the mouth, giving the appearance of a rotating wheel. The phylum includes about 1,800 species which are grouped into three classes: Seisonidea, Bdelloidea and Monogononta.

rotor (L. *rotator*), a revolving part of a machine; especially the revolving head of a → centrifuge in which e.g. cell organelles and macromolecules can be separated in a centrifugal field. Tissue homogenates or any other mixtures to be analysed are transferred into plastic, metal, or glass tubes which are placed into a rotor. **Angle rotors** have diagonal cavities for the tubes, in **swing-out rotors** the tubes are in buckets which rise to horizontal position during centrifugation, and in **vertical rotors** the tubes are in upright cavities. **Zonal rotors** are large-capacity rotors which contain a large cylindrical space divided into sector-shaped compartments. The rotor is filled and emptied while at speed. *See* centrifugation, density gradient centrifugation.

rough endoplasmic reticulum, rough ER, *see* endoplasmic reticulum.

round window (L. *fenestra cochleae*), fenestra

of the cochlea, cochlear fenestra of the vertebrate ear, i.e. a roundish opening covered by a membrane in the wall between the middle and internal ear; closes flexibly the end of the scala tympani in the cochlea.

roundworms, → Nematoda.

row buds, → axillary buds which develop in a row on the leaf axis, i.e. on the place where generally only one bud is developed; e.g in banana.

royal jelly, → queen substance.

R-PCR, → reverse polymerase chain reaction.

R plasmid, → resistance factor.

RQ, → respiratory quotient.

rRNA, → ribosomal RNA.

r-selection (r = intrinsic *r*ate of increase), in ecology, a → selection of life-history traits acting in an environment where changes are rapid and not predictable (**r-strategist environment**); species under r-selection (**r-strategists, r-selected species**) prefer reproduction in allocation of resources. Such species are opportunists, reach their sexual maturity early, and produce a great number of small-sized offspring. Often the species are characterized by → semelparity, quick ontogeny, and small body size. Their population density is usually below the → carrying capacity. R-strategists prevail usually in environments which are in the early stages of → succession, such as in courtyards and fallows; e.g. many weeds, rats, and locusts are r-strategists. *Cf.* K-selection.

r-strategist (species), *see* r-selection.

RT-PCR, → reverse transcriptase polymerase chain reaction.

rubber, natural rubber, caoutchouc, India rubber, an elastic material (elastomer) synthesized in many plants; consists mainly of polyisoprene, $(CH_2CHC(CH_3)CH_2)_n$, a hydrocarbon polymer with an average relative molecular mass about 300,000. About 50 natural plant species produce raw rubber, i.e. latex (milk juice), the rubber tree *Hevea brasiliensis* and its relatives being most important for rubber industry. A shrub, guayule (*Parthenium argentatum*) which grows in northern Mexico produces rubber which is chemically identical to *Hevea* rubber. Latex occurs in the undercortical ducts of the rubber tree, and is tapped from cuts made in the stem. Some water is added and rubber is precipitated with dilute formic or acetic acid to obtain slab raw rubber. **Synthetic rubbers** are e.g. isomers of isoprene, ethylene (ethenyl, vinyl), ethene-propene, butane, and butadiene.

Rubisco, → ribulose 1,5-bisphosphate carboxylase-oxygenase.

rudiment (L. *rudimentum* = first attempt, beginning), **1.** the first appearance of a structure in the course of its development, such as a rudiment of an organ in an embryo; *syn.* primordium, anlage; **2.** an organ whose development is arrested at an early stage, e.g. the teats of male mammals; **3.** an undeveloped structure that has partially or totally lost its former function, either in an individual or phylogenetically; *syn.* vestige; **rudimentation**, the formation of a rudiment. *Adj.* **rudimentary, rudimental.**

rumen, pl. **rumina** or **rumens** (L.), the first compartment of the ruminant stomach, derived from the oesophagus; also called paunch. *See* rumination.

Ruminantia (L. *ruminare* = to chew the cud), **ruminants**, cud-chewing mammals; a suborder in the order Artiodactyla characterized by a high specialization for ingesting cellulose food (*see* rumination); the dentition comprises 32 teeth, canines are reduced or lacking. Ruminants include e.g. the following families: **deer** (Cervidae), **cattle, antelopes, sheep,** and **goats** (Bovidae), **giraffe** and **okapi** (Giraffidae); some authorities include also **camels** and **llamas** (Camelidae), whereas others consider them as a separate suborder, Tylopoda in Artiodactyla, or divide the suborder Ruminantia into two infraorders, Tylopoda and Pecora.

rumination, chewing the cud; a specialized alimentary process of cud-chewing mammals (ruminants) for digestion of food containing mainly cellulose; characterized by regurgitating food from the stomach for rechewing. The stomach of ruminants has four separate compartments the first three of which derive from the oesophagus. The **rumen** and **reticulum** form the anterior division of the fermentation basin where cellulase of bacteria and other microbes begin to digest the cellulose-rich food; the third compartment is the **omasum** with a highly folded mucous membrane efficiently absorbing water. The fourth, **abomasum**, corresponds to the single stomach of other vertebrates. During feeding the grazed food is swallowed into the rumen and partly to the reticulum, where the food later triggers reflectory contractions of the wall muscle, pushing small boli of food from the rumen and reticulum back to the mouth. After chewing the cud flows directly to the omasum where some water is absorbed, and then further to the abomasum, where gastric enzymes begin digestive action. The intestine is very long, and the caecum and ascending colon act as the posterior division of the fermentation basin, containing high concentrations of microbes capable of decomposing cellulose; this polysaccharide cannot be digested at all by the enzymes produced by the animal itself. Dead intestinal bacteria also serve as an important protein source for the animal.

According to their nutrition strategy and structure of the alimentary canal, ruminants are classified into three groups: **roughage feeders,** e.g. cattle and sheep which feed on hay and have the most efficient digestion of cellulose, **concentrate selectors**, e.g. elk, moose, and some deer feeding on fresh leaves and grass, i.e. material with less cellulose, and **intermediate feeders**, e.g. reindeer, caribou, and red deer which are more flexible in their nutritional requirements and are able to change their diet. *See* ruminantia.

runner, a plant shoot which has long internodes and grows horizontally on the ground, or below the ground. It is often important for the dispersion of the plant, and may have → bulbs (e.g. potato) which support reproduction or act as energy store organs. *Syn.* stolon.

runoff, superfluous water from rain and snow flowing to a water body, such as a brook, river, lake, sea. *Syn.* runoff volume.

runoff area, an area from which water flows to a certain water body; e.g. the runoff area of the Baltic Sea is about 4.3 times the area of the sea.

rust fungi, → Uredinales.

S

S, symbol for **1.** sulphur; **2.** serine; **3.** siemens; **4.** substrate; **5.** Svedberg unit; **6.** selection differential.

s, symbol for **1.** second; **2.** sedimentation coefficient; **3.** solubility of substances (soluble); **4.** selection coefficient.

S_1, S_2, S_3 etc. (S = self-fertilization), an abbreviation used to indicate generations of self-fertilization.

sac (L. *saccus* < Gr. *sakkos* = bag), in anatomy, a pouch or a bag-like structure or organ; e.g. the **air sacs** connected with the avian lung, or the **dental sac** that envelopes a developing tooth.

saccharides (Gr. *sakcharon* = sugar), → carbohydrates.

saccharin, benzosulphimide; a synthetic organic compound, 1,2-benzisothiazol-3(2H)-one 1,1-dioxide; its sodium salt is used as a non-caloric sweetening agent, several hundred times sweeter than sucrose.

saccharolytic (Gr. *lysis* = dissolution), having capability of splitting sugars chemically.

saccharomycetes, yeast fungi; unicellular ascomycetes used as bread and brewing yeast (*Saccharomyces*) and as experimental organisms in cell biology and genetics. *Adj.* **saccharomycetic**.

saccharose, → sucrose.

saccule (L. *sacculus,* dim. of *saccus* = bag), a small bag-like structure, a small sac; in vertebrate anatomy e.g. **alveolar saccule** in the lung, or **vestibular saccule** in the internal ear.

saccus, pl. **sacci** (L. < Gr. *sakkos*), a bag, sac, or any bag-like structure in anatomy; e.g. *saccus lacrimalis* (lacrimal sac), the wide upper end of the nasolacrimal duct in many vertebrates.

Sachs' iodine test (*Julius von Sachs*, 1832—1897), a staining method for → starch; amylose is stained blue by iodine in aqueous solution.

sacral nerves, *see* spinal nerves.

sacral vertebra, *see* sacrum.

sacrum (L. *os sacrum* < *sacer* = sacred), a triangular bone situated caudally from the last lumbal vertebra and between the two hip bones in tetrapod vertebrates; usually formed from five fused **sacral vertebrae**. *Adj.* **sacral**.

sagittal (L. *sagittalis* < *sagitta* = arrow), arrow-shaped, straight; e.g. **1.** pertaining to direction from front to back in the median plane or a plane parallel to median; e.g. **sagittal section**, longitudinal section of a bilateral organism; **sagittal level**, median level; **2.** located in the plane of the sagittal suture (or parallel to it) between the parietal bones of the skull.

salamanders, → Caudata.

salamandrin, a poisonous compound secreted by glands in the skin of a tailed amphibian, such as tiger salamander (*Ambystoma tigrinum*).

salicylic acid (L. *Salix* = genus of willow), a phenol derivative, HOC_6H_4COOH, found abundantly in the willow bark; inhibits some enzyme functions in plants and the germination of seeds. Its esters and salts (**salicylates**) act in animals as inhibitors of prostaglandin synthesis and are used as analgesics; *para*-aminosalicylic acid (PAS) has a bactericidal effect and is used in the treatment of tuberculosis.

salinity (L. *sal* = salt), the salt concentration of a solution, especially that of natural waters, e.g. sea water.

saliva (L.), the secretion of special mucous or serous glands, → salivary glands, situated in, or opening into the buccal cavity of vertebrates and many invertebrates; contains inorganic salts, mucin, albumin, globulin, enzymes (e.g. amylase), and epidermal growth factor stimulating the renewal of the alimentary epithelium in vertebrates. Saliva serves often softening, neutralization, dilution, and digestion of food and cleansing the mouth; contains usually also antibacterial compounds (*see* lysozyme), and in some species, poisonous substances for predation and defence.

salivary glands, oral glands; the exocrine glands derived from the epithelium of the anterior part of the alimentary canal, secreting saliva. In mammals, the glands comprise three pairs of large salivary glands (**parotid, submandibular**, and **sublingual glands**) and several smaller glands in the tongue, cheeks, lips, and palate, all opening into the buccal cavity. The parotid and submandibular glands secrete a digestive enzyme, amylase, but the sublingual gland only mucus. The small salivary glands secrete continuously, but the secretion of the large glands is controlled by the autonomic nervous system according to the feeding activity of the animal. In other tetrapods, the glands are usually small, the number, distribution, and detailed structure varying largely; in aquatic tetrapods the glands are reduced. Poison glands of many snakes, some lizards, and shrews are modified salivary glands. Salivary glands exist also in many invertebrates, serving digestion and some other functions, such are e.g. the → spinning glands of some insect larvae.

salivary gland chromosome, a → giant chromosome in the nuclei of salivary gland cells in the larvae of dipteran insects.

S-allele, (S = self-incompatibility), an → allele of those genes which regulate → self-incompatibility in plants.

salmonellosis, any disease in vertebrates caused by bacteria of the genus *Salmonella*. Salmonellae are aerobic, or facultatively anaerobic, Gram-negative bacteria of the family Enterobacteriacae. One species of bacteria may live in several animal species, as *S. enteritidis* especially in rats, mice, and fowl, *S. arizonae* in reptiles, fowl and many domestic mammals, and *S. typhi* mainly in man. Many *Salmonella* species living in animals can be transferred to man e.g. in contaminated food; infection is usually manifested as gastroenteritis with diarrhoea, vomiting, and fever.

Salmonidae (L. *Salmo* = genus of salmon), a family of bony fishes (in the order Salmoniformes) living in the northern hemisphere mainly in fresh waters, but also in oceans; some species are → anadromous; economically important fishes.

Salmoniformes (L. *forma* = shape), an order of bony fishes having many primitive features, often an adipose fin behind the dorsal fin; deep-water species have photophores. The order includes 270 species in five suborders: Esocoidei (pikes, mudminnows), Salmonoidei (salmon, trout, chars), Argentinoidei (e.g. herring smelts, deap-sea smelts), Stomioidei (e.g. scaly dragonfishes) and Giganthuroidei (giganthurids).

salps, → Thaliacea.

saltation (L. *saltare* = to jump, dance), the action of leaping, jumping, or dancing. *Adj.* **saltatory**.

saltatory conduction, the jumping of nerve impulses from one → node of Ranvier to the next in myelinated nerve fibres; saltatory conduction greatly speeds the propagation of a nerve impulse.

saltatory replication, a sudden lateral amplification of a DNA segment producing several copies of this DNA region. *See* amplification.

salt gland, 1. a gland in marine reptiles, birds and some seals excreting excessive salt (NaCl) out of the body by an active ion pump of the glandular cells. The paired salt gland is a modified lacrimal gland located usually inside the edge of the eye ball socket. In marine fish, salt excretion usually takes place through the gills by the action of epidermal cells but some fish species, such as rays, have a salt gland in the hindgut; **2.** a salt-secreting structure on the surface of some plants living in very saline surroundings; e.g. shoots of *Tamarix articulata* may be covered by crystallized salts.

salts, chemical compounds formed when one or more hydrogen atoms of an acid molecule are replaced by a metal or radical acting like a metal; typically formed by the reaction of acids with bases. In aqueous solutions, salts exist in ionic form and are electrolytes like acids and bases. E.g. when sodium hydroxide (NaOH) reacts with hydrochloric acid (HCl), sodium chloride (NaCl, table salt) is formed: $Na^+ OH^- + H^+ Cl^- \longrightarrow NaCl + H_2O$; in water NaCl dissociates completely into Na^+ and Cl^- ions. An example of an organic salt is potassium acetate CH_3COOK that dissociates into K^+ and CH_3COO^- (acetate) ions.

Salvidae, a subclass in the class → Pteropsida, division Pteridophyta, Plant kingdom; in some systems the order Salviniales in the division Pterophyta. All species are water-ferns; those in the family Salviniaceae are free-floating aquatics (e.g. the genera *Salvinia* and *Azolla*), and the species in the other family Marsileaceae grow in marshy places (e.g. *Pilularia* and *Marsilea*).

sampling methods, ecological methods for estimating population densities of animals or structures of vegetation. *See* partial sampling, random sampling, systematic sampling.

sand, a soil type, grain size from 0.2 to 2 mm in diameter.

sandgrouse, → Pteroclidiformes.

sap, the natural fluid or juice of a living organism; e.g. **1.** in the eukaryotic cell; cell sap (hyaloplasm) and nuclear sap (karyoplasm); **2.** a body fluid such as blood, lymph, saliva; **3.** the sugar-rich liquid flowing in the → xylem of foliage trees in spring. The xylem transfers water; the other type of → vascular tissue, phloem, transfers the organic products of photosynthesis, but especially in spring the water in xylem also contains organic compounds, which originate from storage tissues, especially from roots. The best known sap trees are the Canadian maple and the birch species.

saponification (L. *sapo* = soap), any reaction in which a neutral fat reacts with a suitable base, such as sodium hydroxide, NaOH, potassium hydroxide, KOH, or calcium hydroxide, $Ca(OH)_2$, producing a glycerol molecule and three molecules of fatty acid salts (soaps).

saponins, steroid type triterpene glycosides occurring in many plants. They form a foam with water and are used to permeabilize cellular membranes, and as detergents because they consist of lipid-soluble (triterpene) and water-soluble (sugar) elements. Saponins are toxic to cells because of their ability to disrupt cellular membranes, e.g. causing the haemolysis of red blood cells. Saponins obtained from the yam (*Dioscorea*) are used in the synthesis of progesterone-like hormones for contraceptives.

saprobiont (Gr. *sapros* = rotten, *bios* = a life), → saprophage.

saprobe, → saprophage.

saprophage, also **saprophagan** (Gr. *phagein* = to eat), an organism utilizing dead or decaying organic matter; *syn.* **saprobiont**, **saprobe**, **saprotroph**. Most typical saprophages belong to Fungi; a saprophagous (saprotrophic) plant can be called **saprophyte**, an animal **saprozoite** (saprozoon).

saprophyte (Gr. *phyton* = plant), *see* saprophage.

saprotroph (Gr. *trophe* = nutrition), → saprophage.

saprozoite, saprozōon (Gr. *zoon* = animal), *see* saprophage.

sapwood, the outermost region of the xylem of tree trunks, containing the living cells of secondary tissues, functioning in water conduction and storing. *Syn.* splintwood, alburnum. *See* heartwood.

sarc(o)- (Gr. *sarkos* = flesh), denoting relationship to flesh, or muscle.

sarcoblast (Gr. *blastos* = germ), the cell developing into a muscle cell. *Syn.* myoblast.

Sarcodina (Gr. *sarkodes* = like flesh), **sarcodinans**; a protozoan subphylum (or superclass) in the phylum Sarcomastigophora; also classified as a cladss in the subphylum Sarcomastigophora. Sarcodinans include e.g. **Rhizopoda**, such as amoebas (Amoeba) and entamoebas (Entamoeba), and foraminiferans (Foraminifera), and **Actinopoda**, such as heliozoans (Heliozoa) and radiolarians (Radiolaria). Sarcodinans are free-living or parasitic animals with pseudopods or actinopods without flagella. *See* Protozoa.

sarcolemma (Gr. *lemma* = husk), the → plasma membrane of the muscle cell.

sarcoma, pl. **sarcomas** or **sarcomata** (Gr. *sarkoma* = fleshy growth), a tumour type derived from connective tissues, such as fibrosarcoma, lymphosarcoma, chondrosarcoma, osteosarcoma. *Cf.* carcinoma, tumour.

Sarcomastigophora (Gr. *mastix* = whip, *pherein* = to bear), a phylum (or subphylum) in the subkingdom (phylum) → Protozoa; flagellated or pseudopodial protozoans, most are free living, some are parasitic in plants and animals; includes the subphyla Mastigophora, Opalinata, and Sarcodina. *See* Protista.

sarcomere (Gr. *meros* = part), the contractile unit of a **myofibril** in the muscle cell. Successive sarcomeres along the length of the myofibril are separated from each other by **Z-discs**. The length of a sarcomere in a resting cell is usually 2 to 3 μm, and consequently there may be hundreds of successive sarcomeres in one myofibril, especially in skeletal muscle fibres. In muscles having cross striations, as in skeletal and cardiac muscles of vertebrates, the sarcomeres of different myofibrils are situated in a fibre side by side, and thus the distances between the striations correspond to the lengths of myofibrillar sarcomeres.

sarcoplasm (Gr. *plassein* = to mould), the → cytoplasm of a muscle cell.

sarcoplasmic reticulum (L. *reticulum* dim. of *rete* = net), SR,a special type of the → endoplasmic reticulum occurring in the sarcoplasm (cytoplasm) of the muscle cell. It is an agranular form of the reticulum containing calcium stores from which calcium ions, necessary for contraction, are released into the sarcoplasm by → action potentials. The reticulum is best developed in fast-contracting muscles, such as in skeletal muscle cells, and is weakly developed in smooth muscle cells, and in cardiac muscle cells of ectothermic vertebrates. *See* excitation-contraction coupling, calcium-induced calcium release.

sarcoplast (Gr. *plastos* = formed), → satellite cell.

Sarcopterygii (Gr. *pterygion* = fin), **sarcoptergygians, fleshy-finned fishes**; a subclass in the class Osteichthyes; have fleshy fins with a large median lobe, functional lungs, and internal nostrils connected to the pharynx through → choanae like in all terrestrial vertebrates. The subclass is divided into four orders: Coelacanthiformes (**lobe-finned fishes**), Ceratodiformes and Lepidosireniformes (**lungfishes**), and Polypteriformes (**bichirs** and **reedfishes**). The former name for fleshy-finned fishes was Choanichthyes. *See* Dipnoi.

SAT-chromosome (SAT < L. *sine acido thymonucleonico* = without thymonucleic acid, i.e. DNA), a chromosome in which a → secondary constriction separates a → satellite from the rest of the chromosome. The chromosome was named SAT because the constrictive region is not stained similarly to the remaining parts of the chromosome. SAT-chromosomes are fairly common in all groups of organisms.

satellite (L. *satelles,* gen. *satellitis* = bodyguard, escort), companion; one that is secondary or adjacent; e.g. **1.** the chromosomal satellite; a short chromosomal segment separated from the rest of the chromosome by a → secondary constriction; **2.** the centriolar satellite; a special cellular body associated with the → centriole of a cell and serving as a site for polymerization of → tubulin; **3.** a small mass of chromatin found adjacent to the nucleolus in most nerve cells of female mammals; **4.** in astronomy, a celestial body orbiting another of larger size, or a man-made object revolving around a planet or other celestial bodies.

satellite cell, 1. an undifferentiated cell type associated with muscle fibres in the skeletal muscles of vertebrates. When a part of the muscle is destroyed, these cells are able to proliferate new muscle fibres; *syn.* sarcoplast; **2.** an → oligodendrocyte encapsulating a ganglion cell; *see* neuroglia.

satellite DNA, that part of the DNA of the genome of a cell which in centrifugation settles uppermost in the centrifuge tube, i.e. is the lightest layer of DNA and is separable as an independent satellite from the rest of DNA. It consists of highly repetitive DNA, in which short segments or sequences are repeated up to millions of times. In the chromosome, satellite DNA is located in the constitutive → heterochromatin (*see* sequence complexity of DNA).

saturated community, a community in which every niche is occupied by a species adapted to it, i.e. there are equal numbers of species and niches in the community.

saturation, 1. the act or state of a solution containing the maximum amount of solute to be dissolved under given conditions; in this state the solution is in equilibrium with its solute; **2.** the saturation of vapour (gas) under given conditions (temperature, pressure), e.g. the maximum value of relative air humidity (= 100% relative humidity); any increase of the vapour volume in the same conditions causes the vapour to condense into liquid; **3.** the saturation of an organic compound in which all carbon-carbon bonds are single; e.g. butyric acid, $CH_3(CH_2)_2COOH$, and palmitic acid, $CH_3(CH_2)_{14}COOH$; *cf.* unsaturated compounds; **4.** light saturation in → photosynthesis: the light-energetic mechanisms are saturated by strong light; **5.** the treatment of timber (**impregnation**) against rot and blue-stain fungi. *Adj.* **saturable**. *Verb* to **saturate**. *Noun* **saturant**, something that saturates, or is capable of being saturated.

Sauria (Gr. *sauros* = lizard), **Lacertilia, lizards**; a reptilian suborder which together with snakes (Serpentes) and worm lizards (Amphisbaenia) form the order Squamata. Sauria comprises about 3,300 species in 17 families, most of them terrestrial, some semiaquatic. Lizards are → oviparous or ovoviviparous, and usually they have four limbs with five digits on each; some species are legless, such as the slow-worm, *Anguis fragilis*.

Saurischia (Gr. *ischion* = hip), *see* dinosaurs.

Sauropsida, sauropsids, sauropsidans; a collective name for all extant and nearly all extinct reptiles as well as all birds.

savanna(h), a type of → grassland with scattered shrubs and a few or no trees, typical of areas between the desert and tropical rain forests in East Africa.

saxitoxin (L. *Saxidomus giganteus* = a species of butter clam, Gr. *toxikon* = arrow poison), a powerful poison produced by a marine dinoflagellate, *Gonyaulax catenella* ("red-tide"); tends to accumulate in many organisms, especially in mussels. Its action mechanism in cells is similar to that of → tetrodotoxin.

scabies (L. *scabere* = to scratch), scab, itch, mange; a dermatitis caused by mites (e.g. *Sarcoptes*, *Psoroptes*, *Chorioptes*) living in the skin of many wild and domestic animals, mainly mammals. Female mites dig tiny burrows (cuniculi) inside the epidermis where eggs are laid, and in a couple of weeks a new generation has developed. Mites are transmitted by close contact with other hosts, and they may cause an epidemic disease in dense populations of animals (e.g. in foxes) causing severe stress and decreased reproduction. The human itch mite (*Sarcoptes scabiei*) lives mostly in the skin of the joints, and may be stressful, especially if the skin is infected secondarily by bacteria. *Adj.* **scabietic**.

scaffold, a protein structure of any chromosome revealed in the form of a → sister chromatid pair after the histones and DNA have been removed from the chromosome.

scala, pl. **scalae** (L. = staircase), in anatomy, a stair-like structure, especially the canals of the cochlea in the inner ear of vertebrates: *scala vestibuli* (vestibular canal), *scala media* (cochlear duct), *scala tympani* (tympanic canal). *See* cochlea.

scale, a small, lamellar surface structure, which in vertebrates is derived either from the epidermis (**horny scales** or **scutes**) or dermis (**bony scales**). The former are found e.g. covering the bodies of snakes and lizards, on bird's feet, on the tails of rats and mice, and covering the body of armadillos. Bony scales are e.g. the scales of fish, which more specifically are named **cosmoid scales** of crossopterygians and dipnoids, **ganoid scales** of sturgeons, tooth-like **placoid scales** of cartilaginous fishes, and **ctenoid** and **cycloid scales** of bony fishes. Scales exist also in some invertebrate groups, e.g. the **wing scales** of lepidopterans. In plants, different scales may be found, such as appendages of epidermal cells (*see* epidermis).

scale leaves, the first, oldest leaves of a plant, preceding the formation of actual leaves. Their growth is often inhibited and they may be scale-like, especially in → rhizomes, in which they are the only leaves. Scale leaves appear also as bud scales. *Cf.* bracts.

Scandentia (L. *scandentis* = climbing), **tree shrews**; an order of small, squirrel-like mammals; in some classifications a family (Tupaiidae) in the order Insectivora, or in Primates. Tree shrews live in tropical forests in India and southeastern Asia. They are mainly arboreal, some species also terrestrial; some structural characteristics resemble primates owing to → convergent evolution. The order comprises 16 species.

scanning electron microscope, SEM, *see* electron microscope.

scape (L. *scapus* = stalk, shaft of a column), a stalk or shaft; e.g. **1.** the basal joint of an antenna of some insects, especially when it is longer than the other joints; **2.** a leafless flower stalk rising from ground level; e.g. dandelion.

scapula, pl. **scapulae** (L. = shoulder), one of the three bones in the pectoral girdle of tetrapods, the shoulder blade in mammals.

SCAR, → sequence characterized amplified region method.

scattered loading, in environmetal ecology, a chemical loading deposited from many small separate discharge sources, such as vehicles, farms, and other scattered dwellings. The prevention and control of scattered loading is more difficult than that of point-loading (point-pollution). *Syn.* nonpoint-pollution, diffuse pollution.

scavenger, 1. an animal feeding on dead animals and decaying materials, e.g. scavenger beetles; **2.** a chemically active substance in a mixture for removing an undesirable substance, or a substance, such as e.g. vitamin E, influencing the course of a chemical reaction readily combining to free → radicals.

SCE, → sister chromatid exchange.

scent marking (L. *sentire* = to feel, perceive), a behavioural characteristic of many animals to announce themselves by spreading odorous substances into their environment as scent signals. The scents, often produced by special → scent organs or single cells, are spread in air, water, or in faeces or urine; the marking is mainly meant for other individuals of the same species. The functions of scent marking are various: establishing the ownership of territory, attracting sex partners, or warning species companions of danger. → Pheromones are specific scent substances.

scent organs, scent glands or solitary glandular cells producing odorous substances; found in many animal groups usually associated to scent hairs, bristles or scales which spread these substances around. *See* scent marking.

Schiff reagent, Schiff's reagent (*H. Schiff*, 1834—1915), a staining reagent used for determination of aldehydes; a fuchsin solution decolourized with sulphur dioxide producing a purple colour when reacting with aldehyde groups of other compounds. It is used for determination of carbohydrates. It also stains proteins and amino acids with aldehyde groups formed e.g. when heated with a ninhydrin reagent. *See* PAS.

Schistosoma (Gr. *schizein* = to cleave, split, *soma* = body), **schistosomes**, **blood flukes**; the genus of flukes, subclass *Digenea*, includes several blood parasites which cause a severe disease, → schistosomiasis.

schistosomiasis, a disease found in tropical and subtropical zones of the world caused by the blood parasites of the genus *Schistosoma* (digenean trematode); an infection causes haemorrhage and tissue injuries. Aquatic gastropods act as intermediate hosts, from which the parasites pass into warm-blooded vertebrates, three species (*S. haematobium, S. mansoni, S.japonicum*) pass also into man. Schistosomes are widely spread especially in Africa, Asia, and South America. In Africa, the disease is known as **bilharziosis**.

schizocarp (Gr. *karpos* = fruit), a fruit formed from an ovary, the original carpels of which split from each other forming separate **mericarps** (e.g. in the family Apiaceae).

schizocoel, also **schizoc(o)elom** (Gr. *koilos* = hollow), **schizoc(o)elous cavity**; a type of body cavity, → **coelom**, formed by splitting of the embryonic mesoderm; found in some protostomes such as annelids, molluscs, and arthropods. *Cf.* enterocoel.

schizogenic (schizogenous) intercellular space (Gr. *genos* = origin), an intercellular space formed in a plant tissue when some cells are broken and destroyed.

schizogony (Gr. *gone* = birth), a type of asexual reproduction in sporozoans by several mitotic cell divisions; e.g. inside the red blood cells of vertebrates, the schizogony of *Haemosporina* occurs by mitotic division of haploid trophozoites to merozoites. *Cf.* sporogony.

schizopod stage (Gr. *pous* = foot), → mysis.

schizozoite, → merozoite.

Schwann cell (*Theodor Schwann*, 1810—1882), a cell type that enwraps the axons of neurones in the peripheral nervous system of vertebrates, forming the → myelin sheath around an axon; the cells are comparable to → oligodendrocytes in the central nervous system. *See* nerve fibre, neuroglia.

Schwann sheath, neurolemma, neurilemma. *See* nerve fibre.

scintillation counting (L. *scintillare* = to sparkle), a technique for measuring quantities of radioactive isotopes (mainly β-emitters, such as ^{14}C and ^{35}S) present in a biological sample. A scintillation liquid added to a sample converts energy from ionizing radiation to light, and the light flashes are detected as electrical pulses with a **scintillation counter**.

scion, a plant shoot which is grafted to another plant. *See* grafting.

scler(o)-, sclera- (Gr. *skleros* = hard), denoting a hard or dry structure, or the sclera of the eye. *Adj.* **scleroid**, **sclerosal**, **sclerotic**, **sclerous**, pertaining to a hard tissue, having a hard texture.

sclera, pl. **sclerae, scleras** (L.), the fibrous, white, outermost layer of the eyeball in vertebrates, and a similar structure also in the eyes of some invertebrates, e.g. cephalopods; continues anteriorly as the cornea and posteriorly as the outer sheath of the optic nerve. *Adj.* **scleral**.

sclereid (Gr. *eidos* = form), stone cell; a cell type belonging to the → sclerenchyma and further to → supporting tissues of plants. The walls of sclereids are very thick and lignified, and the cells may appear as layers, but mostly they are single, so-called **idioblasts**. The sclereids are classified according to their form: **brachysclereids** are roundish and common e.g. in the flesh of pear, **macrosclereids** are long and column-like (in the seed coat of pea), **trichosclereids** are thin and branching (in some aerial roots), **osteosclereids** are bone-like (in seed coats and leaves of many plants), and **asterosclereids** star-shaped (in many leaves).

sclerenchyma pl. **sclerenchymas** or **sclerenchymata** (Gr. *enchyma* = cast, infusion), a type of plant tissue belonging to the supporting tissues together with the → collenchyma. The cells of the sclerenchyma have mostly a thick, hard, even lignified → secondary wall. The types of sclerenchyma are → sclereids and → fibres. *Syn.* sclerenchyme.

sclerite, a small chitinous plate or spicule, such as the sclerotized plates of the exoskeleton particularly in arthropods. *See* tergum.

scleroblasts (Gr. *blastos* = bud), amoebocyte-like cells of a sponge by which calcareous and siliceous spicules are formed on the body wall of the animal.

sclerocyte (Gr. *kytos* = cavity, cell), a cell type in the body wall of sponges. *See* Porifera.

sclerosis, pl. **scleroses** (Gr. *sklerosis* = hardness), **1.** an abnormal hardening of a tissue or an organ; e.g. hardening of blood vessels (arteriosclerosis) or nervous structures (e.g. multiple sclerosis); **2.** a hardening of plant cell walls by the thickening or deposition of lignin. *Adj.* **sclerotic**.

sclerotium, pl. **sclerotia**, a compact, hard-walled structure formed of fungal hyphae, varying in size from some millimetres to some tens of centimetres. The sclerotium is capable of remaining dormant for long periods allowing the fungus to survive during unfavourable conditions. In good conditions the hyphae continue to grow and the sclerotium may give rise to → fruit bodies, e.g. in the ergot.

sclerotome (Gr. *temnein* = to cut), **1.** the ventromedial part of the → somite in a vertebrate embryo; consists of mesenchymal cells, which migrate towards the notochord to form the skeleton (cartilages and bones); **2.** a surgical instrument for the incision of the sclera.

scopolamine, an alkaloid found in several solanaceous plants (e.g. in the genus *Scopolia*); like another similar alkaloid, atropine, it inhibits the nervous system by blocking acetylcholine receptors in the autonomic nervous system. *Syn.* hyoscine, hyoscyamine.

scorbutus (L.), → scurvy.

Scorpaeniformes (L. *Scorpaena* = the type genus < Gr. *skorpios* = scorpion, L. *forma* = shape), **scorpion fishes, bullheads**; an order of bony fishes characterized by a large head, and spines and bony scales on the body surface; the spines are often associated with poison glands. The order includes about 1,000 species in 21 families.

Scorpionida, Scorpiones, scorpions; an order of Arachnida characterized by strong, pincer-like pedipalps and a venomous sting at the tip of the abdomen. The body comprises the prosoma and a fairly long abdomen. Scorpions are large arachnids, most ranging from 3 to 9 cm in length; ca. 1,200 species which live in tropical and subtropical areas.

Scots pine blister rust, a disease of Scots pine (*Pinus silvestris*), caused by rust fungi *Cronartium* and *Endocronartium* (→ Uredinales). Seedlings usually die soon after infection but full-grown trees may stay alive tens of years. The infected trees are the best material for tar production.

scramble competition, → intraspecific competition.

scrapers, aquatic animals which prey on microscopic periphyton by scraping it from the surface of water plants; scrapers are found e.g. among some gastropods, as well as among larvae of trichopterans, ephemeropterans, and chironomids.

scrapie, a degenerative disease of the central nervous system in sheep and goat, characterized by pruritus, muscular uncoordination, and death; caused by a → prion with a very long incubation time.

scRNA, → small cytoplasmic ribonucleic acid.

scRNP, → small cytoplasmic ribonucleoprotein particle.

scurvy (L. *scorbutus*), a disease of some mammals and birds (e.g. primates, guinea pig), characterized by haemorrhaging in the mucous membranes and skin due to dietary deficiency of ascorbic acid, resulting in disturbance in collagen synthesis.

scute (L. *scutum* = shield), any large scale, or a scale-like structure; e.g. **1.** the tympanic scute, the bony plate separating the upper part of the tympanic cavity from the mastoid cells; **2.** the horny plate on the turtle, or the dermal bony plates on the armadillos. *See* scutum.

scutellum, pl. **scutella** (L. dim. of *scutum* = shield), a small plate in an anatomical structure; e.g. **1.** one of the transverse scales on the tarsi and toes of birds; **2.** the third dorsal → sclerite behind the → scutum of a thoracic segment of an insect; **3.** the shield-like cell layer between the endosperm and the embryo in seeds of grasses (Poaceae); it has an important role in seed germination. *Adj.* **scutellate(d)**, having scutes, or formed into a scutellum. *Cf.* scutum.

scutum, pl. **scuta** (L.), **scute**; a shield, or shield-like structure; e.g **1.** in vertebrate anatomy, the tympanic → scute, or the thyroid cartilage; **2.** a chitinous plate on the dorsal surface of hard-bodied ticks; **3.** the second dorsal sclerite of a thoracic segment behind the praescutum of an insect; **4.** one of the two lower valves of the operculum of a barnacle. *Cf.* scutellum.

scyphistoma, pl. **scyphistomae,** also **scyphistomas** (Gr. *skyphos* = cup, *stoma* = mouth), scyphopolyp; a developmental stage of sessile polyps in the life cycle of scyphozoans; in this stage the body has several tentacles, and it reproduces asexually by budding plate-like ephyrae, and is therefore also called

strobilation stage. *See* strobila.

scyphopolyp (L. *polypus* = polyp), → scyphistoma.

Scyphozoa (Gr. *zoon* = animal), **scyphozoans, jellyfishes**; a class in the phylum Cnidaria. The life cycle of these animals is characterized by a free-swimming, pelagic **medusa stage**, and the sessile **polyp stage** which may be absent or is less significant and smaller in size than the medusa stage; the → coelenteron is often divided into four compartments. Scyphozoans include ca. 200 species classified into four orders: Coronatae, Rhizostomeae, Semaeostomeae, and Stauromedusae; sometimes also Cubomedusae are included in this class.

SDS, → sodium dodecyl sulphate.

SDS-PAGE, sodium dodecyl sulphate-polyacrylamide gel electrophoresis; a technique of → electrophoresis using polyacrylamide gel for the separation of proteins under denaturing conditions. The mixture of proteins is first dissolved in an SDS solution; SDS disrupts nearly all noncovalent interactions in native proteins. Disulphide bonds are reduced by mercaptoethanol or dithiothreitol. SDS forms complexes (about one SDS for two amino acid residues) with denatured proteins and gives a large negative charge which is much greater than the charge of native protein. SDS-PAGE is a rapid and sensitive determination method for protein analyses.

sea cows, → Sirenia.

sea cucumbers, → Holothuroidea.

search(ing) image/model, a hypothetical mental picture evolved in a predator to select such prey species which produce maximum profit in respect to energy used for the preying.

search(ing) time, a factor in assessing predation efficiency, i.e. the time a predator needs for searching of prey. A further factor is → **handling time** used for dealing with the prey after it is caught. The reciprocal relations between the search and handling times, and especially the sum of these times, decide the **optimal feeding behaviour** of the animal. *See* optimum foraging theory.

seasonal migration, a migration of animals occurring in regular rhythms and controlled by the seasonal variation of the environment. Migration is mainly based on annual physiological rhythms of animals influenced by temporary environmental changes. The migration route and goal are usually determined by genetic factors. Seasonal migration is particularly common among birds, some large savanna mammals, bats, fishes, and insects. Birds in particular, have various migration types, as e.g. **route** and **loop migrations**. *See* irruptive species, migration.

seasonal ring, → annual ring.

sea squirts, → Ascidiacea.

sea stars, → Asteroidea.

sea urchins, → Echinoidea.

sebaceous gland (L. *glandula sebacea* < *sebum* = tallow), any exocrine gland in the skin of mammals secreting a thick semifluid fatty substance, **sebum**. It is a tubulous holocrine gland in which the entire cell disperses in the process of secretion forming the material of excretion. The gland opens into the upper part of the hair follicle lubricating the hair and skin. The → oil gland of birds is homologous with the sebaceous glands of mammals.

secalin (L. *secale* = rye), a prolamine type protein that is found as reserve substance in rye seeds.

secalose, a polysaccharide consisting of fructose units found in rye and oat.

secondaries (L. *secundus* = second), the flight feathers attached to the ulna in bird wings. *Cf.* primaries, tertiaries.

secondary cell wall, the secondary part of the cell wall of plants, deposited on the → primary cell wall. The secondary wall is laid down on the primary cell wall after the cell has ceased growing. It is much thicker than the primary wall and is composed mainly of cellulose microfibrils laid down in at least three layers, with the microfibrils in the layers having a different orientation. The secondary cell wall can also be lignified (*see* lignin).

secondary coelom, *see* coelom.

secondary compounds, by-products formed in plant metabolism having no apparent metabolic functions in plant cells. Their evolutionary importance is supported by their allelopathic effects (*see* allelopathy) and usage for defence against herbivores; e.g. alkaloids in Solanaceae plants and nicotine in the tobacco plant. *See* chemical defence.

secondary constriction, any constriction of a chromosome other than the → primary constriction, formed by the centromere. Usually a gene family of the → ribosomal RNA is situated at a secondary constriction, and the constriction is associated with the nucleolus.

secondary cortex, → secondary phloem.

secondary growth, a type of growth due to an increase of thickness in plant shoots and roots, caused by a plant tissue other than the → apical meristem. In practice, secondary growth is caused by the vascular → cambium, which forms → secondary xylem and → secondary phloem. Also the formation of periderm or bark, caused by → phellogen, can be considered as secondary growth, most marked in trees and shrubs. *Cf.* primary growth.

secondary host, → intermediate host.

secondary intergradation zone, an area where formerly geographically isolated populations meet each other, resulting in a zone where variations of individual characteristics of hybrids exceed variations of primary populations.

secondary meristem (Gr. *meristos* = divided), a type of plant → meristem comprising cells which develop from already differentiated tissue and regain the ability to divide; secondary meristems arc the → **cambium** that supports → secondary growth in plants, the → **phellogen** which produces → periderm in roots and shoots, and the wound → **callus** that forms callus tissue protecting damaged areas in plants.

secondary phloem, a → phloem tissue in plants, formed by the vascular → cambium during → secondary growth. It replaces the → primary phloem which belongs to the original structure of shoots and roots.

secondary production, the organic material produced by consumers such as herbivores, carnivores, and detritus feeders in an ecosystem. *Cf.* primary production. *See* productivity.

secondary sex characters, *see* sex characters.

secondary sex ratio, *see* sex ratio.

secondary succession, *see* succession.

secondary tissue, a plant tissue formed secondarily by the vascular → cambium (forms secondary xylem and phloem) or → by → phellogen (forms periderm or bark); causes → secondary growth.

secondary xylem, → xylem tissue in plants, formed by vascular → cambium during secondary growth, replacing the → primary xylem in roots and shoots; forms a cylinder with annual rings especially in tree stems.

second messenger, a substance that transmits the action of the first messenger (e.g. hormones and growth factors) inside its target cell, producing a change in cellular functions, such as enzyme activation or gene function. Examples of second messengers are → cyclic adenosine monophosphate (cAMP), inositol triphosphate (IP_3) and diacylglycerol (*see* inositol), and calcium. *See* signal transduction.

secretin (L. *secernere* = to separate), a polypeptide hormone secreted by glandular cells of the mucosa in the anterior part of the intestine in vertebrates (duodenum and upper jejunum in mammals); it is secreted when acid chyme enters the intestine. Stimulating pancreatic exocrine cells, secretin causes the release of bicarbonate-rich secretion from the pancreas into the gut. To a lesser degree, secretin also stimulates pepsin, bile, and intestinal secretions, and inhibits gastric acid secretion and the activity of intestinal smooth muscle.

secretion (L. *secretio*), **1.** the process of releasing specific products by specialized cells (glandular cells), which often form a whole organ, a → gland; **2.** a substance produced by secretion. *Adj.* **secretory**. *Verb* to **secrete.** *See* endocrine glands, exocrine glands.

section (L. *secare* = to cut), **1.** a part that is cut off, or a surface being cut, as e.g. a slice of tissue or cell for examination under the microscope; **2.** the act of cutting, such as in a medical operation; **3.** a segment of an organ; **4.** a taxonomic group, often a subdivision of a genus, not clearly determined; **5.** a segment of a fruit, as e.g. found in oranges.

security time, a certain minimum time which must pass before plants, treated with → pesticides, can be used by man and domestic animals; varies from one day to several weeks.

sedation (L. *sedatio* < *sedare* = to calm), the process of calming, especially after an administrated drug; the state of being calm.

sedative, 1. calming, allaying activity; **2.** an agent that allays excitement.

sedentary (L. *sedere* = to sit, settle), **1.** stationary, not migratory, staying in the same place, as sedentary birds; **2.** sessile, i.e. permanently attached to a substratum, e.g. sedentary barnacles, hydras, many clams and tunicates; sedentary species usually have motile larvae.

sedimentation, 1. the deposition and stratification of solid matter from air or water to the ground or bottom; the produced matter is called **sediment**; **2.** the act of producing the deposit of a sediment, e.g using a → centrifuge.

sedimentation coefficient, s, the sedimentation

velocity of a particle in the centrifugal field. *See* Appendix 1.

sedoheptulose, a 7-carbon monosaccharide in all photosynthetic plants, first reported to appear in leaves and stems of several species in the genus *Sedum* (stonecrops). Sedoheptulose phosphates function as intermediates in the Calvin cycle (*see* photosynthesis) and in the → pentose phosphate pathway.

seed, a reproductive unit of certain plants (seed plants), i.e. → gymnosperms and → angiosperms; consists of a protecting seed coat, embryo, and usually also the → endosperm as a food store. The seed is formed from the fertilized → ovule. In angiosperms, the ovules are in the ovary which forms the fruit during the development of the seeds, but in gymnosperms the ovules are bare on the carpels and the seeds develop without protection.

seed bank, the storage of seeds of wild plant species and varieties, often of important crop plants. The seeds are collected in order to preserve them for future times and needs; they are stored well protected in dry conditions, or freeze-dried in liquid nitrogen. Seed banks have been established in many countries around the world.

seed coat, the protecting coat of a mature → seed; develops from the integuments of the → ovule.

seed ferns, → Pteridospermae.

seedling, any young plant that develops from the embryo after the germination of the → seed; a typical seedling consists of the → radicle, hypocotyle, cotyledons and the first foliage leaves.

seed plants, Spermatophyta, the most developed plants having a complete structure (i.e. roots, stem, and leaves) and reproduction with seeds.

segment (L. *segmentum*), a part of a larger structure in an organism delimited naturally or arbitrarily from the whole; e.g. spinal segments of vertebrates, body segments of annelids, or leg segments of arthropods. *Adj.* **segmental**. *Cf.* somite.

segmental organ (Gr. *organon* = implement), an organ whose structure repeats in a similar form in each body segment of an animal, e.g. the nephridia of annelids, or an embryonic excretory organ in vertebrates, i.e. the pronephros and mesonephros.

segmentation, 1. division of an organism or part of it into similar parts, such as the structure of an organism that is formed by successive (in bilateral organisms) or radial segments (in radial organisms). Segmentation is typical of most arthropods and worms whose successive segments may be very similar (*see* metamerism). In vertebrates, segmentation occurs most conspicuously in the course of embryonic development (→ somites). Later it is seen especially in the structure of the vertebral column and the positions of spinal nerves, but also in the motor and sensory innervation of the body trunk (*see* dermatomes). Segmentation occurs less clearly in the fully developed musculature and inner organs of higher vertebrates;
2. cleavage, especially series of successive cell divisions after the fertilization of an ovum.

segmentation genes, genes determining the development of body segments in the → zygote; found so far only in insects. They are divided into → **cardinal genes** (gap genes), **pair-rule genes**, and **segment polarity genes**. The segmentation genes work hierarchically in this order regulating the function of → selector genes as well as their own function.

segment polarity genes, → interpretation genes of a zygote reading the positional information created by the → coordinate genes downstream of the → pair-rule genes. The segment polarity genes regulate the polarity of body segments and somites. In insects, mutations of these genes cause developmental disturbance such that the segments have either two anterior or two posterior parts.

segregation (L. *segregatio* = splitting), **1.** the splitting of the homologous chromosomes of a bivalent during → meiosis; **2.** the separation of → allele pairs of → homologous chromosomes from each other and their distribution into different cells during meiosis; → Mendel's laws of inheritance concern the segregation of alleles and chromosomes; **3.** sometimes denoting the distorted behaviour of homologous chromosomes during → mitosis (somatic segregation); **4.** the separation or isolation of individuals or groups from a population.

seiche (Fr.), a stationary oscillation of waves in a small lake, varying from a few minutes to several hours; it is thought to be initiated chiefly by sudden strong winds due to local variations in atmospheric pressure.

seismonasty (Gr. *seismos* = shock, earthquake, *nastos* = pressed), a → nastic movement in plants caused by environmental vibration or mechanical trembling.

Selaginellales, an order in the class Lycopsida, division Pteridophyta (in some systems division Lycophyta), Plant kingdom. The appearance of these small herbaceous species are in many respects like that of the lycopods (→ Lycopodiales), but are characterized by heterospory and extremely reduced prothalli. Only a few species live in Europe but some hundreds in the Tropics.

selectins, a family of carbohydrate-binding adhesive proteins on the cell surface; found in the blood (E and P selectins) and many other tissues of vertebrates, some are attached to the cell membrane of leucocytes (L selectin) regulating their function.

selection, in biology, the consequence of forces which determines the relative reproductive effectiveness of the various genotypes in a population. Selection is the most important evolutionary factor which affects the → gene pool and → gene frequencies of a population. Biological selection may be due to a difference in the survival and reproduction capacity of different genotypes, and has always a positive and negative aspect and effect. The selective success of certain genotypes (**positive selection**) means the elimination of other genotypes (**negative selection**). Selection produced by man in populations of cultivated plants and domestic animals, is called **artificial selection**. Its goals are in plant and animal breeding, and when individuals with certain desired characters are chosen for reproduction, selection usually improves these characters. *See* evolution, natural selection, sexual selection.

selection coefficient, a measure of the effectiveness of → selection; compares the ability of a particular → genotype to transmit gametes to form the next generation with that of the average genotype. The selection coefficient (**s**) indicates the proportion of a given genotype that is eliminated in selection. The selection coefficient is the opposite concept to adaptive value (**w**), so that **w** + **s** = 1. *See* fitness.

selection differential, symbol S; in artificial selection, the difference between the mean genotypic values of individuals selected for breeding and the mean genotypic values of all individuals in the population.

selection limit, a population is at the selection limit when selection no longer has an effect on it.

selection pressure, the effectiveness of → natural selection in a population, usually measured by the magnitude of the change in → gene frequencies due to selection in the population.

selection response, symbol R; in artificial selection, the difference between the mean → genotypic value of the progenies of the individuals selected for breeding, and the mean genotypic value of all individuals in the population.

selective advantage, the profit which the individuals of a given genotype in a population benefit in → natural selection as compared with the other genotypes.

selective disadvantage, a character of a genotype which in the absence of other differences shows selective disability in the population. The genotype is gradually replaced by another genotype with selective advantage if both are present within one habitat. *Cf.* selective advantage.

selective value, → fitness.

selectivity, the quality of being selective, the state characterized by selection; e.g. **1.** the property of the cell membrane for selective ion transport; **2.** the selectivity of a pharmacological agent acting only in one receptor type of the same receptor family.

selector gene, a gene which by regulating the function of other genes guides the realization of whole developmental programmes.

selenium (Gr. *selene* = moon), a yellowish, non-metallic element of the sulphur group, symbol Se, atomic weight 78.96; one of the essential micronutrients (trace elements) of organisms being a cofactor of → glutathione that is an important antioxidant in organisms. Selenium deficiency may cause e.g. the degeneration of the heart muscle. In some geographical areas the soil is so rich in selenium that plants absorb it in concentrations which are toxic for animals feeding on them, resulting in disturbances in the metabolism of the liver and bones, whereas in some other areas selenium must be added artificially to fields.

self-fertile, an individual (→ hermaphrodite) which possesses the sexual organs of both sexes producing gametes capable of forming a → zygote; the phenomenon is called **self-**

fertility.

self-fertilization, the fusion of male and female gametes from the same individual into a → zygote; in plants, self-fertilization is an automatic consequence of → self-pollination.

self-incompatibility (L. *incompatibilis* = not suitable for combination), the suppression of → self-fertilization found in hermaphroditic animals and monoecious plants. It is genetically controlled, and in plants it usually results from the incapability of a pollen grain to fertilize the egg cell which contains the same gene or genes (regulating self-fertilization) as the pollen grain itself.

selfish DNA, DNA located in the chromosome between → genes in all eukaryotic organisms. Its function is not known but according to one hypothesis, it is thought to be parasitic DNA, the only function of which is to replicate and thus propagate itself.

self-pollination, the transfer of pollen grains from anthers to stigma of the same flower (autogamy), or from one flower to another in the same plant (geitonogamy). Continuous self-pollination is often useful, because it keeps plant populations genetically consistent and preserves their adaptation to the environment. *Cf.* cross-pollination.

self-regulation a type of regulation of a biological process in cells, tissues, organs, organisms or communities when the final outcome from the process feeds back to the factors initiating it, and thus the process regulates its own function. The regulation of protein synthesis in a cell, blood circulation in a certain organ, or hormonal secretion are examples of physiological processes in which self-regulation mechanisms are involved. In ecology, the term denotes **self-limitation**, when e.g. in high population densities an intraspecific competition decreases natality or increases mortality.

self-regulation hypothesis, a proposition stating that especially in more developed animals such as vertebrates and arthropods, population densities are regulated at optimum levels in relation to environmental resources. This is achieved e.g. by → epideictic behaviour and conventional competition. *See* population regulation.

sella, pl. **sellae** (L. = saddle), any saddle-shaped anatomical structure; e.g. *sella turcica*, Turkish saddle, a depression on the superior surface of the sphenoid bone of vertebrates, containing the pituitary gland.

SEM, scanning electron microscope. *See* electron microscope.

semelparity (L. *semel* = once, *parere* = to give birth), a type of reproduction in which an organism reproduces only once in its lifetime. Such an organism allocates a large proportion of resources for reproduction, and dies soon thereafter (**big-bang reproduction**), e.g. salmon of the genus *Oncorhynchus*. *Adj.* **semelparous**. *Cf.* iteroparity.

semen, pl. **semina** or **semens** (L. = seed), **1.** any seed or seed-like structure; **2.** the fluid produced by male reproductive organs containing spermatozoa, especially the → sperm of mammals.

semi- (L. = half), denoting half, or partly.

semiaquatic (L. *aquaticus* = pertaining to water), pertaining to an organism living in a wet and humid environment, such as in marsh or at riversides; e.g. adult frogs and the otter.

semicanal, a half canal; a channel or canal that is open on one side, e.g. the semicanal of the auditory tube which is the groove in the bony part of the auditory tube in tetrapod vertebrates.

semicell, one of the two cell halves connected by a narrow isthmus containing the nucleus, found e.g. in some unicellular green algae, especially in placoderm → desmids.

semicircular (L. *circulus* = circle), relating to the form of a half circle, e.g. semicircular canals of the inner ear.

semiconservative, in genetics, describing a type of → replication of the DNA in which the double helix of DNA opens, and both strands build a complementary strand from free → nucleotides in the nucleus. As a result of the replication, two identical DNA molecules are formed, each containing one parental and one new strand.

semidominance, → incomplete dominance.

semilethal (L. *lethalis* = mortal), → sublethal.

semilunar (L. *luna* = moon), a structure resembling a half-moon or crescent, e.g. semilunar valves in the vertebrate heart located at the bases of the aorta and pulmonary artery.

seminal (L. *semen* = seed), pertaining to seed or semen.

seminal receptacle, → spermatheca.

seminal vesicle (L. *vesicula*, dim. of *vesica* = bladder), *vesicula seminalis*; **1.** a paired sacculated gland connected with the male reproductive tract in amniotic vertebrates; comprises several glandular sacs producing a secretion, which is rich in fructose, proteins and prostaglandins. The ducts from the sacs join and open to the ductus deferens releasing secretion in ejaculation; in lower vertebrates the seminal vesicle also stores spermatozoa; *syn.* seminal gland; **2.** the vesicle associated with the reproductive organs in some male or hermaphrodite invertebrates, such as flatworms and earthworms, for maturing and storing the sperm.

seminiferous (L. *ferre* = to bear), producing or conveying semen, or producing or bearing seed; e.g. **seminiferous tubules** and **seminiferous duct** in the vertebrate → testis.

semiparasite, hemiparasite (L. *semi-*, Gr. *hemi-* = half), an organism, which lives partly as a → parasite, but may survive also without a host; e.g. **1.** a microbe which is commonly parasitic but is also capable of living on dead or decaying organic material; **2.** a plant, which develops independently from seeds and is capable of photosynthesis, but has also connections with host plant roots through its own specific roots, called → haustoria; e.g. the mistletoe (*Viscum album*), and numerous species in the family Scrophulariaceae. *Cf.* holoparasite. *See* parasite.

semipermeability (L. *permeabilis* = passing through), partially permeable; a selective → permeability of natural or artificial membranes, freely permeable to water and many other solvents but relatively impermeable or selectively permeable to solutes. *Adj.* **semipermeable**. *See* osmosis.

semispecies, a term used occasionally to designate a taxonomic unit between the species and subspecies; it is a borderline case in speciation and usually concerns populations in geographically isolated districts (*see* isolate).

senescence (L. *senescere* = to grow old), old age; the process of growing old, i.e. the ageing of cells, tissues, organs, and organisms. *Adj.* **senescent**, exhibiting senescence.

senility (L. *senilitas*), **1.** old age; characteristic of old age; **2.** the deterioration associated with old age; **senilism**, premature old age. *Adj.* **senile.**

sensation (L. *sentire* = to perceive, feel), **1.** the function of senses; the awareness or impression conceived through stimuli from the senses; e.g. dermal (cutaneous) sensation, i.e. skin sensation; **2.** a mental (excited) feeling. *Adj.* **sensate**, perceived by the senses.

sense (L. *sensus*), **1.** a faculty by which animals perceive and process stimuli from the outside and inside of the body. Protozoans have special sensory structures in their unicellular body, but multicellular animals have specialized → **sense organs** with specific sensory cells or parts, **sensory receptors**, which are mostly sensitive only to one type of energy (adequate stimulus). In the most developed animals, sensory information is processed in the brain to generate subjective **sensations** (perceptions, experiences). Traditionally it is thought that man has five senses: sight, hearing, smell, taste, and touch, but in fact touch includes different separate senses, such as pain, pressure and thermal senses. In addition, there are senses of equilibrium and muscle or tendon stretch, and several receptors in the internal organs for monitoring the homeostatic state of the body (e.g. receptors for blood pressure, carbon dioxide, and brain temperature); most of these senses do not give any sensation. The sensory world of a species seems to be an adaptation to demands of its special environment, and many animal groups have senses which man does not have, e.g. ultrasound hearing, electric sense, magnetic sense; **2.** in genetics, the term sense pertains to the → sense strand; **3. senses**, the sound mental faculty in humans.

sense organ, the organ in animals specialized to respond to a certain type of energy (adequate stimulus), exceptionally to several types of stimuli. Sense organs contain cells or parts of them, called **sensory receptors**, which are differentiated to respond to certain stimuli and transduce energy into electrical form (→ receptor potential) by changing the ion permeability of the receptor membrane, but sense organs also contain different types of accessory cells and other structures. According to the quality of the adequate stimulus, sense organs are grouped into **mechanic**, **thermal**, **chemical**, and **electromagnetic senses**, and according to the location, into **external** and **internal senses**, the last group further comprising **proprioceptive** (*see* proprioceptors) and **visceral senses**. Some sense organs respond to various types of adequate stimuli, e.g. pain receptors to mechanic, thermal, and chemical stimuli.

sense strand, → coding strand.

sensibility (L. *sensibilitas*), the ability to perceive or feel; responsiveness or susceptibility to sensory stimuli.

sensible, 1. capable of being perceived by senses; perceptible to senses or mind; **2.** having sound judgement.

sensibilization, 1. the act of making an organ or organism more sensitive; **2.** → sensitization.

sensillum, pl. **sensilla** (L. dim. of *sensus* = sense), **1.** a simple sensory organ in many invertebrates often associated with a small hair; consists of only one or a few receptor cells and a sensory nerve fibre; **2.** one of the structural units in a compound sense organ, especially refers to invertebrates.

sensitivity (L. *sentire* = to feel), **1.** the state of (increased) responsiveness to sensory stimulation; **2.** in chemistry, **analytical sensitivity** means the sensitivity of a method to reliably measure the smallest concentration of a substance; **3.** in medicine, **diagnostic sensitivity** means the conditional probability to correctly identify a disease by a clinical test. *Adj.* **sensitive.**

sensitization, the act of rendering an organ or organism more sensitive to a stimulus, e.g. in inducing an immune response to an antigen (immunization). *Verb* to **sensitize**.

sensor (L. *sensorius* < *sensus* = sense), something that senses or recognizes; especially an element in a device that responds to physical or chemical stimuli, such as heat, light, pressure, quality, or concentration of a substance to be measured; **biosensor**, a sensor installed in a living organism to record a certain function *in vivo*.

sensory, denoting relationship to senses, sensation, or neuronal functions associated to senses.

sensory adaptation, the adaptation of the activity of a sense organ to the strength of a stimulus, i.e. sensory receptors become less responsive to repeated or continued stimuli of constant intensity. Sensory adaptation may occur quickly (**phasic receptors**, e.g. olfactory receptors), or slowly, or not at all (**tonic receptors**, e.g. pain receptors). It is often due to an alleviated response of the receptor cell membrane in prolonged stimulation. The adaptation of photic receptors involves the synthesis and decomposition of a photopigment according to the intensity of light. *Cf.* habituation.

sensory epithelium, the epithelium of a sense organ containing sensory cells, which may comprise differentiated epithelial cells, such as hair cells in the inner ear, or neurones or their parts (neuroepithelium), such as olfactory receptors in the mucous membrane in the vertebrate nose.

sensory pathway, a nerve tract from a sense receptor to the brain. In vertebrates, each sensory pathway from all over the body comprises two to five successive neurones. The first neurone leads along a sensory or mixed nerve to the spinal cord or directly to the brain stem; the next neurone, located inside the central nervous system, continues to the thalamus, and the third neurone usually to the sensory cerebral cortex. The olfactory tract has only two successive neurones, the latter ending at the primitive cerebral cortex (rhinencephalon). In auditory and visual tracts, the first two or three neurones are located in the sense organ itself, and the fourth or fifth neurone ends in the special sensory cortex of the cerebrum. The sensory pathways are divided into **specific sensory pathways** (auditory, visual, olfactory, gustatory tracts), and **somatic sensory pathways** (touch, pressure, pain, thermal tracts).

sensory receptor, *see* sense organ.

sepals, the outermost, sterile flower leaves, forming the **calyx**; are often green and protect the flower, and sometimes remain until the ripening of the fruit(s). Sometimes they resemble the → **petals** and may form with them a homochlamydous perianth, as e.g. in many → Monocotyledonae (e.g. tulip).

separation layer, → abscission layer.

separation species, a plant species growing in certain vegetation types but is absent in others; used to differentiate vegetation types, especially of plant → associations. *Cf.* character species.

Sephadex, the trade name for a cross-linked dextran with epichlorohydrin, used as beads for → gel filtration. Related compounds are **Sephacryl**, a cross-linked co-polymer of allyl dextran with N,N'-methylenebisacrylamide, and **Sepharose**, a cross-linked agarose.

sepia (Gr. *sepia* = cuttlefish), **1.** an inky secretion of various cuttlefish containing the pigment melanin (*see* ink sac); **2. sepium**, pl. **sepia**, the cuttlebone, i.e. the calcareous internal shell of the cuttlefish used by man in polishing agents and tooth powders, and for cage birds as a supply of calcium and other salts; *syn.* sepion, sepia bone.

sepsis pl. **sepses** (Gr. = decay), an invasion and persistence of pathogenic microbes or their toxins in the blood and other tissues; the term is in clinical use. *Adj.* **septic**.

septic(a)emia (Gr. *haima* = blood), blood poisoning, a systemic disease with pathogenic microbes in the blood stream.

sept(i)- (L. *septem* = seven), pertaining to seven, or seventh.

sept(o)-, septi- (L. *septum* = wall), pertaining to a wall or membrane of an organ.

septicidal (L. *caedere* = to kill), describes longitudinal dehiscence of a → capsule in plants; the multilocular septicidal capsule splits along septa between the original carpels, and in the syncarpous capsule the walls split along the fused margins of the carpels. *Cf.* loculicidal.

septiferous (L. *ferre* = to bear), a structure having a → septum.

septulum, pl. **septula** (L. dim. of *septum* = wall), a minute septum; a small wall or partition in an anatomical structure; e.g. **septula testis** (also *septa testis*), tissue lamellae inside the vertebrate testis.

septum, pl. **septa** (L.), a dividing wall, partition; e.g. **1.** any of the partitions formed from joined peritoneal membranes between the segments of annelids; **2.** *septum interventriculare*, the wall between the cardiac ventricles; **3.** *septum nasi*, partition of the nose; **4.** partition separating parts in fruits; **5.** partition formed during the division of a plant cell, separating the cell halves, which develop into the two daughter cells.

sequence characterized amplified region method, SCAR (L. *sequi* = to follow), an improved variant of the method of → random amplified polymorphic DNA, used for detecting → genetic polymorphism; the method uses specific → primers instead of random primers.

sequence complexity of DNA, division of the genomic DNA into different classes according to the property of having different number of repeats of similar DNA segments, i.e. → unique copy DNA, → middle repetitive DNA, and → satellite DNA. Unique DNA is present only in one or a few copies, middle repetitive DNA in 100 to 10,000 copies, and satellite DNA (highly repetitive DNA) from 10,000 to one million copies per genome.

sequence-tagged site, STS, a unique DNA sequence, which is identifiable with the → polymerase chain reaction. These tagged sites exist in organisms throughout the whole genome, usually at relatively even intervals.

sequencing, the determination of the monomer order in a macromolecule, e.g. the amino acid order in a polypeptid or protein chain and the nucleotide order in DNA or RNA.

SER, smooth → endoplasmic reticulum.

Ser, ser, → serine.

sere (L. *series* = row, series), in ecology, a successional series in plant communities following one another in the course of biotic development from a pioneer stage to the climax; also a stage in succession. *Adj.* **seral**. *See* succession.

serial homology, a developmental series of transformed structures in any taxon having the same origin and basic structure; e.g. vertebrae in different vertebrate groups, or the various appendages of crustaceans all developed from → biramous appendages.

series (L.), **1.** a group or succession of objects (such as cells, organs or organisms) arranged in regular order and following one another in space or time; e.g. the erythrocytic series, a succession of developmental stages of red blood cells; **2.** in chemistry, a homologous series; a group of compounds related in composition and structure; **3.** in physics, an arrangement of the parts in an electric circuit by connecting them successively in series to form a single path for the current.

serine, abbr. Ser, ser, symbol S; β-hydroxyalanine 2-amino 3-hydroxypropionic acid, $HOCH_2CH(NH_2)COOH$; an amino acid with an aliphatic hydroxyl group; containing this group, serine is very hydrophilic and reactive.

sero- (L. *serum* = whey, blood serum), pertaining to the blood serum or a serum-like fluid.

serosa, pl. **serosae,** also **serosas** (L.), **1.** any serous membrane, especially the membrane (*tunica serosa*) covering the outer free surface of the alimentary canal of vertebrates; **2.** → chorion; **3.** a membrane of blastodermic origin which encloses the embryo of many insects. *Adj.* **serosal**.

serotonin, 3-(2-aminoethyl)-5-hydroxyindole; a substance found in animals, many plants and microbes. In animal tissues it is synthesized from the amino acid tryptophan, and it acts as a messenger substance, i.e. in → synapses of the nervous system as a → neurotransmitter (**serotoninergic transmission**), and in many other tissues as a hormone, e.g.

stimulating contractions of smooth muscles in the alimentary canal and blood vessels. *Syn.* **5-hydroxytryptamine**, **5-HT**, enteramine, thrombocytin, thrombotonin.

serotype, 1. a type or a taxonomic subdivision of microbes distinguished by the kinds of constituent antigens present in the microbe; **2.** as a verb, to determine organisms on the basis of their antigenic properties.

serous (L. *serosus*), pertaining to serum, producing or containing serum, or resembling serum.

Serpentes (L. *serpens* = serpent, snake), **ophidians (Ophidia)**, **snakes**; a polymorphic reptilian suborder in the order Squamata; includes 14 families and about 2,250 species of which about 400 are poisonous. Snakes are distributed all over the world except arctic regions, and they live in all kinds of habitats, e.g. forests, deserts, and shore waters of lakes and oceans. Snakes are scaly, carnivorous animals which lack ear openings, sternum, and secondarily legs; some species have rudiments of hip bones. They have no eyelid but the eyes are covered by a transparent scale (speculum), the left lung is usually reduced or absent, and the skin is usually shed whole. The tongue is slender and forked, and the jaws are held by elastic ligaments rendering the mouth very distensible. In general, snakes are → oviparous or ovoviviparous, some species viviparous.

serpentine receptors, seven-spanning receptors; all the heterotrimeric G protein-binding receptors; large proteins which span the cell membrane seven times; e.g. many **hormone receptors**, such as β_2-adrenergic receptors or muscarinic cholinergic receptors, or **sensory receptors** like rhodopsin in the retina.

Sertoli cell, Sertoli's cell (*Enrico Sertoli*, 1842—1910), a cell type in the walls of the testicular seminiferous tubules of vertebrates; large glycogen-containing cells which support and nourish the maturing spermatids and secrete hormones, such as an androgen-binding protein, → Müllerian regression factor (MRF), and → inhibin. Sertoli cells are also called nurse cells, foot cells, or trophocytes.

serum, pl. **sera** or **serums** (L.= whey), **1.** any clear, watery body fluid, such as the fluid oozing from the serous membrane; **2.** → blood serum.

sessile (L. *sedere* = to sit), **1.** attached by a broad base; not pedunculated; **2.** → sedentary.

Sewall Wright effect (*Sewall Wright*, 1889—1988), → genetic drift. *See* sifting balance theory.

sex (L. *sexus*), the sum of opposing and complementary characteristics and functions of females and males observed within a single species, and the ability to mate and produce → genetic recombination among the progenies. → Sex determination may be due to genetic or environmental factors. Sexuality in the proper sense of the word occurs only in eukaryotic organisms.

sex allocation (L. *ad* = to, *locus* = place), the relative investment of reproductive effort of a parent to offspring of different sexes. According to *R.A. Fisher* (1930), it is evolutionarily stable to invest equally in male and female offspring thus producing the sex ratio of 1:1 (Fisherian sex ratio). In circumstances where the females are in good condition (capable of producing large offspring) they often produce more sons, but those in poor condition produce more daughters. This is common especially among the species whose males are the larger sex, e.g. in birds and mammals. According to the hypothesis of "size-fecundity advantage" the body size and reproductive effort are positively correlated.

sex/sexual characters, characters manifesting the sex; in animals, **1. primary sex/sexual characters**, sexual differences in reproductive organs (genital organs) producing gametes and sex hormones; **2. secondary sex/sexual characters** (characteristics), sexual differences induced by sex hormones in other organs, such as in pelage, plumage, voice, and size.

sex chromatin (Gr. *chroma* = colour), a small mass of → chromatin stained by Feulgen (*see* Feulgen staining) and occurring in the interphase nuclei of most mammalian females; it is usually attached to the nuclear envelope and consists of facultative → heterochromatin. Sex chromatin is derived from one of the two X chromosomes by inactivation and condensation. It is used for determination of genetic femaleness. *Syn.* Barr body. The number of sex chromatin bodies per cell is the number of X chromosomes minus 1.

sex chromosome (Gr. *soma* = body), a chromosome or chromosome group of eukaryotic organisms which distinguishes the sexes and is causally related to genotypic regulation of → sex determination. *Syn.* idiochromosome. *Cf.* autosomes. *See* X, Y, Z, and W chromosome.

sex determination, the determination of the differentiation of sexes in organisms having genuine → sexes, i.e. male and female: **1. phenotypic sex determination**, the sex determined by the influence of internal or external environmental factors (e.g. *Bonellia viridis*); **2. genotypic sex determination**, determined by the → genotype of the → zygote or spores. Sex determination occurring during → haplophase is called **haplogenotypic sex determination**, as compared to **diplogenotypic sex determination** in which also → diplophase is determined to be either male or female.

In diplogenotypic sex determination, special → sex chromosomes play a central role. The XY system and its many varieties are based either on a balance between the number of X chromosomes and the number of the doses of other chromosomes (*see* autosomes), as in many insects and dioecious plants, or on the presence of Y chromosome in the chromosomal complement actively determining the sex, as in mammals. In most turtles and all crocodiles, sex is determined by the temperature of egg incubation. Generally, eggs incubated at low temperatures (22—27°C) produce one sex, whereas eggs incubated at higher temperatures (above 30°C) produce the other sex. Between 27 and 30°C both males and females will be hatched from the same brood of eggs. In many invertebrates, the sex of each cell is determined by the cell's genotype only without any environmental influence (cell autonomous sex determination).

A further genotypic sex determining mechanism is the **haplodiploid mechanism** occurring in many eusocial hymenopterans in which the females are born from fertilized diploid eggs, while the males develop from unfertilized haploid eggs.

sexduction (L. *ductio* = transmission), the transmission of genes in the → conjugation of bacteria from one cell to another with the aid of a sex factor. Also called F-duction.

sex factor, → F factor.

sex hormones, reproductive hormones; steroid hormones regulating the development and maintenance of sexual properties and functions in animals; secreted mostly from the ovaries and testes, but also from other tissues, such as the adrenal cortex in vertebrates. *See* male sex hormones, female sex hormones.

sexine, the outer part of the exine, the outermost layer of the → pollen grain wall.

sex-limited, a genetic characteristic occurring in one sex only.

sex-limited inheritance, the inheritance of genetic characters the genes of which occur in both sexes but can be expressed in one sex only; e.g. many secondary → sex characters.

sex-linked, pertaining to the inheritance of genes located on the → sex chromosomes.

sex organs, → genital organs.

sex ratio, the proportion of males in relation to females in a population; e.g. the number of males divided by the number of females, or the number of males per hundred females; the sex ratio can be determined at conception (**primary sex ratio**), at birth (**secondary sex ratio**), or in a fertile population (**tertiary sex ratio**).

sexual (L. *sexualis*), **1.** pertaining to sex, or to the two sexes; **2.** relating to reproductive organs and processes in them; genital; **3.** erotic; **4.** a person considered as to his or her sexual relations; e.g. **contrary sexual**, a homosexual.

sexual dimorphism (Gr. *dis* = twice, *morphe* = shape), the occurrence of differences in some characteristics between a male and female of the same species, such as shape, size, structure, colours, etc.; e.g. the males of many birds often have brighter coloured plumage than females.

sexuality, 1. the condition of possessing male or female reproductive elements; **2.** having reproductive functions, or sexual functions, such as sexual behaviour; **3.** a degree of sexual activity or sexual attractiveness.

sexual reproduction, reproduction by the fusion of male and female → gametes (bisexual reproduction), or by the development of an unfertilized egg (unisexual reproduction). *Cf.* asexual reproduction. *See* syngamy, parthenogenesis.

sexual selection, a form of → selection in which the choice of the pairing partner is determined by individual characteristics of the partner or competitors. Usually the male competes for the access to females, and the females make the choice. *See* intersexual selection, dominance mating.

sexual success, the ability of an individual to

obtain many, highly reproductive sex partners.

Shannon—Wiener index of diversity (*Claude Shannon*, b. 1913; *Norbert Wiener*, 1894-1964), an index (H) describing the species diversity of a community, presented in the formula: $H = \Sigma^{s}_{i=1} (p_i)(\log_2 p_i)$, where s = the total number of species in the sample; p_i = the number of individuals of one species in relation to the number of individuals in the population. The index is lowest when all the individuals in the community belong to the same species and highest when they represent different species.

sharks, → Lamniformes, Squaliformes.

sheath, a tubular, protective structure around an organ or part; e.g. **1.** the myelin sheath around a nerve fibre, or the tendon sheath; **2.** the lower part of a leaf base surrounding the plant stem.

Shelford's law of tolerance (*Victor E. Shelford*, 1877—1968), a rule stating that the success of an organism in a habitat is controlled by the qualitative or quantitative deficiency or excess of specific environmental factors, such as temperature, moisture, trace elements, etc.; failure is dependent on the tolerance limits of the organism to these factors.

shell, 1. a hard covering of an animal structure; a shell may be formed from calcium, silicon, bone or keratin, such as an eggshell of birds, or a calcareous shell of a mollusc composed of calcium carbonate, aragonite, and an organic substance, conchiolin; **2.** the cover of a fruit, such as a pea (legume) or nut.

shell gland, 1. an organ of invertebrates secreting a shell around an egg or a larva; *see* Mehlis' glands; **2.** an excretory organ (maxillary gland) in branchiopods in which the duct is visibly coiled within the carapace wall; **3.** a glandular part of the oviduct of many animals where the eggshell is formed, as e.g. in birds.

shifting balance theory, a theory proposed by *Sewall Wright* (1889—1988), stating that in → evolution the stochastic forces (like → genetic drift) and directional forces (like mutation and selection pressures), act temporarily in a balanced fashion, stochastic forces dominating at one time and directional forces at another.

shikimic acid, an organic acid in plants and microorganisms synthesized from carbohydrates; an intermediate in the reaction chain, in which → aromatic amino acids are formed from phosphoenolpyruvate and erythrose 4-phosphate (shikimic acid pathway). Aromatic amino acids act e.g. as precursors for the formation of phenolic compounds in plants.

Shine—Dalgarno sequence (*J. Shine, L. Dalgarno*), a purine-rich region in the initiator site of messenger RNA (mRNA) located near the initiating codon AUG; the sequence pairs with the 3' end of 16S ribosomal RNA (rRNA), and thus guides the ribosome to the initiation site (A = adenine, U = uracil, G = guanine).

shivering, 1. trembling from cold or excitement; **2.** muscle shivering; involuntary trembling contractions of skeletal muscles of birds and mammals in the cold. In these homoiothermic animals this is a normal reaction to increase thermogenesis (**shivering thermogenesis**) in a cold environment. The thermoregulatory centre in the hypothalamus controls shivering through motor neurones and → muscle spindles (*see* thermoregulation). Muscle shivering occurs also in some insects, such as honeybees, especially before starting to fly; **3.** a disease of horses characterized by trembling of various muscles.

shock, a condition of a sudden profound disturbance in brain and/or haemodynamic functions, especially in birds and mammals, caused e.g. by a severe stress situation or disease; it is characterized e.g. by paleness, rapid but weak pulse, reduced total blood volume, low blood pressure, decrease of body temperature, rapid but shallow respiration, mental disturbances, nausea, and vomiting. *See* heat shock.

shoot, a plant stem with its leaves; can be above or below the ground.

shorebirds, → Charadriiformes.

short-cycle forest, short rotation coppice; a forest which is cultivated for biomass but not for log and wood production; most common species of foliage trees and shrubs grow rapidly and reproduce vegetatively, such as willows. *See* coppice.

short-day plants, a plant which requires short-day conditions for flowering. *See* photoperiodism.

short interspersed elements, SINES, a class of → retroposons that occurs in short repetitive sequences in the eukaryotic genomes; in the human genome they are repeated approx. one million times.

short shoot, a plant branch, the growth of

which is inhibited and the internodes are very short; e.g. the leaves of many foliage trees develop in short shoots. The significance of the short shoot formation is that it prevents the tree structure from becoming too complicated and it spares material. *Cf.* long shoot.

short tandem repeat, STR, → microsatellite.

shotgun experiment, the cloning of the whole genome of an organism by cleaving the genome with → restriction endonucleases at random into pieces of different length.

shoulder girdle, thoracic girdle, pectoral girdle; the bony or cartilaginous ring-shaped structure (incomplete behind) in tetrapods serving for the attachment of the forelimbs; it comprises the → clavicles, coracoids, scapulae, and anterior segment of the sternum.

shredders, aquatic bottom animals, which cut organic matter into coarse particles for feeding; e.g. the larvae of ephemeropterans, trichopterans, and plecopterans, or the water hog louse (*Asellus*). *Cf.* grazers, gathering-collectors, filter feeders, deposit feeders, scrapers.

shrub, a perennial woody plant, which forms branches near the ground and does not form a trunk.

shrub layer, a vegetation layer of a forest, consisting of shrubs and young trees up to 6 m. *Syn.* shrub stratum. *See* layering.

shufflon, a domain of the genome consisting of a group of tightly clustered → exons, which during evolution have been shuffled together into a single functional unit.

shuttle vector, a → plasmid into which starting points for the → genetic replication of two host organisms (e.g. a bacterium and yeast) have been introduced; thus it may be used as a → vector to transduce foreign DNA back and forth between the two hosts.

sialic acid (Gr. *sialikos* = pertaining to saliva), any of the group of amino sugars containing 9 or more carbon atoms; acyl derivatives of neuraminic acid (e.g. N-acetylneuraminic acid). Sialic acids are components of many glycoproteins, glycolipids and mucopolysaccharides, especially in animals.

sibling species, species separated from each other by means of reproductive → isolation; they are morphologically similar or identical and often appear in the same region.

sibmating, a mating of siblings; the mating system used in inbreeding of animals.

sib method, a method of human genetics aiming to find out the Mendelian ratios (*see* Mendel's laws) of the presence of a certain character in the siblings of a family; using this method, the inheritance of the character can be studied.

sickle-cell an(a)emia, a heritable blood disease of man due to a recessive → point mutation in the gene for β-chain of haemoglobin. In a homozygous condition this mutation produces a severe syndrome with sickle-shaped red blood cells in the blood (**sicklemia**) which often, especially during the first pregnancy, leads to death of the foetus. Since the gene for sickle-cell anaemia in the heterozygous condition offers resistance against malaria, this harmful recessive mutation persists rather frequently (10—20%) in human populations in areas where malaria occurs.

siemens, S (*William Siemens*, 1823—1883), the SI unit of electrical conductance. *See* Appendix 1.

sieve area, *see* sieve tube.

sieve cell, a conducting cell of the → phloem in ferns and gymnosperms; elongated cells are in contact with each other by sieve pores in the cell wall. *See* sieve tubes.

sieve element, a structural unit of plant → phloem; may be formed by one sieve cell or part of the → sieve tube between two sieve plates with → companion cells.

sieve plate, *see* sieve tube.

sievert, Sv, dose equivalent. *See* Appendix 1.

sieve tube, a tube in the → phloem of angiosperms; translocates products of photosynthesis (especially sugars) from production sites to other parts of the plant, such as to the storage roots. The sieve tubes consist of **sieve elements**, which are long, non-nucleated cells attached to one side of living → companion cells. The sieve elements are joined end to end, having **sieve plates** between them. The sieve plates have pores in groups (in **sieve areas**), through which material can pass and which are blocked during an unfavourable time of the year. Sieve areas may also occur in the side walls of the sieve elements.

sigma factor, σ factor, a subunit of the bacterial → RNA polymerase which is necessary in the initiation of → genetic transcription.

sigmoid growth, → logistic growth.

sign (L. *signum* = mark), **1.** a mark or symbol; **2.** an indication of the existence of a condition, e.g. objective evidence of a disease in an organism; *cf.* symptom.

signal, any act, event, or substance which transfers information between cells, organs, or organisms; e.g. **1.** signals inside the organism, such as messenger RNA, hormones, growth factors, neurotransmitters, action potentials, and second messengers; **2.** signals between organisms; may be transmitted by chemical substances, but also by many other elements and energy forms, e.g. in sensory information of animals. In communication between animals the **key stimuli** may act as signals, as well as the behaviour models that produce information between individuals or groups, regardless of their other functions. Signals may be based on sight (**visual signals**), hearing (**acoustic signals**), touch (**tactile signals**), or chemical sensing (**olfactory signals**). Signals may be either intraspecific or interspecific (e.g. warning signals). Functionally, signals are divided into **primer signals** producing changes in activity urges, **trigger signals** expressing the strength of urge, and **metacommunication signals** (*see* metacommunication). *See* allomone, pheromones.

signal recognition particle, SRP, an essential component of protein translocation across the membrane of the → endoplasmic reticulum (ER); contains six polypeptides and a 300-nucleotide RNA. SRP binds to the → signal sequence of a nascent polypeptide chain on the ribosome, and the complex then binds to the **signal recognition particle receptor (SRP receptor)** situated on the ER membrane.

signal sequence (L. *sequens* = following), a part of a nascent secretory protein (usually N-terminus) initiating the transport of this protein through the membrane into the endoplasmic reticulum (ER); it is bound to the **signal sequence receptor** in the ER membrane. *See* signal recognition particle.

signal transduction (L. *transducere* = to lead across), a series of processes by which a ligand, such as a hormone, interacts with its receptor and causes the activation of a → second messenger (e.g. cyclic AMP or Ca^{2+}) that further causes e.g. the activation of intracellular enzymes, kinetic reactions, or genes. Thus the message is transduced inside the cell.

sign stimulus, → releaser.

silencer, a certain DNA sequence in the → promoter of genes in eukaryotic organisms that recognizes → transcription factors and decreases the efficiency of gene function. *Cf.* enhancer.

silent mutation, a gene mutation that does not cause any change in the polypeptide chain encoded by the gene. Silent mutations are possible because of the degeneracy of the → genetic code, and involve the third nucleotide position of the → codon.

silica cells, epidermal cells in graminaceous plants (family Poaceae) containing silicon and alternating with the other epidermal cell types, i.e. the long cells and cork cells.

silicification, also **silicatization,** the impregnation of silicon dioxide (silica, SiO_2) into cell walls of epidermal cells in plants, e.g. in horsetails and *Carex* species and especially in diatom shells. *Verb* to **silicify**.

silicon, a non-metallic element, Si, atomic weight 28.09; is found as silica (SiO_2) and salts of silicon (silicates), the latter being important components of soil, e.g. in clay. Silicon is the second most abundant element on the Earth's crust. It is one of the micronutrients (trace elements) in organisms. *See* silicification.

silicule (L. *silicula,* dim. of *siliqua* = pod, husk), a fruit occurring in the family Brassicaceae; broad, flat and divided by a false septum. *Syn.* silicle. *Cf.* siliqua.

siliqua, pl. **siliquae,** a fruit occurring in the family Brassicaceae; long, thin, and divided by a false septum. *Syn.* silique. *Cf.* silicule.

silk gland, → spinning glands.

silt, a soil type composed of grains from 0.002 to 0.02 mm in size.

Silurian (L. *Silures* = ancient people in southern Wales), a geological period of the Palaeozoic era, between Ordovician and Devonian, about 438 to 408 million years ago. During the period the first terrestrial plants (spore plants), arthropods, and primitive jawed fishes (placoderms) appeared. The period was characterized by flourishing marine invertebrate life with e.g. crinoids and large arthropods. *Adj.* **Silurian**.

Siluriformes (L. *Silurus* = the type genus, *forma* = shape), **catfishes**; an order of bony fishes, most species living in inland waters of Africa, Asia, and South America. Their skin is either bare or covered by bony scales. The order includes about 30 families and over 2,000 species.

simazine, 2-chloro-4,6-bisethylamino-1,3,5-triazine, a herbicide inihibiting the light reactions of photosynthesis and consequently the growth of plants.

simulation (L. *simulatio* < *simulare* = to imitate, represent), **1.** the imitation of structures and characteristics, enabling animals to cheat their predators; e.g. some insects resembling leaves or twigs (camouflage); *see* mimicry; **2.** more generally, means any process that is imitated, such as a biological process by a mathematical model (computer simulation).

SINES, → short interspersed elements.

single-stranded conformational polymorphism, SSCP, a form of DNA polymorphism detected on the basis that single strands of DNA of defined length, isolated from normal and mutant genomes by → polymerase chain reaction, run through a non-denaturating electrophoretic gel at different rates. *See* genetic polymorphism.

single-stranded deoxyribonucleic acid, ssDNA, a type of DNA that consists of only one chain of nucleotides. *See* deoxyribonucleic acid.

singlet oxygen, $^1O_2$ (L. *singulus* = individual, one only), a very reactive but unstable oxygen radical which reacts with double bonds in different organic molecules. It is an excited form of two-atomic oxygen (O_2) in which the electron pair of the outermost orbit has opposite → spins (a spin of zero). In organisms, singlet oxygen is produced normally by the breakdown of peroxides in cellular metabolism, in reactions of ozone and phosphorus, and in photo-oxidation (oxidative reactions caused by visible and UV light). In plants, singlet oxygen is formed from atmospheric oxygen in the presence of light-excited chlorophyll. Singlet oxygen strongly oxidizes organic molecules preventing them from acting; this occurs e.g. in photosynthetic reactions causing photoinhibition in plant cells. Carotenoids in plants and several → antioxidants in animal tissues act as protective substances against singlet oxygen. White blood cells of vertebrates produce singlet oxygen to kill microbes. *See* oxygen radicals.

sinistr(o)- (L. *sinister* = left, on the left side), denoting relationship to the left side; e.g. sinistrocerebral, pertaining to or located in the left cerebral hemisphere.

sino-, sinu- (L. *sinus* = hollow, curve, fold), denoting relationship to the → sinus.

sinoatrial/sinuatrial node, → sinus node.

sinus (L. = curve, gulf), a cavity or recess in tissue; e.g. **1.** a normal cavity or dilated channel usually filled with venous blood, e.g. → sinus venosus in vertebrates, or haemal sinuses in bivalves (*see* haemal system); **2.** a normal cavity filled with air, such as the paranasal sinuses of certain cranial bones; **3.** an abnormal channel or fistula through which pus discharges.

sinus gland, an endocrine gland in the eyestalk of crustaceans, producing hormone which inhibits the moult of the animal, and regulates pigment movements in chromatophores and thus the colour change of the skin. The activity of the sinus gland is controlled by the → X gland.

sinus node, a group of specialized cardiac muscle cells, which in reptiles, birds and mammals is located in the atrial wall at the junction of the superior vena cava and the right atrium (therefore usually called **sinoatrial node**), but in lower vertebrates in the wall of → sinus venosus. The sinus node is the pacemaker of the → conduction system of the vertebrate heart. The cells of the sinus node induce spontaneously action potentials which are conducted to the whole heart thus pacing the normal cardiac rhythm, called **sinus rhythm**.

sinus venosus (L. *vena* vein), venous sinus; **1.** the bulbous membrane chamber attached to the atrium of the embryonic vertebrate heart; receives blood from the large venae and transmits it to the atrium; **2.** the related part in adult fish, amphibians, and reptiles; **3.** the atrial area of the heart of adult birds and mammals derived from the embryonic sinus venosus, i.e. the area where the venae cavae empty; also called *sinus venarum cavarum*.

siphon (Gr. = tube, pipe), funnel; **1.** a tubular duct in many animal groups for transporting fluids; e.g. the excurrent and incurrent siphons of bivalves, transporting water; **2.** the funnel of cephalopods, a flexible, muscular tube passing from the mantle cavity out behind the head; **3.** a bent tube with two branches of unequal length used to transfer liquids by the force of atmospheric pressure from a higher to a lower level.

siphonoglyph(e) (Gr. *glyphein* = to engrave), a smooth ciliated groove for water incurrence to the pharynx and enteron of some anthozoans, e.g. sea anemones.

Siphonophora (Gr. *phorein* = to carry along), **siphonophorans**; an order of pelagic hydrozoans consisting of floating or swimming colonies with hundreds of thousands of individuals. There is great diversity between the members of the colonies, and at least the following types of **polyps** occur: floating, feeding, defensive (fighting), sensitive (feeling), and reproductive polyps. The colonies are buoyant due to gas-filled **pneumatophores**. A well known species is Portuguese man-of-war, *Physalia physalis*.

siphonostele (Gr. *stele* = column), *see* stele.

siphonozo(o)id (Gr. *zoon* = animal, *eidos* = form), a small polyp without tentacles pumping water into the colonies of sea pens (Pennatulacea, *see* Anthozoa) to maintain the turgor pressure.

siphuncle (L. *siphunculus*, dim. of *siphon* = tube, pipe), **1.** a thin, gas-filled membranous tube connecting the internal gas chambers of the *Nautilus*; **2.** the tube with which the animal regulates its gas pressure, as a fish does with the swim bladder.

Sipuncula (L. *Sipunculus* = the type genus < *sipunculus*, *siphunculus* = little pipe), **sipunculids**, **sipunculans**; some species are known as "peanut worms"; a phylum including small, unsegmented marine worms living in sand or mud of shallow seashores. They have a narrow retractile trunk (*proboscis*) and several tentacles around the mouth. Previously sipunculids were grouped as a class (**Sipunculoidea**) in Annelida. The phylum comprises about 325 species in 16 genera.

Sirenia (Gr. *seiren* = siren), **sirenians**, **sea cows** and **manatees**; an order of large aquatic herbivorous mammals including two families: **manatees** (Trichechidae) and **dugong** (Dugongidae). Sea cows resemble whales; their forelimbs are paddle-shaped and the tail is finned; the ear lobes and hindlimbs are absent. They live in shallow shore waters in warm districts; manatees in America and West Africa, sea cows (dugongs) in South Asia and Australia. Steller's sea cow, *Hydrodamalis stelleri*, was exterminated from the North Pacific Ocean by hunters in 1854.

sister chromatids (Gr. *chroma* = colour), the two chromatids which arise as a consequence of the replication of one single → chromosome during → interphase.

sister chromatid exchange, SCE, the result of → crossing-over occurring between → sister chromatids in a replicated, somatic chromosome; can be made microscopically visible using harlequin staining which stains the sister chromatids differently.

site, in genetics, the smallest unit of the gene which can mutate independently (mutational site), or the smallest unit of the gene which can be separated from other such units in intragenic recombination (*see* genetic recombination). Thus the site is the location of the → muton and → recon within the gene.

site-directed mutagenesis, site-specific mutagenesis, → focused mutagenesis.

site-specific recombination, a → recombination occurring between two specific nucleotide sequences of DNA (which are not necessarily homologous); e.g. occurs in the integration of a → prophage into bacterial DNA.

skates, → Rajiformes.

skatole, also **skatol** or **scatole**, (Gr. *skor*, gen. *skatos* = dung, excrement), 3-methyl-1H-indole, C_9H_9N, produced in the intestine by the decomposition of proteins, giving a strong characteristic faecal odour. In some plants it has hormone-like effects. Skatole is obtained from coal tar and used as a fixative in chemical industry.

skeletal muscles (L. *musculi skeleti*), the → striated muscles of vertebrates, attached to bones by tendons or fasciae; serve in locomotion of the animal, and motion of mouth, tongue, pharynx, skin, eyeballs, and ears. The skeletal muscle is composed of numerous muscle cells (**muscle fibres**), which may be from 0.1 mm to several centimeters in length, usually from 10 to 100 μm in width. The cells form bundles of different sizes surrounded by layers of connective tissue, i.e. by the endomysium, perimysium, epimysium. A single muscle cell is derived from many, even hundreds of embryonic muscle cells (**myoblasts**), which are fused together, and the number of nuclei in the mature muscle cell (**myocyte**) corresponds to the number of original myoblasts fused. *Syn.* voluntary muscles. *See* muscle.

skeleton (Gr. = a dried body, mummy), the hard framework of the animal body; in vertebrates, comprises bony or cartilaginous **endoskeleton** (neuroskeleton) and **exoskeleton**, the latter being of dermal origin (dermal bones). The term **splanchnoskeleton** is sometimes used to denote the skeletal structure connected with the viscera (e.g. gills, tongue). In invertebrates the skeleton is mostly of exoskeleton type formed from different substances, such as → chitin, chitin cuticle, spongin, sericin (silk gum), which as a group are called **skeletin**. *See* cytoskeleton.

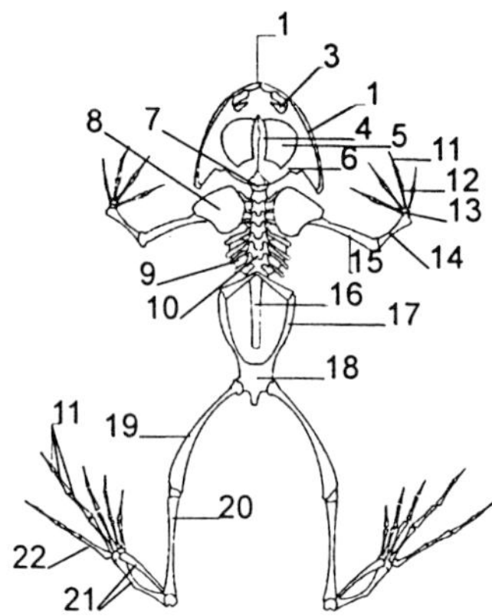

SKELETON OF THE FROG

1. Premaxilla 2. Maxilla 3. Naris 4. Fronto-parietal 5. Orbit 6. Prootic 7. Atlas 8. Supra-scapula 9. Sternum 10. Vertebra 11. Phalanges 12. Metacarpals 13. Carpals 14. Radioulna 15. Humerus 16. Urostyle 17. Ilium 18. Ischium 19. Femur 20. Tibiafibula 21. Tarsus 22. Metatarsus

skin (Gr. *derma*, L. *cutis*), **1.** the external integument or covering of an animal body, or such an integument stripped from the body, i.e. the pelt or fur. The skin of most invertebrates comprises a single layer of cells, → epidermis, with some sort of secreted matter forming the → **cuticle** (*cuticula*) on it. The skin of vertebrates is always multicellular and is formed from the ectodermal epithelial tissue, **epidermis**, which lies upon the mesodermal connective tissue, called → **dermis** (corium); **2.** the outermost covering of a seed or fruit.

skin respiration, cutaneous respiration, the exchange of respiratory gases through the skin; common in invertebrates which live in aquatic or moist environments, such as in different types of worms. The skin functions as an additional respiratory organ also in many vertebrates, e.g. in fish species which rise temporarily onto land, or in amphibians overwintering in the water. Gas exchange occurs also through the skin of higher vertebrates; in many species this is important especially for the skin cells. Respiration through the mucous membrane of the mouth or anal part of the intestine is found in some fish and turtle species.

skull, → cranium.

skull index, pl. **indexes** or **indices** (L.), a number indicating the ratio of skull measurements, such as the ratio of the height and breadth, or of the length and breadth.

slash-and-burn agriculture, a term for field clearing by burning forest; previously a method common e.g. in Nordic countries and still in use in the tropics.

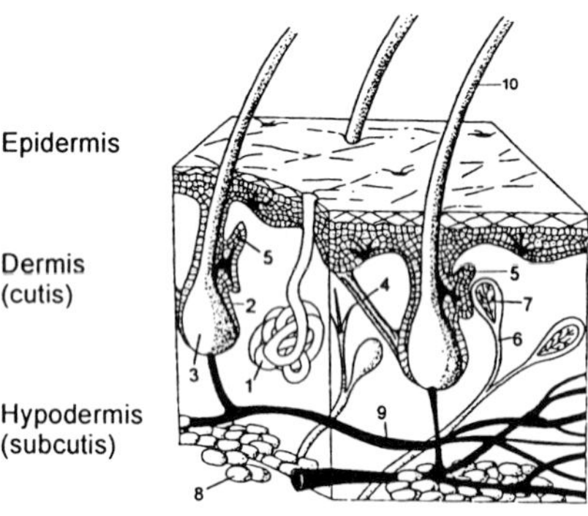

SECTION OF HUMAN SKIN

1. sweat gland 2. hair follicle 3. hair papilla 4. erector muscle 5. sebaceous gland 6. nerve 7. sensory corpuscle 8. fat cell 9. blood vessel 10. hair

sleep, in higher animals, a period of rest during which volition and consciousness are partially or totally lost. Its properties are best known in man, in whom different phases have been found. Sleep begins with drowsiness and then deepens to **orthodox sleep** (deep sleep, slow-wave sleep), which is followed, after 30 to 90 min by **paradoxical sleep** (REM sleep) during which dreaming occurs. Each cycle of paradoxical sleep lasts from 5 to 30 minutes, the longest cycles occurring towards the morning. During paradoxical sleep all skeletal muscles except the eye muscles are totally relaxed, and the electric discharge of the brain cortex shows similar activity as during wakefulness. Probably sleep serves the processing of experiences and seems important for storage of them in the memory. Sleep is controlled especially by certain areas of the hypothalamus (sleep centre) and the limbic system of the cerebrum. Comparisons of behaviour and electric activity of the brain indicate that all mammals have similar sleep patterns, but monotremes have no paradoxical sleep. Like mammals, birds and some reptiles alternate slow-wave and paradoxical sleep.

slime moulds (molds), → Myxomycetes.

small intestine (L. *intestinum tenue*), the anterior and middle portion of the intestine between the stomach and the large intestine of vertebrates where digestion and absorption mainly occur. In mammals, it consists of three histologically different parts, the **duodenum**, **jejunum** and **ileum**, in man representing about 10, 40, and 50%, respectively. The bile and pancreatic ducts open into the duodenum which is anchored by ligaments to the dorsal peritoneal wall; the other two parts move quite freely supported by the mesenterium. In the inner intestinal wall there are many circular permanent folds and numerous **intestinal villi** (→ villus), which increase the intestinal surface area (for absorption). The **intestinal glands** (crypts of Lieberkühn) and **duodenal glands** (Brunner's glands) open between the villi and produce intestinal juice with several enzymes. The free surfaces of the epithelial cells covering the villi are divided into minute **microvilli**, which form a brush border on the surface, important for absorption. Many **lymphatic nodules** in the mucosa serve for immune defence, and several types of endocrine cells for digestive and circulatory functions of the intestine. The tissue layers of the wall are similar to those in other parts of the → alimentary canal.

small cytoplasmic ribonucleic acid, scRNA, ribonucleic acid (RNA) found in the cell cytoplasm consisting of fewer than 300 nucleotides per molecule. *See* ribonucleic acid.

small cytoplasmic ribonucleoprotein particle, scRNP, a complex formed in a cell when → small cytoplasmic ribonucleic acid (scRNA) is associated with a specific protein.

small nuclear ribonucleic acid, snRNA, ribonucleic acid (RNA) found in the cell nucleus consisting of fewer than 300 nucleotides per molecule; six different snRNA types (U1 to U6) are found. *See* small nuclear ribonucleoprotein particle.

small nuclear ribonucleoprotein particle, snRNP, a complex formed in a cell when → small nuclear ribonucleic acid (snRNA) is associated with a specific protein; snRNPs play an important role in the splicing of → heterogeneous nuclear RNA.

smell, sense of smell, → olfaction.

smog (smoke + fog), a mixture of fog and smoke; may be a condition of a local climate with mixtures of various air pollutants, prevailing usually during calm weather periods. Smog includes **industrial smog** consisting mainly of sulphur dioxide and suspended droplets of sulphuric acid as well as a variety of suspended solid particles, and **photochemical smog** that is a mixture of air pollutants produced in the atmosphere by the reaction of hydrocarbons and nitrogen oxides under the influence of sunlight. Smog is harmful to organisms if it contains ozone, peroxyacyl nitrates and various aldehydes.

smolt, a young salmon or sea trout between the parr and grilse stages, usually 2 to 4 years of age, ready to migrate from the river of birth to the sea.

smooth muscle, unstriated muscle, unstriped muscle; a type of muscle without transverse striations in its cellular structure. The smooth muscle functions almost always **involuntarily** and forms contracting tissue sheets (visceral smooth muscle) in the inner organs, such as in the blood and lymph vessels, respiratory pathways, alimentary canal, bladder, glandular and genital ducts, and small separate muscles (multi-unit smooth muscles) e.g. in sensory organs such as the eye. In comparison with → skeletal muscles the contractions are generally much slower. Smooth muscle cells are spindle-shaped mononuclear cells from 20 to 500 μm in length. Usually several → **gap junctions** join adjacent cells to the functional entity, where muscle impulses (action potentials) propagate electrically from cell to cell.

Muscle impulses are mostly generated in muscle cells themselves (**myogenic contraction**), usually in a pacing area of the muscle, and the autonomic nervous system and hormones do not trigger muscle impulses but facilitate or inhibit their generation. In some cases, as in smooth muscles of the iris, contractions are triggered by nerves (neurogenic contraction) as in skeletal muscles, and no gap junctions exist. The smooth muscles (usually called unstriated muscles) of invertebrates are principally similar to those of vertebrates, but structural and functional variations between animal groups are great.

smooth phloem, the functioning tissues and cells of the → phloem, i.e. sieve cells and sieve tubes, companion cells and phloem parenchyma cells. *Cf.* hard phloem.

smut, a disease of cereal grasses (wheat, oats, rye, indian corn) caused by basidiomycetous fungi of the order → Ustilaginales.

smut fungi, → Ustilaginales.

snakes, → Serpentes.

sneeze, a reflective expulsion of air through the mouth and nose to expel irritating material from the nose of air-breathing vertebrates; a reflex excited by an irritation of the nasal mucous membrane or occasionally by sunlight striking the eye. The reflex is controlled by a nucleus of nerve cells in the medulla oblongata.

snow bed, a limited area, on which the snow cover stays longer than in the surroundings, possibly the whole summer. Snow beds are common on mountains and fells, and they support many plant species, which utilize the snow cover and local moisture and which cannot live elsewhere.

snRNA, → small nuclear ribonucleic acid.

snRNP, → small nuclear ribonucleoprotein particle.

S1 nuclease, an enzyme which specifically catalyses the splitting of a single-stranded DNA. *See* nucleases.

social (L. *socialis* < *socius* = companion, associate), pertaining to relationships between individuals living in colonies or communities.

social adaptation (L. *adaptare* = to fit to), all the forms of adaptation that facilitate intraspecific behaviour, usually excluding reproductive activities.

social behaviour, any intraspecific interactions among conspecific individuals, usually excluding interactions in reproductive courtship, mating, parent-offspring, and intersibling behaviour.

social dominance, a phenomenon in a social group where an individual or a group of individuals dominates other members in the group. *See* dominance, hierarchy.

social environment, all interactions between individuals of the same species, excluding reproductive activities.

social facilitation, spreading of a behavioural pattern of an individual among other members of a group, e.g. yawning in a human group.

social group, a group of conspecific individuals which are in mutual interaction; the term excludes mated pairs and dependent offspring.

social imprinting, in ethology, the → imprinting of young to their parents, siblings, or other conspecific individuals, and especially in experimental situations, also to individuals of other species or artificial → models.

social mimicry, evolution of a similar behavioural pattern in two or more social species, facilitating interaction between them; e.g. the similarity of contact sounds of several species in winter tit swarms.

social parasite, a social species (usually an insect) which is parasitically dependent on the society of another species; may be congeneric with the host. The degree of parasitism ranges from mere egg-laying or food-searching behaviour to destruction of eggs, larvae, and nests of the host. Social parasites are common especially among hymenopterans.

social pathology, a physiological or behavioural disturbance or disease among individuals in stress situations, e.g. in overcrowded populations, with the result of decreasing birth rate.

social selection, → genetic selection caused by social behaviour.

social signal, *see* releaser.

society (L. *societas* = company), in biology, **1.** a group of conspecific individuals which are organized in a cooperative manner and usually specialized in different castes with special functions, such as queens, drones, workers, and soldiers in the societies of social hymenopterans; such societies are sometimes called colonies; **2.** in botany, a lower unit in an → association, defined by the predominance of species other than the ones normally dominant.

sociobiology (Gr. *bios* = life, *logos* = word, speech), the comparative study of social organization among animals based on the idea that all behaviour, and consequently social systems, are evolutionarily adaptive. Later, the idea was widened to include also social phenomena of humans, but raised a heated debate.

SOD, → superoxide dismutase.

sodium (L. *natrium* < Arab. *natrun* = ash lye), an alkali metallic element, Na, atomic weight 22.99; occurs abundantly as **sodium chloride**

(NaCl, common salt), e.g. in the sea water and as rock salt. Sodium is an essential element for organisms, particularly for the physiological functions of animals. Sodium ions (Na^+) are mainly extracellular and their transport is controlled by an active transport mechanism (sodium pump, *see* ion pumps). Na^+ is required by cells to maintain their osmotic and electrical conditions. *See* resting potential, action potential.

sodium bicarbonate, sodium hydrogen carbonate, $NaHCO_3$; a salt found e.g. in natural waters; with carbonic acid it constitutes the principal inorganic pH buffer system of cellular and tissue fluids. Also called soda, baking soda, sodium acid carbonate.

sodium dodecyl sulphate, SDS, an anionic detergent that binds to proteins and denatures them by forming a negatively charged SDS-protein complex; it is used as a pH buffer in protein → electrophoresis and as a wetting and emulsifying agent in mixtures. *Syn.* sodium lauryl sulphate.

sodium lauryl sulphate, → sodium dodecyl sulphate.

sodium pump, *see* ion pumps.

soil formation, the processes which change the chemical and physical structure of the soil covering the bedrock. These processes form soil types with characteristic, stable layers (horizons), i.e. the **soil profile**. Factors affecting soil formation are climate, watering and the effects of water, nutrient cycles caused by vegetation, humus formation, and chemical reactions such as oxidation-reduction processes.

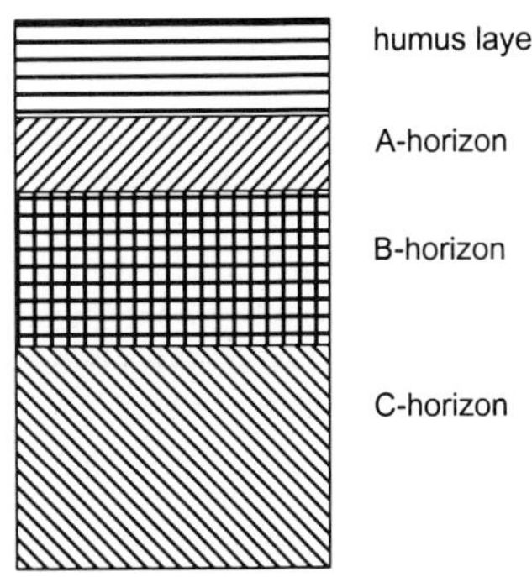

Soil structure: the layers of podzol.

Well known soil types are → **podzol, brown forest soils** (brown earths), **black earth** (chernozem), **tundra soils, desert soils** and tropical **laterites**. Podzol (grey forest soil) is common in cold temperate regions, mostly in coniferous forests; brown earths are typical of areas covered with deciduous forests, e.g. in Central Europe. Tundra soils are common in cold areas; their surfaces are formed from a thin layer of peat or humus. Due to poor plant growth, desert soils contain very little organic matter; in laterites the organic layer is thin because of rapid decomposition and the soil is red due to abundance of iron and aluminium compounds.

soil profile, a series of distinct layers or horizons in vertical sections of the soil. *See* soil formation.

sol (L. *solutio* = solution), **1.** a colloid suspension in a liquid form; *see* colloid, *cf.* gel; **2.** abbreviation for solution.

solanine (L. *Solanum* = the type genus of solanaceous plants), a steroidal glycoalkaloid, $C_{45}H_{73}NO_{15}$, which occurs e.g. in potato and tomato; it is poisonous to most animals and has been used as an insecticide. *Syn.* solatunine.

solar (L. *solaris* < *sol* = sun), **1.** pertaining to the sun, such as to light, heat, or time determined by the sun; **2.** pertaining to the radial anatomical structure, such as the solar plexus (coeliac plexus), an autonomic nerve plexus behind the stomach in vertebrates.

solar constant, the radiation energy of the Sun entering the outer layer of the atmosphere, approx. 1,360 $J/s/m^2$.

solar day, the time interval between two successive returns of the Sun to the meridian. The mean solar day is the average value of the → solar constant.

soldier caste, a caste of some ant and termite species, usually including big individuals with a large head and strongly developed jaws specialized for defending the colony.

solenidion (Gr. dim. of *solen* = channel, tube, pipe), a short and hollow sensory seta in the legs and palps of mites and ticks (Acari); probably a receptor for chemical stimuli.

solenocytes (Gr. *kytos* = hollow, cell), tubular flagellated cells in excretory organs (protonephridia) of many polychaetes and priapulids for filtrating excretions from coelomic fluid to the → protonephridium; they functionally correspond with the **flame cells** of some platyhelminths and rotifers, but each solenocyte includes only one flagellum.

solenostele (Gr. *stele* = column), amphiphloic siphonostele in plants having phloem tissue both outside and inside the xylem. *See* stele.

Solpugida (*Solpuga* = the type genus), **false spiders, sun spiders**; an order of arachnids with a hairy body, very large chelicerae, long pedipalps and legs (first legs tactile, not locomotory), and segmented abdomen; about 900 species, living mainly in arid areas. *Syn.* Solifuga.

solute (L. *solvere* = to dissolve), the dissolved substance in a solution.

solution (L. *solutio*), **1.** a homogeneous mixture of two or more molecular components; the most common type is a solid component dissolved in a liquid, other types are e.g. gases in liquids, liquids in liquids, or solids in solids (solid solutions like some alloys) in which the process of solution takes place in the molten form; **2.** the act of dissolving.

solvate, a solution in which there is a noncovalent, easily reversible combination between the solvent and solute (dissolved substance); the formation of this chemical combination is called **solvation**. When water is the solvent, it is called a → hydrate, and the process is hydration.

solvent, a substance, usually liquid, which has the capability of dissolving other substances in it; the solvent has the same physical state as the solution.

soma, pl. **somata,** also **somas** (Gr. = body), **1.** the body (as distinguished from psyche); **2.** the whole of an organism except the germ cells; **3.** the cell body of branching cells, e.g. of nerve cells.

somat(o)-, somatico-, denoting relationship to a body, bodily, or physical (as distinguished from mental or psychic); also pertains to cells other than reproductive ones.

somatic, 1. pertaining to the animal body, especially to the body wall in contrast to the viscera; **2.** pertaining to the → vegetative as distinguished from the generative, i.e. somatic cells, somatic tissues, which are cells and tissues other than those producing gametes or their progenitor cells.

somatic conjugation, the close juxtaposition and co-orientation of homologous chromosomes during the prophase and metaphase of somatic cell divisions; found e.g. in fruit flies and other dipterans.

somatic hybrid, a → heterokaryon formed by two somatic cells.

somatic mutation, a → mutation occurring in somatic cells of an organism expressed only in the daughter cells of the mutated cell; thus it is not inherited to further generations.

somatic nervous system, that part of the vertebrate nervous system which comprises 1) the sensory nerve tracts from the somatic areas of the body, mainly from the skin and muscles, 2) the motor neurones innervating the skeletal muscles, and 3) the brain areas controlling these functions, especially somatic sensory and motor cortex. *See* sensory pathway, nervous system.

somatocoel (Gr. *koilos* = hollow), a pair of coelomic vesicles in the echinoderm embryo; the left somatocoel gives rise to oral coelom, and the rigt somatocoel developes into aboral coelom. *See* axocoel, hydrocoel.

somatoliberin (L. *liberare* = to set free, liberate), a peptide hormone of vertebrates produced by certain neurosecretory cells of the hypothalamus. When transferred in the blood to the anterior → pituitary gland, it induces the release of → growth hormone. *Syn.* growth hormone-releasing hormone (factor), **GH-RH**, **GRF**, also somatotropin-releasing hormone (factor). *Cf.* somatostatin.

somatomammotropin (L. *mamma* = mammary gland, Gr. *tropos* = turn), a mammalian peptide hormone produced by the placenta, closely related to → growth hormone and → prolactin. Human chorionic somatomammotropin, HCS, (human placental lactogen, HPL) has luteotropic activity, and stimulates milk production and the growth of tissues during pregnancy.

somatomedins (L. *medius* = middle), a group of polypeptide hormones of vertebrates synthesized in many tissues, especially in the liver, and circulating in the blood. → Growth hormone stimulates the tissues to produce and release somatomedins, which then act as → **growth factors** in the body and thus mediate the effects of growth hormone, i.e. the growth of tissues. Somatomedins cause release of → somatostatin from the hypothalamus and thereby decrease their own production and release. The first somatomedin isolated was called **sulphation factor** because it stimulated the incorporation of sulphate into the cartilage.

somatopleure (L. *somatopleura*, Gr. *pleura* = side), a double cell layer of the early vertebrate embryo derived from the association of the lateral mesoderm and ectoderm; develops into the body wall and participates in the formation of amnion. *Cf.* splanchnopleure.

somatostatin (Gr. *stasis* = standing), a polypeptide hormone secreted mainly by certain neurosecretory cells in the median eminence of the → hypothalamus and by delta cells of the pancreas. When transferred via blood to the anterior → pituitary gland it inhibits the secretion of growth hormone and thyrotropin. It also inhibits the secretion of → insulin and other pancreatic hormones; in the intestine it inhibits e.g. the secretion of gastrin, secretin, and in the kidney the secretion of → renin. *Syn.* growth hormone-inhibiting hormone **(GIH)**. *Cf.* somatoliberin.

somatotropin, also **somatotrophin** (Gr. *tropos* = turn, *trophe* = nourishment), → growth hormone.

somite, 1. a segment of a metamerically segmented animal, such as the somites in the abdomen of insects; **2.** one of the paired segments of tissue masses lying alongside the notochord of the vertebrate embryo, derived from the → mesoderm. The number of somites is related to the embryonic stage (increases during development). Later the band of somites is divided into three longitudinal portions, i.e. into medial **sclerotomes** from which the vertebrae and some ventral parts of the skull derive, into intermediary **myotomes** from which the skeletal muscles are formed, and into lateral **dermatomes** which develop into the corium (dermis) of the skin. The embryonic somites are also called mesodermal (mesoblastic) segments.

somni- (L. *somnus* = sleep), denoting relationship to sleep or drowsiness.

somniferous (L. *ferre* = to bring), causing sleep; somnific.

son(i)-, sono- (L. *sonus* = sound), denoting relationship to sound.

sonication, the disruption of cells by using ultrasound waves with frequencies of hundreds of thousands hertz.

sonogram (Gr. *graphein* = to write), an image produced by ultrasonic waves. *See* bioacoustics.

soralium, pl. **soralia** (Gr. *soros* = heap), *see* soredium.

sorbent (L. *sorbere* = to suck), any substance used for sorption. *See* absorption, adsorption.

sorbitol, D-glucitol (L. *Sorbus* = the genus of mountain ash), a white, crystalline polyhydric alcohol, $CH_2OH(CHOH)_4CH_2OH$, first found in berries of the mountain ash. It occurs in many other berries and fruits, and in some seaweeds, and is found also in mammals as an intermediary product of glucose metabolism in the seminal vesicle. Sorbitol is used e.g. as a sweetening agent and osmotic diuretic, and in laboratory studies to maintain the required osmolality in a buffer solution for isolating cell organelles. *See* sugar alcohols.

sorbose, a ketohexose (a monosaccharide), $CH_2OH(CHOH)_3COCH_2OH$, synthesized from xylulose in plants acting as a precursor of sorbitol. Sorbose is manufactured from sorbitol by fermentation with *Acetobacter* species, and used in making ascorbic acid.

sorption, → absorption, → adsorption.

soredium, pl. **soredia** (Gr. *soros* = heap), a → diaspore of lichens, a scale-like structure formed on the thallus; contains both algal or cyanobacterial cells and fungal hyphae, and is able to develop into a new lichen thallus. Soredia may occur scattered on the thallus or in groups called **soralia** (sing. soralium).

sorus, pl. **sori** (L.), **1.** a cluster of sporangia in ferns, found mostly on the underside of fern leaves, i.e. sporophylls; **2.** a cluster of gemmae on the thallus of a lichen; **3.** a mass of spores bursting through the epidermis of the host plant of a parasitic fungus.

SOS response (SOS = common emergency signal), in microbiology, the → induction of several → repair enzymes in bacteria due to DNA damage, causing a significant increase in the quantity of these enzymes. The response results e.g. from the exposure of bacteria to UV radiation or other agents destroying DNA. **SOS box** is a short operator DNA sequence found in some genes involved in the SOS response in *Escherichia coli.*

Southern blotting, a sensitive method for detecting DNA, named after its inventor, *E. Southern*. In the method, DNA restriction fragments to be identified are separated by gel electrophoresis, denatured to form single-stranded DNA, and transferred onto a nitrocellulose filter where DNA fragments are hybridized with a radioactive single-stranded DNA probe. A DNA fragment containing a specific sequence complementary to that of the probe, is then visualized by autoradiography. *See* DNA fingerprint, Northern blotting, Western blotting.

Soxhlet extractor (*Franz von Soxhlet*, 1848—1926), an apparatus for extracting the soluble component of a substance by circulating the boiling solvent through it; e.g. extracting lipids into boiling ether.

spacing behaviour, a behavioural aim of animals to exploit a space; e.g. territorial behaviour, nomadism, etc.; spacing behaviour regulates the dispersion of individuals and pairs.

spadix, pl. **spadices** (Gr. *span* = to pull, draw, tear), **1.** a type of plant inflorescence; a type of succulent spike in which there are numerous sessile flowers in a long, thickened axis. The protecting leaf, **spathe**, is usually associated with the spadix; common in the family Araceae; **2.** in zoology, a conoid organ of the cephalopod mollusc, *Nautilus*, forming the base for the tentacles.

spanandry, spanandria (Gr. *spanos* = scarce, *aner* = man), the scarcity of males in comparison with females in a population; the continuously decreasing proportional number of males in a population.

spasticity (Gr. *spastikos* = drawing in), a state of increased muscular tone, caused by prolonged involuntary muscular contractions (**spasms**). *Adj.* **spastic**.

spathe, a subtending leaf below the → spadix occurring typically in plants whose inflorescence is of a spadix type, or it exists in a modified form in some monocotyledons, e.g. in the family Araceae.

spatium, pl. **spatia** (L.), space; e.g. *spatium intercostale*, the intercostal space between the ribs; *spatium subdurale*, the subdural space or cavity below the → dura on the vertebrate brain.

spawn, 1. a mass of eggs in the water, usually surrounded by mucous liquid, deposited by female clams, crustaceans, fishes, amphibians, and some other aquatic animals; **2.** the mycelium of mushrooms; **3.** to deposit eggs or sperm directly into the water; **4.** to plant with mycelium.

specialist, an individual, population, or species adapted to a particular environmental factor, or ecological niche; e.g. a **food specialist** feeds only on certain prey species or types, a **habitat specialist** lives only in a particular habitat. Specialists are usually unable to adapt to changes in environmental conditions. *Cf.* generalist.

specialization, the evolutionary adaptation of structure and function of an organ, organism, or population to a certain definite life habit, or habitat.

speciation, the evolutionary process by which new species arise. Speciation is possible by separation of one population or population group into two or more reproductive isolates. Speciation has three basic types: **1. sympatric speciation**, development without geographical isolation when reproductive isolation arises in a deme; **2. allopatric speciation**, development as a consequence of geographical isolation, in the course of which genetic differences in isolates increase to the degree that interbreeding with ancient populations is impossible after isolation, and **3. instantaneous speciation**, a sudden birth of a new individual out of its original population into a reproductive isolation (e.g. **allopolyploid**, → polyploid), the isolate being ecologically able to establish a new population.

species, a group of individuals or populations which is actually or potentially capable of interbreeding but is reproductively isolated from all other such groups. Thus the species is a **reproductive unit** that has a common gene supply and an individual is only a temporary **gene carrier** (the conception of **biological species**). Other definitions are: 1) **typological species**, including the species which are unchanged, and intraspecific variation is a consequence of deviations from the basic type; e.g. the species conception of *Carl von Linné* (1707—1778); 2) **multidimensional species**, including two different criteria: the first is **reproductive isolation of species**, and the second is **interbreeding of the species populations** for supporting the unity of species.

Common criteria are established for identifying a species: firstly, the criterion of common descent meaning that members of the species can be traced down to a common ancestral population; secondly, the species must be the smallest distinct grouping of organisms sharing patterns of ancestry and descent. Morphological, chromosomal, and molecular characteristics are important in identifying such grouping. The third criterion is that of reproductive community, as mentioned above.

A species is a taxonomic unit denoted by a **binomen** consisting of its genus and the species' name. In practice, the concept of species is difficult to define because spatial and genetic variations in nature are rich and thus no species is capable of remaining as a static unit. *Cf.* subspecies, race.

species-area relationship, a postulate of the → island theory showing a positive correlation between an island area and the number of species living there. The concept of island may be applied to any area of land which provides an environment that is substantially different from its immediate surroundings. The biodiversity of a community in relation to the area is described by the **species-area curve**: $S = cA^z$, where the number of species is represented by S and the size of the area by A, c is a constant showing the number of species per an areal unit, and z is a key parameter in island models.

species diversity, the number of different species and their relative abundances in a certain community, habitat, or area. The **species diversity indices**, e.g. → Shannon-Wiener index, describe the species diversities in communities or habitats. The diversity can be divided into alpha, beta, and gamma components. The **alpha diversity** is within-area diversity pertaining to the number of species occurring in an area of a given size; it describes the richness of interactive species. The **beta diversity** designates the degree of species change along a given habitat or physiographic gradient thus measuring the between-area diversity (is usually expressed with similarity indices). The **gamma diversity** refers to overall diversity within a large region dealing with species diversity e.g. at the landscape level. *See* biodiversity.

species richness, the total number of species present in a habitat or community at a certain moment.

species selection, the → selection of species during evolution, meaning that the species may be a unit of evolution as well as the individual. It is a form of → group selection in which the number of species with differing characteristics increases (by speciation) or decreases (by extinction) at different rates.

species turnover, a term used in → island biogeography presenting the proportional number of species changed in a particular area during a time period, e.g. 10% species turnover per year.

specific (L. *specificus* < *species*), having a special characteristic, property, or application; e.g. **1.** specific distinctive characteristics of a species; **2.** specific affinity of an → antigen for its antibody, or vice versa; **3.** a specific substance which binds to a certain type of cellular receptors (receptor-specific); **4.** a specific disease caused by a single kind of microbe; **5.** pertaining to a remedy exerting a definitive influence on a particular disease, e.g. quinine is specific to malaria. *Noun* **specificity**.

specific activity, acitivity per unit mass, volume, or molarity; e.g. **1.** the activity (*a*) per unit mass of a pure radioisotope, i.e. the ratio of radioactive to non-radioactive atoms or molecules of the same kind; expressed in disintegrations per second per kg; *see* activity in Appendix 1; **2.** enzyme activity expressed as the quantity of reaction product or substrate used in the reaction; usually μmoles per mg of protein per minute.

specific gravity, the weight of a substance or a body compared with that of an equal volume of another substance regarded as a standard.

specific pathogen free, SPF, pertaining to laboratory animals, which are reared in special conditions behind a barrier so that the animals are free from certain pathogenic germs. *Cf.* gnotobiota.

spectrin, a cytoskeletal protein which binds actin and crosslinks actin filaments, and forms a firm structure (membrane skeleton) under the plasma membrane especially in red blood cells; related actin crosslinking and attachment proteins are α-actinin and dystrophin.

spectrophotometer (L. *specere* = to look, watch, Gr. *phos* = light, *metron* = measure), optical equipment for measuring the intensity of light of a definite wavelength transmitted through a solution. It is used to identify substances and determine their purity and quantities on the basis of their → absorbance. Most spectrophotometers measure the transmission of monochromatic visible and UV light; special equipment uses infra red and fluorescent light.

spectrum, pl. **spectra** or **spectrums** (L. = appearance), an array of entities such as light waves or particles; spectrum is continuous if all its wavelengths or other components are present, and a line type if some of them are lacking; e.g. **1.** light spectrum (visible spectrum), can be seen as different colours when light passes through a prism; **2.** absorption spectrum, i.e. the absorption of electromagnetic radiation (e.g. light) by an agent, such as a chemical compound, having a typical absorption spectrum based on their light → absorbance; **3.** line spectrum, recorded e.g. by a → mass spectrometer; **4.** broadly, the range of any magnitude like that of activity or energy based on the frequency or wave length.

Spemann's organizer (*Hans Spemann*, 1869—1941), that part of the dorsal lip of the → blastopore in the gastrula stage of the vertebrate embryo which causes induction of the neural plate. *See* embryogenesis.

sperm(a)-, spermi-, spermo- (Gr. *sperma* = seed, sperm), pertaining to seed, spermatozoa, sperm, semen.

sperm, pl. **sperm** or **sperms, 1.** a viscid fluid produced by male animals containing spermatozoa and secretions of gonads and accessory sexual glands, such as the → epididymis, seminal vesicles, prostate, and bulbourethral glands, usually released by ejaculation. The sperm of fish is called **milt**, and that of mammals and birds (especially man and domestic animals) **semen**; **2.** → spermatozoon.

spermaceti (Gr. *ketos* = whale), **1.** spermaceti wax; a white, translucent, waxy, fatty substance consisting chiefly of cetyl palmitate (cetin) and other esters of fatty acids obtained from the head of the sperm whale (*Physeter macrocephalus*) and from the oils of related cetaceans; used e.g. in cosmetic creams; earlier, spermaceti wax was believed to be coagulated sperm; **2.** sperm oil, a yellowish oil occurring with spermaceti wax, chemically a liquid wax.

spermagonium, → spermogonium.

spermat(o)- (Gr. *sperma* = seed), denoting relationship to seeds, or to male germ cells (spermatozoa) in animals.

spermatangium, pl. **spermatangia** (Gr. *angeion* = vessel), the male gametangium in red algae; develops into a non-motile gamete (spermatium).

spermateliosis (Gr. *teleiosis* = development), → spermiogenesis.

spermatheca (L. *theca* = case, cover), a saclike organ associated with the reproductive organs of female or hermaphrodite invertebrates, e.g. many insects, oligochaetes, and a few vertebrates such as some amphibians, receiving and storing sperm often for long periods. *Syn.* seminal receptacle.

spermatid (Gr. *eidos* = form), *see* spermatogenesis.

spermatium, pl. **spermatia** (Gr. *spermation* dim. of *sperma* = seed), **1.** a non-motile male gamete in red algae; **2.** a non-motile cell (pycnospore) developed in some fungi by abstriction from a sterigma within a → spermogonium, being able to conjugate with a mycelium of the opposite sex.

spermatocyte (Gr. *kytos* = hollow, cell), *see* spermatogenesis.

spermatogenesis (Gr. *gennan* = to produce), in animals, the development of male gametes in male gonads, testes. The testes contain sperm mother cells, **primary spermatogonia**. They divide, usually rapidly, several times by → mitosis generating **secondary spermatogonia**. These develop without cell division into **primary spermatocytes**, which further undergo the first meiotic division (*see* meiosis), each cell giving rise to two **secondary spermatocytes**, which then undergo the second meiotic division. The formation of these first stages is called spermatocytogenesis. As a consequence of these meiotic divisions four haploid **spermatids** (spermids) arise, which in → **spermiogenesis** develop without any cell division into mature male gametes, called **spermatozoa** (sperm cells, spermia).

In plants, male gametes develop in the male → gametangium.

spermatogonium, pl. **spermatogonia** (Gr. *gone* = birth, origin), *see* spermatogenesis.

spermatophore (Gr. *pherein* = to carry), a sperm capsule, especially of several invertebrate groups, such as annelids, molluscs, and arthropods, transferred by the male to the female in copulation; e.g. using a special organ, **hectocotylus**, the male squid transfers the spermatophore into the mantle cavity of the female. Among vertebrates, the fertilization of newts (Urodela) takes place with the aid of the spermatophore.

Spermatophyta (Gr. *phyton* = plant), **spermatophytes**; seed-bearing plants, the most developed division in the Plant kingdom; divided into subdivisions → Gymnospermae and → Angiospermae. In some systems spermatophytes is a group name without any taxonomical status.

spermatozoid (Gr. *zoon* = animal), **1.** a free-swimming male gamete, present in many algae and mosses, developing in a male gametangium or antheridium; in these organisms, female gametes are sessile and much bigger than male gametes; **2.** sometimes a mature animal sperm cell is called spermatozoid, but more commonly → spermatozoon. *Syn.* antherozoid.

spermatozoon, pl. **spermatozoa**, the mature male gamete of animals, consisting of a head which contains the nucleus and usually the → acrosome at its tip, and the tail that is the motile organelle of the spermatozoon. *Syn.* sperm cell, spermium, sperm. *See* spermatogenesis.

sperm cell, 1. the male gamete of seed plants, formed in the pollen tube; **2.** in animals, → spermatozoon.

sperm competition, the competition between sperms from two or more males to fertilize the eggs; occurs in plants and animals. The competition may occur either outside the body of the female, as in water, or within the female's reproductive tract. Many females have complex sperm storage organs (e.g. → spermatheca) to assortment sperms of different males after copulation; also males have various methods to eliminate sperms of their competitors (e.g. many behavioural patterns). Sperm competition is studied using molecular measures of parentage, e.g. DNA fingerprinting.

spermid, spermatid, *see* spermatogenesis.

spermine, a → polyamine first found in human sperm, but later observed to be associated with nucleic acids in different tissues of animals and plants, in bacteria and some viruses; its precursor is called **spermidine**. These polyamines play an important role in the regulation of cellular proliferation and differentiation. In plants spermidine has hormone-like effects. *Syn.* gerontine, musculamine, neuridine.

spermiogenesis (Gr. *gennan* = to produce), the last stage in → spermatogenesis in which animal spermatids develop into mature male germ cells, spermatozoa. *Syn.* spermateliosis.

spermium, pl. **spermia,** → spermatozoon.

spermogonium, also **spermagonium,** pl. **spermogonia, spermagonia** (Gr. *gone* = birth, origin), a flask-shaped mycelial structure forming special spores (pycnospores) in rust fungi and in fungal partners of lichens.

sperm oil, *see* spermaceti.

SPF, → specific pathogen free.

sphaeridium, also **spheridium,** pl. **sph(a)eridia** (Gr. *sphairidion,* dim. of *sphaira* = ball, sphere, globe), any of the small club-shaped organs on the surface of some echinoderms, e.g. echinoids; probably a sensory organ acting as an equilibrium sense.

Sphagnidae, sphagnum mosses, peat mosses, bog mosses (L. Gr. *sphagnos* = moss), a subclass in the class Musci (Bryopsida), the division Bryophyta; in some systems a class in the division Bryophyta. Comprises only one genus, *Sphagnum.* The growth and structure of sphagnum mosses are characteristic, i.e. the individuals grow continuously in the top whilst the bottom end dies. They grow only in wet, acid areas, where their accumulated remains become compact, forming peat with other plant debris. They are also important swamp formers, because dead cells in their structure effectively absorb surface water.

S phase (S < synthesis), the phase of the mitotic cycle during which DNA is synthesized. *See* mitosis, cell cycle.

Sphenisciformes (Gr. *spheniskos*, dim. of *sphen* = wedge, L. *forma* = form; *Spheniscus* = the type genus), **penguins**; an order comprising the single family Spheniscidae with 18 flightless marine bird species found from Antarctica to the Galápagos Islands. Their body is adapted for swimming; the wings are modified into paddles and the legs with webbed toes locate far back on the body, and the feathers are scale-like. Penguins are adapted to marine life but come on land for breeding.

Sphenodonta (Gr. *sphen* = wedge, *odous* = tooth), → Rhynchocephalia.

Sphenophyllales, an extinct order in the class Sphenopsida, division → Pteridophyta, Plant kingdom. The sphenophylls were herbaceous plants about a metre high and the branching stems had extended internodes and wedge-shaped leaves in whorls. Fossils are known from Upper Devonian to Lower Permian.

Sphenophyta, *see* Sphenopsida.

Sphenopsida (Gr. *opsis* = manifestation), a class in the division Pteridophyta, Plant kingdom; in some systems classified as a separate division, Sphenophyta. The only present order is Equisetales, → horsetails; the orders Sphenophyllales and Protoarticulatales are extinct and found as fossils.

spherosome (Gr. *sphaira* = ball, *soma* = body, piece), a globular organelle in plant cells, 0.4 to 3 µm in diameter; it is enveloped by membrane half of a unit thick, i.e. 3 nm. In the spherosomes, lipids are partially synthesized, stored and hydrolysed. *Syn.* oleosome.

sphincter (Gr. *sphinkter* = band), a circular band of muscle fibres which closes a natural orifice, or constricts a passage of a hollow organ of the animal body; e.g. anal sphincter (*musculus sphincter ani*), pyloric sphincter (*m.s. pyloricus*), or pupillary sphincter (*m.s. pupillae*).

sphingolipid (Gr. *sphingein* = to bind fast), any of a group of lipids, such as → ceramides, cerebrosides, gangliosides, and sphingomyelins, containing → sphingosine or a related base; forms components of animal and plant cell membranes.

sphingomyelin, any of a group of phospholipids containing phosphorylcholine combined with → ceramide; it is one of many → sphingolipids found in animal tissues, especially in nerve, liver, and red blood cells.

sphingosine, a mono-unsaturated aliphatic amino alcohol; $C_{18}H_{37}NO_2$; a long-chain molecule which occurs as a component in → sphingolipids.

spicule (L. *spiculum* = small spike, sting, arrow), a small needle-like structure; e.g. **1.** one of crystal-like calcareous or siliceous needles supporting the soft body wall of many radiolarians, sponges, and holothurians; **2.** copulatory spicule, the organ projecting from the male genital pore of the roundworm, *Ascaris*, or a dart in many terrestrial snails to heighten the excitement before copulation; **3.** a small appendage such as the sterigma in basidiomycetous fungi.

spiders, → Araneae.

spike, 1. an indeterminate → inflorescence with numerous sessile flowers in → spikelets on the main axis; the spike is typical of graminaceous plants (family Poaceae); **2.** a pointed element in a curve of a graph, e.g. a brief → action potential, or an up-and-down deflection in electroencephalograms.

spike leaf, a leaf modified into a → thorn, e.g. in cactus species.

spikelet, the basic unit of the inflorescence of graminaceous plants (family Poaceae) consisting generally of 2—6 perianthless flowers (florets) with bracts called **lodicule(s)**, **palea**, and **lemma**. Two bracts (**glumes**) protect the whole spikelet. The lemma may have a thin hooked projection, called awn. The inflorescence is normally a spike or panicle.

spike stem, a → short plant shoot modified into a → thorn; e.g. hawthorn.

spike stipule, a plant stipule modified into a → thorn.

spin, the rotation of an electron around its axis, or the revolving of the Earth around its axis while orbiting the Sun. *See* electron spin resonance, spin labelling.

spin-, spini-, spino- (L. *spina* = thorn, spine, spinal column), pertaining to the spinal cord, spinal column, or spine. *Adj.* **spinal**.

spinal canal, vertebral canal; the tube inside the successive vertebrae containing the → spinal cord with spinal meninges. *See* vertebra.

spinal column, → vertebral column.

spinal cord (L. *medulla spinalis*, *chorda spinalis*, *chorda dorsalis*), that part of the central nervous system of vertebrates which is located in the spinal canal (vertebral canal); extends through the whole length of the canal only in fish, but e.g. in mammals it ends at the level of the first lumbal vertebrae. → Spinal nerves are separated into five areas: **cervical**, **thoracic**, **lumbal**, **sacral**, and **coccygeal** spinal areas. The nervous tissue of the spinal cord is composed of grey matter consisting of nerve cell bodies and situated in the core area (seen as an H shape in cross section), and of white matter with afferent and efferent, mostly myelinated nerve fibres. The **central canal** forms a long channel in the middle of the spinal cord, and like the brain, the spinal cord is enclosed by three membranes (**spinal meninges**), the dura mater,

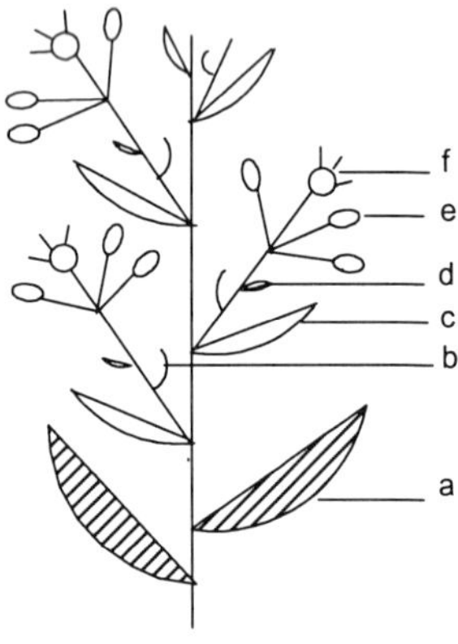

Spikelet. The structure of spikelet: glume (a), palea (b), lemma (c), lodicule (d), stamen (e), pistil (f).

arachnoid, and pia mater (*see* meninx). Both the central canal and the subarachnoidal space are filled with → cerebrospinal fluid.

spinal ganglion, pl. **spinal ganglia** (L. *ganglion spinale*), a ganglion on the dorsal root of each → spinal nerve; also called sensory ganglion.

spinal nerves (L. *nervi spinales*), paired vertebrate nerves passing out from the spinal cord segmentally between the vertebrae. Each nerve (except in fishes) has two roots, the **dorsal root** (sensory root) with sensory nerve fibres and the spinal ganglion, and the **ventral root** (motor root) with somatic motor and some autonomic nerve fibres. The number of spinal nerves varies according to the vertebrate group, e.g. in fish and snakes usually over 100, in man 31 pairs. In tetrapods, they are grouped into **cervical** (8 pairs in man), **thoracic** (12), **lumbal** (5), **sacral** (5), and **coccygeal nerves** (1).

spindle, *see* mitotic spindle, muscle spindle.

spindle attachment site, → centromere.

spine (L. *spina* = backbone), **1.** a sharp-pointed, hard process on the surface of an animal (or plant), or on bones or cartilages, e.g. a spiny fin ray of a fish, or an iliac spine (*spina iliaca*) of the ilium (bone); **2.** spinal column, → vertebral column, backbone. *Cf.* thorn.

spin labelling, a method for labelling biological macromolecules with → free radicals or other paramagnetic probes (e.g. Cu^{2+}) to determine the state of given molecules and their relationship with other molecules by using → electron spin resonance spectroscopy. A nitroxide group (NO) is usually used as a free radical; it is relatively unreactive and contains an unpaired electron which gives the electron a spin resonance signal.

spinneret, spinning organ; **1.** a conical structure, usually occurring in two or three pairs, in front of the anus on the abdominal surface of spiders. The spinnerets are probably the modified 5th and 6th abdominal legs. Many fine **spinning tubes** from the → spinning glands open to the membranous tips of the spinnerets, emitting hardening secretion streams which join together to form a single thread, the silk. The tip is surrounded by hairs and barbs forming the **spinning field**; **2.** an analogous organ of some insect larvae.

spinning glands, silk glands, 1. the glands in the abdomen of some spiders (arachnids) producing proteinaceous secretion that hardens into a silk thread on exposure to air; in some arachnids, as e.g. pseudoscorpions, the glands open to the tips of chelicerae; **2.** the spinning glands of some insect larvae which are not homologous with those of spiders. The glands are located in the pharynx and are usually modified salivary glands whose secretion is used for the spinning of larval and pupal covers, → cocoons. *See* spinneret.

spinothalamic, pertaining to ascending sensory nerve tracts from the spinal cord to the → thalamus in the vertebrate central nervous system.

spiny-headed worms, → Acanthocephala.

spir-, spiri-, spiro- (L. *spira* < Gr. *speira* = coil), **1.** pertaining to a coil or spiral; **2. spiro-** (L. *spirare* = to breath), denoting relationship to respiration or breathing.

spiracle (L. *spiraculum* = air hole), a water or air hole in animals; **1.** the **tracheal spiracle** of terrestrial arthropods, such as insects; *see* trachea; **2.** a small, vestigial gill opening for respiratory water between mandibular and hyoid arches in cartilaginous fishes and some bony fishes, e.g. sturgeon; **3.** lateral gill opening, found in amphibian larvae; **4.** the nostril opening of whales (blowhole), located on the top of the head.

spiral ganglion, the ganglion of the cochlear nerve in the inner ear of tetrapod vertebrates, consisting of bipolar sensory nerve cells whose axons form the cochlear branch of the eighth cranial nerve.

spiral leaf arrangement, *see* phyllotaxy.

spiral organ (L. *organum spirale*), an auditory organ in the inner ear of tetrapod vertebrates. The organ lies on the spiral (less spiral in amphibians) **basilar membrane** in the cochlear duct, with neuroepithelial sensory receptor cells (→ **hair cells**) in the inner and outer rows, and inner and outer pillar and phalangeal cells in several rows supporting the hair cells. The organ is overhung by a shelf-like membrane, **tectorial membrane** (*membrana tectoria*) which, when pressing against the stereocilia of the hair cells, causes irritation. The spiral organ is innervated by the **cochlear nerve**, a branch of the 8th → cranial nerve. *Syn.* organ of Corti, acoustic papilla. *See* cochlea.

spirillum, pl. **spirilla** (L. dim. of *spira* = coil), **1.** a general name for spiral-shaped bacteria; **2.** any bacterium of the genus *Spirillum* (the

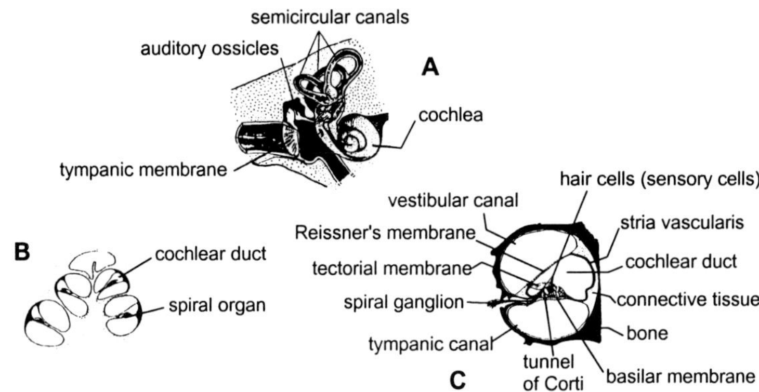

Spiral organ, enlarged in the middle of figure C, seen in the cross section of the cochlea (B) of the mammalian ear (A).

type species *Spirillum volutans* living in fresh water); these bacteria are helically curved, motile microbes, some of which live as parasites in mammals, e.g. *Spirillum minus* in the nasopharyngeal mucosa of rats and mice causing rat-bite fever in man. Many of these species have since been classified in other genera.

spirit (L. *spiritus*), **1.** the principle of conscious life, or a supernatural being; **2.** a volatile or distilled liquid, especially strong ethanol (*spiritus fortis*).

Spirochaetales (Gr. *speira* = coil, *chaite* = hair, mane), an order comprising spiral-shaped, slender, flexuous, motile bacteria, 5 to 500 μm long. The order contains two families: Spirochaetaceae (the type genus *Spirochaeta*) and Treponemataceae, some of which are pathogenic in man, e.g. *Borrelia* (→ borreliosis), and *Treponema pallidum* causing syphilis.

spiroch(a)ete, 1. an organism of the genus *Spirochaeta* of the order → Spirochaetales; **2.** broadly, any spiral microbe of the order Spirochaetales.

spirometry (L. *spirare* = to breathe, *metrum* = measure), the determination of the breathing capacity of human lungs; the instrument used for this measurement is called **spirometer**. *Adj.* **spirometric**.

Spirotrichia (Gr. *thrix* = hair), a protozoan group in the phylum Ciliophora, characterized by external membranelles around the oral groove in their unicellular cell body; e.g. *Stentor*.

splanchn(o)- (Gr. *splanchnon* = viscus, entrail), denoting relationship to the viscera, or splanchnic nerves.

splanchnocoel(e) (Gr. *koilos* = hollow, cavity), that part of the embryonic body cavity (coelom) of vertebrates which develops into pleural, pericardial, and abdominal cavities; also called pleuroperitoneal cavity.

splanchnocranium (Gr. *kranion* = skull), those parts of the skull which support the jaws and → gill arches in fish; in tetrapod vertebrates, those parts of the skull which are derived from the first and second embryonic pharyngeal arches, i.e. the jaw skeleton. *Syn.* viscerocranium. *Cf.* neurocranium.

splanchnopleure (L. *splanchnopleura*, Gr. *pleura* = side), a double tissue layer formed by the union of the mesoderm with the underlying endoderm in an early vertebrate embryo; forms most of the wall tissues of the visceral organs. *Cf.* somatopleure.

spleen (L. *lien*, Gr. *splen* = spleen), a large lymphoid organ in the abdominal cavity of jawed vertebrates; it is linked to the blood circulation rather than the lymphatic circulation as are the lymph nodes; consists of **white pulp**, i.e. the lymphatic tissue producing lymphocytes, and of **red pulp** containing venous sinusoids, and disintegrating red blood cells. The spleen serves also as a store of red blood cells and produces these cells during embryonic development and in the newborn.

splen(o)-, denoting the spleen.

spliceosome (Gr. *soma* =body), a complex in the cell nucleus consisting of several kinds of → small nuclear ribonucleoprotein particles (snRNPs); catalyses → splicing, i.e. the processing of → heterogeneous nuclear RNA into → messenger RNA.

splicing, in genetics, the removal of → introns from the heterogeneous nuclear RNA and the rejoining of → exons; the splicing of precursor mRNA is catalysed by the → spliceosome, → small nuclear RNA molecules (snRNAs) playing an important role in the process. Usually splicing is **constitutive**, i.e. all the exons are joined together. In certain genes also **alternative splicing** occurs in which e.g. some exons can be skipped. Sometimes, especially in preribosomal RNAs and various mitochondrial and chloroplast RNAs, also **self-splicing** occurs, i.e. splicing without the catalytic activity of the spliceosome.

splicing junction, the two DNA segments in a gene which immediately surround the boundaries (junctions) between → exons and → introns of the genes. During splicing the → heterogeneous nuclear RNA is cut at these junctions.

spondyl(o)- (Gr. *spondylos* = vertebra, whorl), pertaining to the vertebra or the vertebral column; e.g. **spondylosis**, a condition with degenerative changes in vertebral joints due to osteoarthritis.

sponges, → Porifera.

spongin (Gr. *spongos* = sponge), a fine, sulphur-containing scleroprotein appearing in the form of fibres which form the network-like (endo)skeleton of some sponges (e.g. bath sponge, *Spongia*). Spongin is secreted by spongioblasts.

spong(i)oblast (Gr. *blastos* = bud), **1.** a cell secreting spongin in some sponges; **2. spongioblast**, a cell type of the embryonic neural tube capable of developing into neuroglia.

spongocoel (Gr. *koilos* = hollow), a central cavity of the system of cavities in sponges; also called spongoc(o)ele, atrium.

spongocyte (Gr. *kytos* = cavity, cell), a cell type in the body wall of sponges. *See* Porifera.

sporangiophore (Gr. *sporos* = seed, *angeion* = vessel, receptacle, *pherein* = to bear), a stalk of the → sporangium.

sporangiospore, *see* spore.

sporangium, pl. **sporangia,** a case within which asexual → spores in algae, mosses, ferns, and fungi are produced. Male sporangia are microsporangia, female ones are macrosporangia (megasporangia).

spore, usually a unicellular structure for maintaining reproduction, dispersal, and survival of some organisms over unfavourable seasons or periods. Pores are the typical main reproductive organs of → spore plants, such as different groups of algae, bryophytes, pteridophytes, and fungi, and of many protists, but actually spores are found in all plant groups, produced by the → sporophyte. Spores are formed in **sporangia** (may thus be called **sporangiospores**) usually after → meiosis (**meiospores**), and develop into a haploid **gametophyte**. In some organisms, spore formation is associated with → mitosis; these spores, **mitospores**, give rise to new sporophytes. Many species of algae and fungi have their own special terms for different spore types. Red algae have diploid **carpospores**, developing into diploid sporophytes which produce **tetraspores** after meiosis and **monospores** in mitosis. Ascomycetes have **ascospores**, and the spores of → Basidiomycetes are called **basidiospores**. Many algae and fungi which live in water have **zoospores**, which have a flagellum and are motile. Rust fungi (→ Uredinales) especially have a very complicated life cycle with numerous different spore types. → Heterosporous plants have separate (male) **macrospores** and (female) **microspores**; seed plants have analogous structures. *See* Sporozoa. *Cf.* conidium.

spore mother cell, a diploid cell giving rise to four haploid cells (spores) in → meiosis. *See* spore.

spore plants, plants which reproduce mainly by → spores; e.g. algae, mosses and ferns. Spore plants have two separate generations in their life cycle, the gametophyte phase and the sporophyte phase. The former produces gametes, and the latter spores, which again give rise to gametophytes. The gametophyte of higher spore plants is reduced and is smaller than the sporophyte. Even → seed plants have structures analogous to spores, but they (and the gametophyte structures formed from them) are much reduced, located inside the the sporophyte structures, i.e. the → ovule and → pollen grain.

sporoblast (Gr. *blastos* = germ), *see* sporogony.

sporocarp (Gr. *karpos* = fruit), a capsule in which one or more sporangia are produced; occurs in some ferns, algae, and fungi.

sporocyst (Gr. *kystis* = bladder), **1.** a sac-like

encysted embryo forming the second generation in the life cycle of parasitic, digenean trematodes; the sporocyst develops from a **miracidium larva**, and produces asexually several → **rediae**; **2.** an encysted stage in the life cycle of some protozoans, e.g. gregarines, during which the animal divides into one or several **sporozoites**. The sporocyst develops from a **sporoblast**; *see* sporogony.

sporoderm, *see* pollen grain.

sporogenesis (Gr. *genesis* = birth, origin), **1.** the formation of → spores; **2.** → sporogony.

sporogenic cells (Gr. *gennan* = to produce), cells which are formed in the anthers and ovules of the carpels during the division of → archesporia. Sporogenic cells develop into → **microspores** (in stamens) or → **macrospores** (in carpels). The other cells formed in the division are **parietal cells**, which form the wall tissue of anthers or the nucellus tissue of carpels.

sporogonium, pl. **sporogonia** (Gr. *gonos* = birth), a → sporangium developed from a fertilized ovum in the archegonium of bryophytes; represents the sporophytic generation.

sporogony, the formation of haploid **sporozoites** in sporozoans (e.g. malaria parasite) following the fusion of gametes in the zygote (**sporont**) formation, i.e. a process of asexual division within the **sporoblast**, which develops into the **sporocyst** within an **oocyst**. Sporogony occurs in the body of blood-sucking invertebrates, such as leech, ticks, and some insects. *Syn.* sporogenesis. *Cf.* gamogony, schizogony.

sporont, *see* sporogony.

sporophore (Gr. *pherein* = to carry), **1.** a structure producing and bearing spores in fungi; **2.** in slime moulds, a part of plasmodium producing spores on its surface.

sporophyll (Gr. *phyllon* = leaf), a structure in which the → sporangia are formed; the structure and size of the sporophyll varies, usually being leaf-like. Also the ordinary leaves may function as sporophylls, e.g. in ferns.

sporophyte (Gr. *phyton* = plant), the diploid stage in the life cycle of a plant; produces haploid **spores** through → meiosis. The spores form the haploid **gametophyte**, which produces gametes, and after fertilization a new sporophyte is formed. The size and degree of independence varies greatly in different plant groups. In most developed groups (e.g. seed plants) the sporophyte (the flowering plant) is the dominant form, the gametophyte being much reduced, forming a small sexual organ inside the ovules and pollen grains of the sporophyte. *See* alternation of generations.

sporopollenin (Gr. *pollen* = flour), a carotenoid polymer of which the outermost layer of pollen grains, i.e. **exine**, is formed. Sporopollenin is one of the most resistant organic substances, and therefore the pollen grains survive well in anaerobic conditions in soil and water; → **pollen analyses** are based on this property.

Sporozoa (Gr. *zoon* = animal), **sporozoans**; a protozoan class in the phylum Apicomplexa; comprises unicellular parasitic animals which have no locomotor organelles or contractile vacuoles in adult stages. Sporozoans have a complicated life cycle usually involving an alternation of a sexual and asexual generation, including the formation of cysts and spores. Some sporozoans, such as plasmodians (e.g. malaria parasites) and coccidians, are the most widely dispersed parasites in the world.

sporozoite, a stage in the life cycle of many sporozoans developed in the process of → sporogony and liberated either from a sporocyst as in gregarines, or from an oocyst as in blood sporozoans (Haemosporidia). The sporozoite is a small, usually motile and elongate stage, which infects a definitive host; e.g. malaria sporozoites are transferred from mosquitoes to man.

spring aspect, *see* aspect.

springwood, wood formed in annual rings of woody plants during spring and early summer, i.e. during the earlier part of the growing season. The cells, vessels and tracheids of springwood are typically larger than those of the later **summerwood**.

sprout, a new plant shoot, formed from a dormant bud on the shoot base. Sprout formation is especially common in stumps of felled foliage trees. Sprouts may also be **root sprouts**, formed from dormant buds in roots, or it may be a shoot grown from a seed or tuber.

squalene (L. *squalus* = shark), a C_{30} hydrocarbon containing six → isoprene units found in liver oil, especially in sharks, but also in various plant oils. It is an intermediate in the biosynthesis of → cholesterol.

Squaliformes (L. *forma* = form, *Squalus* = the

type genus), **dogfish sharks**; an order of cartilaginous fishes (subclass Elasmobranchii) including small or middle-sized sharks, characterized e.g. by spines in front of the two dorsal fins, the caudal fin with a small lower lobe, and by the absence of the anal fin. About 75 species are described.

Squamata (L. *squama* = scale), a polymorphic and large reptilian order in the subclass Lepidosauria including about 5,730 species in three suborders: **lizards** (Lacertilia/Sauria), **snakes** (Serpentes/Ophidia), and **worm lizards** (Amphisbaenia). All of them have skin with horny epidermal scales or shields. Most of the present reptiles belong to Squamata.

squamous epithelium (Gr. *epi* = on, *thele* = nipple, teat), a type of → epithelium in animals composed of plate-like cells; when composed of a single cell layer, it is called pavement epithelium.

squids, *see* Cephalopoda.

SR, → sarcoplasmic reticulum.

SSCP, → single-stranded conformational polymorphism.

ssDNA, → single-stranded deoxyribonucleic acid.

S-shaped growth curve, *see* logistic growth.

stability (L. *stabilis* = firm, stable), the quality or state of being stable or firm; e.g. **1.** the ability of a living system to withstand or recover from perturbations without sudden changes in composition; *see* stable equilibrium; **2.** in chemistry, the resistance to chemical changes.

stabilizing factor, a factor that operates against the extremes and tends to restore a system to its equilibrium state; e.g. the density-dependent factors act to restore a population to an equilibrium within the carrying capacity of the environment.

stabilizing selection, *see* selection.

stable age composition/distribution, *see* age distribution.

stable equilibrium (L. *aequus* = equal, level, *libra* = balance), the functional stable condition to which an open system, like an organ or organ system, attempts to return after disturbances; e.g. in ecology, the equilibrium state of a population or community to which the system has returned after disturbances.

stachyose (L. *Stachys* = the genus of the hedge nettle), a tetrasaccharide, $C_{24}H_{42}O_{21}$, consisting of one saccharose and two galactose residues; present in the phloem of many plants and as a storage carbohydrate in certain seeds and tubers. The concentration of stachyose increases in some plants during cold hardening.

stagnation (L. *stagnum* = standing water, pool), the absence or cessation of growth, movement, or activity; e.g. **1.** the stagnation of the carbon cycle in coal beds; **2.** the stagnation of a water body when water masses are immobile (*see* stratification). *Adj.* **stagnant** (stagnatory), not flowing, not active. *Verb* to **stagnate**, to cease to flow or move, fail to develop.

staining, the artificial colouration of biological material with dyes or stains for examination of cell and tissue structures in a microscope. The staining depends on the chemical property of a dye (may be acid, basic, or neutral) but also on the structure to be stained. In many histochemical stainings the colour is formed in enzyme reactions. **Counterstaining** (double staining) is used in order to stain different parts of the specimen in different ways. The stains used for electron microscopy are usually compounds containing electron-dense heavy metal atoms.

stamen, pl. **stamina** or **stamens** (L. = warp, thread, stamen), the male reproductive, pollen-producing organ in the flower of a seed plant, morphologically a spore-bearing leaf. The stamen consists of a filament bearing the anther where the pollen grains are formed. *Syn.* microsporophyll. *See* pollen grain, archesporium.

staminate, pertaining to a flower having stamens but no carpels. *Cf.* carpellate, pistillate.

staminode, staminodium, pl. **staminodes, staminodia,** a → stamen which does not produce pollen, i.e. is sterile and may lack the anther. Staminodes are normal structures in some plants, e.g. in the flowers of the genus *Parnassia*.

stand, an area of vegetation where the species and their relationships are homogeneous; e.g. in forests, the stands consist of trees of about the same age.

standing crop, the total biomass per unit area at any specified time. The amount of standing crop is often given as energy content (in joules).

stapes, pl. **stapes** or **stapedes** (L.), stirrup; one of the three → auditory ossicles in mammals; represents the → columella of other tetrapod vertebrates, and the hyomandibular bone of fish articulating the lower jaw to the skull.

star activity, in genetics, a reduced specificity of a restriction enzyme under suboptimal conditions; e.g. at low ionic strength the specificity of *Eco* restriction enzyme (*Eco* R1) is reduced so that only the internal tetranucleotide sequence of the hexanucleotide in a substrate DNA is necessary for recognition and cleavage. The activity is called *Eco* RI* i.e. RI-star activity.

starch, L. *amylum*, $(C_6H_{10}O_5)_n$; a polysaccharide consisting of glucose residues bound together with α-1,4 linkages as in amylose (linear polymer), or with both α-1,4 and α-1,6 linkages as in amylopectin (branched polymer). Starch is the end product of photosynthetic carbon assimilation and occurs as the storage carbohydrate in plastids of plant cells, found abundantly in cereal grains and potato. Starch is an important nutritive energy source for many animals. It is degraded in acid hydrolysis, or enzymatically by → amylase, into easily soluble forms such as dextrins, maltose, and finally to → glucose.

starch grain, an energy store structure of the plant cell consisting of → starch; occurs especially in the → storage parenchyma.

starch sheath, the → endodermis of the plant root with numerous → starch grains.

starfishes, → Asteroidea.

Starling's law (of the heart) (*E.H. Starling*, 1866—1927), "the energy of contraction is proportional to the initial length of the cardiac muscle fibre"; it means that the blood output from the heart per beat is directly related to the volume of the diastolic filling, i.e. the extension of the contraction depends on the stretch stage of the cardiac muscle.

start codon, → initiator codon.

stasipatric (Gr. *stasis* = a standing still, *patra* = fatherland), pertaining to a marginal, partly isolated → population.

stasis, pl. **stases** (L.), **1.** a state of equilibrium among prevalent opposing forces; **2.** the local diminution or stoppage of the blood flow, or the flow of other body fluid or semifluid material. *Adj.* **static**, at rest, not dynamic.

static life table (Gr. *statos* = standing), a life table that is constructed on the basis of age structure in a population at a given time.

statins, 1. inhibiting hormones; peptide neurohormones secreted from the hypothalamic area of the vertebrate brain; transferred in blood to the anterior → pituitary gland where they inhibit the release of growth hormone (→ **somatostatin**) and prolactin (→ **prolactostatin**); **2.** inhibiting drugs used to stop a harmful process in tissues, e.g. fluvastatin and provastatin for hypercholesterolaemia in man.

stationary age distribution/composition, *see* age distribution.

stato- (Gr. *statos* = standing), pertaining to rest, equilibrium, or balance.

statoblast (Gr. *blastos* = bud), a chitinous internal winter bud or pupa of some freshwater bryozoans; after the death of the parent organism in spring, the statoblast is activated developing into a new individual. *Cf.* gemmule.

statoconium, pl. **statoconia** (Gr. *konis* = dust, ashes), → statolith.

statocyst (Gr. *kystis* = bladder, pouch), **1.** a small, sac-like, fluid-filled organ in many invertebrates, such as in cnidarians and crustaceans; forms an equilibrium organ in which → **statoliths** lie on the hair-like projections of sensory cells, and thus the organ can sense the position of the animal in space; **2.** rarely, denoting the → utricle and saccule in the inner ear of vertebrates, functioning in the maintenance of static equilibrium.

statocyte (Gr. *kytos* = cavity, hollow vessel), a plant cell supposed to act as a georeceptor, containing one or more → statoliths; e.g. root cap cells.

statolith (Gr. *lithos* = stone), **1.** any of the small, secreted calcareous granules in a → statocyst of vertebrates and many invertebrates (cnidarians, ctenophores, crustaceans); may be a grain of sand, as in crabs; in vertebrates also called statoconium, otolith, otolite, otoconium, otoconite; **2.** in botany, any of the particles in → statocytes, supposed to act in gravity detection; e.g. starch grains in root cap cells.

statolith concept, the theory of detection of gravity in plants by → statocytes and statoliths.

stearic acid (Gr. *stear* = fat, tallow), a saturated fatty acid, $CH_3(CH_2)_{16}COOH$; solid at room temperature, as well as its lipid derivative **stearine**. Stearic acid (together with palmitic acid) is the most common fatty acid in plant and animal fats, especially in homoiothermic animals. Its salts and esters are **stearates**. *Syn.* octadecanoic acid.

stele (Gr. *stele* = column), the system of primary vascular tissues in roots and stem of vascular plants. The stele types are classified according to the reciprocal order of vascular tissues (→ xylem and → phloem) and ground tissues. The most primitive type is the **protostele**, in which the vascular tissues (xylem formed mostly of tracheids, surrounded by primitive phloem) are situated in the middle of ground tissue as a solid, unbroken column. This type exists in some pteridophytes: subtypes are the **haplostele** with a cylinder-shaped column, **actinostele** having the column (vascular cylinder) with proximal spikes, and **plectostele**, in which the column has broken into separate bundles.

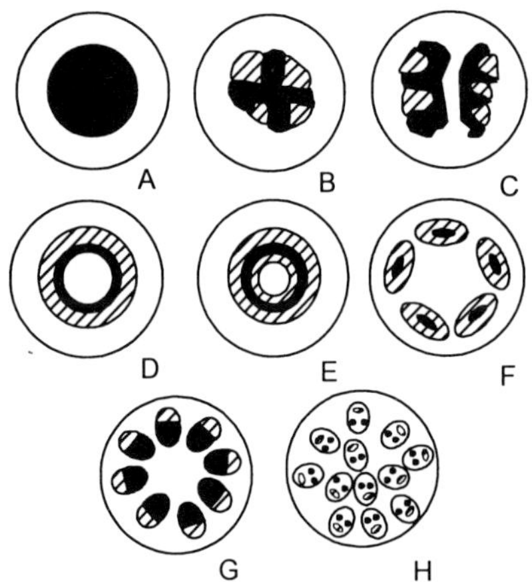

Stele types in plants in cross sections. Xylem is black and phloem hatched: haplostele (A), actinostele (B), plectostele (C), siphonostele (D), amphiphloic siphonostele (E), dictyostele (F), eustele (G), atactostele (H).

If the vascular tissue is ring-like and surrounds the pith of the ground tissue, the type is called **siphonostele** with phloem that may be around (**amphicribral siphonostele**), or on both sides of the xylem (**amphiphloic siphonostele**, **solenostele**). These types are also found in the group Pteridophyta, as well as the **dictyostele** having separate bundles, all with phloem surrounding the xylem. Conifers and vascular plants have a well developed **eustele**, in which vascular tissues form an open net (mostly ring-like) of collateral → vascular bundles with phloem outside and xylem inside. Most monocotyledons have an **atactostele**, in which the vascular bundles are closed and collateral, being surrounded by a bundle sheath, and are scattered randomly in the stem.

stellate hair, *see* hair.

Stelleroidea (L. *stella* = star, Gr. *eidos* = form), **stelleroids**, in some classifications a class in Echinodermata, comprising e.g. the subclasses → Asteroidea and → Ophiuroidea.

STEM, scanning transmission electron microscope. *See* microscope.

stem, 1. the main axis of a vascular plant, branched and bearing leaves, buds, and reproductive parts, e.g. flowers; the stem is normally aerial but may also be a subterranean → rhizome ; **2.** an anatomical structure resembling the stalk of a plant, e.g. the → brain stem in animals.

stem-and-loop structure, the structure of → transfer RNA (tRNA) having four base-paired stems, three of which end as loops which are not base-paired.

stem cell, an undifferentiated cell from which the cells of a certain tissue derive; after the division of the stem cell one daughter cell remains as a stem cell and the other differentiates. Thus the number of stem cells remains constant. *Cf.* progenitor cell.

stem succulent, a type of succulent plant storing water in its stem, e.g cacti.

sten(o)- (Gr. *stenos* = narrow, close, little), denoting narrow, or closed. *Cf.* eury-.

stenohaline (Gr. *hals* = salt), pertaining to an aquatic organism that tolerates only slight variations of salinity. *Cf.* euryhaline.

stenophage (Gr. *phagein* = to eat), an organism or a species subsisting on a narrow food selection. *Adj.* **stenophagous**. *Cf.* monophage, polyphage and euryphage.

stenopodium, pl. **stenopodia** (Gr. *pous* = foot), *see* biramous appendage.

stenothermic (Gr. *therme* = heat), pertaining to an organism which tolerates only slight variations of temperature; also stenothermal, stenothermous. *Cf.* eurythermic.

stenothermophilic (Gr. *philein* = to love), pertaining to organisms, e.g. microbes, growing only at high temperatures, usually at about 60°C; **stenothermophile**, such an organism.

stenotopic (Gr. *topos* = place), pertaining to a species or any other taxon which occurs only in a restricted area; such a species has a narrow tolerance range to changes in environmental conditions. *Noun* **stenotopy**. *Syn.* **stenotropic**. *Cf.* eurytopic.

steppe (Russ. *step* = lowland), a type of → grassland, a vegetation zone which is usually level, dry, treeless, and rich in grasses; the biomass of the vegetation is mainly situated below ground. The largest steppes are in the areas from Hungary (puszta) to the Caspian Sea.

stercoral pocket (L. *stercus* = dung, excrement), a dorsal sac or pocket in the posterior part of the midgut of spiders; collects digestive waste.

stere(o)- (Gr. *stereos* = solid, rigid), **1.** denoting hardness, solidity; **2.** pertaining to three-dimensionality, e.g. stereochemistry or stereoscopic vision.

stereid (Gr. *eidos* = form), a thick-walled supporting plant cell in the stalk or in the middle vein of the leaves of some mosses (Musci).

stereochemistry, the branch of chemistry dealing with the spatial arrangement of atoms in a molecule.

stereocilium, pl. **stereocilia,** a non-motile filamentous cell organelle on the free surface of certain animal cells, especially in mechanoreceptor sensory cells. *Cf.* kinocilium. *See* hair cell, cilium.

stereoscope (Gr. *skopein* = to examine), an optical instrument producing two separate images of the same object into a single three-dimensional image.

stereotactic apparatus (Gr. *taxis* = arrangement), an instrument used in brain research or neurosurgery: exact brain loci can be reached for microprocedures by using coordinates which correspond to exact brain areas in the brain map made of a certain animal species.

stereotaxis, stereotaxy, 1. a three-dimensional arrangement, as e.g. in a → stereotactic apparatus; **2.** an orientation movement (→ taxis) of an animal or locomotive cell in which contact with a solid body is the directive factor. *Adj.* **stereotactic**.

stereotropism (Gr. *trepein* = to turn), the growth of a plant or a movement of a sedentary animal towards (positive) or away (negative) from a solid body.

sterigma, pl. **sterigmata,** also **sterigmas** (Gr. *sterizein* = to prop, support, strengthen), **1.** the stalk-shaped part in the basidium in the group Basidiomycetes, bearing → basidiospores; **2.** peg-shaped projections on twigs of some conifers, to which leaves are attached.

sterile (L. *sterilis*, Gr. *steira*), **1.** an organism incapable of breeding or reproduction; **2.** free from living organisms, especially from microorganisms.

sterility (L. *sterilitas*), **1.** incapability of breeding or reproducing; **2.** the absence of microorganisms.

sterilize, 1. to destroy microbes; **2.** to make sterile by destroying sex organs or inhibiting their functions; **3.** to render land unproductive; **sterilization**, the act of sterilizing, or the condition of being sterilized.

sternum, pl. **sterna** or **sternums** (L. < Gr. *sternon* = breast, chest), **1.** breast bone; a bone or cartilage forming the ventral middle wall of the chest of most tetrapod animals; **2.** the shield covering the ventral thorax in some arthropods, e.g. insects; formed from several hard plates, sclerites. *Cf.* notum, pleuron.

steroids (Gr. *stereos* = solid), a large group of polycyclic lipid compounds with a cyclopentanoperhydrophenanthrene ring as a part of the molecule. Animal steroids are e.g. → sex hormones, corticosteroids, and bile acids. In plants, there are → sterols, sapogenins, cardiac glycosides, and steroid alkaloids, many of which resemble those of animal steroids to the extent that they may have hormonal effects in animals. In plants, steroids have different functions: they may be growth regulators, allelochemicals, and act in the storage of nitrogen and nitrogen wastes (alkaloids).

sterols, one group of steroids containing a side chain of 8 to 10 carbon atoms and at least one -OH group. Plant sterols are triterpenoids with one methyl group which are rather common in plants, such as ergosterol in wheat, cholesterol in many plants, stigmasterol in soybean, and spinasterol in spinach. Heart glycosides of the purple foxglove, *Digitalis purpurea*, belong to the sterol group. The most important sterols in animals are → cholesterol and ergosterol (a precursor of vitamin D_2).

sticky ends, → cohesive ends.

stigma, pl. **stigmata,** also **stigmas** (L.), a mark, spot, or stain; e.g. **1.** a small mark or spot in an anatomical or morphological structure of an animal, such as the follicular stigma in the vertebrate ovary, the eyespot of a protozoan, the opening (spiracle) into the trachea system of insects, a spot on the wings of some butterflies, or one of the openings in the branchial chamber of urochordates; **2.** that part of the seed plant → gynaecium which gathers the pollen grains during pollination, after which → fertilization takes place; **3.** in pathology, a spot in the human skin, bleeding in certain mental states.

stimulus, pl. **stimuli** (L.= goad, incentive), a chemical or physical factor that causes excitation in an organism or one of its parts; e.g. light in the photoreception of animals, or in the phototropism of plants.

stinging hair, a complicated hair in plants excreting poison at a touch, e.g. in nettles.

stipe (L. *stipes* = log, trunk of a tree, branch), **1.** a stem-shaped part (peduncle) in animal morphology, such as the second basal segment of the maxilla of an insect, or of the mandible of a millipede, or a stalk of stalked eyes of crustaceans; in animal morphology, the latin form **stipes** (pl. stipites) is often used; **2.** the stalk of mushrooms or other stalked fungi, or of seaweeds; the petiole of the fern frond; a prolongation of the receptacle beneath the ovary of a seed plant.

stipula, pl. **stipulae** (L. = stalk, straw), → stipule.

stipule, a mostly green, leaf-like outgrowth at the base of a plant leaf; may also be spike-like (*Euphorbia* species) or modified into a tendril (*Gloriosa*). Stipules are often important for the identification of species (*Salix* and *Rosa* species).

stirrup, → stapes.

stochastic (Gr. *stochastikos* = skillful in aiming, able to guess), random; describing a probabilistic event, i.e. a process randomly variable within certain limits (stochastic process); particularly refers to a time series of random variables. In population ecology, a **stochastic model** is a mathematical model based on probabilities; the prediction of the model is not a single fixed number but a range of possible numbers (*cf.* deterministic model). **Stochastic forces** are random processes that affect the community structure (*cf.* deterministic forces).

stock, 1. a group of related animal individuals, which in certain characters differ from other groups of the same species or → breed; also the original group of progenitors (e.g. a race) from which others have descended; **2.** in a modular organism, such as in a colonial hydroid, interconnected zooids form a stock; **3.** the descendants of a single plant individual irrespective of the way of pollination, *syn.* strain; **4.** the main (perennial) stem of a plant, *syn.* trunk; **5.** a plant maintained specifically for the production of slips or cuttings; **6.** a plant or part of it recieving a bud or a scion from another plant in grafting.

stoicheiometry, Am. **stoichiometry** (Gr. *stoicheion* = element, *metrein* = to measure), mathematics of chemistry; the study of relative quantities of reactive substances (atoms, molecules) and products in chemical reactions.

stolon (L. *stolo* = shoot, branch), **1.** hydrorhiza, a root-like branch of the stem in some primitive colonial organisms, e.g. bryozoans and cnidarians, forming new buds; **2.** a plant runner, capable of forming new individuals, e.g. found in strawberry; **3.** a hypha connecting two mycelia of fungi.

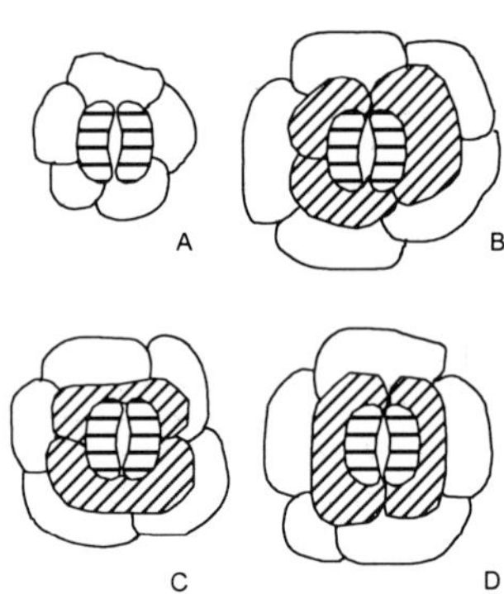

Stoma types in dicotyledonous plants: anomocytic (A), anisocytic (B), paracytic (C), diacytic (D).

stoma, pl. **stomata** or **stomas** (Gr. *stoma* = mouth, opening), any minute opening, pore, or orifice; **1.** in animals, e.g. the openings between epithelial cells and lymph space forming a direct communication between adjacent lymph channels, or a spiracle of the tracheal system of an insect; **2.** in plants, an opening between two **guard cells** in the epidermis of a leaf. Stomata control the transport of water vapour from inner structures of the leaf, and the transport of oxygen and carbon dioxide. Stomata open or close according to the prevailing turgor pressure of the guard cells; usually, they are open in the light and closed in the dark. Around the guard cells there are sometimes **subsidiary cells** (*Syn.* auxiliary cells), which may have a role in the function of the stomata. Different types of stomata are distinguished according to the number and location of subsidiary cells. The main stoma types of dicotyledons (→ Dicotyledonae) are: **anomocytic type**, without special subsidiary cells (Ranunculaceae, Malvaceae); **anisocytic type**, with three subsidiary cells of different sizes (Brassicaceae, Crassulaceae); **paracytic type**, with two or more subsidiary cells parallell to the guard cells (Magnoliaceae, Rubiaceae); **diacytic type**, with two subsidiary cells located transversally to the guard cells. In monocotyledons, the stoma types do not differ as clearly, but four types have been distinguished; in two of them at least four subsidiary cells are found (e.g. *Rhoeo*), in the third type two cells exist (Poaceae), whreas in the fourth type subsidiary cells are totally lacking (e.g. *Allium*).

stomach (L. *stomachus* < Gr. *stomachos*), **ventriculus** (L.), **gaster** (Gr.); the enlargement of the alimentary canal of most multicellular animals storing, diluting and digesting food; **1.** the single stomach of most mammals, situated between the oesophagus and the intestine separated from the latter by the **pylorus** with a strong, circular smooth muscle regulating the passage of food from the stomach. The stomach is anatomically divided into several parts: the **cardia** is situated most cranially followed by the **fundus**, **corpus**, and **antrum**. The wall consists of several muscular and membranous tissue layers (*see* alimentary canal), and in the surface of the mucosa numerous small gastric pits (*foveolae gastricae*) exist, at the bottoms of which → **gastric glands** open. In ruminants, the stomach has four separate compartments of which the first three are derived from the oesophagus (*see* rumination). The stomach of birds is composed of two parts, the **proventriculus**, which secretes gastric juice, and the muscular → **gizzard**, which is lined with horny plates for grinding food. In lower vertebrates, the stomach is usually single as in most mammals. The function of the stomach is regulated by the → autonomic nervous system and some hormones, e.g. → gastrin;
2. the single or multipartite enlargement of the anterior part of the alimentary canal found in many invertebrates; in arthropods the wall of the stomach is lined internally with chitin, which sometimes forms teeth, as in many crustaceans. Some invertebrates, such as spiders, have a **pumping stomach** with which the animals suck the soft tissues of a prey;
3. the ventral part of the trunk, i.e. belly, abdomen.

stomatal resistance, the resistance (R) that controls the flow of water vapour or other gases through stomata in plants. R is analogous to the electric resistance of → Ohm's law. The total resistance of the stomatal → transpiration is expressed as a simple formula: $R = R_s + R_a$, in which R_s is the stomatal resistance based on the opening stage of a stoma, and R_a is the resistance depending on the boundary layer of air outside the stoma. When the stoma is fully open the transpiration of water is dependent on R_a only, because R_s is zero, but when closed, R_s is at its maximum and R_a is insignificant. The total resistance contains the mesophyll resistance (R_m), cuticular resistance (R_c), and internal resistance (R_i).

stomodaeum, Am. **stomodeum,** pl. **stomod(a)ea** or **stomod(a)eums** (Gr. *hodaion* = on the way), **1.** an invagination in the ectoderm which gives rise to the mouth (oral) cavity in the early vertebrate embryo; **2.** the related embryonic part in many invertebrates; e.g. in anthozoans gives rise to the **pharynx**, and in insects to the anterior ectodermal part (foregut) of the alimentary canal, consisting of the mouth, buccal cavity, pharynx, oesophagus, crop, and proventriculus; a stomodeal valve separates the foregut from the midgut.

stone canal, a short, calcareous canal connecting the madreporite to the ring canal of the → water-vascular system in many echinoderms. *Cf.* radial canal.

stone cell, → sclereid.

stoneworts, → Charophyta.

stop codon, → termination codon.

storage parenchyma (Gr. *para* = beside, *enchein* = to pour in, infuse), a type of plant parenchyma tissue for storage and short distance tranport, especially in roots and shoots.

s.t.p. also **stp,** standard temperature and pressure, formerly N.T.P. (NTP); normal temperature and pressure, i.e. a pressure of 1 ATM (= 101.325 kPa) at the temperature of 0°C.

STR, short tamdem repeat, i.e. → microsatellite.

strain, **1.** to over-exercise, e.g. a muscle, tendon or whole body; **2.** strong or excessive muscular effort; **3.** a line descended from a particular ancestral individual, forming a group of organisms within a species characterized by a particular physiological or biochemical quality, e.g. the hypertonic strain of laboratory mice, or a population of microorganisms, containing genetically identical individuals; *syn.* stock.

Strasburger cell (*Eduard Strasburger*, 1844—1912), a cell type in the vascular bundle of coniferous needles, acting as a starch store. *Syn.* albuminous cell.

stratification (L. *stratum* = layer, *facere* = to make), the state, appearance, or process of stratifying or arranging in layers; e.g. **1. stratification of a water body** in temperate and cold climate zones: in the summer the warmer upper water layer (**epilimnion**) is separated from the colder lower layer (**hypolimnion**) by the **thermocline**. The stage is called **summer stagnation**. In the autumn the epilimnion cools down with cold weather resulting in **autumn circulation** (full circulation) followed by **winter stagnation** when the hypolimnion is about +4°C and the epilimnion somewhat colder. In the spring, during **spring circulation**, the stratification changes again resulting in summer stagnation. The described **dimictic circulation system** is typical of cool climate zones, but subtropical water bodies are **monomictic**, summer stagnation being long and dominant, while in the winter only an incomplete circulation is found. Monomictic circulation occurs also in arctic fresh waters, where winter stagnation is the dominant stage. Tropical waters are **oligomictic**, without any circulation. In **polymictic waters** the stratification may be formed and broken off with short intervals, even daily. If the water circulates from the surface to the bottom, the lake is called **holomictic**, but if stagnation continues without any circulation, the lake is **meromictic**, as e.g. in brackish-water lakes in which salt water lies on the bottom (**chemical meromicty**); **2.** a method to improve the germination of plant seeds; seeds are incubated moistened (in sand) and at low temperature (some degrees above 0°C); **3.** in geology, the formation of strata; *see* stratum, defin. 2.

stratosphere (L. *stratum* = layer, bed, *sphaera* = globe), the layer of the → atmosphere between the troposphere and mesosphere, from about 10 to 50 km above the surface of the Earth. The stratosphere includes an important ozone layer that protects organisms against excessive UV-radiation. The temperature in its lower region is stable and that part (or the whole stratosphere) is therefore called the isothermal region, but upwards from 25 km the temperature begins to increase from about -60°C to -10°C, even slightly above 0°C, because ozone absorbs the UV radiation of the sun. *Cf.* ionosphere, mesosphere, troposphere.

stratum, pl. **strata,** Am. also **stratums** (L.), a layer, e.g **1.** a sheet-like layer of a differentiated tissue, such as the *stratum corneum*, the horny layer of the skin epidermis in many vertebrates; **2.** a vertical layer of vegetation, comprising e.g. herbs, shrubs, or trees; **3.** a thin layer of sedimentary rock or earth; **4.** a layer of the ocean or atmosphere.

Strepsirhini (Gr. *strepsis* = twisting), *see* Prosimii.

streptavidin, a protein isolated from the bacterium *Streptomyces avidinii*; an → avidin analogue possessing a high affinity to biotin and is therefore used to detect biotin in tissue preparations.

streptococcal toxins (*Streptococcus* = a genus of Gram-positive bacteria), exotoxins produced by streptococci having haemolytic effects on vertebrates.

streptomycin (*Streptomyces* = a genus of Gram-positive bacteria), a highly basic trisaccharide synthesized by *S. griseus*; inhibits the initiation, elongation, and termination of protein synthesis and causes misreading of RNA in prokaryotes. Commonly used as an antibiotic.

stress, 1. influence, emphasis, strain, pull, tension, pressure; **2.** disturbed physiological balance in an organism. After being exposed to adverse stimuli (**stressors**), which may be physical, chemical, microbial, mental, or emotional, the organism tries to achieve homeostasis in this imbalanced condition. In animals, the reactions elicited (**stress reactions**) are non-specific neuronal and hormonal reactions which improve survival in a stress situation. In vertebrates stress reactions are quite well known. This non-specific response, called also → **general adaptation syndrome**, is induced by various types of stressors and is controlled by brain areas associated with emotionality and the function of the autonomic nervous system. In stress situations the sympathetic nervous system is activated and hormones (**stress hormones**) from the adrenal gland are liberated preparing the animal to "fight or flight" (secretion of adrenaline and noradrenaline), or to tolerate the stress condition for a longer time (secretion of glucocorticoids). In severe or prolonged situations the state of **exhaustion** may occur with several psychosomatic symptoms (e.g. disturbances in sleep, dysfunction of the alimentary canal and the heart, and reduced reproduction). At cellular and molecular level, the stress response occurs as changed enzyme acitivty and in the production of → stress proteins.

Lately the term stress has also been used in plant physiology and molecular biology. Stress reactions in plants are caused by abiotic (e.g. temperature, light, water) or biotic factors (e.g. moulds, viruses). Plants have different means of avoiding or tolerating stress conditions (*see* cold hardening).

stress proteins, proteins the synthesis of which is induced in cells by different environmental factors, as e.g. high or low temperature, pressure, osmotic stress, and many chemicals. The size and properties of stress proteins are known quite well, and they are named by different terms according to the research field, e.g. → heat shock proteins, cold proteins. Their functional mechanisms are less known, but they protect cellular structures from damage caused by stress factors.

stretch receptors, mechanical receptors which sense stretching of tissues, particularly of muscles or tendons, or are sensitive to the bending of joints and different types of → sensilla. The receptors may be single cells or nerve endings, or more complicated multicellular sense organs such as the → muscle spindle and Golgi tendon organ in vertebrates.

striate body, striatum (L. *corpus striatum*), an area of → grey matter inside the forebrain (telencephalon) in vertebrates; it includes the caudate and lentiform nuclei, belonging to the **basal ganglia**, and is a part of the → extrapyramidal system serving locomotor functions of the body.

striated muscle, 1. the → skeletal muscle of vertebrates; **2.** any of similar muscles of invertebrates with cross-striations in the muscle cells (fibres) serving the voluntary locomotions of the body and limbs. *Cf.* smooth muscle, cardiac muscle.

stricture (L. *strictura* = contraction), the decrease of the diameter of a passage, such as a canal or duct in an organ.

stridulation (L. *stridulus* = shrill), the production of high-pitched sounds by some insects, e.g. cicadas and crickets; sounds are generated by rubbing together special organs (**stridulating organs**), such as sclerites, vibrating membranes, or modified edges of the forewings.

Strigiformes (L. *Strix* = the type genus of the family Strigidae, *forma* = form), **owls**; a cosmopolitan order of mostly nocturnal birds of prey which comprises 174 species in two families, Strigidae and Tytonidae; binocular vision and hearing are well developed. Owls live mainly in open grass habitats or forests, some species also live in urban habitats (e.g. barn owl, *Tyto alba*); the snowy owl (*Nyctea scandiaca*) has adapted to the conditions of the arctic tundra.

strobila, pl. **strobilae** (L. < Gr. *strobilos* = anything twisted, a cone), **strobile**, strobilus; a linear series of similar structures; e.g. **1.** an asexual late stage of the → **scyphistoma** in jelly fishes; in the process of **strobilation**, the sessile scyphistoma (now called **strobila**) produces several free-swimming, saucer-shaped buds, **ephyrae**, which grow and develop into adult jellyfishes; **2.** a segmented body of tapeworms composed of the **scolex** (head) and numerous body segments, **proglottides**.

strobilation, 1. the budding of ephyrae from the strobila of scyphozoans; **2.** the detachment of mature proglottides from the strobila of tapeworms.

strobilus, pl. **strobili** (L), **strobile**; **1.** → strobila; **2.** a male or female inflorescence of → Gymnospermae; the male organ is called **microstrobilus** and the female organ **macrostrobilus** (cone); **3.** any of cone-shaped assemblages of sporophylls in horsetails and club mosses.

stroboscope (Gr. *strobos* = whirl, *skopein* = to observe), an instrument by which regular rhythmic motion (e.g. the ciliary rhythm) can be measured. In the apparatus the rhythm of the scintillating light directed to an object can be adjusted to match the rhythm of the object, and thus the movements of the object appear stationary.

stroma, pl. **stromata** (Gr. = anything laid out), **1.** the supporting intercellular material, → matrix, of an animal organ, as distinguished from the specific functional structure, → parenchyma; **2.** the insoluble structural material of a cell, especially of red blood cells, remaining after the lysis of a cell; **3.** in → chloroplasts, the colourless base material; **4.** in fungi, the tissue formed from → hyphae, in which spore-bearing structures are formed; e.g. forms the tissue of a mushroom.

stroma thylakoid, *see* chloroplast.

strong fertilizer, an industrially produced substance used to fertilize soil; contains in a concentrated form one or more plant nutrients, mainly nitrogen, phosphate, and potassium.

strontium (*Strontian* = village in Scotland where strontium was discovered), symbol Sr, an alkaline earth metal, atomic weight 87.62; appears naturally only in a combined form, such as hydroxide and carbonate e.g. in calcium and potassium spars. Strontium is used in industry and may be deposited in minute harmless concentrations in organisms. However, the radioisotope strontium-90 (^{90}Sr, half-life 28 years), present in the fall-out from nuclear explosions and accidents, has become a serious pollutant. In organisms, strontium (also radioactive strontium) attempts to displace calcium and therefore accumulates easily in bones.

strophanthin G, → ouabain.

structural change, in genetics, a change in the structure of the → chromosome leading to → chromosome mutation.

structural diversity, in ecology, diversity of physical structures in a community, regulating the number of habitats and thus eventually the number of species; e.g. the structural diversity for breeding of perching birds is much greater in coppices than in barren coniferous forests.

structural gene, a gene which in the cell codes for the amino acid sequence in the primary structure of a polypeptide (*see* protein synthesis). In addition to the structural genes the chromosome contains sites (regulator genes) which regulate the gene activity. The term is particularly used for prokaryotic organisms; the corresponding genes in eukaryotes are usually called **producer genes**. *Cf.* regulator gene. *See* genetic transcription, translation.

structural heterozygosity (Gr. *heteros* = the other, secondary, different, *zygotos* = yoked together), the structural difference of two chromosomal regions, chromosomes, or sets of chromosomes, resulting from a → chromosome mutation. *Cf.* structural homozygosity.

structural homozygosity (Gr. *homos* = the same, similar), the structural similarity of two chromosomal regions, chromosomes, or sets of chromosomes in the cells of an organism. *Cf.* structural heterozygosity.

struggle for existence, a concept associated with the evolution theory of *Charles Darwin* (1809—1882); according to this, the competition of the members of the natural population living on a limited supply of environmental resources, leads to the elimination of inadequately adapted individuals and to the survival of better adapted individuals, i.e. to → natural selection.

Struthioniformes (L. *struthio* = ostrich, *forma* = form), **ostriches**; an order of flightless, large birds characterized by a long strong neck, long legs and unkeeled sternum. The only species of the order still living is the ostrich (*Struthio camelus*) which inhabits arid areas in Africa; the male is about 2,5 m tall and weighs more than 100 kg. *See* Ratites.

strychnine, a crystalline poisonous alkaloid, $C_{21}H_{22}N_2O_2$, obtained chiefly from the seeds of *Strychnos nux-vomica*; stimulates the nervous system by blocking synaptic inhibition, and is therefore used as an antidote for poisoning caused by depressant drugs, earlier in minute doses also to stimulate appetite.

STS, → sequence-tagged site.

stupor (L. *stupere* = to be benumbed, or astonished), **1.** insensibility to environmental stimuli, or diminution of sensibility of animals, as caused by intoxicants, narcotics, or low body temperature; **2.** a state of apathy.

sturt, Sturtevant unit (*Alfred H. Sturtevant*, 1891—1970), a map distance unit of the → fate map of a fruit fly embryo; based on the analysis of gynandromorphs, i.e. genetically mosaic individuals whose tissues are partly female and partly male. *Cf.* sturtoid.

sturtoid, a generalized modification of → sturt which can be applied to any kinds of → fate maps of fruit fly embryos or → imaginal discs; based on the analysis of different types of genetically mosaic individuals.

style (L. *stilus* = pricker, stake, pole < Gr. *stylos* = pillar), an elongated process or bristle in an anatomical structure; e.g. **1.** a posterior conical region in the stomach of some filter-feeding clams (Bivalvia), releasing enzymes and mixing them with the stomach contents; **2.** → stylet; **3.** in botany, a stalk-like part of the → carpel, connecting the ovary with the stigma.

stylet, 1. style; any of relatively rigid, long and narrow mouthparts, such as those of some insects (e.g. bugs), specialized for stinging and sucking; **2.** needle-like stylets in chelicerae of some parasitic ticks; **3.** in medicine, a slender probe or a wire inside a catheter; *syn.* stilet.

stylostome (Gr. *stoma* = mouth), a duct in the skin of a host animal enzymatically excavated by some parasitic mite larvae (in Trombidiformes); the larvae are capable of sucking body fluids from the host through the ducts.

S-type position effect (S < *stabilis* = stable), *see* position effect.

sub- (L. *sub* = under, below), denoting under, almost, partial, near, less than normally, subordinate.

subalpine zone, a vegetation zone that belongs to the → height zones of mountains and is situated between the treeless alpine zone and the montane forest zone.

subarachnoid cavity/space (L. *cavum subarachnoidea*), the space between the arachnoidea and pia mater on the surface of the central nervous system of vertebrates; traversed by fibrous trabeculae and filled with → cerebrospinal fluid.

subarctic zone (Gr. *arktos* = bear), the vegetation zone between the cool zone (with coniferous forests) and the arctic zone (tundra zone).

subclass (L. *sub* = under, *classis* = division), the systematic taxon between the class and order.

subcutaneous (L. *cutis* = skin), beneath the skin, e.g. a subcutaneous injection (**s.c.**), or the subcutaneous fat.

subcutis (L. *tela subcutanea*), subcutaneous tissue, i.e. the tissue beneath the skin (cutis, dermis); especially denotes the subcutis of vertebrates formed of areolar connective tissue, which in man and aquatic mammals is differentiated largely into adipose tissue (fat tissue). *Syn.* hypodermis, hypoderm.

subdural cavity/space (L. *cavum/spatium subdurale*), the very narrow space between the dura mater and arachnoidea on the surface of the central nervous system of vertebrates.

suberic acid (L. *suber* = cork), octanedioic acid, $HOOC(CH_2)_6COOH$; a crystalline dibasic acid that is present in some plant lipids, usually obtained from castor oil (ricinus oil) and used e.g. for manufacturing of plastics.

suberin, a complex mixture of waxy substances composed of lipid derivatives built up of both saturated and unsaturated fatty acids ranging from 16-carbon to 22-carbon; it is not permeable to water and exists as a protective substance in the plant → periderm, especially in walls of cork cells.

subfossil (L. *sub* = under, below, *fodere* = to dig), any remains of an organism, not yet fossilized but discovered in relatively young deposits, such as in peat and mud, usually originated from Quaternary. *See* fossil.

subgenus, pl. **subgenera** (L. *genus* = race), the systematic taxon between the genus and species; large genera are sometimes divided into subgenera.

subkingdom, a systematic taxon in the → Animal kingdom between the kingdom and phylum.

sublethal (L. *lethalis* = leading to death), nearly lethal, insufficient to cause death; e.g. sublethal dose, temperature, etc.

sublethal factor, a → gene whose expression decreases the viability of individuals in a population so that less than half of the individuals die before the age of reproduction. *Syn.* semilethal factor.

sublimation (L. *sublimare* = to promote, elevate), **1.** the conversion of a substance from a solid state directly into vapour by the action of heat, and subsequent condensation; a substance obtained by sublimation is called **sublimate**, e.g. mercury chloride; **2.** elevating or changing human instinct energy into an action that is socially approved.

sublingual gland (L. *glandula sublingualis*), *see* salivary glands.

sublittoral (L. *li(t)tus* = seashore), *see* littoral.

submandibular gland (L. *glandula submandibularis*), *see* salivary glands.

submaxilla, pl. **submaxillae,** also **submaxillas** (L.), the lower jaw, → mandible, in vertebrates.

submerged water plant, a water plant growing from the bottom upwards but not reaching the surface; e.g. *Elodea*, *Ceratophyllum*. *Syn.* submersed (water) plant.

submissive posture, → appeasement behaviour.

submontane zone, *see* height zones.

submucosa, pl. **submucosae** (*tunica/tela submucosa*), the layer of areolar connective tissue beneath a mucous membrane; e.g. in the alimentary canal and respiratory pathways in vertebrates. *Adj.* **submucous**.

suborder (L. *ordo* = order), the systematic taxon sometimes used between the order and family (or superfamily).

subsidiary cells, *see* stoma.

subspecies (L. *species* = particular kind), the lowest systematic taxon below a species; in general, includes a group of locally isolated populations which are divergent by their taxonomic characteristics and gene frequencies from other groups of the same species. The members of different subspecies are, however, capable of interbreeding succesfully in areas where their distributions overlap. In botany, the subspecies may be → sympatric, but in zoology only → allopatric. In animal classification, the subspecies is a taxon denoted by a **trinomial** name; e.g. *Pusa hispida saimensis*. In botany, the subspecies (if separated from the species) is denoted by adding "ssp." or "subsp." between the names; e.g. *Cladonia gracilis* ssp. *turbinata*, or *C. g.* subsp. *turbinata*. The terms subspecies and → race are sometimes used as synonyms. *Cf.* semispecies.

substance P (P < peptide), a peptide composed of 11 amino acid residues found in vertebrate tissues, e.g. in many neurones throughout the nervous system and in endocrine cells of the alimentary canal. Substance P acts as a neurotransmitter and hormone, e.g. stimulating intestinal movements, causing vasodilatation and sweating, and transmitting impulses from the sensory neurones (especially in slow pain pathways) in the spinal cord. Substance P belongs to tachykinins.

substantia, pl. **substantiae** (L.), material, substance; e.g. material of which tissues and organs are formed, such as *substantia alba*, *substantia grisea*, the white and grey matter of the brain; *substantia spongiosa*, spongy substance of the bone.

substitutional load, *see* genetic load.

substrate (L. *substratum* = the base, underlayer), **1.** any substance upon which an → enzyme specifically acts; the enzyme first attaches to the substrate with a lock-key binding and changes its chemical structure in the reaction; *see* lock-and-key model; **2.** → substratum.

substratum, pl. **substrata** (L.), a stratum or layer lying under another; e.g. **1.** a base or material on which bacteria or other organisms or cells grow; **2.** in agriculture, the subsoil; **3.** → substrate.

subtending leaf (L. *tendere* = to stretch), the leaf in whose axil a bud (axillary bud) or a flower stands.

subtribe (L. *tribus* = one third of the ancient Roman people, tribe), a subdivision of a → tribe.

subtropic(al) zone (Gr. *tropikos* = tropics < *tropos* = turning), the → vegetation zone where the average temperature of the coldest month is above 0°C. Evergreen forests grow in coastal areas, but in inland regions deciduous forests may be typical; also → deserts and → savannas may exist.

subumbrella (L. *umbrella,* dim. of *umbra* = shadow), the concave lower side of the jellyfish, including the mouth.

subunit, a secondary or subordinate unit; in biology, e.g. **1.** one of the polypeptide chains of an oligomeric protein; **2.** one of the two structural parts of a → ribosome.

subvital, pertaining to organisms with impaired ability to live; e.g. denoting → genes or genotypes whose expression substantially decreases viability as compared with normal. The viability of a subvital genotype is 50 to 90% of the normal. *Cf.* sublethal.

succession (L. *successio* < *succedere* = to follow, succeed), the following of one thing or event after another in sequence. In biology, succession is a directional, progressive change in the composition of a community of organisms, achieving gradually a stable state, **climax**. The gradual emergence and changing of vegetation from an initially plantless stage is **primary succession**: **xerarch succession** occurs in terrestrial habitats, such as the succession after a vulcanic eruption, and the **hydrarch succession** in aquatic habitats, in a closing of a lake. **Secondary succession** occurs in places where a preceding vegetation has been destroyed, e.g. in a forest after a fire. There are successions also in fauna, usually closely associated with vegetational changes. The successional series in plant communities following one another in the course of the biotic development from pioneer stage to climax, and also the stages in succession, are called **seres**. *See* allogenic succession, autogenic succession.

succinic acid, a dicarboxylic acid, $(CH_2\text{-}COOH)_2$; one of the intermediates of the → citric acid cycle functioning in cell respiration; its salts are **succinates.** *Syn.* butanedioic acid.

succinyl CoA, succinyl coenzyme A, active succinate formed as a condensation product of succinic acid and → coenzyme A, CoA; it is a high-energy compound of the → citric acid cycle functioning in cell respiration, formed from α-ketoglutarate via an oxidative decarboxylation. Succinyl CoA is converted to succinate in the reaction catalysed by succinyl CoA synthetase, and simultaneously GDP (guanosine diphosphate) is phosphorylated to GTP (guanosine triphosphate). Together with glycine δ-aminolevulinate succinyl CoA acts in the synthesis of → porphyrins.

succulent (L. *succus* = juice), a plant adapted to very dry or desert conditions; typically has thick, swollen, fleshy roots, stem (**succulent stem**), and leaves (**succulent leaves**), chiefly comprising water-storing parenchyma. The leaves may also be reduced to thorns, as in cacti. Due to their water stores, succulents may survive during long, totally dry periods.

sucrase (Fr. *sucre* = sugar), → invertase.

sucrose, saccharose; a non-reducing disaccharide, $C_{12}H_{22}O_{11}$, mol. weight 342.30, comprising one glucose and one fructose moiety; the most important storage and transport sugar in plants, especially in roots and fruits; sucrose is obtained chiefly from sugar cane and sugar beet. Common names: **table sugar**, **cane sugar**, **beet sugar**. *See* invertase.

Suctoria (L. *sugere* = to suck), **suctorians**; a protozoan group (subclass) in the phylum Ciliophora; immature stages are ciliated and free-swimming, but adults are usually sessile and their cilia are lacking; suctorians live both in fresh and salt waters.

Sudan dyes, histochemical dyes for staining lipids. *Adj.* **sudanophilic** (sudanophilous), staining readily with Sudan dyes.

sudoriferous glands (L. *sudor* = sweat, *ferre* = to bear), → sweat glands.

suet (L. *sebum*), the hard fat obtained from the abdominal cavity of ruminant animals; used in cookery, or processed to yield tallow.

sufficient condition, a concept belonging to modal logic describing the causal relations of phenomena. If always when phenomenon A exists, also B exists, phenomenon A is a sufficient condition for phenomenon B. The disjunction of the sufficient conditions (S) (i.e. S_1 or S_2 or ... S_n) is at the same time a → necessary condition of the phenomenon. For example, during the replication of a gene, a → nucleotide substitution is a sufficient condition for a mutation.

sugar alcohols, → alcohols derived from sugars in carbohydrate metabolism, especially in plant cells; e.g. → inositol, mannitol and sorbitol. They act as protective substances in plants under stress conditions, like drought and low temperature. Sugar alcohols are used e.g. in physiological buffer solutions, in → density gradient centrifugation, and as → cryoprotectants.

sugar phosphates, compounds formed from a monosaccharide and phosphate group, usually occurring in metabolic reactions of cell respiration and in the reactions of the Calvin cycle in photosynthesis.

sugars, a group of water-soluble carbohydrates, comprising monosaccharides (e.g. → trioses, pentoses, hexoses), disaccharides (e.g. → sucrose, maltose, and lactose), and oligosaccharides (e.g. → raffinose and stachyose). The skeleton of a monosaccharide molecule $(CH_2O)_n$, has 3 to 7 carbon atoms; disaccharides consist of two monosaccharides, and oligosaccharides have several monosaccharide residues. Sugars are produced in the photosynthesis of plants, first as monosaccharides which are important substances in

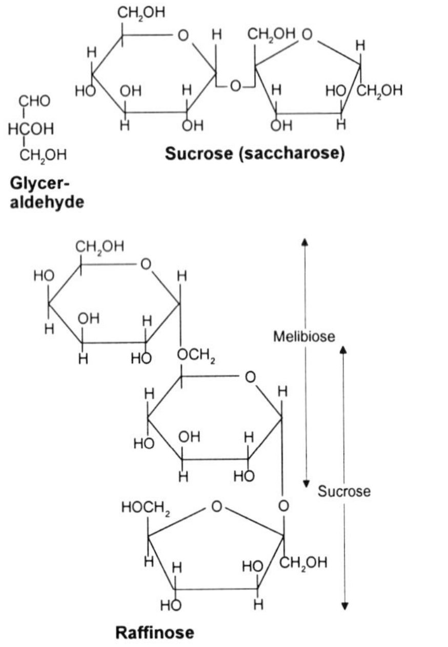

Glyceraldehyde

Sucrose (saccharose)

Raffinose

Examples of sugars: glyceraldehyde (triose), sucrose (disaccharide), raffinose (trisaccharide).

the energy metabolism of cells in all organisms and precursors for syntheses of other saccharides and many other compounds. Sugars produced by plants are important energy sources for animals, di- and oligosaccharides being first hydrolysed in the alimentary canal into monosaccharides, which are then absorbed into the blood or haemolymph through the epithelium of the intestine. *Cf.* polysaccharides.

Suiformes, Suina (*Sus* = the type genus of Suidae), a suborder of even-toed ungulates (order Artiodactyla) including three families, **pigs** (Suidae), **peccaries** (Tayassuidae) and **hippopotamuses** (Hippopotamidae); in general, omnivorous or herbivorous animals whose stomach has two or three parts, the teeth are low-crowned and the legs have four toes (three in peccaries).

sulcus, pl. **sulci** (L.), a groove, furrow, or depression in an anatomical structure; e.g. *sulci cerebri*, *sulci cerebelli*, cerebral and cerebellar sulci between the gyri on the surface of the brain; *sulcus lacrimalis*, lacrimal sulcus. *Cf.* gyrus.

sulph(o)-, Am. **sulf(o)-,** containing a sulphur atom or atoms.

sulphanilamide, Am. **sulfanilamide,** a sulphur-containing aromatic amide, $H_2NC_6H_4SO_2NH_2$, the first sulphonamide discovered; prevents bacterial growth by inhibiting purine synthesis, and was formerly used as an antibacterial drug.

sulpholipid, Am. **sulfolipid,** any sulphur-containing lipid synthesized in the → Golgi complex of cells, occurs e.g. in chloroplasts and chromatophores; yields sulphuric acid in hydrolysis.

sulphur, Am. **sulfur,** a non-metallic element appearing in several allotropic forms; symbol S, atomic weight 32.06; exists naturally either free or combined, especially in **sulphides** and **sulphates**. The most common sulphur mineral is pyrite, FeS_2, i.e. iron sulphide (a salt of hydrogen sulphide, H_2S). When sulphur compounds are burned, sulphur is liberated as dioxide, SO_2. Together with water sulphur dioxide forms an unstable acid, **sulphurous acid** (trioxosulphuric acid), H_2SO_3, the salts and esters of which are **sulphites**. Sulphur trioxide (SO_3) reacts with water and forms **sulphuric acid**, H_2SO_4 (its salts and esters are sulphates). Sulphur is an essential element for cells found in -SH groups of some amino acids, forming -S-S- bridges between peptide chains in proteins. Sulphur oxides have become serious pollutants which are liberated to the atmosphere from the smoke of fossil fuels. *See* acid deposition.

sulphur cycle, a cycle of processes by which → sulphur circulates in the biosphere. The cycle includes assimilation of inorganic sulphur mainly by plants which take sulphur as a sulphate anion (SO_4^{2-}) from the soil, its incorporation into proteins of all organisms, and the liberation of sulphur from dead organic matter by bacteria. Liberated sulphides in the soil are further converted to elemental inorganic sulphur and sulphates, or back to sulphides by the action of sulphur bacteria.

sulphydryl, Am. **sulfhydryl,** the univalent radical of sulphur and hydrogen, -SH; sulphydryl groups are important in organisms forming → disulphide bonds in protein molecules. *Syn.* thiol. *See* protein, cysteine.

summer aspect, *see* aspect.

summer spore, uredospore. *See* Uredinales.

summer stagnation (L. *stagnum* = standing water), stratification of a lake during the summer, when clearly different temperature layers exist in a water body. *See* stratification.

summerwood, a type of wood in annual rings of woody plants developing mainly during the summer, i.e. in the latter part of the growing season. The cells, vessels, and tracheids of summerwood are typically smaller and denser than those of the earlier **springwood**. *Syn.* latewood.

super- (L. *super* = above, over, in addition), denoting above, excessive, more than normal. *Cf.* supra-.

supercoiling, describes the coiling of a double-stranded ring-shaped DNA molecule so that it crosses its own axis.

supercooling, cooling of a liquid to temperatures below its freezing point without ice formation (metastable state of liquid); occurs in water solutions, such as tissue fluids, if there are no ice nucleators present in it. The phenomenon is found in cold-hardened organisms tolerating freezing temperatures because the increased quantity of some organic molecules, such as glucose and glycerol, prevents the formation of ice crystals. *See* cryoprotectants.

supergene, a group of linked genes which always remain together acting as an allelomorphic unit, i.e. → crossing-over does not occur between the genes and hence they are transmitted as a single unit.

supernatant (L. *natare* = to swim), **1.** *adj.* floating on the surface; **2.** a liquid layer above a precipitate in a centrifuge tube.

supernormal stimulus, → supranormal releaser.

superorder (L. *ordo* = order), the systematic taxon ranking between the order and class (or subclass), into which several closely related orders are included.

superorganism, any group of organisms acting as one functional unit; e.g. **1.** the true society of eusocial animals, e.g. termites; **2.** a community including several species with tightly reciprocal functional relations with each other.

superoxide, HO_2; an acid, often presented as $H^+ + O_2^-$, the latter being the → superoxide radical. *Syn.* hyperoxide.

superoxide dismutase, SOD, an enzyme (EC 1.15.1.1) catalysing the breakdown of the harmful → superoxide radical (O_2^-) to hydrogen peroxide and molecular oxygen: $2\ O_2^- + 2\ H^+ \longrightarrow H_2O_2 + O_2$; the active site of the cytosolic enzyme in eukaryotes contains a copper or zinc ion (metalloenzyme). Superoxide dismutase is present in all aerobic organisms in order to protect cells against → oxygen radicals.

superoxide radical, superoxide anion (O_2^-); a reactive intermediate in the reduction of oxygen in which an oxygen molecule has an extra electron. *See* oxygen radicals.

superphosphate, an artificial fertilizer mixture consisting mainly of calcium dihydrogen phosphate, $Ca(H_2PO_4)_2$, but also of calcium sulphate, $CaSO_4$, and other metal sulphates present in raw materials. Common superphosphate contains about 20% and a more concentrated form (triple superphosphate) about 45% of soluble phosphates.

superpoison, a substance that is exceptionally toxic to organisms; may be a natural substance produced by an organism, such as a bacterial toxin, **botulin,** which in a dose of 50 ng may be fatal to man, or a toxic substance produced by man, e.g. → **dioxin**, which is highly poisonous to all organisms.

supersecondary structure, the intermediary structure of proteins between the secondary and tertiary structures; refers to clusters of secondary structures.

supersonic, → ultrasonics.

superspecies (L. *species* = kind, appearance), species group; an informal systematic taxon consisting of several closely related, usually allopatric species with common morphological characteristics; every superspecies is monophyletic, i.e. has a common ancestral form.

supervital, pertaining to organisms, which are more viable than normal; in genetics means genes or → genotypes, the expression of which leads to an increased viability as compared with normal types.

supination (L. *supinatio* < *supinare* = to lay on the back), the act or state of lying face up; especially applied to the hand (or foot) meaning the act of turning the palm forward or upward (lateral rotation, which is the property of primates). *Adj.* **supine**. *Cf.* pronation.

supporting root, a plant root type which develops in the stem above ground and grows into the soil in order to support the plant; e.g. mangrove plants.

supporting tissues, a tissue type in plants made of cells with thickened walls, providing mechanical support to plant structures. The main types are → **collenchyma**, which is common in growing parts of plants, and → **sclerenchyma**, formed from cells which mostly have a very thick, hard, even lignified → secondary wall. Two major types of sclerenchyma are → **sclereids** and **fibres**.

suppression (L. *suppressio* < *supprimere* = to press down), the sudden inhibition or stoppage of a process; e.g. **1.** the inhibition of a biochemical or physiological process, such as secretion or excretion, or activity of the immunological defence; **2.** the reversion of a mutant phenotype into a wild one due to another mutation, the latter mutation located at another mutational site; *see* suppressor mutation; **3.** in psychology, the conscious inhibition of unacceptable impulses.

suppressor, one that brings about → suppression, e.g. any secondary gene mutation that totally or partially restores the function lost in the primary mutation.

suppressor cell, T8 cell, suppressor T cell, having the glycoprotein CD8 on its surface; a type of T lymphocyte inhibiting the function of B lymphocytes; involved in immunotolerance and autoimmunity.

suppressor mutation (L. *mutatio* = change), a secondary → mutation, which partly or totally restores the function damaged in the primary mutation; can occur in the same or a different gene than the primary mutation.The first event is called **intragenic** and the second **intergenic (extragenic) suppression**. Uncommonly, it is a mutation that prevents local or complete → crossing-over in meiotic cells.

supra- (L.), denoting above or over, as **super**, but emphasizing position or location.

supralit(t)oral (L. *lit(t)us* = seashore), relating to, or living in a litoral zone of oceans above the tidal range, but affected by surf.

supranormal releaser, a key stimulus with some characteristic stressed particularly. An animal reacts to the supranormal releaser with an exceptional readiness and the releaser usually elicits an extra strong response. The releaser may be either natural or artificial; e.g. an extra large and bright gape of a cuckoo nestling releases the feeding behaviour in its host parents more easily than the smaller gapes of their own nestlings. Known experimental models of supranormal releasers are e.g. extra large eggs, which elicit the hatching pattern in many bird species more easily than their own eggs (e.g. gulls and oystercatchers). Supranormal releasers are used in many human activities, such as spoon-baits in fishing or in commercial advertisements. *Syn.* supernormal stimulus.

sura, pl. **surae** (L.), sural region, calf; the muscular back part of the leg.

surface frost, frost in the spring when the surface layer of the soil has already thawed but freezes again; may cause injuries (**ice scorch**) in plants if the roots cannot take water from the frozen soil.

surface runner, *see* runner.

surface stem, any plant stem (aerial stem) growing on the surface of the ground. *Syn.* aerial stem.

surface tension, the tension of an open surface of a liquid; symbol σ, given as J/m^2; caused by forces of attraction between the molecules of a liquid, tending to minimize the area of the liquid surface. Capillary action and water uptake by plants against gravity are based on surface tension.

surface water, water resources on the surface of the earth, e.g. brooks, rivers, ponds, lakes. *Cf.* ground water. *See* hydrological cycle.

surfactant, any surface-active agent, such as detergents, emulsifiers, wetting agents, dispersing agents, and surface tension depressants; e.g. **1.** a mixture of certain phospholipids and proteins in a fluid on the surface of the alveolar membrane in the lung preventing the collapse of alveoli; **2.** any surface-active detergent used to improve the penetration of other substances into tissues, e.g. the absorption of herbicides into plants.

surgery (Gr. *cheir* = hand, *ergon* = work), **1.** a branch of medicine and veterinary dealing with manual or other operative methods; **2.** a place where these methods are performed, or a room where the patients are treated; **3.** the work performed by a specialist of surgical operations (surgeon). *See* microsurgery.

survivorship, the likelihood of an organism to survive through a certain time interval.

survivorship curve, a curve describing graphically a probable survival of an age-cohort (or a presumed cohort) in a population as a function of age. The vertical axis describes the number of survivors in the cohort (usually transformed into 1,000 born individuals) in a logarithmic scale, and the horizontal axis presents the age-groups, usually as percentages of the maximal age. Survivorship curves are of three main types: a **convex curve** (type I) describes a population with the mortality concentrated to the end of the maximum lifespan (e.g. human populations in industrial countries, or populations of rotifer species), a **straight line** (type II) describes a population with a constant mortality rate (e.g. many birds), and a **concave curve** (type III) the populations with maximal mortality in initial age-classes, e.g. representing many fish populations.

suspension (L. *suspensio* = hanging, suspending), any mixture of a liquid and very fine solid particles; the solid component can be separated by standing or centrifugation.

suspension feeder, any animal which feeds on material suspended in water.

suspensor, 1. a cell chain developing from the zygote of angiosperms in the beginning of the development of an embryo in the → embryo sac. The suspensor pushes the embryo towards the → endosperm, which serves as an energy store; **2.** one of the two hyphae in fungi of the order Mucorales, bearing gametangia at the tips, later supporting the zygospore.

sustained yield, any renewable environmental source that can be removed without depletion; e.g. the number of individuals possible to be removed from a population (e.g. by hunting) to achieve a planned density as estimated by calculations on a natural input and output of the population.

sutura, pl. **suturae** (L.), seam, suture, suture joint, especially a type of fibrous joint between bony surfaces, so closely united that no movement can occur; particularly found between the skull bones.

Svedberg (unit) (*Theodor Svedberg*, 1884—1971), symbol S; the unit for the sedimentation rate of a particle determined by ultracentrifugation. *See* Appendix 1.

swamp, a tract of wet land with wilt standing water; often periodically flooded, growing shrubs and trees, having no peat accumulation on its surface, and unsuitable for agriculture without artificial drainage.

sweat, 1. the secretion of → sweat glands in mammals; *syn.* perspiration (especially in man); **2.** to secrete sweat.

sweat glands (L. *glandulae sudoriferae*), tubular glands in the skin of mammals secreting salty liquid onto the surface of the body. There are two types of sweat glands: **eccrine sweat glands** which are simple, coiled, tubular glands secreting dilute fluid, **sweat**, containing some sodium chloride, urea and sodium lactate, and **apocrine sweat glands** which are larger, branched tubular glands opening to hair root sheaths and secreting more viscous sweat with many organic compounds, such as proteins and fats.

The eccrine sweat glands are located in the paws of many mammals, and their function is stimulated in exciting situations improving the grasp of the paws. In primates, some glands are found also in other areas of the skin, but in man abundantly all over the skin functioning efficiently in → thermoregulation (*see* aquatic ape theory). The apocrine glands are located mainly in armpits, pubic region and areolae of breasts, and their secretion serves mostly as scent signals. Mammals living in a hot environment, such as odd-toed ungulates, many even-toed ungulates, and many primates, have apocrine glands also in other skin areas where the secretion is more fluid and serves mainly thermoregulation. Also in man, apocrine glands are found all over the skin of an early foetus, but later they degenerate, and are postnatally found only in special areas mentioned above. The sweat glands are also called **sudoriferous glands** (sudoriparous glands). The mammary glands and special scent glands found in many species are differentiated from the sweat glands. In humans, 1 to 2 litres of sweat per day is lost but the amount may increase to 12 litres in hot and dry environments.

sweepstakes dispersal, emigrations of species by chance to districts difficult to reach; e.g. the migration of terrestrial animals to a region without land connections. Usually the attempt to migrate is successful only for a few species and therefore fauna and flora are very different on both sides of the barrier, as e.g. the mammalian fauna in South Africa versus Madagascar.

swim bladder (L. *vesica pneumatica*), the membranous sac filled with gas, situated dorsally in the abdominal cavity of fish; acts as a hydrostatic organ controlling the specific gravity of the animal and thus its buoyancy. The organ is absent in cartilaginous fishes and also in those bony fishes which live deep in the oceans. During embryogenesis the swim bladder derives from the same organ primordium as the oesophagus, and in many species a ductal connection to the mouth remains (**physostomous fishes**, as herrings); in other species the connection disappears during embryogenesis (**physoclistous fishes**, such as perch-like fishes). Gases (CO_2, O_2, N_2) are secreted into the bladder from the blood. In physostomous fishes, gases are removed from the bladder through the duct to the pharynx, but in physoclistous fishes a special organ controls the removal of gases back to the blood. The swim bladder may also act in sound production, e.g. in squirrelfishes (Holocentridae), in sound reception, as in cod (Gadidae), or as an accessory respiratory organ, e.g. in gar (Lepisosteidae). *Syn.* air bladder, gas bladder.

swing-out rotor, → rotor.

sycon grade, syconoid type (Gr. *sykon* = fig), a type of sponge having choanocytes located in sac-like radial canals of the body wall. *Cf.* ascon grade, leucon grade.

sym-, syn- (Gr. *syn* = with, together), pertaining to an association or union.

symbiont (Gr. *symbioun* = to live together), symbiote; an organism living in → symbiosis.

symbiosis, pl. **symbioses** (L. < Gr.), the living of two dissimilar organisms together in a close association; **1.** in a restricted sense means reciprocal actions of two different species beneficial to both; often called →

mutualism; e.g. the symbiosis of the alder and leguminous plants with their root bacteria, and many animals with digestive bacteria in the alimentary canal; **2.** broadly, symbiosis includes also interactive associations from which only one species benefits but causes no harm to the other (**commensalism**), or one benefits at the other's expense (**parasitism**). *Adj.* **symbiotic**. *Cf.* protocooperation.

symmetric competition, *see* competition.

symmetric flower (Gr. *symmetria* < *syn* = together, with, *metron* = measure), a type of flower, the projection of which can be divided into two similar halves in many different ways. *Cf.* asymmetric flower, zygomorphic flower.

symmetry, 1. uniformity, conformity; **2.** the corresponding arrangement in form and size around an axis, such as on each side of a plane of a body. *Adj.* **symmetric(al)**. *See* bilateral, radial symmetry.

sympath(o)-, sympathico- (Gr. *sympatheia* = sharing of compassion, sympathy), denoting relationship to the → sympathetic nervous system. *Adj.* **sympathetic**.

sympathetic nervous system, a portion (*pars sympathica*) of the → autonomic nervous system of vertebrates. Its peripheral nerve tracts comprise two successive neurones. The first one (**preganglionic neurone**), leaves the spinal cord via the ventral root at the thoracolumbal spinal area, and most preganglionic neurones together pass along the **sympathetic nerve trunk** on both sides of the vertebral column. Most preganglionic neurones end on the trunk and synapse with the second neurones (**postganglionic neurones**) forming segmentally **paravertebral ganglia**. Some of the preganglionic neurones continue to other ganglia (**prevertebral ganglia**) located in internal organs outside the sympathetic nerve trunk. The postganglionic neurones reach different peripheral organs and tissues everywhere in the body. In the ganglia the transmission of nerve impulses occurs **cholinergically**, i.e. by → acetylcholine, and in the peripheral tissues mainly **adrenergically**, i.e. by → noradrenaline.

The sympathetic nervous system is highly activated by sudden stress situations eliciting the "fight or flight behaviour" and improving the effort and survival of an animal. The response is due to a direct effect of noradrenaline released from sympathetic nerve endings, but also to hormonal effects transmitted by adrenaline and noradrenaline from the adrenal medulla (embryonally a sympathetic ganglion). These hormones strengthen and prolong the initial nervous action. The brain centres which control functions of the sympathetic nervous system are located mainly in the limbic cortex and hypothalamus, which further can activate the hormonal systems. In many organs the → parasympathetic nervous system acts antagonistically to the sympathetic nervous system.

sympathicotonia (Gr. *tonos* = tone), the increased activity (tonus) of the → sympathetic nervous system. Called also sympatheticotonia. *Cf.* vagotonia.

sympatholytic, sympathicolytic (Gr. *lytikos* = dissolutive), **1.** opposing the activity of the → sympathetic nervous system; **2.** pertaining to an agent (drug) that opposes the activity of the sympathetic nervous system. *Cf.* parasympath(ic)olytic.

sympathomimetic, sympathicomimetic (Gr. *mimetikos* = imitative), **1.** mimicking the action of the → sympathetic nervous system; **2.** pertaining to an agent (drug) that mimicks the effects of the → sympathetic nervous system, i.e. has adrenergic effects. *Cf.* parasympathomimetic.

sympatric (Gr. *patra* = fatherland), pertaining to populations of two or more species living in the same geographic area. *Cf.* allopatric.

sympatric speciation, *see* speciation.

sympetalous, → gamopetalous.

Symphyla (Gr. *phylon* = race), **symphylids**; a class of small, eyeless arthropods living in the soil (e.g. garden centipede). Characteristically, they have a pale-coloured segmented body with one pair of antennae, 12 pairs of jointed legs and a pair of unjointed posterior appendages bearing one pair of → spinnerets. The class includes about 160 species, most are herbivorous, some saprophagous.

symphysis, pl. **symphyses** (Gr. *phyein* = to grow), a site of union; a fibrocartilaginous joint in which the surfaces of two bones are firmly united by a plate of cartilage; e.g. **intervertebral symphyses** between adjacent vertebral bodies, or **pubic symphysis**, the firm joint between the two pubic bones.

symplesiomorphic, symplesiomorphous (Gr. *plesios* = near, *morphe* = form), in cladistic classifications, pertaining to a primitive characteristic shared by several taxa but regarded as inherited from ancestors older than the last common ancestor. *Noun.* **symplesiomorphy**.

sympodial branching (Gr. *podion*, dim. of *pous* = foot), a type of plant branching, in which the development of the parent axis is completely suppressed and growth is continued by one of the side branches which takes the dominant role in the structure, called **sympodium** (pl. sympodia); found e.g. in the grapevine. *Cf.* monopodial branching.

symport, a coupled transport of two different ions or molecules in the same direction across the cell membrane; one of the ions or molecules moves usually down its → electrochemical gradient and with this energy it can pull the other one against its own electrochemical gradient; e.g. sodium-glucose cotransport in animal cells, proton (H^+)-glucose transport in plant cells. *See* cell membrane transport. *Cf.* antiport, uniport.

symptom (L., Gr. *symptoma*), any subjective evidence or indication of disease or disorder in the condition of a patient or diseased animal; also an evident reaction in a plant caused by a pathogen. *Cf.* sign.

syn-, sym- (Gr. *syn* = with, together), pertaining to an association or union.

synanthrophic (Gr. *anthropos* = man), pertaining to an organism which is associated with man or living in human dwellings.

synapomorphy (Gr. *apo* = from, *morphe* = form), in cladistic classification, a common → apomorphic character shared by two or more taxa and therefore believed to have appeared in their common ancestor.

synapse (Gr. *synaptein* = to join together), the junction between two neurones, usually between the terminal button of an axon and a dendrite or the cell body of another neurone. Broadly the term is used to mean also junctions between neurones and effector cells, such as a muscle fibre (*see* neuromuscular junction) or a secretory cell. Certain individual neurones of the central nervous system may get thousands of synaptic contacts from other neurones. Typically the synapse consists of a **synaptic knob** (end-foot, terminal button of axon), the membrane of which is pressed closely against the membrane of the other neurone. The **synaptic cleft** (15 to 30 nm) occurs between the membranes, i.e. between the presynaptic and postsynaptic membranes. Electric excitation (nerve impulses) propagating along the axon to the synaptic knob crosses the cleft chemically by means of special → **neurotransmitters**; only exceptionally electric transmission is possible (*see* ephapse). Nerve impulses cause an influx of calcium ions into the synaptic knob, resulting in the liberation of a neurotransmitter from small vesicles (**synaptic vesicles**) into the synaptic cleft. Neurotransmitter molecules released then bind to their **receptors** (postsynaptic receptors) on the postsynaptic membrane causing an opening or closing of ion channels in the membrane. The process either depolarizes or hyperpolarizes the membrane resulting in the generation of an → **excitatory postsynaptic potential** or **inhibitory postsynaptic potential**, respectively. The spatial and temporal summation of the functions of several synapses determines the generation of a new impulse (or series of impulses) in the **axon hillock** (initial segment) of the postsynaptic neurone. After acting, the neurotransmitters are rapidly destroyed by specific enzymes or taken back up into the presynaptic knobs. Often the synapse also has **presynaptic receptors** inhibiting further liberation of the neurotransmitter, and thus restraining the function of the synapse. More than one transmitter substance may occur in the same synapse.

synapse en passant (Fr.), a type of → synapse in neuromuscular junctions on the surface of smooth and cardiac muscle cells, some also in neurone-neurone junctions; typically comprises many successive varicosities (synaptic expansions) along the terminal branch of an autonomic nerve fibre, forming synapses with many muscle cells. *See* neuromuscular junctions.

synapsid (Gr. *apsis* = arch), pertaining to a skull type with one pair of ventrolateral temporal apertures, found in mammal-like reptiles (*see* Synapsida) and mammals. *Cf.* anapsid, parapsid.

Synapsida, synapsids; a subclass of extinct mammal-like reptiles (e.g. orders Pelycosauria and Therapsida) which are characterized by a synapsid skull. *Cf.* Anapsida, Diapsida.

synapsis, pl. **synapses** (L.), **1.** the tight pairing of homologous chromosomes in meiotic cell division; *see* meiosis; **2.** rarely used to denote → synapse.

synaptic knob, the terminal button of the → axon of a nerve cell; also called synaptic endfoot. *See* synapse.

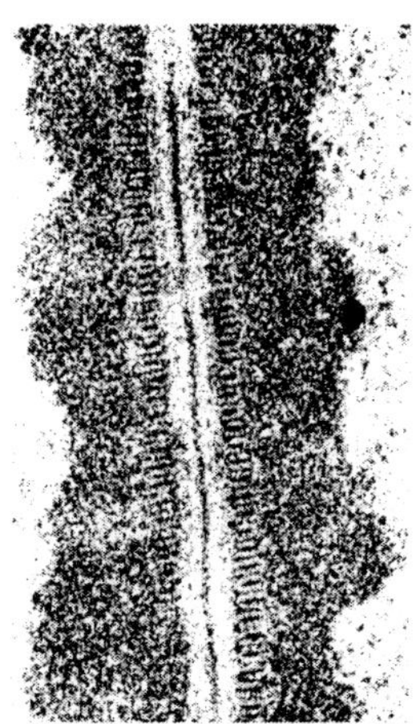

Synaptonemal complex. The ultrastructure of the synaptonemal complex. In the middle the central element. On both sides of it the ladder-like lateral elements, and attached to these the homologous chromosomes. (After v. Wettstein - Cold Spring Harbor Symposia on Quantitative Biology, 23, 1968, with the permission of the publisher).

synaptonemal complex (Gr. *nema* = thread), a complex structure of the cell nucleus, visible in an electron microscope, holding the meiotically paired chromosomes together in zygotene and pachytene, i.e. during the synapsis of → meiosis. The synaptonemal complex is a tripartite protein structure consisting of lateral elements formed by the axes of the chromosomes, and a central element located between these. The complex dissolves during diplotene.

synaptosome (Gr. *soma* = body), any of synaptic knobs (presynaptic endings) pinched off and separated from the brain tissue for the study of synaptic function; each synaptosome contains mitochondria, synaptic vesicles, and axoplasm surrounded by the presynaptic membrane of the knob with some remnants of postsynaptic membrane.

syncarpous (Gr. *karpos* = fruit), pertaining to a plant → gynaecium which has the → carpels united to form a compound gynaecium. *Noun* **syncarpy**. *Cf.* apocarpy.

syncaryon, → synkaryon.

synchronia (Gr. *chronos* = time), **1.** occurrence at the same time; *syn.* **synchronism, synchrony**; **2.** the formation of tissues and organs at the usual time and order; opposed to heterochronia in which the development occurs at an unusual time and order. *Adj.* **synchronous**.

synchronized sleep, the phase of deep sleep. *See* sleep.

syncytium, pl. **syncytia** (Gr. *kytos* = cavity, cell), a multinucleate mass of protoplasm formed by the fusion of individual cells; e.g. **1.** a multinucleated cell produced by the fusion of single cells (e.g. a striated muscle fibre), or by a process in which the cell nuclei have divided without the division of the cells; **2.** syncytiotrophoblast (syntrophoblast), the syncytial outer layer of the → trophoblast; **3.** the functional syncytium of the cardiac muscle, earlier believed to be formed by cell fusions; actually the muscle consists of a vast number of single cells, which, due to several → gap junctions between the cells act electrically as a syncytium; also many smooth muscles form similar functional syncytia. *Adj.* **syncytial**.

syncytium theorem (Gr. *theorema* = theory, sight), a statement presenting that multicellular animals, metazoans, originate from unicellular, polynuclear ciliates by the formation of plasma membranes between the nuclei.

syndactylism, syndactyly (Gr. *dactylos* = digit), **1.** the normal state of many animal species having two or more digits wholly or partly joined, as e.g. in many birds and marsupials; **2.** in man, a congenital and often inheritable anomaly characterized by webbing or fusion of fingers or toes.

syndecan (Gr. *syndein* = to bind together), a cell surface proteoglycan (250 to 300 kD) integrated in the cell membrane of epithelial cells. Its extracellular domain contains sulphates and an N-linked oligosaccharide, the sulphate chains being bound to proteins of intercellular matrix, and the intracellular (cytoplasmic) domain interacting with actin filaments. Thus, syndecan behaves as a receptor for interstitial matrix, serving in the interaction between tissue cells and their surroundings.

syndrome (Gr. = combination, concurrence), a group of symptoms or signs occurring together in a disease or other abnormal physiological conditions (e.g. AIDS, amnestic syndrome, stress syndrome); in genetics, means a pattern of multiple abnormal properties or malformations, which may be inherited together.

synecology (Gr. *oikos* = household, *logos* = word, discourse), a branch of ecology dealing with interactions between organism groups and their environments. *Cf.* autecology.

synergid (Gr. *synergos* = working together), one of two cells in an embryo sac of seed plants, located beside the egg cell at the micropylar end of the → embryo sac.

synergism, the joint effect of different agents on certain biochemical or physiological reactions; may be additive or potentiating, in the latter the combined effect being greater than the sum of their individual effects; e.g. synergistic actions of muscles, or of different environmental toxic agents in organisms, or synergistic actions of alcohol, tobacco, and various drugs in humans. *Cf.* antagonism.

syngamy (Gr. *gamos* = marriage), in sexual reproduction, the union of the two gametes that in fertilization leads to → karyogamy and the formation of a zygote.

syngeneic, *see* isogeneic.

syngraft, → isograft.

synkaryon, also **syncaryon** (Gr. *karyon* = nut, nucleus), zygote nucleus; the cell nucleus of a → zygote formed by fusion of the nuclei of the female and male gametes in fertilization.

syntenic (Gr. *tainia* = band, tape), pertaining to genes which are located in the same chromosome; the genes do not necessarily show → linkage if they are located so far apart that at least one → chiasma is always formed between them.

synthetases, → ligases.

synthetic evolution theory, → neo-Darwinism. *See* evolution theory.

syntopic (Gr. *topos* = place), pertaining to → sympatric populations or species living in the same habitats.

syntype (Gr. *typos* = model, mark), *see* type specimen.

syringyl alcohol (L. *Syringa* = lilac), sinapyl alcohol; a phenylpropanoid alcohol that together with para-coumaryl alcohol and/or → coniferyl alcohol forms → lignin.

syrinx, pl. **syringes,** also **syrinxes** (Gr. = pipe, tube), the vocal organ of birds situated in the lower part of the trachea. It consists of resonant membranes supported by cartilaginous arches and muscles controlling the stretch of the membranes. The sounds themselves are generated when the respiratory air is pressed between the vocal cords inside the syrinx. *Adj.* **syringeal**.

systematics (Gr. *systema* = an organized whole, system), the biological → classification and study dealing with the diversity of organisms and their relationships, especially emphasizing the significance of phylogenetic relations between taxa and dealing with practice of identification and nomenclature. The terms systematics and → taxonomy are often used as synonyms.

systematic sampling, regular sampling; a → sampling method used in ecological research using study areas or points which are situated systematically in the field, e.g. in a straight line. Systematic sampling gives a good illustration e.g. of zonal vegetation. *Cf.* random sampling.

systemic, pertaining to the body as a whole, as distinguished from local tissues.

systemic circulation, greater circulation, major circulation; the circulation of blood in tetrapod vertebrates carrying oxygenated blood from the left ventricle of the heart (in amphibians from the left side of the single ventricle) to various tissues of the body, the venous blood returning to the right atrium. *Cf.* pulmonary circulation. *See* blood circulation.

systole (Gr. *systellein* = to contract), **1.** the contraction phase of the heart; **2.** the contraction of a pulsating contractile vacuole. *Adj.* **systolic**. *Cf.* diastole.

T

T, symbol for **1.** transmission of light; **2.** tritium; **3.** threonine; **4.** thymine, thymidine; **5.** thyroxine (T_4), triiodothyronine (T_3); **6.** *tera* (10^{12}).

***T*,** symbol for temperature.

t, symbol for ton, 1,000 kg.

T3, T4, T7, → bacteriophages of *Escherichia coli*.

tacho- (Gr. *tachos* = speed), pertaining to speed; e.g. **tachography**, the recording of the velocity of blood current.

tachy- (Gr. *tachys* = rapid), denoting rapid.

tachycardia (Gr. *kardia* = heart), an increased heart rate, either physiological or pathological. *Adj.* **tachycardiac**.

tachykinins, short polypeptides found in animal tissues, both in vertebrates and invertebrates; have hormone-like actions in the nervous system and smooth muscles of internal organs. *See* substance P.

tachymetabolic animals (Gr. *metabole* = change), animals with high metabolism; the term is used to describe → endothermic animals, i.e. birds and mammals. *Cf.* bradymetabolic animals.

tachyphylaxis (Gr. *phylaxis* = protection), **1.** the rapid decline in a response of an organ or the whole organism to a physiologically active agent after repeated administration of this agent; e.g. the weakening effect of a drug during repeated administration, or of a drug applied several times to an isolated organ; **2.** the rapid immunization against a high dose of a toxin by previous injections of small doses of the same toxin.

tachypn(o)ea (Gr. *pnoia* = breath), increased rate of respiration.

tactile hairs (L. *tactilis* < *tactus* = touch), sensory hairs (mechanoreceptors) located on the exoskeleton of many invertebrates; each consists of one or more sensory cells sensitive to touch or vibration, such as contact with the body surface and air or water movements. *See* Meissner's corpuscle.

tactile sense, sense of touch; the sense by which an animal receives information from contact with objects; the perception is based on excitation of organs of touch which are mechanical receptors (mechanoreceptors), usually slightly differentiated **free nerve endings** in the skin, often at the base of hairs, → **tactile hairs**. When irritated by touch, the membrane of the receptor bends, changing ion permeability that results in → receptor potential and further in nerve impulses, which through special sensory neurones propagate to the brain. The tactile sense can function also as a vibration organ.

taen-, t(a)en(i)-, t(a)enio-, 1. (L. *taenia* = a flat band, tape), pertaining to an anatomical structure resembling a flat band, **taenia**, e.g. *taeniae coli*, muscle bands in the wall of the colon; **2.** denoting relationship to tapeworms; e.g. **taeniasis**, i.e. infection by tapeworms of the genus *Taenia*.

tagma, pl. **tagmata** (Gr. = corps), a part of the body of an animal formed by the fusing of different segments; e.g. the caput, thorax, and abdomen of insects are tagmata, originally formed by fusion of somites of metameric animals. *Cf.* pseudotagma.

taiga, the coniferous forest zone in the northern hemisphere in the cool zone, south of the arctic (tundra)zone, especially in Siberia.

tail fin, → caudal fin.

tailing, a method by which DNA segments containing identical → nucleotides are added to the ends of a → DNA fragment.

tallow, → suet.

tannins, phenol derivatives synthesized from shikimic acid and acetate; exist abundantly in cortical tissues of plants causing browning of plant parts. Tannins act as natural defensive substances in plants against herbivores, and tannin concentration often increases in plants that animals feed on. Tannins bind tightly to proteins and are therefore used as **tanning** substances for hides.

tapetum, pl. **tapeta** (L. < Gr. *tapetion*, dim. of *tapes* = rug, carpet), **1.** a covering structure such as *tapetum corporis callosi*, a stratum of fibres of the corpus callosum in the cerebrum of vertebrates, or *tapetum nigrum*, the pigment layer behind the retina of the eye, or *tapetum lucidum*, the iridescent pigment epithelium in the eyes of some animals (e.g. cats) which causes shining in the dark; **2.** a layer of cells with an intense secretory activity nourishing the developing pollen grains in the pollen sacs of the stamen in → Angiospermae.

tapeworms, → Cestoda.

tap root, *see* root.

***Taq* polymerase,** a thermostable DNA polymerase isolated from *Thermus aquaticus*, used in the → polymerase chain reaction.

Tardigrada (L. *tardus* = slow, *gradus* = step), **tardigrades, water bears**; a phylum of invertebrates including small animals (less than 1 mm) with four pairs of stubby legs; live in damp soil, some species also in the bottom of lakes and oceans. Tardigrades are known for their good resistance to low temperatures and aridity; about 400 species are described.

tardus (L.), slow, slowly developing.

targeted mutagenesis, → focused mutagenesis.

target theory, a theory stating that a certain process, such as a gene or chromosome mutation, is produced when ionization of macromolecules (e.g. DNA) caused by radiation takes place within a well-defined small area (target), usually in or close to the chromosome.

tarsus (Gr. *tarsos* = flat broad surface), **1.** the vertebrate ankle; the articulation region between the tibia and metatarsus; **2.** the distal segment of the leg of an arthropod; **3.** in the vertebrate eye, the plate of dense fibrous connective tissue giving form and support to the edge of the eyelid, i.e. *tarsus inferior* in the lower eyelid, *tarsus superior* in the upper eyelid. *Adj.* **tarsal**.

tartrate (L. *tartarum* = lees of wine, tartar), a salt, ester, or the ionic form of **tartaric acid**, $HOOC(CHOH)_2COOH$, widely distributed in plants, especially in fruits; tartrates are used e.g. in pharmacy and food preparation (cream of Tartar).

taste, 1. the → chemical sense associated with gustatory receptors, afferent gustatory nerve fibres, and their integrative brain areas; the receptors are stimulated by the contact of soluble substances. In vertebrates, the receptors are located in → taste buds in the mouth cavity, mainly in the epithelium of the tongue; in invertebrates they locate in the gustatory sensilla of various mouthparts, antennae, and appendages of various body segments; **2.** as a verb, to form a sensation of taste.

taste bud, a small flask-shaped structure comprising taste receptors, supporting cells and basal cells in the epithelium of the tongue or mouth cavity of vertebrates, in fish also in the external body surface, especially in the head region.

TATA box (T = thymine, A = adenine), Hogness box; the conserved sequence of six or seven nucleotides in a → promoter acting as a recognizing site for eukaryotic RNA polymerase II in the initiation of → genetic transcription; TATA box is located around position —30 upstream of the transcription initiation site in the → promoter, and recognized by a common transcription factor thus being essential for the initiation of → genetic transcription. TATA box is analogous to → Pribnow box of prokaryotic organisms.

taurine (Gr. *tauros* = bull), 2-aminoethanesulphonic acid, $H_2NCH_2CH_2SO_3H$, synthesized in the liver as a part of the molecule of a common bile salt, **taurocholate**; possibly a neurotransmitter in some invertebrates, virtually absent in plants.

tautomerism (Gr. *tauto* = same, *meros* = part), a form of structural isomerism (*see* isomers) in which a compound appears as a mixture of two isomers, **tautomers** (tautomerides), in equal quantities. The isomers are not directly convertible, but the reversibility of the change is due to the mobility of an atom or atom group in the molecule from one position to another, often with a new arrangement of a double bond. The removal of one of these isomeric forms from the mixture results in the conversion of the other, so that the equilibrium is restored. One commom type is *keto-enol* isomerism. *See* isomers.

taxis, pl. **taxes** (Gr. = arrangement), tactic movement; the movement of animals or locomotive cells towards a source of stimulation (positive taxis), or away from it (negative taxis). According to the type of stimulation, the taxes are grouped into **chemotaxis** (chemical stimulus), **phototaxis** (light stimulus), **heliotaxis** (sunlight stimulus), **geotaxis** (gravity stimulus), **galvanotaxis** (electric stimulus), **barotaxis** (pressure stimulus), and **thigmotaxis** (touch stimulus). In ethology, the taxes existing among invertebrates are grouped on the basis of movement patterns. *See* **klino-**, **meno-**, **mnemo-**, **telo-**, **tropotaxis**. *Cf.* kinesis.

taxodont (Gr. *odous* = tooth), pertaining to a bivalve mollusc with unspecialized hinge teeth.

taxol, a substance obtained from the bark of the Pacific yew tree, *Taxus brevifolia*, and used with its synthetic derivatives (**taxoids**) for treatment of cancer; in cells the substances can bind to → microtubules rendering them static, thus inhibiting cell divisions.

taxon, pl. **taxa**, Am. also **taxons**, a general name for any group of organisms which have been ranked in a hierarchical classification and described with a scientific name; e.g. species, subspecies, genus, class, and phylum are taxa. *Syn.* taxonomic unit.

taxonomic character (Gr. *nomos* = law), any characteristic seen in all members of a taxon but which differs from characteristics of other taxa.

taxonomic hierarchy, → hierarchy.

taxonomic unit, → taxon.

taxonomy, the branch of biology including the description, naming, and classification of extant and extinct organisms. There are different kinds of taxonomy, depending on the method used: **classical taxonomy** deals with morphological and anatomical characteristics, **biochemical taxonomy** is based on biochemical properties of organisms, **numerical taxonomy** estimates similarities and differences between organisms using mathematical procedures; **cytotaxonomy** studies the structure and number of chromosomes of organisms. *See* cladistics. *Cf.* phylogenetics.

2,3,6-TBA, → 2,3,6-trichlorobenzoic acid.

TBSV, tomato bushy stunt virus.

TCA, → trichloroacetic acid.

TCCD, → dioxin.

T cell, T lymphocyte. *See* lymphocytes.

TEA, → tetraethylammonium, triethanolamine.

tectorial membrane (L. *membrana tectoria*), the shelf-like gelatinous membrane of the cochlear duct in the inner ear, i.e. the membrane of the → spiral organ against which sensory cilia of the hair cells will bend under the influence of sound waves.

tectum, pl. **tecta** (L. *tectus* < *tegere* = to cover), an anatomical structure resembling or serving as a roof of an organ; e.g. **1.** *tectum mesencephali*, the roof of the midbrain; **2.** the outermost layer of the → pollen grain wall, i.e. the outer layer of exine.

tegmen, pl. **tegmina** (L.), an integument, covering; e.g. **1.** *tegmen tympani*, a thin bone in the middle ear; **2.** a pair of leathery forewings (elytra) covering the hindwings of orthopterans; **3.** *endopleura*, the inner coating of the seed.

tel(e)-, tel(o)- (Gr. *tele-* = far off, distant), **1.** distant, at or over a distance; **2.** (Gr. *telos* = end, completeness), pertaining to an end, a completeness, or a mature state.

tela, pl. **telae** (L. = web), a layer of web-like tissue, as *tela chorioidea ventriculi (laterali, terti,* and *quarti)*, the vascular tissue structures in the roofs of the four brain ventricles.

telencephalon (Gr. *telos* = end, *enkephalos* = brain), **1.** the most anterior part of the vertebrate brain comprising two halves, **hemispheres**, united by the → corpus callosum; the two lateral brain **ventricles** are located within the hemispheres. Most nerve cell bodies are concentrated to the surface area, → **pallium** (cerebral cortex), and several **nuclei** (*see* striate body) deeper in the basal areas of the hemispheres. The telencephalon is the most developed part of the brain. In ectothermic vertebrates, it serves mainly rhinal functions (→ olfaction, *see* rhinencephalon), and automatic voluntary movements associated with deeper motor nuclei (*see* extrapyramidal system). In homoiothermic vertebrates, particularly mammals, it forms the most dominant part involved in integrative brain functions. The pallium has gradually developed into a very complicated system, → **neocortex**. It controls efficiently muscular and nervous functions, such as accurate voluntary motor functions of the → pyramidal system, emotions associated with the → limbic system, and the integration of sensory and motor areas with other brain parts; **2.** the anterior vesicle of the prosencephalon in the developing embryo of vertebrates. *Cf.* cerebrum.

teleology, the study using purpose or utility of biological functions as an explanation of any natural phenomenon, i.e. a doctrine which explains natural phenomena and events by final causes.

Teleostei (Gr. *teleios* = perfect, *osteon* = bone), **teleosts**; a fish taxon of uncertain category (in some classifications infraclass) comprising higher bony fishes within the subclass Actinopterygii of the class Osteichthyes.

teleut(o)- (Gr. *teleute* < *telos* = end, completeness), denoting relationship to completion.

teleutospore (Gr. *sporos* = seed), an overwintering spore type of rust fungi. *See* Uredinales.

Teliomycetes, *see* Uredinales.

telmatic layer (Gr. *telma* = stagnant water, marsh), an organic layer formed e.g. from peat sedimented in a littoral zone.

teloblast (Gr. *telos* = end, *blastos* = germ), an embryonic cell type of protostomes producing tissue cells from which the mesoderm and later the → schizocoel are derived.

telocentric (Gr. *kentron* = centre), pertaining to chromosomes or chromatids with a terminal → centromere; telocentric chromosomes are usually unstable and are eliminated during a few cell divisions, or will be converted to → isochromosomes.

telodendron, pl. **telodendria** (Gr. *dendron* = tree), the terminal branching of a nerve fibre.

telolecithal (Gr. *lekithos* = yolk), describing an egg type with a large amount of yolk concentrated at one pole (vegetative or vegetal pole), and having the cytoplasm with the cell nucleus at the other pole (animal pole); occurs e.g. in rays, reptiles, and birds.

telomerase, telomere terminal transferase; the ribonucleoprotein enzyme adding nucleotides to a DNA strand at the → telomere.

telomere (Gr. *meros* = part), the end of a → chromosome.

telomere terminal transferase, → telomerase.

telome theory, a theory describing the structure and development of present → vascular plants. According to the theory, the most original, undeveloped terrestrial plants were not differentiated in their aerial parts into shoots and leaves, but formed uniformly forking, undifferentiated shoot systems, telomes. Further, the theory explains that typical organs of the highly differentiated present plants are derived by a series of processes from the primordial telomes.

telophase (Gr. *phasis* = appearance), the last stage of → mitosis or → meiosis in cell division.

telotaxis, pl. **telotaxes** (Gr. *taxis* = arrangement), a tactic movement of an animal directly towards a stimulus (positive) or away from it (negative) without regard to the angle between the direction of movement and the stimulation source, as in → menotaxis. *See* taxis.

telotroch (Gr. *trochos* = wheel), a circular tuft of cilia around the anus in → actinotrocha, trochophore, and tornaria larvae.

telson (Gr. = extremity), **1.** an appendage in the terminal abdominal segment of decapod crustaceans, forming together with two uropods the tail of the crayfish; **2.** a poisonous sting in the terminal abdominal segment of merostomates (horseshoe crabs) and scorpions.

TEM, transmission electron microscope. *See* microscope.

temperate bacteriophage (< L. *temperare* = to control), a → bacteriophage (a type of virus) which, contrary to virulent or lytic bacteriophages, can be integrated into the bacterial chromosome as a → prophage, and in this way cause the lysogenic state of the bacterium (*see* lysogenic).

temperate zone, a vegetation zone, in which the mean temperature is over 10°C at least four months of the year; characterized by foliage trees, steppes, or desert.

temperature sense, a sense by which an animal is able to detect changes in temperature. The receptors (thermal receptors) are slightly differentiated nerve endings situated mainly on the surface of the body, but also inside many organs. Two types of thermal receptors, **cold receptors** and **warm receptors**, are found in the skin of vertebrates; of these types cold receptors are usually manifold in number. The receptors become activated at different temperatures; e.g. the cold receptors of mammals are most active at about 25°C and the warm receptors at about 37°C. Nerve tracts associated with thermal receptors, have two or three successive neurones, the tract ending usually in the thalamus but some continuing further to the cerebral cortex. Homoiothermic animals also have special types of thermal receptors in the thermoregulatory centre of the hypothalamus for measuring blood temperature (*see* thermoregulation). Some snakes have a very specialized type of temperature sense, called → **pit organ**, in which numerous warm receptors are located in a small pit of the face. Among invertebrates, the types and sensitivity of temperature receptors vary according to the environmental temperature they are adapted to. Many invertebrates have a very precise thermal sense; e.g. a bedbug running on the ceiling can in the dark locate a person lying underneath mainly by its thermal receptors, and drop down on him.

temperature-sensitive mutation, a mutation which is expressed only at a certain (usually high) temperature, e.g. many temperature-sensitive lethal mutations of the fruit fly.

template, a macromolecule which acts as a model for the synthesis of other macromolecules; e.g. DNA and RNA templates.

temporal (L. *temporalis*), **1.** (L. *tempus* = time), pertaining to (limited) time, temporary; **2.** (L. *tempora* = the temples), pertaining to the lateral regions of the head.

ten-, teni-, tenio-, → taen-.

tendon (L. *tendo*), a fibrous cord of specialized connective tissue connecting the muscle to the bone or other structures; formed from parallel collagen fibres and cells (fibrocytes). The tendon is surrounded by a layer of looser connective tissue, **tendon sheath**.

tendon organ, → Golgi tendon organ.

tendril, a thin, twining structure in plants; helps especially climbing plants to stick to and climb on other plants and objects; may be formed from stems, leaves, or stipules, in some species from roots. A **tendril stem** may be formed in sympodially growing plants, the tendril being formed at a branching from the smaller branch. In a **tendril leaf** the tip of the leaf is modified into a tendril (e.g. pea); a **tendril root** is a plant root (aerial root), which has changed to a twining tendril (e.g. vanilla).

tension (L. *tensio*, Gr. *tonos*), **1.** the state of being stretched or strained, the act of stretching; **2.** mental or emotional strain; **3.** the pressure of gas or fluid, as the oxygen tension in the blood, or the blood pressure; **4.** electromotive force or potential.

tension wood, the → reaction wood in dicotyledons; develops on the upper side of bent stems and branches and contains less lignin and more cellulose than normal wood. *Cf.* compression wood.

tentacle (L. *tentare* = to touch), any slender flexible organ found mainly around the mouth of many invertebrates, especially in molluscs, polyps, and bryozoans; serves usually as a sense organ of touch as well as an organ for locomotion, respiration, feeding, and defence.

tentorium (L. = tent), an anatomical part that covers something; e.g. **tentorium of cerebellum**, the extension of dura mater covering the cerebellum of the vertebrate brain.

teratogen (Gr. *teras* = monster, *genos* = origin, birth), a chemical or physical agent that causes foetal damage. Radiation, microbes and certain chemicals (e.g. some pollutants) are the best known teratogens of vertebrates. The teratogenic properties of drugs are tested using isolated cells and laboratory animals.

teratogenic, causing developmental malformations; pertaining to a drug or other agents causing malformations during embryonic development.

terbacil, a uracil derivative, i.e. 3-tertbutyl-5-chloro-6-methyluracil; used as a herbicide inhibiting the Hill reaction in → photosynthesis.

tergum, pl. **terga** (L. = back), **1.** the back of an animal; **2.** the dorsal plate (sclerite) of the covering of a metameric segment in some annelids and arthropods; **3.** one of the dorsal plates of the mantle (carapace) of a barnacle.

terminal (L. *terminus* = boundary, end, border), pertaining to, or situated at the end or in the tip, as the → terminal bud.

terminal bud, a bud supporting the longitudinal growth of a plant; may also be differentiated into a flower or inflorescence. *See* bud.

terminal button, → synaptic knob.

terminal nerve system, one of the accessory olfactory systems in vertebrates (*see* olfaction); comprises the **terminal nerve** (*nervus terminalis*) and its sensory nerve endings in the anterior nasal mucous membrane (separate from the main olfactory area and → Jacobson's organ). The nerve is composed of thin, plexiform nerve strands, visible to the naked eye only in some sharks, running to the telencephalon parallel and medial to the tracts of the olfactory nerve. Many neurones of this system contain → luliberin (LHRH), and it is presumed that the system acts in the regulation of reproductive functions via nasal detection of → pheromones.

terminal redundancy, the repetition of the same nucleotide sequence at both ends of the → genome, as occurs e.g. in a bacteriophage.

terminal webb, a cytoskeletal network of filaments in the cell cytoplasm beneath the free surface of certain epithelial cells, especially of absorptive cells with → microvilli, or of the → hairs cells in the inner ear. *See* cytoskeleton.

termination, in genetics, the ending of the synthesis of DNA, RNA, or a → polypeptide; specifically the ligation of the last amino acid residue at the end of the polypeptide molecule in → genetic translation.

termination codon, terminator codon, a → nonsense codon (stop codon) encoding the termination of the formation of a polypeptide chain in → protein synthesis of the cell. In → messenger RNA (mRNA) the termination codons are the triplets UAA, UAG and UGA (U = uracil, A = adenine, G = guanine).

termination factors, terminator factor, proteins that respond to the → termination codons in messenger RNA causing the release of the completed polypeptide chain and the → ribosome from the messenger RNA in → protein synthesis. *Syn.* release factors.

terminology (L. *terminus* = term, *logos* = word, discourse), **1.** the specialist vocabulary in science, art, and any particular subject; **2.** the study dealing with arrangement and construction of terms (nomenclature).

termites, Isoptera.

termone, a chemical compound which evokes the expression of sexual characters, and thus the development of reproductive organs in algae. *Cf.* gamone.

terpenes (L. *terebinthina* = resin), a group of → terpenoids, aliphatic or cyclic hydrocarbons that have the general formula $(C_5H_8)_n$. Terpenes are synthesized from → isoprene and are present in plants, especially in essential oils, resins, and balsams. Terpenes accumulate in cell walls together with suberin and waxes, in vacuoles of plant cells as latex, and in the cytoplasm as oil droplets. **Monoterpenes** $(C_{10}H_{16})$, such as α- and β-pinene, have two isoprene units; **sesquiterpenes** $(C_{15}H_{24})$ have three and **diterpenes** $(C_{20}H_{32})$ four units; plants may also contain larger **tri-**, **tetra-**, and **polyterpenes.** Plant hormones, such as → gibberellins and → abscisic acid, are terpene derivatives; the former are diterpenes and the latter a sesquiterpene.

terpenoids, a group of water-insoluble isoprenoid lipids of plants, either aliphatic or cyclic compounds with five to some hundreds of carbon atoms.

terrestrial (L. *terra* = earth), **1.** pertaining to the earth, land (terrestrial origin), consisting of soil; not → aquatic, or aerial; **2.** describing an organism which lives on land, in the soil, or on the ground.

terricolous, also **terricoline** (L. *colere* = populate), pertaining to an organism which lives on land, or grows on the ground.

terrigenous (Gr. *gennan* = to produce), produced by the earth; pertaining to sedimented material, such as dead organisms or deposits formed by the erosive action of moving water, as e.g. rivers and tides; especially refers to the sediments on the sea bottom.

territorial (L. *territorium* = domain), pertaining to an animal having a → territory.

territory, an area of the home range that an animal or animal group defends against other individuals or groups of the same species; e.g. the breeding territory is usually defended by a male against other conspecific males. The main purpose of the territory is to assure sufficient resources to owners of the territory. According to the use of territories, they are named e.g. **individual, breeding, feeding, roosting,** and **wintering territories.**

tertiaries, tertiary feathers; the flight feathers attached to the humerus in bird wings. *Cf.* primaries, secondaries.

Tertiary (L. *tertius* = third), the earlier of the two periods of the Caenozoic era 66 to 1.6 million years ago (the later is the Quaternary period); includes five epochs, → **Palaeocene, Eocene, Oligocene, Miocene,** and **Pliocene.** The climate was turning cooler, especially at the end of Miocene, leading to a decline of tropical and subtropical vegetation. Tertiary has been called the period of mammals, owing to their rapid speciation (*see* adaptive radiation); also modern bird species appeared during this period. Seed plants developed to the present status but the gymnosperms declined. *Adj.* **Tertiary.**

tertiary butanol, *see* butanol.

tertiary sex ratio, *see* sex ratio.

tertiary structure, the third organization level of macromolecules (e.g. folding and arrangement of helices), particularly in → proteins, to give a three-dimensional steric structure.

TES, N-tris-(hydroxymethyl)-methyl-2-aminoethanesulphonic acid; used as a → buffer substance in biochemical solutions.

test, 1. (L. *testum* = crucible, earthen vessel), an examination; a means of trial; the method for assessing; the process of determining the nature of a substance; **2.** (L. *testa* = shell), the protective covering of certain invertebrates, such as some protozoans, echinoderms, and tunicates.

test cross, crossing of a double or multiple heterozygous individual with the corresponding recessive homozygote to study what kinds of gametes, and in which proportion, are produced by this heterozygous individual.

testis, pl. **testes** (L = witness, testis), **testicle;** the male gonad (reproductive gland) in animals producing spermatozoa, in vertebrates also sex hormones. In mammals, the testes are situated generally in the scrotum, but in other vertebrates in the coelom. The testes are surrounded by two sheaths, the outer serous mesothelial sheath, *tunica vaginalis testis,* derived from the peritoneum with visceral and parietal laminae, and the white tissue layer, *tunica albuginea testis*, immediately covering the testis beneath the visceral laminae of the tunica vaginalis. Connective tissue of the latter extends inside the testis and divides it into several compartments, **lobuli testis**, filled with convoluted **seminiferous tubules** with different stages of generative cells maturing in the process of → spermatogenesis. The spermatozoa move from the tubules through the **rete testis** to the → epididymis and then into the deferent duct (*ductus deferens*, *vas deferens*). In addition to the generative cells the testes comprise supporting cells (→ **Sertoli cells**) and endocrine cells (→ **Leydig cells**).

testosterone, one of the sex hormones of vertebrates, chemically a steroid, 17-β-hydroxyandrost-4-en-3-one, $C_{19}H_{28}O_2$. It is the most potent androgenic hormone produced by → **Leydig cells** of the testes. Secretion of testosterone is stimulated by the luteinizing hormone (LH) produced by the anterior portion of the pituitary gland. Testosterone induces anabolic processes and is responsible for the development of the male sexual organs, spermatogenesis, and affects the secretion of gonadotropins. In peripheral tissues testosterone is converted to **dihydrotestosterone** which is mainly responsible for other male characteristics. Testosterone, like many other **anabolic steroids**, also has anabolic influence on bone and muscle tissues.

testudinal (L. *testudo* = tortoise, turtle), pertaining to, or resembling a tortoise or tortoise shell. *Syn.* testudinarious.

Testudines, Testudinata, → Chelonia.

tetanization (Gr. *teinein* = to stretch), the repeated stimulation of a muscle with continuously increasing frequencies to the state in which successive contractions summate into one large contraction, called summation of contractions (*see* tetanus). *Verb* to **tetanize**.

tetanus (L. < Gr. *tetanos* = rigid, stretched), **1.** summation of contraction; physiological tetanus; a sustained contraction of a skeletal muscle or its motor units caused by repetitive impulses from motor nerves at frequencies so high that individual contractions are summated into one, long-lasting and strong contraction; called also **tetanic contraction** or → tetany; it plays a part in normal muscular work; **2.** pathological tetanus; an infectious disease manifested by spasmic contractions and hyperreflexia due to a toxin (tetanus toxin) produced by the anaerobic bacillus *Clostridium tetani*.

tetany (L. *tetania*), **1.** hyperexcitability of skeletal muscles and nerves due to lowered extracellular Ca^{2+} concentration; **2.** → tetanus, defin. 1.

Tethys Sea (*Tethys* = the genus of the seahare), the sea during Mesozoic about 200 million years ago between the → Gondwana and Laurasia continents. *See* continental drift theory.

tetr(a)- (Gr. *tetras* = four), denoting four.

tetracycline, 1. a yellow broad-spectrum antibiotic, $C_{22}H_{24}N_2O_8$, produced by *Streptomyces viridifacies*; **2.** broadly, any derivative of tetracycline obtained from the genus *Streptomyces*, or prepared synthetically; inhibits the growth of bacteria, protozoans and some viruses. Tetracyclines can cause osteogenetic damage in the foetuses of vertebrates.

tetrad (Gr. *tetras*, gen. *tetrados* = four), a fourpartite (four-sided) structure; e.g. **1.** the four chromatids of a → bivalent observed in the first meiotic division; **2.** the quartet of haploid cells generated in → meiosis. In certain organisms such as some fungi, musci, and algae, all the four cells formed in a single meiosis can be found and analysed genetically (*see* tetrad analysis); **3.** pollen tetrad; during the development of → pollen grains in the anther of a stamen the first meiosis of the pollen mother cell occurs, resulting in the formation of tetrads, from which separate haploid pollen grains develop; **4.** in chemistry, a quadrivalent element.

tetrad analysis, a genetical analysis dealing with all the four cells (or their derivatives) produced by a single → meiosis in cell division. It is possible to use tetrad analysis for organisms, in which these four cells remain together as a group, → tetrad. The term is also used for the analysis of a meiotic stage, pachytene bivalent, to find out how the strands of the bivalent have participated in → crossing-over.

tetradecanoic acid, the systematic name of → myristic acid $CH_3(CH_2)_{12}COO$.

tetraethylammonium, TEA, $(C_2H_5)_4N^+$, a toxic substance that acts as an inhibitor for the function of potassium channels of cell membranes (*see* ion channels), e.g. having a strong effect on the ganglions of the → autonomic nervous system of vertebrates; chlorides and bromides of TEA are used in medicine.

tetraploid (Gr. *tetraploos* = fourfold), an → autopolyploid or → allopolyploid cell, tissue, or organism, which bears four sets of chromosomes (symbol 4x).

Tetrapoda (Gr. *pous*, gen. *podos* = foot), **tetrapods**, vertebrates having four limbs; include amphibians, reptiles, birds, and mammals. The skeletal structure of the limb in all tetrapods derives basically from a five-digit model. *Adj.* **tetrapodic** (quadrupedal).

tetrarch (Gr. *arche* = origin, beginning), pertaining to a root structure in dicotyledons with four-lobed primary xylem (in cross-section) in the central cylinder (stele).

tetrasaccharide (L. *saccharum* = sugar), any of a class of carbohydrates composed of four monosaccharide units, e.g. → stachyose in plants.

tetrasomic (Gr. *soma* = body), pertaining to a → polysomic cell, tissue, or individual, in which one or a few of the chromosomes occur fourfold instead of the normal twofold; the tetrasomic chromosome number is 2n+2 (→ n).

tetrasporangium (Gr. *sporos* = seed, *angeion* = vessel), a type of sporangia in → red algae forming four haploid tetraspores.

tetraspore, a spore type in red algae. *See* spore.

tetrodotoxin, TTX (*Tetraodontidae* = a fish family, Gr. *toxikon* = arrow poison), one of the strongest toxins produced by animals; a selective blocker of sodium channels of the cell membrane (*see* ion channels). It is produced in about 50 fish species, some amphibians and some cephalopods. In Japan, certain restaurants have a special meal made from *fugu* fish (*Fugu exacrum*, puffer-fish) with traces of tetrodotoxin still present, producing a kind of strange mental experience to a brave customer. In nerve and muscle research tetrodotoxin is an important tool.

tetrose, any monosaccharide with four carbon atoms; the general formula is $(CH_2O)_4$, e.g. → erythrose.

TF, → transcription factor.

TGF → transforming growth factor.

TGMV, → tomato golden mosaic virus.

thalamus, pl. **thalami** (L. < Gr. *thalamos* = room), the mediolateral part of the diencephalon of the vertebrate brain forming the lateral walls of the third ventricle; acts especially as a relay area for ascending sensory neurones.

thalassaemia (Gr. *thalassa* = sea, *haima* = blood), a group of hereditary human diseases (haemolytic anaemias) with a decreased rate of synthesis of one or more haemoglobin polypeptide chains (α, β, or δ chains); occurs especially in Mediterranean countries.

thale cress, *Arabidopsis thaliana*, a cruciferous species that has a short life cycle and small genome; it is the most important seed plant in basic genetic research.

Thaliacea (Gr. *thalia* = luxuriance), **thaliaceans, salps**; a class in the subphylum Tunicata, phylum Chordata, comprising free-swimming animals in pelagic oceans. The barrel-shaped body is surrounded by bands of circular muscles, and adults are lacking the notochord and tail; the → alternation of generations is typical in reproduction. Thaliaceans are divided into three orders: Pyrosomida, Doliolida, and Salpida.

thallophyte (Gr. *thallos* = stem, *phyton* = plant), a member of the Plant kingdom having an undifferentiated → thallus. The group includes → algae and → mosses.

thallus, pl. **thalli,** Am. also **thalluses,** a vegetational part of a plant having a structure which is not divided into roots, stem, and leaves, and has no differentiated tissues; e.g. algae have typical thalli. The structure is multicellular but less developed than in vascular plants. *See* thallophyte.

THAM, → tris-(hydroxymethyl)-aminomethane.

theca, pl. **thecae** (Gr. *theke* = box), a protective, covering layer enveloping various biological structures such as organs and organisms; e.g. **1.** → theca folliculi around the vertebrate ovum; **2.** the horny covering of an insect pupa; **3.** a bud of a graptolite; **4.** fungal ascus; **5.** a half of the anther with two pollen sacs.

theca folliculi, an envelope around the → ovarian follicle of vertebrates comprising an external layer (*theca externa)* of fibrous connective tissue, and an internal layer (*theca interna*) of vascular loose connective tissue with ample endocrine cells (thecal cells), which produce female sex hormones.

thecal cells, → theca folliculi.

Thecodontia (Gr. *odous* = tooth), **thecodonts**; an extinct order of archosaurian reptiles who dominated in Triassic; mostly large, carnivorous animals with the teeth in sockets and the limbs with bipedal tendency.

thel(e)- (Gr. *thele* = nipple, teat), pertaining to the nipple or a nipple-like structure.

thelytoky (Gr. *thelys* = female, *-tokos* < *tiktein* = to bear), a type of → parthenogenesis in which female offspring are produced from unfertilized eggs.

theobromine (Gr. *theos* = god, *broma* = food), an alkaloid of the seeds of the cacao tree, *Theobroma cacao*, 3,7-dimethylxanthine, $C_7H_8N_4O_2$; chemical name 3,7-dihydro-3,7-dimethyl-1H-purine-2,6-dione. The physiological effects on animals resemble those of caffeine having a stimulatory effect on the sympathetic nervous system of vertebrates. The alkaloid is used in medicine e.g. as a diuretic, a smooth muscle relaxant, and cardiac stimulant.

theophylline (Gr. *phyllon* = leaf), a basic bitter compound, 1,3-dimethylxanthine, $C_7H_8N_4O_2$, from tea leaves, isomeric with → theobromine; pharmacological effects resemble those of theobromine and → caffeine.

Therapsida (Gr. *ther* = wild beast, *apsis* = arch), **therapsids**; an order of reptiles succeeding pelycosaurs; lived from Permian to Jurassic. Therapsids were the ancestors of mammals and they radiated into several herbivorous and carnivorous lineages. They had upright limbs and thus an efficient erect gait.

therapy, therapeutics (Gr. *therapeia* = attendance), the treatment of a disease.

Theria (Gr. *therion* = wild animal), **therians**; a mammalian subclass that comprises two living infraclasses Metatheria (marsupials) and Eutheria (placental mammals). In some classifications also the infraclass → Ornithodelphia is included in therians.

therm(o)- (Gr. *therme* = heat, *thermos* = hot), pertaining to heat or temperature.

thermal, 1. relating to heat or temperature; *syn.* thermic; **2.** as a noun, a rising body of warm air.

thermal sense, → temperature sense.

thermal sum, a term indicating the ecological significance of temperature in plant development during the growing season; especially trees develop under the control of the thermal sum. It is expressed as **degree days** (d.d.) that is the sum of mean temperatures exceeding a threshold value, e.g. 5°C. The thermal sum is also used in studies dealing with successful development and survival of poikilothermic animals.

thermic, thermal, relating to heat or temperature.

thermocline (Gr. *klinein* = to incline), **1.** a temperature gradient; **2.** a layer in lakes and seas where water temperature decreases rapidly, as e.g. in northern areas in summer. The warm oxygen-rich zone above the thermocline is called → **epilimnion** and the cool oxygen-poor zone, **hypolimnion**. *See* stratification.

thermodynamics (Gr. *dynamis* = force, power), a branch of physics dealing with the relations between heat and other energy types and the conversion of one into the other. The heat content of a thermodynamic system is called **enthalpy**. Systems are open, closed, or isolated depending on whether it is possible to transfer energy or matter from one system to another. All biological systems are open and can accept external matter and energy. The first **law of thermodynamics** confirms that energy can be neither created nor destroyed (the indestructibility of energy). The second law concerns the transferred flow of heat energy that is always in the direction from a higher temperature to a lower; the impairment of energy is → **entropy**. According to the third law, all processes become slower and weaker when approaching the absolute zero of temperature, and absolute zero can never be reached.

thermogenesis, heat generation in organisms; in endothermic animals includes non-shivering and shivering thermogenesis. *See* thermoregulation.

thermogenin, *see* brown fat.

thermonasty (Gr. *nastos* = firm, solid, pressed), a → nastic movement in a plant caused by temperature changes, e.g. the movements of floral leaves. *Adj.* **thermonastic**.

thermoneutral zone, an environmental temperature range for a homoiothermic animal within which the animal does not use induced metabolic heat for the regulation of its body temperature, but heat loss is in balance with basic heat production. In the thermoneutral environment only the **basic metabolism** of a mammal or bird is functioning and can be measured experimentally. For arctic animals the thermoneutral zone is wide (eg. for the arctic fox in winter -30 to +20°C), but for tropical animals or animals of small size, it is narrow (for urban people +28 to 31°C, for mice +28 to 32°C). The upper limit of the thermoneutral zone is called **upper critical temperature**, and the lower limit, **lower critical temperature**.

thermophil(e) (Gr. *philein* = to love), an organism favouring warmth, growing best at high temperature, as e.g. **thermophilic** bacteria.

thermophobic (Gr. *phobos* = fear, flight), avoiding warmth; pertaining to an organism that favours low temperatures.

thermoplegia (Gr. *plege* = stroke) sunstroke, → heat stroke.

thermoreceptors, *see* temperature sense.

thermoregulation, the regulation of body temperature of **homoiothermic** animals (birds and mammals) at a near constant level. As the environmental temperature varies, the regulatory mechanisms maintain the body temperature of mammals at 35 to 37°C and of birds at 39 to 42°C. In thermoregulation the balance between heat production and heat loss is due to the function of **chemical thermoregulation** (i.e. increase or decrease of metabolism), and of **physical thermoregulation** (i.e. changes in body insulation associated with the control of heat loss). These mechanisms involve movements of feathers or hairs, and vasomotor control, sweating, panting, or → wallowing. Chemical thermoregulation has two mechanisms: **shivering thermogenesis**, i.e. shivering of skeletal muscles, and **non-shivering thermogenesis**, which is based on hormonal functions, in mammals mainly on thermogenesis in the → brown fat. The thermoregulatory mechanisms are controlled by the **thermoregulatory centre** of the hypothalamus which has receptors for measuring blood temperature, but the centre also receives information from peripheral receptors. Quick regulatory processes are transmitted to the effectors through motor nerves and the autonomic nervous system, but slower hormonal processes become active in long-term physiological temperature adaptation, called → acclimatization. The physical and chemical thermoregulation of homoiothermic animals is improved by **behavioural thermoregulation**. Animals seek shelter against unfavourable temperatures by digging pits, building nests or taking water or mud baths. Animals living in a cold environment also display **social thermoregulation**. They aggregate in dense groups for reducing heat loss; known examples are the musk-ox in the arctic zone in winter as well as many northern birds especially at night.

Many poikilothermic animals can regulate their body temperature by choosing an environment which is closest to optimal temperature, e.g. basking in sunshine; i.e. these animals have **behavioural** and **social thermoregulation**. Also some physiological thermoregulatory processes are found in some poikilotherms, as in some insect, fish, and reptile species; these animals are called **poikilothermic endotherms**. *See* endothermy.

thermostat (Gr. *statos* = staying), **1.** a device that maintains a desired constant temperature; **2.** hypothalamic thermostat, the control system of body temperature in the hypothalamic thermoregulatory centre of homoiothermic vertebrates, i.e. birds and mammals. *See* thermoregulation.

thermotaxis (Gr. *taxis* = arrangement), an active movement of an organism (e.g. bacterium) in which a temperature gradient acts as the directive factor for the movement. *See* taxis. *Adj.* **thermotactic**.

therophyte (Gr. *theros* = summer, *phyton* = plant), an annual plant that has no winter buds and overwinters as a seed. *See* life form.

thi(o)- (Gr. *theion* = sulphur), a prefix pertaining to sulphur.

thiamin(e), 3-((4-amino-2-methyl-5-pyrimidinyl)-methyl)-5-(2-hydroxyethyl)-4-methylthiazolium; thiamine hydrochloride is vitamin B_1, the active form of which is **thiamine pyrophosphate**. It is synthesized by plants and is found abundantly e.g. in vegetables, beans, corn, and brown rice. *Syn.* aneurin. *See* vitamin B complex.

thigmonasty (Gr. *thigma* = touch, *nastos* = firm, solid, pressed), a → nastic movement of a plant caused by touch. *Adj.* **thigmonastic**.

thigmotaxis (Gr. *taxis* = arrangement), a → taxis caused by touch. *Adj.* **thigmotactic**.

thigmotropism (Gr. *tropos* = turn), a → tropism in which contact serves as the orientating factor; i.e. a turning, bending or growth movement of an organism caused by touch, e.g. the winding of a → tendril around another plant.

thin-layer chromatography, TLC, *see* chromatography.

thin section, an ultrathin section made from a biological material using an ultramicrotome (*see* microtome); sections are 50 to 80 nm in thickness (silver or grey in colour), or sometimes even thinner (down to 10 nm), and they are examined using an electron microscope.

thiocarbamates (Gr. *theion* = sulphur), derivatives of → thiourea (thiocarbamide) used as herbicides; prevent the development and growth of weed seedlings and inhibit photosynthesis. Common thiocarbamate herbicides are e.g. ethyldipropylthiocarbamate (EPTC), diallate, triallate, propham, and chlorpropham. Thiocarbamates are also used to control fungal diseases of plants.

thiocyanate (Gr. *kyanos* = blue), an anion $S{=}C{=}N^-$, salt, or ester of thiocyanic acid; formed in organisms in cysteine metabolism, in animals also in detoxification of cyanides.

thioglycoside (Gr. *glykys* = sweet), a sulphur-containing organic compound that consists of a sugar and aglyconic moiety bound together by a glycosidic bond; thioglycosides containing glucose are also called thioglucosides or glucosinolates. The hydrolysis of thioglucosides is catalysed by thioglucosidases (*see* myrosinase). *See* mustard oils.

thiourea (Gr. *ouron* = urine), thiocarbamide; a synthetic compound, $CS(NH_2)_2$, obtained from urea by replacing the oxygen atom by sulphur; inhibits the function of the thyroid gland; in plants promotes seed germination. Its cyclic derivative, **thiouracil**, is used as an antithyroid drug.

thirst centre, an area in the hypothalamus of vertebrates that controls water intake. Specialized neurones, **osmoreceptors**, located in this area respond to osmotic concentration, especially to an increased Na^+ ion concentration of the plasma. According to the water-balance state of the body, the thirst centre controls thirst and the release of → vasopressin from the posterior pituitary gland (neurohypophysis).

thoracic, thoracal (Gr. *thorax* = chest), pertaining to the chest.

thoracic cavity (L. *cavitas thoracis*), pectoral cavity; the anterior portion of the body cavity of mammals; the respiratory diaphragm separates it from the abdominal cavity.

thoracic duct, *see* lymphatic vessels.

thoracopods, thoracic legs (thoracic appendages) of → malacostracans, such as crabs and lobsters, usually bearing gills.

thorac(ci)otropic hormone, TTH, → prothorac(ic)otropic hormone.

thorax, pl. **thoraces** (Gr.), **1.** the chest of vertebrates; the anterior part of the trunk between the neck and abdomen, supported by the ribs (rib cage); **2.** the middle part of the body of arthropods; in crustaceans and arachnids, the thorax is fused with the head (cephalus) to form the cephalothorax. The thorax of insects is divided into prothorax, mesothorax and metathorax, each possessing a pair of legs.

thorn, 1. a sharp spinose structure on an animal, such as the spines of a sea urchin; **2.** a plant structure, which protects plants from herbivorous animals. Thorns may be formed by epidermal tissue alone, or they may have connections with other tissues situated below the epidermis (e.g. rose thorns). Thorns may also be modified organs, such as leaves (in cacti), branches or short shoots (in hawthorn), or stipules (in *Euphorbia* species).

thorn root, a plant tuber formed in the shoot, the adventitious roots of which have been modified into thorns. Thorn roots occur in tropical epiphyte plants; best known is *Myrmecodia* with a hollow thorn root inhabited by ants.

thorny-headed worms, spiny-headed worms, → Acanthocephala.

threatened species, a species which, as a consequence of direct or indirect human activity, has become so rare, that it risks to disappear completely from a certain area, or to become extinct. The danger may be caused also by natural factors independent of man. The World Nature Conservation Strategy, formulated by International Nature Conservation Association, proposes to conserve the genetic diversity of threatened species in the world. National registers of threatened species have been published in many countries; these are called **red books**, in which the species have been grouped into four categories: **extinct, endangered, vulnerable**, and **watched**.

threat signals, signals of animals in situations where a threatening animal tries to drive away another individual, or cause it to submit without fight. The purpose of the signals is to make the threatener's appearance big and frightening. In vertebrates, threat signals are shown to be associated with physiological changes, such as the activation of the autonomic nervous system, the increase of adrenal secretion, and faster and deeper respiration; hairs of mammals and feathers of birds become erect, the thorax expands, eyes stare and the pupils dilate. Many amphibians, reptiles and some birds threaten by inflating air into air sacs around the face, and so render themselves larger. *Cf.* appeasement behaviour.

threonine, abbr. Thr, thr, symbol T; an amino acid, $CH_3CHOHCH(NH_2)COOH$, found in proteins; one of the → essential amino acids in animals.

threose, a monosaccharide, tetrose, $C_4H_8O_4$; a four-carbon monosaccharide formed in cellular biosynthesis from glyceraldehyde; has two central hydroxyl groups located in *trans* orientation (*cf.* erythrose).

threshold potential, → membrane potential in cells, especially nerve and muscle cells, just at the level where a response in the form of → action potential is generated.

thrombin (Gr. *thrombos* = clot), a proteolytic enzyme found in the vertebrate blood converting fibrinogen to fibrin in the last process of → blood coagulation. Also called fibrinogenase or thrombase.

thrombocyte (Gr. *kytos* = hollow, cell), → platelet.

thromboxanes, TX (Gr. *axane* = ring), hormone-like compounds derived from prostaglandin endoperoxides by the action of thromboxane synthase in vertebrates; they are derivatives of fatty acids belonging to a larger group of → eicosanoids. Best known is thromboxane A_2 (TXA_2), which when released from platelets induces their aggregation and causes constriction of blood vessels, thus acting as an antagonist to prostacyclin. *See* prostaglandin.

thylakoid (Gr. *thylakos* = pocket), a membrane structure in plant cell → chloroplasts.

thym(i)-, thymo-, 1. (Gr. *thymos* = thymus), pertaining to thymus; **2.** (Gr. *thymos* = the mind as the place of strong feelings), denoting relationship to the mind, emotions, or soul.

thymidine, a → nucleoside composed of thymine and deoxyribose; occurs in DNA molecules.

thymine, 5-methyluracil; a pyrimidine base present in deoxyribonucleic acid, DNA.

thymocyte (Gr. *kytos* = hollow, cell), a cell type in the cortex area of the → thymus in vertebrates; thymocytes are probably derived from stem cells of the foetal bone marrow and liver, and develop into T lymphocytes (T cells).

thymol, a phenol derivative having an aromatic odour and antiseptic properties; found e.g. in thyme oil, and is also made synthetically. It is used as a fungicide and preservative, and in cosmetics. Its derivative, **thymol blue** (sulphonaphthalein), is used as an acid-base indicator.

thymosin, any of several polypeptide hormones secreted by thymic epithelial cells in vertebrates; induces the maturation of T lymphocytes (T cells) in the thymus; the most active is thymosin α_1.

thymus, pl. **thymi,** also **thymuses, 1.** (L. < Gr. *thymos* = warty excrescence, sweetbread), **thymus gland**; an organ of the immune system of vertebrates located ventrally in the anterior part of the thorax, in higher vertebrates in the mediastinum of the thoracic cavity between the heart and the breast bone. It is paired, except in mammals, in which it develops secondarily unpaired. The thymus derives as a bud from the third or fourth gill arch, and during embryonic development it moves backwards from the pharyngeal area. It is most active in immature animals and is partly degenerated later in adulthood. Structurally the thymus resembles a lymph gland consisting of lymphocytes and their precursors as the main cell type. Apparently the thymus is the only organ, where T lymphocytes (T cells) are differentiated to serve in the cell-mediated immunity in the body of vertebrates (*see* immunity); **2.** thyme, any menthaceous plant of the genus *Thymus* (Gr. *thymos*) in the family Lamiaceae; the plant contains thyme oil used as a spice.

thyr(o)-, thyreo- (Gr. *thyreos* = shield), denoting relationship to the thyroid gland.

thyrocalcitonin, *see* calcitonin.

thyroglobulin, 1. a protein secreted by the follicle cells in the thyroid gland of vertebrates, usually stored in the colloid of the glandular follicles. It is involved in the synthesis of thyroid hormones with iodination of tyrosine moieties of this protein. Monoiodotyrosines and diiodotyrosines are first formed, two of which then combine to form triiodotyronine and thyroxine (tetraiodotyronine). Secretion of the thyroid hormones (triiodotyronine and thyroxine) needs proteolytic degradation of thyroglobulin; thyroglobulin is sometimes also called thyreoglobulin or iodoglobulin; **2.** a preparation obtained by the fractionation of thyroid glands of domesticated animals; used earlier in medicine.

thyroid (Gr. *thyreoides* < *thyreos* + *eidos* = form), **1.** → thyroid gland; **2.** pertaining to the thyroid gland or thyroid cartilage; **3.** a preparation made from the thyroid glands of domesticated animals to be used in medicine.

thyroid cartilage (L. *cartilago thyr(e)oidea*), the largest, shield-like cartilage of the → larynx.

thyroid gland (L. *glandula thyroidea*), an endocrine gland of vertebrates derived embryonically from the endoderm of the branchial floor; in the adult, it is situated ventrally in the pharyngeal area or in the lower part of the front neck. The thyroid gland is either single with two lobes connected by a bridge (isthmus) as in mammals and reptiles, or paired as in birds and most amphibians. In fish the thyroid tissue is dispersed in the pharyngeal area. The hypobranchial groove of the larvae of cyclostomes is homologous to the thyroid gland. The thyroid gland is made up of multiple small **follicles** (acini) each surrounded by a single layer of cells secreting proteinaceous material (**colloid**) with → thyroid hormones in the follicles. According to the hormonal condition in the body, the hormones are then transferred from the follicles into blood capillaries around them. In mammals, single longitudinal endocrine cells (C cells) are found in the tissue between follicles. They secrete a hormone, called **thyrocalcitonin**, directly into the blood (*see* calcitonin).

thyroid hormones, the vertebrate hormones secreted by the → thyroid gland; two active hormones, **thyroxine (T_4)** and **triiodothyronine (T_3)**, the latter being more active in tissues. The hormones are iodine-containing amino acids derived from tyrosine. In the thyroid follicles they are bound to → thyroglobulin, but secreted as free hormones into the blood circulation, where they are bound largely to albumin and thyroxine-binding globulin (TBG). Hormone synthesis and secretion in the thyroid gland is induced by a pituitary hormone, → **thyrotropin**, which is further controlled by → **thyroliberin**. In mammals, thyroid hormones activate cell metabolism in most tissues, especially in a cold environment (calorigenic action). In all vertebrates they regulate tissue qrowth, especially in the nervous system, and improve the absorption of carbohydrates from the intestine. The deficiency of thyroid hormones (**hypothyroidism** or hypothyreosis) during the ontogenic development results in disturbances especially in the nervous system (cretinism in humans); in adulthood the deficiency causes a complicated syndrome, called myxoedema, with decreased metabolism and skin changes. In frogs the hypothyroidal state inhibits metamorphosis. **Hyperthyroidism** (hyperthyreosis) presents itself in higher vertebrates as increased metabolism, nervousness, hyperphagia, weight loss, increased heart rate and blood pressure, sweating, etc. The mammalian thyroid gland also secretes the third hormone, called → **calcitonin** (thyrocalcitonin), which traditionally is not included in the term thyroid hormones.

thyroid-stimulating hormone, TSH, → thyrotropin.

thyroliberin, a tripeptide hormone of vertebrates secreted from the hypothalamus of the brain; stimulates the release of → thyrotropin and prolactin from the anterior → pituitary gland. *Syn.* thyrotropin-releasing hormone, TRH.

thyrotropin (Gr. *trepein* = to turn), a glycoprotein hormone of vertebrates secreted from the anterior lobe of the → pituitary gland. It activates the production and secretion of thyroxine and triiodotyronine hormones in the → thyroid gland; secretion is controlled by → thyroliberin. *Syn.* thyrotropic hormone, thyroid-stimulating hormone, **TSH**, formerly also thyrotrophic hormone.

thyrotropin-releasing hormone, TRH, → thyroliberin.

thyroxine, also **thyroxin,** tetraiodotyronine (T_4). *See* thyroid hormones.

TIBA, → triiodobenzoic acid.

tibia, pl. **tibiae,** or **tibias** (L.), **1.** the shin bone; the inner and usually larger bone of the hindlimb below the knee in tetrapod vertebrates; together with the fibula, forms the second segment of the hindlimb; **2.** the segment in the legs between the femur and tarsus in insects, and between the patella and tarsus in arachnids.

ticks, → Acari.

tide, the rising and falling of the surface of an ocean and some other water bodies connected to it, caused by the attraction of the moon and to a lesser extent of the sun.

Tiedemann's vesicles (*F. Tiedemann*, 1781—1861), small bodies attached to the ring canal of the → water-vascular system in echinoderms where the → coelomocytes develop.

tight junction, occluding junction; the intercellular junction in which the membranes of two adjacent cells are connected tightly together by integral protein strands produced by these cells; tight junctions are found especially between epithelial cells forming a seal around the cell neck, which is impermeable to the intercellular passage of most molecules, water molecules included.

tillering, the process of forming side shoots (tillers) from lateral buds of grasses.

TIM, → triose phosphate isomerase.

timbal, tymbal (Fr. < Ar. *tabl* = drum), the vibrating, chitinous, abdominal membrane of the shrilling organ in cicadas, producing special sounds.

timberline, the upper limit of arboreal growth in mountains and high latitudes (subarctic or subantarctric areas). *Cf.* tree line.

time budget, a term in population ecology pertaining to time used by individuals in various activities; concerns the allocation of the use of resources in populations.

time delay in a population model, a model of population growth in which the net reproductive rate of individuals at a particular time is determined by the population size some time earlier. Using the model, cyclic variations in population densities are obtained.

time lag, an interval of time between two phenomena related to each other; e.g. in physiology the time between stimulation and the following reaction.

tin (L. *stannum*), a metallic element, symbol Sn, atomic weight 118.69; inorganic tin salts (**stannates**) are observed to be necessary for the normal growth of mammals, and tin is now regarded as one of the essential → trace elements.

Tinamiformes (L. *Tinamus* = the type genus, *forma* = form), **tinamous**; a bird order living in South America and Mexico; comprises 46 species classified into one family, Tinamidae. They resemble small gallinaceous birds and are related to ratites.

tip growth, a type of cell growth occurring in the tip of a thread-like structure; typical of algae, protonemata of mosses, and fungal hyphae.

tissue (Fr. *tissu* = woven), **animal tissue**, an aggregate of similarly differentiated cells and cell products (intercellular substance, matrix) which form the structural material of an animal body and serve a special function, e.g. the → epithelium, connective, muscle, and nerve tissues; **plant tissues** are analogous, similarly differentiated cell systems, of which plants are composed; different types are the → meristem, parenchyma, supporting, conducting, and dermal tissues, as well as excretory and secretory tissues.

tissue culture, the maintenance and growth of tissue samples from animals or plants *in vitro* in cell culture conditions; the term often includes organ culture. *See* cell tissue and organ culture.

tissue fluid, → interstitial fluid.

tissue hormones, local hormones in animals; the term sometimes used to denote hormones or substances with hormone-like properties which act at a district tissue site and are usually rapidly inactivated; e.g. many peptides regulating the function of the alimentary canal. However, the division of hormones into local and general hormones is difficult and arbitrary.

tissue respiration, the respiration of a given tissue; can be measured as oxygen consumption or production of carbon dioxide, and indicates the respiratory activity of the cells in that tissue. *See* cell respiration.

tissue transplantation, → transplantation.

titration (Fr. *titre* = standard), the determination of the quantity (the reactive capacity) of a susbstance in a solution (**titre**) by adding measured amounts of a liquid of a known concentration (**titrant**) until the stoichiometric endpoint is reached. In **neutralization titration** (acid-base titration), the endpoint is obtained as the colour change of an indicator stain or as a rapid change in electric potential. In **redox titration** (reduction-oxidation titration), a compound to be determined is oxidized or reduced from one oxidation state to another. In **precipitation titration** a given substance is precipitated with a titrant; e.g. **agglutination titration** is a method to determine the maximum dilution at which a sample (e.g. serum) gives a reaction in an immunological test with a standard preparation of an antigen or antibody.

Titriplex, → EDTA.

TLC, thin-layer chromatography. *See* chromatography.

T lymphocyte, *see* lymphocytes.

TMV, → tobacco mosaic virus.

TNF, → tumour necrosis factor.

toadstool, commonly used name for poisonous fleshy fungi, having no scientific basis. *Cf.* mushroom.

tobacco, pl. **tobaccos,** also **tobaccoes** (L. *tabacum*), **1.** any solanaceous plant of the genus *Nicotiana*, as *N. tabacum*, sometimes means also similar plants of other genera. Tobacco is cultured mostly for smoking, but the pith tissue of the tobacco stem is good material for → callus cultures; the tobacco plant is also widely used to produce → transgenic plants; **2.** the dried leaves of *N. tabacum* which are rich in → nicotine and are prepared for smoking, chewing, and snuffing.

tobacco mosaic virus, TMV, a simple rod-shaped RNA virus that causes a mosaic disease in tobacco and some other plants; used widely in biological and biochemical studies.

tocopherol (Gr. *tokos* = childbirth, *pherein* = to bear), → vitamin E.

tolerance (L. *tolerantia*), **1.** the act or capacity of enduring, as the ability to endure large doses of a toxin or drug; **2.** acquired drug tolerance, that is observed in an animal as a decreased response to repeated equivalent doses of a drug, e.g. an opiate; **3.** immunological tolerance, in which the immune system does not respond to an antigen; **4.** in ecology, the capacity of an organism to endure extreme conditions. **Shelford's law of tolerance** presents that the minimum and maximum limits of tolerance to different environmental factors determine the distribution of a species. A range of changes of an environmental factor, which an organism or a population of a species can stand, is called **tolerance range**. That part of the tolerance range in which an organism lives well is called **optimum range**. If a change of an environmental factor exceeds the tolerance of a species, the factor becomes limiting (*see* limiting factor), and the success of the species decreases. *Verb* to **tolerate**. *Adj.* **tolerant**.

toluene, a hydrocarbon, methylbenzene, $C_6H_5CH_3$, a colourless liquid obtained from resins and tar; widely used as an organic solvent in industry and in biochemical and chemical studies.

tomato golden mosaic virus, TGMV, a plant virus that contains a single-stranded DNA, causes a yellow mosaic disease in the leaves of a host plant.

Tömösvary's organ, a pair of organs at the base of the antennae of centipedes and some diplopods, probably detecting vibrations or chemical stimuli.

tone (Gr. *tonos*, L. *tonus* = tension, tone, pitch), the resistance to stretch; the normal degree of tension; **1.** in animal physiology, the slight continuous contraction state of a muscle at rest maintained by alternating contractions of small groups of muscle fibres; called also **tonus**; **2.** the quality of voice or sound.

tongue (Gr. *glossa*, L. *lingua*), an organ on the floor of the mouths of vertebrates acting in catching food, mastication and swallowing, cleansing and grooming, articulation of sound, and as the sense of taste. In mammals, the tongue is largely movable due to numerous skeletal muscle fibres oriented in different directions. The surface is covered by the mucosa formed from connective tissue and stratified epithelium with many papillae, mucous glands, and lymphatic tissue (lingual tonsil). In birds the tongue is mostly more cornified with a pointed end, in woodpeckers and hummingbirds forming a long protrusible organ for catching food. In snakes, the tongue is a forked organ with which the animal takes olfactory samples from the air and transfers them into an accessory olfactory organ called → Jacobson's organ. The tongue of amphibians is attached basally to the floor of the mouth, and the prehensile tongue can reach a prey from a long distance. The tongue of fish is just an immovable fold of the mucous membrane. There are also some tongue-like mouthparts in invertebrates which may be called the tongue, as *hypopharynx* and *ligula* in some insects and *radula* of molluscs.

tongue worms, → Pentastomida.

tonic (Gr. *tonikos*), **1.** pertaining to the normal → tone, or slight continuous tension in muscles, i.e. **tonicity**; **2.** pertaining to the quality of voice, sound, or speech; **3.** as a noun, a drug or any agent which increases the tone in the body.

tonoplast (Gr. *plastos* = moulded), the semipermeable protoplasmic membrane surrounding a vacuole in the plant cell; functions like the plasma membrane transporting ions, sugars etc. from the cytoplasm to the vacuole and vice versa.

tonsil (L. *tonsilla*), a rounded mass of animal tissue, especially of lymphoid tissue at the palatoglossal and palatopharyngeal area in the throat of mammals including the **palatine**, **pharyngeal** and **lingual tonsils**, which altogether form the tonsillar ring. They are covered by mucous membrane and contain many lymph follicles and crypts acting in defence against bacteria that invade the mouth.

tonus, → tone.

tooth (L. *dens,* pl. *dentes*), any of the hard bony structures in the jaws of most vertebrates, and/or on other bones in the walls of the mouth or pharynx as found e.g. in many fish species. The teeth are used for catching and killing prey, mastication and swallowing of food, and for offence or defence. The teeth of fishes, amphibians and reptiles are **homodont** (similar to each other) and capable of continuous reneval. The teeth of mammals are **heterodont** being differentiated into **incisors**, **canines** (cuspids), **premolars**, and **molar teeth** (the last two named **cheek teeth**). These may be permanent, or renewable as e.g. the 20 temporary **deciduous teeth** in man. The incisors of rodents are rootless, growing continuously, and the canine teeth of carnivores are very large; in elephants, the tusks are upper incisors which grow continuously. The number and quality of teeth, particularly the cheek teeth vary widely according to the food used: the cusps of cheek teeth are knobby in omnivores (as man and swine), high crowned with enamel ridges in herbivores, and sharp-pointed in insectivores. The cheek teeth of ruminants are separated by a space (**diastema**) from the incisors due to the lack of canines. In carnivores, one pair of molars are specialized into **shearing teeth**. Some vertebrate groups are lacking teeth totally, as lampreys, sturgeon, turtles, birds, monotremes, anteaters, pangolins, and whalebone whales.

Different types of vertebrate teeth are structurally quite similar consisting of a hard bony tissue called dentine (*dentinum* or *substantia eburnea dentis*), which is covered by a very hard inorganic substance, **enamel** (*substantia adamantina*), on the crown, and by bone-like connective tissue, **cementum**, on the root. The soft tissue, **dental pulp** (*pulpa dentis*), in the cavity inside the tooth is composed of richly innervated and vascularized loose connective tissue. The pulp cavity extends into the root (or roots) forming the **root canal** (or canals) through which the nerves and blood vessels enter the tooth. The teeth of lower vertebrates (fishes and most reptiles) have no cementum but the teeth are grown together with the bony or cartilaginous tissue of the jaw.

Also for invertebrates, the term tooth is used denoting some hard, horny, chitinous, and calcareous tooth-like structures around the mouth, as e.g. radula teeth of molluscs.

toothed whales, → Odontoceti.

tooth-like scale, → placoid scale.

tooth shells, → Scaphopoda.

topography (Gr. *topos* = place, *graphein* = to write), **1.** the description of a geographical place such as a tract of land, especially the configuration of a surface; **2.** the description of regional anatomical features, usually presented as a chart or illustration; also called topology.

topoisomerases, DNA topoisomerases, type I and II, EC 5.99.1.2-3; the enzymes that stepwise catalyse the breaking of DNA, the transfer of a DNA segment to this break, and the reseal of the DNA break; thus the enzymes alter the linking number of DNA, which is defined as the number of times one DNA strand winds around the other.

tornaria, pl. **tornariae,** Am. **tornarias** (L. *tornare* = to turn), a planktonic larva of some acorn worms (→ Enteropneusta) in the phylum Hemichordata; resembles the bipinnaria larva of echinoderms, thus indicating the evolutionary relationship between the chordates and echinoderms.

torpid (L. *torpidus* = sluggish, numb), **1.** slow, lethargic, inactive, sluggish, apathetic; **2.** dormant as a hibernating or aestivating animal.

torpor (L.), a suspended responsiveness to normal or ordinary stimuli; dormancy, as e.g. of a hibernating mammal.

tortoises, *see* Chelonia.

torus, pl. **tori** (L. = a rounded moulding or swelling, bulge), **1.** a protuberant part or rounded bony ridge in the anatomy of invertebrates or vertebrates; **2.** in plants, the thickened centre of the membrane which closes the pit cavity in → bordered pits in → tracheids; **3.** the apex of the → flower stalk bearing the flower parts; *syn.* receptacle.

totipotency (L. *totus* = whole, entire, all, *potentia* = ability, power, potency), the ability to generate the whole organism from a part, e.g. from a single cell or a piece of tissue. In animals, fertilized egg cells or early stages of blastomeres are **totipotent** and can develop into a complete individual. In many primitive animals also the cells of more developed embryos and even mature tissues, can form new individuals when separated from the body (*see* regeneration). All living plant cells are totipotent at least in theory, i.e. are capable of regenerating into new plants if cells are separated from their normal environment and cultivated in special media.

touch, *see* tactile sense.

tox(i)-, toxo- 1. (Gr. *toxikon* = arrow poison), denoting relationship to a poison or toxin; **2.** (Gr. *toxon* = bow, arrow), pertaining to bowed or arrow-shaped.

tox(a)emia (Gr. *haima* = blood), **1.** a condition resulting from the presence of bacteria or toxins in the bloodstream; **2.** a condition due to severe metabolic disturbances, as e.g. the toxaemia of pregnancy; **3.** a plant injury caused e.g. by an insect toxin.

toxication, poisoning.

toxicity, the quality or degree of being poisonous; **toxicant**, a poisonous agent; poisonous; **toxic**, poisonous; also manifesting the symptoms of a severe infection. *See* acute toxicity.

toxicology (Gr. *logos* = word, discourse), the scientific study dealing with poisons, their natural existence, detection, pharmacological effects, and the treatment of conditions produced by toxic agents. *Adj.* **toxicologic**.

toxin, any poisonous agent, especially a protein produced by living organisms, such as microbes, and some higher plants and animals; in a more restricted sense, a poisonous substance produced by pathogenic microbes (bacterial toxins).

toxoid, a toxin made non-toxic by a procedure which maintains its antigenic properties.

toxoplasmosis (Gr. *plasma* = anything formed), a widespread disease of mammals caused by a protozoan, *Toxoplasma gondii.* It is transmitted to humans especially from cats (the definitive host) by oocysts in the infected faeces or contaminated soil. In man, the oocysts settle in the brain, eyes, cardiac and skeletal muscles, liver and lungs, causing severe manifestations of disease in immunodeficient individuals but no symptoms in most individuals. The transplacental infection of the foetus results in severe brain and eye damage.

TPN, 1. triphosphopyridine nucleotide, better known by the name → nicotinamide adenine dinucleotide phosphate, **NADP**; **2.** total parental nutrition.

TPNH, reduced triphosphopyridine nucleotide, better known by the name → nicotinamide adenine dinucleotide phosphate, **NADPH**.

trabecula, pl. **trabeculae** or **trabeculas** (L. dim. of *trabs* = beam), in anatomy, a thin supporting strand of connective tissue; e.g. *trabeculae cordis*, muscular ridges in the inner ventricular walls of the vertebrate heart. *Adj.* **trabecular**.

trace elements, chemical elements needed by all organisms in very small amounts; in mammals their concentration is lower than 0.01% of the total body weight. Many trace elements are constituents of enzymes and vitamins acting as regulators of redox conditions in cell metabolism. In most organisms, important trace elements are arsenic, chromium, cobalt, copper, fluorine, iodine, iron, manganese, molybdenum, nickel, selenium, silicon, tin, vanadium, and zinc. These elements are used in ionic forms, formed e.g. in dissociation of metal salts. Trace elements in higher concentrations than needed, are toxic. Plants have some special requirements, such as manganese, chlorine and copper for photosynthetic reactions, molybdenum for nitrate reduction, zinc for auxin biosynthesis, boron for carbohydrate metabolism and iron for chlorophyll biosynthesis. *See* mineral, Hoagland's solution, micronutrients.

trachea, pl. **tracheae** (Gr. *tracheia* = rough), **1.** the windpipe of animals; the unpaired tubular organ of air-breathing vertebrates extending from the larynx and bifurcating to form the right and left main bronchi (in snakes usually only the right bronchus). The trachea is an air passage to and from the lungs. Its inner wall is lined by mucous membrane with ciliated epithelial cells; the submucosa beneath it consists of connective tissue supported by several horseshoe-shaped cartilages, smooth muscle, and lymph follicles. Amphibians have no trachea but the bronchi are branched immediately behind the larynx; **2.** one of the minute air-conveying tubes (called also **tracheal tubules**) of the respiratory system (**tracheal system**) of terrestrial and secondarily aquatic arthropods. The tracheal system usually has openings (**spiracles** or **stigmas**) covered by chitin brushes, situated segmentally on both sides of the body. The spiracles lead to paired, longitudinal main tubes, which branch into many smaller tubes (**tracheoles**), and further form the tracheal capillary system ending close to tissue cells. Except for the smallest capillaries, the walls of tracheae are supported by chitin circles. Animals adapted secondarily to aquatic life, have only one pair of spiracles at the end of the abdomen, or all spiracles are totally degenerated. In some species the tracheal system is developed to **tracheal gills**, in which respiratory gases are exchanged from water through the wall of tracheal capillaries inside the gill appendages; **3.** in plants, trachea is a water-conducting structure in → xylem, usually called **xylem vessel**.

tracheal system, *see* trachea.

tracheid, a water-conducting cell type in the → xylem of plants.

tracheid fibre, *see* libriform cell.

tractus, pl. **tractus** (L. = action of drawing, extension), a tract such as a bundle of nerve fibres which have the same origin, termination, and function.

trade off, generally denoting an even choice or chance; occurs when an increase in some system implies a decrease in some other; **physiological trade off**, occurs when two or more traits compete for materials and energy within a single organism; **microevolutionary trade off**, occurs in a population when the selection of one trait decreases the value of another trait. The trade off is closely connected with the → allocation hypothesis.

trailer (sequence), a rather long sequence of nucleotides at the 3' end of the → messenger RNA following the coding sequence and termination codon and preceding the poly-A end; the trailer sequence is not involved in translation.

trans- (L. *trans* = across, through), **1.** across, through, beyond, or on or to the other side; **2.** having one of the two mutant genes on each homologous chromosome; **3.** in chemistry, denoting the presence of certain radicals or atoms on opposite sides of the molecule. *Cf. cis-*.

transacetylases, → acetyltransferases.

trans-acting factor, a factor, usually a protein, involved in the regulation of gene function affecting the genes which lie either in *trans*-position or *cis*-position with the gene producing a protein. Such proteins bind to *cis*-acting sites of DNA.

transaldolase, a transferase enzyme (EC 2.2.1.2) which catalyses the transfer of a three-carbon unit (aldehyde) from a ketose to an aldose in → pentose phosphate pathway: sedoheptulose 7-phosphate + glyceraldehyde 3-phosphate —> erythrose 4-phosphate + fructose 6-phosphate. *Cf.* transketolase.

transamination, the transfer of an amino group (-NH_2) from an amino acid to an α-keto acid. The reaction is catalysed by aminotransferases (transaminases). Aspartate aminotransferase catalyses e.g. the transfer of the amino group of aspartate to α-ketoglutarate: aspartate + α-ketoglutarate —> oxaloacetate + glutamate.

trans configuration/position, *see* isomers.

transcortin (L. *cortex* = bark, rind), a protein (α-globulin) produced in the liver; specifically binds and transports cortisol in the blood of vertebrates. *Syn.* corticosteroid-binding globulin (CBG), cortisol-binding protein.

transcriptase, a DNA directed RNA polymerase, EC 2.7.7.6, i.e. an enzyme that catalyses RNA synthesis in → genetic transcription.

transcription, → genetic transcription.

transcription complex, a complex formed by interaction of multiple proteins (e.g. → transcription factors) with a DNA region within a gene, called the **internal control region**. The transcription complex is remarkably stable and may account, at least in part, for the control of the varying states of cells expressing one set of genes and repressing another set. Transcription complexes have been found for genes transcribed by all three classes of eukaryotic → RNA polymerases.

transcription factor, TF, any of the enzymes necessary for the initiation and/or prolongation of → genetic transcription. The transcription factors are classified into **general** and **specific** factors. The former are necessary in the transcription of all genes, the latter are specific to given genes. On the other hand, transcription factors are classified into **TF I**, **TF II** and **TF III** groups according to whether they promote the function of RNA polymerase I, II or III, respectively. *See* RNA polymerase.

transcutaneous (L. *cutis* = skin), entering through the skin, e.g. injection of material. *Syn.* percutaneous, transdermic.

transdermic (Gr. *derma* = skin), → transcutaneous.

transdetermination, a change in the determination state of the → imaginal discs of the fruit fly during their cultivation *in vivo*; leads to the development of some other structure from a part of the imaginal disc, than that which the original determination state would have produced. *See* determination.

transducer molecules (L. *transducere* = to lead across), a group of → guanine-nucleotide binding proteins; intermediate molecules, especially → G-protein, within the cell membrane transmitting a hormone-initiated signal from an externally facing hormone receptor to an internal enzyme. One of the molecules is **transducin** that serves in transducing a signal of light in rhodopsin to produce electric potential in retinal rods and cones.

transduction, in genetics, the process in which temperate or virulent → bacteriophages mediate the transfer of genetic material from one bacterial cell to another.

transfection (L. *trans* = on the other side, *facere* = to do), the infection of a cell with a → nucleic acid isolated from a virus; leads to the generation of complete viruses or virus particles in the cell.

transfer cells, plant cells which function in a short-distance transfer between cells and structures in plants, or between plants and the environment. Transfer cells may develop from many different tissues and cell types. Typical of them is the increase of the plasma membrane area and the formation of projections into other cells helping the transport. Transfer cells are important e.g. in association with → conducting tissues and in plant ovules functioning in the nourishment of developing embryos.

transfer efficiency, → ecological efficiency.

transferase (L. *ferre* = to carry), any enzyme of the class EC 2., catalysing the transfer of a chemical group from one compound (donor) to another (acceptor), e.g. aminotransferases (transaminases of EC sub-subclass 2.6.1.) which catalyse the transfer of amino groups, and sulphotransferases (EC sub-subclass 2.8.2.) catalysing the transfer of sulphate groups.

transferrin (L. *ferrum* = iron), a protein of β-globulin type which binds and transports iron in the blood of vertebrates, also called siderophilin. The iron-free form is often called **apotransferrin**, and that with iron, **ferrotransferrin**. *See* ferritin.

transfer RNA, tRNA, a short-stranded → ribonucleic acid having a molecular conformation resembling the shape of a clover leaf. In → protein synthesis, it transfers a particular amino acid to the growing polypeptide chain in the ribosome. With the aid of its → anticodon tRNA fixes the amino acid onto the place of the corresponding → codon in the → messenger RNA, thus adding this amino acid into its correct place in the polypeptide chain under formation. In most cases there are several specific tRNA types for one amino acid. The structure of all tRNA molecules are similar.

transformation (L. *forma* = form), **1.** the transmission of genetic information with the aid of naked DNA fragments from one cell to another within or between species. The transformation, originally found in bacteria, has also been observed in eukaryotic organisms, in which DNA as such can also be transported from one individual to another, even from one species to another; **2.** the artificial genetic manipulation of organisms by transducing foreign DNA into them; **3.** the physiological change of a substance or structure in an organism into another form, e.g. a chemical substance in metabolism, or a larva into the adult stage in → metamorphosis; **4.** the change of a normal cell into a cancer cell.

transforming growth factors, TGFs, a superfamily of dimeric polypeptide growth factors; i.e. **TGF-α** and **TGF-β**, the latter including → **inhibins**, **activins**, and **Müllerian regression factor**. TGFs have effects on cell proliferation and differentiation. TGF-β has a wide range of biological activities being generally stimulatory in cells of mesenchymal origin and inhibitory in cells of epithelial or neuroectodermal origin. Nearly all normal and tumour cells of vertebrates have surface receptors for TGF-β.

transfusion (L. *fusio* = result of melting), transfer from one source to another, especially the artificial transfer of blood or its components, or other fluid (e.g. physiologic saline solution) into a blood vessel. *Cf.* infusion.

transfusion tissue, a tissue type in the vascular bundle system in needles of conifers; it consists of parenchymatous and tracheid-like cells and its task is to transfer water and compounds between vascular tissues and photosynthezising tissue below the epidermis and hypodermis.

transgenic, pertaining to organisms into which foreign DNA has been introduced either from another individual of the same species or from other species.

transgression (L. *transgressus* = having stepped across), the act of crossing, passing over; e.g. in genetics, the appearance of one or more genotypes in a segregating generation (F_2, backcross, etc.) whose character quantities fall outside the limits of variation of the parents and F_1 individuals.

transition, a genetic point mutation in which a pyrimidine base of DNA is substituted by another pyrimidine, or a purine base by another purine.

transition state, an intermediate state of a biochemical reaction where old bonds break and new ones form. The transition state (S^*) is between the substrate (S) and product (P) of any reaction and it has a higher free energy than S or P.

transketolase, an enzyme (EC 2.2.1.1) of the transferase class catalysing the transfer of a two-carbon unit in the carbohydrate metabolism; e.g. in photosynthesis it catalyses the reaction: fructose 6-phosphate + glyceraldehyde 3-phosphate —> xylulose 5-phosphate + erythrose 4-phosphate (i.e. $C_6 + C_3 = C_5 + C_4$, in which C = carbon). Transketolase, together with → transaldolase, forms a reversible link between the → pentose phosphate pathway and → glycolysis.

translation, → genetic translation.

translocation (L. *locatio* = placing), displacement, dislocation; e.g. **1.** translocation of chromosomes (reciprocal translocation), refers to exchange of parts of non-homologous chromosomes without the loss of genetic material; *cf.* crossing-over; **2.** protein translocation, the transfer of a protein across a cellular membrane or to another site in the membrane; **3.** translocation of ribosome, the movement of a ribosome one codon along the → messenger RNA molecule after the addition of each amino acid to the polypeptide chain in → protein synthesis.

transmembrane proteins, *see* integral membrane proteins.

transmission (L. *transmissio* < *missio* = sending), transfer, passage; e.g. **1.** the transfer of a disease from one individual to another; **2.** neurotransmission, the transfer of impulses through the → synapse; **3.** → genetic transmission; **4.** transmitted light, such as light passing through a solution in spectrophotometric analyses. *Verb* to **transmit**.

transmission electron microscope, TEM, *see* microscope.

transmission threshold, in parasitology, the threshold of a net reproduction rate which the parasite population has to overcome, resulting in a disease in the host population.

transmitter, a substance or any other factor that transmits; e.g. → neurotransmitters. *See* transmission.

transpiration (L. *spiratio* = exhalation), the passing of air or vapour through something, such as water evaporating from the skin (insensible perspiration) or lungs, or from the surface of a plant. In plants, transpiration is the driving force of water transport. Most water is transpired via the stomatal structure (stomatal transpiration) and only a few per cent through the cuticular surface of leaves and stems (cuticular transpiration). Big foliage trees may transpire from two to four hundred litres of water in a day. Transpiration is dependent on the relative humidity and temperature of air, wind, light and water supplies of soil. *Verb* to **transpire**. *Adj*. **transpirable**. *See* sweating.

transplacental, crossing the placenta, such as the transplacental passing of immunoglobulins from the mother to the foetus.

transplantation (L. *plantare* = to plant), **1.** the grafting of cells, tissues, or organs taken from one part of an organism or from one individual, and implanted in another; may be **allogeneic**, i.e. between genetically different individuals of the same species, or **syngeneic**, i.e. between individuals of the same pure line; *see* graft; **2.** the removal of an organism or species from one place and setting it to another place, often to an unoccupied area. Man has transplanted many animal and plant species between continents; known examples are rabbit, fox, and dog, transplanted by European emigrants to Australia and New Zealand where they have become dangerous to the native fauna. Usually, however, the transplantations are unsuccessful.

transposase (L. *ponere* = to place), an enzyme required for the replacement of a → transposon.

transposition (L. *positio* = placement), removal from one place to another; displacement to the opposite side; e.g. **1.** the transposition of two atoms in a molecule, or of an organ in the body cavity; **2.** a → chromosome mutation in which a given segment of a chromosome is placed into a new position in the genome so that a reciprocal replacement does not occur; transposition may occur in the same or a non-homologous chromosome.

transposon, Ts (L. *transponere* = to remove), a segment of DNA which has the ability to move in the genome from one location to another. The length of a transposon is usually over 2,000 nucleotide pairs, and at each end it contains genes which are necessary for the integration of this segment into the chromosome. One or several other genes can be situated between the terminal genes. Transposons are found in all organisms from bacteria to man. When a transposon is integrated in the chromosome into the vicinity of or inside a gene, it causes disturbances in the function of this gene.

transsexual, a person with sexual characteristics of one sex, but whose personal identification is with that of the opposite sex; **transsexualism**, the state of being transsexual, or the desire to change to the opposite sex. *Adj.* **transsexual**.

trans-splicing, a splicing reaction in which → exons of heterogeneous nuclear RNA molecules produced by allelic genes, are joined together. *See* splicing.

transvection effect (L. *transvectio* = driving over, going over, going by), in genetics, a → position effect observed in the fruit fly resulting in the enhancement of the effect of a mutation; occurs when the mutation is in the → heterozygous position with a certain structural change of the chromosome.

transvector, an organism which transmits a toxic substance to organisms higher up in the food chain, not produced by itself but accumulated from other sources.

transverse tubules, T tubules, *see* T system.

transversion (L. *transvertere* = to turn), a turning; e.g. **1.** in genetics, a point mutation in which a purine base of DNA is replaced by a pyrimidine or vice versa; **2.** the moving of a tooth in the jawbone into the place of another tooth.

trap flower, a flower that traps entering insects inside it to ensure pollination; e.g. some species in the family Araceae.

tree, a woody, perennial plant with a minimal height of four to six m (the criteria for heights vary in different countries); smaller, young trees are usually classed into shrubs.

tree ferns, tree-like → ferns, growing in tropical and subtropical areas, especially on mountains; evolutionally very old and belong mostly to the family *Cyatheaceae*; well known genera are *Cyathea* and *Dicksonia*.

tree layer, a vegetation layer in the forest, comprising the trees; may consist of several sublayers with different heights.

tree line, the border line against the treeless area in northern or southern cold areas, or on mountains where altitude causes the lack of trees. *Cf.* timberline.

tree shrews, → Scandentia.

tree ring, → growth ring.

Trematoda (Gr. *trematodes* = pierced, with holes), **trematodes, flukes**; a class of parasitic flatworms (Platyhelminthes), most of which live as endoparasites with several larval stages in molluscs (intermediate host) and in vertebrates (definitive host). One or two suckers are situated on the ventral side of the leaf-like body, and the alimentary canal is divided into two main branches. Trematoda comprises about 6,000 species in two (or three) subclasses (orders in some classifications): Aspidogastraea and Digenea, in some systems also Monogenea. The larval stages of the latter are → miracidium, sporocyst, redia, cercaria, and metacercaria.

tremor (L. *tremere* = to shake), an involuntary quivering of the body or an organ, e.g. **kinetic tremor**, occurring in a limb during active movements, or **tremor cordis**, irregularity of the heart.

tri- (L. *tres, tria*, Gr. *treis, tria* = three), three-, treble, triple, triplex.

triacylglycerols, triglycerides, neutral fats; lipids formed in cells from one glycerol and three fatty acid molecules by esterification. Most storage lipids both in animals and plants (e.g. all edible fats) belong to triacylglycerols. In the alimentary canal of animals or in certain intracellular vesicles such as → lysosomes, triacylglycerols are hydrolysed to mono- and diacylglycerols, or to free fatty acids and glycerol.

triad (Gr. *trias* = group of three), any trivalent element or object, or a group of elements or structures comprising three entities.

trial-and-error conditioning (learning), a kind of learning behaviour by experience with trial and error to avoid reactions to stimuli which produce negative experiences, and vice versa to favour stimuli with positive experiences; e.g. birds learn to avoid harmless but vespid-like flower flies (Syrphidae) after the initial encounter with stings of wasps.

triarch (Gr. *arche* = origin), pertaining to a root structure in dicotyledons with three-lobed primary xylem (in cross section) in the central cylinder (stele).

Trias (Triassic), the initial period of the Mesozoic era 245 to 208 million years ago; the age of reptiles when the first dinosaurs, turtles, ichthyosaurs, plesiosaurs, and also therapsids developed, and cycads and conifers dominated among plants.

triazine, a heterocyclic compound having three varying side chains. The triazine group contains common herbicides, such as atrazine, desmetryne, prometryne, and simazine.

triazole, an organic compound composed of one five-ring with three nitrogen atoms; its derivatives, such as amitrole, are used as herbicides.

tribe (L. *tribus*), the taxonomic category in botany between the family and genus, used to indicate groups of closely related genera; also any group of organisms without a taxonomic sense, such as a tribe of sparrows. Sometimes a tribe includes one or more **subtribes**.

tricarboxylic acid cycle, → citric acid cycle.

triceps (L. *tri-* = three, *caput* = head), three-headed; e.g. the triceps muscle in mammals.

trich(o)- (Gr. *thrix* = hair), denoting relationship to hair or hair-like structures.

trichial, thread-like; pertaining e.g. to a primitive structure of the → thallus, common in green algae; also found in some brown and red algae.

trichina, pl. **trichinae** or **trichinas**, a nematode parasite (*Trichinella spiralis*), about 1.5 mm in length. Adult trichinae live mainly in the alimentary canal of omnivorous or carnivorous mammals, such as rat, bear, pig, or man. Its larvae migrate into muscles causing inflammation in several tissues, especially in the heart and nervous system. Man may be infected by eating raw or underdone infected meat. **Trichinosis**, disease caused by trichinae. *Adj.* **trichinous**, infected with trichinae.

trichlor(o)acetic acid, TCA, an aliphatic organic acid, CCl_3COOH. TCA is a caustic substance damaging cells and tissues; even in low concentrations it inhibits cell division and prevents the cuticle synthesis in plant leaves. TCA is an effective precipitant of proteins and is used in biochemical analyses e.g. for quantitative determinations of proteins and as a decalcifier and a fixative in microscopy. TCA has been used as a total herbicide having a very strong contact effect, but as a chlorinated compound, it is a pollutant.

2,3,6-trichlorobenzoic acid, 2,3,6-TBA, an organic compound consisting of a substituted benzoyl ring with three chlorine atoms; used as a hormone herbicide.

trichoblast (Gr. *thrix* = hair, *blastos* = bud), the smaller of the two cell types formed in an unequal cell division of a → meristemoid in the plant root epidermis (rhizodermis); develops into a root hair.

trichobothrium, pl. **trichobothria** (Gr. *bothros* = pit), a sense organ of some arthropods, e.g. some arachnids and myriapods, having a small pit with a sense hair as a mechanoreceptor. The trichobothrium reacts to movements of air, water, and compact objects, also sensing vibrations.

trichocyst (Gr. *kystis* = bladder), a highly specialized cell organelle in the ectoplasm of many ciliates and dinoflagellates. The rod-shaped or oval trichocysts with thread-like shafts are evenly distributed on the surface of the unicellular body; they are either toxic or non-toxic organelles serving defence and adherence.

trichogyne (Gr. *gyne* = woman), **1.** a hair-like elongated structure in the ascogonium of → ascomycetes, functioning in sexual reproduction (gametangiogamy) by joining the female and male gametangia; **2.** the hair-like elongation of the carpogonium of red algae, functioning as a receptive organ for the spermatium.

trichome, 1. an outgrowth of the epidermis in plants; may be a simple or branched hair, vesicle, or stinging hair; **2.** one of the brightly coloured hairs located in bunches on the body surface in some myrmecophilous insects, releasing an attractive secretion to ants.

Trichoptera (Gr. *pteron* = wing), **trichopterans, caddis flies**; an order of holometabolous insects resembling lepidopterans; the wings are hairy and well-veined, and folded roof-like over the hairy body. The adult stage is short-lived. Aquatic larvae live in self-made cases of leaves, sand, or gravel, bound together by secreted silk. About 7,000 species are found in the world.

trichothecenes, mycotoxins formed in the metabolism of *Fusarium* fungi; one of the most toxic subtances is T-2 produced by *F. sporotrichioides* growing on overwintered cereals. Trichothecenes are immunosuppressive and can e.g. cause pulmonary infections and alimentary toxic aleukia (decreased number of leucocytes) if ingested with food.

tricolpate, *see* monocolpate.

tricuspid valve (L. *valva tricuspidalis*, or *valva atrioventricularis dextra*), right atrioventricular valve; the valve between the right atrium and right ventricle of the three- or four-chambered heart in tetrapod vertebrates. The valve comprises three cusps allowing bloodstream only towards the ventricle. *Cf.* bicuspid valve.

trifluralin, a dinitroaniline herbicide (α,α,α-trifluoro - 2,6 - dinitro -N,N- dipropyl-p-toluidine) that inhibits cell division and growth in plant roots; used in biological studies to inactivate the function of → microtubules.

trigeminal nerve (L. *nervus trigeminus*), the 5th of the → cranial nerves.

triglyceride, → triacylglycerol.

trihybrid (Gr. *treis* = three), an individual organism which is heterozygous for three genes. *See* hybrid.

triiodobenzoic acid, TIBA, a benzoic acid derivative having three iodine atoms in the ring; used as a hormone herbicide.

triiodothyronine, *see* thyroid hormones.

Trilobita (Gr. *lobos* = lobe), **trilobites**; an extinct arthropod class mainly from the Cambrian period; trilobites were mainly marine bottom animals with a segmented body moulded into three lobes. About 4,000 fosssil species are described.

trimer (Gr. *meros* = part), a molecule or polymer composed of three simpler molecules; if the constituents are identical the compound is called **homotrimer**, if different, **heterotrimer**.

trimethylamine, a volatile, liquid tertiary amine, a strong base, $(CH_3)_3N$, with a fishy odour; a degradation product of nitrogenous animal and plant substances. Its conjugated forms are widely distributed in animal tissue, especially in many fish species. *See* trimethylamine oxide.

trimethylamine oxide, an end product of nitrogen metabolism in the blood of hyperosmotic fishes living in oceans. Together with its metabolites produced after the death of the fish, it gives a strong taste and smell to the flesh, if not bled immediately after the catch.

trinomial (L. *nomen* = name), a scientific name of a subspecies consisting of three terms; the first term designates the genus, the second the species, and the third the subspecies.

triolein, *see* olein.

triose, the smallest monosaccharide containing three carbon atoms; e.g. glyceraldehyde in → glycolysis, and in the Calvin cycle of → photosynthesis.

triose phosphate isomerase, TIM, the enzyme (EC 5.3.1.1) catalysing the isomerization of triose phosphates (dihydroxyacetone phosphate and glyceraldehyde 3-phosphate) in glycolytic cell metabolism.

triphenyltetrazolium chloride, TTC, an organic cyclic compound rich in nitrogen, $C_{19}H_{15}ClN_4$; colourless in its oxidized form and red in reduced form, called **triphenylformazan**. The **TTC test** is based on the reduction capability of TTC, and is used for testing the viability of seeds, cells, etc. The TTC test is also used for measuring quantitatively the activities of some enzymes like those of the citric acid cycle, e.g. succinic dehydrogenase.

triplet, a set or group of three; e.g. **1.** a group of three contiguous nucleotides in DNA or RNA (*see* genetic code); **2.** a tripartite structure of the microtubule in the → centriole; **3.** one of three offspring born at one birth; **4.** in physics, a group of three elementary particles.

triploblastic (Gr. *blastos* = germ), pertaining to animals whose embryos have three germ layers, i.e. → ectoderm, mesoderm, and endoderm; most animals (except sponges and coelenterates) are triploblastic. *Cf.* diploblastic.

triploid (Gr. *triploos* = threefold, *eidos* = form), an → autopolyploid or allopolyploid cell, tissue, or individual which has a threefold chromosome set in its nuclei (symbol 3x).

tripton (Gr. *triptos* = rubbed, pounded), any of non-living organic remains (as humus), or inorganic microparticles in water.

TRIS, → tris-(hydroxymethyl)-aminomethane.

trisaccharide (Gr. *tri-* = three-, *sakcharon* = sugar), a sugar molecule consisting of three monosaccharides, e.g. → raffinose.

tris-(hydroxymethyl)-aminomethane, TRIS, Tris, an organic compound used as a pH buffer in physiological and biochemical studies.

Syn. tromethamine, THAM, Trizma base.

trisomic (Gr. *soma* = body), pertaining to cells, tissues or individuals in which one or a few chromosomes occur in a threefold instead of the normal twofold dosage. The chromosome number of a trisomic cell is 2n+1. *Noun* **trisomy**; e.g. → Down's syndrome.

tristyly (Gr. *stylos* = column), one type of plant → heterostyly in which there are three types of styles and the stigmata are located at three different levels.

trit(o)- (Gr. *tritos* < *treis* = three), denoting third, tertiary.

tritium, a radioactive isotope of hydrogen, symbol ^{3}H or H-3, and T, half-life about 12.5 years; tritium is used as a radioactive tracer in biological, biochemical, and chemical studies.

triton, the nucleus of the tritium atom.

Triton X-100, alkylpolyphenoxypolyoxy ethanol; a non-ionic detergent which decreases surface tension, and is therefore used to improve the absorption of different substances; also used as an additive in herbicides, homogenization solutions, and emulsions. *Syn*. Octoxynol.

tritonymph (Gr. *tritos* = third, *nymphe* = bride, nymph), the third larval (nymphal) stage in the life cycle of mites and ticks (Acari). *Cf.* protonymph, deutonymph.

triturating stomach (L. *trituratio* = process of threshing), → cardiac stomach.

trivalence, trivalency (L. *valere* = to be strong), the → valence of three.

trivalent (L. *valens* = being of some value), pertaining to something having the value of three, e.g. an atom or antibody molecule; in genetics, the union of three chromosomes in → meiosis.

Trizma base, → tris-(hydroxymethyl)-aminomethane.

tRNA, → transfer RNA.

troch(o)- (Gr. *trochos* = wheel < *trechein* = to run), pertaining to a wheel, or a wheel-like structure; round.

trochanter (Gr. = runner), **1.** L. *trochanter major*, greater throcanter, and *t. minor*, lesser trochanter, two bone processes in the neck of the femur in tetrapod vertebrates; **2.** the second segment between the coxa and femur in the leg of insects and arachnids.

trochophore (Gr. *pherein* = to bear), a free-swimming pelagic larva of some marine annelids, bryozoans, and molluscs; the trochophore has an ovoid body with a preoral and (or) equatorial circlet of cilia. *Syn*. trochosphere.

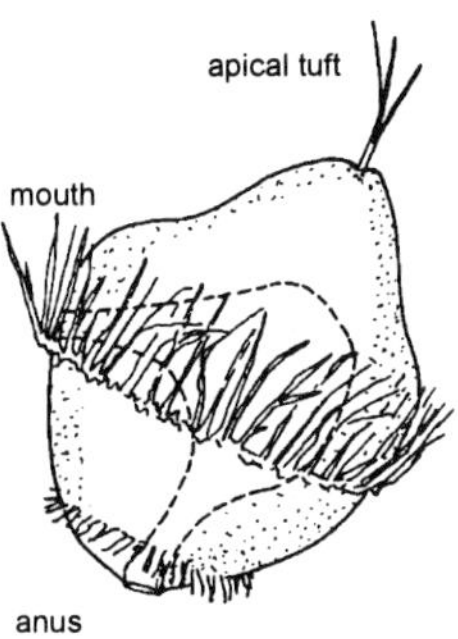

TROCHOPHORE

trochosphere, → trochophore.

Trogoniformes (*Trogon* = the genus of trogons, Gr. *trogein* = to gnaw, L. *forma* = shape), **trogons**; an order of richly coloured, long-tailed birds in tropical forests; includes 39 species in one family, Trogonidae.

tromethamine, → tris-(hydroxymethyl)-aminomethane.

troph(o)- (Gr. *trophe* = nourishment, nutrition), denoting relationship to nourishment, nutrition, food. *Adj*. **trophic**.

trophallaxis (Gr. *allaxis* = change), reciprocal feeding; i.e. a habit of exchanging food between animals, found especially among social insects. Reciprocal feeding often occurs between larvae and adult, but sometimes also between adults, strengthening social bonds in the colony.

trophic, relating to nutrition; **-trophic** or **-trophous**, relating to a specified type of nutrition, such as hypertrophic or hypotrophic.

trophic level, a division of the → food web in which the levels are determined on the basis of feeding methods: **primary producers** (autotrophs), **primary consumers** (herbivores), **secondary**, **tertiary,** etc. **consumers** (carnivores), and **decomposers**.

trophic structure, an organization in a → community based on the feeding relations between the organisms. *See* food web, trophic level.

trophobiosis (Gr. *biosis* = living < *bios* = life), a type of → symbiosis between two species in which one gathers food (trophobiont) and the other defends against predators; e.g. the symbiosis between ants and aphids, in which relationship the aphids are **trophobionts**, secreting honeydew to ants which defend them against predators.

trophoblast (Gr. *blastos* = germ), an extraembryonic tissue layer outside the early mammalian embryo (blastocyst, embryoblast). The trophoblast attaches the embryo to the uterine wall and provides nutrition for its development. It also separates the embryo from the maternal tissue forming the foetal parts of the placenta. The extraembryonic membranes, chorion and amnion, are derived from the trophoblast. The inner cellular layer of the trophoblast is called **cytotrophoblast**, and the outer layer, **syncytiotrophoblast**.

trophosome (Gr. *soma* = body), an organ derived from the midgut of pogonophores; contains symbiotic chemoautotrophic bacteria which oxidize hydrogen sulphide to provide energy for the synthesis of organic compounds from carbon dioxide to be used by the host animal.

trophozoite (Gr. *zoon* = animal), a haploid stage in the life cycle of sporozoans; e.g. in human blood cells, the trophozoites of plasmodians divide actively into → merozoites.

-tropic (Gr. *trope* = turning), **1.** pertaining to a turning, or a change in a turning of an organism or organ as a response to a specified stimulus, e.g. heliotropic or geotropic; **2.** pertaining to certain hormones, tropic hormones; *see* -tropin.

tropics (Gr. *tropikos* = applies turning), the area between the Tropics of Cancer and Capricorn; *adj.* **tropic(al)**; the corresponding → climate zone is the **tropical zone** where the mean temperature of the coldest month is over 18°C; the vegetation zone in the same area is the **tropical vegetation zone**, characterized especially by → rain forests.

-tropin, a hormone that has an effect on the secretion of another hormone; especially refers to pituitary hormones, such as → corticotropin, somatotropin (growth hormone), thyrotropin, and gonadotropins. Also -trophin.

tropism, a bending, turning, or growth movement of an organism (mainly plants) or a part of it, in response to an external stimulus. The movement is either positive, i.e. towards a stimulus, or negative, i.e. away from a stimulus; such stimuli are e.g. light in **phototropism** and gravitation in **gravitropism**. *Cf.* taxis.

tropocollagen, the basic structural unit of → collagen.

tropomyosin, a long protein molecule located attached to actin filaments (thin filaments) of muscle fibres; inhibits contraction by blocking the interaction of actin and myosin. *See* troponin.

troponin, a complex of three proteins (troponin C, I, T) located in the thin filaments (actin filaments) of muscle cells serving in the control of contraction and relaxation in muscle fibres. When troponin C binds calcium ions, troponin C and I move away from → tropomyosin making contraction possible, i.e. the interaction of actin and myosin. *See* muscle contraction.

troposphere (Gr. *sphaira* = ball), the innermost layer of the atmosphere extending about 10 km above the Earth's surface; forms about 95% of the Earth's air mass. *Cf.* ionosphere, mesosphere.

tropotaxis (Gr. *taxis* = arrangement), a tactic movement in which an animal orients itself by making simultaneous comparisons with its bilaterally symmetrical sensory receptors so that the proportion of stimulation to either side of its body regulates the direction of its locomotor activity. *See* taxis.

true breeding, describes genotypes of organisms which according to their particular characteristics produce only progenies of their own kind, since these genotypes are homozygous (*see* homozygote) for the genes regulating the development of these characteristics.

true predator, a functional concept in ecology meaning any predator which kills prey animals almost immediately after the catch; e.g. raptorial birds, carnivorous mammals (as tiger and wolf), as well as insectivorous plants. *Cf.* predator.

truncus, pl. **trunci** (L. = trunk of a tree, shaft of a column); e.g. *truncus sympathicus*, sympathetic nervous trunk, *truncus pulmonalis*, pulmonary trunk in the circulatory system of tetrapod vertebrates.

Trypanosoma (Gr. *trypan* = to bore, *soma* = body), **trypanosomes**; flagellate protozoans belonging to the phylum Sarcomastigophora and subphylum Flagellata; parasites found both among invertebrates and vertebrates. They cause an infectious disease called **trypanosomiasis** in many vertebrate species (e.g. African sleeping sickness in man and nagana disease in domestic livestock). Their → vectors are blood-sucking invertebrates carrying the first developmental stages of trypanosomes (amastigote, promastigote, epimastigote); the last stage, trypomastigote, lives in the blood of the definitive host.

trypsin (Gr. *tryein* = to rub, to wear out), an enzyme (EC 3.4.21.4) of the hydrolase group; initially included several enzymes functioning in the intestine of vertebrates, but later was defined as endopeptidase. It catalyses the cleavage of peptide linkages at the carboxyl group of arginine or lysine. Trypsin is secreted in an inactive form, **trypsinogen**, from the pancreas of vertebrates, and it is converted into the active form in the intestinal fluid by another enzyme, **enterokinase** (enteropeptidase).

tryptamine, an intermediate in the biosynthesis of → auxin, formed via decarboxylation from → tryptophan. In animals 5-hydroxytryptamine, called → serotonin, acts as a neurotransmitter.

tryptophan(e), abbr. Trp, trp, symbol W; an aromatic amino acid, α-amino-3-indolepropionic acid, $C_8H_6N\text{-}CH_2CH(NH_2)COOH$, which exists in plant and animal proteins; it is also a precursor of → serotonin. Tryptophane belongs to the → essential amino acids of animals.

Ts, → transposon.

tsetse flies (*Glossina*), dipteran flies carrying trypanosomes in Africa and causing sleeping sickness in man and nagana disease in domestic livestock. *See* Trypanosoma.

TSH, thyroid-stimulating hormone, → thyrotropin.

T system (of tubules) (T < transverse), an intracellular structure comprising transverse tubules in the skeletal and cardiac muscle cells (fibres), formed as tubulous invaginations of the cell membrane. The T system conducts action potentials from the cell membrane inside the cell close to the → sarcoplasmic reticulum (SR), resulting in the activation of calcium release from SR, the process necessary for the contraction of the muscle cells. *See* excitation contraction coupling.

TTC, → triphenyltetrazolium chloride.

tuba, pl. **tubae** (L.), tube; e.g. *tuba auditiva/Eustachii*, Eustachian tube, or *tuba uterina*, oviduct.

tube feet, organs in the → water-vascular system of echinoderms; each tube foot is a closed cylinder with muscular walls having a sucker at the outer end, and a bulb-like ampulla at its proximal inner end. The feet serve for attaching the animal to the substratum; they also function in the locomotor activity and the handling of food.

tuber (L. = hump, knob, tumour), **1.** a plant structure for storing food and/or functioning in reproduction; mostly swollen structures which may develop in many ways and in various parts of plants, i.e. in the hypocotyl (found e.g. in the radish), in the runners (in potato), in the stem base above the ground (in orchids, called bulbs), or below the ground (in crocus), or in roots formed from the tap root (in rutabaga) or adventitious roots (in some orchids); **2.** in anatomy of animals, a rounded protuberance; e.g. *tuber ischiadicum*, ischial tuberosity; **tuberculum** (pl. tubercula) nodule, small protuberance.

tubocurarine, *see* curare.

Tubulidentata (L. *tubulus* = small tube, *dens* = tooth), a mammalian order including the single extant species, **earth pig** (**aardvark,** erdvark), *Orycteropus afer*, a pig-like mammal living in Africa. The name of the order refers to the radial, thin tubules in the teeth which are rootless and without enamel. Previously the aardvark was located in the order of edentates owing to the lack of the incisor teeth..

tubulin (L. *tubus* = tube), a protein of → microtubules in cells.

tubulus, pl. **tubuli** (L.), **tubule,** small tube; e.g. *tubulus renalis*, renal tubule, *tubulus seminiferus*, seminiferous duct in the vertebrate testis; *cf.* microtubulus.

tubus, pl. **tubi** (L.), **tube**; in anatomy a tube-shaped structure, e.g. *tubus digestorius*, alimentary canal.

tularemia, a zoonotic disease caused by a bacillus *Pasteurella/Francisella tularensis*, found primarily in rodents, but transmitted to man and other animals (rabbits, squirrels, muskrats, etc.) by the bites of fleas, deer flies, and ticks, or by inhalation of their products. The disease is characterized by inflammation of the bitten area, fever, chills, headache, and swollen lymph nodes.

tumour, Am. **tumor** (L. *tumere* = to swell), **1.** a swelling, enlargement; one of the classic signs of → inflammation; **2.** a neoplasmic tissue (neoplasm), a new growth of tissue involving uncontrolled and progressive multiplication of cells.

tumour necrosis factor, TNF, a large polypeptide of a group of → cytokines in vertebrates. **TNF-α** (cachectin) is produced by activated macrophages, **TNF-β** (lymphotoxin) by activated T lymphocytes. They activate a variety of immune defence mechanisms by interactions e.g. with → granulocytes, lymphocytes, and fibroblasts, and thereby destroy tumour cells.

tumour suppressor gene, a gene normally regulating the cell cycle, but through a mutation becomes tumourigenic; in contrast to → oncogenes, the tumourigenic mutations of the tumour suppressor genes are recessive.

tundra (Russ., of Finno-Ugric origin = arctic mountain), a treeless vegetation zone in cold areas, between the nival and cool zones. In the tundra zone the mean temperature of the warmest month is mostly 10°C, often below 5°C; for nine months the mean temperature is below 0°C, and snow covers the ground which is frozen throughout the year. Because of the ice, the surface water will not be absorbed in the soil and bog development occurs commonly. In the northern hemisphere the southern border of the tundra zone is generally 70 °N, in the southern hemisphere tundra is found between 45 and 61 °S.

tundra soil, *see* soil formation.

tunica, pl. **tunicae** (L. = tunic, integument), a coat, covering, or tunic; usually a membrane covering or lining an organ, e.g. **1.** *tunica mucosa*, a mucous membrane lining tubular structures, such as the alimentary canal and respiratory pathways of vertebrates; **2.** in plants, the outer part of the → apical meristem consisting of 1—3 cell layers on the surface, being responsible for the formation of e.g. leaf primordia; *cf.* corpus, defin. 4.

Tunicata, tunicates, a subphylum in the phylum Chordata; the animals have features of chordates, such as the notochord and spinal cord in the larval stage, but these structures are mostly reduced in adults. About 1,500 tunicate species in three classes are described: **sea squirts** (Ascidiacea), **thaliaceans** (Thaliacea) and **larvaceans** (Larvacea/ Appendicularia). *Syn.* Urochorda(ta).

Tupaiidae, tree shrews. *See* Scandentia.

Turbellaria (L. *turbella,* dim. of *turba* = tumult, disorder, confusion), turbellarians; a class of mostly free-living flatworms (Platyhelminthes) widely distributed in aquatic and humid environments. Their body is leaf-like, covered with cilia, having no hooks and suckers; the ventral mouth and strong proboscis are followed by a widely branched intestine. The animals have no circulatory or respiratory organs. The class includes about 3,000 species divided into 12 orders.

turbid, pertaining to a cloudy or muddy liquid; **turbidimetry**, an optical method for the determination of the concentration of solutions, based on **turbidity**.

turgor (L. *turgor* = swelling, expansion), an internal pressure in cells and tissues caused by water uptake. *Adj.* **turgid**. *See* osmosis, water potential.

turion (L. *turio* = sprout, young branch), a scaly shoot developed from a bud on a subterranean rootstock.

Turner's syndrome (*H.H. Turner*, 1892—1970), a human developmental disturbance due to the → aneuploidy (monosomy) of the sex chromosomes; is characterized by short stature, sterility due to undifferentiated gonads, and sometimes mental retardation. The chromosome number in Turner's syndrome is 45, X0, and the phenotype is female (X = X chromosome, 0 indicates the lack of the homologous chromosome). The syndrome can appear also in mosaic forms, the tissues being partly of XO and partly of XX chromosome constitution.

turtles, *see* Chelonia.

tusk shells, → Scaphopoda.

Tween 80, polyoxyethylene sorbitan monooleate; a detergent and lipid solvent used to decrease the surface tension in dispersion and isolation of biological material such as cell organelles and lipids. Related compounds are e.g. Tween 20, Tween 40. *Syn.* Polysorbate 80. *Cf.* Triton X-100.

twin method, a method in human genetics for studying the influence of genes and the environment on the development of traits by comparing monozygotic and dizygotic twins.

twins, two individuals who in mammals have developed simultaneously in the same uterus, in birds within the same vitelline membrane (likewise triplets, quadruplets etc.). The twins may be → monozygotic or → dizygotic, the former developing from a single zygote, the latter from two fertilized eggs. Monozygotic twins are genetically identical.

two-winged flies, → Diptera.

Tylopoda (Gr. *tylos* = pad, lump, callus, knob, *pous* = foot), **tylopods, camels**; a suborder of artiodactyls including two genera, humpbacked camels (*Camelus*) in the Old World, and llamas and alpacas (*Llama*) without humps in South America; both of them have a tripartite stomach and feet with two toes and well-developed, soft soles.

tylosis, pl. **tyloses,** an intrusion of a parenchyma cell through a pit into a xylem vessel in foliage trees; the formation of tylosis may block the vessels in old parts of trees, leading to the formation of dead heartwood in the middle of the trunk. *Cf.* tylosoid.

tylosoid (Gr. *eidos* = form), an intrusion of a secretory epithelial cell into a resin duct of coniferous trees. The tylosoids may block the resin ducts leading to the formation of dead heartwood in the middle of the trunk. *Cf.* tylosis.

tymbal, → timbal.

tympanal organ (L. *tympanum* = drum), a specialized auditory organ of some insect groups, such as noctuid moths, grasshoppers and crickets; consists of a tympanic membrane (**tympanum**) attached to a few mechanoreceptor cells within an air-filled cavity.

tympanic cavity (L. *cavitas tympanica*, *cavum tympani*), **1.** the air-filled cavity of the → middle ear of tetrapod vertebrates; **2.** an analogous structure of the → tympanal organ in some insects.

tympanic membrane (L. *membrana tympani*), eardrum; **1.** the membrane of an auditory canal separating the outer ear from the middle ear; **2.** an analogous structure of the → tympanal organ in some insects.

tympanum, 1. → tympanic membrane, tympanal organ; **2.** tracheal tympanum, a partially ossified segment in the sound-producing organ of some birds acting as a resonator.

Tyndall effect (*John Tyndall*, 1820—1893), the scattering of light passing through a medium containing minute suspended particles, e.g. seen in a colloidal suspension, or dusty or smoky air, thereby making a visible beam.

type specimen, a specimen or individual according to which a new → taxon is described and designated. Type specimens may be type individuals, type species, or type genera. **Holotype** is an individual or part of it which is used in describing a new species or lesser taxon, and which has been named a type specimen by the author. The duplicate specimen of the holotype taken simultaneously e.g. from the same plant is named **isotype**; **syntype** is any specimen of a type series when no holotype was designated. **Paratype** is a specimen of a type series other than the holotype, usually used in animal taxonomy. **Lectotype** is chosen afterwards as the type specimen from the group of syntypes, **paralectotype** being any of the type series remaining after the designation of the lectotype. **Neotype** is chosen as a type specimen from the original type locality if the original type material is damaged or lost. **Topotype** is a specimen collected at the locality of the original type.

typhlosole (Gr. *typhlos* = blind, *solen* = channel, pipe), a dorsal longitudinal fold of the wall of the digestive canal of some invertebrates (e.g. earthworms, bivalve molluscs) and of cyclostomes. The main function of the typhlosole is to increase the absorptive surface area.

typologic species, *see* species.

tyramine (Gr. *tyros* = cheese), a decarboxylation product of → tyrosine found in ergot, decaying organisms, and also in ripe cheese; it is a sympathomimetic amine affecting animals in some respects like → adrenaline.

tyrosine, Tyr, tyr, symbol Y; β-(*p*-hydroxyphenyl)alanine, $C_9H_{11}NO_3$; an aromatic amino acid found in many proteins. In animals, tyrosine is synthesized from phenylalanine and is so a nutritionally non-essential amino acid. It acts as a precursor of catecholamines, thyroid hormones, and melanin. In plants, tyrosine is e.g. the precursor in the biosynthesis of phenolic compounds. The first step in the use of tyrosine is catalysed by an enzyme, **tyrosine amino transferase** (EC 2.6.1.5): L-tyrosine + 2-ketoglutarate —> 4-hydroxyphenylpyruvate + L-glutamate. **Tyrosyl** is the acyl radical of tyrosine.

U

U, symbol for **1**. unit; **2**. uridine, uracil; **3**. uranium.

U, symbol for voltage (expressed in volts).

ubiquinone, coenzyme Q; 2,3-dimethoxy-5-methyl-1,4-benzoquinone (coenzyme Q_0) with different multiprenyl side chains (coenzymes Q_1, Q_2, Q_6, Q_7, Q_9, and Q_{10}). Ubiquinones act as electron carriers in the → electron transfer chain of cell respiration in mitochondria, occurring in the majority of aerobic organisms from bacteria to higher plants and animals. Their reduced forms are called ubiquinoles. Ubiquinone preparates are called vitamin Q.

ubiquitin, Ub, a small acidic protein present in all prokaryotic and eukaryotic cells; plays an important role in tagging proteins for destruction. When activated, attaches covalently to a protein targeted for degradation and forms a multiprotein complex of proteolytic enzymes (**proteosome**), causing a rapid hydrolysis of e.g. abnormal intracellular proteins. Ubiquitin plays an important role in normal processes during the **cell cycle**; it may function in chromatin organization and regulation, and in cellular response to stress. It is highly conserved in evolution, e.g. differing only by three amino acid residues of the 76-residue protein in yeast and man.

UDP, uridine diphosphate. *See* uridine phosphates.

UDP-glucose, glucose activated by the binding energy of uridine diphosphate, UDP; occurs in many metabolic reactions of plant cells, e.g. in the synthesis of → sucrose and → cellulose.

ulcer (L. *ulcus* = wound), a local inflammatory excavation of the surface of an organ or tissue; e.g. **peptic ulcer** in the alimentary canal (oesophageal, gastric, or duodenal ulcer), caused by the action of acid gastric juice, especially in mammals.

ulcerosis (L. *ulcerosus* = full of sores), common name for several inflammatory skin diseases in fish.

ulna (L.), usually the larger of the two forearm bones in tetrapod vertebrates. *Cf.* radius.

ultimate factor (L. *ultimus* = last, most distant), any factor considered to act as a fundamental cause (**ultimate cause**) of a biological phenomenon during the evolutionary history (the term coined by *Ernst Mayr*). For example, when the coat colour of the hare turns white in the autumn, the ultimate factor is the protective colour, increasing the fitness of the species. Thus, the ultimate factors or causes are teleological, not mechanistic like the → proximate factors.

ultimobranchial organ/body (Gr. *branchia* = gills), an endocrine gland in vertebrates (except mammals), located near the carotid arteries in the pharyngeal region. The gland secretes → calcitonin, which regulates calcium and phosphate levels in the blood and other tissues; it is thus homologous with the C cells in the → thyroid gland of mammals.

ultra- (L.), exceeding or beyond.

ultracentrifuge (L. *centrum* = centre, *fugere* = to flee), a type of → centrifuge equipped with a rotor capable of up to 120,000 rpm, corresponding to approx. 700,000 times *g* (gravity). An analytical ultracentrifuge includes photographic equipment and a fluorescent plate for viewing and saving of sedimentation profiles.

ultradian (Gr. *dias* = day), pertaining to rhythmic phenomena in organisms occurring in greater frequency than once a day. *See* biorhythms.

ultrafiltration (Ger. *Filz* = felt), a filtration through pores of microscopical size, e.g. the filtration of plasma through capillary walls; in laboratory work, the separation of minute particles, such as microbes, from a solution using special filters.

ultramicrotome (Gr. *mikros* = small, *tome* = section), an instrument used for the preparation of ultrathin (< 1 μm) tissue sections for electron microscopy.

ultrasonics (L. *sonus* = sound), the study of acoustics dealing with → ultrasounds. Ultrasonic radiation is used for cell and tissue fractionation, therapeutically e.g. for the treatment of arthritis, and as a diagnostic aid using the visual display of echoes from irradiated tissues (e.g. in foetal diagnostics).

ultrasound, a high-frequency sound; a mechanical radiant energy with frequencies above the range of human hearing, i.e. above 20,000 Hz. *Syn.* supersound. *See* hearing.

ultrastructure (L. *structura* = composition), any submicroscopic cell structure visible only in an → electron microscope.

ultraviolet microscope, *see* microscope.

ultraviolet radiation, → UV radiation.

umbel, an → inflorescence composed of numerous flowers, the stalks of which are equally long and rise from the same point on the tip of the flower shoot (e.g. in Apiacae).

umbilical cord, a flexible foetal organ in mammals with a discoidal placenta, in which blood vessels pass to the placenta; develops from mesodermal tissue that extrudes from the embryo and forms a stem-like link with the chorion, developing from the → trophoblast. The umbilical cord is found e.g. in primates.

umbilicus (L.), the navel; the point of attachment of the → umbilical cord.

umbo, pl. **umbones** (L. = shield boss), **1.** the oldest, anterior portion on each valve of the bivalve shell; **2.** the posterior projecting peak on the ventral valve of the brachiopod shell.

umbrella, pl. **umbrellae,** or **umbrellas** (L. = sunshade), an umbrella-like body of jellyfishes, fringed by a row of marginal tentacles.

unconscious, 1. not conscious, insensible, without awareness, not responding to sensory stimuli; **2.** in *Freud's* model the part of the mind of which the ego is not aware. *Noun* **unconsciousness**.

uncompetitive inhibition, a type of inhibition of a biochemical reaction in which an inhibitory compound (antagonist) does not compete for the same active site (receptor site) as does the actual substrate. Also called non-competitive inhibition or antagonism. *Cf.* competitive inhibition.

uncoupler, a toxic compound that separates reactions normally linked with each other in cellular metabolism; e.g. dinitrophenol (DNP) uncouples → electron transfer from → oxidative phosphorylation (ATP production) in cell respiration.

unequal crossing-over, an event of → crossing-over producing one chromatid that contains a gene doubled and another chromatid in which that gene is lacking (*see* deletion); also refers to an event of crossing-over occurring between tandem duplications so that a triplication and a reversion are produced.

unguis (L.), → nail.

Ungulata (L. *ungula* = hoof), **ungulates**, **hoofed mammals**; a common name for odd-toed (→ Perissodactyla) and even-toed (→ Artiodactyla) mammals.

uni- (L. *unus* = one), denoting one, single.

unicellular (L. *cellula* = cell), pertaining to organisms consisting of a single cell only, such as bacteria and protozoans. *Cf.* acellular.

unifacial structure, a structure of a plant organ, usually a petiole or a leaf, with only one face, the upper face being undeveloped.

unilateral (L. *latus* = side), pertaining to one side only.

uniparental, an individual originated from a single parent, e.g. by budding. *Cf.* parthenogenesis.

uniparental disomy (Gr. *di-* = two, *soma* = body), a chromosomal constitution in which an individual has inherited two members of a → homologous chromosome pair from one parent and none from the other parent. **Uniparental isodisomy** and **uniparental heterodisomy** are distinguished, i.e. in the former the individual has inherited two copies of one member of the chromosome pair from one parent, and in the latter both members of the chromosome pair from one parent.

uniparous (L. *parere* = to beget), pertaining to an organism that produces only one offspring at a time or in the course of its lifetime.

uniport, *see* cell membrane transport.

unipotent, unipotential (L. *potentia* = power), having power in one way only, e.g. a cell that is capable of developing only in one direction or into one end product.

unique copy DNA (L. *unicum* = the only one), a sequence of → deoxyribonucleic acid, DNA, that occurs in a genome as a single copy or a few copies (*see* sequence complexity of DNA). Structural genes encoding proteins typically consist of unique copy DNA.

Uniramia (L. *ramus* = branch), **uniramians**; a subphylum in the phylum Arthropoda (in some classifications a phylum in the superphylum Arthropoda); includes **insects, chilopods, diplopods, symphylans**, and **pauropods**, in all, more than 1 million described species. Common characteristics are uniramous and jointed appendages, a single pair of antennae, and unsegmented mandibles. Some taxonomists divide uniramians into two superclasses, Insecta and Myriapoda.

uniserial, uniseriate (L. *series* = row), arranged in one row or series; e.g. **uniserial ray**, a plant → ray with one row of parenchyma cells. *Cf.* multiserial.

unisexual organism, an organism in which only one sex is represented; e.g. a plant whose flowers bear only the → stamens or the → pistils but not both. Most animals are unisexual. *Cf.* hermaphrodite. *See* gonochorism.

unit membrane, a lipid bilayer membrane of all cells and cell organelles.

unitunicate (L. *unus* = one, *tunica* = coat), pertaining to a single-layer structure of the → tunica in the plant → apical meristem.

univalent (L. *valens* = to be of some value), **1.** having a → valence of one; **2.** chromosomes which do not pair during the first division of → meiosis. Univalents either lack a homologous chromosome or are results from disturbances in chromosome pairing; this may be due to either genetic or environmental factors. *Cf.* bivalent, multivalent.

universal veil, *see* veil.

unmyelinated neurones, neurones in which the → myelin sheath is absent.

unsaturated, in chemistry, **1.** describes a solution with ability to still dissolve more of a substance; **2.** unsaturated compound, i.e. an organic compound with double or triple bonds especially between two carbon atoms, e.g. oleic acid, $CH_3(CH_2)_7CH{=}CH(CH_2)_7$-COOH, or arachidonic acid, $CH_3(CH_2)_4$-$(CH{=}CHCH_2)_4CH_2CH_2COOH$. A compound with a number of double or triple bonds is termed a polyunsaturated compound; **3.** in ecology, an unsaturated community, having unoccupied → niches.

unstable equilibrium, in ecology, refers to the instability of a population, community, or environmental resources, in which even a minor change brings about ever-increasing disturbances.

unstriated/unstriped muscle, → smooth muscle.

upwelling, a vertical movement of deeper water towards the sea surface, caused by offshore currents; occurs especially in certain coastal areas. As a result from upwelling, the deeper water, rich in nutrients, reaches the → euphotic zone thus allowing a high productivity of phytoplankton.

ur(o)-, ure(o)-, uron(o)- (Gr. *ouron* = urine), pertaining to urine or urea.

uracil, a pyrimidine base occurring in ribonucleic acid, RNA.

urate, a salt of → uric acid.

urate oxidase, uricase, EC 1.7.3.3; an enzyme catalysing the oxidation of uric acid to allantoin; belongs to the oxidoreductase class, liberating CO_2 and H_2O in the reaction. It is found commonly in mammals but not in primates.

urban ecology, a research field dealing with ecological problems in settled areas, especially in urban environments. As a result of the great increase in the urbanization of human populations, urban ecology has become a steadily growing branch within ecology.

urea (Gr. *ouron* = urine), the diamine of carbonic acid, $CO(NH_2)_2$; a nitrogenous end product of protein metabolism in → ureotelic animals (especially in amphibians and mammals). Urea is formed in the liver via the → urea cycle and is excreted by the kidneys into urine. In the environment urea is decomposed into ammonia by microbes. Urea is widely used as a fertilizer in forestry. *Syn.* carbamide.

urea cycle, a cyclic reaction chain whereby urea is formed in the liver of ureotelic vertebrates. Ornithine is the carrier molecule which is converted to citrulline and further to arginine. Arginine is hydrolysed into urea and ornithine, which restarts the cycle. Nitrogen-containing molecules enter the cycle, using ATP as energy source. *Syn.* ornithine cycle.

urea-nitrogen recycling, the circulation of urea from the urinary bladder into the blood and its excretion into the gut, where intestinal microbes decompose it into ammonium ions. In this form nitrogen is reabsorbed into the blood; after combining with glycerol in the liver, urea is used for the biosynthesis of new amino acids. Urea-nitrogen recycling is found in mammals who are seasonally exposed to a low-protein diet, e.g. the reindeer in the winter, or the bears during winter sleep.

urease, an aminohydrolase enzyme, EC 3.5.1.5, catalysing the decomposition of → urea into carbon dioxide (CO_2) and ammonia (NH_3), and the hydrolysis of hippuric acid into benzoic acid and glycine. The enzyme is found in microorganisms as well as in many plants and animals. Urease was the first enzyme to be crystallized.

Uredinales, rust fungi; an order in the class Basidiomycetes, subdivision Basidiomycotina, division Amastigomycota, kingdom Fungi; in some systems an order in the class Teliomycetes, division Basidiomycota. Rust fungi are parasitic, causing diseases in plants. They do not form → fruit bodies and a typical life cycle is very complicated, normally with two host species. Rust fungi pass the cold season as **winter spores (teleutospores)**, the other spore types being **basidiospores**, **pycnospores**, **aecidiospores**, and **uredospores (summer spores)**. The order includes approx. 5,000 species causing damage e.g. among graminids, cereal plants included.

uredospore (L. *uredo* = smut, Gr. *sporos* = seed), summer spore, a spore type of → Uredinales.

ureotelic (Gr. *ouron* = urine, *telos* = end), pertaining to animals having urea as their main nitrogenous end product, i.e. amphibians, some reptiles, and mammals. *Cf.* ammonotelic, uricotelic.

uresis, urination, the flow of urine; as a suffix, **-uresis**, the excretion of a substance in the urine; e.g. **saluresis**, the urinary excretion of sodium chloride.

ureter (Gr. *oureter*), the duct conveying urine from the kidney to the bladder (in mammals) or to the cloaca (in birds and reptiles). The development of the ureter is related to the formation of the metanephros type of kidney in the Amniota group of vertebrates. Its walls are composed of connective tissue with several layers of smooth muscles, and the transitional epithelium of the mucous membrane lines the lumen.

urethra (Gr. *ourethra*), the tube in mammals extending from the urinary bladder to the exterior of the body, i.e. in females to the ventral area of the vulva and in males to the tip of the penis.

urge, a force or readiness of animals for a particular, relatively restricted instinctive behaviour or activity, controlled by hormones; e.g. urge for migration, or urge for courtship. The concept of urge is not easily distinguished from the concepts of activity level and motivation. *See* drive.

uric acid (Gr. *ouron* = urine), 2,6,8-trihydroxypurine, $C_5H_4N_4O_3$, an end product of the nitrogen metabolism of uricotelic animals (most reptiles, all birds, and most terrestrial invertebrates) and of the purine metabolism of some mammals, e.g. primates. Uric acid is the least toxic compound of all nitrogenous end products and can occur in high, even prepicitating concentrations in excretory organs, also inside the egg shell during embryogenesis. In most mammals it is an intermetabolite which is converted into allantoin that is excreted into urine. The water-solubility both of uric acid and its salts, **urates**, is weak, and they are easily precipitated in tissue fluids (causing gout in humans) and excretions (the excrements of birds and reptiles).

uricase, → urate oxidase.

uricotelic (Gr. *telos* = end), pertaining to animals having → uric acid as the main end product of nitrogen metabolism; most reptiles, all birds, and most terrestrial invertebrates are uricotelic. *Cf.* ammonotelic, ureotelic.

uridine, a → nucleoside composed of uracil and ribose.

uridine phosphates, nucleotides formed from uridine and phosphate group(s). These compounds are uridine mono-, di- or triphosphates (UMP, UDP, UTP) and they occur e.g. in RNA or in metabolism as sugar activating compounds; an example of the latter is → UDP-glucose in the synthesis of glycogen.

urinary bladder, the membranous sac in the caudal portion of the abdominal cavity in vertebrates, serving as a reservoir for urine. In fish it is formed as an extension of the Wolffian duct, in amphibians as an extension of the cloaca. In amniotes it is generally formed from the main part of allantois, while in birds and most reptiles it does not develop at all. Histologically the urinary bladder is composed of connective tissue with several layers of smooth muscle; the transitional epithelium of the mucous membrane lines the lumen. Also some invertebrates, such as the crustaceans, have a urinary bladder.

urine, liquid or semisolid excretion from the kidneys in most multicellular animals containing especially end products of nitrogen metabolism, such as → urea, uric acid, ammonium ions, and salts (especially sodium chloride).

urobilin (L. *bilis* = bile), a brown pigment which is an oxidized form of urobilinogen in the bile; excreted in the faeces of vertebrates.

Urochordata (Gr. *oura* = tail, *chorde* = cord), **urochordates**; → Tunicata.

Urodela (Gr. *delos* = visible), **urodeles,** → Caudata.

urogenital organs (Gr. *ouron* = urine, L. *gignere* = to beget), a common name for excretory and reproductive organs, the development of which are closely linked during embryogenesis.

uro(hypo)physis (Gr. *oura* = tail, *phyein* = to expand), an endocrine gland close to the caudal part of the spinal cord in most fishes; secretes peptide hormones, **urotensines,** which regulate the salt content of the blood.

uropodium, pl. **uropodia,** Am. also **uropodiums** (Gr. *pous* = foot), **uropod**; the last pair of abdominal appendages in malacostracan crustraceans, forming a tail fin with a flattened terminal telson.

uropyge, also **uropygium** (Gr. *pyge* = rump), the rump and lower back of birds supporting tail feathers. *See* oil gland.

uropygial gland, → oil gland.

urostyle (Gr. *stylos* = pillar), an unsegmented, rod-like bone in the vertebral column of anurans (frogs), formed by the fusion of the caudal vertebrae.

Ustilaginales, smut fungi, an order in the class Basidiomycetes, subdivision Basidiomycotina, division Amastigomycotina, kingdom Fungi; in some systems an order in the class Ustomycetes, division Basidiomycota. The species produce both basidiospores and special chlamydospores but form no fructifications. They are parasitic having only one host in higher plants, causing smut diseases which in cereals cause heavy economic loss.

Ustomycetes, *see* Ustilaginales.

uterine tube (L. *uterus* = womb, *tubus* = pipe), → oviduct.

uterus, pl. **uteri** (L.), (Gr. *hystera*), **1. womb**; a hollow muscular organ between the oviduct and the vagina in female mammals (except Monotremata), where the embryo fastens to the uterine wall by the placenta. The uterus is divided into **uterine body** (*corpus uteri*) and **uterine cervix** (*cervix uteri*). Primarily the uterus is a paired organ, and in the most primitive mammals the two uteri are separate (**uterus duplex**). In most mammalian species, however, the distal ends of the two are fused (**uterus bicornus**), while in higher primates a complete union (**uterus simplex**) is formed. The walls of the uterus consist of three layers: the mucous membrane (*endometrium*), the muscle layer (*myometrium*) and the external membrane (*perimetrium*). In particular the mucous membrane undergoes great changes during the oestrus cycle and gravidity; *see* progesterone; **2.** a modified section of an oviduct in many invertebrates and vertebrates (other than mammals), serving as a place for the secretion of the eggshell, or the development of eggs or offspring. *Adj.* **uterine**.

utricle (L. *utriculus* = small bag), a bladder-like structure; **1.** the upper of the two sacs in the labyrinth of the internal ear in vertebrates; *cf.* saccule; *see* equilibrium sense; **2.** the bract subtending the female flower in *Carex* and related species.

uvea (L. *uva* = grape), the portion of the vertebrate eye composed of the iris, ciliary body, and choroid.

uvomorulin, E-cadherin. *See* cadherins.

UV radiation, ultraviolet radiation, electromagnetic radiation between the visible and X-ray regions of the spectrum: UV-A 315—400 nm, UV-B 280—315 nm and UV-C shorter than 280 nm. UV light of the Sun is mainly radiation with wavelenghts below 295 nm, but the ozone layer of the atmosphere absorbs the greatest part of this wavelength. UV radiation kills unprotected organisms, highly irritating unpigmented skin and eyes of animals. In tissues it causes the formation of → oxygen radicals which damage cell structures. Because UV radiation kills bacteria, it is used to sterilize laboratories, operating rooms, and other working spaces. UV radiation with wavelength approx. 260 nm is absorbed by → nucleic acids and it can therefore cause mutagenic changes.

V

V, symbol for **1**. vanadium; **2**. valine; **3**. volt, voltage; **4**. velocity (V or v).

V, symbol for volume.

v, abbreviation for vein (L. *vena*).

vaccine (L. *vaccinus* < *vacca* = cow), preparation of killed or attenuated microbes or their parts used for preventive inoculation or for treatment of infectious diseases. The immunization procedure in which the vaccine is injected is called **vaccination**. *Adj.* **vaccinal** or **vaccinial**. *Verb* to **vaccinate**.

vacuole (L. *vacuus* = empty), **1.** a minute space or cavity in any tissue; **2.** a cavity in the protoplasm of a cell surrounded by a membrane; different types of vacuoles exist both in animal and plant cells, e.g. **autophagic vacuole**, (autoplasmosome); **condensing vacuole**, spherical vacuoles of → Golgi complex in secretory cells, **digestive vacuole** (secondary lysosomes), **contractile vacuole**, an organelle that ejects excess water from the cell typical of protozoans, **food vacuole**, for suspension of food material in holozoic protozoans, and **heterophagic vacuole** (heterophagosome). *See* lysosome, phagocytosis, pinocytosis.

vacuum activity, in ethology, an instinctive behaviour which is activated in the apparent absence of any external stimulus. This is because the animal has not been able to perform a certain behaviour for a long time, e.g. owing to the lack of the releasing stimuli. As a result the threshold value for the releasing stimuli is lowered.

vagina, pl. **vaginae** (L.), **1**. a sheath-like structure in anatomical nomenclature; **2**. the canal from the vulva to cervix uteri in female mammals receiving the penis in copulation, or any of the analogous organs in many invertebrates. *Adj.* **vaginal**.

vagotonia (vagus nerve, Gr. *tonos* = tension), hyperexcitability of the parasympathetic vagus nerve. *Syn.* parasympathicotonia.

vagus (nerve) (*nervus vagus*), the tenth cranial nerve of vertebrates. It is an important parasympathetic nerve of the autonomic nervous system and innervates widely organs in front and middle parts of the body (eg. the heart, lungs, liver, pancreas, kidneys, and the main part of the alimentary duct). It also has some somatic neurones, such as motor neurones to the throat and sensory neurones to internal organs. *Adj.* **vagal**.

valence (L. *valere* = to be strong), the combining power of an atom or a radical, i.e. the number of bonds an atom can form with other atoms in chemical reactions. Valence depends on the structure of electron shells around the atom nucleus (*see* octet), and the valence of an atom is equal to the ionic charge in ionic compounds and equal to the number of bonds formed in covalent compounds.

valine, Val, val, symbol V, α-aminoisovaleric acid, $(CH_3)_2CHCH(NH_2)COOH$; an amino acid synthesized by plants, and common in plant proteins; in animals, it is an → essential amino acid for protein synthesis. Also called 2-aminoisovaleric acid, or isopropyl-aminoacetic acid, or 2-amino-3-methylbutanoic acid.

valinomycin, an antibiotic produced by bacteria of the genus *Streptomyces* (e.g. species *S. fulvissimus*); acts as an → ionophore transporting K^+ ions through cell membranes and uncouples → oxidative phosphorylation in mitochondria. Valinomycin as well as other ionophores are used to study ion transport properties of cells, but also as an insecticide and nematocide.

value, a central concept of → quantitative genetics, referring to the measure for a given character in an individual; e.g. the → phenotypic value, genotypic value, breeding value, or environmental deviation value.

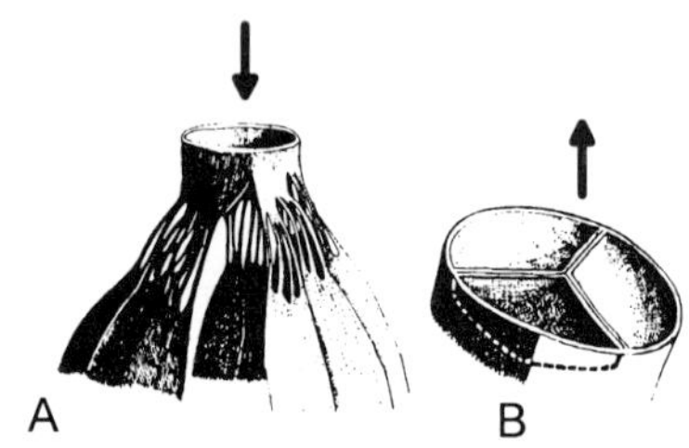

Cardiac valves in vertebrates. A: the atrioventricular or tricuspid valve between the right atrium and ventricle; B: semilunar valve; ↑ direction of the blood flow.

valve (L. *valva*), any anatomical structure that permits flow in one direction only; e.g. **1.** cardiac valves, i.e. atrioventricular valves betveen the atria and ventricles, and semilunar valves at the bases of the aorta and pulmonary artery; **2.** one of the two calcareous hinged halves of the shells of clams and lamp shells;

dim. **valvule** (*valvula*), e.g. one part of the semilunar valve, or dorsal and ventral valvules forming the → ovipositor in insects.

van der Waals' forces (*J.D. van der Waals*, 1837—1923), weak, nonspecific, attractive forces between molecules based upon permanent or fluctuating electric charges; e.g. the pressure of any gas is lower than expected according to the equation of state ($pV = RT$), because the attractive force between molecules diminishes the expansive tendency of gas. Van der Waals' forces operate between all types of molecules, polar and nonpolar. These forces rapidly weaken when the distance of molecules lengthens, i.e. the forces are inversely proportional to the sixth power of the distance between the molecules or atoms.

vanadium (*Vanadis* = a Norse deity), a rare metallic element, symbol V, atomic weight 50.94; occurs in nature as **vanadates** (salts of vanadic acid). It is a part of haemovanadin pigment (haemovanadium) in the blood of tunicates. Nanomolar concentrations of vanadates inhibit P-type ATPase enzymes in cells. Vanadates have been used in treating various diseases.

vanadocyte (Gr. *kytos* = hollow, cell), a green blood cell type of some ascidians (Ascidiidae and Perophoridae). Vanadocytes contain haemovanadium and they take part in the synthesis of tunica.

vanilmandelic acid (VMA), one of the principal metabolites of adrenaline and noradrenaline occurring in the urine of vertebrates.

van't Hoff's law (*Jacobus van't Hoff*, 1852—1911), **1**. the equation $pV = nRT$ is proved to apply to any solution that has n moles of solute in V litres of solution; p = pressure, V = volume, n = the number of moles, R = gas constant (8.3144 J mol^{-1} K^{-1}), and T = temperature (K). From the equation one can get the equation of osmotic pressure: $p = nRT/V$; if the moles in one litre (n/V) is expressed as c (= concentration), the equation takes the form: $p = cRT$. Osmotic pressure is usually expressed as π, then π = cRT. Pressure is proportional to the concentration of any solution and also proportional to temperature; *see* osmosis, osmotic potential, water potential; **2**. the rule describing that the velocity of chemical and biochemical reactions increases twofold or more for each 10°C rise of temperature; *see* Q_{10}; **3**. in stereochemistry, van't Hoff's law explains that in all optically active compounds, one or more polyvalent atoms are bound to four different atoms forming an asymmetric and steric structure.

variability, the amount, state or capability of → variation, i.e. the inclination to vary.

variable number of tandem repeats, VNTR, a form of → genetic polymorphism in which varying numbers of short, identical, contiguous DNA segments are found in the genomes of different individuals. The repeats occur in different positions on the chromosome set. In man, this polymorphism is so wide that every individual, with the exception of identical twins, differs from each other in respect to the number of these repeats, and hence an individual can be identified even on the basis of a small DNA sample.

variance (L. *varians* = varying), a statistical concept which indicates the distribution of the material on both sides of the mean (symbol s^2), i.e. the square of standard deviation.

variation, the occurrence of hereditary or nonhereditary differences in the stable structures of cells (intraindividual variation), between different individuals of a → population (individual variation) or between populations (group variation). The main sources of variation are genetic differences and differences caused by the environment. Biological variation can be divided into three classes: 1) **Phenotypic variation** is the total observable variation in a given character of an organism. It can be qualitative or quantitative. The symbol of quantitative phenotypic variation is V_P. It can be partitioned into genetic (V_G) and environmental (V_E) variation. By definition $V_P = V_G + V_E$. 2) **Environmental variation** is the variation caused by all non-genetic internal and external factors of the cell or organims appearing in their → phenotype. This part of the variation is indicated as V_E. 3) **Genetic variation** is that part of the variation which is purely heritable (symbol V_G). It is caused by genes and their interactions. This part of variation is indicated as → heritability the symbol of which is H. By definition H = V_G/V_P.

variety, also **varietas,** abbr. var; a unit of classification, specifically of plants, which is lower than species or subspecies. A variety differs from the main type of the species more than a → form which is the lowest unit of classification.

Varroa mite (*Varroa Jacobsoni*), a mite species causing damage to bee colonies; it is spread progressively in the world through transplantation of honey bees.

vas, pl. **vasa** (L.), a vessel or duct in animal anatomy; **1.** *vasa afferentia*, vessels that transport fluid to an organ or part of it, e.g. *vasa afferentia nodi lymphatici*, afferent vessels of a lymph node; *vas deferens* (→ ductus deferens, deferent duct); **2.** *vasa efferentia*, vessels that transport fluid from an organ or a part of it, e.g. *vasa efferentia nodi lymphatici*, efferent vessels of a lymph node.

vascular, 1. pertaining to, or composed of vessels or ducts which convay blood, lymph, or sap; **2.** relating to circulation of fluids in xylem and phloem in plants.

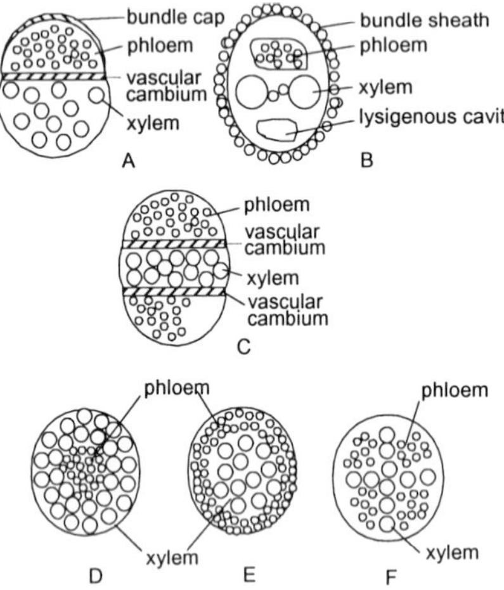

Types of vascular bundles in plants: collateral, open type (A), collaretal, closed type (B), bicollateral (C), concentric, leptocentric (D), concentric, hadrocentric (E), radial (F).

vascular bundle, a tubular structure in vascular plants, formed from vascular tissues (xylem and phloem), conducting water, nutrients and organic compounds. Vascular bundles run from roots through the shoot to the leaves, where they form leaf veins. The most common vascular bundle type is **collateral**, which may be **open** or **closed**. The open type is typical of dicotyledons. It contains xylem and phloem layers, between which the **vascular cambium** develops and causes the later secondary growth. The closed collateral type is typical of monocotyledons; no cambium and no secondary growth occurs, but a well developed **bundle sheath** is developed around the structure. In the **bicollateral** type there are two cambium-layers (many climbing plants), and in the **concentric** type either xylem or phloem tissue surrounds the other. If xylem is the outermost layer, the vascular bundle is **leptocentric**; this type is found in many rhizomes of monocotyledons; the other is a **hadrocentric** type, common in ferns.

vascular cambium, *see* cambium.

vascular parenchyma (Gr. *enchyma* = poured in), a → parenchyma tissue of plants which occurs associated with vascular tissues and rays.

vascular plant, a plant with a vascular system of xylem and phloem forming → vascular bundles, which conduct water, nutrients and organic compounds. The group of vascular plants comprises the division Pteridophyta (e.g. ferns), and the division Spermatophyta (seed plants). In total, about 260,000 species are described.

vascular tissues, conductive plant tissues; → **xylem** conducts water and nutrients and → **phloem** transports organic compounds. Vascular tissues form → **vascular bundles** in higher plants.

vasectomy (Gr. *ektome* = excision), surgical removal of the ductus deferens (vas deferens) or a part of it, performed e.g. to induce sterility in domestic animals or man; the same result is achieved by ligation of the ducts.

vasoactive intestinal peptide, VIP, a polypeptide of 28 amino acid residues found in mammals, functioning as a tissue hormone and probably as a transmitter in the nervous system; e.g. relaxes the smooth muscle of blood vessels and increases hormonal secretion from the pancreas, gut, and hypothalamus.

vasomotor (L. *motor* = mover), **1.** affecting the diameter of a blood vessel; **2.** an agent that causes a change in the diameter of blood vessels, i.e. results in **vasomotion**, which can be **vasodilatation** or **vasoconstriction** (vasocontraction). The part of the nervous system that controls vasomotor muscles is called **vasomotor system** (vasomotorium).

vasopressin, an octapeptide hormone of mammals synthesized by neurosecretory cells of the hypothalamus and secreted from the posterior lobe of the → pituitary gland (neurohypophysis). In high concentrations it elevates blood pressure by stimulating the constriction of arterioles and capillaries, and increases intestinal peristalsis. Because one of its principal physiological effects is the retention of water by the kidneys, vasopressin is often called **antidiuretic hormone** (**ADH**). Related hormones (slightly different in structure) are also found in other vertebrates. *See* pituitary gland. *Cf.* oxytocin.

vasotocin, an octapeptide hormone of vertebrates (except mammals) secreted from the posterior lobe of the pituitary gland; the corresponding hormone in mammals is called → vasopressin. *See* pituitary gland.

vasotonia (Gr. *tonos* = tone), angiotonia; the tension or tone of blood vessels, especially in arteries and arterioles. *Syn.* angiotonia.

Vater—Pacini corpuscle (*Abraham Vater*, 1684—1751; *Filippo Pacini*, 1812—1883), → Pacinian corpuscle.

vector (L. *vehere* = to carry), **1.** in gene technology, a plasmid, virus, liposome, or transposon used to introduce foreign DNA into a cell; **2.** a carrier, intermediate host, such as an arthropod that transfers infective agents from one host to another (e.g. anopheline mosquitoes as vectors of malaria), or an animal, wind, water, etc. carrying pollen from stamen to stigma; **3.** a quantity possessing magnitude and direction which is represented by a straight line resembling an arrow; **4.** the electrical axis of the heart represented by an arrow.

vegetal (L. *vegetare* = to grow), **1.** pertaining to a plant; **2.** → vegetative.

vegetal pole, vegetative pole, vitelline pole; that pole of an animal ovum (or early embryo) at which the main part of yolk is deposited; opposite to animal pole. In freely floating eggs the vegetal pole turns downward.

vegetation analysis, the study of the structure of vegetation including ground, field, tree, and bush layers. Common parameters used are species composition, abundance, and spatial patterns. The chosen study area is mostly analysed using squares (e.g. 1 m^2), which are placed in the area at random, or by a systematic manner, e.g. a line transect (line analysis). The species, their → coverage and frequency are determined in every square or along the line, and the results are generalized, and usually analysed statistically.

vegetation map, a map showing the characteristics of the vegetation in some area, i.e. the existence of different → plant communities or → associations. In some countries the whole area has been mapped this way (e.g. in Austria), and the maps are used in community planning.

vegetation zones, clearly separate zones in the vegetation; may be **local**, when the zonation is caused by an alteration of some local external factor (e.g. the effect of water in littoral areas), or **global**, caused by changes of temperature conditions between the poles and the equator. The **nival zone** is the area around the poles, north of the tundra zone, and is characterized by permanent snow and ice. The **tundra** area is a treeless vegetation zone in cold areas between the nival and cool zones characterized by mosses, lichens, dwarf herbs, and shrubs. In the tundra zone the mean temperature of the warmest month is mostly 10°C, often below 5°C. The mean temperature is below 0°C during nine months when snow covers the soil which is frozen throughout the year. Because of the ice, water is not absorbed into the soil and bog development is common. In the northern hemisphere the southern border of the tundra zone is located generally at 70 °N, and in the southern hemisphere between 45 and 61 °S. In the **cool zone** the average temperature of the warmest month is over 10°C and coldest below —3°C. In the northern hemisphere, the cool zone is located south of the tundra zone and is characterized by coniferous forests. It is more difficult to demarcate the cool zone in the southern hemisphere. In the **temperate zone** the mean temperature is over 10°C during at least four months a year. The zone is characterized by foliage trees, steppes, or deserts. In the **subtropic zone** the average temperature of the coldest month is over 0°C. Evergreen forests grow in coast areas but inland also deciduous forests are typical. In the subtropic zone there are also → deserts and → savannas. The **tropics** or tropical area is the region lying between the tropics of Cancer and Capricorn, the mean temperature of the coldest month being over 18°C. The area is characterized especially by → rain forests.

These zones are often divided into subzones determined by humidity, or by altitude and local temperature in mountain areas. *See* height zones.

vegetative, 1. pertaining to growth and nutrition; **2.** functioning involuntarily such as the vegetative or → autonomic nervous system; **3.** pertaining to asexual reproduction, e.g. budding; **4.** characteristics of plants. *See* vegetative cell, vegetal pole.

vegetative cell, 1. an early cell type of the animal embryo developing from a megalecithal egg; vegetative cells form a → **vegetal pole** (vegetative pole) of an embryo; **2.** in plants, a cell formed in pollen grain (microspore) during microsporogenesis together with **a generative cell**. A vegetative cell contains a **vegetative nucleus**. *See* fertilization.

vegetative nervous system, → autonomic nervous system.

vegetative nucleus, *see* vegetative cell.

vegetative organs, the involuntarily functioning organs of animals, such as the alimentary canal, respiratory, circulatory, and excretory organs; better known as **autonomic organs**.

vegetative propagation, *see* vegetative reproduction.

vegetative reproduction, asexual reproduction in plants with the help of different vegetatively formed structures, which may be → runners, bulbils, tubers, bulbs, or shoot parts. The new offspring are thus genetically similar to the parent plants. Vegetative reproduction is utilized in artificial production of plants (**vegetative propagation**), particularly in horticulture, using e.g. cuttings and → layering, recently also → cell and tissue cultures (micropropagation).

vegetative segregation, the segregation of organelle genes (mitochondrial and plastid genes) during mitotic division.

vegetative shoot, a plant shoot without flowers.

vehicle (L. *vehiculum*), a means by which something is carried or conveyed; e.g. **1.** any substance (usually fluid) in which an active drug is administered to an animal or which is used to bind active ingredients or pigments in tablets (an excipient); **2.** a carrier, e.g. of infection.

veil (L. *velum*), **1.** → velum; **2.** a sheet formed by the marginal tissue of the young pileus or cap in some fungi (mushroom), being joined to the stipe or stalk (**partial veil**), as in many basidiomycetes, or enclosing the young pileus and stalk totally, as in some *Amanita* species and related genera (**universal veil**). The partial veil may totally disappear during maturation (*Boletus* species), or its remains may form a ring (annulus) on the stalk. The universal veil forms a cap-shaped remnant around the base of the stalk. *Syn.* volva.

vein (L. *vena*), any vessel conveying blood from various organs back to the heart; **2.** in insects, a nervure as a part of the framework of the wing; **3.** in plants, a vascular bundle and supporting tissues in a leaf; **4.** a natural water channel beneath the surface of the earth; **5.** a crack or stratum in a rock containing mineral matter deposited from a solution. *Adj.* **venous**.

velamen, pl. **velamina** (L.), a covering, membrane, tegument, meninx; e.g. **1.** velamen vulvae; **2.** a type of corky epidermis in aerial roots in certain plants, absorbing water; found e.g. in epiphytic orchids.

velarium (L. = awning), a velum-like organ in Cubomedusae.

velar tentacles (L. *tentare* = to probe), → tentacles around the mouth of the lancelet (*Amphioxus*) for drawing water with tiny food particles into the mouth.

veld, open temperate → grassland in southern Africa.

veliger (L. *gerere* = to carry), the second larval stage (after trochophora) in the life cycle of some molluscs; the veliger has a head, foot, and mantle, as well as a ciliated swimming structure, called **velum**. *See* larva.

velum, pl. **vela** (L.), a veil or veil-like structure; e.g. **1.** *velum palatinum* (the soft palate) in the mouth of vertebrates; **2.** a ciliated swimming structure in the → veliger; **3.** a membrane on the subumbrella surface of jellyfish (Hydrozoa); **4.** a sheath-like structure protecting the cap (**velum parietale**) or the whole fruiting body (**velum universale**) in some mushrooms during development; *see* veil. *Adj.* **velar**.

ven(o)- (L. *vena* = vein), pertaining to the vein; e.g. **venoatrial**, pertaining to the vena cava and the right atrium; **venomotor**, pertaining to the constriction or dilatation of veins.

vena, pl. **venae** (L.), → vein; e.g. *vena pulmonalis*, pulmonary vein.

venation types, patterns formed by the leaf → veins in plants. Most primitive and simple is the **dichotomous** type found in *Ginkgo biloba* and some ferns; in **pinnate venation**, a clear middle vein with smaller lateral veins are found on both sides of the leaf (typical of dicotyledons), and in **parallel venation** the large veins run parallel through the leaf (in most monocotyledons). Special types of pinnate venation are **digitate** and **reticulate venation**; in the former there are some large veins beginning from the same point and forming a digitate arrangement in the leaf (e.g. maple); in the latter the veins form a clear reticulate structure.

venom (L. *venenum* = poison), **1.** a poisonous (toxic) substance which some animals, e.g. many snakes, spiders and insects, secrete and introduce into their victims; **2.** broadly, any poison. *Adj.* **venomous**, poisonous, toxic, secreting venom.

venose (L. *vena* = vein), **1.** provided with veins; **2.** venous.

venous, pertaining to the veins; also **venosinal**.

ventilation (L. *ventilatio* = blowing), **1.** providing a space with fresh air; **2.** air exchange in the lungs, or water flow around the gills in respiration; **3.** in psychiatry, verbalization of emotional problems.

ventral (L. *ventralis* < *venter* = belly), pertaining to the belly or venter, or to the abdominal side of the body.

ventral leaves, *see* lateral leaves.

ventral nerve cord, the pair of longitudinal nerves with their segmental ganglia running along the length of the body in the ventral midline, characteristic of many higher invertebrates, e.g. annelids and arthropods. *See* nervous system.

ventral root, the ventral of the two roots of each → spinal nerve in tetrapod vertebrates; comprises motor and autonomic nerve fibres leaving the central nervous system.

ventricle (L. *ventriculus*), a cavity or a hollow organ in an animal body; e.g. **1.** heart ventricle (*ventriculus cordis*), a chamber of the heart which pumps blood out from the heart; **2.** one of the four cavities of the vertebrate brain. *Adj.* **ventricular**.

ventriculus (L. dim. of *venter* = belly), **1.** stomach, gaster; **2.** → ventricle.

venule (L. *venula,* dim. of *vena* = vein), a small vessel drawing blood from capillaries; venules join to form the veins.

vermiculite (L. *vermiculus* = small worm), a net silicate, manufactured from biotite mineral; when heated, it expands strongly into worm-like threads; used in seed planting as inert growth support for hydroponics of plants, and as an insulating material. *See* water culture.

vermin, pl. **vermin** (L. *vermis* = worm), a general term for noxious or repulsive animals, especially those of small size that are difficult to control, as fleas, louse, bedbugs, cockroaches, mice, rats, etc.; also birds and mammals which prey on game competing with man, such as wild mink; **vermination**, being infested with vermin, the breeding of vermin. *Adj.* **verminous**.

verminal (L. *verminalis*), pertaining to worms or vermins.

vermis (L. = worm), **1.** a worm-like anatomical structure, especially the *vermis cerebelli* that is the narrow median part of the cerebellum between the lateral hemispheres; **2.** an old-fashioned term for annelids and some other worm-like invertebrates.

vernalin, a hypothetic plant hormone said to promote flowering in the absence of → vernalization.

vernalization, the hastening of the flowering and fruiting of plants by treating seeds, bulbs, or seedlings to shorten the vegetative period; e.g. biennials can be induced to flower in the first summer, using a hormone treatment or low temperature.

vernation (L. *vernatio* = moult), an arrangement, in which young leaves are folded in a bud. *Cf.* aestivation, defin. 2.

vertebra (L.), any of the short bones (cartilage in some fishes) of the spinal column of vertebrates. In tetrapods, it typically comprises the cylindrical **centrum** (*corpus vertebrae*) with paired dorsal processes that close together forming the **arch** (neural arch), leaving the vertebral foramen (*foramen vertebrae*) inside. The **spinal canal** (neural/vertebral canal, *tubus vertebralis*) is formed by the successive vertebral foramina, containing the → spinal cord with spinal meninges. The unpaired, long **spinous process** (*processus spinalis*), and paired **transverse processes** push out from the arch. In some lower vertebrates there is also a pair of ventral processes often closed together forming the arch (haemal arch, haemal canal). Vertebra types, found in different vertebrate groups, are: **amphic(o)elous**, concave on both surfaces, **opisthoc(o)elous**, concave behind, convex in front, **proc(o)elous**, concave in front, convex behind, **saddle-shaped** (cervicales in birds), and **ac(o)elous** vertebra, with a flat-ended centre. Vertebrae are formed around the notochord from mesenchyme of mesodermal origin. *See* vertebral column.

vertebral canal, spinal canal. *See* vertebra.

vertebral column, spinal column, the columnal assemblage of the → vertebrae from the skull dorsally along the body to the caudal end of the skeleton; in fish includes the thoracic and caudal vertebrae, and in tetrapods the cervical (neck), thoracic (chest), lumbar (back), sacral (pelvic), and caudal (tail) vertebrae. In mammals, sacral vertebrae are fused to form the → sacral bone, in birds some of the lumbar and caudal vertebrae are fused with sacrales, forming the synsacrum. In tailless species, the caudal vertebrae are reduced in number and size, and/or fused, forming the → urostyle (pygostyle, coccyx). *See* vertebra.

Vertebrata, vertebrates; a subphylum of Chordata, segmented animals characterized by a *cranium* (skull) and the *vertebral column* constructed by bony (in some taxa cartilaginous) vertebrae. Vertebrata comprise two superclasses, **Agnatha** (jawless fishes) and **Gnathostomata** (jawed vertebrates). The former includes the classes Cephalaspidomorphi (**lampreys**), Pteraspidomorphi (e.g. **hagfishes**) and Ostracodermi (extinct), the latter are represented by seven classes: extinct **placoderms** (Placodermi), **cartilaginous fishes** (Chondrichthyes), **bony fishes** (Osteichthyes), **amphibians** (Amphibia), **reptiles** (Reptilia), **birds** (Aves), and **mammals** (Mammalia). *Syn.* Craniata, craniates.

vertical rhizome, *see* rhizome.

vesicle (L. *vesiculus*), a small blister or sac in the cytoplasm surrounded by a membrane; in animal cells e.g. **synaptic vesicles** in the axon terminal button containing transmitter substances (*see* synapse), **secretory vesicles** containing e.g. hormones, or **endocytotic vesicles** containing material taken outside the cell (→ endocytosis).

vesicular-arbuscular mycorrhiza, *see* mycorrhiza.

vessel, 1. in animals, a tube or duct, such as a → blood vessel, lymphatic vessel, or bile vessel; **2.** in plants, a water-conducting structure in the → xylem.

vestibular organ/apparatus (L. *vestibulum* = entrance hall), *see* equilibrium sense.

vestibule, a cavity in an organ at the entrance to a canal; e.g. in vertebrate anatomy, **auricular vestibule** (*vestibulum auris*), an oval cavity in the middle of the bony labyrinth where the saccule and utricle are located in the inner ear; **nasal vestibule**, the vestibule of the nose; *Adj.* **vestibular**.

vestigial organ (L. *vestigium* = trace), a rudimentary organ (relict).

veterinary, pertaining to the medical treatment of mainly domestic animals (**veterinary medicine**).

viability (Fr. *vie* < L. *vita* = life), ability to live. *Adj.* **viable**.

vibration sense, a sense with mechanoreceptors in the skin of vertebrates (e.g. → Pacinian corpuscles, vibrissae), or in different → sensilla on the antennae or skin of invertebrates e.g. the → Johnston's organs in insects, or the → lyriform organs (slit sense) in spiders.

vibrissa, pl. **vibrissae** (L.), **1.** any of sensory hairs located near the mouth region of many mammals (e.g. cats and dogs), or some motile sensilla in insects; sensitive to contacts, also to air vibration; **2.** hairs growing in the nasal cavity of man.

vicarious species (L. *vicarius* = substituting), a pair of closely related taxa of organisms living in similar but spatially isolated habitats, such as the reindeer in Lapland and the caribou in Alaska.

vicillin (L. *Vicia* = the genus of vetch), a reserve protein in seeds of leguminous plants.

vicious circle (L. *circulus vitiosus*), a chain of abnormal processes in an organ or organ system produced by positive feedback reactions in which a certain harmful factor or disorder leads to another, aggravating it; e.g. decreased coronary circulation as a result of bleeding weakens the contraction of the heart muscle which further decreases coronary circulation; also the mutually accelerating action of two independent diseases.

vigilance (L. *vigilantia*), a state of being wakeful, watchful. *Adj.* **vigilant**.

villikinin (L. *villus* = tuft of hair, Gr. *kinein* = to move), a hypothetic hormone that is secreted from the mucosa of duodenum; supposed to increase the motility of epithelial villi.

villin, a cellular protein that binds actin in the framework of → microvilli.

villus, pl. **villi,** a small protrusion from the surface of the membrane; e.g. **amniotic villi**, in some places on the surface of the amnion; **intestinal villi** (*villi intestinales*), mostly fingerlike projections which cover the surface of the small intestine of vertebrates, functionally differentiated for absorption; **arachnoid villi**, in the central nervous system, projections of the arachnoid into the venous sinuses; **chorionic villi**, on the surface of the chorion; **pericardial villi**, on the surface of the pericardium; **pleural villi**, on the surface of the pleura; **synovial villi**, projections from the free surface into the joint cavity. *Adj.* **villous, villiform**.

vimentin, a polypeptide that through polymerization forms intermediate filaments in animal cells, especially in mesenchymal cells.

vinblastine (L. *Vinca*, a plant genus, *blast* = developing cell), an alkaloid, $C_{46}H_{58}N_4O_9$, produced by some plants; because of its property to inhibit cell division it is used in the treatment of human neoplastic diseases; also a related compound, **vincristine**, is used as an anticancer drug. Both substances can be produced in plant cell cultures.

vinculin, a protein of the cytoskeleton of animal cells; anchors intracellular → microfilaments to the cell membrane.

violaxanthin (L. *Viola* = violet, Gr. *xanthos* = yellow), a common → xanthophyll in plants.

VIP, → vasoactive intestinal peptide.

viral, pertaining to a → virus.

virgin (L. *virgo* = maid), an animal never fertilized, e.g. an unfertilized insect.

viril, Am. **virile, 1.** characteristic of men or the male sex; **2.** possessing masculine traits or energy.

virilism, the development or presence of masculine characteristics in females, often due to abnormal secretion of male hormones.

virion, a complete virus particle comprising nucleic acid and enclosed by a protein shell.

viroids, small naked RNA particles which can cause diseases in plants and animals; capable of → replication by an unknown mechanism.

virulence (L. *virus* = poison), **1.** the ability of microbes to cause disease. *Adj.* **virulent**, characterized by virulence, pathogenic, deleterious.

virus (L.), a particle containing DNA or RNA that can grow, multiply and express its genes in a host cell only. Viruses (15—300 nm in size) are visible only in an electron microscope and vary morphologically from spherical, disc-shaped, or polyhedral, to rod-shaped and tadpole-shaped forms. They are customarily separated into three subgroups: bacterial viruses, animal viruses, and plant viruses, and classified further on the basis of their origin, host specificity, transmission, or manifestation in the host. Bacterial viruses are called **phages** (bacteriophages), e.g. T-phage of *Escherichia coli*. The head of the bacteriophage contains the DNA core of genetic material enclosed by a protein coat, **nucleocapsid**. It has also a **collar** and a **tail** with tail fibres. Some viruses have a **mantle** consisting of lipids and proteins as the outermost layer. Small plant viruses, **viroids**, have single-stranded RNA of low molecular weight.

viscer(o)- (L. *viscus*, pl. *viscera* = internal organs), pertaining to internal organs, especially organs in the abdominal cavity of vertebrates, to bowels. *Adj.* **visceral**.

visceral arch, *see* gill arch.

visceral cavity, → coelom.

visceral connectives (L. *connectere* = to bind together), the nerve cords in molluscs joining together the cerebral and visceral ganglia. *Syn.* visceral cords, visceral commissures. *See* pedal connectives.

visceral ganglion (Gr. *ganglion* = knot, little tumour), one of the three pairs of nerve ganglia of molluscs, situated in the visceral sac. *See* ganglion. *Cf.* cerebral ganglion, pedal ganglia.

visceroceptors (L. *recipere* = to receive), sensory receptors of the visceral organs.

viscosity (L. *viscosus* = sticky), the physical property of a liquid to resist flow; liquids become less fluid with decreasing temperature. The unit of viscosity is poise (*see* Appendix 1). *Adj.* **viscous.**

visible light/spectrum, → electromagnetic radiation within the wavelength range of 390—770 nm.

vision, sight, sensing with the eyes, act of seeing (*see* eye, photoreception); e.g. **binocular vision**, vision of the same object with both eyes; **monocular vision**, vision with one eye; **photopic vision**, cone vision with light-adapted eye; **scotopic vision**, rod or night vision; **colour vision**.

visual, pertaining to sight or vision.

visual pigments, photosensitive pigments of the eye; e.g. visual purple, i.e. → **rhodopsin**, and visual violet, → **iodopsin**. *See* photoreception.

vital (L. *vitalis* < *vita* = life), **1.** pertaining to life; **2.** necessary to life; **3.** being the necessary source of life, e.g. vital organs; **4.** having liveliness.

vital capacity, VC, respiratory capacity (of

lungs), i.e. the maximum volume of expired air after maximal inspiration.

vitalism, a scientifically false doctrine claiming that functions of an organism are due to the vital principle (*vis vitalis*) and not explainable by chemical and physical processes.

vitality, capacity for living, vigorousness of an organism; reduction in vitality is observable e.g. in the retardation of growth, and in the reduction of the production of offspring or fruit. Reduction in the vitality of dominant species in a plant community is an indication of harmful changes in that environment.

vital staining, staining of living cells or tissues by non-toxic dyes.

vitamin, an organic compound that is necessary in small quantities for the metabolism of organisms. Animals are not usually able to synthesize vitamins, but have to obtain them from their diet. Vitamins function in cell metabolism as cofactors of enzymes or as antioxidants; some act like hormones. Many of them are water-soluble (→ vitamin B complex and ascorbic acid), some are fat-soluble, such as → vitamin A, D, E and K.

vitamin A, a fat-soluble compound of 20-C alcohol found in organisms in two forms: retinol (vitamin A_1) and dehydroretinol (A_2). In animals they are synthesized from carotenes produced by plants. Vitamin A is needed for differentiation of epithelial tissue and for a part of photopigments in the eye. In vertebrates, the deficiency causes disturbances in the epithelial tissue and night blindness. Dietary sources of vitamin A are e.g. vegetables, whole grain cereals, egg yolk, liver, and meat.

vitamin B complex, a group of water-soluble compounds most of which are coenzymes of cell metabolism. Generally only plants and microbes can synthesize these vitamins, and animals get them directly or indirectly from those sources. The group includes several substances. Vitamin B_1 or **thiamine**, is a cofactor for some oxidative decarboxylases the deficiency of which results in disturbances in neural functions (beriberi disease). Vitamin B_2 or **riboflavin** is a cofactor of flavoprotein enzymes; deficiency shows as disturbances e.g. in mouth mucosa. **Niacin** or nicotinic acid and **niacinamide** or nicotinamide are constituents of redox coenzymes NAD and NADP. Deficiency causes disturbances in neural functions, mucosal membranes, and skin (pellagra). Vitamin B_6 or **pyridoxine** (and its derivatives) is a coenzyme of some decarboxylases and transaminases. Deficiency results in spasmic contractions of muscles. **Pantothenic acid** is a constituent of → coenzyme A. Deficiency, not occurring in humans, produces in laboratory animals disturbances especially in functions of the skin, alimentary canal and adrenal gland. **Biotin** or vitamin H functions as a cofactor in fat synthesis (coenzyme R). Deficiency (very rare in humans) causes disturbances especially in the skin and alimentary canal. **Folates** (salts of folic acid) work as coenzymes of carbotransferase in methylation reactions. Deficiency appears as anaemia and disturbances in intestinal absorption, and is globally quite common in humans, especially in small children, elderly people and during pregnancy. Vitamin B_{12} or **cyanocobalamin** is a coenzyme for enzymes functioning in amino acid metabolism. It must be combined with the intrinsic factor of the stomach mucosa to be absorbed in the small intestine, and the absence of this factor leads to malabsorption of B_{12}. Cyanocobalamin, also called **external factor**, is needed for maturation of red blood cells, and the deficiency causes pernicious anaemia.

vitamin C, → ascorbic acid.

vitamin D, one of several fat-soluble compounds (sterols) that have antirachitic properties, collectively called calciferols, including **ergocalciferol** (vitamin D_2) and **cholecalciferol** (vitamin D_3). Vitamin D acts like a hormone in the regulation of calcium concentration in blood by facilitating the intestinal absorption of calcium and phosphates. It is stored in the liver and therefore the deficiency will develop slowly causing disturbances in ossification (rickets). Carnivores obtain vitamin D in animal fat and herbivores synthesize it from plant sterols. Some mammals, like man, can synthesize vitamin D from 7-dehydrocholesterol in the skin cells using ultraviolet radiation as energy source. In medication, vitamin D can be overdosed with harmful consequences.

vitamin E, alpha-tocopherol; a fat-soluble compound acting as a coenzyme for cytochrome enzymes in cell respiration and also as an antioxidant in tissues; e.g. it inhibits oxidation of cell membrane lipids. Its deficiency, although rare in humans, may cause liver degeneration and cell damage in various tissues resulting in muscle dystrophy, anaemia, and disturbances in gamete production (therefore called fertility vitamin). Dietary sources for humans are e.g. green vegetables, wheat germ oil, and egg yolk.

vitamin F, a former term for → essential fatty acids.

vitamin H, biotin; *see* vitamin B complex.

vitamin K, one of several fat-soluble thermostable compounds, including naturally occurring K_1, i.e. phylloquinone (phytonadione), K_2, i.e. menaquinone, and their synthetic derivatives K_3—K_6. Vitamin K is necessary for blood coagulation because of its role in the synthesis of prothrombin in the liver. It is synthesized by intestinal bacteria in higher animals and therefore its deficiency rarely occurs in them.

vitellaria, pl. **vitellariae** (L. *vitellus* = yolk), a non-feeding, barrel-shaped larva of crinoids and some ophiuroids and holothuroids with an apical tuft and a number of transverse ciliated bands. *Cf.* auricularia, doliolaria.

vitellarium, pl. **vitellaria,** → yolk gland.

vitellin, a phosphoprotein of the egg → yolk.

vitelline, pertaining to the yolk of an egg or ovum.

vitelline membrane, *see* egg membranes.

vitellogenin (Gr. *gennan* = to produce), a large protein that is synthesized in the liver of many vertebrates. It is secreted into the bloodstream where it combines with lipids and carbohydrates and is then endocytotically taken up by developing oocytes. It supplies the oocyte with lipids, carbohydrates, phosphates, calcium, and storage yolk proteins such as vitellin and phosphovitin.

vitreum (L. *corpus vitreum*), the transparent hyaloid gel that fills the eyeball behind the lens; called also **vitreous body** (hyaloid body).

vitrification, 1. transformation from a liquid phase to a non-crystallized stage (e.g. water); **2.** plant tissues forming a watery and transparent mass in tissue cultures due to too low agar content or excess of water in the medium; **3.** the glassy or vitrified hairs of epidermal tissue of plants (e.g. birch) produced in tissue cultures.

viviparous (L. *vivus* = living, *parere* = to give birth), **1.** giving birth to living young; *cf.* oviparous, ovoviviparous; **2.** pertaining to plants having shoots which produce small plantlets or → bulbils, which separate from the parent plant and develop into separate individuals; the reproduction is fully vegetative. Sexual viviparous reproduction occurs if an embryo begins to develop inside the seed before it is shed from the parent.

vivisection (L. *secare* = to cut), cutting of, or operation on a living animal body usually for physiological or pathological purposes.

V_{max}, maximal velocity, the maximal velocity of an enzyme reaction. *See* Michaelis—Menten kinetics.

VNTR, → variable number of tandem repeats.

vocal cords, two pairs of folds of mucous membrane in the cavity of the larynx in tetrapod vertebrates; **false vocal cords**, the folds of the laryngeal membrane forming the boundary toward the vestibule; **true vocal cords** (*plicae vocales*), the folds that cover the vocal muscles in the larynx and form the inferior boundary toward the trachea; the vocal cords and the opening between them (*rima glottidis*) form the vocal apparatus in the glottis.

volant (L. *volare* = to fly), pertaining to flight.

volar, 1. (L. *volaris* < *vola* = a concave surface), pertaining to the sole (plantar), or to the palm; **2.** (L. *volare* = to fly), pertaining to flight; also volant.

Volkmann's canal (*A.W. Volkmann*, 1800—1877), *see* bone.

volt (*Alessandro Volta*, 1745—1827), **1.** the SI unit of electric potential, symbol V; *see* Ohm's law, Appendix 1; also the unit of electromotive force; **2.** electron volt; *see* Appendix 1.

voltage, the electric potential, potential difference, or electromotive force expressed in volts.

voltage clamp, a method used in electrophysiology for the measurement of ionic currents in cell or tissue preparations; the voltage level is clamped to certain values and ionic currents are measured at these voltage levels. *Cf.* patch clamp.

voltine (It. *volta* = time), pertaining to a breeding rhythm, voltinism; **univoltine, bivoltine, multivoltine** producing one, two, or many sets of offspring in a season.

voltinism, 1. a type of polymorphism in insects in which a part, but not all, of a population enter a diapause (rest period); **2.** breeding rhythm, brood frequency.

voluntary nervous system, that part of the nervous system that innervates skeletal muscles (**voluntary muscles**); consists of motor nerves to skeletal muscles and brain areas involved in motor functions. *See* pyramidal system, extrapyramidal system, reflex movement.

volva (L.), *see* veil.

volvent (L. *volvere* = to roll), one of the three types of nematocysts in cnidarians; a small, pear-shaped bladder containing a short, thick thread, which when discharged coils around a prey. *Syn.* desmodeme. *Cf.* penetrant, glutinant.

vomeronasal organ (L. *vomer* = plowshare, the bone of the nasal septum, *nasus* = nose), → Jacobson's organ.

vomit ball (L. *vomitare* = to discharge, vomit), undigested food remains in a ball-shaped clot (pellet) ejected through the mouth. Birds of prey (e.g. eagles, hawks, owls), and also birds feeding on fish or insects (e.g. gulls, terns, swallows) eject vomit balls.

vomiting reflex, in many mammals and birds, a series of reflex movements that causes the expulsion of the contents of the stomach or crop to the mouth. Usually after an inspiration the abdominal muscles (in mammals also the diaphragm) contract strongly and cause the expulsion. The reflex is usually activated by irritation of the stomach, but starts in many mammals also by irritation of the balance sense (travel sickness). The reflex is controlled by a group of nerve cells in the brain (**vomiting centre**) located in the area postrema of medulla oblongata. Toxic substances in blood can also directly irritate the centre resulting in vomiting. For some mammals and birds vomiting is a pattern of normal behaviour e.g. eliminating pellets of undigested material or for feeding offspring.

vulva (L.), **1.** the external genital organs of female mammals, including the clitoris, labia, and vestibular glands; **2.** also used to refer to the analogous genital organs in invertebrates.

W

W, symbol for **1.** watt; **2.** tryptophan; **3.** W chromosome.

waggle dance, a dance of bees in the figure of 8 in which the bee waggles the abdomen sidewards giving information about the distance and direction of a food source in relation to the sun; described in detail by *Karl von Frisch* (1886—1982).

Wallacea (*A.R. Wallace*, 1823—1913), a separate transitional zoogeographic region between the Oriental and Australian regions, marked by Weber's line in the east and Wallace's line in the west; comprises e.g. the islands of Sulawesi, Lombok, Flores, and Timor.

Wallace's line, an imaginary zoogeographical boundary between the Oriental and Australian regions running between islands of Bali and Lombok and then between Sulawesi and Borneo, passing the Philippines on the east side and continuing northwards (modified by E. Mayr in 1944).

wallowing, the thermoregulatory behaviour found in many mammals, increasing evaporative heat loss by spreading an aqueous fluid such as urine, saliva, water, or mud on the body surface, or rolling in mud or dust.

Warburg—Dickens pathway (*Otto Warburg*, 1883—1970, *Frank Dickens*, b. 1899), → pentose phosphate pathway.

warm-blooded animals, homoiotherms. *See* homoiothermy.

warm receptors, *see* temperature sense.

water, colourless liquid composed of hydrogen and oxygen (H_2O) by covalent bonds. Water is a polar compound having (—)-charge at the oxygen side and (+)-charge at the side of hydrogen atoms. Hydrogen bridges bind water molecules together forming groups of two to several molecules. Water is heaviest at +4°C, a property which prevents natural lakes and ponds from freezing to the bottom.

Water is a good universal solvent and is present in all organisms, and it moves osmotically through cell membranes. The seeds, spores, and other resting forms of many organisms can be very dry (water < 10%), but most fresh tissues of plants and animals contain 70—90% water. In cells and tissues, water is partly bound to proteins and other macromolecules, a bigger portion, however, is free water. Colloids (e.g. clay colloids) bind water very tightly, and many compounds contain crystallized water. *See* water circulation.

water balance, maintenance of osmotic concentration in organisms, based on water regulation and on regulation of concentration of salts and other compounds in cells and tissues. The mechanisms and ability to maintain water balance vary in animals. Most aquatic (especially marine) invertebrates cannot regulate the osmotic concentration of their tissue fluids but can tolerate moderate osmotic changes in their tissues (osmoconformers, poikilosmotic animals). On the other hand, many invertebrates like almost all vertebrates can regulate their tissue fluids (osmoregulators or homoiosmotic animals); *see* osmoregulation.

A sufficient → water potential maintains the turgor in plant cells as well as their metabolic activity, e.g. photosynthesis. The roots of plants can take water from the soil if the concentration of cell sap is higher than that of the surrounding solution. Some lower plants can take water through their entire epidermis. A strong → transpiration can cause the imbalance of the water economy. Some herbicides inhibit the synthesis of → cuticle, and treated weeds are therefore killed because of an uncontrolled transpiration. *See* turgor.

water bears, → Tardigrada.

water circulation, 1. physical circulation of water on the Earth and in the atmosphere; *see* hydrological cycle; **2.** chemical circulation of water (H_2O); includes the **synthesis of water** in burning of organic material and in the respiration of organisms when hydrogen derived from organic compounds and atmospheric oxygen react with each other (*see* cell respiration), and the chemical **decomposition of water** that occurs in → photosynthesis in green plants using light energy, and spontaneously in very high temperatures, e.g. during a lightning.

water culture, the cultivation of plants without soil. Roots are grown in a liquid nutrient solution, or seedlings are planted into a chemically inert material such as quartz sand, → perlite, → vermiculite or styrox mass watered with nutrient solutions. This technique is much used for experimental purposes, but also some vegetables and other crops are produced by this method, also called **hydropon-**

ics.

water expulsion vesicle, → contractile vacuole.

water flowering, algal → bloom; occurs in waters in temperate zones during late summer due to mass occurrence of colonies of → cyanobacteria. May cause harmful, even toxic changes in water.

water luminescence, a type of bioluminescence, i.e. blue-white glow occurring in tropical seas, caused by some dinoflagellates (e.g. *Noctiluca miliaris*). *See* luminescence.

water overturning, vertical circulation pattern of water in lakes due to temperature changes in the spring and autumn; an important process for oxygenation of natural waters especially in arctic and subarctic areas. *See* stratification.

water plants, plants (hydrophytes, aquatic plants) living in water; grouped according to their living forms: **1. isoetids** living on the bottom, totally under water (e.g. quillworts, *Isoëtes*); **2. elodeids** rising from the bottom but not reaching the surface (e.g. Canadian pondweed, *Elodea canadensis*; **3. nympheids** with leaves floating on the water (yellow water lily, *Nuphar*, and water lily, *Nymphaea*); **4. helophytes** with their shoots and leaves mainly above the water surface (reed); **5. lemnids** floating on the water without roots (duckweed-species).

water potential, symbol ψ, the water balance of a cell or any other system. *See* osmosis.

water storage parenchyma, → parenchyma tissue composed of large, thin-walled, highly vacuolate cells in plants, storing water; found e.g. in cacti.

water stress, a stress condition in plants caused by water deficit.

water-vascular system, a circulatory system of echinoderms for locomotion, clinging, food handling, and respiration. The system comprises water-filled canals, i.e. a → **stone canal**, a **ring canal**, and five **radial canals** connected with each other. The system is in connection with surrounding water, i.e. the stone canal opens to the body surface through narrow pores of the → madreporite (plate), and many short **lateral canals** connect the radial canal to → tube feet. *See* Tiedemann's vesicles.

Watson—Crick helix (*James Watson*, b. 1928; *Francis Crick*, b. 1916), the molecular structure of DNA. *See* deoxyribonucleic acid.

watt (*James Watt*, 1736—1819), unit of power. *See* Appendix 1.

waxes, compounds formed from esters of fatty acids and polyalcohols. Waxes cover and protect the integument of many invertebrates, the fur of mammals, and different parts of plants, serving e.g. in preventing water evaporation or protecting the surface of fruits. Some plant and insect waxes have economic value, among others carnauba wax obtained from the carnauba wax palm in South America is used e.g. in car wax. Beeswax, a substance secreted by bees for constructing the honeycomb, is used for several purposes, e.g. in cosmetics.

W chromosome, a type of sex chromosome in the females of some animal groups (e.g. birds, reptiles, and butterflies) in which the female is the heterogametic sex, i.e. the sex chromosomes are different; occurs as a pair of the → Z chromosome, and corresponds to the Y chromosome of mammalian males. *See* sex determination.

Weberian organ (*E.H. Weber*, 1795—1878), an auditory organ of bony fishes in the orders Cypriniformes and Siluriformes. It includes three small ossicles which have originated from the processes of the vertebral column, acting as a lever arm that transfers water vibrations from the swim bladder to the internal ear. The Weberian organ is analogous to the auditory ossicles of mammals. *Syn.* Weberian ossicles, Weberian apparatus.

Weber's line, a zoogeographical line, more eastern than the → Wallace's line, running to the north between the Molucca Islands and Sulawesi and to the south between Timor and the Kai Islands. The line separates the Oriental and Australian regions from each other.

weeds, plants growing in undesirable places or causing economic damage or doing harm to useful plants. Weeds are divided into herbs, woody plants, and other weeds (mosses, lichens, and algae).

Wegener's continental drift theory (*Alfred Wegener*, 1880—1930), → continental drift theory.

Wernicke's area/field/centre (*Karl Wernicke*, 1848—1905), sensory speech centre, located in the cerebral cortex near the lateral sulcus of the left hemisphere in man; corresponds to → Brodmann's areas 40, 39, and 22. The area is concerned with comprehension of auditory and visual information and controls the formulation of coherent speech associating with the motor speech centre, → Broca's area.

Western blotting (etym. *see* Southern blotting), a technique for the analysis and identification of protein antigens. Proteins are transferred (blotted) from an electrophoresis gel to a nylon or other membrane where they are labelled with an → antibody or another probe. *Cf.* Northern blotting, Southern blotting.

wet deposition, *see* acid deposition.

whalebone, any of the long parallel bony plates with fringy margins hanging from the palate in the mouth cavity of whalebone whales, → Mysticeti. The animals feed by gathering small marine organisms with the fringes of the whale-bones. *Syn.* baleen.

whalebone whales, baleen whales, → Mysticeti.

whales, → Cetacea.

wheat germ extract, an extract obtained from wheat embryos and used for → *in vitro* translation of proteins. *See* genetic translation.

wheel animals, → Rotifera.

white (blood) cells, leucocytes, Am. leukocytes; colourless nucleate cells of the blood in vertebrates, intruding from blood vessels to other tissues; include → **lymphocytes**, different types of → **granulocytes**, and → **monocytes**. The number of leucocytes varies greatly depending on the animal group, developmental stage, and physiological condition of an animal, usually comprising about 0.1—0.2% of the number of red blood cells. White blood cells function in the physiological defence of animals. *See* immunity, inflammation, phagocytosis, macrophages.

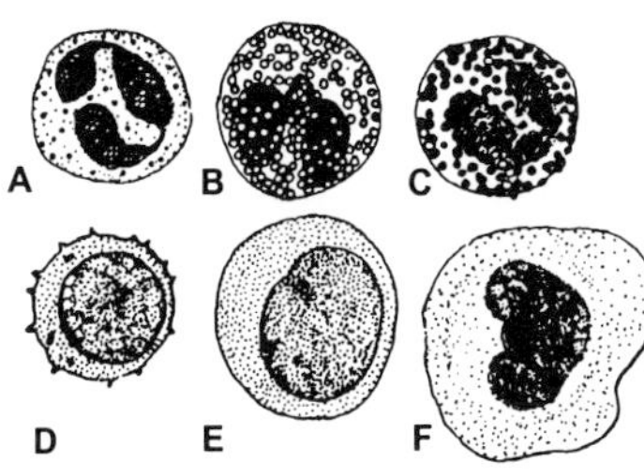

Different types of white blood cells: neutrophilic (A), basophilic (B), and eosinophilic (C) granulocytes; small (D) and large (E) lymphocytes; and a monocyte (F).

white fibres, anaerobic (glycolytic) muscle fibres of vertebrate skeletal muscles having a highly active glycolytic enzyme system (*see* glycolysis), and less effective aerobic enzyme system, i.e. less mitochondria and myoglobin than in → red fibres; therefore they look white. The contractions of white fibres are faster than those of red, but because of the accumulation of lactic acid they become fatigued quickly. Those muscles which contain mainly white muscle fibres are called **white muscles**, e.g. leg muscles in many animals. Most muscles are of mixed type.

white matter/substance (L. *substantia alba*), the tissue of white fatty areas in the vertebrate brain composed of parallel axons covered by → myelin sheaths. The white substance forms superficial tracts along the brain stem and spinal cord, and deeper tracts in the cerebral and cerebellar tissues. *Cf.* grey matter.

white medulla, → spleen.

white muscle, *see* white fibres.

white rot, *see* rot.

whole arm translocation, a → translocation involving whole chromosome arms. *Syn.* Robertsonian translocation.

whorled leaf arrangement, *see* leaf arrangement.

wild allele, an → allele of a gene which is common in natural → populations.

wild type, a type or form of an organism or a gene which is prevalent in natural → populations.

wilting, *see* permanent wilting point.

wind net trap, an insect trap rotating with the wind, constructed especially to catch cicadas.

windpipe, → trachea.

wind pollination, transport of pollen grains from a flower to another with the help of the wind.

winter bud, an overwintering plant → bud protected by hard, thick scales.

winter seeders, plants having seeds which develop just before winter and loosen during winter; this is useful for dispersal of plants, such as spruce and pine, elms and maples.

winter sleep, a special strategy of some mammals (e.g. bear, badger) to pass the winter in cold regions; the animal spends the coldest time in deep sleep at which state the body temperature decreases 4—5°C. Tissue metabolism decreases 30—50% and nitrogen metabolism changes so that urea is not secreted (*see* urea nitrogen recycling). When disturbed, the animal wakes up immediately and reacts usually with fight or flight. Winter sleep is also called facultative → hibernation, or winter torpor.

winter spore, a spore type in → Uredinales (rust fungi). *Syn.* teleutospore.

winter stagnation (L. *stagnum* = standing water), a situation in a water body during winter when the temperature at the bottom is 4°C and the surface water is cooler; the water body is then stationary and unmixed. *See* summer stagnation, stratification.

winter territory, a territory where an animal overwinters in an active condition. *See* overwintering.

witches' broom, an abnormally grown part of a shoot with very short internodes and dense group of branches; typical of birch branches in which it is caused by the saprophyte fungus *Taphrina betulina*; in conifers, witches' brooms are mostly caused by → bud mutations.

wobble hypothesis, a hypothesis which explains the pairing of the → codon in → messenger RNA (mRNA) and the → anticodon in → transfer RNA (tRNA) in the third position of the codon. According to the hypothesis, the same tRNA molecule is able to recognize more than one codon in which case the base pairing in the first two positions is strictly consistent with the → base pair rule, but wobbly in the third position.

Wolffian duct (*Kaspar Wolff*, 1733—1794), mesonephric duct; the primary duct that conducts urine from the → kidney (mesonephros type) in cyclostomes, fishes, and amphibians. In amniote embryos (reptiles, birds and mammals), when the metanephros develops, the Wolffian duct degenerates in females but forms the ductus deferens in males.

wood, the hard, lignified, mostly non-living part of a tree, formed from secondary → xylem.

wooded meadow, a type of vegetation formed from a → coppice by man's activities, especially by grazing of cattle; tree and bush groups alternate with open meadow areas.

woodpeckers, → Piciformes.

wrist (L. *carpus*), a joint or articulation between the forearm and metacarpus in tetrapod vertebrates. Primarily it consists of several small bones that form ventrally a wrist canal, where tendons of flexor muscles run under the ligament. Dorsally the tendons of extensor muscles lie under a similar ligament. There is much variation in wrist structures in different tetrapod classes.

X

x, symbol for **1.** the basic haploid (monoploid) chromosome number of a species or group of species; the smallest number of chromosomes in the chromosome set; *cf.* n; **2.** xanthosine (in nucleotides).

xanthine (Gr. *xanthos* = yellow), a white → purine base (its nitrate is yellow) in animal and plant tissues; formed by oxidation of hypoxanthine and may be oxidized to uric acid. Xanthine and its derivatives have pharmacological effects on animals. *See* caffeine, theobromine, theophylline.

Xanthophyceae (Gr. *phykos* = seaweed), a class in the division → Xanthophyta, yellow-green algae.

xanthophyll cycle (Gr. *phyllon* = leaf), the xanthophyll reactions in the → thylakoids of chloroplasts. Violaxanthin is an epoxide that is transformed to zeaxanthin by a de-epoxidase on the inner surface of a thylakoid and an epoxidase of the outer surface converting zeaxanthin to violaxanthin. The concentration of violaxanthin decreases in light and under anaerobic conditions, and is restored in the dark in the presence of oxygen. The cycle is controlled by the proton gradient formed by light.

xanthophylls, yellow carotenoid pigments of plant tissues, synthesized via oxidation of carotenes; found also in some animal tissues, such as corpus luteum and egg yolk, in many animals also in the skin (e.g. crayfish). The most common xanthophylls are lutein, viola-, neo-, zeaxanthin in plants, and lutein and astaxanthin in animals.

Xanthophyta (Gr. *phyton* = plant), **yellow-green algae**, division in Plant kingdom; unicellular and coenocytic plants (*see* coenocyte) living mostly in fresh waters or moist soil. The cells contain chlorophylls a and e, → xanthophylls, and beta-carotene, which give the cells the characteristic yellow-green colour. The best known genus is *Vaucheria*, which tends to form a layer on floors of greenhouses.

xanthopsin, a visual pigment with 3-hydroxyretinal and → opsin, called also visual yellow; results from the bleaching of → rhodopsin.

X chromosome (Gr. *chroma* = colour, *soma* = body), one of the two sex chromosomes existing in both sexes in which case the females are XX and the males XY (e.g. in mammals), or the males are XO if the → Y chromosome is lacking (e.g. in many insects). *See* sex determination.

X-chromosome inactivation, X inactivation; a mechanism of → dosage compensation in female mammals whereby one of the two X chromosomes becomes inactivated and changes into a → Barr body in the → somatic cells at an early embryonic stage. The inactivation is a random and irreversible event, i.e. in one cell the Barr body may be maternal and in another cell it may be paternal. It occurs in each cell independent of the other cells, but once it has occurred it is inherited in cell divisions to the daughter cells. *Syn.* Lyonization.

xeno- (Gr. *xenos* = strange, foreign), pertaining to strange or odd, or denoting relationship to foreign material.

xenobiotics (Gr. *bios* = life), chemicals foreign to organisms, e.g. pollutants.

xenogamy (Gr. *gamos* = marriage), cross-pollination, the pollination of a flower with pollen of some other, genetically non-identical plant individual of the same species; leads to cross-fertilization.

xenogenous (Gr. *gennan* = to produce), originated outside the organism.

xenograft, → heterograft.

xerarch (Gr. *xeros* = dry, *arche* = beginning), originating in a dry habitat.

xerarch succession (L. *succedere* = to follow), development of vegetation on dry soil towards more mesic conditions; occurs e.g. on the ground rising from the sea after the Ice Age (in Scandinavia). *See* succession.

xeromorphic (Gr. *morphe* = form), structurally modified to succeed in dry (xeric) conditions. *See* xerophyte.

xerophyte (Gr. *phyton* = plant), a plant adapted to a dry habitat, i.e. to xeric conditions; such plants have xeromorphic structural features, such as hairs, low surface/volume ratio, leaf reduction into thorns, which prevent transpiration of water. Many xerophytes, as e.g. cacti, can store water during moist seasons for dry periods in a special tissue, called water parenchyma.

xerosere (L. *serere* = sow), → succession of vegetation on dry land. *Cf.* hydrosere.

X-gland, X-organ, a paired endocrine gland

associated with the anterior cerebral ganglia (*medulla terminalis*) in crustaceans. It contains neurosecretory cells whose axon endings form another gland (**sinus gland**) near the eye, in decapods at the bottom of the eyestalk. The X-gland controls the secretion of the sinus gland, stimulating the release of peptide hormones which inhibit moulting (*see* ecdysone) and regulate colour change, osmotic concentration, and cell metabolism. The effect on moulting is transmitted by its inhibitory action on the → Y-organ.

xiphisternum (Gr. *xiphos* = sword, *sternon* = sternum), the most caudal of the three segments of the sternum in tetrapod vertebrates.

X-organ, → X-gland

X-ray crystallography, a method to study the three-dimensional structures of molecules using diffraction patterns produced by X-rays.

X-ray micronalysis, a method used for determination of elements in samples made for electron-microscopical studies; the analyser identifies the elements according to their characteristic emission of X-ray irradiation. *Syn.* roentgen microanalysis.

X-rays, Roentgen rays, Röntgen rays; a penetrating electromagnetic radiation (wavelength 0.1—10 nm) that causes ionization in organisms. X-rays are produced when highly accelerated electrons hit atoms of heavy metals (electric potentials more than 10,000 V are used). To man, the lethal whole-body single dose of radiation is about 500 roentgens (*see* Appendix 1). *See* electromagnetic radiation.

xyl(o)- (Gr. *xylon* = wood), pertaing to wood.

xylan, hemicellulose present in deciduous trees, herbs, and seeds.

xylem, a water-conducting plant tissue, belonging to → vascular tissues. Xylem consists of different cell types: **tracheids** and **vessels**, **parenchyma cells** and **xylem fibres**. The water-conducting parts are the tracheids and vessels; the former are separate cells connected with each other by bordered pits in the cell walls, vessels being unbroken tubes. Tracheids are more primitive and typical of → gymnosperms, vessels are more developed and typical of → angiosperms. The parenchyma cells (vascular parenchyma) function in the xylem as storage tissue, and the xylem fibres as supporting structures. *See* phloem.

xylem fibre, Am. **xylem fiber,** a type of thick-walled supporting cell in → xylem.

xylem ray, a band formed from parenchyma cells penetrating through the secondary xylem in the direction of shoot or root radius. *See* ray.

xylem sap, *see* sap.

xylitol, a sweet natural alcohol, CH_2OH-$(CHOH)_3CH_2OH$, formed from xylose. Xylitol is used in chewing gums and sweets instead of glucose; mouth bacteria cannot use xylitol and the increased secretion of saliva caused by xylitol has an antibacterial effect.

xylol, dimethylbenzene, $C_6H_4(CH_3)_2$, a hydrocarbon synthesized from methanol; used in microscopical techniques as a solvent and clarifier. *Syn.* xylene.

xylose, a pentose sugar, $CH_2OH(CHOH)_3CHO$, present in mucopolysaccharides of animal tissues and in plants as a partial structure of xylan.

xylulose, a pentose, $C_5H_{10}O_5$, occurring e.g. as an intermediate in the Calvin cycle of → photosynthesis.

XYY-syndrome, an abnormality of the human sex chromosomes due to → aneuploidy which is characterized by tall stature and often mental lability. The chromosome set in this syndrome is 47, XYY, and accordingly these individuals are men.

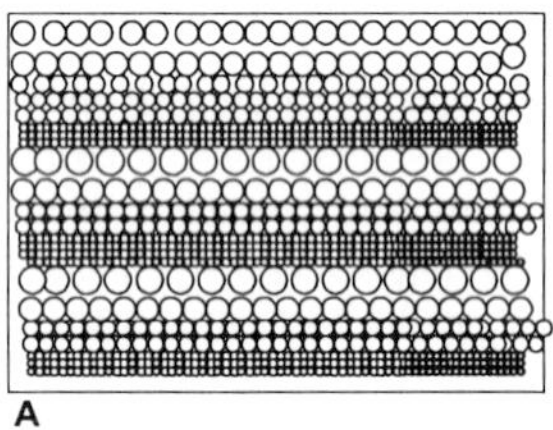

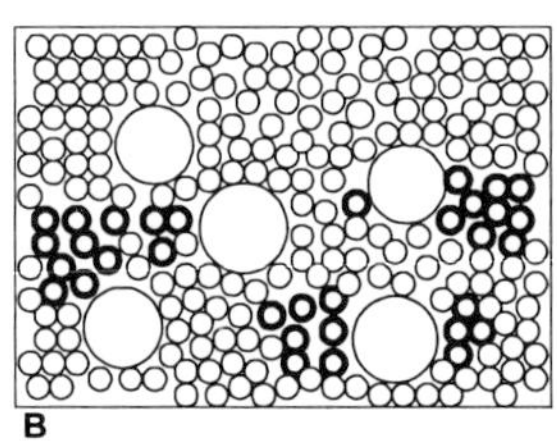

Xylem. Structures of secondary xylem in gymnosperms (A), composed of only tracheids; the growth rings formed by tracheids of different sizes are clearly seen. The secondary xylem of dicotyledonous trees (B) is composed of large tracheas, parenchyma cells (light small circles) and xylem fibres (dark small circles).

Y

Y, symbol for **1.** tyrosine; **2.** yttrium.

YAC, → yeast artificial chromosome.

Y chromosome (Gr. *chroma* = colour, *soma* = body), the male sex chromosome in those animal species in which the male is the heterogametic sex; in these species the two sex chromosomes are different, as e.g. in mammals and dipterans, the males being of the XY type and females of the XX type. *See* X chromosome, sex determination.

YCp, → yeast centromeric plasmid.

yeast, any species in the heterogeneous group of unicellular or thread-like fungi, used in brewing and baking industry. Most yeasts belong to the division Amastigomycota (Ascomycota). Some of them are placed in division Deuteromycota. There are about 350 yeast species.

yeast artificial chromosome, YAC, any artificial chromosome which replicates in yeast cells; consists of the yeast → centromere and → telomeres, between which any fragment of DNA (e.g. human DNA) can be packed using methods of genetic engineering. One YAC can carry 5,000 kb of DNA. These chromosomes behave normally during the division of the yeast cell in → mitosis.

yeast centromeric plasmid, YCp, an artificial → plasmid construction of yeast containing the → centromere and is thus stable and capable of attaching to the → mitotic spindle, but which in contrast to linear → yeast artificial chromosomes, is circular.

yeast episomal plasmid, YEp, a natural → plasmid of yeast which can be used as a → vector in producing transgenic yeast cells, and which can exist either as a free plasmid or integrated into the chromosomal complement of the host cell.

yeast expression plasmid, YXp, an artificial → plasmid construction of yeast containing a → yeast episomal plasmid and in addition, an → expression complex; usually contains many different restriction cleavage sites.

yeast integrative plasmid, YIp, a natural → plasmid of yeast which can be used as a → vector in producing transgenic yeast cells, and which must be integrated into the chromosomal complement of the yeast.

yeast replicative plasmid, YRp, a natural → plasmid of yeast which can be used as a → vector in producing transgenic yeast cells; replicates only in yeast cells and cannot be integrated into its chromosomal complement.

yeast two-hybrid screen, a method for elucidating interactions between proteins. Its basic principle is, that the protein of interest is fused with the heterologous DNA-binding → domain. This hybrid protein is then used as a "bait" to screen a library where cDNA fragments are expressed as fusions to an activation domain. Both types of hybrids are expressed in a *Saccharomyces cerevisiae* strain that contains two integrated reporter constructs, the yeast *HIS3* gene and the bacterial *lacZ* gene. Individually none of the hybrid proteins are expected to be able to activate transcription from the reporter constructs, but when the DNA-binding domain of the bait and the activation domain are brought together via an interaction between the bait and one of its putative partners, this results in → transactivation. The transactivation can be easily detected, since it enables the host cells to grow in the absence of histidine, and makes them blue in the presence of suitable substrates due to the production of beta-galactosidase encoded by the *lacZ* gene.

yellow fever, an infectious disease caused by a flavivirus transmitted to man by mosquitoes which get the virus either from man (urban type, *Aedes aegypti* as a vector) or from other mammals (jungle type, *Aedes africanus* and *Haemagogus* species as vectors). It is marked by fever, in severe forms leading to liver and renal damage. The disease is most common in tropical regions of Africa and South America.

yellow spot (L. *macula lutea retinae*), a yellowish area at the optic axis in the retina; the central fovea is located in the centre of it. The fovea is the area of colour vision where the visual acuity is greatest.

YEp, → yeast episomal plasmid.

Yersinia (*A. Yersin*, 1862—1943), a genus of bacteria; many species are zoonotic causing diseases in animals, transmissible to man; e.g. *Y.enterocolitica* causes yersiniosis, transmitted through pork, *Y. pestis* (*Pasteurella pestis*) causes plague, and *Y. pseudotuberculosis* causes pseudotuberculosis in birds and rodents and a type of mesenteric lymphadenitis in man.

Y-gland, Y-organ, a paired endocrine gland in the antennal or maxillary segment of crustaceans, secreting moulting hormones. *See* ecdysone. *Cf.* X-gland.

YIp, → yeast integrative plasmid.

Yoldia Sea, a cool, marine stage of the Baltic Sea after the Ice Age about 12,000—9,000 years ago, succeeded the Baltic Ice Sea stage; a species typical of this stage was the clam *Yoldia arctica.*

yolk, the part of the contents of the egg necessary for nourishing the animal embryo during development; rich in protein, fat, and micronutrients. The yolk of egg-laying vertebrates, such as fishes, amphibians, reptiles and birds, nearly fills the whole egg, or a big part of it (**mega(lo)lecithal egg** or **mesolecithal egg**, respectively); in mammals the yolk comprises not more than 5% of the egg volume (**oligolecithal egg**). The yolk is synthesized within the egg from raw materials supplied by the surrounding follicle cells, in many invertebrates by the action of the → yolk gland.

yolk gland, a gland in many invertebrate females secreting yolk to nourish a developing embryo. *Syn.* vitellarium.

yolk sac, the first extraembryonic membrane; an organ present during the embryonic development of vertebrates, derived from endoderm; forms an energy store for the development of the embryo, in mammals only for initial stages, but in many animals (e.g. birds and reptiles) for the entire embryonic life. The yolk sac of mammals with oligolecithal eggs have two developmental stages; the **primary yolk sac** serves as an energy storage only for initial development of the embryo. Later it degenerates and the endoderm and mesoderm together form the **secondary yolk sac** which gets nutrients from the uterus. The first blood cells are derived from the yolk sac. Later the blood-forming stem cells move to other tissues, such as to the liver, spleen, and bone marrow. Also the stem cells of gametes, i.e. cells of the germinal epithelium of the gonads, are derived from yolk sac cells.

Y-organ, *see* Y-gland.

YRp, → yeast replicative plasmid.

YXp, → yeast expression plasmid.

Z

Z, symbol for **1.** ionic charge number; **2.** glutamine or glutamic acid.

***Z*,** symbol for atomic number.

Z chromosome, a type of sex chromosome in some animal groups (e.g. birds, reptiles, and butterflies) in which the female is the → heterogametic sex, i.e. the sex chromosomes are different. In this system the females are of WZ and the males of ZZ type. *See* W chromosome, sex determination.

Z disc/line/band, a thin intracellular disc in the centre of the → I band of a muscle fibril inside the muscle cell; contains α-actinin that anchors actin filaments. The segment between two adjacent Z lines is called **sarcomere**.

Z-DNA, 1. a rare form of deoxyribonucleic acid (DNA) conformation in which the double helix is a left-handed coil, forming a zigzag-like structure; occurs in living cells of both prokaryotic and eukaryotic organisms where it is believed to recognize signals of gene regulatory proteins; **2.** a small fraction (0.3%) of the total amount of DNA in a cell to be synthesized during zygotene of → meiosis.

zeatin (L. *Zea mays* = maize, corn), a natural → cytokinin first found in maize; an adenine derivative like most other cytokinins.

zeaxanthin (Gr. *xanthos* = yellow), a yellow pigment in plants; belongs to → xanthophylls.

zein, a storage protein found in the maize kernel; belongs to alcohol-soluble proteins of the prolamine group. *See* prolamines.

Zeitgeber (Ger.), an external timer in biological rhythms, e.g. the intensity of light.

zigzag dance, a fixed-action pattern in the courtship behaviour of the male three-spined stickleback to attract a female to lay eggs into the male's nest.

zinc (L. *zincum*), a bluish-white metal, symbol Zn, atomic weight 65.38; exists naturally e.g. as zinc blende i.e. zinc sulphide, ZnS, as zinc oxide, ZnO, and in silicates and carbonates. It is a trace element in organisms and an essential part of several enzymes, e.g. in carbonic anhydrase. The deficiency in vertebrates results in anaemia, impaired healing of wounds, and underdevelopment of gonads. Plants require zinc for the biosynthesis of → auxin.

zinc finger protein, a protein, the tertiary structure of which is formed from finger-like loops (zinc finger motifs) when zinc atoms in the secondary structure couple cysteine residues together. The protein is encoded by regulator genes, and it recognizes certain DNA-sequences, fixes to them, and thus regulates the function of other genes; it is commonly found in organisms.

Z-line, → Z-disc.

zoaea, zoea, pl. **zoeae, zoeas** (Gr. *zoe* = life), an early larval stage of some decapods, like shrimps, with an easily distinguishable cephalothorax, abdomen, and compound eyes.

Zoantharia (Gr. *zoon* = animal, *anthos* = flower), **zooantharian corals**; a cnidarian subclass in the class → Anthozoa; polyp-like, colonial or solitary marine animals, usually having unbranched tentacles. The subclass is divided into five orders. Called earlier Hexacorallia (hexacorals) because of their hexamerous symmetry. *Cf.* Alcyonaria.

zona, pl. **zonae** (L.), zone, e.g. **1.** *zona pellucida*, the pellucid zone surrounding the vertebrate ovum; **2.** *zona fasciculata*, *zona glomerulosa*, *zona reticularis*, the layers of the cortex of the adrenal gland in vertebrates. *Cf.* zonula.

zonal rotor, *see* rotor.

zonation, 1. the formation of zones, bands, or concentric layers, as e.g. the zonate structure of a growing plant tissue; **2.** a pattern of distribution of species determined primarily by regular and parallel changes in environmental factors, e.g. climate. Zonation may be stratified as is the case in water bodies and mountains, or it is successive as found on the Earth's surface; e.g. the climatic and vegetation zones from the Equator to the Arctic, or those from coasts to inland regions. The environmental change may be steep in some regions, resulting in narrow zones such as littoral vegetation zones of sea shores.

zonula (L. dim. of *zona* = zone), zonule, a small zone in a histological structure; e.g. **1.** *zonula adherens* and *z. occludens*, junctional complexes between the epithelial cells; *see* tight junction; **2.** *z. ciliaris Zinni*, the ciliary zonule between the ciliary body and the equator of the lens in the vertebrate eye, a system of fibres holding the lens in place.

zoobenthos (Gr. *zoon* = animal, *benthos* = depth of the sea), animals living at the bottom of a water body.

zoochlorella, any chlorella (unicellular alga) living in symbiosis with animals such as rhizopods, radiolarians, sponges, or hydrozoans.

zoochore (Gr. *chorein* = to withdraw, go), a plant distributed by animals. *See* anemochore, antropochore, hydrochore.

zoochrome (Gr. *chroma* = colour), any animal pigment, such as melanotic, haematogenous, and hepatogenous pigments.

zooecium, Am. **zoecium** (Gr. *oikos* = house), a capsule or sac (body wall) that encloses each individual (zooid) in a colony of bryozoans.

zooflagellates, → Zoomastigophora.

zoogeography (Gr. *ge* = earth, *graphein* = to write), the field of science dealing with the geographical distribution of animals.

zoogeographical regions/realms, the wide geographical areas characterized by unified faunal features. Usually the regions are separated from each other by oceans or mountain ranges. The world is globally divided into the **Australian**, **Ethiopian**, **Nearctic**, **Neotropic**, **Oriental**, and **Palaearctic regions.** Every region consists of several smaller zoogeographical areas, **subregions**. *See* floristic kingdoms.

zoogloea, Am. **zooglea** (Gr. *glia* = glue), a jelly-like mass of microorganisms. *Adj.* **zoogloeal, zoogleal**.

zooid, 1. an individual in a combined colony of animals (as cnidarian and bryozoan colonies); **2.** an animal-like object or form. *See* blastozooid.

zoology (Gr. *logos* = word, discourse), **1.** the biology of animals; the branch of biology dealing with the structure, function, development, classification, evolution, and environmental relationship of animals; **2.** the study of animal life in a particular region. *Adj.* **zoological**.

Zoomastigina (Gr. *mastix* = whip), zooflagellates; in some classifications presented as a phylum, → Zoomastigophora.

Zoomastigophora (Gr. *pherein* = to bear), **zoomastigophorans,** a class in the phylum Sarcomastigophora, subphylum Mastigophora; unicellular flagellates, which are heterotrophic without chloroplasts and pigment organelles (*cf.* Phytomastigophora). The number of flagella varies from one to many hundred. Many zoomastigophorans are parasitic (e.g. the genera *Trichomonas* and *Trypanosoma*), but also free-living species are found.

zoometry, the measurement of proportional sizes of body parts of animals. *Adj.* **zoometric**.

zoonosis, pl. **zoonoses** (Gr. *nosos* = disease), an animal disease communicable to man. The infection often spreads from a symptomless animal **carrier** e.g. by faeces, urine and saliva, but is often transmitted by **vectors**, such as some insects and ticks. Zoonoses may be caused by **viruses** (e.g. epidemic nephropathy, foot and mouth disease, rabies and many viral diseases caused by arboviruses), by **bacteria** (borreliosis, plague, tularemia, salmonellosis, listeriosis, yersiniosis) or **parasites** (e.g. toxoplasmosis, malaria, echinococcosis, trichinosis).

zooparasites, parasitic animals. *See* parasite.

zoophage, → carnivore.

zoophilous (Gr. *philein* = to love), **1.** pertaining to plants pollinated by animals other than insects; **2.** having an attraction for animals, also **zoophilic**.

zooplankton (Gr. *planktos* = wandering), small animal organisms living in → plankton.

zoospore (Gr. *sporos* = seed), a motile flagellated spore of an aquatic alga or a fungus. *See* planospore.

zootoxin (Gr. *toxikon* = arrow poison), a poisonous (toxic) substance of animal origin, e.g. venom of snakes. *See* toxin.

zootype, a genetically determined, general ground plan for a developing structure of an animal embryo, typical of different phyla. *Cf.* phylotype.

zooxanthellae (Gr. *xanthos* = yellow), → dinoflagellates living in symbiosis with animals.

Z scheme, electron transport via the light reactions of → photosynthesis, presented according to oxidation-reduction potentials of reacting metabolites, e.g. plastoquinone, cytochrome complexes, and plastocyanin.

Zuckerkandl's body, *see* paraganglion.

zwitterion (Ger.), a dipolar ion carrying both a positive and negative electric charge; e.g. amino acids with COOH and NH_2 groups in their molecules form zwitterions (COO^- and NH_3^+) in aquatic solutions.

zygodactylous (Gr. *zygotos* = yoked together, *daktylos* = digit), pertaining to a foot type, usually found in birds, having two toes pointing forwards and two backwards; e.g. in the climbing feet of woodpeckers.

zygomorphic flower, a flower with only one plane of symmetry. *Syn.* dorsiventral flower, monosymmetric flower. *Cf.* asymmetric flower, bilateral flower, symmetric flower.

Zygomycetes (Gr. *mykes* = fungus), a class in the subdivision Zygomycotina, division Amastigomycota (in some systems division Zygomycota), kingdom Fungi. Typical of the species (ca. 500) is the formation of thick-walled **zygospores** generated when the tips of two hyphae are joined. The zygospores develop into sporangia. The group includes some well-known moulds, e.g. *Mucor* species.

Zygomycotina, a subdivision in the division Amastigomycota, kingdom Fungi. The group is divided into classes → Zygomycetes and Trichomycetes. In some systems Zygomycotina is classified as a separate division, Zygomycota.

zygospore (Gr. *sporos* = seed), *see* Zygomycetes.

zygote, 1. a cell formed by the union of two gametes; i.e. the fertilized ovum, egg; **2.** broadly, an individual derived from such a cell.

zygotene (Gr. *tainia* = string), the second stage of the first meiotic prophase in cell division, during which the two homologous chromosomes undergo pairing to form a bivalent structure. *Syn.* amphitene. *See* meiosis.

zymogen (Gr. *zyme* = leaven), proenzyme; any inactive enzyme protein which is activated through proteolytic cleavage. *Adj.* **zymogenous** (zymogenic).

SYMBOLS, UNITS AND NOMENCLATURE IN PHYSICS

SI system

SI is an abbreviation for Système International d'Unités (International System of Units). It is an international metric system of units recommended by the Conférence Générale des Poids et Mesures in 1960.

Physical concepts and units

Concepts include e.g. time, length, mass, force, frequency, resistance and the corresponding units are second, metre, kilogramme, newton, hertz and ohm, respectively.

There are several basic units and derived units. Furthermore, there are a number of supplementary units.

Basic units

concept	symbol	name of unit	abbr. of unit name
length	l	metre	m
mass	m	kilogramme	kg
time	t	second	s
electric current	I	ampere	A
thermodynamic temperature	T	kelvin	K
amount of substance	n	mole	mol
luminous intensity	I	candela	cd

Derived units

A derived unit is defined by a combination of basic units, which makes it possible to use both basic and derived units in a formula without any conversions. For example, the velocity is defined as the relation of length (distance) to time (m/s). Altogether 21 derived units have been given a special name. In the explanations below the equivalent units or alternative definitions are given e.g. 1 N = 1 $kgms^{-2}$ or = 1 kgm/s^2.

concept	symbol	name of unit	abbr. of unit name
absorbed dose	D	gray	Gy = J/kg
activity	A	becquerel	Bq = 1/s
dose equivalent	H	sievert	Sv = J/kg
electric capacitance	C	farad	F = As/V
electric charge	Q	coulomb	C = As
electric conductance	G	siemens	S = 1/W
electric potential	V	volt	V = W/A
electric resistance	R	ohm	W = V/A

concept	symbol	name of unit	abbr. of unit name
force	F	newton	N = kgm/s^2
frequency	f	herz	Hz = 1/s
illumination	E	lux	lx = lm/m^2
inductance	L	henry	H = Vs/A
luminous flux	Φ	lumen	lm = cdsr
magnetic flux	Φ	weber	Wb = Vs
magnetic flux density	B	tesla	T = Wb/m^2
plane angle etc.	α,β etc.	radian	rad
power	P	watt	W = J/s
pressure (stress)	p	pascal	Pa = N/m^2
solid angle	ω	steradian	sr
temperature	t	degree Celsius	^{o}C
work (energy, heat)	W	joule	J = Nm

For various purposes the decimal multiples and sub-multiples can be used with basic and derived units. For example the prefix kilo (k) indicates a unit multiplied by 1000 as in 1 km = 1000 m = 10^3 m. The following Greek derived prefixes and symbols are in use:

prefix	symbol	multiple	prefix	symbol	sub-multiple
exa	E	10^{18}	deci	d	10^{-1}
peta	P	10^{15}	centi	c	10^{-2}
tera	T	10^{12}	milli	m	10^{-3}
giga	G	10^9	micro	μ	10^{-6}
mega	M	10^6	nano	n	10^{-9}
kilo	k	10^3	pico	p	10^{-12}
hecto	h	10^2	femto	f	10^{-15}
deca	da	10^1	atto	a	10^{-18}

The multiples hecta, deca and centi are used with only some units, such as metre and litre. Generally, it is recommended that the prefixes are used so that the numeric value is bigger than 0.1 and smaller than 1000, for example 2.5 kJ but not 2500 J. A prefix representing 10 raised to a power that is a multiple of three should be used, for example, 10 mN (not 1 cN) or 100 g (not 1 hg).

The decimal multiples of the unit kilogramme (kg) are made by adding the prefix to the unit gram (g), for example, mg (milligram) and not mkg (microkilogramme). The prefix μ is always read as micro and not as micron as it is not its own special unit.

Supplementary units

Besides the SI units there are numerous units of measurements which are commonly used or used as alternatives in certain special fields.

Time units: minute (1 min = 60 s), hour (1 h = 60 min), day (1 d = 24 h), year (1 a = 365 d).

Plane angle units: degree ($1^\circ = 2\pi/360$ rad); degrees are converted to radians by multiplying by $2\pi/360$, and radians to degrees by multiplying by $360/2\pi$.

Volume units: litre (l or L = 1 dm^3). Decimal multiples of litre are, for example, microlitre, millilitre, decilitre and hectolitre.

Mass units: ton (1 ton = 1000 kg) and atomic mass unit (1 *u* = $1.66053 \cdot 10^{-27}$kg).

Energy units: electronvolt (1 eV = $1.602 \cdot 10^{-19}$ J).

Pressure units: bar (1 bar = 10^5 Pa).

Units are indicated with abbreviations or written in full with lower case letters. The abbreviations are unpunctuated. When talking about a specific value of a concept, the name of the concept must be used, for example, the power of an electric bulb is 40 W (but the amount of watts is not 40 W). Units and numeric values are not to be used as combined units, for example, the energy content of milk is approximately 2.3 MJ/kg instead of 230 kJ/100 g.

Biologically important concepts and units

absorbed dose, *D*, the dose of ionizing radiation when the average energy per unit mass imparted to matter by ionizing radiation is 1 gray (Gy), equal to 1 J/kg.

acceleration, *a*; gravity, *g*, the rate of change of velocity in m/s^2, the standard acceleration of gravity (at sea level, at 45° latitude) is 1 g = 9.80665 m/s^2. The former cgs unit of acceleration is gal, 1 gal = 1 cm/s^2.

activity, *A*, the number of disintegrations of a radfionuclide per unit time, measured in becquerels (Bq). Thus 1 Bq refers to one spontaneous nuclear disintegration per second. Former unit, the curie (Ci) is equal to $3.7 \cdot 10^{10}$ Bq.

(a)eon, a unit of time equal to 10^9 (a milliard, a billion) years.

ampere, A, the SI unit of electric current; defined as the constant current that, if maintained in two parallel conductors of infinite length placed one meter apart in a vacuum, would produce a force between the conductors of $2 \cdot 10^{-7}$ Nm^{-1}. If the current is 1 A there are $6.24 \cdot 10^{18}$ electrons per second going through the conductor. Multiple units kiloampere (10^3 A), milliampere (1 mA = 10^{-3} A) etc.; 1 A = 1 V/1 Ω.

ampere second, AS, the unit of electric charge; is equal to the charge transferred by a current of one ampere in one second. 1 As = 1 C (coulomb). The capacity of an accumulator battery and of other power sources is usually expressed in ampere hours (Ah); 1 Ah = 3600 As = 3600 C = 3.6 kC.

angström, angstroem unit, ångström, Å, 10^{-10} m = 10^{-1} nm; an old unit of length, superseded by nanometre (nm).

aperture, → numerical aperture.

ASA units, system of units developed by American Standards Association; e.g. in → photographic speed rating.

atmosphere, atm, a former unit of pressure; 1 atm = the pressure of 760 mm of mercury at 0°C at sea level at 45° latitude (1 atm = 101.325 kPa = 760 torr).

atomic mass unit, → relative atomic mass.

atomic weight, → relative atomic mass.

atomic number, *Z*, the number of protons in the nucleus or the number of electrons orbiting the nucleus of an atom.

Avena unit, the amount of auxin that causes an *Avena* (oat) coleoptile tip to bend 10° under certain circumstances within two hours.

Avogadro's number, N_A, $6.022 \cdot 10^{23}$, the number of molecules in one mole of a compound or the number of atoms in one gram atomic weight of an element.

bar, a unit of pressure; 1 bar = 100 kPa; *see* pascal.

becquerel, Bq, the unit of → activity.

calorie, cal, the cgs unit of heat (work); 1 cal_{15} = 4.1855 J; often in nutritional work calorie is misleadingly used instead of a kilocalorie.

candela, cd, the SI unit of luminous intensity; 1 cd = 1/60 of the luminous intensity of one square centimetre of a black body surface at 1773.5°C (the solidification temperature of platinum). A more accurate unit used in radiometry is W/cm^2.

candle, a former unit of luminous intensity, 1 candle = 1.01 candela.

capacitance, *see* farad.

Celsius, *t*, *θ*, a derived unit of temperature; $t = T - T_o$, in which T_o = 273.15 K (melting point of ice), measured in Celsius (°C).

cgs (c.g.s.) unit, based on the centimetre, the gram and the second as the fundamental units of length, mass and time. Today replaced for most purposes by SI units.

colour temperature, the temperature of a black body which radiates the same dominant wavelengths as those emitted from a given source, expressed usually in → kelvins.

concentration, *c*, → molarity.

conductance, *G*, → siemens.

coulomb, C, the SI unit of electric charge, 1 C = 1 As.

curie, Ci, the former unit of radioactivity, now replaced by → becquerel; 1 Ci = $3.7 \cdot 10^{10}$ Bq.

dalton, D, Da, → relative atomic mass.

darwin, a unit describing the rate of evolution. When a given trait of a species (see: phenotypic value) increases or decreases by 2.7-fold in a million years, the rate of evolution of that particular trait is one darwin.

decibel, dB, originally, the unit used to compare two power levels. The ratio between power levels P_1 and P_2 in decibels is: $10 \log_{10} P_1/P_2$; e.g. a 10-fold increase of power = 10 dB, 100-fold increase = 20 dB. Used to measure the sound intensity by taking the sound pressure (10^{-12} W/m^2) of the threshold value of hearing as a reference level.

degree, °, 1. the unit of plane angle equal to 1/360th of a complete revolution (*see* grade); **2**. the unit of temperature difference; *see* Celsius (°C), Fahrenheit (°F), Réaumur (°R); **3**. the measure of distance in geography: → latitude, → longitude.

density, ρ, the mass of unit volume, $\rho = m/V$; expressed as kg/dm^3, in gases g/dm^3.

DIN, Deutsche Industrie-Norm; the German national standards organization.

dioptre, the unit of power of a lens, generally it is the reciprocal of its focal length in metres, 1/m.

dose equivalent, describes the biological efficiency measured in sieverts (Sv). 1 Sv = 1 J/kg. The dose equivalent (short: dose) is quality factor Q multiplied by modifying factor N multiplied by absorbed dose *D*.

dyne, the unit of force in the cgs system; 1 dyne equals the force required to give a mass of one gram an acceleration of 1 cm/s^2; 1 dyne = 10^{-5} newton.

e, the base of natural, Napierian logarithms, defined as the limiting value of $(1 + 1/n)^n$ as n approaches infinity; $e = \lim_{n\to\infty}(1 + 1/n)^n = 1 + 1/1! + 1/2! + 1/3!...... \cong 2.7182818284$.

einstein, *E*, the amount of energy in a photochemical process bound in one mole of substance = $6.022 \cdot 10^{23} \cdot 6.6256 \cdot 10^{-34}$ Js · frequency of radiation.

electric charge, *see* coulomb.

electron charge, e, 1 $e = 1.602 \cdot 10^{-19}$ C.

electron rest mass, m_e, 1 $m_e = 9.1095 \cdot 10^{-31}$ kg.

electronvolt, eV, a unit of energy equal to the energy gained by an electron in moving through a potential difference of one volt; 1 eV = $1.602 \cdot 10^{-19}$ J.

energy units, the SI unit is the joule, others include the erg, the calorie, the kilowatt-hour.

equivalent dose, *H*, the concept that designates the biological effect of radiation; the unit is sievert (Sv), 1 Sv = 1 J/kg. The equivalent dose (i.e. dose) = *Q* (the quality coefficient of radiation) · N (modification coefficient) · *D* (absorbed dose).

equivalent weight, eq, the relative atomic or molecular mass divided by the valence; e.g. the atomic weight of oxygen is 16 and valence 2, thus eq is 8. The equivalent weight of a substance expressed in grams is the gram equivalent.

erg, a unit of energy or work in system; 1 erg = 10^{-7} J.

Fahrenheit scale, °F, a unit of temperature: In Fahrenheit scale the °F freezing point of water is 32°F and boiling point 212°F; 1°C = 1.8°F, and 1°F = approx. 0.56°C .

farad, F, the SI unit of capacitance; a capacitor with a capacitance of one farad has a potential difference of one volt if charged with one coulomb; 1 F = 1 C/V = 1 As/V

Faraday constant, *F*, 9.648456 · 10^4C/kg.

flow, *q*, streamflow, measured in m^3/s.

f-number, *see* luminous intensity.

gravitation (force of gravity), *F*, the force with which the Earth attracts a body of mass; $F = Gm_1m_2/r^2$ (G = the constant of gravitation, m_1 = Earth's mass, m_2 = the mass of the particle, r = the distance between Earth's centre and the particle).

gas constant, *R*, 8.31434 J/K · mol.

grade,^{g}, the unit of plane angle equal to 1/400 of a complete revolution; 1° = approx. 1.111^g, $1^g = 0.9^o$.

gram equivalent, *see* equivalent weight.

gram percent, g-%, the amount of a substance in grams per 100 grams (w/w) or per 100 millilitres (w/v); w = weight, v = volume.

gravity, *g*, *see* acceleration.

gray, Gy, *see* absorbed dose.

half-life, the time in which half of the atoms of a given radioactive substance undergo disintegration and activity falls to half of its initial value.

hardness, there are several scales of hardness for minerals; in Mohs' scale 1 = talc, 2 = gypsum, 3 = calcite, 4 = fluorite, 5 = apatite, 6 = orthoclase, 7 = quartz, 8 = topaz, 9 = corundum, 10 = diamond. For water, the quantity of calcium carbonate is used as a measure of hardness expressed in ppm (parts per million). 0—55 ppm is soft, 56—100 is hardish, 101—200 hard and over 200 is very hard water.

henry, H, the SI unit of inductance. A circuit has an inductance of 1 henry if an electro-magnetic force of 1 volt is produced by a current variation of 1 ampere per second; 1 H = 1 Wb/A.

herz, Hz, the SI unit of frequency; 1 Hz = 1 cycle /s.

international unit, *IU*, a quantity of a substance, as a vitamin, hormone, drug, that produces a given reaction or effect as agreed upon. Examples of absolute quantities in international units:

penicillin	0.6 μg benzylpenicillin
vitamin B_1	8.0 μg thiamine hydrochloride
vitamin A	0.3 μg as alcohol
	0.344 μg as acetate
vitamin E	1.0 mg DL-tocopherol acetate
insulin	0.125 mg standard preparation

joule, J, the SI unit of work and energy; 1 joule is equal to the work done when a force of 1 newton moves its point of application 1 metre in the direction of the force = 1 Nm. 1J = 1 kgm^2s^{-2} = 10^7 erg = 1 Ws; 1 J = 0.23901 cal = $2.78 \cdot 10^{-7}$ kWh = $6.2418 \cdot 10^{18}$ eV.

kelvin, K, the SI unit of temperature; the temperature interval of 1 kelvin equals that of the celsius; the absolute zero of temperature is 0 K = -273.15°C.

kilowatt-hour, kWh, a unit of energy, used especially as a commercial unit of electrical energy, 1 kWh = 3.6 MJ.

knot, the speed used in navigation, 1 knot = 1 nmph (1 nautical mile/h *i.e.* 1852 m/h = 0.514 m/s).

latitude, °, angular distance north or south from the equator measured through 90 degrees. At equator parallels of latitude 1° apart are separated on the Earth´s surface by 111.137 km.

LET, linear energy transfer; the linear rate of energy dissipated by radiation whilst penetrating absorbing media.

light year, the distance travelled by light in a year, appr. $9.4605 \cdot 10^{15}$ m.

longitude, °, meridians of longitude are half great circles passing through both poles of the Earth; at the equator meridians 1 ° apart are separated by about 111.12 km.

lumen, lm, the unit of luminous flux, 1 lm is the amount of light emitted from a point source of 1 candela in a solid angle of 1 steradian (= 1 cd · sr)

luminous intensity, *I*, a physical quantity measured in → candela; in a lens, the luminous intensity is expressed by f-number which is the reciprocal of the relative aperture of a lens expressing the ratio of the effective diameter of the lens to the focal length.

lux, lx, a unit of illumination in photometry; 1 lux is equal to the direct illumination on a surface that is one metre apart from a point source of 1 candela; 1 lx = 1 lm/m^2.

mass, *m*, a measure of a body's resistance to acceleration or a measure of the attraction of one body to another. The mass of a body can be measured against a known reference mass e.g. by weighing. The SI unit of mass is the kilogramme, kg.

mass number, *M*, the total number of nucleons in the nucleus of an atom, i.e. $M = N + Z$ where N is the number of neutrons and Z that of protons.

mesh, the mesh size designates the screen size as the number of openings per linear inch in a grid, net or sieve.

mm Hg, a former unit of pressure, mercury barometre; *see* pascal, torr.

Mohs' scale, *see* hardness.

molality, *m*, the number of moles of dissolved substance per kilogramme of solvent, mol/kg.

molar volume, V_m, the volume occupied by a mole of ideal gas at standard temperature and pressure, i.e. at 0°C and 101.3 kPa; $[V_m] = dm^3/mol$.

molarity, M, the number of moles of dissolved substance per one litre of solution, mol/l i.e. mol/dm^3.

mole, mol, the amount of substance corresponds to the relative molecular weight in grams for any chemical compound; one mole of substance contains $6.0220 \cdot 10^{23}$ molecules (→ Avogadro constant).

molecular mass, molecular weight, → relative molecular mass.

Napierian logarithm, → e.

natural logarithm, → e.

neutron number, *N*, the number of neutrons in the nucleus of an atom.

newton, N, the SI unit of force that gives a body of 1 kg an acceleration of 1 m/s^2; 1 N = 10^5 dyn.

normal temperature and pressure, NTP, the state of gas at 0°C and 101.3 kPa. The molar volume of an ideal gas at NTP is 22.4141 dm^3. Also abbreviated STP.

normality, *N*, the number of gram equivalents of a substance per one litre of solution; *see* equivalent weight.

numerical aperture, the refractive index of the object space multiplied by the semiaperture of the objective lens. The resolution of a microscope is usually proportional to the wave lenght of the light used and to the numerical aperture.

ohm, Ω, the SI unit of electrical resistance; 1 ohm equals to a resistance in a circuit in which a current of 1 ampere causes a potential difference of 1 volt; 1 Ω = 1 V/A.

optical density, *D*, the ratio of the radiant flux intensity passing through the medium to that incident on the medium; $D = \log 1/T$ (T = transmission ratio); also known as absorbance or extinction; *T* is sometimes expressed as transmsission percentage 100 · *T*. The reciprocal of transmission quotient 1/*T* is called opacity in photography.

ounce, oz, a unit of mass, 28.3495 g; in precious metals (troy ounce, oz tr) and apothecaries (oz ap) = 31.1035 g.

pascal, Pa, the SI unit of pressure; 1 Pa = 1N/m^2 = 10 dyne/cm^2 = 10^{-5} bar = 0.987 · 10^{-5} atm = 7.50 · 10^{-3} torr.

pH value, a measure of acidity; *see* acidity in the glossary.

photographic speed rating, The speeds of photographic emulsions; most important scales used are DIN, ASA and Schneider:

Relative speed	1	4	16	64	128
DIN	7	13	19	25	28
ASA	4	16	64	250	500
°Schneider (Eur.)	17	23	29	35	38

In practice 18°DIN = 50 ASA, 21°DIN = 100 ASA.

pi, π, the ratio of the circumference of a circle to its diameter ≅ 3.141592654.

Planck's constant, *h*, 6.626 · 10^{-34} Js.

poise, P, a cgs unit of viscosity; 1 P = 1 dyn/cm^2 = 1 g/cm/s.

pond, p, a cgs unit of force; equals the weight of a mass of gramme under normal conditions; 1 kgf = 9.80665 N.

pound, lb, the unit of mass in the old UK system; 1 lb = 0.454 kg (= 16 oz, → ounce).

pound-force, lbf, 1 lbf = 4.448 N.

ppb or p.p.b., parts per billion.

ppm or p.p.m., parts per million.

pressure units, → pascal, torr, bar.

proton rest mass, m_p, 1.673 · 10^{-27} kg; *see* relative atomic mass.

psi, → pound-force per square inch; 1 psi = 1 lbf/in^2 = 6.895 kPa.

quality factor, QF, → relative biological efficiency.

quantum, a term for an increment of energy, especially radiation energy; the amount of energy of a quantum E = $h\nu$, in which *h* = Planck's constant, ν = frequency); e.g. a light bulb of 100 W radiates visible light approximately 10^{19} quanta per second.

radian, rad, the SI unit of plane angular measure; 1 rad = the angle subtended at the centre of a circle by an arc equal in lenght to the radius = 57.29578° = 57°17'45".

Rankine temperature scale, °R, An absolute temperature scale based on the Fahrenheit scale; $0^{o}C = 32^{o}F = 273.15\ K = 492.69^{o}R$; *see* Kelvin.

Réaumur scale, °R, the temperature scale in which the melting point of ice is $0^{o}R$ and the boiling point of water is $80^{o}R$.

refraction ratio, $\sin\alpha/\sin\beta$, in which α is the angle of the incident beam to the normal of the reflecting interface and β is the angle of the refracted beam to the normal of the reflecting interface; almost identical to refractive index (in case of a beam entering a medium from vacuum).

relative atomic mass (weight), the ratio of the mass of an element to 1/12 of the mass of ^{12}C, which equals to the atomic mass unit *u*: 1 *u* = 1 D(a) (dalton) = $1.660 \cdot 10^{-27\ kg}$; e.g. for oxygen, the relative atomic mass is (former atomic weight) is 16 *u*, but the *u* is usually omitted.

relative biological efficiency, QF, measures the effectiveness of ionizing radiation that is harmful to tissues; is the inverse relation of the absorbed dose to the absorbed X-ray dose caused by a 200 kV potential. The absorbed dose is measured in grays (Gy) and multiplied by QF gives the effective biological dose in → sieverts. The QF of roentgen- and gamma-radiation is approx. 1 and that of alpha emission approx. 10. Now replaced by the effective dose equivalent.

relative molecular mass (weight), the sum of the relative atomic masses of all the atoms of a molecule; e.g. the relative molecular mass of O_2 is 32; in proteins the atomic mass may be hundreds of thousands, in which cases → daltons (D, Da) are usually used.

resistance, *R*, a property of a conductor by the virtue of which the passage of current is opposed; equal to the voltage across the conductor divided by the current in the conductor. Resistance causes electric energy to be transformed into heat. Resistance is usually measured in → ohms (Ω).

rhe, the unit of fluidity, the reciprocal unit of → poise in cgs system.

R_f- (rf-) value, relative fastness; in chromatography and electrophoresis the ratio of the distance moved by a substance (seen as a dot or band) to that moved by the solvent front.

roentgen, R, an old unit expressing the dose of ionizing radiation; 1 R = a dose of radiation that produces in one cubic centimetre of dry air both positive and negative ions equal to one electrostatic unit of charge = $0.258 \cdot 10^{-3}$ C/kg.

RPM, rpm, r.p.m. (r/m), the velocity of rotation, revolutions per minute; also used revolutions per second, **rps, r.p.s. (r/s)**.

sedimentation coefficient s, S, the sedimentation velocity of a particle divided by acceleration; 1 s = $v/\omega^2 r$, v = the speed of the interface between the fluid with particles and the solvent in a centrifuge, ω = angular velocity of rotor in rad/s, and r = the distance of the particle from the axis of rotation. Sedimentation units usually fall below 1 picosecond, therefore a Svedberg unit has been taken into practical use, defined as 1 · 10^{-13} s.

siemens, S, the unit of electrical conductance; reciprocal of → ohm; 1 S = 1/Ω = 1 A/V.

sievert, Sv, *see* dose equivalent.

specific activity, α, activity divided by the mass of the sample; expressed in Bq/kg (*see* bequerel).

steradian, sr, the unit of solid angular measure. 1 sr = the solid angle subtended at the centre of a sphere by an area on its surface equal to the square of the radius.

stilb, sb, the unit of luminance, 1 sb = lx/cm^2.

Svedberg unit, S, 10^{-13} s = 10^{-1} ps; *see* sedimentation coefficient.

torr, a former unit of pressure; 1 torr = 1 mm mercury at °C temperature = 133.322 Pa (→ pascal, the SI unit of pressure); 760 torr = 1 atm.

velocity of light, c, in vacuo approx. 299,792.5 km/s.

viscosity η, the property of a fluid that resists the fluid to flow; the viscous force is $F = \eta A \cdot \Delta v/\Delta s$, in which η is the viscocity coefficient, A the area of adjacent fluid layers, Δv velocity gradient, and Δs the distance between the layers. Viscosity is measured in pascal seconds, Pa · s (SI unit) or → poise, P (cgs unit).

volt, V, The SI unit of electrical potential (potential difference); 1 V = 1 W/A = 1 J/C = 1 A · Ω (a potential difference of 1 V produces a current of one ampere in a circuit that has a resistance of one ohm).

volume flow rate, q_v, V/t (V = volume), usually expressed in m^3/s.

watt, W, the SI unit of power; 1 W = 1 J/s = 1 Nm/s = 1 VA = 1 $A^2\Omega$ = 10^7 erg/s = 59.91549 J/min.

weber, Wb, the SI unit of magnetic flux; 1 Wb = 1 Vs.

weight, F_g, = mg (m = mass of the body, g = local gravity).

ELEMENTS

Name, symbol	Atomic number, Z	Relative atomic mass (longest-lived isotope)	Number of stable and	radioactive isotopes	Valences	Density	Melting point, °C	Boiling point, °C
actinium, Ac	89	(227)	0	25	3	10.07	1050	3300
aluminium, Al	13	26.98	1	10	3	2.70	660	2327
americium, Am	95	(243)	0	12	3,4,5,6	13.7	1173	2067
antimony, Sb	51	121.75	2	22	3,5	6.68	630	1635
argon, Ar	18	39.95	3	12	-	1.78	-189.2	-185.7
arsenic, As	33	74.92	1	21	3,5	5.7	818	613
astatine, At	85	(219)	0	20	1,3,5,7	-	302	337
barium, Ba	56	137.33	7	15	2	3.6	710	1600
berkelium, Bk	97	(247)	0	11	3,4	14	985	
beryllium, Be	4	9.01	1	6	2	1.85	1287	2500
bismuth, Bi	83	208.98	1	16	3,5	9.78	271	1560
bohrium, Bh	107	(262)	0	3	-	-	-	-
boron, B	5	10.81	2	3	3	2.35	2200	2550
bromine, Br	35	79.90	2	15	1-7	3.10	-7.25	59.47
cadmium, Cd	48	112.41	7	21	2	8.65	321	765
calcium, Ca	20	40.08	6	7	2	1.55	850	1460
californium, Cf	98	(251)	0	18	2,3,4	8.7-15	900	-
carbon, C	6	12.01	2	6	4			
diamond						3.51	>3500	4827
graphite						2.15	>3500	4200
cerium, Ce	58	140.12	3	25	3,4	6.77	798	3433
cesium, Cs	55	132.91	1	20	1	1.87	28.5	705
chlorine,Cl	17	35.45	2	9	1-7	3.21	-101	-34.6
chromium,Cr	24	52.00	4	9	1-6	7.14	1903	2642
cobalt, Co	27	58.93	1	15	1,2,3,4,5	8.92	1493	3100
copper, Cu	29	63.55	2	9	1,2	8.94	1083	2595
curium, Cm	96	(247)	0	14	3,4	13	1345	3110
dubnium, Db	104	(261)	0	10	-	-	-	-
dysprosium, Dy	66	162.50	7	16	2,3,4	8.5	1400	2562
einsteinium, Es	99	(252)	0	14	2,3	-	860	-
erbium, Er	68	167.26	6	20	3	9.1	1529	2868
europium,Eu	63	151.96	2	21	2,3	5.2	826	1429
fermium, Fm	100	(257)	0	18	-	-	-	-
fluorine, F	9	19.00	1	1	1	1.70	-219.6	-188.1
francium, Fr	87	(223)	0	9	1	-	-	-
gadolinium, Gd	64	157.25	6	15	3	7.87	1300	3266
gallium, Ga	31	69.72	2	12	1,2,3	5.91	29.8	2400
germanium, Ge	32	72.59	5	9	2,4	5.36	938	2800
gold, Au	79	196.97	1	21	1,3	19.3	1064.8	2700
hafnium, Hf	72	178.49	6	14	2,3,4	13.3	2150	5400
hahnium, Hn	108	(263)	0	3	-	-	-	-
helium, He	2	4.003	2	5-8	0	0.18	-272.2	-268.6
holmium, Ho	67	164.93	1	24	3	8.8	1460	2695
hydrogen, H	1	1.008	2	1	1	0.09	-259.2	-252.8
indium, In	49	114.82	2	15	3	7.3	156.2	2080
iodine, J	53	126.90	1	22	1-7	4.93	113.6	185.2
iridium, Ir	77	192.22	2	16	1-6	22.65	2450	4500
iron, Fe	26	55.85	4	6	2,3	7.86	1535	3000

Name, symbol	Atomic number, Z	Relative atomic mass (longest-lived isotope)	Number of stable and	radioactive isotopes	Valences	Density	Melting point, °C	Boiling point, °C
joliotium, Jl	105	(262)	0	3	-	-	--	-
krypton, Kr	36	83.80	6	20	0	3.73	- 156.6	-152.3
lanthanum, La	57	138.91	2	26	3	6.15	920	3464
lawrencium, Lw	103	(262)	0	10	3	-	--	
lead, Pb	82	207.19	4	16	2,4	11.34	327.3	1750
lithium, Li	3	6.94	2	3	1	0.53	179	1330
lutetium, Lu	71	174.97	2	28	3	9.87	1650	3327
magnesium, Mg	12	24.31	3	3	2	1.74	651	1107
manganese, Mn	25	54.94	1	9	1,2,3,4,6,7	7.3	1244	2095
meitnerium, Mt	109	(266)	0	1	-	-	-	-
mendelevium, Md	101	(258)	0	12	2,3	-	-	-
mercury, Hg	80	200.59	7	19	1,2	13.55	-38.87	356.7
molybdenum, Mo	42	95.94	7	11	2-6	10.22	2610	5560
neodymium, Nd	60	144.24	7	19	2,3,4	7.0	1024	3100
neon, Ne	10	20.18	3	7	-	0.90	- 248.7	-245.9
neptunium, Np	93	(237)	0	13	4,5,6	20.45	640	3902
nickel, Ni	28	58.69	5	9	1-4	8.90	1453	2732
niobium, Nb	41	92.91	1	13	2-5	8.57	2468	4900
nitrogen, N	7	14.01	2	5	3,5	1.25	- 210.01	-195.8
nobelium, No	102	(259)	0	10	2,3	-	-	-
osmium, Os	76	190.2	7	18	1-8	22.57	3000	5000
oxygen, O	8	15.999	3	5	2	1.43	-218.4	-183.0
palladium, Pd	46	106.4	6	12	2,4,6	12.0	1550	2900
phosphorus, P	15	30.97	1	6	3,5	1.82	44.1	280
platinum, Pt	78	195.08	6	23	1,2,4,5,6	21.45	1773	3827
plutonium, Pu	94	(242)	0	15	3-7	19.8	640	3235
polonium, Po	84	(210)	0	26	2,4,6	9.3	254	962
potassium, K	19	39.10	2	8	1	0.86	63.4	765
praseodymium, Pr	59	140.91	1	24	3,4	6.77	935	3020
promethium, Pm	61	(145)	0	24	3	-	1035	2730
protactinium, Pa	91	(231)	0	13	3,4,5	15.4	1570	4227
radium, Ra	88	(226)	0	13	2	5	700	1700
radon, Rn	86	(222)	0	18	0	9.73	- 71	- 61.8
rhenium, Re	75	186.2	1	15	1-7	21.0	3180	5900
rhodium, Rh	45	102.91	1	11	1-6	12.4	1966	3700
rubidium, Rb	37	85.47	1	17	1	1.53	38.8	700
ruthenium, Ru	44	101.07	7	9	1-8	12.4	2450	4150
rutherfordium, Rf	106	(263)	0	4	-	-	-	-
samarium, Sm	62	150.36	7	29	2,3	7.5	1070	1900
scandium, Sc	21	44.96	1	11	3	3.0	1540	2836
selenium, Se	34	78.96	6	11	2,4,6	4.8	217	685
silicon, Si	14	28.09	3	5	4	2.4	1410	2355
silver, Ag	47	107.87	2	16	1,2	10.50	960.5	2200
sodium, Na	11	22.99	1	6	1	0.97	97.8	881
strontium, Sr	38	87.62	4	14	2	2.6	757	1366
sulphur, S	16	32.06	4	6	2,4,6	2.07	112.8	444.6
tantalum, Ta	73	180.95	2	13	2-5	16.6	2996	5429
technetium, Tc	43	(97)	0	21	4,7	11.5	2172	4877
tellurium, Te	52	127.60	8	13	2,4,6	6.2	450	990
terbium, Tb	65	158.93	1	20	3,4	8.3	1356	3230
thallium, Tl	81	204.37	2	25	1,3	11.85	303.5	1457
thorium, Th	90	232.04	0	23	4	11.7	1750	4790

Name, symbol	Atomic number, Z	Relative atomic mass (longest-lived isotope)	Number of stable and	radioactive isotopes	Valences	Density	Melting point, °C	Boiling point, °C
thulium, Tm	69	168.93	1	28	3	9.3	1545	1725
tin, Sn	50	118.69	10	15	2,4	7.31	231.9	2507
titanium, Ti	22	47.88	5	4	2,3,4	4.51	1677	3277
tungsten, W	74	183.85	5	13	2-6	19.3	3410	5900
uranium, U	92	238.03	0	15	3-6	18.95	1133	3818
vanadium, V	23	50.94	1	8	3,5	6.11	1917	3380
xenon, Xe	54	131.29	9	26	2,4,6,8	5.89	-111.9	-107.1
ytterbium, Yb	70	173.04	7	20	2,3	7.0	819	1196
yttrium, Y	39	88.91	1	21	3	4.5	1500	2927
zinc, Zn	30	65.38	5	8	2	7.14	419.5	908
zirconium, Zr	40	91.22	5	14	3,4	6.5	1857	3577

PROKARYOTES

System 1:

division Bacteria

subdivision Archaebacteria

subdivision Eubacteria

class Bacteriobionta
class Cyanobionta (Cyanobacteria)

System 2:

EUBACTERIA

division Gracilicutes

class Scotophobia	all heterotrophic Gram-negative bacteria and some other groups, e.g. Myxobacteria, spirochaetes (e.g. *Borrelia*), enterobacteria (e.g. *Escherichia*), nitrogen-fixing bacteria (e.g. *Rhizobium*, *Azotobacter*), chemoautotrophic bacteria: nitrifying and denitrifying bacteria (e.g. *Nitrobacter*, *Nitrosomonas*), sulphur-oxidizing bacteria (e.g. *Thiobacillus*) and methane-oxidizing bacteria (e.g. *Methylomonas*)
class Anoxyphotobacteria	photosynthetic bacteria other than cyanobacteria, e.g. green and purple photosynthetic sulphur bacteria
class Oxyphotobacteria	cyanobacteria, e.g. *Nostoc*, *Anabaena*, *Rivularia*
class Prochlorophyta	prochlorophytes, e.g. *Prochloron*, *Prochlorothrix*

division Firmicutes

class Firmibacteria	Gram-positive cocci and rods, e.g. *Streptococcus*, *Staphylococcus*, *Bacillus*, *Clostridium*
class Thallobacteria	actinobacteria (e.g. *Actinomyces*), mycobacteria (e.g. *Mycobacterium tuberculosis*)

division Tenericutes

class Mollicutes — wall-less prokaryotes with questionable origin and relationships: rickettsiae, mycoplasmas, chlamydiae

ARCHAEBACTERIA

division Mendosicutes

class Archaebacteria — prokaryotes with unusual cell wall structure and membrane lipids: extreme halophiles (e.g. *Halobacterium*) and extreme thermophiles (*Thermococcus*)

THE PLANT KINGDOM

System 1:

division Chlorophyta (green algae)

class Chlorophyceae
class Prasinophyceae

division Charophyta (stoneworts)

class Charophyceae
order Charales

division Phaeophyta (brown algae)

class Phaeophyceae

division Rhodophyta (red algae)

class Rhodophyceae
subclass Bangiophycidae
subclass Florideophycidae

division Chrysophyta

subdivision Bacillariophyceae (diatoms)

subdivision Chrysophyceae (golden algae)

division Xanthophyta (yellow-green algae)

class Xanthophyceae
class Haptophyceae

division Pyrrophyta (Dinophyta)(dinoflagellates)

division Euglenophyta (euglenoids)

class Euglenophyceae

division Cryptophyta (cryptophytes)

class Cryptophyceae

division Bryophyta (mosses)

class Hepaticopsida (liverworts)
order Marchantiales
order Metzgeriales
order Jungermanniales
class Anthocerotae (hornworts)
class Bryopsida (Musci)
subclass Sphagnidae (peat mosses)
subclass Bryidae (true mosses)
subclass Andreaeidae (granite mosses)

division Pteridophyta (seedless vascular plants)

class Psilotopsida (whisk ferns)
class Lycopsida (lycophytes)
order Lycopodiales (club mosses)
order Selaginellales) (*Selaginella*)
order Lepidodendrales + (tree lycophytes)
order Isoëtales (quillworts)

class Sphenopsida (horsetails)
order Equisetales (horsetail, *Equisetum*)
order Sphenophyllales +
order Protoarticulatales

class Pteropsida
order Archaeopteridae +
order Ophioglossidae (adder's tongues, moonworts)
order Polypodidae
order Marsileidae (water ferns)
order Salvidae (water ferns)

division Spermatophyta (seed plants)

subdivision Gymnospermae (gymnosperms)

class Pteridospermae +
class Cycadinae (cycads)
class Bennettitinae +
class Cordaitinae +
class Ginkgoinae (*Ginkgo*)

+ = an extinct taxon

class Coniferae (conifers)
order Pinales
class Gnetinae

subdivision Angiospermae (angiosperms)

class Magnoliatae (Dicotyledonae)(dicotyledons)
subclass Magnoliidae
subclass Ranunculidae
subclass Caryophyllidae
subclass Hamamelididae
subclass Rosidae
subclass Dillenidae
subclass Lamiidae
subclass Asteridae

class Liliatae (Monocotyledonae)(monocotyledons)
subclass Alismatidae
subclass Aridae
subclass Liliidae
subclass Zingiberidae
subclass Commelinidae
subclass Arecidae

System 2:

ALGAE

division Chlorophyta (green algae)

class Chlorophyceae

division Charophyta (stoneworts)

class Charophyceae

division Phaeophyta (brown algae)

class Phaeophyceae

division Rhodophyta (red algae)

class Rhodophyceae

division Chrysophyta

subdivision Bacillariophyceae (diatoms)

subdivision Chrysophyceae (golden algae)

division Xanthophyta (yellow-green algae)

class Xanthophyceae
class Haptophyceae

division Euglenophyta (euglenoids)

class Euglenophyceac

division Pyrrophyta (Dinophyta)(dinoflagellates)

division Cryptophyta (cryptophytes)

BRYOPHYTES

division Hepatophyta (liverworts)

order Marchantiales
order Metzgeriales
order Jungermanniales

division Anthocerophyta (hornworts)

division Bryophyta

class Sphagnidae (peat mosses)
class Andreaeidae (granite mosses)
class Bryidae (true mosses)

PTERIDOPHYTES, SEEDLESS VASCULAR PLANTS

division Psilophyta (Psilotophyta)(whisk ferns)

division Lycophyta (lycophytes)

order Lycopodiales (club mosses)
order Lepidodendrales + (tree lycophytes)
order Selaginellales (*Selaginella*)
order Isoëtales (quillworts)

division Sphenophyta (horsetails)

order Equisetales (horsetail, *Equisetum*)

order Sphenophyllales +
order Protoarticulatales +

division Pterophyta (Filicophyta)(ferns)

order Marattiales
order Ophioglossales (adder's tongues, moonworts)
order Osmundales
order Filicales (most living ferns)
order Marsileales (water ferns)
order Salviniales (water ferns)

division Progymnophyta (progymnosperms)

SPERMATOPHYTES, SEED PLANTS

GYMNOSPERMS

division Pteridospermophyta + (seed ferns)

division Cycadeoidophyta (Bennettitales) + (cycadeoids)

division Cycadophyta (cycads)

division Ginkgophyta (*Ginkgo*)

division Coniferophyta (conifers and allies)

order Coniferales (most living conifers)
order Cordaitales +
order Voltziales +
order Taxales

division Gnetophyta

order Welwitschiales
order Ephedrales
order Gnetales

ANGIOSPERMS

division Anthophyta (Magnoliophyta)

class Dicotyledonaes (Magnoliopsida)
subclass Magnoliidae
subclass Ranunculidae
subclass Hamamelididae

subclass Caryophyllidae
subclass Dillenidae
subclass Rosidae
subclass Asteridae

class Monocotyledonaes (Liliatae)
subclass Alismidae
subclass Liliidae
subclass Commelinidae
subclass Arecidae

THE FUNGI

System 1:

division Gymnomycota (slime moulds)

subdivision Acrasiogymnomycotina (acrasiomycetes)

class Acrasiomycetes

subdivision Plasmodiogymnomycotina

class Protosteliomycetes
class Myxomycetes (plasmoidal slime moulds)

division Mastigomycota (algal fungi)

subdivision Haplomastigomycotina

class Hyphochytridiomycetes
class Chytridiomycetyes
class Plasmodiophoromycetes

subdivision Diplomastigomycotina

class Oomycetes

division Amastigomycota

subdivision Zygomycotina

class Zygomycetes
class Trichomycetes

subdivision Ascomycotina (sac fungi)

class Ascomycetes

subclass Hemiascomycetidae
subclass Plectomycetidae
subclass Hymenoascomycetidae
subclass Laboulbeniomycetidae
subclass Loculoascomycetidae

subdivision Basidiomycotina

class Basidiomycetes
subclass Holobasidiomycetidae
subclass Phragmobasidiomycetidae
subclass Teliomycetidae

division Deuteromycota

class Deuteromycetes (Fungi imperfecti)
subclass Blastomycetidae
subclass Coelomycetidae
subclass Hyphomycetidae

System 2:

division Acrasiomycota (acrasiomycetes)

division Dictyosteliomycota (dictyostelids)

division Ascomycota (sac fungi)

division Basidiomycota

class Basidiomycetes
class Teliomycetes
class Ustomycetes

division Hyphochytridiomycota (hyphochytridiomycetes)

division Chythridiomycota (chythridiomycetes)

division Plasmodiophoromycota (plasmodiophoromycetes)

division Oomycota (oomycetes)

division Zygomycota

class Zygomycetes
class Trichomycetes

division Labyrinthulomycota (labyrinthulomycetes)
order Labyrinthulales
order Thraustochtriales

division Myxomycota (plasmoidal slime moulds)

class Myxomycetes
class Protosteliomycetes

THE ANIMAL KINGDOM

The following list draws in outline of the taxonomic classification of living animals. Only part of the taxa have been considered; the invertebrates usually to the level of class. In some essential groups such as molluscs, arthropods and all vertebrates also subclasses or orders are given. The arrangement and nomenclature of the taxa follow with some exceptions the book "Synopsis and Classification of Living Organisms" by Sybil P. Parker (McGraw-Hill Company, N.Y. 1982). Systematics of fishes is based on the classification by Joseph S. Nelson ("Fishes of the World" - John Wiley & Sons, N.Y. 1976).

subkingdom Protozoa

phylum Sarcomastigophora

subphylum Mastigophora (flagellated protozoans)

class Dinoflagellata (dinoflagellates)
class Phytomastigophora (plant-like flagellates)
class Zoomastigophora (zooflagellates)

subphylum Opalinata

subphylum Sarcodina

superclass Rhizopoda (e.g. amoebae)
class Lobosa
class Filosa
class Acarpomyxa
class Acrasea
class Eumycetozoa
class Plasmodiophorea
class Xenophyophorea
class Granuloreticulosa

superclass Actinopoda (e.g. heliozoans, radiolarians)
class Acantharia
class Polycystina
class Phaeodaria
class Heliozoa

phylum Labyrinthulata

class Labyrinthulea

phylum Apicomplexa (Sporozoa, sporozoans)

class Perkinsea
class Sporozoea

phylum Microspora

class Rudimicrosporea
class Microsporea

phylum Myxozoa (Cnidospora)

class Myxosporea
class Actinomyxea

phylum Ascetospora (Haplospora)

class Stellatosporea
class Paramyxea

phylum Ciliophora (ciliates)

class Kinetofragminophora
(e.g. subclass Suctoria)
class Oligohymenophora
(e.g. subclass Peritrichia)
class Polyhymenophora
subclass Spirotrichia

subkingdom Phagocytellozoa

phylum Placozoa (placozoans)

subkingdom Parazoa

phylum Porifera

class Calcarea (calcareous sponges)
class Hexactinellida (glass sponges)
class Demospongiae (e.g. horny sponges)
class Sclerospongiae (coralline sponges)

subkingdom Eumetazoa

phylum Cnidaria

class Hydrozoa (e.g. hydrae, colonial hydroids)
class Scyphozoa ("true" jellyfishes)
class Cubozoa (cubomedusae)

class Anthozoa ("flower animals")
subclass Alcyonaria (soft and horny corals)
subclass Zoantharia (e.g. sea anemones, hard corals)
subclass Ceriantipatharia (e.g. thorny corals)

phylum Ctenophora ("sea walnuts", "comb jellies")

class Tentaculata
class Nuda

PROTOSTOMIA; ACOELOMATA

phylum Platyhelminthes (flatworms)

class Turbellaria (free-living flatworms)

class Trematoda (flukes)
subclass Monogenea (monogenean flukes)
subclass Aspidogastrea
subclass Digenea (digenean flukes)

class Cestoda (tapeworms)
subclass Cestodaria
subclass Eucestoda

phylum Mesozoa

class Rhombozoa
class Orthonectida

phylum Nemertea (Rhynchocoela) (ribbon worms)

class Anopla
class Enopla

phylum Gnathostomulida

PROTOSTOMIA; PSEUDOCOELOMATA

phylum Rotifera (rotifers)

class Seisonidea
class Bdelloidea
class Monogononta

phylum Kinorhyncha

phylum Loricifera

phylum Nematoda (roundworms)

class Phasmidia (Secernentea)
class Aphasmidia (Adenophorea)

phylum Gastrotricha
class Macrodasy(o)ida
class Chaetonot(o)ida

phylum Nematomorpha (horsehair worms)

phylum Acanthocephala (spiny-headed worms)

phylum Entoprocta (Kamptozoa)

phylum Priapulida (priapus worms)

PROTOSTOMIA; COELOMATA

phylum Ectoprocta (Bryozoa) (moss animals)

class Phylactolaemata
class Stenolaemata
class Gymnolaemata

phylum Phoronida (horseshoe worms)

phylum Brachiopoda (lamp shells)

class Articulata
class Inarticulata

phylum Mollusca

class Caudofoveata
class Solenogastres (solenogasters)
class Monoplacophora
class Polyplacophora (chitons)

class Gastropoda (snails, slugs)
subclass Prosobranchia
subclass Opistobranchia (e.g. sea slugs)
subclass Pulmonata (e.g. garden snails)

class Scaphopoda (tusk shells)

class Bivalvia (Pelecypoda) (mussels, clams, oysters)
subclass Protobranchia
subclass Pteriomorpha
subclass Palaeoheterodonta
subclass Heterodonta
subclass Anomalodesmata

class Cephalopoda
subclass Nautiloidea (pearly nautilus)
subclass Coleoidea
- order Sepioidea (cuttlefish)
- order Teuthoidea (squids)
- order Octopoda (octopuses)
- order Vampyromorpha

phylum Annelida (segmented worms)

class Polychaeta (bristleworms)
class Oligochaeta (earthworms, pot worms, etc.)
class Hirudinea (leeches)

phylum Sipuncula (peanut worms)

phylum Echiura (spoonworms)

phylum Pogonopohora (beardworms)

phylum Onychophora (velvet worms, walking worms)

phylum Arthropoda

subphylum Trilobita (trilobites, extinct)

subphylum Chelicerata

class Merostomata
subclass Eurypterida (water scorpions)
subclass Xiphosura (horseshoe crabs)

class Arachnida, e.g.:
- order Araneae (spiders)
- order Scorpiones (scorpions)
- order Uropygi (e.g. whipscorpions)
- order Pseudoscorpiones (false scorpions)
- order Palpigradi (microwhipscorpions)
- order Solpugida (e.g. sun spiders)
- order Amblypygi (whipspiders)
- order Phalangida (Opiliones) (harvestmen)
- order Acari (mites, ticks)

class Pycnogonida (Pantopoda, sea spiders)

subphylum Crustacea

class Branchiopoda
order Notostraca (tadpole shrimps)
order Anostraca (fairy shrimps)
order Conchostraca (clam shrimps)
order Cladocera (water fleas)

class Cephalocarida
class Ostracoda (mussel shrimps)
class Remipedia
class Mystacocarida

class Copepoda, e.g.:
order Calanoida
order Harpacticoida
order Cyclopoida

class Tantulocarida
class Branchiura (fish lice)
class Cirripedia (barnacles)

class Malacostraca, e.g.:
order Stomatopoda (mantis shrimps)
order Mysidacea (opossum shrimps)
order Cumacea
order Tanaidacea
order Isopoda (e.g. woodlice, sow bugs)
order Amphipoda (sandhoppers)
order Euphausiacea (krill)
order Decapoda (crabs, lobsters, shrimps)

subphylum Uniramia

class Chilopoda (centipedes)
class Diplopoda (millipedes)
class Symphyla
class Pauropoda

class Insecta (insects)
subclass Apterygota (wingless insects)
order Protura (bark lice)
order Diplura (japygids)
order Collembola (springtails, snow fleas)
order Thysanura (silverfish, bristletails)

subclass Pterygota (winged insects)
superorder Exopterygota (Hemimetabola)
order Ephemeroptera (mayflies)

order Odonata (dragonflies, damselflies)
order Blattodea (cockroaches)
order Mantodea (preying mantis, mantises)
order Isoptera (termites)
order Zoraptera (zorapterans)
order Grylloblattaria (rock crawlers)
order Dermaptera (earwigs)
order Orthoptera (e.g. grasshoppers, locusts)
order Phasmida (Phasmatoptera) (e.g.stick insects, walking sticks)
order Embioptera (webspinners)
order Plecoptera (stoneflies)
order Hemiptera (true bugs)
order Psocoptera (book lice, bark lice)
order Mallophaga (biting lice)
order Anoplura (sucking lice)
order Thysanoptera (thrips)
order Homoptera (cicadas, leaf/tree hoppers)

superorder Endopterygota (Holometabola)
order Neuroptera (lacewings, ant lions, etc.)
order Coleoptera (beetles, weevils)
order Strepsiptera (stylops)
order Mecoptera (scorpionflies)
order Siphonaptera (fleas)
order Diptera (true flies; e.g. mosquito, tsetse fly)
order Trichoptera (caddis flies)
order Lepidoptera (butterflies, moths)
order Hymenoptera (e.g. ants, bees, wasps)

phylum Pentastomida (tongue worms)

phylum Tardigrada (water bears)

DEUTEROSTOMIA

phylum Echinodermata (echinoderms)
class Crinoidea (sea lilies, feather stars)
class Asteroidea (sea stars)
class Ophiuroidea (brittle stars)
class Echinoidea (sea urchins, sand dollars)
class Holothuroidea (sea cucumbers)
class Concentricycloidea (sea daisies)

Phylum Chaetognatha (arrow worms)

Phylum Hemichordata (hemichordates)
class Enteropneusta (acorn worms)
class Pterobranchia (pterobranchs)
class Planctosphaeroidea

Phylum Chordata (chordates)

Subphylum Urochordata (Tunicata, tunicates)
class Thaliacea (salps)
class Larvacea
class Ascidiacea (sea squirts)

Subphylum Cephalochordata (lancelets)

Subphylum Vertebrata

superclass Agnatha (vertebrates without jaws)
class Cephalaspidomorphi (lampreys)
order Petromyzoniformes

class Pteraspidomorphi (hagfishes)
order Myxiniformes

superclass Gnathostomata (jawed vertebrates)
class Chondrichthyes (cartilaginous fishes)
subclass Elasmobranchii (sharks, skates, rays)
order Heterodontiformes
order Hexanchiformes
order Lamniformes
order Squaliformes
order Rajiformes
subclass Holocephali (chimaeras, or ghostfish)
order Chimaeriformes

class Osteichthyes (bony fishes) *
subclass Sarcopterygii (fleshy-finned fishes)
order Coelacanthiformes (lobe-finned fishes)
order Ceratodiformes (Australian lungfishes)
order Lepidosireniformes (lungfishes)
order Polypteriformes (bichirs)

subclass Actinopterygii (ray-finned fishes)

** in some classifications the class Osteichthyes is divided:*

subclass Dipneusti (Dipnoi, lungfishes)
order Ceratodiformes
order Lepidosireniformes

subclass Crossopterygii
order Coelacanthiformes

subclass Brachiopterygii
order Polypteriformes

subclass Actinopterygii

---

infraclass Chondrostei
  order Acipenseriformes (sturgeons)
infraclass Holostei
  order Semionotiformes
  order Amiiformes (Amia)
infraclass Teleostei
  order Osteoglossiformes (mooneyes, knifefishes)
  order Mormyriformes (elephant fishes)
  order Clupeiformes (herrings, anchovies, etc)
  order Elopiformes (tarpons, bonefishes)
  order Anguilliformes (eels)
  order Notacanthiformes (halosaurs, spiny eels)
  order Salmoniformes (salmon, smelts, etc.)
  order Gonorynchiformes (milkfishes)
  order Cypriniformes (carps, minnows, etc.)
  order Siluriformes (catfishes)
  order Myctophiformes (lanternfishes, etc.)
  order Polymixiiformes (beardfishes)
  order Percopsiformes (troutperches, etc.)
  order Gadiformes (cods, etc.)
  order Batrachoidiformes (toadfishes)
  order Lophiiformes (anglerfishes)
  order Indostomiformes
  order Atheriniformes (toothcarps, etc.)
  order Lampridiformes (crestfishes, etc.)
  order Beryciformes (slimeheads, etc.)
  order Zeiformes (boarfishes, dories, etc.)
  order Syngnathiformes (pipefishes, sea-horses, etc.)
  order Gasterosteiformes (sticklebacks, etc.)
  order Synbranchiformes (singleslit eels)
  order Scorpaeniformes (scorpionfishes etc.)
  order Dactylopteriformes (flying gurnards)
  order Pegasiformes (sea-moths)
  order Perciformes (perches, mackerel, etc.)
  order Pleuronectiformes (flatfishes)
  order Tetraodontiformes (puffers, molas, etc.)

class Amphibia
subclass Lissamphibia
  order Caudata (Urodela) (newts, salamanders)
  order Anura (Salientia) (frogs, toads)
  order Gymnophiona (Apoda) (caecilians)

class Reptilia
subclass Anapsida
  order Testudines (Chelonia) (turtles, tortoises)
subclass Diapsida
  superorder Lepidosauria
  order Rhynchocephalia (tuatara)

  order Squamata
  suborder Sauria (Lacertilia) (lizards)

suborder Serpentes (Ophidia) (snakes)

subclass Archosauria
order Crocodylia (crocodilians)

class Aves (birds)
subclass Neornithes

superorder Palaeognathae
order Struthioniformes (ostriches)
order Rheiformes (rheas)
order Casuariiformes (cassowaries, emus)
order Apterygiformes (kiwis)
order Tinamiformes (tinamous)

superorder Neognathae
order Sphenisciformes (penguins)
order Gaviiformes (loans, divers)
order Podicipediformes (grebes)
order Procellariiformes (albatrosses, etc.)
order Pelecaniformes (pelicans, cormorants, gannets, etc.)
order Ciconiiformes (storks, herons, etc.)
order Phoenicopteriformes (flamingoes)
order Anseriformes (swans, geese, ducks)
order Falconiformes (eagles, hawks, etc.)
order Galliformes (grouse, pheasants, etc.)
order Opisthocomiformes (hoatzin)
order Gruiformes (cranes, rails, coots)
order Charadriiformes (shorebirds, waders, gulls, etc.)
order Columbiformes (pigeons, doves)
order Psittaciformes (parrots, parakeets)
order Cuculiformes (cuckoos, roadrunners)
order Strigiformes (owls)
order Caprimulgiformes (nighthawks, etc.)
order Apodiformes (swifts, hummingbirds)
order Coliiformes (mousebirds)
order Trogoniformes (trogons)
order Coraciiformes (kingfishers, bee-eaters, hornbills, etc.)
order Piciformes (woodpeckers, etc.)
order Passeriformes (perching songbirds)

class Mammalia
subclass Prototheria
order Monotremata (egg-laying mammals)

subclass Theria
infraclass Metatheria
order Marsupialia (marsupials)

infraclass Eutheria (Placentalia) (placental mammals)
order Insectivora (shrews, moles, etc.)
order Dermoptera (flying lemurs)

order Chiroptera (bats)
suborder Megachiroptera
suborder Microchiroptera

order Primates
suborder Prosimii (lemurs, tarsiers, etc.)
suborder Anthropoidea
superfamily Ceboidea (New World monkeys)
superfamily Cercopithecoidea (Old World monkeys)
superfamily Hominoidea
    family Hylobatidae (gibbons)
    family Hominidae (great apes, human)

order Edentata (anteaters, sloths, armadillos)
order Pholidota (pangolins)

order Cetacea (whales)
suborder Mysticeti (baleen whales)
suborder Odontoceti (toothed whales, dolphins)

order Carnivora (flesh-eating mammals)
order Pinnipedia (seals, walruses)
order Lagomorpha (rabbits, hares, pikas)
order Rodentia (gnawing mammals)
order Tubulidentata (aardwark)
order Proboscidea (elephants)
order Hyracoidea (hyraxes)
order Sirenia (manatees, sea cows)
order Perissodactyla (odd-toed ungulates)

order Artiodactyla (even-toed ungulates)
suborder Suina (Suiformes) (pigs, peccaries, hippopotamuses, etc.)
suborder Tylopoda (camels, llamas)
suborder Ruminantia (deer, cattle, etc.)

Geologic Timetable

Era	Period	Epoch	Years before present (B.P.)
CAENOZOIC	Quaternary	Recent	10 thousand
		Pleistocene	1.6 million
	Tertiary	Pliocene	5 million
		Miocene	24 million
		Oligocene	38 million
		Eocene	55 million
		Palaeocene	66 million
MESOZOIC	Cretaceous		144 million
	Jurassic		208 million
	Triassic		245 million
PALAEOZOIC	Permian		285 million
	Carboniferous		360 million
	Devonian		408 million
	Silurian		438 million
	Ordovician		505 million
	Cambrian		570 million
PRECAMBRIAN	Precambrian		

NEW ENTRIES

betaine (L. *beta* = beet), trimethylglycine anhydride, glycine betaine, trimethyl glycocoll, oxyneurine $(CH_3)_3NCH_2COOH$; oxidation product of choline in plants and animals regulating osmotic balance (osmoprotective effect). In animals it also has lipotropic activity.

candidate gene, a gene which is supposed to be the cause of a given trait.

catechol-O-methyltransferase, COMT, the enzyme (EC 2.1.1.6) inactivating → catecholamine neurotransmitters in the liver and kidney by catalysing the methylation of 3-hydroxy group of the catechol ring. The enzyme uses S-adenosylmethionine as the methyl donor. The enzyme inactivating catecholamines in the nervous system is → monoamine oxidase.

COMT, → catechol-O-methyltransferase.

cybernetics (Gr. *kybernetos* = mate), a general theory of → regulation; in biology the branch of science connecting biology and technology which is specifically involved in the study of self-regulation (autoregulation) mechanisms and systems of living organisms like the regulation of the function of the genes and the balance of ecosystems.

DNA chip, a microarray technique of studying large samples of DNA. In this array even tens of thousands of different DNA samples, derived e.g. from different cells, tissues, individuals, populations or species, are loaded on a silylated microscope slide, and then analysed by hybridizing → cDNA or → RNA probes (→ probe) to them. By this technique e.g. the expression of different genes in different cells or tissues can be studied.

lignans (L. *lignum* = wood), in plants occurring phenol compounds having the carbon skeleton $(C_6\text{-}C_3)_2$. Lignans are found in → lignin and they occur in many different plant groups, e.g. pinoresinol in pine and futoquinol in pepper. Many lignans occur as water-soluble glycosides in plant tissues. Some lignans have antimicrobial and antifungal activities, and some of them may prevent the development of breast cancer.

molecular anthropology (Gr. *anthropos* = man, *logos* = discourse), the study of human populations and their genetic differences by comparing biological macromolecules like → DNA and → proteins of different human individuals.

Musci (L. pl. of *muscus* = moss), → Bryopsida.

myocardium (Gr. *kardia* = heart), → cardiac muscle.

nerve ganglion, *see* ganglion.

nuclear magnetic resonance, NMR (Gr. *magnes* = magnet, L. *resonantia* = echo), a resonance phenomenon observed in a magnetic field if the magnetic moment of an atomic nucleus differs from zero. When the intensity of the magnetic field is changed a strong resonance is found in a certain area and is seen as loss of energy. NMR technique is a method to determine energy of atomic nuclei and it is used e.g. in organic chemistry to study the molecular structure of compounds, and in medicine an application is nuclear magnetic imaging (NMI).

r_m, *see* innate capacity for increase, Malthusian parameter.

translation complex, a molecule aggregate formed of → messenger RNA and → translation factors necessary for the initiation of → genetic translation. The translation factors bind both to the → poly-A tail and to the → leader sequence of the messenger RNA.

translation factors, proteins which act in the initiation of → genetic translation in eukaryotic organisms, and which together with the → messenger RNA form the → translation complex.

tunnelling electron microscope, *see* microscope.

vitamin C, ascorbic acid, antiscorbutic vitamin, $C_6H_8O_6$; a water-soluble vitamin synthesized by plants and most animals; especially abundant in fruits, berries and vegetables. Ascorbic acid functions in oxidation-reduction reactions of cells, acting as an important → antioxidant in many enzymatic reactions. Many animal species (like man and other primates) who have got ascorbic acid enough from their food sources (e.g. fruits), have lost their ability to synthesize this vitamin. In man deficiency causes → scurvy and poor wound repair. Suggestions for daily need varies from 50 to several hundreds of milligrams. Ascorbic acid is used as antimicrobial agent and antioxidant in foodstuffs.

ERRATA

p. 22, **aldose**: or glyceralhehyde; should be: or glyceraldehyde
p. 27, **alpha diversity**, *Cf.* beta diversity. should be: *See* species diversity.
p. 33, **amylases**: starch) by the action; should be: starch) begins by the action
p. 39, **Anoplura**: About 2,400 species; should be: About 500 species
p. 41, **antiauxin** (L. ; should be: **antiauxin** (Gr.
p. 44, **apes**, *see* Primates; should be: *see* Anthropoidea.
p. 45, **Apodiformes**: A Chinese species, **edible-nest swiftlet;** should be: A Malesian and Indonesian species, **edible-nest swiftlet**
p. 51, **Arthropoda**: **myriapods** (Uniramia), and **pentastomids** (Pentastomida); previously; should be: **myriapods** (Uniramia); previously
p. 54, **atmospheric pressure**: Many organisms are adpated; should be: Many organisms are adapted
p. 60, **azo-**: Lavoiser); should be: Lavoisier),
p. 61, **bacteriocin**: by *Eschericia coli*; should be: by *Escherichia coli*
p. 69, **bioacoustics**: a **sound spectrogram;** should be: a **sound spectrograph**
p. 92, **CAM,** should be moved to the right alphabetical order to page 91.
p. 95, **carbon**: 5,760 years; should be 5,730 years
p. 103, **Celcius, °C** (*Anders Celcius*; should be: **Celsius, °C** (*Anders Celsius*...; and should be moved to the right alphabetical order.
p. 114, **Chelicerata**: (Pygnogonida); should be: (Pycnogonida)
p. 115, **chiasma**: letter chi), a crossing; should be: letter chi), chiasm, a crossing and: **2.** optic chiasma; should be: **2.** optic chiasm
p. 126, **chromosome pairing**: 2) **nons-pecific;** should be: non-specific
p. 129, **Cirripedia**, the second line; should be: **barnacles**; a crustacean class (or a subclass in the class Maxillopoda),
p. 131, **clathrin**: **chlathrin-coated vesicle;** should be: **clathrin-coated vesicle**
p. 135, **codon**: (the regeneracy of the → genetic code); should be: (the degeneracy of the → genetic code).
p. 185, **Dictyoptera**, the end of the paragraph: should be: in the order of Orthoptera, or they are divided into two orders, Blattodea and Mantodea.
p. 185, **diestrus,** → dioestrus; should be: **diestrus, dioestrus,** *see* oestrus cycle.
p. 191, **dithiothreitol**: *Syn.* Cleland´s reagent, dierythritol; should be: *Syn.* Cleland´s reagent.
p. 202, **ecophenotype**: a habitat form of an organism; should be: a form of an organism modified by the habitat;
p. 202, **ecosphere**: *Cf.* biosphere; should be: *Syn.* biosphere.
p. 205, **eicosanoids**: delete the last sentence: Also called...
p. 205, **electr(o)** (Gr. *electron*; should be: (Gr. *elektron*
p. 210, **embryogenesis**: **neurulation** (<- neurula); should be: **neurulation** (→ neurula); the **hypocotyle;** should be: the **hypocotyl**
p. 218, **environmental resource**: delete: without disadvantage to other organisms; and replace it by: and so take away from other organisms;
p. 226, **erythromycin**: *Streptomyces erythraeus*; should be: *Streptomyces erythreus*
p. 226, **ESS-theory**: evolutionary; should be: evolutionarily
p. 248, **flavin**: delete: **1.**; and delete: **2.** a yellow acridine dye, **quercetin**, $C_6H_7(OH)_5$, obtained from oak bark.

p. 249, **floral formula**: K4C4A4+2*G*(2); should be: K4C4A4+2<u>*G*</u>(2)
p. 249, **flower**: **flower leaves;** should be: **floral leaves**
p. 251, **fluorouracil**: inhibiting the RNA synthesis; should be: inhibiting as deoxyuridylate the DNA synthesis;
p. 262, **gastric caecum/cecum**: located at the anterior end of the foregut; should be: located at the posterior end of the foregut.
p. 264, **Gay-Lussac´s law**: degree Celcius; should be: degree Celsius
p. 287, **gonadoliberin**: RHRF; should be: LHRF
p. 288, **G protein**: guanine triphosphate; should be: guanosine triphosphate
p. 293, **guanosine**: constituent of RNA; should be: constituent of DNA and RNA
p. 298, **Haemosporina**: in the class Coccidia, class Sporozoa, comprising; should be: in the class Sporozoea (phylum Apicomplexa), comprising
p. 301, **haploidization**: should be moved to the right alphabetical order.
p. 312, **hexadecanoid acid,** should be: **hexadecanoic acid**
p. 317, **Hominidae**: the gorilla (*Gorilla*) are incorporated; should be: the gorilla (*Gorilla*) with the man (*Homo sapiens*) are incorporated
p. 317, **Homo**: 1 to 2,5 million; should be: 2.5 to 1 million
p. 329, **hypodermis**: *dermis* ; should be: *derma*
p. 331, **hypoxanthine**: *ksanthos*; should be: *xanthos*
p. 333, **-ia**: as a verb; should be: as a word
p. 374, **Laurer´s canal**: the oviducts and yolk gland; should be: the oviducts and shell gland
p. 377, **Lepidosireniformes**: delete: *lepidus* = pretty,
p. 378, **leptome**: compagnion cells; should be: companion cells
p. 387, **loop migration,** → seasonal migration; should be: → seasonal migration, circular migration.
p. 400, **mandibular arch**: Merkel´s cartilage; should be: Meckel´s cartilage
p. 402, **Marsupialia**: Australia, but one species (**opossum**) is found in South America where; should be: Australia, but some species of opossums are found in North and South America where
p. 402, **martelism**: marten; should be: ermine; wolverine; should be: glutton
p. 453, **naphthyl**: $C_{10}C_7$-; should be: $C_{10}H_7$-
p. 458, **Neopterygii**: neoptertygians represent; should be: neopterygians represent
p. 458, **neoxanthin**: *ksanthos*; should be: *xanthos*
p. 496, **Ostracoda**: ca. 2,000 species; should be: ca. 5,700 species
p. 505, **Paneth´s cell**: secrete lysosome; should be: secrete lysozyme
p. 647, **spore**: separate (male) **macrospores** and (female) **microspores**; should be: separate (female) **macrospores** and (male) **microspores**.
p. 684, **tooth**: continuous reneval; should be: continuous renewal.
p. 689, **Trematoda**: The larval stages of the latter are; should be: The larval stages of the Digenea are
p. 707, **vertebra**: **spinous process** (*processus spinalis*); should be: **spinous process** (*processus spinosus*)